Air Conditioning and Refrigeration

Air Conditioning and Refrigeration

Rex Miller
Professor Emeritus
State University College at Buffalo
Buffalo, New York

Mark R. Miller
Professor, Industrial Technology
The University of Texas at Tyler
Tyler, Texas

McGraw-Hill

New York Chicago San Francisco Lisbon London
Madrid Mexico City Milan New Delhi San Juan
Seoul Singapore Sydney Toronto

The McGraw-Hill Companies

Library of Congress Cataloging-in-Publication Data

Miller, Rex, date.
Air conditioning and refrigeration / Rex Miller, Mark R. Miller.
p. cm.
Includes index.
ISBN 0-07-146788-2
1. Air conditioning. 2. Refrigeration and refrigerating machinery. I. Miller, Mark R. II. Title.

TH7687.M4855 2006
621.5′6—dc22 2005058099

1 2 3 4 5 6 7 8 9 0 QPD/QPD 0 1 2 1 0 9 8 7 6

ISBN 0-07-146788-2

The sponsoring editor for this book was Larry S. Hager and the production supervisor was Pamela A. Pelton. It was set in Times by International Typesetting and Composition. The art director for the cover was Handel Low.

Printed and bound by Quebecor/Dubuque.

Contents

7 Refrigeration Compressors

8 Condensers, Chillers, and Cooling Towers

9 Working with Water-Cooling Problems

10 Evaporators

11 Refrigerant: Flow Control

12 Servicing and Safety

14 Temperature, Psychrometrics, and Air Control

15 Comfort Air Conditioning

16 Commercial Air-Conditioning Systems

20 Air-Conditioning and Refrigeration Careers

Appendices

Preface

An introduction to the basic principles and practices of the air-conditioning and refrigeration industry is more than just a review of the facts and figures. It requires a complete look at the industry. This text presents the basics of all types of refrigeration. It explains the equipment that makes it possible for us to live comfortably in air-conditioned spaces and enjoy a wide variety of foods.

Up-to-date methods of equipment maintenance are stressed. The latest tools are shown. The applications of the newer types of units are emphasized. The field of air-conditioning technology is still growing and will continue to grow far into the future. New technicians will need to be aware of the fact that change is inevitable. They will have to continue to keep up with the latest developments as long as they stay in the field.

This textbook has been prepared to aid in instructional programs in high schools, technical schools, trade schools, and community colleges. Adult evening classes and apprenticeship programs may also find it useful. This book provides a thorough knowledge of the basics and a sound foundation for anyone entering the air-conditioning and refrigeration field.

The authors would like to give a special thanks to Mr. Burt Wallace who is an instructor in the air conditioning and refrigeration program in Tyler Junior College and Mr. Andy Bugg an AC Applications Engineer for one of the largest air conditioning manufacturers for their most valuable contributions to the book. Both live in Tyler, Texas.

REX MILLER
MARK R. MILLER

Acknowledgments

No author works without being influenced and aided by others. Every book reflects this fact. This book is no exception. A number of people cooperated in providing technical data and illustrations. For this we are grateful.

We would like to thank those organizations that so generously contributed information and illustrations. The following have been particularly helpful:

Admiral Group of Rockwell International
Air Conditioning and Refrigeration Institute
Air Temp Division of Chrysler Corp.
Americold Compressor Corporation
Amprobe Instrument Division of SOS Consolidated, Inc.
Arkla Industries, Inc.
Bryant Manufacturing Company
Buffalo News
Calgon Corporation
Carrier Air Conditioning Company
E.I. DuPont de Nemours & Co., Inc.
Dwyer Instruments, Inc.
Ernst Instruments, Inc.
General Controls Division of ITT
General Electric Co. (Appliance Division)
Haws Drinking Faucet Company
Hubbell Corporation
Hussman Refrigeration, Inc.
Johnson Controls, Inc.
Karl-Kold, Inc.
Kodak Corporation
Lennox Industries, Inc.
Lima Register Co.
Marley Company
Marsh Instrument Company, Division of General Signal
Mitsubishi Electric, HVAC Advanced Products Division
Mueller Brass Company
National Refrigerants
Packless Industries, Inc.
Parker-Hannifin Corporation
Penn Controls, Inc.
Rheem Manufacturing Company
Schaefer Corporation
Sears, Roebuck and Company
Snap-on Tools, Inc.
Sporlan Valve Company
Superior Electric Company
Tecumseh Products Company
Thermal Engineering Company
Trane Company
Turner Division of Clean-weld Products, Inc.
Tuttle & Bailey Division of Allied Thermal Corporation
Tyler Refrigeration Company
Union Carbide Company, Linde Division
Universal-Nolin Division of UMC Industries, Inc.
Virginia Chemicals, Inc.
Wagner Electric Motors
Weksler Instrument Corporation
Westinghouse Electric Corp.
Worthington Compressors

ABOUT THE AUTHORS

Rex Miller is Professor Emeritus of Industrial Technology at State University College at Buffalo and has taught technical curriculum at the college level for more than 40 years. He is the coauthor of the best-selling *Carpentry & Construction*, now in its fourth edition, and the author of more than 80 texts for vocational and industrial arts programs. He lives in Round Rock, Texas.

Mark R. Miller is Professor of Industrial Technology at the University of Texas at Tyler. He teaches construction courses for future middle managers in the trade. He is coauthor of several technical books, including the best-selling *Carpentry & Construction*, now in its fourth edition. He lives in Tyler, Texas.

Air Conditioning and Refrigeration

CHAPTER

Air-Conditioning and Refrigeration Tools and Instruments

PERFORMANCE OBJECTIVES

After studying this chapter the reader should be able to:

1. Understand how tools and instruments make it possible to install, operate, and troubleshoot air-conditioning and refrigeration equipment.
2. Know how electricity is measured.
3. Know how to use various tools specially made for air-conditioning and refrigeration work.
4. Know how to identify by name the tools used in the trade.
5. Know the difference between volt, ampere, and ohm and how to measure each.
6. Know how to work with air-conditioning and refrigeration equipment safely.

TOOLS AND EQUIPMENT

The air-conditioning technician must work with electricity. Equipment that has been wired may have to be replaced or rewired. In any case, it is necessary to identify and use safely the various tools and pieces of equipment. Special tools are needed to install and maintain electrical service to air-conditioning units. Wires and wiring should be installed according to the *National Electrical Code* (NEC). However, it is possible that this will not have been done. In such a case, the electrician will have to be called to update the wiring to carry the extra load of the installation of new air-conditioning or refrigeration equipment.

This section deals only with interior wiring. Following is a brief discussion of the more important tools used by the electrician in the installation of air-conditioning and refrigeration equipment.

Pliers and Clippers

Pliers come in a number of sizes and shapes designed for special applications. Pliers are available with either insulated or uninsulated handles. Although pliers with insulated handles are always used when working on or near "hot" wires, they must not be considered sufficient protection alone. Other precautions must be taken. Long-nose pliers are used for close work in panels or boxes. Slip-joint, or gas, pliers are used to tighten locknuts or small nuts. See Fig. 1-1. Wire cutters are used to cut wire to size.

Fig. 1-1 *Pliers.*

Fuse Puller

The fuse puller is designed to eliminate the danger of pulling and replacing cartridge fuses by hand, Fig. 1-2.

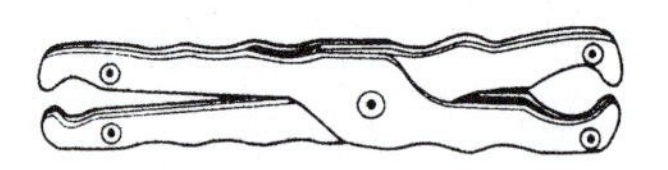

Fig. 1-2 *A fuse puller.*

It is also used for bending fuse clips, adjusting loose cutout clips, and handling live electrical parts. It is made of a phenolic material, which is an insulator. Both ends of the puller are used. Keep in mind that one end is for large-diameter fuses; the other is for small-diameter fuses.

Screwdrivers

Screwdrivers come in many sizes and tip shapes. Those used by electricians and refrigeration technicians should have insulated handles. One variation of the screwdriver is the screwdriver bit. It is held in a brace and used for heavy-duty work. For safe and efficient use, screwdriver tips should be kept square and sharp. They should be selected to match the screw slot. See Fig. 1-3.

The Phillips-head screwdriver has a tip pointed like a star and is used with a Phillips screw. These

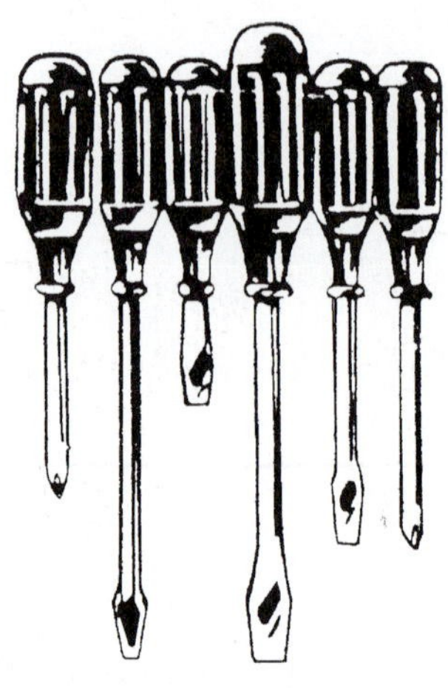

Fig. 1-3 *Screwdrivers.*

screws are commonly found in production equipment. The presence of four slots, rather than two, assures that the screwdriver will not slip in the head of the screw. There are a number of sizes of Phillips-head screwdrivers. They are designated as No. 1, No. 2, and so on. The proper point size must be used to prevent damage to the slot in the head of the screw. See Fig. 1-4.

Fig. 1-4 *A Phillips-head screwdriver.*

Wrenches

Three types of wrenches used by the air-conditioning and refrigeration trade are shown in Fig. 1-5.

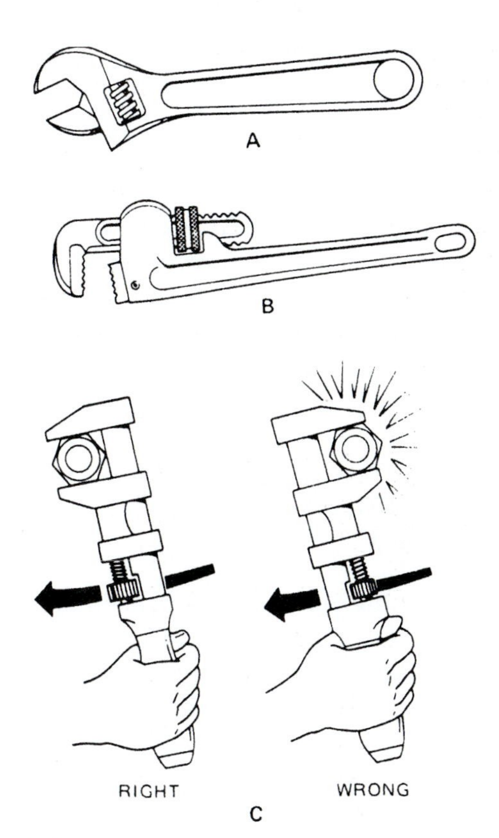

Fig. 1-5 *Wrenches. (A) Crescent wrench. (B) Pipe wrench. (C) Using a monkey wrench.*

- The adjustable open-end wrenches are commonly called *crescent wrenches.*
- *Monkey wrenches* are used on hexagonal and square fittings such as machine bolts, hexagonal nuts, or conduit unions.
- *Pipe wrenches* are used for pipe and conduit work. They should not be used where crescent or monkey wrenches can be used. Their construction will not permit the application of heavy pressure on square or hexagonal material. Continued misuse of the tool in this manner will deform the teeth on the jaw face and mar the surfaces of the material being worked.

Soldering Equipment

The standard soldering kit used by electricians consists of the same equipment that the refrigeration mechanics use. See Fig. 1-6. It consists of a nonelectric soldering device in the form of a torch with propane fuel cylinder or an electric soldering iron, or both.

The torch can be used for heating the solid-copper soldering iron or for making solder joints in copper tubing. A spool of solid tin-lead wire solder or flux-core

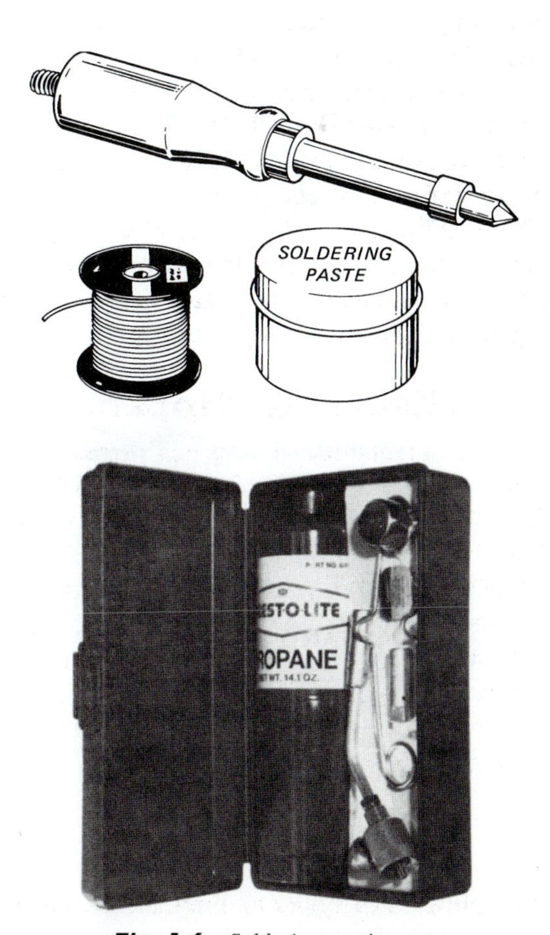

Fig. 1-6 *Soldering equipment.*

solder is used. Flux-core solder with a rosin core is used for electrical soldering.

Solid-core solder is used for soldering metals. It is strongly recommended that acid-core solder not be used with electrical equipment. Soldering paste is used with the solid wire solder for soldering joints on copper pipe or solid material. It is usually applied with a small stiff-haired brush.

Drilling Equipment

Drilling equipment consists of a brace, a joint-drilling fixture, an extension bit to allow for drilling into and through thick material, an adjustable bit, and a standard wood bit. These are required in electrical work to drill holes in building structures for the passage of conduit or wire in new or modified construction.

Similar equipment is required for drilling holes in sheet-metal cabinets and boxes. In this case, high-speed or carbide-tipped drills should be used in place of the carbon-steel drills that are used in wood drilling. Electric power drills are also used. See Fig. 1-7.

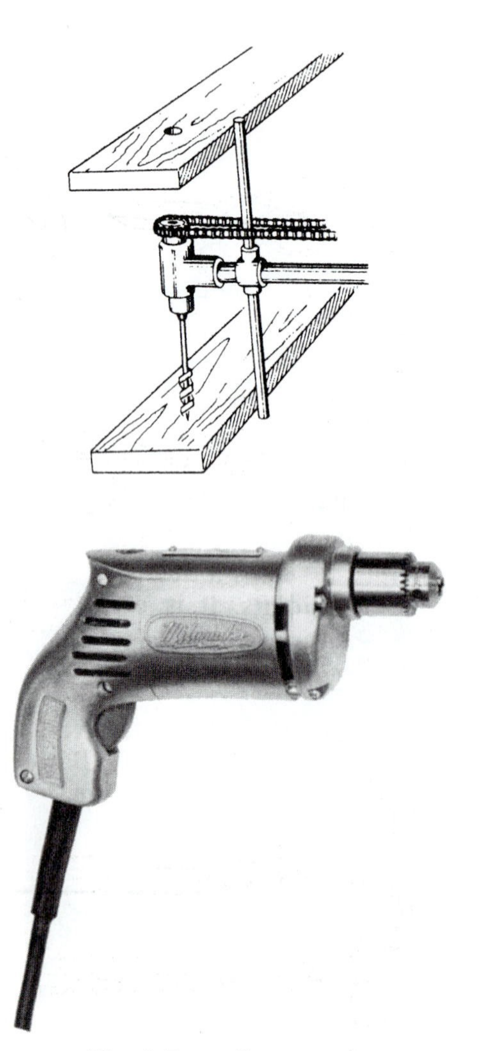

Fig. 1-7 *Drilling equipment.*

Woodworking Tools Crosscut saws, keyhole saws, and wood chisels are used by electricians and refrigeration and air-conditioning technicians. See Fig. 1-8. They are used to remove wooden structural members, obstructing a wire or conduit run, and to notch studs and joists to take conduit, cable, box-mounting brackets, or tubing.

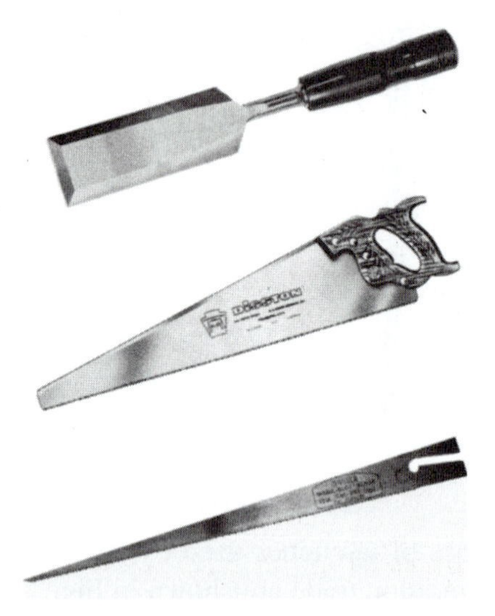

Fig. 1-8 *Woodworking tools.*

They are also used in the construction of wood-panel mounting brackets. The keyhole saw will again be used when cutting an opening in a wall of existing buildings where boxes are to be added or tubing is to be inserted for a refrigeration unit.

Metalworking Tools The cold chisel and center punch are used when working on steel panels. See Fig. 1-9. The knockout punch is used either in making or in enlarging a hole in a steel cabinet or outlet box.

The hacksaw is usually used when cutting conduit, cable, or wire that is too large for wire cutters. It is also a handy device for cutting copper tubing or pipe. The mill file is used to file the sharp ends of such cutoffs. This is a precaution against short circuits or poor connections in tubing.

Masonry Working Tools The air-conditioning technician should have several sizes of masonry drills in the tool kit. These drills normally are carbide-tipped. They are used to drill holes in brick or concrete walls. These holes are used for anchoring apparatus with expansion screws or for allowing the passage of conduit, cable, or tubing. Figure 1-10 shows the carbide-tipped bit used with a power drill and a hand-operated masonry drill.

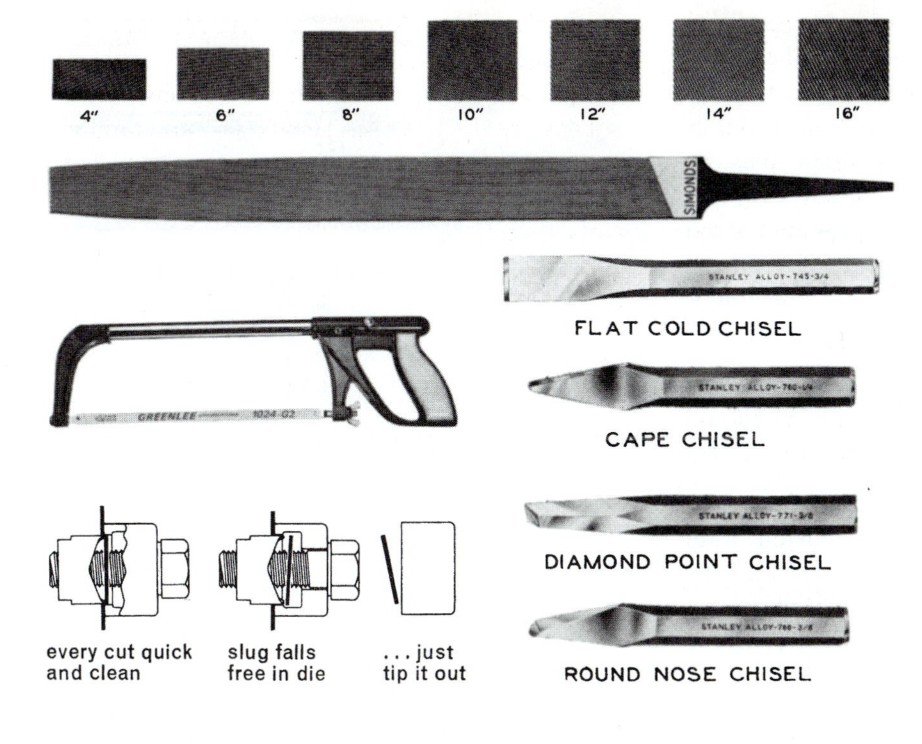

Fig. 1-9 *Metalworking tools.*

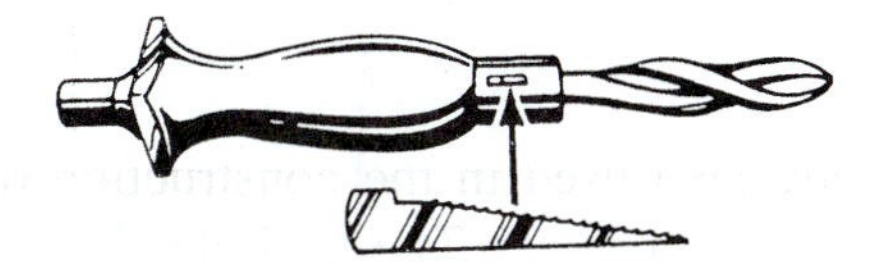

Fig. 1-10 *Masonry drills.*

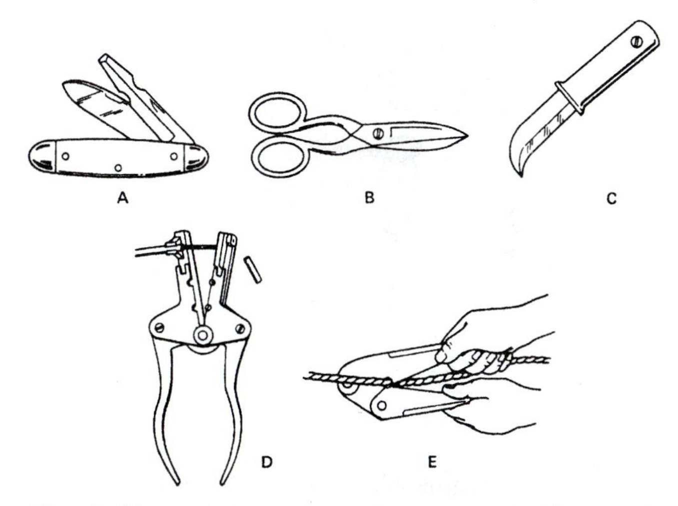

Fig. 1-11 *Tools for cutting and stripping. (A) Electrician's knife. (B) Electrician's scissors. (C) Skinning knife. D) Stripper. (E) Cable cutter.*

Knives and Other Insulation-Stripping Tools

The stripping or removing of wire and cable insulation is accomplished by the use of tools shown in Fig. 1-11. The knives and patented wire strippers are used to bare the wire of insulation before making connections. The scissors are used to cut insulation and tape.

The armored cable cutter may be used instead of a hacksaw to remove the armor from the electrical conductors at box entry or when cutting the cable to length.

Hammers Hammers are used either in combination with other tools, such as chisels, or in nailing equipment to building supports. See Fig. 1-12. The figure shows a carpenter's claw hammer and a machinist's ball-peen hammer.

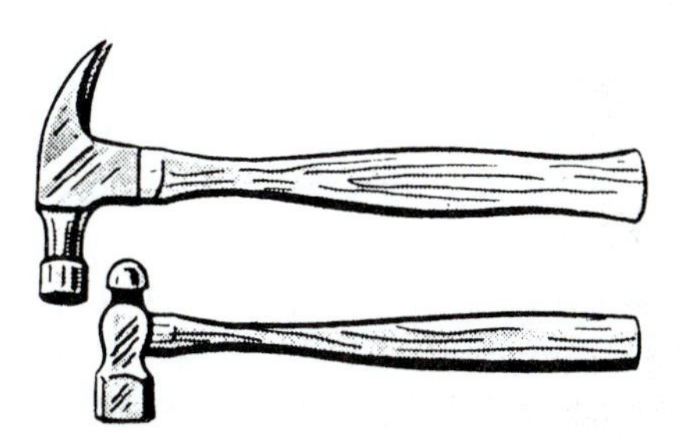

Fig. 1-12 *Hammers.*

Tape Various tapes are available. They are used for replacing removed insulation and wire coverings.

Friction tape is a cotton tape impregnated with an insulating adhesive compound. It provides weather resistance and limited mechanical protection to a splice already insulated.

Rubber tape or varnished cambric tape may be used as an insulator when replacing wire covering.

Plastic electrical tape is made of a plastic material with an adhesive on one side of the tape. It has replaced friction and rubber tape in the field for 120- and 208-V circuits. It serves a dual purpose in taping joints. It is preferred over the former tapes.

Ruler and Measuring Tape The technician should have a folding rule and a steel tape. Both of these are aids to cutting to exact size.

Extension Cord and Light The extension light shown in Fig. 1-13, is normally supplied with a long extension cord. It is used by the technician when normal building lighting has not been installed and where the lighting system is not functioning.

Fig. 1-13 *Extension light.*

Wire Code Markers Tapes with identifying numbers or nomenclature are available for permanently identifying wires and equipment. The wire code markers are particularly valuable for identifying wires in complicated wiring circuits, in fuse boxes, and circuit breaker panels, or in junction boxes. See Fig. 1-14

Meters and Test Prods

An indicating voltmeter or test lamp is used when determining the system voltage. It is also used in locating the ground lead and for testing circuit continuity through the power source. They both have a light that glows in the presence of voltage. See Fig. 1-15.

A modern method of measuring current flow in a circuit uses the hook-on voltammeter. See Fig. 1-16. This instrument does not have to be hooked into the

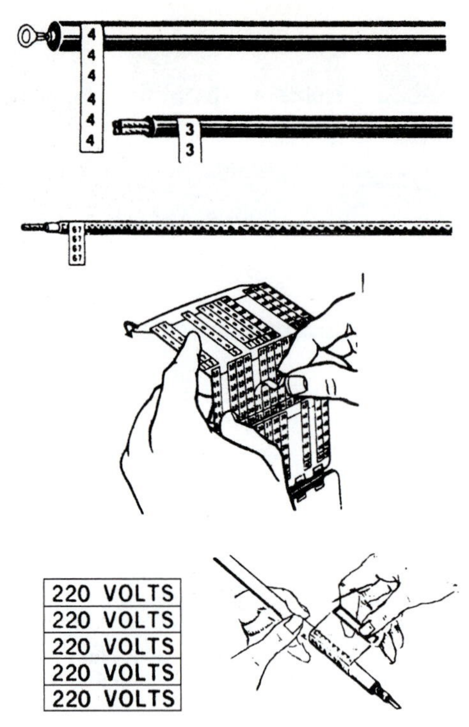

Fig. 1-14 *Wire code markers.*

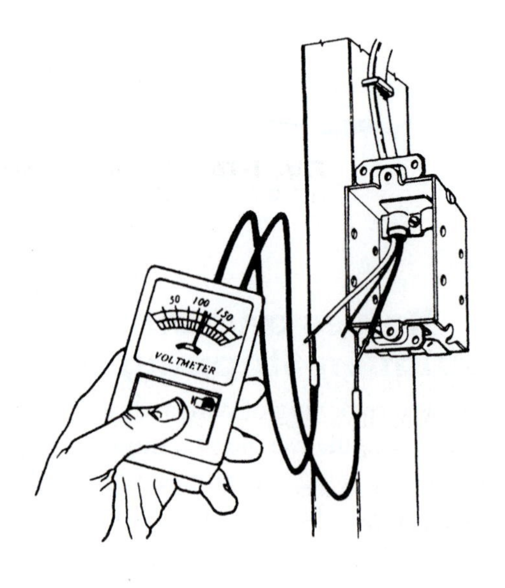

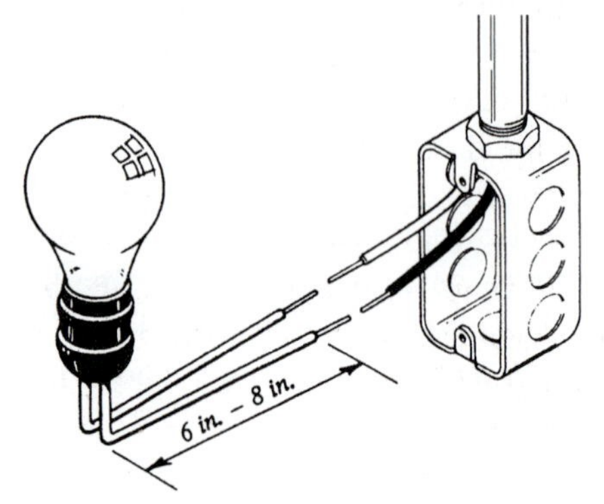

Fig. 1-15 *Test devices.*

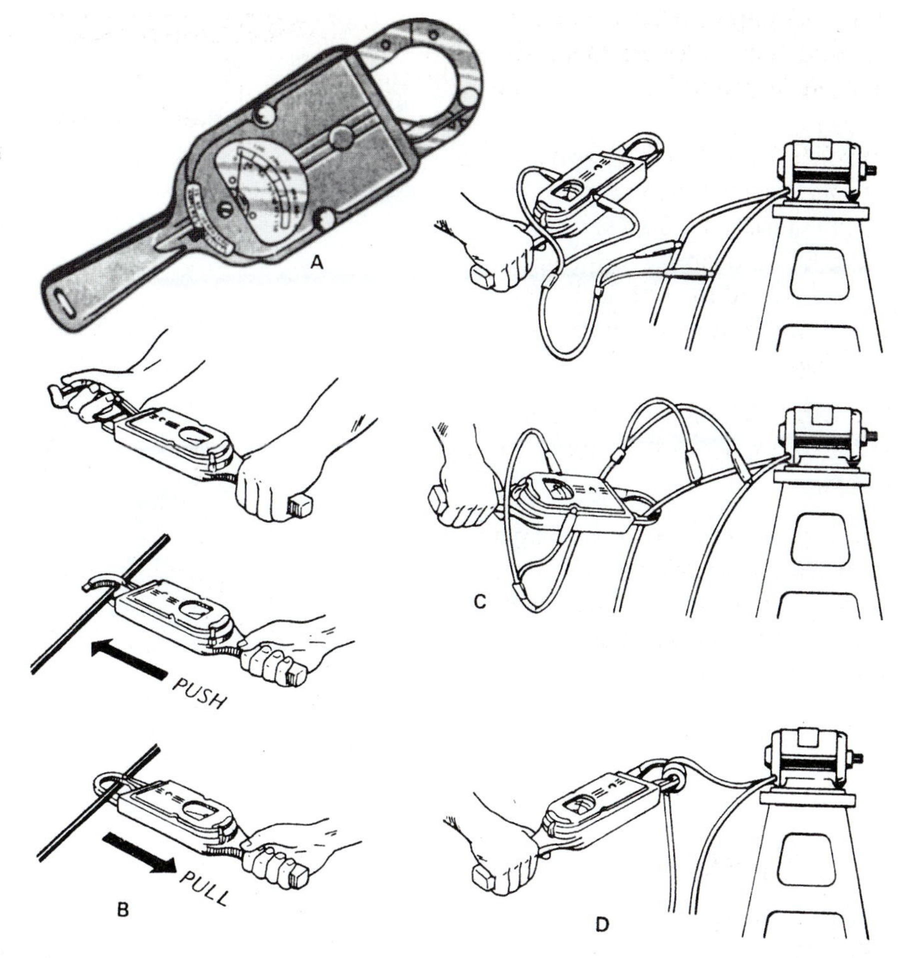

Fig. 1-16 *Hook-on volt-ammeter. (A) The volt-ammeter. (B) Correct operation. (C) Measuring alternating current and voltage with a single setup. (D) Looping conductor to extend current range of transformer.*

circuit. It can be operated with comparative ease. Just remember that it measures only one wire. Do not clamp it over a cord running from the consuming device to the power source. In addition, this meter is used only on *alternating current* (AC) circuits. The AC current will cancel the reading if two wires are covered by the clamping circle. Note how the clamp-on part of the meter is used on one wire of the motor.

To make a measurement, the hook-on section is opened by hand and the meter is placed against the conductor. A slight push on the handle snaps the section shut. A slight pull on the handle springs open the tool on the C-shaped current transformer and releases a conductor. Applications of this meter are shown in Fig. 1-16. Figure 1-16B shows current being measured by using the hook-on section. Figure 1-16C shows the voltage being measured using the meter leads. An ohmmeter is included in some of the newer models. However, power in the circuit must be off when the ohmmeter is used. The ohmmeter uses leads to complete the circuit to the device under test.

Use of the voltammeter is a quick way of testing the air-conditioning or refrigeration unit motor that is drawing too much current. A motor that is drawing too much current will overheat and burn out.

Tool Kits

Some tool manufacturers make up tool kits for the refrigeration and appliance trade. See Fig. 1-17 for a good example. In the Snap-on tool kit, the leak detector is part of the kit. The gages are also included. An adjustable wrench, tubing cutter, hacksaw, flaring tool, and ball-peen hammer can be hung on the wall and replaced when not in use. One of the problems for any repairperson is keeping track of tools. Markings on a board will help locate at a glance when one is missing.

Figure 1-18 shows a portable tool kit. Figure 1-18J shows a pulley puller. This tool is used to remove the

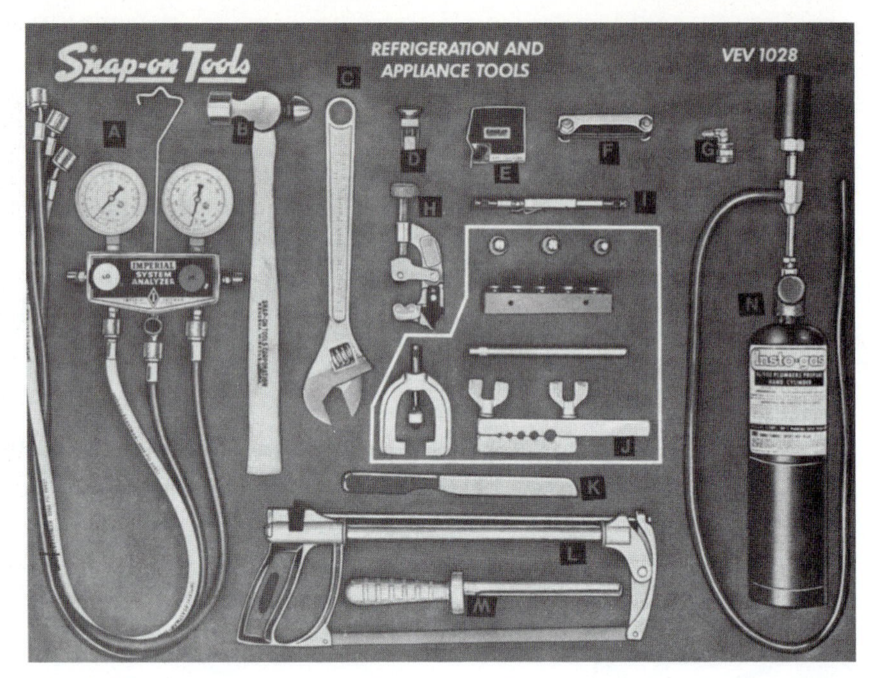

Fig. 1-17 *Refrigeration and appliance tools. (A) Servicing manifold. (B) Ball-peen hammer. (C) Adjustable wrench. (D) Tubing tapper. (E) Tape measure. (F) Allen wrench set. (G) 90° adapter service part. (H) Tubing cutter. (I) Thermometer. (J) Flaring tool kit. (K) Knife. (L) Hacksaw. (M) Jab saw. (N) Halide leak detector.* (Snap-On Tools)

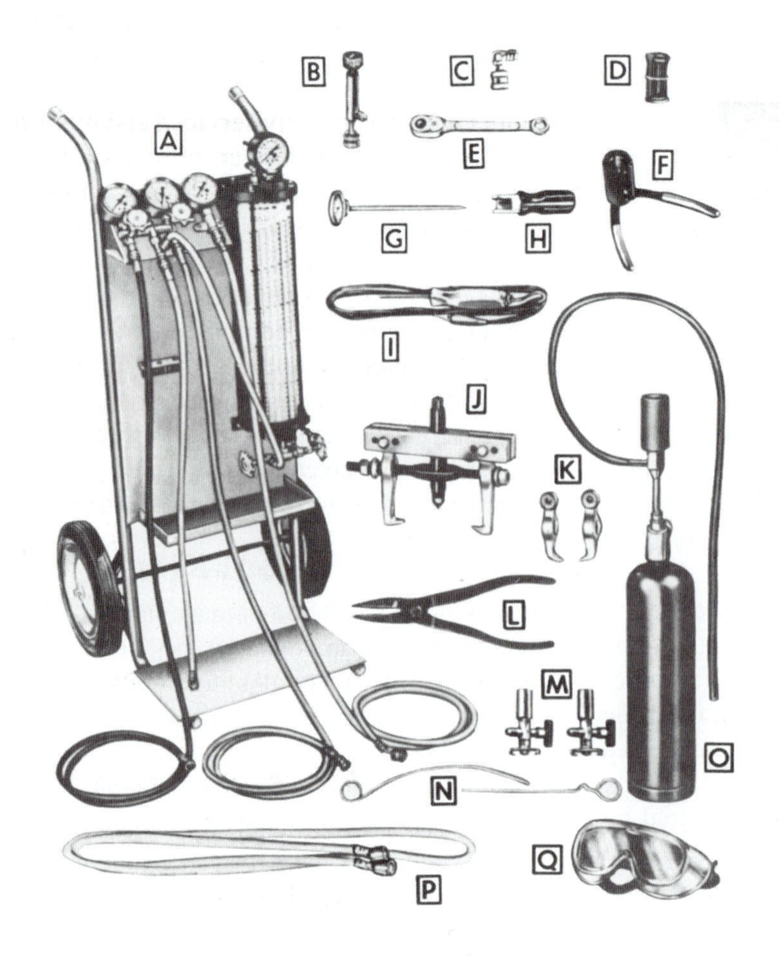

Fig. 1-18 *Air-conditioning and refrigeration portable tool kit. (A) Air-conditioning charging station. (B) Excavating/charging valve. (C) 90 adapter service port. (D) O-ring installer. (E) Refrigeration ratchet. (F) Snap-ring pliers. (G) Stem thermometer. (H) Seal remover and installer. (I) Test light. (J) Puller. (K) Puller jaws. (L) Retainer ring pliers. (M) Refrigerant can tapper. (N) Dipsticks for checking oil level. (O) Halide leak detector. (P) Flexible charging hose. (Q) Goggles.* (Snap-On Tools)

pulley if necessary to get to the seals. A cart (A) is included so that the refrigerant and vacuum pump can be easily handled in large quantities. The goggles (Q) protect the eyes from escaping refrigerant.

Figure 1-19 shows a voltmeter probe. It detects the presence of 115 to 750 V. The handheld meter is used to find whether the voltage is AC or DC and what the potential difference is. It is rugged and easy to handle. This meter is useful when working around unknown power sources in refrigeration units.

Fig. 1-19 *AC and DC voltage probe voltmeter.* (Amprobe)

Figure 1-20 shows a voltage and current recorder. It can be left hooked to the line for an extended period. Use of this instrument can be used to determine the exact cause of a problem, since voltage and current changes can affect the operation of air-conditioning and refrigeration units.

Fig. 1-20 *Voltage and current recorder.* (Amprobe)

GAGES AND INSTRUMENTS

It is impossible to install or service air-conditioning and refrigeration units and systems without using gages and instruments.

A number of values must be measured accurately if air-conditioning and refrigeration equipment is to be operated properly. Refrigeration and air-conditioning units must be properly serviced and monitored if they are to give the maximum efficiency for the energy expended. Here, the use of gages and instruments becomes important. It is not possible to analyze a system's operation without the proper equipment and procedures. In some cases, it takes thousands of dollars worth of equipment to troubleshoot or maintain modern refrigeration and air-conditioning system.

Instruments are used to measure and record such values as temperature, humidity, pressure, airflow, electrical quantities, and weight. Instruments and monitoring tools can be used to detect incorrectly operating equipment. They can also be used to check efficiency. Instruments can be used on a job, in the shop, or in the laboratory. If properly cared for and correctly used, modern instruments are highly accurate.

Pressure Gages

Pressure gages are relatively simple in function. See Fig. 1-21. They read positive pressure or negative pressure, or both. See Fig. 1-22. Gage components are

Fig. 1-21 *Pressure gage.* (Weksler)

Fig. 1-22 *This gage measures up to 150 psi pressure and also reads from 0 to 30 for vacuum. The temperature scaled runs from –40° to 115°F (–40° to 46.1°C).*

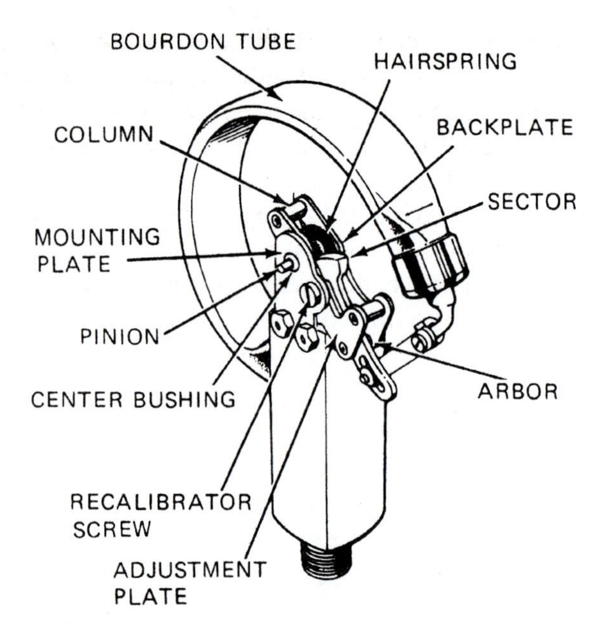

Fig. 1-23 *Bourdon tube arrangement and parts of a gage.* (Marsh)

relatively few. However, different combinations of gage components can produce literally millions of design variations. See Fig. 1-23. One gage buyer may use a gage with 0 to 250 psi range, while another person with the same basic measurement requirements will order a gage with a range of 0 to 300 psi. High-pressure gages can be purchased with scales of 0 to 1000, 2000, 3000, 4000, or 5000 psi.

There are, of course, many applications that will continue to require custom instruments, specially designed and manufactured. Most gage manufacturers have both stock items and specially manufactured gages.

Gage Selection

Since 1939, gages used for pressure measurements have been standardized by the *American National Standards Institute* (ANSI). Most gage manufacturers are consistent in face patterns, scale ranges, and grades of accuracy. Industry specifications are revised and updated periodically.

Gage accuracy is stated as the limit that error must not exceed when the gage is used within any combination of rated operating conditions. It is expressed as a percentage of the total pressure (dial) span.

Classification of gages by ANSI standards has a significant bearing on other phases of gage design and specification. As an example, a test gage with ±0.25 percent accuracy would not be offered in a 2 in. dial size. Readability of smaller dials is not sufficient to permit the precision indication necessary for this degree of accuracy. Most gages with accuracy of ±0.5 percent and better have dials that are at least 4.5 in. Readability can be improved still further by increasing the dial size.

Accuracy How much accuracy is enough? This is a question only the application engineer can answer. However, from the gage manufacturer's point of view, increased accuracy represents a proportionate increase in the cost of building a gage. Tolerances of every component must be more exacting as gage accuracy increases.

Time is needed for technicians to calibrate the gage correctly. A broad selection of precision instruments is available and grades A (±1 percent), 2A (±0.5 percent), and 3A (±0.25 percent) are examples of tolerances available.

With the advent of modern electronic gages and more sophisticated equipment it is possible to obtain heretofore undreamed of accuracy automatically with equipment used in the field.

Medium In every gage selection, the medium to be measured must be evaluated for potential corrosiveness to the Bourdon tube of the gage.

There is no ideal material for Bourdon tubes. No single material adapts to all applications. Bourdon tube materials are chosen for their elasticity, repeatability, ability to resist "set" and corrosion resistance to the fluid mediums.

Ammonia refrigerants are commonly used in refrigeration. All-steel internal construction is required. Ammonia gages have corresponding temperature scales. A restriction screw protects the gage against sudden impact, shock, or pulsating pressure. A heavy-duty movement of stainless steel and Monel steel prevents corrosion and gives extra-long life. The inner arc on the dial shows pressure. The other arc shows the corresponding temperature. See Fig. 1-24.

Fig. 1-24 *Ammonia gage.* (Marsh)

Line Pressure

The important consideration regarding line pressures is to determine whether the pressure reading will be constant or whether it will fluctuate. The maximum pressure at which a gage is continuously operated should not exceed 75 percent of the full-scale range. For the best performance, gages should be graduated to twice the normal system-operating pressure.

This extra margin provides a safety factor in preventing overpressure damage. It also helps avoid a permanent set of the Bourdon tube. For applications with substantial pressure fluctuations, this extra margin is especially important. In general, the lower the Bourdon tube pressure, the greater the overpressure percentage it will absorb without damage. The higher the Bourdon tube pressure, the less overpressure it will safely absorb.

Pulsation causes pointer flutter, which makes gage reading difficult. Pulsation also can drastically shorten gage life by causing excessive wear of the movement gear teeth. A pulsating pressure is defined as a pressure variation of more than 0.1 percent full-scale per second. Following are conditions often encountered and suggested means of handling them.

The restrictor is a low-cost means of combating pulsation problems. This device reduces the pressure opening. The reduction of the opening allows less of the pressure change to reach the Bourdon tube in a given time interval. This dampening device protects the Bourdon tube by the retarding overpressure surges. It also improves gage readability by reducing pointer flutter. When specifying gages with restrictors, indicate whether the pressure medium is liquid or gas. The medium determines the size of the orifice. In addition, restrictors are not recommended for dirty line fluids. Dirty materials in the line can easily clog the orifice. For such conditions, diaphragm seals should be specified.

The needle valve is another means of handling pulsation if used between the line and the gage. See Fig. 1-25. The valve is throttled down to a point where pulsation ceases to register on the gage.

In addition, to the advantage of precise throttling, needle valves also offer complete shutoff, an important safety factor in many applications. Use of a needle valve can greatly extend the life of the gage by allowing it to be used only when a reading is needed.

Liquid-filled gages are another very effective way to handle line pulsation problems. Because the movement is constantly submerged in lubricating fluid, reaction to pulsating pressure is dampened and the pointer flutter is practically eliminated.

Silicone-oil-treated movements dampen oscillations caused by line pressure pulsations and/or mechanical oscillation. The silicone oil, applied to the movement, bearings, and gears, acts as a shock absorber.

SOFT SEAT NEEDLE VALVES

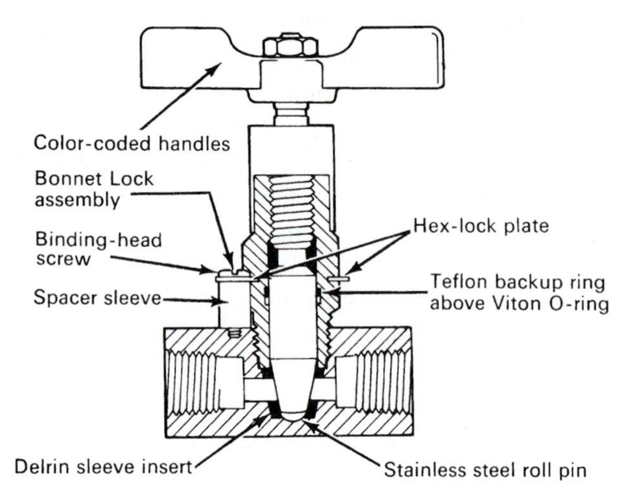

STAINLESS STEEL NEEDLE VALVES

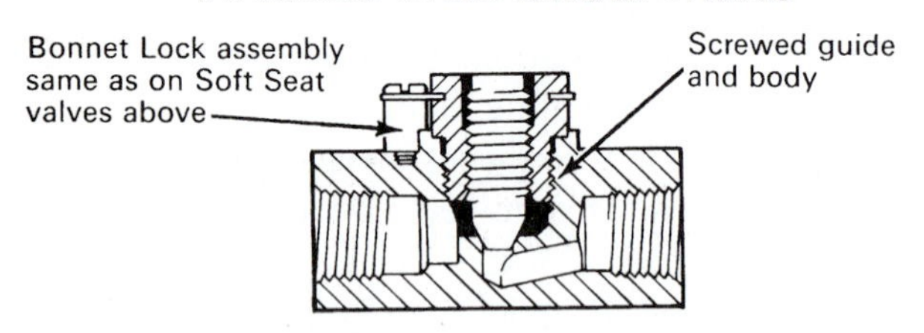

ALLOY STEEL NEEDLE VALVES

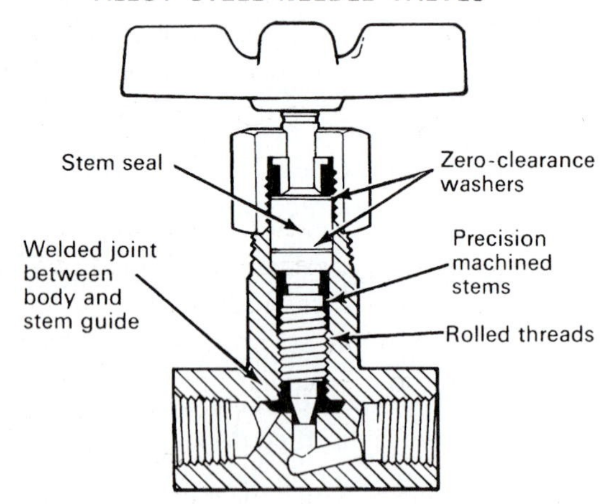

Fig. 1-25 *Different types of needle valves.* (Marsh)

This extends the gage life while helping to maintain accuracy and readability.

Effects of Temperature on Gage Performance

Because of the effects of temperature on the elasticity of the tube material, the accuracy may change. Gages calibrated at 75°F (23.9°C) may change by more than 2 percent at:

- *Full scale* (FS) below –30°F (–34°C)
- Above 150° F (65.6°C)

Care of Gages

The pressure gage is one of the service person's most valuable tools. Thus, the quality of the work depends on the accuracy of the gages used. Most are precision-made instruments that will give many years of dependable service if properly treated.

The test gage set should be used primarily to check pressures at the low and high side of the compressor. The ammonia gage should be used with a steel Bourdon tube tip and socket to prevent damage.

Once you become familiar with the construction of your gages, you will be able to handle them more efficiently. The internal mechanism of a typical gage is shown in Fig. 1-23. The internal parts of a vapor tension thermometer are very similar.

Drawn brass is usually used for case material. It does not corrode. However, some gages now use high-impact plastics. A copper alloy Bourdon tube with a brass tip and socket is used for most refrigerants. Stainless steel is used for ammonia. Engineers have found that moving parts involved in rolling contact will last longer if made of unlike metals. That is why many top-grade refrigeration gages have bronze-bushed movements with a stainless steel pinion and arbor.

The socket is the only support for the entire gage. It extends beyond the case. The extension is long enough to provide a wrench flat enough for use in attaching the gage to the pressure source. Never twist the case when threading the gage into the outlet. This could cause misalignment or permanent damage to the mechanism.

NOTE: Keep gages and thermometers separate from other tools in your service kit. They can be knocked out of alignment by a jolt from a heavy tool.

Most pressure gages for refrigeration testing have a small orifice restriction screw. The screw is placed in the pressure inlet hole of the socket. It reduces the effects of pulsations without throwing off pressure readings. If the orifice becomes clogged, the screw can be easily removed for cleaning.

Gage Recalibration

Most gages retain a good degree of accuracy in spite of daily usage and constant handling. Since they are precision instruments, however, you should set up a regular program for checking them. If you have a regular program, you can be sure that you are working with accurate instruments.

Gages will develop reading errors if they are dropped or subjected to excessive pulsation, vibration, or a violent surge of overpressure. You can restore a gage to accuracy by adjusting the recalibration screw. See Fig. 1-26. If the gage does not have a recalibration screw, remove the ring and glass. Connect the gage you are testing and a gage of known accuracy to the same pressure source. Compare readings at midscale. If the gage under test is not reading the same as the test gage, remove the pointer and reset.

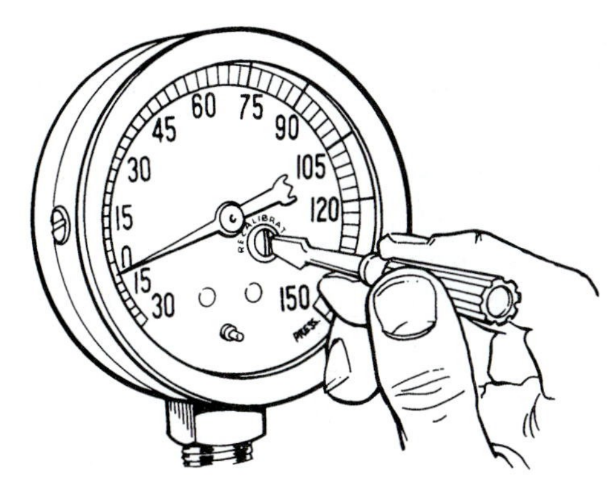

Fig. 1-26 *Recalibrating a gage. (Marsh)*

This type of adjustments on the pointer acts merely as a pointer-setting device. It does not reestablish the original even increment (linearity) of pointer travel. This becomes more apparent as the correction requirement becomes greater.

If your gage has a recalibrator screw on the face of the dial, as in Fig. 1-26, remove the ring and glass. Relieve all pressure to the gage. Turn the recalibration screw until the pointer rests at zero.

The gage will be as accurate as when it left the factory if it has a screw recalibration adjustment. Resetting the dial to zero restores accuracy throughout the entire range of dial readings.

If you cannot calibrate the gage by either of these methods, take it to a qualified specialist for repair.

THERMOMETERS

Thermometers are used to measure heat. A thermometer should be chosen according to its application. Consider first the kind of installation—direct mounting or remote reading.

If remote readings are necessary, then the vapor tension thermometer is best. It has a closed, filled Bourdon tube. A bulb is at one end for temperature sensing. Changes in the temperature at the bulb result in pressure changes in the fill medium. Remote reading thermometers are equipped with 6 ft of capillary tubing as standard. Other lengths are available on special order.

The location of direct or remote reading is important when choosing a thermometer. Four common types of thermometers are used to measure temperature:

- Pocket thermometer
- Bimetallic thermometer
- Thermocouple thermometer
- Resistance thermometer

Pocket Thermometer

The pocket thermometer depends upon the even expansion of a liquid. The liquid may be mercury or colored alcohol. This type of thermometer is versatile. It can be used to measure temperatures of liquids, air, gas, and solids. It can be strapped to the suction line during a superheat measurement. For practical purposes, it can operate wet or dry. This type of thermometer can withstand extremely corrosive solutions and atmospheres.

When the glass thermometer is read in place, temperatures are accurate if proper contact is made between the stem and the medium being measured. Refrigeration service persons are familiar with the need to attach the thermometer firmly to the suction line when taking superheat readings. See Fig. 1-27A and B. Clamps are available for this purpose. One thing should be kept in mind, that is, the depth at which the thermometer is to be immersed in the medium being measured. Most instruction sheets point out that for liquid measurements the thermometer should be immersed so many inches. When used in a duct, a specified length of stem should be in the airflow. Dipping only the bulb into a glass of water does not give the same readings as immersing to the prescribed length.

Shielding is frequently overlooked in the application of the simple glass thermometer. The instrument should be shielded from radiated heat. Heating repairpersons often measure air temperature in the furnace bonnet. Do not place the thermometer in a position where it receives direct radiation from the heat exchanger surfaces. This causes erroneous readings.

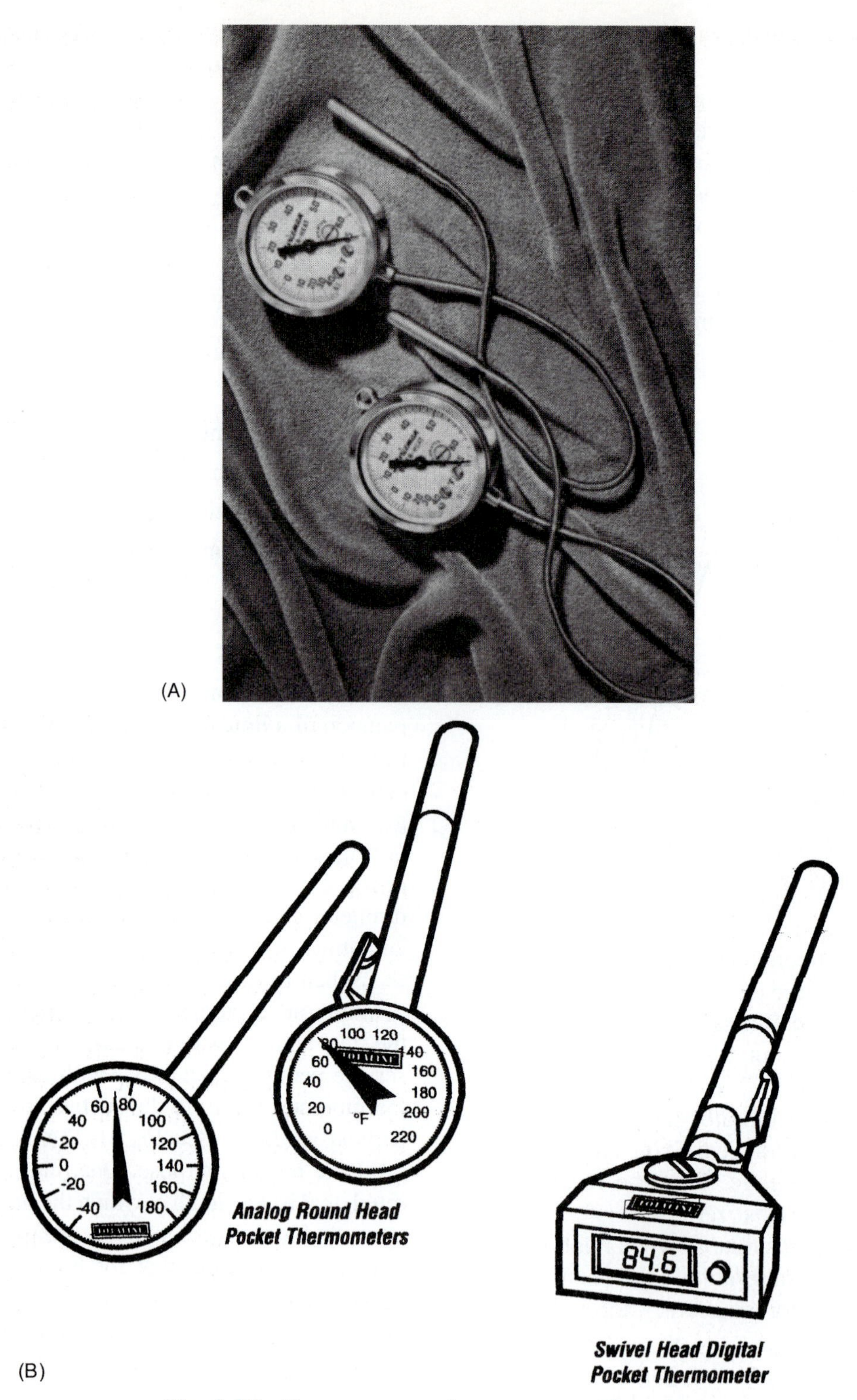

Fig. 1-27 *Thermometers used to measure superheat.* *(Marsh)*

The greatest error in the use of the glass thermometer is that it is often not read in place. It is removed from the outlet grille of a packaged air conditioner. Then it is carried to eye level in the room at ambient temperatures. Here it is read a few seconds to a minute later. It is read in a temperature different from that in which it was measured.

A liquid bath temperature reading is taken with the bulb in the bath. It is then left for a few minutes, immersed, and raised so that it can be read.

A simple rule helps eliminate incorrect readings:

- Read glass thermometers while they are actually in contact with the medium being measured.
- If a thermometer must be handled, do so with as little hand contact as possible. Read the thermometer immediately!

A recurring problem with mercury-filled glass thermometers is separation of the mercury column. See Fig. 1-28. This results in what is frequently termed as a *split thermometer*. The cause of the column's splitting is always rough handling. Such handling cannot be avoided at all times in service work. Splitting does not occur in thermometers that do not have a gas atmosphere over the mercury. Such thermometers allow the mercury to move back and forth by gravity, as well as temperature change. Such thermometers may not be used in other than vertical positions.

A split thermometer can be repaired. Most service thermometers have the mercury reservoir at the bottom of the tube. In this case, cool the thermometer bulb in shaved ice. This draws the mercury to the lower part of the reservoir. Add more ice or salt to lower the temperature, if necessary. With the thermometer in an upright position, tap the bottom of the bulb on a padded piece of paper or cloth. The entrapped gas causing the split column should then rise to the top of the mercury. After the column has been joined, test the service thermometer against a standard thermometer. Do this at several service temperatures.

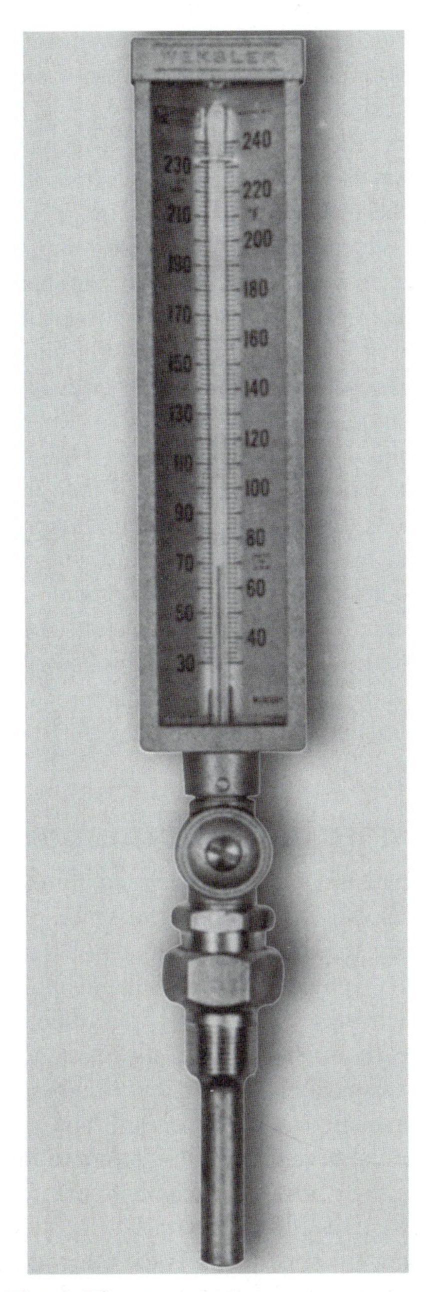

Fig. 1-28 *Mercury thermometer.* (Weksler)

Bimetallic Thermometers

Dial thermometers are actuated by bimetallic coils, mercury, vapor pressure, or gas. They are available in varied forms that allow the dial to be used in a number of locations. See Fig. 1-29. The sensing portion of the instrument may be located somewhere else. The dial can be read in a convenient location.

Bimetallic thermometers have a linear dial face. There are equal increments throughout any given dial ranges. Dial ranges are also available to meet higher temperature measuring needs. Ranges up to 1000°F (537.8°C) are available. In four selected ranges, dials giving both Celsius and Fahrenheit readings are available. Bimetallic thermometers are economical. There is no need for a machined movement or gearing. The temperature-sensitive bimetallic element is connected directly to the pointer. This type of thermometry is well adapted to measuring the temperature of a surface. Dome-mounted thermal protectors actually react to the surface temperature of the compressor skin. These thermometers are used where direct readings need to be taken, such as on:

- Pipelines
- Tanks
- Ovens
- Ducts
- Sterilizers
- Heat exchangers
- Laboratory temperature baths

Fig. 1-29 *Dial-type thermometer.* (Weksler)

The simplest type of dial thermometer is a stem. The stem is inserted into the medium to be measured. With the stem immersed 2 in. in liquids and 4 in. in gases, this thermometer gives reasonably accurate readings.

Although dial thermometers have many uses, there are some limitations. They are not as universally applicable as the simple glass thermometer. When ordering a dial thermometer, specify the stem length, scale range, and medium in which it will be used.

One of the advantages of bimetallic thermometry is that the thermometer can be applied directly to surfaces. It can be designed to take temperatures of pipes from 0.5 through 2 in.

In operation, the bimetallic spiral is closely coupled to the heated surface that is to be measured. The thermometer is held fast by two permanent magnets. One manufacturer claims their type of thermometer reaches stability within 3 min. Its accuracy is said to be plus or minus 2 percent in working ranges.

A simple and inexpensive type of bimetallic thermometer scribes temperature travel on a load of food in transit. It can be used also to check temperature variations in controlled industrial areas. The replacement chart gives a permanent record of temperature variations during the test period.

Bimetallic drives are also used in control devices. For example, thermal overload sensors for motors and other electrical devices use bimetallic elements. Other examples will be discussed later.

Thermocouple Thermometers

Thermocouples are made of two dissimilar metals. Once the metals are heated, they give off an EMF (*electromotive force* or voltage). This electrical energy can be measured with a standard type of meter designed to measure small amounts of current. The meter can be calibrated in degrees, instead of amperes, milliamperes, or microamperes.

In use, the thermocouples are placed in the medium that is to be measured. Extension wires run from the thermocouple to the meter. The meter then gives the temperature reading at the remote location.

The extension wires may be run outside closed chests and rooms. There is no difficulty in closing a door, and the wires will not be pinches. On air-conditioning work, one thermocouple may be placed in the supply grille and another in the return grille. Readings can be taken seconds apart without handling a thermometer.

Thermocouples are easily taped onto the surface of pipes to check the inside temperature. It is a good idea to insulate the thermocouple from ambient and radiated heat. Although this type of thermometer is rugged, it should be handled with care. It should not be handled roughly. Thermocouples should be protected form corrosive chemicals and fumes. Manufacturer's instructions for protection and use are supplied with the instrument.

Resistance Thermometers

One of the newer ways to check temperature is with a thermometer that uses a resistance- sending element. An electrical sensing unit may be made of a thermistor. A *thermistor* is a piece of material that changes resistance rapidly when subjected to temperature changes. When heated, the thermistor lowers its resistance. This decrease in resistance makes a circuit increase its current. A meter can be inserted in the circuit. The change in current can be calibrated against a standard thermometer. The scale can be marked to read temperature in degrees Celsius or degrees Fahrenheit.

Another type of resistance thermometer indicates the temperature by an indicating light. The resistance-sensing bulb is placed in the medium to be measured. The bridge circuit is adjusted until the light comes on. The knob that adjusts the bridge circuit is calibrated in degrees Celsius or degrees Fahrenheit. The knob then shows the temperature. The sensing element is just one of the resistors in the bridge circuit. The bridge circuit is described in detail in Chap. 3.

There is the possibility of having practical precision of ±1°F (0.5°C). In this type of measurement, the range covered is –325 to 250°F (–198 to 121°C). A unit may be used for deep freezer testing, for air-conditioning units, and for other work. Response is rapid. Special bulbs are available for use in rooms, outdoors, immersion, on surfaces, and in ducts.

Superheat Thermometer

The superheat thermometer is used to check for correct temperature differential of the refrigerator gas. The inlet and outlet side of the evaporator coil have to be measured to obtain the two temperatures. The difference is obtained by subtracting.

Test thermometers are available in boxes. See Fig. 1-30. The box protects the thermometer. It is important to keep the thermometer in operating condition. Several guidelines must be followed. Figure 1-31 illustrates how to keep the test thermometer in good working condition.

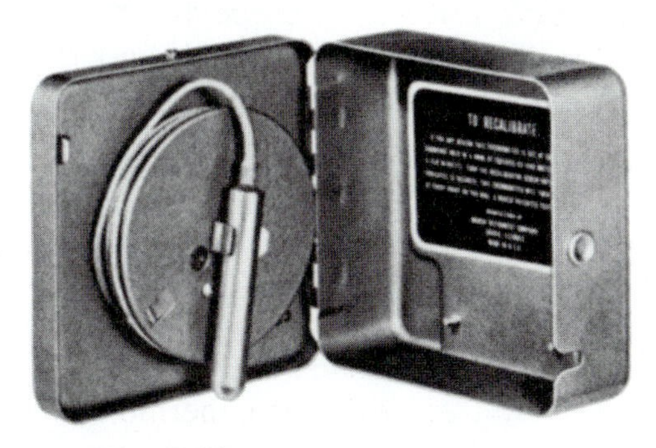

Fig. 1-30 *Test thermometer. (Marsh)*

Preventing kinks in the capillary is important. Keep the capillary clean by removing grease and oil. Clean the case and crystal with a mild detergent.

SUPERHEAT MEASUREMENT INSTRUMENTS

Superheat plays an important role in refrigeration and air-conditioning service. For example, the thermostatic expansion valve operates on the principle of superheat. In charging capillary tube systems, the superheat measurement must be carefully watched. The suction line superheat is an indication of whether the liquid refrigerant is flooding the compressor from the suction side. A measurement of zero superheat is a definite indicator that liquid is reaching the compressor. A measurement of 6 to 10°F (–14.4 to –12.2°C) for the expansion valve system and 20°F (6.7°C) for capillary tube system indicates that all refrigerant is vaporized before entering the compressor.

The superheat at any point in a refrigeration system is found by first measuring the actual refrigerant temperature at that point using an electronic thermometer. Then the boiling point temperature of the refrigerant is found by connecting a compound pressure gage to the system and reading the boiling temperature from the center of the pressure gage. The difference between the actual temperature and the boiling point temperature is superheat. If the superheat is zero, the refrigerant must be boiling inside. Then, there is a good chance that some of the refrigerant is still liquid. If the superheat is greater than zero, by at least 5°F or better, then the refrigerant is probably past the boiling point stage and is all vapor.

The method of measuring superheat described here has obvious faults. If there is no attachment for a pressure gage at the point in the system where you are measuring superheat, the hypothetical boiling temperature cannot be found. To determine the superheat at such a point, the following method can be used. This method is particularly useful for measuring the refrigerant superheat in the suction line.

Instead of using a pressure gage, the boiling point of the refrigerant in the evaporator can be determined by measuring the temperature in the line just after the expansion valve where the boiling is vigorous. This can be done with any electronic thermometer. See Fig. 1-32. As the refrigerant heats up through the evaporator and the suction line, the actual temperature of the refrigerant can be measured at any point along the suction line. Comparison of these two temperatures gives a superheat measurement sufficient for field service

KEEP YOUR THERMOMETERS WORKING BY FOLLOWING THESE STEPS

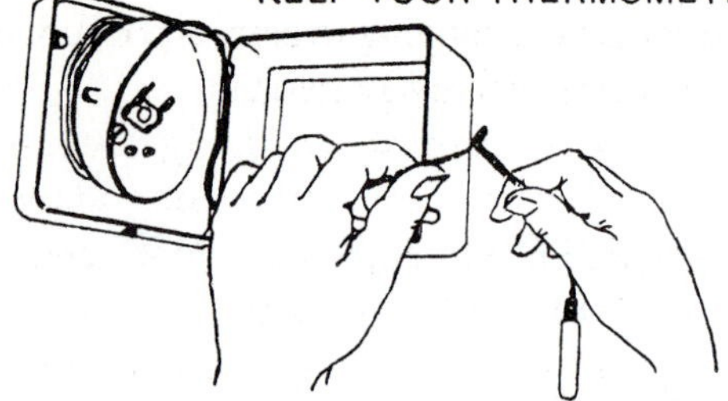

1. DO NOT CUT, TWIST, OR KINK CAPILLARY.

   When capillary becomes kinked, remove the kink by carefully bending the capillary in a direction opposite to the kink.

   To straighten twisted capillary, grasp the tubing in both hands and untwist short sections at a time, being careful not to break the fine wire armor.

   Cutting the capillary will release the charge and render the instrument useless.

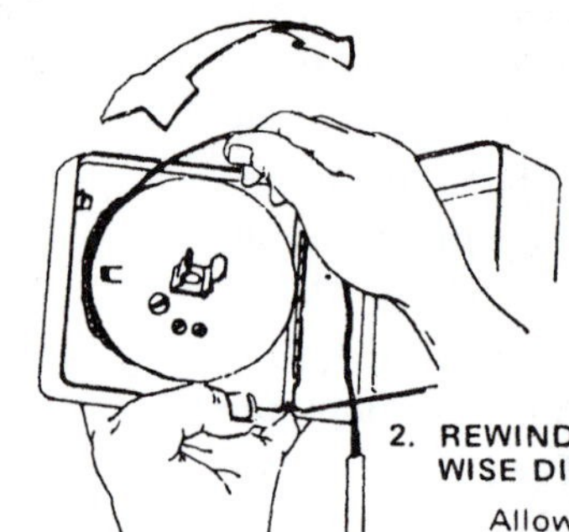

2. REWIND CAPILLARY CAREFULLY IN CLOCKWISE DIRECTION.

   Allow bulb to hang free and turn with winding.

   Keep bulb in holding clip when thermometer is not in use. Clip will turn in any direction to receive bulb.

3. UNREEL CAPILLARY CAREFULLY AND PLACE IN SLOT AT SIDE OF CASE BEFORE CLOSING.

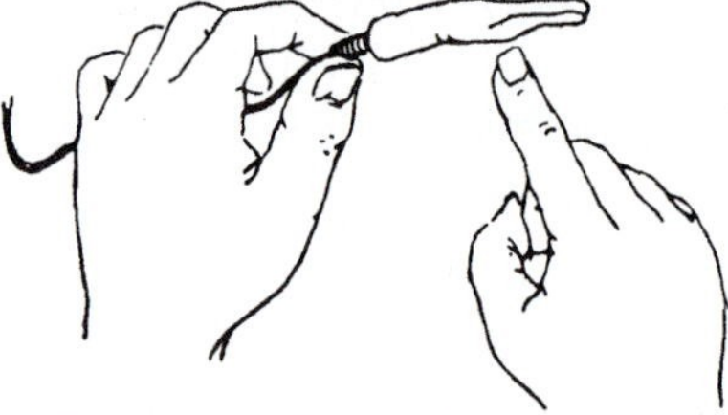

4. DO NOT BEND OR FLATTEN BULB.

   Distortion of the bulb will result in false reading.

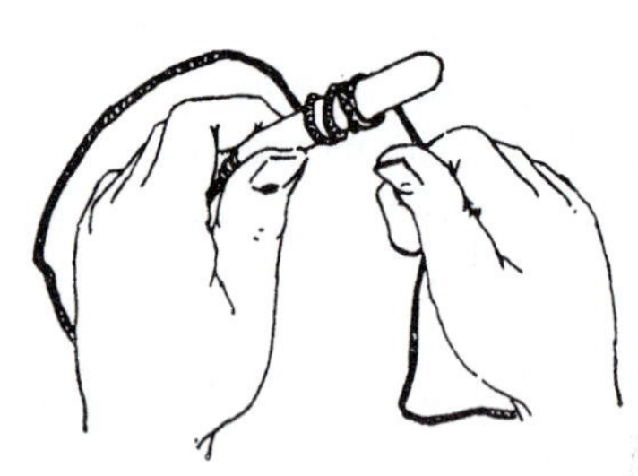

5. DO NOT TWIST CAPILLARY AROUND BULB TO HOLD IN POSITION.

   A small piece of tape will usually be adequate to hold bulb in place.

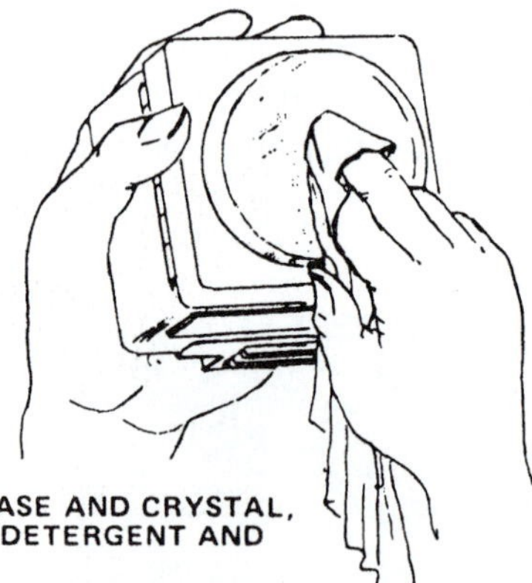

6. TO CLEAN CASE AND CRYSTAL, USE A MILD DETERGENT AND SOFT RAG.

7. TO CLEAN OIL OR GREASE FROM CAPILLARY OR BULB, DIP IN CARBON TETRACHLORIDE AND WIPE WITH SOFT RAG.

8. MAGNETIC BASE UNIT USED FOR CONVENIENT POSITION MOUNTING OF THERMOMETER.

Fig. 1-31 *How to take care of the thermometer?* (Marsh)

Fig. 1-32 *Hand-held electronic thermometer.* (Amprobe)

unless a distributor-metering device is used or the evaporator is very large with a great amount of pressure drop across the evaporator.

By using the meter shown in Fig. 1-33, it is possible to read superheat directly, using the temperature differential feature. Strap one end of the differential probe to the outlet of the metering device. Strap the other end to the point on the suction line where the superheat measure is to be taken. Turn the meter to temperature differential and the superheat will be directly read on the meter.

Figure 1-34 illustrates the way superheat works. The bulb "opening" force (F-l) is caused by bulb temperature. This force is balanced against the system back-pressure (F-2) and the valve spring force (F-3). The force holds the evaporator pressure within a range that will vaporize the entire refrigerant just before it reaches the upper part or end of the evaporator.

The method of checking superheat is shown in Fig. 1-35. The procedure is as follows:

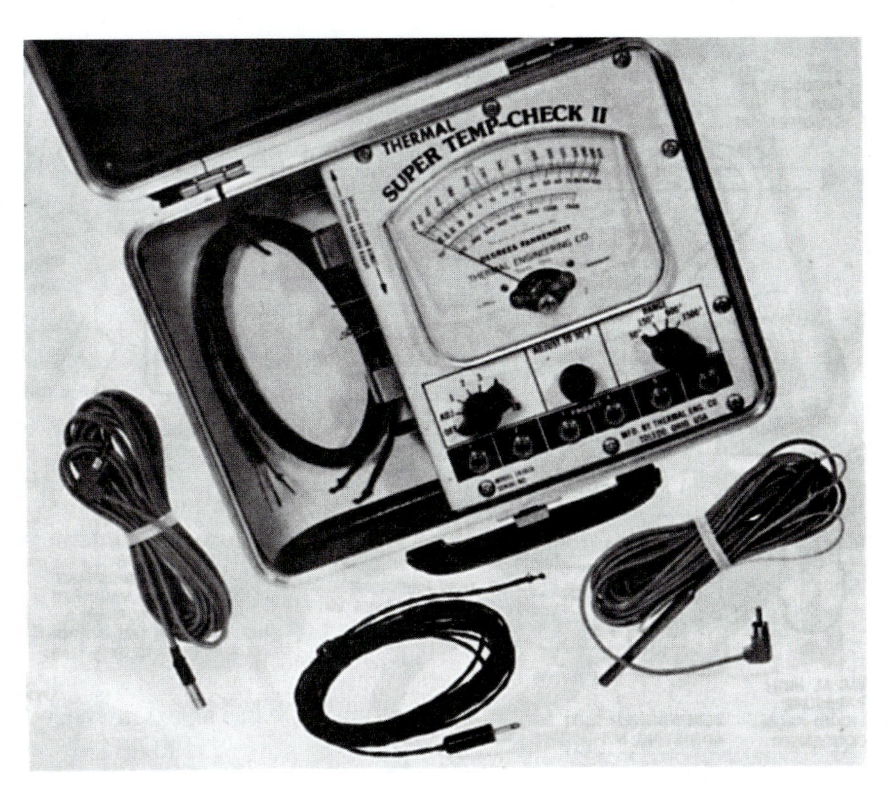

Fig. 1-33 *Electronic thermometer for measuring superheat. The probes are made of thermo-couple wire. They can be strapped on anywhere with total contact with the surface. This thermometer covers temperatures from –50° to 1500°F on four scales. The temperature difference between any two points directly means it can read superheat directly. It is battery operated and has a ±2 percent accuracy on all ranges. Celsius scales are available.* (Thermal Engineering)

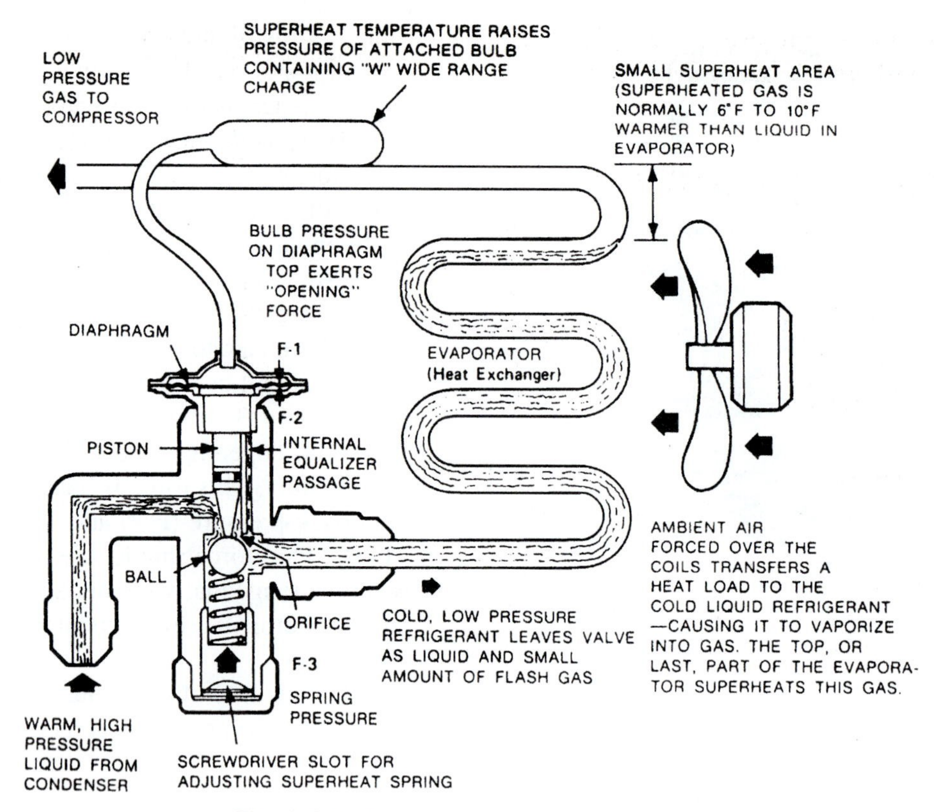

Fig. 1-34 *How superheat works.* (Parker-Hannefin)

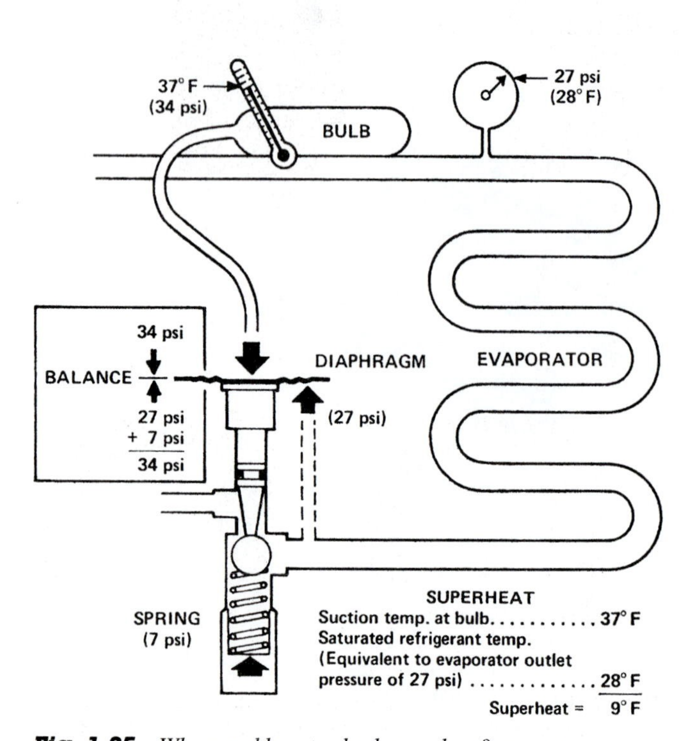

Fig. 1-35 *Where and how to check superheat?* (Parker-Hannefin)

1. Measure the temperature of the suction line at the bulb location. In the example, the temperature is 37°F.
2. Measure the suction line pressure. In the example, the suction line pressure is 27 psi.
3. Convert the suction line pressure to the equivalent saturated (or liquid) evaporator temperature by using a standard temperature-pressure chart (27 psi = 28°F).
4. Subtract the two temperatures. The difference is superheat. In this case, superheat is found by the formula: 37°F – 28°F = 9°F

Suction pressure at the bulb may be obtained by either of the following methods:

- If the valve has an external equalizer line, the gage in this line may be read directly.
- If the valve is internally equalized, take a pressure gage reading at the compressor base valve. Add to this the estimated pressure drop between the gage and the bulb location. The sum will approximate the pressure at the bulb.

The system should be operating normally when the superheat is between 6 and 10°F (−14.4 and −12.2°C).

HALIDE LEAK DETECTORS

Not too long ago leaks were detected by using soap bubbles and water. If possible, the area of the suspected leak was submerged in soap water. Bubbles pinpointed the leak area. If the unit or suspected area was not easily submerged in water then it was coated with soap solution. In addition, where the leak was covered with soap, bubbles would be produced. These indicated the location of the leak. These methods are still used today in some cases. However, it is now possible to obtain better indications of leaks with electronic equipment with halide leak detectors.

Halide leak detectors are used in the refrigeration and air-conditioning industry. They are designed for locating leaks and noncombustible halide refrigerant gases. See Figs. 1-36 and 1-37.

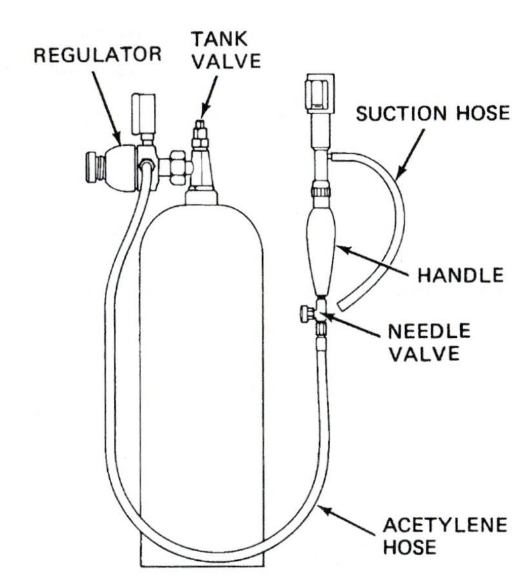

Fig. 1-37 *Halide leak detector for use with an MC tank.* (Union Carbide)

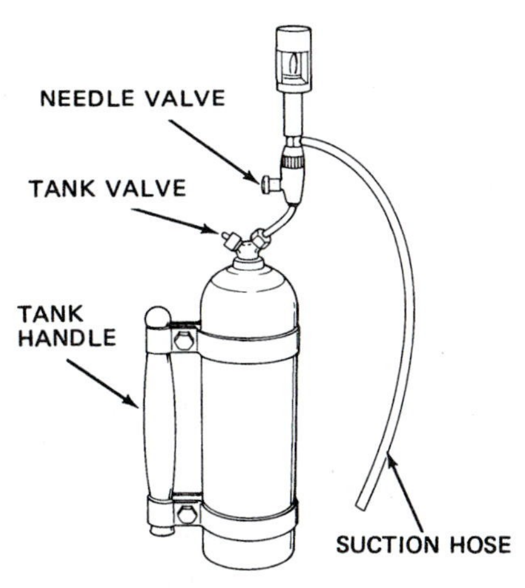

Fig. 1-36 *Halide leak detector for use with a B tank.* (Union Carbide)

The supersensitive detector will detect the presence of as little as 20 parts per million of refrigerant gases. See Fig. 1-38. Another model will detect 100 parts of halide gas per million parts of air.

Setting Up

The leak detector is normally used with a standard torch handle. The torch handle has a shut-off valve. Acetylene can be supplied by a "B" tank (40 ft^3) or MC tank (10 ft^3). In either case, the tank must be equipped with a pressure-reducing regulator; the torch handle is connected to the regulator by a suitable length of fitted acetylene hose. See Fig. 1-36.

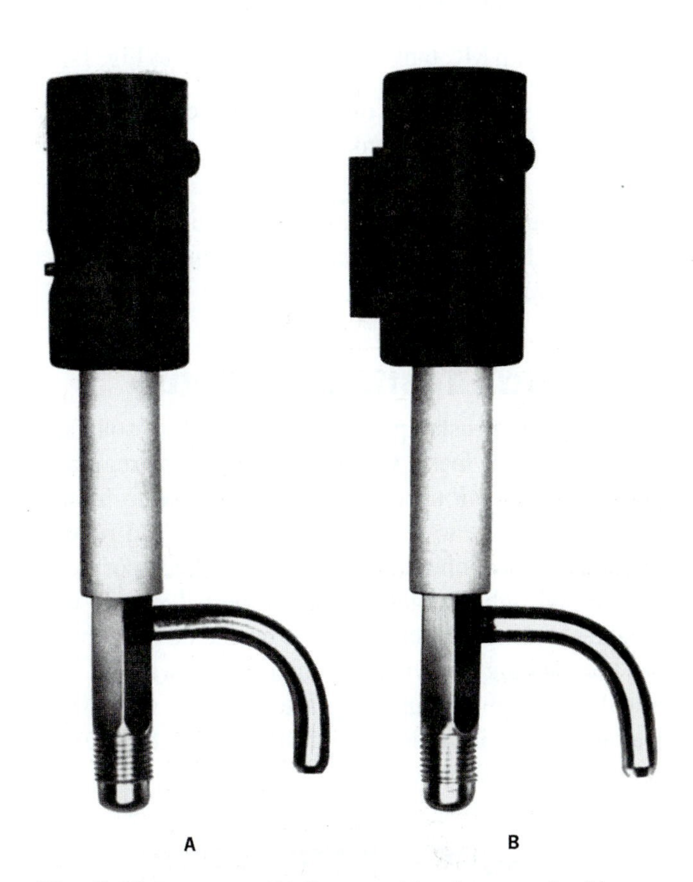

Fig. 1-38 Detectors. *(A) Supersensitive detector of refrigerant gases. This detects 20 parts per million. (B) Standard model detector torch. This detects 100 parts per million.* (Union Carbide)

An alternate setup uses an adapter to connect the leak detector stem to an MC tank. No regulator is required. The tank must be fitted with a handle. See Fig. 1-37.

In making either setup, be sure all seating sources are clean before assembling. Tighten all connections securely. Use a wrench to tighten hose and regulator connections. If you use the "B" tank setup, be sure to follow the instructions supplied with the torch handle and regulator.

Lighting

Setup with tank, regulator, and torch handle. Refer to Fig. 1-36.

- Open the tank valve one-quarter turn, using a P-O-L tank key.
- Be sure the shut-off valve on the torch handle is closed. Then, adjust the regulator to deliver 10 psi. Do this by turning in the pressure-adjusting screw until the "C" marking on the flat surfaces of the screw is opposite the face of the front cap. Test for leaks.
- Open the torch handle shut-off valve and light the gas above the reaction plate. Use a match or taper.
- Adjust the torch until a steady flame is obtained.

Setup with MC tank and adaptor. Refer to Fig. 1-37.

- With the needle valve on the adaptor closed tightly, just barely open the tank valve, using a P-O-L tank key. Test for leaks.
- Open the adapter needle valve about one-quarter turn. Light the gas above the reaction plate. Use a match or taper.

Leak Testing the Setup

Using a small brush, apply a thick solution of soap and water to test for leaks. Check for leaks at the regulator and any connection point. Check the hose to handle connection, hose to regulator connection, and regulator or adaptor connection. If you find a leak, correct it before you light the gas. A leak at the valve stem of a small acetylene tank can often be corrected by tightening the packing nut with a wrench. If this will not stop a leak, remove the tank. Tag it to indicate valve stem leakage. Place it outdoors in a safe spot until you can return it to the supplier.

Adjusting the Flame

Place the inlet end of the suction hose so that it is unlikely to draw in air to contaminate the refrigerant vapor. Adjust the needle valve on the adapter or torch handle until the pale blue outer envelope of the flame extends about 1 in. above the reaction plate. The inner cone of the flame, which should also be visible above the reaction plate, should be clear and sharply defined.

If the outer envelope of the flame, when of proper length, is yellow, not pale blue, the hose is picking up refrigerant vapors. There may also be some obstruction in the suction hose. Make sure the suction tube is not clogged or bent sharply. If the suction tube is clear, shut off the flame. Close the tank valve. Disconnect the leak detector from the handle or adaptor. Check for dirt in the filter screw or mixer disc. See Fig. 1-39. Use a $^{1}/_{8}$ in. socket key (Allen wrench) to remove or replace the filter screw. This screw retains the mixer disc.

Detecting Leaks

To explore the leaks, move the end of the suction hose around all points where there might be leaks. Be careful not to kink the suction hose.

Watch for color changes in the flame as you move the end of the suction hose:

- With the model that has a large opening in the flame shield (wings on each side), a small leak will change the color of the outer flame to a yellow or an orange-yellow hue. As the concentration of halide gas increases, the yellow will disappear. The lower part of the flame will become a bright, light blue. The top of the flame will become a vivid purplish blue.
- With the model that has no wings alongside the flame shield opening, small concentrations of halide gas will change the color. A bright blue-green outer flame indicates a leak. As the concentration of the halide gas increases, the lower part of the flame will lose its greenish tinge. The upper portion will become a vivid purplish blue.
- Watch for color intensity changes. The location of small color leaks can be pinpointed rapidly. Color in the flame will disappear almost instantly after the intake end of the hose has passed the point of leakage. With larger leaks, you will have to judge the point of leakage. Note the color change from yellow to purple-blue or blue-green to blue-purple, depending on the model used.

Maintenance

With intensive usage, an oxide scale may form on the surface of the reaction plate. Thus, sensitivity is reduced. Usually this scale can be easily broken away from the late surface. If you suspect a loss in sensitivity,

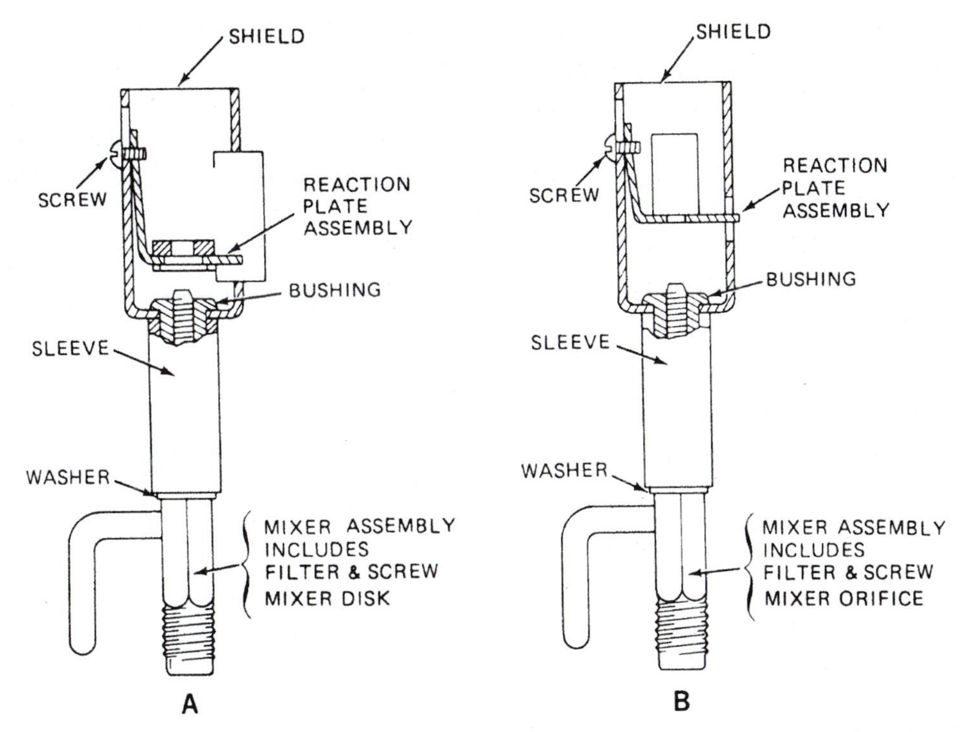

Fig. 1-39 *Position of filter screw and mixer disc on Prest-O-Lite halide leak detector (A) Standard model. (B) Supersensitive model.*

remove the reaction plate. Scrape its surface with a knife or screwdriver blade, or install a new plate.

ELECTRICAL INSTRUMENTS

Several electrical instruments are used by the air-conditioning serviceperson to see if the equipment is working properly. Studies show that the most trouble calls on heating and cooling equipment are electrical in nature.

The most frequently measured quantities are volts, amperes, and ohms. In some cases, wattage is measured to check for shorts and other malfunctions. A wattage meter is available. However, it must be used to measure *volt-amperes* (VA) instead of watts. To measure watts, it is necessary to use DC only or convert the VA to watts by using the power factor. The power factor times the volt-amperes produces the actual power consumed in watts. Since most cooling equipment use AC, it is necessary to convert to watts by this method.

A number of factors can be checked with electrical instruments. For example, electrical instruments can be used to check the flow rate from a centrifugal water pump, the condition of a capacitor, or the character of a start or run winding of an electric motor.

Ammeter

The ammeter is used to measure current. It can measure the amount of current flowing in a circuit. It may use one of a number of different basic meter movements to accomplish this. The most frequently used of the basic meter movements is the D'Arsonval type. See Fig. 1-40. It uses a permanent magnet and an electromagnet to determine circuit current. The permanent magnet is used as a standard basic source of magnetism. As the current flows through the coil of wire, it creates a magnetic field around it. This magnetic field is strong or weak, depending upon the amount of current flowing through it. The stronger the magnetic field created by the moving coil, the more it is repelled by the permanent magnet. This repelling motion is calibrated to read amperes, milliamperes (0.001 A), or microamperes (0.000001 A).

The D'Arsonval meter movement may also be used on AC when a diode is placed in series with the moving coil winding. The diode changes the AC to DC and the meter works as on DC. See Fig. 1-41. The dial or face of the instrument is calibrated to indicate the AC readings.

There are other types of AC ammeters. They are not always as accurate as the D'Arsonval, but they are effective. In some moving magnet meters, the coil

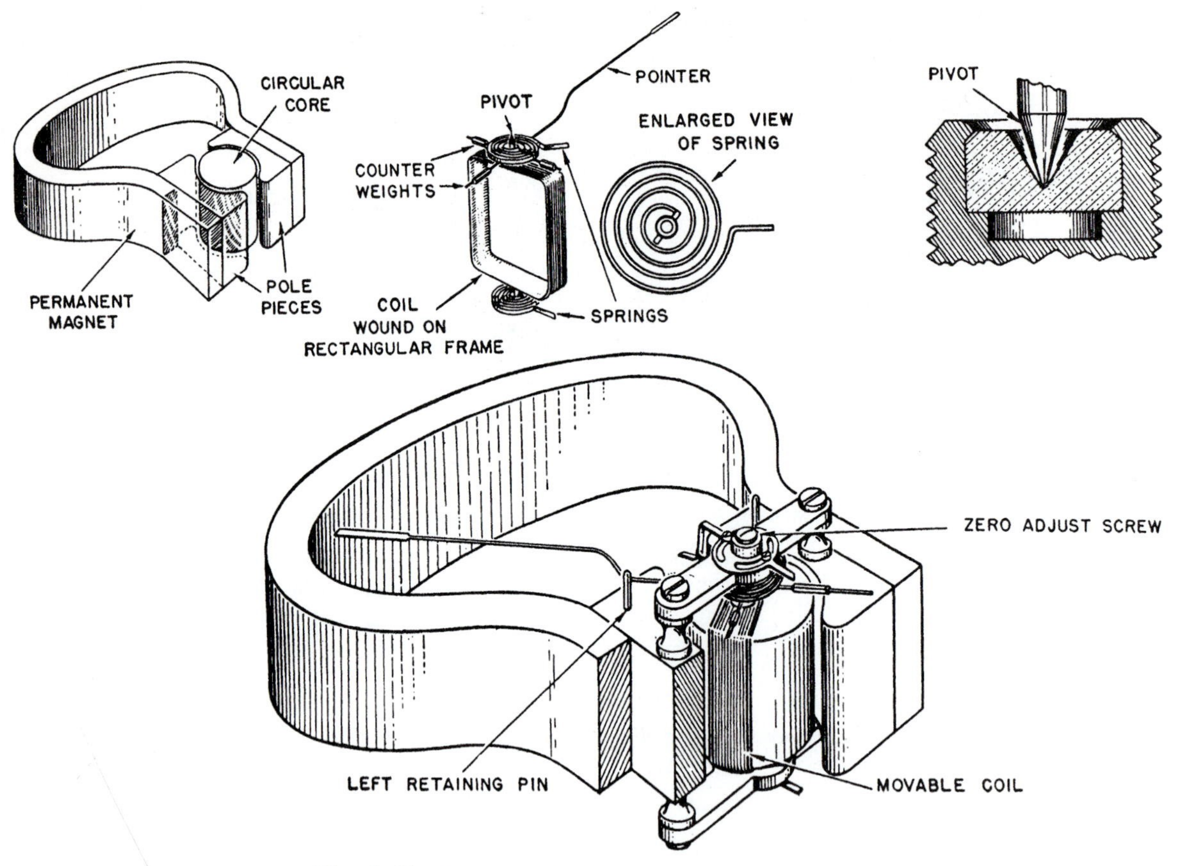

Fig. 1-40 *Moving coil (D'Arsonval) meter movement.*

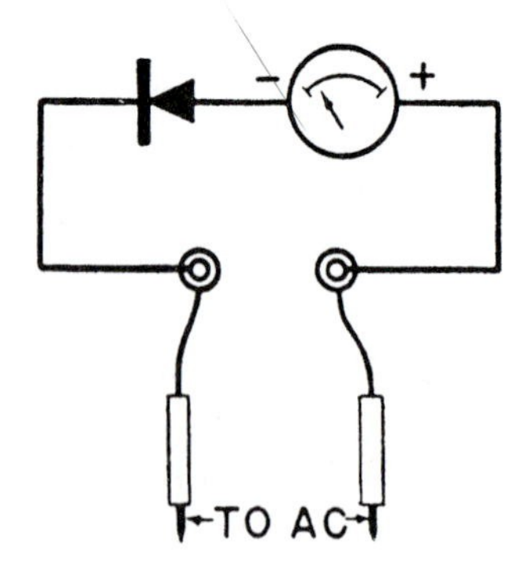

Fig. 1-41 *Diode inserted in the circuit with a D'Arsonval movement to produce an AC ammeter.*

is stationary and the magnet moves. Although rugged, this type is not as accurate as the D'Arsonval type meter.

The moving vane meter is useful in measuring current when AC is used. See Fig. 1-42.

The clamp-on ammeter has already been discussed. It has some limitations. However, it does have one advantage in that it can be used without having to break the line to insert it. Most ammeters must be connected in series with the consuming device. That means one line has to be broken or disconnected to insert the meter into the circuit.

The ampere reading can be used to determine if the unit is drawing too much current or insufficient current. The correct current amount is usually stamped on the nameplate of the motor or the compressor.

Starting and running amperes may be checked to see if the motor is operating with too much load or it is shorted. The flow rate of some pumps can be determined by reading the current the motor pulls. The load on the entire line can be checked by inserting the ammeter in the line. This is done by taking out the fuse and completing the circuit with the meter. Be careful.

If the ammeter has more than one range, it is best to start on the highest range and work down. The reading should be in or near the center of the meter scale for a more accurate reading. Make sure you have some idea what the current in the circuit should be before inserting the meter. Thus, the correct range—or, in some instances, the correct meter—can be selected.

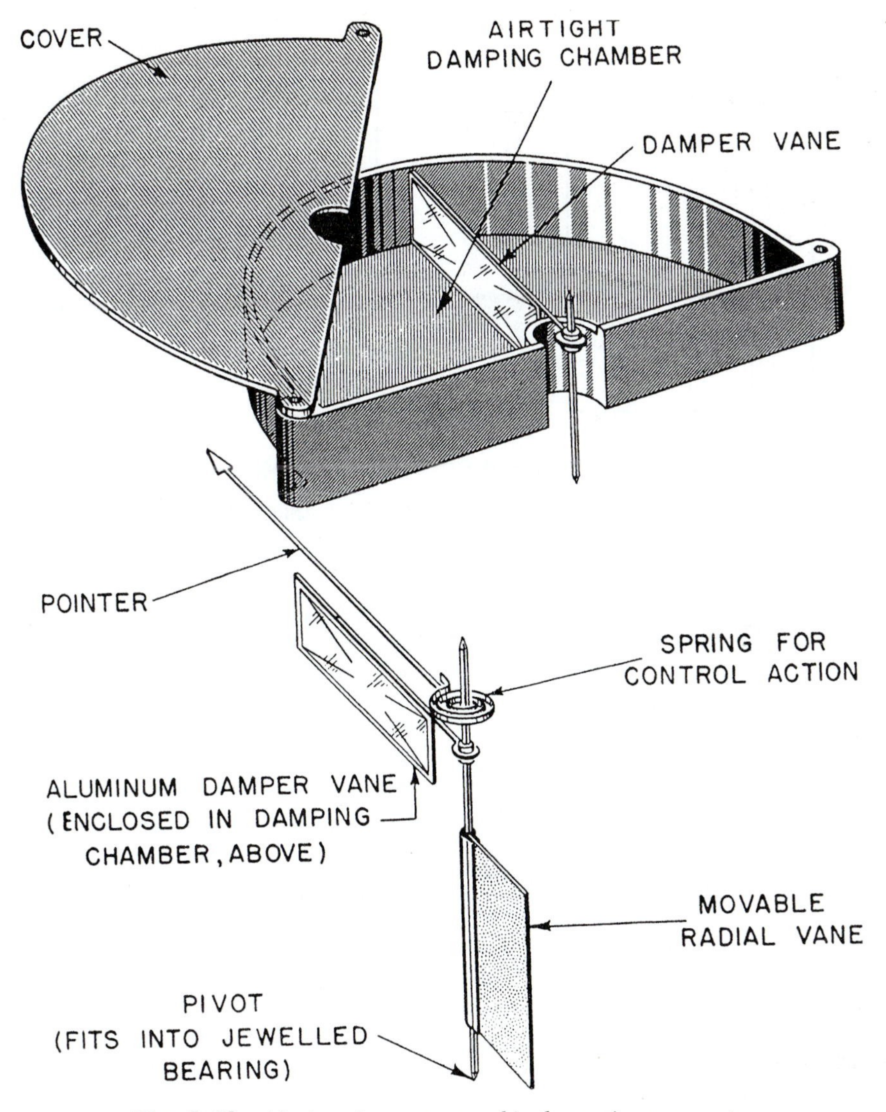

Fig. 1-42 *Air-damping system used in the moving-vane meter.*

Voltmeter

The *voltmeter* is used to measure voltage. Voltage is the electrical pressure needed to cause current to flow. The voltmeter is used across the line or across a motor or whatever is being used as a consuming device.

Voltmeters are nothing more than ammeters that are calibrated to read volts. There is, however, an important difference. The voltmeter has a very high internal resistance. That means very small amounts of current flow through its coil. See Fig. 1-43. This high resistance is produced by multipliers. Each range on the voltmeter has a different resistor to increase the resistance so the line current will not be diverted through it. See Fig. 1-44. The voltmeter is placed across the line, whereas the ammeter is placed in series. You do not have to break the line to use the voltmeter. The voltmeter has two leads. If you are measuring DC, you have to observe polarity. The red lead is the positive (+)

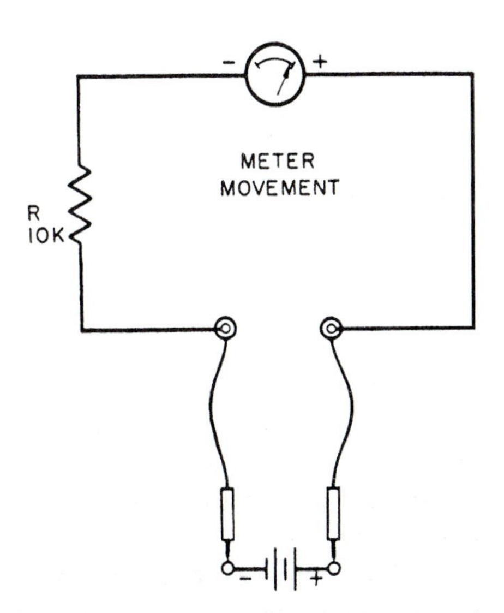

Fig. 1-43 *An ammeter with high resistance in series with the meter movement allow it to measure voltage.*

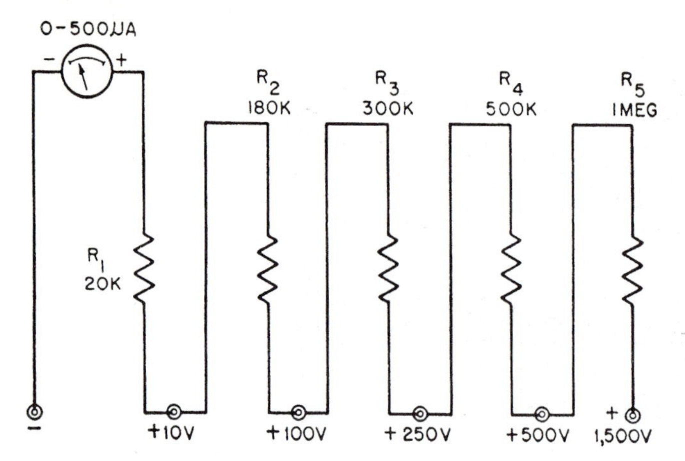

Fig. 1-44 *Different types of multirange voltmeters. This view shows the interior of the meter box or unit.*

and the black lead is the negative (–). However, when AC is used, it does not matter which lead is placed on which terminal. Using a D'Arsonval meter movement, voltmeters can be made with the proper diode to change AC to DC. Voltmeters can be made with a stationary coil and a moving magnet. Others types of voltmeters are available. They use various means of registering voltage.

If the voltage is not known, use the highest scale on the meter. Turn the range switch to appoint where the reading is in the midrange of the meter movement.

Normal line voltage in most locations is 120 V. When the line voltage is lower than normal, it is possible for the equipment to draw excessive current. This will cause overheating and eventual failure and/or burnout. The correct voltage is needed for the equipment to operate according to its designed specifications. The voltage range is usually stamped on the nameplate of the device. Some will state 208 V. This voltage is obtained from a three-phase connection. Most home or residential voltage is supplied at 120 V or 230 V. The range is 220 to 240 V for normal residential service. The size of the wire used to connect the equipment to the line is important. If the wire is too small, voltage will drop. There will be low voltage at the consuming device. For this reason a certified electrician with knowledge of the NEC should wire a new installation.

Ohmmeter

The *ohmmeter* measures resistance. The basic unit of resistance is the ohm (Ω). Every device has resistance. That is why it is necessary to know the proper resistance before trying to troubleshoot a device by using an ohmmeter. The ohmmeter has its own power supply. See Fig. 1-45. Do not use an ohmmeter on a line that

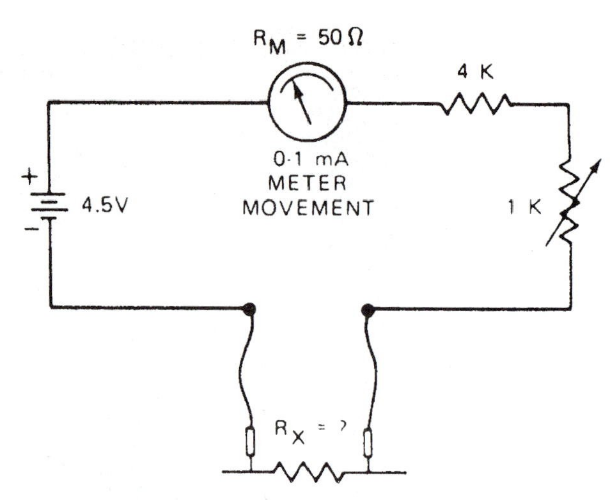

Fig. 1-45 *Internal circuit of an ohmmeter.*

is energized or connected to a power source of any voltage.

An ohmmeter can read the resistance of the windings of a motor. If the correct reading has been given by the manufacturer, it is then possible to see if the reading has changed. If the reading is much lower, it may indicate a shorted winding. If the reading is infinite (ω), it may mean there is a loose connection or an open circuit.

Ohmmeters have ranges. Figure 1-46 shows a meter scale. The $R \times 1$ range means the scale is read as is. If the $R \times 10$ range is used, it means that the scale reading must be multiplied by 10. If the $R \times 1000$ range is selected, then the scale reading must be multiplied by 1000. If the meter has a $R \times 1$ meg range, the scale reading must be multiplied by one million. A meg is one million.

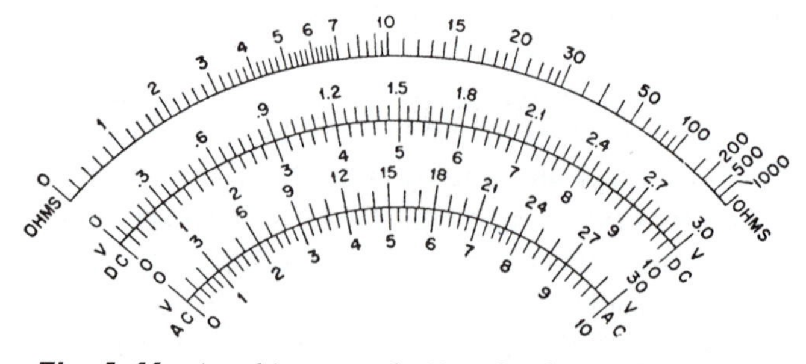

Fig. 1-46 *A multimeter scale. Note the ohms and volt scales.*

Multimeter

The *multimeter* is a combination of meters. See Fig. 1-47. It may have a voltmeter, an ammeter, and an ohmmeter

Fig. 1-47 *Two types of multimeters.*

in the same case. This is the usual arrangement for fieldwork. This way it is possible to have all three meters in one portable combination. It should be checked for each of the functions.

The snap-around meter uses its scale for a number of applications. It can be read current by snapping around the current carrying wire. If the leads are used, it can be used as a voltmeter or an ohmmeter. Remember that the power must be off to use the ohmmeter. This meter is mounted in its own case. It should be protected from shock and vibration just as any other sensitive instrument.

Wattmeter

The *wattmeter* is used to measure watts. However, when used on an AC line it measures volt-amperes. If watts are to be measured, the reading must be converted to watts mathematically. Multiply the reading on the wattmeter by the power factor (usually available on the nameplate) to obtain the reading in watts.

Wattmeters use the current and the voltage connections as with individual meters. See Fig. 1-48. One coil is heavy wire and is connected in series. It measures the current. The other connection is made in the same way as with the voltmeter and connected across the line. This coil is made of many turns of fine wire. It measures the voltage. By the action of the two magnetic fields, the current is multiplied by the voltage. Wattage is read on the meter scale.

The volt-ampere is the unit used to measure volts time amperes in an AC circuit. If a device has inductance (as in a motor) or capacitance (some motors have run-capacitors), the true wattage is not given on a wattmeter. The reading is in volt-amperes instead of watts. It is converted to watts by multiplying the reading by the power factor. A wattmeter reads watts only when it is connected to a DC circuit or to an AC circuit with resistance only.

The power factor is the ratio of true power to apparent power. Apparent power is what is read on a wattmeter on an AC line. True power is the wattage reading of DC. The two can be used to find the power factor. The power factor is the cosine of the phase angle. The power factor can be found by using a mathematical computation or a very delicate meter designed for the purpose. However, the power factor of equipment using alternating current is usually stamped on the nameplate of the compressor, the motor, or the unit itself.

Wattmeters are also used to test capacitors. Some companies provide charts to convert wattage ratings to microfarad ratings. The wattmeter can test the actual connection of the capacitor. The ohmmeter tells if the capacitor is good or bad. However, it is hard to indicate how a capacitor will function in a circuit with the

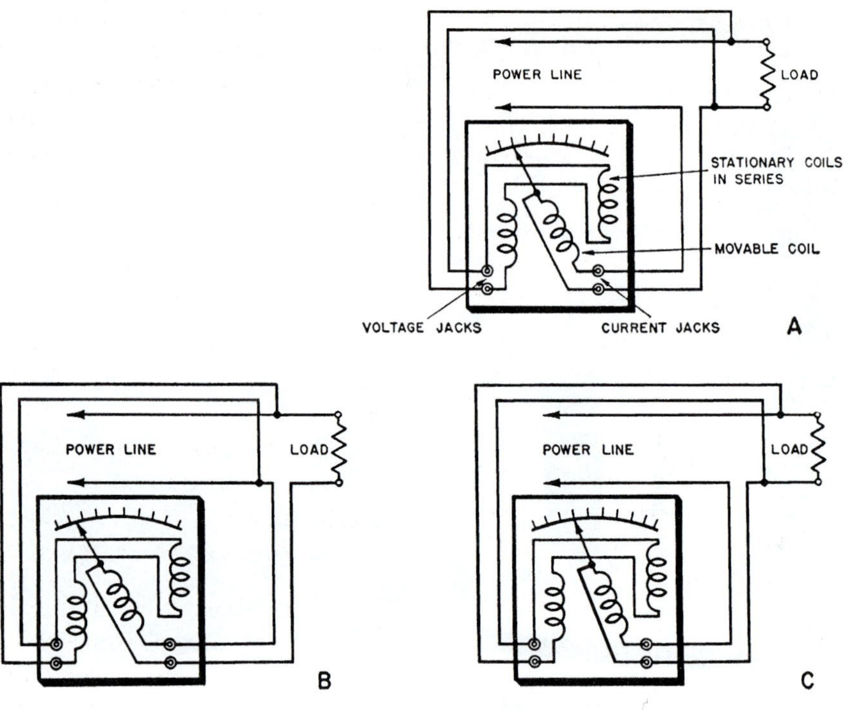

Fig. 1-48 *(A) Wattmeter connection for measuring input power. (B) Alternate wattmeter connection. (C) With load disconnected, uncompensated wattmeter measures its own power loss.*

voltage applied. This is why testing with the wattmeter is preferred.

OTHER INSTRUMENTS

Many types of meters and gages are available to test almost any quantity or condition. For example:

- Air–filter efficiency gages
- Air-measurement gages
- Humidity-measuring devices
- Moisture analyzers
- *British thermal unit* (Btu) meters

Vibration and sound meters and recorders are also available.

Air–Filter Efficiency Gages

Air measurements are taken in an air-distribution system. They often reveal the existence and location of unintentionally closed or open dampers and obstructions. Leaks in the ductwork and sharp bends are located this way.

Air measurements frequently show the existence of a blocked filter. Dirty and blocked filters can upset the balance of either a heating or cooling system. This is important whether it is in the home or in a large building.

Certain indicators and gages can be mounted in air plenums. They can be used to show that the filter has reached a point where it is restricting the airflow. An *air plenum* is a large space above the furnace heating or cooling unit.

Air-Measurement Instruments

The volume and velocity of air are important measurements in the temperature control industries. Proper amounts of air are indispensable to the best functioning of refrigeration cycles, regardless of the size of the system. Air-conditioning units and systems also rely upon volume and velocity for proper distribution of conditioned air.

Only a small number of contractors are equipped to measure volume and velocity correctly. The companies that are doing the job properly are in great demand. Professional handling of air volume and velocity ensures the efficient use of equipment. Large buildings are very much in need of the skills of air-balancing teams.

Some people attempt to obtain proper airflow by measuring air temperature. They adjust dampers and blowers speeds. However, they usually fail in their attempts to balance the airflow properly.

There are instruments available to measure air velocity and volume. Such instruments can accurately measure the low pressures and differentials involved in air distribution.

Draft gages do measure pressure. However, their specific application to air control makes it more appropriate to discuss them here, rather than under pressure gages. They measure pressure in inches of water. They come in several styles. The most familiar is the slanted type. It may be used either in the field or in the shop.

Meter type draft gages are better for fieldwork. They can be carried easily. They can sample air at various locations, with the meter box in one location.

Besides air pressure, it is frequently necessary to measure air volume, which is measured in cubic feet per minute or cfm. Air velocity is measured in feet per minute or fpm. The measure of airflow is still somewhat difficult. However, newer instruments are making accurate measurements possible.

Humidity-Measurement Instruments

Many hygroscopic (moisture absorbing) materials can be used as relative-humidity sensors. Such materials absorb or lose moisture until a balance is reached with the surrounding air. A change in material moisture content causes a dimensional change, and this change can be used as an input signal to a controller. Commonly used materials include:

- Human hair
- Wood
- Biwood combinations similar in action to a bimetallic temperature sensor
- Organic films
- Some fabrics, especially certain synthetic fabrics

All these have the drawbacks of slow response and large hysteresis effects. Accuracy also tends to be questionable unless they are frequently calibrated. Field calibration of humidity sensors is difficult.

Humidity is read in rh or relative humidity. To obtain the rh, it is necessary to use two thermometers. One thermometer is a dry bulb, the other is a wet bulb. The device used to measure rh is the sling psychrometer. It has two glass-stem thermometers. The wet bulb thermometer is moistened by a wick attached to the bulb. As the dual thermometers are whirled, air passes over them. The dry and wet bulb temperatures are recorded. Relative humidity is determined by:

- Graphs
- Slide rules
- Similar devices

Thin-film sensors are now available, which use an absorbent deposited on a silicon substrate such that the resistance or capacitance varies with relative humidity. They are quite accurate in the range of ±3 to 5 percent. They also have low maintenance requirements.

Stationary Psychrometers Stationary psychrometers take the same measurements as sling psychrometers. They do not move. However, they use a blower or fan to move the air over the thermometer bulbs.

For approximate rh readings, there are metered devices. They are used on desks and walls. They are not accurate enough to be used in engineering work

Humidistats, which are humidity controls, are used to control humidifiers. They operate the same way as thermometers in closing contacts to complete a circuit. They do not use the same sensing element, however.

Moisture Analyzers It is sometimes necessary to know the percentage of water in a refrigerant. The water vapor or moisture is measured in parts per million. The necessary measuring instrument is still used primarily in the laboratory. Instruments for measuring humidity are not used here.

Btu Meters The Btu is used to indicate the amount of heat present. Meters are especially designed to indicate the Btu in a chilled water line, a hot water line, or a natural gas line. Specially designed, they are used by skilled laboratory personnel at present.

Vibration and Sound Meters

More cities are now prohibiting conditioning units that make too much noise. In most cases, vibration is the main problem. However, it is not an easy task to locate the source of vibration. However, special meters have been designed to aid in the search for vibration noise.

Portable noise meters are available. The dB, or decibel, is the unit for the measurement of sound. There are a couple of bands on the noise meters. The dB-A scale corresponds roughly to the human hearing range. Other scales are available for special applications.

More emphasis is now being placed on noise levels in factories, offices, and schools. The *Occupational Safety and Hazards Act* (OSHA) lays down strict guidelines regarding noise levels. There are penalties for noncompliance. Thus, it will be necessary for all new and previously installed units to be checked for noise.

High-velocity air systems—used in large buildings—are engineered to reduce noise to levels set by the OSHA. For example, there are chambers to lower the noise in the ducts. Air engineers are constantly working on high-velocity systems to try to solve some of the problems associated with them.

SERVICE TOOLS

Service personnel use some special devices to help them with repair jobs in the field. One of them is the chaser kit. See Fig. 1-49. It is used for cleaning partially plugged capillary tubes. The unit includes 10 spools of lead alloy wire. These wires can be used as chasers for the 10 most popular sizes of capillary tubes. In addition to the wire, a cap tube gage, a set of sizing tools, and a combination file/reamer are included in the metal case. This kit is used in conjunction with the Cap-Check. The Cap-Check is a portable, self-contained hydraulic power unit with auxiliary equipment especially adapted to cleansing refrigeration capillary tubes. See Fig. 1-50. A small plug of wire from the chaser kit is inserted into the capillary tube. The wire is a few thousandths of an inch smaller than the internal diameter of the capillary tube. This wire is pushed like a piston through the capillary tube with hydraulic pressure from the Cap-Check. A 0 to 5000 psi gage shows pressure buildup if the capillary tube is restricted. It also shows when the chaser has passed through the tube. A trigger-operated gage shutoff is provided so the gage will not be damaged if pressure greater than 5000 psi is desired. When the piston stops against a partial restriction, high-velocity oil is directed around the piston and against the wall, washing the restriction away and allowing the wire to move through the tube. The lead wire eventually ends up in the bottom of the evaporator, where it remains. The capillary tube is then as clean as when it was originally installed.

A 30 in. high-pressure hydraulic hose with a $^1/_4$ in. *Society of Automotive Engineers* (SAE) male flare outlet connects the cap tube to the Cap-Check for simple handling. An adapter comes with the Cap-Check for simple handling. Another adapter comes with the unit to connect the cap tube directly to the hose outlet without a flared fitting.

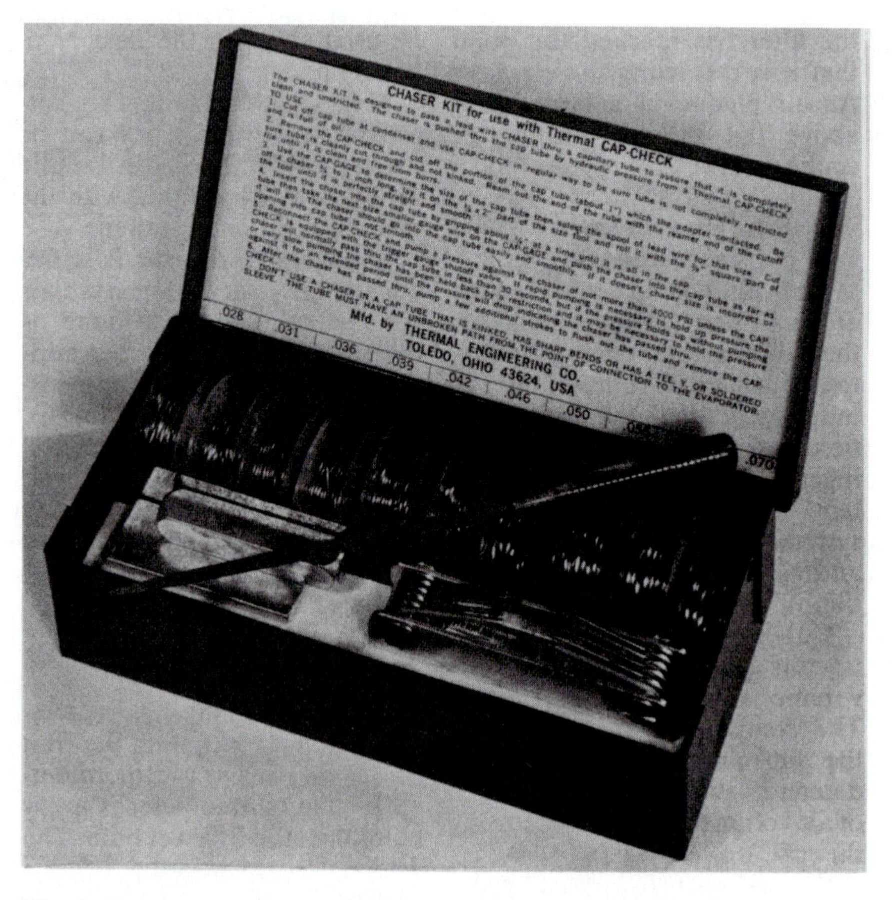

Fig. 1-49 *Cap-Check chaser kit. This is a means to clean partially plugged capillary tubes. It has 10 spools of lead alloy wire. These wires can be used as a chaser for the 10 most popular sizes of cap tubes. A cap tube gage, set of sizing tools, and a combination file/reamer are included in the kit.*

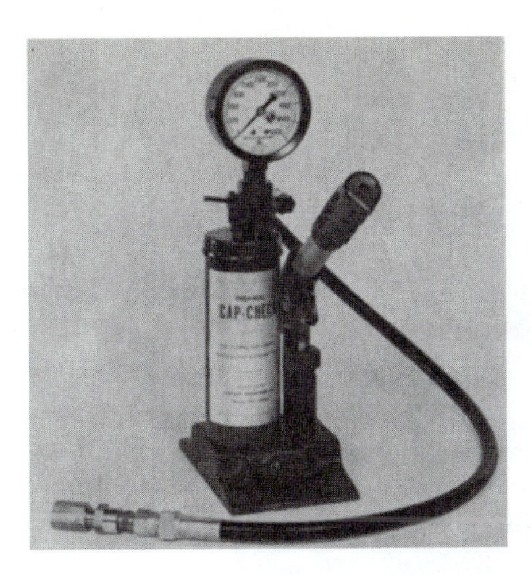

Fig. 1-50 *Cap-Check is a portable self-contained hydraulic power unit with auxiliary equipment that is especially adapted to cleaning refrigeration capillary tubes. It is hand operated.*

The Cap-Gage is a capillary tube gage. It has 10 stainless steel gages to measure the most popular sizes of capillary tubes. See Fig. 1-51.

More up to date tools and test equipment are shown in the Appendices. Go online to find the latest available tools and instruments. One source for tools and test equipment is yellowjacket.com or the Ritchie Engineering Company in Minnesota. Another is Mastercool.com in New Jersey.

SPECIAL TOOLS

Eventually, almost every refrigerant-charging job turns into a vapor-charging job. Unless the compressor is turned on, liquid can be charged into the high side only so long before the system and cylinder pressures become unfavorable. Once that happens, all refrigerant must be taken in the low side in the form of vapor.

Vapor charging is much slower than liquid charging. To create a vapor inside the refrigerant cylinder, the liquid refrigerant must be boiling. Boiling refrigerant absorbs heat. This is the principle on which refrigeration operates.

The boiling refrigerant absorbs heat from the refrigerant surrounding it in the cylinder. The net effect is that the cylinder temperature begins to drop soon after you begin charging with vapor. As the temperature drops, the remaining refrigerant will not vaporize as readily. Charging will be slower.

To speed charging, service personnel add heat to the cylinder by immersing part of it in hot water. The cylinder temperature rises. The boiling refrigerant becomes vigorous and charging returns to a rapid rate. It is not long, though, before all the heat has been taken from the water and more hot water must be added.

The Vizi-Vapr is an example of how a device can remove liquid from a cylinder and apply it to the system in the form of a vapor. See Fig. 1-52. No heat is required. This eliminates the hazards of using a torch and hot water. The change from a liquid to a gas or vapor

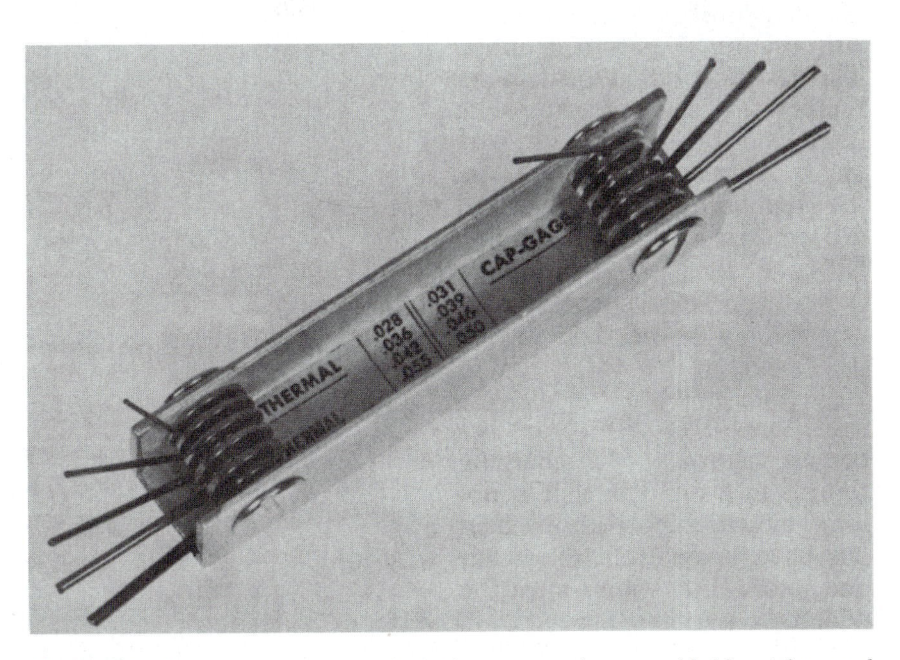

Fig. 1-51 *The Cap-Gage is a pocketknife-type cap tube gage with 10 stainless steel gages to measure the most popular sizes of cap tubes.* (Thermal Engineering)

Fig. 1-52 *The Vizi-Vapr is a device that allows rapid charging of a compressor without heating the cylinder of refrigerant.* (Thermal Engineering)

takes place in the Vizi-Vapr. It restricts the charging line between the cylinder and compressor. This restriction is much like an expansion valve in that it maintains high cylinder pressure behind it to hold the refrigerant as a liquid.

However, it has a large pressure drop across it to start evaporation. Heat required to vaporize refrigerant is taken from the air surrounding the unit, not from the remaining refrigerant. This produces a dense, saturated vapor.

The amount of restriction in the unit is very critical. Too much restriction will slow charging considerably. It also will allow liquid to go through and cause liquid slugging in the compressor. The restriction setting is different for each size system, for different types of refrigerants and even for different ambient temperatures.

VACUUM PUMPS

Use of the vacuum pump may be the single most important development in refrigeration and air-conditioning servicing. The purpose of a vacuum pump is to remove the undesirable materials that create pressure in a refrigeration system. These include:

- Moisture
- Air (oxygen)
- Hydrochloric acid

In addition, there are other materials that will vaporize at low micron range. These, along with a wide variety of solid materials, are pulled into the vacuum pump in the same way a vacuum cleaner sucks up dirt.

Evacuation is being routinely performed on almost every service call on which recharging is required.

NOTE: It is no longer permitted to simply add refrigerant to the system with one end open for evacuation into the atmosphere. This shortcut was a favorite of many service technicians over the years since it was quick and the refrigerant was inexpensive.

Vacuum levels formerly unheard of for field evacuations are being accomplished daily by service persons who are knowledgeable regarding vacuum equipment. These service persons have found through experience that the two-stage pump is much better than the single-stage pump for deep evacuations. See Figs. 1-53 and 1-54. It was devised as a laboratory instrument and with minor alterations; it has been adapted to the refrigeration field. It is the proper tool for vacuum evacuations in the field. The latest in vacuum pumps is shown in Appendix 4.

Fig. 1-53 *Single-stage portable vacuum pump.* (Thermal Engineering)

Fig. 1-54 *Dual or two-stage portable vacuum pump.* (Thermal Engineering)

To understand the advantages of a two-stage pump over a single-stage pump refer to Fig. 1-55. This shows the interior of a two-stage vacuum pump. This is a simplified version of a vacuum stage. It is built on the principle of a Wankel engine.

There is a stationary chamber with an eccentric rotor revolving inside. The sliding vanes pull gases through the intake. They compress them and force them into the atmosphere through the exhaust. The vanes create a vacuum section and a pressure section inside the pump. The seal between the vacuum and the pressure sections is made by the vacuum pump oil. These seals are the critical factor in the depth that a vacuum pump can pull. If the seals leak, the pump will not be able to draw a deep vacuum. Consequently, less gas can be processed. A pump with high leakage across the seal will be able to pull a deep vacuum on a small system, but the leakage will decrease the pumping speed (cfm) in the

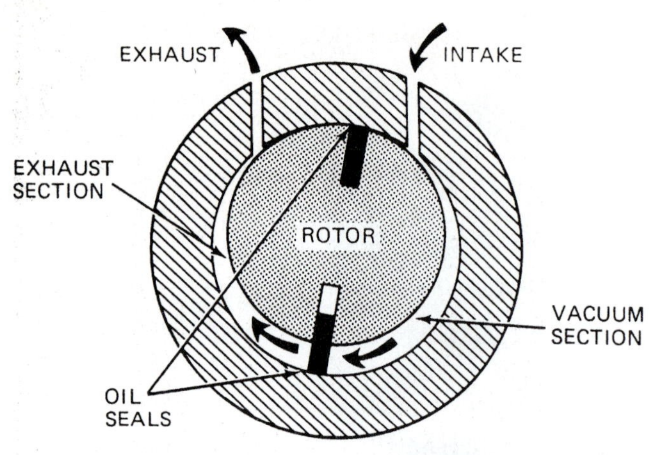

Fig. 1-55 *Two-stage vacuum pump showing seals and intake, exhaust and vacuum section.* (Thermal Engineering)

deep vacuum region. Long pull-down times will result. There are three oil seals in a single-stage vacuum pump. Each seal must hold against a high pressure on one side and a deep vacuum on the other side. This places a great deal of strain on the oil seal. A two-stage vacuum pump cuts the pressure strain on the oil seal in half. Such a pump uses two chambers instead of one to evacuate a system. The first chamber is called the *deep vacuum chamber*. It pulls in the vacuum gases from the deep vacuum and exhausts them into the second chamber at a moderate vacuum. The second chamber, or stage, brings in these gases at a moderate vacuum and exhausts them into the atmosphere. By doing this, the work of a single chamber is split between two chambers. This, in turn, cuts in half the strain on each oil seal, which reduces the leakage up to 90 percent.

A two-stage vacuum pump is more effective than a single-stage vacuum pump. For example, a single-stage vacuum pump rated for 1.5 cfm capacity will take one and one-half hours to evacuate one drop of water. A two-stage vacuum pump with the same rating will evacuate the drop in 12 min.

For evacuation of a 5-ton system saturated with moisture, a minimum of 15 h evacuation time is required in using a single-stage vacuum pump. A two-stage pump with the same cfm rating could do the job in as little as 2 h.

Another advantage of the two-stage pump is reliability. As you can see, if the oil seal is to be effective, the tolerances in these vacuum pumps must be very close between rotor and stator. If the tolerances are not correct, the oil seal will not be effective. Slippage of tolerance due to wear is the major cause of vacuum pump failure. With a single-stage pump, when the tolerance is in the stage slips, the pump loses effectiveness. With a two-stage pump, if one stage loses tolerance, the other one will still pull the vacuum of a single-stage pump.

Larger cfm, two-stage vacuum pumps are preferred to the single-stage vacuum pumps. The cost difference between the two is not great. In addition, the time saved by using the two-stage pump is evident on the first evacuation.

Vacuum Pump Maintenance

The purpose of vacuum pump oil is to lubricate the pump and act as a seal. To perform this function the oil must have:

- A low vapor pressure that does not materially increase up to 125°F (51.7°C)
- A viscosity sufficiently low for use at 60°F (15.6°C) yet constant up to 125°F (51.7°C)

These requirements are easily met by using a low vapor pressure, paraffin-based oil having a viscosity of approximately 300 SSU (shearing stress units) at 100°F (37.8°C) and a viscosity index in the range of 95 to 100. This type of uninhibited oil is readily obtainable. It is the material provided by virtually all sellers of vacuum pump oil to the refrigeration trade.

Vacuum Pump Oil Problems

The oils used in vacuum pumps are designed to lubricate and seal. Many of the oils available for other jobs are not designed to clean as they lubricate. Neither are they designed to keep in suspension the solids freed by the cleaning action of the oil. In addition, the oil is not usually heavily inhibited against the action of oxygen. Therefore, the vacuum pump must be run with flushing oil periodically to clean it. Otherwise, its efficiency will be reduced. The use of flushing oils is recommended by pump manufacturers.

If hydrochloric acid has been pulled into the pump, water, solids, and oil will bond together to form sludge or slime that may be acidic. The oil also may deteriorate due to oxidation (action on the oil by oxygen in air pulled through the pump). This results in a pump that will not pull a proper vacuum, may wear excessively, seriously corrode, or rust internally.

Operating Instructions

Use vacuum-pump oil in the pump when new. After 5 to 10 h of running time, change the oil. Make sure all of the original oil is removed from the pump. Thereafter, change the oil after every 30 h of operation when the oil becomes dark due to suspended solids drawn into the pump. Such maintenance will ensure peak efficiency in the pump operation.

If the pump has been operated for a considerable time on regular pump oil, drain the oil and replace with dual-purpose vacuum-pump oil. Drain the oil and replace with dual purpose after 10 h of operation. The oil will probably be quite dark due to sludge removed from the pump. Operate the second charge of oil for 10 h and drain again. The second charge of oil may still be dark. However, it will probably be lighter in color than the oil drained after the first 10 h.

Change the oil at 30-h intervals. After that, change the oil before such intervals if it becomes dark due to suspended solids pulled into the pump. Be sure to change the oil every 30 h thereafter to keep the vacuum pump in peak condition.

Evacuating a System

How long should it take? Some techniques of evacuation will clean refrigeration and air-conditioning system to a degree never reached before. Properly used, a good vacuum pump will eliminate 99.99 percent of the air and virtually all of the moisture in a system. There is no firm answer regarding the time it will take a pump to accomplish this level of cleanliness. The time required for evacuation depends on many things. Some factors that must be considered are:

- The size of the vacuum pump
- The type of vacuum pump—single or two-stage
- The size of the hose connections
- The size of the system
- The contamination in the system
- The application for the system

Evacuations sometimes take fifteen minutes. Then, again, they may take weeks. The only way to know when evacuation is complete is to take micron vacuum readings, using a good electronic vacuum gage. A number of electronic meters are available. See Figs. 1-56 and 1-57.

Evacuating down to 29 in. eliminates 97 percent of all air. Moisture removal, however, does not begin until a vacuum below 29 in. is reached. This is the micron level of vacuum. It can be measured only with an electronic vacuum gage. Dehydration of system does not certainly begin until the vacuum gage reads below 5000 microns. If the system will not pump down to this level, something is wrong. There may be a leak in the vacuum connections. The vacuum-pump oil may be contaminated. There may be a leak in the system. Vacuum gage readings between 500 and 1000 microns

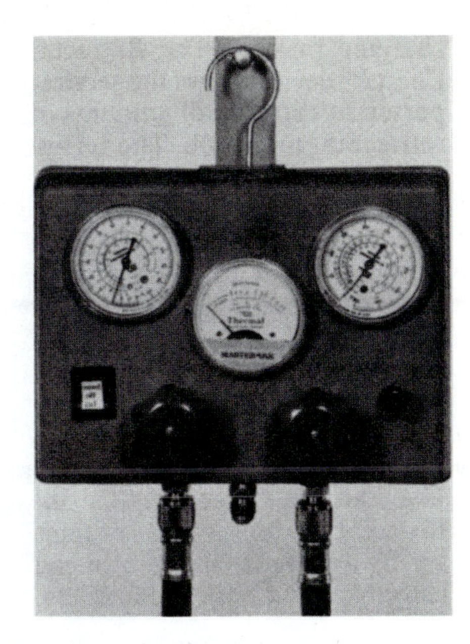

Fig. 1-56 *This vacuum check gage is designed to be as handy as a charging manifold.* (Thermal Engineering)

assure that dehydration is proceeding. When all moisture is removed, the micron gage will pull down below 1000 microns.

Pulling a system down below 1000 microns is not a perfect test for cleanliness. If the vacuum pump is too large for the system, it may pull down this level before all of the moisture is removed. Another test is preferred. Once the system is pulled down below 1000 microns it will not go any further. The system should be valved off from the vacuum pump and the pump turned off. If the vacuum in the system does not rise over 2000 microns in the next 5 min, evacuation has been completed. If it goes over this level, either the moisture is not completely removed or the system has a slight leak. To find which, reevacuate the system to its lowest level. Valve it off again and shut off the vacuums pump. If the vacuum leaks back to the same level as before, there is a leak in the system. If, the rise is much slower than before, small amounts of moisture are probably left in the system. Reevacuate until the vacuum will hold.

CHARGING CYLINDER

The charging cylinder lets you charge with heat to speed up the charging process. This unit, with its heater assembly, allows up to 50 W of heat to be used in charging. Refrigerant is removed rapidly from the cylinder as liquid, but injected into the system as a gas with the Vizi-Vapr. It requires no heat

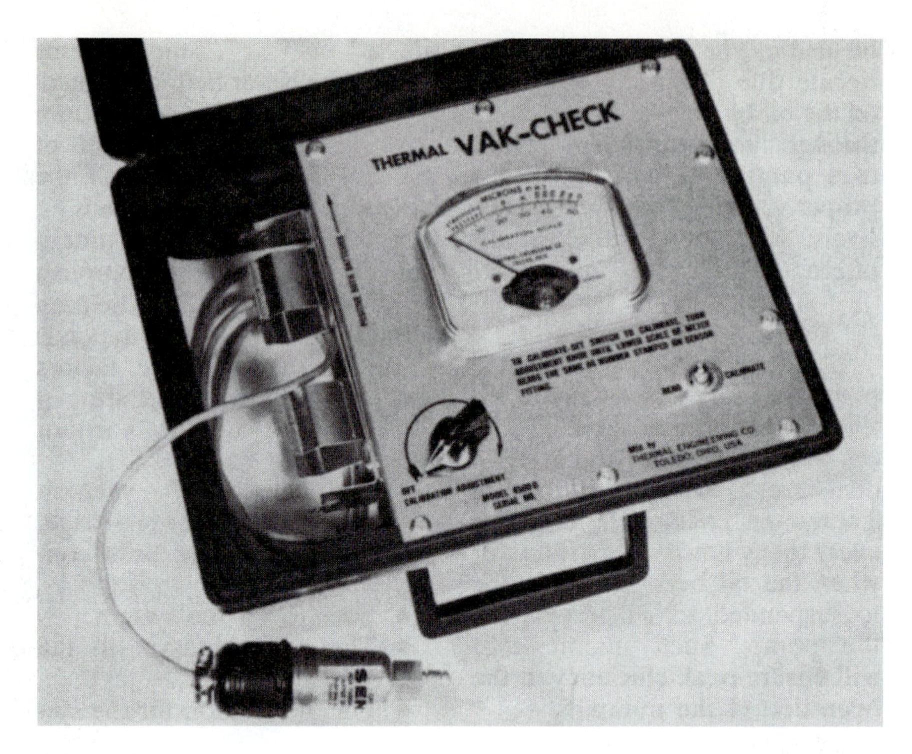

Fig. 1-57 *An electronic high-vacuum gage that reads directly in microns.* (Thermal Engineering)

during the charging process. The Extracta-Charge device allows the serviceperson to carry small amounts of refrigerant to the job. The refrigerant can be bought in large drums and stored at the shop. The Extracta-Charge comes in a rugged, steel carrying case to protect it from tough use. It provides a method for draining refrigerant even from capillary tube, sealed systems.

It is now mandatory to capture the escaping refrigerant. The Extracta-Charge is the instrument to use. When systems are overcharged, the excess can be transported back to the drum. The amount removed can be measured also. A leak found after the charging operation usually means the loss of the full charge. Using this device, the serviceperson can extract the charge and save it for use after the leak has been found and repaired.

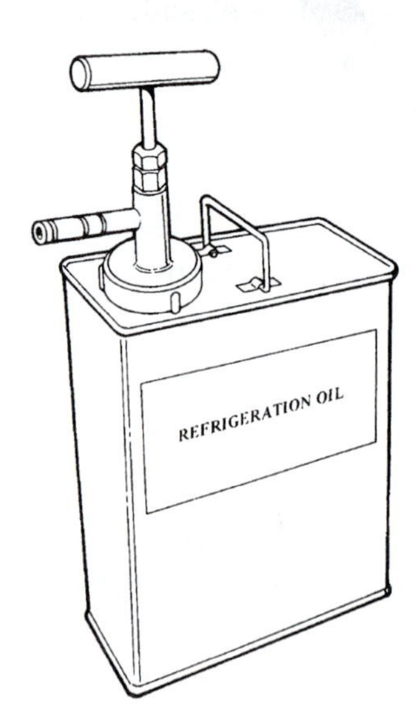

Fig. 1-58 *Oil charging pump.* (Thermal Engineering)

CHARGING OIL

In charging a compressor with oil, there is danger of drawing air and moisture into the refrigeration system. Use of the pump shown in Fig. 1-58 eliminates this danger. This pump reduces charging time by over 70 percent without pumping down the compressor. The pump fits the can with a cap seal, so the pump need not be removed until the can is empty. It is a piston-type high-pressure pump designed to operate at pressures up to 250 psi. It pumps one quart in 20 full strokes of the piston. The pump can be connected to the compressor by a refrigerant charging line or copper tubing from a $\frac{1}{2}$ in. male flare fitting.

CHANGING OIL

Whenever it is impossible to drain oil in the conventional manner, it becomes necessary to hook up a pump. Removing oil from refrigeration compressors before dehydrating with a vacuum is a necessity. The pump shown in Fig. 1-59 has the ability to remove one quart of oil with about 10 strokes. It is designed for use in pumping oil from refrigeration compressors, marine engines, and other equipment.

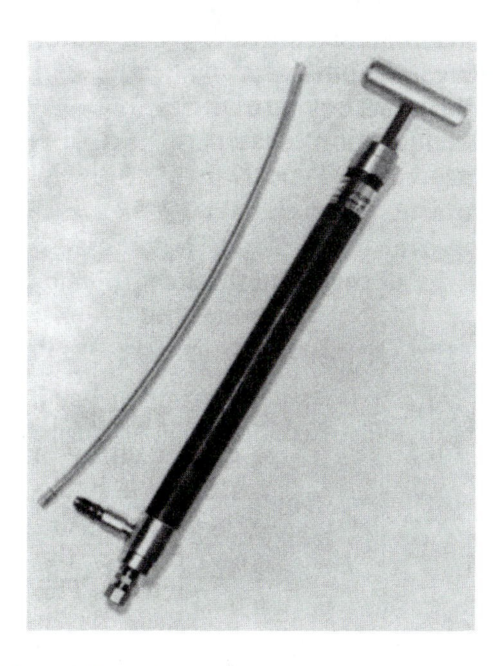

Fig. 1-59 *Oil changing pump.* (Thermal Engineering)

MOBILE CHARGING STATIONS

Mobile charging stations can be easily loaded into a pickup truck, van, or station wagon. They take little space. See Fig. 1-60. Stations come complete with manifold gage set, charging cylinder, instrument and tool sack, and vacuum pump. The refrigerant tank can also be mounted on the mobile charging station.

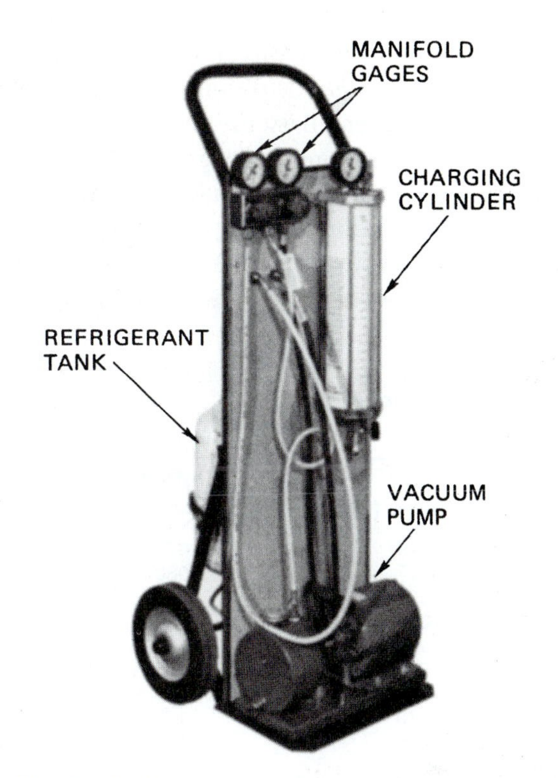

Fig. 1-60 *Mobile charging station.* (Thermal Engineering)

TUBING

Several types of tubing are used in plumbing, refrigeration, and air-conditioning work. Air conditioning and refrigeration, however, use special tubing types. Copper, aluminum, and stainless steel are used for tubing materials. They ensure that refrigerants do not react with the tubing. Each type of tubing has a special application. Most of the tubing used in refrigeration and air conditioning is made of copper. This tubing is especially processed to make sure it is clean and dry inside. It is sealed at the ends to make sure the cleanliness is maintained.

- Stainless steel tubing is used with R-717 or ammonia refrigerant.
- Brass or copper tubing should *not* be used in ammonia refrigerant systems.
- Aluminum tubing is used in condensers in air-conditioning systems for the home and automobile.

This calls for a special type of treatment for soldering or welding. Copper tubing is the type most often used in refrigeration systems. There are two types of copper tubing—hard-drawn and soft copper tubing. Each has a particular use in refrigeration.

Soft Copper Tubing

Some commercial refrigeration systems use soft copper tubing. However, such tubing is most commonly found in domestic systems. Soft copper is annealed. *Annealing* is the process whereby the copper is heated to a blue surface color and allowed to cool gradually to room temperature. If copper is hammered or bent repeatedly, it will become hard. Hard copper tubing is subject to cracks and breaking.

Soft copper comes in rolls and is usually under $^1/_2$ in. in *outside diameter* (OD). Small-diameter copper tubing is made for capillary use. It is soft drawn and flexible. It comes in random lengths of 90 to 140 ft. Table 1-1 gives the available inside and outside diameters. This type of tubing usually fits in a $^1/_4$ in. (OD) solder fitting that takes a $^1/_8$ in. (OD) diameter tubing.

Table 1-1 *Inside and Outside Diameter of Small Capillary Tubing**

Inside Diameter (ID), in.	Outside Diameter (OD), in.
.026	.072
.31	.083
.036	.087
.044	.109
.050	.114
.055	.125
.064	.125
.070	.125
.075	.125
.080	.145
.085	.145

*Reducing bushing fits in $^3/_8$ in. OD solder fitting and takes $^3/_8$ in. OD tubing.

There are three types of copper tubing—types K, L, and M.

- Type-K tubing is heavy duty. It is used for refrigeration, general plumbing, and heating. It can also be used for underground applications.
- Type-L tubing is used for interior plumbing and heating. Type-M tubing is used for light duty waste vents, water, and drainage purposes.
- Type-K soft copper tubing that comes in 60-ft rolls is available in outside diameters of $^5/_8$, $^3/_4$, $^7/_8$, and $1^1/_8$ in. It is used for underground water lines. Wall thickness and weight per foot are the same as for hard copper tubing.

Copper tubing used for air-conditioning and refrigeration purposes is marked "ACR." It is deoxidized and dehydrated to ensure that there is no moisture in it. In most cases, the copper tubing is capped after it is cleaned and filled with nitrogen. Nitrogen keeps it dry and helps prevent oxides from forming inside when it is heated during soldering.

Refrigeration dehydrated and sealed soft copper tubing must meet standard sizes for wall thickness and outside diameter. These sizes are shown in Table 1-2.

Hard and soft copper tubings are available in two wall thicknesses—K and L. The L thickness is used most frequently in air-conditioning and refrigeration systems.

Hard-Drawn Copper Tubing

Hard-drawn copper tubing is most frequently used in refrigeration and air-conditioning systems. Since it is hard and stiff, it does not need the supports required by soft copper tubing. This type of tubing is not easily bent. In fact, it should not be bent for refrigeration work. That is why there are several tubing fittings available for this type of tubing.

Hard-drawn tubing comes in 10 or 20 ft lengths. See Table 1-3. Remember, there is a difference between hard copper sizes and nominal pipe sizes. Table 1-4 shows the differences. Nominal sizes are used in water lines, home plumbing, and drains. They are *never* used in refrigeration systems. Keep in mind that Type K is

Table 1-2 *Dehydrated and Sealed Copper Tubing Outside Diameters, Wall Thicknesses, and Weights**

50-Foot Coils		
Outside Diameter (in.)	**Wall Thickness (in.)**	**Approximate Weight (lbs)**
1/8	.030	1.74
3/16	.030	2.88
1/4	.030	4.02
5/16	.032	5.45
3/8	.032	6.70
1/2	.032	9.10
5/8	.035	12.55
3/4	.035	15.20
7/8	.045	22.75
1 1/8	.050	44.20
1 3/8	.055	44.20

*The standard soft dehydrated copper tubing is made in the wall thickness recommended by the Copper and Brass Research Association to the National Bureau of Standards. Each size has ample strength for its capacity.

Table 1-3 *Outside Diameter, Wall Thickness, and Weight per Foot of Hard Copper Refrigeration Tubing*

Outside Diameter (in.)	Wall Thickness	Weight Per Foot
	Type-K Tubing	
3/8	0.035	0.145
1/2	0.049	0.269
5/8	0.049	0.344
3/4	0.049	0.418
7/8	0.065	0.641
1 1/8	0.065	0.839
1 3/8	0.065	1.040
1 5/8	0.072	1.360
2 1/8	0.083	2.060
2 5/8	0.095	2.930
3 1/8	0.109	4.000
4 1/8	0.134	6.510
	Type-L Tubing	
3/8	0.030	0.126
1/2	0.035	0.198
5/8	0.040	0.285
3/4	0.042	0.362
7/8	0.045	0.445
1 1/8	0.050	0.655
1 3/8	0.055	0.884
1 5/8	0.060	1.114
2 1/8	0.070	1.750
2 5/8	0.080	2.480
	Type-M Tubing	
1/2	0.025	0.145
5/8	0.028	0.204
7/8	0.032	0.328
1 1/8	0.035	0.465
1 3/8	0.042	0.682
1 5/8	0.049	0.940

Table 1-4 *Comparison of Outside Diameter and Nominal Pipe Size*

Outside Diameter (in.)	Nominal Pipe Size (in.)
3/8	1/4
1/2	3/8
5/8	1/2
3/4	—
7/8	3/4
1 1/8	1

heavy-wall tubing, Type L is medium-wall tubing, and Type M is thin-wall tubing. The thickness determines the pressure the tubing will safely handle.

Cutting Copper Tubing

Copper tubing can be cut with a copper tube cutter or a hacksaw. ACR tubing is cleaned, degreased, and dried before the end is sealed at the factory. The sealing plugs are reusable.

To provide further dryness and cleanliness, nitrogen, an inert gas, is used to fill the tube. It materially reduces the oxide formation during brazing. The remaining nitrogen limits excess oxides during succeeding brazing operations. Where tubing will be exposed inside food compartments, tinned copper is recommended.

To uncoil the tube without kinks, hold one free end against the floor or on a bench. Uncoil along the floor or bench to the desired length. The tube may be cut to length with a hacksaw or a tube cutler. In either case, deburr the end before flaring. Bending is accomplished by use of an internal or external bending spring. Lever-type bending tools may also be used. These tools will be shown and explained later.

The hacksaw should have a 32-tooth blade. The blade should have a wave set. No filings or chips can be allowed to enter the tubing. Hold the tubing so that when it is cut the scraps will fall out of the usable end.

Figure 1-61 shows some of the tubing cutters available. The tubing cutter is moved over the spot to be cut. The cutting wheel is adjusted so it touches the

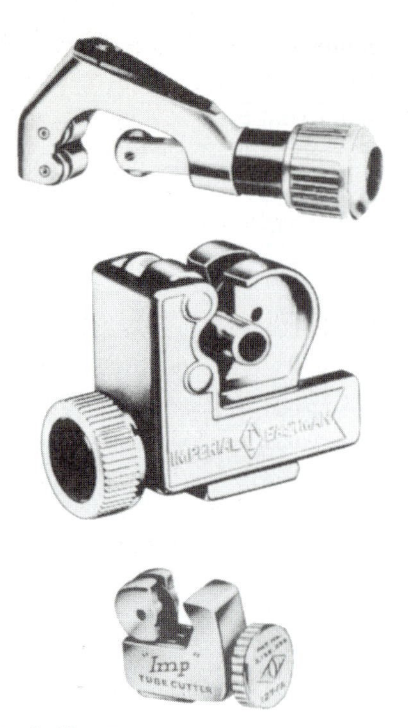

Fig. 1-61 *Three types of tubing cutters.* (Mueller Brass)

copper. A slight pressure is applied to the tightening knob on the cutter to penetrate the copper slightly. Then the knob is rotated around the tubing. Once around, it is tightened again to make a deeper cut. Rotate again to make a deeper cut. Do this by degrees so that the tubing is not crushed during the cutting operation.

After the tubing is cut through, it will have a crushed end. The crushed end is prepared for flaring by filing and reaming. See Fig. 1-62. A file and the

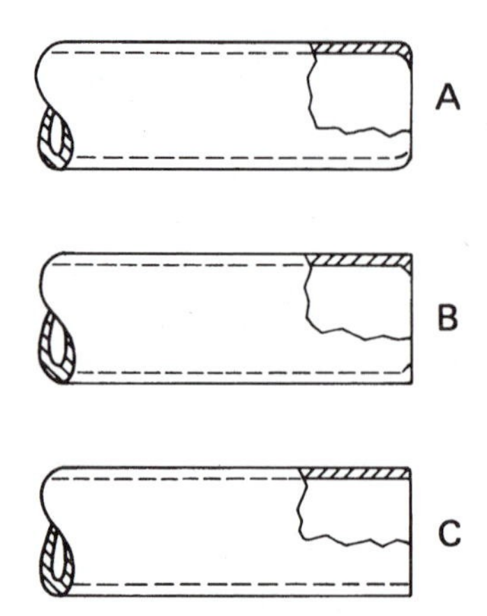

Fig. 1-62 *The three steps in removing a burr after the tubing has been cut with a tubing cutter. (A) The end of the cut tubing. (B) Squaring with a file produces a flat end. (C) The tube has been filed and reamed. It can now be flared.*

deburring attachment on the cutting tool can also be used. After the tubing is cut to length, it probably will require flaring or soldering.

Flaring Copper Tubing

A flaring tool is used to spread the end of the cut copper tubing outward. Two types of tools are designed for this operation. See Fig. 1-63. The flaring process is shown in Fig. 1-64. Note that the flaring is done by holding the end of the tubing rigid at a point slightly below the protruding part of the tube. This protruding part allows for the stretching of the copper.

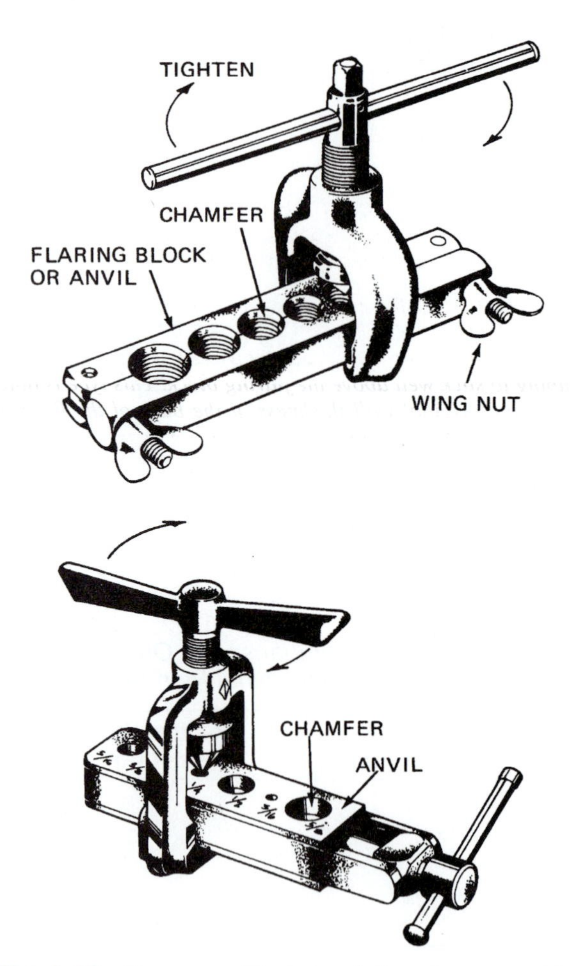

Fig. 1-63 *Two types of flaring tools for soft copper tubing.*

A flare is important for a strong, solid, leak-proof joint. The flares shown in Fig. 1-64 are single flares. These are used in most refrigeration systems. The other type of flare is the double flare. Here the metal is doubled over to make a stronger joint. They are used in commercial refrigeration and automobile air conditioners. Figure 1-65 shows how the double flare is made. The tool used is called a *block-and-punch.*

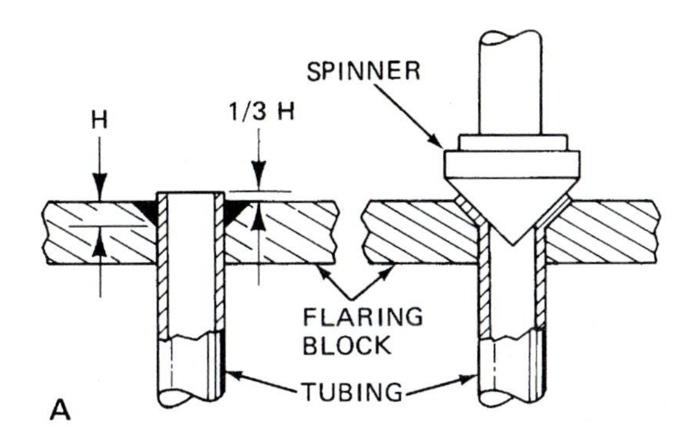

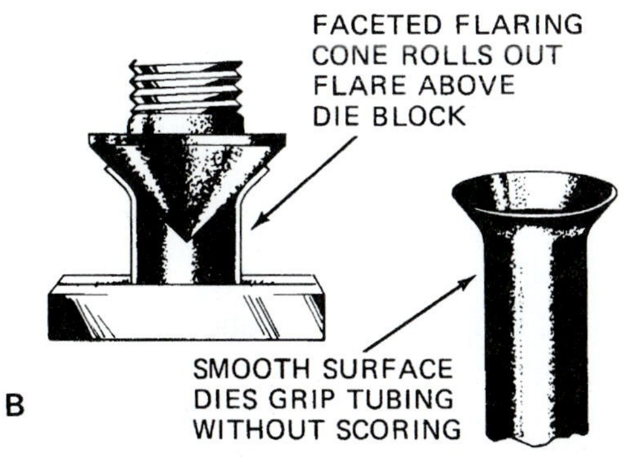

Fig. 1-64 *Flaring tools. (A) This type of tool calls for the tubing to be inserted into the proper size hole with a small amount of the tube sticking above the flaring block. (B) This type of tool calls for the tubing to stick well above the flaring block. This type is able to maintain the original wall thickness at the base of the flare. The faceted flaring cone smoothes out any surface imperfections.*

Adapters can be used with a single-flare tool to produce a double flare. See Fig. 1-66.

Figure 1-67 shows joints that use the flare. The flared tubing fits over the beveled ends. The flare tee uses the flare connection on all three ends. The half-union elbow uses the flare at one end and a *male pipe thread* (MPT) on the other end. A *female pipe thread* is designated by the abbreviation FPT.

Double flaring is recommended for copper tubing $^{5}/_{16}$ in. and over. Double flares are not easily formed on smaller sizes of tubing.

Constricting Tubing

A tubing cutter adapted with a roller wheel is used to constrict a tubing joint. Two tubes are placed so that one is inserted inside the other. They should be within 0.003 in. when inserted. This space is then constricted by a special wheel on the tube cutter. See Fig. 1-68. The one shown is a combination tube cutter and constrictor. The wheel tightens the outside tube around the inside tube. The space between the two is then filled with solder. Of course, proper cleanliness for the solder joint must be observed before attempting to fill the space with solder.

Both pieces of tubing must be hot enough to melt the solder. Flux must be used to prevent oxidation during the heating cycle. Place flux only on the tube to be inserted. No flux should be allowed to penetrate the inside of the tubing. It can clog filters and restrict refrigerant flow.

Swaging Copper Tubing

Swaging joins two pieces of copper without a coupling. This makes only one joint, instead of the two that would be formed if a coupling were used. With

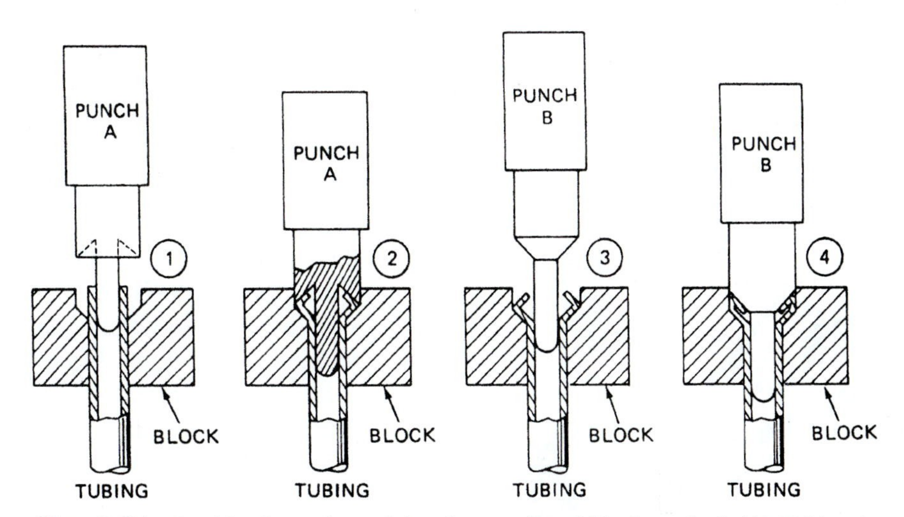

Fig. 1-65 *Double flares formed by the punch-and-block method. (1) Tubing is clamped into the block opening of the proper size. The female punch, Punch A, is inserted into the tubing. (2) Punch A is tapped to bend the tubing inward. (3) The male punch, Punch B, is tapped to bend the tubing inward. (4) The male punch is tapped to create the final double flare.*

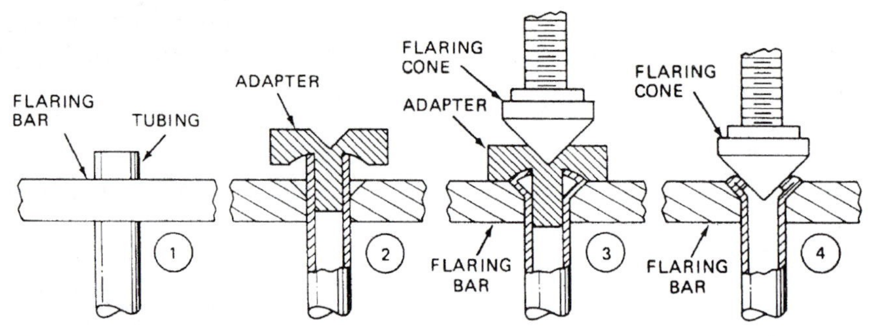

Fig. 1-66 *Making a double flare with an adapter for the single-flare tool. (1) Insert the tubing into the proper size hole in the flaring bar. (2) Place the adapter over the tubing. (3) Place the adapter inside the tubing. Apply pressure with the flaring cone to push the tubing into a doubled-over configuration. (4) Remove the adapter and use the flaring cone to form a double-thickness flare.*

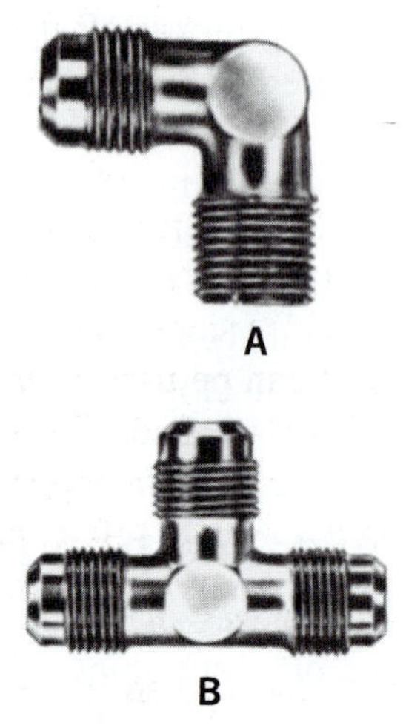

Fig. 1-67 *A half-union elbow (A) and a flare tee (B). Note the 45° angle on the end of the half-union elbow fitted for a flare. Also, note the 45° angles on both ends of the flare tee. Note that the flared end does not have threads to the end of the fitting.*

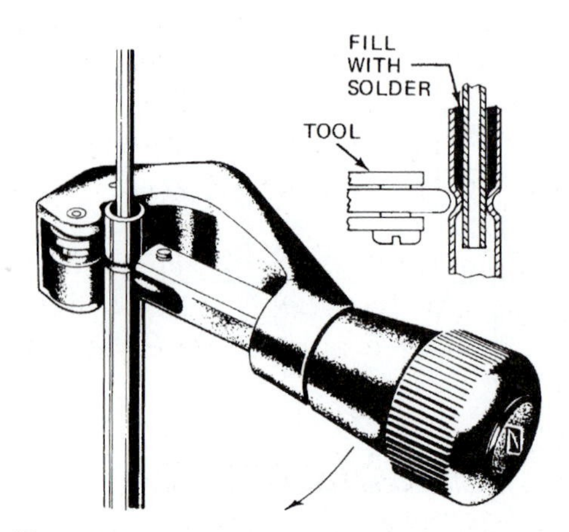

Fig. 1-68 *Tubing cutter adapter with a roller wheel to work as a tubing constrictor.*

fewer joints, there are fewer chances of leaks. Punch-type swaging tools and screw-type swaging tools are used in refrigeration work. The screw-type swaging tool works the same as the flaring tool.

Tubing is swaged so that one piece of tubing is enlarged to the outside diameter of the other tube. The two pieces of soft copper are arranged so that the inserted end of the tubing is inside the enlarged end by the same amount as the diameter of the tubing used. See Fig. 1-69. Once the areas have been properly prepared for soldering, the connection is soldered. Today, most mechanics use fittings, rather than take the time to prepare the swaged end.

Forming Refrigerant Tubing

There are two types of bending tools made of springs. One fits inside the tubing. The other fits outside and over the tubing being bent. See Fig. 1-70. Tubing must be bent so that it does not collapse and flatten. To prevent this, it is necessary to place some device over the tubing to make sure that the bending pressure is applied evenly. A tube bending spring may be fitted either inside or outside the copper tube while it is being bent. See Fig. 1-71. Keep in mind that the minimum safe distance for bending small tubing is five times its diameter. On larger tubing, the minimum safe distance is ten times the diameter. This prevents the tubing from flattening or buckling.

Make sure the bending is done slowly and carefully. Make a large radius bend first, then go on to the smaller bends. Do not try to make the whole bend at one time. A number of small bends will equalize the applied pressure and prevent tubing collapse. When using the internal bending spring, make sure part of it

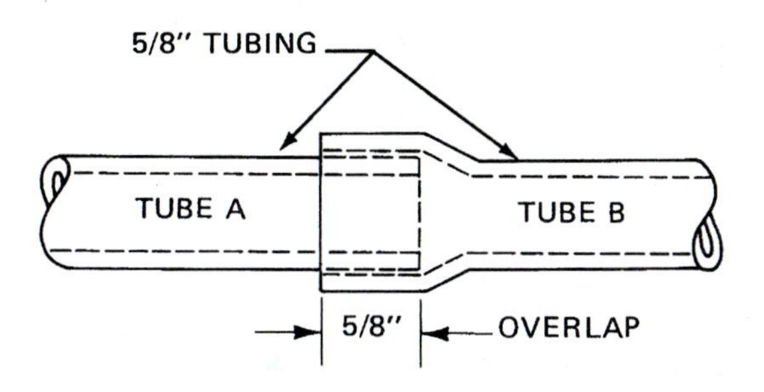

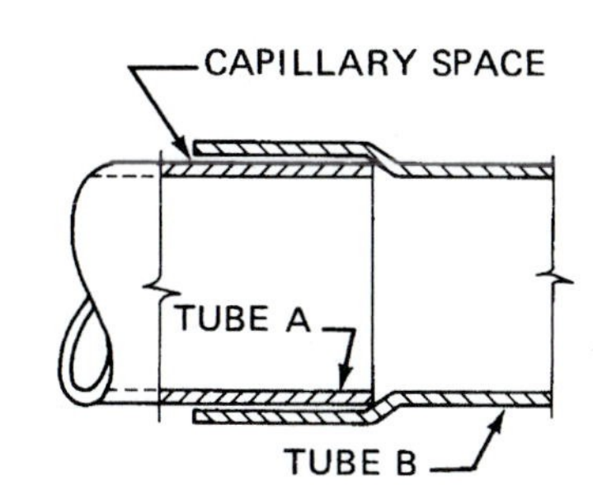

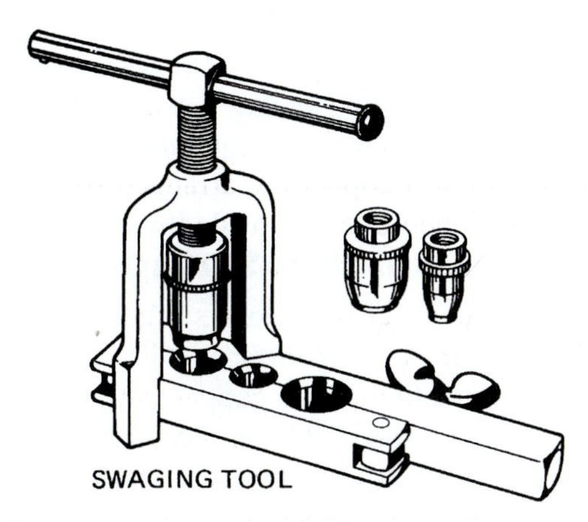

Fig. 1-69 *Swaging tool and swaging techniques. The swaging punches screw into the yoke and are changed for each size of tubing. Swages are available in 1/2, 5/8, and 7/8 in. OD, or 3/8, 1/2, and 3/4 in. nominal copper and aluminum tubing sizes.*

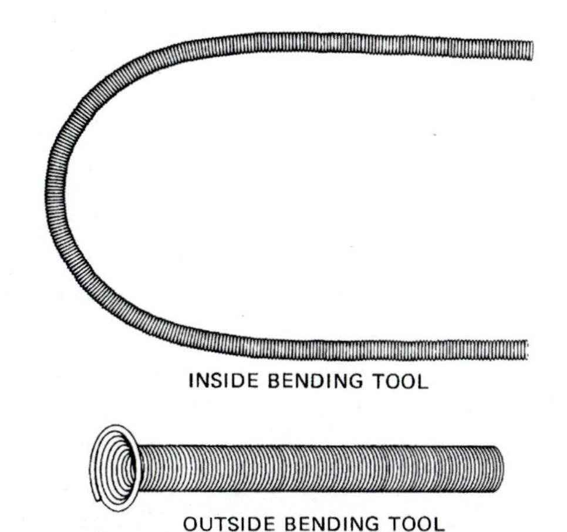

Fig. 1-70 *Bending tools for soft copper tubing.*

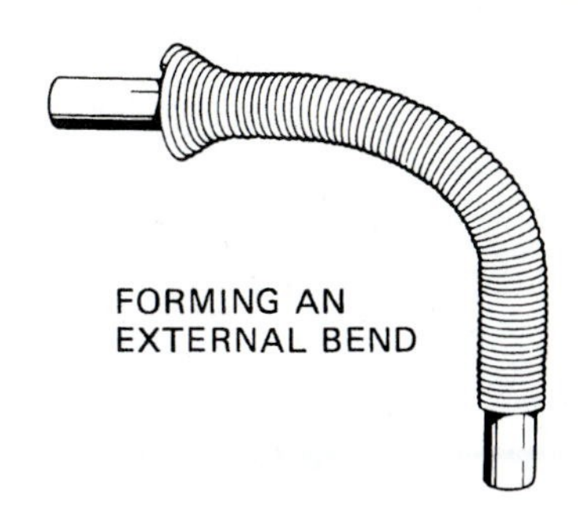

Fig. 1-71 *Using a spring-type tool to bend tubing.*

is outside the tubing. This gives you a handle on it when it is time to remove it after the bending. You may have to twist the spring to release it after the bend. By bending it so the spring compresses, it will become smaller in diameter, and pull out easily. The external spring is usually used in bending tubing along the midpoint. It is best to use the internal spring when a bend comes near the end of the tubing or close to a flared end.

The lever-type tube bender is also used for bending copper tubing. See Fig. 1-72. This one-piece open-side bender makes a neat, accurate bend since it is calibrated in degrees. It can be used to make bends up to 180°. A 180° bend is U-shaped. This tool is to be used when working with hard-drawn copper or steel tubing. It can also be used to bend soft copper tubing. The springs are used only for soft copper, since the hard-drawn copper would be difficult to bend by hand. Hard-drawn copper tubing can be bent, if necessary, using tools that electricians use to bend conduit.

Fitting Copper Tubing by Compression

Making leak-proof and vibration-proof connections can be difficult. A capillary tube connection can be used. See Fig. 1-73. This compression fitting is used with a capillary tube. The tube extends through the nut and into the connector fitting. The nose section is forced tightly against the connector fitting as the nut is tightened. The tip of the nose is squeezed against the tubing.

If you service this type of fitting, you must cut back the tubing at the end and replace the soft nose nut. If the nut is reused, it will probably cause a leaky connection.

SOLDERING

Much refrigeration work requires soldering. Brass parts, copper tubing, and fittings are soldered. The

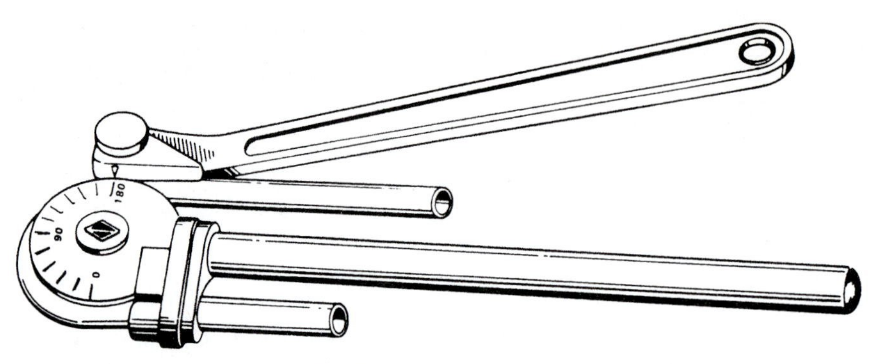

Fig. 1-72 *A tube bender.*

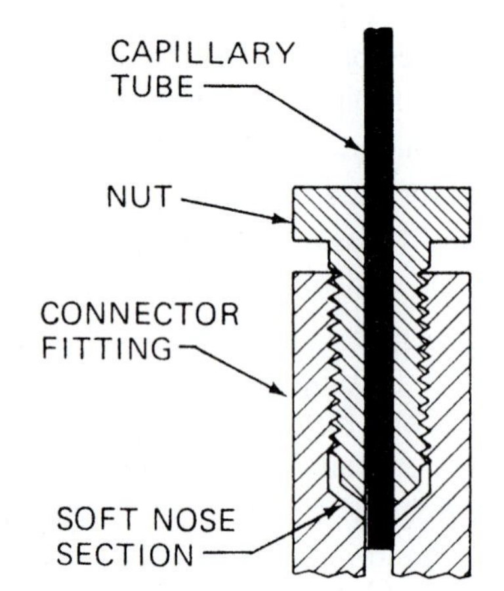

Fig. 1-73 *A capillary tube connection.*

cooling unit is also soldered. Thus, the air-conditioning and refrigeration mechanic should be able to solder properly.

Two types of solders are used in refrigeration and air-conditioning work. Soft solder and silver solder are most commonly used for making good joints. Brazing is actually silver soldering. Brazing requires careful preparation of the products prior to heating for brazing or soldering. This preparation must include steps to prevent contaminants such as dirt, chips, flux residue, and oxides from entering and remaining in an installation. A general-purpose solder for cold water lines and hot water lines with temperatures below 250°F (121.1°C) is 50-50. The solder is made of 50 percent tin and 50 percent lead. The 50-50 solder flows at 414°F (212.2°C).

Another low-temperature solder is 95-5. It flows at 465°F (240.5°C). It has a higher resistance to corrosion. It will result in a joint shear strength approximately two and a half times that of a 50-50 joint at 250°F (121.1°C).

A higher temperature solder is No. 122. It is 45 percent silver brazing alloy. This solder flows at 1145°F (618.2°C). It provides a joining material that is suitable for a joint strength greater than the other two solders. It is recommended for use on ACR copper tubing.

Number 50 solder is 50-50 lead and tin. Number 95 solder is 95 percent tin and 5 percent antimony. Silver solder is really brazing rod, instead of solder. The higher temperature requires a torch to melt it.

Soft Soldering

Soldering calls for a very clean surface. Sand-cloth is used to clean the copper surfaces. Flux must be added to prevent oxidation of the copper during the heating process. A no-corrode solder is necessary. See Fig. 1-6. Acid-core solder must not be used. The acid in the solder will corrode the copper and cause leaks.

Soldering is nothing more than applying a molten metal to join two pieces of tubing or a tubing end and a fitting. It is important that both pieces of metal being joined are at the flow point of the solder being used. Never use the torch to melt the solder. The torch is used to heat the tubing or fitting until it is hot enough to melt the solder.

The steps in making a good solder joint are shown in Fig. 1-74. Cleanliness is essential. Flux can damage any system. It is very important to keep flux out of the lines being soldered. The use of excessive amounts of solder paste affects the operation of a refrigeration system. This is especially true of R-22 systems. Solder paste will dissolve in the refrigerant at the high liquid line temperature. It is then carried through a drier or strainer and separated out at the colder expansion valve temperature. Generally, R-22 systems will be more seriously affected than those carrying R-12. This is because the solid materials separate out at a higher

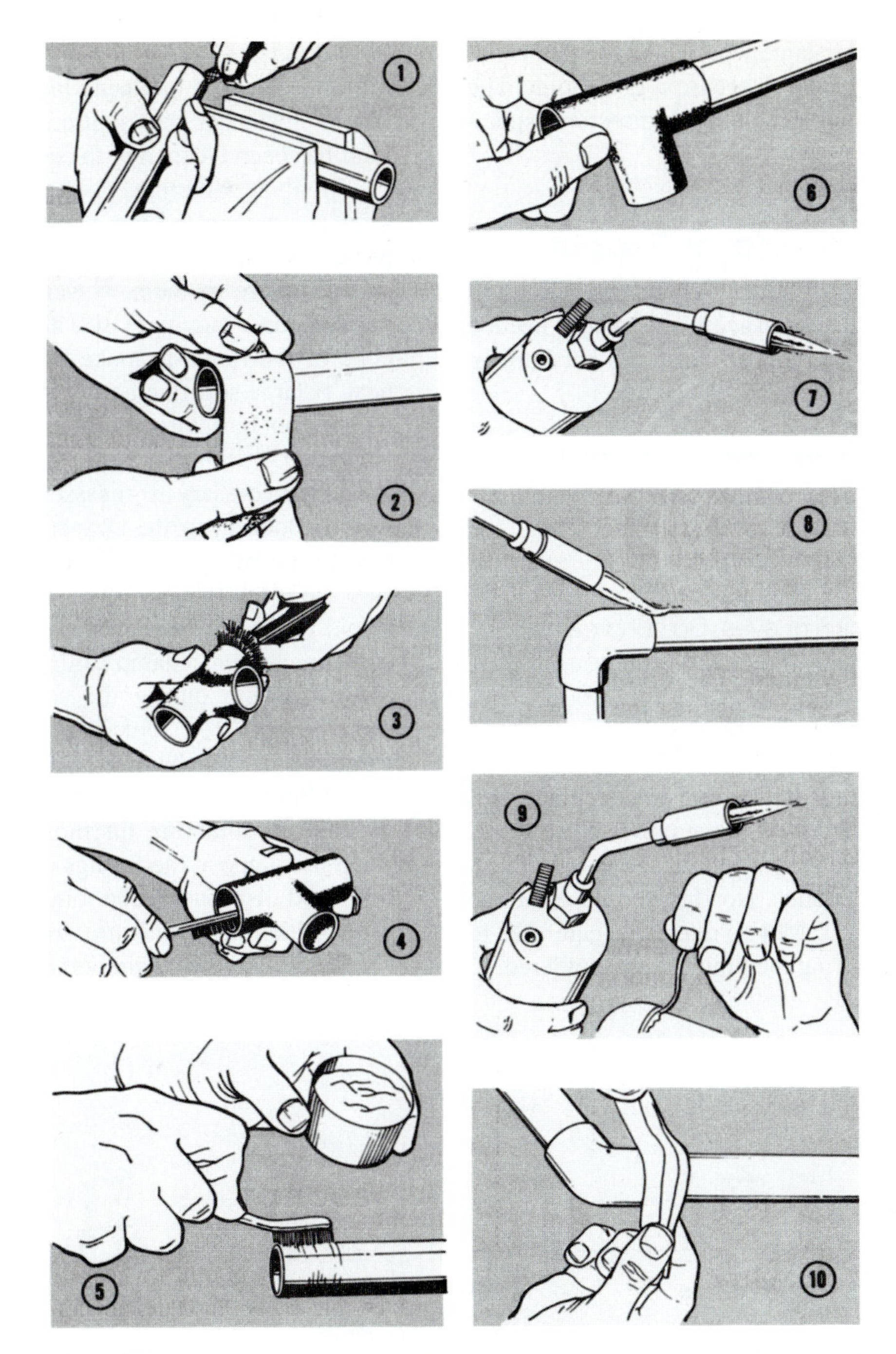

Fig. 1-74 *Soldering procedures. (1) Cut the tubing to length and remove the burrs. (2) Clean the joint area with sandpaper or sand-cloth. (3) Clean inside the fitting. Use sandpaper, sand-cloth, or wire brush. (4) Apply flux to the inside of the fitting. (5) Apply flux to the outside of the tubing. (6) Assemble the fitting onto the tubing. (7) Obtain proper tip for the torch and light it. Adjust the flame for the soldering being done. (8) Apply heat to the joint. (9) When solder can be melted by the heat of the copper (not the torch), simply apply solder so it flows around the joint. (10) Clean the joint of excess solder and cool it quickly with a damp rag.*

temperature. Sound practice would indicate the use of only enough solder paste to secure a good joint. The paste should be applied according to directions specified by the manufacturer.

Silver Soldering or Brazing

Silver solder melts at about 1120°F (604.4°C) and flows at 1145°F (618°C). An acetylene torch is needed for the high heat. It is used primarily on hard-drawn copper tubing.

> CAUTION: Before using silver solder, make sure it does not contain cadmium. Cadmium fumes are very poisonous. Make sure you work in a very well-ventilated room. The fumes should not contact your skin or eyes. Do not breathe the fumes from the cadmium type of silver solder. Most manufacturers will list the contents on the container.

Silver soldering also calls for a clean joint area. Use the same procedures as shown previously for soldering. See Fig. 1-74. Figure 1-75 shows good and poor design characteristics. No flux should enter the system being soldered. Make your plans carefully to prevent any flux entering the tubing being soldered.

Nitrogen or carbon dioxide can be used to fill the refrigeration system during brazing. This will prevent any explosion or the creation of phosgene when the joint has been cleaned with carbon tetrachloride.

In silver soldering, you need a tip that is several sizes larger than the one used for soft soldering. The pieces should be heated sufficiently to have the silver solder adhere to them. Never hold the torch in one place. Keep it moving. Use a slight feather on the inner cone of the flame to make sure you have the proper heat. A large soft flame may be used to make sure the tip does not burn through the fitting or the tubing being soldered.

It is necessary to disassemble sweat-type valves when soldering to the connecting lines. In soldering sweat-type valves where they connect to a line, make sure the torch flame is directed away from the valve. Avoid excessive heat on the valve diaphragm. As an extra precaution, a damp cloth may be wrapped around the diaphragm during the soldering operation. The same is true for soldering thermostatic expansion valves to the distributor.

Either soft or hard solder or silver brazing is acceptable in soldering thermostatic expansion valves. Keep the flame at the fittings and away from the valve body and distributor tube joints. *Do not overheat.* Always solder the *outside diameter* (OD) of the distributor, never the *inside diameter* (ID).

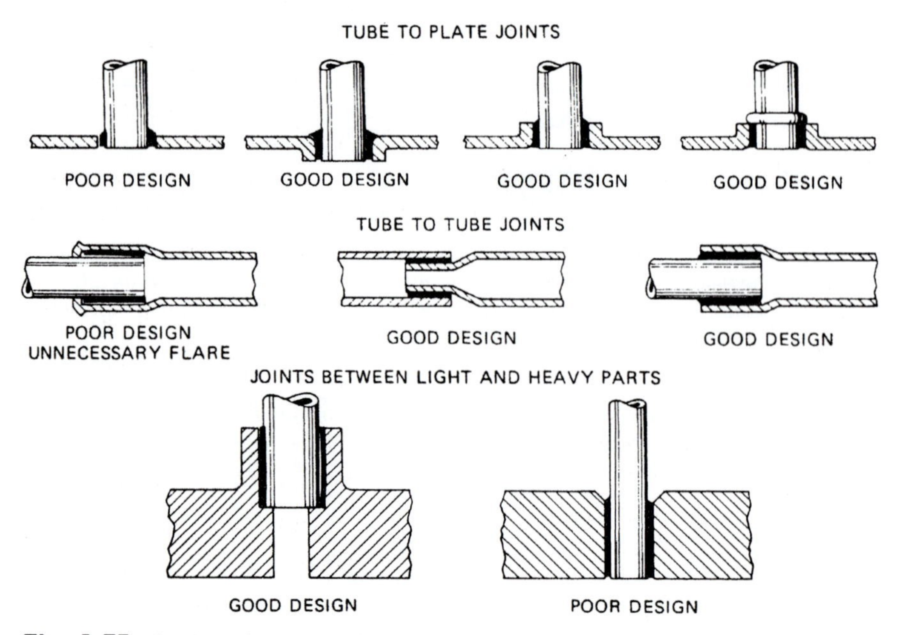

Fig. 1-75 *Designs that are useful in silver soldering copper tubing. Here, the clearances between the copper tubing are exaggerated for the sake of illustration. They should be much less than shown here.* (Handy and Harmon)

TESTING FOR LEAKS

Never use oxygen to test a joint for leaks. Any oil in contact with oxygen under pressure will form an explosive mixture.

Do not use emery cloth to clean a copper joint. Emery cloth contains oil. This may hinder the making of a good soldering joint. Emery cloth is made of silicon carbide, which is a very hard substance. Any grains of this abrasive in the refrigeration mechanism or lines can damage a compressor. Use a brush to help clean the area after sanding.

CLEANING AND DEGREASING SOLVENTS

Solvents, including carbon tetrachloride (CCl_4), are frequently used in the refrigeration industry for cleaning and degreasing equipment. No solvent is absolutely safe. There are several which may be used with relative safety. Carbon tetrachloride is *not* one of them. Use of one of the safer solvents will reduce the likelihood of serious illness developing in the course of daily use. Some of these solvents are stabilized methyl chloroform, methylene chloride, trichlorethylene, and perchloroethylene. Some petroleum solvents are available. These are flammable in varying degrees.

Most solvents may be used safely if certain rules are followed.

- Use no more solvent than the job requires. This helps keep solvent vapor concentrations low in the work area.
- Use the solvent in a well-ventilated area and avoid breathing the vapors as much as possible. If the solvents are used in shop degreasing, it is wise to have a ventilated degreasing unit to keep the level of solvent vapors as low as possible.
- Keep the solvents off the skin as much as possible. All solvents are capable of removing the oils and waxes that keep the skin soft and moist. When these oils and waxes are removed, the skin becomes irritated, dry, and cracked. A skin rash may develop more easily.

> CAUTION: While commonly used solvent, carbon tetrachloride has many virtues as a solvent, it has caused much illness among those who use it. Each year several deaths result from its use. Usually, these occur in the small shop or the home. Most large industries have discontinued its use. It is used only with extreme caution. A measure of its harmful nature is indicated by the fact that it bears a poison label. It should never be placed in a container that is not labeled "poison." It is for industrial use only.

While occasional deaths result from swallowing carbon tetrachloride, the vast majority of deaths are caused by breathing its vapors. When exposure is very great, the symptoms will be headache, dizziness, nausea, vomiting, and abdominal cramping. The person may lose consciousness. While the person seems to recover from breathing too much of the vapor, a day or two later he or she again becomes ill. Now there is evidence of severe injury to the liver and kidneys. In many cases, this delayed injury may develop after repeated small exposures or after a single exposure not sufficient to cause illness at the time of exposure. The delayed illness is much more common and more severe among those who drink alcoholic beverages. In some episodes where several persons were equally exposed to carbon tetrachloride, the only one who became ill or the one who became most seriously ill was the person who stopped for a drink or two on the way home. When overexposure to carbon tetrachloride results in liver and kidney damage, the patient begins a fight for life without the benefit of an antidote. The only sure protection against such serious illness is not to breathe the vapors or allow contact with the skin.

Human response to carbon tetrachloride is not predictable. A person may occasionally use carbon tetrachloride in the same job in the same way without apparent harm. Then, one day severe illness may result. This unpredictability of response is one factor that makes the use of "carbon tet" so dangerous.

Other solvents will do a good job of cleaning and degreasing. It is much safer to select one of those solvents for regular use rather than to expose yourself to the potential dangers of carbon tetrachloride.

REVIEW QUESTIONS

1. What does NEC stand for?
2. What type of solder core is preferred for electrical work?
3. What type of tips must masonry drill bits have?
4. What is a thermocouple?
5. What is a thermistor?
6. What is superheat?
7. What symbol identifies infinite resistance on an ohmmeter?

8. What is a draft gage?
9. What is the difference between a sling psychrometer and a stationary psychrometer?
10. Where are humidistats used?
11. What is a British thermal unit (Btu)?
12. What is a capillary tube?
13. Why is vapor charging slower than liquid charging?
14. What is the purpose of a vacuum pump?
15. What is a micron?
16. What type of tubing is needed with R-717 or ammonia refrigerant?
17. Name the three types of copper tubing and describe each.
18. What does ACR on a piece of copper tubing signify?
19. How do you shape or form copper tubing without collapsing it?
20. What is swaging?
21. At what temperature does silver solder melt?

2
CHAPTER

Development of Refrigeration

PERFORMANCE OBJECTIVES

After studying this chapter, you should:

1. Know various types of refrigeration systems.
2. Know how pressure operated devices function.
3. Know how compression ratios are figured.
4. Know how various heat related factors influence refrigeration systems.
5. Know how specific heat is figured.
6. Know how to troubleshoot systems using what you have learned so far.

Refrigeration is the process of removing heat from where it is not wanted. Heat is removed from food to preserve its quality and flavor. It is removed from room air to establish human comfort. There are innumerable applications in industry in which heat is removed from a certain place or material to accomplish a desired effect.

During refrigeration, unwanted heat is transferred mechanically to an area where it is not objectionable. A practical example of this is the window air conditioner that cools air in a room and exhausts hot air to the outdoors.

The liquid called the refrigerant is fundamental to the heat transfer accomplished by a refrigeration machine. Practically speaking, a commercial refrigerant is any liquid that will evaporate and boil at relatively low temperatures. During evaporation or boiling, the refrigerant absorbs the heat. The cooling effect felt when alcohol is poured over the back of your hand illustrates this principle.

In operation, a refrigeration unit allows the refrigerant to boil in tubes that are in contact, directly or indirectly, with the medium to be cooled. The controls and engineering design determine the temperatures reached by a specific machine.

Fig. 2-1 *One of the first commercial home refrigerators.* (General Electric)

HISTORICAL DEVELOPMENT

Natural ice was shipped from the New England states throughout the western world from 1806 until the early 1900s. Although ice machines were patented in the early 1800s, they could not compete with the natural ice industry. Artificial ice was first commercially manufactured in the southern United States in the 1880s.

Domestic refrigerators were not commercially available until about 1920. See Fig. 2-1. During the 1920s, the air-conditioning industry also got its start with a few commercial and home installations. The refrigeration industry has now expanded to touch most of our lives. There is refrigeration in our homes, and air conditioning in our place of work, and even in our automobiles. Refrigeration is used in many industries, from the manufacture of instant coffee to the latest hospital surgical techniques.

STRUCTURE OF MATTER

To be fully acquainted with the principles of refrigeration, it is necessary to know something about the structure of matter. *Matter* is anything that takes up space and has weight. Thus, matter includes everything but a perfect vacuum.

There are three familiar physical states of matter: solid, liquid, and gas or vapor. A *solid* occupies a definite amount of space. It has a definite shape. The solid does not change in size or shape under normal conditions.

A *liquid* takes up a definite amount of space, but does not have any definite shape. The shape of a liquid is the same as the shape of its container.

A *gas* does not occupy a definite amount of space and has no definite shape. A gas that fills a small container will expand to fill a large container.

Matter can be described in terms of our five senses. We use our senses of touch, taste, smell, sound, and sight to tell us what a substance is. Scientists have accurate methods of detecting matter.

Elements

Scientists have discovered 105 building blocks for all matter. These building blocks are referred to as elements. Elements are the most basic materials in the universe. Ninety-four elements, such as iron, copper, and nitrogen, have been found in nature. Scientists have made 11 others in laboratories. Every known substance, solid, liquid, or gas, is composed of elements. It is very rare for an element to exist in a pure state. Elements are nearly always found in combinations called *compounds*. Compounds contain more than one element. Even such a common substance as water is a compound, rather than an element. See Fig. 2-2.

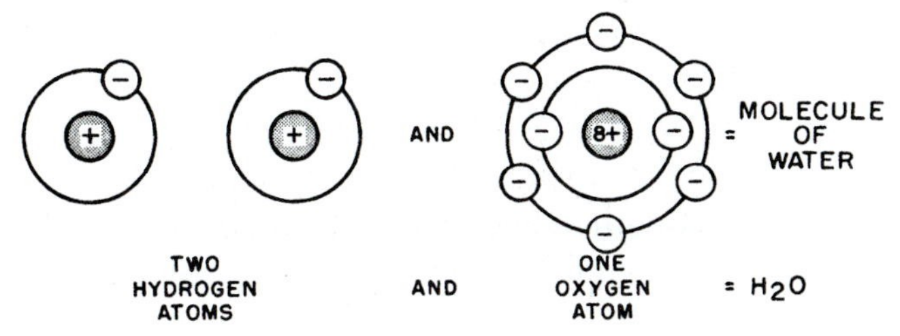

Fig. 2-2 *Two or more atoms linked are called a molecule. Here two hydrogen atoms and one oxygen atom form a molecule of the compound water H_2O.*

Atom

An atom is the smallest particle of an element that retains all the properties of that atom, that is, all hydrogen atoms are alike. They are different from the atoms of all other elements. However, all atoms have certain things in common: They all have an inner part—the nucleus. This is composed of tiny particles called protons and neutrons. An atom also has an outer part. It consists of other tiny particles, called electrons, which orbit around the nucleus. See Figs. 2-3 and 2-4.

Neutrons have no electrical charge, but protons have a positive charge. Electrons are particles of energy and have a negative charge. Because of these charges, protons and electrons are particles of energy. That is, these charges form an electric field of force within the atom. Stated very simply, these charges are always pulling and pushing each other. This makes energy in the form of movement.

The atoms of each element have a definite number of electrons, and they have the same number of protons. A hydrogen atom has one electron and one proton. An aluminum atom has 13 of each. The opposite charges—negative electrons and positive protons—attract each other and tend to hold electrons in orbit. As long as this arrangement is not changed, an atom is electrically balanced. When chemical engineers know the properties of atoms and elements they can then engineer a substance with the properties needed for a specific job. Refrigerants are manufactured in this way.

PROPERTIES OF MATTER

It is important for a refrigeration technician to understand the structure of matter. With this knowledge, the person can understand those factors that affect this structure. These factors can be called the properties of matter. These properties are chemical, electrical, mechanical, or thermal (related to heat). Some of these properties are force, weight, mass, density, specific gravity, and pressure.

Force is described as a push or a pull on anything. Force is applied to a given area. Weight is the force of gravity pulling all matter toward the center of earth. The unit of weight in the English system is the pound. The unit of mass in the metric system is the gram. Mass

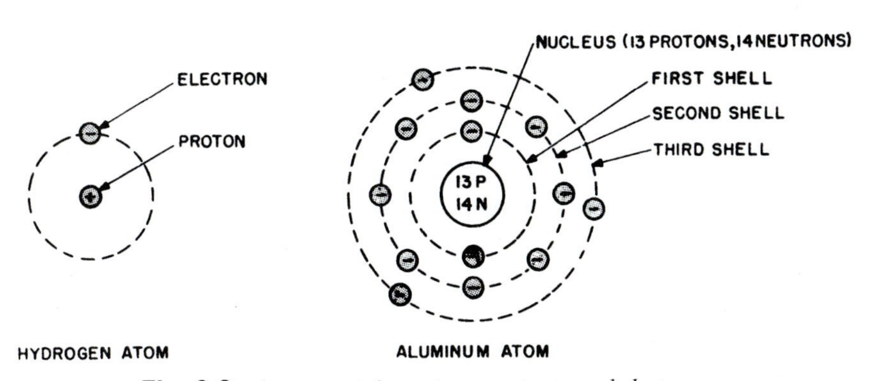

Fig. 2-3 *Atoms contain protons, neutrons, and electrons.*

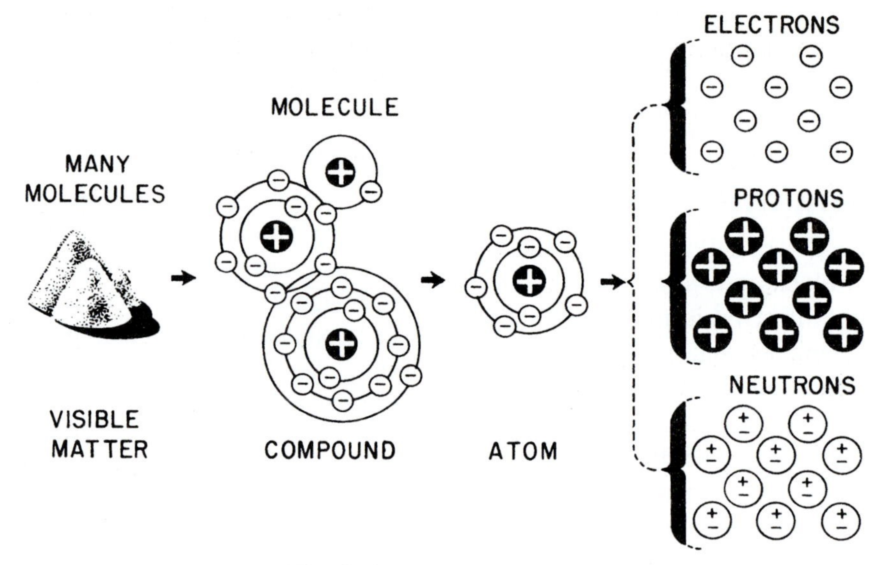

Fig. 2-4 *Molecular structure.*

is the amount of matter present in a quantity of any substance. Mass is not dependent on location. A body has the same mass whether here on earth, on the moon, or anywhere else. The weight does change at other locations. In the metric system, the kilogram (symbol kg) is the unit of mass. In the English system, the *slug* is the unit of mass.

Density is the mass per unit of volume. Densities are comparative figures, that is, the density of water is used as a base and is set at 1.00. All other substances are either more or less dense than water.

The densities of gases are determined by a comparison of volumes. The volume of 1 lb of air is compared to the volume of 1 lb of another gas. Both gases are under standard conditions of temperature and pressure.

The specific gravity of a substance is its density compared to the density of water. Specific gravity has many uses. It can be used as an indicator of the amount of water in a refrigeration system. Testing methods are discussed in later chapters.

PRESSURE

Pressure is a force that acts on an area. Stated in a formula, it becomes:

$$P = \frac{F}{A}$$

where F = force
A = area
P = pressure

The unit of measurement of pressure in the English system is the pound per square foot or pounds per square inch (psi). The metric unit of pressure is the kilopascal (kPa). Pressure measuring elements translate changes or differences in pressure into motion. The three types most commonly used are the diaphragm, the bellows, and the Bourbon spring tube.

Pressure Indicating Devices

Pressure indicating devices are most important in the refrigeration field. It is necessary to know the pressures in certain parts of a system to locate trouble spots.

The diaphragm is a flexible sheet of material held firmly around its perimeter so there can be no leakage from one side to the other. See Fig. 2-5. Force applied to one side of the diaphragm will cause it to move or flex. Diaphragms, in some cases, are a made of a flat sheet of material with a limited range of motion. Other

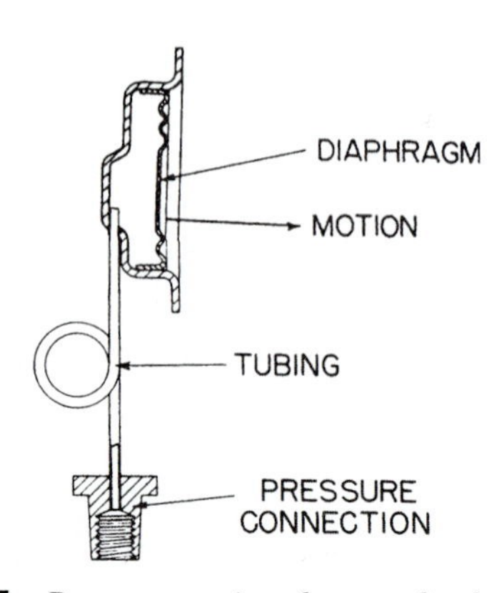

Fig. 2-5 *Pressure sensing element, diaphragm type.*

diaphragms are at least one corrugation or fold. This allows more movement at the point where work is produced.

Some types of pressure of pressure controllers require more motion per unit of force applied. To accomplish the desired result, the diaphragm is joined to the housing by a section with several convolutions or folds called bellows. Thus, the diaphragm moves in response to pressure changes. Each holds only a small amount. See Fig. 2-6. The bellow element may be assembled to extend or to compress as pressure is applied. The bellows itself act as a spring to return the diaphragm section to the original position when the pressure

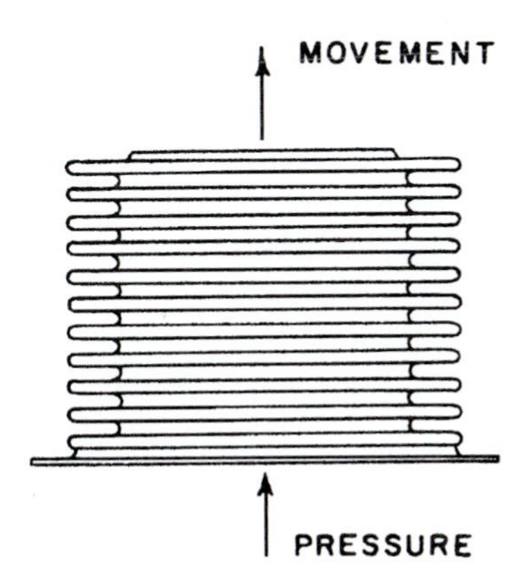

Fig. 2-6 *Pressure sensing element, bellows type.* (Johnson)

differential is reduced to zero. If a higher spring return rate is required, to match or define the measured pressure range, then an appropriate spring is added.

One of the most widely used types of pressure measuring elements is the Bourdon spring tube, discussed in Chap. 1. It is readily adaptable to many types of instruments. See Fig. 2-7. The Bourdon tube is a

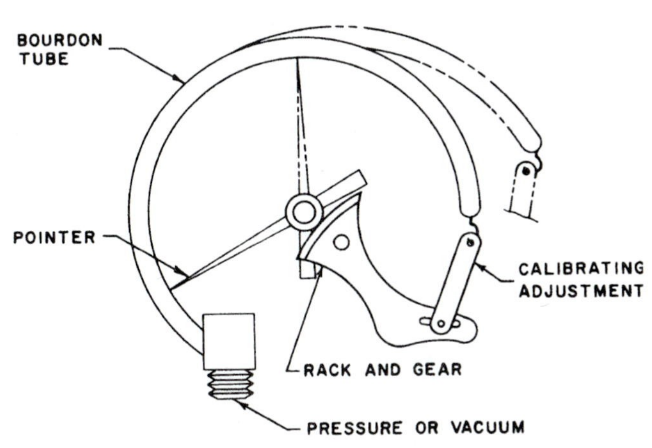

Fig. 2-7 *Pressure sensing element, Bourdon spring tube type.* (Johnson)

flattened tube bent into a spiral or circular form closed at one end. When fluid pressure is applied within the tube, the tube tends to straighten or unwind. This produces motion, which may be applied to position an indicator or actuate a controller.

Pressure of Liquids and Gases

Pascal's law states that when a fluid is confined in a container that is completely filled, the pressure on the fluid is transmitted at equal pressure on all surfaces of the container. The pressure of a gas is the same on all areas of its container.

Atmospheric Pressure

The layer of air that surrounds the earth is several miles deep. The weight of the air above exerts pressure in all directions. This pressure is called, atmospheric pressure. Atmospheric pressure at sea level is 14.7 psi. On converting, it is 1.013×10^5 N/m^2.

The instrument used to measure atmospheric pressure is called a *barometer*. Two common barometers are the aneroid barometer and the mercury barometer. The aneroid barometer has a sealed chamber containing a partial vacuum. As the atmospheric pressure increases, the chamber is compressed causing the needle to move. As the atmospheric pressure decrease, the chamber expands, causing the needle to move in the other direction. A dial on the meter is calibrated to indicate the correct pressure.

The mercury barometer has a glass tube about 34 in. long. The tube holds a column of mercury. The height of this column reflects the atmospheric pressure. Standard atmospheric pressure at sea level is indicated by 29.92 in. of mercury. That converts to 759.96 mm.

Gage Pressure

Gage pressure is the pressure above or below atmospheric pressure. This is the pressure measured with most gages. A gage that measures both pressure and vacuum is called a *compound gage*. Vacuum is pressure that is below atmospheric pressure. A gage indicates zero pressure before you start to measure. It does not take the pressure of the atmosphere into account. In the customary system, gage pressure is measured in pounds per square inch (psi).

Absolute Pressure

Absolute pressure is the sum of the gage pressure and atmospheric pressure. This is abbreviated as *psia*. A good example of this is the pressure in a car tire. This is usually 28 psi. That would be 42.7 psia. For example:

$$\begin{aligned} \text{psi (gage)} &= 28 \text{ psi} \\ \text{Atmospheric pressure} &= 14.7 \text{ psi} \\ \text{Absolute pressure} &= 42.7 \text{ psi} \end{aligned}$$

The abbreviation for pounds per square inch gage is *psig*. The abbreviation for pounds per square inch absolute is psia. Absolute is found by adding 14.7 to the psig. However, the atmospheric pressure does vary with altitude. In some cases, it is necessary to convert to the atmospheric pressure at the altitude where the pressure is being measured. This small difference can make a tremendous difference in correct readings of psia. To convert psi to kPa (kilopascals), the metric unit of pressure, multiply psi by 6.9.

Compression Ratio

Compression ratio is defined as the absolute head pressure divided by the absolute suction pressure.

$$\text{Compression ratio} = \frac{\text{absolute head pressure}}{\text{absolute suction pressure}}$$

Example 1

When the gage reading is 0 or above.

Absolute head pressure = gage reading + 15 lbs (14.7 actually)

Absolute suction pressure = gage reading + 15 lbs (14.7 actually)

Example 2

When the low side reading is in vacuum range.

Absolute head pressure = gage reading + 15 lbs (14.7 actually)

$$\text{Absolute suction pressure} = \frac{30 - \text{gage reading in inches}}{2}$$

The calculation of compression ratio can be illustrated by the following.

Example 1

Head pressure = 160 lbs

Suction pressure = 10 lbs

$$\text{Compression ratio} = \frac{\text{absolute head pressure}}{\text{absolute suction pressure}} = \frac{160+15}{10+15} = \frac{175}{25} = 7{:}1$$

Example 2

Head pressure = 160 lbs

Suction pressure = 10 in. of vacuum

Absolute head pressure = 160 + 15 = 175 lbs

$$\text{Absolute suction pressure} = \frac{30-10}{2} = \frac{20}{2} = 10$$

$$\text{Compression ratio} = \frac{\text{absolute head pressure}}{\text{absolute suction pressure}} = \frac{175}{10} = 17.5{:}1$$

The preceding examples show the influence of back pressure on the compression ratio. A change in the head pressure does not produce such a dramatic effect. If the head pressure in both cases were 185 lb, the compression ratio in Example 1 would be 8:1, and in Example 2 it would be 20:1.

A high compression ratio will make a refrigeration system run hot. A system with a very high compression ratio may show a discharge temperature as much as 150°F [65.6°C] above normal. The rate of a chemical reaction approximately doubles with each 18°F (7.8°C) rise in temperature. Thus, a system running an abnormally high head temperature will develop more problems than, a properly adjusted system. The relationship between head pressure and back (suction) pressure, wherever possible, should be well within the accepted industry bounds of a 10:1 compression ratio.

It is interesting to compare, assuming a 175-lb heat pressure in both cases: *Refrigerant* 12 (R-12) versus *Refrigerant* 22 (R-22) operating at –35°F (–37°C) coil. At a –35°F (–37°C) coil, as described, the R-22 system would show a 10.9:1 compression ratio while the R-12 system would be at 17.4:1. The R-22 system is a borderline case. The R-12 system is not in the safe range and it would run very hot with all of the accompanying problems.

A number of other factors will produce serious high-temperature conditions. However, high compression ratio alone is enough to cause serious trouble. The thermometer shown in Fig. 2-8 reads temperature as a function of pressure. This device reads the pressure of R-22 and R-12. It also indicates the temperature in degrees Fahrenheit on the outside scale.

TEMPERATURE AND HEAT

The production of excess heat in a system will cause problems. Normally, matter expands when heated. This is the principle of thermal expansion. The linear

Fig. 2-8 *Thermometer and pressure gage.* (Marsh)

dimensions increase, as does the volume. Removing heat from a substance causes it to contract in linear dimensions and in volume. This is the principle of the liquid in a glass thermometer.

Temperature is the measure of hotness or coldness on a definite scale. Every substance has temperature.

Molecules are always in motion. They move faster with a temperature increase, and more slowly with a temperature decrease. In theory, molecules stop moving at the lowest temperature possible. This temperature is called *absolute zero.* It is approximately –460°F (–273°C).

The amount of heat in a substance is directly related to the amount of molecular motion. The absence of heat would occur only at absolute zero. Above that temperature, there is molecular motion. The amount of molecular motion corresponds to the amount of heat.

The addition of heat causes a temperature *increase.* The removal of heat causes a temperature *decrease.* This is true except when matter is going through a change of state.

Heat is often confused with temperature. Temperature is the measurement of heat intensity. It is not a direct measure of heat content. Heat content is not dependent on temperature. Heat content depends on the type of material, the volume of the material, and the amount of heat that has been put into or taken from the material. For example, one cup of coffee at 200ºF (93.3°C) contains less heat than one gallon of coffee at 200ºF (93.3°C). The cup at 200ºF (93.3°C) can also contain less heat than the gallon at a lower temperature of 180ºF (82.2°C).

Specific Heat

Every substance has a characteristic called *specific heat.* This is the measure of the temperature change in a substance when a given amount of heat is applied to it.

One Btu (British thermal unit) is *the amount of heat required to raise* 1 *lb of water* 1ºF *at* 39º *F.* With a few exceptions, such as ammonia gas and helium, all substances require less heat per pound than water to raise the temperature 1° F.

Thus, the specific heat scale is based on water, which has a specific heat of 1.0. The specific heat of aluminum is 0.2. This means that 0.2 Btu will raise the temperature of 1 lb of aluminum by 1° F. One Btu will raise the temperature of 5 lb of aluminum by 1° F, or of 1 lb, 5° F.

Heat Content

Every substance theoretically contains an amount of heat equal to the heat energy required to raise its temperature from absolute zero to its temperature at a given time. This is referred to as *heat content,* which consists of *sensible heat* and *latent heat.* Sensible heat can be felt because it changes the temperature of the substance. Latent heat, which is not felt, is seen as it changes the state of substance from solid to liquid or liquid to gas.

Sensible Heat

Heat that changes the temperature of a substance, without changing its state, when added or removed is called sensible heat. Its effect can be measured with a thermometer in degrees as the difference in temperatures of a substance (Delta T, or ΔT).

If the weight and specific heat of a medium are known, the amount of heat added or removed in Btu can be computed by multiplying the sensible change (ΔT) by the weight of the medium and by its specific heat. Thus, the amount of heat required for raising the temperature of one gallon of water (8.34 pounds) from 140ºF to 160ºF is:

$$\text{Sensible heat} = \Delta T \times \text{weight} \times \text{specific heat} =$$
$$(160 - 140) \times 8.34 \times 1 = 20 \times 8.34 = 166.8 \text{ Btu}$$

Latent Heat

The heat required to change the state of a substance without changing its temperature is called latent heat, or hidden heat. Theoretically, any substance can be a gas, liquid, or solid, depending on its temperature and pressure. It takes heat to change a substance from a solid to a liquid or, from a liquid to a gas.

For example, it takes 144 Btu of latent heat to change 1 lb of ice at 32ºF to 1 lb of water at 32ºF. It takes 180 Btu of sensible heat to raise the temperature of 1 lb of water 180ºF from 32ºF to 212ºF. It takes 970 Btu of latent heat to change 1 lb of water to steam at 212ºF. When the opposite change is affected, equal amounts of heat are taken out or given up by the substance.

This exchange of heat, or the capability of a medium, such as water to take and give up heat, is the basis for most of the heating and air-conditioning industry. Most of the functions of the industry are concerned with adding or removing heat at a central point and distributing the heated or cooled medium throughout a structure to warm or cool the space.

Other Sources of Heat

Other heat in buildings comes principally from four sources: electrical energy, the sun, outdoor air temperatures and the building's occupants. Every kilowatt of electrical energy in use produces 3413 Btu/h, whether it is used in lights, the heating elements of kitchen ranges, toasters, or irons.

The sun is a source of heat. At noon, a square foot of surface directly facing the sun may receive 300 Btu/h, on a clear day. When outdoor air temperatures exceed the indoor space temperature. The outdoors become a source of heat. The amount of heat communicated depends on the size and number of windows, among other factors.

The occupants of a building are a source of heat, since body temperatures are higher than normal room temperatures. An individual, seated and at rest, will give off about 400 Btu/h in a 74°F (23.3°C) room. If the person becomes active, this amount of heat may be increased two or three times, depending upon the activity involved. Some of this heat is sensible heat, which the body gives off by convection and radiation. The remainder is latent heat, resulting from the evaporation of visible or invisible perspiration. The sensible heat increases the temperature of the room. The latent heat increases the humidity. Both add to the total heat in the room.

REFRIGERATION SYSTEMS

The refrigerator was not manufactured until the 1920s. Before that time, ice was the primary source of refrigeration. A block of ice was kept in the icebox. The icebox was similar to the modern refrigerator in construction. It was well insulated and had shelves to store perishables. The main difference was the method of cooling.

The iceman came about once a week to put a new 50- or 100-lb block of ice in the icebox. How much cooling effect does a 50-lb block of ice produce? The latent heat of melting for 1 lb of ice is 144 Btu. The latent heat of melting for a 50-lb block is 50 × 144, or 7200 Btu. The latent heat of melting for the 100-lb block was 14,400 Btu. The refrigeration was accomplished by convection in the icebox.

One of the first refrigerators is shown in Fig. 2-9. The unit on the top identified it as a refrigerator instead of an icebox. Some of these units, made in the 1920s, are still operating today.

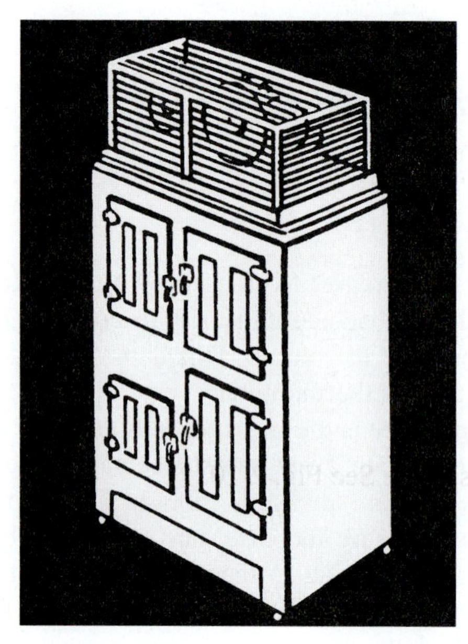

Fig. 2-9 *Early modification of the icebox to make it a refrigerator unit.*

Refrigeration from Vaporization (Open System)

The perspiration on your body evaporates and cools your body. Water kept in a porous container is cooled on a hot day. The water seeps from the inside. There is a small amount of water on the outside surface. The surface water is vaporized—it evaporates.

Much of the heat required for vaporization comes from the liquid in the container. When heat is removed this way the liquid is cooled. The heat is carried away with the vapor.

Basic Refrigeration Cycle

A substance changes state when the inherent amount of heat is varied. Ice is water in a solid state and steam is a vapor state of water. A solid is changed to liquid and a liquid to a vapor by applying heat. Heat must be added to vaporize or boil a substance. It must, be taken away to liquefy or solidify a substance. The amount of heat necessary will depend on the substance and the pressure changes in the substance.

Consider, for example, an open pan of boiling water heated by a gas flame. The boiling temperature of water at sea level is 212ºF (100°C). Increase the temperature of the flame and the water will boil away more rapidly, although the temperature of the water will not change. To heat or boil a substance, heat must be removed from another substance. In this case, heat is removed from the gas flame. Increasing the temperature of the flame merely speeds the transfer of heat. It does not increase the temperature of the water.

A change in pressure will affect the boiling point of a substance. As the altitude increases above sea level, the atmospheric pressure and the boiling temperature drop. For example, water will boil at 193°F (89.4°C) at an altitude of 10,000 feet. At pressures below 100 psi, water has a boiling point of 338°F (170°C).

The relationship of pressure to refrigeration is shown in the following example. A tank contains a substance that is vaporized at atmospheric pressure. However, it condenses to a liquid when 100 lb of pressure are applied. The liquid is discharged from the tank through a hose and nozzle into a long coil of tubing to the atmosphere. See Fig. 2-10.

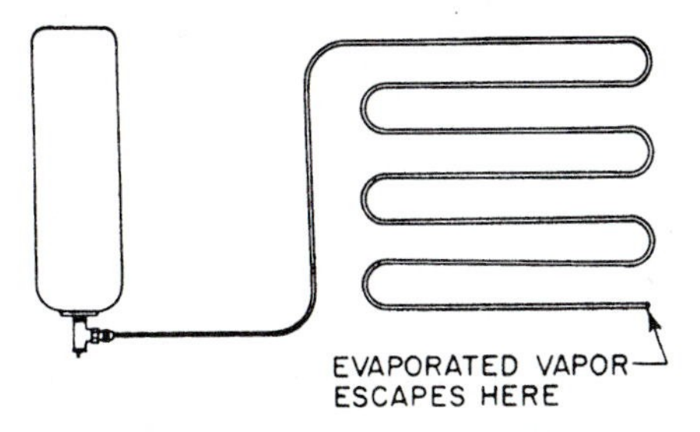

Fig. 2-10 *Basic step of refrigeration.* (Johnson)

As the liquid enters the nozzle, its pressure is reduced to that of the atmosphere. This lowers its vaporization or boiling point. Part of the liquid vaporizes or boils, using its own heat. The unevaporated liquid is immediately cooled as its heat is taken away. The remaining liquid takes heat from the metal coil or tank and vaporizes, cooling the coil. The coil takes heat from the space around it, cooling the space. This unit would continue to provide cooling or refrigeration for as long as the substance remained under pressure in the tank.

All of the other components of a refrigeration system are merely for reclaiming the refrigeration medium after it has done its job of cooling. The other parts of a refrigeration system, in order of assembly, are tank, or liquid receiver, expansion valve, evaporator coil, compressor, and condenser.

Figure 2-11 illustrates a typical refrigeration system cycle. The refrigerant is in a tank or liquid receiver under high pressure and in a liquid state. When the refrigerant enters the expansion valve, the pressure is lowered, and the liquid begins to vaporize. Complete evaporation takes place when the refrigerant moves into the evaporator coil. With evaporation, heat must be added to the refrigerant. In this case, the heat comes from the evaporator coil. As heat is removed from the coil, the coil is cooled. The refrigerant is now a vapor under low pressure. The evaporator section of the system is often called the low pressure, back pressure, or suction side. The warmer the coil, the more rapidly evaporation takes place and the higher the suction pressure becomes.

The compressor then takes the low-pressure vapor and builds up the pressure sufficiently to condense the refrigerant. This starts the high side of the system. To return the refrigerant to a liquid state (to condense it), heat picked up in the evaporator coil and the compressor must be removed. This is the function of the condenser used with an air- or water-cooled coil. Being cooler than the refrigerant, the air or water absorbs its heat. As it cools, the refrigerant condenses into a liquid and flows into the liquid receiver or tank. Since the pressure of the refrigerant has been increased, it will condense at a lower temperature.

In some systems, the liquid receiver may be part of another unit such as the evaporator or condenser.

Capacity

Refrigeration machines are rated in tons of refrigeration. This rating indicates the size and ability to produce cooling energy in a given period. One ton of refrigeration has cooling energy equal to that produced by one ton of ice melting in 24 hours. Since it takes 288,000 Btu of heat to melt 1 ton of ice, a 1-ton machine will absorb 288,000 Btu in a 24 hour period.

Refrigerants

Theoretically, any gas that can be alternately liquefied and vaporized within mechanical equipment can serve as a refrigerant. Thus, carbon dioxide serves as a refrigerant on many ships. However, the piping and machinery handling it must be very heavy-duty.

Practical considerations have led to the use of several refrigerants that can be safely handled at moderate pressures by equipment having reasonable mechanical

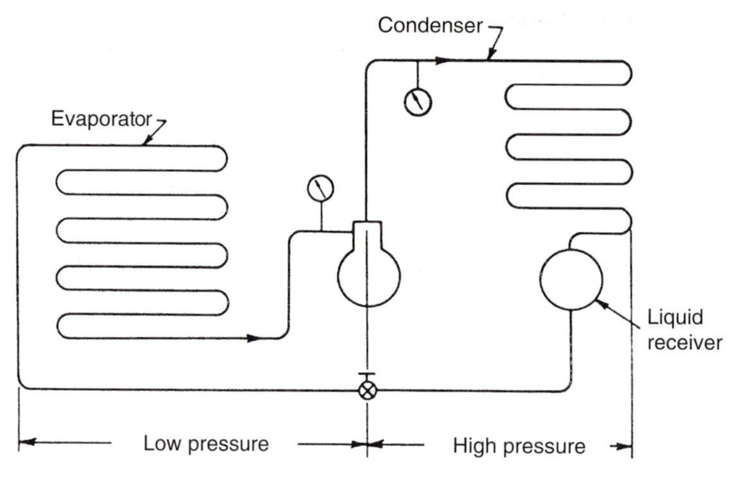

Fig. 2-11 *High and low sides of a refrigeration system.*

strength and with lines of normal size and wall thickness. While no substance possesses all the properties of an ideal refrigerant, the hydrocarbon (Freon) refrigerants come quite close.

Refrigerant 12 is made of carbon (C), chlorine (Cl), and fluorine (F). Its formula is CCl_2F_2. It is made of a combination of elements. Refrigerant 22 is made of carbon (C), hydrogen (H), chlorine (Cl), and fluorine (F). Its formula $CHClF_2$ is slightly different from that of R-12.

Each of these manufactured refrigerants has its own characteristics, such as odor and boiling pressure.

Refrigerants are the vital working fluids in refrigeration systems. They transfer heat from one place to another for cooling air or water in air-conditioning installations.

Many substances can be used as refrigerants, including water under certain conditions. The following are some common refrigerants:

- *Ammonia.* The oldest commonly used refrigerant, still used in some systems. It is very toxic.
- *Sulphur dioxide.* First to replace ammonia and to be used in small domestic machines. It is very toxic.
- *Refrigerant* 12 . The first synthetic refrigerant to be used commonly. Used in a large number of reciprocating machines operating in the air-conditioning range. It is nontoxic.
- *Refrigerant* 22. Used in many of the same applications as R-12. Its lower boiling point and higher latent heat permit the use of smaller compressors and refrigerant lines. It is nontoxic.
- *Refrigerant* 40. Methylchloride is used in the commercial refrigeration field, particularly in small installations. Today it is no longer used. It will explode when allowed to combine with air. It is nontoxic.

Refrigerant Replacements and the Atmosphere

Refrigerants such as ammonia are used for low-temperature systems. These include food and process cooling, ice rinks, and so forth. Propane has been used for some special applications. Now that chlorinated hydrocarbons have been determined to be harmful to the earth's ozone layer, R-11 (CCl_3F), R-12 (CCl_2F_2), and other similar compounds that were in common use along with the less harmful refrigerant R-22 ($CHClF_2$) have had much attention in the press. Recent international protocols (standards) have set schedules for the elimination of damaging refrigerants from commercial use. Replacements have been, and are being, developed. Part of the challenge is technical and part is economic. First, to find a fluid that has optimal characteristics and is safe is a challenge. Second, to encourage manufacture in sufficient quantities to produce and distribute the fluid at an affordable price is another. R-123 ($CHCl_2CF_3$) has been developed as a near-equivalent replacement for R-11, with R-134a (CH_2FCF_3) replacing R-12. R-123 still comes under criticism for having some chlorine in it. R-134a can be bought at auto supplies stores for automobile air conditioners. Most new cars are required to have R-134a in their A/C systems.

R-22 is used widely in residential and commercial air-conditioning scroll compressor systems. It too will be phased out someday (probably during the period 2020–2030). However, finding a suitable, widely accepted replacement has not come as quickly as first thought.

REVIEW QUESTIONS

1. Define refrigeration.
2. Define refrigerant.
3. What is a compound?
4. What is an atom?
5. What is a Bourdon spring?
6. What is the difference between an aneroid barometer and a mercury barometer?
7. Describe gage pressure.
8. Describe Pascal's law.
9. What is absolute zero?
10. What is specific heat?
11. What is sensible heat?
12. What is latent heat?
13. How much heat is produced by a kilowatt-hour of electrical energy?
14. What amount of heat is needed to melt 1 ton of ice in 24 hours?
15. What is absolute pressure?

3

CHAPTER

Voltage, Current, and Resistance

PERFORMANCE OBJECTIVES

After studying this chapter, you should:

1. Understand the five ways electricity is produced.
2. Understand how the units of measurements for electricity were developed and are used.
3. Be able to work with Ohm's law problems.
4. Understand how volts, ohms, and amps are related in an electrical circuit.
5. Understand how to work electrical-power problems.

Every electrical circuit has current, voltage, and resistance. The movement of electrons along a wire or conductor is referred to as current.

Voltage is the electrical pressure that pushes electrons through a resistance. Voltage is measured in volts (V). Electrical pressure, *electromotive force* (EMF), difference of potential, and voltage are all terms used to designate the difference in electrical pressure or potential. For example, a battery is a common power source. It furnishes the energy needed to cause electrical devices to function. A battery has a difference of potential between its terminals. This difference of potential is called voltage.

Current is the flow of electrons. Current is measured in amperes (A). A coulomb is 6.28×10^{18} electrons. When a coulomb is standing still, or static, it is referred to as static electricity. Once the coulomb is in motion, it is referred to as current electricity. The movement of 1 C past a given point in 1 s is 1 A. At times, it is necessary to refer to smaller units of ampere. The milli-ampere is one-thousandth of an ampere (0.001 A or 1 mA). A microampere is one-millionth of an ampere (0.000001A or 1 μA). These smaller units are commonly used in working with transistorized circuits.

Resistance is the opposition offered to the passage of electrical current. Resistance is measured in ohms (Ω). The ohm is the amount of opposition presented by a substance when a pressure of 1 V is applied and 1 A of current flows through it.

OHM'S LAW

Ohm's law states the relationship among the three factors of an electrical circuit. A circuit is a path for the flow of electrons from one side of a power source or potential difference to the other side. See Fig. 3-1.

Ohm's law states that the current (I) in a circuit is equal to the voltage (E) divided by the resistance (R). Ohm's law is expressed by the following three formulas:

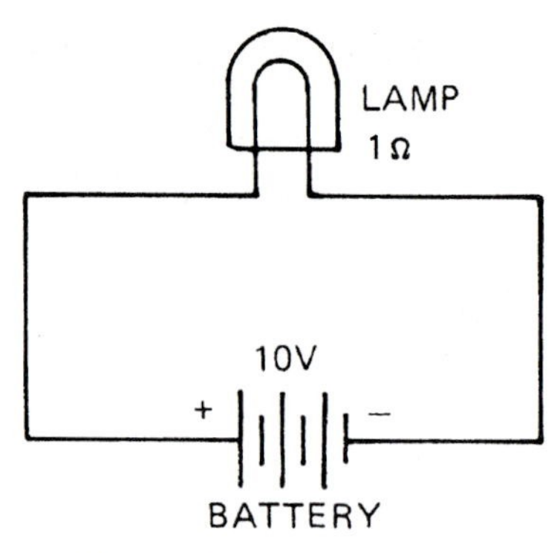

Fig. 3-1 *A simple circuit.*

$$I = \frac{E}{R}$$

$$R = \frac{E}{I}$$

$$E = I \times R$$

The best way to become familiar with Ohm's law is to do a few problems. If two of the factors or quantities are known, it is easy to find the unknown. Since the size of the wire used in a circuit is determined by the amount of current it is to handle, it is necessary to find the current and check a chart to see what size the wire should be. See Tables 3-1 and 3-2.

Problem: The voltage available is 120 V. The resistance of the circuit is 60 Ω. What is the current? What size of wire will handle this amount of current? See Fig. 3-2 and Tables 3-1 and 3-2.

$$I = \frac{E}{R}$$

$$I = \frac{120}{60}$$

$$I = 2\text{ A}$$

Now that you know the amount of current: 2 A refer to Table 3-2 to find the size of wire that would be used to handle 2 A. The table says that a No. 1 8 wire handles 2.32 A. Thus, there is a safety factor of 0.32 A, or 320 mA.

SERIES CIRCUITS

A series circuit consists of two or more consuming devices connected with one terminal after the other. Figure 3-3 shows that the current through the circuit is the same in all devices. However, the total resistance is

Table 3-1 Current-Carrying Ability of Copper Wire with Different Types of Insulation Coating

	In Conduit or Cable		In Free Air		
Wire size	Type RHW* THW*	Type TW, R*	Type RHW* THW*	Type TW, R*	Weather-Proof Wire
14	15A	15	20	20	30
12	20A	20	25	25	40
10	30A	30	40	40	55
8	45A	40	65	55	70
6	65A	55	95	80	100
4	85A	70	125	105	130
3	100A	80	145	120	150
2	115A	95	170	140	175
1	130A	110	195	165	205
0	150A	125	230	195	235
00	175A	145	265	225	275
000	200A	165	310	260	320

*Types "RHW," "THW," "TW," or "R" are identified by markings on outer cover.

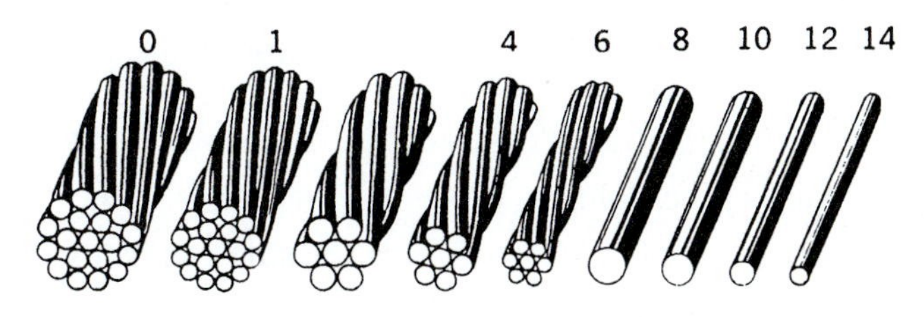

Actual size of copper conductors. Note the larger the gage number, the smaller the diameter of the wire.

Table 3-2 Wire Size and Current-Carrying Capacity

Wire Size A.W.G. (B & S)	Current-Carrying Capacity at 700 cm/A
8	23.6
10	14.8
12	9.33
14	5.87
16	3.69
18	2.32
20	1.46
22	.918
24	.577
26	.363
28	.228
30	.144
32	.090
34	.057
36	.036
38	.022
40	.014

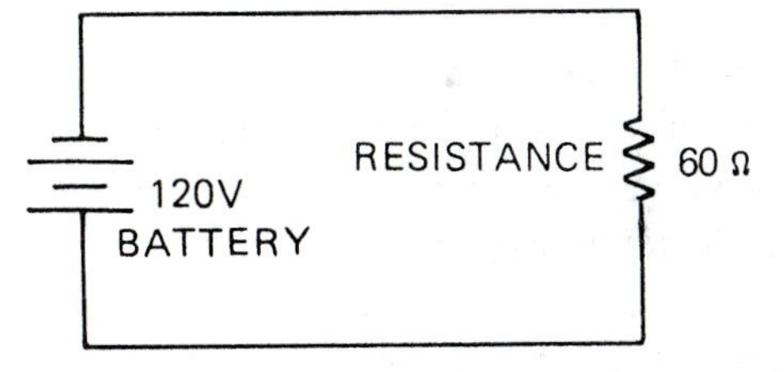

Fig. 3-2 *A circuit with one resistor.*

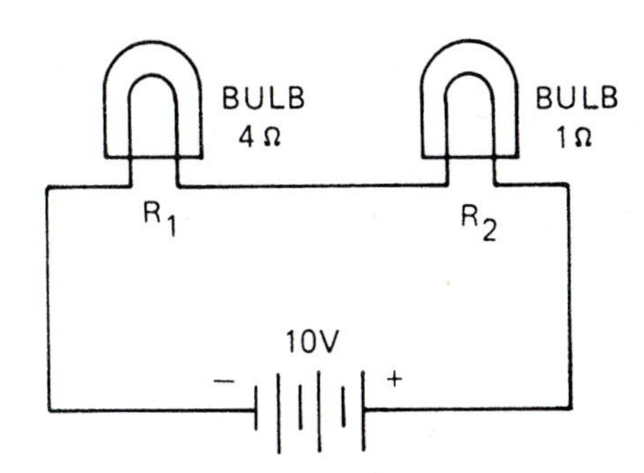

Fig. 3-3 *A series circuit with two bulbs.*

found by adding the resistances. Thus, $R_T = R_1 + R_2 + R_3 + ...$Therefore, if a resistance of 1 Ω and a resistance of 4 Ω are connected in series, the total resistance is 5 Ω. To find the total current, divide the total resistance into the voltage (in this case 10 V). That gives Ohm's law another use—finding the total current in the circuit.

Since the total current in a series circuit is the current through each resistance, the individual light bulbs will have the same current through them. Or,

$$I = \frac{10\text{ V}}{5\,\Omega} = 2\text{ A}$$

There are three basic laws regarding series circuits:

- Current is the same in all parts of the circuit.
- Voltage drop across each resistance varies according to the resistance of the individual device.
- Resistance is added to equal the total. Or, $R_T = R_1 + R_2 + R_3 + \cdots$

Another example of how series circuit laws and Ohm's law can be of assistance follows:

In a circuit with 120 V and a current of 5 A, what is the resistance?

$$R = \frac{E}{I}$$

$$R = \frac{120}{5}$$

$$R = 24\ \Omega$$

In a circuit with 20 A and 40 Ω, what is the voltage needed for normal operation?

$$E = I \times R$$

$$E = 20 \times 40$$

$$E = 800\ \text{V}$$

Suppose you have a series circuit for which you know the voltage (120 V), the current (4 A), and the resistance of one of the two resistors (20 Ω). How do you find the value of the other resistor in the circuit?

Use Ohm's law and the laws of a series circuit:

$$R = \frac{E}{I}$$

$$R = \frac{120}{4}$$

$$R = 30\ \Omega$$

Subtract the known resistance of 20 Ω from the total of 30 Ω. This gives 10 Ω for the missing resistor value.

PARALLEL CIRCUITS

Parallel circuits are the most common type of circuit. They are used for wiring lights in a house or for connecting equipment that must operate on the same voltage as the power source.

A parallel circuit consists of two or more resistors connected as in Fig. 3-4. Both resistors have the same voltage available as furnished by the battery. Thus, if the battery puts out 12 V, the resistors will have 12 V across them.

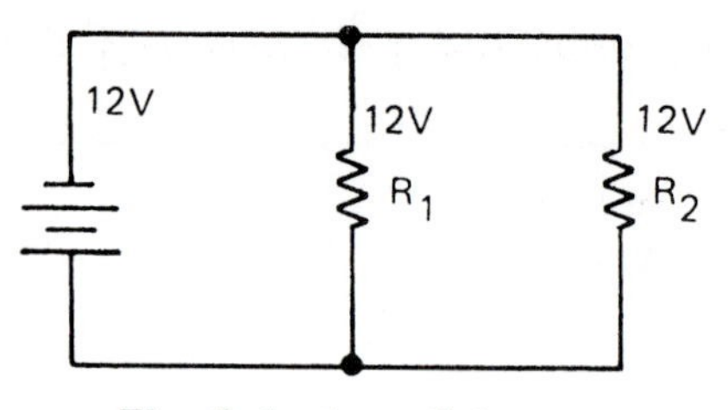

Fig. 3-4 *A parallel circuit.*

There are three basic laws regarding parallel circuits:

1. The voltage is the same across each resistor.
2. The current divides according to the resistance.
3. There are two ways of finding total resistance.

This formula can be used for only two resistors:

$$R_T = \frac{R_1 \times R_2}{R_1 + R_2}$$

This formula can be used for any number of resistors:

$$\frac{1}{R_T} = \frac{1}{R_1} + \frac{1}{R_2} + \frac{1}{R_3} \cdots$$

Current in a Parallel Circuit

The current divides according to the resistance. For example:

$$R_1 = 10\ \Omega$$

$$R_2 = 20\ \Omega$$

$$R_3 = 30\ \Omega$$

If the voltage is 60 V, then the following method is used to determine the current through each resistor.

Voltage across each resistor is the same (60 V). Therefore, the resistance and the voltage are known. Use Ohm's law and find the current through each resistor:

$$I_{R_1} = \frac{E_{R_1}}{R_1} \quad \text{or} \quad I_{R_1} = \frac{60}{10} = 6\ \text{A}$$

$$I_{R_2} = \frac{E_{R_2}}{R_2} \quad \text{or} \quad I_{R_2} = \frac{60}{20} = 3\ \text{A}$$

$$I_{R_3} = \frac{E_{R_3}}{R_3} \quad \text{or} \quad I_{R_3} = \frac{60}{30} = 2\ \text{A}$$

Since current is divided according to the resistance of the individual resistor, the total current is found by adding the individual currents:

$$I_{R_T} = I_{R_1} + I_{R_2} + I_{R_3}$$

Resistance in a Parallel Circuit

As has already been stated, the total resistance of a parallel circuit can be found by two methods. For instance, find a common denominator for the fractions: Add the numerators:

$$R_1 = 10\ \Omega$$
$$R_2 = 15\ \Omega$$
$$R_3 = 20\ \Omega$$

Add the numerators:

$$\frac{1}{R_T} = \frac{1}{R_1} + \frac{1}{R_2} + \frac{1}{R_3} \qquad \frac{1}{R_T} = \frac{1}{10} + \frac{1}{15} + \frac{1}{20}$$

Find a common denominator for the fractions:

$$\frac{1}{R_T} = \frac{6+4+3}{60}$$

Add the denominators for the fractions:

$$\frac{1}{R_T} = \frac{13}{60}$$

Add the numerators:

$$\frac{R_I}{1} = \frac{60}{13}$$

Invert:

$$R_T = 4.6153\,846$$

Notice that the total resistance is always smaller than the resistance of the smallest resistor. If this circuit with three resistors of 10, 15, and 20 Ω in parallel has 120 V applied, what is the current through each resistor? See Fig. 3-5.

Total resistance is 4.6153846 Ω. Applied voltage is 120 V. Therefore, using Ohm's law you can find the total current in the circuit:

$$I_T = \frac{120}{4{,}6153846} = 26.0 \text{ A}$$

This means that the total current in the circuit divides three ways through each resistor. Now, check to see if each resistor has the proper amount of current so that the total of the individual currents is 26.

$$I_{R_1} = \frac{120}{10} = 12 \text{ A}$$
$$I_{R_2} = \frac{120}{15} = 8 \text{ A}$$
$$I_{R_3} = \frac{120}{20} = 6 \text{ A}$$

Add these individual currents. The result is 26 A.

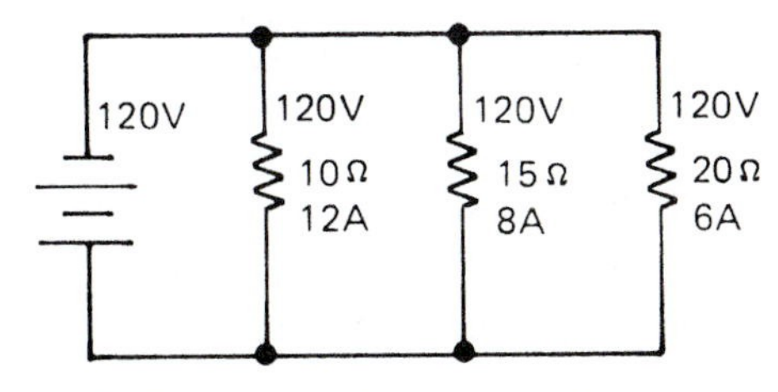

Fig. 3-5 *A parallel circuit with three resistors.*

AC AND DC POWER

Electrical power can be supplied in two different forms—AC and DC. The difference is in the characteristics of the current flow. A power source that causes current to flow in only one direction is referred to as a *direct current* (DC) source. A power source that causes current to flow alternately in one direction and then in the other is referred to as an *alternating current* (AC) source.

Batteries and automobile DC generators are common examples of DC electrical power sources. Normally, a DC current flow is thought of as a continuous unidirectional (*uni* means one) flow that is constant in magnitude. However, a pulsating current flow that changes in magnitude, but not direction, is also considered DC.

The power supplied by power companies in the United States is the most common example of AC power. If the magnitude of the current is recorded as it varies with time, the shape of the resultant curve is called the waveform. The waveform produced by the power companies' generators is a sine wave as, shown in Fig. 3-6.

When the waveform of an AC voltage or current passes through a complete set of positive and negative values, it completes a cycle (now called a *hertz*). The frequency of an AC voltage or current is the number of hertz (cycles) that occur in 1 s. The frequency of the voltage supplied by U. S. power companies is 60 Hz. In Europe, it is 50 Hz. Hertz is abbreviated as Hz.

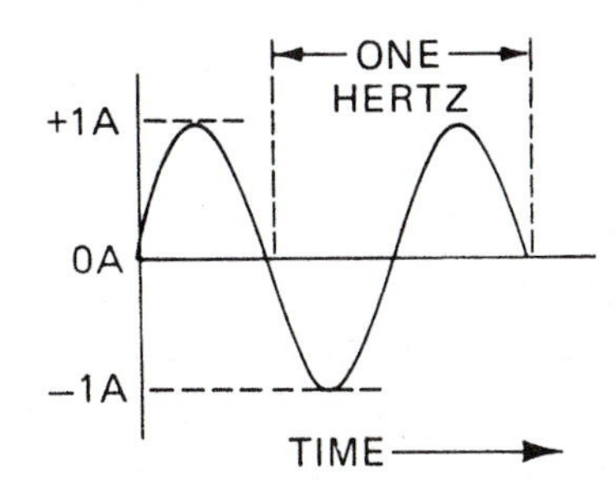

Fig. 3-6 *One Hertz of alternating current.*

All calculations using AC voltage are based on sine waves. Four values of sine waves are of particular importance.

Instantaneous value. The voltage or current in an AC circuit is continuously changing. The value varies from zero to maximum and back to zero. If you measure the value at any given instant, you will obtain the instantaneous voltage or current value.

Maximum value. For two brief instants in each Hertz the sine wave reaches a maximum value. One is a positive maximum and the other is a negative maximum. Maximum value is often referred to as peak value. The two terms have identical meanings and are interchangeable.

Average value. The positive and negative halves of a sine wave are identical. Thus, the average value can be found by determining the area below the wave, and calculating what DC value would enclose the same area over the same amount of time. For either the positive or negative, half of the sine wave, the average value is 0.636 times the maximum value.

Effective. The effective value is often referred to as rms (*root-mean-square*). The effective value is the same as a DC voltage or current required providing the same average power or heating effect. The heating effect is independent of the direction of electron flow. The effective value of an AC sine wave is equal to 0.7071 times the maximum value. Thus, the alternating current of a sine wave, having a maximum value of 10 A, produces in a circuit the amount of heat produced by a DC current of 7.071 A.

The heating effect varies with the square of the voltage or current. If we square the instantaneous values of a voltage or current sine wave, we obtain a wave that is proportional to the instantaneous power, or heating effect of the original sine wave. The average of this new waveform represents the average power that will be supplied. The square root of this average value is the voltage or current that represents the heating effect of the original sine wave of voltage or current. This is the effective value of the wave, or the rms value, the square root of the average of the squared waveform.

The effective values of voltage and current are more important than instantaneous, maximum, or average values. Most AC voltmeters and ammeters are calibrated to read in rms values.

Phase

The phase of an AC voltage refers to the relationship of its instantaneous polarity to that of another AC voltage. Figure 3-7 shows two AC sine waves in phase, but unequal in amplitude. Figure 3-8 shows the two sine waves 45° out-of-phase. Figure 3-9 shows the two sine waves 180° out-of-phase.

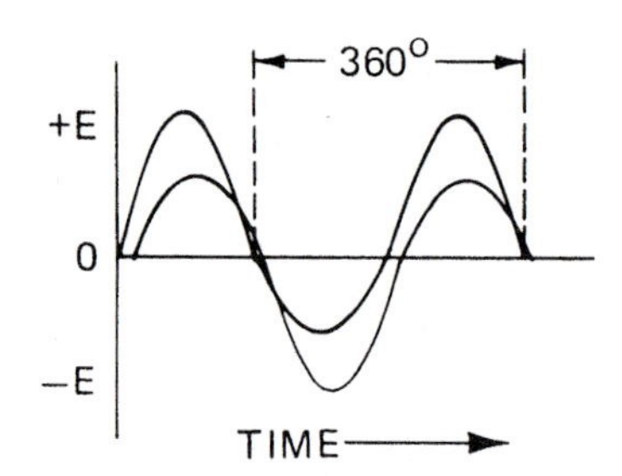

Fig. 3-7 *Two AC waveforms, unequal in amplitude, but in-phase.*

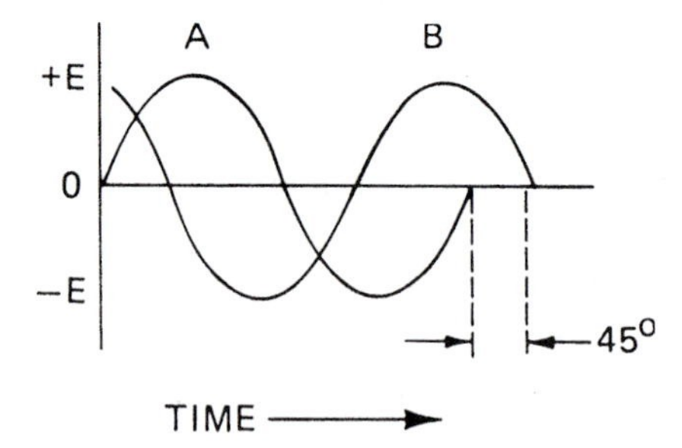

Fig. 3-8 *Two AC waveforms, out-of-phase by 45°.*

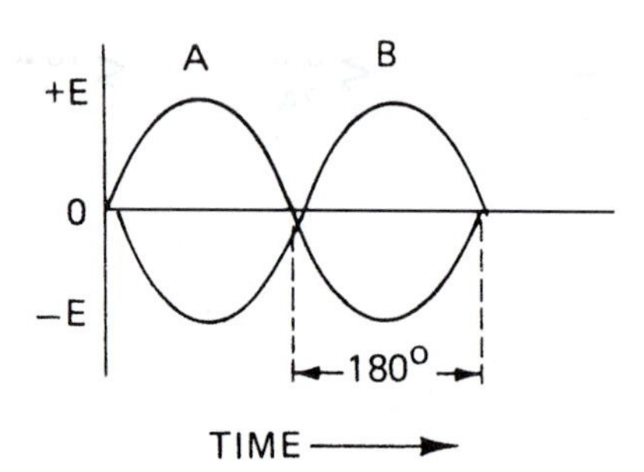

Fig. 3-9 *Two AC waveforms, out-of-phase by 180°.*

The length of a sine wave can be measured in angular degrees because each hertz is a repetition of the previous one. One complete cycle (hertz) of a sine wave is 360.

Power in DC Circuits

Whenever a force causes motion, work is performed. Electrical force is expressed as voltage. When voltage causes a movement of electrons (current) from one point to another, energy is expended. The rate of work, or the rate of producing, transforming, or expending energy, is generally expressed in watts or kilowatts. A kilowatt (kW) is 1000 W. In a DC circuit, 1 V forcing a current of 1 A through a 1-Ω resistance results in 1 W of power being expended. The formula for this is:

$$P \text{ (watts)} = E \text{ (volts)} \times I \text{ (amperes)}$$

Find the power used by the light bulbs in a series circuit. Use the values given in Fig. 3-3.

$$I = \frac{E}{R_T}$$

$$I = \frac{10\text{ V}}{5\,\Omega}$$

$$I = 2\text{ A}$$

Then, because $P = E \times I$, $P = 10\text{ V} \times 2\text{ A} = 20\text{ W}$. The same calculations can be made for the bulbs in a parallel connection. See Fig. 3-10. Total current is 3 A. Applied voltage is 10 V. Since it is a parallel circuit, each resistor will have a 10-V potential across it. Now, use the power formula:

$$P = E \times I$$

$$P = 10\text{ V} \times 3\text{ A}$$

$$P = 30\text{ W}$$

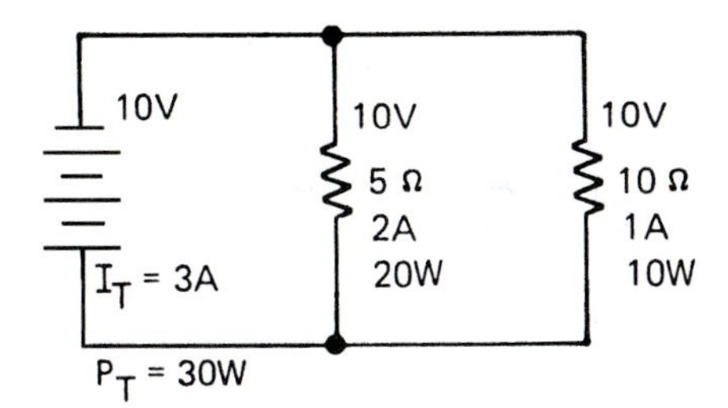

Fig. 3-10 *Total current is found by adding the individual currents.*

Power can be computed if any two of the three values of current, voltage, and resistance are known.

When resistance is unknown:

$$P = E \times I$$

When voltage is unknown:

$$P = I^2 \times R$$

When current is unknown:

$$P = \frac{E^2}{R}$$

POWER RATING OF EQUIPMENT

Most electrical equipment is rated for both voltage and power. Electric lamps rated at 120 V are also rated in watts. Then, they are commonly identified by their wattage rating, rather than by voltage. The voltage is usually 120 V in the United States, and 240 V in Europe and certain other countries.

The wattage rating of a light bulb or other electrical device indicates the rate at which electrical energy is changed into other forms of energy, such as light and heat. The greater the amount of electrical power, the brighter the lamp will be. Therefore, a 100-W bulb furnishes more light than a 75-W bulb.

Similarly, the power ratings of motors, resistors, and other electrical devices indicate the rate at which the devices are designed to change electrical energy into some other form of energy. If the rated wattage is exceeded, the excess energy is usually converted to heat. Then, the equipment will overheat and this can lead to it being damaged. Some devices will have maximum DC voltage and current ratings instead of wattage. Multiplied, these values give the effective wattage.

Resistors are rated in watts dissipated, in addition to ohms of resistance. Resistors of the same resistance value are available with different wattage ratings. Usually, carbon composition resistors are rated from about 0.1 to 2 W. Carbon composition resistors have color bands to indicate their resistance. The physical size determines their wattage rating. The fourth band determines the tolerance of the resistor. See Fig. 3-11.

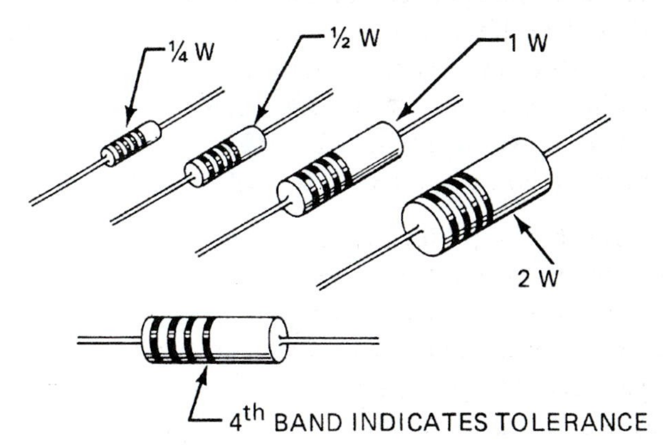

Fig. 3-11 *Wattage rating of carbon composition resistors varies from one-fourth to 2 W. Color bands indicate their ohmic value and tolerance.*

Wire-wound resistors are used when a higher wattage is needed. Generally the larger the physical size of the resistor, the higher the wattage rating, since a larger amount of surface area exposed to the air is capable of dissipating more heat. See Fig. 3-12.

CAPACITORS

A capacitor is a device that opposes any change in the circuit voltage. It may be used in AC or DC circuits.

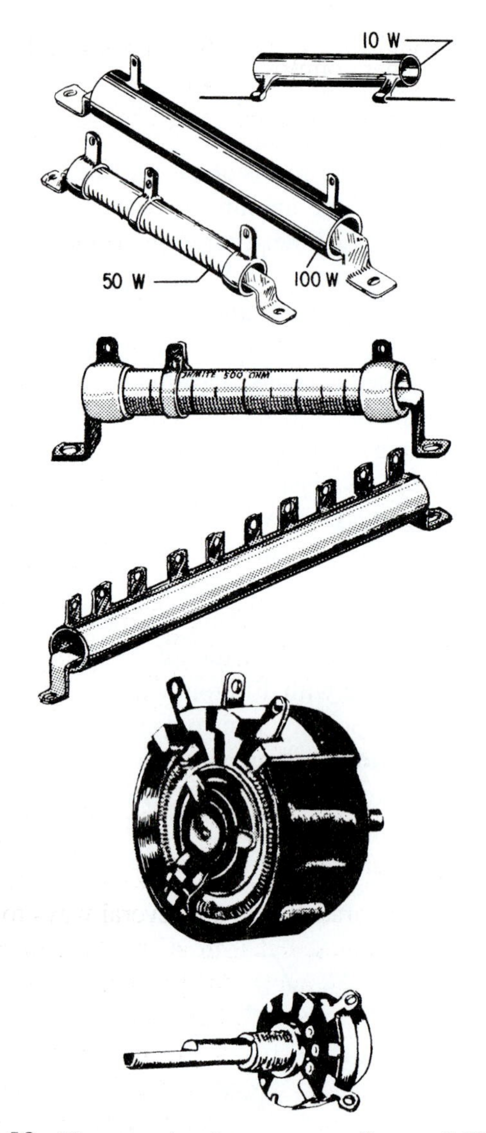

Fig. 3-12 *Wire-wound resistors are usually over 2 W. Shown above are various shapes of wire-wound resistors.*

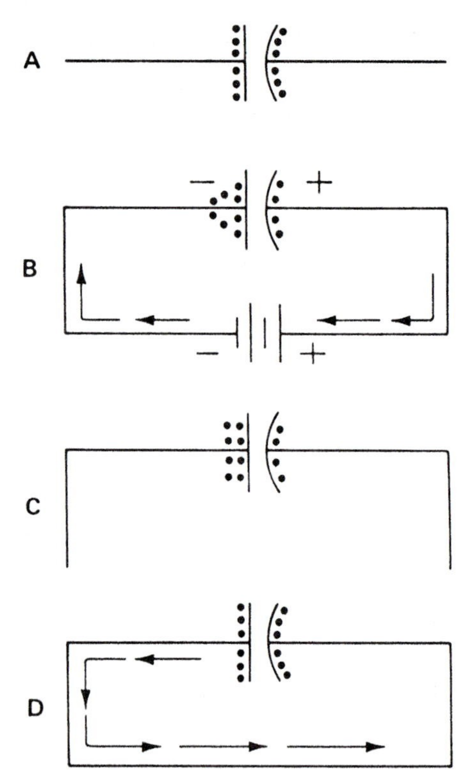

Fig. 3-13 *Capacitor charges. (A) A capacitor with no charge. (B) A capacitor charged by a battery. (C) A capacitor holding its charge after battery is removed. (D) A capacitor discharging, since it has been shorted.*

It does, however, have different uses for different types of current—AC or DC.

Capacitance is that property of a capacitor that opposes any change in circuit voltage. In a capacitor, a device used to obtain capacitance, is made of two plates of a conductor material that are isolated from each other by a dielectric. A dielectric is a material that does not conduct electrons easily. Electrons are stored on the surface of the two plates. If the surface area is made larger, there is more room to store electrons and more capacitance is produced.

How a Capacitor Works

If a capacitor has no electron charge, it is neutral, or uncharged. See Fig. 3-13A. This is the condition when no applied voltage has been connected to the plates. When a source of voltage is connected to the two leads of the capacitor, the difference in potential created by the voltage source causes electrons to be transferred from the positive plate and placed on the negative plate. See Fig. 3-13B.

This transfer continues until the accumulated charge equals the potential difference of the applied voltage. Once the voltage source is removed, Fig. 3-13C, the potential difference remains until a conductor for discharging the excess electrons on the negative plate is connected to the positive or deficient plate. See Fig. 3-13D.

The discharge path for electrons, from one plate to the other, is in the opposite direction of the charge path. This indicates that any change in circuit voltage also results in a minor change in the capacitor charge. Some electrons leave the excess negative plate to try to keep the voltage in the circuit constant. This capability of a capacitor to oppose a change in circuit voltage by placing stored electrons back into circulation is called capacitance.

Capacitance tries to hold down circuit voltage when it increases, and tries to hold it up when circuit voltage decreases.

Since DC voltage varies only when turned on and off, there is little capacitance effect other than at these times. However, AC is continuously changing. Thus, the capacitance effect is continuous in an AC circuit. The symbols used for devices placed in circuits to produce capacitance are shown in Fig. 3-14.

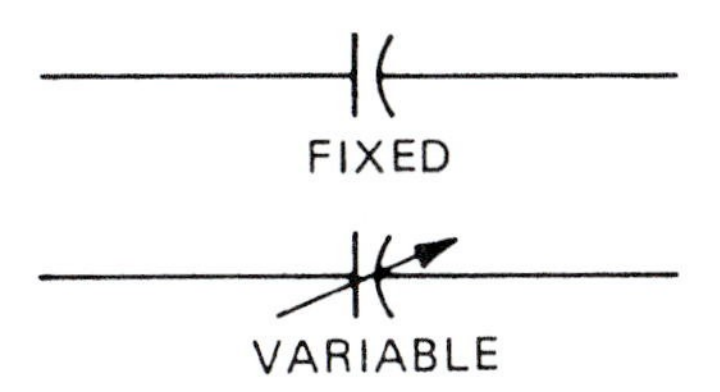

Fig. 3-14 *Capacitor symbols.*

Capacity of a Capacitor

The plates of a capacitor may be made of any material. A dielectric is made of an insulator type of material such as air, a vacuum, wood, mica, plastic, rubber, Bakelite, paper, or oil. If electrons accumulate on a surface, it has capacitance. The larger the surface area, the larger the capacity of the capacitor.

The following three factors determine the capacity of a capacitor:

1. Area of the plate
2. Distance between the plates
3. Material used for the dielectric

The area of the plates determines the ability of the capacitor to hold electrons. The larger the plate area, the more electrons it can hold. The distance between the plates determines the effect of the electron charges on one another.

The electrostatic field between opposite plates can store a greater charge when the plates are close together. They also can produce more electron interaction than plates that are farther apart. The capacitance between two plates increases as the plates are brought closer together. Capacitance decreases when plates are separated.

The thinner the dielectric, the closer together are the plates of a capacitor. This insures a greater effect of the stored charges on the plates and a greater capacitance. Some dielectrics have better insulating properties than others. This insulating property is referred to as the dielectric constant.

Dielectric Failure

The break down voltage is the voltage at which the dielectric will break down and allow a path of electrons to flow through it. The dielectric, at this point, is no longer an insulator.

It will short the capacitor. In some cases, the dielectric allows small amounts of electrons to flow at different times. A capacitor in which this happens is referred to as a leaky capacitor.

Basic Units of Capacitance

The farad is the basic unit of capacitance. It was named after the English physicist Michael Faraday. The farad is equal to the capacitance of a capacitor that has stored in its dielectric 1 C of electrons. One coulomb is 6.28×10^{18} electrons, or 6280000000000000000 electrons. Thus, 1 C of electrons on one plate and no electrons on the other would produce a difference in capacitance of 1 F.

As can be seen by the number of electrons, the farad is a large unit. For practical purposes, it is broken down into smaller quantities. The microfarad is one millionth of a farad (0.000001 F) and the micromicrofarad is one-millionth of one-millionth of a farad (0.000000000 001F). Micromicrofarad is an old term, but can still be found on older capacitors. The term micromicro has been replaced by pico. The symbol for micro is the Greek letter mu, or μ. The symbol for picofarad is pF. The symbol MMF was used for micromicrofarad or picofarad. There are several ways to represent capacitor values, but then all use microfarads or picofarads. For instance, MMF, mmf, UUF, uuf, UUFD, and MMFD were all formerly used to designate micromicrofarads. Today, pF is used as the prefix for the symbol for micromicrofarad. The letters MM and UU were used to symbolize micromicro. This made it unnecessary to buy a separate font of Greek letters and use just one of them. The symbols MFD or mfd may also be found on equipment with older component parts.

Today, MF is used almost exclusively. Occasionally, the Greek letter mu (μ) will be used with the F to represent microfarads.

Working with Capacitive Values

Sometimes it is necessary to convert the farad to smaller units. It may also be necessary to change the smaller units to larger units. For example, it may become necessary to convert 10, 000 pF to microfarads or farads.

This would mean moving the decimal place. For example, 10,000 pF equals 0.01 μ F or 0.00000001 F. A schematic may be marked 10K pF, meaning 10,000 pF. However, some schematics may call for a 0.01-M F capacitor. This would be equivalent to a 10,000-pF capacitor. The 10,000 is sometimes abbreviated as 10K on disc ceramic capacitors, and no pF follows. It is assumed that such a large value could only be in the pF range.

Table 3-3 lists the methods by which capacitive values can be converted.

Table 3-3 *Capacitive Value Conversion Table*

To Convert	Move Decimal Point
pF to MF	Six places to the left
MF to F	Six places to the left
F to MF	Six places to the right
MF to pF	Six places to the right
pF to F	Twelve places to the left
F to pF	Twelve places to the right

Capacitor Types

The following five types of capacitors are available for commercial applications:

- Air
- Mica
- Paper
- Ceramic
- Electrolytic

The capacitor with polarity markings (+ or −) is called an electrolytic. The other four types are not polarized and are not marked + or −. Some of the other capacitors will have a black band around one end. This indicates the terminal or lead that is connected to the outside foil of the capacitor. See Fig. 3-15.

- *Air capacitors.* Air capacitors have air as the dielectric separating the plates. These capacitors are usually variable capacitors.
- *Mica capacitors.* Mica is the dielectric separating aluminum foil plates. Mica capacitors are not common today. Many other materials are less expensive. Mica capacitors are usually contained in Bakelite. They usually have capacitances of 50 to 500 pF.
- *Paper capacitors.* In paper capacitors, paper is the dielectric separating the two plates of aluminum foil. The materials, aluminum foil and paper separators, are rolled in a cylinder. Leads are attached to each foil layer. See Fig. 3-15. The cylindrical roll is placed in a container tube made of cardboard and sealed with wax. Paper capacitors usually have a capacitance

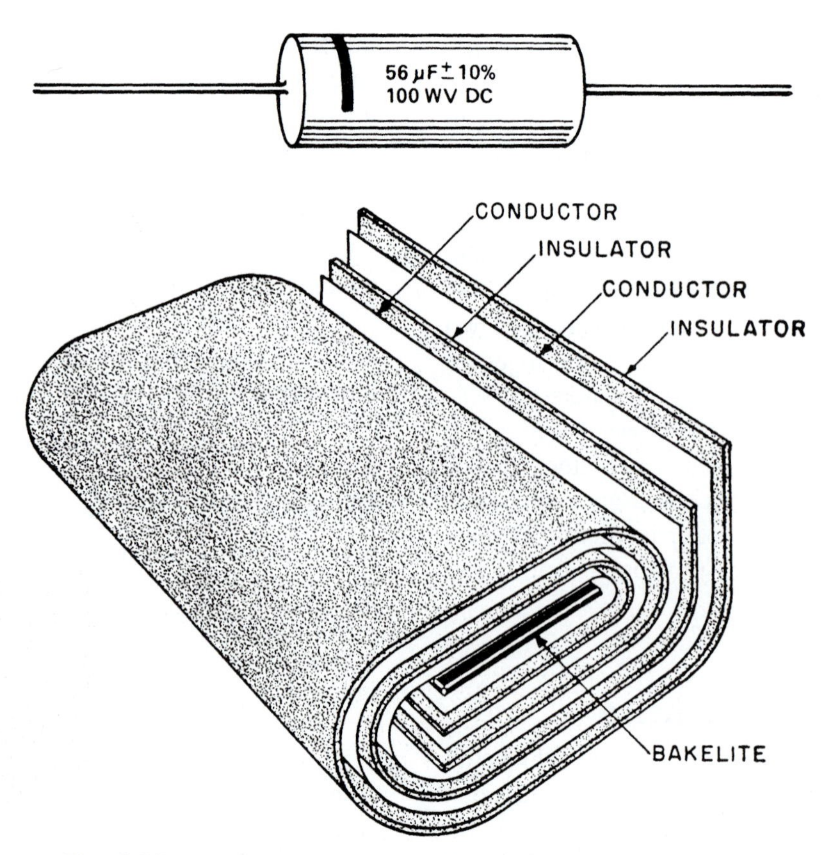

Fig. 3-15 *A paper capacitor, encased, and with the top cover removed.*

of 0.001 to 1.0 MF. Some use Teflon or Mylar, instead of paper, as a dielectric. These capacitors have the added advantage of high breakdown voltage and low losses. They operate efficiently over a longer period than the regular paper capacitors.

- *Oil-filled capacitors.* Oil-filled capacitors are paper capacitors encased in oil. Usually mounted in a metal case, they are referred to as bathtub capacitors. Their values are not over 1 MF. Their main advantage is a higher breakdown voltage and ruggedness.
- *Ceramic capacitors.* A ceramic capacitor has a high-voltage rating, since ceramic is a good insulator. They are usually small and rugged. They consist of a ceramic disc with a coating of silver on both sides. Leads are soldered to the coatings. The whole assembly is then covered with a ceramic glaze and fired.

They are made with values from 1 pF to 0.05 MF. Breakdown voltages can be as high as 10,000 V. The value and the code for voltage are stamped on the capacitor. Small ceramic capacitors are now made for transistor circuits with very low voltage breakdown capability. Such low voltages are common in transistor circuits.

- *Tubular ceramic capacitors.* Tubular ceramic capacitors are used in electronic circuits where stability of capacitance is required, as in control circuits. Such a capacitor is nothing more than a ceramic tube coated on the inside and outside with a metallic substance, small wires soldered to them and then coated with epoxy or some other coating to protect the plate surface. Their value is applied by color code. They have been designed to replace mica capacitors. Values of 1 to 500 pF are common. Their physical size is their greatest advantage.
- *Electrolytic capacitors.* An electrolytic capacitor is easily identified, since it will have a – or a + at one end of the tubular case. There are two types of electrolytic capacitors, wet and dry. The dry type is the most common. The wet type is used in heavy-duty electronic equipment, such as transmitters.

Electrolytic Capacitors Dry electrolytic capacitors are not really dry. They have an electrolyte that is damp. Once the electrolyte dries up, the capacitor is defective. Such drying can occur under a number of conditions. For example, an electrolytic capacitor will dry up if allowed to sit without use for a period of time. In some cases they will become leaky, shorted, or open due to age.

Electrolytic capacitors are often called merely electrolytics. They are available in sizes starting at 1 MF and going up to 1 F. Of course, the working voltage—the point where there is a difference of potential across the plates—is very small at the higher values. The method used to construct electrolytic capacitors is shown in Fig. 3-16.

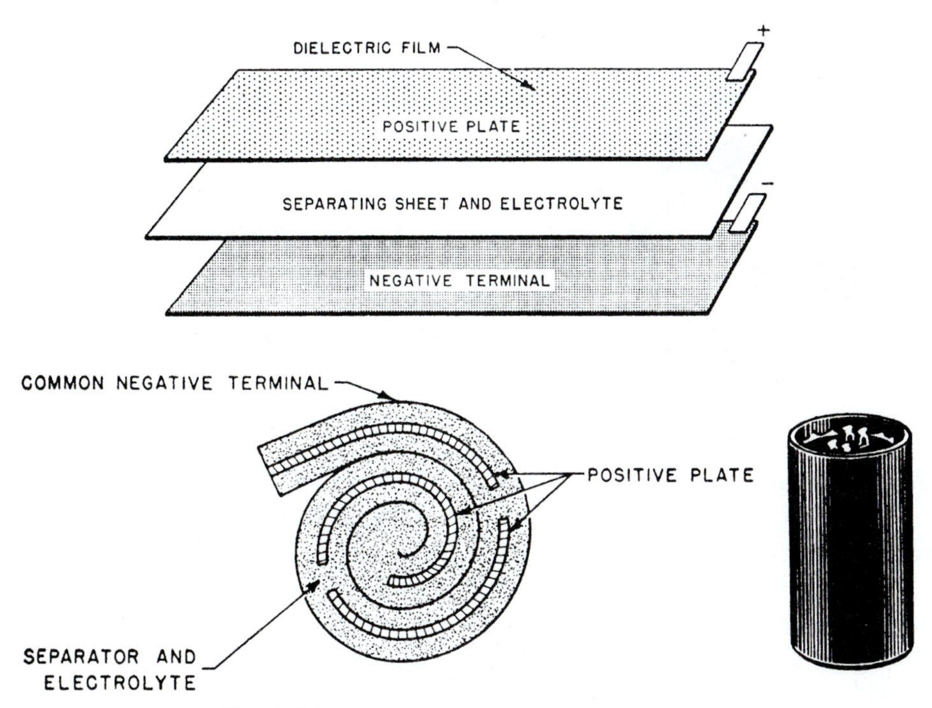

Fig. 3-16 *How an electrolytic capacitor is made?*

Making an Electrolytic In manufacturing, a DC voltage is applied to the electrolytic. An electrolytic action creates a molecule-sized film of aluminum oxide with a thin layer of gas at the junction between the positive plate and the electrolyte.

The oxide film is a dielectric. There is capacitance between the positive plate and the electrolyte through the film. The negative plate provides a connection to the electrolyte. This thin film allows many layers of foil to be placed into a can or cardboard cover. Larger capacitance values can thus be produced by having plates closer together. Some electrolytic capacitors have more than one capacitor in a case. Such capacitors are either labeled on the cover or at the bottom of the can. See Fig. 3-17.

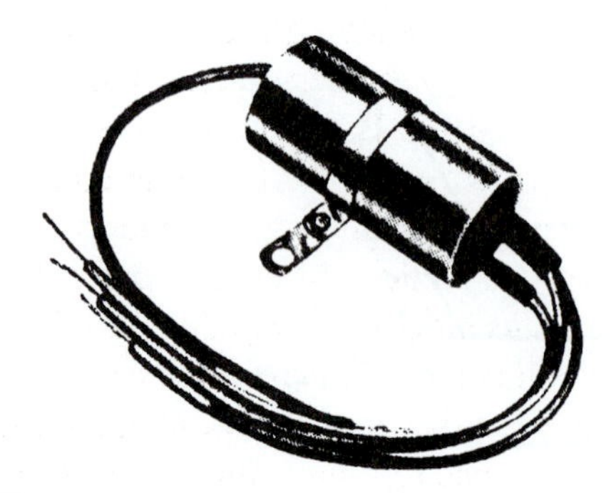

Fig. 3-17 *The can capacitor has more than one electrolytic. The tubular capacitor has more than one electrolytic.*

Connecting Electrolytics in Circuits The polarity of electrolytics must be observed when they are connected in a circuit. If they are not connected properly, that is, – to – and + to +, the oxide film that was formed during manufacture will break down and form large amounts of gas under pressure. This can rupture the can or container and cause an explosion. Thus, it is best to make sure polarity is properly connected.

Electrolytics should be used in circuits where at least 75 percent of their *working voltage* (WVDC) is available. This will keep the capacitor formed to its rating.

AC and Electrolytics AC electrolytics are found in air-conditioning units in connection with the motors that power the compressors. These electrolytics are not polarized. Nonpolarized electrolytics are made by connecting them in series, but back-to-back. Thus, two capacitors of 50 MF can be used to produce an AC nonpolarized capacitor by placing them in series and connecting the two negative (–) terminals and using the two positive (+) terminals for connections in the circuit. It can also be done by connecting the two positive terminals and using the negative terminals for connection purposes. Using this arrangement, it is possible to arrange two standard electrolytics to substitute for a capacitor. Remember—the placing of the two capacitors in series lowers the capacitance of the combination.

Series Capacitors Capacitors placed in series effectively separate the plates. This reduces the total capacitance of the capacitors placed in series. The working voltage, however, is increased by placing the plates farther apart.

$$C_T = \frac{1}{\frac{1}{C_1} + \frac{1}{C_2} + \frac{1}{C_3} \cdots} \quad \text{or} \quad \frac{1}{C_T} = \frac{1}{C_1} + \frac{1}{C_2} + \frac{1}{C_3}$$

or, for only two capacitors:

$$C_T = \frac{C_1 \times C_2}{C_1 + C_2}$$

Therefore, two 50-MF capacitors would make a series combination of:

$$\frac{50 \times 50}{50 + 50} = \frac{2500}{100} = 25 \text{ MF}$$

When connecting capacitors in series, consider the *working voltage DC* rating (WVDC). If two capacitors are connected in series, the outside plates are farther apart. This increase in distance between the plates increases the WVDC rating of the capacitor. For example, if one capacitor has a 100-WVDC rating and the other a 50-WVDC rating, then the total WVDC rating would be 150. Just add the two WVDC ratings.

Parallel Capacitors Connecting capacitors in parallel increases the capacitance. This is primarily since the plate area is increased by parallel connections. The area for electron storage is increased. Total capacitance is found by adding the individual capacitances. With a parallel connection and with the working volts DC, total working voltage equals the working voltage of the smallest capacitor. For instance:

$C_1 = 50$ MF at 400 WVDC

$C_2 = 25$ MF at 200WVDC

$C_3 = 75$ MF at 200 WVDC

The weakest point in the connection is the 200-WVDC capacitor. That would be the one used to protect the combination from voltage breakdown.

Capacitor Tolerances

Capacitors have a tolerance of ±20 percent, unless otherwise noted. The manufacturer's specifications must be checked to make sure. In some cases, a ±10 percent capacitor is available. However, this is not the case with electrolytic capacitors. The electrolytic may have a tolerance of –20 and ±100 percent. For instance, the capacitor marked 50 MF may be somewhere between 40 and 100 MF

In the case of AC electrolytics made for use on AC circuits (as opposed to one made from DC electrolytics), the capacitance range will be given on the capacitor. For example, it may read 40 to 100 MF at 200 V AC, 60 Hz.

If you are working with close tolerance control equipment, you may encounter the mica or tubular ceramic capacitor. These capacitors have extremely small tolerances. Their tolerance may be ±2 to ±20 percent. The closer the tolerance, the more expensive the capacitor. If very close tolerances are required, silver-plated mica may be specified with a ±1 percent tolerance.

THE AC CIRCUIT AND THE CAPACITOR

DC and AC affect a capacitor differently. When DC is applied to a capacitor, the capacitor charges to the voltage of the source. Once the voltage source is removed from the capacitor, the capacitor will discharge through the resistor in the opposite direction than that from which it was charged. See Fig. 3-18. No current flow takes place once the capacitor is charged to the source voltage level.

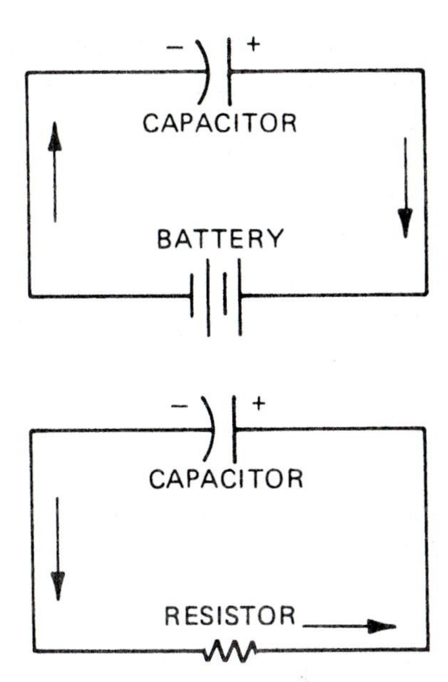

Fig. 3-18 *Note direction of charge and discharge of a capacitor.*

In an AC circuit with a capacitor, capacitive reactance (X_c) must be considered. Capacitive reactance is the opposition to current flow presented by a given capacitance. Capacitive reactance is determined by the frequency of the AC and the capacity of the capacitor. Capacitive reactance is found by the following formula:

$$X_c = \frac{1}{2\pi FC}$$

where X_c = capacitive reactance, measured in ohms

$\pi = 3.14$

F = frequency (usually 60 Hz)

C = capacity in Farads

Alternating current appears to pass through a capacitor. However, it is blocked. The capacitor is charged first in one direction and then the other as the current alternates. See Fig. 3-19. Note that the circuit allows current to flow when the capacitor is charging and discharging. The AC source voltage increases to a maximum, decreases to zero, then increases to a minimum in the opposite direction. Then it drops to zero again. Since the current is alternating, the charging and discharging current moves through the lamp as quickly as the source can change its direction. At 60 Hz, the bulb increases and decreases its intensity so rapidly (120 times per second) that the human eye is unable to detect the change. However, the bulb appears to glow continuously.

A small capacitor will cause the lamp to glow dimly. A larger capacitor will cause the lamp to glow brightly. This change indicates that the same amount of current is not available to make the lamp glow brightly in the dimmer circuit. That means something must have caused the difference to the bulb brightness. Since nothing was changed except the size of the capacitor, it must be surmised that the size of the capacitor affects the brightness of the bulb's glow.

The following problem illustrates the exactness with which this phenomenon can be checked mathematically.

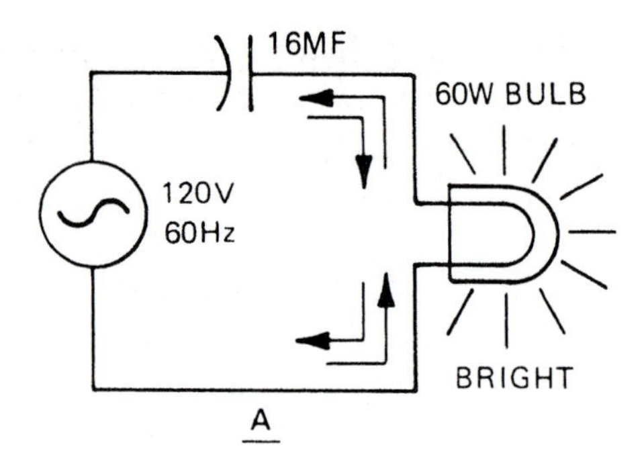

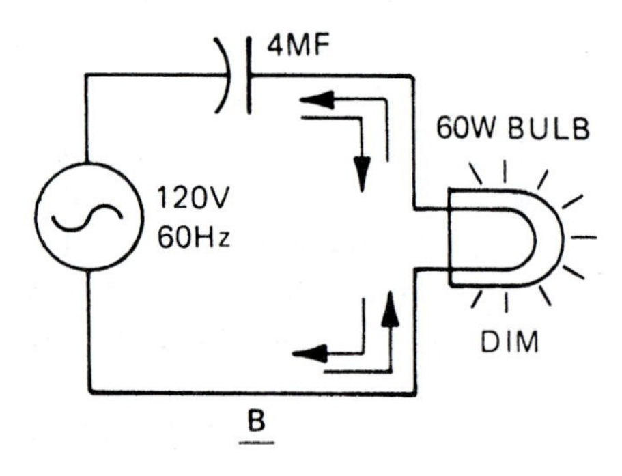

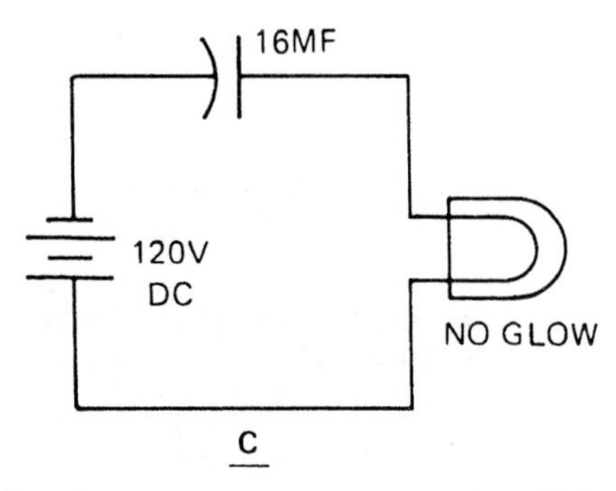

Fig. 3-19 *Alternating current in a capacitor. (A) Large capacitor (16 MF) allows the bulb to glow brightly. (B) Small capacitor (4 MF) allows the bulb to glow dimly. (C) Capacitor in DC circuit will not allow the bulb to glow.*

Problem: A circuit has 120 V, 60 Hz, AC applied to a 40-W light bulb in series with a 10-MF capacitor. What will be the current flow through the bulb?

Solution: The capacitive reactance (X_c) is the opposition. Use it where the resistance is called for in the Ohm's law formula.

$$X_c = \frac{1}{2\pi \text{FC}}$$

or

$$\frac{1}{6.28 \times 60 \times 0.00001}$$

Note: The capacitance must be measured in farads.

$$X_c = \frac{1}{0.003768} = 265.39\ \Omega$$

Since the voltage for the whole circuit is 120 V, the wattage rating of the bulb tells what the current should be, or:

$$I = \frac{E}{R} = \frac{120}{360} = 0.3333\text{ A}$$

Resistance of the bulb is found by:

$$P = \frac{E^2}{R} \quad \text{or} \quad R = \frac{E^2}{P}$$

In this case,

$$R = \frac{14400}{40} = 360\ \Omega$$

$E = 120$ V
$E^2 = 14{,}400$
R = resistance of filament
P = watts (40 W in this case)
I, then, is equal to 0.3333 A

To find the impedance, or total opposition (Z), made up of the capacitive reactance and the resistance of the lamp bulb filament, use the following formula:

$$Z = \sqrt{R^2 + X_c^2}$$
$$Z = \sqrt{360^2 + 265.39^2} \quad \text{or}$$
$$\sqrt{129{,}600 + 70{,}431.85} = 447.2\ \Omega$$

Now that the impedance (Z), has been found, the problem of finding the total current in the circuit with the capacitor and bulb can be found using the following formula:

$$I_T = \frac{E_{\text{Applied}}}{Z} \quad \text{or} \quad \frac{120}{447.2} = 0.2683\text{ A}$$

The answer, 0.2683 A, is less than the 0.3333 A needed to give the bulb full brightness. Thus, the bulb glows dimmer than it would without the capacitor in the circuit.

The same procedure can be followed with the larger capacitor. If the capacity of the capacitor is increased, it means the capacitive reactance is lower. If the X_c is lower, a larger current value is obtained when the voltage is divided by the capacitive reactance. Thus,

$$\uparrow I_T = \frac{120}{X_c} \downarrow$$

The smaller the capacitive reactance (X_c), the larger the total current (I_T). Thus, the brighter the bulb glows.

Uses of Capacitors

Capacitors are used in electronic circuits for one of the three basic purposes:

- To couple an AC signal from one section of a circuit to another.
- To block out and/or stabilize any DC potential from a component.
- To bypass or filter out the AC component of a complex wave.

Capacitors are also used as part of a circuit in an electric motor. They improve the operating characteristics of some motors. It is possible to start a motor under load if it is a capacitor-start type. This is very important when an air-conditioning unit must start under load. A capacitor-start, capacitor-run type of motor is also used in air-conditioning units. This type of motor will be discussed later.

INDUCTANCE

Inductors have inductance. *Inductance* is that property of a coil that opposes any change in circuit current. Inductance is measured in henries (H). The symbol for inductance is L. Inductance is sometimes measured in millihenries. (Milli means 1/1000 or 0.001 H). The symbol for millihenry is mH. There are occasions when even smaller units of the henry are used, such as microhenry. (Micro means one-millionth or 0.000001.)

Inductors are used in circuits containing audio frequencies (those that can be heard) and in circuits containing radio frequencies (those that cannot be heard). The symbol for a coil is xx. If the coil has application in a circuit with audio frequencies it will have an iron core. The symbol will be .

The symbol for an inductor used in a circuit with radio frequencies is . Notice there is no core in a radio frequency choke. In some cases, a ferrite core is used and then the symbol will be .

Four Methods of Changing Inductance

The following four factors affect the inductance of a coil:

1. The number of turns.
2. The diameter of the coil.
3. The permeability of the core material.
4. The length of the coil.

Changing any of these factors will change the inductance of the coil. In a coil with an air core having the turns close-wound, inductance is increased four times by doubling the number of turns. Doubling the diameter of the coil also quadruples the inductance. The length of the coil directly increases the inductance.

Self-Inductance

The capability of a conductor to induce voltage in itself when the current changes is called self-inductance, or inductance. When a current that is changing in value (such as AC) passes through a coil, the moving magnetic field around the windings of the coil produces electromagnetic induction. The magnetic field around each turn of the coil cuts across the remaining turns and a voltage is generated across the coil. Because this induced voltage is generated by the moving magnetic field produced by an increasing or decreasing current, it is generated in the opposite direction to the voltage that caused it. This is referred to as a *counter-electromotive force* (CEMF). See Fig. 3-20.

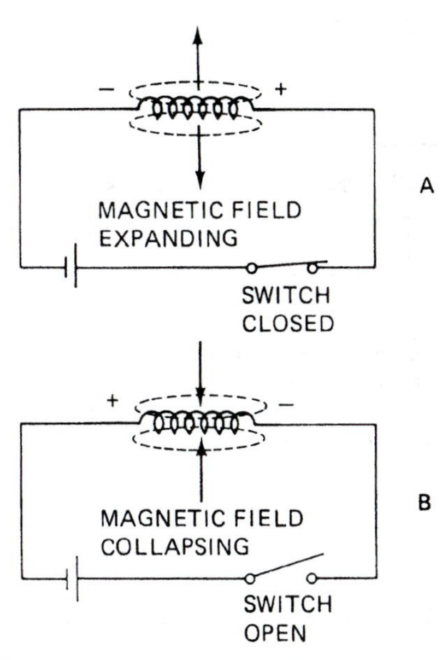

Fig. 3-20 *Counter-electromotive force. (A) Magnetic field builds up and expands when switch is closed. (B) Magnetic field collapses when the switch is opened.*

One henry is the amount of inductance present when a current variation of 1 A/s results in an induced EMF of 1 V.

In Fig. 3-20A the current is shown rising from zero to a maximum rather quickly. This causes the magnetic field around the coil to expand. A CEMF is produced by the expanding magnetic field, cutting the windings of the coil ahead of the current. The windings are usually alongside or on top of the energized part of the coil. In Fig. 3-20B the circuit is shown opened by a switch. The magnetic field collapses and the current in the circuit changes from its maximum value to zero. As the field collapses, it induces a voltage across the coil. This opposes the decrease in current and prevents the current from dropping to zero as quickly as it would in a straight wire. Note that the time lag shown in Fig. 3-21 is produced by a coil.

Mutual Inductance

Mutual inductance is concerned with two or more coils. *Mutual inductance* refers to the condition in which two circuits share the energy of one circuit. The energy in one circuit is transferred to the other circuit. The coupling that takes place between the circuits is done by means of magnetic flux. See Fig. 3-22. When two coils have a mutual inductance of 1 H it means that

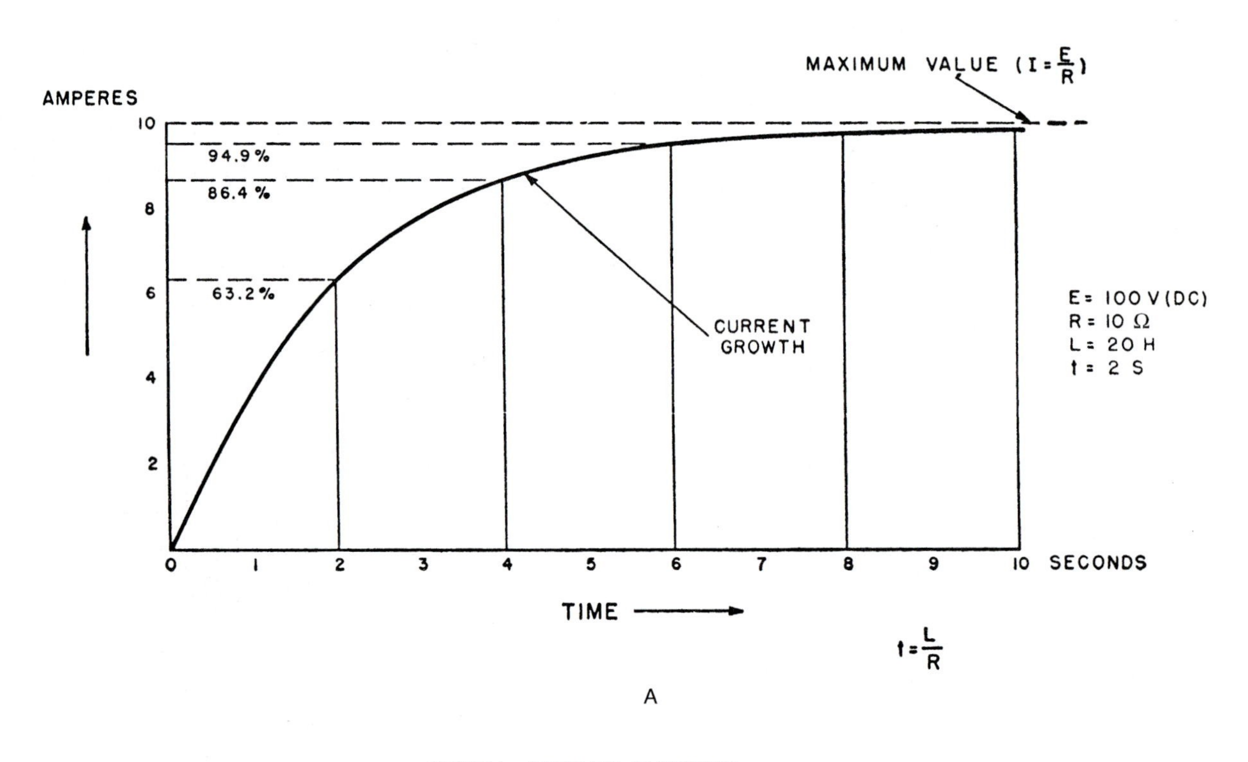

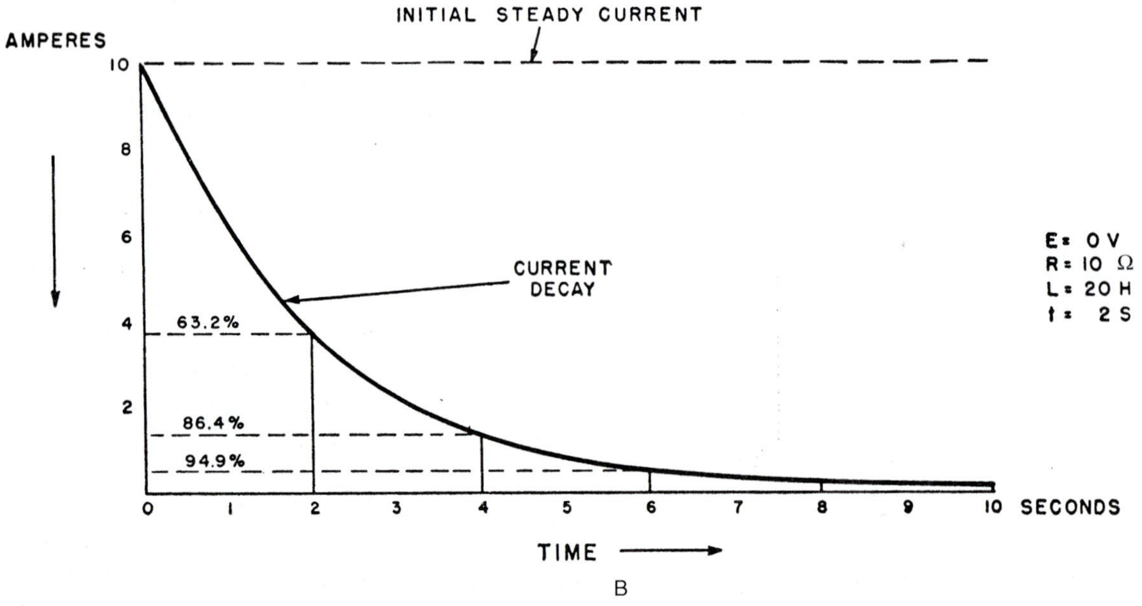

Fig. 3-21 *The time lag produced by a coil. (A) The way in which a time lag is introduced in a circuit by an inductor. (B) It takes time for the magnetic field to collapse.*

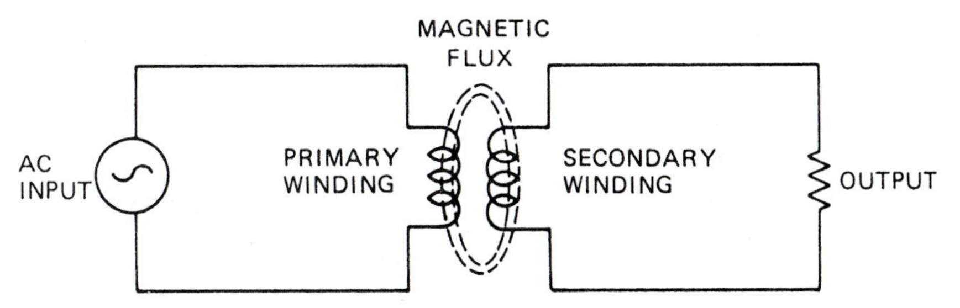

Fig. 3-22 *Magnetic flux is the coupling between the primary and secondary transformer circuits.*

a change in current of 1 A takes place in 1 s. One coil induces 1 V in the other coil.

Inductive Reactance

When alternating current flows through an inductor it creates a certain amount of opposition to its flow. This opposition is called *inductive reactance* (X_L). Inductive reactance is measured in ohms. This type of reactance is not present in a coil when energized by DC. The only opposition encountered by a DC current passing through a coil is the resistance of the copper wire used to wind the coil.

A number of factors determine inductive reactance. Frequency and inductance are the major factors. The formula for inductive reactance is:

$$X_L = 2\pi FL$$

where $2\pi = 6.28$
F = frequency of alternating current
L = inductance (in henries)

Uses of Inductive Reactances

Inductive reactances are very important in filter circuits. It is sometimes necessary to smooth out the variations in a power source current. The inductor can help make the fluctuations less severe.

Inductive reactance (X_L) becomes very useful when dealing with electronic circuits. When combined with capacitive reactance (X_C), it is possible to obtain a resonant frequency. Inductive reactance and capacitive reactance can have the same value. Under such conditions, they can cause a circuit to resonate at a given frequency and no other. Thus, it is possible to pick out one frequency from a number present. This is helpful in tuning in a radio or television station.

TRANSFORMERS

A *transformer* is a device that transfers energy from one circuit to another without being physically connected to both circuits. A transformer operates on the principle of mutual inductance. Magnetic lines of force (or a flux field) are created by the primary side of the transformer. These lines of force (or the force field) change, as the AC changes in polarity. The changing magnetic field creates an induced emf in the secondary side of the transformer. The amount of current available is determined by the size of the wire and the amount of iron in the core of the transformer. The symbols for the transformers are shown in Fig. 3-23.

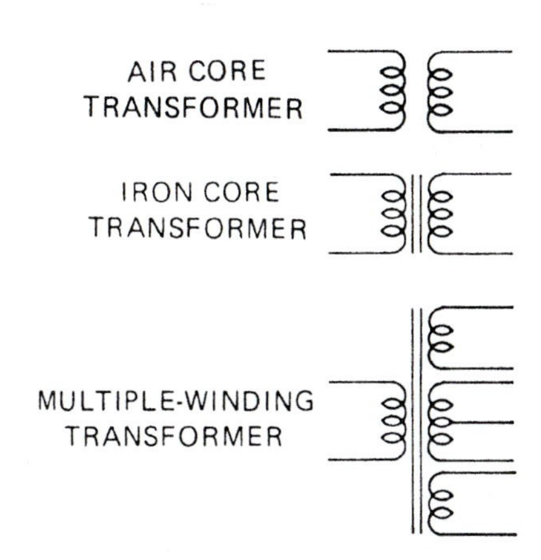

Fig. 3-23 *Transformer symbols.*

Transformer Construction

Transformers are constructed with a coil in the primary winding and a coil in the secondary winding. The coil in the primary winding is connected to the power source. The secondary coil is connected to the circuit needing its particular voltage and current. The primary coil is the input. The secondary coil is the output. There is some power loss in the transfer of energy from the input to the output coils. Nevertheless, transformers are very close to being 100 percent efficient. This is due partly to the fact that there are no moving parts—only the current varies. The core of the transformer may be air (no core) or iron. Air cores are used in radio frequency applications.

Iron cores are used in power line frequencies and audio-frequency applications. The magnetic path is usually through the iron core. The core makes a difference in the capability of the transformer to transfer large amounts of energy from one coil to the other. The core also represents a power loss potential.

The following three types of losses are encountered in transformers:

- Hysteresis losses are caused by the reluctance of the iron core to change polarity with changes in current direction and resultant changes in magnetic field polarity.
- Eddy current losses are created by small currents induced into the core material by changing magnetic fields.
- Copper losses are due to the copper content of the wire. This copper has resistance as an inherent factor.

Losses can be reduced by the following methods:

- Hysteresis losses are reduced by using silicon steel.
- Eddy current losses are reduced by using laminations.
- Copper losses are reduced by using the correct size of wire.

Turns Ratio

A transformer's output voltage is determined by its number of turns as compared to those of the input primary. For example, if the primary has 100 turns and the secondary has 10 turns, then the turns ratio is 10:1. Thus, if 100 V are applied to the primary, the secondary will put out 10 V. However, if the input current is 1 A, then the available output current would be 10 A. The power in must equal the power out, less any inefficiency. Power in (or, $P = E \times I$) is equal to power out (or, $P = E \times I$). Therefore, a step-up transformer refers to the voltage because the current will be the opposite of voltage. The example just mentioned is a step-down transformer. In such a transformer, the output voltage is less than the input.

Transformer Applications

Most heating and cooling devices use transformers to step down the voltage for control circuits. A transformer means that you can have the proper voltage for use by any type of equipment. It makes operating various equipment from one voltage source possible. Transformers are used on AC only, since DC does not have a moving magnetic field.

Transformers are used in electronic air cleaners to step up the voltage sufficiently to operate the equipment and trap dust particles.

SEMICONDUCTORS

Semiconductors are used in making diodes and transistors. These devices are made primarily of germanium and silicon crystals. Controlled amounts of impurities are placed into a 99.999999 percent pure silicon wafer or germanium wafer. When arsenic or antimony are added, the N-type semiconductor material is formed. This means the material has an excess of electrons. Electrons have a negative charge.

When gallium or indium is used as the impurity, a P-type semiconductor material is produced. This means it has a positive charge, or is missing an electron.

Diodes

When N- and P-type materials are joined, they form a diode, also called a *rectifier*. This device is used to change AC to DC. The PN junction (diode) acts as a one-way valve to control the current flow. The forward, or low-resistance direction through the junction, allows current to flow through it. The high-resistance direction does not allow current to flow. This means that only one-half of an AC hertz is allowed to flow in a circuit with a diode. Figure 3-24 shows how a diode is used in the forward biased direction that allows current to flow. Figure 3-25 indicates the arrangement in the reverse bias configuration. No current is allowed to flow under these conditions. Note the polarity of the battery.

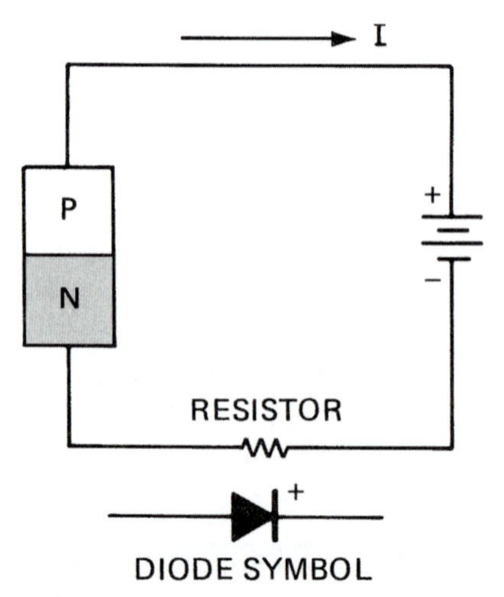

Fig. 3-24 *Diode placed in a circuit. The symbol and the silicon wafer are represented in a circuit.*

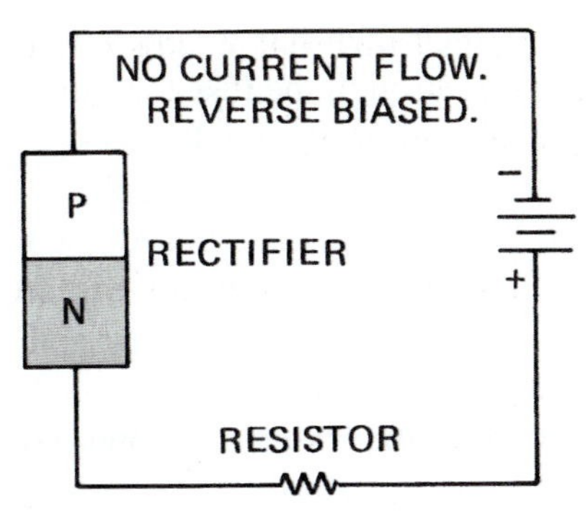

Fig. 3-25 *Reverse-biased diode circuit.*

Diodes are also used in isolating one circuit from another. A simple rectifier circuit is shown in Fig. 3-26. The output from the transformer is an AC voltage, as shown in Fig. 3-27. However, the rectifier action of the diode blocks current flow in one-half of the sine wave and produces a pulsating DC across the resistor. See Fig. 3-27B.

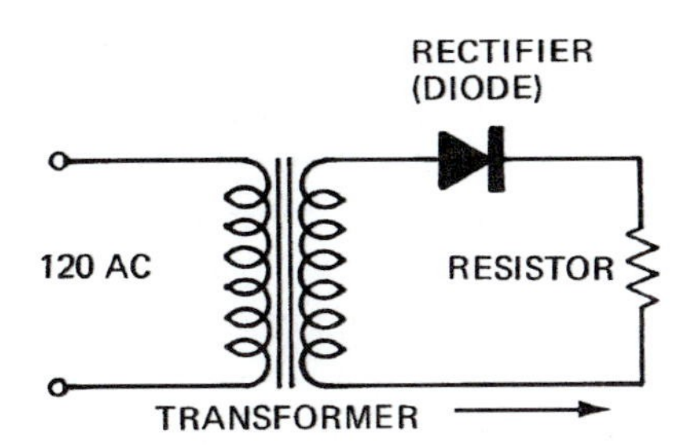

Fig. 3-26 *Rectifier circuit using a diode to produce DC from AC.*

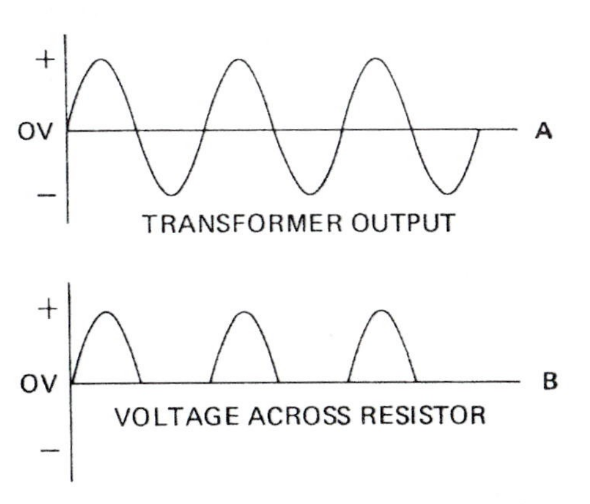

Fig. 3-27 *Results of the rectifier circuit. The transformer output is changed to pulsating DC across the resistor.*

Zener Diode When one polarity of voltage is applied to a rectifier (diode), it blocks the current flow. However, if the voltage is raised high enough the diode breaks down. This allows current to flow. Normal diodes would be destroyed by this breakdown. However, a zener diode is designed to operate in the breakdown region.

Figure 3-28 shows how a zener diode is connected in a circuit. The breakdown voltage on the diode is 8.2 V. As long as the battery voltage is 8.2 V or lower, the output across the diode will be 8.2 V or lower. However, if the battery voltage is more than 8.2 V the voltage drop across the diode is still 8.2 V. If the battery voltage is 10 V, the voltage drop is 1.8 across the series resistor and 8.2 across the diode. If the battery voltage reaches 12 V, the voltage is 3.8 across the resistor and 8.2 across the diode. As can be seen in Fig. 3-28, the zener diode can be used in a circuit to regulate the voltage and keep it constant, or at least no higher than its rating. That is why the circuit is called a voltage regulator circuit. Such a circuit is very useful when a constant voltage is necessary for sensing equipment to operate accurately.

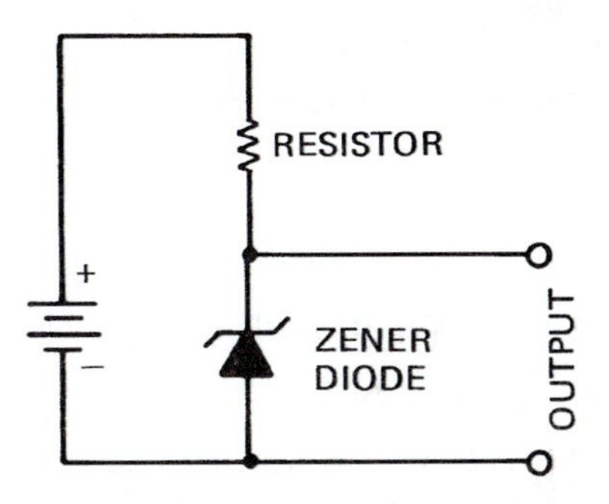

Fig. 3-28 *Zener diode in a circuit. Resistor is necessary to the proper operation of the circuit.*

Transistors

In 1948, the Bell Telephone research laboratories announced that a crystal could amplify. Such a crystal was called a transistor–meaning transfer resistor. The transistor has replaced the vacuum tube in almost all applications. It is made up of three layers of P- and N-type semiconductor material arranged in either of two ways. See Fig. 3-29.

The transistor is used as a switching device or an amplifier. The advantages of the transistor are well known, since it is used in the transistor radio and the semiconductor television receiver.

Figure 3-30 shows a simple transistor amplifier. Battery-1 (B_1) and adjustable resistor R, determine the input current to the transistor. When R, is high in resistance, the current flowing from the base to the emitter is very small. When the base-to-emitter current is small,

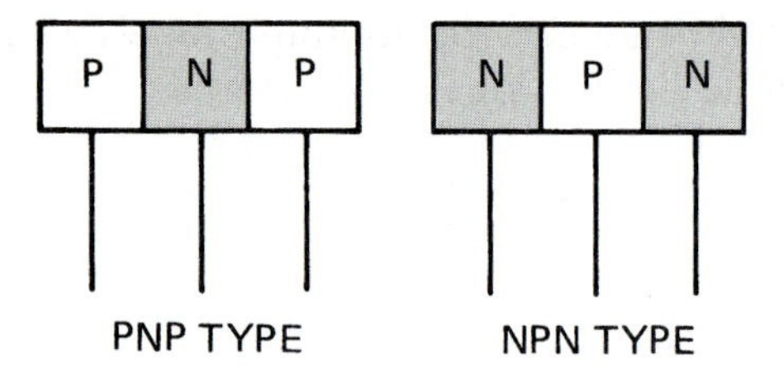

Fig. 3-29 *Arrangement of wafers of silicon or germanium to produce a PNP or NPN transistor.*

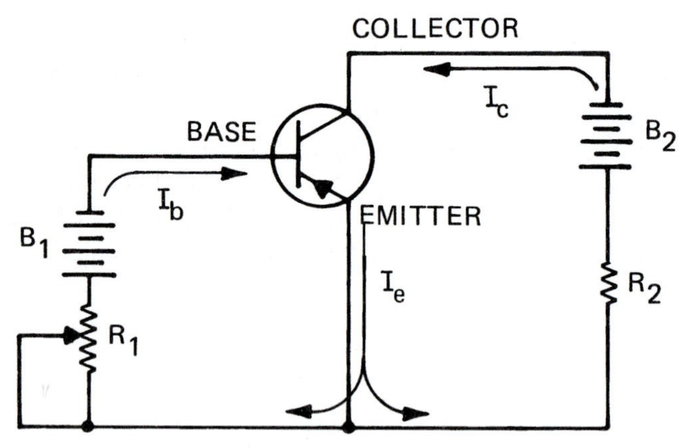

Fig. 3-30 *A simple transistor amplifier.*

the collector-to-emitter resistance appears as a very high resistance. This limits the current flow from battery 2(B_2). The result is a limiting of the voltage drop across R_2.

As the resistance of R_1 is lowered, the current flowing through the base-to-emitter junction increases. As the base-to-emitter current is increased, the resistance of the transistor from collector to emitter is decreased. More current is flowing from B_2 through R_2 and the voltage drop across R_2 is increased. A very small change in the current from B_1 causes a large change in the current from B_2. The ratio of the large change to the small change is defined as the gain of the transistor.

Silicon-controlled Rectifier (SCR)

The *silicon-controlled rectifier* (SCR) is a four-layered PNPN device. The SCR can be defined as a high-speed semiconductor switch. It requires only a short pulse to turn it on. It remains on as long as current is flowing through it.

Look at the circuit shown in Fig. 3-31. Assume that the SCR is off. (It would then have a very high resistance.) No current would be flowing through the resistor. When switch S_2 is closed just long enough to turn on the SCR (which then has a very low resistance) a current will flow through the resistor and the SCR. The SCR will remain on until switch S_2 stops the flow of current through the resistor and SCR. Then the SCR will turn off. When S_2 is again closed, the resistance of the SCR remains high. No current will flow through the resistor until S_1 is reclosed. Figure 3-32A shows an SCR represented schematically. Figure 3-32B shows the arrangement of the layers of P- and N-type materials that produce the SCR effect. The anode is the positive terminal. The gate is the terminal used to turn on the SCR. The cathode is the negative terminal.

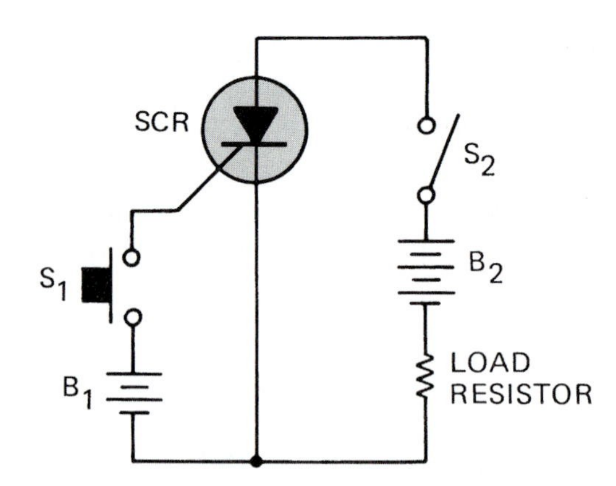

Fig. 3-31 *A silicon controlled rectifier (SCR) circuit.*

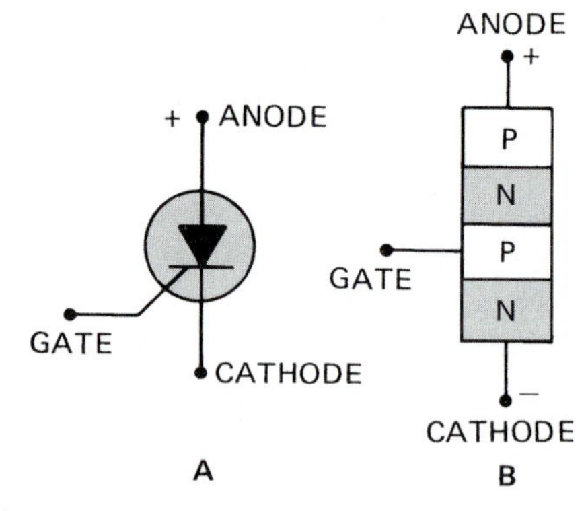

Fig. 3-32 *A silicon controlled rectifier. (A) Schematic representation of an SCR. (B) Wafer arrangement needed to produce a SCR.*

BRIDGE CIRCUITS

Wheatstone Bridges

A bridge circuit is a network of resistances and capacitive or inductive impedances. The bridge circuit is usually used to make precise measurements. The most common bridge circuit is the Wheatstone bridge. This consists of variable and fixed resistances. Simply, it is a series-parallel circuit. Redrawn, as shown in Fig. 3-33, is a Wheatstone bridge circuit. The branches of the circuit forming the diamond shape are called "legs."

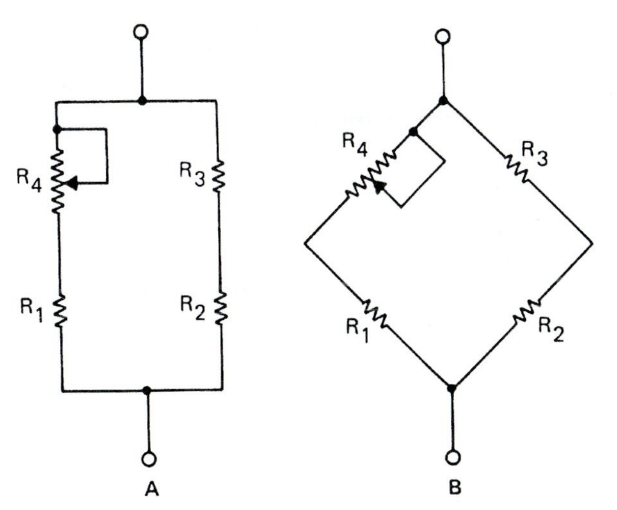

Fig. 3-33 *Two ways of drawing a bridge circuit.*

If 10 V DC were applied to the bridge shown in Fig. 3-34, one current would flow through R_1, and R_2, and another through R_3 and R_4. Since R_1, and R_2 are both fixed 1000-Ω resistors, the current through them is constant. Each resistor will drop one-half of the battery voltage, or 5 V. Five volts is dropped across each resistor. The voltmeter senses the sum of the voltage drops across R_2 and R_3. Both are 5 V. However, the R_2 voltage drop is a positive (+) to negative (–) drop. The R_3 drop is a negative to positive drop. They are opposite in polarity and cancel each other. This is called a

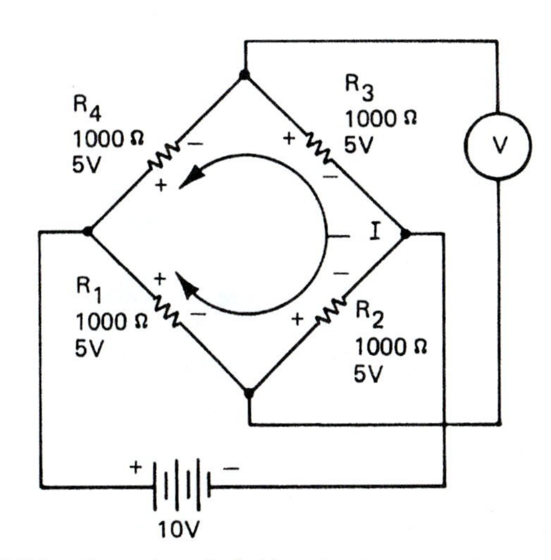

Fig. 3-34 *Operation of a bridge circuit.*

balanced bridge. The relationship is usually expressed as a ratio of

$$\frac{R_1}{R_2} = \frac{R_3}{R_4}$$

The actual resistance values are not important. What is important is that this ratio is maintained and the bridge is balanced.

Variable Resistor

In Fig. 3-35, the value of the variable resistor R_4 is 950 Ω. The other resistors have the same value. Using Ohm's law, the voltage drop across R_4 is found to be 4.9 V. The remaining voltage, 5.1 V, is dropped across R_3. As shown in Fig. 3-35, the voltmeter measures the sum of the voltage drops across R_2 and R_3 as 5 V (+ to –) and 5.1 V (– to +). It registers a total of –0.1 V.

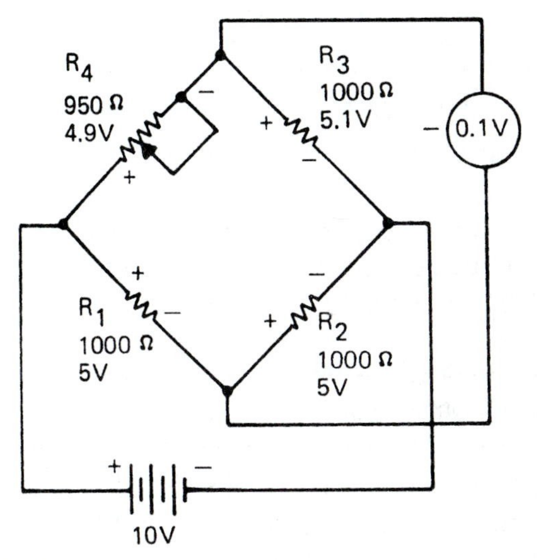

Fig. 3-35 *Operation of a bridge circuit.*

In Fig. 3-36, the converse is true. The value of R_4 is 1050 Ω. The voltage drop across R_3 is 4.9 V. The voltmeter senses the sum of 5 V (+ to –) and 4.9 V (– to +), or + 0.1 V.

When R_4 changes the same amount, above or below the balanced bridge resistance, the magnitude of the DC output, measured by the voltmeter, is the same. However, the polarity is reversed.

SENSORS

The sensor in a control system is a resistance element that varies in resistance value with changes in the variable it is measuring. These resistance changes are

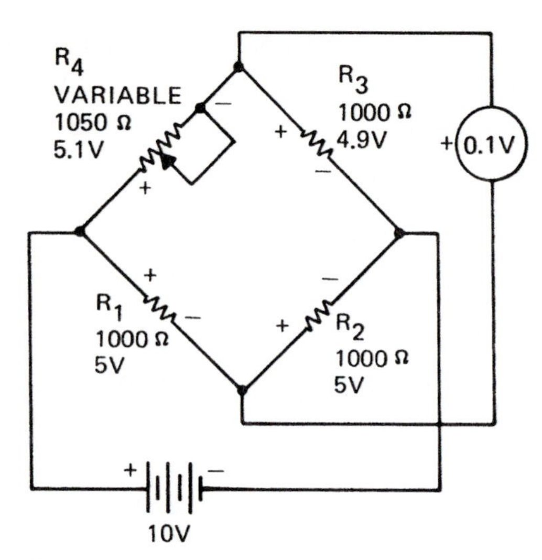

Fig. 3-36 *Operation of a bridge circuit.*

converted into proportional amounts of voltage by a bridge circuit. The voltage is amplified and used to position actuators that regulate the controlled variable.

Temperature Elements

The temperature element used in cybertronic devices is a nickel wire winding. (A *cybertronic device* is an electronic control system.) This wire is very sensitive to temperature changes. It increases its resistance to current flow at the rate of approximately 3 Ω for every degree Fahrenheit increase in temperature. This is called a positive temperature coefficient. The length and type of wire give the winding a reference resistance of 1000 Ω at 70°F. A temperature drop decreases the resistance and a temperature rise increases the resistance. The winding is accurate over a range of –40 to 250°F.

Humidity Elements

There are many moisture-absorbing materials used as relative-humidity sensors. Such materials absorb or lose moisture until a balance is reached with the surrounding air. A change in material moisture content causes a dimensional change. This change can be used as an input signal to a controller. Commonly used materials include:

- Human hair
- Wood
- Biwood combinations, similar in action to a bimetal temperature sensor
- Organic films
- Some fabrics, especially certain synthetic fabrics

All these have the drawbacks of slow response and large hysteresis effects. Their accuracy tends to be questionable unless they are frequently calibrated. Field calibration of humidity sensors is difficult.

Thin-film sensors are now available. They use an absorbent deposited on a silicon substrate such that the resistance or capacitance varies with relative humidity. These are quite accurate ±3 to 5 percent—and have low-maintenance requirements.

Improvements in the design of humidity-sensing elements and the materials used in their construction have minimized many of the past limitations of humidity sensors. One of the humidity sensors used with the electronic controls is a resistance *cellulose acetate butyrate* (CAB) element. This resistance element is an improvement over other resistance elements. It has greater contamination resistance, stability, and durability. The humidity CAB element is a multilayered humidity-sensitive polymeric film. It consists of an electrically conductive core and insulating outer layers. These layers are partially hydrolyzed. The element has a nominal resistance of 2500 Ω. It has a sensitivity of 2 Ω per 1 percent relative humidity (rh) at 50 percent rh. Humidity sensing range is rated at 0 to 100 percent rh.

The CAB element consists of a conductive humidity-sensitive film, mounting components, and a protective cover. See Fig. 3-37. The principle component of this humidity sensor is the film. The film has five layers of CAB in the form of a ribbonlike strip. The CAB material is used because of its good chemical and mechanical stability and high sensitivity to humidity. It also has excellent film-forming characteristics. See Fig. 3-38.

The CAB resistance element is a carbon element having the resistance/humidity tolerances shown in Fig. 3-39. With an increase in relative humidity, water is absorbed by the CAB, causing it to swell. This

CAB RESISTIVE ELEMENT

Fig. 3-37 *CAB-resistive element.* (Johnson Controls)

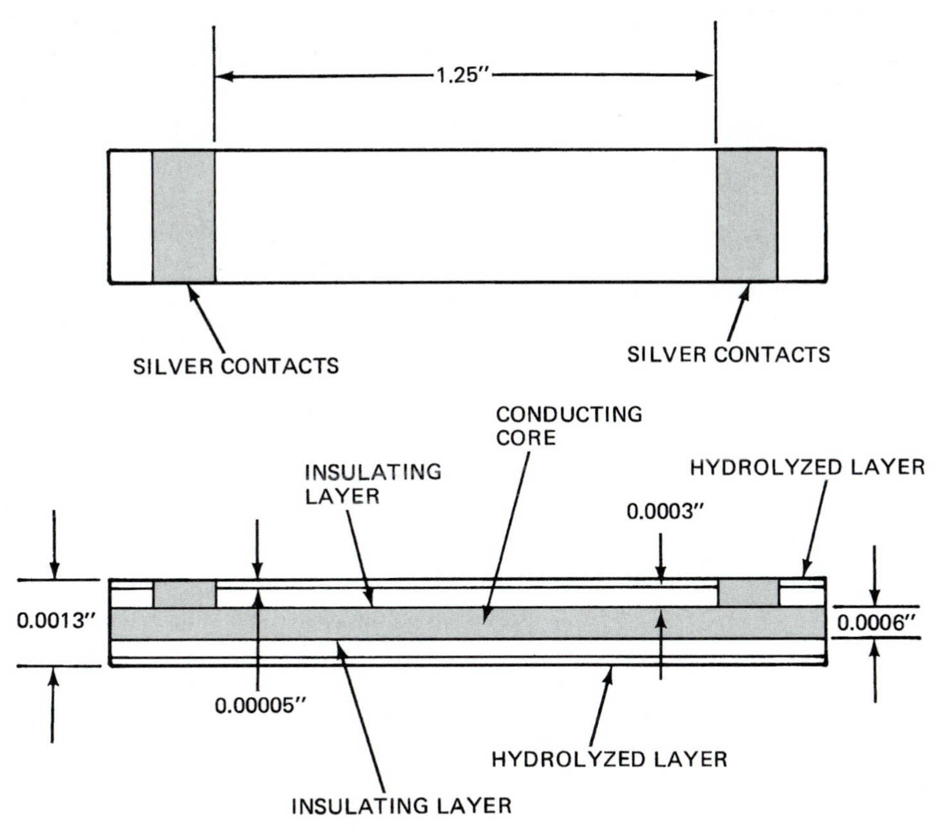

Fig. 3-38 *A hydrolyzed humidity element.*

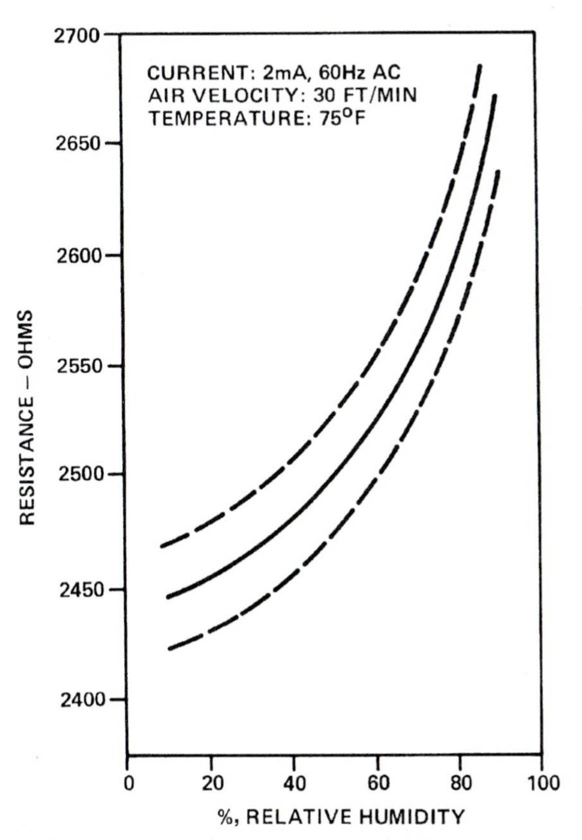

Fig. 3-39 *Operational characteristics of a humidity element.* *(Johnson Controls)*

swelling of the polymer matrix causes the suspended carbon particles to move farther apart from each other. This results in an increased element resistance.

When relative humidity decreases, water is given up by the CAB. The contraction of the polymer causes the carbon particles to come closer together. This, in turn, makes the element more conductive, or less resistive.

CONTROLLERS

The sensing bridge is the section of the controller circuit that contains the temperature-sensitive element or elements. The potentiometer for establishing the "set point" is also part of the control system. The bridges are energized with a DC voltage. This permits long wire runs in sensing circuits without the need for compensating wires or for other capacitive compensating schemes.

Both integral (room) and remote-sensing element controllers produce a proportional 0- to 16-V DC output signal in response to a measured temperature change. Controllers can be wired to provide direct or reverse action. Direct-acting operation provides an increasing output signal in response to an increase in temperature. Reverse-acting operation provides an increasing output signal in response to a decrease in temperature.

Single-Element Controllers

Electronic controllers have three basic parts:

- The bridge
- The amplifier
- The output circuit.

Bridge theory has been covered previously. Two legs of the bridge are variable resistances. See Fig. 3-40. The sensor and the set point potentiometer are shown in the bridge circuit. If temperature changes, or if the set point is changed, the bridge is in an unbalanced state. This gives a corresponding output result. The output signal, however, lacks power to position actuators. Therefore, this signal is amplified.

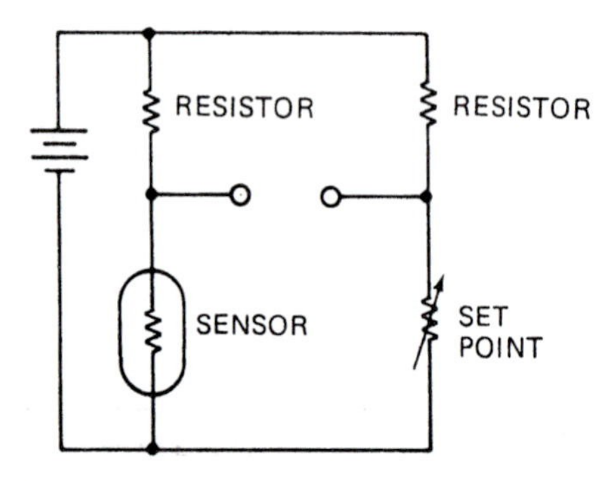

Fig. 3-40 *A bridge arrangement with a sensor and set point.*

Differential Amplifiers Controllers utilize direct-coupled DC differential amplifiers to increase the millivolt signal from the bridge to the necessary 0- to 16-V level for the actuators. There are two amplifiers—one for direct reading and one for reversing signals. Each amplifier has two stages of amplification. This arrangement is shown in block form in Fig. 3-41.

The differential transistor circuits provide gain and good temperature stability. Figure 3-42 compares a single transistor amplifier stage with a differential amplifier. Transistors are temperature sensitive. That is, the current they allow to pass depends upon the voltage at the transistor and its ambient temperature. An increase in ambient temperature in the circuit shown in Fig. 3-42A would cause the current through the transistor to increase. The output voltage would, therefore, decrease. The emitter resistor R_E reduces this temperature effect. It also reduces the available voltage gain in the circuit because the signal voltage across the resistor amounts to a negative feedback voltage. That is, it causes a decrease in the voltage difference which was originally produced by the change in temperature at the sensing element.

Since it is desirable for the output voltage of the controller to correspond only to the temperature of the sensing elements and not the ambient temperature of the amplifier, the circuit shown in Fig. 3-42B is used. Here, any ambient temperature changes affect both transistors simultaneously. The useful output is taken as the difference in output levels of each transistor and the effects of temperature changes are cancelled. The voltage gain of the circuit shown in Fig. 3-42B is much higher than that shown in Fig. 3-42A. This is because

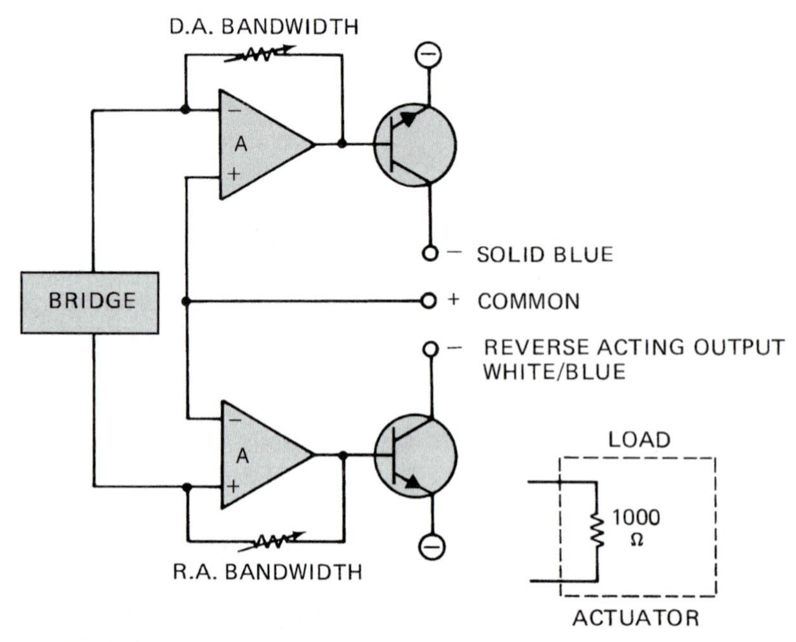

Fig. 3-41 *DC differential amplifiers for use in a controller circuit.*

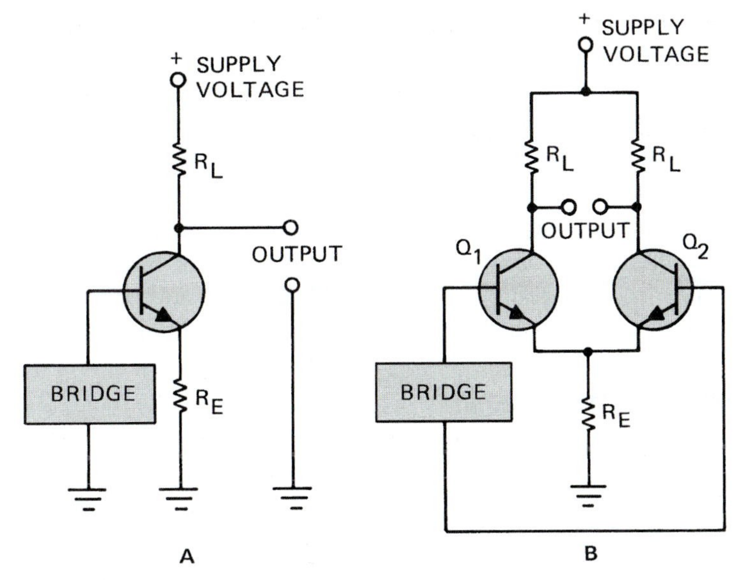

Fig. 3-42 *Amplifier stages. (A) Single transistor amplifier stage. (B) Two-transistor amplifier stage.*

the current variations in the two transistors produced by the bridge signal are equal and opposite. An increase in current through Q_1 is accompanied by a decrease in current through Q_2. The sum of these currents through R_E is constant. No signal voltage appears at the emitters to cause negative feedback as in Fig. 3-42A.

Output Circuit Connections The output circuit of the controller has three connections:

- Common positive (+), solid red wire
- Direct acting negative (–), solid blue wire
- Reverse acting negative, white/blue wire

A load in the form of an actuator, which is equivalent to 1000 Ω, can be connected to either set of wires or terminals. This depends upon the controller action desired.

The controller's amplifier and output circuits are also designed to provide sequential operation of two actuators. This is accomplished by connecting an actuator to the direct output, as well as to the reverse acting output.

The result is sequentially varying DC signals in response to temperature change at the sensing element. See Fig. 3-43. When sequential operation is used, the controller is calibrated so the set point and sensing element provide the bridge with a balanced condition at set point. This means both the direct and reverse acting outputs are zero.

When the temperature is significantly below the set point, a 16-V DC output is present on the reverse

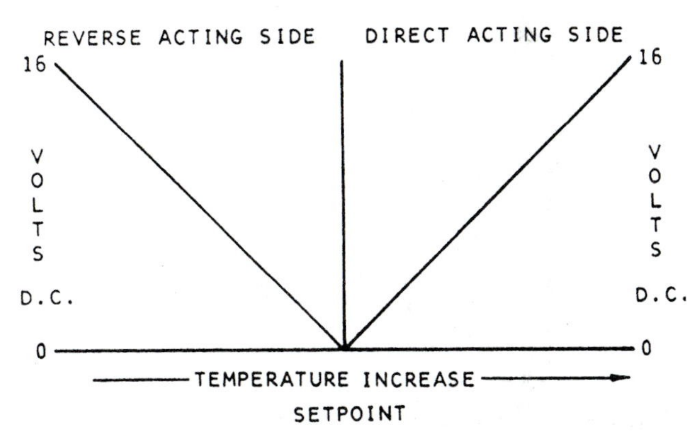

Fig. 3-43 *Result of sequentially varying DC signals in response to temperature change at the sensing element.*

acting side and a 0-V DC signal is present on the direct acting side. As temperature increases, the reverse acting signal decreases. When the temperature reaches set point, both outputs are 0-V DC or at "null." On a further increase in temperature, the direct acting signal increases from 0 to 16-V DC. When the temperature is such that operation is on the reverse acting side of null, only the actuator connected to that side is operating. Similarly, when the temperature is above the set point, operation is on the direct acting side of null and only that actuator is operating. In other words, the actuators operate in sequence, not simultaneously.

Bandwidths Bandwidths in these controllers are adjusted separately for direct- and reverse-acting signals. This permits optimum settings for both heating and cooling systems. See Fig. 3-43.

Bandwidth adjustment of an electronic controller is defined as "the number of desired degree changes at the element needed to cause a full 0- to 16-V DC change in the output signal." When sequential operation is utilized, the total temperature change at the element, which caused the outputs of both sides of null to vary, must be considered in the evaluation of system control.

Since there are two bandwidth settings, consider individually the temperature change from set point to where the full 16-V output on each side of null should occur.

Dual-Element Controllers

Dual-element controllers function the same way as single-element controllers with one exception. In place of one bridge, two bridges are used. Two bridge controllers are used where temperature effects on one element are to be used to readjust the set point of another element to provide greater accuracy of control and improved comfort for occupants.

Dual Bridge A dual-bridge arrangement is shown in Fig. 3-44. Bridge output is proportional to the algebraic sum of the temperature effects on both elements. This algebraic sum is expressed in terms of percentage of authority. An authority of 100 percent simply means that a temperature change (Δt) on the auxiliary element has the same effect as a temperature change (Δt) on the main element, except that the temperature change at each element is the opposite in direction. This is referred to as reverse adjustment.

Main Element Determining main and auxiliary assignments is dependent upon the measured temperature span at each element. The main element is always the element having the least measured temperature change of the two elements. The auxiliary element is always the element having the greatest measured temperature change. This arrangement is essential, since authority settings are always between 0 and 100 percent.

A typical system might have a ratio of main to auxiliary sensor effects of 20 to 1. This corresponds to a 5 percent authority setting. This means that a 20°F change in temperature at the auxiliary element produces a bridge output equal to that of' a 1°F change at the main element. For a 2 to 1 ratio, authority is 50 percent. This means a 2°F change at the auxiliary element has the effect of a 1°F change at the main element.

Dual-element controllers differ from single-element controllers only in regard to bridge configurations. There is an interacting effect within the bridge circuitry caused by the two elements and the authority setting. The amplifier circuitry and output circuitry cause the signals on both sides of null to be identical to those encountered with single-element controllers.

ACTUATORS

Electro-Hydraulic Actuators

Cybertronic actuators perform the work in an electronic system. They accept a control signal and translate that signal into mechanical movement to position valves or dampers. The electro-hydraulic actuators are so called because they convert an electric signal into a fluid movement and force. Damper actuators, equipped with linkage for connection to dampers and valve

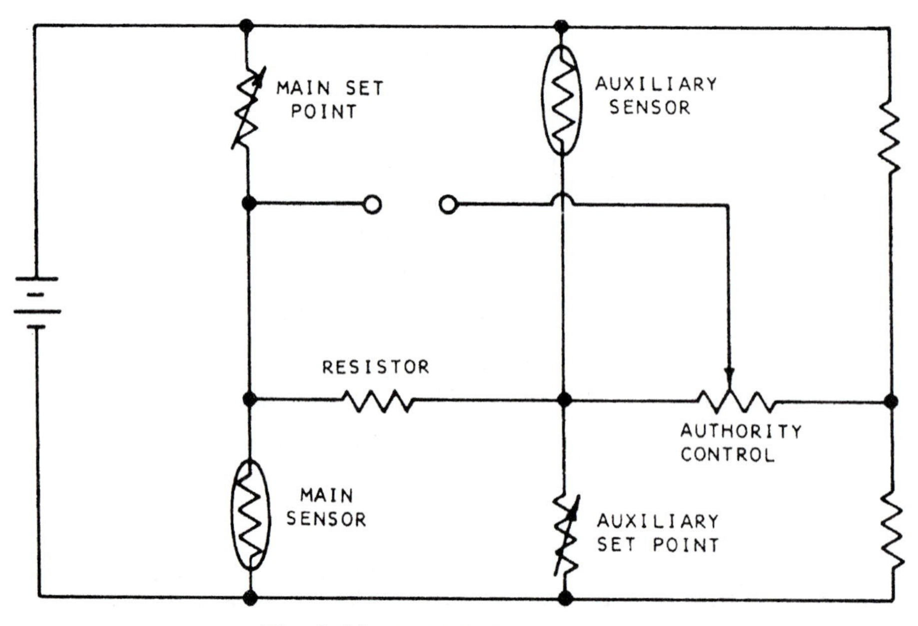

Fig. 3-44 *A dual-bridge arrangement.*

actuators, having a yoke and linkage to facilitate mounting on a valve body, are available.

Operation of Actuators Two voltages are applied to the actuator. See Fig. 3-45. A 0- to 16-V DC control signal regulates or controls the servo valve. Then a 24- or 120-V AC signal, depending on the unit, operates the oil pump. The oil pump moves oil from the upper chamber to the lower chamber. The servo valve controls pressure at the diaphragm by varying the return flow from the lower to the upper chamber.

When there is no DC voltage applied to the servo valve, the flapper is pushed off the servo port by way of the hydraulic pressure developed by the pump. The open servo port allows the pump to move all the oil through the lower chamber back into the upper chamber.

When the voltage on the servo increases, a magnetic force is developed. This magnetic force holds the flapper down over the servo port. The pump continues to pump oil into the lower chamber. But, the return flow to the upper chamber is stopped by the blocked servo port. Pressure is built up in the lower chamber until the magnetic force on the flapper is overcome and the flapper is pushed away from the servo port. This equalizes the flow through the pump and servo valve, while maintaining a pressure in the lower chamber.

Each increase in DC voltage results in a hydraulic pressure increase in the lower chamber. The increased pressure begins to overcome the opposing pressure from the return spring, and forces out the actuator shaft. Each further increase in DC voltage causes an increased extension of the actuator shaft.

The servo valve represents the load of 1000 Ω required by the controller to cause a variation of 0- to 16-V DC output signal. Two actuators can be connected in parallel across the output terminals of an electronic controller. However, this will provide only 500 Ω resistance, which the controller also can handle.

Thermal Actuators

Thermal actuators should more properly be called electro-thermal actuators. This is because they take a 0- to 16-V DC signal and convert the signal into heat. The thermal damper actuator has linkage for connection to a damper. The thermal-valve actuator is directly connected to the valve body.

Operation of a Thermal Actuator A thermal actuator is shown in Fig. 3-46. A small electrical control circuit is encapsulated in the electrical cable about 12 in. from the thermal unit. The 0- to 16-V DC signal from the controller and the 24-V AC supply voltage are fed into the control circuit. The circuit allows the 0- to 16-V DC signal to control the amount of current from the 24-V supply to the actuator.

Inside the actuator, the controlled current from the 24-V source heats up a small heater that is embedded in wax. When the wax reaches approximately 180°F it changes from solid to liquid. During this change, wax expands. This is the point at which the motion of the device is controlled.

As the wax expands, the power element shaft is forced out to move the piston. This, in turn compresses

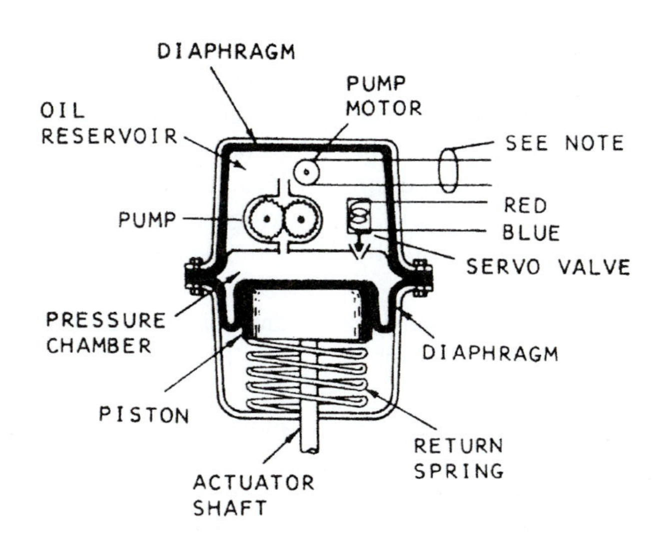

Fig. 3-45 *An elctro-hydraulic actuator.*

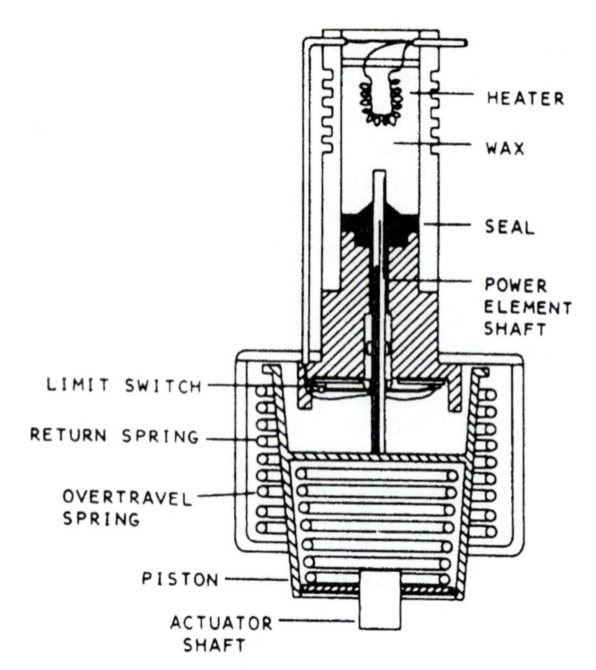

Fig. 3-46 *A thermal actuator.*

the return spring and moves the actuator shaft. After the power element shaft has traveled the full stroke, a limit switch is opened to stop the flow of current to the heater. The wax begins to cool and contract.

The power element shaft is forced to retract by the return spring. This closes the limit switch and the sequence is repeated. However, only when the control signal is high enough to hold the actuator at its fully extended position does it take place.

AUXILIARY DEVICES

Low- and high-signal selectors accept several control signals. Such selectors then compare them and pass the lowest or highest. For example, a high-signal selector can be used on a multizone unit to control the cooling coil. The zone requiring the most cooling transmits the highest control signal. This, in turn, will be passed by the high-signal selector to energize the cooling.

Minimum position networks are used to ensure that the outdoor air dampers are positioned to admit a minimum amount of air for ventilation, regardless of the controller demand. Reversing networks change the action of a controller output signal from direct to reverse or reverse to direct acting. Sequencing networks amplify a selected portion of an input voltage from a controller. A common application is where two actuators function in sequence.

Two-position power supplies permit two-position override of a proportional control system. A unison amplifier allows a controller to operate up to eight actuators, where a controller alone will operate only two.

ELECTRONIC COMPRESSOR MOTOR PROTECTION

Solid-state circuitry for air-conditioning units has been in use for some time. The following is an illustration of how some of the circuitry has been incorporated into the protection of compressor motors. This module is manufactured by Robertshaw Controls Co. of Milford, Connecticut.

Solid-state motor protection prevents motor damage caused by excessive temperature in the stator windings. These solid-state devices provide excellent phase-leg protection by means of separate sensors for each phase winding. The principal advantage of this solid-state system is its speed and sensitivity to motor temperature and its automatic reset provision.

There are two major components to the protection system:

1. The protector sensors are embedded in the motor windings at the time the motor is manufactured.
2. The control module is a sealed enclosure containing a transformer and switch. Figure 3-47 shows two models.

Operation

Leads from the internal motor sensors are connected to the compressor terminals as shown in Fig. 3-48. Leads from the compressor terminals to the control module are connected as shown in Fig. 3-49. Figure 3-49A shows the older model and Fig. 3-49B the newer model. While the exact internal circuitry is quite complicated, basically the modules sense resistance change through the sensors as the result of motor-temperature changes in the motor windings. This resistance change triggers, the action of the control circuit relay at predetermined opening and closing settings, which causes the line voltage circuit to the compressor to be broken and completed, respectively.

The modules are available for either 208/240- or 120-V circuits. The module is plainly marked as to the input voltage. The sensors operate at any of the stated because an internal transformer provides the proper power for the solid-state components.

The two terminals on the module marked "power supply" (T1 and T2) are connected to a power source of the proper voltage, normally the line terminals on the compressor motor contact, or the control-circuit transformer as required.

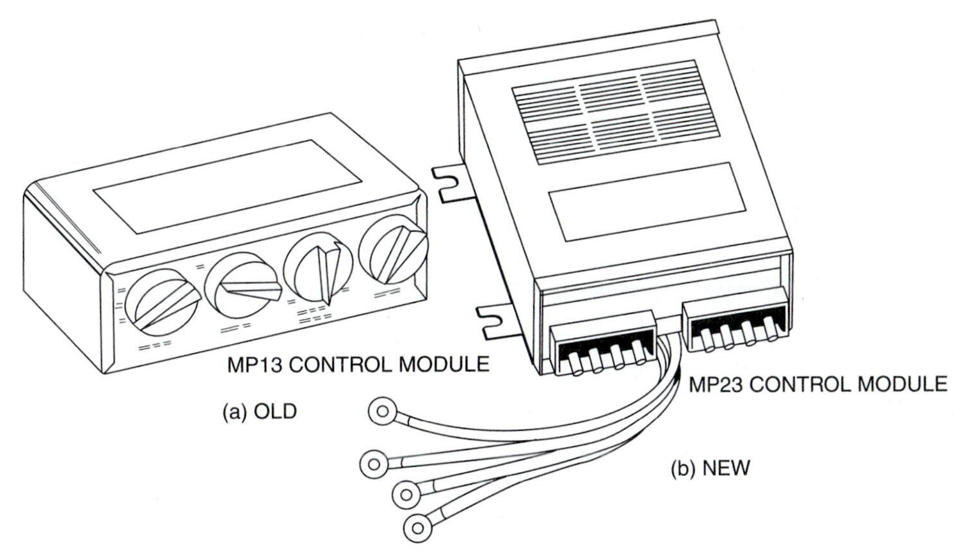

Fig. 3-47 *Solid-state control modules. (A) Older unit. (B) Newer unit.* (Robertshaw)

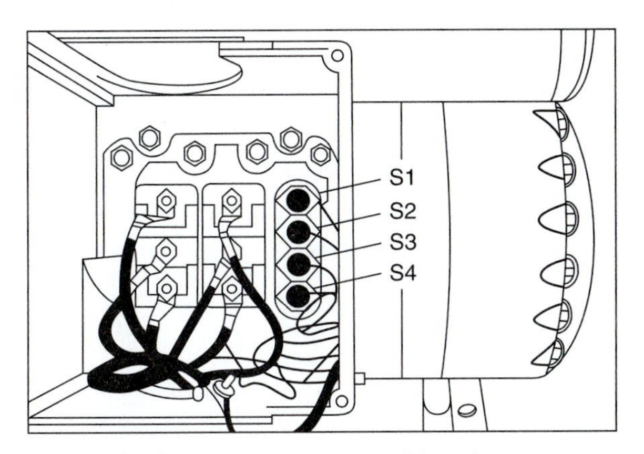

Fig. 3-48 *Compressor terminal board.* (Robertshaw)

Troubleshooting the Control

The solid-state module cannot be repaired in the field, and if the cover is opened or the module physically damaged, the warranty on the module is voided. No attempt should be made to adjust or repair this module, and if it becomes defective, it must be returned intact for replacement. This is the usual procedure for most solid-state units. However, if the unit becomes defective, you should be able to recognize that fact and replace it.

If the compressor motor is inoperable or is not operating properly, the solid-state control circuit may be checked as follows:

1. If the compressor has been operating and has tripped on the protector, allow the compressor to cool for at least 1 h before checking to allow time for the motor to cool and the control circuit to reset.
2. Connect a jumper wire across the control-circuit terminals on the terminal board. See Fig. 3-49. This will bypass the relay in the module. If the compressor will not operate with the jumper installed, then the problem is external to the solid-state protection system. If the compressor operates with the module bypassed, but will not operate when the jumper wire is removed, then the control-circuit relay is open.
3. If, after allowing time for motor cooling, the protector still remains open, the motor sensors may be checked as follows:
   - Remove the wiring connections from the sensor and common terminals on the compressor board. See Figs. 3-48 and 3-49.
   - *Warning*. Use an ohmmeter with a 3-V maximum battery power supply. The sensors are sensitive and easily damaged, and no attempt should be made to check continuity through them. Any external voltage or current applied to the sensors may cause damage, necessitating compressor replacement.
   - Measure the resistance from each sensor terminal to the common terminal. The resistance should be in the following range: 75 (cold) to 125 Ω(hot). Resistance readings in this range indicate the sensors are good. A resistance approaching zero indicates a short. A resistance approaching infinity indicates an open connection. If the sensors are damaged, they cannot be repaired or replaced in the field, and the compressor must be replaced to restore motor protection.

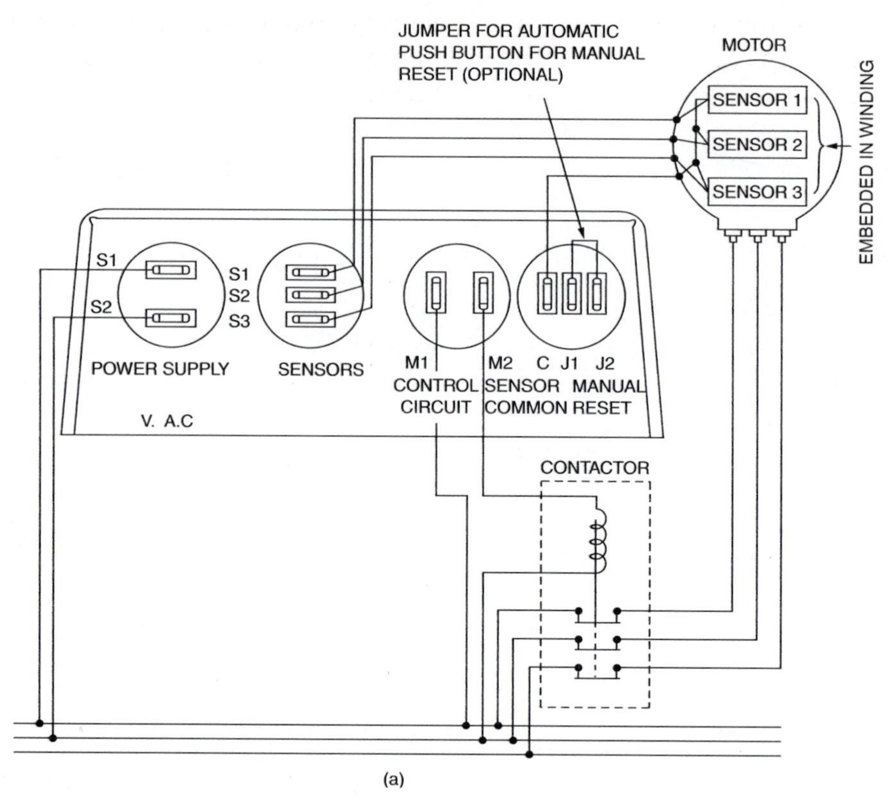

(a)

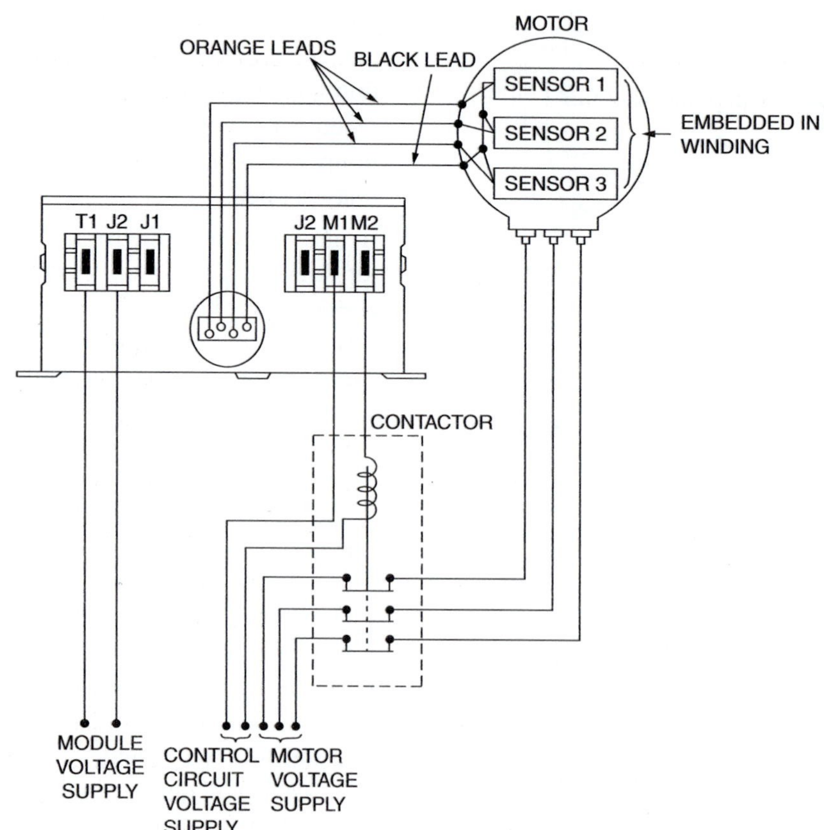

Note: Control is automatic reset when terminals J1 and J2 are not included. The control is manual reset when terminals J1 and J2 are included.

(b)

Fig. 3-49 *(A) Solid-state control modules (Older unit wiring details). (Robertshaw) (B) Continuing schematic for control modules (Newer unit wiring details). (Robertshaw)*

If the sensors have proper resistance and the compressor will run with the control circuit bypassed, but will not run when connected properly, the solid-state module is defective and must be replaced. The replacement module must be the same voltage and made by the same manufacturer as the original module on the compressor.

Restoring Service

In the unlikely event that one sensor is damaged and has an open circuit, the control module will prevent compressor operation even though the motor may be in perfect condition. If such a situation should be encountered in the field, as an *emergency* means of operating the compressor until such time as a replacement can be made, a properly sized resistor can be added between the terminal of the open sensor and the common sensor terminal in the compressor terminal box. See Figs. 3-48 and 3-50. This, then indicates to the control module an acceptable resistance in the damaged sensor circuit, and compressor operation can be restored. The emergency resistor should be a 2 W, 82-Ω, wire wound with a tolerance of ±5 percent.

In effect, the compressor will continue operation with two-leg protection rather than three-leg protection. While this obviously does not provide the same high degree of protection, it does provide a means of continuing compressor operation with a reasonable degree of safety.

REVIEW QUESTIONS

1. What is Ohm's law?
2. Describe a parallel circuit.
3. What is the formula for finding resistance in a parallel circuit?
4. What is the basic unit of measurement for electrical power?
5. What is a capacitor?
6. What is a dielectric?
7. What three factors determine the capacitance of a capacitor?
8. What is a microfarad?
9. What makes an electrolytic capacitor different from a standard paper-type?
10. What is the unit of measurement for inductance?
11. What is the symbol for an audio-frequency inductor?
12. What is inductive reactance?
13. What effect does the turns ratio have on the output voltage of a transformer?

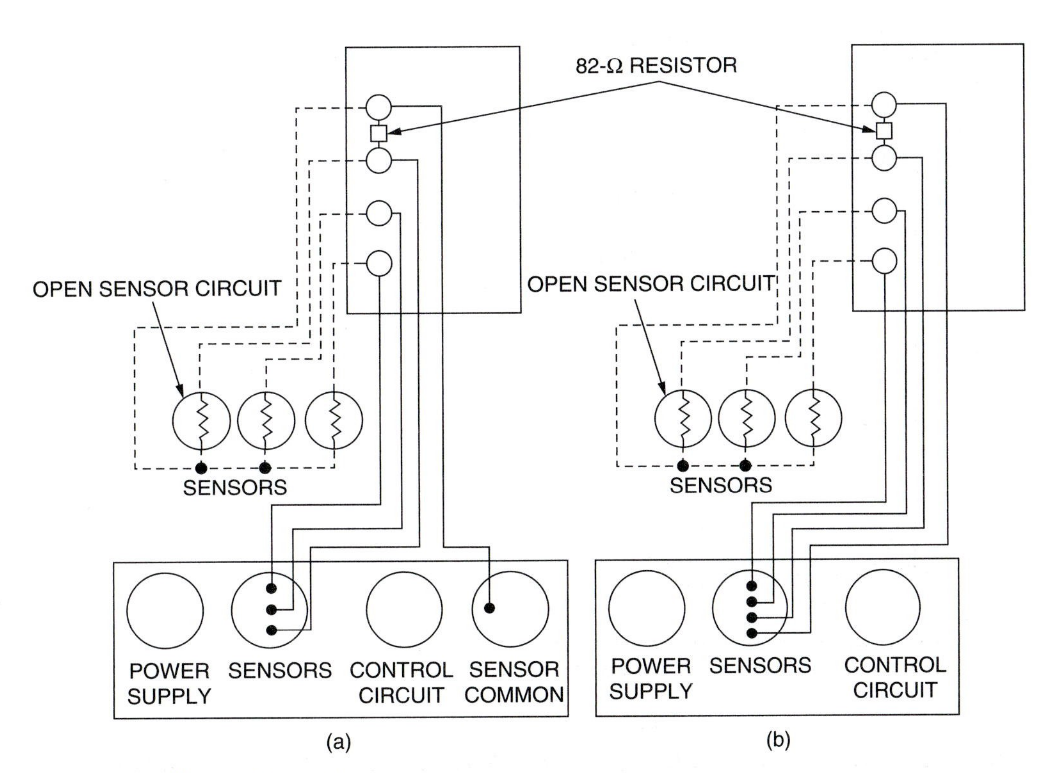

Fig. 3-50 *Adding a resistor to compensate for an open sensor.* (Robertshaw)

14. What is a zener diode?
15. What do the letters SCR stand for?
16. What is a bridge circuit?
17. What are the three parts of electronic controllers?
18. What is a thermal actuator?
19. What does the word semiconductor mean?
20. What are the two materials used for semiconductor devices?
21. What is a diode?
22. What is the PN junction diode used for?
23. What are the uses for diodes?
24. What are the two main uses for transistors?
25. What is an SCR? Where is it used?
26. What are the two main uses for a thermister?
27. What is a PNP transistor?
28. What is an integrated circuit?
29. What does CAB stand for in a humidity circuit?
30. What is a bridge circuit?
31. How do balanced and unbalanced bridge circuits differ?
32. What is a sensor?
33. How is a sensing bridge connected?
34. What is an actuator?
35. What is a differential amplifier used for?

4

CHAPTER

Solenoids and Valves

PERFORMANCE OBJECTIVES

After studying this chapter, you should:

1. Know the basis of magnetic induction.
2. Know how electromagnets are made.
3. Know how solenoids differ from relays.
4. Know the primary purpose of an electrically-operated solenoid valve.
5. Know the difference between NC and NO.
6. Know what happens when a valve leaks in a hot-gas defrost system.
7. Know the usual voltage ratings of solenoid valves.
8. Know what VA stands for and why you need to know it.

A *solenoid*, where the length is greater than the diameter is one of the most common types of coil construction used in electricity and electronics. The field intensity is the highest at the center in an iron-core solenoid. At the ends of the air-core coil, the field strength falls to a lower value.

A solenoid that is long, compared to the diameter, has a field intensity at the ends approximately one-half of that at the center. If the solenoid has a ferromagnetic core, the magnetic lines pass uniformly through the core.

Mechanical motion can be produced by the action of a solenoid or it can generate a voltage that is a result of some mechanical movement. The term solenoid has commonly come to mean a coil of wire with a moving iron core that can center itself lengthwise within the coil when current is applied to the coil. Then if a ferromagnetic core is properly suspended and under suitable tension, it can be moved in and out of a solenoid coil form with the application of coil current. This is the operating basis of some relays and a number of other electromechanical devices. If an outside force is used to move the ferromagnetic core physically, it is possible to induce a voltage in the solenoid coil.

There is a tendency in a solenoid for the core to move so that it encloses a maximum number of magnetic lines of force. Each line of force has the shortest possible length (Fig. 4-1). In the illustration the core is outside the coil. Because it is a ferromagnetic material, the coil presents a low-reluctance path to the magnetic lines of force at the north end of the coil. These lines of force concentrate on the soft-iron core and then complete their paths back to the south pole of the electromagnet.

Electromagnetic lines of force that pass through the core magnetize it. This means that the induced magnetic field in the core has a south pole near the coil's north pole. Inasmuch as unlike poles attract, the

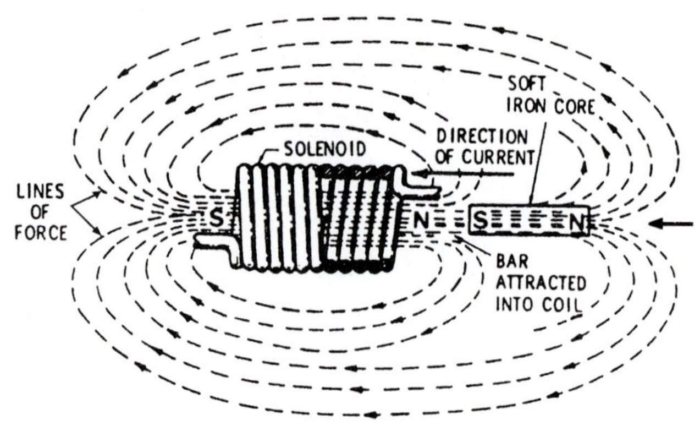

Fig. 4-1 *Solenoid pulls core into the coil. Sucking effect of a coil.*

core is attracted toward the hole in the solenoid coil. This attraction tends to pull the core into the coil. As the iron core is pulled into the coil, the magnetic field becomes increasingly shorter and the magnetic lines of force travel the shortest possible distance when the core centers itself in the coil.

By attaching a spring to the core, it is possible to have the core return to its outside position once the power is interrupted to the coil. When the power is then turned on again, it pulls the core back into the coil. It is this type of movement that is utilized in the construction of industrial solenoids that operate switch contacts in relays and motor starters, and valves in gas, air, and liquid lines of various types.

INDUSTRIAL SOLENOIDS

Industry has many uses for solenoids. They are electrically operated and can be controlled from a distance by low voltage and small currents. They come in many sizes and shapes. There are two classifications, which may be of interest to the air-conditioning and refrigeration personnel in any residential or industrial as well as commercial field.

Tubular Solenoids

There are various uses for solenoids. Figure 4-2 shows tubular solenoids. Notice the type, voltage rating, coil resistance, and the minimum and maximum lifts and strokes. Some are pull types and others are push types. They are also specified as to intermittent and continuous duty.

Frame Solenoids

The frame-mounted types of solenoids (Fig. 4-3) are available in intermittent and continuous duty as well as usable on either AC or DC. Types 11 and 28 can operate

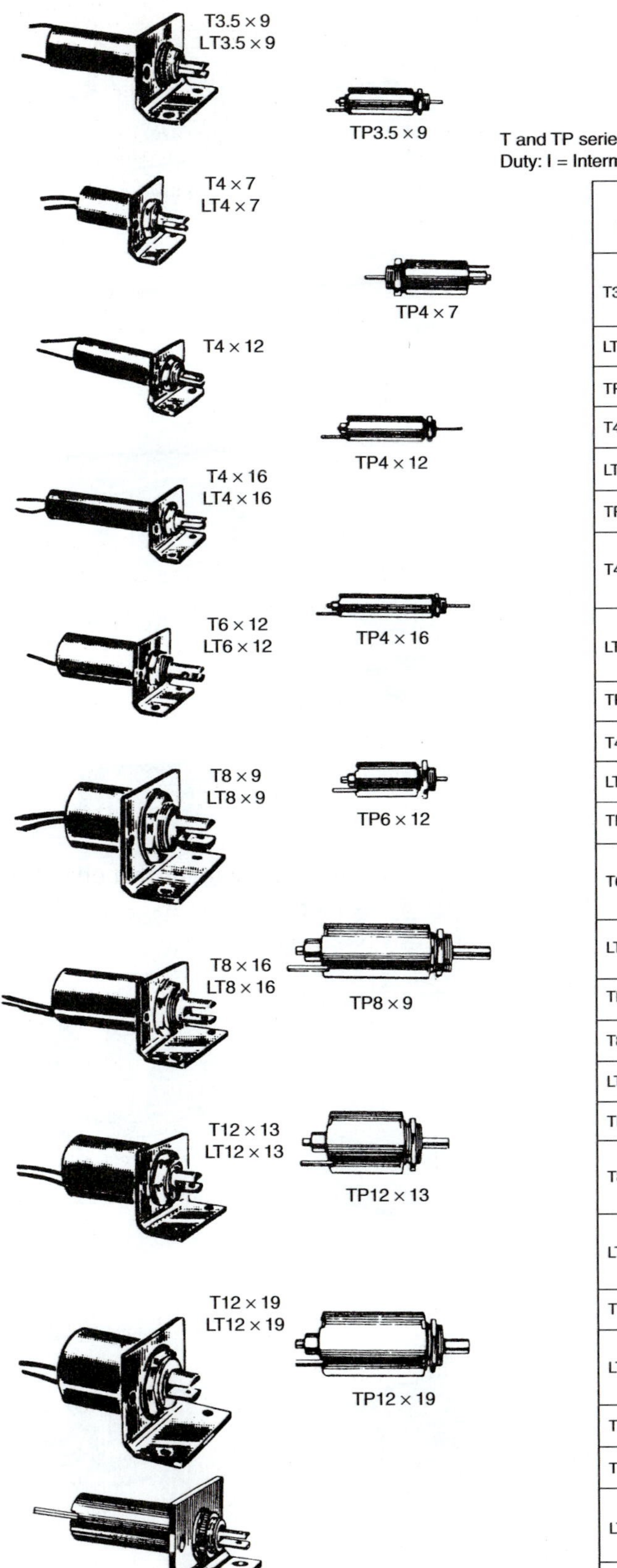

T, TP AND LT SERIES SOLENOIDS

T and TP series are UL recognized.
Duty: I = Intermittent. C = Continuous

Type	Duty-volt.	Coil Resis. Ohms	Lifts and Strokes			
			Min.		Max.	
			Oz.	@in	Oz.	@in
T3.5 × 9	C-12D	60.2	4 1/2	1/32	0.3	5/16
	I-12D	31.1	7	1/32	0.7	1/2
	I-24D	122	7	1/32	0.7	1/2
	C-24D	254	4 1/2	1/32	0.3	5/16
LT3.5 × 9	C-12D	52.4	2.5	1/32	0.2	3/8
	C-24D	221	2.5	1/32	0.2	3/8
TP3.5 × 9	I-24D	122	6	1/32	0.5	1/2
	C-24D	254	4	1/32	0.3	5/16
T4 × 7	I-24D	131	7	1/32	1	1/4
	C-24D	270	5	1/32	0.5	1/4
LT4 × 7	C-12D	63.3	2.5	1/32	1	0.15
	C-24D	264	2.5	1/32	1	0.15
TP4 × 7	I-24D	131	6.5	1/32	0.75	0.15
	C-24D	270	3.5	1/32	0.50	1.5
T4 × 16	C-12D	45.1	4.5	1/16	1.2	1/2
	I-12D	17.7	7	1/16	2.5	1
	C-24D	173	4.5	1/16	1.2	1/2
	I-24D	72.7	7	1/16	2.5	1
LT4 × 16	C-12D	42.5	4	1/16	2	1/2
	I-12D	14	8	1/16	2.5	1/2
	C-24D	168	4	1/16	2	1/2
	I-24D	69.1	8	1/16	2.5	1/2
TP4 × 16	C-24D	173	3.8	1/16	1	1/2
	I-24D	72.7	5.5	1/16	2	1/2
T4 × 12	C-24D	195	6	1/32	1.5	0.15
	I-24D	96.7	9	1/32	2	0.15
LT4 × 12	C-12D	49.3	7.5	1/32	0.5	1/2
	C-24D	19.2	7.5	1/32	0.5	1/2
TP4 × 12	I-24D	96.7	7	1/32	1.5	0.15
	C-24D	195	5	1/32	1.0	0.15
T6 × 12	C-12D	31.7	15	1/16	1	3/4
	I-12D	12.1	23	1/16	3	3/4
	C-24D	121	15	1/16	1	3/4
	I-24D	60.6	23	1/16	3	3/4
LT6 × 12	C-12D	35	18	1/16	1	1/2
	I-12D	13.8	35	1/16	2.5	1/2
	C-24D	138	18	1/16	1	1/2
TP6 × 12	C-24D	121	12	1/16	0.8	3/4
	I-24D	60.6	18	1/16	2.5	3/4
T8 × 9	C-24D	135	18	1/16	2.5	7/16
	I-24D	44	45	1/16	5	7/16
LT8 × 9	C-24D	109	18	1/16	1	1/2
	I-24D	44.6	35	1/16	3	1/2
TP8 × 9	I-24D	44	36	1/16	4	7/16
	C-24D	135	14	1/16	2	7/16
T8 × 16	I-12D	9.3	58	1/16	5	0.60
	C-12D	28.3	30	1/16	2.5	0.60
	I-24D	36.1	58	1/16	5	0.60
	C-24D	110	30	1/16	2.5	0.60
LT8 × 16	I-12D	6.2	67	1/16	12	1/2
	C-12D	19.3	33	1/16	2.5	1/2
	I-24D	29.7	62	1/16	10	1/2
	C-24D	77.2	33	1/16	2.5	1/2
T12 × 13	I-24D	28.4	100	1/8	15	0.60
	C-24D	90.4	40	1/8	5	0.60
LT12 × 13	I-12D	5.9	110	1/8	10	0.5
	C-12D	18.3	35	1/8	1	0.5
	I-24D	22	110	1/8	10	0.5
	C-24D	71.2	35	1/8	1	0.5
TP12 × 13	I-24D	28.4	80	1/8	12	0.60
	C-24D	90.4	32	1/8	4	0.60
T12 × 19	I-24D	22.1	130	1/8	40	3/4
	C-24D	68	70	1/8	10	3/4
LT12 × 19	I-12D	4.66	130	1/8	15	3/4
	C-12D	14.8	70	1/8	5	3/4
	I-24D	18.6	130	1/8	15	3/4
	C-24D	71.8	70	1/8	3	3/4
TP12 × 19	I-24D	22.1	100	1/8	30	3/4
	C-24D	68	55	1/8	8	3/4

Fig. 4-2 *Tubular solenoids.*

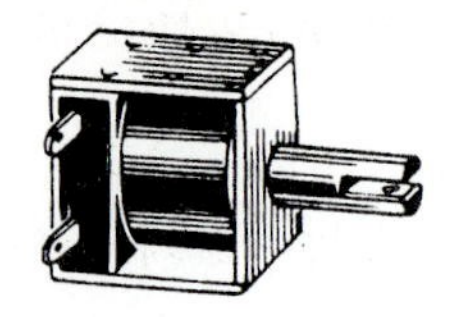

Type 2

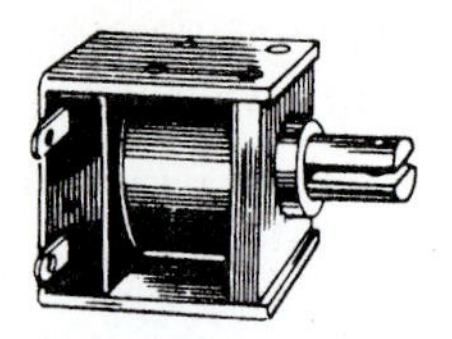

Type 2HD

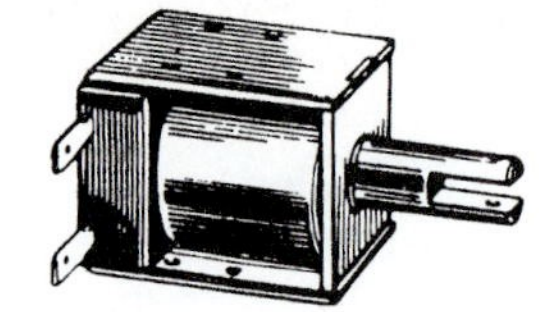

Type 4

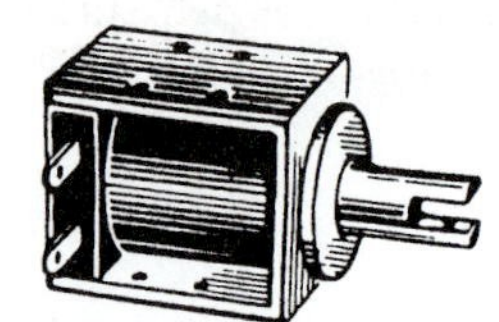

Type 4HD

Type 11

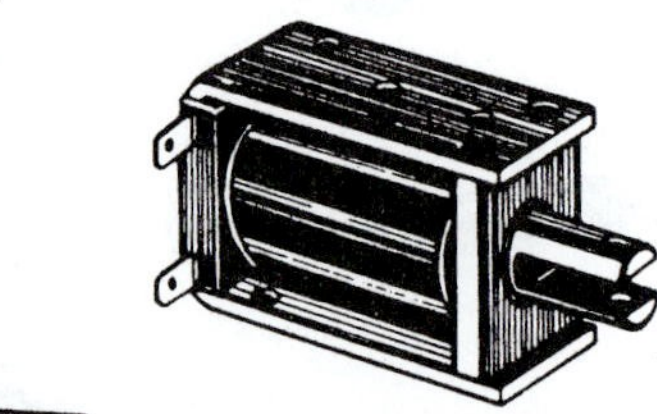

Type 11HD

DC Voltage Model Solenoids

Type	Duty	Volts	Ohms	Oz. @inch of stroke	
				Minimum	Maximum
2HD	Intermittent	24	22.6	96@1/8"	15@1/2"
2HD	Continuous	24	71	48@1/8"	5@1/2"
4	Intermittent	24	15.8	100@1/8"	20@1"
4	Continuous	24	61.3	60@1/8"	7@4/5"
4	Intermittent	110	296	100@1/8"	20@1"
4	Continuous	110	1215	60@1/8"	7@4/5"
4HD	Intermittent	24	18.9	130@1/8"	25@3/4"
4HD	Continuous	24	57.5	80@1/8"	5@3/4"
4HD	Intermittent	110	354	130@1/8"	25@3/4"
4HD	Continuous	110	1140	80@1/8"	5@3/4"
11	Intermittent	6	1.88	45@1/8"	10@1/2"
11	Continuous	6	4.69	30@1/8"	4@1/2"
11	Intermittent	24	29.1	45@1/8"	10@1/2"
11	Continuous	24	93.1	30@1/8"	4@1/2"
11HD	Intermittent	24	29.3	70@1/8"	5@3/4"
11HD	Continuous	24	76.3	30@1/8"	2@3/4"
11P	Continuous	24	93.1	24@1/8"	3.2@1/2"
22	Intermittent	6	5.8	17@1/16"	2@0.3"
22	Continuous	6	11.5	11@1/16"	1@0.3"
22	Intermittent	24	93.2	17@1/16"	2@0.3"
22	Continuous	24	182	11@1/16"	1@0.3"
28	Intermittent	6	3.03	40@1/16"	3@1/2"
28	Continuous	6	7.5	23@1/16"	2@1/2"
28	Intermittent	12	11.9	40@1/16"	3@1/2"
28	Continuous	12	29.8	23@1/16"	2@1/2"
28	Intermittent	24	47.4	40@1/16"	3@1/2"
28	Continuous	24	116	23@1/16"	2@1/2"

Type 12

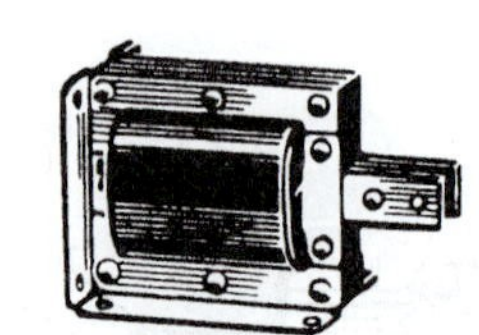

Type 14

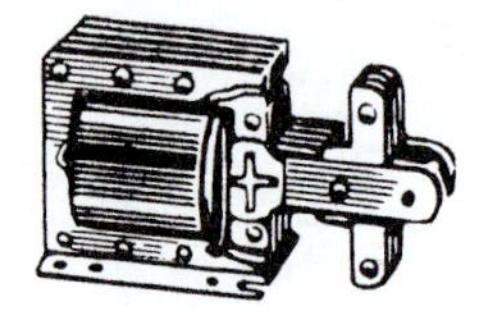

Type 16

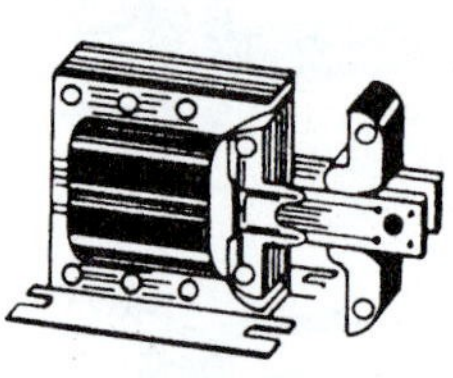

Type 18

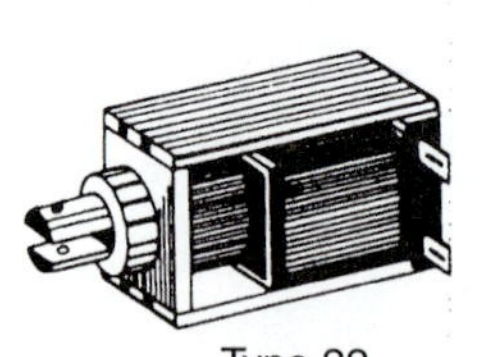

Type 22

AC Voltage Model Solenoids

2	Intermittent	120	60	45@1/8"	11@7/8"
2	Continuous	120	166	14@1/8"	3@7/8"
2HD	Intermittent	120	36	70@1/8"	16@3/4"
2HD	Continuous	120	113	25@1/8"	6@3/4"
4	Intermittent	120	37	36@1/8"	26@1"
4	Continuous	120	133	8@1/8"	7@1"
11	Intermittent	120	85	21@1/8"	11@3/4"
11	Continuous	120	200	12@1/8"	6@3/4"
11HD	Continuous	120	165	12@1/8"	31/2@1"
11P	Continuous	120	200	9.6@1/8"	4.8@3/4"
12	Intermittent	120	100	48@1/8"	9@7/8"
12	Continuous	120	150	28@1/8"	6@7/8"
14	Intermittent	120	11	108@1/8"	56@11/2"
14	Continuous	120	18	75@1/8"	40@11/2"
16	Intermittent	120	41	110@1/8"	28@3/4"
16	Continuous	120	85	63@1/8"	15@3/4"
16	Continuous	240	350	63@1/8"	15@3/4"
16P	Intermittent	120	41	88@1/8"	22.5@3/4"
16P	Continuous	120	85	50.5@1/8"	12@3/4"
18	Intermittent	120	8.8	350@1/8"	208@7/8"
18	Continuous	120	19.7	152@1/8"	100@7/8"
18	Intermittent	240	45	350@1/8"	208@7/8"
18	Continuous	240	78	150@1/8"	100@7/8"
18P	Intermittent	120	8.8	315@1/8"	187@7/8"
18P	Continuous	120	19.7	137@1/8"	90@7/8"
24	Continuous	120	500	10@1/16"	2@5/8"
28	Continuous	24	17.4	24@1/16"	5@5/8"
28	Continuous	120	400	24@1/16"	5@1/2"

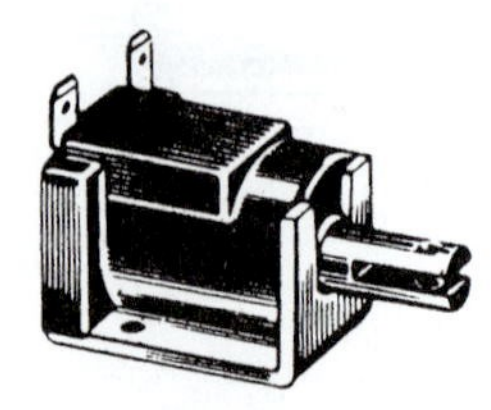

Type 24

Type 28

- **All models are UL recognized**

These intermittent and continuous duty solenoids are available in AC and DC versions, and in three constructions: box frame, U-frame and laminated. **Types 2, 2HD, 4, 4HD, 11, 11HD, 11P, 22** and **28** are box frame. **Types 12, 14, 16,16P, 18** and **18P** are laminated. **Type 24** is U-frame. Suffix **P** indicates a push type model. Suffix **HD** indicates a heavy duty model. All box frame models have quick connect terminals.

Fig. 4-3 *Frame solenoids.*

on AC/DC. The other types are identified as to whether they operate best on AC or DC.

APPLICATIONS

Solenoids are devices that turn electricity, gas, oil, or water on and off. Solenoids can be used, for example, to turn the cold water on, and the hot water off, to get the proper mix of warm water in a washing machine. To control the hot water solenoid, a thermostat is inserted in the circuit.

Figure 4-4 shows a solenoid for controlling natural gas flow in a hot-air furnace. Note how the coil is wound around the plunger. The plunger is the core of the solenoid. It has a tendency to be sucked into the coil whenever the coil is energized by current flowing through it. The electromagnetic effect causes the plunger to be attracted upward into the coil area. When the plunger is moved upward by the pull of the electromagnet, the soft disk (10) is also pulled upward, allowing gas to flow through the valve. This basic technique is used to control water, oil, gasoline, or any other liquid or gas.

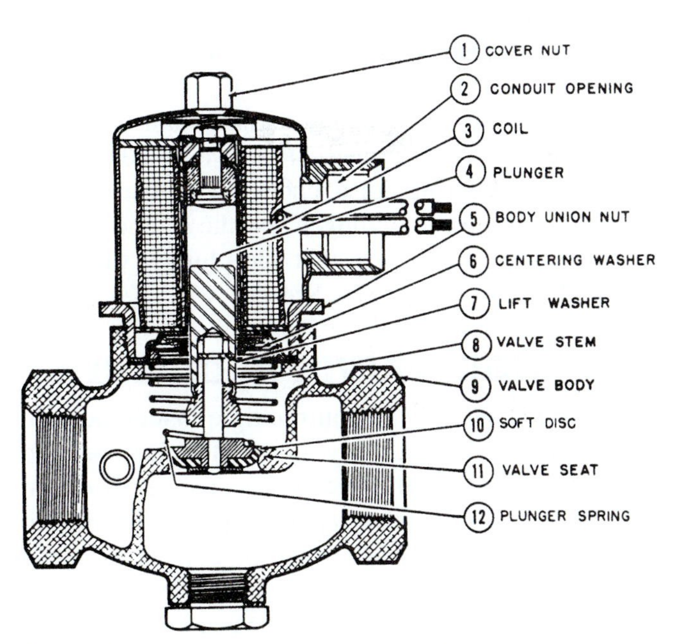

Fig. 4-4 *Solenoid for controlling natural gas flow to a hot-air furnace.* (Honeywell)

The starter solenoid on an automobile uses a similar procedure except that the plunger has electrical contacts on the end that complete the circuit from the battery to the starter. The solenoid uses low voltage (12 V) and low current to energize the coil. The coil in turn sucks the plunger upward. The plunger, with a heavy-duty copper washer attached, then touches heavy-duty contacts that are designed to handle the 300 A needed to start a cold engine. In this way, low voltage and low current are used, from a remote location, to control low voltage and high current.

Solenoids as Electromagnets

An electromagnet is composed of a coil of wire wound around a core of soft iron. A solenoid is an electromagnet. When current flows through the coil, the core becomes magnetized. The magnetized core can be used to attract an armature and act as a magnetic–circuit breaker (Fig. 4-5). Note how the magnetic–circuit breaker is connected in series with both the load circuit to be protected and with the switch contact points. When excessive current flows in the circuit, a strong magnetic field in the electromagnet causes the armature to be attracted to the core. A spring attached to the armature causes the switch contacts to open and break the circuit. The circuit breaker must be reset by hand to allow the circuit to operate properly again. If the overload is still present, the circuit breaker will "trip" again. It will continue to do so until the cause of the short circuit or overload is found and corrected.

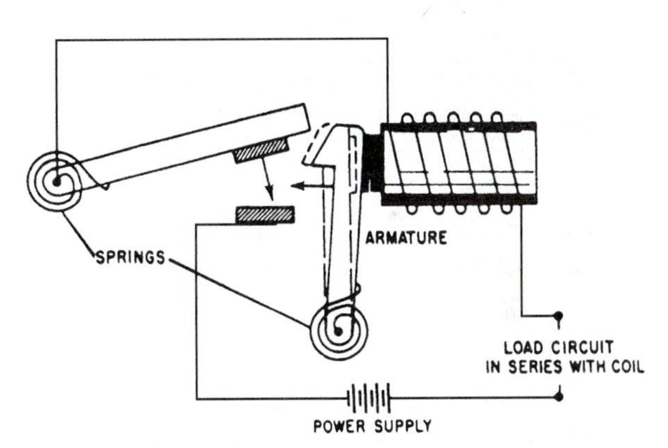

Fig. 4-5 *Magnetic–circuit breaker.*

Solenoid Coils

The coil is the most important part of the solenoid, as the valve or switch contacts, that it operates, cannot work unless the coil is capable of being energized. There are at least three types of coils you should be aware of in solenoids used in air conditioning, refrigeration, and heating circuits. For various applications they are divided into classes, as outlined in Table 4-1. See Fig. 4-6.

Servicing Coils

Coils can be replaced when they malfunction. Excessive heat causes coil malfunction. Make sure that the

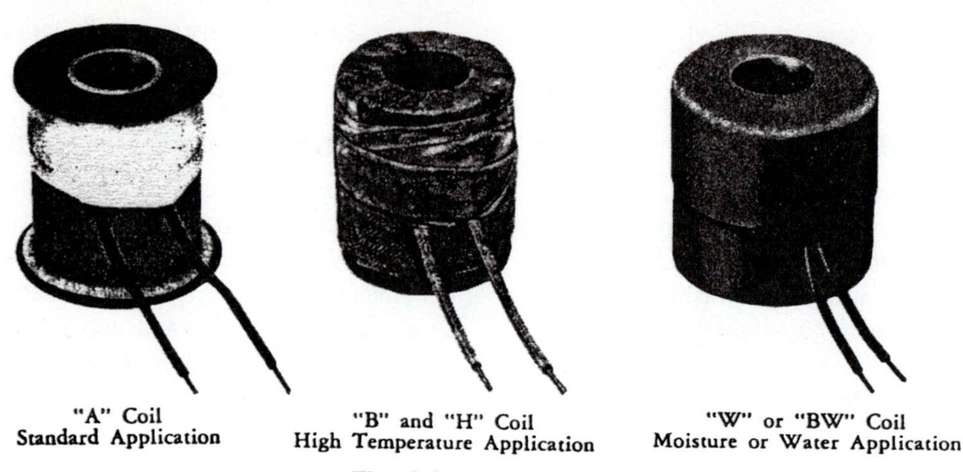

Fig. 4-6 *Solenoids coils.*

Table 4-1 *Classes of Solenoid Coils*

Class	Application
A	Moisture-resistant coil for normal use of gas or fluid up to 175°F.
B	Ambient and fluid temperature up to 200°F.
H	Temperatures up to 365°F, high-steam pressure, rapid-valve cycling, high voltage, fungusproof.
BW	Same as coil B, and waterproof, fungusproof, plastic-encapsulated for temperatures up to 200°F.
W	Same as coil A, and waterproof, fungusproof, plastic-encapsulated for temperatures up to 175°F.

valve is not heated to a temperature above the coil rating. When replacing a coil, reassemble the solenoid correctly. A missing part or improper reassemble causes excessive coil heat. See the exploded view in Fig. 4-7.

Applied voltage must be at the coil-rated frequency and voltage. A damaged plunger tube or tube sleeve causes heat and can prevent the solenoid front operating. For applications requiring greater resistance or different electrical requirements, use the proper coil in the solenoids. Do not change from AC to DC, or DC to AC, without changing the entire solenoid assembly (coil, plunger, plunger tube, and base fitting).

When replacing a coil, first be sure to turn off the electric power to the solenoid. It will not be necessary in most instances to remove the valve from the pipeline. Disconnect the coil leads. Disassemble the solenoid carefully and reassemble in reverse order. Failure to reassemble the solenoid properly can cause coil burn out.

Surge suppressors are available to protect the coil from unusual line surges. Figure 4-8 shows how the coil leads can be connected to allow for 120- or 240-V operation. These are referred to as dual-voltage coils.

The valve shown in Fig. 4-7 is a series–balanced diaphragm solenoid valve that provides on-off control for domestic and industrial furnaces, boilers, conversion burners, and similar units using thermostats, limit controls, or similar control devices. The valve uses a balanced diaphragm (or high-operating pressure with low electrical power consumption). It is suitable for use with all gases and comes in a variety of sizes, capacities, and pressures.

Presence of a low, barely audible hum is normal when the coil is energized. If the valve develops a buzzing or chattering noise, check for proper voltage. Thoroughly clean the plunger and the interior of the plunger tube. Make sure that the plunger tube and solenoid assembly are tight. See Fig. 4-9.

SOLENOID VALVES IN CIRCUITS

Solenoid valves are used on multiple installations in refrigeration systems. They are electrically operated as shown in Fig. 4-10. When connected, as shown in the illustration, the valve remains open when current is supplied to it. It closes when the current is turned off. In general, solenoid valves are used to control the liquid refrigerant flow into the expansion valve or the refrigerant gas flow from the evaporator when it, or the fixture it is controlling, reaches the desired temperature.

The most common application of the solenoid valve is in the liquid line and operates with a thermostat. With this hookup, the thermostat is set for the desired temperature in the fixture. When this temperature is reached, the thermostat opens the electrical circuit and shuts off the current to the valve. The solenoid valve closes and shuts off the refrigerant supply to the expansion valve. The condensing unit operation is controlled by the low-pressure switch. In other applications,

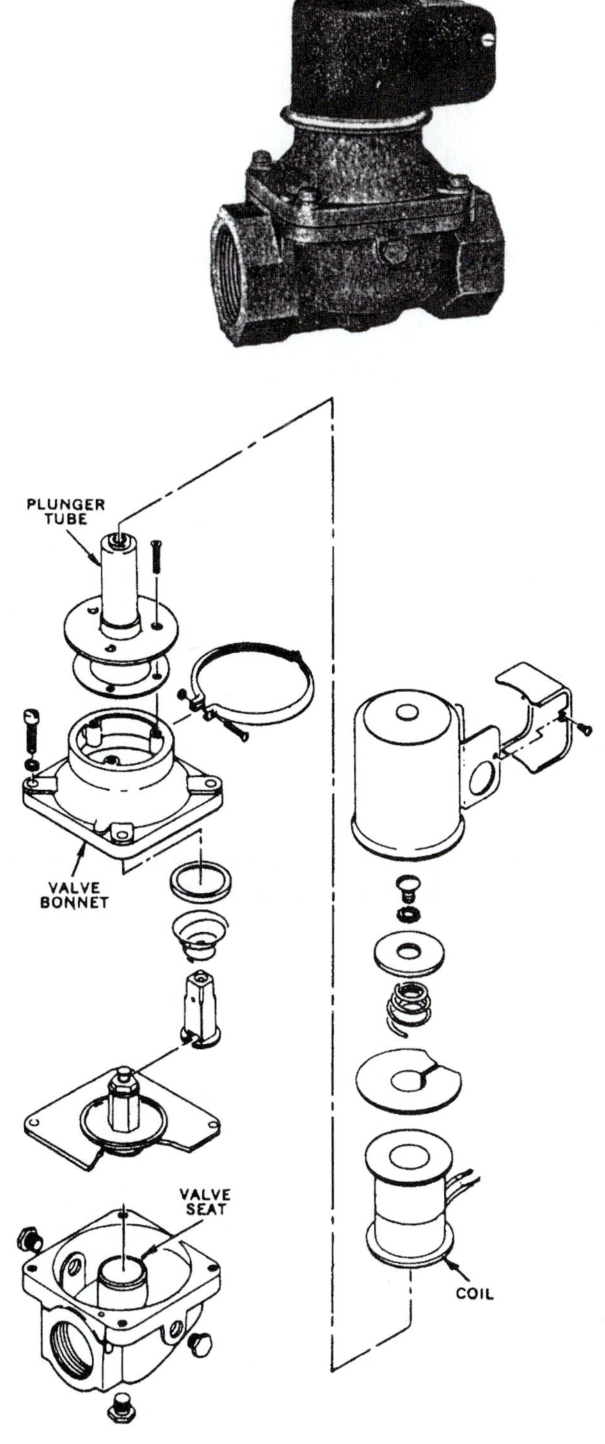

Fig. 4-7 *Exploded view of balanced diaphragm valve*

where the evaporator is in operation for only a few hours each day, a manually-operated snap switch is used to open and close the solenoid valve.

Refrigeration Valve

The solenoid valve, shown in Fig. 4-11, is operated with a normally closed status. A direct-acting metal

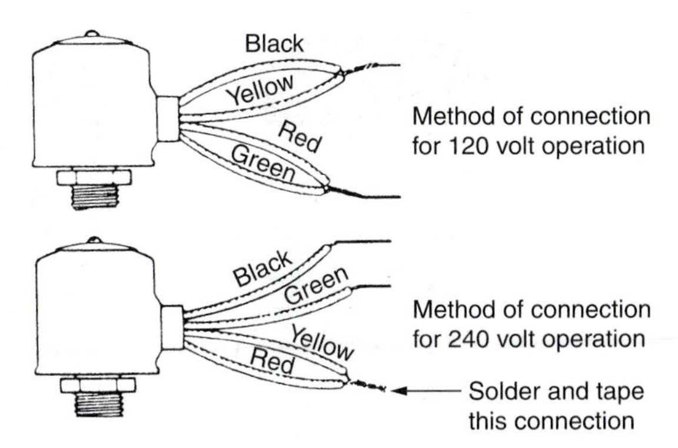

Fig. 4-8 *Dual-voltage coil wiring diagrams.*

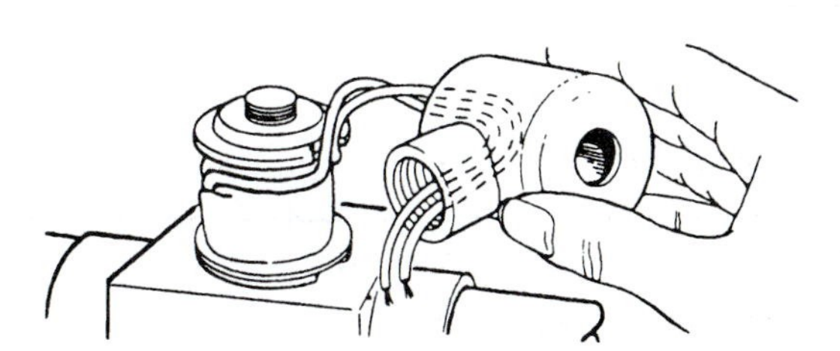

Fig. 4-9 *Solenoid coil with cover removed.*

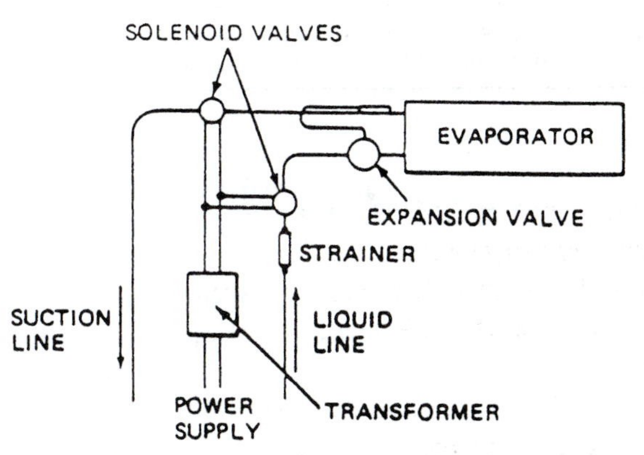

Fig. 4-10 *Solenoid valves connected in the suction and liquid evaporator lines of a refrigeration system.*

ball and seat assure tight closing. The two-wire, class W coil is supplied standard for long life in low-temperature service or sweating conditions. Current failure or interruption causes the valve to fail-safe in the closed position. Explosion-proof models are available for use in hazardous areas.

This solenoid valve is usable with all refrigerants except ammonia. It can also be used for air, oil, water, detergents, butane or propane gas, and other noncorrosive liquids or gases.

A variety of temperature control installations can be accomplished with these valves. Such installations include bypass, defrost, suction line, hot gas service,

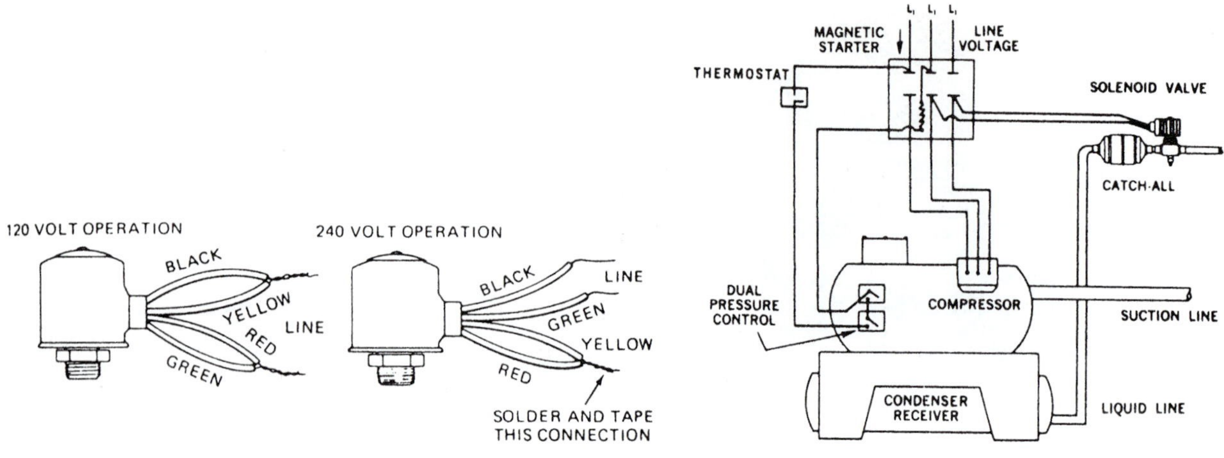

Fig. 4-11 *Schematic illustration of a refrigeration installation.*

humidity control, alcohols, unloading, reverse cycle, chilled water, cooling tower, brine, and liquid–line stop installations and ice makers.

The valves are held in the normally closed position by the weight of the plunger assembly and fluid pressure on top of the valve ball. The valve is opened by energizing the coil. This magnetically lifts the plunger and allows full flow by the valve ball. Deenergizing the coil permits the plunger and valve ball to return to the closed position.

REVIEW QUESTIONS

1. Define solenoid.
2. What is meant by the term sucking effect of a solenoid?
3. What does the armature of an electromagnet do?
4. What is the most important part of the solenoid?
5. List the five classes of solenoids.
6. What are dual-voltage coils? How are they wired?
7. Where are series-balanced diaphragm solenoid valves used?
8. What does a barely audible hum indicate when a coil is energized?
9. Where are solenoid valves used?
10. How are valves made fail-safe?

5
CHAPTER

Electric Motors: Selection, Operational Characteristics, and Problems

PERFORMANCE OBJECTIVES

1. Know the principle of operation of a DC motor.
2. Know how motors operate.
3. Know how to start a motor.
4. Know different types of motors.
5. Know how AC/DC motors differ one from the other.
6. Know how to select the proper type and size of motor for a specific job.
7. Know how to troubleshoot motors using the probable cause-remedy chart.
8. Know how to use a V-O-M to test a refrigeration motor circuit.
9. Know how contactors, starters, and relays operate.
10. Know how motor-overload protectors operate.
11. Know how thermostats work electrically.
12. Know how hot-gas defrosting works.
13. Know how to read a "ladder" schematic.

It is often necessary for the air conditioning and refrigeration repair or service individual to work on the electric motors that move the air and refrigerant in a system. In some cases it is the movement of water in the chiller system.

The individual who services these machines must be able to understand how they work and why they are used where they are in the system. And, of course, it is very necessary to be able to recognize any symptom that indicates trouble and correct it at once.

CONSTRUCTION OF AN INDUCTION MOTOR

In an induction motor, the stationary portion of the machine is called a *stator*. The rotating member is called a *rotor*. Instead of salient poles in the stator, distributed windings are used. These are placed in slots around the periphery of the stator. See Fig. 5-1.

It is not usually possible to determine the number of poles from visual inspection of an induction motor. A look at the nameplate will usually tell the number of poles. The nameplate also gives the rpm, the voltage required, and the current needed. This rated speed is usually less than the synchronous speed because of the slip. Slip is due to the inability of the rotor to keep up with the rotating field. To determine the number of poles per phase of the motor, divide 120 times the frequency by the rated speed:

$$P = \frac{120 \times f}{N}$$

P = number of poles per phase
f = frequency in hertz (Hz)
N = rated speed in rpm
120 = constant

The result is very nearly the number of poles per phase. For example, consider a 60-Hz, three-phase machine rated with a speed of 1750 rpm. In this case:

$$P = \frac{120 \times 60}{1750} = \frac{7200}{1750} = 4.1$$

Therefore, the motor has four poles per phase. If the number of poles per phase is given on the nameplate, the synchronous speed can be determined. Divide 120 times the frequency by the number of poles per phase. In the example just given, the synchronous speed is equal to 7200 divided by 4, or 1800 rpm.

The rotor of an induction motor consists of an iron core with longitudinal slots around its circumference, in which heavy copper or aluminum bars are embedded in the slots. These bars are welded to a heavy ring of high conductivity on either end. This composite structure is sometimes called a "squirrel cage." Motors containing

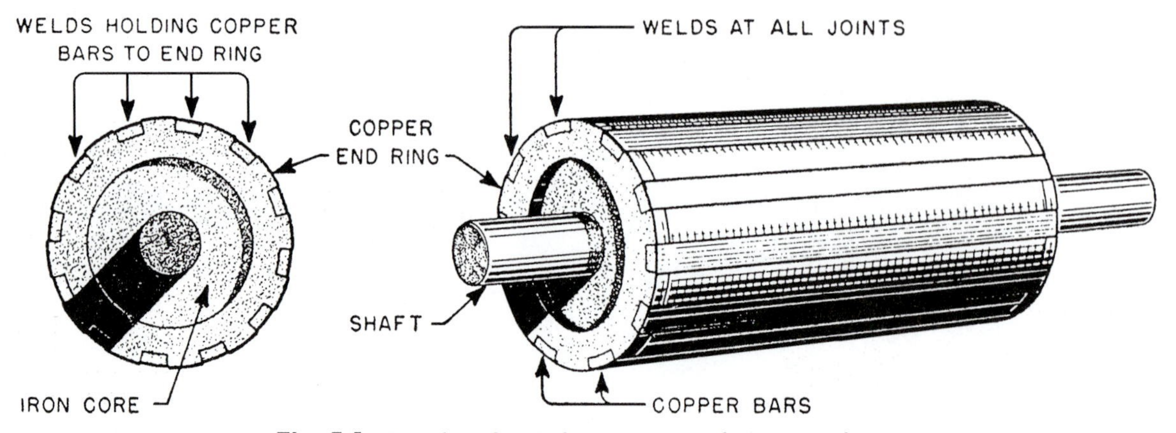

Fig. 5-1 *Note how the windings are inserted in a motor frame.*

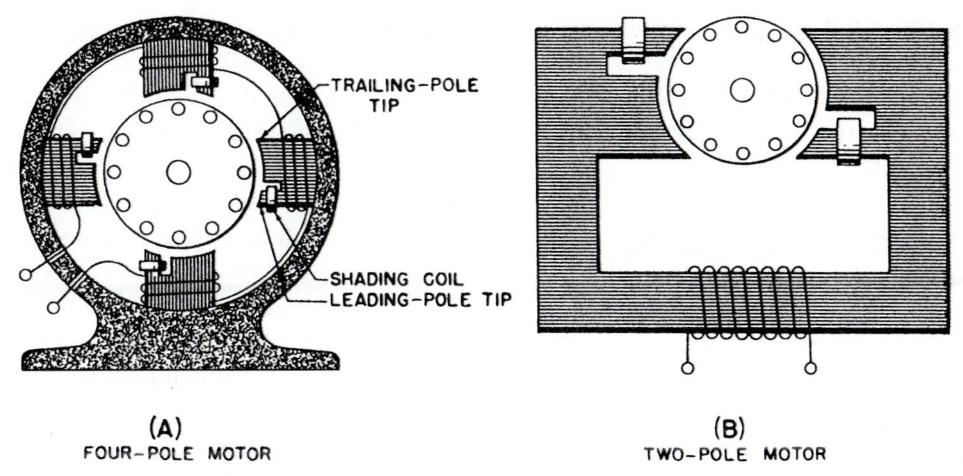

Fig. 5-2 *Shaded-pole motor.*

such a rotor are called squirrel-cage induction motors. See Fig. 5-1.

Single-Phase Motors

The field of a single-phase motor, instead of rotating merely pulsates. No rotation of the rotor takes place. A single-phase pulsating field may be visualized as two rotating fields revolving at the same speed, but in opposite directions. It follows, therefore, that the rotor will revolve in either direction at nearly synchronous speed—if, it is given an initial impetus in either one direction or the other. The exact value of this initial rotational velocity varies widely with different machines. A velocity higher than 15 percent of the synchronous speed is usually sufficient to cause the rotor to accelerate to the rated or running speed. A single-phase motor can be made self-starting if means can be provided to give the effect of a rotating field.

Shaded-Pole Motor

The shaded-pole motor resulted from one of the first efforts to make a self-starting single-phase self-starting motor. See Fig. 5-2. This motor has salient poles. A portion of each pole is encircled by a heavy copper ring. The presence of the ring causes the magnetic field through the ringed portion of the pole face to lag behind that through the other portion of the pole face. See Fig. 5-3.

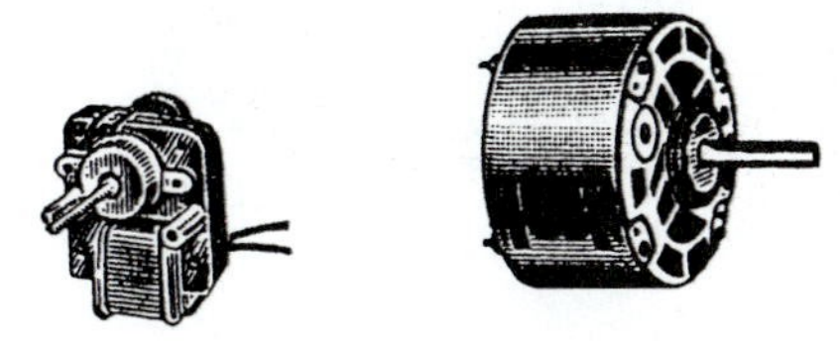

Fig. 5-3 *Shaded-pole motors used for fans and clocks.*

The effect is the production of a slight component of rotation of the field. That slight component of rotation is sufficient to cause the rotor to revolve. As the rotor accelerates, the torque increases until the rated speed is obtained. Such motors have low-starting torque. Their greatest use is in small fans where the initial torque is low. They are also used in clocks, inexpensive record players, and some electric typewriters. See Fig. 5-3.

Split-Phase Motor

Many types of split-phase motors have been made. Such motors have a start winding that is displaced 90 electrical degrees from the main or run winding. In some types, the start winding has a fairly high resistance. This causes the current in it to be out-of-phase with the current in the run winding.

This condition produces, in effect, a rotating field and the rotor revolves. A centrifugal switch is used to disconnect the start winding automatically after the rotor has attained approximately 75 percent of its rated speed. See Fig. 5-4.

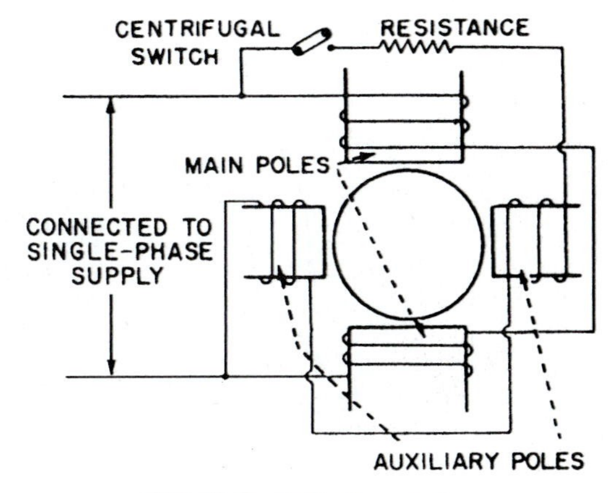

Fig. 5-4 *Split-phase motor.*

Split-phase motors are used where there is no need to start under load. They are used on grinders, buffers, and other similar devices. They are available in fractional horsepower sizes with various speeds, and are wound to operate on 120 V AC or 240 V AC.

Capacitor-Start Motor

With the development of high-quality and high-capacity electrolytic capacitors, a variation of the split-phase motor, known as the capacitor-start motor, has been made. Almost all fractional-horsepower motors in use today on refrigerators, oil burners, washing machines, table saws, drill presses, and similar devices, are capacitor-start motors.

A capacitor motor has a high starting current and has the ability to develop about four times its rated horsepower if it is suddenly overloaded. In this adaptation of the split-phase motor, the start winding and the run winding have the same size and resistance value. The phase shift between currents of the two windings is obtained by means of capacitors connected in series with the start winding.

Capacitor-start motors have a starting torque comparable to their torque at rated speed and can be used in places where the initial load is heavy. A centrifugal switch is required for disconnecting the start winding when the rotor speed is up to about 25 percent of the rated speed. Figure 5-5 shows a disassembled capacitor motor.

Note in Fig. 5-6, also a capacitor-start motor, the centrifugal-switch arrangement with the governor mechanism. Figure 5-7 shows the windings, the rotor, and the capacitor housing on top of the motor. Note that the windings overlap.

One of the advantages of the single-value capacitor-start motor is its ability to be reversed easily and frequently. See Figs. 5-8 and 5-9. The motor is quiet and smooth running. If a 5- to 20-hp capacitor-start motor is called for, the two-value capacitor motor is used. See Fig. 5-10.

This motor has two sets of field windings in the stator—an auxiliary winding, called a phase winding, and the main winding. The phase winding is designed for continuous duty; a capacitor remains in series with the winding at all times. A start capacitor is added to the phase current to increase starting torque.

However, it is disconnected by a centrifugal switch during acceleration. This type of motor is used in many air-conditioning applications where the unit is 2 to 4 tons.

In general, the single-phase motor is more expensive to purchase and to maintain than the three-phase motor. It is less efficient, and its starting currents are relatively high. All run at essentially constant speed. Nonetheless, most machines using electric motors around the home, on the farm, or in small commercial plants are equipped with single-phase motors.

Those who select a single-phase motor usually do so because three-phase power is not available to them. See Fig. 5-11 where it shows the simple methods used in the construction of a three-phase motor. Note this is a half-etched, squirrel-cage rotor. The bearings are not sealed ball bearings. They are a sleeve-type with the oil caps placed so that oil may be added occasionally to keep the bearings lubricated.

Figure 5-12 is a cutaway view of a three-phase motor. Note the simple rotor and fan blades. The windings and the sealed ball bearings make it simple for maintenance. This is an almost maintenance-free motor. Figure 5-13 shows a polyphase motor that has been made explosion-proof.

SIZES OF MOTORS

Some single-phase induction motors are rated as high as 2 hp. The major field of use is 1 hp, or less, at 120 or 240 V for the smaller sizes. For larger power ratings,

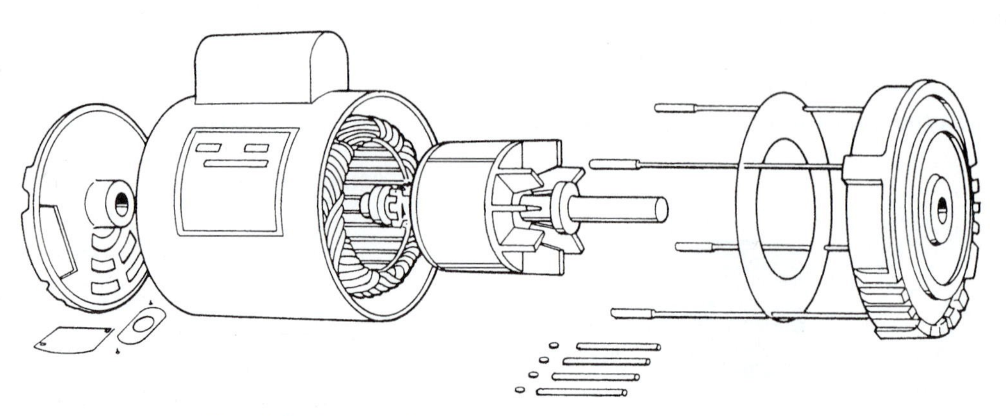

Fig. 5-5 *Disassembled single-phase, capacitor-start motor.*

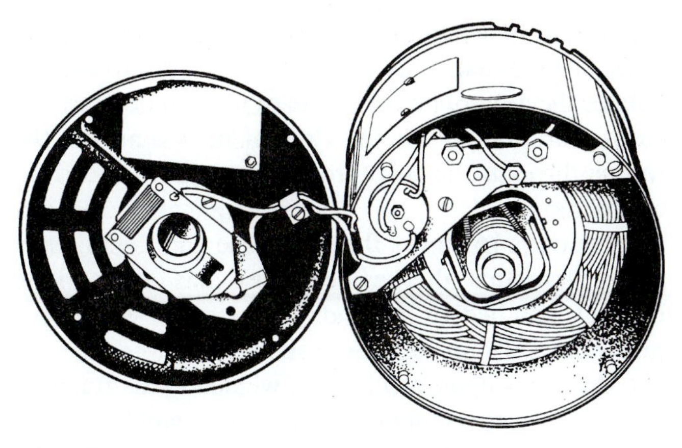

Fig. 5-6 *Single-phase starting switch and governor mechanism.*

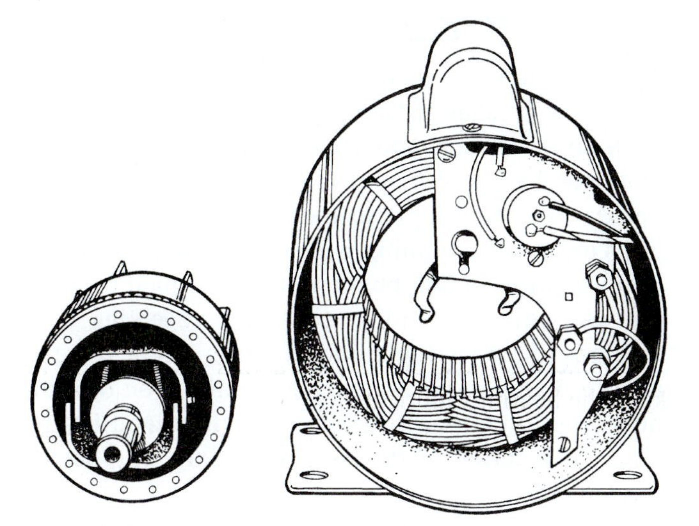

Fig. 5-7 *Single-phase stator and rotor to the left of the frame.*

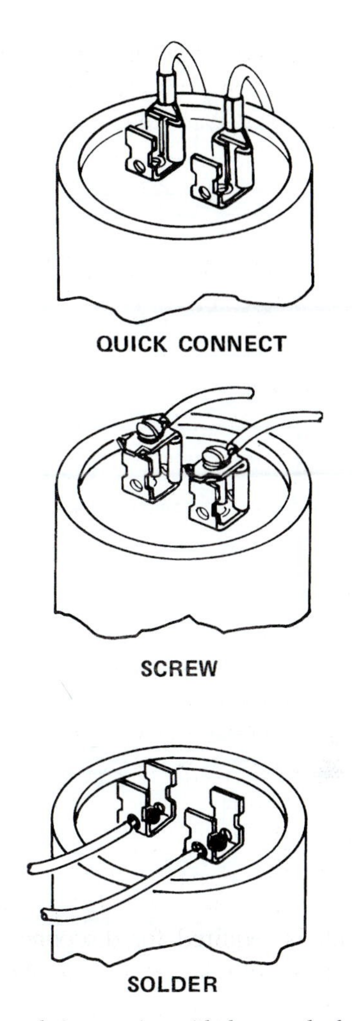

Fig. 5-8 *Electrolytic capacitor with three methods of connection.*

polyphase—two-phase, three-phase, and so forth—are generally specified, since they have excellent starting torque and are practically maintenance free.

Figure 5-14 is a brush-lifting, repulsion-start, induction-run, single-phase motor. The following should be noted about this type of motor. The rotor is wound (just like that in a DC motor). The brushes can be lifted by centrifugal force once the rotor comes up to speed. This means the rotor can then act as a squirrel-cage type. This type pulls a lot of current in starting, but is capable of starting under full-load conditions.

COOLING AND MOUNTING MOTORS

Figure 5-15 shows an improved motor-ventilating system. A large volume of air is directed through the motor to reduce temperatures. The large blower on the right is located behind a baffle that controls air movement to

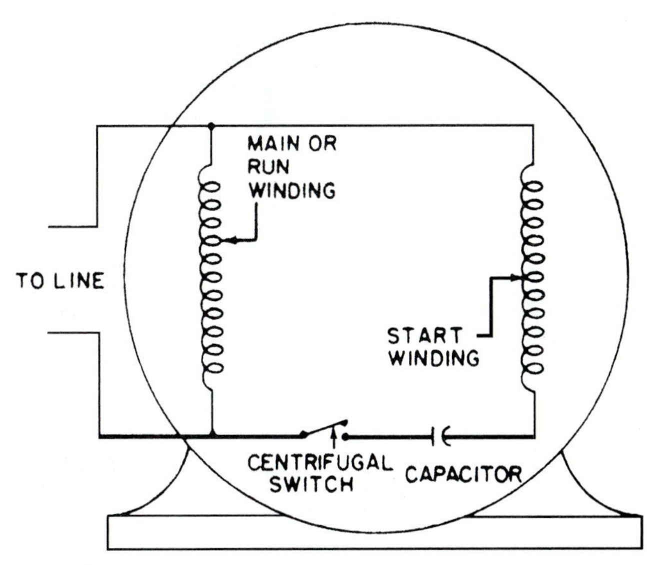

Fig. 5-9 *Note how the switch takes out the start winding when it gets up to speed.*

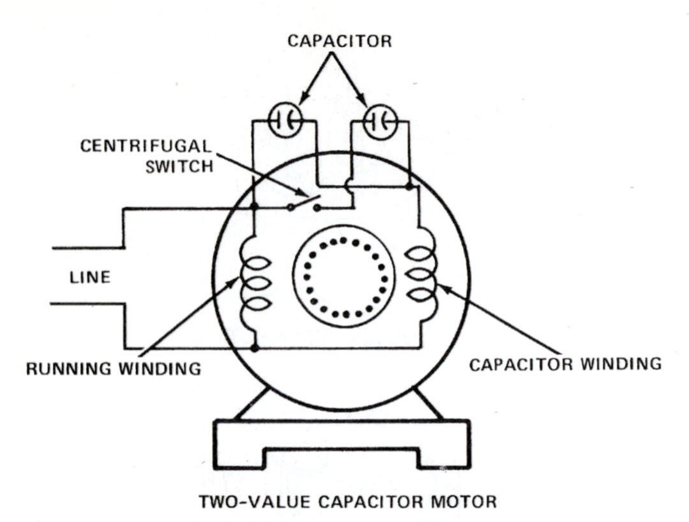

Fig. 5-10 *Two-value capacitor-start motor.*

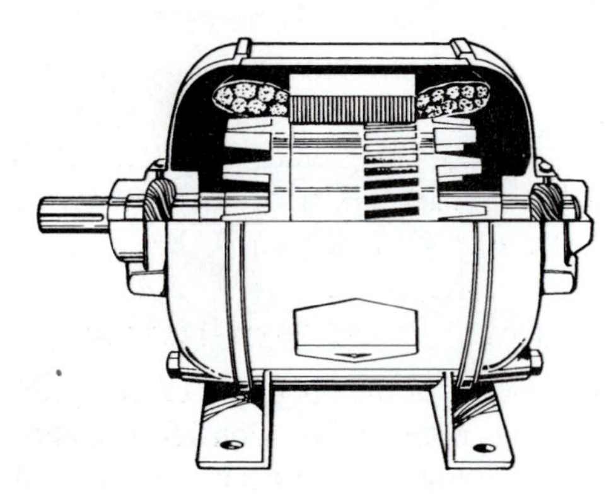

Fig. 5-11 *Cutaway view of a three-phase motor with a half-etched squirrel-cage rotor.* (Wagner)

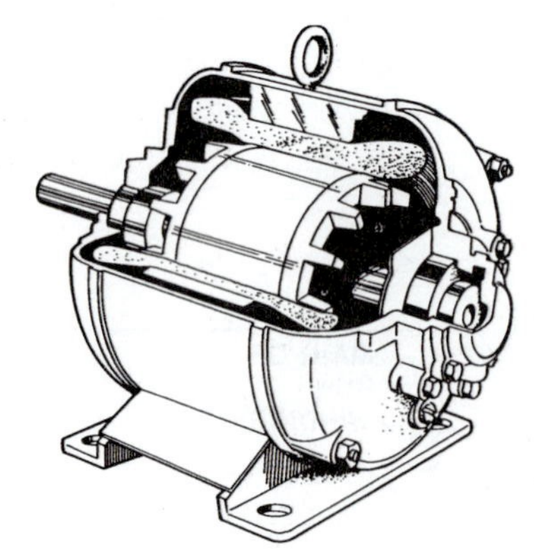

Fig. 5-12 *Cutaway view of a three-phase motor showing the cast rotor.*

the blower blades. The blower draws outside air through the large drip-proof openings in the back end plate.

It then forces the cooling air around the back coil extension, through the rotor vent holes, the air gap, and through the passages between the stator core and the frame. A second blower on the front end of the rotor at left, cast as an integral part of the rotor, circulates the air around the inside of the front coil extensions and then speeds the flow of heated air out the motor through the drip-proof openings in the front end plate.

Figure 5-16 shows the rigid base and the resilient base. Note that the resilient base has a mounting bracket attached to the ends of the rotor with some material used to make it more silent. However, the rigid base has its support mechanism welded to the frame of the motor. If the support mechanism is welded, it can transmit the noise of the running motor to whatever it is attached to in operation.

DIRECTION OF ROTATION

The direction of rotation of a three-phase induction motor can be changed simply by reversing two of the leads to the motor. The same effect can be obtained in a two-phase motor by reversing connections to one phase. In a single-phase motor, reversing connections to the start winding will reverse the direction of rotation. Most single-phase motors designed for general use have provisions for readily reversing connections to the start winding. Nothing can be done to a shaded-pole motor to reverse the direction of rotation. The direction of rotation is determined by the physical location of the copper shading ring on the shaded pole.

If, after starting, one connection to a three-phase motor is broken, the motor will continue to run but will deliver only one-third of the rated power. Also, a two-phase motor will run at one-half its rated power if one phase is disconnected. Neither motor will start under these conditions. They can be started by hand in either direction, manually. Once started by hand, they do run. Incidentally, the only place that a two-phase motor will be found is in Europe where some two-phase power is distributed for local use. In the United States only single-phase power is available to residential customers. Three-phase power is usually available to industry and commercial establishments. Schools usually have three-phase power located within easy connection, if needed.

Some parts of the southwestern United States, now have three-phase power distributed to homes. This is primarily due to the requirements of the air-conditioning units they need. The three-phase motor requires fewer service calls and has a long life. Therefore, it is often worth the extra expense to have three-phase power brought in by the power company. It is less expensive if a whole subdivision uses three-phase power.

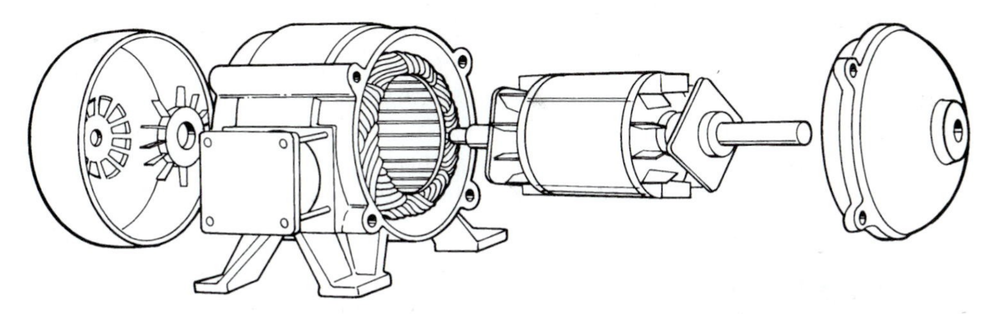

Fig. 5-13 *This is a polyphase motor with explosion-proof construction.*

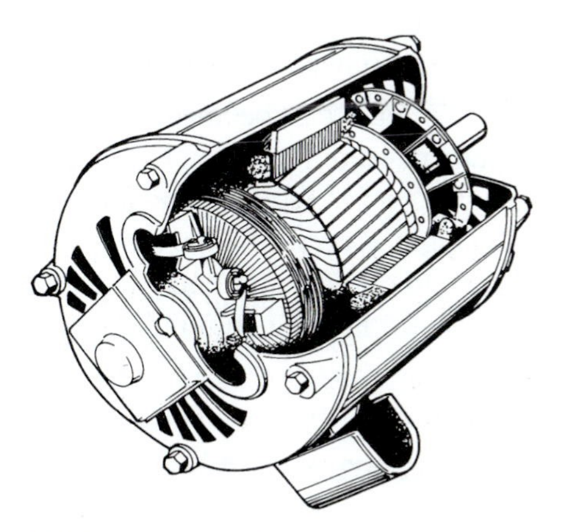

Fig. 5-14 *A brush-lifting, repulsion-start, induction-run, single-phase motor. Note the brushes and the wound rotor.* (Wagner)

SYNCHRONOUS MOTOR

A synchronous motor is one of the principal types of AC motors. Like the induction motor, the synchronous motor is designed to take advantage of a rotating magnetic field. Unlike the induction motor, however, the torque developed does not depend upon the induction of currents in the rotor. Briefly, the principle of operation of the synchronous motor is as follows.

A multiphase source of AC is applied to the stator windings and a rotating magnetic field is produced. A direct current is applied to the rotor windings and another magnetic field is produced. The synchronous motor is so designed and constructed that these two fields react upon each other. They act in such a manner that the rotor is dragged along. It rotates at the same speed as the rotating magnetic field produced by the stator windings.

Theory of Operation

An understanding of the operation of the synchronous motor may be obtained by considering the simple motor shown in Fig. 5-17. Assume that poles A and B are being rotated clockwise by some mechanical means to produce a rotating magnetic field.

The rotating poles induce poles of opposite polarity, as shown in the illustration of the soft-iron rotor, and forces of attraction exist between corresponding north and south poles. Consequently, as poles A and B rotate, the rotor is dragged along at the same speed.

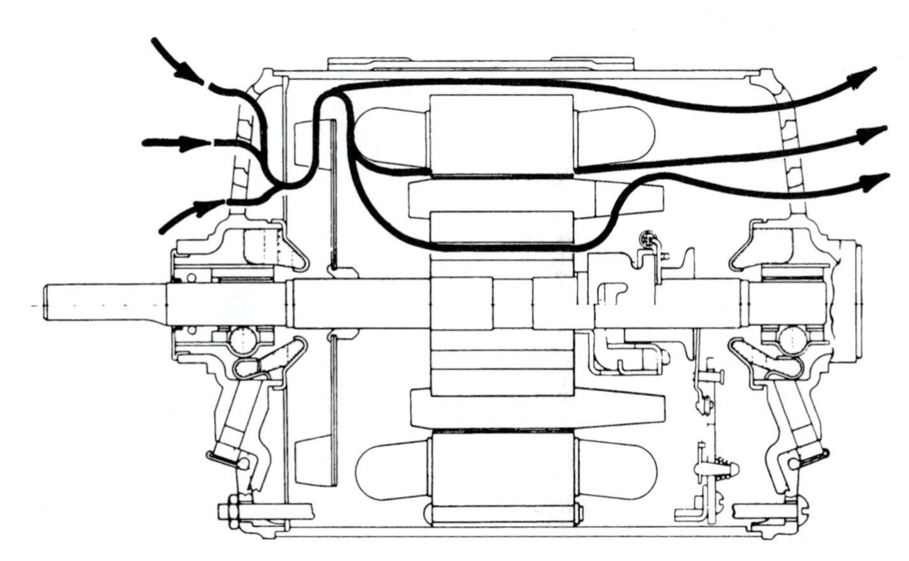

Fig. 5-15 *Cooling system using two fans to keep the air moving inside an electric motor.* (Wagner)

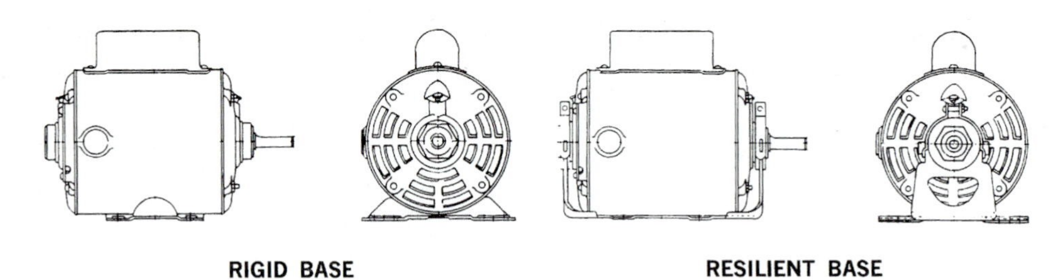

Fig. 5-16 *Rigid-base and resilient-base mountings for electric motors.* (Wagner)

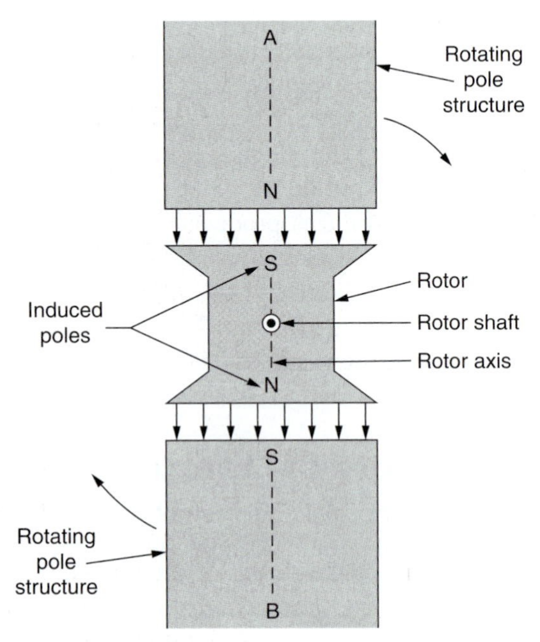

Fig. 5-17 *Simple synchronous motor.*

However, if a load is applied to the rotor shaft, the rotor axis will momentarily fall behind that of the rotating field, but will thereafter continue to rotate with the field at the same speed, as long as the load remains constant. If the load is too large, the rotor will pull out of synchronization with the rotating field. As a result, it will no longer rotate with the field at the same speed. The motor is then said to be overloaded.

Synchronous Motor Advantages

Some advantages of the synchronous motor are as follows:

- When used as a synchronous capacitor, the motor is connected on the AC line in parallel with the other motors on the line. It is run either without load or with a very light load. The rotor field is overexcited just enough to produce a leading current that offsets the lagging current of the line with the motors operating. A unity power factor (1.00) can usually be achieved. This means the load on the generator is the same as though only resistance made up the load.
- The synchronous motor can be made to produce as much as 80 percent leading power factor. However, because a leading power factor on a line is just as detrimental as a lagging power factor, the synchronous motor is regulated to produce just enough, leading current to compensate for lagging current in the line.

Properties of the Synchronous Motor

The synchronous motor is not a self-starting motor in most cases. The rotor is heavy. From a dead stop, it is impossible to bring the rotor into magnetic lock with the rotating magnetic field. For this reason, all synchronous motors have a starting device. Such a simple starter is another motor, either AC or DC, which can bring the rotor up to approximately 90 percent, of the synchronous speed. The starting motor is then disconnected and the rotor locks in step with the rotating field.

Another starting method is a second winding of the squirrel-cage type on the rotor. This induction winding brings the rotor almost into synchronous speed. When the DC is connected to the rotor windings, the rotor pulls into step with the field. The latter method is the more commonly used.

Figure 5-18 shows a small synchronous motor that has a number of applications. Because of their

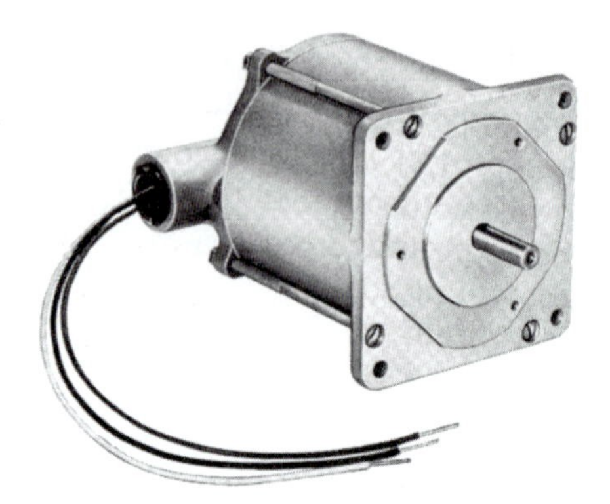

Fig. 5-18 *Synchronous motor.*

stable speed, synchronous motors were used for turntables in stereo equipment. This type is also used in timing devices.

ELECTRIC MOTORS

Electric motors are designed to deliver their best overall performance when operated at the design voltage shown on the nameplate. However, this voltage is often not maintained. Instead, it varies between minimum and maximum limits over what is termed voltage spread. The voltage spread is usually due to the wiring and transformers of the electrical distribution system and varies in proportion to motor or load currents.

In most modern plants using load-center power-distribution systems, variations in voltage normally will be within recommended limits of 110 to 220, 220 to 240, 440 to 480, and 550 to 600 for single-phase and three-phase squirrel-cage and synchronous motors. However, there are older plants throughout the country with large low-voltage systems. Long low-voltage feeders often cause voltage drops that result in below standard voltages at the motor terminals, especially during motor starting, when currents may be up to six times normal full load. Table 5-1 shows the effect of voltage variations on the performance of polyphase-induction motors.

Table 5-1 *Voltage Variations*

Rated Voltage	Lower Limit	Upper Limit
220	210	240
440	420	480
550	525	600
2300	2250	2480
4000	3920	4320
4600	4500	5000
6600	6470	7130

Single-phase and polyphase motors call for different approaches or methods to start them under various conditions of operation. Most single-phase motors are started by the turning on of an on-off switch or a magnetic starter.

Starting the Motor

One of the most important parts of the electric motor is the start mechanism. A special type is needed for use with single-phase motors. A centrifugal switch is used to take a start winding out of the circuit once the motor has come up to within 75 percent of its run speed. The split phase, capacitor start, and other variations of these types need the start mechanism to get them running.

The stator of a split-phase motor has two types of coils, one called the run winding and other the start winding. The *run winding* is made by winding the enamel-coated copper wire through the slots in the stator *punchings*. The *start winding* is made in the same way except that the wire is smaller. Coils that form the start windings are positioned in pairs in the stator directly opposite each other and between the run windings. When you look at the end of the stator, you see alternating run windings and start windings. See Fig. 5-19.

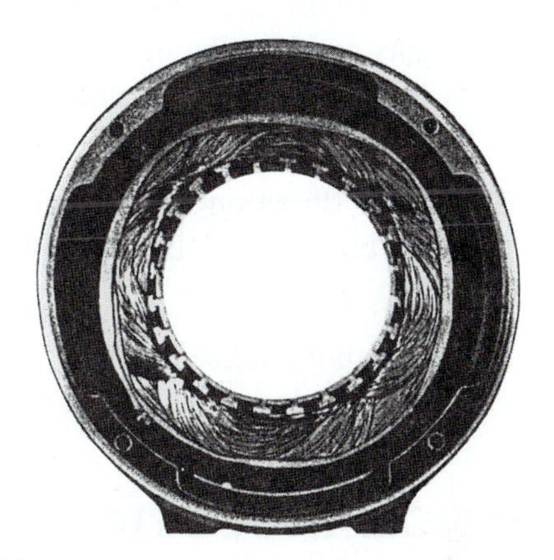

Fig. 5-19 *Split-phase motor windings.* (Bodine Electric Co.)

The run windings are all connected together, so the electrical current must pass through one coil completely before it enters the next coil, and so on through all the run windings in the stator. The start windings are connected together in the same way and the current must pass through each in turn. See Fig. 5-20.

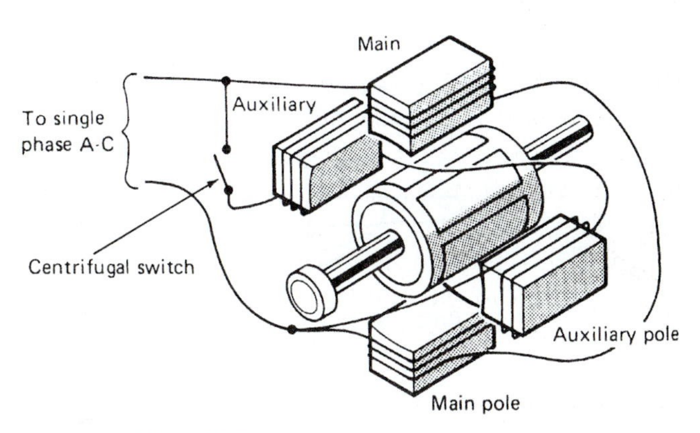

Fig. 5-20 *Single-phase induction motor.*

The two wires from the run windings in the stator are connected to terminals on an insulated terminal block in one end bell where the power cord is attached to the same terminals. One wire from the start winding is tied to one of these terminals also. However, the other wire from the start winding is connected to the stationary switch mounted in the end bell. Another wire then connects this switch to the opposite terminal on the insulated block. The stationary switch does not revolve but is placed so that the weights in the rotating portion of the switch, located on the rotor shaft, will move outward when the motor is up to speed and open the switch to stop electrical current from passing through the start winding.

The motor then runs only on the main winding until such times as it is shut off. Then, as the rotor decreases in speed, the weights on the rotating switch again move inward to close the stationary switch and engage the start winding for the next time it is started.

Reversibility The direction of rotation of the split-phase motor can be changed by reversing the start-winding leads.

Uses This type of motor is used for fans, furnace blowers, oil burners, office appliances, and unit heaters.

REPULSION-INDUCTION MOTOR

The repulsion-induction motor starts on one principle of operation and, when almost up to speed, changes over to another type of operation. Very high twisting forces are produced during starting by the repulsion between the magnetic pole in the armature and the same kind of pole in the adjacent stator field winding. The repulsing force is controlled and changed so that the armature rotational speed increases rapidly, and if not stopped, would continue to increase beyond a practical operating speed. It is prevented by a speed-actuated mechanical switch that causes the armature to act as a rotor that is electrically the same as the rotor in single-phase induction motors. That is why the motor is called a *repulsion-induction motor*.

The stator of this motor is constructed very much like that of a split-phase or capacitor-start motor, but there are only run or field windings mounted inside. End bells keep the armature and shaft in position and hold the shaft bearings.

The armature consists of many separate coils of wire connected to segments of the commutator. Mounted on the other end of the armature are governor weights which move pushrods that pass through the armature core. These rods push against a short-circuiting ring mounted on the shaft on the commutator end of the armature. Brush holders and brushes are mounted in the commutator end bell, and the brushes, connected by a heavy wire, press against segments on opposite sides of the commutator. See Fig. 5-21.

When the motor is stopped, the action of the governor weights keeps the short-circuiting ring from touching the commutator. When the power is turned on and current flows through the stator field windings, a current is induced in the armature coils. The two brushes connected together form an electromagnetic coil that produces a north and south pole in the armature, positioned so that the north pole in the armature is

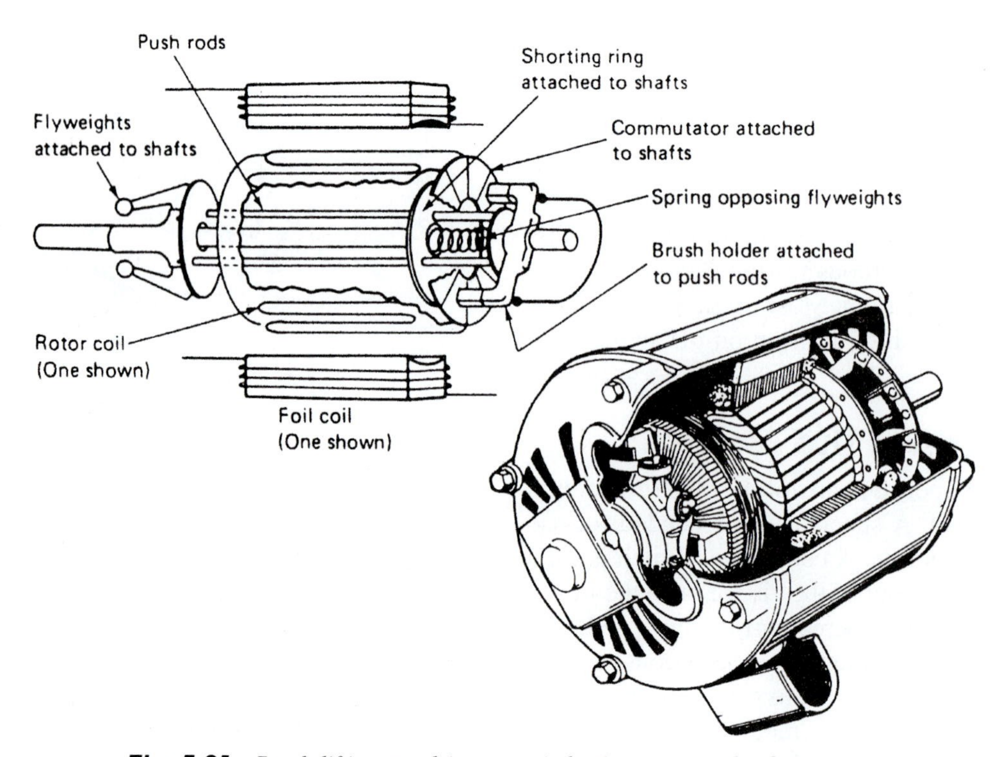

Fig. 5-21 *Brush-lifting, repulsion-start, induction- run, single-phase motor.*

next to a north pole in the stator field windings. Since like poles try to move apart, the repulsion produced in this case can be satisfied in only one way, by the armature turning and moving the armature coil away from the field windings.

The armature turns faster and faster, accelerating until it reaches what is approximately 80 percent of the run speed. At this speed the governor weights fly outward and allow the pushrods to move. The pushrods, which are parallel to the armature shaft, have been holding the short-circuiting ring away from the commutator. Now that the governor has reached its designed speed, the rods can move together electrically in the same manner that the cast aluminum disks did in the cage of the induction-motor rotor. This means that the motor runs as an induction motor.

Uses The repulsion-induction motor can start very heavy, hard-to-turn loads without drawing too much current. They are made from 0.5 to 20 hp. This type of motor is used for such applications as large air compressors, refrigeration equipment, large hoists, and are particularly useful in locations where low line voltage is a problem. This type of motor is no longer used in the refrigeration industry. Some older operating units may be found with this type of motor still in use.

CAPACITOR-START MOTOR

The capacitor motor is slightly different from a split-phase motor. A capacitor is placed in the path of the electrical current in the start winding. See Fig. 5-22. Except for the capacitor, which is an electrical component that slows any rapid change in current, the two motors are the same electrically. A capacitor motor can usually be recognized by the capacitor can or housing that is mounted on the stator. See Fig. 5-23.

By adding the capacitor to the start winding, it increases the effect of the two-phase field described in connection with the split-phase motor. The capacitor means that the motor can produce a much greater twisting force when it is started. It also reduces the amount of electrical current required during starting to about 1.5 times the current required after the motor is up to speed. Split-phase motors require three or four times the current in starting that they do in running.

Reversibility An induction motor will not always reverse while running. It may continue to run in the

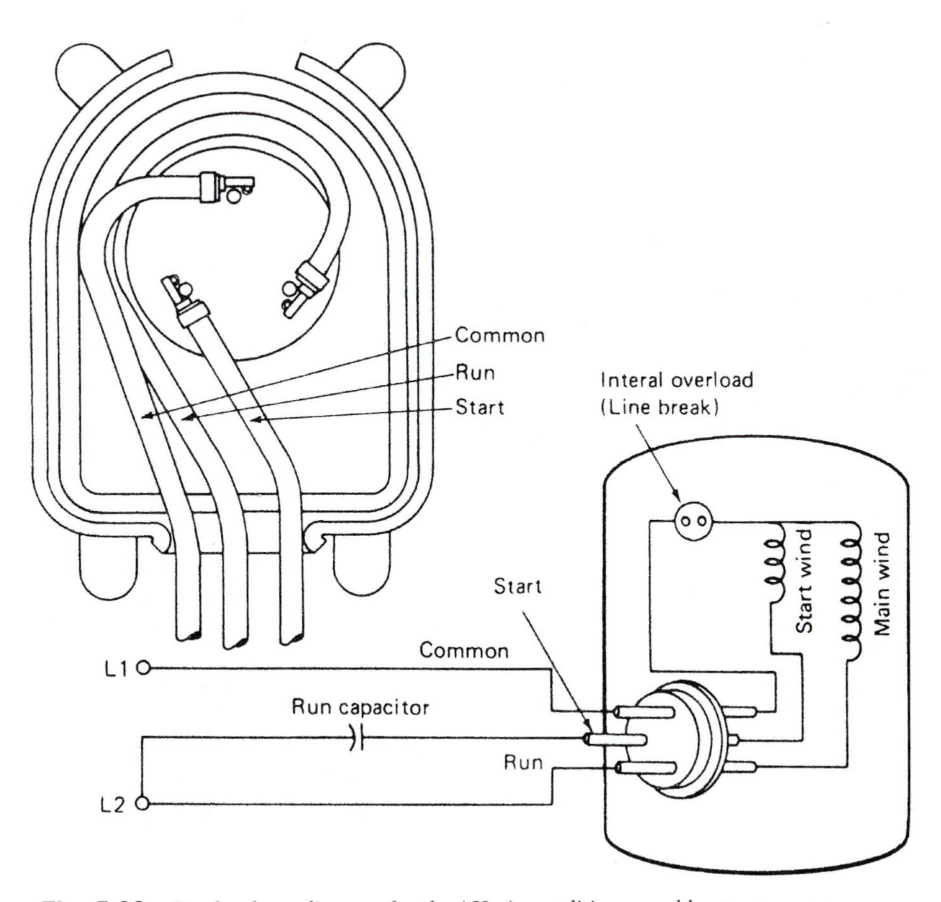

Fig. 5-22 *Single-phase diagram for the AH air conditioner and heat-pump compressor.*

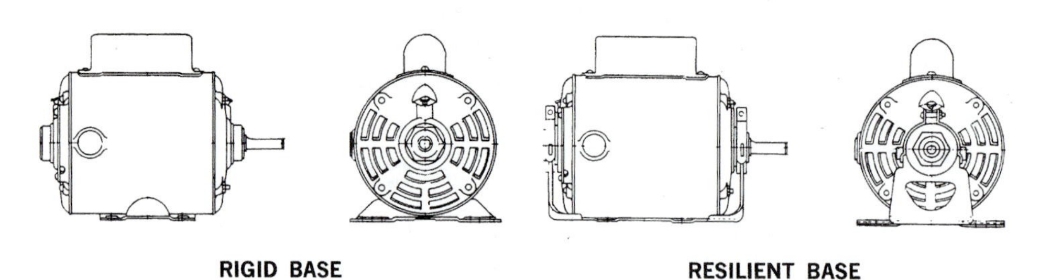

Fig. 5-23 *Capacitor-start motor.*

same direction but at a reduced efficiency. An inertia-type load is difficult to reverse. Most motors that are classified as reversible while running will reverse with a noninertial-type load. They may not reverse if they are under no-load conditions or have a light load or an inertial load.

One of the problems related to the reversing of a motor while it is still running is the damage done to the transmission system connected to the load. In some cases it is possible to damage a load. One of the ways to avoid this is to make sure that the right motor is connected to a load.

Reversing (while standing still) the capacitor-start motor can be done by reversing its start-winding connections. This is usually the only time that will work on a motor. The available replacement motor may not be rotating in the direction desired, so the electrician will have to locate the start-winding terminals and reverse them in order to have the motor start in the desired direction.

Figure 5-24A shows a capacitor-start, induction-run motor used in a compressor. This type uses a relay to place the capacitor in and out of the circuit. More details regarding this type of relay will be given later. Figure 5-24B shows how the capacitor is located outside the compressor.

Uses Capacitor motors are available in sizes from 6 to 20 hp. They are used for fairly hard-starting loads that can be brought up to run speed in under 3 s. They may be used in industrial-machine tools, pumps, air conditioners, air compressors, conveyors, and hoists.

PERMANENT SPLIT-CAPACITOR MOTOR

The *permanent split-capacitor* (PSC) motor is used in compressors for air-conditioning and refrigeration units. It has an advantage over the capacitor-start motor inasmuch as it does not need the centrifugal switch and its associated problems.

The PSC motor has a run capacitor in series with the start winding. Both run capacitor and start winding remain in the circuit during start and after the motor is up to speed. Motor torque is sufficient for capillary and other self-equalizing systems. No start capacitor or relay is necessary. The PSC motor is basically an air-conditioner compressor motor. It is very common through 3 hp. It is also available in the 4- and 5-hp sizes. See Fig. 5-25.

SHADED-POLE MOTOR

The shaded-pole induction motor is a single-phase motor. It uses a unique method to start the rotor turning. The effect of a moving magnetic field is produced by constructing the stator in a special way. See Fig. 5-26.

Portions of the pole-piece surfaces are surrounded by a copper strap called a shading coil. The strap causes the field to move back and forth across the face of the pole piece. In Fig. 5-27 the numbered sequence and points on the magnetization curve are shown. As the alternating stator field starts increasing from zero (1), the lines of force expand across the face of the pole piece and cut through the strap. A voltage is induced in the strap. The current that results generates a field that opposes the cutting action (and decreases the strength) of the main field. This action causes certain actions: As the field increases from zero to a maximum of 90°, a large portion of the magnetic lines of force are concentrated in the unshaded portion of the pole (1). At 90° the field reaches its maximum value. Since the lines of force have stopped expanding, no EMF is induced in the strap, and no opposite magnetic field is generated. As a result, the main field is uniformly distributed across the poles as shown in (2).

From 90° to 180° the main field starts decreasing or collapsing inward. The field generated in the strap opposes the collapsing field. The effect is to concentrate the lines of force in the shaded portion of the poles as shown in (3).

Note that from 0° to 180°, the main field has shifted across the pole face from the unshaded to the shaded portion. From 180° to 360°, the main field goes through the same change as it did from 0° to 180°.

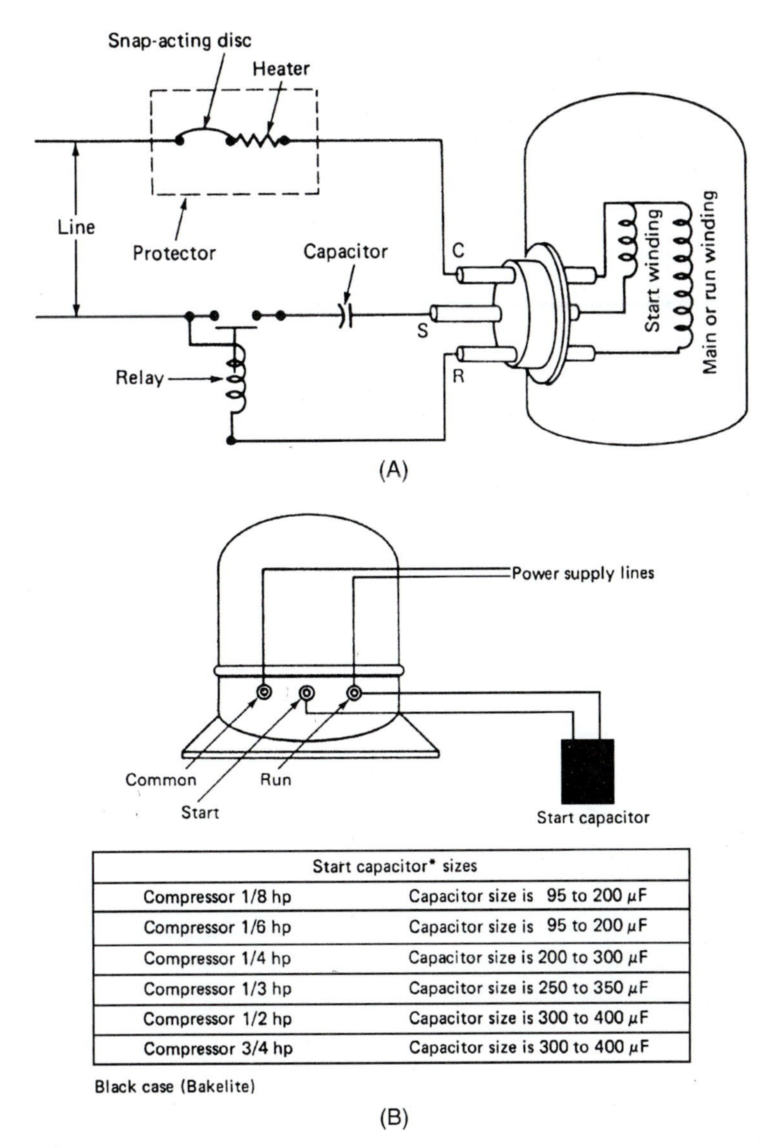

Start capacitor* sizes	
Compressor 1/8 hp	Capacitor size is 95 to 200 μF
Compressor 1/6 hp	Capacitor size is 95 to 200 μF
Compressor 1/4 hp	Capacitor size is 200 to 300 μF
Compressor 1/3 hp	Capacitor size is 250 to 350 μF
Compressor 1/2 hp	Capacitor size is 300 to 400 μF
Compressor 3/4 hp	Capacitor size is 300 to 400 μF

Fig. 5-24 *(A) Capacitor-start, induction-run motor used for a compressor. (B) Location of the start capacitor in a compressor circuit.*

However, it is now in the opposite direction (4). The direction of the field does not affect the way the shaded pole works. The motion of the field is the same during the second half-hertz as it was during the first half-hertz.

The motion of the field back and forth between shaded and unshaded portions produces a weak torque. This torque is used to start the motor. Because of the weak starting torque, shaded-pole motors are built in only small sizes. They drive such devices as fans, clocks, and blowers.

Reversibility Shaded-pole motors can be reversed mechanically. Turn the stator housing and shaded poles end-for-end. These motors are available from 1/250th to ½ hp.

Uses As mentioned previously, this type of motor is used as a fan motor in refrigerators and freezers. They can also be used as fan motors in some types of air-conditioning equipment where the demand is not too great. They can also be used as part of the timing devices used for defrost timers and other sequenced operations.

The fan and motor assembly are located behind the provisions compartment in a refrigerator, directly above the evaporator in the freezer compartment. The suction fan pulls air through the evaporator and blows it through the provisions compartment air duct and freezer compartment fan grille. See Fig. 5-28. This is a shaded-pole motor with a molded plastic fan blade. For maximum air circulation the location of the fan on the motor shaft is most important. Mounting the fan blade

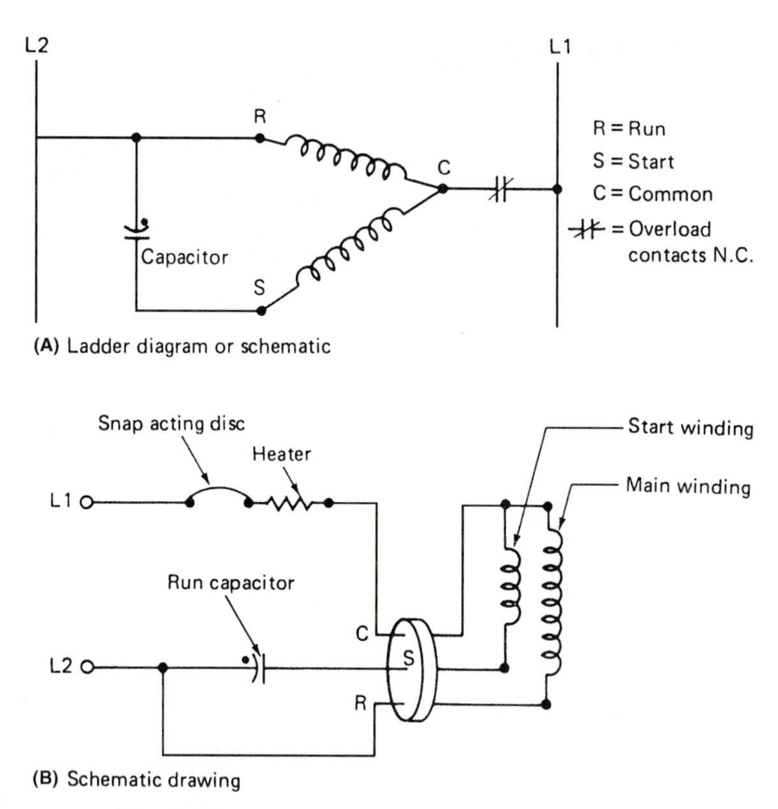

Fig. 5-25 *Permanent split-capacitor motor schematic.*

too far back or too far forward on the motor shaft, in relation to the evaporator cover, will result in improper air circulation. The freezer compartment fan must be positioned with the lead edge of the fan 1/4 in. in front of the evaporator cover.

The fan assembly shown in Fig. 5-29 is used on the top-freezer, no-frost, fiberglass–insulated model refrigerators. The freezer fan and motor assembly is located in the divider partition directly under the freezer air duct.

SPLIT-PHASE MOTOR

Instead of rotating, the field of a single-phase motor merely pulsates. No rotation of the rotor takes place. A single-phase pulsating field may be visualized as two rotating fields revolving at the same speed but in opposite directions. It therefore follows that the rotor will revolve in either direction at nearly synchronous speed—if it is given an initial impetus in either one direction or the other. The exact value of this initial rotational velocity varies widely with different machines.

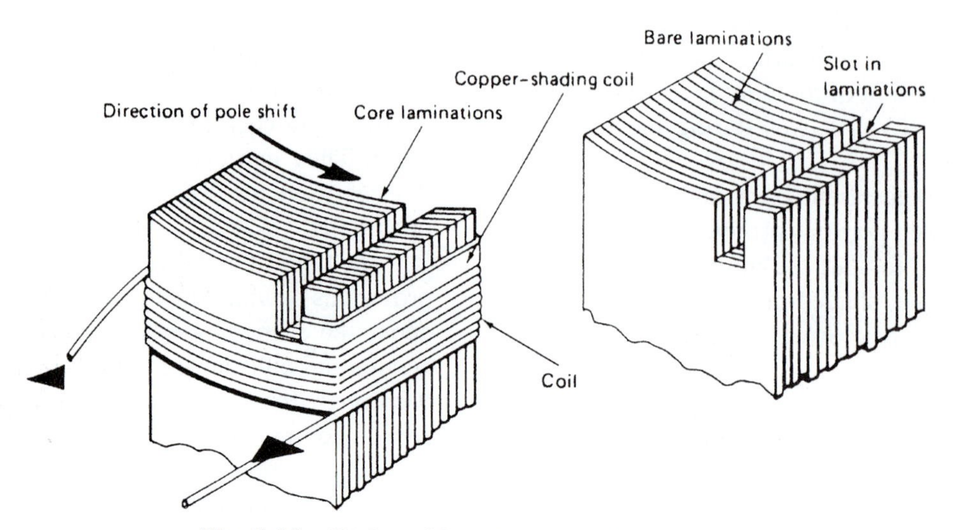

Fig. 5-26 *Shading of the poles of a shaded-pole motor.*

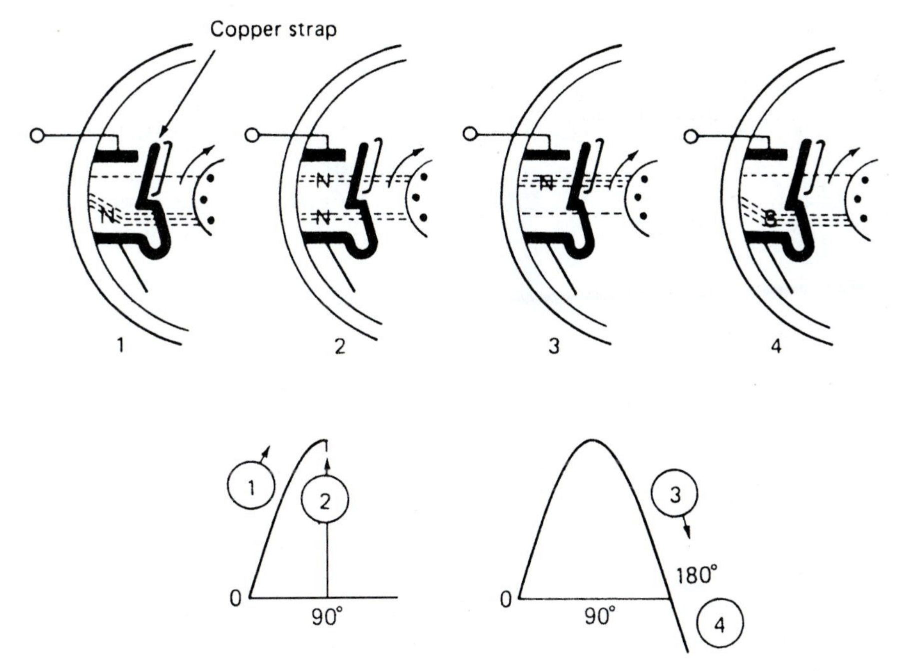

Fig. 5-27 *Shaded poles as used in shaded-pole motors.*

A velocity higher than 15 percent of the synchronous speed is usually sufficient to cause the rotor to accelerate to the rated or running speed. A single-phase motor can be made self-starting if means can be provided to give the effect of a rotating field.

To get the split-phase motor running, a run winding and a start winding are incorporated into the stator of the motor. Figure 5-30 shows the split-phase motor with the end cap removed so that you can see the starting switch and governor mechanism.

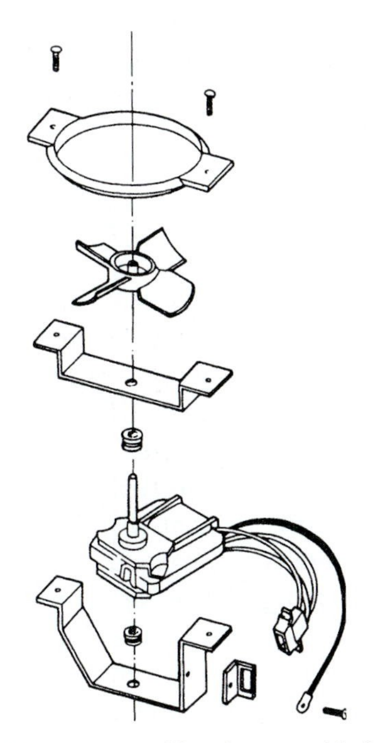

Fig. 5-28 *Fan, motor, and bracket assembly for refrigerator.*

This type of motor is difficult to use with air-conditioning and refrigeration equipment inasmuch as it has very little starting torque and will not be able to start a compressor since it presents a load to the motor immediately upon starting. This type of motor, however, is very useful in heating equipment. See Fig. 5-31.

POLYPHASE-MOTOR STARTERS

The simple manual starter works for single-phase motors and also, in some instances, for polyphase motors. Most of the polyphase manual starters consisting of an on-off switching arrangement are designed for motors of 1 hp or less. Figure 5-32 shows the magnetic-motor starter designed for across-the-line control of squirrel-cage

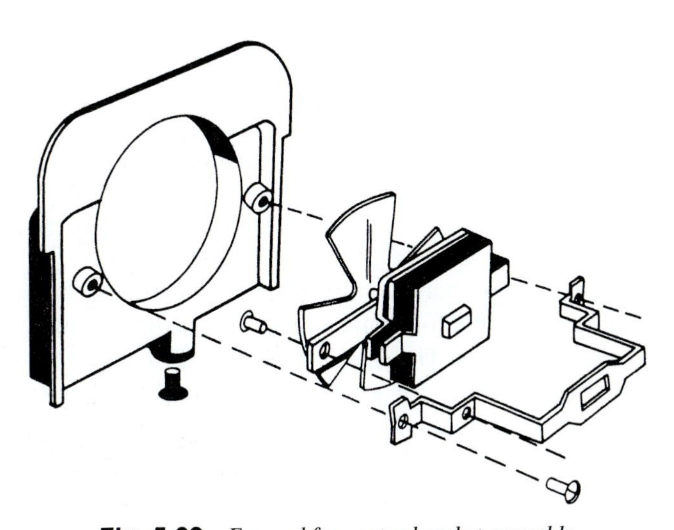

Fig. 5-29 *Fan and fan-motor bracket assembly.*

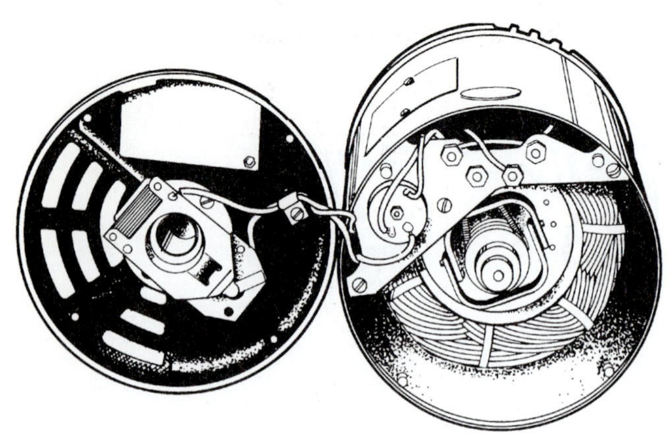

Fig. 5-30 *Single-phase starting switch and governor mechanism.*

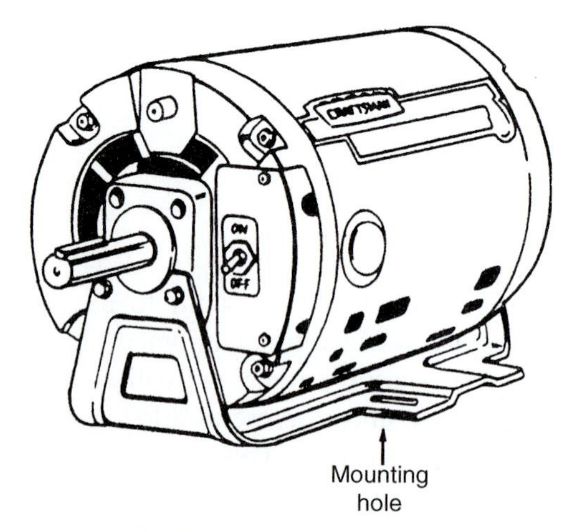

Fig. 5-31 *Single-phase, split-phase motor.*

Fig. 5-32 *Noncombination magnetic motor starter.* (Westinghouse)

motors or as primary control for wound-rotor motors. They are available for nonreversing, reversing, and two-speed applications. The drawing in Fig. 5-33A shows the difference between the single- and three-phase nonreversing type of starter. Figure 5-33B shows the reversing drawing; Figure 5-33C is the two-speed, one winding starter; and Figure 5-33D is the two-speed, two-winding starter for motors up to 100 hp.

During across-the-line starting, motor input current is five to eight times normal full-load current. This can cause an excessive temporary voltage drop on power lines that causes lights to flicker or may even interrupt the service.

To control these temporary voltage drops, power companies have restrictions such as:

- A specific maximum starting current (or kVA)
- A specific limit on kVA/hp
- A maximum horsepower motor size which can be started across-the-line
- A specific maximum line current that can be drawn in steps (increment starting)

The specified restrictions vary considerably between power companies, even within one company's service area. It is wise to check local power company restrictions before making a larger motor installation.

REDUCED-VOLTAGE STARTING METHODS

Reduced-voltage starters operate such that input current and, consequently, torque are reduced during starting. Table 5-2 briefly describes the various methods of starting and gives features and limitations of each.

When motors are started at reduced voltage, the current at the motor terminals is reduced in direct proportion to the voltage reduction, while the torque is reduced as the square of the voltage reduction. For example, if the "typical" motor were started at 65 percent of line voltage, the starting current would be 42 percent and the torque would be 42 percent of full-voltage values. Thus, reduced-voltage starting provides an effective means of reducing both current and torque. See Fig. 5-37.

Primary-Resistor Starting

In primary-resistor starting, a resistor is connected in each motor line (in one line only in single-phase starters) to produce a voltage drop due to the motor starting current. A timing relay shorts out the resistors after the motor has accelerated. Thus, the motor started at reduced voltage but operates at line voltage

Figure 5-34 shows two types of motor-start resistors. The resistance element will retain its mechanical and electrical properties, both during and after repeated heating and cooling. All metal parts are either plated with or fabricated of corrosion-resin material for overall corrosion protection. Under certain conditions operating temperatures may reach 600°C and not change the resistance value. These are 11, 14, 17, and 20 in. long

Non-Reversing Single Phase

To Line
2 1-L1 L2
M
Lead "A"
3
OL
"X2"
(When Used)
T1 T2
To Load.

Non-Reversing Three Phase

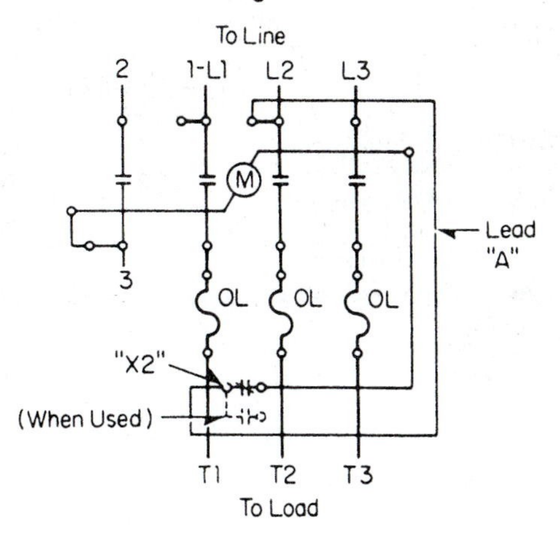

(A)

Reversing

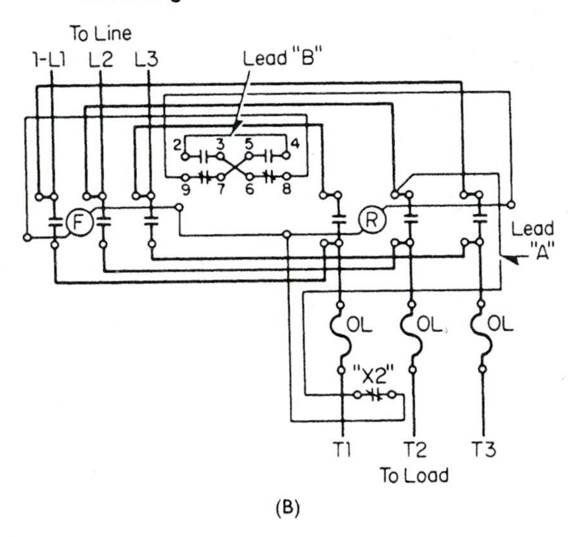

(B)

Two Speed, One Winding

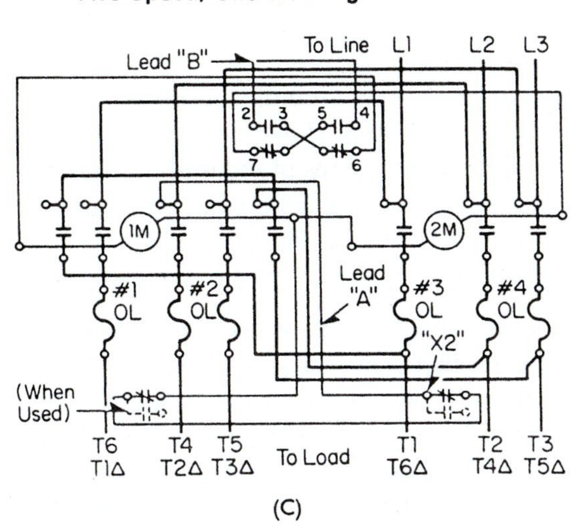

(C)

Two Speed, Two Winding

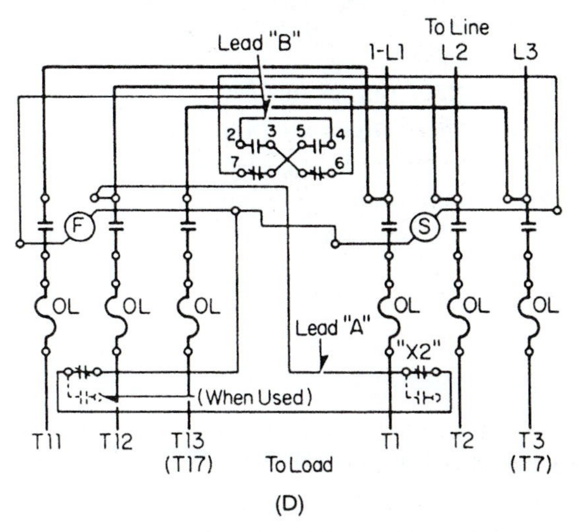

(D)

Fig. 5-33 *(A) Nonreversing single-phase and nonreversing three-phase wiring diagrams. (B) Reversing three-phase wiring diagram. (C) Two-speed, one winding, three-phase delta. (D) Two-speed, two-winding, three-phase starter diagram.* (Westinghouse)

Table 5-2 Starting Method Characteristics

Starting Method	Operation	Starting Current (% of Locked Rotor Current)	Starting Torque (% of Locked Rotor Torque)	Open or Closed Transition	Basic Characteristics: Advantages	Basic Characteristics: Limitations
Across-the-line	Initially connects motor directly to power lines.	100%	100%	None	1. Lowest cost 2. Highest starting torque 3. Used with any standard motor 4. Least maintenance	1. High-starting current 2. High-starting torque 3. May shock driven machine
Primary resistance reduced voltage	Inserts resistance units in series with motor during first step(s).	50–80%	25–64%	Closed	1. Smoothest starting 2. Least shock to driven machine 3. Most flexible in application 4. Used with any standard motor	1. High power loss because of heating resistors 2. Heat must be dissipated 3. Low torque per ampere input 4. Highest cost
Autotransformer reduced voltage	Uses autotransformer to reduce voltage applied to motor. **Tap** 50% 65% 80%	 25% 42% 64%	 25% 42% 64%	Closed	1. Best for hard to start loads 2. Adjustable starting torque 3. Used with any standard motor 4. Less strain on motor	1. May shock driven machine 2. High cost
Wye-Delta	Starts motor with windings wye connected, then reconnects them in delta connection for running.	33%	33%	Open or closed	1. Medium cost 2. Low-starting current 3. Low-starting torque 4. Less strain on motor	1. Low-starting torque 2. Requires delta-wound motor
Part Winding	Starts motor with only part of windings connected, then adds remainder for running.	70–80%	50–60% Minimum pull-up torque 35% of full-load torque.	Closed	1. Low cost 2. Popular method for medium-starting torque applications 3. Low maintenance	1. Not good for frequent starts 2. May require special wound motor 3. Low pull-up torque 4. May not come up to speed on first step when started with load applied

NOTE: The reduced-starting torque (LRT) indicated in this table for the various reduced-starting methods can prevent starting high-inertia loads and must be considered when sizing motors and choosing starters.

and come in wattage ratings of 450 to 1320. Table 5-3 shows the resistance ranges and other factors. Note the current-handling ability of the resistors.

Primary resistor starters are sometimes known as "cushion starters." The main reason for the name is their ability to produce a smooth, cushioned acceleration with closed transition. However, this method is not as efficient as other methods of reduced-voltage start but it is ideally suited for applications such as conveyors, textile machines, or other delicate machinery where reduction of starting torque is of prime consideration.

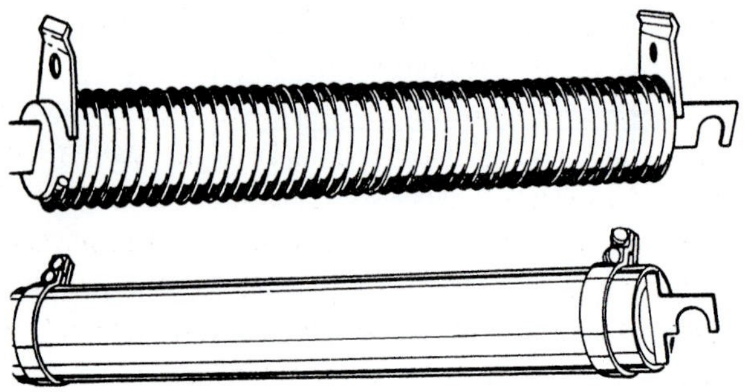

Fig. 5-34 *Wire-wound resistors used in primary resistor starter circuits.* (Westinghouse)

Operation Figure 5-35 is the reduced-voltage magnetic starter that uses resistors to operate a three-phase motor properly at start. Closing the START button or other pilot device energizes the start contactor (S) shown in Fig. 5-36. This connects the motor in series with the starting resistors for a reduced-voltage start. The contactor (S) is now sealed in through its interlock (Sa). Timing relay (TR) is energized, and after a preset time interval its contacts (TR_{TC}) close. This energizes the run contactor, RUN, which seals through its interlock (RUN_a,). The contacts (RUN) close, bypassing the starting

Table 5-3 Resistor Ranges and Properties

	Low R-High Current			High R-Low Current		
Unit Length (in.)	Resistance Range (Ω)	Current Range (A)	Heat Dissipation (Watts Per Unit*)	Resistance Range (Ω)	Current Range (A)	Heat Dissipation (Watts Per Unit*)
11	0.051–4.3	11–104	450–630	4.0–2000	0.46–10.3	426
14	0.069–5.7	11–104	620–820	5.0–2500	0.48–10.8	575
17	0.085–7.1	11–104	770–1080	5.0–2500	0.53–12.0	700
20	0.10–8.6	11–104	900–1320	6.4–4000	0.47–11.8	900

*Approximate only.

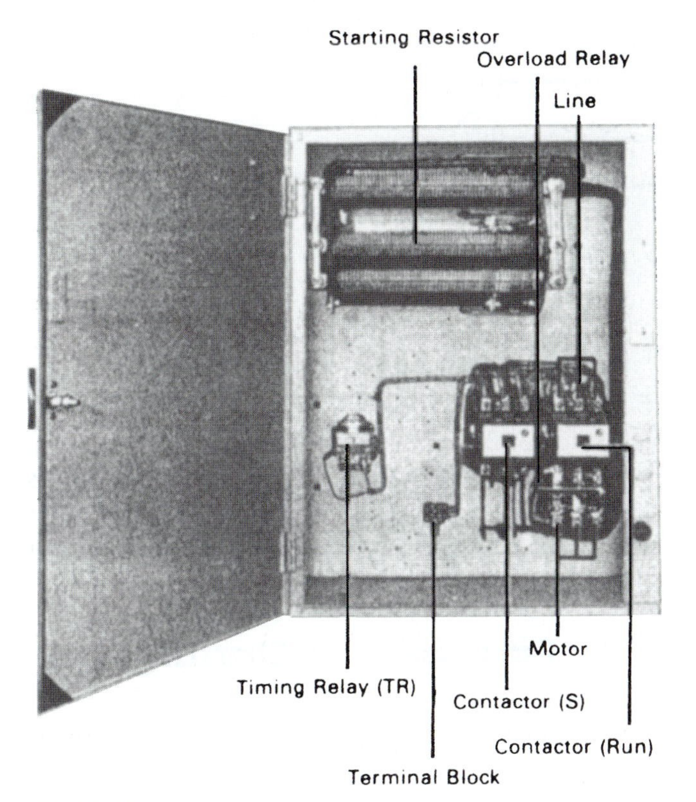

Fig. 5-35 *Primary resistor type of magnetic starter.* (Westinghouse)

resistors, and the motor will now be running at full voltage. The contactor (S) and timing relay (TR) are deenergized when the interlock (RUN_a) opens.

An overload, which opens the STOP button or other pilot device, deenergizes the (RUN) contactor. This removes the motor from the line.

Primary-resistor starters provide extremely smooth starting due to the increasing voltage across the motor terminals as the motor accelerates. Since motor current decreases with increasing speed, the voltage drop across the resistor decreases as the motor accelerates and the motor terminal voltage increases. Thus, if a resistor is shorted out as the motor reaches maximum speed, there is little or no increase in current or torque.

Autotransformer Starting

Autotransformer starters provide reduced-voltage starting at the motor terminals through the use of a tapped, three-phase autotransformer. Upon initiation of the controller pilot device, a two- and a three-pole contactor close to connect the motor to the preselected autotransformer taps. A timing relay causes the transfer of the motor from the reduced-voltage start to line-voltage operation without disconnecting the motor from the power source. This is known as closed-transition starting.

Taps on the autotransformer provide selection of 50, 65, or 80 percent of line voltage as a starting voltage. Starting torque will be 25, 42, or 64 percent, respectively, of line-voltage values. However, because of transformer action, the controller line current will be less than motor current, being 25, 42, or 64 percent of full-voltage values. This autotransformer starting may be used to provide maximum torque available with minimum line current, together with taps to permit both of these factors to be varied. Figure 5-37 shows torque and voltage tap points.

Manual autotransformer starters are used to start squirrel-cage polyphase motors when the characteristics of the driven load or power company limitations require starting at reduced voltage. See Fig. 5-38.

National Electrical Manufacturers Association (NEMA) permits one start every 4 min, for a total of four starts followed by a rest period (2 h). Each starting period is not to exceed 15 s. Figure 5-39 shows an autotransformer type of starter. Note the location of the taps on the starting transformer.

The autotransformer provides the highest starting torque per ampere of line current. Thus, it is an effective means of motor starting for applications where the inrush current must be reduced with a minimum sacrifice of starting torque. This type of starter arrangement features closed-circuit transition, an arrangement that maintains a continuous power connection to the motor during the transition from reduced to full voltage. This avoids the high-transient switching currents characteristic

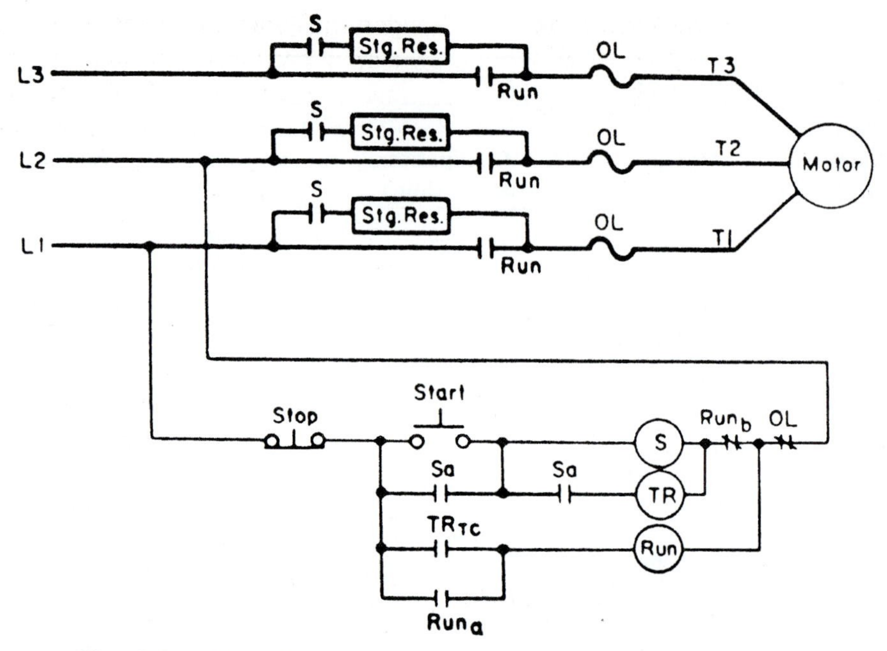

Fig. 5-36 *Wiring diagram for a primary resistor type starter.* (Westinghouse)

of starters using open-circuit transition. It provides smoother acceleration as well.

Operation Operating an external START button, or pilot device, closes the neutral and start contactors, applying reduced voltage to the motor through the autotransformer. After a preset interval, the timer contacts drop out the neutral contactor, breaking the autotransformer connection but leaving part of the windings connected to the motor as a series reactor. The RUN contactor then closes to short out this reactance and apply full voltage to the motor. Transition from reduced to full voltage is accomplished without opening the motor circuit.

For starters rated up to 200 hp you should allow a 15-s operation out of every 4 min for 1 h followed by a rest period of 2 h. For starters rated above 200 hp, you should allow three 30-s operations separated by 30-s intervals followed by a rest period of 1 h. The major disadvantages of this type of starter are its expense for lower horsepower ratings and its low power factor.

Part-winding Starting

Part-winding motors have two sets of identical windings—intended to be operated in parallel—which can be energized in sequence to provide reduced starting current and reduced starting torque. Most (but not all)

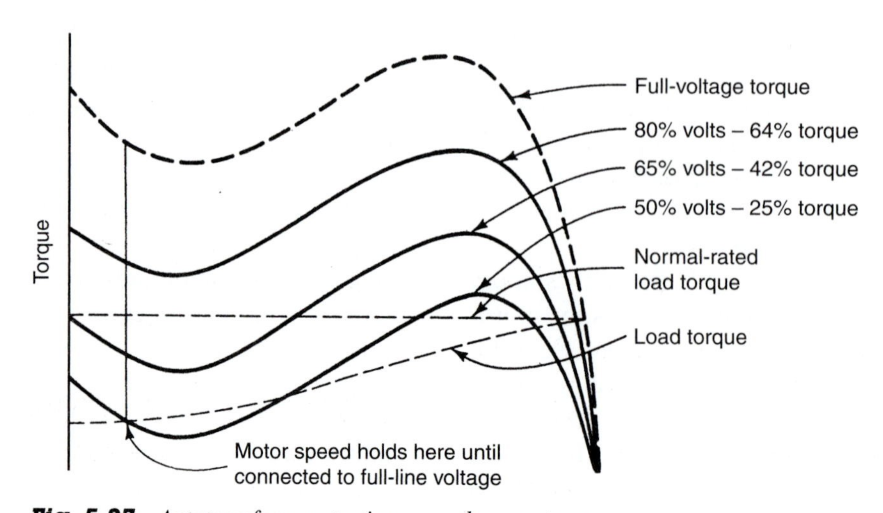

Fig. 5-37 *Autotransformer starting—speed versus torque.* (Lincoln Electric Co.)

(A)

(B)

Fig. 5-38 *(A) Autotransformer type of magnetic starter. (B) Corresponding wiring diagram.* (Allen-Bradley)

dual-voltage 230/460 V motors are suitable for part-winding starting at 230 V.

When one winding of a part-winding motor is energized, the torque produced is about 50 percent of "both winding" torque, and line current is 60 to 70 percent (depending on motor design) of comparable line-voltage values. Thus, although part-winding starting is not truly a reduced voltage means, it is usually also classified as such because of its reduced current and torque.

When a dual-voltage delta-connected motor is operated at 230 V from a part-winding starter having a three-pole start and a three-pole run contactor, an unequal current division occurs during normal operation resulting in overloading of the starting contactor. To overcome this defect, some part-winding starters use a four-pole starting contactor and a two-pole run contactor. This arrangement eliminates the unequal current division obtained with a delta-wound motor, and it enables wye-connected part-winding motors to be given either a one-half or two-thirds part-winding start.

The class 8640 starters have a start contactor, a timing relay, a run contactor, and necessary overload relays. Closing the pilot device contact causes the start contactor to close to connect the start winding and to initiate the time cycle. After expiration of the preset timing, the run contactor closes to connect the balance of the motor windings. A time setting of 1 s is recommended. Most motor manufacturers do not permit energizing of the start winding alone for longer than 3 s. Part-winding starters provide closed-transition starting.

Operation The part-winding type of starter is shown in Fig. 5-40. The parts are located for ease in understanding the operation. By taking a look at the schematic in Fig. 5-41 you can see how the starter operates. Closing the START button or other pilot device energizes the start contactor (1M) that seals in through its interlock ($1M_a$) and energizes the timer (TR). The (1M) contacts connect the first half-winding of the motor across the line. After a preset time interval, timer (TR_{TC}) contacts close the energizing contactor (2M). The (2M) contacts connect the second half-winding of the motor across-the-line.

An overload, which opens the STOP button or other pilot device, deenergizes contacts 1M, 2M, and timer TR, removing the motor from the line. The three-pole contactor (1M) connects only the first half-winding of the motor for reduced inrush current on starting. A three-pole contactor (2M) connects the second half-winding of the motor for running.

Advantages and Disadvantages Part-winding starters are the least expensive reduced voltage controller. They use closed-transition starting and are small in size.

The disadvantages are that they are unsuitable for long acceleration or frequent starting, require special motor design, and that there is no flexibility in selecting starting characteristics.

Wye-delta or Star-delta Starters

Wye-delta (Y-Δ) or star-delta starters are used with delta-wound squirrel-cage motors that have all leads brought out to facilitate a wye connection for reduced-voltage

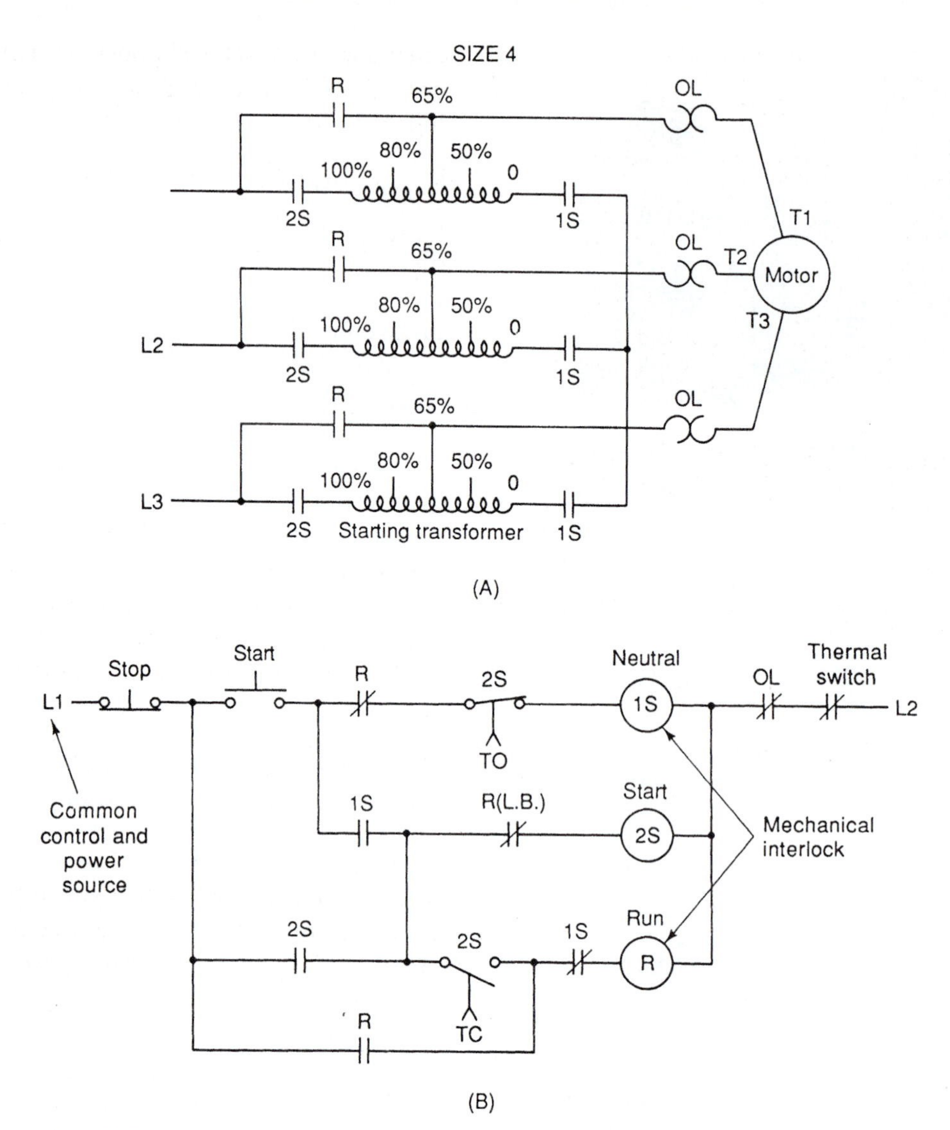

Fig. 5-39 *Typical wiring diagram for an autotransformer type of reduced-voltage starter.*
(Allen-Bradley)

starting. This starting method is particularly suitable for applications involving long accelerating times or frequent starts. Wye-delta starters are typically used for high-inertia loads such as centrifugal air-conditioning units, although they are applicable in cases where low-starting torque is necessary or where low-starting current is necessary and low-starting torque is permissible.

When 6- or 12-lead delta-connected motors are started star-connected, approximately 58 percent of full-line voltage is applied to each winding and the motor develops 33 percent of full-voltage starting torque and draws 33 percent of normal locked-rotor current from the line. When the motor has accelerated, it is reconnected for normal delta operation.

Operation Operating an external START button energizes the motor in the wye connection. See Fig. 5-42. This applies approximately 58 percent of full-line voltage to the windings. At this reduced voltage, the motor will develop about 33 percent of its full-voltage starting torque and will draw about 33 percent of its normal locked-rotor current.

After an adjustable time interval, the motor is automatically connected in delta, applying full-line voltage to the windings. In starters with open-circuit transition the motor is momentarily disconnected from the line during the transition from the wye to delta. With closed transition (Fig. 5-43) the motor remains connected to the line through the resistors. This avoids the current surges associated with open-circuit transition.

Advantages and Disadvantages The advantages are moderate cost and its suitability for high-inertial, long-acceleration loads. It does have torque efficiency. However, the disadvantages are that it requires special motor design, starting torque is low, and it is inherently

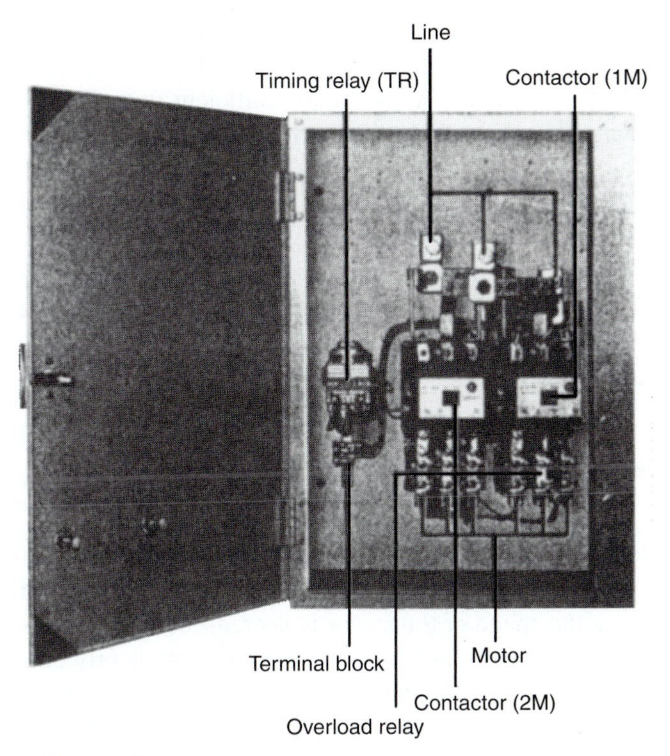

Fig. 5-40 *Part-winding type of magnetic starter.* (Westinghouse)

open transition—closed transition is available at added cost. There is no flexibility in selecting starting characteristics.

Star-Delta (Wye-Delta) Connections There is the 12-lead motor wound for Y-Δ starting operation on either low voltage or a higher voltage. See Fig. 5-44. There is also a six-lead single-voltage motor suitable for Y-Δ starting. Figure 5-44B shows the connection to the lines for the six-lead motor. Keep in mind that overload relay protection is required by the National Electrical Code. The size of the protection is determined by the manufacturer of the motor (Table 5-4).

Multispeed Starters

Multispeed starters are designed for the automatic control of two-speed squirrel-cage motors of either the consequent pole or separate winding types. These starters are available for constant-horsepower, constant-torque, or variable-torque three-phase motors. Multispeed motor starters are commonly used on machine tools, fans, blowers, refrigeration compressors, and many other types of equipment.

Low-Speed Compelling Relay When added to a standard starter, the low-speed compelling relay compels the operator always to start the motor in low speed before switching to a higher speed. This is a safety

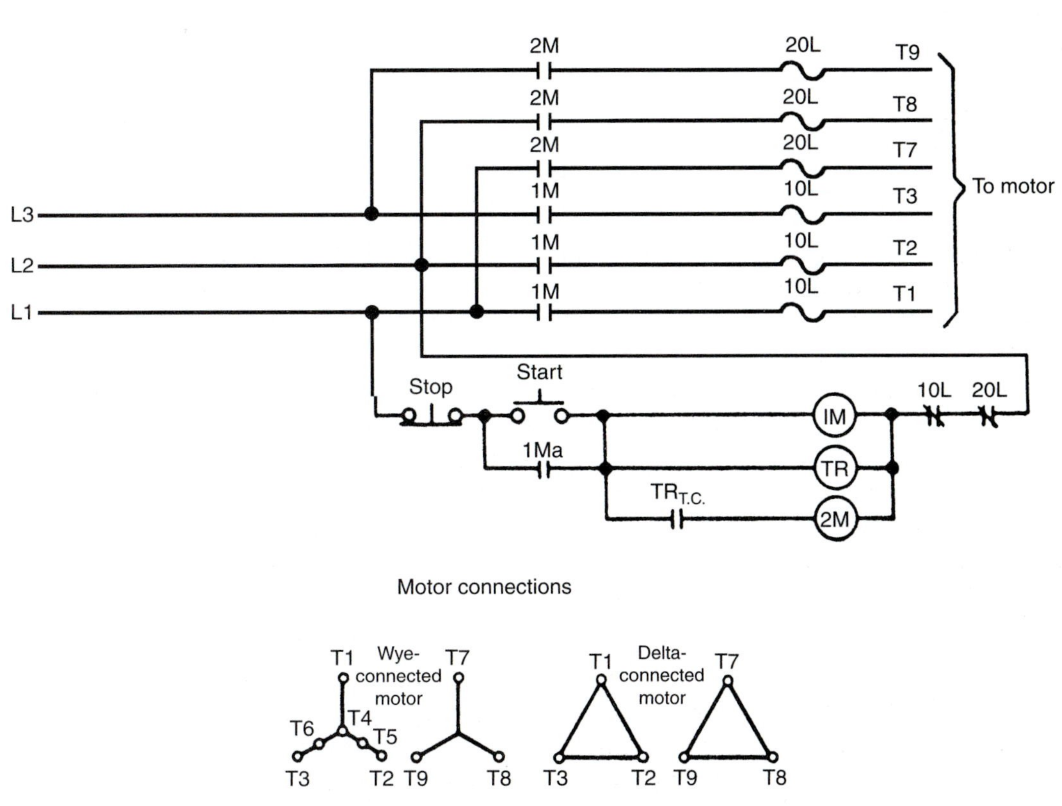

Fig. 5-41 *Typical wiring diagram for part-winding type of starter.* (Westinghouse)

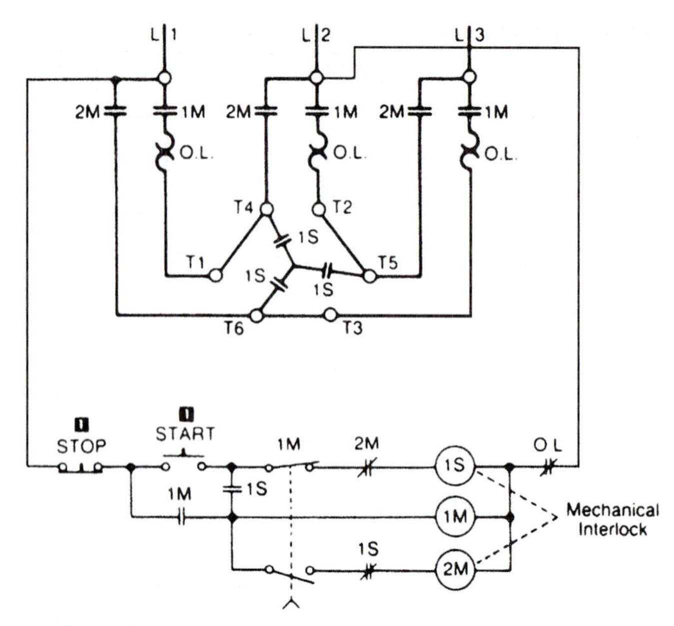

Fig. 5-42 *Typical wiring diagram for wye-delta starter, open-circuit transition.* (Allen-Bradley)

feature where damage to equipment may result when the motor is started at high speed. See Fig. 5-45.

Automatic Sequence Accelerating Relay The automatic sequence accelerating relay will control the sequence of acceleration from low speed up to high speed.

Automatic Sequence Decelerating Relay The automatic sequence decelerating relay is used with large-inertia loads. The braking effect caused by a sudden change from high to low speed may cause damage to the motor or to the driven machine. To avoid this danger, the operation should give the motor sufficient time to slow down by pushing the STOP button and then waiting for a short interval before pushing the button for a lower speed.

To help provide correct operation, multispeed starters can be equipped with an automatic sequence decelerating relay for each lower-speed step. This relay automatically interposes a time delay between the speed steps and makes it unnecessary to press the STOP button when switching to a lower speed.

CONSEQUENT-POLE MOTOR CONTROLLER

By increasing the number of poles of a motor it is possible to change its speed. By increasing the number of poles, the speed of the motor is decreased. Inasmuch as a motor is wound and mounted rather permanently on a frame, it is not easily possible to take out or put in poles or the associated windings. Therefore, an electrical means must be found if the speed of the motor is to be changed by using the number of poles method. One method of doing this is the consequent-pole arrangement. This method can be used for two-speed, one-winding motors or four-speed, two-winding motors.

The reversal of some of the currents in the windings has the same effect as physically increasing or decreasing the number of poles. Three-phase motors are wound, in some cases, with six leads brought out for connection

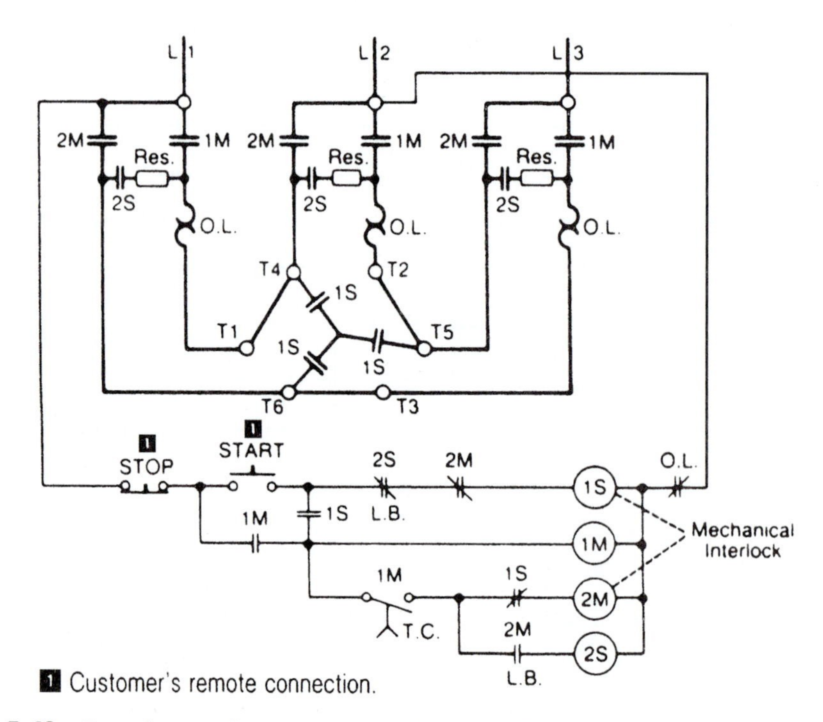

Fig. 5-43 *Typical wiring diagram for wye-delta starter, closed-circuit transition.* (Allen-Bradley)

STAR-DELTA (YΔ) CONNECTIONS

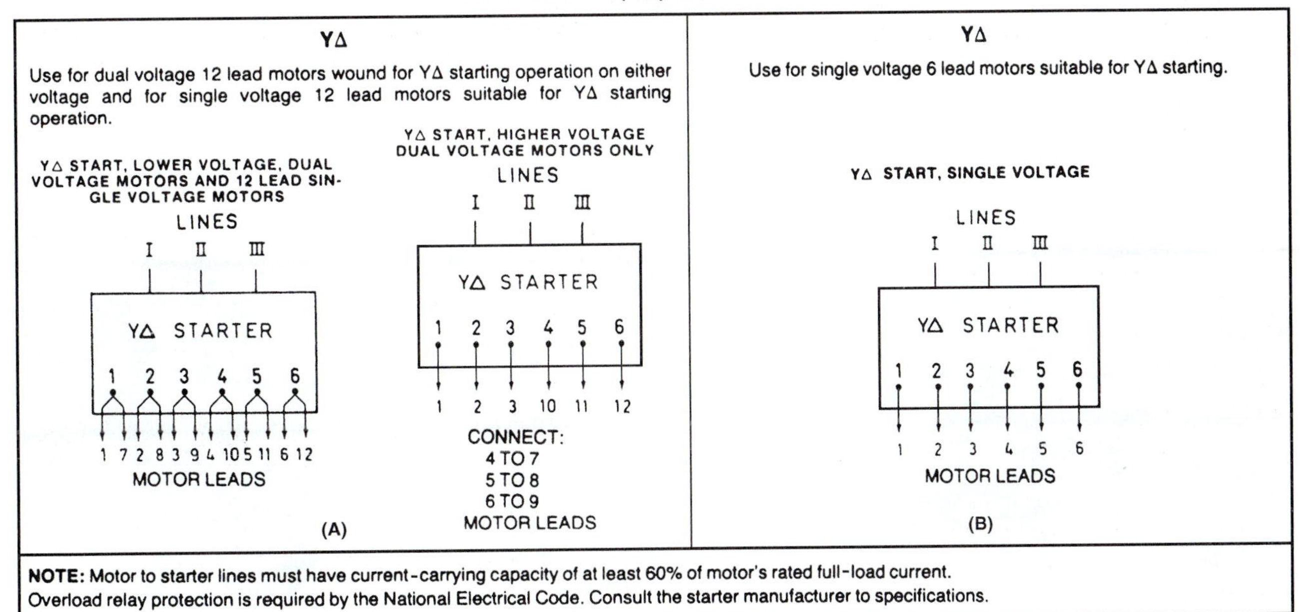

Fig. 5-44 *Star-delta connections.* (Lincoln Electric Co.)

Table 5-4 *Selection of a Controller Best Suited for a Particular Characteristic*

Characteristic Wanted	Type of Starter to use (Listed in Order of Desirability)	Comments
Smooth acceleration	1. Solid state (class 8660) 2. Primary resistor (class 8647) 3. Wye-delta (class 8630) 4. Autotransformer (class 8606) 5. Part-winding (class 8640)	Little choice between 3 and 4.
Minimum line current	1. Autotransformer (class 8606) 2. Solid state (class 8660) 3. Wye-delta (class 8630) 4. Part winding (class 8640) 5. Primary resistor (class 8647)	
High-starting torque	1. Autotransformer (class 8606) 2. Solid state (class 8660) 3. Primary resistor (class 8647) 4. Part winding (class 8640) 5. Wye-delta (class 8630)	
High-torque efficiency (torque vs. line current)	1. Autotransformer (class 8606) 2. Wye-delta (class 8630) 3. Part winding (class 8640) 4. Solid state (class 8660) 5. Primary resistor (class 8647)	Little choice between 3, 4, and 5.
Suitability for long acceleration	1. Wye-delta (class 8630) 2. Autotransformer (class 8606) 3. Solid state (class 8660) 4. Primary resistor (class 8647)	For acceleration time greater than 5s, primary resistor requires nonstandard resistors. Part-winding controllers are unsuitable for acceleration time greater than 2s.
Suitability for frequent starting	1. Wye-delta (class 8630) 2. Solid state (class 8660) 3. Primary resistor (class 8647) 4. Autotransformer (class 8606)	Partwinding is unsuitable for frequent starts.
Flexibility in selecting starting characteristics	1. Solid state (class 8660) 2. Autotransformer (class 8606) 3. Primary resistor (class 8647)	For primary resistor, resistor change required to change starting characteristics. Starting characteristics cannot be changed for wye-delta or part-winding controllers.

Source: Courtesy of Square D.

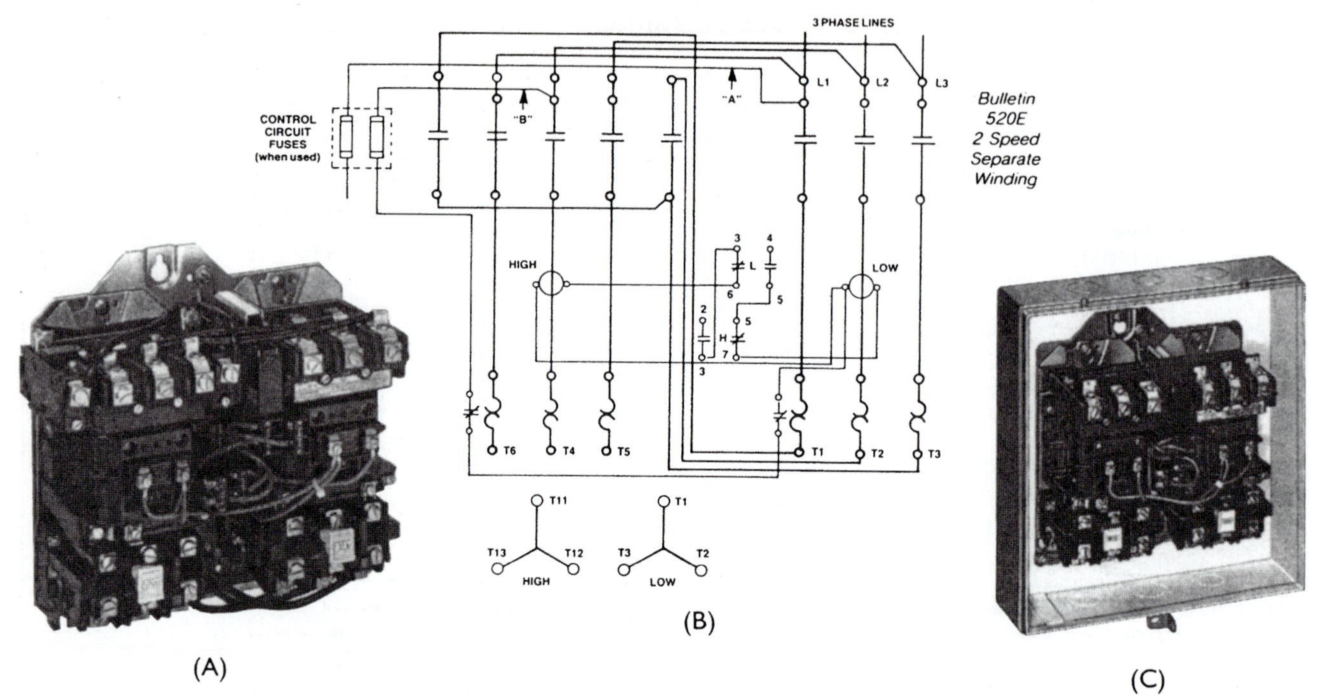

Fig. 5-45 *(A) Multispeed starter and two-speed consequent pole starter without enclosure. (B) Typical wiring diagram for two-speed separate winding-motor starter. (C) General purpose enclosure with cover removed.* (Allen-Bradley)

purposes. It is possible to connect the windings, using combinations of the terminals for connection purposes, either in series delta or in parallel wye. See Fig. 5-46. By tapping the windings it is possible to send current in two different directions, effectively creating more poles and decreasing the speed of the motor. The number of poles is doubled by reversing through half a phase. Two speeds are obtained by producing twice as many consequent poles for low-speed operation as for high speed.

Figure 5-47 shows how the controller is wired to produce consequent poles for constant torque or variable torque. The wiring diagram and the line drawing (Fig. 5-48) illustrate connections for the following method of operation: The motor can be started in either HIGH or LOW speed. The change from LOW to HIGH or from HIGH to LOW can be made without first pressing the STOP button. Figure 5-49 shows pilot devices with connections that can be made to obtain different sequences and methods of operation. The series delta arrangement produces high speed. It also produces the same horsepower rating at high and low speeds.

The torque rating is the same for both speeds if the winding is such that the series delta connection gives the low speed and the parallel wye connection gives the high speed. Consequent-pole motors that have a single winding for two speeds have the extra tap at the midpoint of the winding. This permits the various connection possibilities. However, the speed range is limited to a 1:2 ratio of or 600/1200 or 900/1800 rpm.

Figure 5-50 shows the motor terminal markings and connections for a constant-horsepower delta. The wiring diagram (Fig. 5-51) and the line drawing (Fig. 5-52) illustrate connections for the following method of operation: Motor can be started in either HIGH or LOW speed. The change from LOW to HIGH can be made without first pressing the STOP button. When changing from HIGH to LOW, the STOP button must be pressed between speeds. The pilot devices shown in Fig. 5-53 show the other connections that can be made to obtain different sequences and methods of operation.

CONNECTIONS MADE BY STARTER			
Speed	Supply Lines L1 L2 L3	Open	Together
Low	T1 T2 T3	T4, 5, 6	None
High	T6 T4 T5	None	T1, 2 3

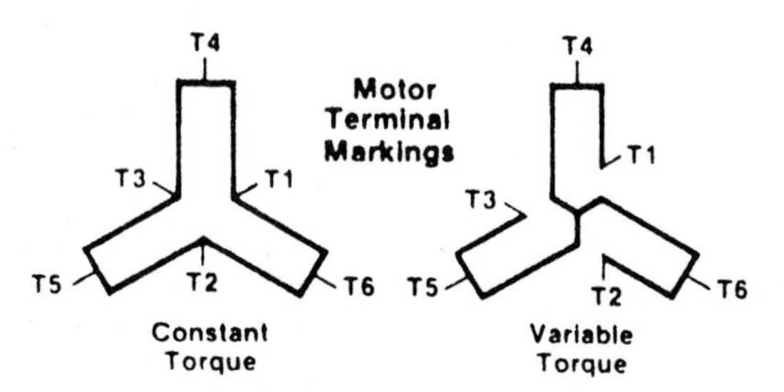

Fig. 5-46 *Connections made by the consequent-pole starter for constant torque or variable torque.* (Allen-Bradley)

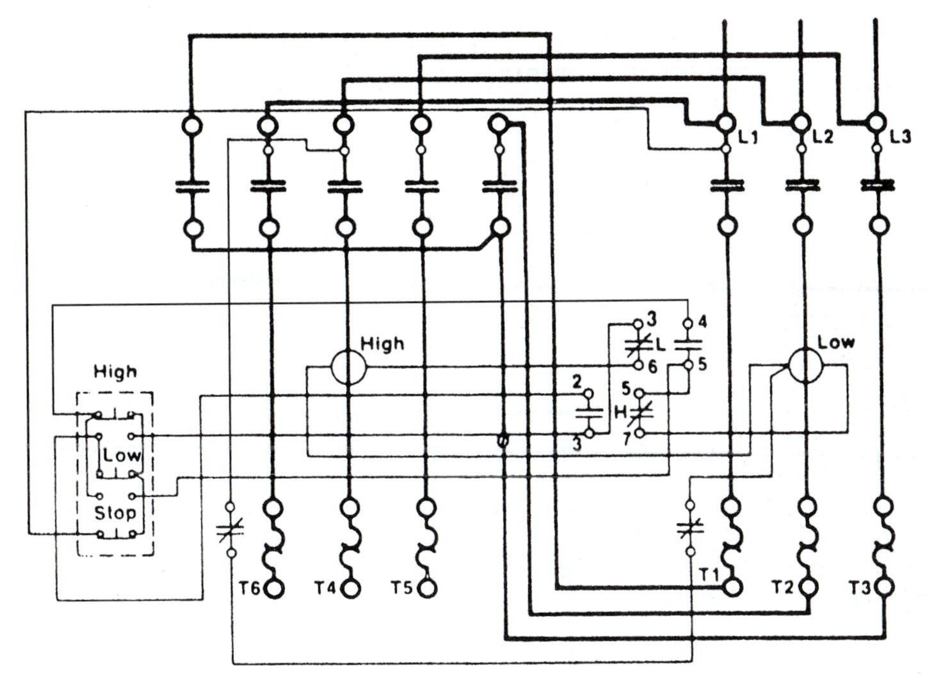

Fig. 5-47 *Wiring diagram for a two-speed, consequent-pole, constant-horsepower motor, NEMA size 0-4.* (Square D)

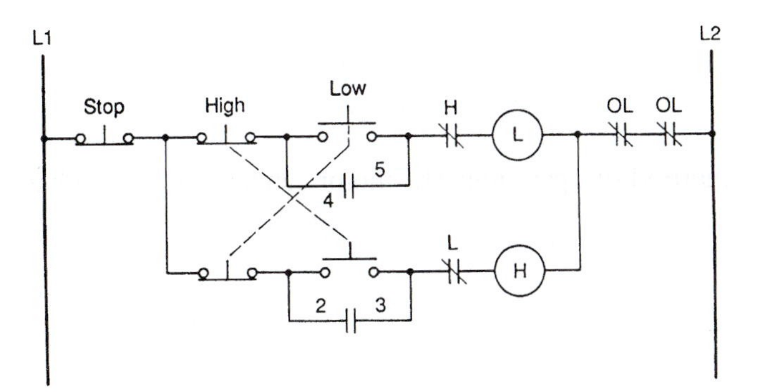

Fig. 5-48 *Line diagram for a two-speed motor.* (Allen-Bradley)

Four-speed, two-winding consequent-pole motor controllers can be used on squirrel-cage motors that have two reconnectable windings and two speeds for each winding. This type of motor does need a special type of starting sequence. This means that it must use the properties of the compelling relay, accelerating relay, and decelerating relay to operate correctly.

Figure 5-54 shows the two-speed consequent-pole starter with variable-torque and constant-torque connections. Figure 5-55 shows how the four-speed, two-winding controller is connected for the possible arrangements using this type of motor.

FULL-VOLTAGE CONTROLLERS

The least expensive of the starters is the full-voltage type. There is no limit to the horsepower, size, voltage rating, or type of motor that can be started on full voltage when the power is available.

Full-voltage starters are always the first choice when the power system can supply initial inrush current, and the motor and the driven machine can withstand the

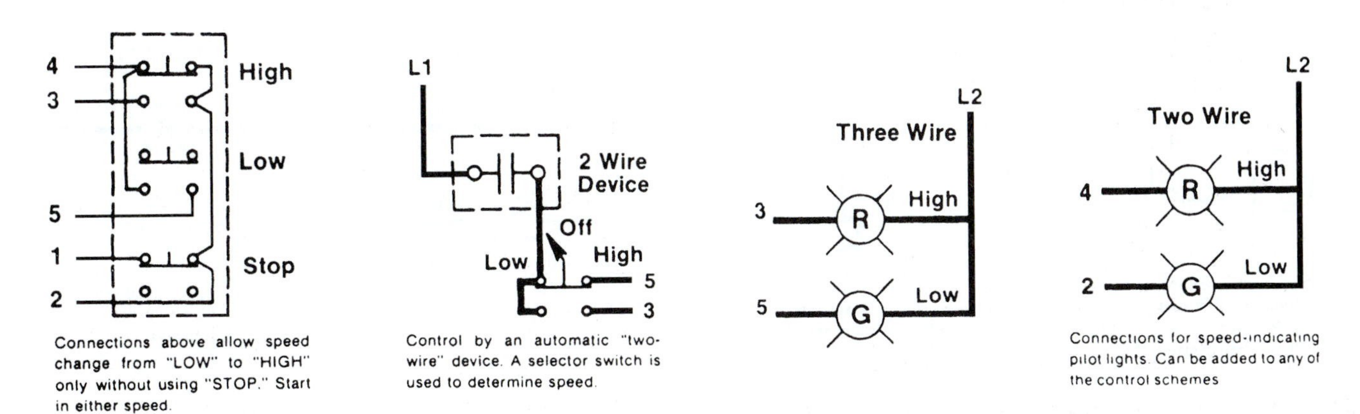

Fig. 5-49 *Pilot-device diagrams show connections that can be made to obtain different sequences and methods of operation.*

CONNECTIONS MADE BY STARTER						
Speed	Supply Lines L1	L2	L3	Open	Together
Low	T1	T2	T3	None	T4, 5, 6
High	T6	T4	T5	T1, 2, 3	None

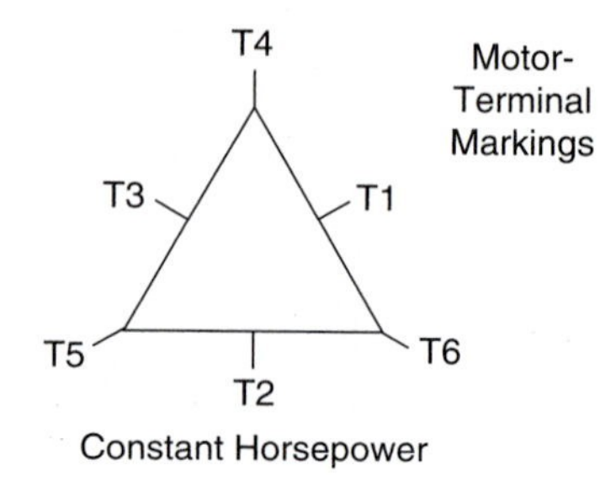

Fig. 5-50 *Connections made by the starter for constant horsepower.* (Allen-Bradley)

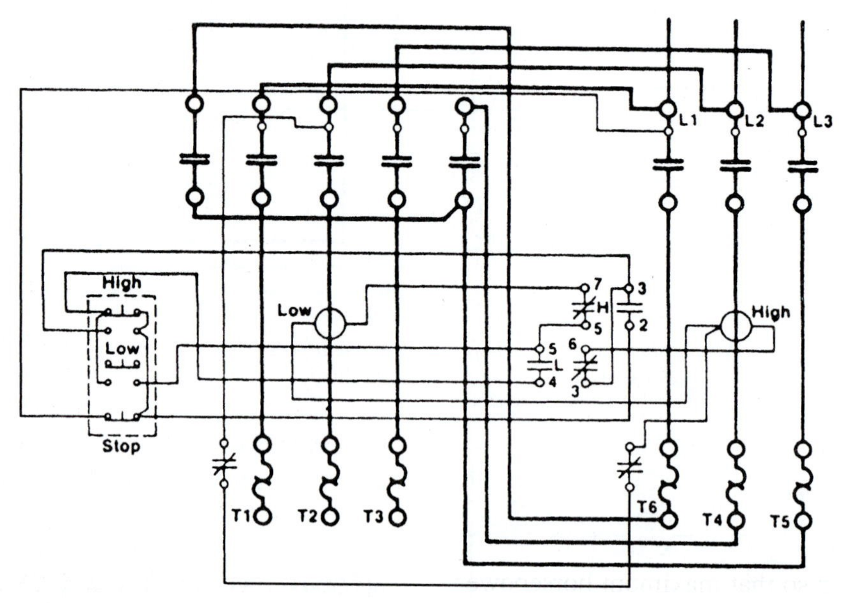

Fig. 5-51 *Wiring diagram for two-speed, consequent-pole, constant-or variable-torque starter, NEMA size 0-4.* (Square D)

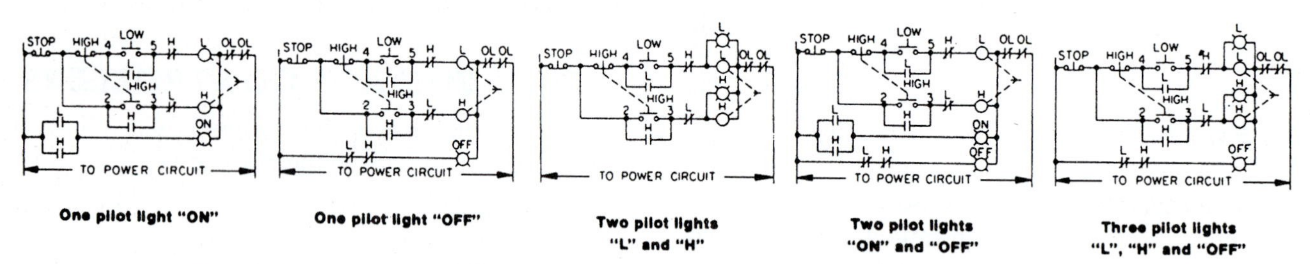

Fig. 5-52 *Elementary drawing of the control circuit for a consequent-pole starter.* (Allen-Bradley)

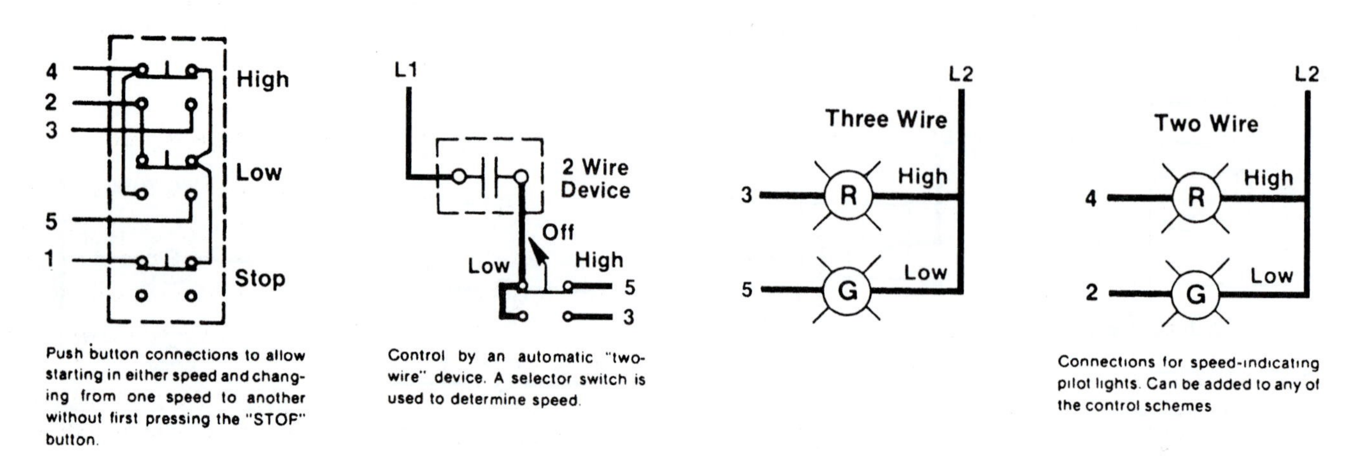

Fig. 5-53 *Connections for different sequences and methods of operation.* (Allen-Bradley)

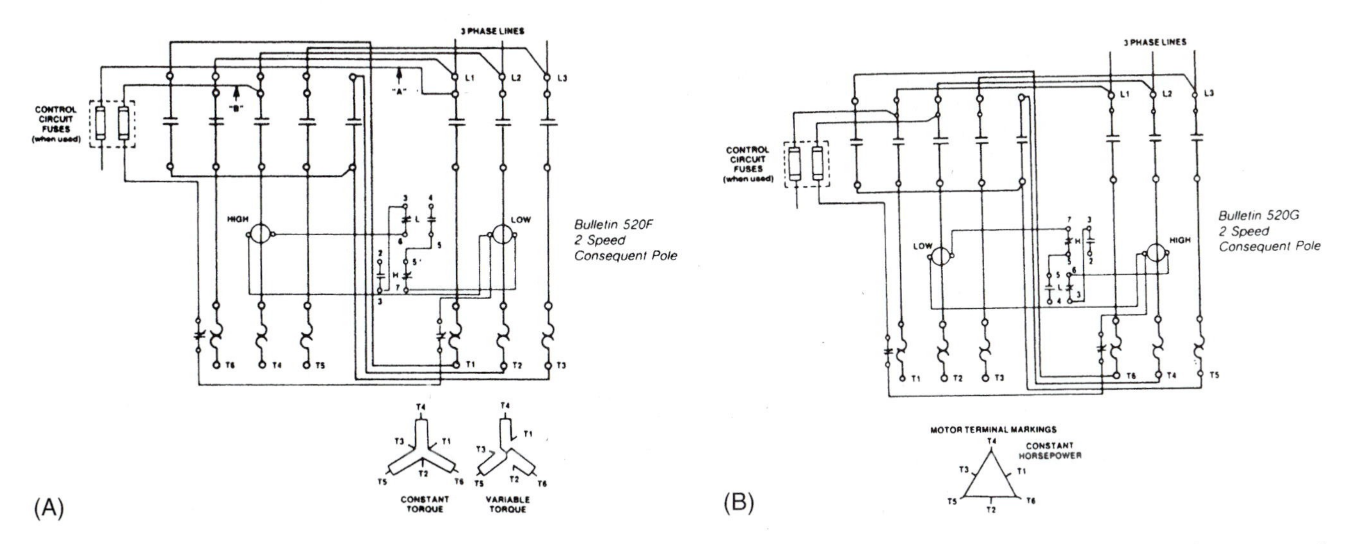

Fig. 5-54 *(A) Typical wiring diagram for two-speed consequent-pole starter* (Wye). *(B) Typical wiring diagram for two-speed consequent-pole starter* (Delta).

sudden starting shock. Examples of this are machines that start unloaded, as well as those that require little torque; or machines may be equipped with some form of unloading device to reduce starting torque, as in the use of an unloader valve in a compressor. A clutch may be inserted between a machine and motor so that the motor may be started unloaded. When the motor is up to speed, the clutch is engaged. Clutches are sometimes used on large machines so that maximum horsepower can be exerted during breakaway without serious power-system disturbance. Use of clutches also permits using motors with lower torque and locked-rotor currents. In most instances, up-to-date installations use solid-state motor controllers to better advantage. Many of the older types of starters are still in use and will continue to provide good service for many more years. As they deteriorate, they are usually replaced by a solid-state type of starter so that the clutch arrangements are unnecessary.

Figure 5-56 shows the general-purpose enclosure for a full-voltage starter. This type of starter is designed for full-voltage starting of polyphase squirrel-cage motors and primary control of slip-ring motors. This type of starter may be operated by remote control with push buttons, float switches, thermostats, pressure switches, snap switches, limit switches, or any other suitable two- or three-wire pilot device.

Starting Sequence

If full-voltage starting produces excessive current demands on the distribution system, motors should be started individually or in blocks of permissible size by using some method of time delay, such as motor driver, pneumatic, or mercury plunger timing relays. When large and small motors are to be started on a common power system, best results are obtained by starting the largest sizes first. This gives larger motors the advantage of full-line capacity. If synchronous motors are on the system with other types of AC motors, the synchronous units should always be started first since they provide voltage stability for starting the induction motors.

Protection Against Low Voltage

Low-voltage protection is needed while the motors are running even though systematic starting permits all motors to be started without excessive line-voltage drop. When three-wire control circuits are used, a severe dip in line voltage or a momentary complete outage breaks the control-sealing circuits, and the controller drops out and stops the motor. This provides low-voltage protection and prevents simultaneous acceleration of all motors to full speed after being slowed down by a voltage dip. However, all motors are disconnected from the line during the voltage dip, and each must be restarted.

Time-Delay Protection

It is possible to wire the circuitry so that a time-delay under-voltage arrangement can be used. This permits dropout of the controllers on low-voltage dips but allows restarting automatically if normal voltage is restored within a preset time delay. The usual time delay is 2 s or less.

Time-delay under-voltage protection on controllers will prevent some complete shutdowns but should be applied with caution. If used on all motor controllers, restoration of voltage within the time-delay setting

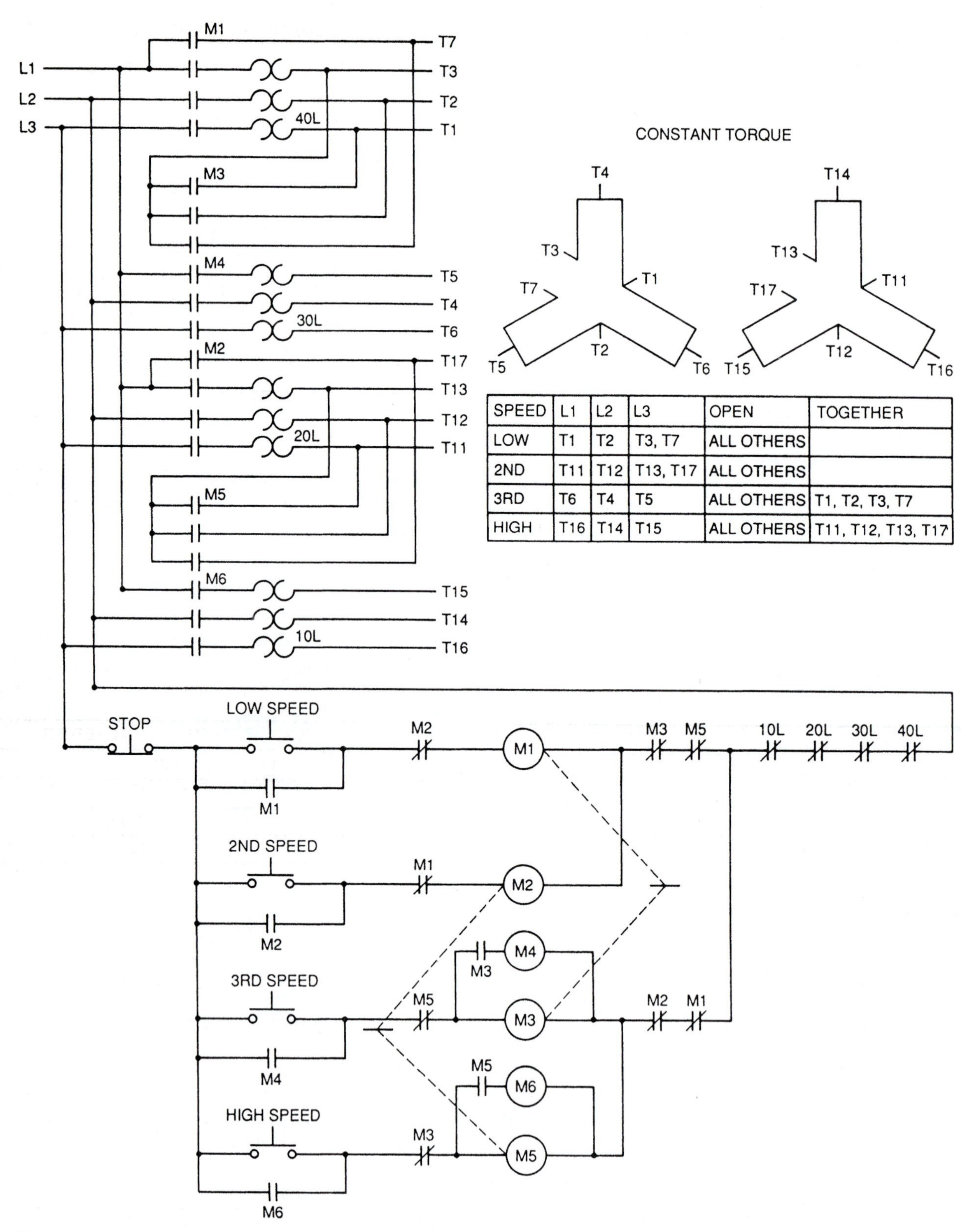

SPEED	L1	L2	L3	OPEN	TOGETHER
LOW	T1	T2	T3, T7	ALL OTHERS	
2ND	T11	T12	T13, T17	ALL OTHERS	
3RD	T6	T4	T5	ALL OTHERS	T1, T2, T3, T7
HIGH	T16	T14	T15	ALL OTHERS	T11, T12, T13, T17

Fig. 5-55 *Elementary diagram of a four-speed, two-winding controller and the possible arrangements for motor connections.* (Allen-Bradley)

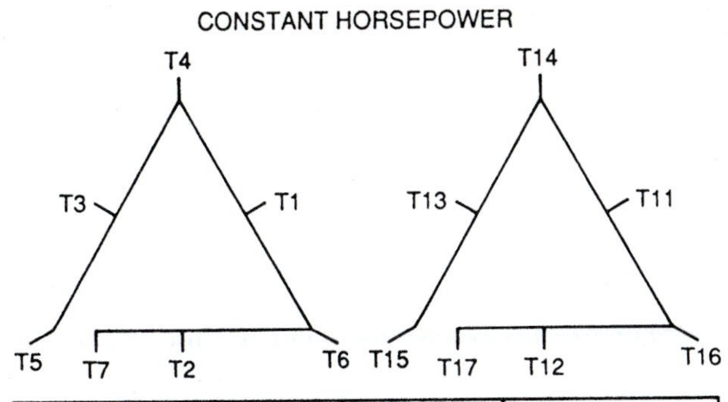

SPEED	L1	L2	L3	OPEN	TOGETHER
LOW	T1	T2	T3	ALL OTHERS	T4, T5, T6, T7
2ND	T6	T4	T5, T7	ALL OTHERS	
3RD	T11	T12	T13	ALL OTHERS	T14, T15, T16, T17
HIGH	T16	T14	T15, T17	ALL OTHERS	

(A)

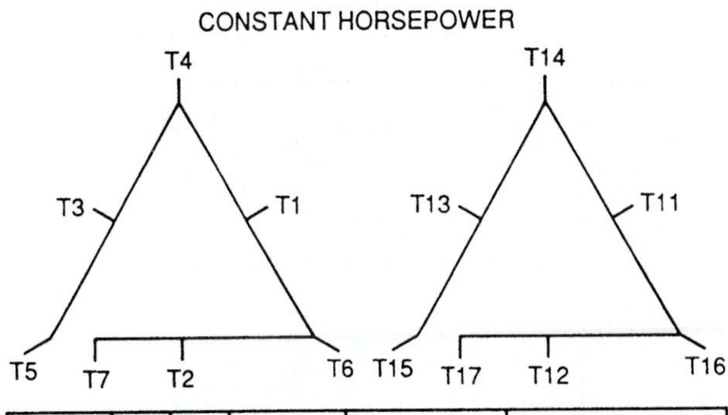

SPEED	L1	L2	L3	OPEN	TOGETHER
LOW	T1	T2	T3	ALL OTHERS	T4, T5, T6, T7
2ND	T11	T12	T13	ALL OTHERS	T14, T15, T16, T17
3RD	T6	T4	T5, T7	ALL OTHERS	
HIGH	T16	T14	T15, T17	ALL OTHERS	

(B)

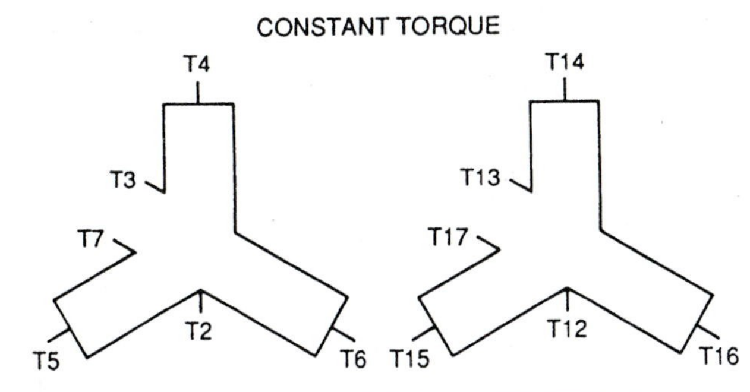

SPEED	L1	L2	L3	OPEN	TOGETHER
LOW	T1	T2	T3, T7	ALL OTHERS	
2ND	T6	T4	T5	ALL OTHERS	T1, T2, T3, T7
3RD	T11	T12	T13, T17	ALL OTHERS	
HIGH	T16	T14	T15	ALL OTHERS	T11, T12, T13, T17

(C)

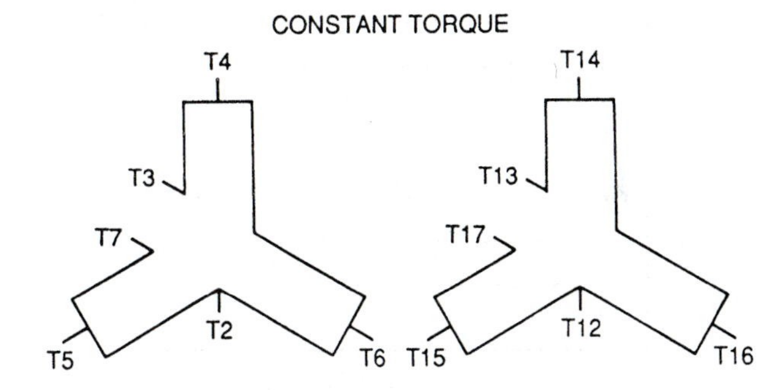

SPEED	L1	L2	L3	OPEN	TOGETHER
LOW	T1	T2	T3, T7	ALL OTHERS	
2ND	T11	T12	T13, T17	ALL OTHERS	
3RD	T6	T4	T5	ALL OTHERS	T1, T2, T3, T7
HIGH	T16	T14	T15	ALL OTHERS	T11, T12, T13, T17

(D)

VARIABLE TORQUE

T4 T14 T1 T11 T3 T13 T2 T12 T5 T6 T15 T16

SPEED	L1	L2	L3	OPEN	TOGETHER
LOW	T1	T2	T3	ALL OTHERS	
2ND	T6	T4	T5	ALL OTHERS	T1, T2, T3
3RD	T11	T12	T13	ALL OTHERS	
HIGH	T16	T14	T15	ALL OTHERS	T11, T12, T13

(E)

VARIABLE TORQUE

T4 T14 T1 T11 T3 T13 T2 T12 T5 T6 T15 T16

SPEED	L1	L2	L3	OPEN	TOGETHER
LOW	T1	T2	T3	ALL OTHERS	
2ND	T11	T12	T13	ALL OTHERS	
3RD	T6	T4	T5	ALL OTHERS	T1, T2, T3
HIGH	T16	T14	T15	ALL OTHERS	T11, T12, T13

(F)

Fig. 5-55 *(Continued)*

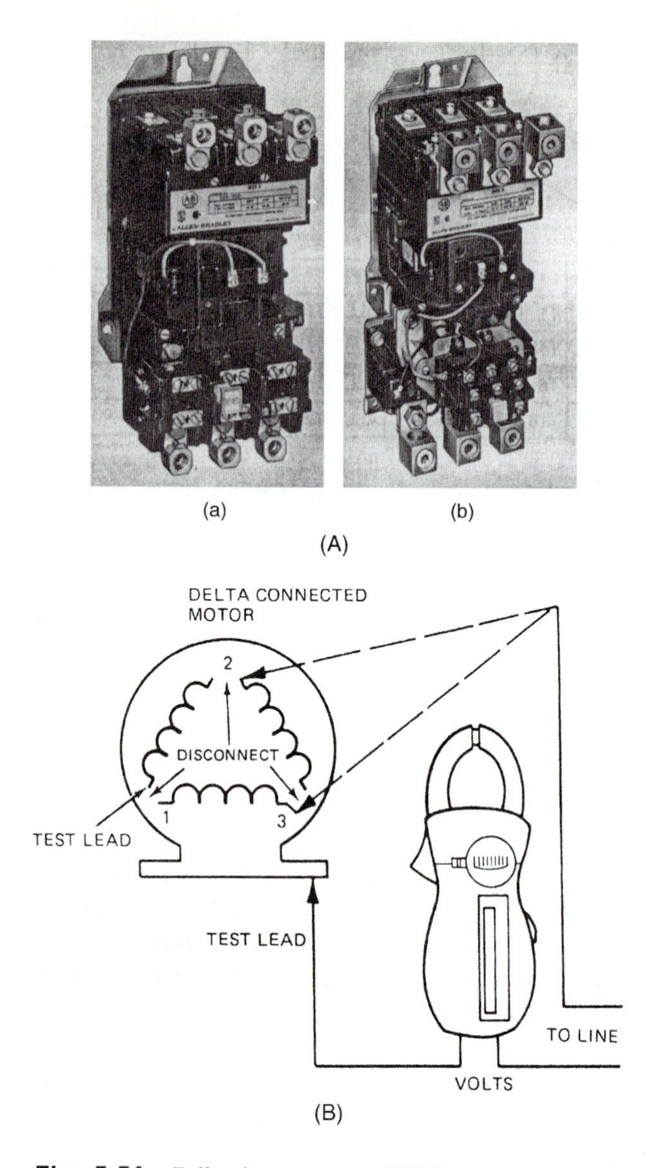

Fig. 5-56 *Full-voltage starters (NEMA), open type, without enclosure: (A) Size 3; (B) Size 5. (A-B)* *(Amprobe)*

after a voltage dip causes each motor to attempt to accelerate simultaneously, thus producing excessive currents that may operate backup protection and starter-overload devices, and disconnect the motors.

Pilot devices such as pressure, float, or temperature switches automatically start and stop motors as the demand arises. On severe voltage dips or voltage failure, motor controllers drop open even though the demand switch is closed. Upon restoration of full voltage all units attempt to restart at the same time. This operating hazard can be overcome by adding a time delay in the starting circuit of each motor and timing the demand for starting at slightly different intervals. Time delays of various units can then be staggered so that at the restoration of voltage only one unit at a time will be started.

ELECTRIC MOTORS: THEIR USES, OPERATION, AND CHARACTERISTICS

Electric motors are devices which convert electric energy to kinetic energy, usually in the form of a rotating shaft that can be used to drive a fan, pump, compressor, and so forth. Single-phase motors are commonly used up to 3 hp, occasionally larger. Three-phase motors are preferred in electrical design for $^3/_4$ hp motors and larger, since they are self-balancing on the three-phase service.

Motors come in various styles and with different efficiency ratings. The efficiency is typically related to the amount of iron and copper in the windings. That is, the more iron for magnetic flux the more copper for reduced resistance means the more efficient the motor. Words such as standard and premium efficiency are common.

- Inverter duty implies a motor built to withstand the negative impacts of variable-frequency drive.
- *Open drip-proof* (ODP) motors are used in general applications.
- *Totally enclosed fan-cooled* (TEFL) motors are used in severe-duty environments.
- Explosion-proof motors may be needed in hazardous environments.

Motors are typically selected to operate at or below their motor nameplate rating, although ODP motors often have a service factor of 1.15, which implies that the motor will tolerate a slight overload, even on a continuous basis.

Since motors are susceptible to failure when they are operated above the rated temperature, care must be taken in motor selection for hot environments such as downstream from a heating coil. For altitudes above 3300 feet, motor manufacturers typically discount the service factor to 1.0.

Motor windings are protected by overload devices which open the power circuit if more than the rated amperage passes for more than a predetermined time. This raises an interesting issue for a motor assigned to drive a fan that has a disproportionately high moment of rotational inertia. On start-up, a motor draws much more than the full-speed operating current. The time required to bring a fan up to speed may be too long if the motor does not have enough torque to both meet the load and accelerate the fan wheel.

If the motor does not come up to speed within 10 to 15 s, it is likely that the motor protector will cut out based on the starting amperage. A motor sized tightly to a fan load may never get started. Therefore, it is important to size a motor for, both load and fan-wheel inertia. Fan vendors can help with this concern. This problem is particularly common on large boiler induced-draft fans where the dense-air, cold-start-up condition requires much more driver power than the hot-operating condition.

Motor Rotation

In single-phase motors, the direction of motor rotation is determined by the factory-established internal wiring characteristics of the motor. Changing the connection of leads to the power source may have no effect on the direction of rotation. To make a change requires a change in an internal connection as directed by the manufacturer.

In polyphase motors, a lead sequence is established at the power plant. The motor presents three sets of lead wires, which are connected to the three phases of the service. If a three-phase motor is found running backward, all that is needed to change the direction is to exchange any two leads.

Variable-Speed Drives

One of the most useful electrical developments in recent years has been the AC *variable frequency drive* (VFD) for motor speed control. Electric speed control of motors is not a new concept—dc drives have been used for decades in the industrial environment—but low cost AC drives suitable for the HVAC market are a relatively new product. These new drives typically use electronic circuitry to vary the output frequency which in turn varies the speed of the motor.

Since the power required to drive a centrifugal fan or centrifugal pump is proportional to the cube of the fan or pump speed, large reductions in power consumption are obtained at reduced speed. These savings are used to pay for the added cost of the VFD on a life cycle cost basis. A quality VFD usually obtains greater energy savings than does a variable-pitch inlet vane or other mechanical-flow volume control. In low-budget projects, the owner may have elected to forgo the higher-quality VFD service in favor of the lower-first-cost inlet vane damper for fans, or modulating-valve differential pressure control for pumps.

In applying a VFD to a duty, several factors have to be considered:

- The VFD needs to be in a relatively clean, air-conditioned environment. Since it is a sophisticated electronic device, particulates in the ambient air, wide swings in ambient air conditions, temperatures above 90°F, and humid condensing environments are all threatening to drive life expectancy.
- The drive should be matched to the driven motor. Reduced motor speeds relate to reduced motor cooling while internal motor energy losses may be high in an inappropriately configured motor. High efficiency or inverter duty motors are typically preferred for VFD service.
- Drives and motors may be altitude-sensitive or may be affected by other local conditions. Drive and motor selection should be confirmed in every case by the drive vendor.
- Some drives use a carrier frequency in the audible range, whirl may be emitted at the drive and/or at the motor. The noise may be objectionable. This is a difficult problem to abate in some applications. Some newer drives allow the carrier frequency to be set above the normal hearing range, which eliminates the noise problem, but shorten motor life expectancy.
- Some variable-speed drives impose "garbage" waveforms on the incoming utility lines or create harmonic distortions which affect the current flow in the neutral conductor of a three-phase power supply. Isolation transformers are not always effective in eliminating harmonic distortion back to the line.

  Harmonic distortions are also implicated in premature fan shaft bearing failures, where vagrant currents overwhelm the insulating qualities of bearing grease to arc from inner to outer bearing races, violating the normally smooth rolling surfaces with metal deposits.
- If VFDs are applied to critical loads, it may be helpful to have bypass circuitry to run the motor at full speed in the event of a drive outage. This creates a concern for pressure control since the full-speed operation will develop a maximum pressure condition whether needed or not. Relief dampers may be considered.
- Most VFDs can accept a remote input signal of 4 to 20 mA, or 0 to 10 V DC, derived from pressure transducers or flow meters. The drives typically have a manual speed-selection option if an occasional or seasonal speed change is all that is needed. The manual setting is also useful in a test-and-balance period.

TROUBLESHOOTING ELECTRIC MOTORS WITH A VOLT-AMMETER

Electrical equipment is designed to operate at a specific voltage and current. Usually the equipment will work satisfactorily if the line voltage differs plus or minus 10 percent from the actual nameplate rating. In a few cases, however, a 10 percent voltage drop may result in a breakdown. Such may be the case with an induction motor that is being loaded to its fullest capacity both on start and run. A 10 percent loss in line voltage will result in a 20 percent loss in torque.

The full load current rating on the nameplate is an approximate value based on the average unit coming off the manufacturers' production line. The actual current for any one unit may vary as much as plus or minus 10 percent at rated output. However, a motor whose load current exceeds the rated value by 20 percent or more will reduce the life of the motor due to higher-operating temperatures and the reason for excessive current should be determined. In many cases it may simply be an overloaded motor. The percentage increase in load will not correspond with percentage increase in load current. For example, in the case of a single-phase induction motor, a 35 percent increase in current may correspond to an 80 percent increase in output torque.

The operating conditions and behavior of electrical equipment can be analyzed only by actual measurement. A comparison of the measured terminal voltage and current will check whether the equipment is operating within electrical specifications.

The measurement of voltage and current requires the use of two basic instruments—a voltmeter and an ammeter. To measure voltage, the test leads of the voltmeter are in contact with the terminals of the line under test. To measure current, the conventional ammeter must be connected in series with the line so that the current will flow through the ammeter.

The insertion of the ammeter means shutting down the equipment, breaking open the line, connecting the ammeter, starting up the equipment, reading the meter and then going through as much work to remove the ammeter from the line. Additional time-consuming work may be involved if the connections at the ammeter have to be shifted to a higher- or lower-range terminal.

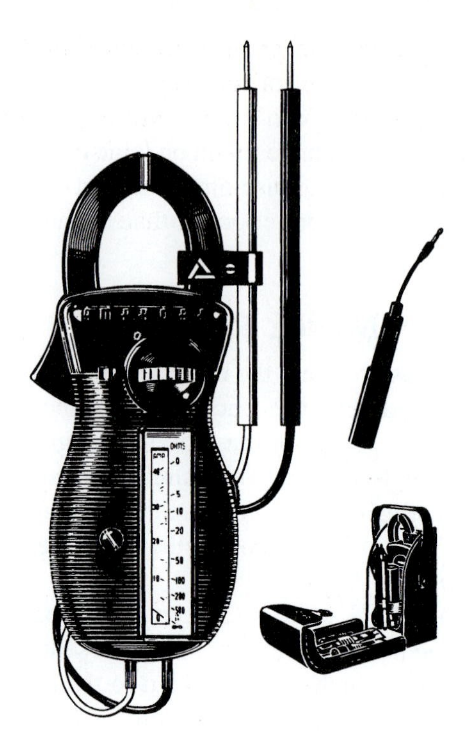

Fig. 5-57 *Clamp-on volt-ampere-ohmmeter with rotary scale.* (Amprobe)

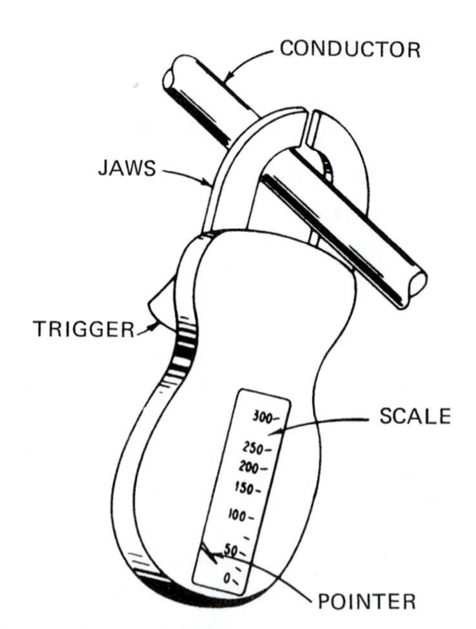

Fig. 5-58 *The clamp-on volt-ampere-ohmmeter with parts labeled.*

SPLIT-CORE AC VOLT-AMMETER

These disadvantages are practically eliminated by use of the split-core AC volt-ammeter. See Fig. 5-57. This instrument combines an AC voltmeter and AC split-core ammeter into a single pocket-size unit with a convenient range switch to select any of tile–multiple voltage ranges or current ranges. See Fig. 5-58. With the split-core ammeter, the line to be tested does not have to be disconnected from its power source.

This type of ohmmeter uses the transformer principle to connect the instrument into the line. Since any conductor carrying alternating current will set up a changing magnetic field around itself, that conductor can be used as the primary winding of the transformer. The split-core ammeter carries the remaining parts of the transformer, which are the laminated steel core and the secondary coil. To get transformer action, the line to be tested is encircled with the split-type core by simply pressing the trigger button. See Fig. 5-59. Aside from measuring terminal voltages and load currents, the split-core ammeter-voltmeter can be used to track down electrical difficulties in electric-motor repair.

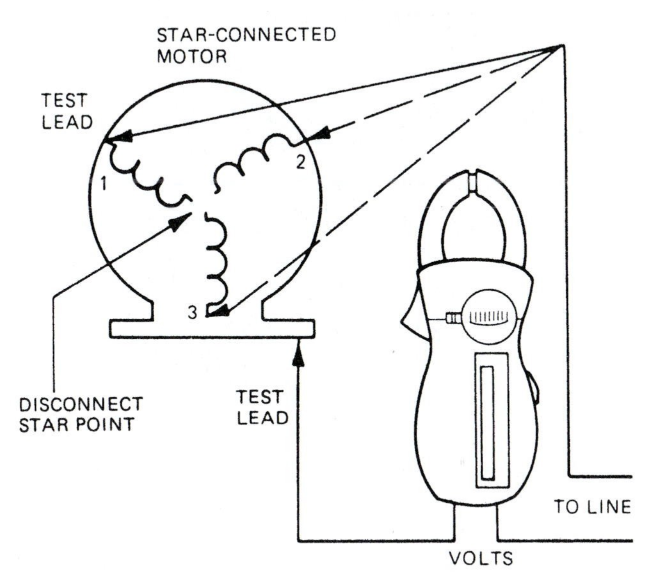

Fig. 5-59 *Find the location of a grounded phase of a motor.* (Amprobe)

Testing for Grounds

To determine whether a winding is grounded or has a very low value of insulation resistance, connect the unit and test leads as shown in Fig. 5-60. Assuming the available line voltage is approximately 120 V, use the unit's lowest voltage range. If the winding is grounded to the frame, the test will indicate full-line voltage. A high-resistance ground is simply a case of low-insulation resistance. The indicated reading for a high-resistance ground will be a little less than line voltage. A winding that is not grounded will he evidenced by a small or negligible reading. This is mainly due to the capacitive effect between the winding and the steel lamination.

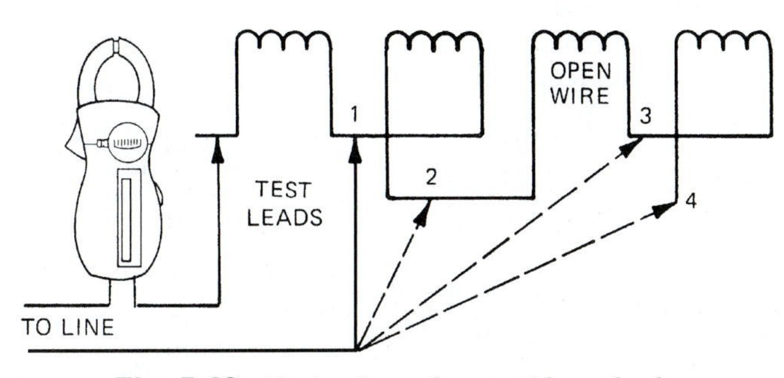

Fig. 5-60 *Testing for and open with test leads.*

To locate the grounded portion of the winding, disconnect the necessary connection jumpers and test. Grounded sections will be detected by a full-line voltage indication.

Testing for Opens

To determine whether a winding is open, connect test leads as shown in Figs. 5-61 and 5-62. If the winding is open, there will be no voltage indication. If the circuit is not open, the voltmeter indication will read full-line voltage.

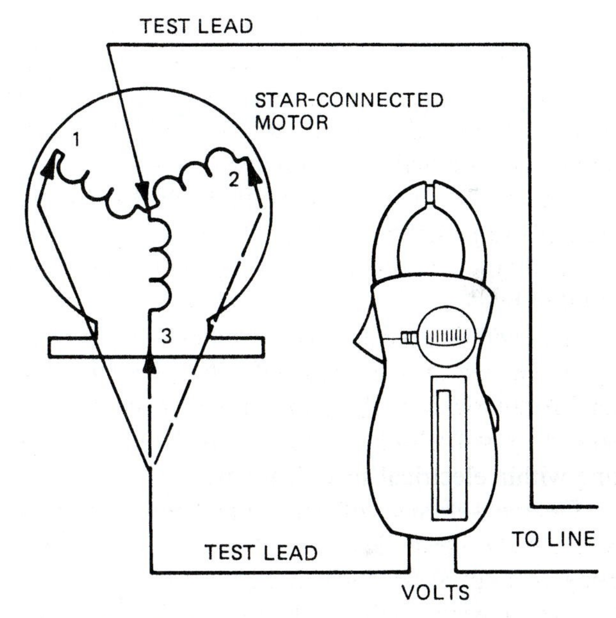

Fig. 5-61 *Isolating an open phase.* (Amprobe)

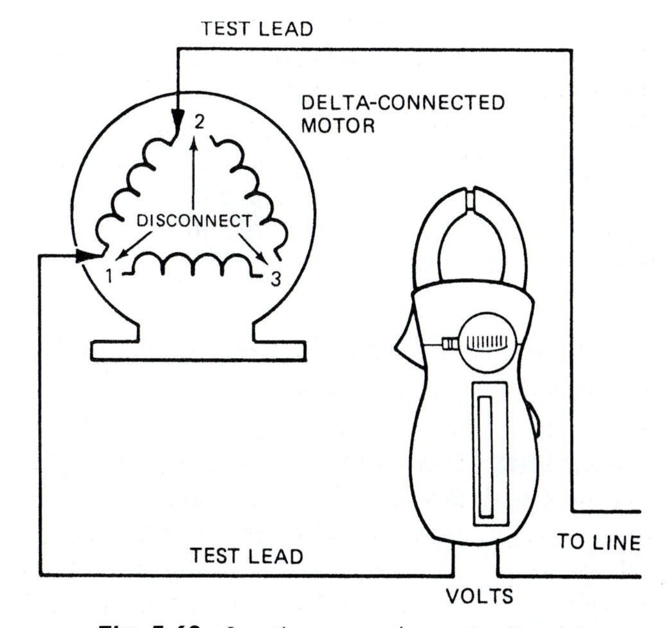

Fig. 5-62 *Locating an open in a motor.* (Amprobe)

Checking for Shorts

Shorted turns in the winding of a motor behave like a shorted secondary of a transformer. A motor with a shorted winding will draw excessive current while running at no load. Measurement of the current can be made without disconnecting lines. This means you engage one of the lines with the split-core transformer of the tester. If the ampere reading is much higher than the full-load ampere rating on the nameplate, the motor is probably shorted.

In a two- or three-phase motor, a partially shorted winding produces a higher current reading in the shorted phase. This becomes evident when the current in each phase is measured.

Testing Squirrel-Cage Rotors

In some cases, loss in output torque at rated speed in an induction motor may be due to opens in the squirrel-cage rotor. To test the rotor and determine which rotor bars are loose or open, place the rotor in a growler as shown in Fig. 5-63. Set the switch to the highest current range. Switch on the growler and then set the test unit to the appropriate current range. Rotate the rotor in the growler and take note of the current indication whenever the growler is energized. The bars and end rings in the rotor behave similarly to a shorted secondary of a transformer. The growler winding acts as the primary. A good rotor will produce approximately the same current indications for all positions of the rotor. A defective rotor will exhibit a drop in the current reading when the open bars move into the growler field.

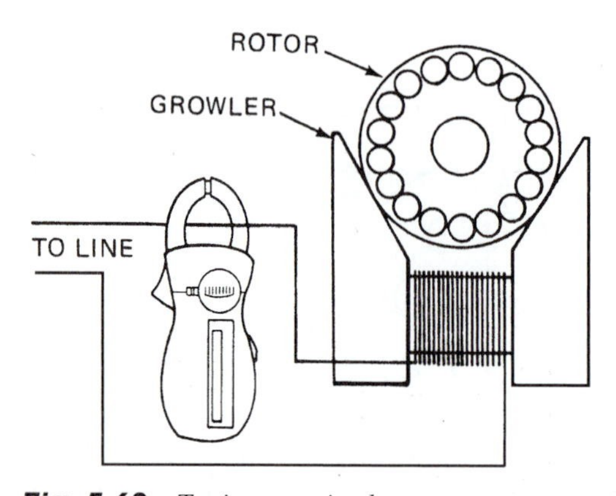

Fig. 5-63 *Testing a squirrel-cage rotor.* (Amprobe)

Testing the Centrifugal Switch in a Split-Phase Motor

A faulty centrifugal switch may not disconnect the start winding at the proper time. To determine conclusively that the start-winding remains in the circuit, place the split-core ammeter around one of the start-winding leads. Set the instrument to the highest current range. Turn on the motor switch. Select the appropriate current range. Observe if there is any current in the start-winding circuit. A current indication signifies that the centrifugal switch did not open when the motor came up to speed. See Fig. 5-64.

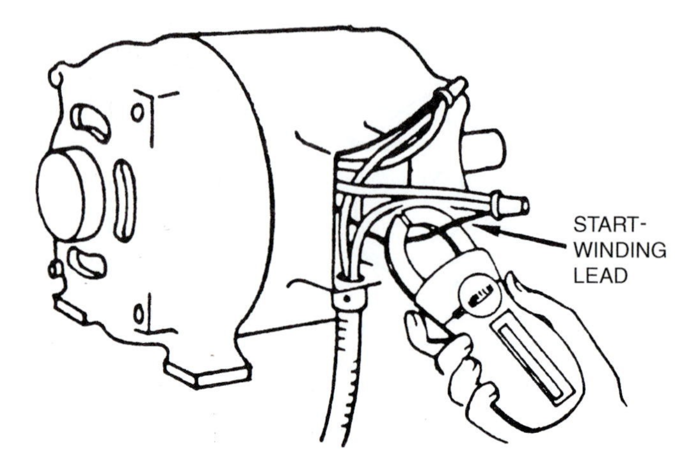

Fig. 5-64 *Testing a centrifugal switch on a motor.*

Test for Short Circuit Between Run and StartWindings

A short between run and start windings may be determined by using the ammeter and line voltage to check for continuity between the two separate circuits. Disconnect the run- and start-winding leads and connect the instrument as shown in Fig. 5-65. Set the meter on voltage. A full-line voltage reading will be obtained if the windings are shorted to one another.

Test for Capacitors

Defective capacitors are very often the cause of trouble in capacitor-type motors. Shorts, opens, grounds, and insufficient capacity in microfarads are conditions for which capacitors should be tested to determine whether they are good.

To determine a grounded capacitor, set the instrument on the proper voltage range and connect the instrument and capacitor to the line as shown in Fig. 5-66. A full-line voltage indication on the meter signifies that the capacitor is grounded to the can. A high-resistance ground will be evident by a voltage reading that is somewhat below line voltage. A negligible reading or a reading of no voltage will indicate that the capacitor is not grounded.

To measure the capacity of the capacitor, set the test unit's switch to the proper voltage range and read the line-voltage indication. Then set to the appropriate

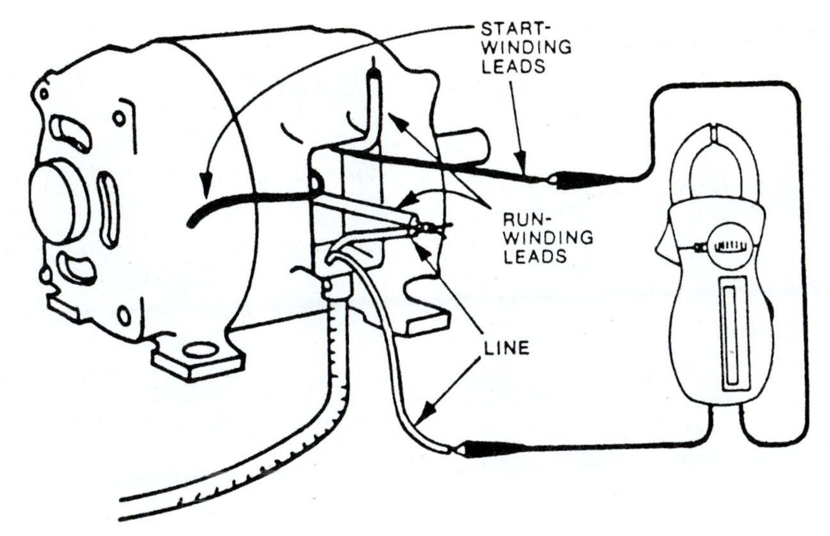

Fig. 5-65 *Test for finding a winding short circuit.* (Amprobe)

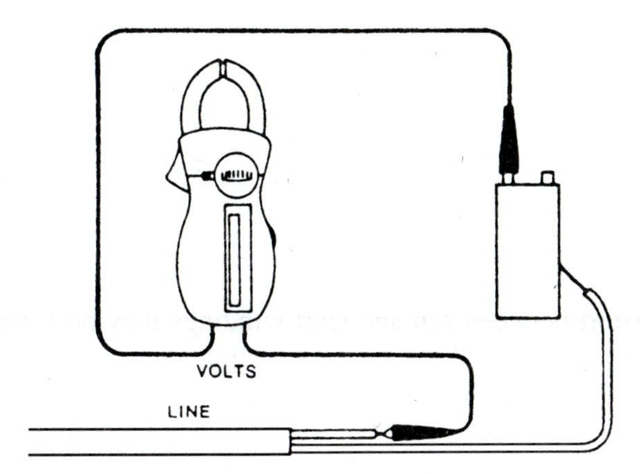

Fig. 5-66 *Test for finding a grounded capacitor.* (Amprobe)

current range and read the capacitor-current indication. During the test, keep the capacitor on the line for a very short period of time, because motor-starting electrolytic capacitors are rated for intermittent duty. See Fig. 5-67. The capacity in microfarads is then computed by substituting the voltage and current readings in the following formula, assuming that a full 60-Hz line was used:

$$\text{Microfarads} = \frac{2650 \times \text{amperes}}{\text{volts}}$$

An open capacitor will be evident if there is no current indication in the test. A shorted capacitor is easily detected. It will blow the fuse when the line switch is turned on to measure line voltage.

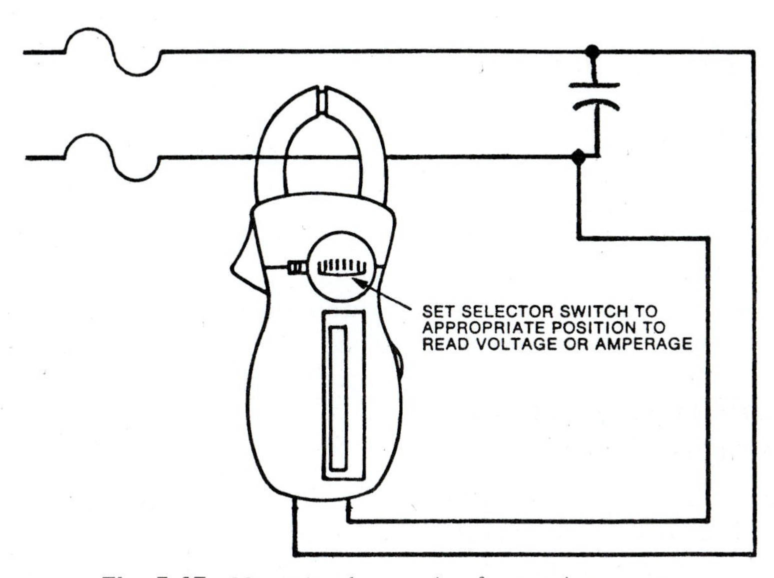

Fig. 5-67 *Measuring the capacity of a capacitor.* (Amprobe)

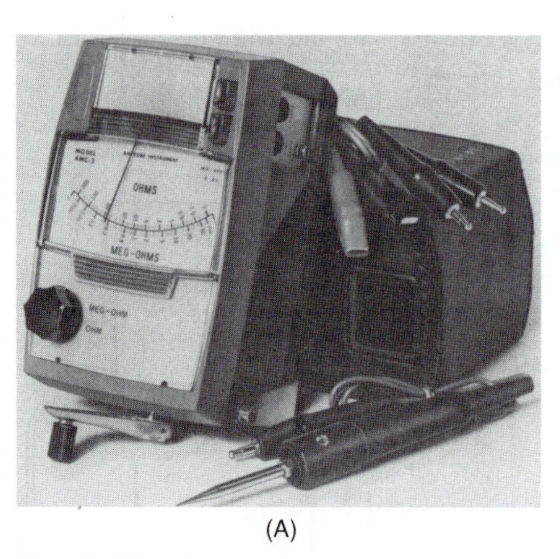

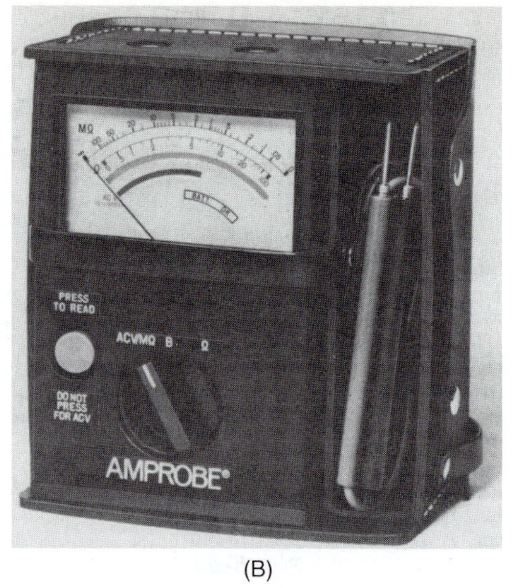

(A) (B)

Fig. 5-68 *Megohmmeters. (A) Megger or megohmmeter with a hand crank. (B) Megger or megohmmeter with a battery for power.* (Amprobe)

USING THE MEGOHMMETER FOR TROUBLESHOOTING

The megohmmeter (sometimes called a *megger*) is a device that can be used to measure millions of ohms. See Fig. 5-68. Meg means "million." The equipment usually uses high voltage to push a small amount of' current through the insulator being measured. The insulation resistance is very important in the proper operation of motors, compressors, and other electrical equipment.

Some meggers use batteries. Others use a crank that turns a small coil of wire in a magnetic field. Turning the crank handle causes the coil of wire to generate an EMF. The EMF is usually of high voltage. Thus, the megger can shock you if you touch the lead ends when the handle is cranked. There is very low current, so there may be little actual damage caused by the electrical energy through your body. Needless to say, read the instructions and follow them closely. Do not use a megger in an explosive atmosphere.

Equipment under test with the megohmmeter may build up a capacitive charge from the testing. One model has a "press to read" button. When it is released it automatically discharges the capacitive charge. With other models you must wait a few minutes for the charge to dissipate or remove the test lead from the earth (ground) jack on the tester and touch to the equipment terminal that the other test lead, line, is connected to. Never use the megger on a live circuit. Since it has a self-contained power, it is not necessary to draw current from the line.

There are two possible conducting or leakage paths in the insulation of all electrical apparatus-one through the insulating material and the other over its surface. By using the guard terminal, the surface leakage can be separated and a direct measurement made of the insulation itself. See Fig. 5-69.

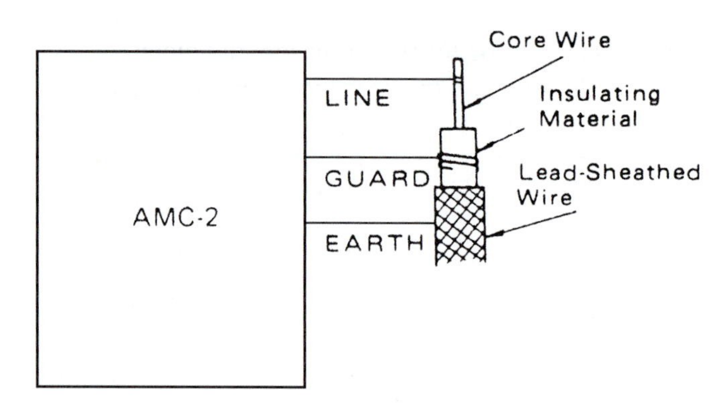

Fig. 5-69 *Hand-cranked model used to test insulation of a cable.*

INSULATION-RESISTANCE TESTING

The primary purpose of insulation is to keep electricity flowing in the desired path. The perfect insulation would have infinite resistance, which would prevent the flow of any current through the insulation to ground. However, there is no perfect insulation material. Thus, there is always some current flow. Good insulation is one that has and keeps a high-resistance value to minimize the current flow.

Unless there is an accidental damage, insulation failure is generally gradual, rather than sudden. This is because failure is generally the result of repeated heating and cooling, the related expansion and contraction, and dirt, physical abrasion, vibration, moisture, and chemicals.

When insulation starts to fail, its resistance decreases. This allows more current to flow through the insulation. If the resistance continues to decrease, the condition of the insulation may reach a point where it may permit through the insulation a current flow, large enough to cause the blowing of a fuse, equipment damage, or fatal shock.

Measuring Insulation Resistance

Insulation-resistance measurements are affected by a number of factors. Temperature and the duration of the measurement are two primary ones. Humidity may also affect readings. Thus, it is a good idea to make a note as to whether the air is dry or humid at the time of the measurement. You may find that insulation resistance readings are lower on humid days and higher on dry days. Wet or flooded equipment should be dried and cleaned as much as possible before measurements are taken. Lastly, dirt and other contaminants (corrosion, chemicals, and so forth) can also affect readings. You should be certain that the contact points at which measurements are to be taken are reasonably clean.

The duration of the resistance measurement also affects the reading. If the insulation is good, the reading will continually increase as long as the megohmmeter is connected to the insulation. The most common megger measurement is taken at the end of a 60-s interval. This time period generally gives a satisfactory measurement of the insulation resistance.

A second test involves taking a reading after 30 s and 60 ss. The 60-s reading divided by the 30-s reading is known as the dielectric absorption ratio. Comparing periodic ratios may prove more useful than comparing 1 minmin readings.

Generally speaking, a ratio of 1.25 is the bottom limit for borderline insulation. An extension of this test involves readings taken after 60 s and 10 min. The ratio of the 10-min reading to the 60-s reading is referred to as the polarization index. The resistance measurement taken at the end of 10 min should be considerably higher than that taken at 60 s. The measured insulation resistance of a dry winding in good condition should reach a relatively steady value in 10 min. If the winding is wet or dirty, the steady value will usually be reached in 1 or 2 min. The index is helpful in evaluating the winding dryness and fitness for over-potential testing.

As a guide, the recommended minimum value of the polarization index for AC and DC rotating machines is 1.5 for 221° F [105°C] (class A) insulation systems and 2.0 for 266°F [130°C] (class B) insulation systems.

Power Tools And Small Appliances

For double-insulated power tools, the megohmmeter lead shown connected to the housing would be connected to some metal part of the tool (such as the chuck or blade). See Fig. 5-70. The switch of the power tool must be in the "on" position.

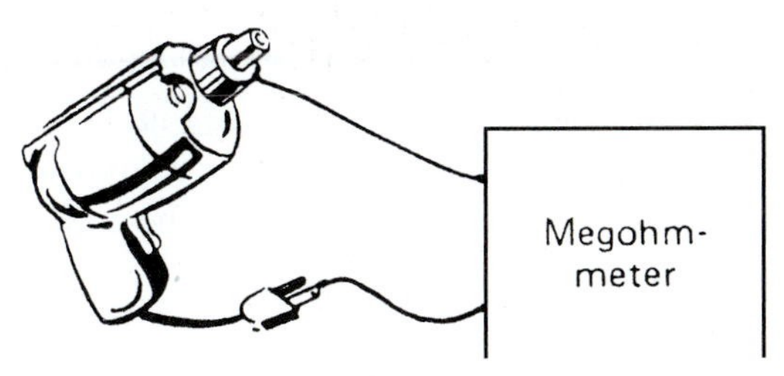

Fig. 5-70 *Using a megohmmeter to check insulation of a small hand drill.*

Motors For testing (AC), disconnect the motor from the line by disconnecting the wires at the motor terminals or by opening the main switch. If the main switch is used and the motor also has a starter, then the starter must be held in the on position. In the latter case, the measured resistance will include the resistances of the motor, wire, and all other components between the motor and the main switch. If a weakness is indicated, the motor and other components should be checked individually.

If the motor is disconnected at the motor terminals, connect one megohmmeter lead to the grounded motor housing. Connect the other lead to one of the motor leads. See Fig. 5-71. For testing (DC), disconnect the motor from the line. To test the brush rigging, field

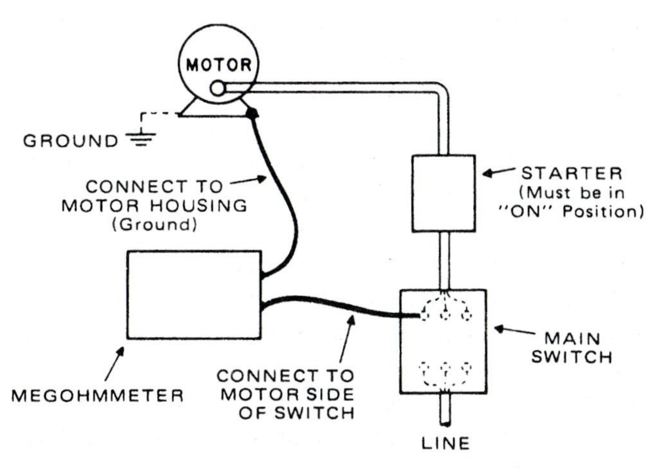

Fig. 5-71 *Using a megohmmeter to check the insulation of the windings of a motor.*

coils, and armature, connect one megohmmeter lead to the brush on the commutator. If the resistance measurement indicates a weakness, raise the brushes off the commutator and separately test the armature, field coils, and brush rigging. Do this by connecting one megohmmeter lead to each of them individually, leaving the other lead connected to the grounded motor housing.

Cables Disconnect the cable from the line. Also disconnect the opposite end to avoid errors due to leakage from other equipment. Check each conductor to ground and/or lead sheath by connecting one megohmmeter lead to each of the conductors in turn. Check insulation resistance between conductors by connecting megohmmeter leads to conductors in pairs. See Fig. 5-72.

To test a relay, connect one megger lead to the relay contact. The other megger lead goes to the coil. Then it goes to the core.

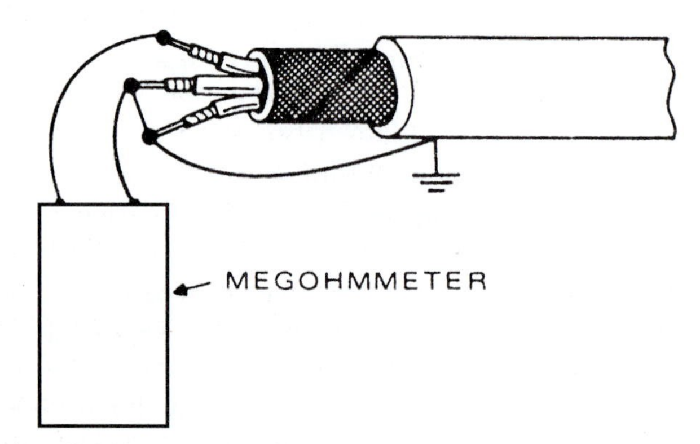

Fig. 5-72 *Using a megohmmeter to check the insulation qualities of wires between conductors.*

AC Motor Control Wound-rotor motors and AC-commutator motors have only a limited application. The squirrel-cage induction motor is the most widely used motor. The use of high voltages (2400 V and higher) introduces requirements that are additional to those needed for 600-V equipment. However, the basic principles are unchanged.

The motor, machine, and motor controller are interrelated and need to be considered as a package when choosing a specific device for a particular application. In general, three basic factors are considered when selecting a controller for a motor:

- The electrical service
- The motor
- The operating characteristics of the controller

HERMETIC COMPRESSOR SYSTEMS

Table 5-5 may be used as a guide to determine the extent to which a system may be contaminated by moisture.

Circuit Breakers and Switches

Circuit breakers and switches to be tested should be disconnected from the line. To test each terminal around, connect one megger lead to the frame or ground. Connect the other megger lead to each terminal of the circuit breaker or switch, one after the other. To test between terminals, connect the megger leads to pairs of terminals.

Coils and Relays

Disconnect from the line relays and coils to be tested. To test the coil, connect one megger lead to one of the coil leads. The other megger lead goes to ground. Then connect the megger between one coil lead and the core.

To test a relay, connect one megger lead to the relay contact. The other megger lead goes to the coil. Then it goes to the core.

AC MOTOR CONTROL

In general, five basic factors were considered when the controller for the motor was selected. The electrical service, the motor, the operating characteristics of the controller environment, and the *National Electrical*

Table 5-5 *Moisture in Hermetic Compressor Systems*

Megger Reading	Compressor Condition	Suggest Preventive Maintenance
100 megohms–infinity.	Good	None necessary
50–100 megohms.	Moisture present	Change drier
20–50 megohms.	Severe moisture and possible contaminated oil	Change numerous driers. Change oil. Acid present.
0–20 megohms.	Severe contamination	Dump oil and entire refrigeration charge. Evacuate system. Install liquid and suction line driers. Recharge system with new oil and refrigerant.

Code (NEC) have to be included in the selection and maintenance of the motor.

Motor Controller

A motor controller will perform some or all of' the following functions: starting, stopping, overload protection, over-current protection, reversing, changing speed, jogging, plugging, sequence control, and pilot-light indication. The controller can also provide the control for auxiliary equipment such as brakes, clutches, solenoids, heaters, and signals. A motor controller may be used to control a single motor or a group of motors.

The terms *starter* and *controller* mean practically the same thing. Strictly speaking, a starter is the simplest form of controller. It is capable of starting and stopping the motor and providing it with overload protection.

AC SQUIRREL-CAGE MOTOR

The workhorse of industry is the AC squirrel-cage motor. The vast majority of the thousands of motors used today in general applications is of the squirrel cage type. Squirrel-cage motors are simple in construction and operation.

The squirrel-cage motor gets its name because of its rotor construction. The rotor resembles a squirrel cage and has no wire winding. A number of terms need to be explained to understand motor control. *Full load current* (FLC) is the current required to produce full-load torque at rated speed. *Locked rotor current* (LRC) is the inrush current when the motor is connected directly to the line. The LRC can be from four to ten times the motor's full-load current. The vast majority of motors have an LRC of about six times FLC. Therefore, this figure is generally used. The "six-times" value is expressed as 600 percent of FLC.

Motor speed depends on the number of poles in the motor's winding. On 60 Hz, a two-pole motor runs about 3450 rpm, a four-pole motor runs at 1725 rpm, and a six-pole motor runs at 1150 rpm. Motor nameplates are usually marked with actual full-load speeds. However, frequently motors are referred to by their synchronous speed. Synchronous speeds are 3600 for the 3450-rpm, 1800 for the 1725-rpm, and 1200 for the 1150-rpm motor.

Torque is the "turning" or "twisting" force of the motor. It is usually measured in foot pounds. Except when the motor is accelerating to speed, the torque is related to the motor horsepower by the formula:

$$\text{Torque (in pound-feet)} = \frac{\text{hp} \times 5252}{\text{rpm}}$$

The torque of a 25-hp motor running at 1725 rpm would be computed as follows:

$$\text{Torque} = \frac{25 \times 5252}{1725} \text{ or, approximately 76 lb/ft}$$

If 90 lb/ft were required to drive a particular load, this motor would be overloaded and would draw a current in excess of full-load current.

Temperature rise is the difference between the winding temperature of the motor when running and the ambient temperature. Current passing through the motor windings results in an increase in motor temperature. The temperature rise produced at full load is not harmful, provided the ambient temperature does not exceed 104°F [40°C].

Higher temperature, caused by increased current or higher ambient temperatures, has a deteriorating effect on motor insulation and lubrication. One rule states that for each increase of 10°F [5.5°C] above the rated temperature, motor life is cut by one-half.

Duty rating is the rating of the motor for continuous or intermittent operation. Most motors have a continuous duty rating, permitting indefinite operation at rated load. Intermittent duty ratings are based on a fixed operating time (such as 5, 15, 30, or 60 mins) after which the motor must be allowed to cool.

Motor service factor is given by the motor's manufacturer. It means that the motor can be allowed to develop more than its rated or nameplate hp without causing undue deterioration of the insulation. The service factor is a margin of safety. If, for example, a 10-hp motor has a service factor of 1.15, the motor can be allowed to develop 11.5 hp. The service factor depends on the motor design.

Jogging describes the repeated starting and stopping of a motor at frequent intervals for a short period of time. A motor would be jogged when a piece of driven equipment has to be positioned fairly closely. Thus, jogging might occur when positioning the table of a horizontal boring mill during setup or aligning any motor-driven device. If jogging is to occur more frequently than five times per minute, NEMA standards require that the starter be derated. For instance, a size 1 starter has a normal duty rating of 7 $^1/_2$ hp at 230 V, polyphase. On jogging applications, this same starter has a maximum rating of 3 hp.

Plugging occurs when a motor running in one direction is momentarily reconnected to reverse the direction. It will be brought to rest very rapidly. If a motor is plugged more than five times per minute, derating of the controller is necessary. The contacts of the controller overheat. Plugging may be used only if the driven machine and its load will not be damaged by the reversal of the motor torque.

Enclosures

NEMA and other organizations have established standards of enclosure construction for control equipment in general; equipment would be enclosed for one or more of the following reasons:

1. To prevent accidental contact with live parts.
2. To protect the control from harmful environmental conditions.
3. To prevent explosion or fires that might result from the electrical are caused by the controller.

Code

The NEC deals with the installation of electrical equipment. It is primarily concerned with safety. It is adopted on a local basis, sometimes incorporating minor changes. NEC rules and provisions are enforced by governmental bodies exercising legal jurisdiction over electrical installations.

The code is used by insurance inspectors. Minimum safety standards are thus assured if the NEC is followed.

Protection of the Motor

Motors can be damaged, or their effective life reduced, when subjected to a continuous current only slightly higher than their full-load current rating times the service factor.

Damage to insulation and windings of the motor can also be sustained on extremely high currents of short duration. These occur when there are grounds and shorts.

All currents in excess of full-load current can be classified as overcurrents. In general, a distinction is made based on the magnitude of the overcurrent and equipment to be protected. Overcurrent up to locked rotor current is usually the result of a mechanical overload on the motor. The NEC covers this in one of its Articles.

Overcurrents due to short circuits or grounds are much higher than locked rotor currents. Equipment used to protect against damage due to this type of overcurrent must protect not only the motor, but also the branch circuit conductors and the motor controller.

The function of the overcurrent-protective device is to protect the motor branch circuit conductors, control apparatus, and motor from short circuits or grounds. The protective devices commonly used to sense and clear overcurrents are thermal magnetic circuit breakers and fuses. The short-circuit device shall be capable of carrying the starting current of the motor, but the device setting shall not exceed 250 percent of full-load current depending upon the code letter of the motor. Where the value is not sufficient to carry the starting current, it may be increased. However, it shall not exceed 400 percent of the motor full-load current. The NEC (with a few exceptions) requires a means to disconnect the motor and controller from the line, in addition to an overcurrent-protective device to clear short-circuit faults.

CONTACTORS, STARTERS, AND RELAYS

If the condensing unit has a motor larger than $1^1/_2$ hp, it will have a starter or contactor. They are usually furnished with the unit.

Relays are a necessary part of many control and pilot-light circuits. They are similar in design to contactors, but are generally lighter in construction, so they carry smaller currents.

Magnetic contactors are normally used for starting polyphase motors, either squirrel cage or single phase. Contactors may be connected at any convenient point in the main circuit between the fuses and the motor. Small control wires may be run between the contactor and the point of control.

Protection of the motor against prolonged overload is accomplished by time-limit overload relays that are operative during the starting period and running period. Relay action is delayed long enough to take care of heavy starting currents and momentary overloads without tripping.

Motor-Overload Protector

Motors for commercial condensing units are normally protected by a bimetallic switch operating on the thermo, or heating, principle. This is a built-in motor-overload protector. It limits the motor-winding temperature to a safe value. In its simplest form, the switch or motor protector consists essentially of a bimetal switch mechanism that is permanently mounted and connected in series with the motor circuit. See Fig. 5-73.

When the motor becomes overloaded or stalled, excessive heat is generated in the motor winding due to the heavy current produced by this condition. The protector located inside the motor is controlled by the motor current passing through it and the motor temperature. The bimetal element is calibrated to open the motor circuit when the temperature, as a result of an excessive current, rises above a predetermined value. When the temperature decreases, the

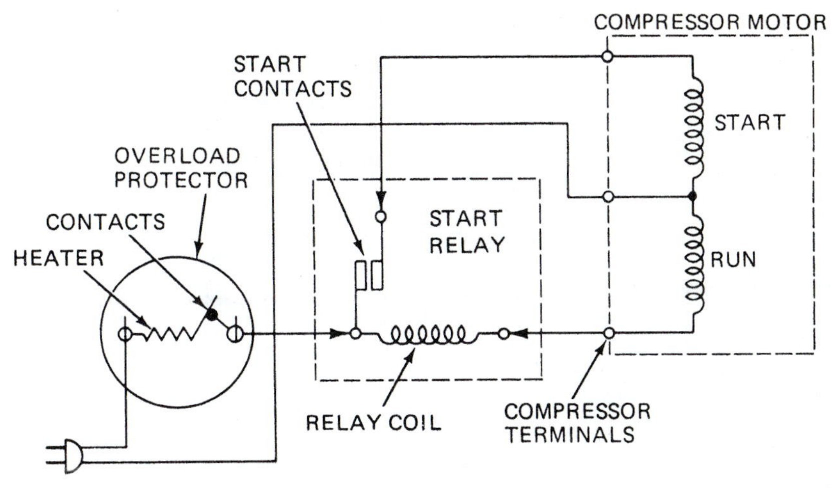

Fig. 5-73 *Circuit for a domestic refrigerator.*

protector automatically resets and restores the motor circuit.

This device reduces service calls due to temporary overloads. The device stops the motor until it cools off and then allows it to start again when needed.

Servicing of motors with built-in overload devices must be handled with care. The compressor may be idle due to an overload. Hence, it will start as soon as the motor cools off. This could result in a serious mishap to the operator or repairperson. To avoid such difficulties, open the electrical circuit by pulling the line plug or switch prior to any repair or servicing operation.

Motor-Winding Relays

A motor-winding relay is usually incorporated in single-phase motor-compressor units. This relay is an electromagnetic device for making and breaking the electrical circuit to the start winding. A set of normally closed contacts is in series with the motor start winding. See Fig. 5-74.

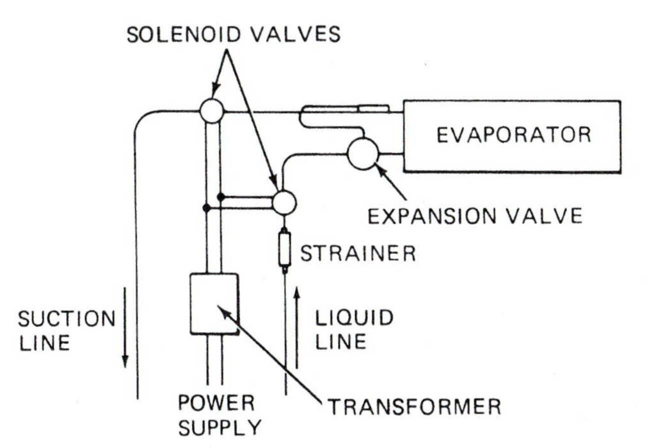

Fig. 5-74 *Solenoid valve connected in the suction and liquid evaporator lines.*

The electromagnetic coil is in series with the auxiliary winding of the motor. When the control contacts close, the motor start and run windings are energized. A fraction of a second later the motor comes up to speed and sufficient voltage is induced in the auxiliary winding to cause current to flow through the relay coil. The magnetic force is sufficient to attract the spring-loaded armature, which mechanically opens the relay starting contacts. With the starting contacts open, the start winding is out of the circuit. The motor continues to run on only the run winding. When the control contacts open, power to the motor is interrupted. This allows the relay armature to close the starting contacts. The motor is now ready to start a new cycle when the control contacts again close.

SOLENOID VALVES

Solenoid valves are used on multiple installations. They are electrically operated. A solenoid valve, when connected as in Fig. 5-75, remains open when current is supplied to it. It closes when the current is turned off. In general, solenoid valves are used to control the liquid refrigerant flow into the expansion valve, or the refrigerant gas flow from the evaporator when it or the fixture it is controlling reaches the desired temperature. The most common application of the solenoid valve is in the liquid line and operates with a thermostat. With this hookup, the thermostat may be set for the desired temperature in the fixture. When this temperature is reached, the thermostat will open the electrical circuit and shut off the current to the valve. The solenoid valve then closes and shuts off the refrigerant supply to the expansion valve. The condensing unit operation should be controlled by the low-pressure switch. In other applications, where the evaporator is to be in operation

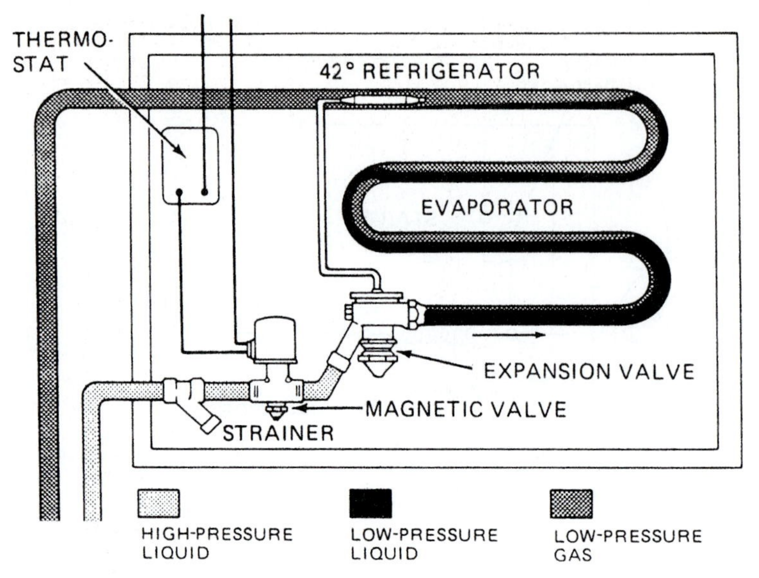

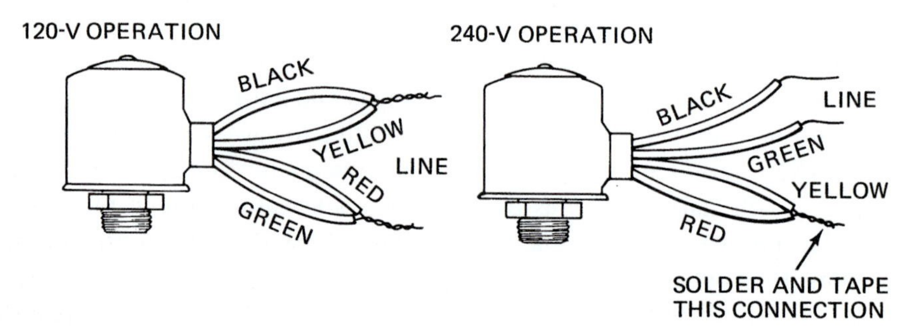

Fig. 5-75 *Solenoid valve leads identification and refrigeration installation.* (General Controls)

for only a few hours each day, a manually operated snap switch may be used to open and close the solenoid valve.

REFRIGERATION VALVE

The solenoid valve in Fig. 5-76 is operated with a normally closed status. A direct-acting metal ball and seat assure tight closing, The two-wire, class-W, coil is supplied standard for long life on low-temperature service or sweating conditions. Current failure or interruption will cause the valve to fail-safe in the closed position. The solenoid cover can be rotated 360° for easy installation. Explosion-proof models are available for use in hazardous areas.

Application

This solenoid valve is usable with all refrigerants except ammonia. Also it can be used for air, oil, water, detergents, butane or propane gas, and other noncorrosive liquids or gases.

A variety of temperature control installations can be accomplished with these valves. Such installations include bypass, defrosting, suction line, hot-gas service, humidity control, alcohols, unloading, reverse cycle, chilled water, cooling tower, brine, and liquid line stop installations and ice makers.

Operation

The valves are held in the normally closed position by the weight of the plunger assembly and the fluid pressure on top of the valve ball. The valve is opened by energizing the coil and magnetically lifting the plunger and allowing full flow by the valve ball. Deenergizing the coil permits the plunger and valve ball to return to the closed position.

The piloted piston solenoid valve is somewhat different. See Fig. 5-77. It too, is normally closed. It can be used on all refrigerants except R-717.

When the solenoid is energized the plunger rises, lifting the pilot valve to allow pressure to bleed from

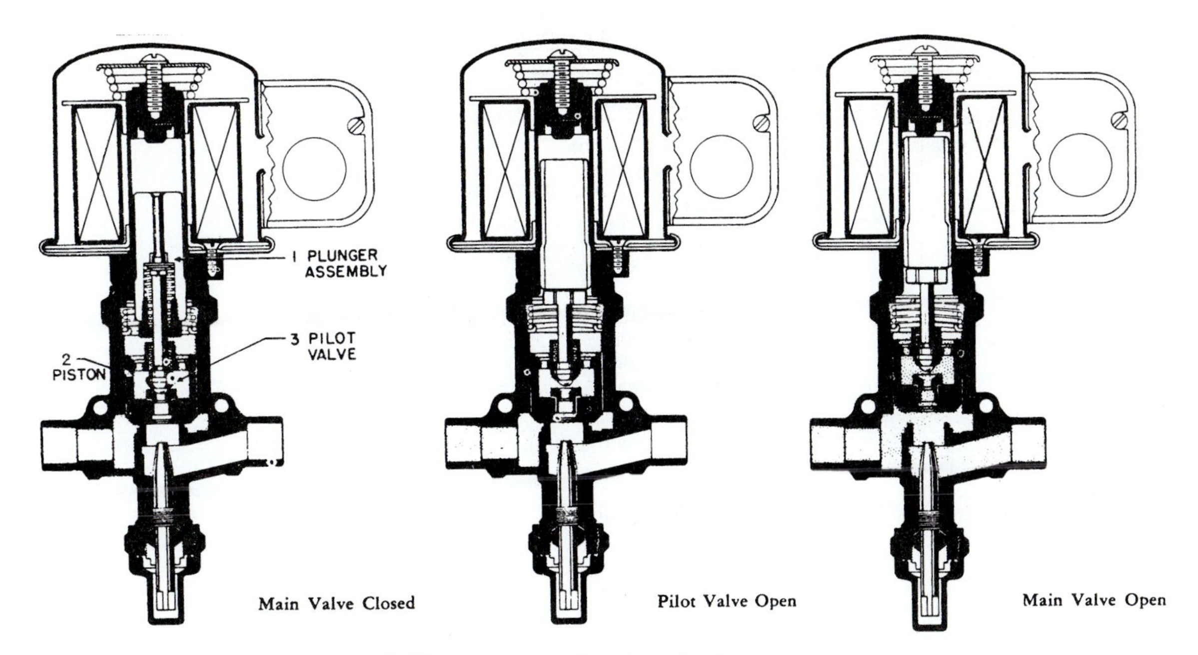

Fig. 5-76 *Operation of a solenoid valve.* (General Controls)

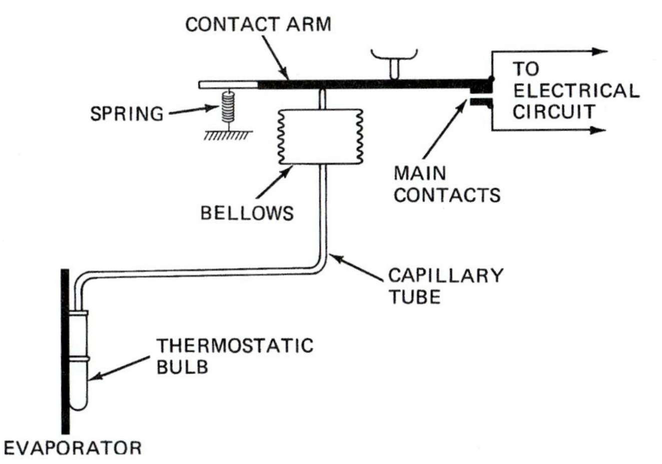

Fig. 5-77 *Thermostatic control switch using bellows.*

above the piston. The pilot valve continues its rise and the piston follows due to the lower pressure affected above the piston. The piston is then held in a fully open position by the plunger and pilot stem to allow full flow through the valve with minimum pressure drop. When the solenoid is deenergized, the plunger drops and allows the pilot valve to seat. The pressure above the piston balances with that on the underside. The combined weight of the piston and plunger assembly causes the valve to return to the closed position.

Installation

Install in a horizontal line with the solenoid upright. With the threaded type do not use the solenoid cover to turn the valve. Provide enough clearance for solenoid removal. On the solder type, remove the solenoid coil before installing the valve. Do not remove plunger tube. Wrap the valve with wet asbestos or a wet cloth while making up fittings. Improper handling may distort the cylinder and cause the piston to bind.

Table 5-6 lists service suggestions for the solenoid valve.

Temperature Controls

In modern condensing units, low-pressure control switches are being largely superseded by thermostatic control switches. A thermostatic control consists of three main parts:

- A bulb
- A capillary tube
- A power element (switch)

The bulb is attached to the evaporator in a manner that assures contact with the evaporator. It may contain a volatile liquid, such as a refrigerant. The bulb is connected to the power element by means of a small capillary tube. See Fig. 5-78.

Operation of the thermostatic control switch is such that, as the evaporator temperature increases, the bulb temperature also increases. This raises the pressure of the thermostatic liquid vapor. This, in turn, causes the bellows to expand and actuate an electrical contact. The contact closes the motor circuit, and the motor and compressor start operating. As the evaporator

Table 5-6 *Service Suggestions*

Trouble	Possible Cause	Remedy
Valve fails to open	Timers, limit controls or other devices holding circuit open	Check circuit for limit control operation, blown fuses, short circuit, and loose wiring
	Solenoid coil shorted, burned out, or wrong voltage	Replace with solenoid coil of correct voltage
	Dirt, pipe compound, or other foreign matter restricting operation of piston or pilot valve	Disassemble and clean internal parts with carbon tetrachloride, install strainer ahead of valve
Valve will not close	Manual-opening device holding valve open	Turn manual opening stem counter-clockwise until stem backseats
	Dirt, pipe compound, or other foreign matter restricting operation of piston or pilot valve	Disassemble and clean internal parts with carbon tetrachloride, install strainer ahead of valve
	Damaged plunger tube preventing plunger operation	Replace plunger tube
Low leakage	Foreign matter in valve interior or damaged seat or seat disc	Clean valve interior with carbon tetrachloride, check condition of seat and seat disc, and replace if necessary

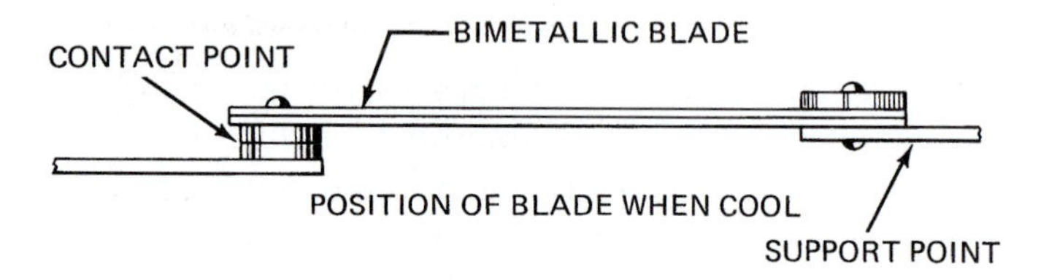

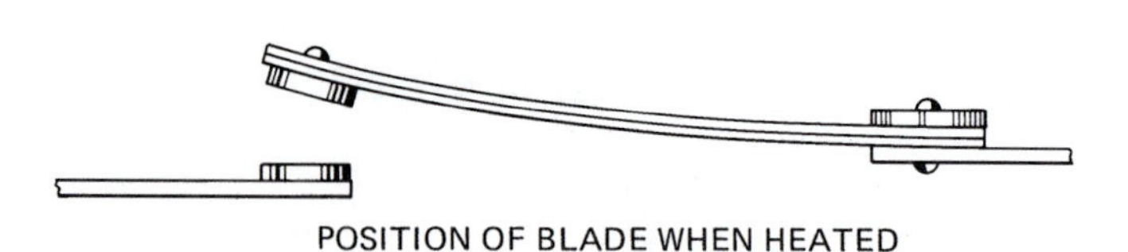

Fig. 5-78 *Working principles of a simple bimetallic thermostat.*

temperature decreases, the bulb becomes colder and the pressure decreases to the point where the bellows contract sufficiently to open the electrical contacts controlling the motor circuit. In this manner, the condensing unit is entirely automatic. Thus, it is able to produce exactly the amount of refrigeration to meet any normal operating condition.

An automatic temperature control system is generally operated by making and breaking an electric circuit or by opening and closing a compressed-air line. When using the electric thermostat, the temperature is regulated by controlling the operation of an electric motor or valve. When using the compressed-air thermostat, temperature regulation is obtained by actuating a compressed-air operated motor or drive. Electrically-operated temperature-control systems are used generally by manufacturers for practically all installations. However, compressed-air temperature-control systems have applications in extremely large central and multiple installations in close temperature work. This is where a large amount of power is required for small control devices.

BIMETALLIC THERMOSTATS

The bimetallic thermostat operates as a function of expansion or contraction of metals due to temperature changes. Bimetallic thermostats are designed for the control of heating and cooling in air-conditioning units, refrigeration storage rooms, greenhouses, fan coils, blast coils, and similar units.

The working principle of such a thermostat is shown in Fig. 5-79. As noted, two metals, each having a different coefficient of expansion, are welded together to form a bimetallic unit or blade. With the blade securely anchored at one end, a circuit is formed and the two contact points are closed to the passage of an electric current. Because an electric current provides heat in its passage through the bimetallic blade, the metals in the blade begin to expand, but at a different rate. The metals in the blade are so arranged that the one with a greater coefficient of expansion is placed at the bottom of the unit. After a certain time, the operating temperature is reached and the contact points become separated, thus disconnecting the appliance from its power source.

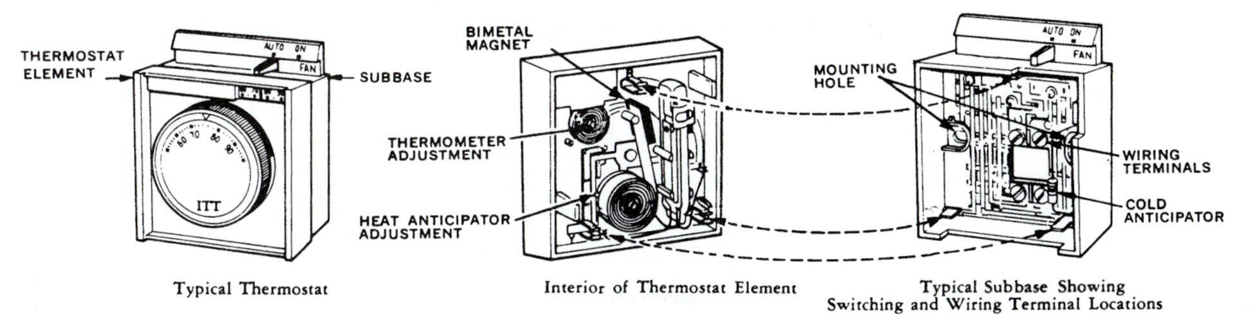

Fig. 5-79 *Modern thermostat for both cooling and heating using a metallic strip's expansion ability to move a magnet close to a magnetic switch. (A) Typical thermostat. (B) Interior of thermostat element. (C) Typical subbase showing switching and wiring terminal locations.* (General Controls)

After a short period, the contact blade will again become sufficiently cooled to cause the contact points to join, thus reestablishing the circuit and permitting the current again to actuate the circuit leading to the appliance. The foregoing cycle is repeated over and over again. In this way, the bimetallic thermostat prevents the temperature from rising too high or dropping too low.

Thermostat Construction and Wiring

Some thermostats are designed for use on both heating and cooling equipment. The thermostat shown in Fig. 5-80 is such a device. The basic thermostat element has a permanently sealed, magnetic *single-pole double-throw* (SPDT) switch. The thermostat element plugs into the subbase and contains the heat anticipation, the magnetic switching, and a room temperature thermometer. The subbase unit contains fixed cold anticipation and circuitry. This thermostat is for use with 24-V equipment. In this case, the thermostatic element (bimetal) does not make direct contact with the electrical circuit. The expansion of the bimetal causes a magnet to move. This, in turn, causes a switch to close or open. Figure 5-80 illustrates the fact that the bimetal is not in the electrical circuit at all.

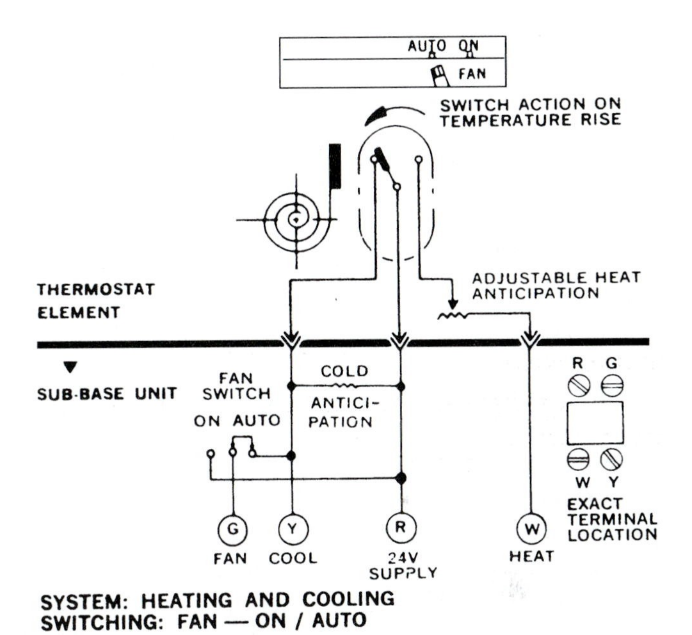

Fig. 5-80 *Wiring diagram showing how the thermostat is wired and hooked into a circuit.*

DEFROST CONTROLS

Automatic defrost is common in domestic refrigeration, it is accomplished in several ways. The control method used depends on the type of refrigeration system, the size and number of condensing units, and other factors.

Defrost Timer Operation

In small and medium-size domestic refrigerators, an automatic defrost-control clock may be set for a defrost cycle once every 24 h, or as often as deemed necessary. The defrosting is usually accomplished by providing one or more electric heaters. They are energized by the action of the electric clock and provide the heating action necessary for complete defrosting.

The defrost controls, as usually employed, are essentially SPDT switching devices in which the switch arm is moved to the defrost position by an electric clock. The switch arm is returned to the normal position by a power element that is responsive to changes in temperature. As the evaporator is warmed during a defrost period, the feeler tube of the defrost control is also warmed until it reaches the defrost cutout point of approximately 45°F [7°C]. The defrost-control bellows then force the switch arm to snap from the defrost position to the normal position. See Fig. 5-81. This starts the motor compressor.

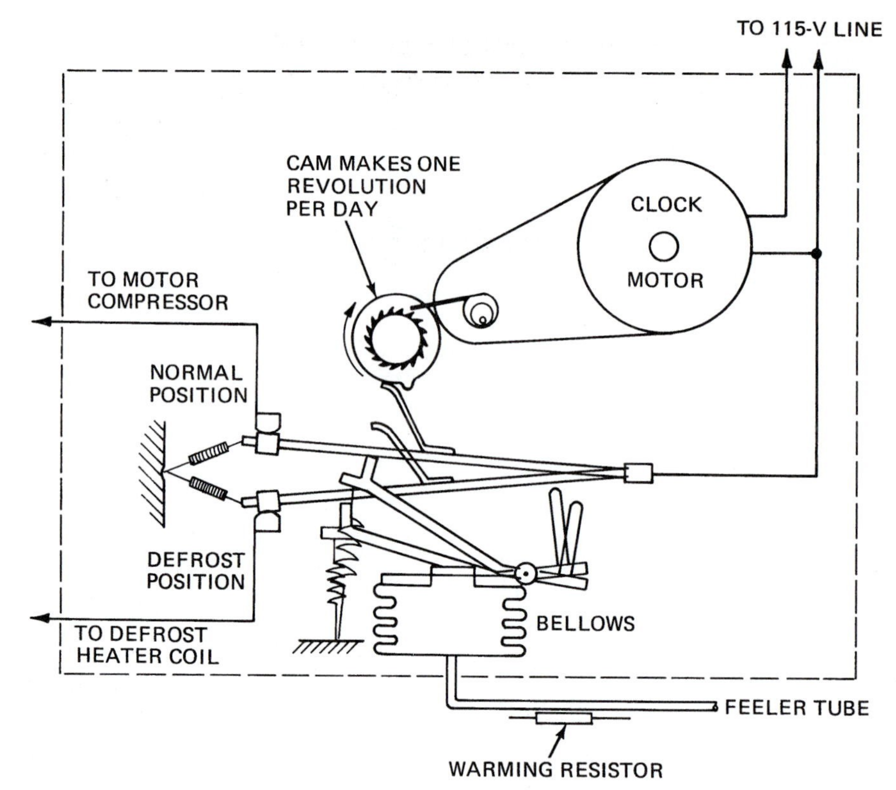

Fig. 5-81 *Twenty–four hour clock used to activate the defrost cycle.*

Another common method of automatic defrosting is the so-called defrost cycle method. In this, the defrost cycle occurs during each compressor off-cycle. In a defrost system of this type, the defrost heaters are connected across the thermostat-switch terminals. When the thermostat switch is closed, the heaters are shunted out of the circuit. When the thermostat opens, the heaters are energized, completing the circuit through the overload relay and compressor. Figure 5-83 illustrates this type of defrosting. Note that the serpentine resistor is shorted out by the temperature control. This occurs when the temperature control is closed and the compressor motor is running normally.

Hot-Gas Defrosting

In any low-temperature room, where the air is to be maintained below freezing, some adequate means for removing accumulated frost from the cooling surface should be provided. An improved hot-gas method of quick defrosting for direct-expansion low-temperature evaporators is now available. To apply this method, two or more evaporators are needed in the system. This is because the hot gas required to defrost part of the system must be provided by the heat absorbed from the other cooling surface in a given system. Defrosting a plate bank with hot gas can be accomplished automatically by installing the proper controls.

MOTOR BURNOUT CLEANUP

The following cleanup methods are simple, rapid, and economical. They represent a drastic reduction in labor requirements over the obsolete flushing methods.

Procedure for Small Tonnage Systems

In systems up to 40 tons, the refrigerant charge is relatively small. Motor burnout contaminants are not diluted to the extent that they are in large tonnage systems. As a result, there is a greater need to isolate the motor compressor from all harmful soluble and insoluble materials that might cause another burnout.

Driers should be installed in both the liquid and suction lines. See Fig. 5-82. The desiccant in the driers removes all harmful soluble chemicals that cause corrosion and attack motor-winding insulation. Suction-line filtration should be employed to prevent harmful solids above 5 μm (0.0002 in.) from returning to the compressor. Through abrasion, foreign particles such as casting dust, copper and aluminum dust,

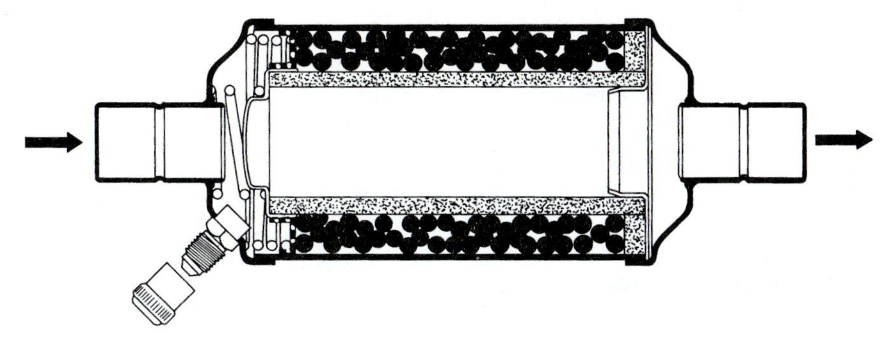

Fig. 5-82 *Low-side filter-drier.* (Virginia Chemicals)

and flux contribute to motor burnouts and compressor damage.

The type of drier used in the suction line is of great importance. Throwaway-type liquid-line driers, when used in the suction line, are usually too small. They may create a dangerously high pressure drop. This may cause overheating of the motor compressor and a repeat burnout.

Until recently, it has been necessary to use large replaceable cartridge-type driers for this purpose. Even these have a limited range. They are costly, heavy, and difficult to mount. Their filtering ability is very questionable. In addition, since the system must be opened to remove them, replacement of the liquid-line drier and reevacuation of the system are needed. These operations add to the cost.

New low-side filter-driers eliminate these difficulties. Two essential components are combined into one. Thus, the filter-drier is both a suction-line filter and a suction-line drier. It is designed for permanent installation. The blended mixture of activated alumina and molecular sieves provides an enormous capacity for adsorbing moisture and other harmful soluble contaminants. It can adsorb inorganic and organic acids and oil breakdown materials.

Evidence indicates that these soluble contaminants are most easily removed in the suction line for the following reasons:

Field experience shows success with soluble contaminants removal when properly-sized driers have been used in the suction line following hermetic motor burnouts.

Modern drying materials have a substantially higher capacity for moisture and acids at lower temperatures. Suction-line temperatures are normally from 20 to 60°F (11 to 33°C) lower than liquid-line temperatures, depending upon the application and ambient conditions.

Since oil breakdown materials dissolve readily in oil, higher concentrations of contaminants are in contact with the desiccant. This results in conditions more favorable to maximum pickup.

Liquid refrigerant does not compete with the desiccant to accept soluble contaminants. In the liquid line it does by greatly diluting soluble materials. This greatly reduces contact time and, consequently, reduces the rate of pickup.

The low-side filter drier has an access valve on the inlet side for checking pressure drop and charge adjustment.

There is no method of cleanup after a burnout that does not carry some risk. No cleanup procedure will guarantee 100 percent success. The procedures that follow have been generally successful. They are practical at the field level and economical enough to be used by the equipment owner.

1. Discharge oil refrigerant mixture in liquid phase. If water cooled, drain all water containing areas first.
2. Remove burned-out compressor, taking care not to touch oil or sludge with bare hands.
3. Blow out coils and condenser with clean, dry-liquid refrigerant.
4. Install new motor compressor.
5. Install a moisture indicator and an oversize high-side filter-drier in the liquid line.
6. Install a low-side filter-drier in the suction line as close to the compressor as possible. If the system is larger than 20 actual tons, install two low-side filter-driers in parallel.
7. Triple evacuate to 500 μm, or as low as practical, and charge.

Optional Check back in two weeks and perform an acid test on the oil. Use the acid-test kit. If the oil is acidic or discolored, chance the oil, both driers, and again evacuate. Another twoweek checkup is desirable.

This method, due to line sizing and refrigerant cost, is applicable up to 40 tons. Consideration may also be given to saving the refrigerant if the charge is above 100 lb.

Procedure for Large Tonnage Systems

In systems above 40 tons, the large refrigerant charge so dilutes the motor burnout contaminants that discarding the refrigerant is unnecessary. It cannot be justified from cost considerations. Oil breakdown materials and organic acids are more soluble in the oil than the refrigerant. They tend to concentrate in the oil. By repeated oil changes and drier changes, with the oil and drier extracting the contaminants, such systems can be cleaned up. The following procedure has been used over an extended period of time by many large contractors with successful cleanups from 40 to over 500 tons.

Due to design variations, the mechanics of carrying out the following procedures must be adapted to the system involved. The basic procedure is as follows:

1. If possible, wash out coil and condenser with clean refrigerant. In some designs, this is possible, but with others, completely impractical.
2. Reinstall the rebuilt compressor with a fresh, clean charge of oil.
3. Install the largest possible drier in the liquid phase of the system.
4. Operate 24 hours.
5. Change oil and drier or drier cores.
6. Operate 24 hours.
7. Change oil and drier or drier cores.
8. Operate 24 hours.
9. Change oil and drier or drier cores.
10. Triple evacuate to 500 μm or as low as practical, and charge.
11. Operate two weeks and check oil color. Perform an acid test on the oil. If it is neutral and the color normal, consider the job done. If the oil is acidic or discolored, repeat the above steps until neutrality is secure and the oil color is normal.

READING A SCHEMATIC

It is often difficult to read a schematic at first glance. Figure 5-83 shows the schematic for a home appliance. The voltage being used is 115 V. Follow the brown wires and see how they control the freezer light, cabinet light, and mullion heater. The brown wire on the right is spliced to an orange wire. This orange wire connects to one side of the freezer door switch, one side of the refrigerator door switch, and one side of the mullion heater. The brown wire from the left side of the schematic connects to two orange wires that attach to one side of the freezer light and one side of the cabinet light. The brown wire on the left connects to the other side of the mullion heater. There is a wire connecting the freezer light and the freezer door switch. Likewise, there is a wire connecting the cabinet light and the cabinet door switch.

Now, trace the schematic. Start at the top of the schematic at the 115 V lead. Trace the left side first. The brown wire on the left side goes down to the orange wires that connect to the freezer light and cabinet light. The brown wire also connects to the mullion heater. Now, take the brown wire leading from the right side of the 115 V plug. It is spliced to the orange wire that connects to the door switch and the mullion heater. This means that the mullion heater is on when the plug is inserted into a power source. It also means that the freezer light does not come on until the door is open and the refrigerator door switch is closed. Likewise, the cabinet light does not come on until the door is open and the door switch is closed.

Referring again to Fig. 5-81, note the way in which the defrost controls are wired. Note in this case that the brown wire on the left side of the schematic—coming from the 115 V plug—has the temperature control inserted in series with the rest of the wiring and devices. Tracing from the left to right you will find that a black wire runs from the temperature control to the door switch. An ivory wire runs from the door switch to the freezer fan. Another ivory wire runs from the freezer fan to the defrost control (point 4). If the defrost control switch is up, it completes the path from point 3 to the brown wire that leads back to the 115-V plug. Thus, if the temperature control (refrigerator thermostat) and the freezer door switch are closed and the defrost control switch is up, the circuit is complete for the freezer fan to operate.

Note that the defrost control is operated by a timer motor. The timer motor is in the circuit at all times when the temperature control switch is closed. This means the defrost control timer will operate and complete its cycle faster if the thermostat is closed. Therefore, the more the refrigerator compressor runs, the faster the defrost control advances to its predetermined point of operation.

To trace the defrost control's source of power, start at the 115-V plug. Trace from the left side through the temperature control and down the black wire to the

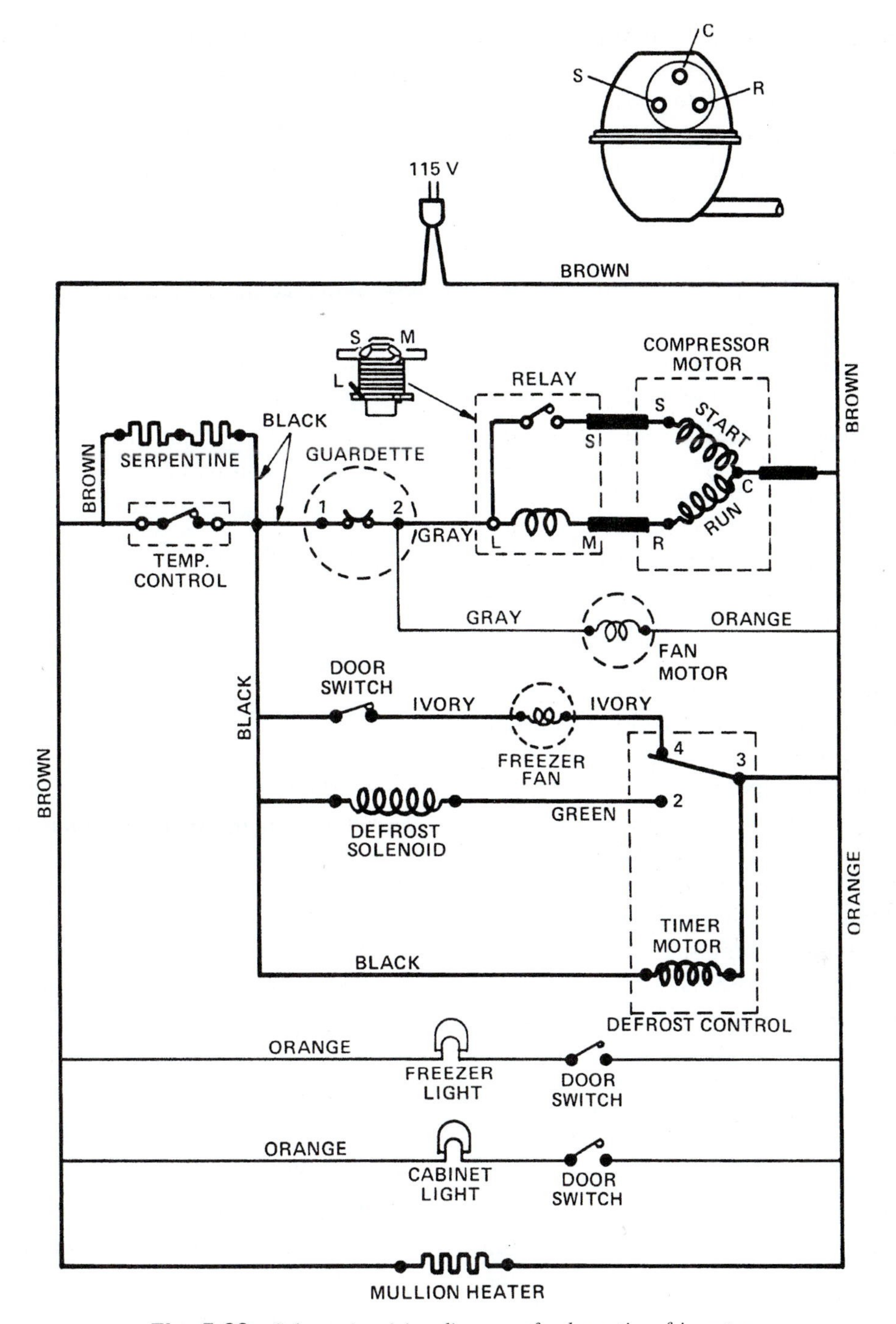

Fig. 5-83 *Schematic-wiring diagram of a domestic refrigerator.*

defrost timer motor and through it to point 3, then to the brown wire from the other side of the power supply. The defrost timer motor is operating anytime that the temperature control (refrigerator thermostat) is closed.

The defrost solenoid in the circuit between the freezer fan and the timer motor is activated as follows. When the defrost control has its switch in the downward direction (from point 3 to point 2) the circuit is completed from the 115-V plug through the temperature control and defrost solenoid to point 2 on the defrost control and through the switch in the downward position to point 3. Point 3 is connected to the other side of the power line through the brown wire on the right side of the schematic. This completes the circuit for the defrost solenoid. As you can see, the defrost control must be in the downward position (connecting points 2 and 3) to complete the circuit and cause the defrosting cycle to begin. Note that the freezer fan motor is not in the circuit. Thus, the fan in the freezer is not running at this time.

The refrigerator motor is controlled as follows. Starting at the left side of the 115-V plug, trace the brown wire to the junction of the serpentine and temperature control. This temperature control switch shorts out the serpentine when the switch is closed. Thus, the serpentine is not in the circuit when the refrigerator is running. A black wire runs from the temperature control switch to the guardette (circuit breaker). A gray wire leads from the guardette to point L of the relay. From point L to point M on the relay is the relay's coil. This coil (point M to point R on the compressor motor) is in series with the run-winding of the compressor motor. Point R to point C of the compressor motor represents the run-winding of the compressor motor. Point C is common to start and run-winding. Note the drawing of the compressor above the schematic. Here, the S, C, and R points are shown relative to their true location within the refrigerator. It can be seen that the temperature control and guardette must be closed for the run-winding to have a complete circuit to the power source lines.

The relay is in series with the run-winding. When the motor starts, the relay contacts are closed. Current through the contacts also completes its path to the common side of the power line (point C). Once the motor comes up to speed, the run-winding draws more current and causes the relay to be energized.

Once energized, the relay opens the contact points and takes the start-winding out of the circuit. When the motor stops again (when the thermostat opens), the relay deenergizes and the contacts close. This means the relay is ready for the next starting sequence. If the relay contacts stick, the start-winding stays in the circuit and draws current. The guardette is brought into action and opens the circuit to protect the motor windings from overheating.

For the refrigerator fan motor to operate, it must have power. It runs when the temperature control and the guardette are closed. To trace the circuit for the fan motor, start at the left side of the 115-V plug. Trace the brown wire through the temperature control, the closed switch, and the guardette. From the number 2 position on the guardette, a gray wire is connected to one side of the fan motor. The other side of the fan motor is connected by an orange wire to the brown wire leading to the other side of the 115-V plug. Thus, the temperature control switch and the guardette must be closed before the fan switch can run. Also, the fan motor runs whenever the compressor motor runs.

The serpentine heater is in the circuit whenever the temperature control is off or the refrigerator is not operating. It is a heating element wrapped around the evaporator coil. It prevents frost build-up between defrosting cycles.

Look again at Fig. 5-83. See if you can more easily read the schematic.

REVIEW QUESTIONS

1. State the left-hand rule for current in a conductor.
2. State the right-hand rule for motors.
3. What is the main advantage of a DC series motor?
4. What is the difference between a single-phase motor and a three-phase motor?
5. How does the capacitor-start motor differ from a split-phase motor?
6. What is the advantage of a three-phase motor over a single-phase motor?
7. What is a squirrel-cage rotor?
8. What is a megger?
9. What is synchronous speed?
10. What is meant by the service factor of a motor?
11. What does NEMA stand for?
12. Describe the operation of a bimetallic thermostat.
13. How is automatic defrost accomplished in today's refrigerators?
14. Where are driers located in a refrigeration system?
15. What is a schematic?
16. What is a serpentine heater?
17. What is voltage spread?
18. What is the purpose of a centrifugal switch on a single-phase motor?
19. How can direction of rotation be reversed on a split-phase motor?
20. What type of motor uses pushrods and a wound armature?
21. Where are capacitor-start motors used?
22. How are capacitor-start motors reversed when standing still?
23. What advantage does the permanent split-capacitor motor have?
24. What are shaded-pole motors most likely to be used for?
25. What is needed to get a split-phase motor to run?
26. How much current does an across-the-line motor draw when it starts?
27. What is another name for primary resistor starters?
28. What is the major disadvantage of the autotransformer starter?
29. What type of starting does part-winding starters provide?

30. What is the least expensive method of motor starting?
31. Where are wye-delta starters typically used?
32. Why are wye-delta starters used with delta-wound squirrel-cage motors?
33. Why are compelling relays needed?
34. What happens to motor speed when more poles are added?
35. How do consequence pole motors obtain two speeds?
36. What is the advantage of reduced-voltage motor starting?

Refrigerants: New and Old

PERFORMANCE OBJECTIVES

After studying this chapter, you should:

1. Know the classifications of refrigerants.
2. Know some of the physical properties of Freon.
3. Know the potential hazards of fluorocarbons.
4. Know operating pressures of refrigerants.
5. Know about moisture and refrigerants.
6. Know some of the problems with older refrigerants.
7. Know why new refrigerants are needed.

Refrigerants are used in the process of refrigeration. Refrigeration is a process whereby heat is removed from a substance or a space.

A refrigerant is a substance that picks up latent heat when the substance evaporates from a liquid to a gas. This is done at a low temperature and pressure. A refrigerant expels latent heat when it condenses from a gas to a liquid at a high pressure and temperature. The refrigerant cools by absorbing heat in one place and discharging it in another area.

The desirable properties of a good refrigerant for commercial use are:

- Low boiling point
- Safe nontoxic
- Easy to liquefy and moderate pressure and temperature
- High latent-heat value
- Operation on a positive pressure
- Not affected by moisture
- Mixes well with oil
- Noncorrosive to metal.

There are other qualities that all refrigerants have. These qualities are molecular weight, density, compression ratio, heat value, and temperature of compression. These qualities will vary with the refrigerants. The compressor displacement and compressor type or design will also influence the choice of refrigerant.

CLASSIFICATION OF REFRIGERANTS

Refrigerants are classified according to their manner of absorption or extraction of heat from substances to be refrigerated. The classifications can be broken down into class 1, class 2, and class 3.

Class 1 refrigerants are used in the standard compression type of refrigeration systems. Class 2 refrigerants are used as immediate cooling agents between class 1 and the substance to be refrigerated. They do the same work for class 3. Class 3 refrigerants are used in the standard absorption-type systems of refrigerating systems.

Class 1. This class includes those refrigerants that cool by absorption or extraction of heat from the substances to be refrigerated by the absorption of their latent heats. Table 6-1 lists the characteristics of typical refrigerants.

Table 6-1 *Characteristics of Typical Refrigerants*

Name	Boiling Point (°F)	Heat of Vaporization at Boiling Point Btu/lb. 1 At.
Sulfur dioxide	14.0	172.3
Methyl chloride	−10.6	177.8
Ethyl chloride	55.6	177.0
Ammonia	−28.0	554.7
Carbon dioxide	−110.5	116.0
Freezol (isobutane)	10.0	173.5
Freon 11	74.8	78.31
Freon 12	−21.7	71.04
Freon 13	−114.6	63.85
Freon 21	48.0	104.15
Freon 22	−41.4	100.45
Freon 113	117.6	63.12
Freon 114	38.4	58.53
Freon 115	−37.7	54.20
Freon 502	−50.1	76.46

Class 2. The refrigerants in this class are those that cool substances by absorbing their sensible heats. They are air, calcium-chloride brine, sodium-chloride (salt) brine, alcohol, and similar nonfreezing solutions.

Class 3. This group consists of solutions that contain absorbed vapors of liquefiable agents or refrigerating media. These solutions function through their ability to carry the liquefiable vapors. The vapors produce a cooling effect by the absorption of their latent heat. An example is aqua ammonia, which is a solution composed of distilled water and pure ammonia.

Common Refrigerants

Following are some of the more common refrigerants. Table 6-1 summarizes the characteristics to a selected few of the many refrigerants available for home, commercial, and industrial use.

Sulfur Dioxide Sulfur dioxide (SO_2) is a colorless gas or liquid. It is toxic, with a very pungent odor. When sulfur is burned in air, sulfur dioxide is formed. When sulfur dioxide combines with water it produces sulfuric and sulfurous acids. These acids are very corrosive to metal. They have an adverse effect on most materials. Sulfur dioxide is not considered a safe refrigerant.

Sulfur dioxide is not considered safe when used in large quantities. As a refrigerant, sulfur dioxide operates on a vacuum to give the temperatures required. Moisture in the air will be drawn into the system when a leak occurs. This means the metal parts will eventually corrode, causing the compressor to seize.

Sulfur dioxide (SO_2) boils at 14°F (−10°C) and has a heat of vaporization at boiling point (1 atmosphere) of 172.3 Btu/1b. It has a latent-heat value of 166 Btu/lb.

To produce the same amount of refrigeration, sulfur dioxide requires about one-third more vapor than Freon and methyl chloride. This means the condensing unit has to operate at a higher speed or the compressor cylinders must be larger. Since sulfur dioxide does not mix well with oil, the suction line must be on a steady slant to the machine. Otherwise, the oil will trap out, constricting the suction line. This refrigerant is not feasible for use in some locations.

Methyl Chloride Methyl chloride (CH_3Cl) has a boiling point of −10.6° F (−23.3°C). It also has heat of vaporization at boiling point (at 1 atmosphere) of 177.8 Btu/lb. It is a good refrigerant. However, because it will burn under some conditions, some cities will not allow it to be used. It is easy to liquefy and has a comparatively high latent-heat value. It does not corrode metal when in its dry state.

However, in the presence of moisture it damages the compressor. A sticky black sludge is formed when excess moisture combines with the chemical. Methyl chloride mixes well with oil. It will operate on a positive pressure as low as −10°F (−23°C). The amount of vapor needed to cause discomfort in a person is in proportion to the following numbers:

Carbon dioxide	100
Methyl chloride	70
Ammonia	2
Sulfur dioxide	1

That means methyl chloride is 35 times safer than ammonia and 70 times safer than sulfur dioxide.

Methyl chloride is hard to detect with the nose or eyes. It does not produce irritating effects. Therefore, some manufacturers add a 1 percent amount of *acrolein*, a colorless liquid with a pungent odor, as a warning agent. It is produced by destructive distillation of fats.

Ammonia Ammonia (NH_3) is used most frequently in large industrial plants. Freezers for packing houses usually employ ammonia as a refrigerant. It is a gas with a very noticeable odor. Even a small leak can be detected with the nose. Its boiling point at normal atmospheric pressure is −28°F (−33°C). Its freezing point is −107.86°F (−77.7°C). It is very soluble in water. Large refrigeration capacity is possible with small machines. It has high latent heat [555 Btu at 18°F (−7.7°C)]. It can be used with steel fittings. Water-cooled units are commonly used to cool down the refrigerant. High pressures are used in the lines (125 to 200 $lb/in.^2$). Anyone inside the refrigeration unit when it springs a leak is rapidly overcome by the fumes. Fresh air is necessary to reduce the toxic effects of ammonia fumes. Ammonia is combustible when combined with certain amounts of air (about one volume of ammonia to two volumes of air). It is even more combustible when combined with oxygen. It is very toxic. Heavy steel fittings are required since pressures of 125 to 200 $lb/in.^2$ are common. The units must be water cooled.

Carbon Dioxide Carbon dioxide (CO_2) is a colorless gas at ordinary temperatures. It has a slight odor and an acid taste. Carbon dioxide is nonexplosive and nonflammable. It has a boiling point of 5°F (−15°C). A pressure of over 300 $lb/in.^2$ is required to keep it from evaporation. To liquefy the gas, a condenser temperature of 80°F (26.6°C) and a pressure of approximately 1000 $lb/in.^2$ are needed. Its critical temperature is 87.8°F (31°C). It is harmless to breathe except in extremely large concentrations. The lack of oxygen can cause suffocation under certain conditions of carbon dioxide concentration.

Carbon dioxide is used aboard ships and in industrial installations. It is not used in household applications. The main advantage of using carbon dioxide for a refrigerant is that a small compressor can be used. The compressor is very small since a high pressure is required for the refrigerant. Carbon dioxide is, however, very inefficient, compared to other refrigerants. Thus, it is not used in household units.

Calcium Chloride Calcium chloride ($CaC1_2$) is used only in commercial refrigeration plants. Calcium chloride is used as a simple carrying medium for refrigeration.

Brine systems are used in large installations where there is danger of leakage. They are used also where the temperature fluctuates in the space to be refrigerated. Brine is cooled down by the direct expansion of the refrigerant. It is then pumped through the material or space to be cooled. Here, it absorbs sensible heat.

Most modern plants operate with the brine at low temperature. This permits the use of less brine, less piping or smaller diameter pipe, and smaller pumps. It also lowers pumping costs. Instead of cooling a large volume of brine to a given temperature, the same number of refrigeration units are used to cool a smaller volume of brine to a lower temperature. This results in greater economy. The use of extremely low-freezing brine, such as calcium chloride, is desirable in the case of the shell-type cooler.

Salt brine with a minimum possible freezing point of −6°F (−20.9°C) may solidify under excess vacuum on the cold side of the refrigerating unit. This can cause considerable damage and loss of operating time. There are some cases, in which the cooler has been ruined.

Ethyl Chloride Ethyl chloride (C_2H_5Cl) is not commonly used in domestic refrigeration units. It is similar to methyl chloride in many ways. It has a boiling point of 55.6°F (13.1°C) at atmospheric pressure. Critical temperature is 360.5°F (182.5°C) at a pressure of 784 lb absolute. It is a colorless liquid or gas with a pungent ethereal odor and a sweetish taste. It is neutral toward all metals. This means that iron, copper, and even tin and lead can be used in the construction of the refrigeration unit. It does, however, soften all rubber compounds and gasket material. Thus, it is best to use only lead for gaskets.

FREON REFRIGERANTS

The Freon refrigerants have been one of the major factors responsible for the tremendous growth of the home refrigeration and air-conditioning industries. The safe properties of these products have permitted their use under conditions where flammable or more toxic refrigerants would be hazardous to use. There is a Freon refrigerant for every application—from home and industrial air conditioning to special low-temperature requirements.

The unusual combination of properties found in the Freon compounds is the basis for the wide application and usefulness. Table 6-2 presents a summary of the specific properties of some of the fluorinated products. Figure 6-1 gives the absolute pressure and gage pressure of Freon refrigerants at various temperatures.

Molecular Weights

Compounds containing fluorine in place of hydrogen have higher molecular weights and often have unusually low boiling points. For example, methane (CH_4) with a molecular weight of 16 has a boiling point of −258.5°F (−161.4°C) and is nonflammable. Freon 14 (CF_4) has a molecular weight of 88 and a boiling point of −198.4°F (−128°C) and is nonflammable. The effect is even more pronounced when chlorine is also present. Methylene chloride (CH_2Cl_2) has a molecular weight of 85 and boils at 105.2°F (40.7°C) while Freon 12 (CCl_2F_2, molecular weight 121) boils at −21.6°F (−29.8°C). It can be seen that Freon compounds are high-density materials with low boiling points, low viscosity, and low surface tension. Freon includes products with boiling points covering a wide range of temperatures. See Table 6-3.

The high molecular weight of the Freon compounds also contributes to low vapor, specific-heat values, and fairly low latent heats of vaporization. Tables of thermodynamic properties including enthalpy, entropy, pressure, density, and volume for the liquid and vapor are available from manufacturers.

Freon compounds are poor conductors of electricity. In general, they have good dielectric properties.

Flammability

None of the Freon compounds are flammable or explosive. However, mixtures with flammable liquids or gases may be flammable and should be handled with caution. Partially halogenated compounds may also be flammable and must be individually examined.

Toxicity

Toxicity means intoxicating or poisonous. One of the most important qualities of the Freon fluorocarbon compounds is their low toxicity under normal conditions of handling and usage. However, the possibility of serious injury or death exists under unusual or uncontrolled exposures or in deliberate abuse by inhalation of concentrated vapors. The potential hazards of fluorocarbons are summarized in Table 6-4.

Skin Effects

Liquid fluorocarbons, with boiling points below 32°F (0°C), may freeze the skin, causing frostbite on contact. Suitable protective gloves and clothing give insulation protection. Eye protection should be used. In the event of frostbite, warm the affected area quickly to body temperature. Eyes should be flushed copiously with water. Hands may be held under armpits or immersed in warm water. Get medical attention immediately. Fluorocarbons with boiling points at or above ambient temperature tend to dissolve protective fat from the skin. This leads to skin dryness and irritation, particularly after prolonged or repeated contact. Such contact should be avoided by using rubber gloves or plastic gloves. Eye protection and face shields should be used if splashing is possible. If irritation occurs following contact, seek medical attention.

Oral Toxicity

Fluorocarbons are low in oral toxicity as judged by single-dose administration or repeated dosing over long periods.

However, direct contact of liquid fluorocarbons with lung tissue can result in chemical pneumonitis,

Table 6-2 Physical Properties of Freon Products*

		Freon 11	Freon 12	Freon 13	Freon 13B1	Freon 14
Chemical formula		CCl_3F	CCl_2F_2	$CClF_3$	$CBrF_3$	CF_4
Molecular weight		137.37	120.92	104.46	148.92	88.00
Boiling point at 1 atm	**°C** **°F.**	23.82 74.87	–29.79 –21.62	–81.4 –114.6	–57.75 –71.95	–127.96 –198.32
Freezing point	**°C** **°F.**	–111 –168	–158 –252	–181[1] –294	–168 –270	–184[2] –299
Critical temperature	**°C** **°F.**	198.0 388.4	112.0 233.6	28.9 83.9	67.0 152.6	–45.67 –50.2
Critical pressure	**atm** **lbs/sq in abs**	43.5 639.5	40.6 596.9	38.2 561	39.1 575	36.96 543.2
Critical volume	**cc/mol** **cu ft/lb**	247 0.0289	217 0.0287	181 0.0277	200 0.0215	141 0.0256
Critical density	**g/cc** **lbs/cu ft**	0.554 34.6	0.588 34.8	0.578 36.1	0.745 46.5	0.626 39.06
Density, liquid at 25°C (77°F.)	**g/cc** **lbs/cu ft**	1.476 92.14	1.311 81.84	1.298 81.05 } @ –30°C (–22°F.)	1.538 96.01	1.317 82.21 } @ –80°C (–112°F.)
Density, sat'd vapor at boiling point	**g/l** **lbs/cu ft**	5.86 0.367	6.33 0.395	7.01 0.438	8.71 0.544	7.62 0.476
Specific heat, liquid (Heat capacity) at 25°C (77°F.)	**cal/(g)(°C) or Btu/(lb)(°F.)**	0.208	0.232	0.247 @ –30°C (–22°F.)	0.208	0.294 @ –80°C (–112°F.)
Specific heat, vapor, at const pressure (1 atm) at 25°C (77°F.)	**cal/(g)(°C) or Btu/(lb)(°F.)**	0.142 @ 38°C (100°F.)	0.145	0.158	0.112	0.169
Specific heat ratio at 25°C and 1 atm	C_p/C_v	1.137 @ 38°C (100°F.)	1.137	1.145	1.144	1.159
Heat of vaporization at boiling point	**cal/g** **Btu/lb**	43.10 77.51	39.47 71.04	35.47 63.85	28.38 51.08	32.49 58.48
Thermal conductivity at 25°C (77°F.) Btu/(hr) (ft) (°F.) **liquid** **vapor (1 atm)**		0.0506 0.00451	0.0405 0.00557	0.0378 0.00501 } @ –30°C (–22°F.)	0.0234 0.00534	0.0361 0.00463 } @ –80°C (–112°F.)
Viscosity[7] at 25°C (77°F.) **liquid** **vapor (1 atm)**	**centipoise** **centipoise**	0.415 0.0107	0.214 0.0123	0.170 0.0119 } @ (–30°C) (–22°F.)	0.157 0.0154	0.23 0.0116 } @ (–80°C) (–112°F.)
Surface tension at 25°C (77°F.) dynes/cm		18	9	14 @ –73°C –100°F.	4	4 @ –73°C (–100°F.)
Refractive index of liquid at 25°C (77°F.)		1.374	1.287	1.199 @ –73°C (–100°F.)	1.238	1.151 @ –73°C (–100°F.)
Relative dielectric strength[8] at 1 atm and 25°C (77°F.) (nitrogen = 1)		3.71	2.46	1.65	1.83	1.06
Dielectric constant **liquid** **vapor (1 atm)[9a]**		2.28 @ 29°C 1.0036 @ 24°C[9b]	2.13 1.0032 } @ 29°C (84°F.)	1.0024 29°C @(84°F.)		1.0012 @ 24.5°C (76°F.)
Solubility of "Freon" in water at 1 atm and 25°C (77°F.)	**wt %**	0.11	0.028	0.009	0.03	0.0015
Solubility of water in "Freon" at 25°C (77°F.)	**wt %**	0.011	0.009		0.0095 21°C (70°F.)	
Toxicity		Group 5a[12]	Group 6[12]	Probably Group 6[13]	Group 6[12]	Probably Group 6[13]

pulmonary edema, and hemorrhage. Fluorocarbons 11 and 113, like many petroleum distillates, are fat solvents and can produce such an effect. If products containing these fluorocarbons were accidentally or purposely ingested, induction of vomiting would be contraindicated (medically wrong). In other words, *do NOT induce vomiting.*

Central Nervous System (CNS) Effects

Inhalation of concentrated fluorocarbon vapors can lead to *central nervous system* (CNS) effects comparable to the effects of general anesthesia. The first symptom is a feeling of intoxication. This is followed by a

Table 6-2 Physical Properties of Freon Products (Continued)*

		Freon 21	Freon 22	Freon 23	Freon 112	Freon 113	Freon 114
Chemical formula		$CHCl_2F$	$CHClF_2$	CHF_3	$CCl_2F—CCl_2F$	$CCl_2F—CClF_2$	$CClF_2—CClF_2$
Molecular weight		102.93	86.47	70.01	203.84	187.38	170.93
Boiling point at 1 atm	**°C** **°F.**	8.92 48.06	–40.75 –41.36	–82.03 –115.66	92.8 199.0	47.57 117.63	3.77 38.78
Freezing point	**°C** **°F.**	–135 –211	–160 –256	–155.2 –247.4	26 79	–35 –31	–94 –137
Critical temperature	**°C** **°F.**	178.5 353.3	96.0 204.8	25.9 78.6	278 532	214.1 417.4	145.7 294.3
Critical pressure	**atm** **lbs/sq in abs**	51.0 750	49.12 721.9	47.7 701.4	34[3] 500	33.7 495	32.2 473.2
Critical volume	**cc/mol** **cu ft/lb**	197 0.0307	165 0.0305	133 0.0305	370[3] 0.029	325 0.0278	293 0.0275
Critical density	**g/cc** **lbs/cu ft**	0.522 32.6	0.525 32.76	0.525 32.78	0.55[3] 34	0.576 36.0	0.582 36.32
Density, liquid at 25°C (77°F.)	**g/cc** **lbs/cu ft**	1.366 85.28	1.194 74.53	0.670 41.82	1.634[b] 102.1 } @ 30°C (86°F)	1.565 97.69	1.456 90.91
Density, sat'd vapor at Boiling Point	**g/l** **lbs/cu ft**	4.57 0.285	4.72 0.295	4.66 0.291	7.02[5] 0.438	7.38 0.461	7.83 0.489
Specific heat, liquid (heat capacity) at 25°C (77°F.)	**cal/(g)(°C) or Btu/(lb)(°F.)**	0.256	0.300	0.345 @ –30°C –22°F		0.218	0.243
Specific heat, vapor, at const pressure (1 atm) at 25°C (77°F.)	**cal/(g)(°C) or Btu/(lb)(°F.)**	0.140	0.157	0.176		0.161 @ 60°C (140°F.)	0.170
Specific-heat ratio at 25°C and 1 atm	C_p/C_v	1.175	1.184	@ 1.191 0 pressure		1.080 @ 60°C (140°F.)	1.084
Heat of vaporization at boiling point	**cal/g** **Btu/b**	57.86 104.15	55.81 100.45	57.23 103.02	37 (est) 67	35.07 63.12	32.51 58.53
Thermal conductivity[1] at 25°C (77°F.) Btu/(hr) (ft) (°F.) **liquid** **vapor (1 atm)**		0.0592 0.00506	0.0507 0.00609	0.0569 0.0060 } @ –30°C (–22°F.)	0.040	0.0434 0.0044 (0.5 atm)	0.0372 0.0060
Viscosity[1] at 25°C (77°F.) **liquid** **vapor (1 atm)**	**centipoise** **centipose**	0.313 0.0114	0.198 0.0127	0.167 0.0118 } @ –30°C (–22°F.)	1.21[6]	0.68 0.010 (0.1 atm)	0.36 0.0112
Surface tension at 25°C (77°F.) dynes/cm		18	8	15 @ –73°C (–100°F.)	23 @ 30°C (86°F.)	17.3	12
Refractive index of liquid at 25°C (77°F.)		1.354	1.256	1.215 @ –73°C (–100°F.)	1.413	1.354	1.288
Relative-dielectric strength[8] at 1 atm and 25°C (77°F.) (nitrogen = 1)		1.85	1.27	1.04	5 (est)	3.9 (0.44 atm)	3.34
Dielectric constant **liquid** **Vapor (1 atm)[9a]**		5.34 @ 28°C 1.0070 @ 30°C	6.11 @ 24°C 1.0071 @ 25.4°C	1.0073 @ 25°C[9b]	2.54 @ 25°C (77°F.)	2.41 @ 25°C (77°F.)	2.26 @ 25°C 1.0043 @26.8°C
Solubility of "Freon" in water at 1 atm and 25°C (77°F.)	**wt %**	0.95	0.30	0.10	0.012 (Sat'n Pres)	0.017 (Sat'n Pres)	0.013
Solubility of water in "Freon" at 25°C (77°F.)	**wt %**	0.13	0.13			0.011	0.009
Toxicity		much less than Group 4, somewhat more than Group 5[12]	Group 5a[12]	probably Group 6[13]	probably less than Group 4, more than Group 5[13]	much less than Group 4, somewhat more than Group 5[12]	Group 6[12]

Table 6-2 *Physical Properties of Freon* Products*

FC 114B2	Freon 115	Freon 116	Freon 500	Freon 502	Freon 503
$CBrF_2—CBrF_2$	$CClF_2—CF_3$	$CF_3—CF_3$	a	b	c
259.85	154.47	138.01	99.31	111.64	87.28
47.26	−39.1	−78.2	−33.5	−45.42	−87.9
117.06	−38.4	−108.8	−28.3	−49.76	−126.2
−110.5	−106[10]	−100.6	−159		
−166.8	−159	−149.1	−254		
214.5	80.0	19.7[4]	105.5	82.2	19.5
418.1	175.9	67.5	221.9	179.9	67.1
34.4	30.8	29.4[4]	43.67	40.2	43.0
506.1	453	432	641.9	591.0	632.2
329	259	225	200.0	199	155
0.0203	0.0269	0.0262	0.03226	0.02857	0.0284
0.790	0.596	0.612	0.4966	0.561	0.564
49.32	37.2	38.21	31.0	35.0	35.21
2.163 135.0	1.291 80.60	1.587 99.08 }[4] @ −73°C (−100°F.)	1.156 72.16	1.217 75.95	1.233 76.95 } @ −30°C (−22°F.)[7]
	8.37 0.522	9.01[4] 0.562	5.278 0.3295	6.22 0.388	6.02 0.374
0.166	0.285	0.232 @ −73°C (−100°F.)[4]	0.258	0.293	0.287 @ −30°C (−22°F.)
	0.164	0.182[11] @ 0 pressure	0.175	0.164	0.16
	1.091	1.085 (est) @ 0 pressure	1.143	1.132	1.21 @ −34°C (−30°F.)
25 (est)	30.11	27.97	48.04	41.21	42.86
45 (est)	54.20	50.35	86.47	74.18	77.15
0.027	0.0302 0.00724	0.045 0.0098 } @ −73°C (−100°F.)	0.0432	0.0373 0.00670	0.0430 @ −30°C (−22°F.)[7]
0.72	0.193 0.0125	0.30 0.0148	0.192 0.0120	0.180 0.0126	0.144 @ −30°C (−22°F.)
18	5	16 @ −73°C (−100°F.)	8.4	5.9	6.1 @ −30°C (−22°F.)
1.367	1.214	1.206 @ −73°C (−100°F.)	1.273	1.234	1.209 @ −30°C (−22°F.)
4.02 (0.44 atm)	2.54	2.02		1.3	
2.34 @ 25°C (77°F.)	1.0035 @ 27.4°C	1.0021 @ 23°C (73°F.)		6.11 @ 25°C 1.0035 (0.5 atm)	
	0.006				0.042
Group 5a[12]	Group 6[12]	probably Group 6[13]	0.056 Group 5a	0.056 Group 5a[12]	probably Group 6[13]

*FREON is Du pont's registered trademark for its fluorocarbon products

a. CCl_2F_2/CH_3CHF_2 (73.8/26.2% by wt.)
b. $CHClF_2/CClF_2CF_3$ (48.8/51.2% by wt.)
c. $CHF_3/CClF_3$ (40/60% by wt.)

loss of coordination and unconsciousness. Under severe conditions, death can result. If these symptoms are felt, the exposed individual should immediately go or be moved to fresh air. Medical attention should be sought promptly. *Individuals exposed to fluorocarbons should NOT be treated with adrenalin (epinephrine).*

Cardiac Sensitization

Fluorocarbons can, in sufficient vapor concentration, produce cardiac sensitization. This is a sensitization of the heart to adrenaline brought about by exposure to high concentrations of organic vapors. Under severe exposure, cardiac arrhythmias may result from sensitization of the heart to the body's own levels of adrenaline. This is particularly so under conditions of emotional or physical stress, fright, panic, and so forth. Such cardiac arrhythmias may result in ventricular fibrillation and death. Exposed individuals should immediately go or be removed to fresh air. There, the hazard of cardiac effects will rapidly decrease. Prompt medical attention and observation

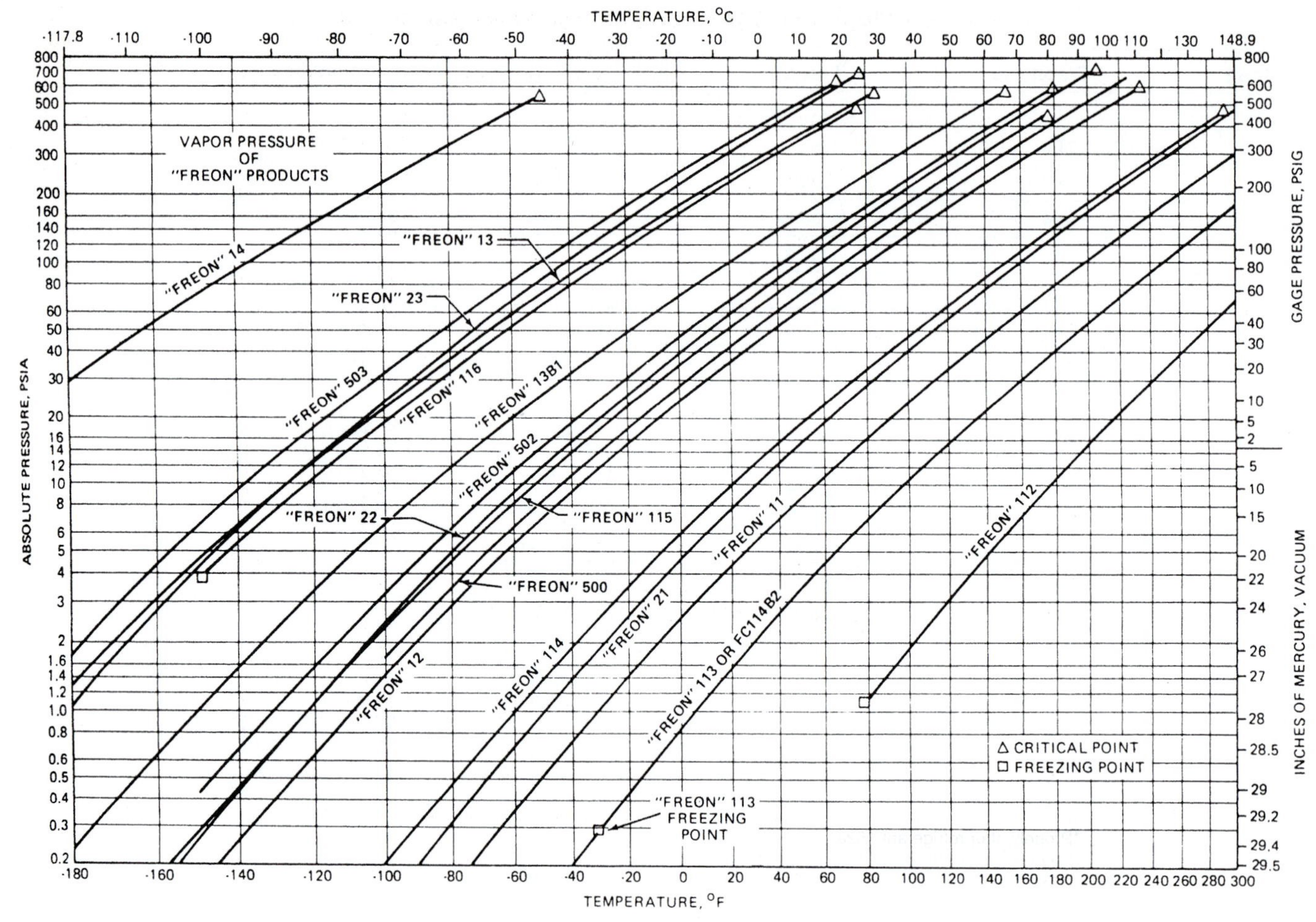

Fig. 6-1 *The absolute and gage pressures of Freon refrigerants.*

should be provided following accidental exposures. *A worker adversely affected by fluorocarbon vapors should NOT be treated with adrenalin (epinephrine)* or similar heart stimulants since these would increase the risk of cardiac arrhythmias.

Thermal Decomposition

Fluorocarbons decompose when exposed directly to high temperatures. Flames and electrical-resistance heaters, for example, will chemically decompose fluorocarbon vapors. Products of this decomposition in air include halogens and halogen acids (hydrochloric, hydrofluoric, and hydrobromic), as well as other irritating compounds. Although much more toxic than the parent fluorocarbon, these decomposition products tend to irritate the nose, eyes, and upper respiratory system. This provides a warning of their presence. The practical hazard is relatively slight. It is difficult for a person to remain voluntarily in the presence of' decomposition products at concentrations where physiological damage occurs.

When such irritating decomposition products are detected, the area should be evacuated and ventilated. The source of the problem should be corrected.

APPLICATIONS OF FREON REFRIGERANTS

There is a Freon refrigerant for every application from home and industrial air conditioning to special low-temperature requirements. Following are a few of the Freon refrigerants.

Freon 11 ($CC1_3F$) has a boiling point of 74.9°F (23.8°C) and is widely used in centrifugal compressors for industrial and commercial air-conditioning systems, it is also used for industrial process water and brine cooling. Its low viscosity and freezing point have also led to its use as a low-temperature brine.

Freon 12 (CCl_2F_2) has a boiling point of –21.6°F (–29.8°C) and is the most widely known and used of the Freon refrigerants. It is used principally in household and commercial refrigeration and air conditioning. It is used for refrigerators, frozen food locker plants, water coolers, room and window air-conditioning units and similar equipment. It is generally used in reciprocating compressors ranging in size from fractional to 800 horsepower. It is also used in the smaller–rotary type compressors, Fig. 6-2.

Table 6-3 *Fluorinated Products and Their Molecular Weight and Boiling Point*

Product	Formula	Molecular Weight	Boiling Point °F	Boiling Point °C
Freon Products				
Freon 14	CF_4	88.0	−198.3	−128.0
Freon 503	$CHF_3/CClF_3$	87.3	−127.6	−88.7
Freon 23	CHF_3	70.0	−115.7	−82.0
Freon 13	$CClF_3$	104.5	−114.6	−81.4
Freon 116	$CF_3—CF_3$	138.0	−108.8	−78.2
Freon 13B1	$CBrF_3$	148.9	−72.0	−57.8
Freon 502	$CHClF_2/CClF_2—CF_3$	111.6	−49.8	−45.4
Freon 22	$CHClF_2$	86.5	−41.4	−40.8
Freon 115	$CClF_2—CF_3$	154.5	−37.7	−38.7
Freon 500	CCl_2F_2/CH_3CHF_2	99.3	−28.3	−33.5
Freon 12	CCl_2F_2	120.9	−21.6	−29.8
Freon 114	$CClF_2—CClF_2$	170.9	38.8	3.8
Freon 21	$CHCl_2F$	102.9	48.1	8.9
Freon 11	CCl_3F	137.4	74.9	23.8
Freon 113	$CCl_2F—CClF_2$	187.4	117.6	47.6
Freon 112	CCl_2F-CCl_2F	203.9	199.0	92.8
Other Fluorinated Compounds				
FC 114B2	$CBrF_2—CBrF_2$	259.9	117.1	47.3
1,1-Difluoroethane*	$CH_3—CHF_2$	66.1	−13.0	−25.0
1,1,1-Chlorodifluoroethane†	$CH_3—CClF_2$	100.5	14.5	−9.7
Vinyl fluoride	$CH_2═CHF$	46.0	−97.5	−72.0
Vinylidene fluoride	$CH_2═CF_2$	64.0	−122.3	−85.7
Hexafluoroacetone	CF_3COCF_3	166.0	−18.4	−28.0
Hexafluoroisopropanol	$(CF_3)_2CHOH$	168.1	136.8	58.2

*Propellant or refrigerant 152a
†Propellant or refrigerant 142b

Table 6-4 *Potential Hazards of Fluorocarbons*

Condition	Potential Hazard	Safeguard
Vapors may decompose in flames or in contact with hot surfaces.	Inhalation of toxic decomposition products.	Good ventilation. Toxic decomposition products serve as warning agents.
Vapors are four to five times heavier than air. High concentrations may tend to accumulate in low places.	Inhalation of concentrated vapors can be fatal.	Avoid misuse. Forced-air ventilation at the level of vapor concentration. Individual breathing devices with air supply. Lifelines when entering tanks or other confined areas.
Deliberate inhalation to produce intoxication.	Can be fatal.	Do not administer epinephrine or other similar drugs.
Some fluorocarbon liquids tend to remove natural oils from the skin.	Irritation of dry, sensitive skin.	Gloves and protective clothing.
Lower boiling liquids may be splashed on skin.	Freezing.	Gloves and protective clothing.
Liquids may be splashed into eyes.	Lower boiling liquids may cause freezing. Higher boiling liquids may cause temporary irritation and if other chemicals are dissolved, may cause serious damage.	Wear eye protection. Get medical attention. Flush eyes for several minutes with running water.
Contact with highly reactive metals.	Violent explosion may occur.	Test the proposed system and take appropriate safety precautions.

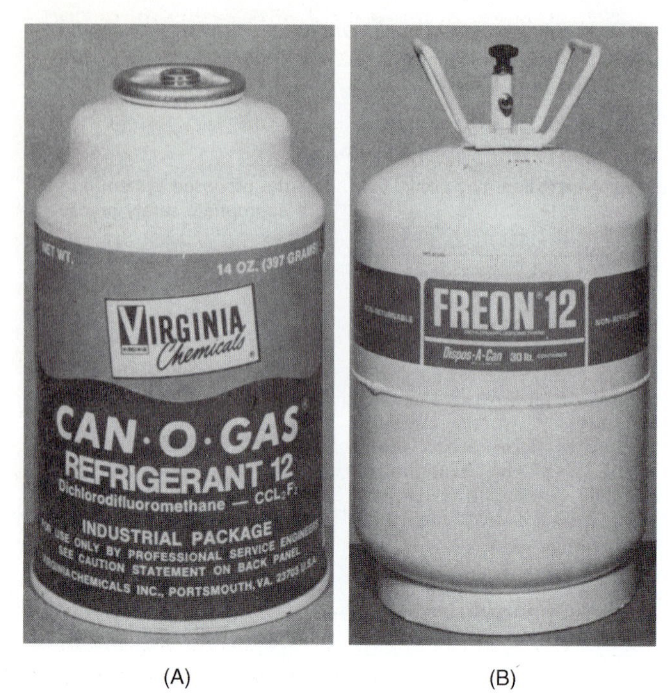

(A) (B)

Fig. 6-2 *Freon can be purchased in a number of sizes.* (Virginia Chemical)

Freon 13 ($CClF_3$) has a boiling point of −114.6°F (−81.4°C) and is used in low-temperature specialty applications using reciprocating compressors and generally in cascade with Freon 12, Freon 22, or Freon 522.

Freon 22 ($CHClF_2$) has a boiling point of −41.4°F (−40.8°C) and is used in all types of household and commercial refrigeration and air-conditioning applications with reciprocating compressors. The outstanding thermodynamic properties of Freon 22 permit the use of smaller equipment than is possible with similar refrigerants. This makes it especially attractive for uses where size is a problem. See Fig. 6-3.

Freon 113 ($CCl_2F{\cdot}CClF_2$) has a boiling point of 117.6°F (47.6°C). It is used in commercial and industrial air conditioning and process water and brine cooling with centrifugal compression. It is especially useful in small-tonnage applications.

Freon 114 ($CClF_2{\cdot}CClF_2$) has a boiling point of 38.8°F (3.8°C). It is used in small refrigeration systems with rotary-type compressors. It is used in large industrial process cooling and air-conditioning systems using multistage centrifugal compressors.

Freon 500 (CCl_2F_2/CH_3CHF_2) is an azeotropic mixture. *Azeotropic* means that a mixture is liquid, maintains a constant boiling point, and produces a vapor of the same composition as the mixture with CH_3CHF_2. It is composed of 73.8 percent Freon 12 (CCl_2F_2) and 26.2 percent CH_3CHF_2. It boils at −28.3°F (−33.5°C). It is used in home and commercial air conditioning in small and medium-size equipment and in some refrigeration applications.

Fig. 6-3 *Freon 22 is marketed in containers of various sizes, such as a 1-lb, 2-lb, and 15-lb cans.* (Virginia Chemical)

Freon 502 is an azeotropic mixture also. It consists of 48.8 percent of Freon 22 and 51.2 percent of Freon 115, by weight. It boils at −49.8°F (−45.4°C). With Freon 502, refrigeration capacity is greater than with Freon 22. Note the pressure differences on the pressure gage in Fig. 6-4. Discharge temperatures are comparable to those found with Freon 12. Freon 502 is finding new applications in low and medium-temperature cabinets for the display and storage of foodstuffs, in food freezing, and in heat pumps.

Freon 503 is an azeotropic mixture of CHF_3 and $CClF_3$. The weight ratio is 40 percent CHF_3 and 60 percent $CClF_3$. The boiling point of this mixture is −127.6°F (−88.7°C). It is used in low-temperature cascade systems.

Freon 13B1 ($CBrF_3$) boils at −72°F (−57.8°C). It serves the temperature range between Freon 502 and Freon 13.

Fig. 6-4 *Pressure gage for R-12, R-22, and R-502.* (Marsh)

These are some of the refrigerants that are now under close scrutiny because of their chlorine content and their effect on the environment. Some have been banned and cannot be manufactured anywhere in the world. Others are being phased out gradually and replaced by a new combination of chemicals.

REACTION OF FREON TO VARIOUS MATERIALS FOUND IN REFRIGERATION SYSTEMS

Metals

Most of the commonly used construction metals—such as steel, cast iron, brass, copper, tin, lead, and aluminum—can be used satisfactorily with the Freon compounds under normal conditions of use. At high temperatures some of the metals may act as catalysts for the breakdown of the compound. The tendency of metals to promote thermal decomposition of the Freon compounds is in the following general order. Those metals that least promote thermal decomposition are listed first.

- Inconel®
- Stainless steel
- Nickel
- 1340 steel
- Aluminum
- Copper
- Bronze
- Brass
- Silver

The above order is only approximate. Exceptions may be found for individual Freon compounds or for special conditions of use.

Magnesium alloys and aluminum containing more than 2 percent magnesium are not recommended for use in systems containing Freon compounds where water may be present. Zinc is not recommended for use with Freon 113. Experience with zinc and other Freon compounds has been limited and no unusual reactivity has been observed. However, it is more chemically reactive than other common construction metals. Thus, it would seem wise to avoid its use with the Freon compounds unless adequate testing is carried out.

Some metals may be questionable for use in applications requiring contact with Freon compounds for long periods of time or unusual conditions of exposure. These metals, however, can be cleaned safely with Freon solvents. Cleaning applications are usually for short exposures at moderate temperatures.

Most halocarbons may react violently with highly reactive materials, such as sodium, potassium, and barium in their free metallic form. Materials become more reactive when finely ground or powdered. In this state, magnesium and aluminum may react with fluorocarbons, especially at higher temperatures. Highly reactive materials should not be brought into contact with fluorocarbons until a careful study is made and appropriate safety precautions are taken.

Plastics

A brief summary of the effect of Freon compounds on various plastic materials follows. However, compatibility should be tested for specific applications. Differences in polymer structure and molecular weight, plasticizers, temperature, and pressure may alter the resistance of the plastic toward the Freon compound.

Teflon - TFE - fluorocarbon resin. No swelling observed when submerged in Freon liquids, but some diffusion found with Freon 12 and Freon 22.

Polychluorotrifluoroethylene. Slight swelling, but generally suitable for use with Freon compounds.

Polyvinyl alcohol. Not affected by the Freon compounds, but very sensitive to water. Used especially in tubing with an outer protective coating.

Vinyl. Resistance to the Freon compounds depends on vinyl type and plasticizer. Considerable variation is found. Samples should be tested before use.

Orlon-acrylic fiber. Generally suitable for use with the Freon compounds.

Nylon. Generally suitable for use with Freon compounds, but may tend to become brittle at high temperatures in the presence of air or water. Tests at 250°F (121°C) with Freon 12 and Freon 22 showed the presence of water or alcohol to be undesirable. Adequate testing should be carried out.

Polyethylene. May be suitable for some applications at room temperatures. However, it should be thoroughly tested since greatly different results have been found with different samples.

Lucite-acrylic resin (methacrylate polymers). Dissolved by Freon 22. However, it is generally suitable for use with Freon 12 and Freon 114 for short exposure. On long exposure, it tends to crack, craze, and become cloudy. Use with Freon 113 may be questionable. It probably should not be used with Freon 11.

Cast Lucite acrylic resin. Much more resistant to the effect of solvents than extruded resin. It can probably be used with most of the Freon compounds.

Polystyrene. Considerable variation found in individual samples. However, it is generally not suited for use with Freon compounds. Some applications might be all right with Freon 114.

Phenolic resins. Usually not affected by the Freon compounds. However, composition of resins of this type may be quite different. Samples should be tested before use.

Epoxy resins. Resistant to most solvents and entirely suitable for use with the Freon compounds.

Cellulose acetate or nitrate. Suitable for use with Freon compounds.

Delrin-acetal resin. Suitable for use with Freon compounds under most conditions.

Elastomers. Considerable variation is found in the effect of the Freon compounds on elastomers. The effect depends on the particular compound and elastomer type. In nearly all cases a satisfactory combination can be found. In some instances the presence of other materials, such as oils, may give unexpected results. Thus, preliminary testing of the system involved is recommended.

REFRIGERANT PROPERTIES

Refrigerants can be characterized by a number of properties. These properties are pressure, temperature, volume, density, and enthalpy. Also, flammability, ability to mix with oil, moisture reaction, odor, toxicity, leakage tendency, and leakage detection are important properties that characterize refrigerants.

Freon refrigerants R-11, R-12, R-22, plus ammonia and water will be used to show their properties in relationship to the mentioned categories. Freon R-11, R-12, and R-22 are common Freon refrigerants. The number assigned to ammonia is R-717, while water has the number R-718.

Pressure

The pressure of a refrigeration system is important. It determines how sturdy the equipment must be to hold the refrigerant. The refrigerant must be compressed and sent to various parts of the system under pressure. The main concern is keeping the pressure as low as possible. The ideal low-side pressure or evaporating pressure should be as near atmospheric pressure (14.7 lb/in.^{2}) as possible. This keeps down the price of the equipment. It also puts positive pressure on the system at all points. By having a small pressure, it is possible to prevent air and moisture from entering the system. In the case of a vacuum or a low pressure, it is possible for a leak to suck in air and moisture. Note the five refrigerants and their pressures in Table 6-5.

Table 6-5 *Operating Pressures*

Refrigerant	Evaporating Pressure (PSIG) at 5°F	Condensing Pressure (PSIG) at 86°F
R-11	24.0 in. Hg	3.6
R-12	11.8	93.2
R-22	28.3	159.8
R-717	19.6	154.5
R-718	29.7	28.6

Freon R-11 is used in very large systems because it requires more refrigerant than others, even though it has the best pressure characteristics of the group. Several factors must be considered before a suitable refrigerant is found. There is no ideal refrigerant for all applications.

Temperature

Temperature is important in selecting a refrigerant for a particular job. The boiling temperature is that point at which a liquid is vaporized upon the addition of heat. This, of course, depends upon the refrigerant and the absolute pressure at the surface of the liquid and vapor. Note that in Table 6-6, R-22 has the lowest boiling temperature. Water (R-718) has the highest boiling temperature. Atmospheric pressure is 14.7 lb/in.^{2}.

Once again, there is no ideal atmospheric boiling temperature for a refrigerant. However, temperature-pressure relationships are important in choosing a refrigerant for a particular job.

Volume

Specific volume is defined as the definite weight of a material. Usually expressed in terms of cubic feet per

Table 6-6 Refrigerants in Order of Boiling Point

ASHRAE number	Type of Refrigerant	Class of Refrigerant	Boiling Point °F (°C)
123	Single component	HCFC	82.2 (27.9)
11	Single component	CFC	74.9 (23.8)
245fa	Single component	HFC	59.5 (15.3)
236fa	Single component	HFC	29.5 (–1.4)
134a	Single component	HFC	–15.1 (–26.2)
12	Single component	CFC	–21.6 (–29.8)
401A	Zeotrope	HCFC	–27.7 (–32.2)
500	Azeotrope	CFC	–28.3 (–33.5)
409A	Zeotrope	HCFC	–29.6 (–34.2)
22	Single component	HCFC	–41.5 (–40.8)
407C	Zeotrope	HFC	–46.4 (–43.6)
502	Azeotrope	CFC	–49.8 (–45.4)
408A	Zeotrope	HCFC	–49.8 (–45.4)
404A	Zeotrope	HFC	–51.0 (–46.1)
507	Azetrope	HFC	–52.1 (–46.7)
402A	Zeotrope	HCFC	–54.8 (–48.2)
410A	Zeotrope	HFC	–62.9 (–52.7)
13	Single component	CFC	–114.6 (–81.4)
23	Single component	HFC	–115.7 (–82.1)
508B	Azeotrope	HFC	–125.3 (–87.4)
503	Azeotrope	CFC	–126.1 (–87.8)

pound, the volume is the reciprocal of density. The specific volume of a refrigerant is the number of cubic feet of gas that is formed when 1 lb of the refrigerant is vaporized. This is an important factor to be considered when choosing the size of refrigeration-system components. Compare the specific volumes (at 5°F) of the five refrigerants we have chosen. Freon R-12 and R-22 (the most often used refrigerants) have the lowest specific volumes as vapors. Refer to Table 6-7.

Table 6-7 Specific Volumes at 5°F

Refrigerant	Liquid Volume (cubic feet/lb)	Vapor Volume (cubic feet/lb)
R-11	0.010	12.27
R-12	0.011	1.49
R-22	0.012	1.25
R-717	0.024	8.15
R-718 (water)	0.016	12 444.40

Density

Density is defined as the mass or weight per unit of volume. In the case of a refrigerant, it is the weight in terms of volume given in pounds per cubic foot (lb/cu ft). Note in Table 6-8 that the density of R-11 is the greatest. The density of R-717 (ammonia) is the least.

Table 6-8 Liquid Density at 86°F

Refrigerant	Liquid Density (lb/ft^3)
R-11	91.4
R-12	80.7
R-22	73.4
R-717	37.2
R-718	62.4

Enthalpy

Enthalpy is the total heat in a refrigerant. The sensible heat plus the latent heat makes up the total heat. Latent heat is the amount of heat required to change the refrigerant from a liquid to a gas. The latent heat of vaporization is a measure of the heat per pound that the refrigerant can absorb from an area to be cooled. It is, therefore, a measure of the cooling potential of the refrigerant circulated through a refrigeration system. See Table 6-9. Latent heat is expressed in Btu per pound.

Table 6-9 Enthalpy (Btu/lb. at 5°F [–15°C])

Refrigerant	Liquid Enthalpy	+	Latent Heat of Vaporization	=	Vapor Enthalpy
R-11	8.88	+	84.00	=	92.88
R-12	9.32	+	60.47	=	78.79
R-22	11.97	+	93.59	=	105.56
R-717	48.30	+	565.00	=	613.30
R-718 (at 40°F)	8.05	+	1071.30	=	1079.35

Flammability

Of the five refrigerants mentioned so far, the only one that is flammable is ammonia. None of the Freon compounds is flammable or explosive. However, mixtures with flammable liquids or gases may be flammable and should be handled with caution. Partially halogenated compounds may also be flammable and must be individually examined. If the refrigerant is used around fire, its flammability should be carefully considered. Some city codes specify which refrigerants cannot be used within city limits.

Capability of Mixing with Oil

Some refrigerants mix well with oil. Others, such as ammonia and water, do not. The ability to mix with oil has advantages and disadvantages. If the refrigerant mixes easily, parts of the system can be lubricated easily by the refrigerant and its oil mixture. The refrigerant will bring the oil back to the compressor and moving parts for lubrication.

There is a disadvantage to the mixing of refrigerant and oil. If it is easily mixed, the refrigerant can mix with the oil during the off cycle and then carry off the oil once the unit begins to operate again. This means that the oil needed for lubrication is drawn off with the refrigerant. This can cause damage to the compressor and moving parts. With this condition, there is foaming in the compressor crankcase and loss of lubrication. In some cases, the compressor is burned out. Procedures for cleaning up a burned-out motor will be given later.

Moisture and Refrigerants

Moisture should be kept out of refrigeration systems. It can corrode parts of the system. Whenever low temperatures are produced, the water or moisture can freeze. If freezing of the metering device occurs, then refrigerant flow is restricted or cut of. The system will have a low efficiency or none at all. The degree of efficiency will depend upon the amount of icing or the part affected by the frozen moisture.

All refrigerants will absorb water to some degree. Those that absorb very little water permit free water to collect and freeze at low-temperature points. Those that absorb a high amount of moisture will form corrosive acids and corrode the system. Some systems will allow water to be absorbed and frozen. This causes corrosion.

Hydrolysis is the reaction of a material, such as Freon 12 or methyl chloride, with water. Acid materials are formed. The hydrolysis rate for the Freon compounds as a group is low compared with other halogenated compounds. Within the Freon group, however, there is considerable variation. Temperature, pressure, and the presence of other materials also greatly affect the rate. Typical hydrolysis rates for the Freon compounds and other halogenated compounds are given in Table 6-10.

Table 6-10 *Hydrolysis Rate in Water Grams/Litre of Water/Year*

	1 atm Pressure 86°F		Saturation Pressure 122°F
Compound	Water Alone	With Steel	With Steel
CH_3Cl	*	*	110
CH_2Cl_2	*	*	55
Freon 113	<0.005	ca. 50†	40
Freon 11	<0.005	ca. 10†	28
Freon 12	<0.005	0.8	10
Freon 21	<0.01	5.2	9
Freon 114	<0.005	1.4	3
Freon 22	<0.01	0.1	*
Freon 502	<0.01††	<0.01††	

*Not measured
† Observed rates vary
†† Estimated

With water alone at atmospheric pressure, the rate is too low to be determined by the analytical method used. When catalyzed by the presence of steel, the hydrolysis rates are detectable but still quite low. At saturation pressures and a higher temperature, the rates are further increased.

Under neutral or acidic conditions, the presence of hydrogen in the molecule has little effect on the hydrolytic stability. However, under alkaline conditions compounds containing hydrogen, such as Freon 22 and Freon 21, tend to be hydrolyzed more rapidly.

Odor

The five refrigerants are characterized by their distinct odor or the absence of it. Freon R-11, R-12, and R-22 have a slight odor. Ammonia (R-717) has a very acrid odor and can be detected even in small amounts. Water (R-718), of course, has no odor.

A slight odor is needed in a refrigerant so that its leakage can be detected. A strong odor may make it impossible to service equipment. Special gas masks may be needed. Some refrigerated materials may be ruined if the odor is too strong. About the only time that an odor is preferred in a refrigerant is when a toxic material is used for a refrigerant. A refrigerant that may be very inflammable should have an odor so that its leakage can be detected easily to prevent fire or explosions.

Toxicity

Toxicity is the characteristic of a material that makes it intoxicating or poisonous. Some refrigerants can be very toxic to humans, Others may not be toxic at all, The halogen refrigerants (R11, R-12, and R-22) are harmless in their normal condition or state. However, they form a highly toxic gas when an open name is used around them.

Water, of course, is not toxic. However, ammonia can be toxic if present in sufficient quantities. Make sure the manufacturer's recommended procedures for handling are followed when working with refrigerants.

Tendency to Leak

The size of the molecule makes a difference in the tendency of a refrigerant to leak. The greater the molecular weight, the larger the hole must be for the refrigerant to escape. A check of the molecular weight of a refrigerant will indicate the problem it may present to a sealed refrigeration system. Table 6-11 shows that R-11 has the least tendency to leak, whereas ammonia is more likely to leak.

Table 6-11 *Molecular Weight of Selected Refrigerants*

Refrigerant	Molecular Weight
R-11	137.4
R-12	120.9
R-22	86.5
R-717 (ammonia)	17.0
R-718 (water)	18.0

DETECTING LEAKS

There are several tests used to check for leaks in a closed refrigeration system. Most of them are simple. Following are some useful procedures:

- Hold the joint or suspected leakage point under water and watch for bubbles.
- Coat the area suspected of leakage with a strong solution of soap. If a leak is present, soap bubbles will be produced.

Sulfur Dioxide

To detect sulfur dioxide leaks, an ammonia swab may be used. The swab is made by soaking a sponge or cloth—tied onto a stick or piece of wire—in aqua ammonia. Household ammonia may also be used. A dense white smoke forms when the ammonia comes in contact with the sulfur dioxide. The usual soap bubble or oil test may be used when no ammonia is available.

If ammonia is used, check for leakage in the following ways:

- Burn a sulfur stick in the area of the leak. If there is a leak, a dense white smoke will be produced. The stronger the leak, the denser the white smoke.
- Hold a wet litmus paper close to the suspected leak area. If there is a leak, the ammonia will cause the litmus paper to change color.

Refrigerants that are halogenated hydrocarbons (Freon compounds) can be checked for leakage with a halide leak test. This involves holding a torch or flame close to the leak area. If there is a refrigerant leak, the flame will turn green. In every instance, the room should be well ventilated when the torch test is made.

An electronic detector for such refrigerant leaks is presently available. The detector gives off a series of rapid clicks if the refrigerant is present. The higher the concentration of the refrigerant, the more rapid the clicks. See Fig. 6-5.

Carbon Dioxide

Leaks can be detected with a soap solution if there is internal pressure on the part to be tested. When carbon dioxide is present in the condenser water, the water will turn yellow with the addition of bromthymol blue.

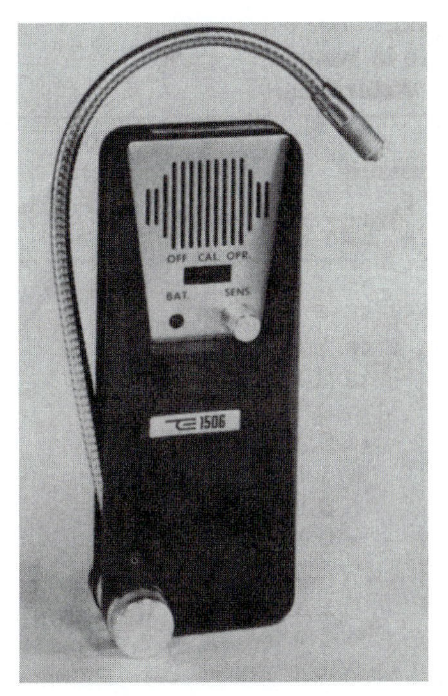

Fig. 6-5 *A hand-held electronic leak detector.* (Thermal Engineering)

Ammonia

Leaks are detected (in small amounts of ammonia) when a lit sulfur candle is used. The candle will give off a very thick, white smoke when it contacts the ammonia leak. The use of phenolphthalein paper is also considered a good test. The smallest trace of ammonia will *cause* the moistened paper strip to turn pink. A large ammonia will cause the phenolphthalein paper to turn a vivid scarlet.

Methyl Chloride

Leaks are detected by a leak-detecting halide torch. See Fig. 6-6. Some torches use alcohol for fuel and produce a colorless flange. When a methyl chloride leak is detected, the flame turns green. A brilliant blue flame is produced when large or stronger concentrations are present. In every instance, the room should be well ventilated when the torch test is made. The combustion of the refrigerant and the flame produces harmful chemicals. If a safe atmosphere is not present, the soap-bubble test or oil test should be used to check for leaks.

Fig. 6-6 *A halide gas leak detector.* (Turner)

As mentioned, methyl chloride is hard to detect with the nose or eyes. It does not produce irritating effects. Therefore, some manufacturers add a 1 percent amount of *acrolein* as a warning agent. Acrolein is a colorless liquid (C_3H_4O) with a pungent odor.

BAN ON PRODUCTION AND IMPORTS OF OZONE-DEPLETING REFRIGERANTS

In 1987 the Montreal Protocol, an international environmental agreement, established requirements that began the worldwide phase out of ozone-depleting *chlorofluorocarbons* (CFCs). These requirements were later modified. This led to the phase out, in 1996, of CFC production in all developed nations. In 1992 an amendment to the Montreal Protocol established a schedule for the phase out of refrigerants, hydrochlorofluorocarbons (HCFCs).

HCFCs are substantially less damaging to the ozone layer than CFCs. However, they still contain ozone-destroying chlorine. The Montreal Protocol, as amended, is carried out in the United States through Title VI of the Clean Air Act. This Act is implemented by the EPA or *Environmental Protection Agency*.

An HCFC, known as R-22, has been the refrigerant of choice for residential heat pump and air-conditioning systems for more than four decades. Unfortunately for the environment, releases of R-22 that result from system leaks contribute to ozone depletion. In addition, the manufacture of R-22 results in a by-product that contributes significantly to global warming.

As the manufacture of R-22 is phased out over the coming years as part of the agreement to end production of HCFCs, manufacturers of residential air-conditioning systems are beginning to offer equipment that use ozone-friendly refrigerants. Many homeowners may be misinformed about how much longer R-22 will be available to service their central air-conditioning systems and heat pumps. The future availability of R-22, and the new refrigerants that are replacing R-22 will be covered here. The EPA document assists consumers in deciding what to consider when purchasing a new air-conditioning system or heat pump, or when having an existing system repaired.

Phase-out Schedule for HCFCs, Including R-22

Under the terms of the Montreal Protocol, the United States agreed to meet certain obligations by specific dates. That will affect the residential heat pump and air-conditioning industry.

January 1, 2004 In accordance with the terms of the Protocol, the amount of all the HCFCs that can be produced nationwide must be reduced by 35 percent by 2004. In order to achieve this goal, the United States has ceased production of HCFC-141b, the most ozone damaging of this class of chemicals, on January 1, 2003. This production ban should greatly reduce nationwide use of HCFCs as a group and make it likely that the 2004 deadline will have a minimal effect on R-22 supplies.

January 1, 2010 After 2010, chemical manufacturers may still produce R-22. But this is to service existing equipment and not for use in new equipment. As a result, *heating, ventilation and air-conditioning* (HVAC) system manufacturers will only be able to use preexisting supplies of R-22 in the production of new air conditioners and heat pumps. These existing supplies will include R-22 recovered from existing equipment and recycled by licensed reclaimers.

January 1, 2020 Use of existing refrigerant, including refrigerant that has been recovered and recycled, will be allowed beyond 2020 to service existing systems. However, chemical manufacturers will no longer be able to produce R-22 to service existing air conditioners and heat pumps.

What does the R-22 phase out mean for consumers? The following paragraphs are an attempt to answer this question.

Availability of R-22

The Clean Air Act does not allow any refrigerant to be vented into the atmosphere during installation, service, or retirement of equipment. Therefore, R-22 must be:

- Recovered and recycled (for reuse in the same system)
- Reclaimed (reprocessed to the same purity levels as new R-22)
- Destroyed

After 2020, the servicing of R-22-based systems will rely on recycled refrigerants. It is expected that reclamation and recycling will ensure that existing supplies of R-22 will last longer and be available to service a greater number of systems. As noted earlier, chemical manufacturers will be able to produce R-22 for use in new air-conditioning equipment until 2010, and they can continue production of R-22 until 2020 for use in servicing that equipment. Given this schedule, the transition away from R-22 to the use of ozone-friendly refrigerants should be smooth. For the next 20 years or more, R-22 should continue to be available for all systems that require R-22 for servicing.

Cost of R-22

While consumers should be aware that prices of R-22 may increase as supplies dwindle over the next 20 or 30 years, EPA believes that consumers are not likely to be subjected to major price increases within a short time period. Although there is no guarantee that service costs of R-22 will not increase, the lengthy phase-out period for R-22 means that market conditions should not be greatly affected by the volatility and resulting refrigerant price hikes that have characterized the phase out of R-12, the refrigerant used in automotive air-conditioning systems and replaced by R-134a.

ALTERNATIVES TO R-22

Alternatives for residential air conditioning will be needed as R-22 is gradually phased out. Nonozone-depleting alternative refrigerants are being introduced. Under the Clean Air Act, EPA reviews alternatives to ozone-depleting substances like R-22 in order to evaluate their effects on human health and the environment. The EPA has reviewed several of these alternatives to R-22 and has compiled a list of substitutes that the EPA has determined are acceptable. One of these substitutes is R-410A, a blend of HFCs, substances that do not contribute to depletion of the ozone layer, but, like R-22, contribute to global warming. R-410A is manufactured and sold under various trade names, including Genetron AZ 20, SUVA 410A, and Puron. Additional refrigerants on the list of acceptable substitutes include R-134a and R-407C. These two refrigerants are not yet available for residential applications in the United States, but are commonly found in residential air-conditioning systems and heat pumps in Europe. EPA will continue to review new nonozone-depleting refrigerants as they are developed.

Servicing existing units

Existing units using R-22 can continue to be serviced with R-22. There is no EPA requirement to change or convert R-22 units for use with a nonozone-depleting substitute refrigerant. In addition, the new substitute refrigerants cannot be used without making some changes to system components. As a result, service technicians who repair leaks to the system will continue to charge R-22 into the system as part of that repair.

Installing new units

The transition away from ozone-depleting R-22 to systems that rely on replacement refrigerants, like R-410A, has required redesign of heat pump and air-conditioning systems. New systems incorporate compressors and other components specifically designed for use with specific replacement refrigerants. With these significant product and production process changes, testing and training must also change. Consumers should be aware that dealers of systems that use substitute refrigerants should be schooled in installation and service techniques required for use of that substitute refrigerant.

Servicing Your System

Along with prohibiting the production of ozone-depleting refrigerants, the Clean Air Act also mandates the use of common sense in handling refrigerants. By containing and using refrigerants responsibly, that is, by recovering, recycling, and reclaiming, and by reducing leaks, their ozone depletion and global-warming consequences are minimized. The Clean Air Act outlines specific refrigerant containment and management practices for HVAC manufacturers, distributors, dealers, and technicians. Properly installed home comfort systems rarely develop refrigerant leaks, and with proper servicing, a system using R-22, R-410A, or another refrigerant will minimize its impact on the environment. While EPA does not mandate repairing or replacing small systems because of leaks, system leaks can not only harm the environment, but also result in increased maintenance costs.

One important thing a homeowner can do for the environment, regardless of the refrigerant used, is to select a reputable dealer that employs service technicians who are EPA-certified to handle refrigerants. Technicians often call this certification "Section 608 certification," referring to the part of the Clean Air Act that requires minimizing releases of ozone-depleting chemicals from HVAC equipment.

Purchasing New Systems

Another important thing a homeowner can do for the environment is to purchase a highly energy-efficient system. Energy-efficient systems result in cost savings for the homeowner. Today's best air conditioners use much less energy to produce the same amount of cooling as air conditioners made in the mid-1970s. Even if your air conditioner is only 10 years old, you may save significantly on your cooling energy costs by replacing it with a newer, more efficient model. Products with EPA's Energy Star label can save homeowners 10 to 40 percent on their heating and cooling bills every year. These products are made by most major manufacturers and have the same features as standard products, but also incorporate energy-saving technology. Both R-22 and R-410A systems may have the Energy Star label. Equipment that displays the Energy Star label must have a minimum *seasonal energy efficiency ratio* (SEER). The higher the SEER specification, the more efficient the equipment.

Energy efficiency, along with performance, reliability, and cost, should be considered in making a decision. And do not forget that when purchasing a new system, you can also speed the transition away from ozone-depleting R-22 by choosing a system that uses ozone-friendly refrigerants.

AIR CONDITIONING AND WORKING WITH HALON

Several regulations have been issued under Section 608 of the Clean Air Act to govern the recycling of refrigerants in stationary systems and to end the practice of venting refrigerants to the air. These regulations also govern the handling of halon fire-extinguishing agents. A Web site and both the regulations themselves and fact sheets are available from the EPA Stratospheric Ozone Hotline at 1-800-296-1996.

NOTE: The handling and recycling of refrigerants used in motor vehicle air-conditioning systems are governed under section 609 of the Clean Air Act.

General Information

April 13, 2005 EPA is finalizing a rulemaking amending the definition of refrigerant to make certain that it only includes substitutes that consist of a class I or class II *ozone-depleting substance* (ODS). This rulemaking also amends the venting prohibition to make certain that it remains illegal to knowingly vent nonexempt substitutes that do not consist of a class I or class II ODS, such as R-134a and R-410A.

January 11, 2005 EPA has published a final rule extending the leak repair required practices and the associated reporting and record-keeping requirements to owners and/or operators of comfort cooling, commercial refrigeration, or industrial process refrigeration appliances containing more than 50 pounds of a substitute refrigerant, if the substitute contains a class I or class II ozone-depleting substance (ODS). In addition, EPA has defined leak rate in terms of the percentage of the appliance's full charge that would be lost over a consecutive 12-month period, if the current rate of loss were to continue over that period. EPA now requires calculation of the leak rate every time that refrigerant is added to an appliance.

March 12, 2004 EPA finalizes rulemaking sustaining the Clean Air Act prohibition against venting hydrofluorocarbon (HFC) and perfluorocarbon (PFC) refrigerants. This rulemaking finds that the knowing venting of HFC and PFC refrigerants during the maintenance, service, repair, and disposal of air-conditioning and refrigeration equipment (i.e., appliances) remains illegal under Section 608 of the Clean Air Act. The ruling also restricts the sale of HFC refrigerants that consist of an ODS to EPA-certified technicians. However, HFC refrigerants and HFC refrigerant blends that do not consist of an ODS are not covered under *The Refrigerant Sales Restriction*, a brochure that documents the environmental and financial reasons to replace CFC chillers with new, energy-efficient equipment. A partnership of governments, manufacturers, *nongovernmental organizations* (NGOs) and others have endorsed the brochure to eliminate uncertainty and underscore the wisdom of replacing CFC chillers.

LEAK REPAIR

The leak-repair requirements, promulgated under Section 608 of the Clean Air Act Amendments of 1990, require that when an owner or operator of an appliance that normally contains a refrigerant charge of more than 50 lb discovers that refrigerant is leaking at a rate that would exceed the applicable trigger rate during a 12-month period, the owner or operator must take corrective action.

Trigger Rates

For all appliances that have a refrigerant charge of more than 50 lb, the following leak rates for a 12-month period are applicable (Table 6-12).

Table 6-12 *Trigger Leak Rates*

Appliance Type	Trigger Leak Rate
Commercial refrigeration	35%
Industrial process refrigeration	35%
Comfort cooling	15%
All other appliances	15%

In general, owners or operators must either repair leaks within 30 days from the date the leak was discovered, or develop a dated retrofit/retirement plan within 30 days and complete actions under that plan within 1 year from the plan's date. However, for industrial process refrigeration equipment and some federally-owned chillers, additional time may be available.

Industrial process refrigeration is defined as complex customized appliances used in the chemical, pharmaceutical, petrochemical, and manufacturing industries. These appliances are directly linked to the industrial process. This sector also includes industrial ice machines, appliances used directly in the generation of electricity, and in ice rinks. If at least 50 percent of an appliance's capacity is used in an industrial process refrigeration application, the appliance is considered industrial process refrigeration equipment and the trigger rate is 35 percent.

Industrial process refrigeration equipment and federally-owned chillers must conduct initial and follow-up verification tests at the conclusion of any repair efforts. These tests are essential to ensure that the repairs have been successful. In cases where an industrial process shutdown is required, a repair period of 120 days is substituted for the normal 30-day repair period. Any appliance that requires additional time may be subject to record keeping/reporting requirements.

When Additional Time is Necessary

Additional time is permitted for conducting leak repairs where the necessary repair parts are unavailable or if other applicable federal, state, or local regulations make a repair within 30/120 days impossible. If owners or operators choose to retrofit or retire appliances, a retrofit or retirement plan must be developed within 30 days of detecting a leak rate that exceeds the trigger rates. A copy of the plan must be kept on site. The original plan must be made available to EPA upon request. Activities under the plan must be completed within 12 months (from the date of the plan). If a request is made within 6 months from the expiration of the initial 30-day period, additional time beyond the 12-month period is available for owners or operators of industrial process refrigeration equipment and federally-owned chillers in the following cases: EPA will permit additional time to the extent reasonably necessary where a delay is caused by the requirements of other applicable federal, state, or local regulations; or where a suitable replacement refrigerant, in accordance with the regulations promulgated under Section 612, is not available; and EPA will permit one additional 12-month period where an appliance is custom-built and the supplier of the appliance or a critical component has quoted a delivery time of more than 30 weeks from when the order was placed (assuming the order was placed in a timely manner). In some cases, EPA may provide additional time beyond this extra year where a request is made by the end of the ninth month of the extra year.

Relief from Retrofit/Retirement

The owners or operators of industrial process refrigeration equipment or federally-owned chillers may be relieved from the retrofit or repair requirements if:

- Second efforts to repair the same leaks that were subject to the first repair efforts are successful
- Within 180 days of the failed follow-up verification test, the owners or operators determine the leak rate is below 35 percent. In this case, the owners or operators must notify EPA as to how this determination will be made, and must submit the information within 30 days of the failed verification test.

System Mothballing

For all appliances subject to the leak-repair requirements, the timelines may be suspended if the appliance has undergone system mothballing. System mothballing means the intentional shutting down of a refrigeration appliance undertaken for an extended period of time where the refrigerant has been evacuated from the appliance or the affected isolated section of the appliance to at least atmospheric pressure. However, the timelines pick up again as soon as the system is brought back on line.

EPA-CERTIFIED REFRIGERANT RECLAIMERS

The EPA listing of reclaimers is updated when additional refrigerant reclaimers are approved. Reclaimers appearing on this list are approved to reprocess used refrigerant to at least the purity specified in appendix A to 40 CFR part 82, subpart F (based on ARI Standard 700, "Specifications for Fluorocarbon and Other Refrigerants"). Reclamation of used refrigerant by an EPA-certified reclaimer is required in order to sell used refrigerant not originating from and intended for use with motor vehicle air conditioners.

The EPA encourages reclaimers to participate in a voluntary third-party reclaimer certification program operated by the *Air-Conditioning and Refrigeration Institute* (ARI). The volunteer program offered by the ARI involves quarterly testing of random samples of reclaimed refrigerant. Third-party certification can enhance the attractiveness of a reclaimer's program by providing an objective assessment of its purity.

NEWER REFRIGERANTS

Since the world has become aware of the damage the Freon refrigerants can do to the ozone layer, there has been a mad scramble to obtain new refrigerants that can replace all those now in use. There are some problems with adjusting the new and especially existing equipment to the properties of new refrigerant blends.

It is difficult to directly replace R-12 for instance. It has been the mainstay in refrigeration equipment for years. However, the automobile air-conditioning industry has been able to reformulate R-12 to produce an acceptable substitute, R-134a. There are others now available to substitute in the more sophisticated equipment with large amounts of refrigerants. Some of these will be covered here.

FREON REFRIGERANTS

The Freon family of refrigerants has been one of the major factors responsible for the impressive growth of not only the home refrigeration and air-conditioning industry, but also of the commercial refrigeration industry. The safe properties of these products have permitted their use under conditions where flammable or more toxic refrigerants would be hazardous.

Classifications

Following were commonly used Freon refrigerants:

- *Freon-11.* Freon-11 has a boiling point of 74.87°F and has wide usage as a refrigerant in indirect industrial and commercial air-conditioning systems employing single or multistage centrifugal compressors with capacities of 100 tons and above. Freon-11 is also employed as brine for low-temperature applications. It provides relatively low operating pressures with moderate displacement requirements.
- *Freon-12.* The boiling point of Freon-12 is –21.7°F. It is the most widely known and used of the Freon refrigerants. It is used principally in household and commercial refrigeration and air-conditioning units, for refrigerators, frozen-food cabinets, ice-cream cabinets, food-locker plants, water coolers, room and window air-conditioning units, and similar equipment. It is generally used in reciprocating compressors, ranging in size from fractional to 800 hp. Rotary compressors are useful in small units. The use of centrifugal compressors with Freon-12 for large air-conditioning and process-cooling applications is increasing.
- *Freon-13.* The boiling point of Freon-13 is –144.6°F. It is used in low-temperature specialty applications employing reciprocating compressors and generally in cascade with Freon-12 or Freon-22.
- *Freon-21.* Freon-21 has a boiling point of 48°F. It is used in fractional-horsepower household refrigerating systems and drinking-water coolers employing rotary vane-type compressors. Freon-21 is also used in comfort-cooling air-conditioning systems of the

absorption type where dimethyl ether or tetraethylene glycol is used as the absorbent.

- *Freon-22.* The boiling point of Freon-22 is –41.4°F. It is used in all types of household and commercial refrigeration and air-conditioning applications with reciprocating compressors. The outstanding thermodynamic properties of Freon-22 permit the use of smaller equipment than is possible with similar refrigerants, making it especially suitable where size is a problem.
- *Freon-113.* The boiling point of Freon-113 is 117.6°F. It is used in commercial and industrial air conditioning and process water and brine cooling with centrifugal compression. It is especially useful in small-tonnage applications.
- *Freon-114.* The boiling point of Freon-114 is 38.78°F. It is used as a refrigerant in fractional-horsepower household refrigerating systems and drinking-water coolers employing rotary vane-type compressors. It is also used in indirect industrial and commercial air-conditioning systems and in industrial process water and brine cooling to –70°F employing multistage centrifugal-type compressors in cascade of 100-tons refrigerating capacity and larger.
- *Freon-115.* The boiling point of Freon-115 is –37.7°F. It is especially stable, offering a particularly low discharge temperature in reciprocating compressors. Its capacity exceeds that of Freon-12 by as much as 50 percent in low-temperature systems. Its potential applications include household refrigerators and automobile air conditioning.
- *Freon-502.* Freon-502 is an azeotropic mixture composed of 48.8 percent Freon-22 and 51.2 percent Freon-115 by weight. It boils at –50.1°F. Because it permits achieving the capacity of Freon-22 with discharge temperatures comparable to Freon-12, it is finding new reciprocating compressor applications in low-temperature display cabinets and in storing and freezing of food.

Properties of Freons

The Freon refrigerants are colorless and almost odorless, and their boiling points vary over a wide range of temperatures. Those Freon refrigerants that are produced are nontoxic, non-corrosive, nonirritating, and nonflammable under all conditions of usage. They are generally prepared by replacing chlorine or hydrogen with fluorine. Chemically, Freon refrigerants are inert and thermally stable up to temperatures far beyond conditions found in actual operation. However, Freon is harmful when allowed to escape into the atmosphere. It can deplete the ozone layer around earth and cause more harmful ultraviolet rays to reach the surface of the earth.

Physical Properties

The pressures required in liquefying the refrigerant vapor affect the design of the system. The refrigerating effect and specific volume of the refrigerant vapor determine the compressor displacement. The heat of vaporization and specific volume of the liquid refrigerant affect the quantity of refrigerant to be circulated through the pressure-regulating valve or other system device.

Flammability Freon is nonflammable and noncombustible under conditions where appreciable quantities contact flame or hot metal surfaces. It requires an open flame at 1382°F to decompose the vapor. Even at this temperature, only the vapor decomposes to form hydrogen chloride and hydrogen fluoride, which are irritating but are readily dissolved in water. Air mixtures are not capable of burning and contain no elements that will support combustion. For this reason, Freon is considered nonflammable.

Amount of Liquid Refrigerant Circulated It should be noted that the Freon refrigerants have relatively low-heat values, but this must not be considered a disadvantage. It simply means that a greater volume of liquid must be circulated per unit of time to produce the desired amount of refrigeration. It does not concern the amount of refrigerant in the system. Actually, it is a decided advantage (especially in the smaller- or low-tonnage systems) to have a refrigerant with low-heat values. This is because the larger quantity of liquid refrigerant to be metered through the liquid-regulating device will permit the use of more accurate and more positive operating and regulating mechanisms of less sensitive and less critical adjustments. Table 6-13 lists the quantities of liquid refrigerant metered or circulated per minute under standard ton conditions.

Volume (Piston) Displacement For reason of compactness, cost of equipment, reduction of friction, and compressor speed, the volume of gas that must be compressed per unit of time for a given refrigerating effect, in general, should be as low as possible. Freon-12 has a relatively low-volume displacement, which makes it suitable for use in reciprocating compressors, ranging from the smallest size to those of up to 800-ton capacity, including compressors for household and commercial refrigeration. Freon-12 also permits the construction of compact rotary compressors in the commercial sizes. Generally, low-volume displacement (high-pressure) refrigerants are used in reciprocating compressors; high-volume displacement (low-pressure) refrigerants are used in large-tonnage

***Table 6-13** Quantities of Refrigerant Circulated per Minute Under Standard Ton Conditions*

Refrigerant	Pounds Expanded per Minute	Ft³/lb Liquid 86°F	In.³ Liquid Expanded per Minute	Specific Gravity Liquid 86°F (water-1)
Freon-22	2.887	0.01367	67.97	1.177
Freon-12	3.916	0.0124	83.9	1.297
Freon-114	4.64	0.01112	89.16	1.443
Freon-21	2.237	0.01183	45.73	1.360
Freon-11	2.961	0.01094	55.976	1.468
Freon-113	3.726	0.01031	66.48	1.555

centrifugal compressors; intermediate-volume (intermediate-pressure) refrigerants are used in rotary compressors. There is no standard rule governing this usage.

Condensing Pressure Condensing (high-side) pressure should be low to allow construction of lightweight equipment, which affects power consumption, compactness, and installation. High pressure increases the tendency toward leakage on the low side as well as the high side when pressure is built up during idle periods. In addition, pressure is very important from the standpoint of toxicity and fire hazard.

In general, a low-volume displacement accompanies a high-condensing pressure, and a compromise must usually be drawn between the two in selecting a refrigerant. Freon-12 presents a balance between volume displacement and condensing pressure. Extra-heavy construction is not required for this type of refrigerant, and so there is little or nothing to be gained from the standpoint of weight of equipment in using a lower-pressure refrigerant.

Evaporating Pressure Evaporating (low-side) pressures above atmospheric are desirable to avoid leakage of moisture-laden air into the refrigerating systems and permit easier detection of leaks. This is especially important with open-type units. Air in the system will increase the head pressures, resulting in inefficient operations, and may adversely affect the lubricant. Moisture in the system will cause corrosion and, in addition, may freeze out and stop operation of the equipment.

In general, the higher the evaporating pressure, the higher the condensing pressure under a given set of temperatures. Therefore, to keep head pressures at a minimum and still have positive low-side pressures, the refrigerant selected should have a boiling point at atmospheric pressure as close as possible to the lowest temperature to be produced under ordinary operating conditions. Freon-12, with a boiling point of –21.7°F, is close to ideal in this respect for most refrigeration applications. A still lower boiling point is of some advantage only when lower-operating temperatures are required.

REFRIGERANT CHARACTERISTICS

The freezing point of a refrigerant should be below any temperature that might be encountered in the system. The freezing point of all refrigerants, except water (32°F) and carbon dioxide (–69.9°F, triple point), are far below the temperatures that might be encountered in their use. Freon-12 has a freezing point of –252°F. See App. 1 for more details on refrigerants.

Critical Temperature

The critical temperature of a refrigerant is the highest temperature at which it can be condensed to a liquid, regardless of a higher pressure. It should be above the highest condensing temperature that might be encountered. With air-cooled condensers, in general, this would be above 130°F. Loss of efficiency caused by superheating of the refrigerant vapor on compression and by throttling expansion of the liquid is greater when the critical temperature is low.

All common refrigerants have satisfactorily high critical temperatures, except carbon dioxide (87.8°F) and ethane (89.8°F). These two refrigerants require condensers cooled to temperatures below their respective critical temperatures, thus generally requiring water.

Hydrofluorocarbons (HFCs) There are some HFC refrigerants (such as R-134a) that are made to eliminate the problems with refrigerants in the atmosphere caused by leaks in systems. The R-134a is a nonozone-depleting refrigerant used in vehicle air-conditioning systems. DuPont's brand name is Suva, and the product is produced in a plant located in Corpus Christi, Texas, as well in Chiba, Japan. According to DuPont's Web site, R-134a was globally adopted by all vehicle manufacturers in the early 1990s as a replacement for CFC-12. The transition to R-134a was completed by the mid-1990s for most major automobile manufacturers. Today, there are more than 300 million cars with air conditioners using the newer refrigerant.

Latent Heat of Evaporation

A refrigerant should have a high latent heat of evaporation per unit of weight so that the amount of refrigerant circulated to produce a given refrigeration effect may be small. Latent heat is important when considering its relationship to the volume of liquid required to be circulated. The net result is the refrigerating effect. Since other factors enter into the determination of these, they are discussed separately.

The refrigerant effect per pound of refrigerant under standard ton conditions determines the amount of refrigerant to be evaporated per minute. The refrigerating effect per pound is the difference in Btu content of the saturated vapor leaving the evaporator (5°F) and the liquid refrigerant just before passing through the regulating valve (86°F). While the Btu refrigerating effect per pound directly determines the number of pounds of refrigerant to be evaporated in a given length of time to produce the required results, it is much more important to consider the volume of the refrigerant vapor required rather than the weight of the liquid refrigerant. By considering the volume of refrigerant necessary to produce standard ton conditions, it is possible to make a comparison between Freon-12 and other refrigerants so as to provide for the reproportioning of the liquid orifice sizes in the regulating valves, sizes of liquid refrigerant lines, and so on.

A refrigerant must not be judged only by its refrigerating effect per pound, but the volume per pound of the liquid refrigerant must also be taken into account to arrive at the volume of refrigerant to be vaporized. Although Freon-12 has relatively low refrigerating effect, this is not a disadvantage, because it merely indicates that more liquid refrigerant must be circulated to produce the desired amount of refrigeration. Actually, it is a decided advantage to circulate large quantities of liquid refrigerant because the greater volumes required will permit the use of less sensitive operating and regulating mechanisms with less critical adjustment.

Refrigerants with high Btu refrigerating effects are not always desirable, especially for household and small commercial installations, because of the small amount of liquid refrigerant in the system and the difficulty encountered in accurately controlling its flow through the regulating valve. For household and small commercial systems, the adjustment of the regulating-valve orifice is most critical for refrigerants with high Btu values.

Specific Heat

A low specific heat of the liquid is desirable in a refrigerant. If the ratio of the latent heat to the specific heat of a liquid is low, a relatively high proportion of the latent heat may be used in lowering the temperature of the liquid from the condenser temperature to the evaporator temperature. This results in a small net-refrigerating effect per pound of refrigerant circulated and, assuming other factors remain the same, reduces the capacity and lowers the efficiency. When the ratio is low, it is advantageous to precool the liquid before evaporation by heat interchange with the cool gases leaving the evaporator.

In the common type of refrigerating systems, expansion of the high-pressure liquid to a lower-pressure, lower-temperature vapor and liquid take place through a throttling device such as an expansion valve. In this process, energy available from the expansion is not recovered as useful work. Since it performs no external work, it reduces the net-refrigerating effect.

Power Consumption

In a perfect system operating between 5 and –86°F conditions, 5.74 Btu is the maximum refrigeration obtainable per Btu of energy used to operate the refrigerating system. This is the theoretical maximum *coefficient of performance* on cycles of maximum efficiency (for example, the Carnet cycle). The minimum horsepower would be 0.821 hp/ton of refrigeration. The theoretical coefficient of performance would be the same for all refrigerants if they could be used on cycles of maximum efficiency.

However, because of engineering limitations, refrigerants are used on cycles with a theoretical maximum coefficient of performance of less than 5.74. The cycle most commonly used differs in its basic form from (1) the Carnet cycle, as already explained in employing expansion without loss or gain of heat from an outside source, and (2) in compressing adiabatically (compression without gaining or losing heat to an outside source) until the gas is superheated above the condensing medium temperature. These two factors, both of which increase the power requirement, vary in importance with different refrigerants. But, it so happens that when expansion loss is high, compression loss is generally low, and vice versa. All common refrigerants (except carbon dioxide and water) show about the same overall theoretical power requirement on a 5 to –86°F cycle. At least the theoretical differences are so small that other factors are more important in determining the actual differences in efficiency.

The amount of work required to produce a given refrigerating effect increases as the temperature level to which the heat is pumped from the cold body is increased. Therefore, on a 5 to –86°F cycle, when gas is superheated above 86°F temperature on compression, efficiency is decreased and the power requirement increased unless the refrigerating effect caused by superheating is salvaged through the proper use of a heat interchanger.

Volume of Liquid Circulated

Volumes of liquid required to circulate for a given refrigerant effect should be low. This is to avoid fluid-flow (pressure-drop) problems and to keep down the size of the required refrigerant change. In small-capacity machines, the volume of liquid circulated should not be so low as to present difficult problems in accurately controlling its flow through expansion valves or other types of liquid-metering devices.

With a given net-refrigerating effect per pound, a high density of liquid is preferable to a low volume. However, a high density tends to increase the volume circulated by lowering the net-refrigerating effect.

HANDLING REFRIGERANTS

One of the requirements of an ideal refrigerant is that it must be nontoxic. In reality, however, all gases (with the exception of pure air) are more or less toxic or asphyxiating. It is therefore important that wherever gases or highly volatile liquids are used, adequate ventilation be provided, because even nontoxic gases in air produce a suffocating effect.

Vaporized refrigerants (especially ammonia and sulfur dioxide) bring about irritation and congestion of the lungs and bronchial organs, accompanied by violent coughing, vomiting, and, when breathed in sufficient quantity, suffocation. It is of the utmost importance, therefore, that the serviceman subjected to a refrigerant gas find access to fresh air at frequent intervals to clear his lungs. When engaged in the repair of ammonia and sulfur dioxide machines, approved gas masks and goggles should be used. Carrene, Freon (R-12), and carbon dioxide fumes are not irritating and can be inhaled in considerable concentrations for short periods without serious consequences.

It should be remembered that liquid refrigerant would refrigerate or remove heat from anything it meets when released from a container. In the case of contact with refrigerant, the affected or injured area should be treated as if it has been frozen or frostbitten.

Storing and Handling Refrigerant Cylinders

Refrigerant cylinders should be stored in a dry, sheltered, and well-ventilated area. The cylinders should be placed in a horizontal position, if possible, and held by blocks or saddles to prevent rolling. It is of utmost importance to handle refrigerant cylinders with care and to observe the following precautions:

- Never drop the cylinders, or permit them to strike each other violently.
- Never use a lifting magnet or a sling (rope or chain) when handling cylinders. A crane may be used when a safe cradle or platform is provided to hold the cylinders.
- Caps provided for valve protection should be kept on the cylinders at all times except when the cylinders are actually in use.
- Never overfill the cylinders. Whenever refrigerant is discharged from or into a cylinder, weigh the cylinder and record the weight of the refrigerant remaining in it.
- Never mix gases in a cylinder.
- Never use cylinders for rollers, supports, or for any purpose other than to carry gas.
- Never tamper with the safety devices in valves or on the cylinders.
- Open the cylinder valves slowly. Never use wrenches or tools except those provided or approved by the gas manufacturer.
- Make sure that the threads on regulators or other unions are the same as those on the cylinder-valve outlets. Never force a connection that does not fit.
- Regulators and gages provided for use with one gas must not be used on cylinders containing a different gas.
- Never attempt to repair or alter the cylinders or valves.
- Never store the cylinders near highly flammable substances (such as oil, gasoline, or waste).
- Cylinders should not be exposed to continuous dampness, salt water, or salt spray.
- Store full and empty cylinders apart to avoid confusion.
- Protect the cylinders from any object that will produce a cut or other abrasion on the surface of the metal.

LUBRICANTS*

Lubricant properties can be evaluated. It can be determined if the product is right for the job. Three basic properties are:

- Viscosity
- Lubricity
- Chemical stability

They must be satisfactory to protect the compressor. The correct viscosity is needed to fill the gaps between parts and flow correctly where it is supposed to go. Generally speaking, smaller equipment with smaller

*Courtesy of National Refrigerants.

gaps between moving parts requires a lighter viscosity, and larger equipment with bigger parts needs heavier viscosity oils.

Lubricity refers to the lubricant's ability to protect the metal surfaces from wear. Good chemical stability means the lubricant will not react to form harmful chemicals such as acids, sludges, and so forth that may block tubing or there may be carbon deposits. The interaction of lubricant and refrigerant can cause potential problems as well.

Miscibility defines the temperature region where refrigerant and oil will mix or separate. If there is separation of the oil from the refrigerant in the compressor it is possible that the oil is not getting to metal parts that need it. If there is separation in the evaporator or other parts of the system it is possible that the oil does not return to the compressor and eventually there is not enough oil to protect it.

Solubility determines if the refrigerant will thin the oil too much. That would cause it to lose its ability to protect the compressor. The thinning effect also influences oil return.

Once you mix a blend at a given composition, the pressure-temperature relationships follow the same general rules as for pure components. For example, the pressure goes up when the temperature goes up. For three blends containing different amounts of A and B, the pressure curve is similarly shaped, but the resulting pressure will be higher for the blend which contains more of the A or higher pressure component.

Some refrigerant blends that are intended to match some other product. R-12 is a good example. It will rarely match the pressure at all points in the desired temperature range. What is more common is the blend will match in one region and the pressures will be different elsewhere.

In the example, the blend with concentration C1 matches the CFC at cold-evaporator temperatures, but the pressures run higher at condenser conditions. The blend with composition C2 matches closer to room temperature. And, it may show the same pressure in a cylinder being stored, for example. The operation pressures at evaporator and condenser temperatures, however, will be somewhat different. Finally, the blend at C3 will generate the same pressures at hot condenser conditions, but the evaporator must run at lower pressures to get the same temperature. See Fig. 6-7.

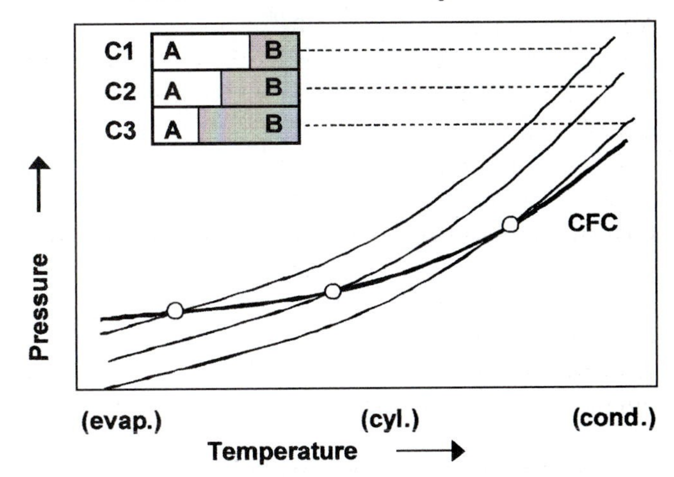

Fig. 6-7 *New refrigerants variable: composition.* (National Refrigerants)

It can be seen later that the choice of where the blend matches the pressure relationship can solve, or cause, certain retrofit-related problems.

Generally speaking, the R-12 retrofit blends have higher temperature glide. They do not match the pressure/temperature/capacity of R-12 across the wide temperature application range which R-12 once did. In other words, one blend does not fit all. Blends which match R-12 at colder evaporator temperatures may generate higher pressures and discharge temperatures when used in warmer applications or in high-ambient temperatures. These are called *refrigeration blends*.

In refrigeration it is often an easier, and cheaper, retrofit job if you can match evaporator pressures to R-12 and split the glide. That is because you can get similar box temperatures in similar run times. And, you would probably not need to change controls or the TXVs, which are sensitive to pressure.

Blends, which match R-12 properties in hot conditions, such as in automotive AC condensers, may lose capacity or require lower suction pressures when applied at colder evaporator temperatures. These are called *Automotive Blends.*

For automotive air conditioning many of the controls and safety switches are related to the high-side pressure. If the blend generates higher discharge pressures you could short cycle more often and lose capacity in general. It is better to pick the high side to match R-12 and let the low side run a little lower pressure.

R-134a REFRIGERANT

The blended refrigerant R-134a is a long-term, HFC alternative with similar properties to R-12. It has become the new industry-standard refrigerant for automotive air-conditioning and refrigerator/freezer appliances.

R-134a refrigerating performance will suffer at lower temperatures (below −10°F). Some traditional R-12 applications have used alternatives other than R-134a for lower temperatures.

R-134a requires *polyolester* (POE) lubricants. Traditional mineral oils and alkyl benzenes do not mix with HFC refrigerants and their use with R-134a may cause operation problems or compressor failures. In addition, automotive AC systems may use poly *alkaline glycols* (PAGs), which are typically not seen in stationary equipment.

Both POEs and PAGs will absorb moisture, and hold onto it, to a much greater extent than traditional lubricants. The moisture will promote reactions in the lubricant as well as the usual problems associated with water corrosion and acid formation. The best way to dry a wet HFC system is to rely on the filter drier. Deep vacuum will remove "free" water, but not the water that has been absorbed into the lubricant.

R-134a Applications

Appliances, refrigeration both commercial and self-contained equipment, centrifugal chillers, and automotive air conditioning utilize R-134a. Retrofitting equipment with a substitute for R-12 is sometimes difficult as there are a number of considerations to be examined before undertaking the task.

R-12 SYSTEMS—GENERAL CONSIDERATIONS

1. For centrifugal compressors it is recommended that the manufacturer's engineering staff become involved in the project; special parts or procedures may be required. This will ensure proper capacity and reliable operation after the retrofit.
2. Most older, direct expansion systems can be retrofit to R-401A, R-409A, R-414B or R-416A (R-500 to R-401 B or R-409A), so long as there are no components that will cause fractionation within the system to occur.
3. Filter driers should be changed at the time of conversion.
4. System should be properly labeled with refrigerant and lubricant type.

R-12 Medium/High Temperature Refrigeration (>0°F evap)

1. See "Recommendation Table" (this can be found on National Refrigerants Web site—click on "Technical Manual") for blends that work better in high-ambient heat conditions.
2. Review the properties of the new refrigerant you will use, and compare them to R-12. Prepare for any adjustments to system components based on pressure difference or temperature glide.
3. Filter driers should be changed at the time of conversion.
4. System should be properly labeled with refrigerant and lubricant type.

R-12 Low Temperature Refrigeration (<20°F evap)

1. See "Recommendation Table" for blends that have better low-temperature capacity.
2. Review the properties of the new refrigerant you will use, and compare them to R-12. Prepare for any adjustments to system components based on pressure difference or temperature glide.
3. Filter driers should be changed at the time of conversion.
4. System should be properly labeled with refrigerant and lubricant type.

Another blended refrigerant that can be used to substitute for R-12 is 401A . It is a blend of R-22, 152a, and 124. The pressure and system capacity match R-12 when the blend is running an average evaporator temperature of 10 to 20°F.

Applications for this refrigerant is a direct expansion refrigerate for R-12 in air-conditioning systems and in R-500 systems.

R-401B

This blend refrigerant is similar to R-401A except, it is higher in R-22 content. This blend has higher capacity at lower temperatures and matches R-12 at –20°F. It also provides a closer match to R-500 at air-conditioning temperatures.

Applications for R-401B are in normally lower temperature R-12 refrigeration locations and in transport refrigeration, and in R-500 as a direct expansion refrigerant in air-conditioning systems.

R-402A

This is a blend of R-22 and R-125 with hydrocarbon R-290 (propane) added to improve mineral oil circulation. This blend is formulated to match R-502 evaporator pressures, yet it has higher discharge pressure than 502. Although the propane helps with oil return, it is still recommended that some mineral oil be replaced with alkyl benzene.

Applications are in low-temperature (R-502) refrigeration locations. Retrofitting—it is used for R-502 substituting.

R-402B

Similar to R-402A, but with less R-125, and more R-22. This blend will generate higher discharge temperatures, which makes it work particularly well in ice machines.

Applications are in ice machines where R-502 was used extensively.

RECLAIMING REFRIGERANT

One of the means available for reclaiming refrigerant is called the TOTALCLAIM system. It is furnished to the trade by Carrier, long known for its dominance in the field of refrigeration and air conditioning.

The information in this section is designed to aid the service technician in understanding the construction and operation of the TOTALCLAIM system. A thorough understanding of the system is the most effective tool for troubleshooting.

Description

TOTALCLAIM extracts refrigerant from an air-conditioning or refrigeration system, removes contaminants from the refrigerant, and stores the charge until it is returned to the original system, or another system. TOTALCLAIM can determine the level of acid and moisture contamination in the refrigerant through the use of TOTALTEST.

In recovery operations (Fig. 6-8) refrigerant is extracted from an air-conditioning or refrigeration system and temporarily stored in the TOTALCLAIM storage cylinder. In *recovery* mode, the target system is evacuated to a pressure less than zero (0) psig. In *recovery plus* mode, the target is evacuated to a negative pressure of approximately 20 in. Hg (4 psia).

In the *recycle* mode (Fig. 6-9) refrigerant already stored in the storage cylinder is reprocessed through the TOTALCLAIM unit to remove additional contaminants.

In the *recharge* mode (Fig. 6-10), the refrigerant stored in the TOTALCLAIM storage cylinder is returned to the target air-conditioning or refrigeration system.

In the *service* mode, the internal solenoid valves are positioned so that the TOTALCLAIM system can be

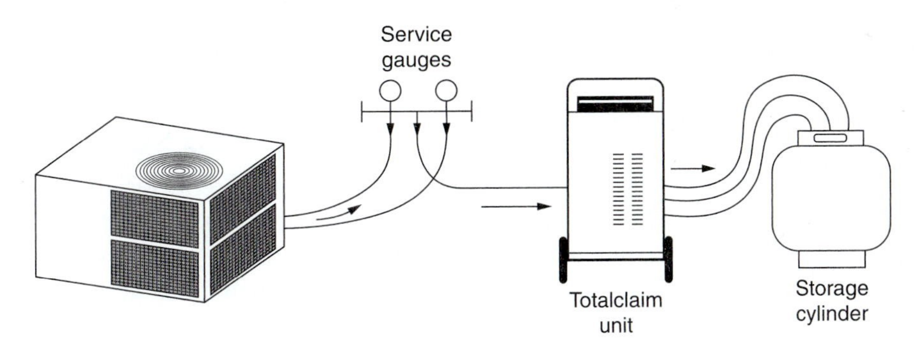

Fig. 6-8 *Recovery operations.* (Carrier)

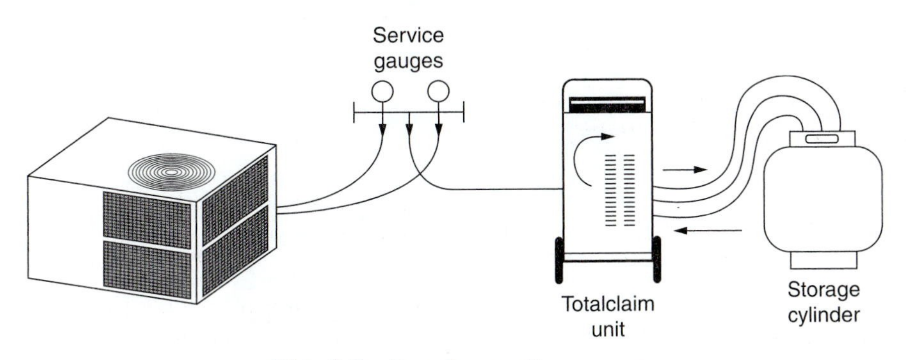

Fig. 6-9 *Recycle operations.* (Carrier)

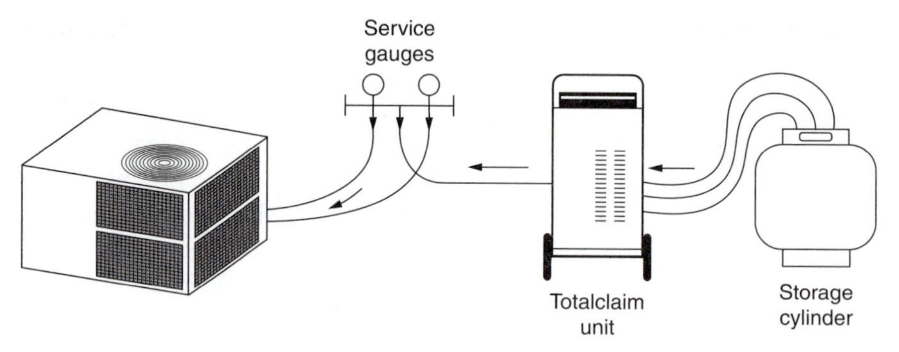

Fig. 6-10 *Recharge operations.* (Carrier)

evacuated. Service mode would be used when a different refrigerant is to be recovered, or when piping connections within TOTALCLAIM must be opened to permit repair.

The *test* mode permits the service technician to energize individual solenoid valves for the purpose of checking out the energizing paths. This mode is intended solely for control-circuit troubleshooting.

In all modes, the pattern of refrigerant flow is determined by solenoid valves, which is controlled by the microprocessor-based control.

Description and Component Location The TOTALCLAIM unit is approximately 35-in. (90 cm) high, including the handle. It is 16-in. (40.7 cm) wide and 10.5-in. (26.5 cm) deep. The TOTALCLAIM unit weighs about 75 lb (34 kg). It is accompanied by a 50-lb (22.7 kg) capacity D.O.T.—approved refrigerant storage cylinder—modified for the TOTALCLAIM application. The hoses required to connect the storage cylinder to the unit are also provided. An external filter-drier is available as an accessory.

In Fig. 6-11 all electrical and electronic controls, except for the solenoid valves, are located in the upper section. This section contains the control panel display board, microprocessor control (standard control module), and a relay board. The only replaceable discrete components in the electronics section are the power switch, transformer, compressor/fan motor contactor, and circuit breaker. If a malfunction is traced to the electronic controls, the entire control module, display board, or relay board must be replaced.

Compressor

In Fig. 6-12 the TOTALCLAIM uses a rotary compressor to pump refrigerant. The compressor is equipped with an external, automatic-reset overload device that trips on excess current or temperature. A discharge temperature thermistor (T_{DIS}) senses the compressor discharge temperature. From here the data are sent to the microprocessor. Both the suction and discharge sides of the compressor are monitored by pressure transducers. These transducers send pressure data to the microprocessor.

Oil Separator

The oil separator collects lubricating oil that escapes with the compressor-discharge gas. A float-valve arrangement inside the oil separator returns oil to the compressor when it reaches a predetermined level.

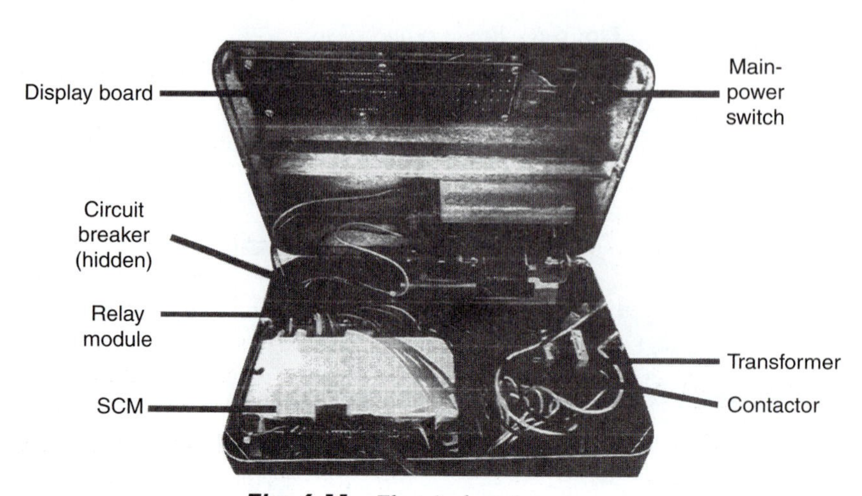

Fig. 6-11 *Electrical section.* (Carrier)

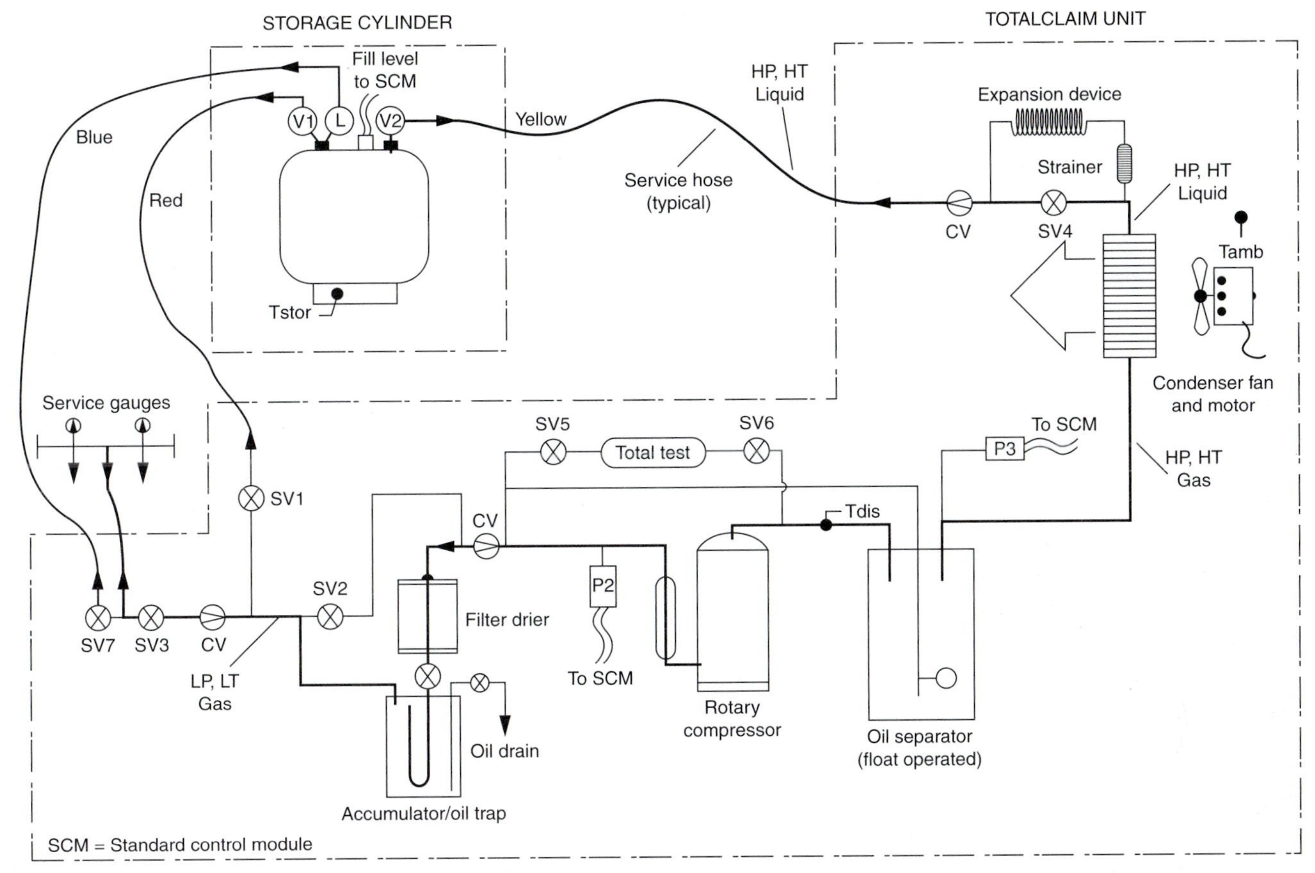

Fig. 6-12 *Liquid recovery—functional flow.* (Carrier)

Condenser

The condenser fan blows air across the condenser, which is mounted to the rear wall. It is a copper-tube/aluminum-fin condenser. The ambient air thermistor (T_{AMB}) senses the temperature at the condenser and sends that data to the microprocessor.

Filter Drier

The primary filter-drier is located behind the access door on the side of the unit. The reset switch for the circuit breaker is also located in this section. Knurled, quick-connect fittings permit the filter drier to be removed and installed without the need for tools. The filter drier shutoff valve must be turned off to allow the filter drier to be replaced while the system is under pressure. See Fig. 6-13.

Accumulator/Oil Trap

The suction-line accumulator/oil trap intercepts oil coming from the unit being evacuated. An oil drain with a valve and an oil measurement bottle are provided so that the trapped oil may be removed. Oil must be drained after each use. The oil-drain valve should be opened slowly to prevent excessive release of refrigerant.

The refrigerant hose connections are equipped with caps, which must be in place when the hoses are disconnected. The hoses have positive shutoff connections at the tank end. The end that connects to the unit is equipped with standard service fittings.

OPERATION OF THE UNIT

The flow of refrigerant through the TOTALCLAIM system, is determined by the mode or submode in which the unit is operating. The state of the refrigerant at various points in the refrigerant cycle is determined by the mode or submode. A key to understanding the system is to know that the storage cylinder plays an important role in the refrigerant cycle. It sometimes acts as a collector, sometimes as an evaporator, other times as a condenser, and still other times as an ambient-temperature charging bottle. Another key to understanding the system is knowing when the various solenoid valves are open and closed. This information is provided

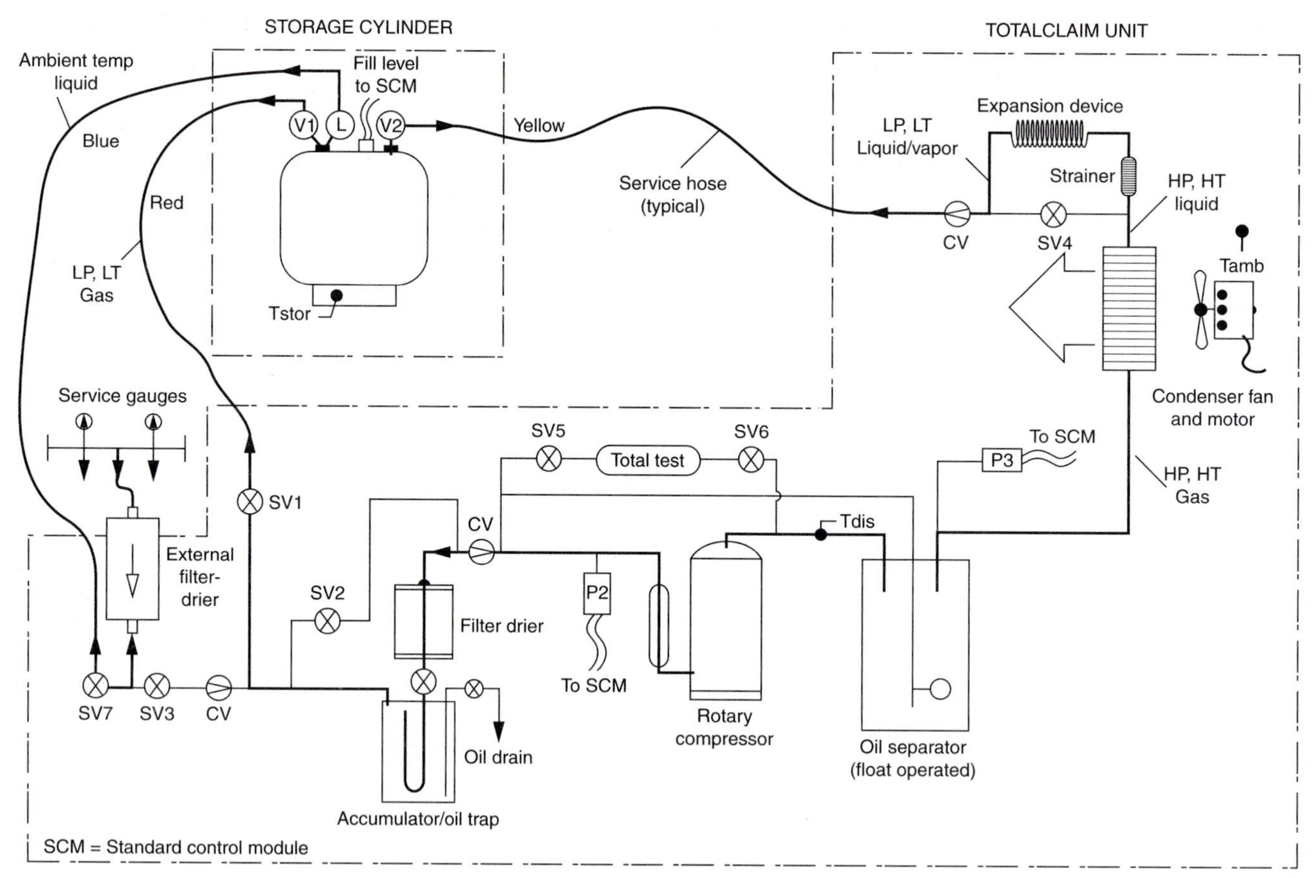

Fig. 6-13 *Vapor recovery—functional flow.* (Carrier)

in Table 6-13, which should be used in conjunction with the mode descriptions that follow.

Recovery Plus/Recovery Operations

Figure 6-12 shows the arrangement of refrigeration cycle components in the TOTALCLAIM system. In a normal recovery operation, the unit will extract liquid refrigerant first. Table 6-13 shows that SV-7 is open and SV-3 is closed during liquid recovery. These conditions would exist if the operator selected the "LIQUID" option at the keyboard.

Given the solenoid-valve conditions shown, liquid at about ambient temperature is extracted from the target system and flows into the storage cylinder. Low-pressure, low-temperature gas is drawn from the cylinder, through SV-1. The high-temperature, high-pressure gas leaving the compressor discharge is condensed to a high-temperature, high-pressure liquid. Note that SV-4 is closed in liquid recovery mode, while in vapor recovery mode it is open. With SV-4 closed, the refrigerant flows through the expansion device. Because of the pressure drop, the refrigerant returns to the storage cylinder as a low-temperature, low-pressure liquid/vapor mixture. This process cools the storage cylinder. In the *recovery plus* and *recovery* modes, the system will automatically enter a storage "cylinder cooling" cycle after 2 min of liquid recovery. Storage cylinder cooling is described later.

When R-22 or R-502 is being processed, the microprocessor will open SV-4, which acts as a parallel expansion device for these refrigerants at higher ambient temperatures.

In vapor recovery operations, the flow is changed significantly, as shown in Fig. 6-12. SV-7 and SV-1 are closed, and SV-3 is open, bypassing the storage cylinder. Therefore, ambient temperature, low-pressure vapor extracted from the target system, is pulled directly into the TOTALCLAIM unit. The other significant difference from liquid recovery operations is that SV-4 is open. Thus, the relatively high-temperature, high-pressure liquid leaving the condenser will enter the storage cylinder in that state.

In the *recovery* mode, one complete recovery cycle is performed. The cycle ends when the pressure in the system being evacuated reaches 0 psig or below.

In the *recovery plus* mode, multiple recovery cycles are performed. Refrigerant is extracted as shown in Figs. 6-12 and 6-13. First it is liquid, then vapor, unless *vapor* recovery is selected at the control panel. During both the liquid and vapor recovery cycles, storage cylinder cooling will be initiated as determined by time, temperature, and/or pressure conditions.

If TOTALTEST is selected (Fig. 6-14), the microprocessor will open SV-5 and SV-6 for 1 to 4 min, depending on the refrigerant type, at the end of the recovery or recycle operation. The refrigerant will be sampled during that period.

Storage Cylinder Cooling

Low-suction pressures are created at the TOTALCLAIM compressor. As the target system approaches a vacuum it causes reduced refrigerant flow. This, in turn, causes high temperatures that would eventually damage the TOTALCLAIM compressor.

To avoid compressor damage, the microprocessor automatically switches TOTALCLAIM into the "storage cylinder cooling" submode, as needed to maintain proper cooling of the compressor. In this submode, connections to the target unit are closed and SV-1 is opened. See Fig. 6-15 and Table 6-13. Now the TOTALCLAIM functions in a closed loop, SV-4 is closed, placing the capillary tube expansion device in the loop. The storage cylinder now acts as a flooded evaporator to cool the refrigerant. The cooling period lasts from 90 s to 15 min, depending on the recovery mode and the state of the refrigerant being processed (liquid or vapor). Over a period of 10 to 15 min, the cylinder temperature is reduced from 60 to 70°F (15.6–21.1°C) below surrounding ambient temperature. Thus, during subsequent recovery cycles, the storage cylinder acts as a low-temperature condenser in addition to the higher temperature air-cooled condenser.

Recycle Operation

In the *recycle* mode, shown in Fig. 6-16, the microprocessor closes SV-3 and SV-7, and opens SV-1. Under those conditions, refrigerant vapor is drawn from the storage cylinder and cycled through the TOTALCLAIM unit to remove additional contaminants. The operator sets the recycle time at the control panel and the microprocessor stops the cycle at the end of that time. If the operator does not select a run time, a default time of 1 h is automatically selected.

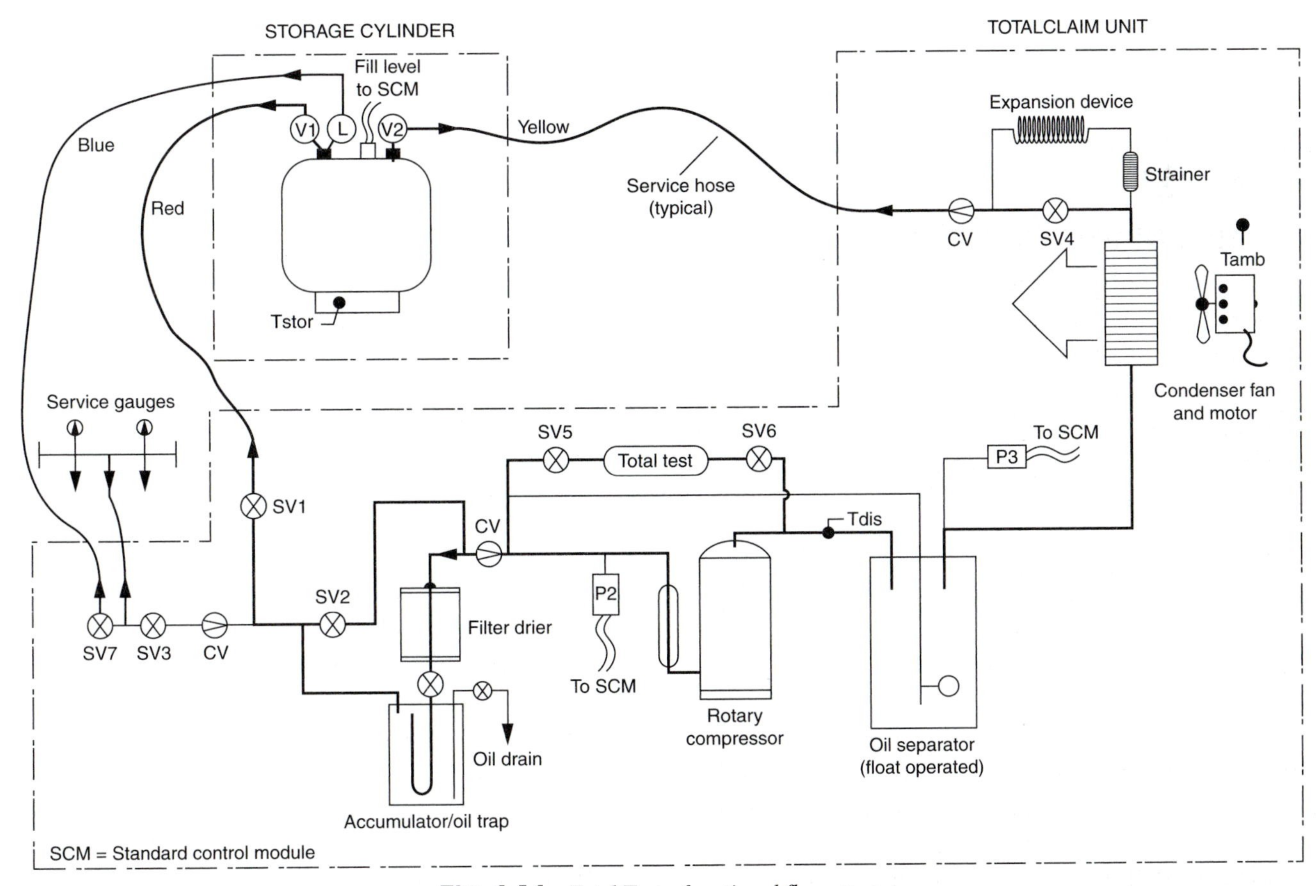

Fig. 6-14 *Total Test—functional flow.* (Carrier)

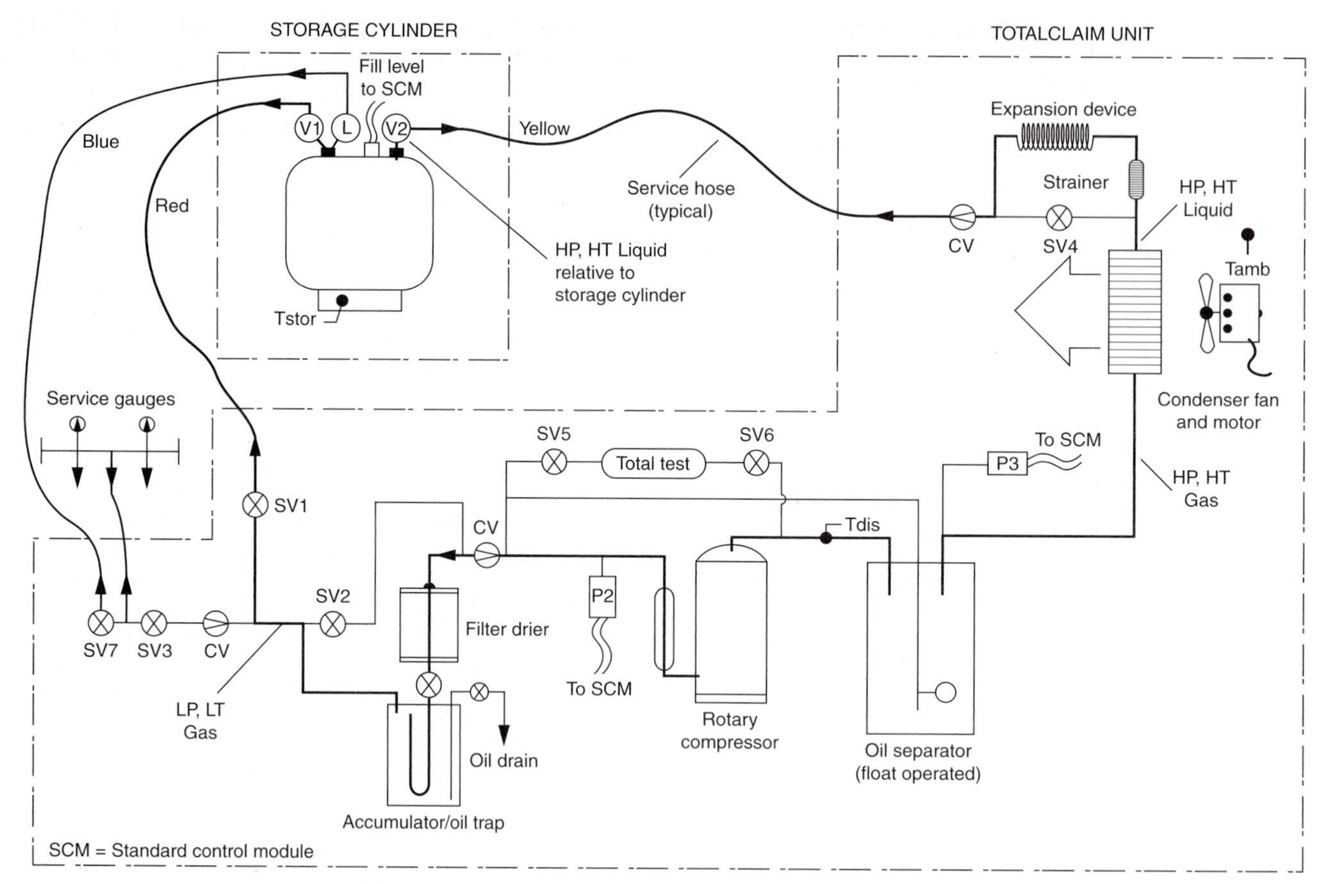

Fig. 6-15 *Storage cylinder cooling—functional flow.* (Carrier)

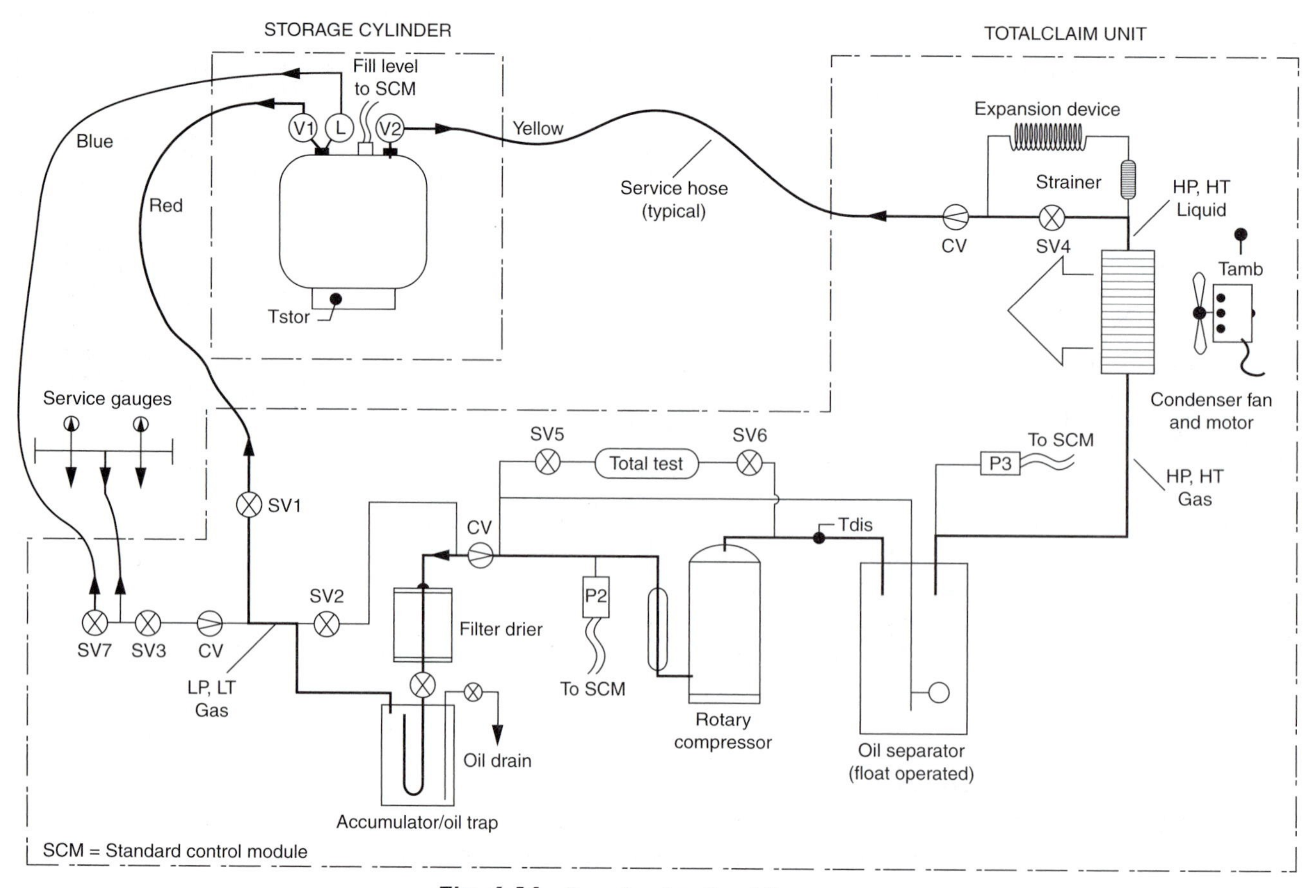

Fig. 6-16 *Recycle—functional flow.* (Carrier)

Recharge Operation

In the *recharge* mode, shown in Fig. 6-17, TOTALCLAIM is basically a charging cylinder. All solenoid valves, except SV-7, are closed. The compressor and condenser fan are turned off. The target unit draws liquid refrigerant from the TOTALCLAIM storage cylinder. In applications where vapor recharging is required, the blue hose must be moved from valve L (blue-handled) to valve V 1 (red-handled) on the storage cylinder.

Service Operation

The *service* mode is selected when it is necessary to evacuate the TOTALCLAIM system. The compressor and condenser fan are shut off, and all solenoid valves, except SV-5, SV6, and SV-7, are open to permit refrigerant to be drawn from the TOTALCLAIM system.

Test Operation

The *test* mode permits the technician to energize individual solenoid valves in order to simplify troubleshooting of the control circuits. This mode is selected by pressing the "RESET" and "MODE" keys for 5 s. The *test* mode takes priority over all other modes. When the *test* mode is turned on, all solenoids are deenergized. Then, using the arrow keys, the technician can energize individual solenoids and trace the energizing signal along the path. If there is a malfunction in the path, it can be isolated to the solenoid valve, relay module, or *standard control module* (SCM). SV-5 and SV-6 are energized at the same time. All the others are operated individually. The "START/STOP" key is used to exit the *test* mode.

Control Circuits

The compressor and fan, both operate from 115-V, single-phase power. The contactor, C, which is energized by the SCM, controls both components. The relay is energized under the '"compressor/condenser fan on" conditions shown in Table 6-13. An external overload device disables the compressor in the event of a current overload or excessively high temperature. The device will reset automatically when internal temperature drops to a safe level. See Fig. 6.18.

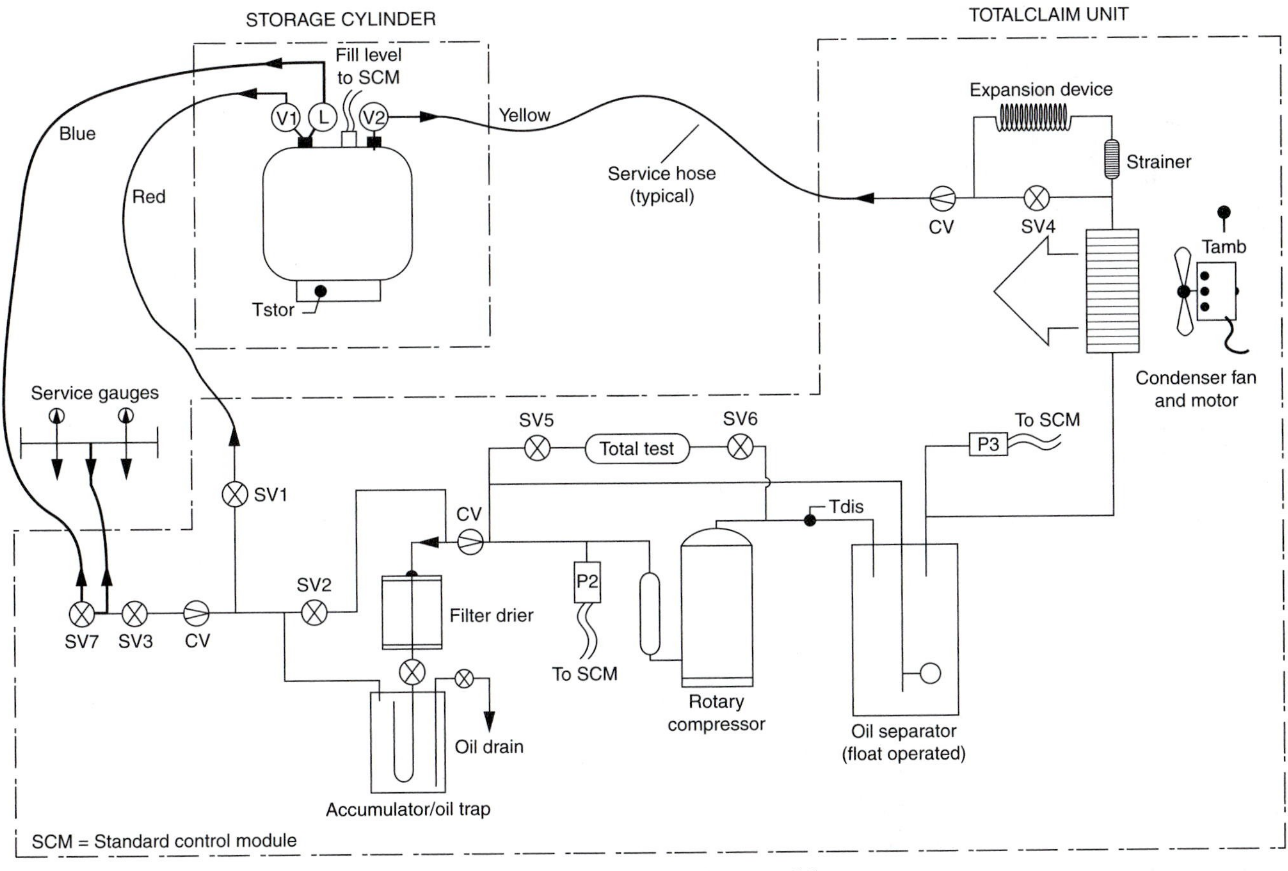

Fig. 6-17 *Recharge (liquid)—functional flow.* (Carrier)

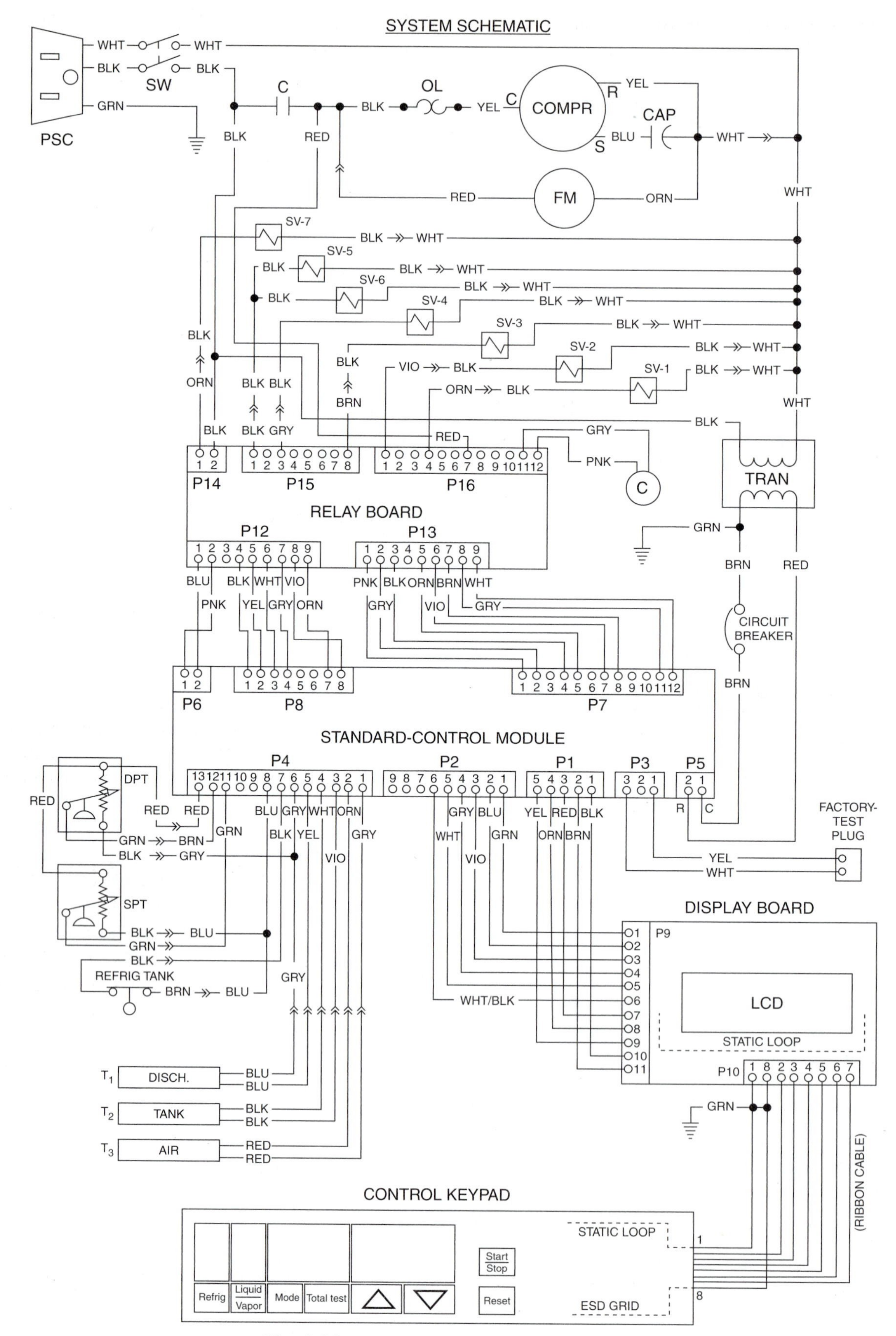

Fig. 6-18 *TOTALCLAIM wiring diagram.* (Carrier)

The SCM receives data from several sources within the unit. Temperature data is supplied by thermistors located in the compressor discharge (T_{DIS}), the storage cylinder (T_{STOR}), and the ambient air intake (T_{AMB}). These are identified as T_1, T_2, and T_3 on the schematic diagram. The unit will shut down, and an error code will appear on the display, if any of these thermistors fail, either open or shorted.

Error Code Thermistor

E03	Ambient air
E04	Compressor discharge
E05	Storage cylinder

A level-sensing device inside the storage cylinder allows the SCM to monitor the contents of the tank. When the tank reaches 80 percent full, 50 lb or 22.7 kg, the SCM will stop the recovery process, and error code E09 will appear on the display.

The SCM monitors suction and discharge pressures from the pressure transducers, SPT and DPT, respectively. In the refrigerant flow diagram, Figs. 6- 12 through 6-17, they are designated as P2 and P3, respectively. The SCM will not allow the compressor to start unless the pressures in the unit reach the correct levels within 3 min of start. The flow chart in Fig. 6-19 shows the sequencing of this process. If the pressure differential is greater than 30 psi, the SCM will energize SV-1, SV-2, and SV-4 to allow the pressures to equalize. The probable causes of an "A11" alarm are storage cylinder refrigerant hoses not being connected, storage cylinder valves not being open, or service manifold valves or target unit valves being closed.

The SCM prevents overloads by sequencing its outputs one at a time. This is known as a "soft start." Once the "START" button is pressed, the selected solenoid valves will cycle on or off in their numerical sequence (SV-1, SV-2, and so forth) at no less than 1-s intervals. The compressor and condenser fan are sequenced on, after all the selected solenoid valves have been energized or deenergized. See Table 6-14.

The solenoid valves and contactor (C) are controlled by the SCM through relays on the relay board. Display functions are controlled by the SCM through the display board.

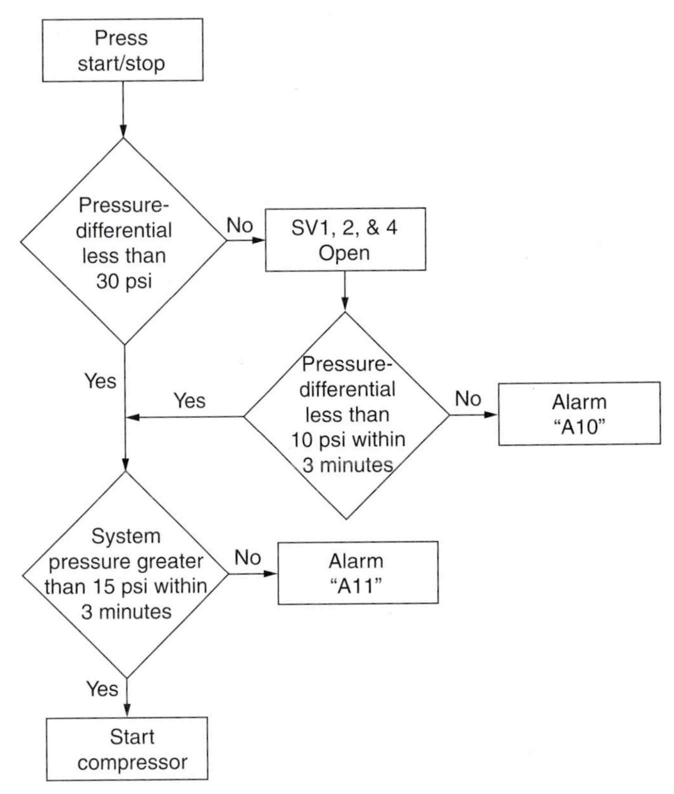

Fig. 6-19 *Pressure equalization at startup.* (Carrier)

TROUBLESHOOTING

The use of modular, solid-state electronics, and built-in diagnostic testing reduces the amount of trouble analysis that must be performed in order to isolate malfunctions. Many malfunction conditions will be diagnosed by the system, and an error message will be displayed to tell the service technician what component has failed. In some cases, however, it will be necessary to isolate the fault by using standard troubleshooting methods, supported by the built-in test capability.

Troubleshooting Approach

The troubleshooting diagram at the front of the unit's manual contains most of the information needed to troubleshoot the TOTALCLAIM system. The flow diagram provides a good starting point if you have no idea what the problem is. If you have an error message on the display panel, you should go directly to the errors and alarms table and perform the indicated action or troubleshoot the failed component.

A *test* mode is provided as an aid in troubleshooting the relay-/solenoid-valve logic. This approach is discussed in detail in the manual that is available on the internet. For more information on this unit contact Carrier at www.carrier.com.

REVIEW QUESTIONS

1. List five desirable properties of a good refrigerant for commercial use.
2. How are refrigerants classified?

3. Where is calcium chloride used in refrigeration systems?
4. What type of refrigerant is used in home refrigerators?
5. Are Freon compounds flammable?
6. What does toxicity mean?
7. What is a fluorocarbon?
8. What is a halocarbon?
9. Why is the pressure of a refrigeration system important?
10. Define specific volume.
11. How are refrigerant leaks detected?
12. What is the Montreal Protocol?
13. What is R-134a used for?
14. Why are R-22 and R-12 so harmful?
15. Where will technicians get their Freon after 2020?
16. What is a trigger leak rate?

7

CHAPTER

Refrigeration Compressors

PERFORMANCE OBJECTIVES

1. Know how to identify refrigeration compressors.
2. Know types of condensers.
3. Be able to explain how evaporative condensers work.
4. Know the types of compressors.
5. Know how to read serial-plate information on a hermetic compressor.
6. Know how the internal line-break motor protector works.
7. Know how hermetic compressors work.
8. Know how PSC motor hook-ups are made.
9. Know different types of hermetic compressors.
10. Know how relays operate and malfunction.
11. Know how and why crankcase heaters are needed in certain compressors.
12. Know how CSIR or high-starting torque motors operate.
13. Know how rotary compressors operate.

Refrigeration compressors can be classified according to the following:

- Number of cylinders
- Method of compression
- Type of drive
- Location of the driving force or motor

The method of compression may be *reciprocating, centrifugal,* or *rotary.* The location of the power source also classifies compressors. Independent compressors are belt driven. Semihermetic compressors have direct drive, with the motor and compressor in separate housings. The hermetic compressor has direct drive, with the motor and compressor in the same housing.

Reciprocating units have a piston in a cylinder. The piston acts as a pump to increase the pressure of the refrigerant from the low side to the high side of the system. A reciprocating compressor can have twelve or more cylinders. See Fig. 7-1.

Fig. 7-1 *Reciprocating compressor.* (Trane)

The most commonly used reciprocating compressor is made for refrigerants R-22 and R-134a. These are for heating, ventilating and air conditioning, and process cooling. The most practical refrigerants used today are R-134a and R-22. However, R-134a, is gaining acceptance, in view of the CFC regulations worldwide. As a matter of fact, some countries only accept R-134a today. Other environmentally acceptable refrigerants are R-404A and R-507. They are for low- and medium-temperature applications. R-470C is for medium temperatures and air-conditioning applications. Recently, R-410A has gained acceptance as an environmentally acceptable substitute for R-22, but only for residential and small equipment. R-410A is not a drop-in refrigerant for R-22.

There are three types of reciprocating compressors: open drive, hermetic, and semihermetic.

Two methods of capacity control are generally applied to the reciprocating refrigeration compressor used on commercial air-conditioning systems. Both methods involve mechanical means of unloading cylinders by holding open the suction valve.

The most common method of capacity control uses an internal multiple-step valve. This applies compressor-oil pressure or high–side refrigerant pressure to a bellow or piston that actuates the unloader.

The second method of capacity control uses an external solenoid valve for each cylinder. The solenoid valve allows compressor oil or high-side refrigerant to pass to the unloader.

A centrifugal compressor is basically a fan or blower that builds refrigerant pressure by forcing the gas through a funnel-shaped opening at high speed. See Fig. 7-2.

Compressor capacity is controlled when the vanes are opened and closed. These vanes regulate the amount of refrigerant gas allowed to enter the fan or turbine. See Fig. 7-3. When the vanes restrict the flow of refrigerant, the turbine cannot do its full amount of work on the refrigerant. Thus, its capacity is limited. Most centrifugal machines can be limited to 10 to 25 percent of full capacity by this method. Some will operate at almost zero capacity. However, another, though less common, method is to control the speed of the motor that is turning the turbine.

CONDENSERS

A condenser must take the superheated vapor from the compressor, cool it to its condensing temperature, and

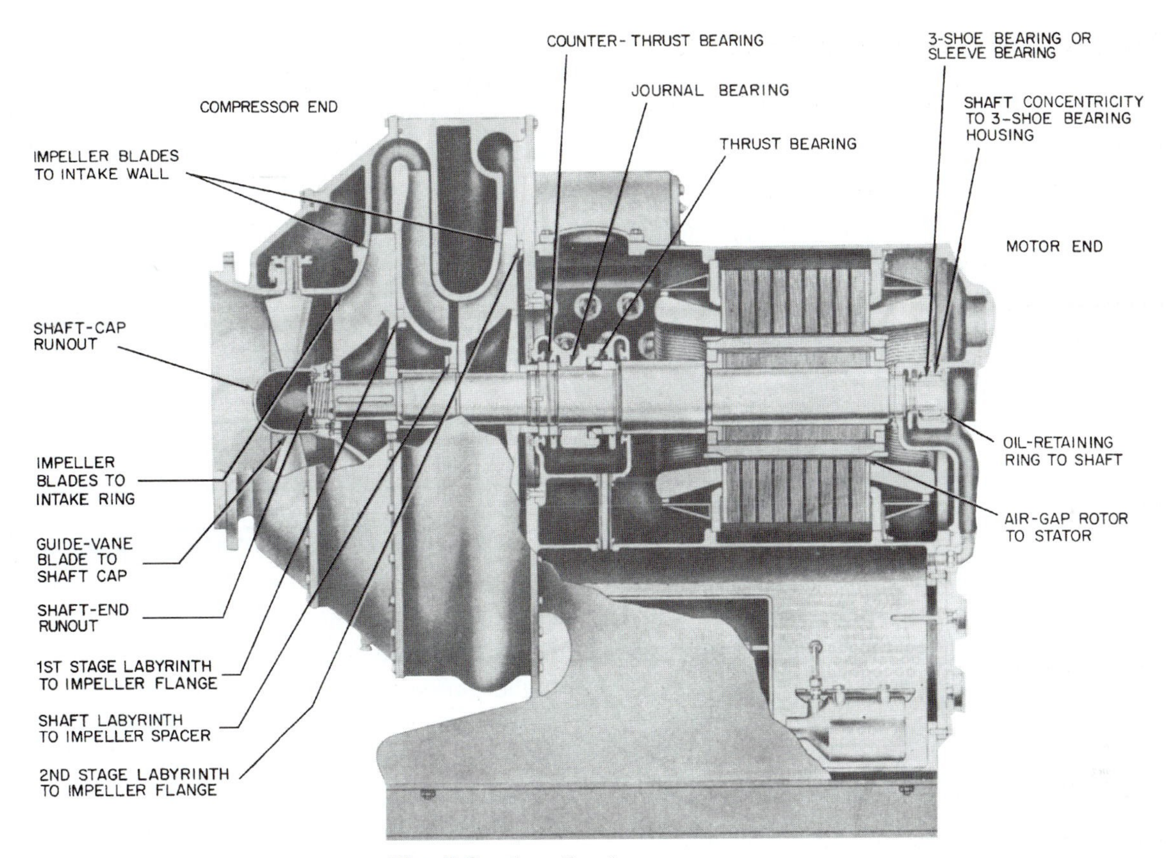

Fig. 7-2 *Centrifugal compressor.* (Carrier)

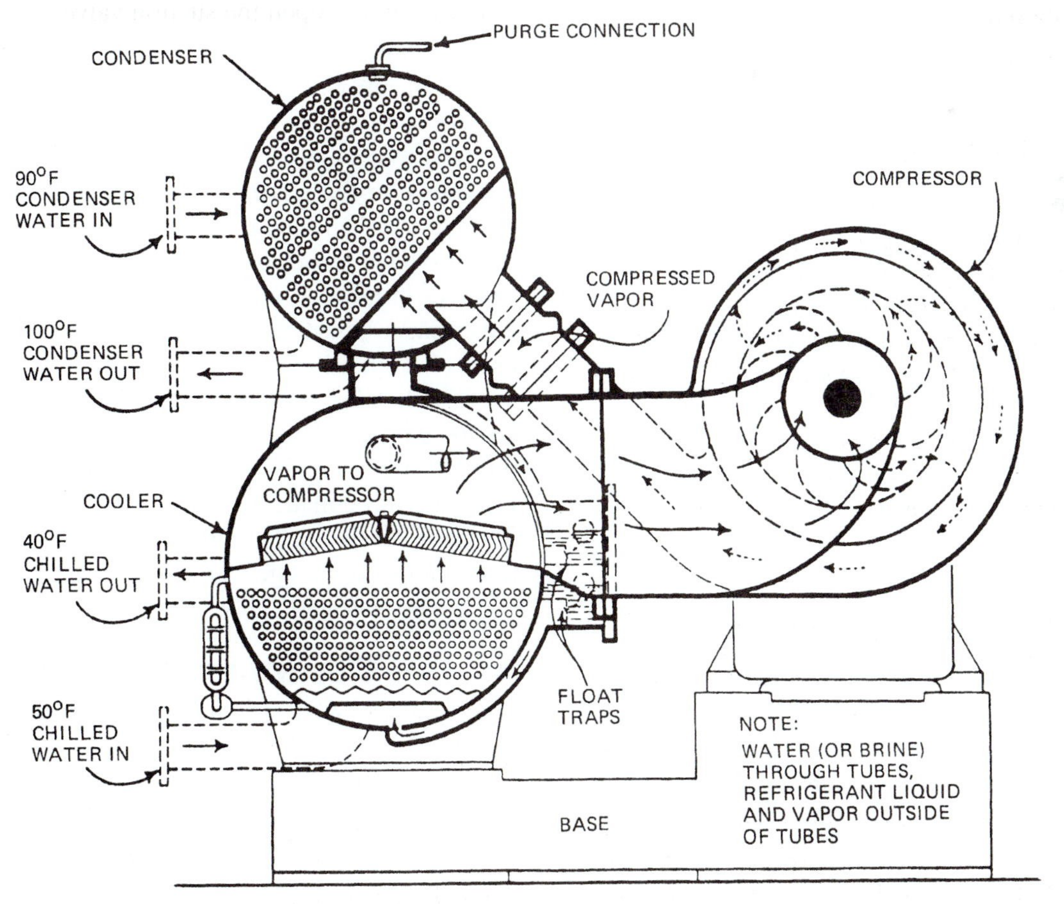

Fig. 7-3 *Centrifugal compressor system.* (Carrier)

then condense it. This action is opposite to that of an evaporator. Generally, two types of condensers are used—air cooled and water cooled.

Air-Cooled Condensers

Air-cooled condensers are usually of the fin and tube type, with the refrigerant inside the tubes and air flowing in direct contact over the outside. Usually, a fan forces the air over the coil. This increases its cooling capabilities. See Figs. 7-4 and 7-5.

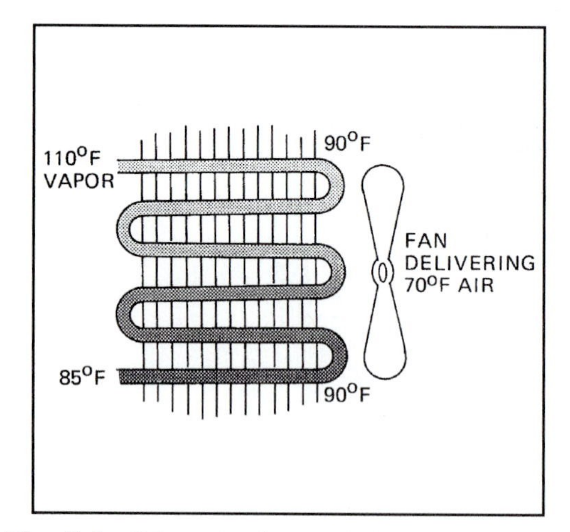

Fig. 7-4 *Schematic of air-cooled condenser.* (Johnson)

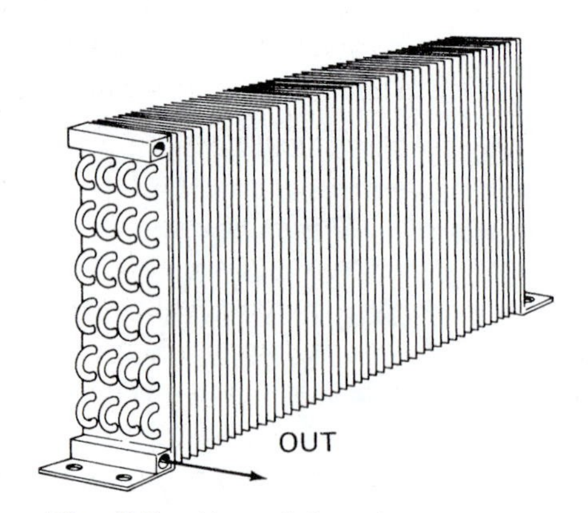

Fig. 7-5 *Air-cooled condenser.* (Johnson)

Water-Cooled Condensers

In the water-cooled condenser, the refrigerant is cooled with water within pipes. See Fig. 7-6. The tubing containing water is placed inside a pipe or housing containing the warm refrigerant. The heat is then transferred from the refrigerant through the tubing to the water. Water-cooled condensers are more efficient than air-cooled condensers. However, they must be supplied with large quantities of water. This water must be either discharged or reclaimed by cooling it to a temperature that makes it reusable.

A cooling tower usually accomplishes reclaiming. See Figs. 7-7 and 7-8.

The tower chills the water by spraying it into a closed chamber. Air is forced over the spray. Cooling towers may be equipped with fans to force the air over the sprayed water.

Another device used to cool refrigerant is an evaporative condenser. See Fig. 7-9. Here, the gas-filled condenser is placed in an enclosure. Water is sprayed on it and air forced over it to cool the condenser by evaporation.

HERMETIC COMPRESSORS

A hermetic compressor is a direct-connected motor compressor assembly enclosed within a steel housing. It is designed to pump low-pressure refrigerant gas to a higher pressure.

A hermetic container is one that is tightly sealed so no gas or liquid can enter or escape. Welding seals the container.

Tecumseh hermetic compressors have a low-pressure shell, or housing. This means that the interior of the compressor housing is subjected only to suction pressure. It is not subjected to the discharge created by the piston stroke. This point is emphasized to stress the hazard of introducing high-pressure gas into the compressor shell at pressures above 150 psig.

The major internal parts of a hermetic compressor are shown in Fig. 7-10. The suction is drawing into the compressor shell then to and through the electric motor that provides power to the crankshaft. The crankshaft revolves in its bearings, driving the piston or pistons in the cylinder or cylinders. The crankshaft is designed to carry oil from the oil pump in the bottom of the compressor to all bearing surfaces. Refrigerant gas surrounds the compressor crankcase and the motor as it is drawn through the compressor shell and into the cylinder or cylinders, through the suction muffler and suction valves. The gas is compressed by the moving piston and is released through the discharge valves, discharge muffler, and compressor-discharge tube.

Compressor Types

Hermetic compressors have different functions. Some are used for home refrigeration. Some are used to produce air conditioning. Others are used in home or commercial freezers. Hermetic compressors are also used for food-display cases.

The serial-number plate on the compressor tells several things about the compressor. See Fig. 7-11.

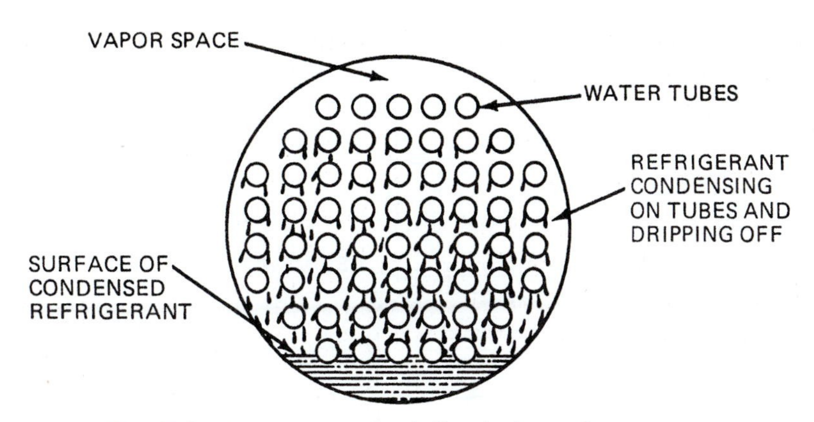

Fig. 7-6 *Cross-section of a shell and tube condenser.* (Johnson)

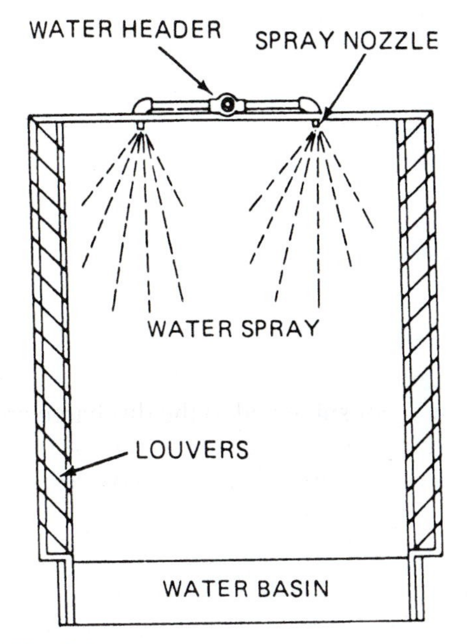

Fig. 7-7 *Spray-type cooling tower.* (Johnson)

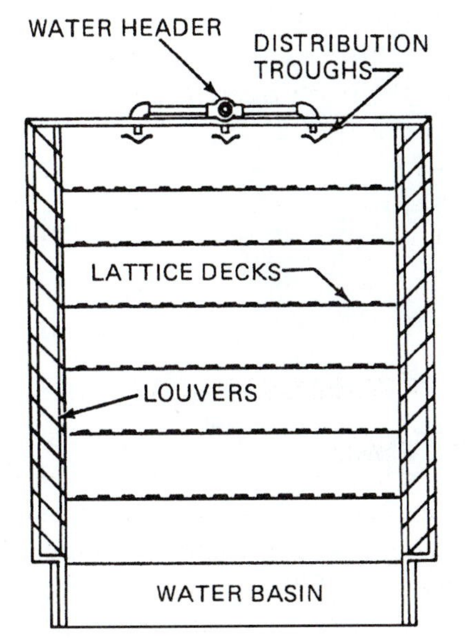

Fig. 7-8 *Deck-type cooling tower.* (Johnson)

Also notice, that several manufacturers made the motor for the compressor:

- A.O. Smith
- Aichi
- Delco
- Emerson
- General Electric
- Ranco
- Wagner
- Westinghouse

Makers of electric motors for compressors usually mark them for the compressor manufacturers. Newer models are rated in Hertz (Hz) and very old models may be marked in cycles per second (cps) instead of hertz.

The following is a brief outline of some of the points to be remembered in servicing.

Pancake models are designated with a P as the first letter of serial number. They are made with 1/20th to $^1/_3$ hp motors. All of them use an oil charge of 22 oz and R-134a or the latest replacement as a refrigerant. They have a temperature range of 20 (–6°C) to 55°F (13°C). The smaller horsepower models are used where –30 (–34°C) to 10°F (–12°C) is required.

The T and AT compressor models have $^1/_6$, $^1/_5$, $^1/_4$, and $^1/_3$ hp motors. All of these models use R-134a refrigerant or its equivalent replacement. The smaller sizes use a 38-oz oil charge, while the larger horsepower models use 32 oz. They have temperature ranges of –30 (–34°C) to 10°F (–12°C) and 20 (–6°C) to 55°F (13°C).

The AE compressors are used for household refrigerators, freezers, dehumidifiers, vending machines, and water coolers. See Fig. 7-12. They are made in $^1/_{20}$, $^1/_{12}$, $^1/_8$, $^1/_6$, $^1/_5$, and $^1/_4$-hp units. The oil charge may be 10, 16, 20, or 23 oz. This AE-compressor model line uses

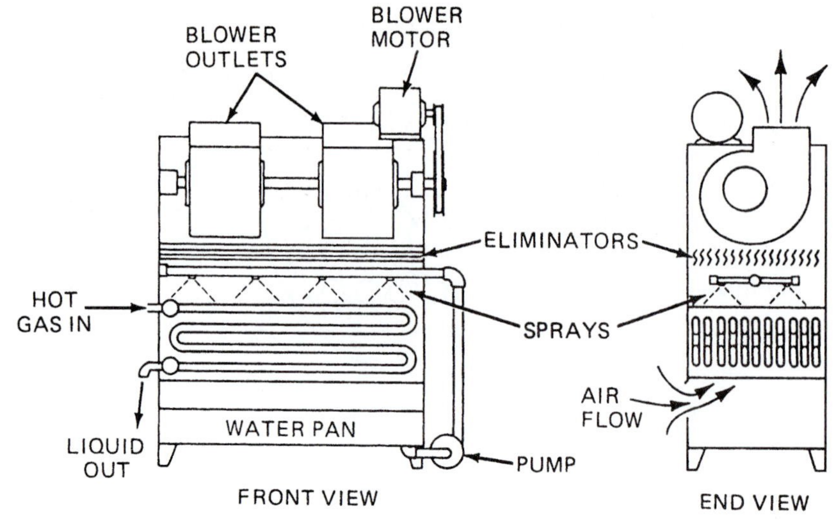

Fig. 7-9 *Evaporative condenser.* (Johnson)

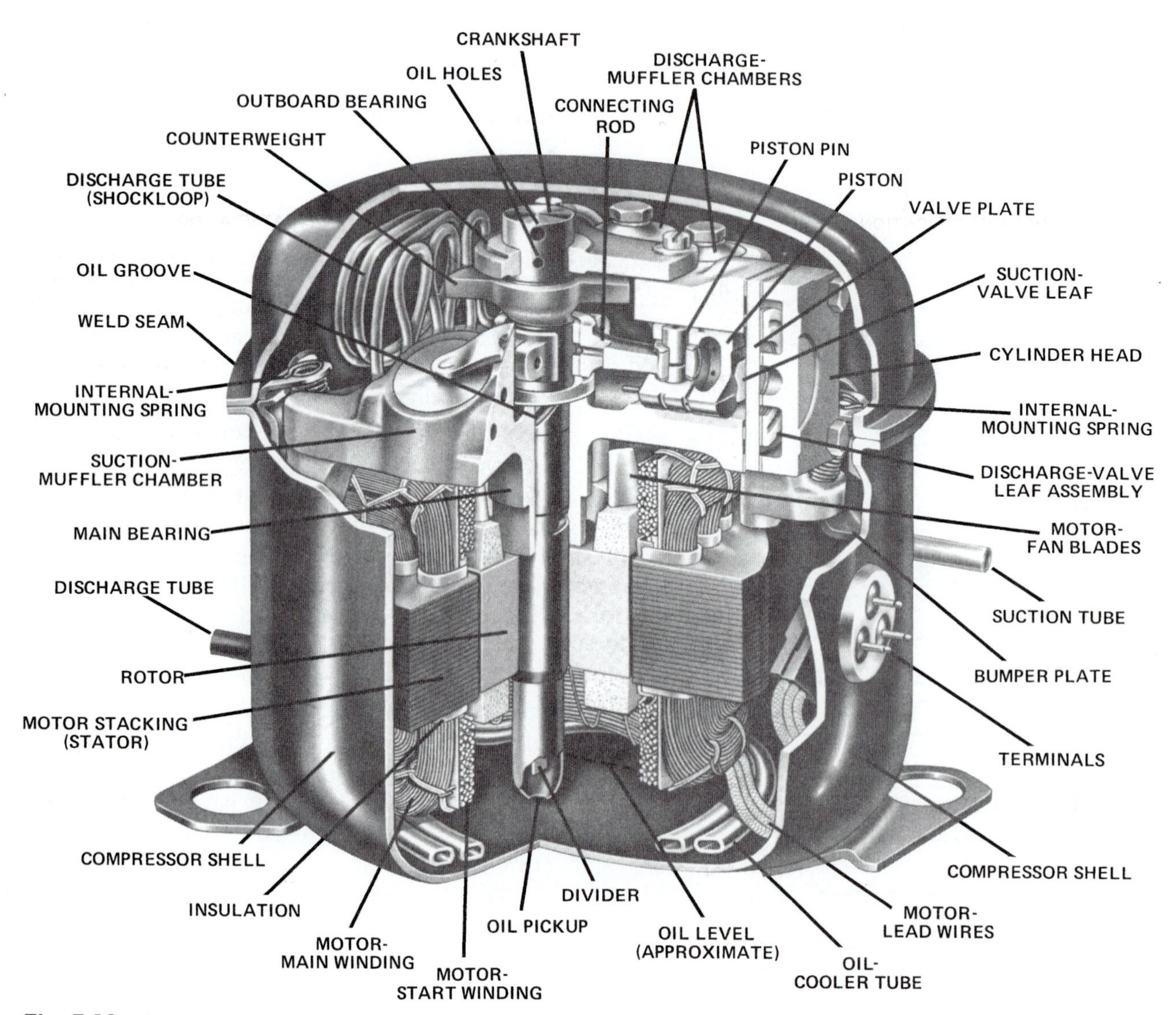

Fig. 7-10 *Cutaway view of a compressor. Note the motor is on the bottom of the compressor and the piston is on the top.* (Tecumseh)

TYPICAL SERIAL PLATE 1953-1958

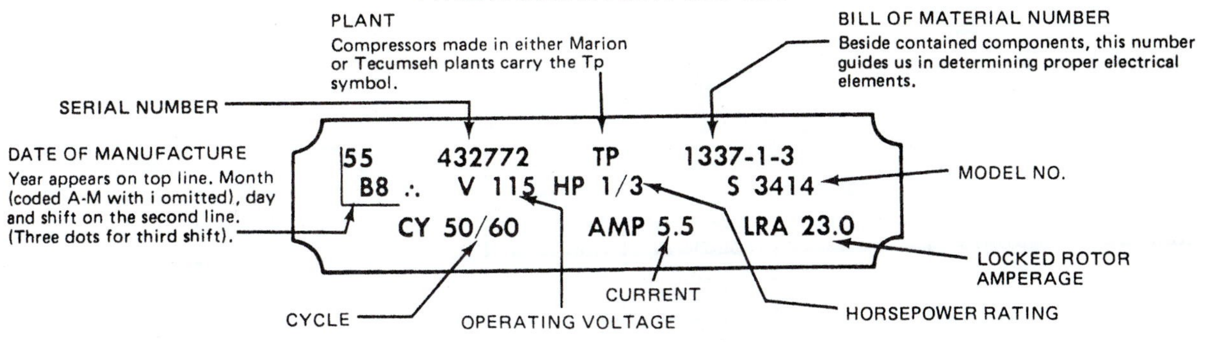

TYPICAL SERIAL PLATE 1958-1964 (AUGUST)

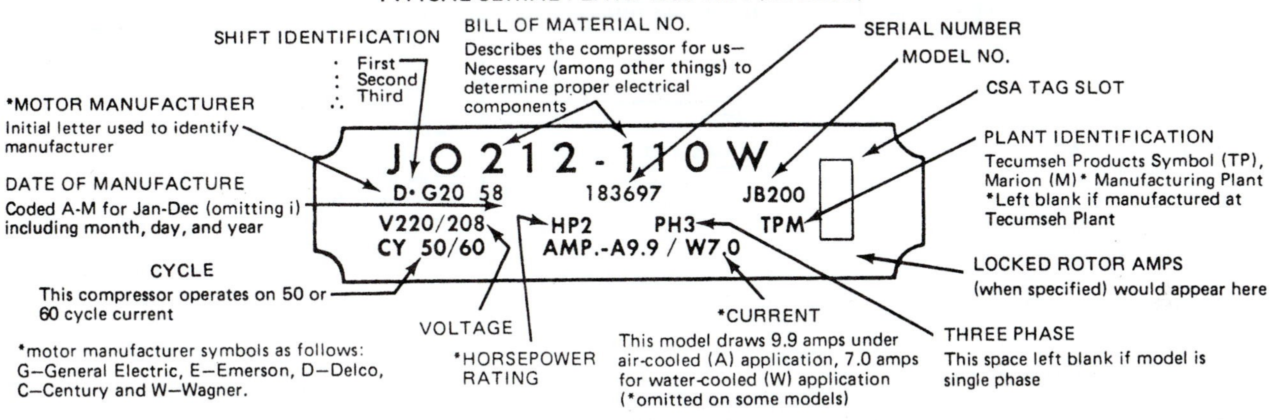

TYPICAL CURRENT SERIAL PLATE

DATE OF MANUFACTURE

The date of manufacture is determined by a code on the serial plate or unit nameplate. This code is as follows:

Starting in January 1940 the date designation on all hermetic compressors was simplified to one letter and one figure. The months are lettered as follows:

January—A	March—C	May—E	July—G	September—J	November—L
February—B	April—D	June—F	August—H	October—K	December—M

Preceding this letter is a numeral indicating the year this compressor was built. For example, 1A would indicate the compressor was built January 1941, 7C would indicate the compressor was built March 1947. This system will hold for compressors manufactured from 1940 through 1949.

For compressors manufactured from 1950 to 1952, the year precedes the letter designating the month. For example, 51L is a compressor manufactured in November 1951. From 1953 to 1958 the year is the first numeral and the month and day are on the second line.

From 1958 on, the second line reading from left to right is: Letter indicating motor, dots to identify shift, letter for month, 2 digits for day and 2 digits for year of manufacture.

(A)

Fig. 7-11 *(A) Serial plate information on Tecumseh's hermetic compressors. (B) Model numbers. (C) Nomenclature explained. (D) Compressor application categories.* (Tecumseh)

MODEL NUMBERS
Nomenclature Explained

EXAMPLE:

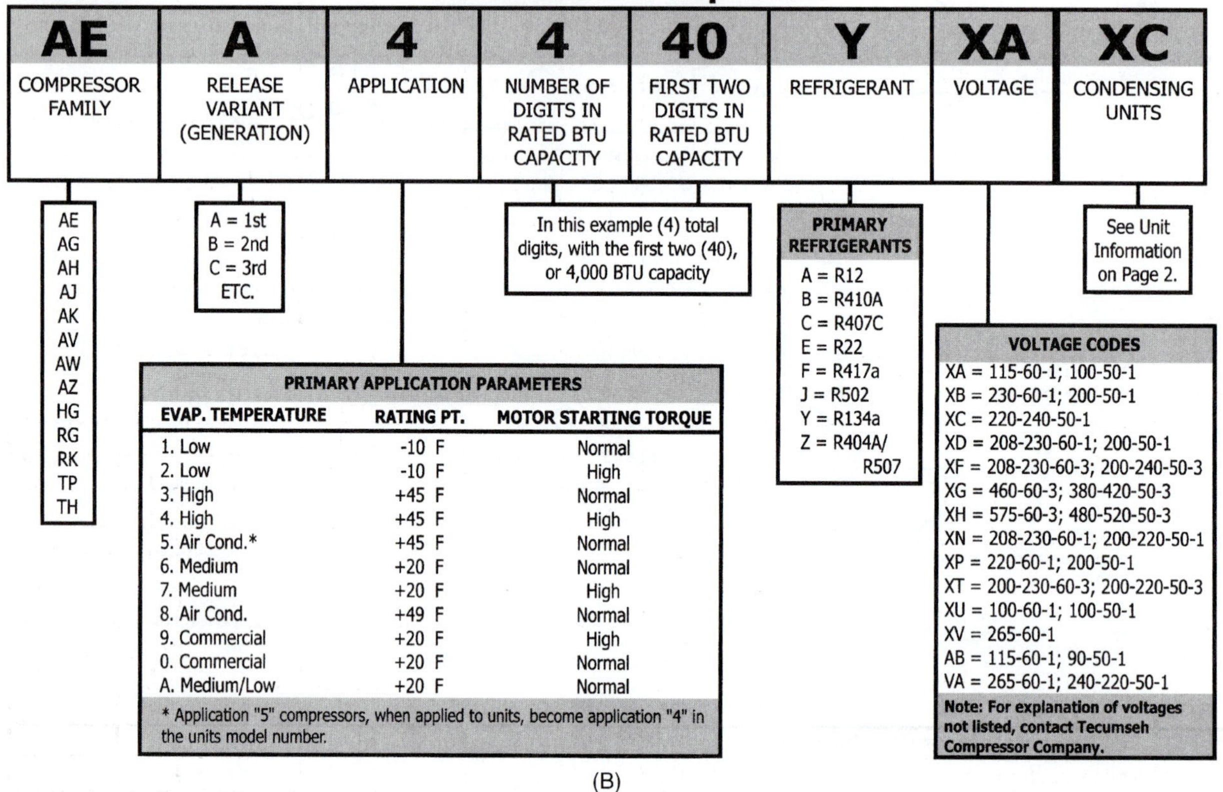

(B)

Fig. 7-11 *(Continued)*

R-134a, or an acceptable substitute, and, in some cases, R-22 as a refrigerant. The older models still in use may have R-12 refrigerant. When servicing older equipment it is best to remember some of the older charges.

Figure 7-13 shows the overload relay in its proper location with the cover removed to indicate the proper positioning. Figure 7-14 shows all parts assembled under the cover. The cover is secured to the fence with a bale strap.

This type of compressor may have a resistance-start induction-run motor. See Fig. 7-15. It may have a capacitor-start induction-run motor. See Fig. 7-16.

Model AK compressors are rated in Btu per hour. They have a 7,000 to 12,000-Btu rating range. All model AK compressors are used for air-conditioning units. The refrigerant used is R-22. A 17-oz charge of oil is used on all models.

The AB compressors are also used for air-conditioning units. However, they are larger, starting with the 19,000-Btu/h rating and extending up to 24,000 Btu, or a 2-ton limit. Keep in mind that 12,000 Btu equals 1 ton. Refrigerant-22 is used with a 36-oz charge of oil.

The AU and AR compressors are made in $^{1}/_{2}$, $^{3}/_{4}$, 1-, and $1^{1}/_{4}$ hp sizes. They are used primarily for air-conditioning units. Most of the models use R-22, except for a few models that use R-134a or its equivalent substitute. A 30-oz charge of oil is standard, except in one of the $^{1}/_{2}$-hp models. Because of such exceptions, you must refer to the manufacturer's specifications chart to obtain the information related to a specific model number within a series.

The *internal spring mount* (ISM) series of compressors range in sizes from $^{1}/_{8}$ to 1 hp. Their temperature range is from −30 (−34°C) to 10°F (−12°C) and from 20 (−6°C) to 55°F (13°C). The oil charge is either 40 or 45 oz, depending upon the particular model.

The AH compressors are designed for residential and commercial air-conditioning and heat-pump applications. See Fig. 7-17. They can be obtained with either three- or four-point mountings. See Fig. 7-18. The internal line-break motor protector is used. It is located precisely in the center of the heat sink position of the motor windings. Thus, it detects excessive motor-winding

AEA4440YXA XC
1.

1. E and G = Evaporative Condensate Units.
X = A holding character, reserved for future use.
Condensing Unit Features, see chart below.

The Letters I, O, and Q are eliminated.		Fan Cooled	Water Cooled	Air Water Cooled	Receiver Tank	BX Cable	Interconnect Compressor	See B/M	Accumulator
A	Standard Unit	●							
B	Std. Unit w/Receiver Tank	●			●	●			
C	Std. Unit w/ Tank & BX Cable	●			●	●			
D	Std. Unit w/ BX Cable	●				●			
E,F,K	Physical Design Variant (Conduit)	●						●	
G,H,J,L,P	Physical Design Variant (Standard)	●						●	●
M	Advanced Commercial Design	●			●	●			●
N	Advanced Commercial Design	●				●			
S	Customer Special						●	●	
T	Interconnect Compressor								●
U	Water Cooled - Adv. Commercial Design		●		●	●			
V	Electrical Special (Conduit Design)	●				●		●	
W	Water-Cooled Unit		●		●	●	●		
X	Interconnect Unit	●			●	●			
Y	Air Water Cooled Unit			●		●			
Z	Electrical Special (Standard Unit)	●						●	
Evaporative Condensate Units									
EC	Large Evaporative Condensate Units Black Plastic Base	●							
ED	Large Evaporative Condensate Units Black Plastic Base	●							
EE	Large Evaporative Condensate Units Black Plastic Base	●							
GA	Small Evaporative Condensate Units Gray Plastic Base	●							
GB	Small Evaporative Condensate Units Gray Plastic Base	●			●				
GK	Small Evaporative Condensate Units Gray Plastic Base	●							
GL	Small Evaporative Condensate Units Gray Plastic Base	●							

(C)

Fig. 7-11 *(Continued)*

temperature and safely protects the compressor from excessive heat and/or current flow. See Fig. 7-19.

The snap on terminal-cover assembly is shown in Fig. 7-20. It is designed for assembly without tools. The molded fiberglass terminal cover may be secured or held in place by a bale strap.

This AH compressor series has a run capacitor in the circuit, as shown in Fig. 7-21. This compressor is designed for single-phase operation. Figure 7-22 shows the terminal box with the position of the terminals, and the ways in which they are connected for RUN, START, and COMMON.

The AH compressors are rated in Btu/h. They range from 3500 to 40,000 Btu/h. These models use 45 oz of oil for the charge. They are used as air-conditioning units and for almost any other temperature-range applications. They use R-134a or suitable substitute or R-22 for refrigerant.

The B and model compressors are available in $^{1}/_{3}$-, $^{1}/_{2}$-, $^{3}/_{4}$-, 1, $1^{1}/_{2}$-, $1^{3}/_{4}$-, and 2-hp units. All of them use a 45-oz charge of oil. They have a wide variety of temperature and air-conditioning applications. These models may have a B or a C preceding the model serial number. This is to indicate the series of compressors.

The AJ series of air-conditioning compressors range in sizes from 1100 to 19,500 Btu. See Fig. 7-23. An oil charge of 26 or 30 oz is standard, depending upon the model. They are mounted on three or four points. See Fig. 7-24. A snap-on terminal cover allows quick access to the connections under the cover. See Fig. 7-25. This particular model has an antislug feature that is standard on all AJM 12 and larger models. See Fig. 7-26 (An antislug feature keeps the liquid refrigerant moving.).

This type of compressor relies upon the permanent split-capacitor motor. In this instance, the need for both start and run capacitor is not presented. The start relay and the start capacitor are eliminated in this arrangement. See Fig. 7-27. With the permanent split-capacitor (PSC) motor, the run capacitor acts as both a start and run capacitor. It is never disconnected. Both motor

Compressor Application Categories

Application	LT	MT	Commercial Temp	HT	Air Conditioning	
Rating Point Temp Range	-10°F -40°F to +10°F	+20°F -10°F to +30°F	+20°F -10°F to +45°F	+45°F +20°F to +55°F	Cooling +45°F +32°F to 55°F	Heating -10°F -15°F to 57°F
	Domestic Refrigerators Domestic Freezers Ice-Cream Cabinets Soft Ice Cream Mashines Sluch Machines Beverage Vendor w/Ice Reach-In Freezers Walk-In Freezers Bakery Freezers Ice Storage Cabinets Baeverage Glass Chillers	Draft-Beer Coolers Ice Machines Commercial Refrigerators Soft Ice-Cream Machines Beverage Vendors w/Ice Reach-In Refrigerators Walk-In Coolers Dough Retarders Ice-Storage Cabinets Beverage Dispensers Bottle Coolers Egg Coolers Air Drier Open Display Cases Fur-Storage Refrigeration Meat Cases Salad Cabinets Mile and Dairy Cases Liquid Chillers	Draft-Beer Coolers Commercial Refrigerators Dehumidifiers Walk-In Coolers Reach-In Refrigerators Beverage Vendors w/Ice Dough Retarders Ice-Storage Cabinets Beverage Dispensers Bottle Coolers Air Drier Open Display Cases Fur-Storage Refrigeration Mortuary Refrigeration Meat Cases Salad Cabinets Milk and Dairy Cases Liquid Chillers	Draft-Beer Coolers Ice Machines Commercial Refrigerators Dehumidifiers Bulk Milk Coolers Beverage Vendor - Ice Reach-In Refrigerators Walk-In Coolers Dough Retarders Florist Refrigerators Beverage Dispensers Bottle Coolers Egg Coolers Air Drier Open Display Cases Fur-Storage Refrigeration Mortuary Refrigeration Drinking Water Coolers Salad Cabinets Milk and Dairy Cases Liquid Chillers	Air Conditioning	Heat Pumps

Notes:

1. On high temperature systems, subtract 20°F from the desired temperature of the product to obtain approximate evaporating temperature of the compressor application, on medium temperature systems subtract 10 to 12°F, and on low temperature systems subtract 6 to 8°F.
2. For air conditioning, sutract 15°F from the desired room temperature, then apply paragragh one.
3. If in doubt, use the higher temperature pump.

(D)

Fig. 7-11 *(Continued)*

windings are always engaged while the compressor is starting and running.

PSC motors provide good running performance and adequate starting torque for low-line voltage conditions. They reduce potential motor trouble since the electrical circuit is simplified. See Fig. 7-28.

The figure shows a run capacitor designed for continuous duty. It increases the motor efficiency while improving power and reducing current drain from the line. Do not operate the compressor without the designated run capacitor. Otherwise, an overload results in the loss of start and run performance. Adequate motor-overload protection is not available either. A run capacitor in the circuit causes the motor to have some rather unique characteristics. Such motors have better pullout characteristics when a sudden load is applied.

Figure 7-29 shows how this particular series of compressors is wired for using the capacitor in the run and start circuit. Note the overload is an external line breaker. This motor-overload device is firmly attached to the compressor housing. It quickly senses any unusual temperature rise or excess current draw. The bimetal disc reacts to either excess temperature and/or excess current draw. It flexes downward, thereby disconnecting the compressor from the power source. See Fig. 7-30.

The CL compressor series is designed for residential and commercial air conditioning and heat pumps. These compressors are made in $2^1/_2$-, 3, $3^1/_2$-, 4, and 5-hp sizes. They can be operated on three phase or single phase. See Fig. 7-31. Since this is one of the larger compressors, it has two cylinders and pistons. It needs a good protection system for the motor. This one has an internal thermostat to interrupt the control circuit to the motor contactor. The contactor then disconnects the compressor from the power source. Figure 7-32 shows the location of the internal thermostat.

There is a supplementary overload in the compressor terminal box so that it can be reached for service. See Fig. 7-33. A locked rotor, or another condition producing excessive current draw, causes the bimetal disc to flex upward. This opens the pilot circuit to the motor contactor.

The contactor then disconnects the compressor from the power source. Single-phase power requires one supplementary overload. See Fig. 7-34. Three-phase power

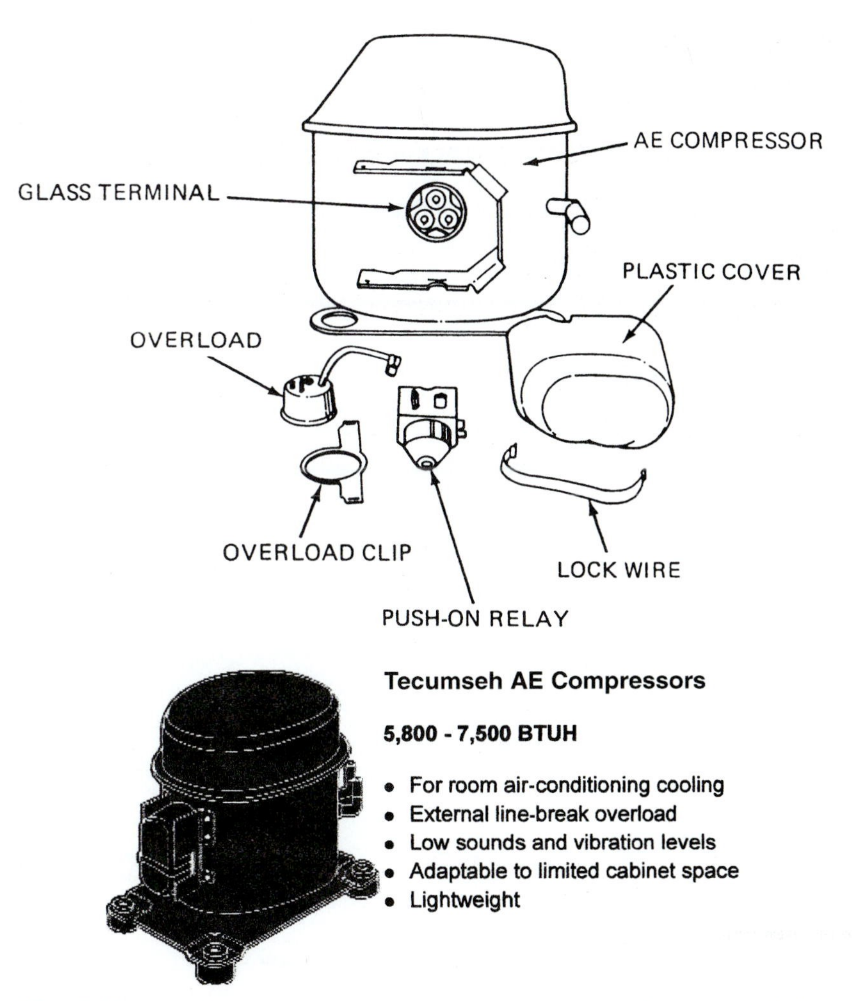

Fig. 7-12 *An AE compressor showing the glass-terminal, overload, overload clip, push-on relay, plastic cover, and lock wire.* (Tecumseh)

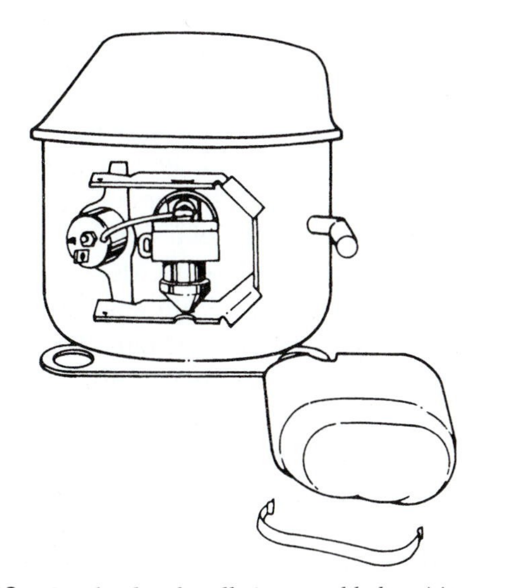

Fig. 7-13 *Overload and really in assembled positions.* (Tecumseh)

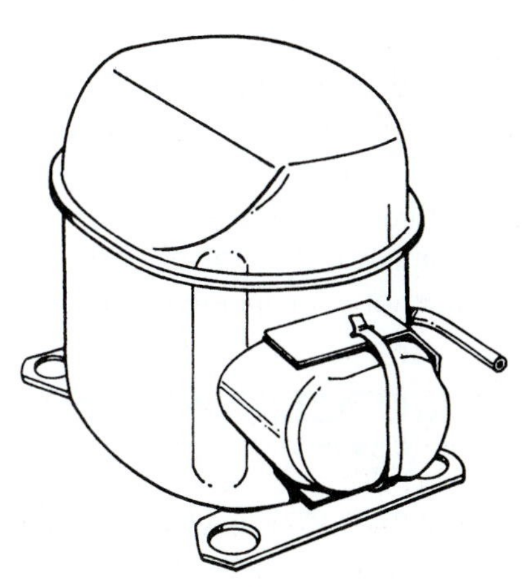

Fig. 7-14 *Completely assembled compressor.* (Tecumseh)

Resistance Start - Induction Run (RSIR)

- Motor Type Can Be Determined By Looking At Model Nameplate Information Located On The AS400 Bill Of Material (Shift F5)
- Normal-Starting Ability (Approved For Self Equalizing Systems)
- Model Numbers Begin With: 0, 1, 3, 6
- Accessories Include Current Relay
- Connections To Relay: Three
- Relay Removes Start Winding From Circuit During Operation

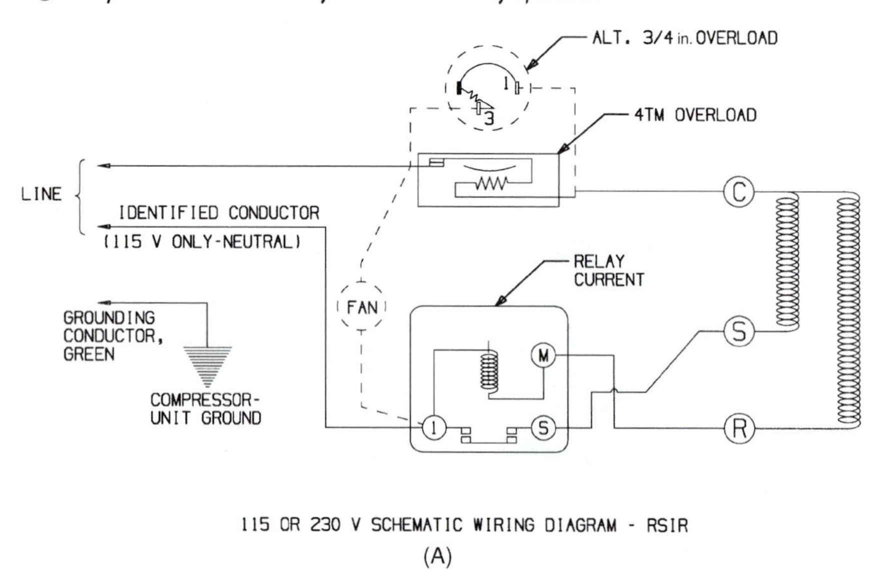

Fig. 7-15 *(A) Resistance-start induction-run motor for a compressor. (B) 115 or 230 V schematic diagram—PTCS-IR.* (Tecumseh)

requires two supplementary overloads. See Fig. 7-35. This CL line of compressors uses R-22 and R-12 (R-134a) or its suitable substitute refrigerants. They also use an oil charge of either 45 or 55 oz. In some cases, when the units are interconnected, they use 65 oz.

The H, J, and PJ compressors vary from $^3/_4$ to 3 hp. They have a wide temperature range. All of these models use a 55-oz oil charge. R-22 and R-12 refrigerants or, R-12 substitutes like R-134a that meet environmental requirements, are used.

The F and PF compressors are 2-, 3-, 4-, and 5-hp units. They use either 115 or 165 oz of oil. They, too, are used for a number of temperature ranges.

NEWER MODELS DESIGNATIONS AND CODING

The newer Tecumseh models are classified according to back pressure. For instance, the *commercial back pressure* (CBP) models start with No. 0 for no starting capacitor or No. 9 when the starting capacitor is required. These CBP models have an evaporator temperature range of −10 to +45°F. Compressor capacity is measured at +20°F evaporator temperature.

High back pressure (HBP) model numbers start with No. 3 for no starting capacitor or No. 4 when starting capacitor is required. The evaporator temperature range for these models is +20 to +55°F. Compressor capacity is measured at +45°F evaporator temperature.

Air-conditioning models of a single cylinder reciprocating type are designated with AE, AK, AJ. Model numbers starting with No. 5 are standard and those with No. 8 are high-efficiency models. The evaporator temperature range of these models is +32 to +55°F. Compressor capacity is measured at 45°F for standard models or +49°F evaporator temperature for high-efficiency models.

Air-conditioning and heat-pump (AC/HP) compressors come in two and three cylinder-reciprocating AC models (AW, AV, AG) plus rotary AC models.

PTC Start - Induction Run (PTCS/IR)

- Normal-Starting Ability (Aprroved For Self-Equalizing Systems)
- Model-Numbers Begin With 1 (AZ, TH, AE, TW, TP)
- Accessories Include PTC Relay
- Relay Removes Start Winding From Circuit During Operation

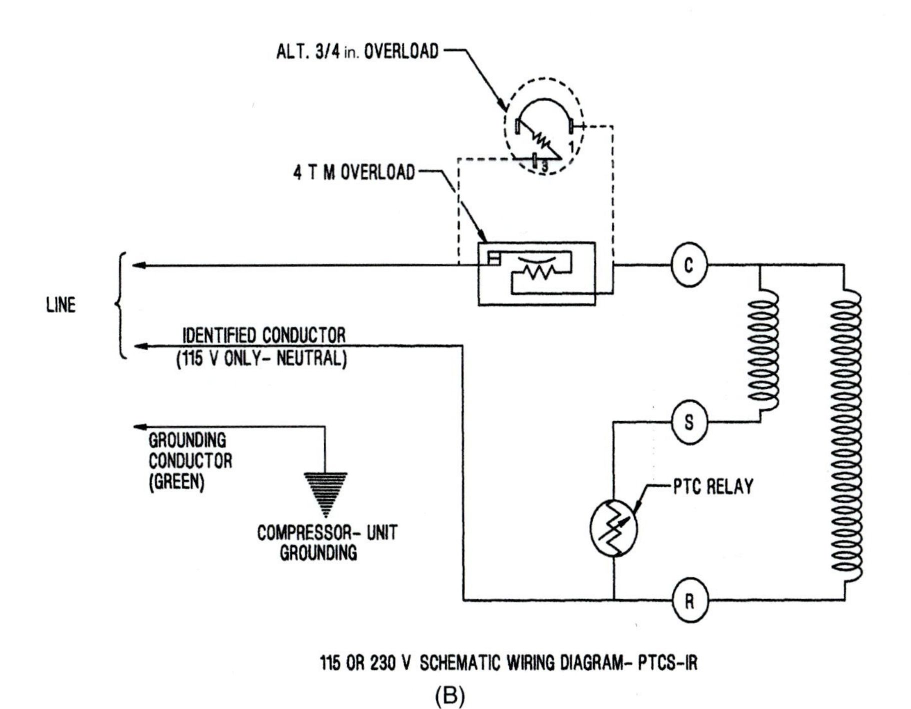

(B)

Fig. 7-15 *(Continued)*

The AC/HP model number starts with No. 5. The evaporator temperature range is −15 to +55°F. Compressor capacity evaporator temperature is measured at +45°F (AC) and −10°F for heat-pump models.

Low back pressure models start with No. 1 for no starting capacitor or No. 2 with starting capacitor required models. The evaporator temperature range is −40 to +10°F. Compressor capacity is measured at −10°F evaporator temperature.

Medium back pressure (MBP) model numbers start with No. 6 for no starting capacitor and No. 7 with starting capacitor required. The evaporator temperature range is −10 to +30°F. Compressor capacity is measured at +20°F evaporator temperature.

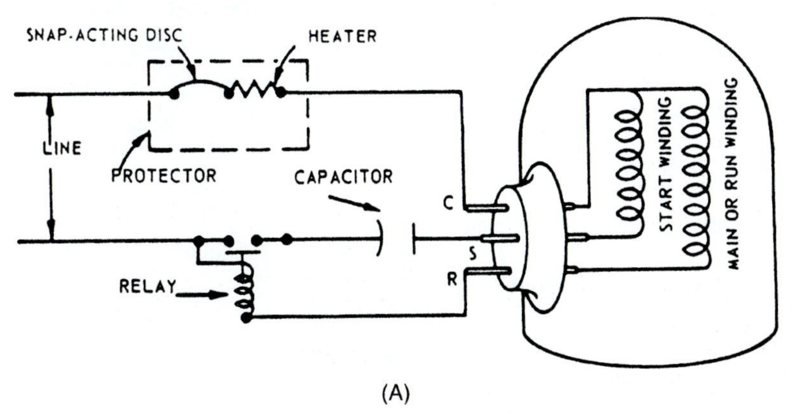

(A)

Fig. 7-16 *(A) Capacitor-start induction-run motor for a compressor. (B) PTC/capacitor run (PTCS/CR). (C) PTC relay.* (Tecumseh)

PTC Start - Capacitor Run (PTCS/CR)

- Normal-Starting Ability (Aprroved For Self-Equalizing Systems)
- Model Numbers Begin With 1 (AZ, TH, TP)
- Accessories Include PTC Relay And Run Capacitor
- Relay Removes Start Winding From Circuit During Operation
- Run Capacitor Remains In The Circuit During Operation

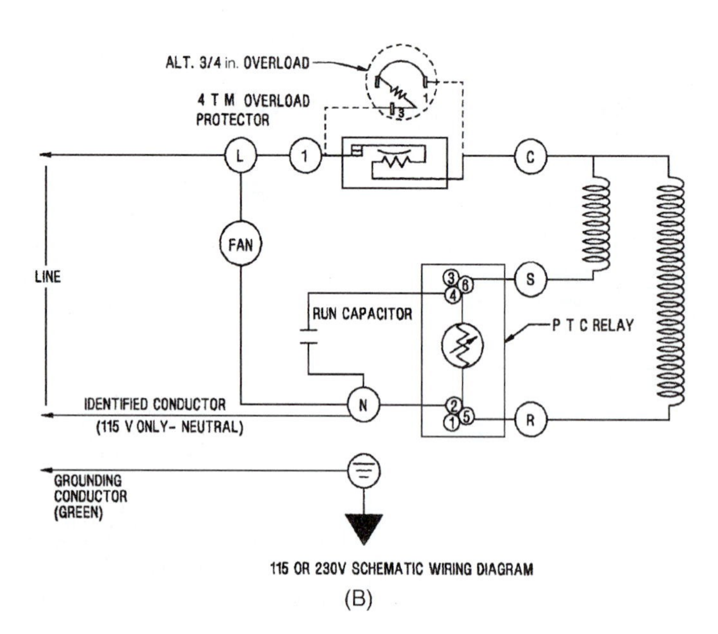

(B)

PTC Relay

- PTC = Positive Temperature Coefficent
- Resistance of PTC Material Increases with Increasing Temperature
- PTC Relay Sometimes used in Place of Current Relay
- Senses Current in the Motor-Start Winding
- Normally has Very Low Resistance
- When Power is Applied, Relay Resistance Increases, Removing Start Winding from Curcuit
- Relay Will Reset After a Cooldown Period of 3 to 10 S
- Usually Limited to use on Domestic Refrigerators and Freezers

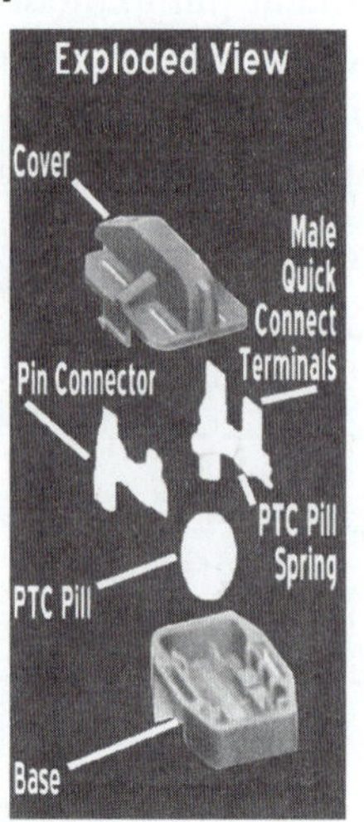

(C)

Fig. 7-16 *(Continued)*

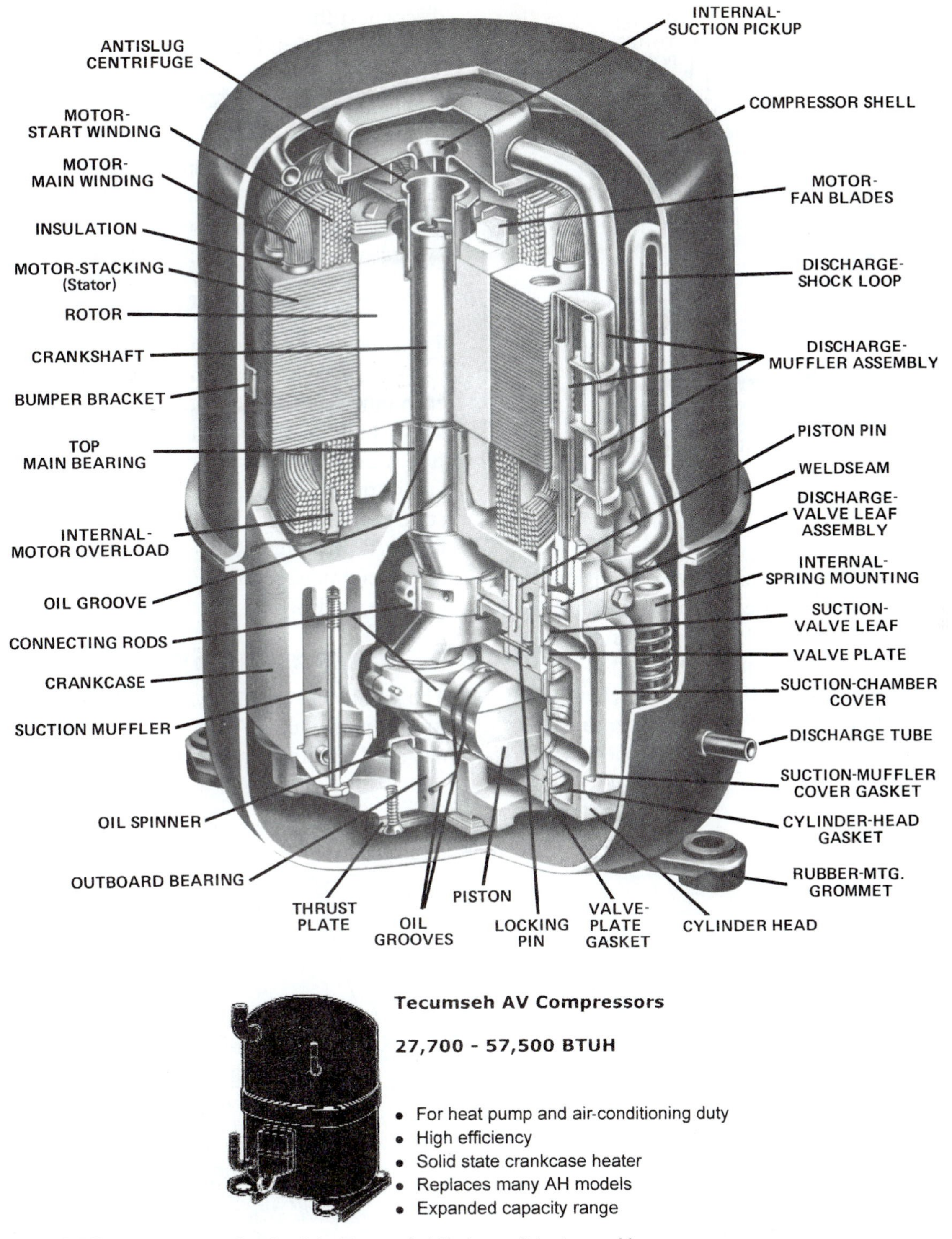

Fig. 7-17 *Construction details of the Tecumseh AH air-conditioning and heat-pump compressors.* (Tecumseh)

HERMETIC COMPRESSOR MOTOR TYPES

There are four general types of single-phase motors. Each has distinctly different characteristics. Compressor motors are designed for specific requirements regarding starting torque and running efficiency. These are two of the reasons why different types of motors are required to meet the various demands.

Resistance Start-Induction Run

The *resistance start-induction run* (RSIR) motor is used on many small hermetic compressors through

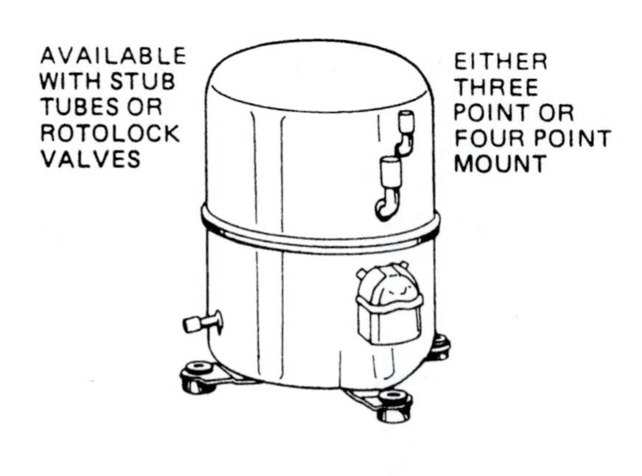

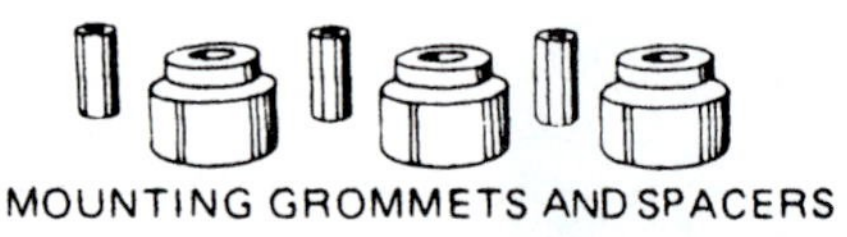

Fig. 7-18 *External view of the Tecumseh AH airconditioning and heat-pump compressor with its grommets and spacers.* (Tecumseh)

1/3 hp. The motor has low-starting torque. It must be applied to completely self-equalizing capillary-tube systems such as household refrigerators, freezers, small water coolers, and dehumidifiers.

This motor has a high-resistance start winding that is not designed to remain in the circuit after the motor has come up to speed. A current relay is necessary to disconnect the start winding as the motor comes up to design speed. See Fig. 7-36.

Capacitor Start-induction Run

The *capacitor start-induction run* (CSIR) motor is similar to the RSIR. However, a start capacitor is included in series with the start winding to produce a higher starting torque. This motor is commonly used on commercial refrigeration systems with a rating through 3/4 hp. See Fig. 7-37.

Capacitor Start and Run

The *capacitor start and run* (CSR) motor arrangement uses a start capacitor and a run capacitor in parallel with each other and in series with the motor start winding. This motor has high-starting torque and runs efficiently. It is used on many refrigeration and air conditioning applications through 5 hp. A potential relay removes the start capacitor from the circuit after the motor is up to speed. Potential relays must be accurately matched to the compressor. See Fig. 7-38. Efficient operation depends on this.

Permanent Split Capacitor

The *permanent split capacitor* (PSC) has a run capacitor in series with the start winding. Both run capacitor and start winding remain in the circuit during start and

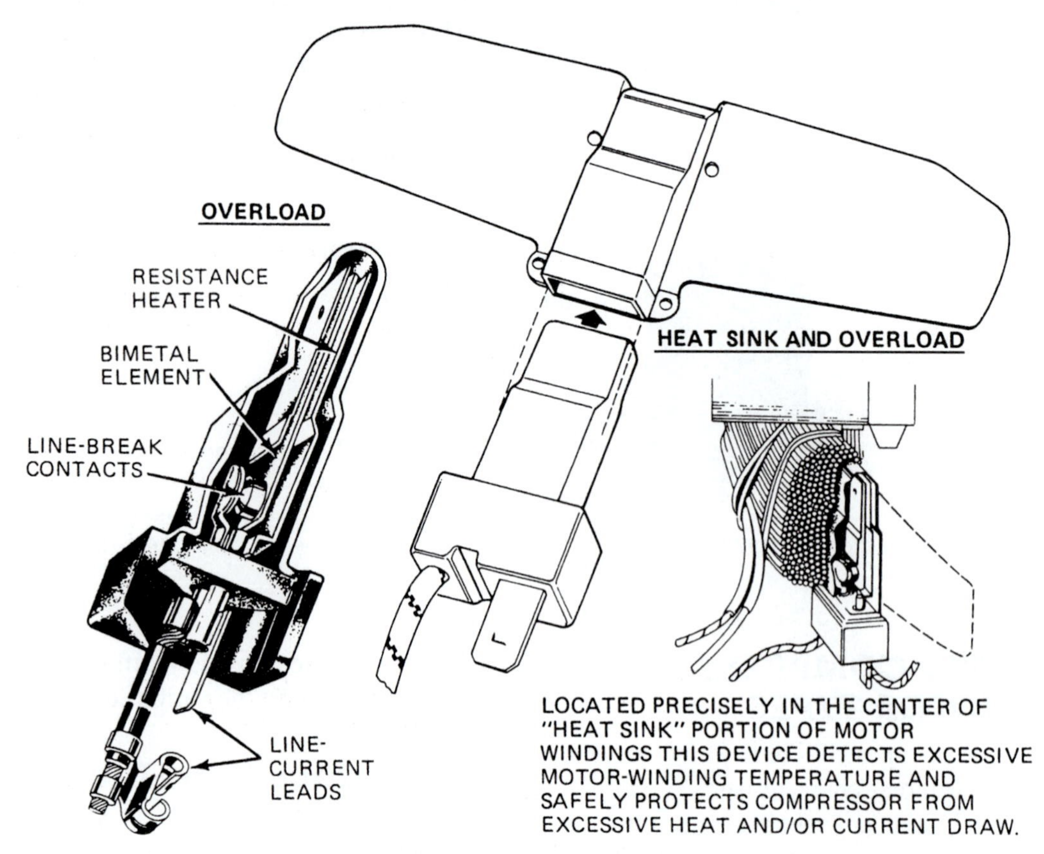

Fig. 7-19 *Internal line-break motor protector.* (Tecumseh)

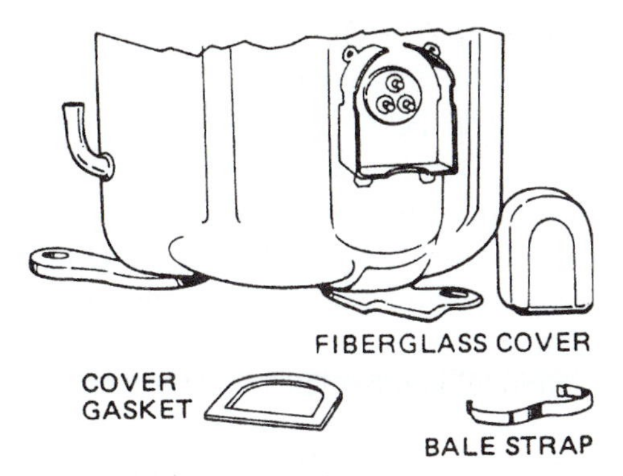

Fig. 7-20 *Snap-on terminal cover assembly. (Tecumseh)*

after the motor is up to speed. Motor torque is sufficient for capillary and other self-equalizing systems. No start capacitor or relay is necessary. The PSC motor is basically an air-conditioning compressor motor. It is very common through 3 hp. It is also available in 4- and 5-hp sizes. See Fig. 7-39.

COMPRESSOR MOTOR RELAYS

A hermetic compressor motor relay is an automatic switching device designed to disconnect the motor start winding after the motor has attained a running speed.

There are two types of motor relays used in the refrigeration and air-conditioning compressors:

- The current-type relay
- The potential-type relay

Current-type Relay

The current-type relay is generally used with small refrigeration compressors up to $^3/_4$ hp. When power is applied to the compressor motor, the relay-solenoid coil attracts the relay armature upward, causing bridging contact and stationary contact to engage. See Fig. 7-40. This energizes the motor start winding. When the compressor motor attains running speed, the motor main winding current is such that the relay-solenoid coil deenergizes and allows the relay contacts to drop open. This disconnects the motor start winding.

The relay must be mounted in true vertical position so that the armature and bridging contact will drop free when the relay solenoid is deenergized.

Potential-type Relay

This relay is generally used with large commercial and air-conditioning compressors. The motors may be capacitor start capacitor run types up to 5 hp. Relay contacts are normally closed. The relay coil is wired across the start winding. It senses voltage change. Start winding voltage increases with motor speed. As the voltage increases to the specific pick-up value, the armature pulls up, opening the relay contacts and deenergizing the start winding. After switching, there is still sufficient voltage induced in the start winding to keep the relay coil energized and the relay-starting contacts open. When power is shut off to the motor, the voltage drops to zero. The coil is deenergized and the start contacts reset. See Fig. 7-41.

Many of these relays are extremely position sensitive. When changing a compressor relay, care should be taken to install the replacement in the same position as the original. Never select a replacement relay solely by horsepower or other generalized rating. Select the correct relay from the parts guidebook furnished by the manufacturer.

COMPRESSOR TERMINALS

For the compressor motor to run properly it must have the power correctly connected to its terminals outside

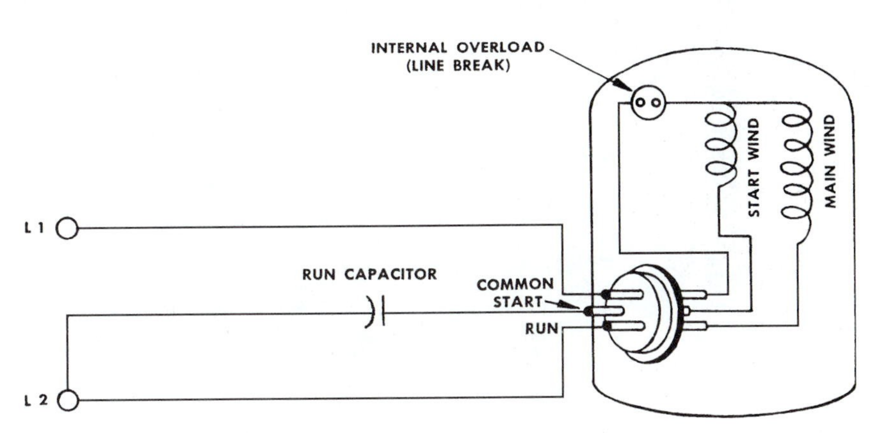

Fig. 7-21 *Single-phase diagram for the AH air-conditioner and heat-pump compressor. (Tecumseh)*

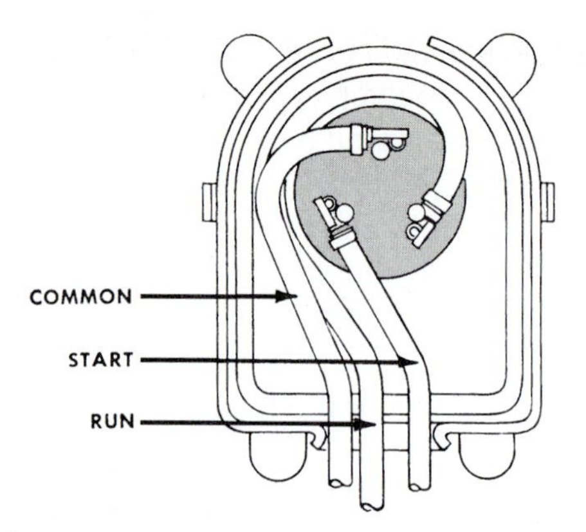

Fig. 7-22 *Terminal box showing the position of the terminals on the AH series of compressors.* (Tecumseh)

of the hermetic shell. There are several different types of terminals used on the various models of Tecumseh compressors.

Tecumseh terminals are always thought of in the order of common, start, and run. Read the terminals in the same way you would read the sentences on a book's page. Start at the top left-hand corner and read across the first line from left to right. Then, read the second line from left to right. In some cases three lines must be "read" to complete the identification process. Figure 7-42 shows the different arrangements of terminals. All Tecumseh compressors, except one model, follow one of these patterns. The exception is the old twin-cylinder, internal-mount compressor built at Marion. This was a 90° piston model designated with an "H" at the beginning of the model number (that is, HA 100). The terminals were reversed on the H models and read run, start, and common. See Fig. 7-43. These compressors were replaced by the J-model series in 1955. All J models follow the usual pattern for common, start, and run.

BUILT-UP TERMINALS

Some built-up terminals have screw- and nut-type terminals for the attaching of wires. See Fig. 7-44.

Others may have different arrangements. The pancake compressors built in 1953, and after, have glass terminals that look something like those shown in Fig. 7-45. The terminal arrangement for S and C

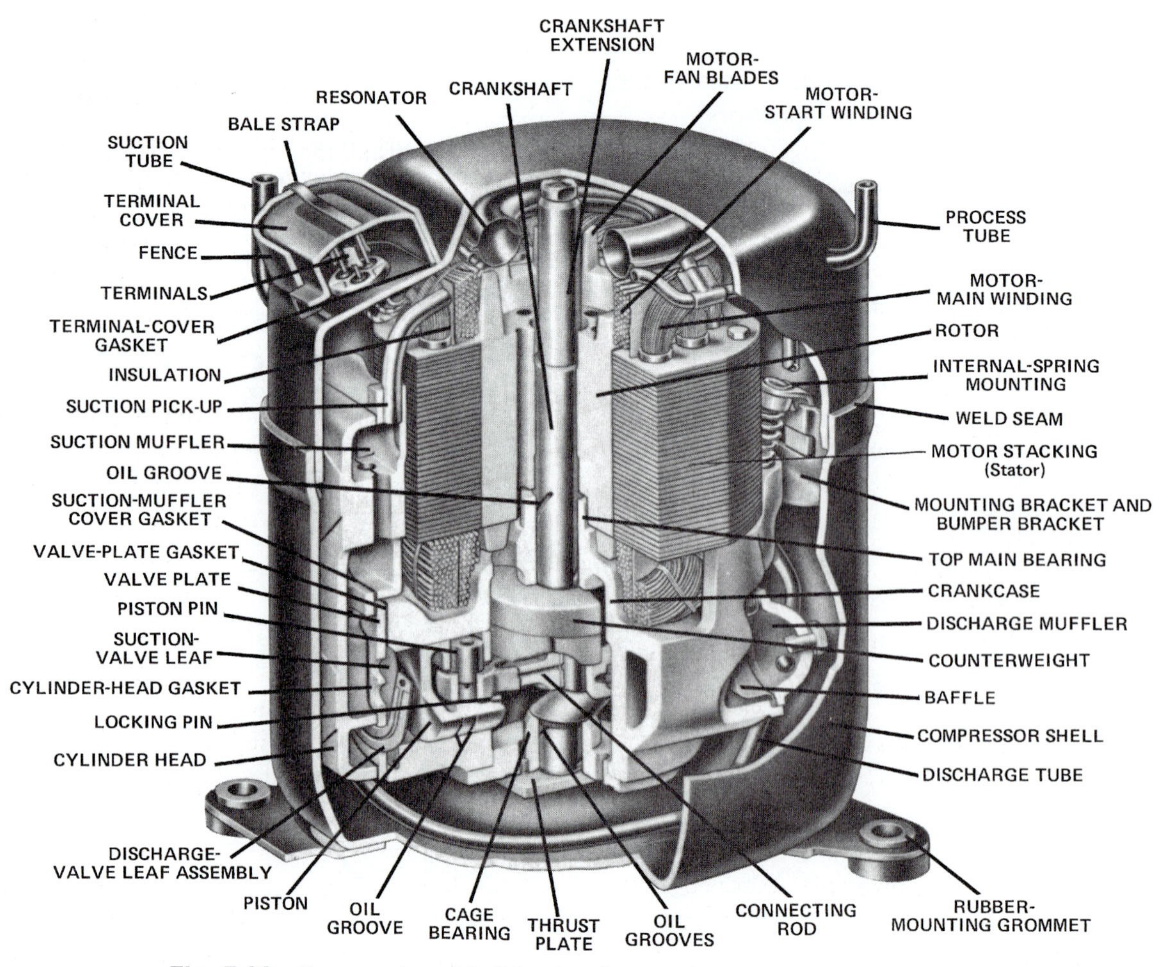

Fig. 7-23 *Cutaway view of the AJ series of air-conditioning compressors.* (Tecumseh)

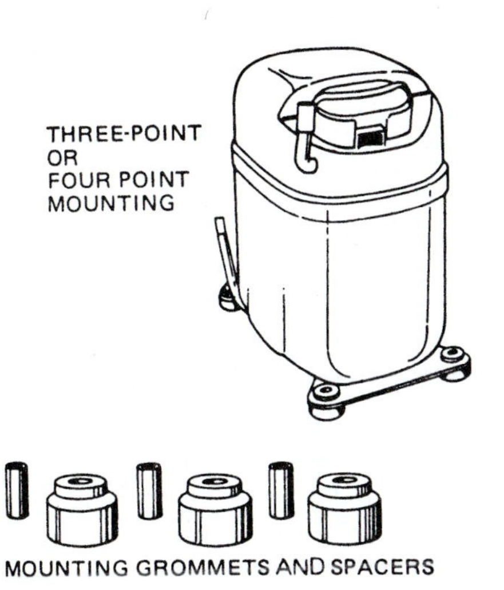

Fig. 7-24 *External view of the AJ compressor.* (Tecumseh)

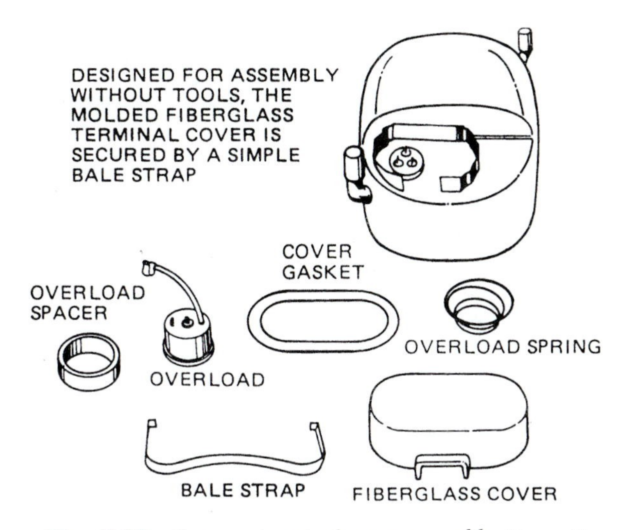

Fig. 7-25 *Snap-on terminal cover assembly.* (Tecumseh)

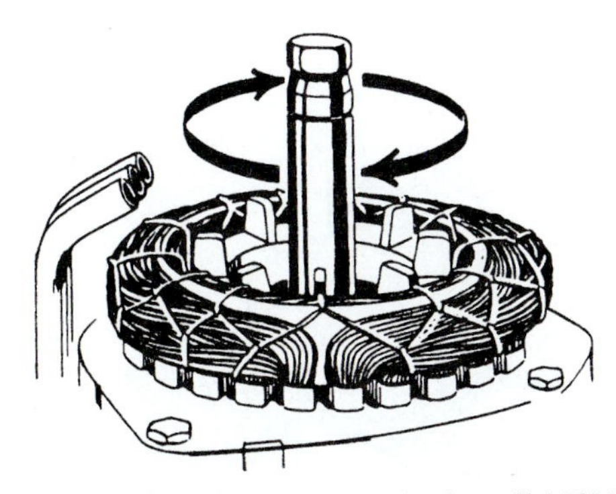

Fig. 7-26 *An antislug feature is standard on all AJ1M12 models and on larger models of the AJ series.* (Tecumseh)

single-cylinder ISM models resembles that shown in Fig. 7-46. Models J and PJ with twin-cylinder internal mount have a different terminal arrangement. It looks like that shown in Fig. 7-47.

GLASS QUICK-CONNECT TERMINALS

Figure 7-48 shows the quick-connect terminals used on S and C single-cylinder ISM models. The AK and CL models also use this type of arrangement. Many of the CL models have the internal thermostat terminals located close by.

Quick-connect glass terminals are also used on AU and AR air-conditioning models. The AE air-conditioning models also use glass quick connects. Models AB, AJ, and AH also use glass quick connects, but notice how their arrangement of common, start, and run vary from that shown in Fig. 7-48. Figure 7-49 shows how the AU, AR, AE, AB, AJ, and AH models terminate.

Glass terminals are also used on pancake-type compressors with P, R, AP, and AR designations. See Fig. 7-50. The T and AT models, as well as the AE-refrigeration models, also use the glass terminals, but without the quick connect.

Keep in mind that you should never solder any wire or wire termination to a compressor terminal. Heat applied to a terminal is liable to crack the glass-terminal base or loosen the built-up terminals. This will, in turn, cause a refrigerant leak at the compressor.

MOTOR MOUNTS

To dampen vibration, hold the compressor while in shipment, and cushion horizontal thrust when the compressor starts or stops, some type of mounting is necessary. Several different arrangements are used. However, each of them uses a base plate. Also, some space is allowed between the rubber grommet and the washer on the nut. The rubber grommet absorbs most of the vibration. See Fig. 7-51.

In Fig. 7-52, you can see the use of a spring to prevent damage to the rubber grommet. This is used for the heavier compressors. There are usually three, but sometimes four of these rubber motor mounts on each compressor model. One of the greatest uses of this type of mount is to make sure that vibrations are not transferred to other parts of the refrigeration system or passed on to the pipes. There, they would weaken the soldered joints.

CRANKCASE HEATERS

Most compressors have crankcase heaters. This is because most air-conditioning and commercial systems are started up with a large part of the system refrigerant charge in the compressor. This is especially when the unit has been idle for some time or when the compressor is being started for the first time. On start-up, the refrigerant boils off, taking the oil charge with it. This

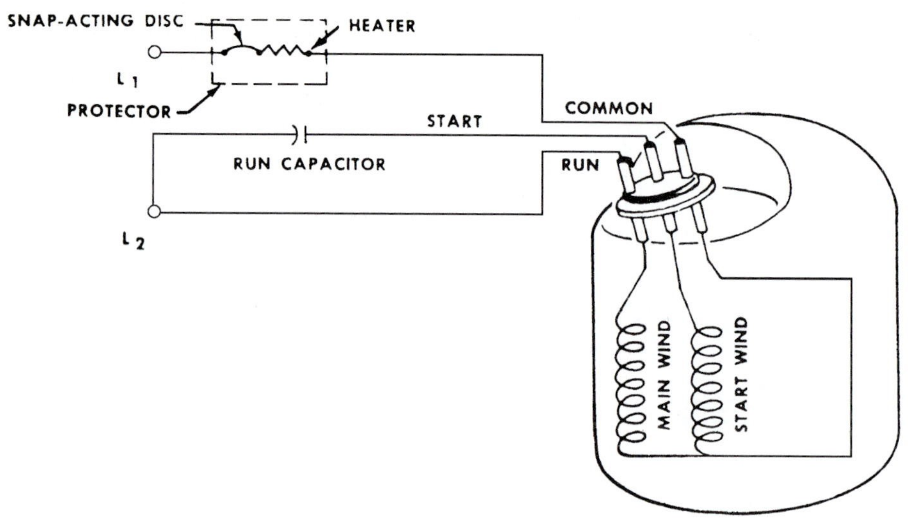

Fig. 7-27 *Permanent split-capacitor schematic.* (Tecumseh)

means the compressor is forced to run for as long as 3 or 4 min. until the oil charge circulates through the system and returns to the crankcase. Obviously, this shortens the service life of the compressor.

The solution is to charge the system so that little or no refrigerant collects in the crankcase and to operate the crankcase heater at least 12 h before start-up or after a prolonged down time.

Two types of crankcase heaters are in common use on compressors. The wrap-around type is usually referred to as the "belly band." The other type is the run capacitance off-cycle heat method.

The wrap-around heater should be strapped to the housing below the oil level and in close contact with the housing. A good heater will maintain the oil at least 10°F (5°C) above the temperature of any other system

Run Capacitors

- Defined by Microfarad and Voltage Ratings
- When Substituting, Microfarad Rating Must be the Same, Voltage Rating can be Higher
- Aides in Starting and Improving the Efficiency of the Motor
- Continuous Duty - Always in Motor Circuit

Fig. 7-28 *Run capacitors.* (Tecumseh)

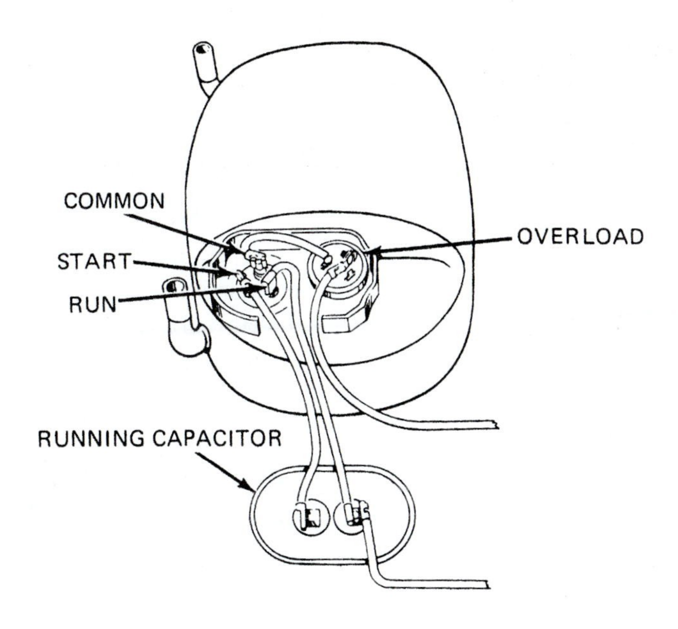

Permanent Split Capacitor (PSC)

- Normal-Starting Ability (Aprroved For Self-Equalizing Systems)
- Model Numbers Begin With: 5, 8
- Accessories Include Run Capacitor
- Run Capacitor Remains In The Circuit During Operation

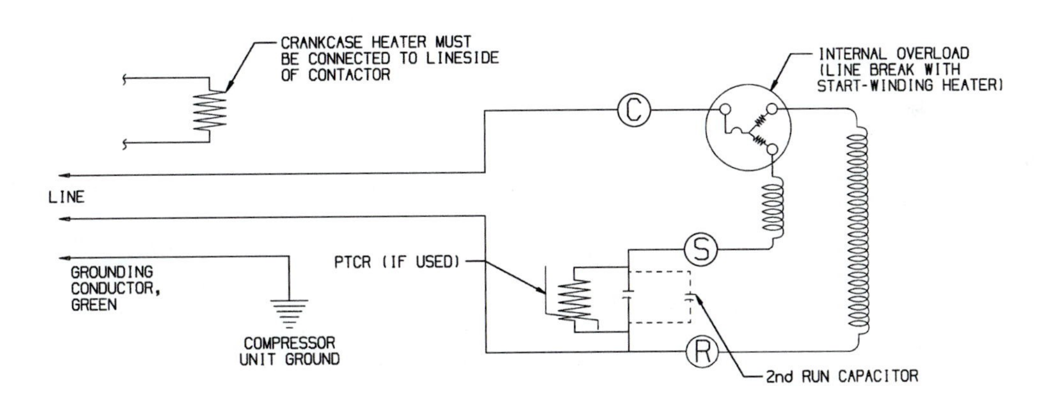

230 V-SCHEMATIC WIRING DIAGRAM
PSC - INTERNAL OVERLOAD (LINE BREAK) - START-WINDING HEATER - CRANKCASE HEATER

Fig. 7-29 *A PSC-motor hookup.* (Tecumseh)

component. When the compressor is stopped it will maintain it at or above a minimum temperature of 80°F (27°C).

The run capacitance, off-cycle heat method, single-phase compressors are stopped by opening only one leg (L_1). Thus, the other leg to the power supply (L_2) of the run capacitor remains "hot." A trickle current through the start windings results, thereby warming the motor windings. Thus, the oil is warmed on the "off-cycle."

Make sure you pull the switch that disconnects the whole unit from the power source before working on such a system.

Capacitance crankcase heat systems, can be recognized by one or more of the following:

- Contactor or thermostat breaks only one leg to the compressor and condenser fan.
- Equipment carries a notice that power is on at the compressor when it is not running and that the main breaker should be opened before servicing.
- Run capacitor is sometimes split (it has three terminals) so that only part of the capacitance is used for off-cycle heating.

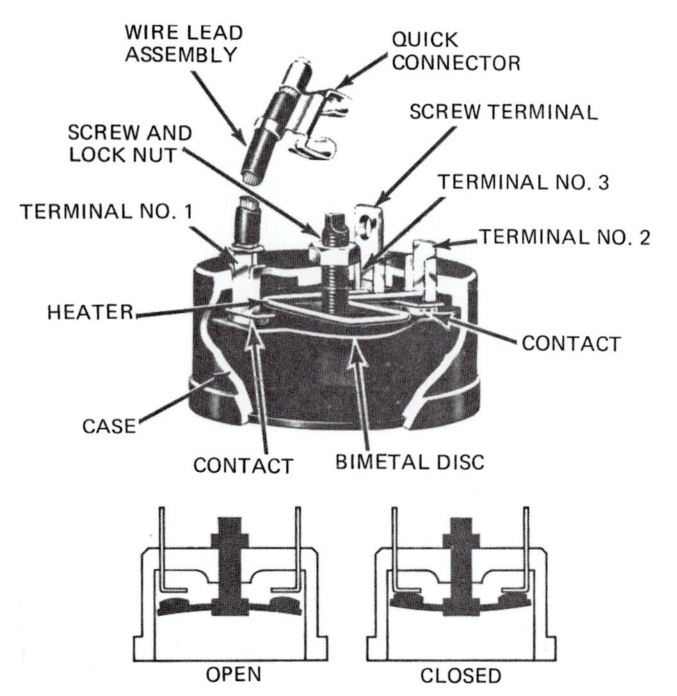

Fig. 7-30 *External line-break overload.* (Tecumseh)

CAUTION: Make sure you use an exact replacement when changing such dual-purpose run capacitors. The capacitor must be fused and carry a bleed resistor across the terminals.

The basic wiring diagram for a PSC compressor with a run capacitance off-cycle heat is shown in Fig. 7-53.

ELECTRICAL SYSTEMS FOR COMPRESSOR MOTORS

Most of the problems associated with hermetic compressors are electrical. Most of the malfunctions are in the current relay, potential relay, circuit breaker, or loose connections. In most cases, internal parts of the compressor housing can be checked with an ohmmeter.

Normal-Starting Torque Motors (RSIR) with a Current-Type Relay

Normal starting torque motors (RSIR) with a current-type relay mounted on the compressor terminals require several tests that must be performed in the listed sequence. Figure 7-54 shows a two-terminal external overload device in series with the start and run windings.

The fan motor runs from point 1 on the current relay to point 3 on the overload device. L_2 has the relay coil inserted in series with the run winding. When the winding draws current, the solenoid is energized. This is done by the initial surge of current through the run winding. When the relay energizes with sufficient current, it closes the contacts (points 1 and S) and places the start winding in the circuit. The start winding stays

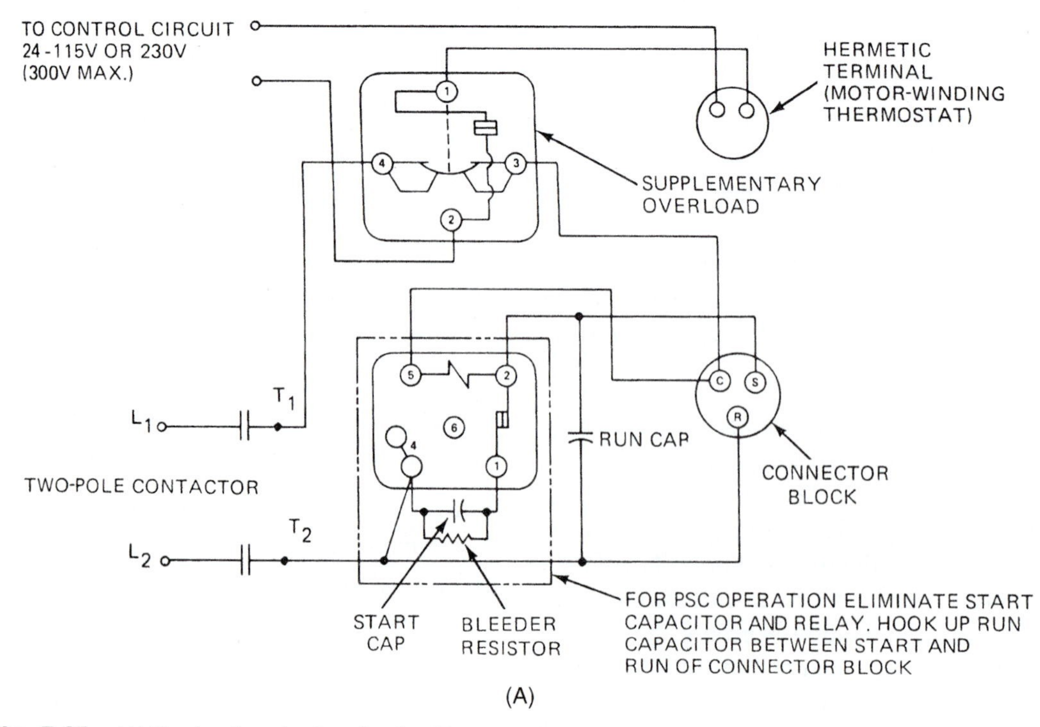

Fig. 7-31 *(A) Single-phase hookup for the CL air-conditioning and heat-pump compressors. (B) Three-phase hookup.* (Tecumseh)

Three Phase

- Increased Starting Ability (Approved For NonSelf-Equalizing Systems)
- Available In Two And Three Cylinder Reciprocating Designs (AW, AV, AG)
- No Accessories Required For Starting

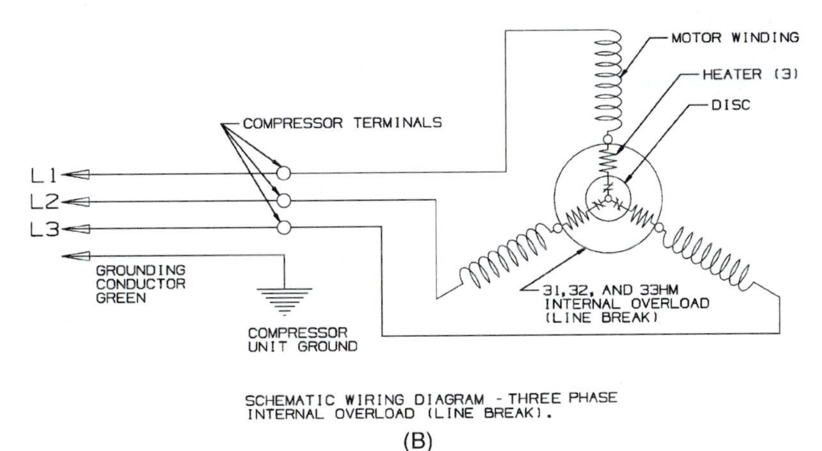

(B)

Fig. 7-31 *(Continued)*

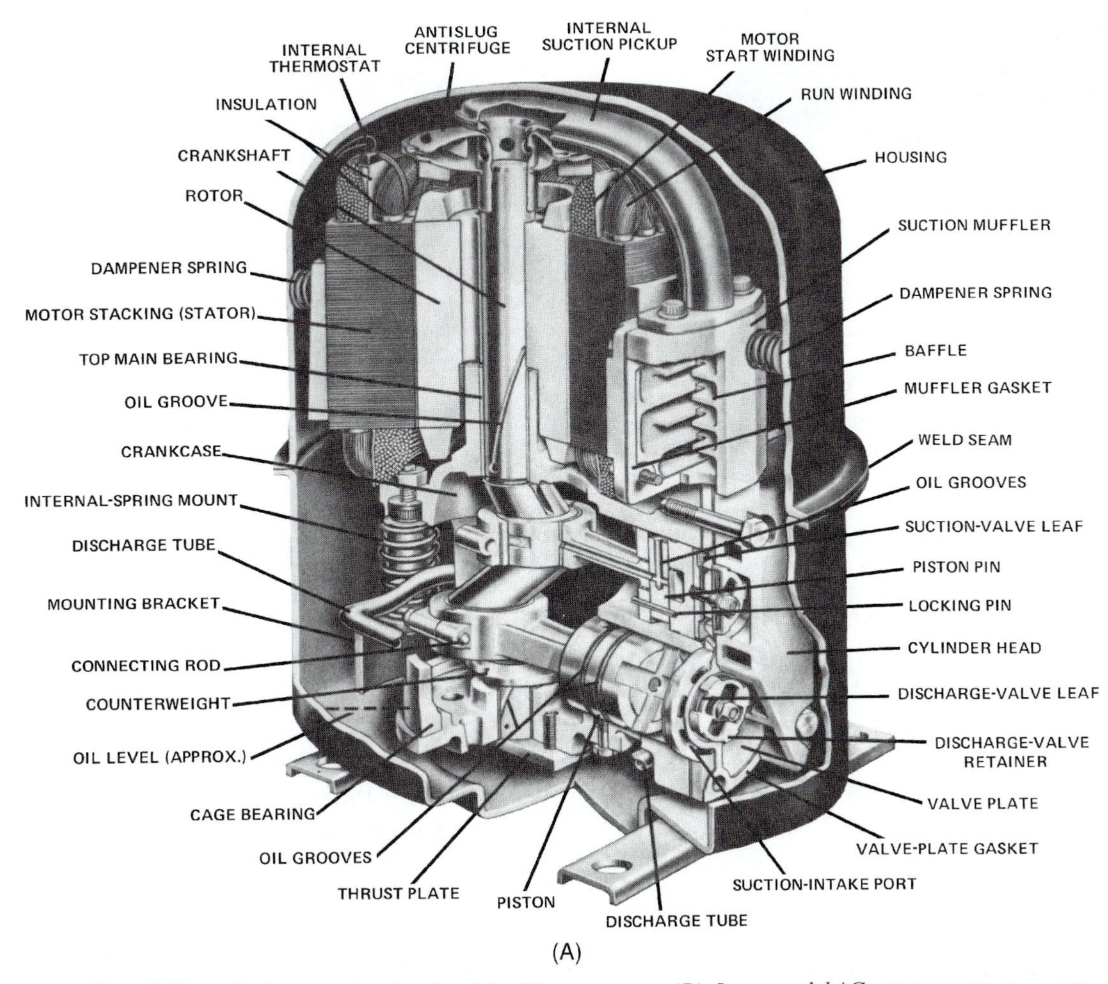

(A)

Fig. 7-32 *(A) Construction details of the CL compressor. (B) Later model AG compressors.* (Tecumseh)

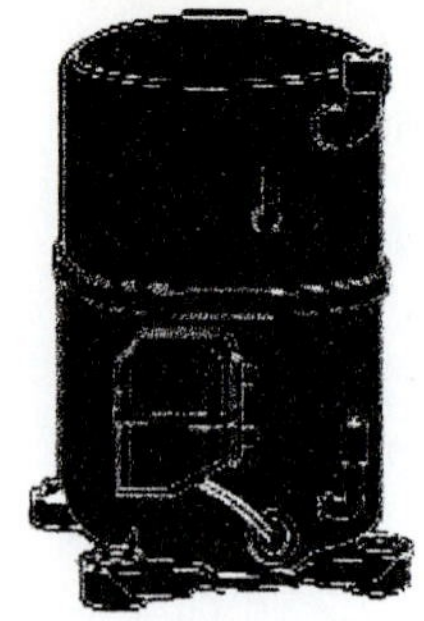

Tecumseh AG Compressors

46,000 - 71,280 BTUH

- For heat pump or central air-conditioning duty
- Internal line-break overload
- Internal pressure-relief valve
- Low sounds and vibration levels
- Replaces CL models and other brands

(B)

Fig. 7-32 *(Continued)*

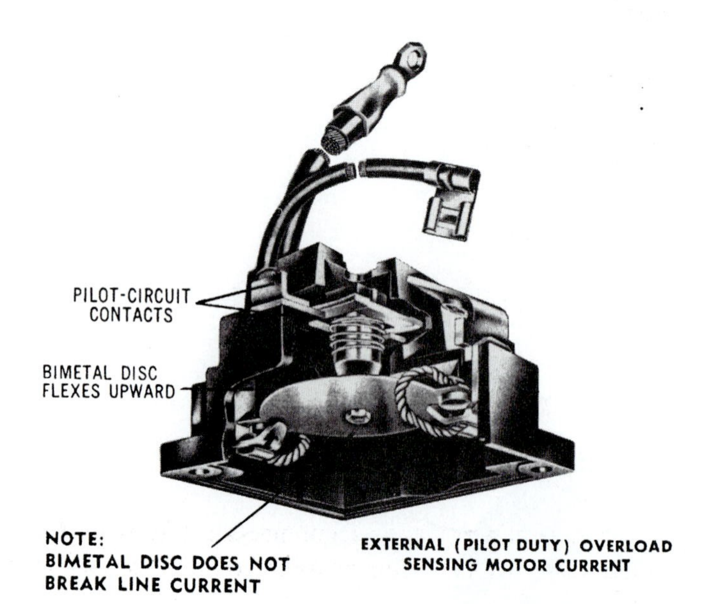

Fig. 7-33 *Cutaway view of the supplementary overload.* (Tecumseh)

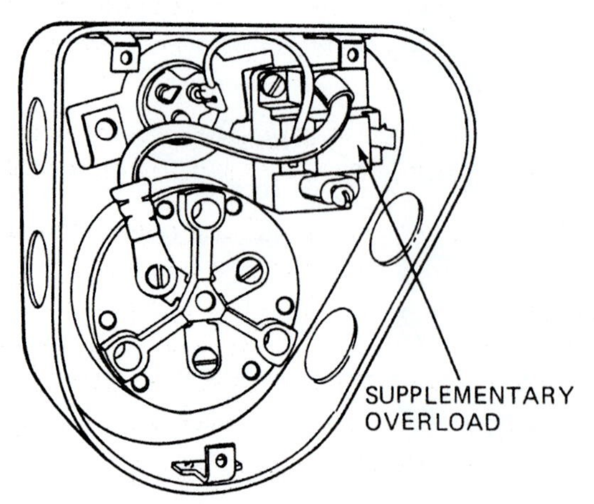

Fig. 7-34 *Location of the supplementary overload on the CL-compressor series.* (Tecumseh)

in the circuit until the relay deenergizes. When the motor comes up to about 75 percent of its run speed, the relay deenergizes since the current through the run winding drops off. This change in current makes it a very sensitive circuit. The sensing relay must be in good operating condition. Otherwise, it will not energize or deenergize at the proper times.

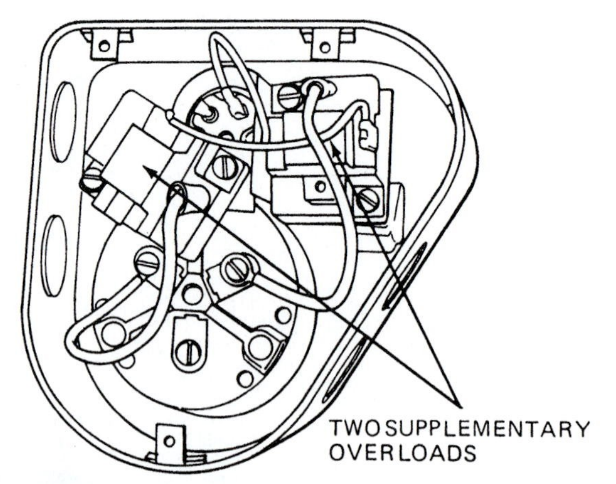

Fig. 7-35 *Location of two supplementary overloads in the terminal box makes it applicable for three-phase power connections.* (Tecumseh)

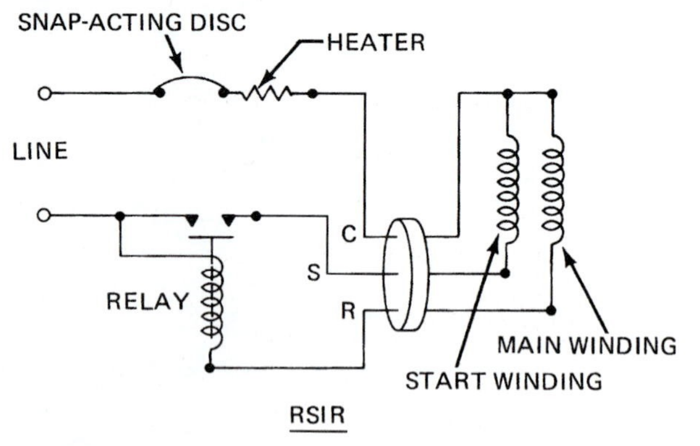

Fig. 7-36 *Resistance-start induction-run motor schematic.* (Tecumseh)

The start contacts on the current-type relay are normally open. See Fig. 7-54. Check the electrical system on this type of compressor system by using a voltmeter for obtaining line-voltage reading. Then, use an ohmmeter

PTC Start Device

- PTC = Positive Temperature Coefficent
- Resistance of PTC Material Increases With Increasing Temperature
- PTC Device By-Passes Run Capacitor Initially, Which Increases Starting Ability
- PTC Resistance Increases, Putting The Run Capacitor Back Into Operation
- Device Resets When The Compresor Is Shut Off

Fig. 7-36 *(Continued)*

to check continuity. That means the power must be off. Make sure the circuit breaker is off at the main power supply for this unit. If a fan is used (as shown in the dotted lines of Fig. 7-54) make sure lead is disconnected from the line. Next, check continuity across the following:

1. Check continuity across L_1 and point 3 of the overload. There is no continuity. Close control contacts by hand. If there is still no continuity, replace the control.
2. Check continuity across No. 3 and No. 1 on the overload. If there is no continuity, the protector may be tripped. Wait 10 min and check again. If there is still no continuity, the protector is defective. Replace it.
3. Pull the relay off the compressor terminals. Be sure to keep it in an upright position.
4. Check continuity across relay-terminal 1 (or L) and S. If there is continuity, relay contacts are closed, when they should be open. Replace the relay.
5. Check continuity across No. 1 and M. If there is no continuity, replace the relay. The solenoid is open.
6. Check continuity across compressor terminals C and R. If there is no continuity, there is an open run winding. Replace the compressor.
7. Check continuity across compressor terminals C and S. If there is no continuity there is an open start winding. Replace the compressor.
8. Check continuity across compressor terminal C and the shell of the compressor. There is no continuity. This means the motor is grounded. Replace the compressor.
9. Check the winding resistance values against those published by the manufacturer of the particular model.

If all the tests prove satisfactory, and there is no capillary restriction, plus, the unit continues to fail to operate properly, change the relay. The new relay will eliminate any electrical problems, such as improper pickup or dropout, that cannot be determined by the tests listed. If a good relay fails to correct the difficulty, the compressor is inoperative due to internal defects. It must be replaced.

High-Starting Torque Motors (CSIR) with a Current-Type Relay

High-starting torque motors (CSIR) with a current-type relay mounted on the compressor terminals can be easily checked for proper operation. Remember from

Capacitor Start - Induction Run (CSIR)

- Increased Starting Ability (Approved For NonSelf-Equalizing Systems)
- Model Numbers Can Begin With: 2, 4, 7, 9
- Accessories Include Current Relay And Start Capacitor
- Connections To Relay: Five
- Relay Removes Start Winding And Start Capacitor From Circuit During Operation

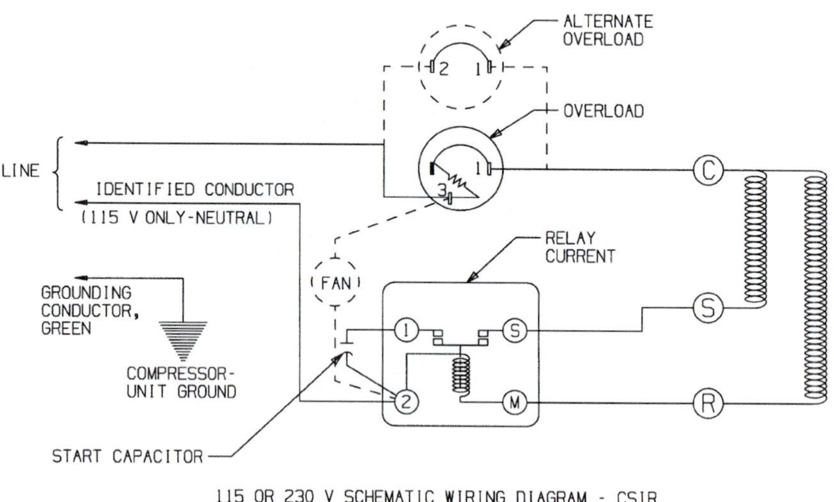

(A)

Start Capacitors

- Defined by Microfarad and Voltage Ratings
- When Substituing, Microfarad Rating Must be the Same, Voltage Rating Can be Higher
- Used to Increase Starting Ability of the Motor
- Intermittent Duty - Must be Removed from Circuit Once CompressorStarts
- If Used with Run Capacitor, Start Capacitor Should Have Resistor Attached

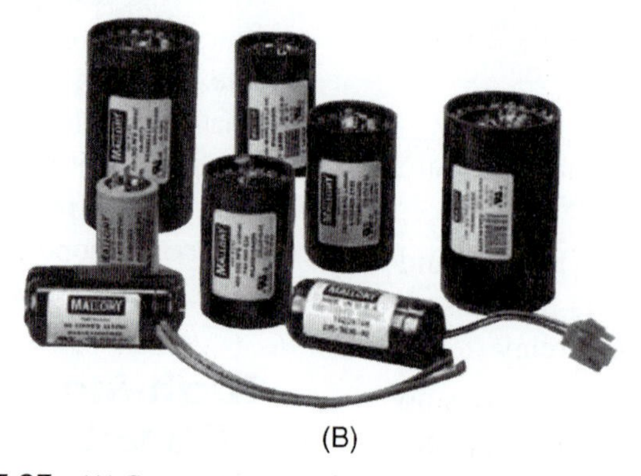

(B)

Fig. 7-37 *(A) Capacitor-start induction-run motor schematic. (B) Start capacitors—PTC-start device.* (Tecumseh)

Capacitor Start - Run (CSR)

- Increased Starting Ability (Approved For NonSelf-Equalizing Systems)
- Model Numbers Can Begin With: 2, 4, 7, 9
- Accessories Include Potential Relay, Start, And Run Capacitors
- Connections To Relay: Five
- Relay Removes Start Winding And Start Capacitor From Circuit During Operation
- Run Capacitor Remains In The Circuit For PSC Operation (See Below)

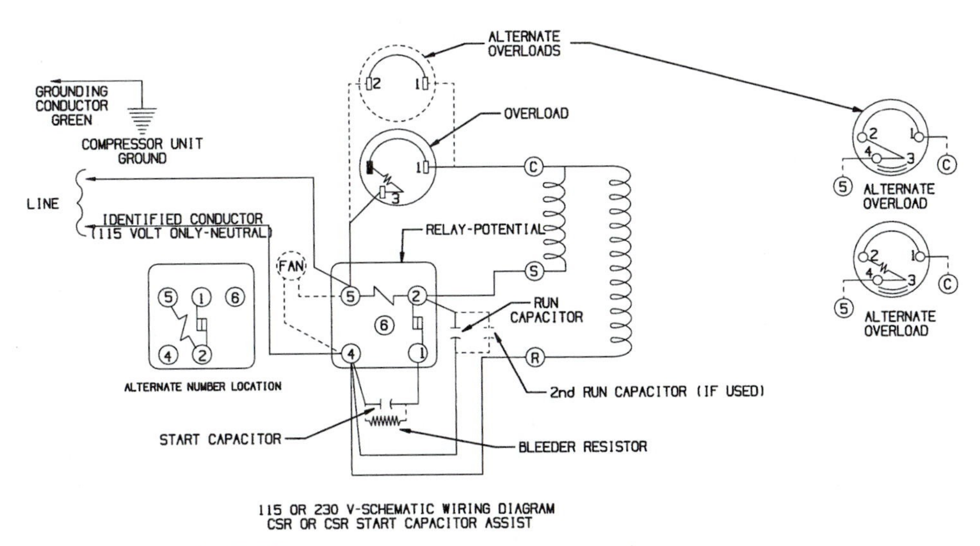

Fig. 7-38 *Capacitor start and run motor schematic.* (Tecumseh)

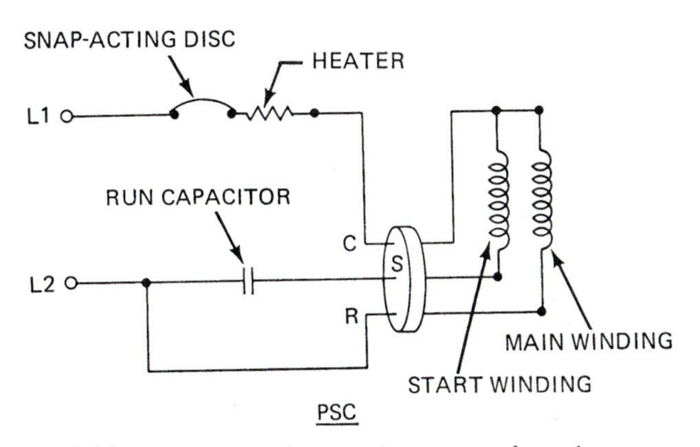

Fig. 7-39 *Permanent split-capacitor motor schematic.* (Tecumseh)

the previous type that the current-type relay normally has its contacts open.

Use a voltmeter first to check the power source. Use an ohmmeter to check continuity. Make sure the power is off and the fan-motor circuit is open. The electrical system on this type of hermetic system can be checked as follows. See Fig. 7-55.

1. Check continuity across L_1 and 3. If there is no continuity, close the control contacts. If there is still no continuity, replace the control.
2. Check continuity across No. 3 and No. 1 on the overload. If there is no continuity, the protector may be tripped. Wait for 10 min and check again. If there is still no continuity, the protector is defective. Replace it.
3. Pull relay off compressor terminals. *Keep it upright!*
4. Check continuity across relay terminal 1 and S. If there is continuity, the relay contacts are closed when they should be open. Replace the relay.
5. Check continuity across relay terminals 2 and M. If there is no continuity, replace the relay.
6. Check continuity across compressor terminals C and R. If there is no continuity, there is an open run winding. Replace the compressor.
7. Check continuity across compressor C and S. If there is no continuity, there is an open start winding. Replace the compressor.

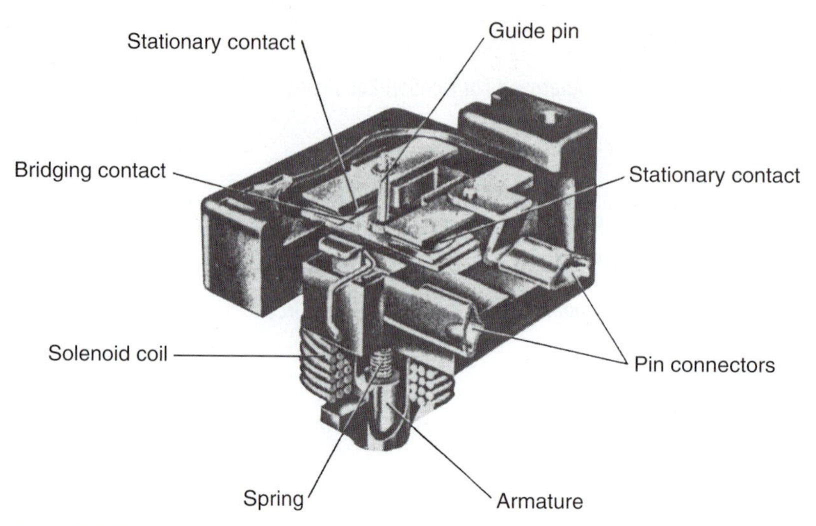

Fig. 7-40 *Current-type relay. This is generally used with small refrigeration compressors up to ³/₄ hp.* (Tecumseh)

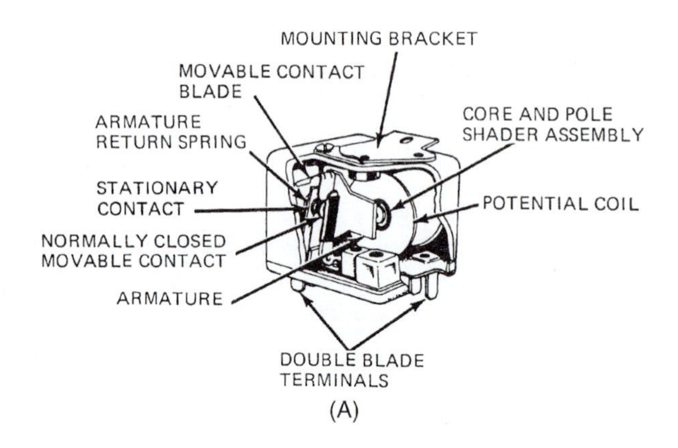

(A)

Potential Relay

- **Senses Voltage in the Motor Start Winding**
- **Relay Contacts are Normally Closed**
- **When Power is Applied, Relay Contacts Eventually Open, Removing Start Winding from Circuit**
- **Relay Contacts Close When Power to the Compressor is Shut Off**

(B)

Fig. 7-41 *(A) Potential-type relay. Usually found on large commercial and air-conditioning compressors to 5 hp. (B) Potential relay.* (Tecumseh)

C S R
C
C S C
C
S R
R
S
S
R
R

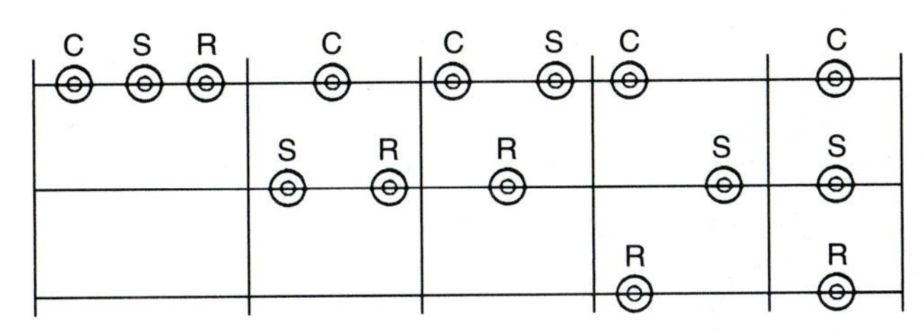

Fig. 7-42 *Identification of compressor terminals.* (Tecumseh)

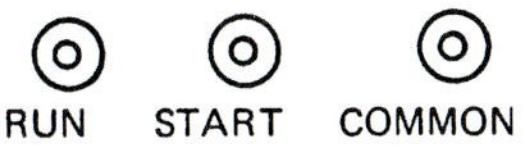

RUN START COMMON

Fig. 7-43 *Built-up terminals. These are on the obsolete twin-cylinder internal mount H models.* (Tecumseh)

COMMON START RUN

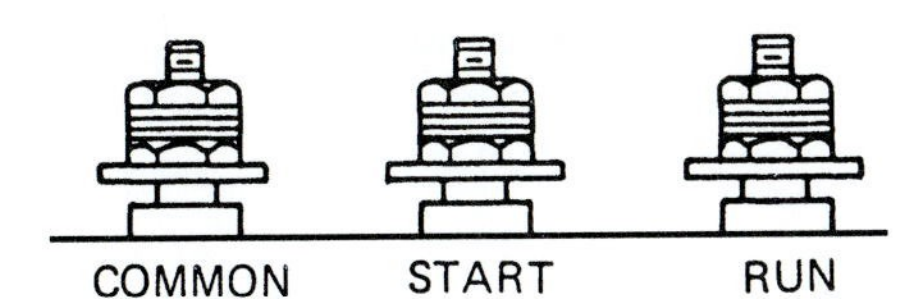

Fig. 7-44 *Built-up terminals. These are on all external-mount B and C twin-cylinder models and on F, PF, and CF four-cylinder external-mount models.* (Tecumseh)

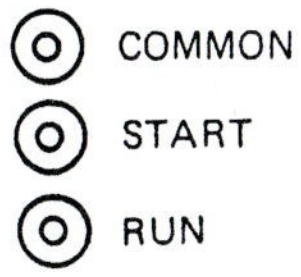

COMMON
START
RUN

Fig. 7-45 *Built-up terminals on pancake compressors manufactured before 1952.* (Tecumseh)

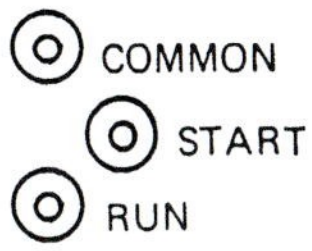

COMMON
START
RUN

Fig. 7-46 *Built-up terminals on S and C single-cylinder ISM models.* (Tecumseh)

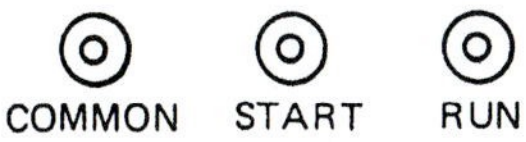

COMMON START RUN

Fig. 7-47 *Built-up terminals on twin-cylinder internal-mount J and PJ models.* (Tecumseh)

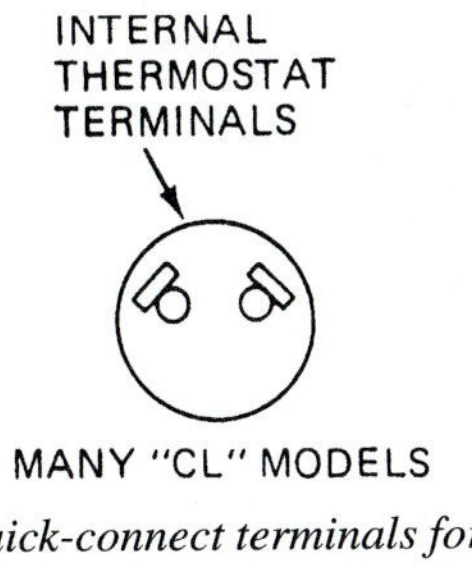

INTERNAL
THERMOSTAT
TERMINALS
MANY "CL" MODELS

Fig. 7-48 *Glass quick-connect terminals for Au and AR models.* (Tecumseh)

COMMON
START
RUN

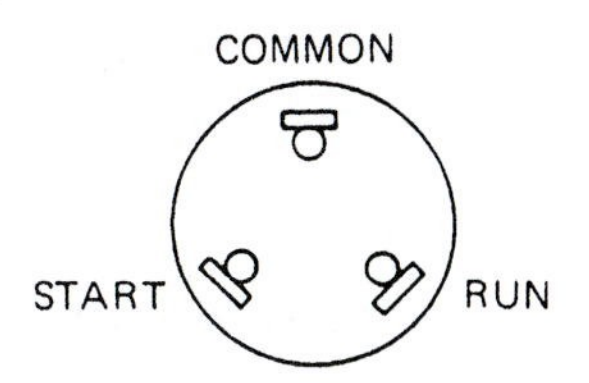

Fig. 7-49 *Glass quick-connect terminals for pancake-type models.* (Tecumseh)

COMMON
START
RUN

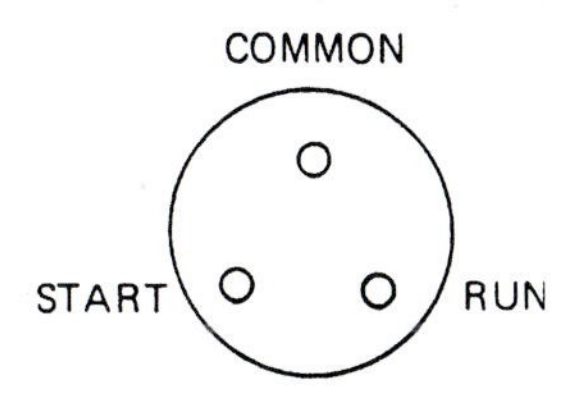

Fig. 7-50 *Glass terminals for pancake-type compressors.* (Tecumseh)

8. Check continuity across C and shell of the compressor. If there is continuity, there is a grounded motor. Replace the compressor.
9. Check the winding resistance against the values given in the manufacturer's resistance tables.
10. Check continuity across relay terminals 1 and 2. Place the meter on the R × 1 scale. If there is continuity, there is a shorted capacitor. Replace the start capacitor. Place the meter on the R × 100,000 scale. If there is no needle deflection, there is an open capacitor. Replace the start capacitor.

If all the tests prove satisfactory, there is no capillary restriction, and the unit still fails to operate properly, change the relay. The new relay will eliminate electrical problems such as improper pickup and dropout. These cannot be determined with the tests listed here. If a good relay fails to correct the difficulty, the compressor is inoperative due to internal defects, and must be replaced.

High-Starting Torque Motors (CSIR) with a Two-Terminal External Overload and a Remote-Mounted Potential Relay

High-starting torque motors (CSIR) with a two-terminal external overload and a remote- mounted potential relay represent another type that must be checked. These are used in compressors for light air-conditioning units and also for commercial and residential refrigeration units.

In this type of motor the starting contacts on the potential-type relay are normally closed. The electrical system on this type of hermetic system can be seen in

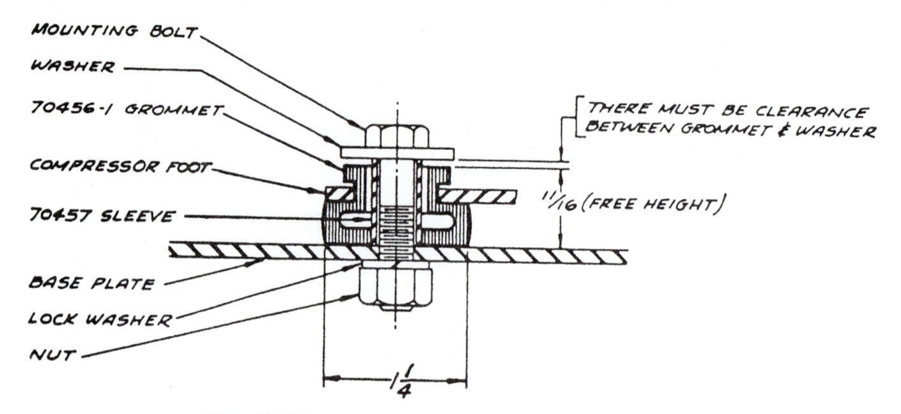

Fig. 7-51 *Mounting grommet assembly.* (Tecumseh)

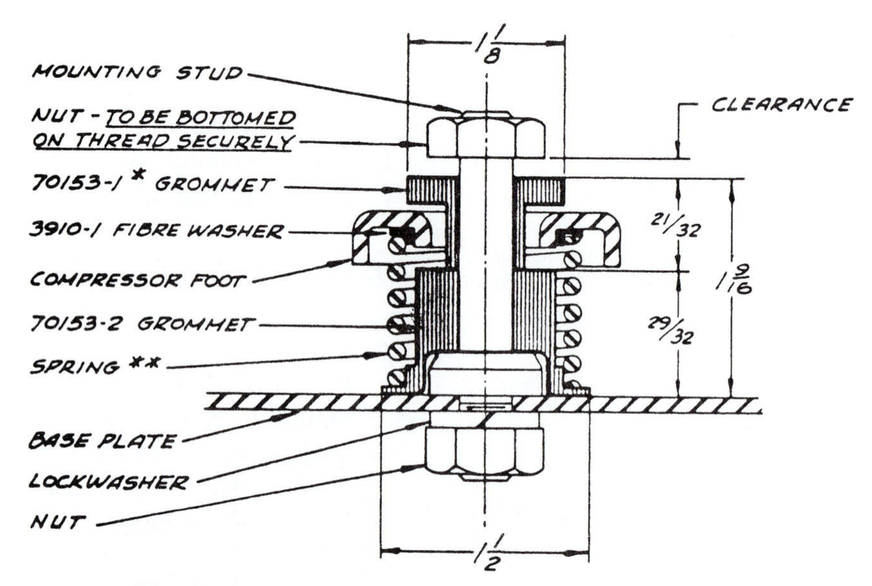

Fig. 7-52 *Mounting spring and grommet assembly.* (Tecumseh)

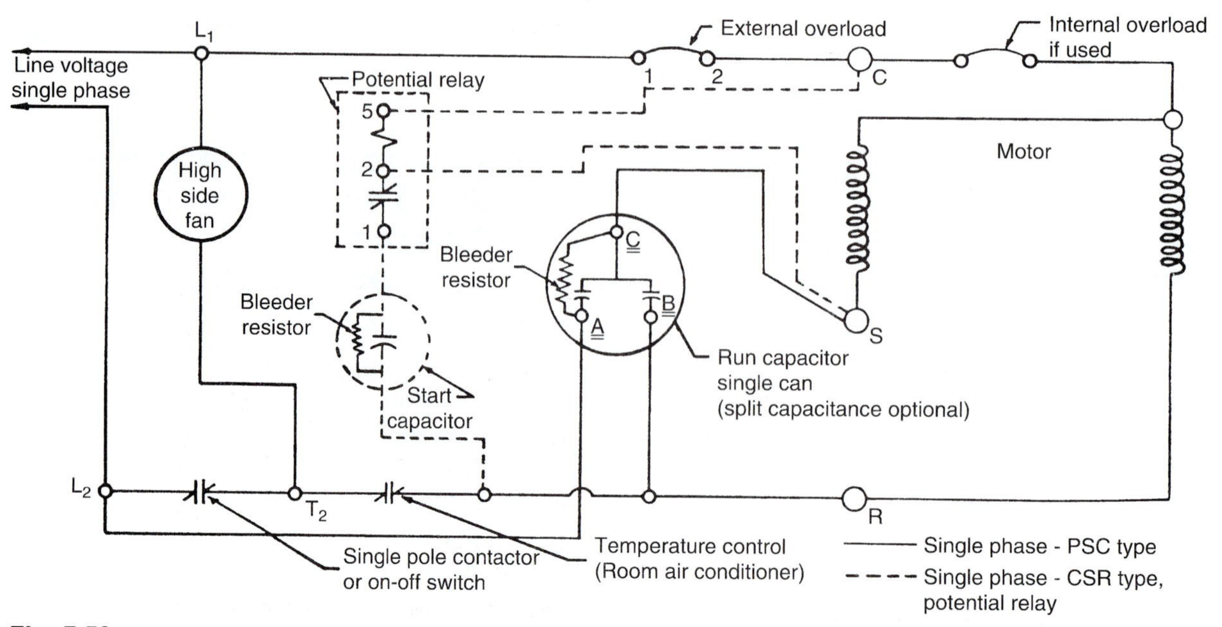

Fig. 7-53 *Single-phase CSR- or PSC-type compressor motor hookup with internal or external line-break overloads.* (Tecumseh)

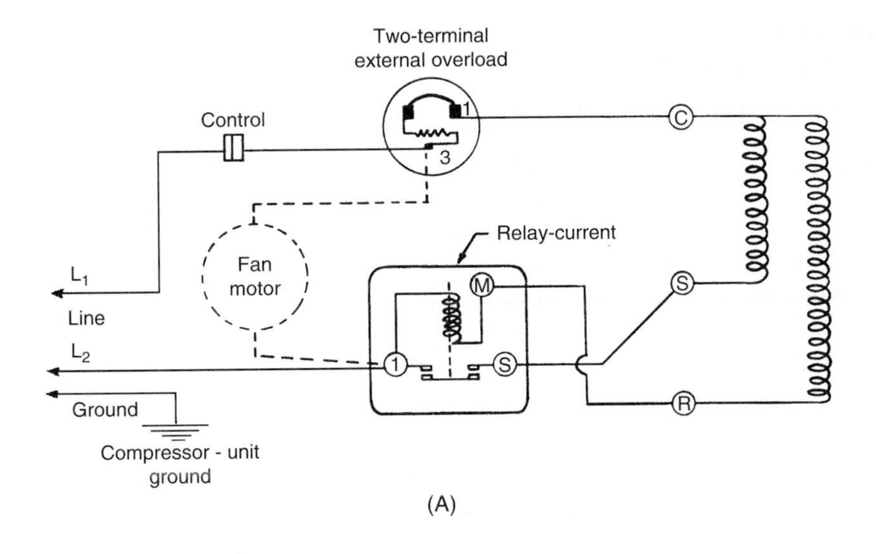

Current Relay

- Senses Current in the Motor-Main Winding
- Relay Contacts are Normally Open
- When Power is Applied, Relay Contacts Close, Energizing the Start Winding
- Relay Contacts Open when the Motor Reaches Running Speed

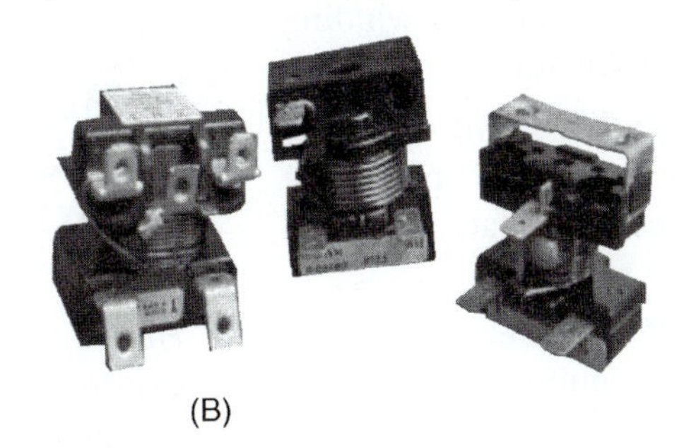

(B)

Fig. 7-54 *(A) Normal-starting motors (RSIR) with current relay mounted on the compressor terminals. (B) Current relays.* (Tecumseh)

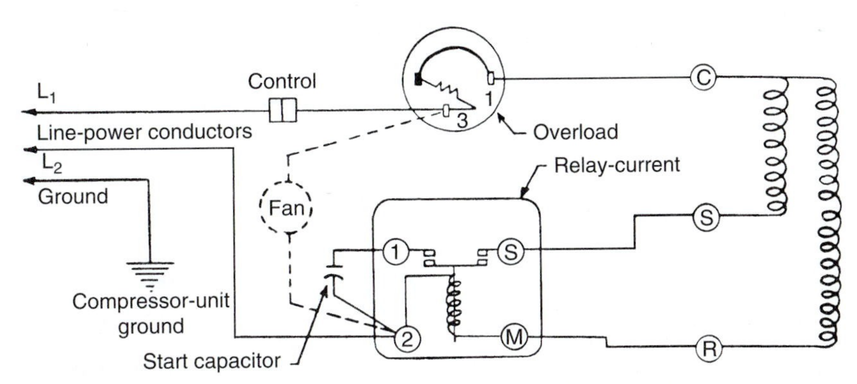

Fig. 7-55 *High-torque motors (CSIR) with current relay mounted on the compressor terminals.* (Tecumseh)

Fig. 7-56. Use a voltmeter to check the power source. Then use an ohmmeter, with the power turned off, to check continuity. Make sure leads 2S and 4R are disconnected. Open the fan circuit, if there is one.

Now, using the ohmmeter, check continuity across the following:

1. Check continuity across L and 3. If there is no continuity, close the control contacts. If there is still no continuity, replace the control.
2. Check continuity across No. 3 and No. 1 on the overload. If there is no continuity, the protector may be tripped. Wait 10 min and try again. If there is still no continuity, the protector is defective. Replace the protector.
3. Check continuity across No. 3 on the overload and No. 5 on the relay. If there is no continuity, check the leads between No. 3 on the protector and No. 5 on the relay.
4. Check continuity across No. 1 on the overload and C on the compressor. If there is no continuity, check the leads between No. 1 on the overload and C on the compressor.
5. Check continuity across C and S on the compressor. If there is no continuity, an open start winding is indicated. Replace the compressor.
6. Check continuity across C and R on the compressor. If there is no continuity, there is an open in the run winding of the compressor. Replace the compressor.
7. Check continuity across No. 5 on the relay and No. 2 on the relay. If there is no continuity, the solenoid's coil is open. The relay is defective. Replace the relay.
8. Check continuity across No. 2 and No. 1 on the relay. If there is no continuity, the contacts are open when they should be closed. Replace the relay.
9. Check continuity across No. 1 on the relay and No. 4 on the relay with the meter on the R × 1 scale. If there is continuity, the capacitor is shorted. Replace the start capacitor. No needle deflection on the meter when it is on the R × 100,000 scale means the capacitor is open. Replace the capacitor.
10. Check between C and the shell of the compressor. If there is continuity, there is a short. The motor is grounded. Replace the compressor.
11. Check the motor-winding resistances against the manufacturer's specification sheet.
12. Check continuity between leads 2S and 4R and reconnect the unit. If all the tests prove satisfactory and the unit still does not operate properly, change the relay. The new relay will eliminate any electrical problems, such as improper pickup and dropout, which cannot be determined with the checks just performed. If a good relay fails to correct the difficulty, the compressor is inoperative due to internal defects. It must be replaced.

High-Starting Torque Motors (CSR) with Three-Terminal Overloads and Remote-Mounted Relays

High-starting torque motors (CSR) with three-terminal overloads and remote-mounted potential relays are another type of motor used in the hermetic compressor systems. See Fig. 7-57.

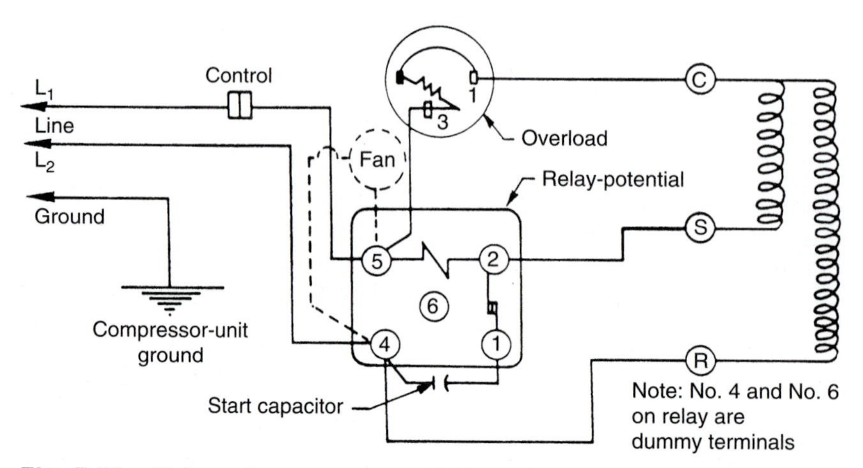

Fig. 7-56 *High-starting torque motors (CSIR) with two-terminal external overload and potential relay mounted remote.* (Tecumseh)

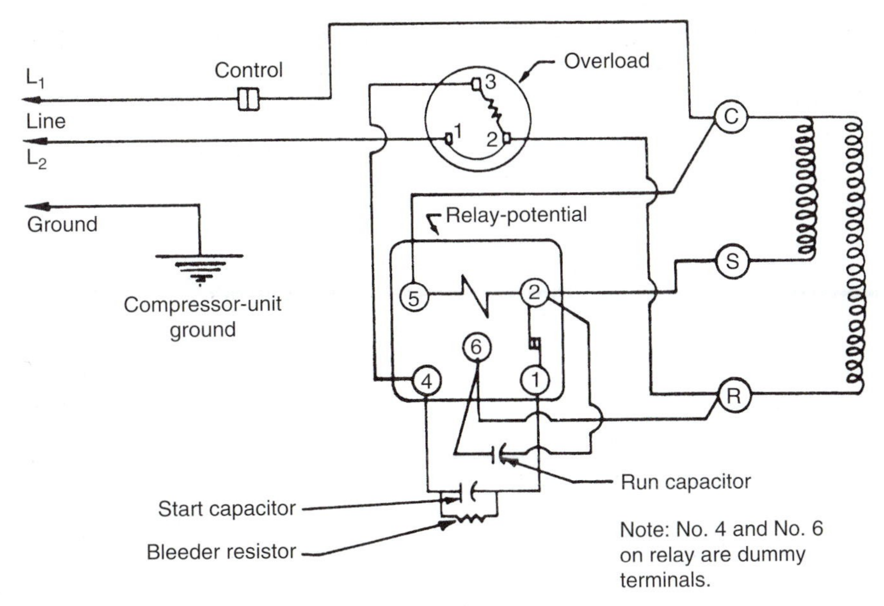

Fig. 7-57 *High-starting torque motors (CSR) with a three-terminal external overload and potential relay mounted remote.* (Tecumseh)

Starting contacts on the potential type of relay are normally closed. The electrical system power supply of the compressor can be checked. Use a voltmeter to check the power source. First, disconnect the leads so that no external wiring connects terminals 5-C, S-2 on the relay, and R-2 on the overload.

Using the ohmmeter, check continuity across the following locations:

1. Check continuity across the control contacts—L_1 and C—on the compressor. The control contacts must be closed. If they are open, replace the compressor.
2. Check continuity across No. 5 on the relay and No. 2 on the relay. No continuity indicates an open potential coil. Replace the relay.
3. Check continuity across No. 2 and No. 1 on the relay. No continuity indicates an open contact situation. Replace the relay.
4. Check continuity across terminals C and S on the compressor. No continuity indicates an open start winding. Replace the compressor.
5. Check continuity across terminals C and R on the compressor. No continuity indicates an open run winding. Replace the compressor.
6. Check continuity across No. 6 and No. 2 on the relay with the meter on the R × 1 scale. Continuity shows a shorted capacitor. Replace the run capacitor. Set the meter on the R × 100,000 scale. If there is no needle deflection, the capacitor is open. Replace the run capacitor.
7. Check continuity across No. 1 on the relay and No. 3 on the overload. Check as in the preceding step.
8. Check continuity across No. 1 and No. 3 on the overload. No continuity indicates the overload is open and should be replaced. However, it should have been given at least 10 min to replace itself properly.
9. Check continuity across C terminal on the compressor and the other ohmmeter lead to the shell of the compressor. Continuity indicates the motor has become grounded to the shell. Replace the compressor.
10. Check the resistance of the motor windings against the values given in the manufacturer's resistance tables.
11. Check continuity of the leads removed above and reconnect terminals 5 to C, S to 2 on the relay, and R and 2 on the overload.

If the tests prove satisfactory and the unit still does not operate properly, replace the relay. The new relay will eliminate any electrical problems, such as improper pickup and dropout, which cannot be determined with the checks just performed. If a good relay fails to correct the difficulty, the compressor is inoperative due to internal defects. It must be replaced.

PSC Motor with a Two-Terminal External Overload and Run Capacitor

Another type of motor used on compressors is the PSC. See Fig. 7-58. It has a two-terminal external overload and a run capacitor. It does not have a start capacitor or relay.

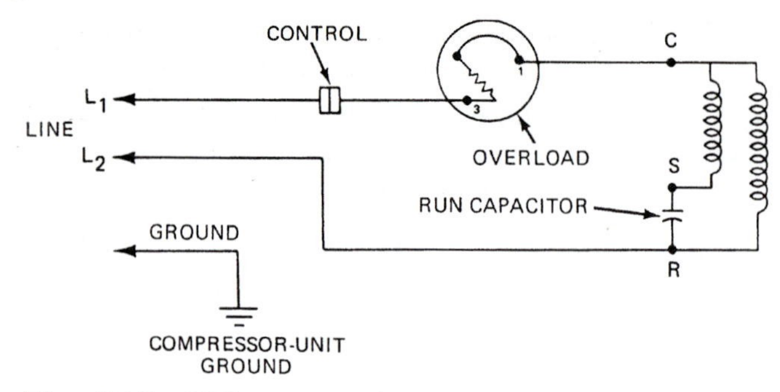

Fig. 7-58 *PSC motors with two-terminal external overload.* (Tecumseh)

Use a voltmeter to check the source voltage. Then, using an ohmmeter, perform the following checks. Disconnect the run capacitor from terminals S and R before starting the tests.

1. L_1 and No. 3 on the overload show no continuity. Close the control contacts. If there is still no continuity, replace the control.
2. C and S terminals on the compressor show no continuity. This means the start winding is open. Replace the compressor.
3. C and R terminals on the compressor show no continuity. This means the run winding is open. A replacement compressor is needed to correct the problem.
4. C and 1 on the overload show no continuity. A defective lead from C to 1 is the probable cause.
5. No. 1 and No. 3 on the overload indicate no continuity. The protector may be tripped. Wait 10 min before checking again. If there is still no continuity, the protector is defective. Replace the overload protector.
6. C and the shell of the compressor show continuity. The motor is shorted to the shell or ground. Replace the compressor.
7. Check the motor windings against the manufacturer's tables.
8. Check across the run capacitor with the meter on the R × 1 scale. If it shows continuity, the capacitor is shorted and must be replaced. Set the meter on R × 100,000 scale. No needle deflection indicates that the capacitor is open and needs to be replaced.
9. Reconnect the capacitor to the circuit at terminals S and R. The marked terminal should go to R.

If the PSC tests reveal no difficulties, but the compressor does not operate properly, add the proper relay and start capacitor to provide additional starting torque. Figure 7-59 gives the proper wiring for a field-installed relay and capacitor. If the unit still fails to operate, the compressor is inoperative due to internal defects. It must be replaced.

PSC Motor with an Internal Overload (Line Breaker)

Those PSC motors with an internal overload (line breaker) are a little different from those just checked. Thus, the testing sequence varies somewhat. This compressor has an internal line break overload and a run

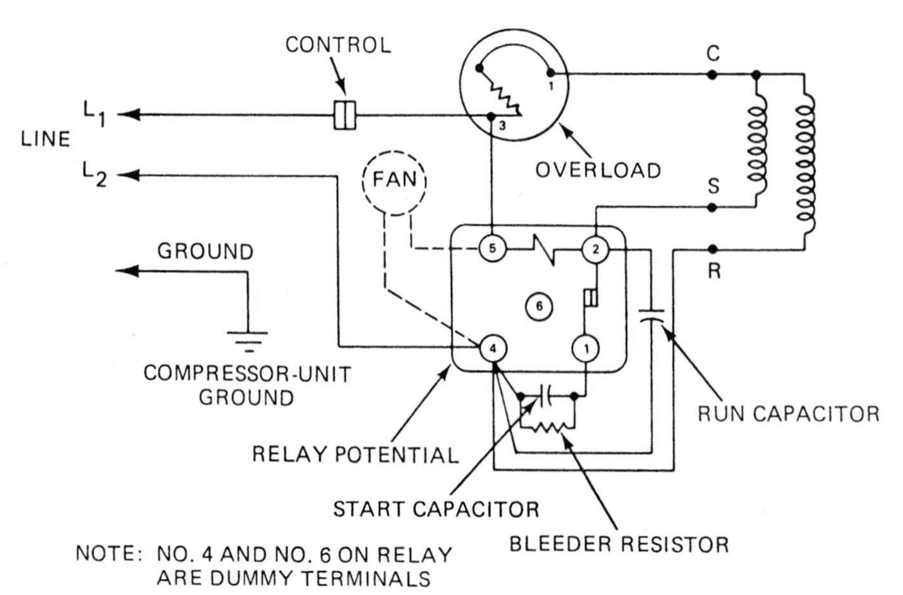

Fig. 7-59 *PSC motors with two-terminal external overload with start components field installed.* (Tecumseh)

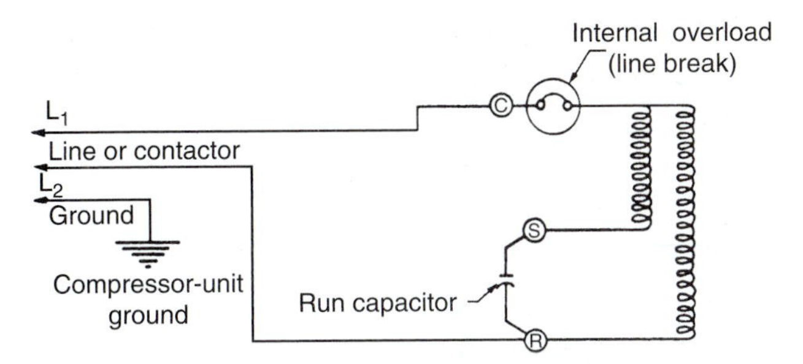

Fig. 7-60 *PSC motors with internal overload or line breaker.*
(Tecumseh)

capacitor. It does not have a start capacitor or relay. See Fig. 7-60.

1. Use a voltmeter to check the power source. Check the voltage at compressor terminals C and R. If there is no voltage, the control circuit is open.
2. Unplug the unit and check continuity across, the thermostat and/or contactor. Check the contactor-holding coil.
3. If the line voltage is present between terminals C and R and the compressor does not operate, unplug the unit and disconnect the run capacitor from S and R.

    NOTE: The compressor shell must be at 130°F (54°C) or less for the following checks. This temperature can be read by a method termed as Tempstik. However, using the hand provides a less reliable guide. If it can remain in contact with the compressor shell without discomfort at a temperature of 130°F (54°C) or less, the motor is not overheated.

4. Using the ohmmeter, check the following:
    a. Check continuity between R and S. If there is continuity, it can be assumed that both windings are intact. If there is no continuity, it can be assumed that one or both of the windings are open and the compressor should be replaced.
    b. Check continuity between R and C. If there is no continuity, the internal overload is tripped. Wait for it to cool off and close. It sometimes takes more than an hour.
    c. There is continuity between R and S, but no continuity between R and C (or S and C). If the motor is cool enough [below 130°F (54°C)] to have closed the overload, then it can be assumed that the overload is defective. The compressor should be replaced.
    d. Check continuity between the S terminal and the compressor shell, and between the R terminal and the compressor shell. If there is continuity in either or both cases, the motor is grounded. The compressor should be replaced.
    e. Check the motor-winding resistance against the values given in the manufacturer's charts.
    f. Check across the run capacitor with the meter on the R × 1 scale. If there is continuity, the capacitor is shorted and should be replaced.
    g. Check across the run capacitor with the meter on the R × 100,000 scale. If there is no needle deflection on the meter, the capacitor is open and should be replaced.
5. Reconnect the run capacitor into the circuit at S and R.

CSR or PSC Motor with the Start Components and an Internal Overload or Line breaker

The next combination is the CSR or PSC motor with the start components and an internal overload or line breaker. The run capacitor, start capacitor, and potential relay are the major components outside the compressor. See Fig. 7-61.

1. Using the voltmeter, check the power source. Check voltage at the compressor terminals C and R. If there is no voltage, the control circuit is open.
2. Unplug the unit and check continuity across the thermostat and/or contactor. Check the contactor-holding coil.
3. If the line voltage is present between terminals C and R and the compressor does not operate, unplug the unit and disconnect the connections to the compressor terminals.

    NOTE: The compressor shell must be at 130°F (54°C) or less for the following checks. A Tempstik can read this temperature. A less-reliable guide is that at this temperature the hand can remain in contact with the compressor shell without discomfort.

4. Using the ohmmeter, check the following:
    a. Check continuity between R and S. If there is continuity, it can be assumed that both windings are intact. If there is no continuity, it can be assumed that one or both of the windings are open. The compressor should be replaced.

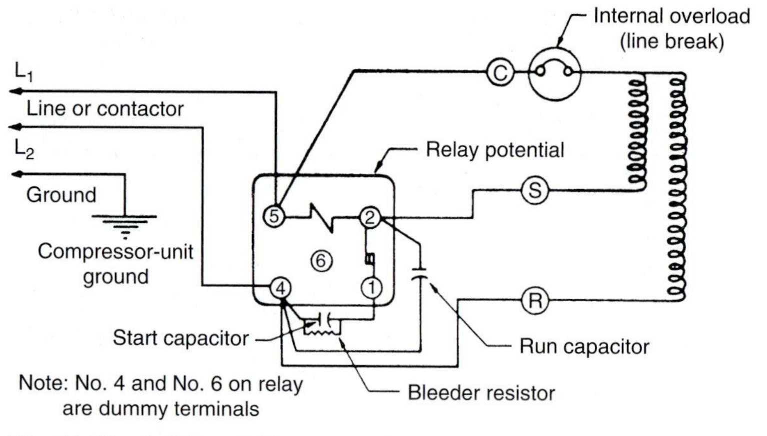

Fig. 7-61 *A CSR or PSC motor with start components and internal overload (line breaker).* (Tecumseh)

b. Check continuity between R and C. If there is no continuity, the internal overload is tripped. Wait for it to cool off and close. It sometimes takes more than an hour.

c. There is continuity between R and S, but no continuity between R and C (or S and C). If the motor is cool enough [130°F (54°C)] to have closed the overload, then it can be assumed that the overload is defective. The compressor must be replaced.

d. Check continuity between the S terminal and the compressor shell and between R and the compressor shell. If there is continuity in either or both cases, the motor is grounded and the compressor should be replaced.

e. Check the motor-winding resistance against the values given in the manufacturer's tables for the specific model being tested.

f. Check continuity across the run capacitor with the meter on the R × 1 scale. If there is continuity, the capacitor is shorted and should be replaced.

g. Check continuity across the run capacitor with the meter on the R × 100,000 scale. If there is no needle deflection, the capacitor is open and should be replaced.

5. Check continuity across No. 5 and No. 2 on the relay. No continuity indicates an open potential coil. Replace the relay. The electrolytic capacitor used for the start usually has its contents on the outside of the compressor housing. If the coil did not energize properly, it leaves the start capacitor in the circuit too long (only a few seconds). This means the capacitor will get too hot. When this happens, the capacitor will spew its contents outside the container.

6. Check continuity across No. 2 and No. 1 on the relay. No continuity shows an open-contacts condition. Replace the relay.

7. Check continuity across No. 4 and No. 1 on the relay with the meter on the R × 1 scale. Continuity indicates a shorted capacitor. Replace it. With the meter on the R × 100,000 scale, if there is no needle deflection, the start capacitor is open. Replace the capacitor.

If all of the tests prove satisfactory and the unit still fails to operate properly, change the relay. If a new relay does not solve the problem, then it is fairly safe to assume that the compressor is defective and should be replaced.

Compressors with Internal Thermostat, Run Capacitor, and Supplementary Overload

Some compressors have an internal thermostat, a run capacitor, and a supplementary overload. However, they do not have a start capacitor or relay. The schematic for such a compressor is shown in See Fig. 7-62.

The supplementary overload has normally closed contacts connected in series with the normally closed contacts of the internal thermostat in the motor. Operation of either of these devices will open the control circuit to drop out the contactor. Make sure the control thermostat and the system-safety controls are closed. Using a voltmeter, check the power source at L_1, L_2, and the control-circuit power supply. If the contactor is not energized, the contactor holding coil is defective or the control circuit is open in either the supplementary overload or the motor thermostat. Unplug the unit and disconnect the run capacitor from terminals S and R.

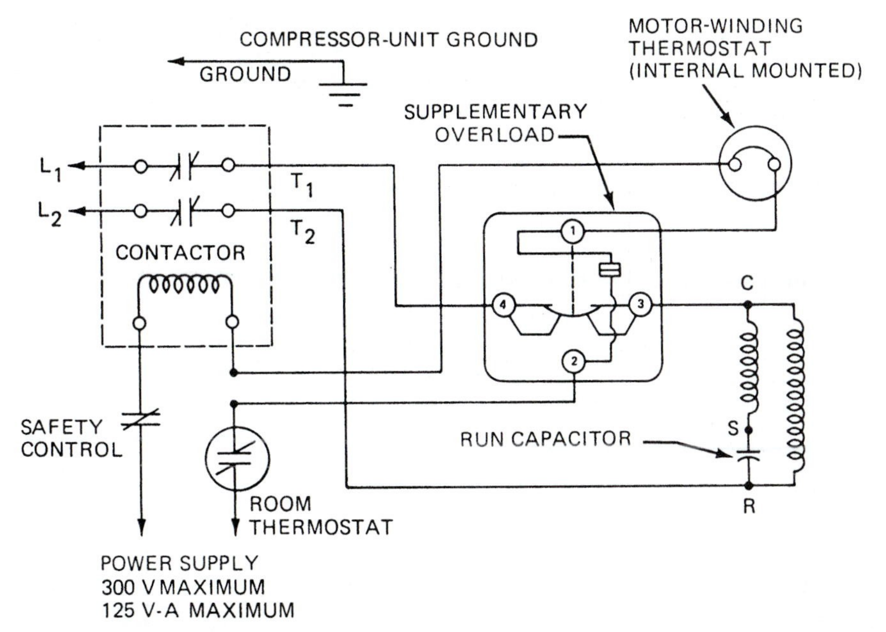

Fig. 7-62 *PSC motors with internal thermostat and supplementary external overload. (Tecumseh)*

Using the ohmmeter, check the continuity across the following:

1. Check continuity across No. 3 and No. 4 on the overload. No continuity means the supplementary overload is defective. Replace it.
2. Check continuity across No. 1 and No. 2 on the overload. No continuity can mean the overload may be tripped. Wait 10 min. Test again. If there is still no continuity, the overload is defective. Replace the overload.
3. Check continuity across the internal (motor winding) thermostat terminals at the compressor. Check Fig. 7-48 for the location of the internal-thermostat terminals. If there is no continuity, the internal thermostat may be tripped. Wait for it to cool down and close. It sometimes takes an hour. If the compressor is cool to the touch [below 130°F (54°C)] and there is still no continuity, internal-thermostat circuitry is open and the compressor must be replaced.
4. Check continuity across terminals C and S on the compressor. No continuity indicates an open start winding. Replace the compressor.
5. Check continuity across terminals C and R. No continuity indicates an open run winding. Replace the compressor.
6. Check continuity across terminal C and the shell of the compressor. Continuity shows a grounded compressor. Replace the compressor.
7. Check the motor-winding resistance with the chart given by the manufacturer.
8. Check continuity across the run capacitor with a meter on the R × 1 scale. If there is continuity, the capacitor is shorted. Replace it. Place the meter on the R × 100,000 scale. If there is no needle deflection, the capacitor is open. Replace the capacitor.
9. Reconnect the capacitor to the circuit at terminals S and R.

CSR or PSC Motor with Start Components, Internal Thermostat, and Supplementary External Overload

Another arrangement for single-phase compressors is the CSR or PSC motor with start components, internal thermostat, and supplementary external overload. See Fig. 7-63.

This type of compressor is equipped with an internal thermostat, run capacitor, start capacitor, potential relay, and supplemental overload.

The supplemental overload has normally closed contacts connected in series with the normally closed contacts of the internal thermostat located in the motor. Operation of either of these devices will open the control circuit to drop out the contactor. See Fig. 7-64 where it shows the details of the internal thermostat.

Make sure the control thermostat and system-safety controls are closed. Using the voltmeter, check the power source at L_1 and L_2. Also check the control-circuit power supply with the voltmeter.

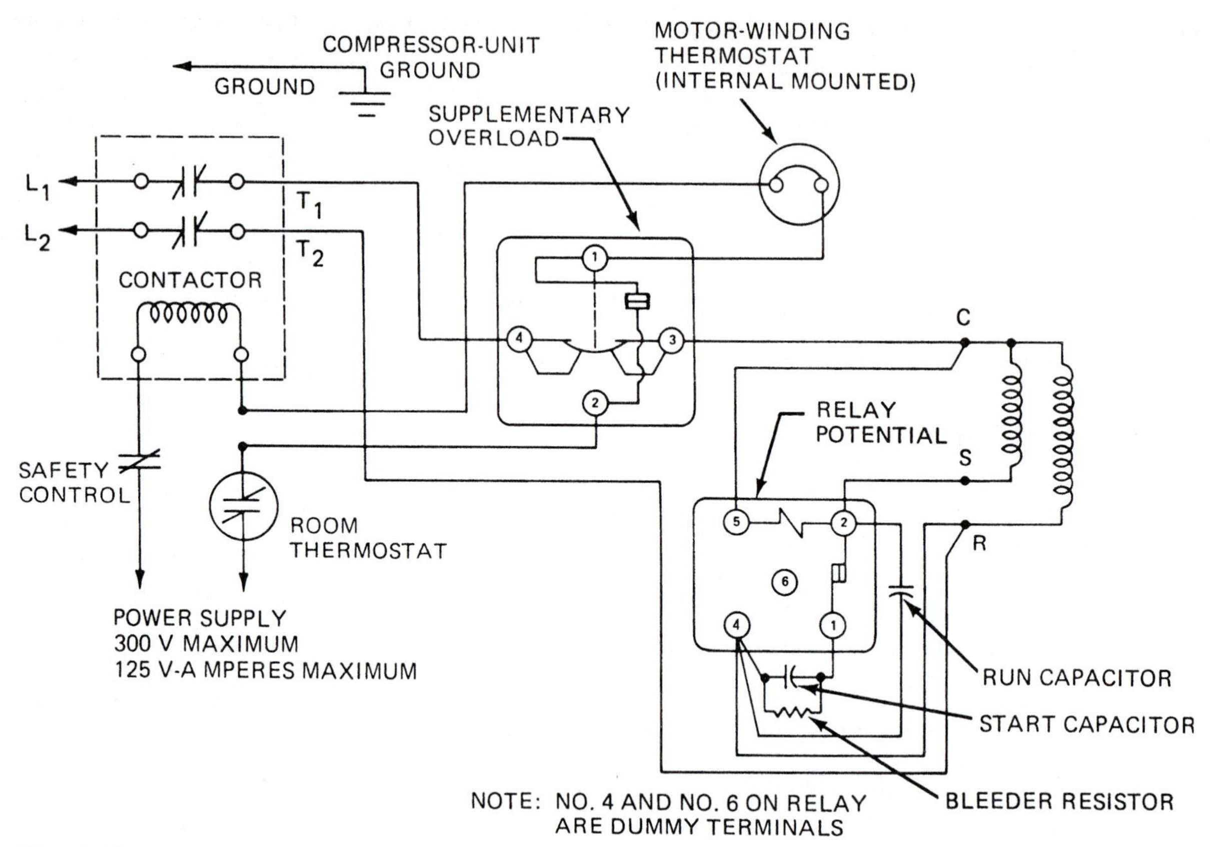

Fig. 7-63 *A CSR or PSC motor with start components and internal thermostat and supplementary external overload.* (Tecumseh)

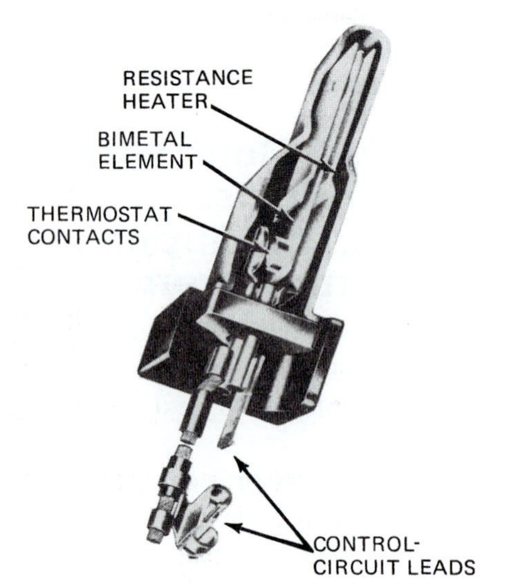

Fig. 7-64 *Internal thermostat embedded in the motor winding.* (Tecumseh)

If the contactor is not energized, the contactor-holding coil is defective or the control circuit is open in either the supplemental overload or the motor thermostat. Unplug the unit and disconnect the connections to the compressor terminals.

Use the ohmmeter, check for continuity across the following:

1. With the control-circuit power supply off, check the continuity of the contactor-holding coil.
2. Check continuity across No. 4 and No. 3 of the supplemental overload. No continuity means the overload is defective. Replace the overload.
3. Check continuity across No. 1 and No. 2 of the overload. No continuity means the overload may be tripped. Wait at least 10 min and test again. If there is still no continuity, the overload is defective. Replace the defective overload.
4. Check continuity across the internal-thermostat terminals at the compressor. See Fig. 7-48 for the location of the terminals of the internal thermostat. If there is no continuity, the internal thermostat may be tripped. Wait for it to cool off and close. It sometimes takes more than 1 h to cool. If the compressor is cool to the touch [below 130°F (54°C)], and there is still no continuity, the internal-thermostat circuitry is open and the compressor must be replaced.
5. Check for continuity across terminals R and S on the compressor. If there is no continuity, one or both of the windings are open. Replace the compressor.

6. Check for continuity across terminal S and the compressor shell. Check for continuity across terminal R and the compressor shell. If there is continuity in either or both cases, the motor is grounded and the compressor should be replaced.
7. Check the motor-winding resistances with the chart furnished by the compressor manufacturer.
8. Check for continuity across terminals 5 and 2 on the relay. No continuity indicates an open potential coil. Replace the relay.
9. Check for continuity across terminals 2 and 1 on the relay. No continuity indicates open contacts. Replace the relay.
10. Check for continuity across terminals 4 and 1 on the relay with the meter on the R × 1 scale. If continuity is read, it indicates a shorted capacitor. Replace the capacitor. Repeat with the meter on the R × 100,000 scale. No needle deflection indicates the start capacitor is open. Replace the start capacitor.
11. Discharge the run capacitor by placing a screwdriver across the terminals. Remove the leads from the run capacitor. With the meter set on the R × 1 scale, continuity across the capacitor terminals indicates a shorted capacitor. Replace the capacitor. Repeat the same test with the meter set on the R × 100,000 scale. No needle deflection indicates an open capacitor. Replace the run capacitor. If all the tests prove satisfactory and the unit still fails to operate properly, change the relay. The new relay will eliminate electrical problems such as improper pickup or dropout, which cannot be determined by the given tests. If a good relay fails to correct the difficulty, the compressor is inoperative due to internal defects. It must be replaced.

One other arrangement for a compressor using single-phase current is a CSR or PSC motor with start components, internal thermostat, supplemental external overload, and start-winding overload. See Fig. 7-65.

The diagnosis is identical to that described for the previous type of motor circuit. However, there is an additional start-winding overload in the control circuit in series with the internal thermostat and supplemental overload.

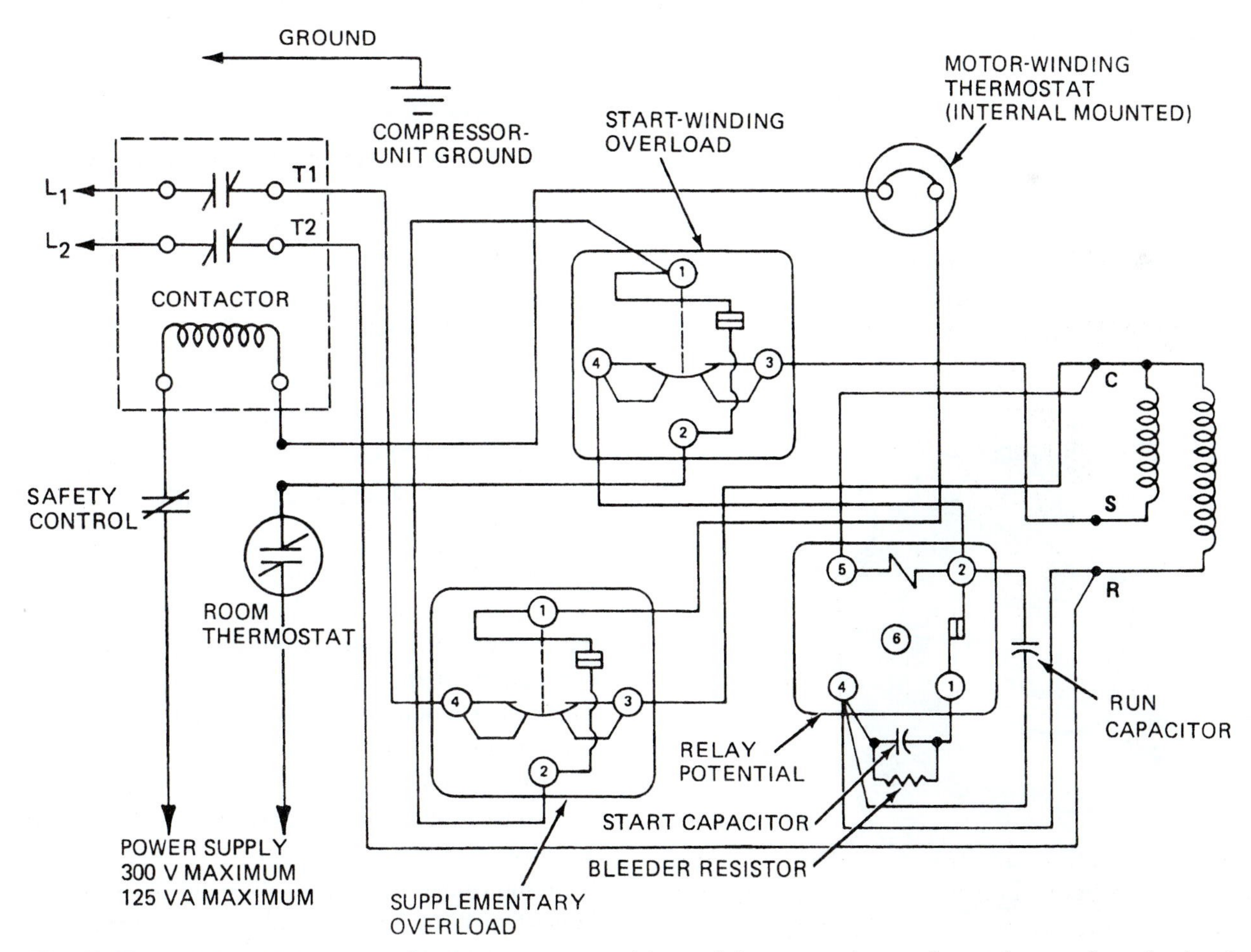

Fig. 7-65 *A CSR or PSC motor with start components and internal thermostat plus supplemental external overload and start-winding overload. Note No. 4 and No. 6 on the relay are dummy terminals.* (Tecumseh)

Check the start-winding overload in the same way the supplemental overload is checked.

COMPRESSOR CONNECTIONS AND TUBES

Tecumseh, like other compressor manufacturers, made compressors for many manufacturers of refrigerators, air-conditioning systems, and coolers. Because of this, the same compressor model may be found in the field in many suction and discharge variations. Each variation depends upon the specific application for which the compressor was designed.

Suction connections can usually be identified as the stub tube with the largest diameter in the housing. If two stubs have the same outside diameter, then the one with the heavier wall will be the suction connection. If both of the largest stub tubes have the same outside diameter and wall thickness, then either can be used as the suction connection. However, the one farthest from the terminals is preferred.

The stub tube not chosen for the suction connection may be used for processing the system. Compressor connections can usually be easily identified. However, occasionally some question arises concerning oil-cooler tubes and process tubes.

Oil-cooler tubes are found only in low-temperature refrigeration models. These tubes connect to a coil or hairpin bend within the compressor oil sump. See Fig. 7-66. This coil or hairpin bend is not open inside the compressor. Its only function is to cool the compressor sump oil. The oil-cooler tubes are generally connected to an individually separated tubing circuit in the air-cooled condenser.

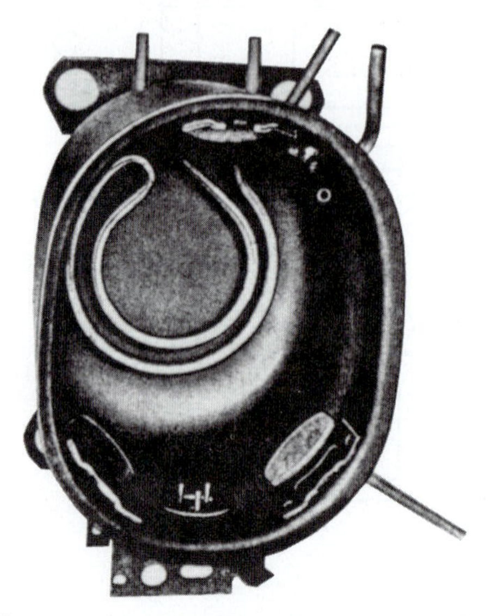

Fig. 7-66 *Location of the oil-cooling tubes inside the compressor shell.* (Tecumseh)

Process Tubes

Process tubes are installed in compressor housings at the factory as an aid in factory dehydration and charging. These can be used in place of the suction tube if they are of the same diameter and wall thickness as the suction tube.

Standard discharge tubing arrangements for Tecumseh hermetic compressors are shown in Fig. 7-67. Discharge tubes are generally in the same position within any model family. Suction and process tube positions may vary.

Other Manufacturers of Compressors

Besides Tecumseh, there are other manufacturers of compressors for the air-conditioning and refrigeration trade. One is Americold Compressor Corporation. Two of the models made by Americold are the M series and the A series. See Fig. 7-68. Both use the same overload relay and current relay connections. See Fig. 7-69. All of these models use R-12, or suitable substitute, as the refrigerant. They are made in sizes ranging from $^1/_{10}$ (0.10) through $^1/_4$ hp. They weigh 21 to 25 lb. Figure 7-70 shows the location of the suction and discharge stubs as well as the process tube.

ROTARY COMPRESSORS

The rotary compressor is made in two different configurations—the *stationary blade* rotary compressor and the *rotating blade* rotary compressor. The stationary blade rotary compressor is the type that has just been described. Both of these compressors have problems regarding lubrication. This problem has been partly solved.

Stationary Blade Rotary Compressors

The only moving parts in a stationary blade rotary compressor are a steel ring, an eccentric or cam, and a sliding barrier. See Fig. 7-71. Figure 7-72 shows how the rotation of the off-center cam compresses the gas refrigerant in the cylinder of the rotary compressor. The cam is rotated by an electric motor. As the cam spins it carries the ring with it. The ring rolls on its outer rim around the wall of the cylinder.

To be brought into the chamber, the gas must have a pathway. Note that in Fig. 7-73 the vapor comes in from the freezer and goes out to the condenser through holes that have been drilled in the compressor frame. Note that an offset rotating ring compresses the gas. Figure 7-74 shows how the refrigerant vapor in the

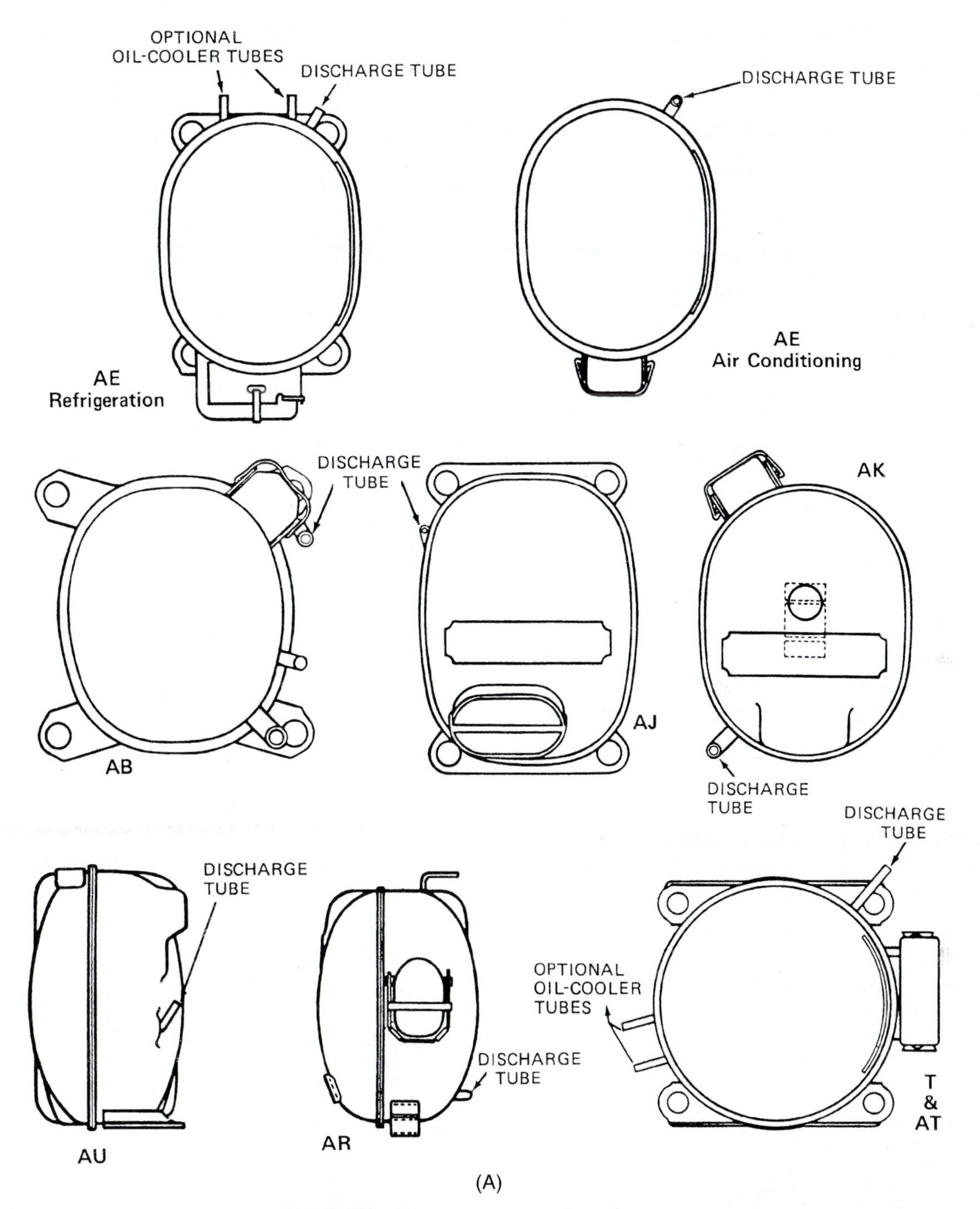

Fig. 7-67 *Compressor-connection tubes.* (Tecumseh)

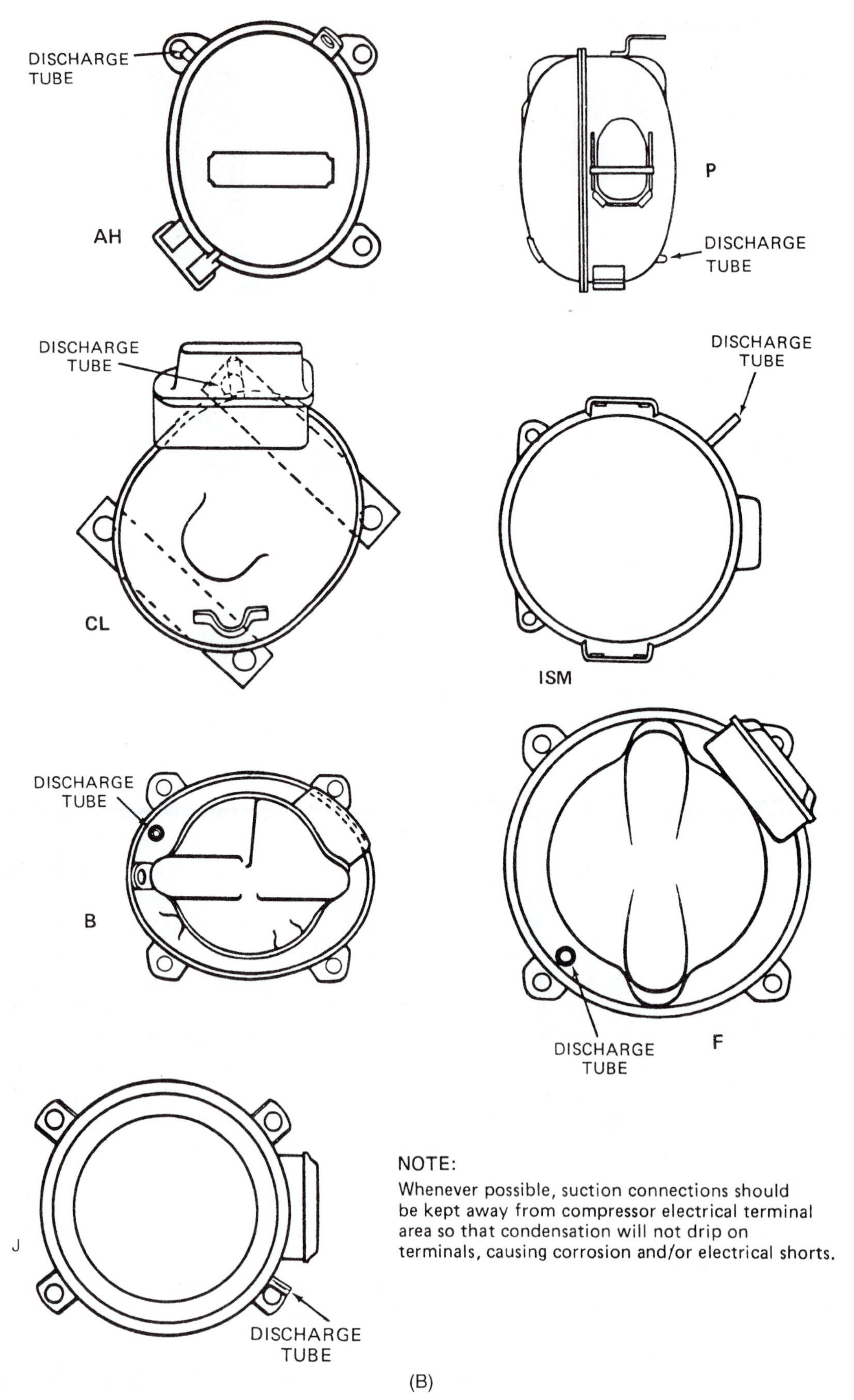

(B)

Fig. 7-67 *(Continued)*

Tecumseh AJ Compressors

12,500 – 19,000 BTUH

- RAC and small central air conditioning only
- External line-break overload
- Low sounds and vibration levels
- Adaptable to limited cabinet space

Tecumseh AW Compressors

13,600 – 32,000 BTUH

- For heat pump and air-conditioning duty
- Internal line-break overload
- Internal pressure-relief valve

17,000 – 31,500 BTUH (60 H) 17,300 – 27,000 BTUH (50 H)

- For heat pump and air-conditioning duty
- Internal line-break overload
- Internal pressure-relief valve

(C)

Fig. 7-67 *(Continued)*

compressor is brought from the freezer. Then, the exit port is opening. When the compressor starts to draw in the vapor from the freezer the barrier is held against the ring by a spring.

This barrier separates the intake and exhaust ports. As the ring rolls around the cylinder it compresses the gas and passes it on to the condenser. See Fig. 7-75. The finish of the compression portion of the stroke or operation is shown in Fig. 7-76. The ring rotates around the cylinder wall. The spring tension of the barrier's spring and the pressure of the cam being driven by the electric motor hold it in place. This type of compressor is not used as much as the reciprocating hermetic type of compressor.

Rotating Blade Rotary Compressors

The rotating blade rotary compressor has its roller centered on a shaft that is eccentric to the center of the cylinder. Two spring-loaded roller blades are mounted 180° apart. They sweep the sides of the cylinder. The roller is mounted so that it touches the cylinder at a point between the intake and the discharge ports. The roller rotates. In rotating, it pulls the vapor into the cylinder through the intake port. Here, the vapor is trapped in the space between the cylinder wall, the blade, and the point of contact between the roller and the cylinder. As the next blade passes the contact point, the vapor is compressed. The space or the vapor becomes smaller and smaller as the blade rotates.

Once the vapor has reached the pressure determined by the compressor manufacturer, it exits through the discharge port to the condenser.

On this type of rotating blade rotary compressor the seals on the blades present a particular problem. There also are lubrication problems. However, a number of rotary compressors are still in operation in home refrigerators.

Some manufacturers make rotary blade compressors for commercial applications. They are used primarily with ammonia. Thus, there is no copper or copper-alloy tubing or parts. Most of the ammonia tubing and working metal is stainless steel.

SCREW COMPRESSORS

Screw compressors operate more or less like pumps, and have continuous flow refrigerant compared to reciprocals. Reciprocal have pulsations. This results in smooth

Tecumseh AK Compressors

6,900 – 15,000 BTUH

- For room air-conditioning cooling and special heat-pump applications as determined by the original equipment manufacturer
- External line-break overload
- Low sound and vibration levels
- Adaptable to limited cabinet space

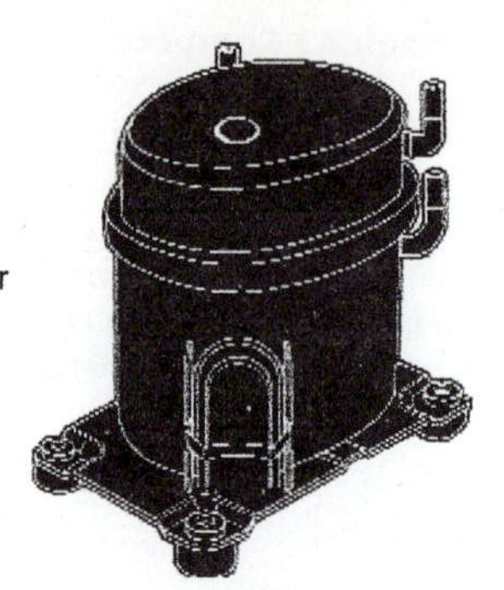

Tecumseh RK Compressors

8,100 – 17,700 BTUH (Rotary)

- Rotary design means low noise and vibration levels
- Air-conditioning and heat-pump application
- High efficiency
- Compact Design

Tecumseh SF Compressors

72,000 – 150,000 BTUH

- "Quadro-flex model"
- High-pressure housing
- No crankcase heater required
- Direct suction flow into cylinder adds to high efficiency
- Factory-installed suction screen

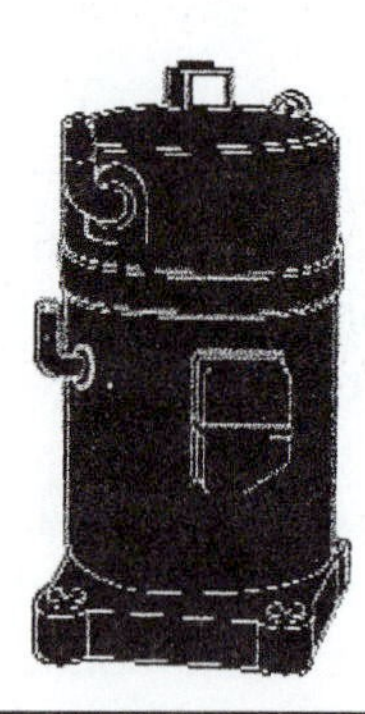

(D)

Fig. 7-67 *(Continued)*

compression with little vibration. Reciprocals, on the other hand, make pulsating sounds and vibrate. They can be very noisy.

Screw compressors have almost linear capacity-control mechanisms. That results in excellent part-load performance. Due to its smooth operation, low-vibration screw compressors tend to have longer life than reciprocals.

Centrifugals are constant-speed machines. These machines surge under certain operating conditions. This results in poor performance and high-power consumption at part load. Screw compressors have proven themselves in tough refrigeration applications including on-board ships. Today, screw compressors practically dominate refrigerated ships, transporting fruits, vegetables, meats, and frozen foods across the ocean with good reliability. These compressors have replaced the traditional shipboard centrifugals.

Screw compressors were developed in Germany in the 1800s. They were patented in 1883 in Italy, but not in the United States until 1905. This type of compressor is a positive-displacement compressor. That means it uses a rotor driving another rotor (twin) or gate rotors (single) to provide the compression cycle. Both methods

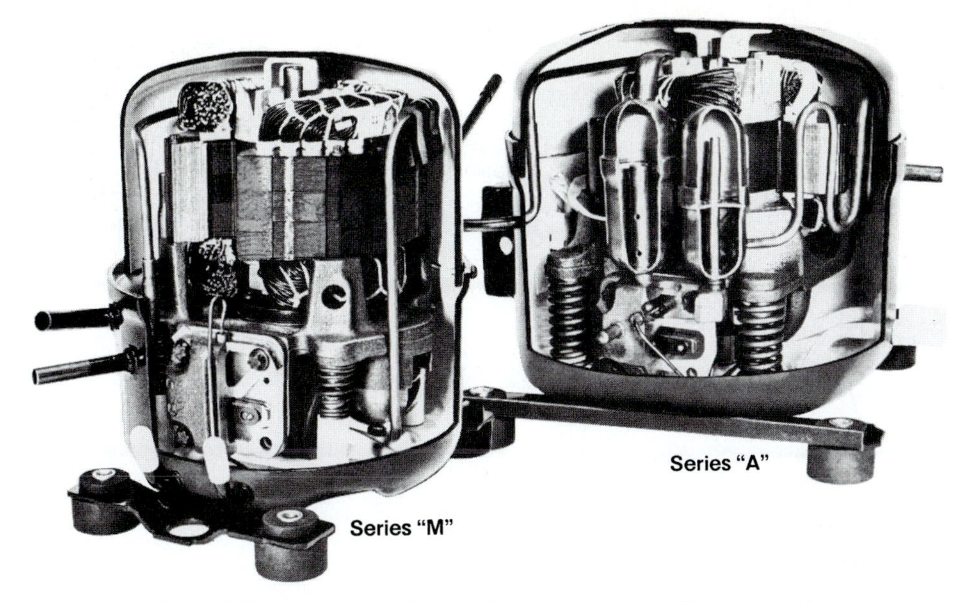

Fig. 7-68 *Series M and series A compressors made by Americold.*

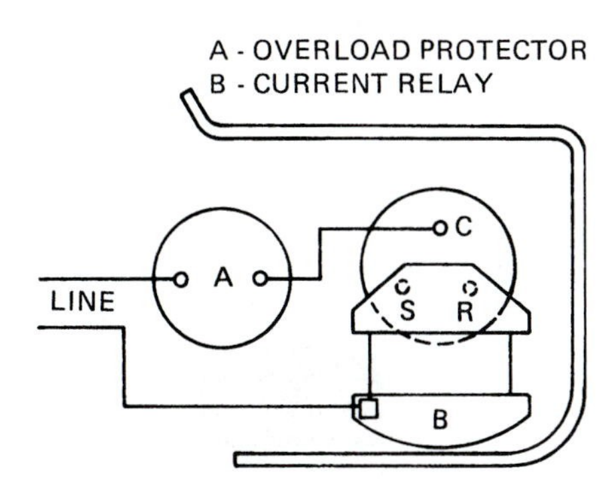

Fig. 7-69 *Location of the terminals for the compressors and electrical connections on the Americold compressors.* (Americold)

use injected fluids to cool the compressed gas, seal the rotor or rotors, and lubricate the bearings.

Single Screw

A single screw compressor is shown in Fig. 7-77. The compression process starts with the rotors meshed at the inlet port of the compressor. The rotors turn. The lobes separate at the inlet port, increasing the volume between the lobes. This increased volume causes a reduction in pressure. Thus, drawing in the refrigerant gas. The intake cycle is completed when the lobe has turned far enough to be sealed off from the inlet port.

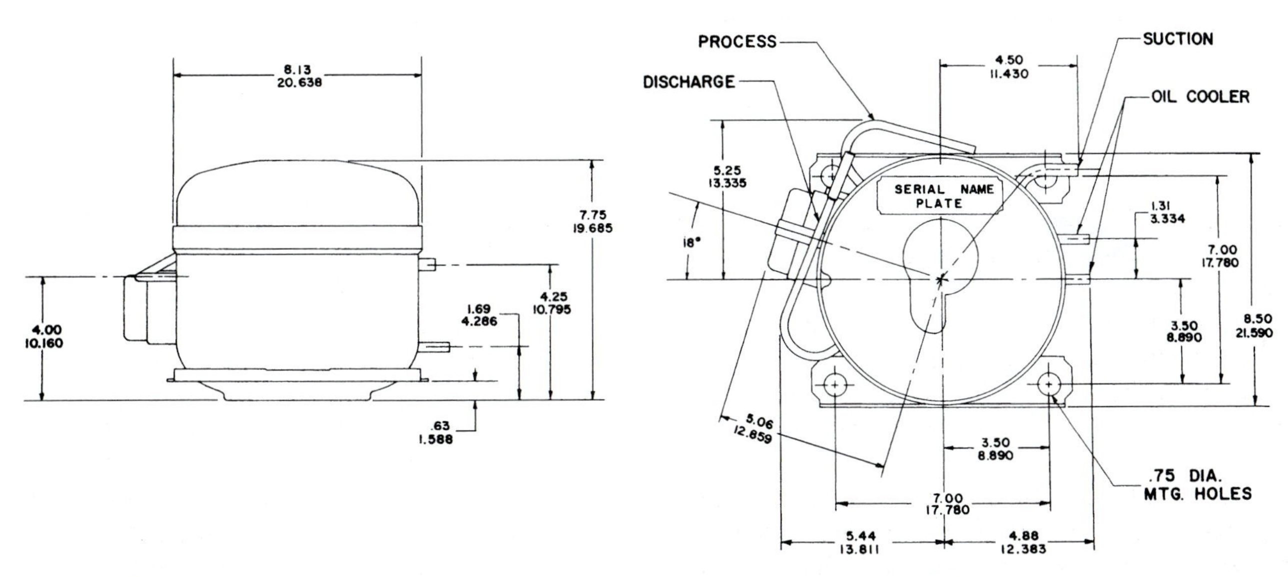

Fig. 7-70 *Location of process, discharge, suction, and oil-cooler stubs on Americold compressors.* (Americold)

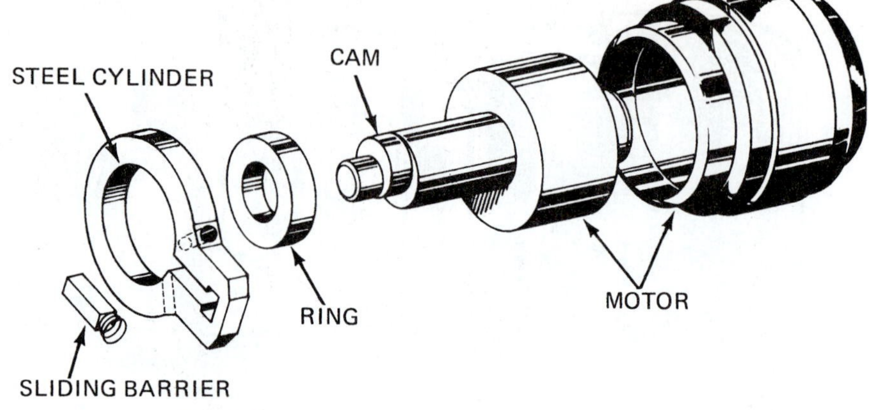

Fig. 7-71 *Parts of a rotary compressor.* (General Motors)

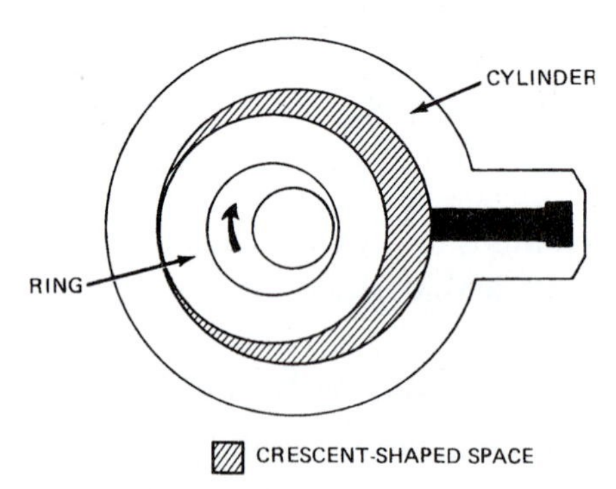

Fig. 7-72 *Operation of a rotary compressor.* (General Motors)

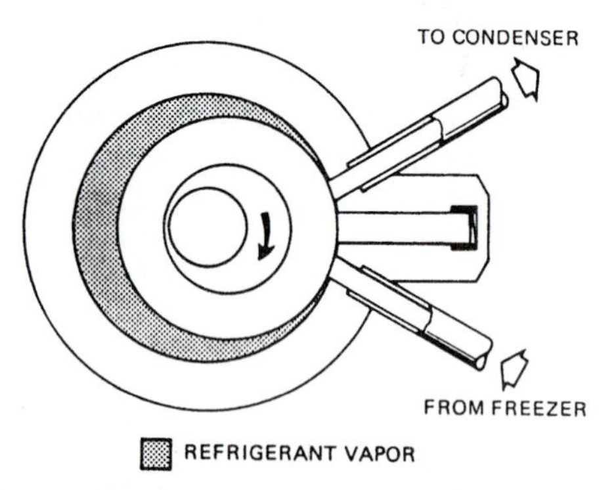

Fig. 7-73 *Beginning of the compression phase of a rotary compressor.* (General Motors)

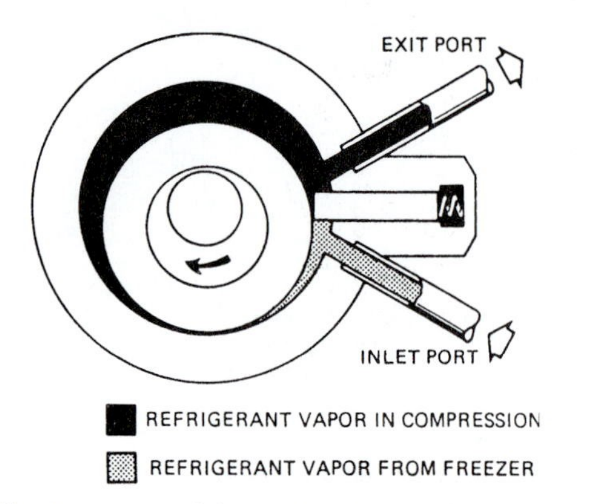

Fig. 7-74 *Beginning of the intake phase in a rotary compressor.* (General Motors)

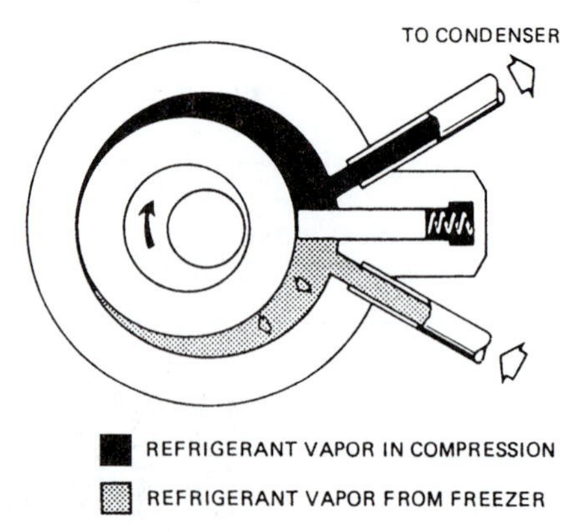

Fig. 7-75 *Compression and intake phases half completed in a rotary compressor.* (General Motors)

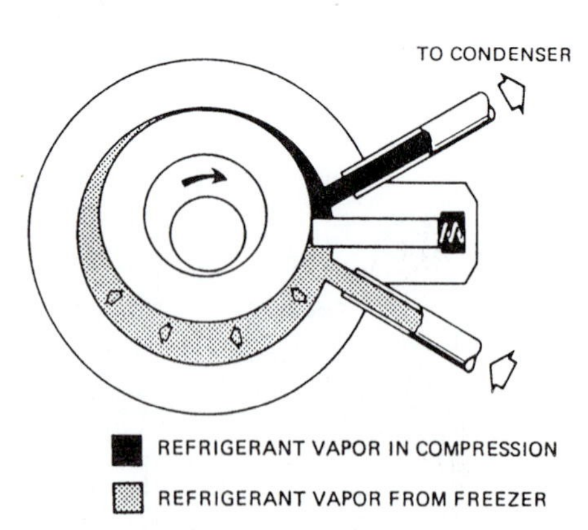

Fig. 7-76 *Finish of the compression phase of the rotary compressor.* (General Motors)

As the lobe continues to turn, the volume trapped in the lobe between the meshing point of the rotors, the discharge housing, and the stator and rotors, is continuously decreased. When the rotor turns far enough, the lobe opens to the discharge port, allowing the gas to leave the compressor. See Fig. 7-78.

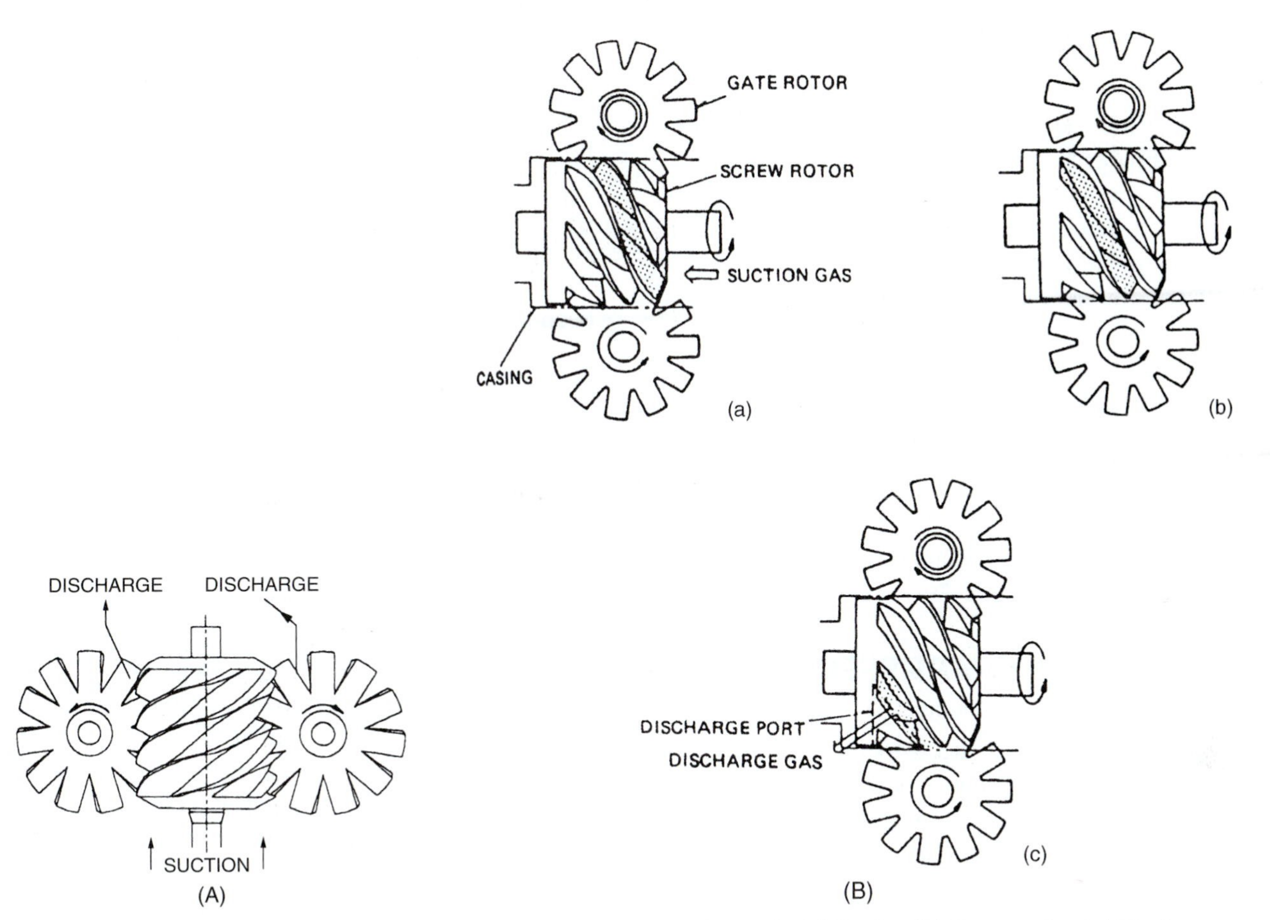

Fig. 7-77 *(A) Single-screw compressor. (B) Mono-screw compression cycle (a) Suction (b) Compression (c) Discharge or exhaust* (Single Screw Compressor, Inc.)

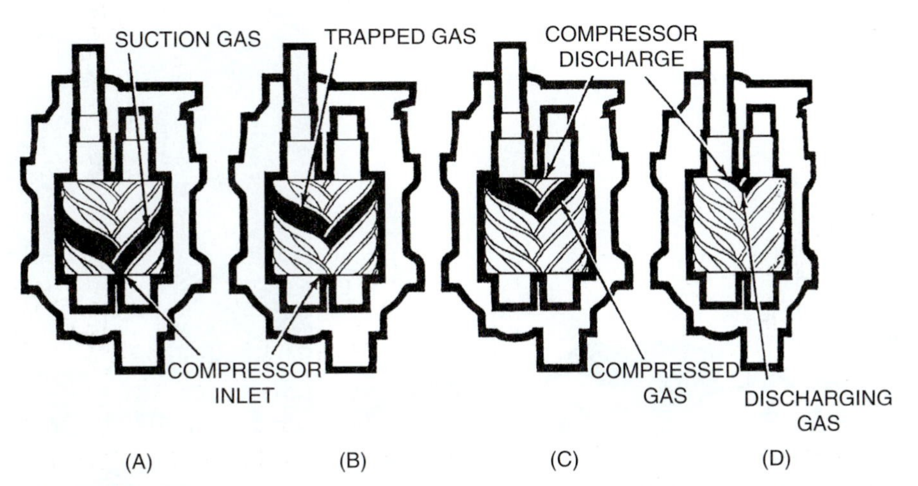

Fig. 7-78 *Twin-screw compression cycle. (A) Intake of gas. (B) Gas trapped in compressor housing and rotor cavities. (C) Compression cycle. (D) Compressed gas is discharged through the discharge port.* (Sullair Refrigeration)

Twin Screw

The twin screw is the most common type of screw compressor used today. It uses a double set of rotors (male and female) to the compress the refrigerant gas. The male rotor usually has four lobes. The female rotor consists of six lobes. Normally, this is referred to as a 4 + 6 arrangement. However, some compressors, especially air conditioners are using other variations, such as 5 + 7.

MAKING THE ROTORS

Not until the mid-1960s were the rotors cut using a symmetrical or circular profile. This was in turn replaced by the asymmetrical profile. This is a line-generated profile that improved the adiabatic efficiency of the screw compressor. See Fig. 7-79.

SCROLL COMPRESSORS

The scroll compressor (Fig. 7-80) is being used by the industry in response to the need to increase the efficiency of air-conditioning equipment. This is done in order to meet the U.S. Department of Energy Standards of 1992. The standards apply to all air conditioners. All equipment must have a *seasonal energy efficiency ratio* (SEER) of 10 or better. The higher the number, the more efficient the unit is. The scroll compressor seems to be the answer to more efficient compressor operation.

Scroll-Compression Process

Figure 7-81 shows how spiral-shaped members fit together. A better view is shown in Fig. 7-82. The two members fit together forming crescent-shaped gas pockets. One member remains stationary, while the second member is allowed to orbit relative to the stationary member.

This movement draws gas into the outer pocket created by the two members, sealing off the open passage. As the spiral motion continues, the gas is forced toward the center of the scroll form. As the pocket continuously becomes smaller in volume it creates increasingly higher gas pressures. At the center of the pocket, the high-pressure gas is discharged from the port of the fixed scroll member. During the cycle, several pockets of gas are compressed simultaneously. This provides a smooth, nearly continuous compression cycle.

This results in a 10 to 15 percent more efficient operation than with the piston compressors. A smooth, continuous compression process means very low flow losses. No valves are required. This eliminates all valve losses. Suction and discharge locations are separate. This substantially reduces heat transfer between suction and discharge gas. There is no reexpansion volume. This increases the compressor's heat pump capacity in low-ambient operation. Increased heat pump capacity in low-ambient temperatures reduces the need for supplemental heat when temperatures drop.

During summer, this means less cycling at moderate temperatures. It also allows better dehumidification to keep the comfort level high. When temperatures rise, the scroll compressor provides increased capacity for more cooling.

During the winter, the scroll compressor heat pumps deliver more warm air to the conditioned space than conventional models.

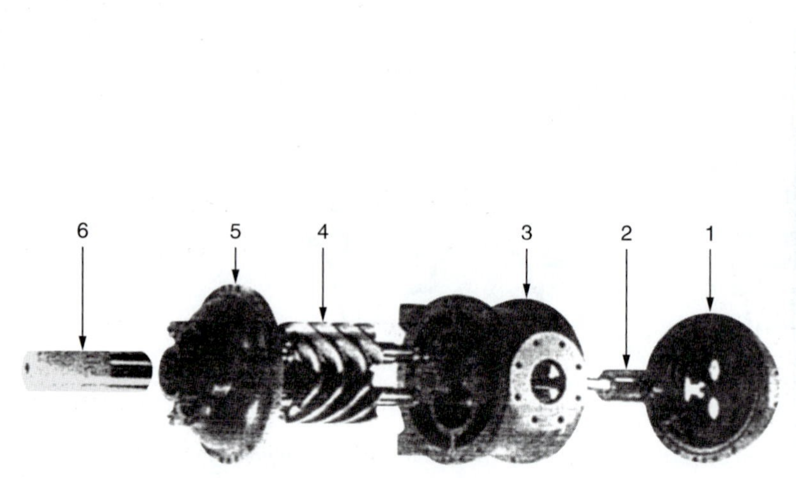

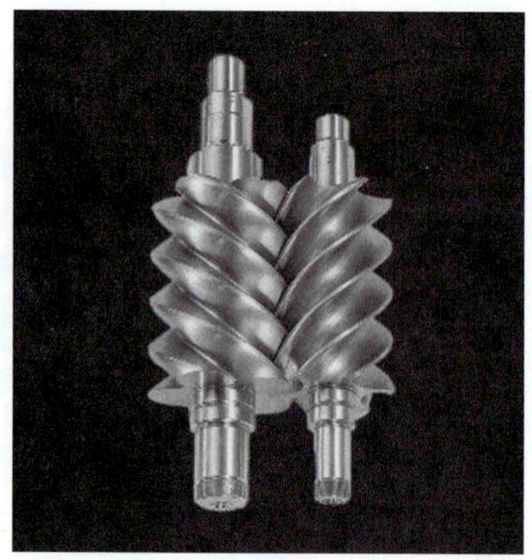

Fig. 7-79 *Twin-screw compressor parts. 1, discharge housing; 2, slide valve; 3, stator; 4, male and female rotors; 5, inlet housing; 6, hydraulic capacity control cylinder.* (Sullair Refrigeration)

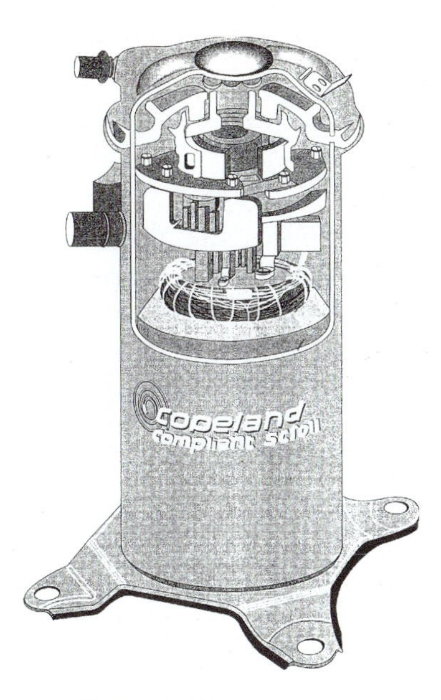

Components:

- OrbitingScroll Set
- StationaryScroll Set
- Crankshaft
- Motor
- Housing

Fig. 7-80 *A Copeland scroll compressor.* (Lennox)

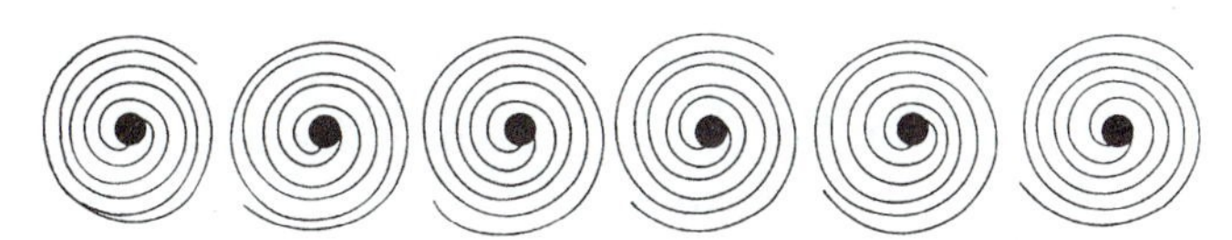

Fig. 7-81 *Scroll-compression process.* (Lennox)

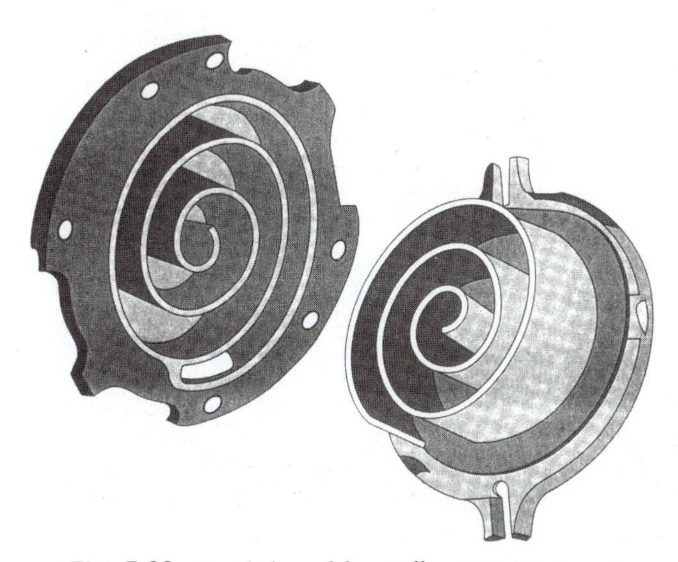

Fig. 7-82 *Two halves of the scroll compressor.* (Lennox)

Operation

The scroll compressor has no valves and low-gas pulses. No valves and low-gas pulses allow for smooth and quiet operation. It has fewer moving parts (only 2 gas compression parts as compared to 15 components in piston-type compressors) and no compressor start components are required. There is no accumulator or crankcase heater required. And, there is not a high- pressure cutout needed.

The radial compliance design features a superior liquid-handling capacity. This allows small amounts of liquid and dirt to pass through without damaging the compressor. At the same time, this eliminates high stress on the motor and provides high reliability. Axial compliance allows the scroll tips to remain in continuous contact, ensuring minimal leakage. Performance actually gets better over the time because there are no seals to wear and causes gas leakage.

Scroll Compressor Models

Examples of air-conditioner units with the scroll compressor are the Lennox HP-20 and HS-22.

HP-20 Model The HP 20 is also designed for efficient use in heat-pump installations. See Fig. 7-83. It has a large coil surface area to deliver more comfort per watt of electricity. A copper-tube coil with aluminum fins provides effective heat transfer for efficient heating and cooling. See Fig. 7-84. The scroll-compressor technology has been around for a long time. However, this was one of the first to make use of it in heat pumps.

Model HS-22 The HS-22 model also uses a Copeland compliant scroll compressor. See Fig. 7-85. The insulated cabinet allows it to operate without disturbing the neighbors in closely arranged housing developments. The cabinet, with vertical air discharge, creates a unit that has sound ratings as low as 7.2 bells.

The condensing unit has a SEER rating as high as 13.5. The scroll compressor is highly efficient with a large double-row condenser. This highly efficient coil increases the efficient use of energy even more.

REVIEW QUESTIONS

1. List three classifications of refrigeration compressors.
2. List three types of hermetic compressor motors.
3. What are the two types of motor relays used in refrigeration and air-conditioning compressors?
4. Describe a glass quick-connect terminal.
5. What is the purpose of the rubber grommet used to hold a compressor in place?

Scroll compressor

Strengths:

- Efficient gas compression
- Low sound and vibration levels
- Fewer parts, smaller size, lighter weight, more per pallet
- No internal suspension system

Weaknesses:

- Orbiting and stationary scrolls must match perfectly
- Used in air conditioning/commercial compressors under development
- Need to reduce vibration to unit using generous amounts of tubing

Fig. 7-83 *HP-20 Model with a heavy-duty scroll compressor. 1, cabinet; 2, coil area; 3, copper tubing; 4, fan; 5, scroll compressor.* (Lennox)

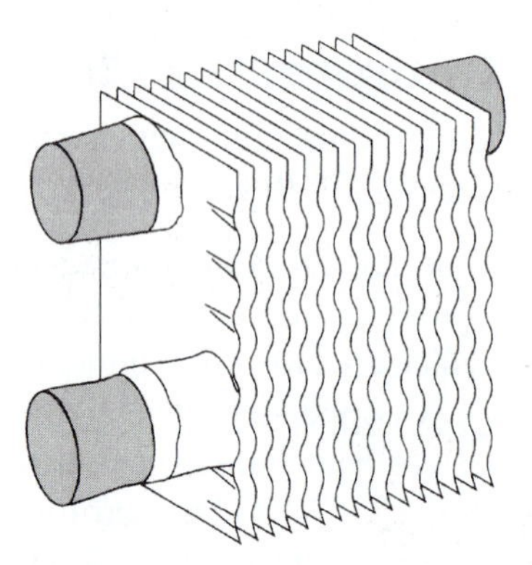

Fig. 7-84 *Copper tubing in the condenser with aluminum fins.* (Lennox)

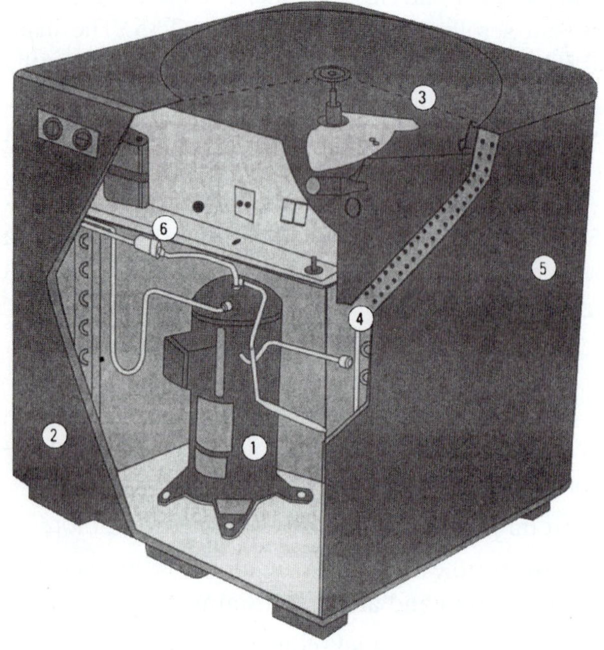

Fig. 7-85 *HS-22 Model. 1, scroll compressor; 2, cabinet; 3, fan; 4, copper tubing in condenser; 5, coil area in compressor; 6, filter drier.* (Lennox)

6. What are the two types of crankcase heaters?
7. What does RSIR stand for in motor terminology?
8. What does CSIR stand for in motor terminology?
9. What is a stub tube?
10. Why are oil-cooler tubes needed?
11. What is a process tube?
12. What are the two types of rotary compressors?
13. How does the scroll compressor compress the refrigerant gas?
14. Why is the scroll compressor so efficient?
15. Why is scroll compressor particularly useful as a heat-pump compressor?

8
CHAPTER

Condensers, Chillers, and Cooling Towers

PERFORMANCE OBJECTIVES

After studying this chapter you should:

1. Understand the function and operation of various types of condensers.
2. Know the different types of condensers.
3. Know the purpose and operation principles of chillers.
4. Know how cooling towers operate and why they are employed.
5. Know different types of cooling towers and how to clean them properly.

The condenser is a heat-transfer device. It is used to remove heat from hot refrigerant vapor. Using some method of cooling, the condenser changes the vapor to a liquid. There are three basic methods of cooling the condenser's hot gases. The method used to cool the refrigerant and return it to the liquid state serves to categorize the two types of condensers. Thus, there are two types of condensers: air cooled and water cooled. Cooling towers are also used to cool the refrigerant.

Most commercial or residential home air-conditioning units are air cooled. Water is also used to cool the refrigerant. This is usually done where there is an adequate supply of fairly clean water. Industrial applications rely upon water to cool the condenser gases. The evaporative process is also used to return the condenser gases to the liquid state. Cooling towers use the evaporative process.

CONDENSERS

Air-Cooled Condensers

Figure 8-1 illustrates the refrigeration process within an air-cooled condenser. Figure 8-2 shows some of the various types of compressors and condensers mounted as a unit. These units may be located outside the cooled space. Such a location makes it possible to exhaust the heated air from the cooled space. Note that the condenser has a large-bladed fan that pushes air through the condenser fins. The fins are attached to coils of copper or aluminum tubing. The tubing houses the liquid and the gaseous vapors. When the blown air contacts the fins, it cools them. The heat from the compressed gas in the tubing is thus transferred to the cooler fin.

Heat given up by the refrigerant vapor to the condensing medium includes both the heat absorbed in the evaporator and the heat of compression. Thus, the

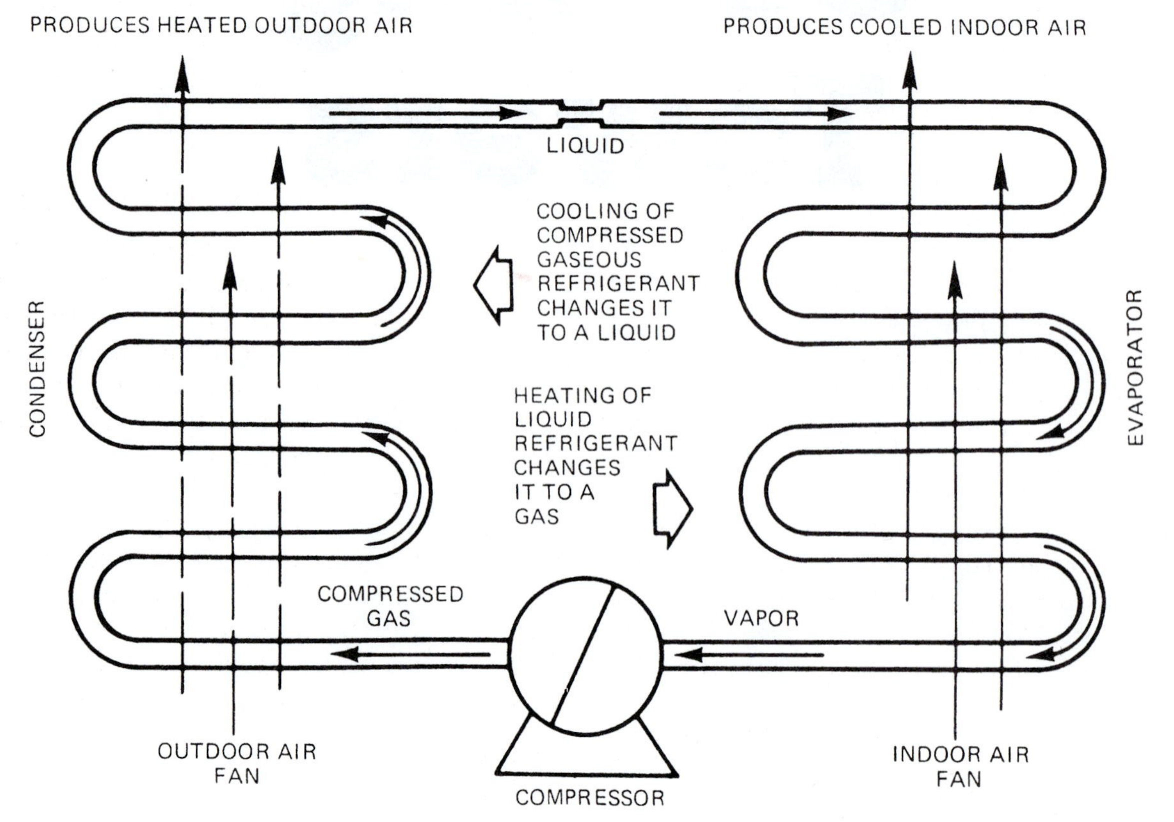

Fig. 8-1 *Refrigeration cycle.*

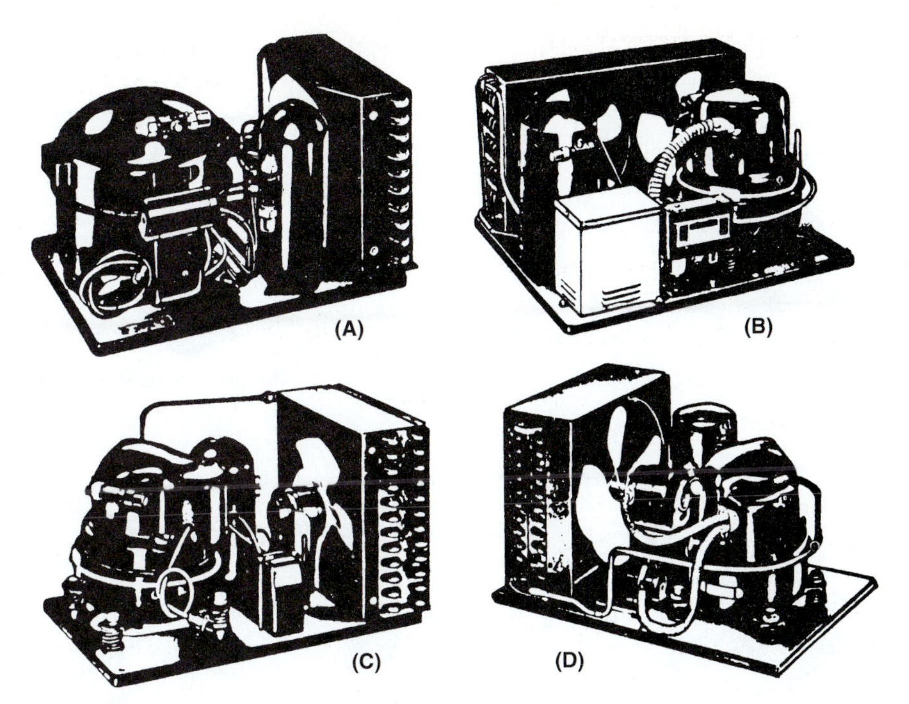

Fig. 8-2 *A condenser, fan, and compressor. Self-contained, in one unit.* (Tecumseh)

condenser always has a load that is the sum of these two heats. This means the compressor must handle more heat than that generated by the evaporator. The quantity of heat (in Btu) given off by the condenser is rated in heat per minute per ton of evaporator capacity. These condensers are rated at various suction and condensing temperatures.

The larger the condenser area exposed to the moving air stream, the lower will be the temperature of the refrigerant when it leaves the condenser. The temperature of the air leaving the vicinity of the condenser will vary with the load inside the area being cooled. If the evaporator picks up the additional heat and transfers it to the condenser, then the condenser must transmit this heat to the air passing over the surface of the fins. The temperature rise in the condensing medium passing through the condenser is directly proportional to the condenser load. It is inversely proportional to the quantity and specific heat of the condensing medium.

To exhaust the heat without causing the area being cooled to heat up again, it is common practice to locate the condenser outside of the area being conditioned. For example, for an air-conditioned building, the condenser is located on the rooftop or on an outside slab at grade level. See Fig. 8-3.

Some condensers are cooled by natural airflow. This is the case in domestic refrigerators. Such natural convection condensers use either plate surface or finned tubing. See Fig. 8-4.

Air-cooled condensers that use fans are classified as chassis mounted and remote. The chassis-mounted type is shown in Fig. 8-2. Here the compressor, fan, and condenser are mounted as one unit. The remote type is shown in Fig. 8-3. Remote air-cooled condensers can be obtained in sizes that range from 1 to 100 tons. The chassis-mounted type is usually limited to 1 ton or less.

Water-Cooled Condensers

Water is used to cool condensers. One method is to cool condensers with water from the city water supply and then exhaust the water into the sewer after it has been used to cool the refrigerant. This method can be expensive and, in some instances, is not allowed by law. When there is a sewer problem, a limited sewer treatment plant capacity, or drought, it is impractical to use this cooling method.

The use of recirculation to cool the water for reuse is more practical. However, in recirculation, the power required to pump the water to the cooling location is part of the expense of operating the unit.

There are three types of water-cooled condensers. They are:

- The double tube
- The shell and coil
- The shell and tube types

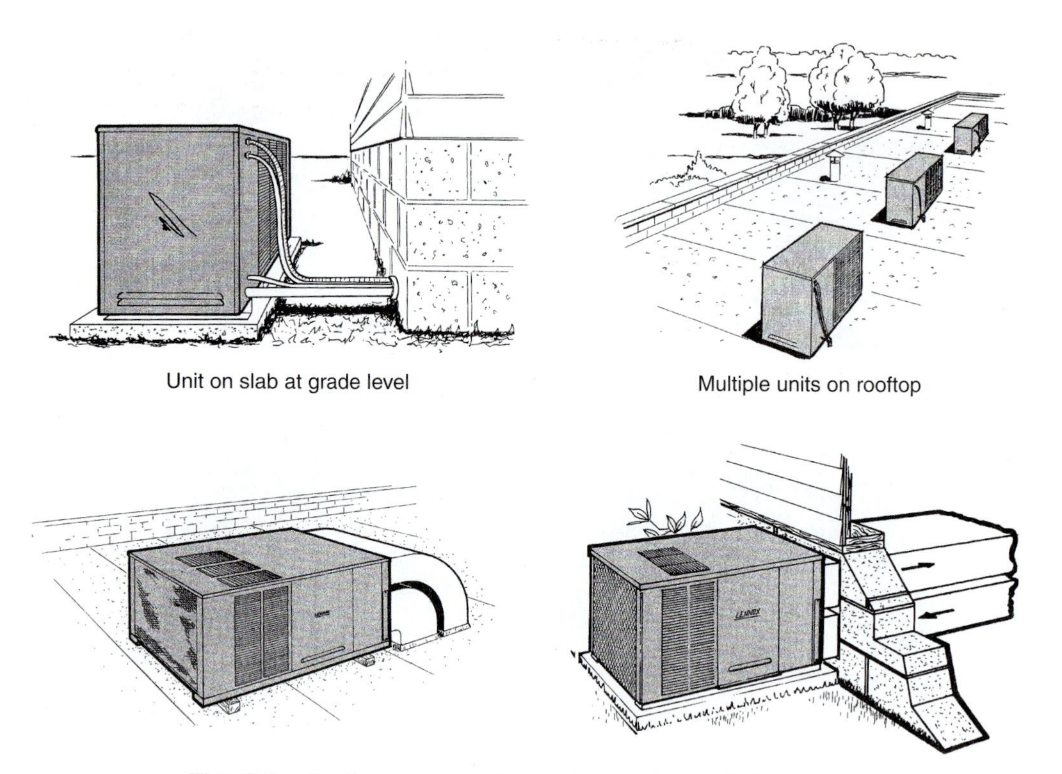

Fig. 8-3 *Condensers mounted on rooftops and at grade level.* (Lennox)

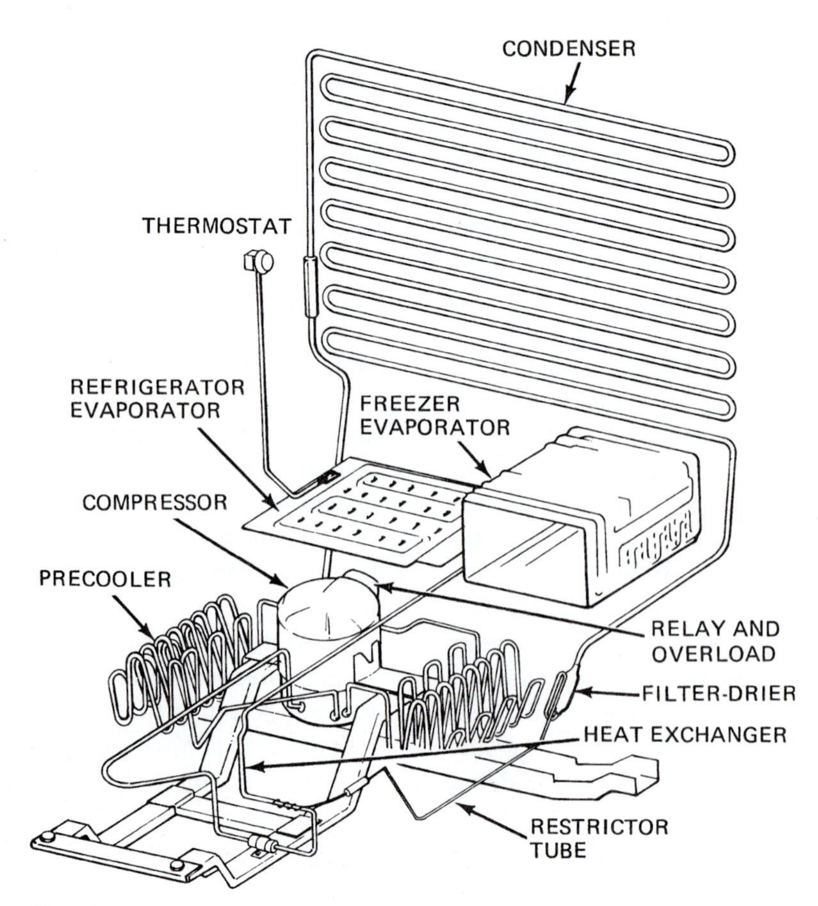

Fig. 8-4 *Flat, coil-type condenser, with natural air circulation. Used in refrigeration in the home.* (Sears)

The double-tube type consists of two tubes, one inside the other. See Fig. 8-5. Water is piped through the inner tube. Refrigerant is piped through the tube that encloses the inner tube. The refrigerant flows in the opposite direction than the water. See Fig. 8-6.

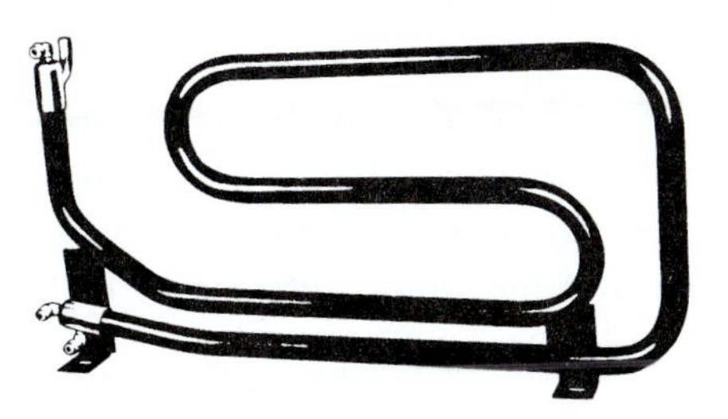

Fig. 8-5 *Coaxial, water-cooled condenser. Used with refrigeration and air-conditioning units where space is limited.*

This type of coaxial water-cooled condenser is designed for use with refrigeration and air–conditioning condensing units where space is limited. These condensers can be mounted vertically, horizontally, or at any angle.

They can be used with cooling towers also. They perform at peak heat of rejection with water pressure drop of not more than 5 $lb/in.^2$, utilizing flow rates of 3 gal/min/ton.

The typical counter-flow path shows the refrigerant going in a 105°F (41°C) and the water going in at 85°F (30°C) and leaving at 95°F (35°C). See Fig. 8-7.

The counter-swirl design, shown in Fig. 8-6, gives heat-transfer performance of superior quality.

The tube construction provides for excellent mechanical stability. The water-flow path is turbulent. This provides a scrubbing action that maintains cleaner surfaces. The construction method shown also has very high system pressure resistance.

The water-cooled condenser shown in Fig. 8-5 can be obtained in a number of combinations. Some of these combinations are listed in Table 8-1. Copper tubing is suggested for use with fresh water and with cooling towers. The use of *cupronickel* is suggested when salt water is used for cooling purposes.

Convolutions to the water tube result in a spinning, swirling water flow that inhibits the accumulation of deposits on the inside of the tube. This contributes to the antifouling characteristics in this type of condenser. Figure 8-8 shows the various types of constructions for the condenser.

This type of condenser may be added as a booster to standard air-cooled units. Figure 8-9 shows some of' the configurations of this type of condenser:

- The spiral
- The helix
- The trombone

Note the input for the water and the input for the refrigerant. Using a cooling tower to furnish water to contact the outside tube can further cool the condensers. Also, a water tower can be used to cool the water sent through the inside tube for cooling purposes. This type of' condenser is usable where refrigeration or air-conditioning requirements are $^1/_3$ to 3 tons.

Placing a bare tube or a finned tube inside a steel shell makes the shell and coil condenser. See Fig. 8-10. Water circulates through the coils. Refrigerant vapor is injected into the shell. The hot vapor contacts the cooler tubes and condenses. The condensed vapor drains from the coils and drops to the bottom of the tank or shell. From there it is recirculated through the refrigerated area by way of the evaporator. In most cases, placing chemicals into the water cleans the unit. The chemicals have a tendency to remove the deposits that build up on the tubing walls.

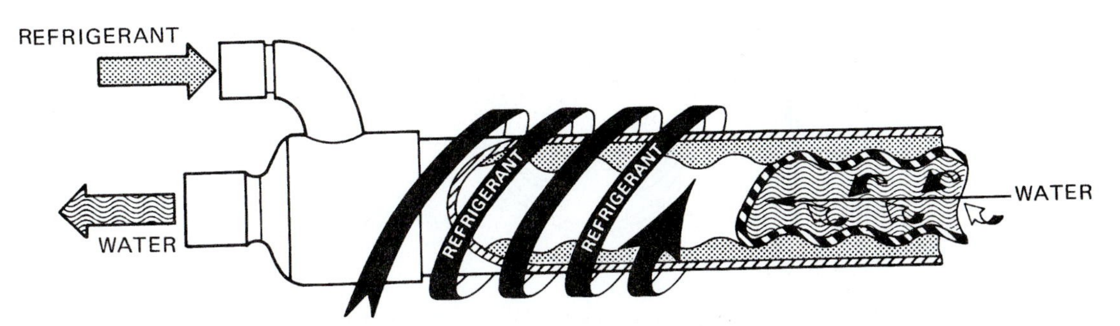

Fig. 8-6 *A typical counter-flow path inside a coaxial water-cooled condenser.* (Packless)

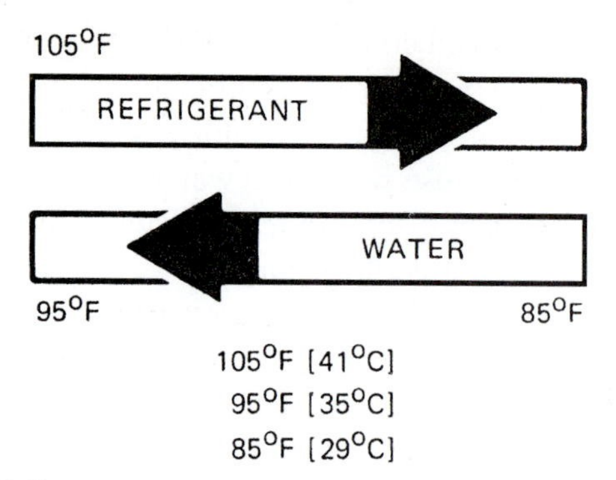

Fig. 8-7 *Water and refrigerant temperatures in a counter-flow, water-cooled condenser.* (Packless)

Table 8-1 *Some Possible Metal Combinations in Water-Cooled Condensers*

Shell Metal	Tubing Metal
Steel	Copper
Copper	Copper
Steel	Cupronickel
Copper	Cupronickel
Steel	Stainless steel
Stainless steel	Stainless steel

CHILLERS

A chiller is part of a condenser. Chillers are used to cool water or brine solutions. The cooled (chilled) water or brine is then fed through pipes to evaporators. This cools the area in which the evaporators are located. This type of cooling, using chilled water or brine, can be used in large air-conditioning units. It can also be used for industrial processes where cooling is required for a particular operation.

Figure 8-11 illustrates such an operation. Note how the compressor sits atop the condenser. Chillers are the answer to requirements of 200 to 1600 tons of refrigeration. They are used for process cooling, comfort air conditioning, and nuclear power plant cooling. In some cases, they are used to provide ice for ice-skating rinks. The arrows in Fig. 8-11 indicate the refrigerant flow and the water or brine flow through the large pipes. Figure 8-12 shows the machine in a cutaway view. The following explanation of the various cycles will provide a better understanding of the operation of this type of equipment.

Refrigeration Cycle

The machine compressor continuously draws large quantities of refrigerant vapor from the cooler, at a rate determined by the size of the guide-vane opening. This compressor suction reduces the pressure within the cooler, allowing the liquid refrigerant to boil vigorously at a fairly low temperature [typically 30 to 35°F (−1 to 2°C)].

Liquid refrigerant obtains the energy needed, for the change to vapor, by removing heat from the water in the cooler tubes. The cold water can then be used in the air-conditioning process.

After removing heat from the water, the refrigerant vapor enters the first stage of the compressor. There, it is compressed and flows into the second stage of the compressor. Here it is mixed with flash-economizer gas and further compressed.

Compression raises the refrigerant temperature above that of the water flowing through the condenser tubes. When the warm [typically 100 to 105°F (38 to 41°C)] refrigerant vapor contacts the condenser tubes, the relatively cool condensing water [typically 85 to 95°F (29 to 35°C)] removes some of the heat and the vapor condenses into a liquid.

Further heat removal occurs in the group of condenser tubes that form the thermal economizer. Here, the condensed liquid refrigerant is subcooled by contact with the coolest condenser tubes. These are the tubes that contain the entering water.

The subcooled liquid refrigerant drains into a high-side valve chamber. This chamber maintains the proper fluid level in the thermal economizer and meters the refrigerant liquid into a flash economizer chamber. Pressure in this chamber is intermediate between condenser and cooler pressures. At this lower pressure,

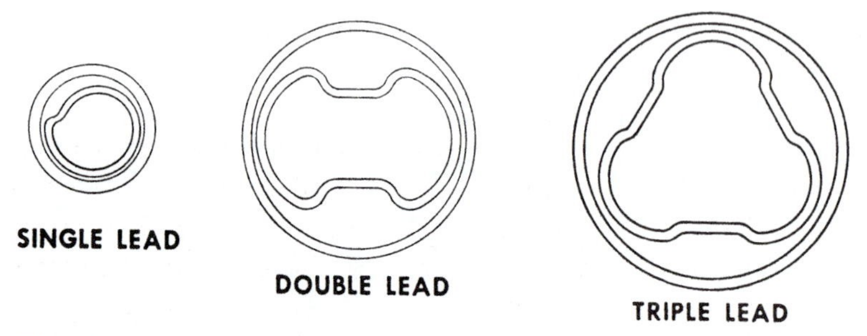

Fig. 8-8 *Different types of tubing fabrication, located inside the coaxial type water-cooled condenser.* (Packless)

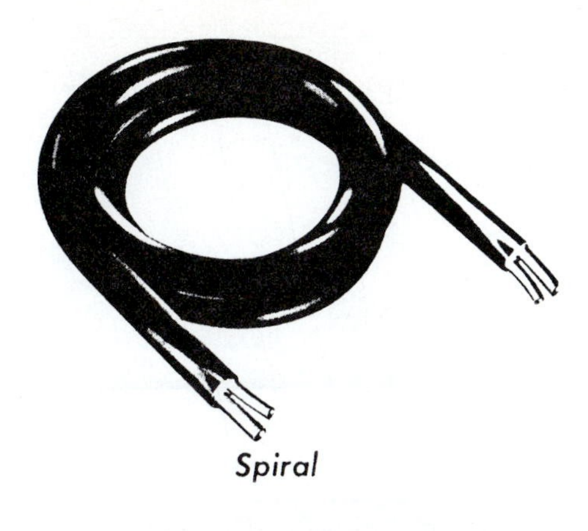

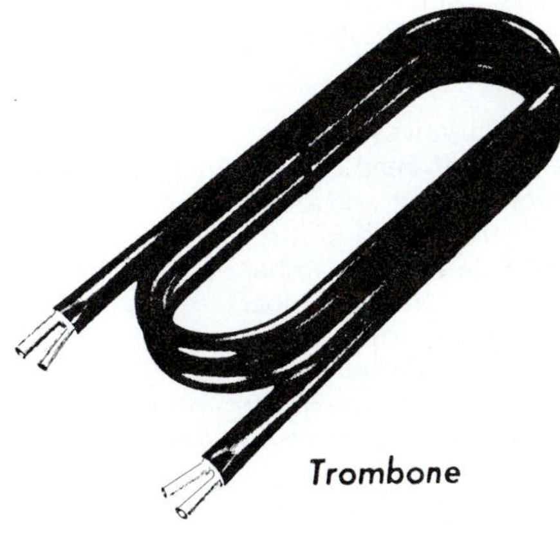

Fig. 8-9 *Three configurations of coaxial, water-cooled condensers.* (Packless)

some of the liquid refrigerant flashes to gas, cooling the remaining liquid. The flash gas, having absorbed heat, is returned directly to the compressor's second stage. Here, it is mixed with gas already compressed by the first-stage impeller. Since the flash gas must pass through only half the compression cycle to reach condenser pressure, there is a savings in power.

The cooled liquid refrigerant in the economizer is metered through the low-side valve chamber into the cooler. Because pressure in the cooler is lower than economizer pressure, some of the liquid flashes and cools the remainder to evaporator (cooler) temperature. The cycle is now complete.

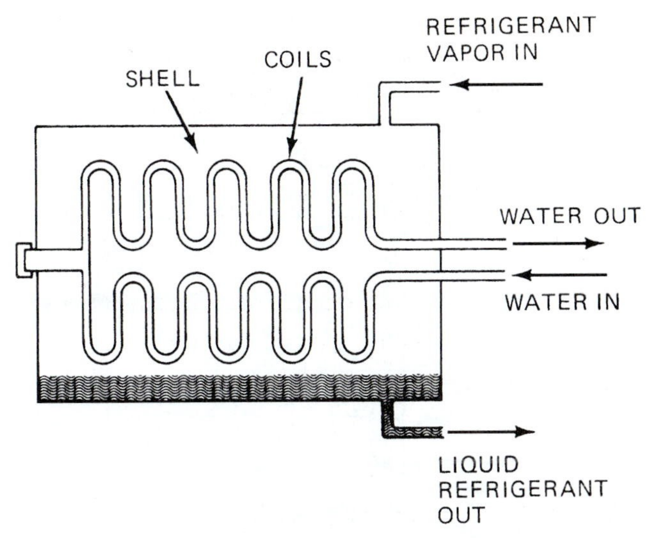

Fig. 8-10 *The shell and coil condenser.*

Motor-Cooling Cycle

Refrigerant liquid from a sump in the condenser (No. 24 in Fig. 8-11) is subcooled by passage through a line in the cooler (No. 27 in Fig. 8-11). The refrigerant then flows externally through a strainer and variable orifice (No. 11 in Fig. 8-11) and enters the compressor motor end. Here it sprays and cools the compressor rotor and stator. It then collects in the base of the motor casing. Here, it drains into the cooler. Differential pressure between the condenser and cooler maintains the refrigerant flow.

Dehydrator Cycle

The dehydrator removes water and noncondensable gases. It indicates any water leakage into the refrigerant. See No. 6 in Fig. 8-11.

This system includes a refrigerant condensing coil and chamber, water-drain valve, purging valve, pressure gage, refrigerant-float valve, and refrigerant piping.

A dehydrator sampling line continuously picks up refrigerant vapor and contaminants, if any, from the condenser. Vapor is condensed into a liquid by the dehydrator-condensing coil. Water, if present, separates and floats on the refrigerant liquid. The water level can be observed through a sight glass.

Water may be withdrawn manually at the water-drain valve. Air and other noncondensable gases collect

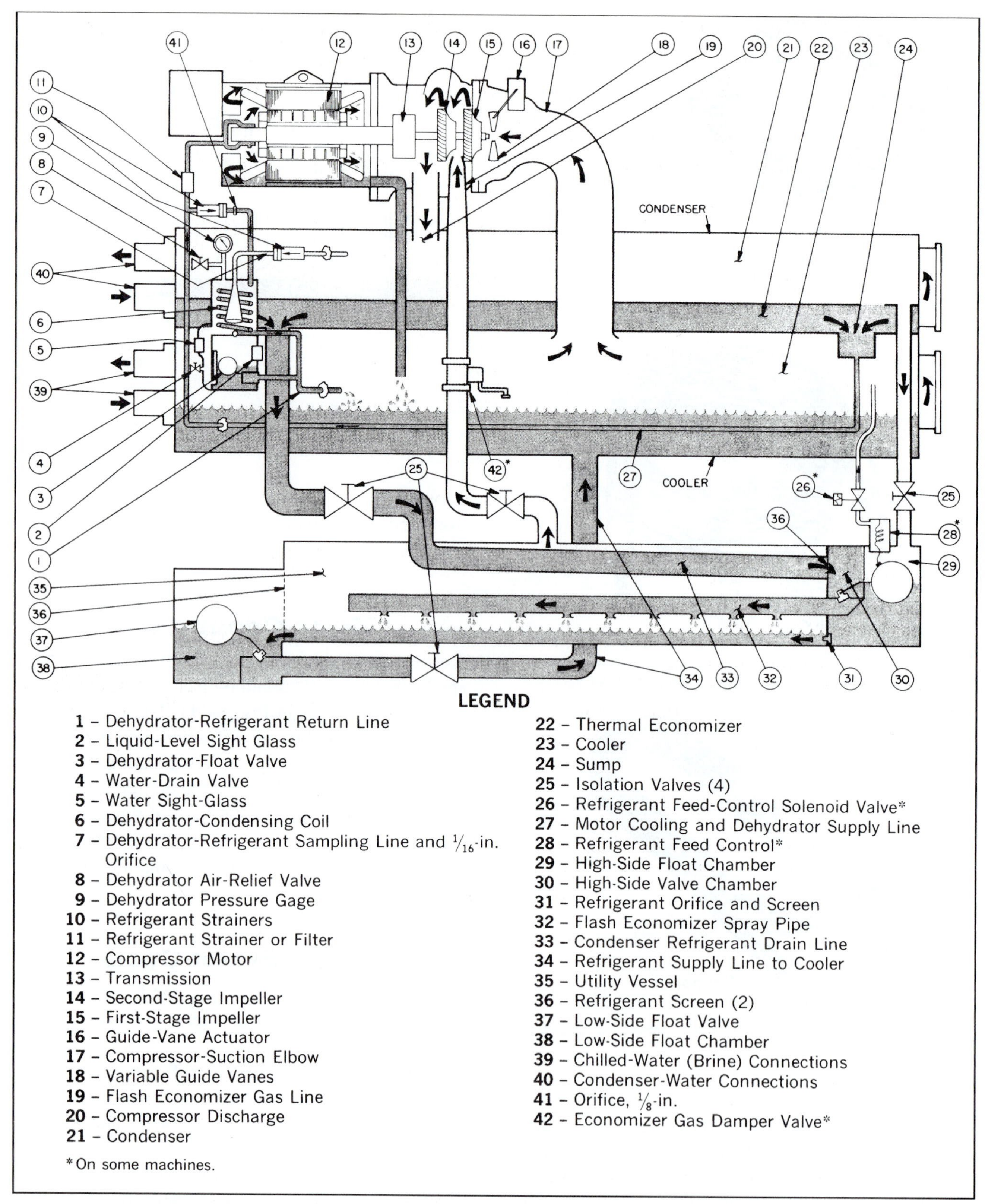

Fig. 8-11 *The chiller, compressor, condenser, and cooler are combined in one unit.* (Carrier)

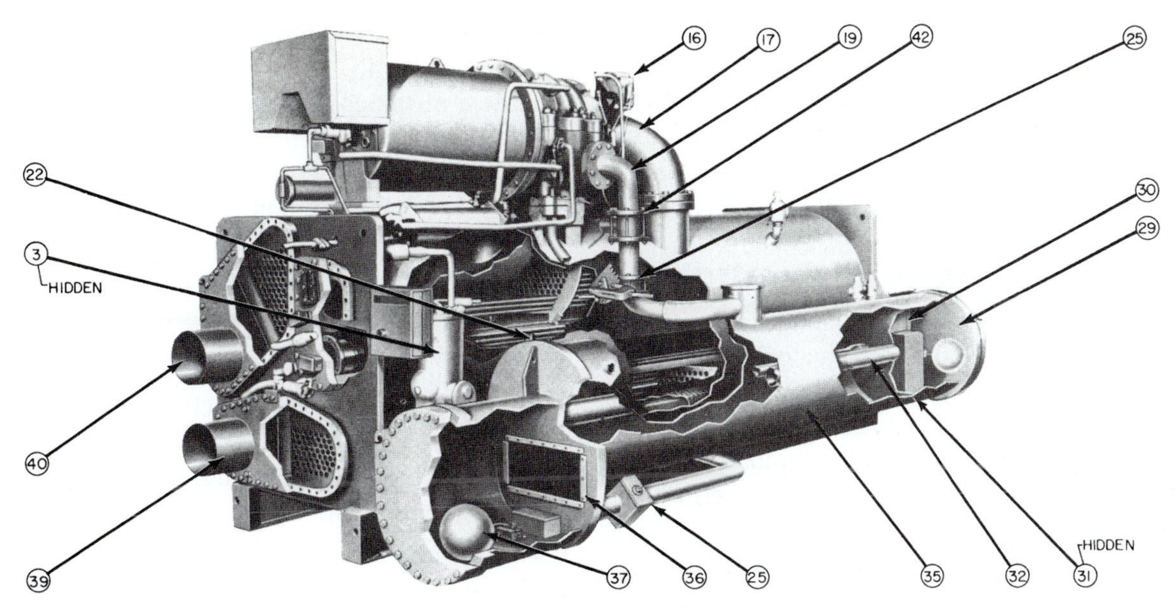

Fig. 8-12 *Cutaway view of a chiller.*

in the upper portion of the dehydrator-condensing chamber. The dehydrator gage indicates the presence of air or other gases through a rise in pressure. These gases may be manually vented through the purging valve.

A float valve maintains the refrigerant liquid level and pressure difference necessary for the refrigerant-condensing action. Purified refrigerant is returned to the cooler from the dehydrator-float chamber.

Lubrication Cycle

The oil pump and oil reservoir are contained within the unishell. Oil is pumped through an oil-filter cooler that removes heat and foreign particles. A portion of the oil is then fed to the compressor motor-end bearings and seal. The remaining oil lubricates the compressor transmission, compressor thrust and journal bearings, and seal. Oil is then returned to the reservoir to complete the cycle.

CONTROLS

The cooling capacity of the machine is automatically adjusted to match the cooling load by changes in the position of the compressor inlet guide vanes. See Fig. 8-13.

A temperature-sensing device in the circuit of the chilled water leaving the machine cooler continuously transmits signals to a solid-state module in the machine control center. The module, in turn, transmits the amplified and modulated temperature signals to an automatic guide-vane actuator.

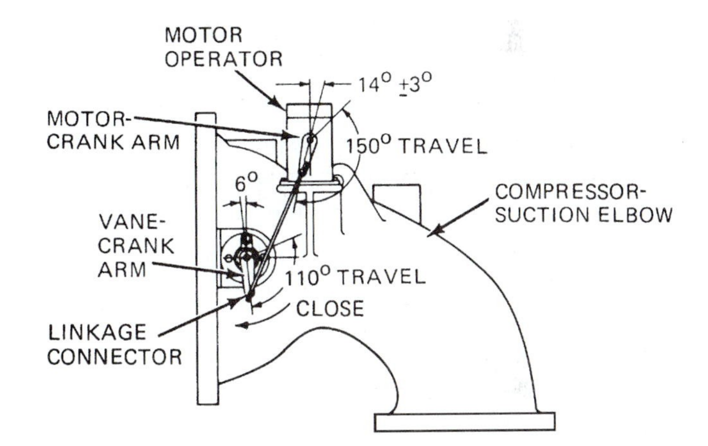

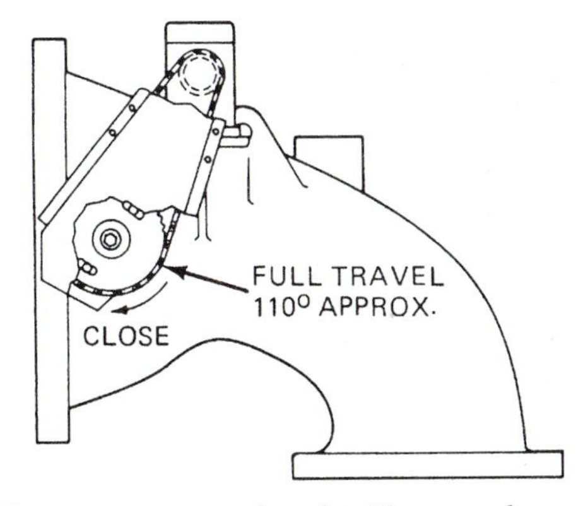

Fig. 8-13 *Vane motor-crank angles. These are shown as No. 16 and No. 17 in Fig. 8-11.*

A drop in the temperature of the chilled water leaving the circuit causes the guide vanes to move towards the closed position. This reduces the rate of refrigerant evaporation and vapor flow into the compressor. Machine capacity decreases. A rise in chilled water temperature opens the vanes. More refrigerant vapor moves through the compressor and the capacity increases.

The modulation of the temperature signals in the control center allows precise control of guide-vane response, regardless of the system load.

Solid-State Capacity Control

In addition to amplifying and modulating the signals from chilled water sensor to vane actuator, the solid-state module in the control center provides a means for preventing the compressor from exceeding full-load amperes. It also provides a means for limiting motor current down to 40 percent of full-load amperes to reduce electrical demand rates.

A throttle-adjustment screw eliminates guide-vane hunting. A manual capacity-control knob allows the operator to open, close, or hold the guide-vane position when desired.

COOLING TOWERS

Cooling towers are used to conserve or recover water. In one design the hot water from the condenser is pumped to the tower. There, it is sprayed into the tower basin. The temperature of the water decreases as it gives up heat to the air circulating through the tower. Some of the towers are rather large, since they work with condensers yielding 1600 tons of cooling capacity. See Fig. 8-14.

Most of the cooling that takes place in the tower results from the evaporation of part of the water as it falls through the tower.

The lower the wet-bulb temperature of the incoming air, the more efficient the air is in decreasing the temperature of the water being fed into the tower.

The following factors influence the efficiency of the cooling tower.

- Mean difference between vapor pressure of the air and pressure in the tower water
- Length of exposure time and amount of water surface exposed to air
- Velocity of air through the tower
- Direction of airflow relative to the exposed water surface (parallel, transverse, or counter)

Theoretically, the lowest temperature to which the water can be cooled is the temperature of the air (wet bulb) entering the tower. However, in practical terms, it is impossible to reach the temperature of the air. In most instances, the temperature of the water leaving the tower will be no lower than 7 to 10°F (4 to 6°C) above the air temperature.

The range of the tower is the temperature of the water going into the tower and the temperature of the

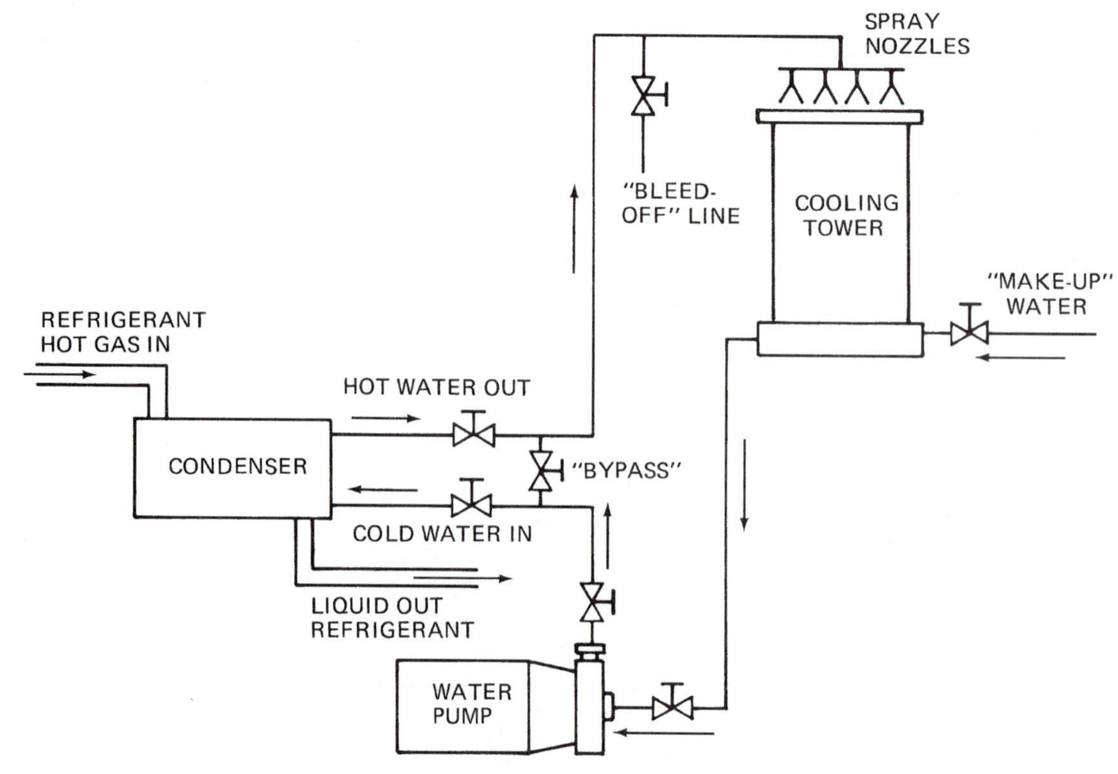

Fig. 8-14 *Recirculating water system using a tower.*

water coming out of the tower. This range should be matched to the operation of the condenser for maximum efficiency.

Cooling Systems Terms

The following terms apply to cooling-tower systems.

Cooling range is the number of degrees Fahrenheit through which the water is cooled in the tower. It is the difference between the temperature of the hot water entering the tower and the temperature of the cold water leaving the tower.

Approach is the difference in degrees Fahrenheit between the temperature of the cold water leaving the cooling tower and the wet-bulb temperature of the surrounding air.

Heat load is the amount of heat "thrown away" by the cooling tower in Btu per hour (or per minute). It is equal to the pounds of water circulated multiplied by the cooling range.

Cooling-tower pump head is the pressure required to lift the returning hot water from a point level with the base of the tower to the top of the tower and force it through the distribution system.

Drift is the small amount of water lost in the form of fine droplets retained by the circulating air. It is independent of, and in addition to, evaporation loss.

Bleed off is the continuous or intermittent wasting of a small fraction of circulating water to prevent the build up and concentration of scale-forming chemicals in the water.

Makeup is the water required to replace the water that is lost by evaporation, drift and bleed off.

Design of Cooling Towers

Classified by the air-circulation method used, there are two types of cooling towers. They are either natural-draft or mechanical-draft towers. Figure 8-15 shows the operation of the natural-draft cooling tower. Figure 8-16 shows the operation of the mechanical-draft cooling tower. The forced-draft cooling tower, shown in Fig. 8-17, is just one example of the mechanical-draft designs available today.

Cooling-tower ratings are given in tons. This is based on heat-transfer capacity of 250 Btu/min/ ton. The normal wind velocity taken into consideration for tower design is 3 mi/h. The wet bulb temperature is usually 80°F (27°C) for design purposes. The usual flow of water over the tower is 4 gal/min for each ton of cooling desired. Several charts are available with current design technology. Manufacturers supply the specifications for their towers. However, there are some important points to remember when use of a tower is being considered:

1. In tons of cooling, the tower should be rated at the same capacity as the condenser.
2. The wet-bulb temperature must be known.
3. The temperature of the water leaving the tower should be known. This would be the temperature of the water entering the condenser.

Towers present some maintenance problems. These stem primarily from the water used in the cooling system. Chemicals are employed to control the growth of bacteria and other substances. Scale in the pipes and on parts of the tower also must be controlled.

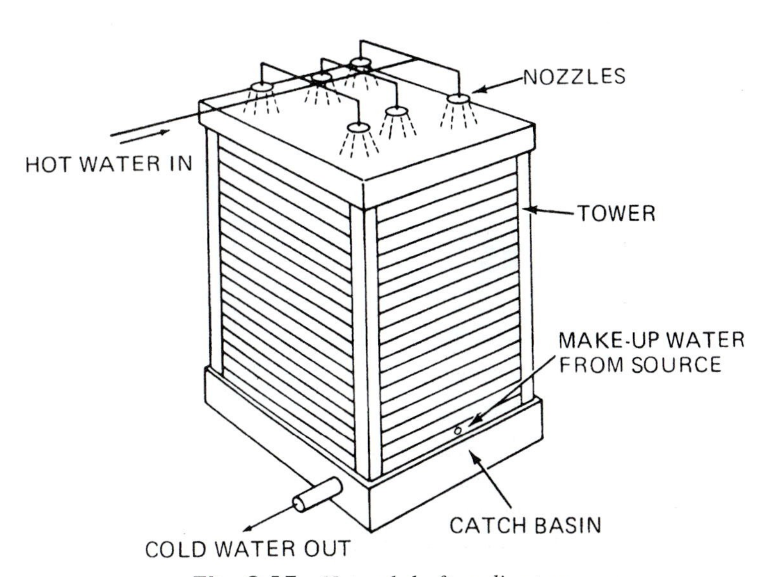

Fig. 8-15 *Natural-draft cooling tower.*

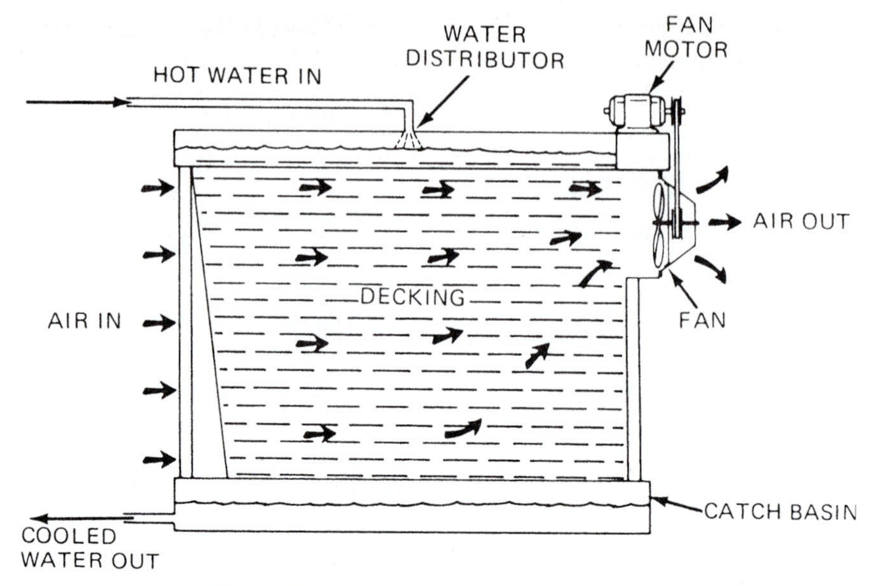

Fig. 8-16 *Small induced-draft cooling tower.*

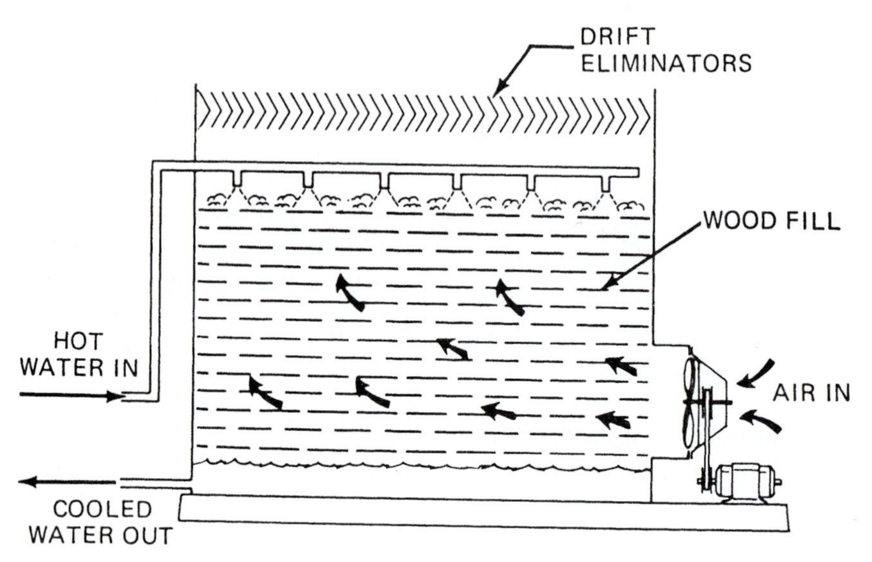

Fig. 8-17 *Forced-draft cooling tower.*

Chemicals are used for each of these controls. This problem will be discussed in the next chapter.

EVAPORATIVE CONDENSERS

The evaporative condenser is a condenser and a cooling tower combined. Figure 8-18 illustrates how the nozzles spray water over the cooling coil to cool the fluid or gas in the pipes. This is a very good water-conservation tower. In the future, this system will probably become more popular. The closed-circuit cooler should see increased use because of dwindling water supplies and more expensive treatment problems. The function of this cooler is to process the fluid in the pipes.

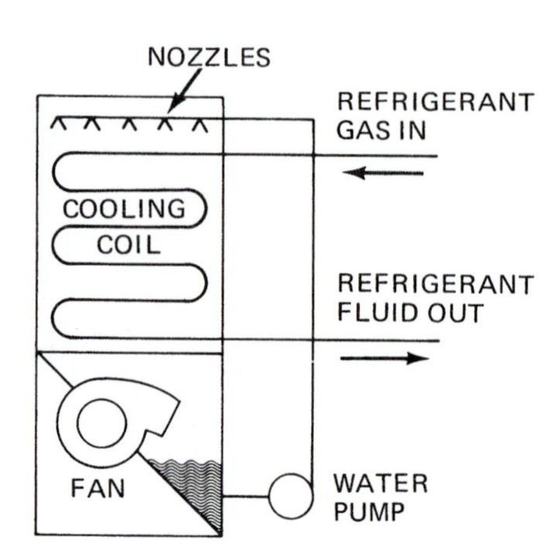

Fig. 8-18 *Evaporative cooler has no fill deck. The water-cooling process fluid directly.* (Marley)

This is a sealed contamination-free system. Instead of allowing the water to drop onto slats or other deflectors, this unit sprays the water directly onto the cooling coil.

NEW DEVELOPMENTS

All-metal towers with housing, fans, fill, piping, and structural members made of galvanized or stainless steel are now being built. Some local building codes are becoming more restrictive with respect to fire safety. Low maintenance is another factor in the use of all-metal towers.

Engineers are beginning to specify towers less subject to deterioration due to environmental conditions. Thus, all-steel or all-metal towers are called for. Already, galvanized-steel towers have made inroads into the air-conditioning and refrigeration market. Stainless-steel towers are being specified in New York City, northern New Jersey, and Los Angeles. This is primarily due to a polluted atmosphere, which can lead to early deterioration of nonmetallic towers and, in some cases, metals.

Figure 8-19 shows a no-fans design for a cooling tower. Large quantities of air are drawn into the tower by cooling water as it is injected through spray nozzles at one end of a venturi plenum. No fans are needed. Effective mixing of air and water in the plenum permits evaporative heat transfer to take place without the fill required in conventional towers.

The cooled water falls into the sump and is pumped through a cooling-water circuit to return for another cycle. The name applied to this design is Baltimore Aircoil. In 1981, towers rated at 10 to 640 tons with 30 to 1920 gal/min were standard. Using pre-strainers in the high-pressure flow has minimized the nozzle-clogging problem. There are no moving parts in the tower. This results in very low maintenance costs.

Air-cooled condensers are reaching 1000 tons in capacity. Air coolers and air condensers are quite attractive for use in refineries and natural gas compressor stations. They are also used for cooling in industry, as well as for commercial air-conditioning purposes. Figure 8-19 shows how the air-cooled condensers are used in a circuit system that is completely closed. These are very popular where there is little or no water supply.

TEMPERATURE CONVERSION

A cooling tower is a device for cooling a stream of water. Evaporating a portion of the circulating stream does this. Such cooled water may be used for many purposes, but the main concern here is its utilization as a heat sink for a refrigeration-system condenser. A number of types of cooling towers are used for industrial and commercial purposes. They are usually regarded as a necessity for large buildings or manufacturing processes. Some of these types have already been mentioned, but the following will bring you more details on the workings of cooling towers and their differences.

Cooling water concerns must be addressed for the health of those who operate and maintain the systems. There is the potential for harboring and for the growth of pathogens in the water basin or related surface. This may occur during many the summer and also during idle periods. When the temperature falls in the 70 to 120°F range, there are periods when the unit will not be operational and will sit idle. Dust from the air will settle in the water and create an organic medium for the culture of bacteria and pathogens. Algae will grow in the water—some need sunlight, others grow without.

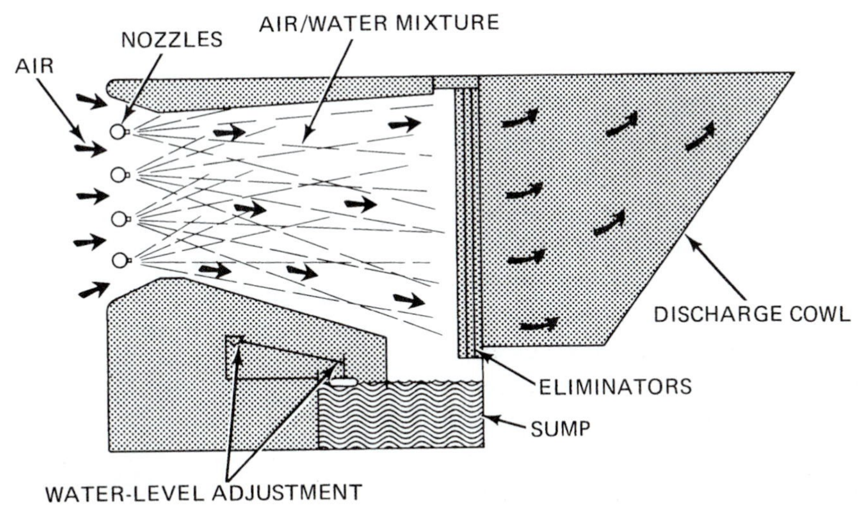

Fig. 8-19 *Cooling tower with natural-draft properties. There are no moving parts in the cooling tower.* (Marley)

Some bacteria feed on iron. The potential for pathogenic culture is there, and cooling-tower design should include some kind of filtration and/or chemical sterilization of the water.

TYPES OF TOWERS

The atmospheric type of tower does not use a mechalnical device, such as a fan, to create airflow through the tower. There are two main types of atmospherictowers—large and small. The large hyperbolic towers are equipped with "fill" since their primary applications are with electric-power plants. The steam-driven alternator has very high temperature steam to reduce to water or liquid state.

Atmospheric towers are relatively inexpensive. They are usually applied in very small sizes. They tend to be energy intensive because of the high spray pressures required. The atmospheric towers are far more affected by adverse wind conditions than are other types. Their use on systems requiring accurate, dependable cold-water temperatures is not reocmmended. See Fig. 8-20.

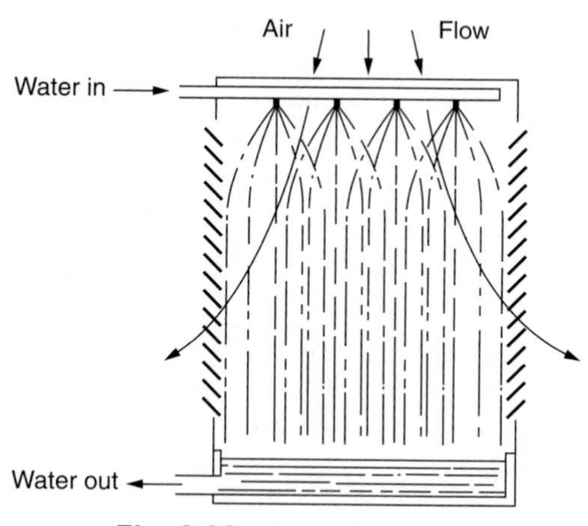

Fig. 8-20 *Atmospheric tower.*

Mechanical-draft towers, such as in Fig. 8-21, are categorized as either forced-draft towers or induced draft. In the forced-draft type the fan is located in the ambient air stream entering the tower. The air is also brought through or induced to enter the tower by a fan above in Fig. 8-22. In the later type the induced draft draws air through the tower by an induced draft.

Forced-draft towers have high air-entrance velocities and low-exit velocities. They are extremely susceptible to recirculation and are therefore considered to have less performance stability than induced-draft towers. There is concern in northern climates as the forced-draft fans located in the cold entering ambient air stream can become subject to severe icing. The resultant imbalance comes when the moving air, laden with either natural or recirculated moisture, becomes ice.

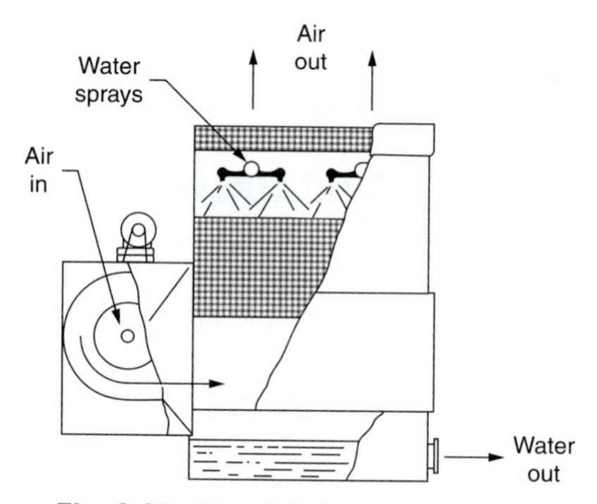

Fig. 8-21 *Forced-draft counter flow tower.*

Usually forced-draft towers are equipped with centrifugal blower-type fans. These fans require approximately twice the operating horsepower of propeller-type fans. They have the advantage of being able to operate against the high static pressures generated with ductwork. So equipped, they can be installed either indoors or within a specifically designed enclosure that provides sufficient separation between the air intake and discharge locations to minimize recirculation. See Fig. 8-23.

Crossflow Towers

Crossflow towers, as seen in Fig. 8-24, have a fill configuration through which the air flows horizontally. That means it is across the downward fall of the water. The water being cooled is delivered to hot-water inlet basins. The basins are located above the fill areas. The water is distributed to the fill by gravity through metering orifices in the basins' floor. This removes the need for a pressure-spray distribution system. And, it places the resultant gravity system in full view for maintenance.

A cooling tower is a specialized heat exchanger. See Fig. 8-25. The two fluids, air and water, are brought into direct contact with each other. This is to effect the transfer of heat. In the spray-filled tower, such as Fig. 8-25, this is accomplished by spraying a flowing mass of water into a rainlike pattern. Then an upward-moving mass flow of cool air is induced by the action of the fan.

Fluid Cooler

The fluid cooler is one of the most efficient systems for industrial and HVAC applications. See Fig. 8-26. By

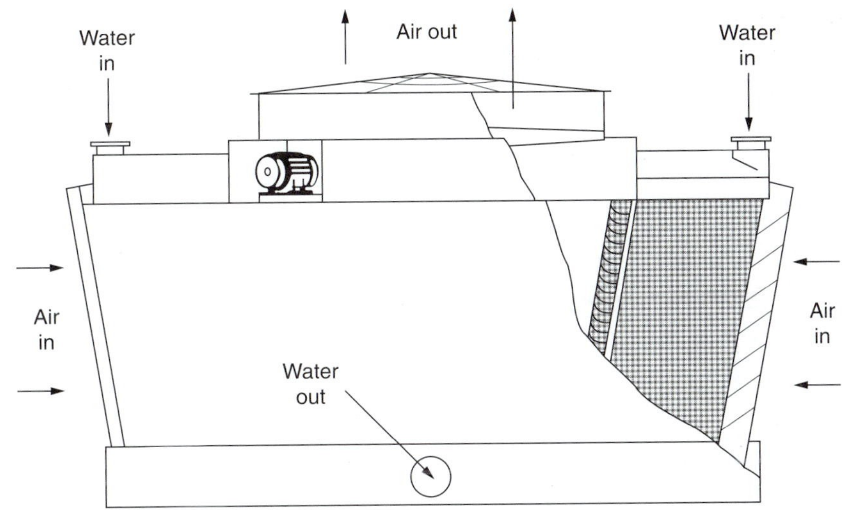

Fig. 8-22 *Induced-draft cross flow tower.*

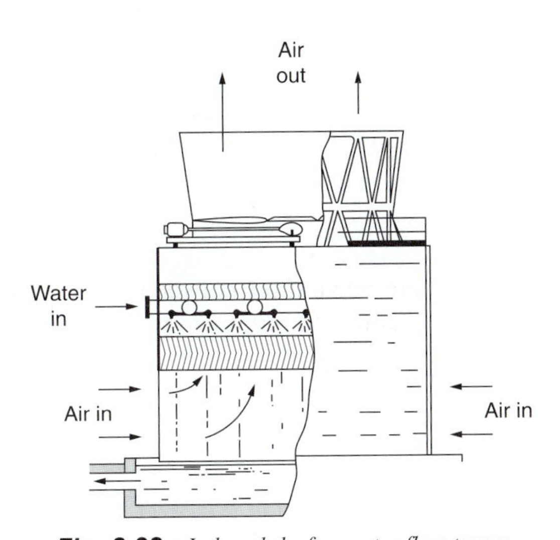

Fig. 8-23 *Induced-draft counter flow tower.*

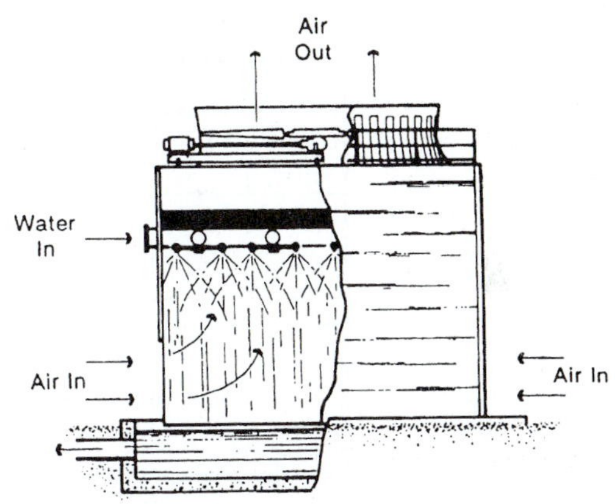

Fig. 8-25 *Spray-filled counter flow tower.*

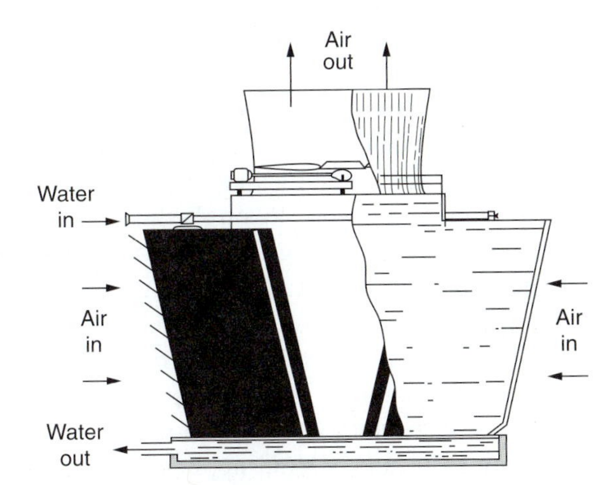

Fig. 8-24 *Double-flow cross flow tower.*

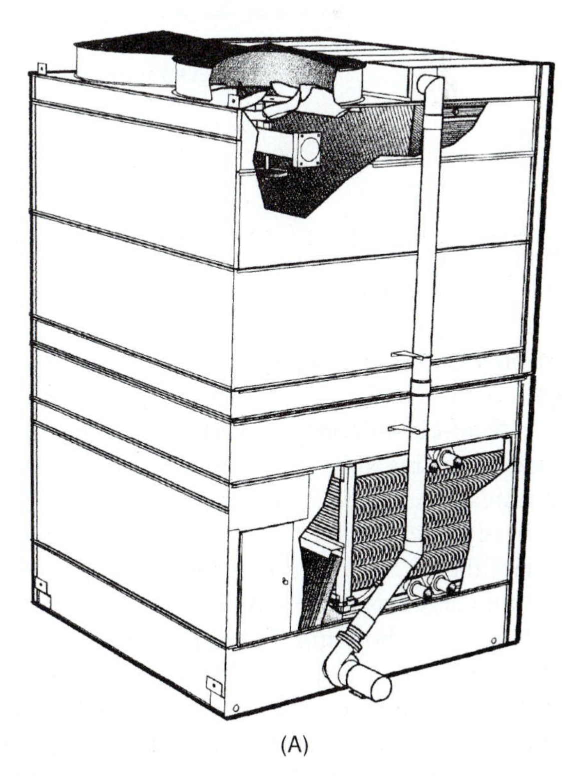

(A)

Fig. 8-26 *MH-fluid cooler. (A) Rear view. (B) Various views. (C) Front view.* (Marley)

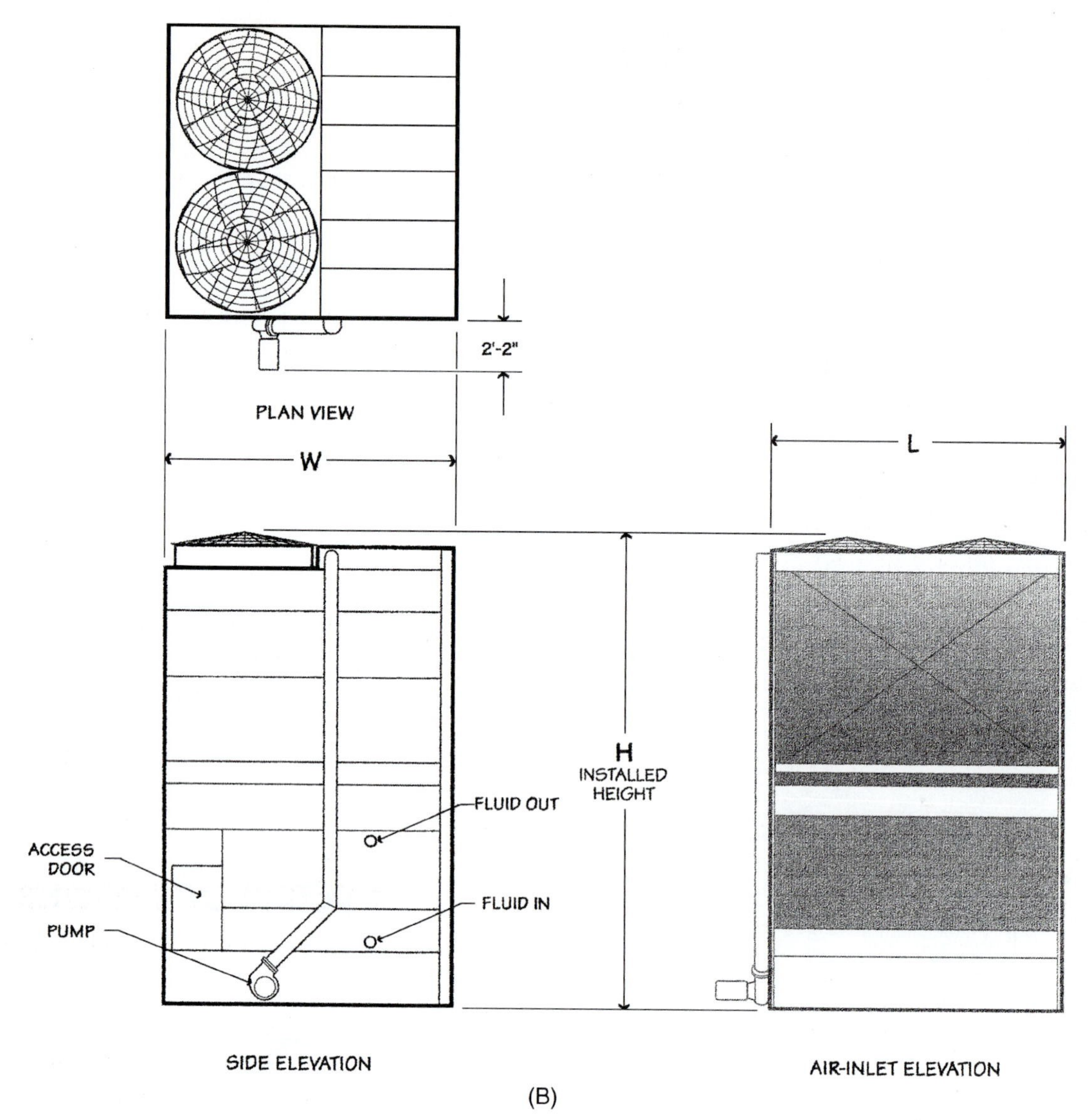

Fig. 8-26 *(Continued)*

keeping the cooling process fluid and in a clean, closed loop it combines the function of a cooling tower and heat exchanger into one system. It is possible to provide superior operational and maintenance benefits.

The fluid-cooler coil is suitable for cooling water, oils, and other fluids. It is compatible to most oils and other fluids when the carbon-steel coil in a closed, pressurized system. Each coil is constructed of continuous steel tubing, formed into a serpentine shape and welded into an assembly. See Fig. 8-27. The complete asssembly is then hot dipped in liquid tin to galvanize it after fabrication. The galvanized-steel coil has proven itself through the years. Paints and electrostatically applied coatings ca not seem to approach galvanization for inceasing coil longevity. The coils can also be made of stainless steel.

Operation of the Fluid Cooler The fluid cooler uses a mechanically induced draft, crossflow technology. And, the fill media is located above the coil. The process fluid is pumped internally through the coil. Recirculating water is cooled as it passes over the fill media, Fig. 8-28. The process fluid is thermally equalized and redistributed over the outside of the coil. A small portion of recirculating water is evaporatd by the air drawn that is passing through the coil and fill media. This cools the process fluid. The coil section rejects heat through evaporative cooling. This process uses the fresh air stream and precooled recirculating spray water. Recirculated water falls from the coil into a collection basin. From the base it is then pumped back up to be distributed over the fill media.

For industrial and HVAC applications this is an ideal type of system. The process fluid is kept in a

(C)

Fig. 8-26 *(Continued)*

Fig. 8-27 *MH-fluid cooler coil.* (Marley)

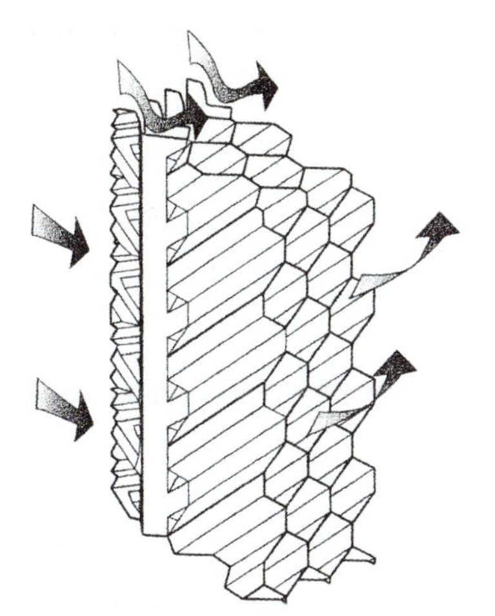

Fig. 8-28 *MH-fluid cooler fill media.* (Marley)

clean, closed loop. It combines the function of a cooling tower and heat exchanger into one system. This improves efficiency and has many maintenance benefits. The unit shown here has a capacity ranging from 100 to 650 tons in a compact enclosure. It is suitable for cooling a wide range of fluids from water and glycols, to quench oils and plating solutions.

REVIEW QUESTIONS

1. What is the purpose of a condenser in a refrigeration system?
2. List the three basic methods for cooling hot gases.
3. How does a chiller serve as a cooling system?
4. Describe the dehydrator cycle in a chiller operation.
5. What is the purpose of the solid-state module?
6. Why are cooling towers necessary?
7. How are cooling towers rated?
8. Describe the term "make-up water."
9. Why are stainless-steel towers needed?
10. What does the word *venturi* mean?

9

CHAPTER

Working with Water-Cooling Problems

PERFORMANCE OBJECTIVES

After studying this chapter, you should:

1. Be able to understand why water fit for human consumption is not necessarily acceptable for use in boilers and cooling equipment.
2. Know how to treat cooling water for corrosion, algae, slime, and fungi.
3. Know how to clean cooling towers and evaporative condensers.
4. Know how to use cleaning chemicals safely.
5. Know how to use solvents and detergents.

Three-fourths of the earth's surface is covered with water. The earth is blanketed with water vapor, which is an indispensable part of the atmosphere. Heat from the sun shining on oceans, rivers, and lakes evaporate some water into the atmosphere. Warm, moisture-laden air rises and cools. The cooling vapor condenses to form clouds. Wind currents carry clouds over landmasses where the precipitation may occur in the form of rain, snow, or sleet. Because of the sun and upper air currents, this process is repeated again and again. Pure water has no taste and no odor. Pure water, however, is actually a rarity.

All water found in oceans, rivers, lakes, streams, and wells contains various amounts of minerals picked up from the earth. Even rainwater is not completely pure. As rain falls to earth, it washes from the air various gases and solids such as oxygen, carbon dioxide, industrial gases, dust, and even bacteria. Some of this water sinks into the earth and collects in wells or forms underground streams. The remainder runs over the ground and finds its way back into various surface-water supplies.

Water is often referred to as the *universal solvent.* Water runs over and through the earth mixes with many minerals. Some of these mineral solids are dissolved or disintegrated by water.

PURE WATER

Pure water and sanitary water are the same as far as municipalities are concerned. *Pure* in this case means that the water is free from excessive quantities of germs and will not cause disease. Mineral salts or other substances in water do not have to be removed by water-treatment plants unless they affect sanitary conditions. Mineral salts are objectionable in water used for many other purposes. These uses include generating power, heating buildings, processing materials, and manufacturing. *Water fit for human consumption is not necessarily acceptable for use in boilers or cooling equipment.*

Water is used in many types of cooling systems. Heat removal is the main use of water in air-conditioning or refrigeration equipment. Typical uses include once-through condensers, open recirculating cooling systems employing cooling towers, evaporative condensers, chilled-water systems, and air washers. In evaporative condensers, once-through systems, and cooling towers, water removes heat from refrigerant and then is either wasted or cooled by partial evaporation in air. Knowledge of impurities in water used in any of these systems aids in predicting possible problems and methods of preventing them.

Cooling towers are usually remotely located; it becomes necessary to regularly inspect and clean the tower according to the manufacturer's recommendation. The few hours, each month, spent on inspecting the cooling tower and maintaining it will pay dividends. The life of a tower varies according to:

- Materials of construction
- Location within the system
- The location of the city or country

Generally, the premium materials of construction are:

- Wood
- Concrete
- Stainless steel
- Fiberglass

These units are expected to last from 20 to 30 years if properly cared for. The less-expensive units, made of galvanized steel, will operate for 8 to 20 years. Of course, tower life will vary due to the extremes of weather, number of hours used each year, and type of water treatment. It is sufficient to say that in order to get the most of the tower, cooling-tower manufacturers want to make that tower last as long as possible.

FOULING, SCALING, AND CORROSION

Fouling reduces water flow and heat transfer. Fouling can be caused by the collection of loose debris over pump-suction screens in sumps, growth of algae in sunlit areas, and slime in shade or dark sections of water systems. Material can clog pipes, or other parts of a system, after it has broken loose and been carried into the system by the water stream. *Scaling* also reduces water flow and heat transfer. The depositing of dissolved minerals on equipment surfaces causes scaling. This is particularly so in hot areas, where heat transfer is most important.

Corrosion is caused by impurities in the water. In addition to reducing water flow and heat transfer it also damages equipment. Eventually, corrosion will reduce operational efficiency. It may lead to expensive repairs or even equipment replacement.

Impurities have at least five confirmed sources. One is the earth's atmosphere. Water falling through the air, whether it be natural precipitation or water showering through a cooling tower, picks up dust, as well as oxygen and carbon dioxide. Similarly, synthetic atmospheric gases and dust affect the purity of water. Heavily industrialized areas are susceptible to such impurities being introduced into their water systems.

Decaying plant life is a source of water impurity. Decaying plants produce carbon dioxide. Other products of vegetable decay cause bad odor and taste. The by-products of plant decay provide a nutrient for slime growth.

These three sources of impurities contaminate water with material that makes it possible for water to pick up more impurities from a fourth source, *minerals*. Minerals found in the soil beneath the earth's surface are probably the major source of impurities in water. Many minerals are present in subsurface soil. They are more soluble in the presence of the impurities from the first three sources, mentioned earlier.

Industrial and municipal wastes are a fifth major source of water impurities. Municipal waste affects bacterial count. Therefore, it is of interest to health officials, but is not of primary concern from a scale or corrosion standpoint. Industrial waste, however, can add greatly to the corrosive nature of water. It can indirectly cause a higher than normal mineral content.

The correction or generation of finely divided material that has the appearance of mud or silt causes fouling. This sludge is normally composed of dirt and trash from the air. Silt is introduced with make-up water. Leaves and dust are blown in by wind and washed from the air by rain. This debris settles in sumps or other parts of cooling systems. Plant growth also causes fouling. Bacteria or algae in water will result in the formation of large masses of algae and slime. These may clog system water pipes and filters. Paper, bottles, and other trash also cause fouling.

Prevention of Scaling

There are two ways to prevent scaling.

- Eliminate or reduce the hardness minerals from the feed water. Control of factors that cause hardness salts to become less soluble is important. Hardness minerals are defined as water-soluble compounds of calcium and magnesium. Most calcium and magnesium compounds are much less soluble than are corresponding sodium compounds. By replacing the calcium and magnesium portion of these minerals with sodium, solubility of the sulfates and carbonates is improved to such a degree that scaling no longer is a problem. This is the function of the water softener.
- The second method of preventing scale is by controlling water conditions that affect the solubility of scale-forming minerals. The five factors that affect the rate of scale formation are:
    - Temperature
    - *Total dissolved solids* (TDS)
    - Hardness
    - Alkalinity
    - pH

These factors can, to some extent, be regulated by proper design and operation of water-cooled equipment. Proper temperature levels are maintained by ensuring a good water flow rate and adequate cooling in the tower. Water flow in recirculating systems should be approximately 3 gal/min/ton. Lower flow levels allow the water to remain in contact with hot surfaces of the condenser for a longer time and pick up more heat. Temperature drop across the tower should be 8 to 10°F (4.5 to 5.5°C) for a compression-refrigeration system, and 18 to 20°F (10 to 11°C) in most absorption systems. This cooling effect, due to evaporation, is dependent on tower characteristics and uncontrollable atmospheric conditions.

Airflow through the tower and the degree of water breakup are two factors that determine the amount of evaporation that will occur. Since heat energy is required for evaporation, the amount of water that is changed into vapor and lost from the system determines the amount of heat. That is, the number of Btu to be dissipated is the heat factor. One pound of water, at cooling-tower temperatures, requires 1050 Btu to be converted from liquid to vapor. Therefore, the greater the weight of water evaporated from the system, the greater the cooling effect or temperature drop across the tower.

Total dissolved solids, hardness, and alkalinity are affected by three interrelated factors. These are evaporation, makeup, and bleed or blow-down rates. Water, when it evaporates, leaves the system in a pure state, leaving behind all dissolved matter. Water volume of evaporative cooling systems is held at a relatively constant figure through the use of float valves.

Fresh make-up water brings with it dissolved material. This is added to that already left behind by the evaporated water. Theoretically, assuming that all the water leaves the system by evaporation and the system volume stays constant, the concentration of dissolved material will continue to increase indefinitely. For this reason, a bleed or blow down is used.

There is a limit to the amount of any material that can be dissolved in water. When this limit is reached, the introduction of additional material will cause either sludge or scale to form. Controlling the rate at which dissolved material is removed controls the degree to which this material is concentrated in circulating water.

Scale Identification

Scale removal depends on the chemical reaction between scale and the cleaning chemical. Scale identification is important. Of the four scales most commonly found, only carbonate is highly reactive with cleaning chemicals generally regarded as safe for use in cooling equipment. The other scales require a pretreatment that renders them more reactive. This pretreatment depends on the type of scale to be removed. Attempting to remove a problem scale without proper pretreatment can waste time and money.

Scale identification can be accomplished in one of the three ways:

- Experience
- Field-tests
- Laboratory analysis

With the exception of iron scale, which is orange, it is very difficult, if not impossible, to identify scale by appearance. Experience is gained by cleaning systems in a given area over an extended period of time. In this way, the pretreatment procedure and the amount of scale remover required to remove the type of scale most often found in this area become common knowledge. Unless radical changes in feed-water quality occur, the type of scale encountered remains fairly constant. Experience is further developed through the use of the two other methods. Figure 9-1 shows a water-analysis kit.

Field Testing

Field tests, which are quite simple to perform, determine the reactivity of scale with the cleaning solution. Adding 1 tablespoon of liquid scale remover, or 1 teaspoon of solid scale remover, to $^1/_2$ pint of water, prepares a small sample of cleaning solution. A small piece of scale is then dropped into the cleaning solution.

The reaction rate usually will determine the type of scale. The reaction between scale remover and carbonate scale results in vigorous bubbling. The scale eventually dissolves or disintegrates. However, if the scale sample is of hard or flinty composition, and little or no bubbling in the acid solution is observed, heat should be applied. Sulfate scale will dissolve at 140°F (60°C). The small-scale sample should be consumed in about an

Fig. 9-1 *Field kit for testing pH, phosphate, chromate, total hardness, calcium hardness, alkalinity, and chloride.* (Virginia Chemicals)

hour. If the scale sample contains a high percentage of silica, little or no reaction will be observed. Iron scale is easily identified by appearance. Testing with a clean solution usually is not required.

Since this identification procedure is quite elementary, and combinations of all types of scale are often encountered, it is obvious that more precise methods may be required. Such methods are most easily carried out in a laboratory. Many chemical manufacturers provide this service. Scale samples that cannot be identified in the field may be mailed to these laboratories. Here a complete breakdown and analysis of the problem scale will be performed. Detailed cleaning recommendations will be given to the sender.

Most scales are predominately carbonate, but they may also contain varying amounts of sulfate, iron, or silica. Thus, the quantities of scale remover required for cleaning should be calculated specifically for the type of scale present. The presence of sulfate, iron, or silica also affects other cleaning procedures.

Corrosion

There are four basic causes of corrosion:

- Corrosive acids
- Oxygen
- Galvanic action
- Biological organisms

Corrosive Acids Aggressive or strong acids, such as sulfurous, sulfuric, hydrochloric, and nitric, are found in most industrial areas. These acids are formed when certain industrial waste gases are washed out of the atmosphere by water showering through a cooling tower. The presence of any of these acids will cause a drop in circulating water pH. Water and carbon dioxide are found everywhere. When carbon dioxide is dissolved in water, carbonic acid is formed. This acid is less aggressive than the acids already mentioned. Because it is always present, however, serious damage to equipment can result from the corrosive effects of this acid.

Corrosion by Oxygen Corrosion by oxygen is another problem. Water that is sprayed into the air picks up oxygen. This oxygen then is carried into the system. Oxygen reacts with any iron in equipment. It forms iron oxide, which is a porous material. Flaking or blistering of oxidized metal allows corrosion of the freshly exposed metal. Blistering also restricts water flow and reduces heat transfer. Reaction rates between oxygen and iron increase rapidly as temperatures increase. Thus, the most severe corrosion takes place in hot areas of equipment with iron parts.

Oxygen also affects copper and zinc. Zinc is the outer coating of galvanized material. Here, damage is much less severe because oxidation of zinc and copper forms an inert metal oxide. This sets up a protective film between the metal and the attacking oxygen.

Galvanic Action Galvanic corrosion is the third cause of corrosion. Galvanic corrosion is basically a reaction between two different metals in electrical contact. This reaction is both electrical and chemical in nature. The following three conditions are necessary to produce galvanic action.

1. Two dissimilar metals possessing different electrochemical properties must be present.
2. An electrolyte, a solution through which an electrical current can flow, must be present.
3. An electron path to connect these two metals is also required.

Many different metals are used to fabricate air-conditioning and refrigeration systems. Copper and iron are two dissimilar metals. Add a solution containing ions, and an electrolyte is produced. Unless the two metals are placed in contact, no galvanic action will take place. A coupling is made when two dissimilar metals, such as iron and copper, are brought into contact with one another. This sets up an electrical path or a path for electron movement. This allows electrons to pass from the copper to the iron. As current leaves the iron and reenters the solution to return to the copper, corrosion of the iron takes place. Copper-iron connections are common in cooling systems.

Greater separation of metals in the galvanic series results in their increased tendency to corrode. For example, if platinum is joined with magnesium, with a proper electrolyte, then platinum would be protected and magnesium would corrode. Since they are so far apart on the scale, the corrosion would be rapid. See Table 9-1. If iron and copper are joined, we can tell by their relative positions in the series that iron would corrode, but to a lesser degree than the magnesium mentioned in the previous example. Nevertheless, corrosion would be extensive enough to be very damaging. However, if copper and silver were joined together, then the copper would corrode. Consequently, the degree of corrosion is determined by the relative positions of the two metals in the galvanic series.

Improperly grounded electrical equipment or poor insulation can also initiate or accelerate galvanic action. Stray electrical currents cause a similar type of corrosion, usually referred to as *electrolytic corrosion.* This generally results in the formation of deep pits in metal surfaces.

Table 9-1 *Galvanic Series*

Anodic (Corroded End)
Magnesium
Magnesium alloy
Zinc
Aluminum
Mild steels
Alloy steels
Wrought iron
Cast iron
Soft solders
Lead
Tin
Brass
Copper
Bronze
Copper-nickel alloys
Nickel
Silver
Gold
Platinum
Cathodic (Protected End)

Biological Organisms Another cause of corrosion is biological organisms. These are algae, slime, and fungi. Slimes thrive in complete absence of light. Some slimes cling to pipes and will actually digest iron. This localized attack results in the formation of small pits, which, over a period of time, will expand to form holes.

Other slimes live on mineral impurities, especially sulfates, in water. When doing so, they give off hydrogen-sulfide gas. The gas forms weak hydro-sulfuric acid. (Do not confuse this with strong sulfuric acid.) This acid slowly, but steadily, deteriorates pipes and other metal parts of the system. Slime and algae release oxygen into the water. Small oxygen bubbles form and cling to pipes. This oxygen may act in the same manner as a dissimilar metal and cause corrosion by galvanic action. This type of corrosion is commonly referred to as *oxygen cell* corrosion.

Algae Algae are a very primitive form of plant life. They are found almost everywhere in the world. The giant Pacific kelp are algae. Pond scums and the green matter that grows in cooling towers are also algae. Live algae range in color from yellow, red, and green to brown and gray. Like bacterial slime, they need a wet or moist environment and prefer a temperature between 40 and 80°F (4 and 27°C). Given these conditions, they will find mineral nourishment for growth in virtually any water supply.

Slime Bacteria cause slime. Slime bacteria can grow and reproduce at temperatures from well below freezing [32°F (0°C)] to the temperature of boiling water [212°F (100°C)]. However, they prefer temperatures between 40 and 80°F (4 and 27°C). They usually grow in dark places. Some types of slime also grow when exposed to light in cooling towers. The exposure of the dark-growing organisms to daylight will not necessarily stop their growth. The only condition essential to slime propagation is a wet or moist environment.

Fungi Fungi are a third biological form of corrosion. Fungi attack and destroy the cellulose fibers of wood. They cause what is known as brown rot or white rot. If fungal decay proceeds unchecked, serious structural damage will occur in a tower.

CONTROL OF ALGAE, SLIME, AND FUNGI

It is essential that a cooling system be kept free of biological growths as well as scale. Fortunately, several effective chemicals are available for controlling algae and slime. Modern algaecides and slimeicides fall into three basic groups: the chlorinated phenols (penta-chloro-phen-ates), quaternary-ammonium compounds, and various organo-metallic compounds.

A broad range of slime and algae control agents is required to meet the various conditions that exist in water-cooled equipment. Product selection is dependent on the following:

- The biological organism present
- The extent of the infestation
- The resistance of the existing growths to chemical treatment
- The type and specific location of the equipment to be treated

There is considerable difference of opinion in the trade as to how often algaecides should be added and whether "slug" or continuous feeding is the better method.

In treating heavy biological growths, remember that when these organisms die they break loose and circulate through the system. Large masses can easily block screens, strainers, and condenser tubes. Some provision should be made for preventing them from blocking internal parts of the system. The best way to do this is to remove the thick, heavy growths *before* adding treatment. The day after treatment is completed, thoroughly drain and flush the system and clean all strainers.

BACTERIA

One of the most critical areas of concern about cleanliness is bacteria-breeding grounds. The most difficult issue to deal with is stagnant water. A system's piping should be free of "dead legs," and tower flow should be maintained. When dirt accumulates in the collection basin of a tower, it provides the right combination of supplies for the creation of *Legionella* bacteria:

- Moisture
- Oxygen
- Warm water
- Food supply

These bacteria can be found in water supplies as well as around rivers and/or streams. They are contained in water droplets and can become airborne. This bacteria makes humans susceptible to it by breathing in the contaminated air. No chemicals can positively eliminate all bacteria from the water supply in a cooling tower. However, evidence exists to suggest that good maintenance along with comprehensive treatment can dramatically minimize the risk.

THE PROBLEM OF SCALE

Air conditioning or refrigeration is basically the controlled removal of heat from a specific area. The refrigerant that carries heat from the cooled space must be cooled before it can be reused. Cooling and condensation of refrigerant require the use of a cooling medium that, in many systems, is water.

There are two types of water-cooled systems. The first type uses once-through operation. The water picks up heat and is then discarded or wasted. In effect, this is 100 percent, or total, bleed. Little, if any, mineral concentration occurs. The scale that forms is due to the break down of bicarbonates by heat. These form carbonates, which are less soluble at high temperatures than at low. Such scale can be prevented through use of a treatment chemical.

The other type of water-cooled system is the type in which heat is removed from water by partial evaporation. The water is then recirculated. Water volume lost by evaporation is replaced. This type of system is more economical from the standpoint of water use. However, the concentration of dissolved minerals leads to conditions which, if not controlled and chemically treated, may result in heavy scale formation.

Evaporative Systems

One method of operating evaporative recirculating systems involves 100 percent evaporation of the water with no bleed. This, of course, causes excessive mineral concentration. Without a bleed on the system, water conditions will soon exceed the capability of any treatment chemical. A second method of operation employs a high bleed rate without chemical treatment. Scale will form and water is wasted. The third method is the reuse of water, with a bleed to control concentration of scale-forming minerals. Thus, by the addition of minimum amounts of chemical treatment, good water economy can be realized. This last approach is the most logical and least expensive. Figure 9-2 shows how connections for bleed lines are made on evaporative condensers and cooling towers.

Scale Formation

Scale is formed as a direct result of mineral insolubility. This in turn, is a direct function of temperature, hardness, alkalinity, pH, and total-dissolved solids. Generally speaking, as these factors increase, solubility or stability of scale-forming minerals decreases. Unlike most minerals, scale-forming salts are less soluble at high temperatures. For this reason, scale forms most rapidly on heat-exchanger surfaces.

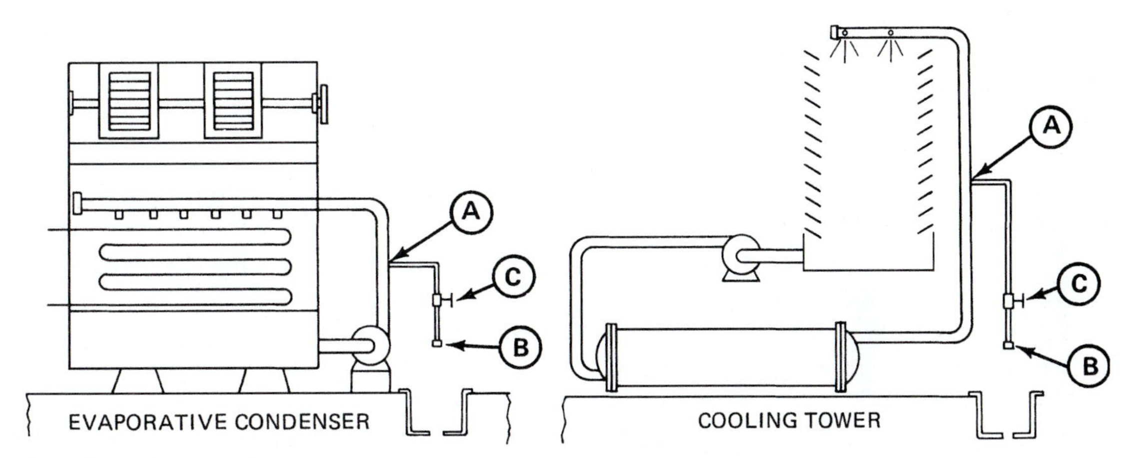

Fig. 9-2 *Connections for bleed lines for evaporative condensers and cooling towers.* (Virginia Chemical)

HOW TO CLEAN COOLING TOWERS AND EVAPORATIVE CONDENSERS

To clean cooling towers and evaporative condensers, first determine the amount of water in the system. This is done by the following procedures.

Determining the Amount of Water in the Sump

Measure the length, width, and water depth in feet. See Fig. 9-3.

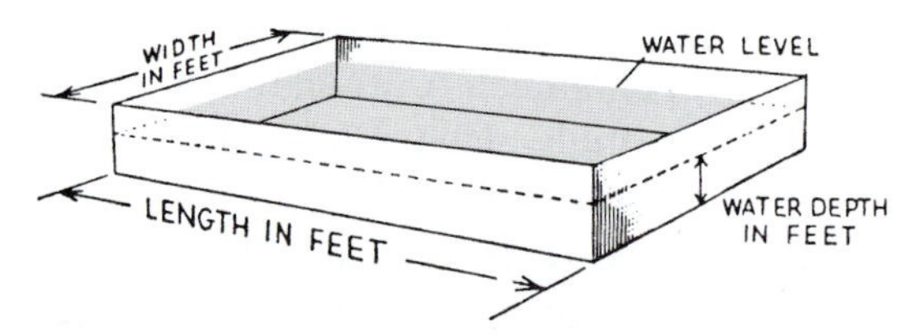

Fig. 9-3 *Method of calculating the amount of water in a rectangular tank or sump.*

Use the following formula—length × width × water depth × 7.5 = gallons of water in the sump.

Example: A sump is 5 ft long and 4 ft wide, with a water depth of 6 in.

Solution:

$$5 \times 4 \times 0.5 \times 7.5 = 75 \text{ gal of water in the sump}$$

Determining the Amount of Water In the Tank

1. Measure the diameter of the tank and the depth of the water in feet. See Fig. 9-4.
2. Use the following formula: diameter^2 × water depth × 6 = gallons of water in tank.

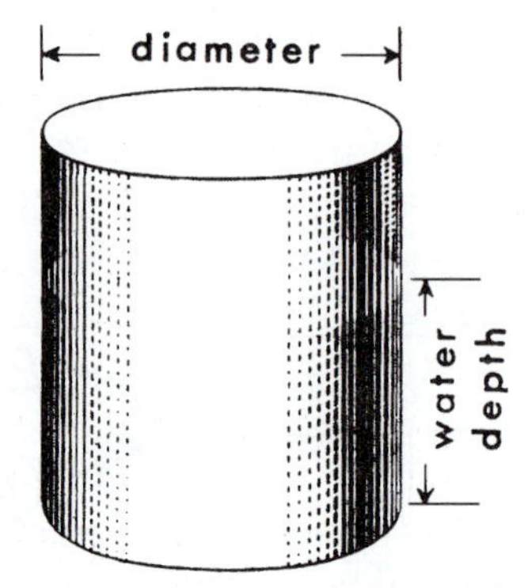

Fig. 9-4 *Calculating the amount of water in a round tank.* (Virginia Chemicals)

Example: A tank has a diameter of 3 ft and the water is 3 ft deep.

Solution:

$$3^2 \times 3 \times 6, \text{ or } 9 \times 3 \times 6 = 162 \text{ gal of water in the tank}$$

Total Water Volume

The preceding two formulas will give you the water volume in either the tank or sump. Each is figured separately since they are both part of the system's circulating water supply. There is also water in the connecting lines. These lines must be measured for total footage. Once you find the pipe footage connecting the system you can figure its volume of water too. Simply take 10 percent of the water volume in the sump for each 50 ft of pipe run. This is added to the water in the sump and the water in the tank to find the *total* system water volume.

For example, a system has 75 gal of water in the sump and 162 gal of water in the tank. The system has 160 ft of pipe.

$$75 \text{ gal} + 162 \text{ gal} = 237 \text{ gal}$$

$$160 \text{ ft} \div 50 \text{ ft} = 3.2 \text{ ft}$$

$$75 \text{ gal in sump} \div 10 = 7.5 \text{ gal for every 50 ft in the pipes}$$

$$7.5 \text{ gal} \times 3.2 = 24 \text{ gal in the total-pipe system}$$

237 gal (in tank and sump) + 24 gal (in pipes) = 261 gal in the total system. This is the amount of water that must be treated to keep the system operating properly.

Now that you have determined the volume of water in the system, you can calculate the amount of chemicals needed.

1. Drain the sump. Flush out, or remove manually, all loose sludge and dirt. This is important because they waste the chemicals.
2. Close the bleed line and refill the sump with fresh water to the lowest level at which the circulating pump will operate. See Fig. 9-5.
3. Calculate the total gallons of water in the system. Next, while the water is circulating, add starting amounts of either chemical slowly, as follows.

> NOTE: These amounts are for *hot water systems.*

For cold water systems, see the section "Chilled Water Systems."

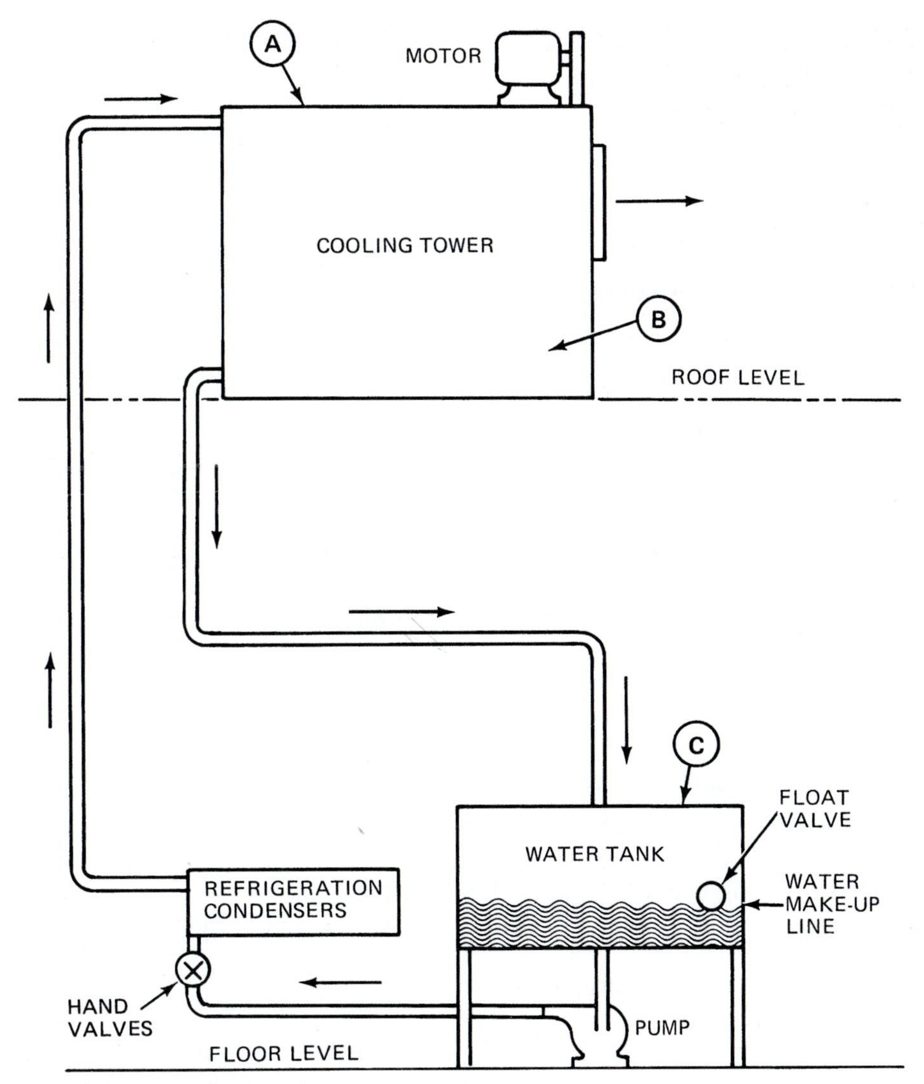

Fig. 9-5 *Forced- and natural-draft towers. (A) Water-tower distribution plate. (B) Sump. (C) Water tank.* (Virginia Chemicals)

- Solid-scale remover: 5 lb per 10 gal of water.
- Regular liquid-scale remover: 1 gal per 15 gal of water.
- Concentrated liquid-scale remover: 1 gal per 20 gal of water.

Refer to Fig. 9-5. The scale removers can be introduced at the water-tower distribution plate (A), the sump (B), or the water tank (C). Convenience is the keyword here. The preferred addition point is directly into the pump-suction area.

When using liquid-scale remover, add 1 ampoule of antifoam reagent per gallon of chemical. This will usually prevent excessive foaming if added before the scale remover. Extra antifoam is available in 1-pint bottles. When using the solid-scale remover, stir the crystals in a plastic pail or drum until completely dissolved. Then pour slowly as a liquid. Loose crystals, if allowed to fall to the bottom of the sump, will not dissolve without much stirring. If not dissolved, they might damage the bottom of the sump.

Figure 9-6 shows how to prepare the crystals in the drum. Use a 55-gal drum. Install a drain or spigot about 6 to 8 in. from the bottom. Set the drum in an upright position. Fill the drum with fresh water within 6 in. from the top; preferably warm water at about 80°F (27°C). Since the fine particles of the water-treatment crystals are quite irritating to the nose and eyes, immerse each plastic bag in the water. Cut the bag below the surface of the water. See Fig. 9-6. Stir the crystals until dissolved. About 6 to 8 lb of crystals will dissolve in each gallon of water.

Drain or pump this strong solution into the system. Repeat this procedure until the required weight of the chemical has been added in concentrated solution. Then, add fresh water to fill the system. The treatment should be repeated once each year for best results.

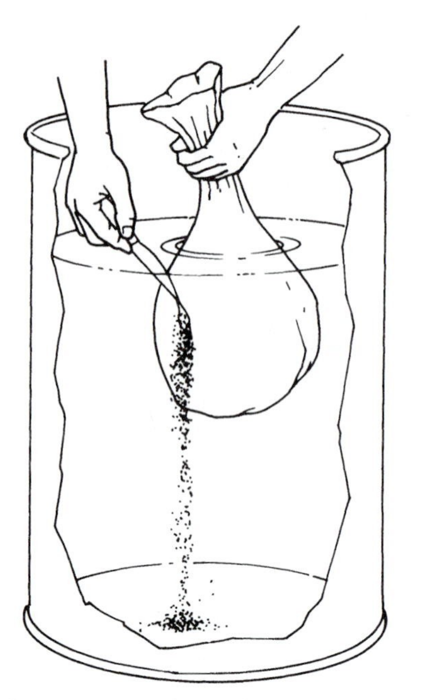

Fig. 9-6 *Preparing crystals in a drum or tank.* (Virginia Chemicals)

If make-up water is needed during the year, be sure to treat this water also at the rate of 6 lb per 100 gal of water added.

Chilled Water Systems

For chilled water systems, follow the instructions just outlined for hot water systems. However, use only 3 lb of circulating water-treatment solution for each 100 gal of water in the system. This treatment solution should be compatible with antifreeze solutions.

Feed through a Bypass Feeder For easy feeding of initial and repeat doses of water-treatment solution, install a crystal feeder in a bypass line. The crystals will dissolve as water flows through the feeder. See Fig. 9-7.

Install the feeder in either a by-pass or in-line arrangement, depending upon the application. Place the feeder on a solid, level floor, or foundation. Connection with standard pipe unions is recommended.

Connecting pipe threads should be carefully cut and cleaned to remove all burrs or metal fragments. Apply a good grade of pipe dope. Use the dope liberally. Always install valves in the inlet and the outlet lines. Install a drain line with the valve in the bottom (inlet) line.

Before opening the feeder, always close the inlet and the outlet valves. Open the drain valve to relieve the pressure and drain as much water as necessary. When adding crystals or chemicals molded in shapes (balls, briquettes, and the like), it is advisable to have the feeder about one-half filled with water.

If stirring in the feeder is necessary, use only a soft wood. Stir gently to avoid damaging the epoxy lining.

Fill the feeder to the level above the outlet line. Coat the top opening, the gasket, and the locking grooves of the cap with petroleum jelly or a heavier lubricant.

Open the outlet valve fully. Then slowly open the inlet valve. If throttled flow is desired for control of treatment feed rate, throttle with inlet valve only.

Normal Operation Once the chemicals have been properly introduced, operate the system in the normal manner. Check the scale remover strength in the sump

CLOSED-SYSTEM INSTALLATIONS

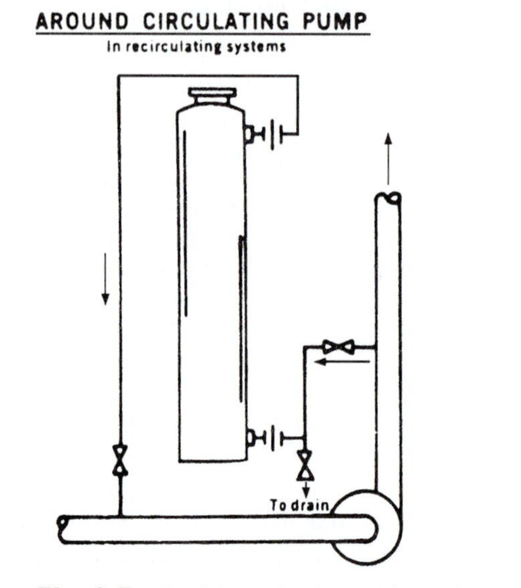

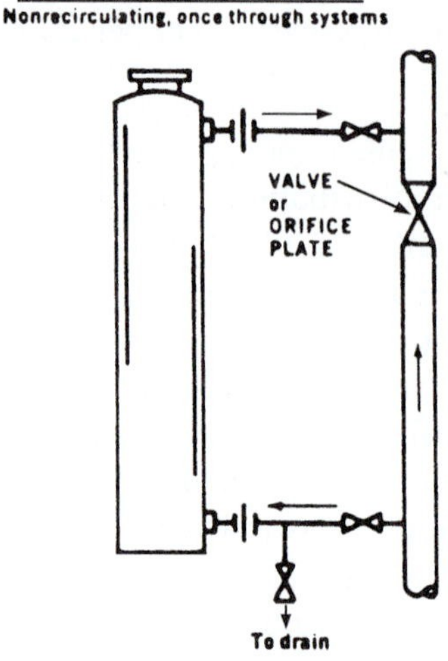

Fig. 9-7 *Feed through a bypass feeder.* (Virginia Chemicals)

by observing the color of the solution when using the solid-scale remover or using test papers. When chemical removers are used, a green solution indicates a very strong cleaner. A blue solution indicates normal cleaning strength. A purple solution indicates that more cleaner is needed. If necessary, dip a sample of the sump solution in a glass to aid color check. Check with the maker of the chemicals and their suggested color chart for accurate work.

If, for instance, Virginia Chemicals scale remover is used, in either solid or liquid form, use test papers to check for proper mixture and solution strength. Red test paper indicates there is enough cleaner. Inspection of the evaporative condenser tubes or lowering of head pressure to normal will indicate when the unit is clean. With shell and tube condensers, inspection of the inside of the water outlet pipe of the condenser will indicate the amount of scale in the unit.

After scale removal is completed, drain the spent solution to the sewer. Thoroughly rinse out the system with at least two fillings of water. Do not drain spent solutions to lawns or near valuable plants. The solution will cause plant damage, just as will any other strong salt solution. Do not drain to a septic tank. Refill the sump with fresh water and resume normal operation.

HOW TO CLEAN SHELL (TUBE OR COIL) CONDENSERS

Isolate the condenser to be cleaned from the cooling-tower system by an appropriate valve arrangement or by disconnecting the condenser piping. Pump in at the lowest point of the condenser. Venting the high points with tubing returning to the solution drum is necessary in some units to assure complete liquid filling of the waterside.

As shown in Fig. 9-8, start circulating from a plastic pail or drum the minimum volume of water necessary to maintain circulation. After adding antifoam reagent or solid-scale remover, slowly add liquid-scale remover until the test strips indicate the proper strength for cleaning. Test frequently and observe the sputtering in the foam caused by carbon dioxide in the return line. Add scale remover as necessary to maintain strength. Never add more than 1 lb of solid-scale remover per gallon of solution. Most condensers with moderate amounts of carbonate scale can be cleaned in about 1 h. Circulation for 30 to 40 min without having to add cleaner to maintain cleaning strength usually indicates that action has stopped and that the condenser is clean.

Empty and flush the condenser after cleaning with at least two complete fillings of water. Reconnect the condenser in the line.

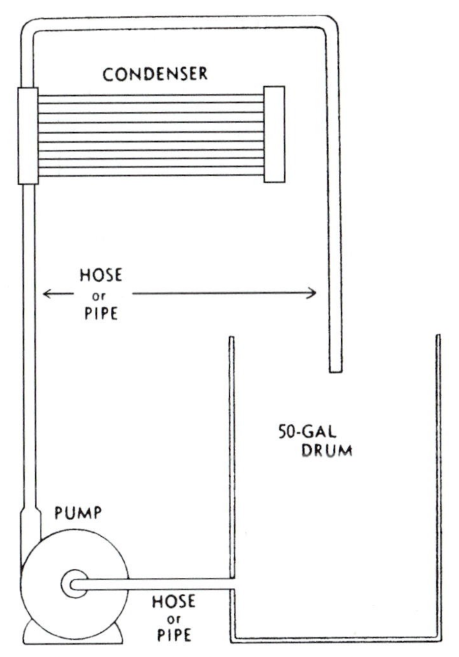

Fig. 9-8 *Cleansing a shell (tube or coil) condenser.* (Virginia Chemicals)

If a condenser is completely clogged with scale, it is sometimes possible to open a passageway for the cleaning solution by using the standpipe method, as shown in Fig. 9-9. Enough scale-remover liquid is mixed with an equal volume of water to fill the two vertical pipes to a level slightly above the condenser.

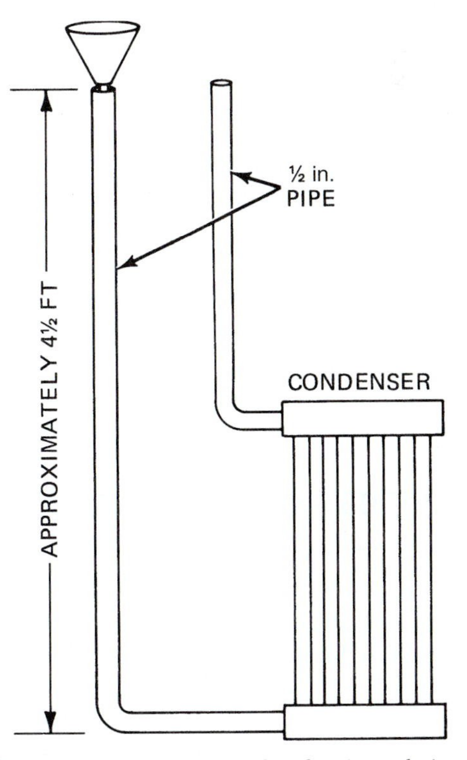

Fig. 9-9 *Opening a passageway for cleaning solution by using the stand-pipe method.*

Some foaming will result from the action of the cleaner solution on the scale. Thus, some protective measures should be taken to prevent foam from injuring surrounding objects. The antifoam reagent supplied with each package will help control this nuisance. When the cleaning operation has been completed drain the spent solution to the sewer. Rinse the condenser with at least two fillings of fresh water.

SAFETY

Most areas of the tower must be inspected for safe working and operating conditions. A number of items should be inspected yearly and repaired immediately in order to guarantee the safety of maintenance personnel.

- Scale remover contains acid and can cause skin irritation. Avoid contact with your eyes, skin, and clothing. In case of contact, flush the skin or eyes with plenty of water for at least 15 min. If the eyes are affected, get medical attention.
- Keep scale remover and other chemicals out of the reach of children.
- Do not drain the spent solution to the roof or to a septic tank. Always drain the spent solution in an environmentally safe way, *not to the storm sewer.*
- Safety is all-important. All chemicals, especially acids, should be treated with great respect and handled with care. Rubber gloves, acid-proof coveralls, and safety goggles should be worn when working with chemicals. Cleaning a system through the tower, although easier and faster than some of the other methods, presents one unique hazard—*wind drift.*

Wind drift, even with the tower fan off, is a definite possibility. Wind drift will carry tiny droplets of acid that can burn eyes and skin. These acid droplets will also damage automobile finishes and buildings. Should cleaning solution contact any part of the person, it should be washed off immediately with soap and water.

Using forethought and reasonable precaution can prevent grief and expense.

SOLVENTS AND DETERGENTS

There are several uses for solvents and detergents in the ordinary maintenance schedule of air-cooled fin coil condensers, evaporator coils, permanent-type air filters, and fan blades. In most instances, a high-pressure spray washer is used to clean the equipment with detergent. Then a high-pressure spray rinse is used to clean the unit being scrubbed. The pump is usually rated at 2 gal/min at 500 lb/in.^{2} of pressure. The main function is to remove dirt and grease from fans and cooling surfaces. It takes about 10 to 15 min for the cleaning solution to do its job. It is then rinsed with clean water.

Using the dipping method cleans permanent-type filters. Prepare a cleaning solution-one part detergent to one part water. Use this solution as a bath in which the filters may be immersed briefly. After dipping, set the filter aside for 10 to 15 min. Flush with a stream of water. If water is not available, good results may be obtained by brisk agitation in a tank filled with fresh water.

When draining the solution used for cleaning purposes, be sure to follow the local codes on the use of the storm sewers for disposal purposes. Proper disposal of the spent solution is critical for legal operation of this type of air-conditioning unit.

Another more recent requirement is the use of asbestos in the construction of the tower fill. If discovered when inspections are conducted, make sure it is replaced with the latest materials. The older towers can become more efficient with newer fill of more modern design.

Cooling tower manufacturers design their units for:

- A given performance standard
- Conditions such as:
  - The type of chiller used
  - Ambient temperatures
  - Location
  - Specifications

As a system ages, it may lose efficiency. The cleanliness of the tower and its components are crucial to the success of the system. An unattended cold-water temperature will rise. This will send warmer water to the chiller. When the chiller kicks out on high head pressure, the system may shut down. Certain precautions should be taken to prevent shutdown from occurring.

REVIEW QUESTIONS

1. Why is rainwater not pure?
2. Why is water that is fit for human consumption not usable in boilers and cooling equipment?
3. How can scaling be prevented?
4. List some aggressive or strong acids.
5. In what way does oxygen cause corrosion?
6. What is meant by galvanic action?
7. What are algae?
8. What is slime?

9. What damage does fungi cause?
10. How are shell condensers cleaned?
11. What safety precautions should be taken when using scale remover?
12. How does wind drift affect tower operation?
13. How is detergent used to clean condensers and evaporator coils?
14. How are permanent-type filters cleaned?
15. How do you dispose off spent chemicals?

10
CHAPTER

Evaporators

PERFORMANCE OBJECTIVES

After studying this chapter you should:

1. Be able to identify various types of evaporators.
2. Know how a shell-and-tube chiller operates.
3. Know how hot-gas defrost of ammonia evaporators controls operate.
4. Know how various types of evaporators operate.
5. Know the value of control valves in the proper operation of evaporators.
6. Know how to troubleshoot a differential pressure-relief regulator.
7. Know how the differential pressure-relief regulator works.

The evaporator removes heat from the space being cooled. As the air is cooled, it condenses water vapor. This must be drained. If the water condensing on the evaporator coil freezes when the temperature is below 32°F (0°C), the refrigerator or freezer must work harder. Frozen water or ice acts as an insulator. It reduces the efficiency of the evaporator. When evaporators are operated below 32°F, they must be defrosted periodically. This eliminates frost buildup on the coils or the evaporator plates.

There are several types of evaporators. The *coiled evaporator is* used in warehouses for refrigerating large areas. The *fin evaporator is* used in the air-conditioning system that is part of the furnace in a house. See Fig. 10-1. The finned evaporator has a fan that blows air over its thin metal surfaces. *Plate evaporators* use flat surfaces for their cooling surface. See Fig. 10-2. They are commonly used in freezers. If the object to be cooled or frozen is placed directly in contact with the evaporator plate, the cold is transferred more efficiently.

Figure 10-3 shows a home refrigerator cooling system. Note the evaporator.

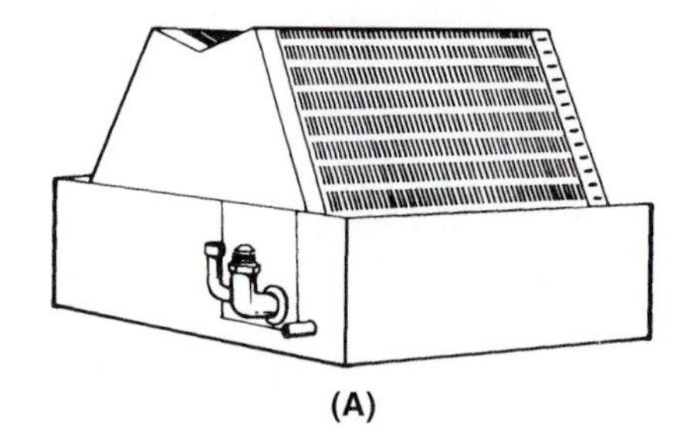

(A)

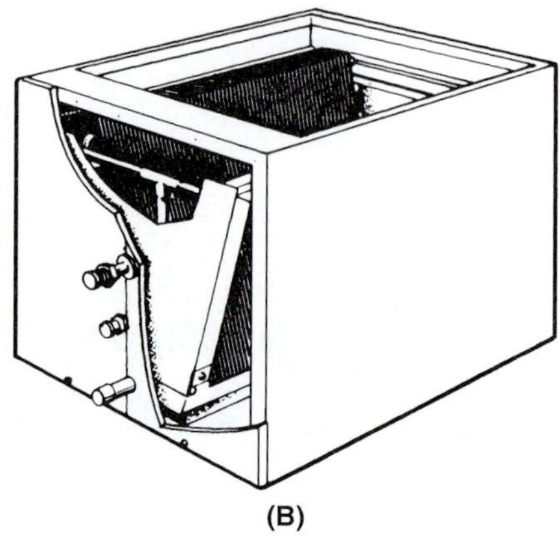

(B)

Fig. 10-1 *Evaporators. (A) Evaporator used in home air-conditioning systems where the unit is placed in the bonnet of the hot-air furnace. (B) Slanted evaporator used in home air conditioning.* (Lennex)

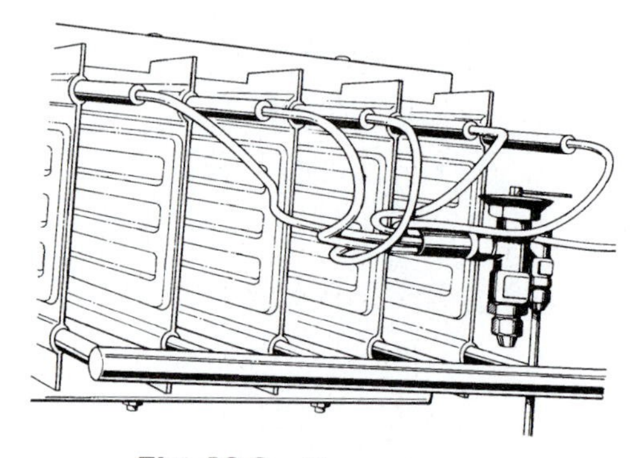

Fig. 10-2 *Plate evaporator.*

COILED EVAPORATOR

Evaporator coils on air-conditioning units fall into two categories:

- *Finned-tube coil.* The finned-tube coil is placed in the air stream of the unit. Refrigerant vaporizes in it. The refrigerant in the tubes and the air flowing around the fins attached to the tubes draw heat from the air. This is commonly referred to as a direct expansion cooling system. See Fig. 10-4.
- *Shell-and-tube chiller.* Shell-and-tube units are used to chill water for air-cooling purposes. Usually, the refrigerant is to tubes mounted inside a tank or shell containing the water or liquid to be cooled. The refrigerant in the tubes draws the heat through the tube wall and from the liquid as it flows around the tubes in the shell. This system can be reversed. Thus, the water would be in the tubes and the refrigerant would be in the tank. As the gas passes through the tank over the tubes, it would draw the heat from the water in the tubes. See Fig. 10-5.

Figure 10-5 shows how K-12 is used in a standard vapor-compression refrigeration cycle. System water for air conditioning and other uses is cooled as it flows through the evaporator tubes. Heat is transferred from the water to the low-temperature, low-pressure refrigerant. The heat removed from the water causes the refrigerant to evaporate. The refrigerant vapor is drawn into

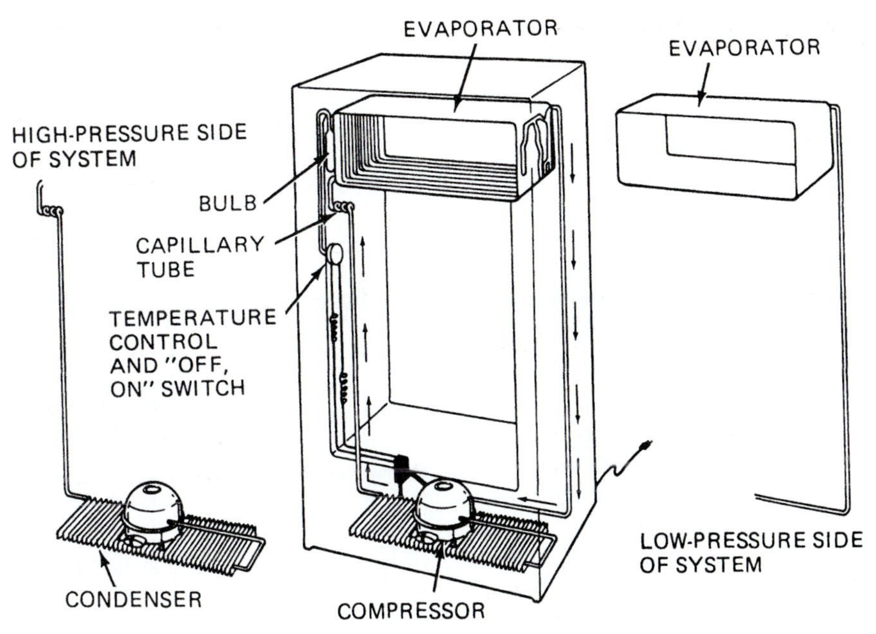

Fig. 10-3 *A home refrigerator's cooling system.*

Fig. 10-4 *Finned-coil evaporator.* (Johnson Controls)

the first stage of the compressor at a rate controlled by the size of the guide-vane opening. The first stage of the compressor raises the temperature and pressure of the vapor. This vapor, plus vapor from the flash economizer, flows into the second stage of the compressor. There, the saturation temperature of the refrigerant is raised above that of the condenser water.

This vapor mixture is discharged directly into the condenser. There, relatively cool condenser water removes heat from the vapor, causing it to condense again to liquid. The heated water leaves the system, returning to a cooling tower or other heat-rejection device.

A thermal economizer in the bottom section of the condenser brings warm condensed refrigerant into contact with the inlet water tubes. These are the coldest water tubes. They may hold water with a temperature as low as 55°F (13°C). This subcools the refrigerant so that when it moves on in the cycle, it has greater cooling potential. This improves cycle efficiency and reduces power per ton requirements. The liquefied refrigerant leaves the condenser through a plate-type control. It flows into the flash economizer or utility vessel. Here, the normal flashing of part of the refrigerant into vapor cools the remaining refrigerant. This flash vapor is diverted directly to the second stage of the compressor. Thus, it does not need to be pumped through the full compression cycle. The net effect of the flash economizer is energy savings and lower operating costs. A second plate-type control meters the flow of liquid refrigerant from the utility vessel back to the cooler, where the cycle begins again. See Fig. 10-6.

APPLICATION OF CONTROLS FOR HOT-GAS DEFROST OF AMMONIA EVAPORATORS

To defrost ammonia evaporators, it is sometimes necessary to check the plumbing arrangement and the valves used to accomplish the task. To enable hot-gas defrost systems to operate successfully, several factors must be considered. There must be an adequate supply of hot gas. The gas should be at a minimum of 100 psig. The defrost cycle should be accurately timed. Condensate removal or storage must be provided. An automatic suction

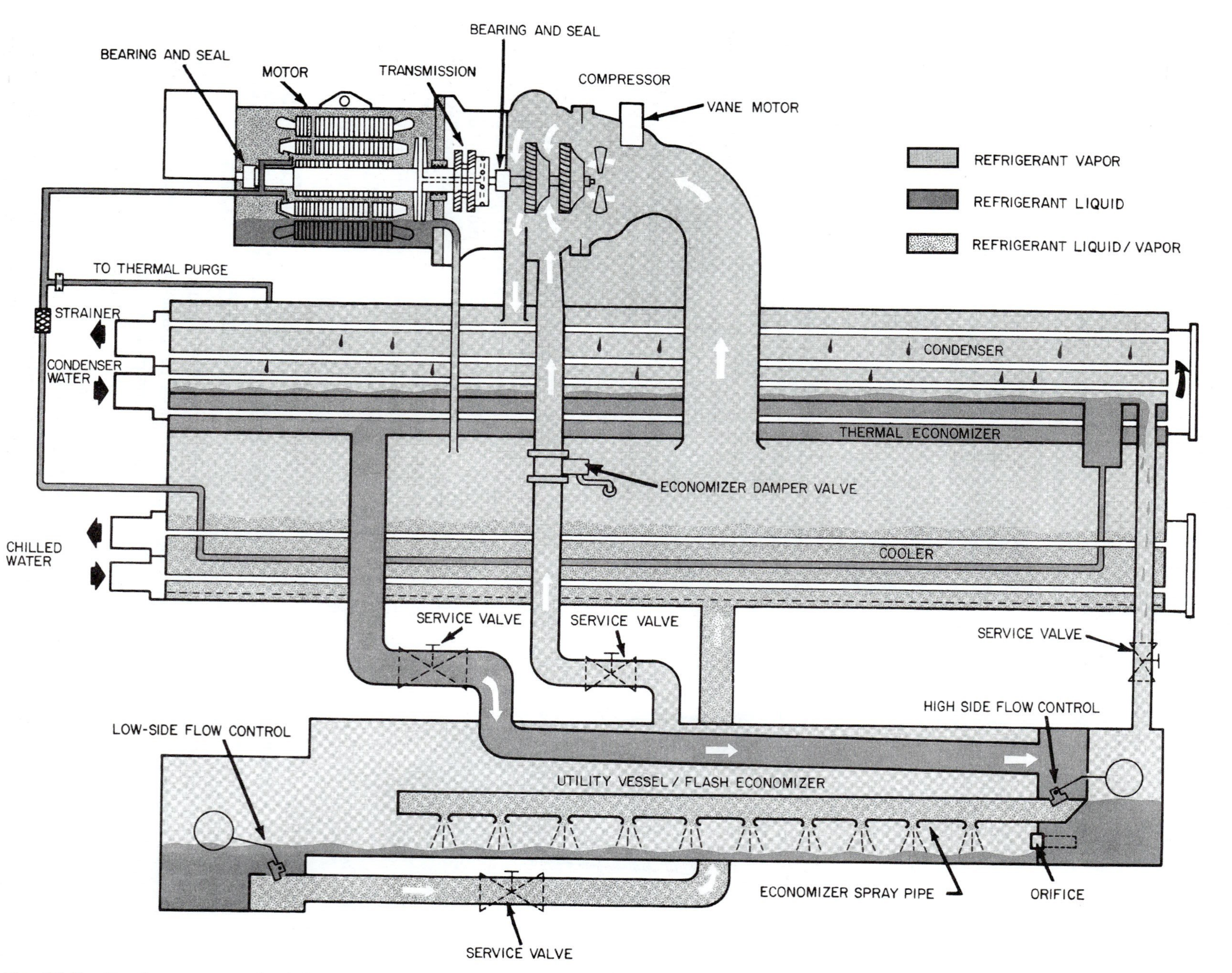

Fig. 10-5 *Complete operation of a shell-and-tube chiller.* (Carrier Corporation)

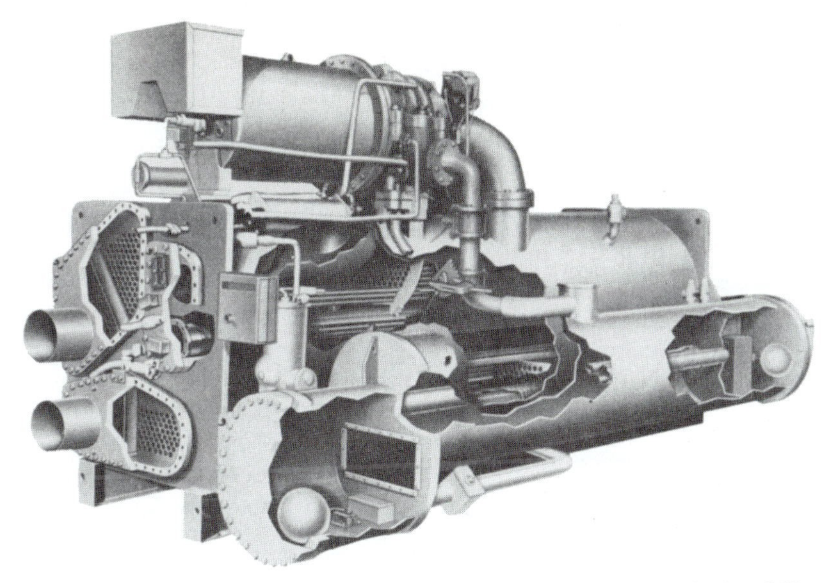

Fig. 10-6 *Cutaway view of the chiller portion of the shell-and-tube chiller shown in Fig. 10-5.*

accumulator or heat reservoir should be used to protect compressors from liquid-refrigerant slugs if surge drums or other evaporators are not adequate to handle the excess gas and condensates. See Fig. 10-7.

Controls must be used to direct and regulate the pressure and flow of ammonia and hot gas during refrigeration and defrost cycles.

Direct-Expansion Systems

Figure 10-7 shows a high-temperature system [above 32°F (0°C)] with no drip-pan defrost. During the normal cooling cycle, controlled by a thermostat, the room temperature may rise above the high setting of the thermostat. This indicates a need for refrigeration. The liquid solenoid (valve A), pilot solenoid (valve B), and the dual-pressure regulator (valve D) open, allowing refrigerant to flow. When solenoid (valve D) is energized. The low-pressure adjusting bonnet controls the regulator. The regulator maintains the predetermined suction pressure in the evaporator.

When the room temperature reaches the low setting on the thermostat, there is no longer a need for refrigeration. At this time, solenoid valve A and solenoid valve D close and remain closed until further refrigeration is required.

The hot-gas solenoid (valve C) remains closed during the normal refrigeration cycle. When the three-position selector switch is turned to DEFROST, liquid-solenoid valve A and valve D with a built-in pilot solenoid close. This allows valve D to operate as a defrost pressure regulator on the high setting. The hot-gas solenoid (valve C) opens to allow hot gas to enter the evaporator. When the defrost is complete, the system is switched back to the normal cooling cycle.

The system may be made completely automatic by replacing the manual switch with an electric time clock. Table 10-1 shows the valve sizes needed for this system.

Valves Used in Direct-Expansion Systems The pilot-solenoid valve (B) is a $^{1}/_{8}$ in. ported-solenoid valve that is direct- operated and suitable as a liquid, suction, hot gas, or pilot valve at pressures up to 300 lb.

Solenoid valve A is a one-piston, pilot-operated valve suitable for suction, liquid, or gas lines at pressures up to 300 lb. It is available with a $^{9}/_{16}$ or $^{3}/_{4}$ in. port.

Solenoid valve C is a rugged, pilot-operated, two-piston valve with spring return for positive closing under the most adverse conditions. It is used for compressor unloader, and for liquid, and hot-gas applications.

The dual-pressure regulator valve (D) is designed to operate at two predetermined pressures without resetting or adjustment. By merely opening and closing a pilot solenoid, it is capable of maintaining either the low- or high-pressure setting.

Figure 10-8 shows a pilot-light assembly. It is placed on valves when it is essential to know their condition for troubleshooting procedures.

A low-temperature defrost system with water being used to defrost the drain pan is shown in Fig. 10-9.

Cooling Cycle

During the normal cooling cycle, controlled by a thermostat, as room temperature rises above the high setting on the thermostate there is a need for refrigeration. Liquid solenoid (A) and the built-in pilot valve (D) open,

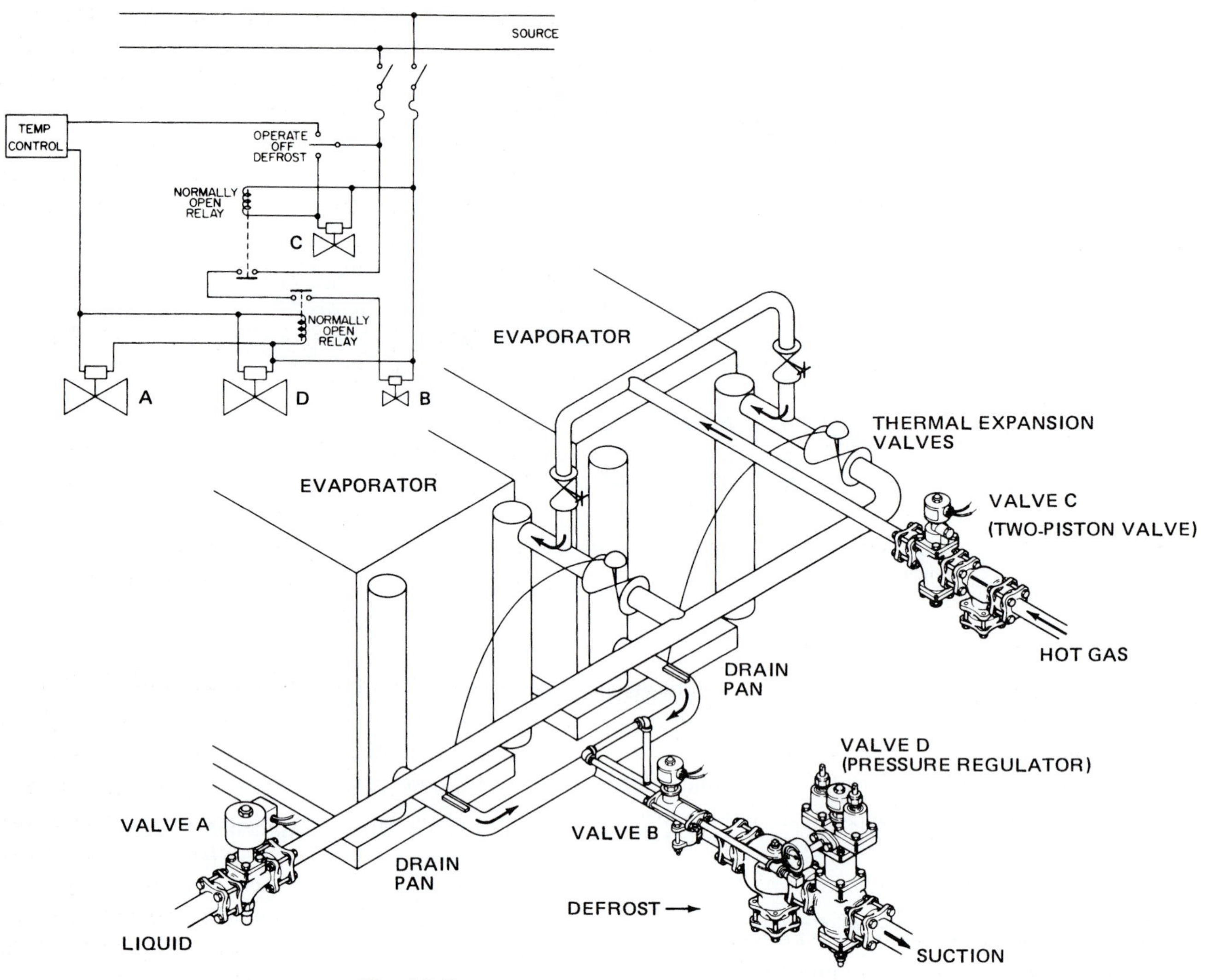

Fig. 10-7 *High-temperature defrost system.* (Hubbell)

Table 10-1 *Valve Sizing for High-Temperature System*

Tons Refrigerant	Liquid Solenoid (in.)	Hot-Gas Solenoid (in.)	Pilot Solenoid (in.)	Dual-Pressure Regulator (in.)
3	1/2	1/2	1/4	3/4
5	1/2	1/2	1/4	3/4
7	1/2	1/2	1/4	1
10	1/2	1/2	1/4	1 1/4
12	1/2	1/2	1/4	1 1/4
15	1/2	1/2	1/4	1 1/2
20	1/2	1/2	1/4	1 1/2
25	1/2	1/2	1/4	2
30	1/2	1/2	1/4	2
35	1/2	1/2	1/4	2
40	3/4	3/4	1/4	2
45	3/4	3/4	1/4	2 1/2
50	3/4	3/4	1/4	2 1/2

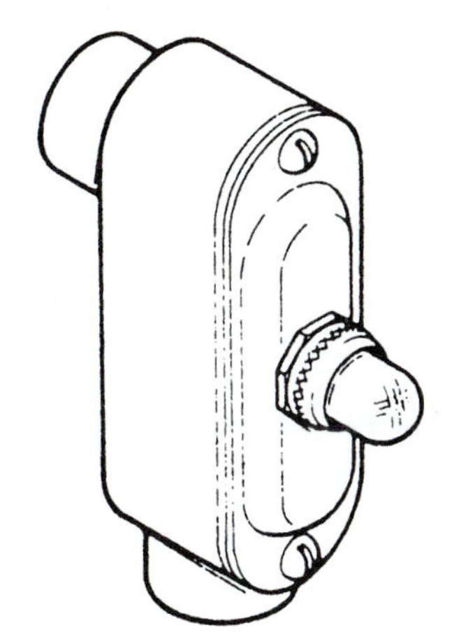

Fig. 10-8 Pilot light assembly. (Hubbell)

allowing refrigerant to flow. The opening of the built-in pilot allows the presure to bypass the sensing chamber of valve D. This forces it to remain wide open with resultant minimum pressure drop through the valve.

When the room temperature drops to the low setting on the thermostat, there is no longer a need for referation. Solenoid valve A and pilot vlave D close. They remain closed until refrigeration is again required. Hot-gas valve C and defrost-water solenoid valve E remain closed during the cooling cycle.

Defrost Cycle When the three-position selector switch is turned to defrost, solenoid valve A and pilot-solenoid valve D close as hot-gas valve C and evaporator-pilot valve B open. This allows hot gas to enter the evaporator. Valve D now acts as a back-pressure regulator, maintaining a predetermined pressure above the freezing point. After a regulated delay, preferably toward the end of the defrost cycle, the time delay allows the water solenoid to open. This causes water to spray over the evaporator, melting ice that may be lodged between coils and flushing the drain pan.

When the evaporator is defrosted, the system is returned to the cooling cycle by turning the three-position selector switch. The hot-gas solenoid (valve C) and built-in pilot valve E) close as the liquid solenoid (valve A) opens.

This system can be made completely automatic by replacing the manual selector with an electric time clock. Table 10-2 shows some of the valve sizings for the low-temperature system.

DIRECT EXPANSION WITH TOP HOT-GAS FEED

In the evaporator shown in Fig. 10-10, when the defrost cycle is initiated, the hot gas is introduced through the hot-gas solenoid valve to the manifold. It then passes through the balancing-glove valve and the pan coil to a check vlalve that prevents liquid crossover. From the check valve, hot gas is directed to the top of the evaporator. Here, it forces the refrigerant and accumulated oil from the relief regulator (valve A). This regulator has been deenergized to convert it to a relief regulator set at about 70 psig. It meters defrost condensate to the suction line and acculmulator.

DIRECT EXPANSION WITH BOTTOM HOT-GAS FEED

Compare the systems shown in Figs. 10-10 and 10-11. In the system shown in Fig. 10-11, the defrost hot gas is introduced into the bottom of the evaporator through the drain pan. The system operates similarly to that shown in Fig. 10-10. However, most of the liquid refrigerant is retained in the evaporators as defrost proceeds from the bottom to the top.

FLOODED LIQUID SYSTEMS

Figure 10-12 shows a flood-gas and liquid-leg shutoff (top hot-gas feed) system. Here, the gas-powered valve is used in both ends of the evaporator. It is a gas-powered check valve. At defrost, the normally closed type-A pilot solenoid is energized. Hot-gas pressure closes the gas-powered check valves. Hot gas flows through the solenoid, globe valves, pan coil, and in-line check valve into the top of the evaporator. Here, it purges the evaporator of fluids. The evaporator is discharged at the metered rate through valve B that has been deenergized and acts as a regulator during defrost.

At the end of the defrost cycle, excess pressure will bleed from the relief line at a safe rate through the energized valve B. The gas-powered valves will not open the evaporator to the surge drum until the gas pressure is nearly down to the system pressure.

Flooded-gas Leg Shutoff (Bottom Hot-Gas Feed)

The system shown in Fig. 10-13 is similar to that shown in Fig. 10-12. However, the liquid leg of the evaporator dumps directly into the surge drum without a relief valve. In this system, valve C is a defrost regulator. It is placed in the suction line, where it is normally open. During defrost, valve C is deenergized, converting to a defrost regulator. In such a system, it is recommended that a large-capacity surge drum or valve A be used as a bypass valve. This will bleed defrost pressure gradually around valve C into the suction line. Note how the in-line check valve is used to prevent cross flow.

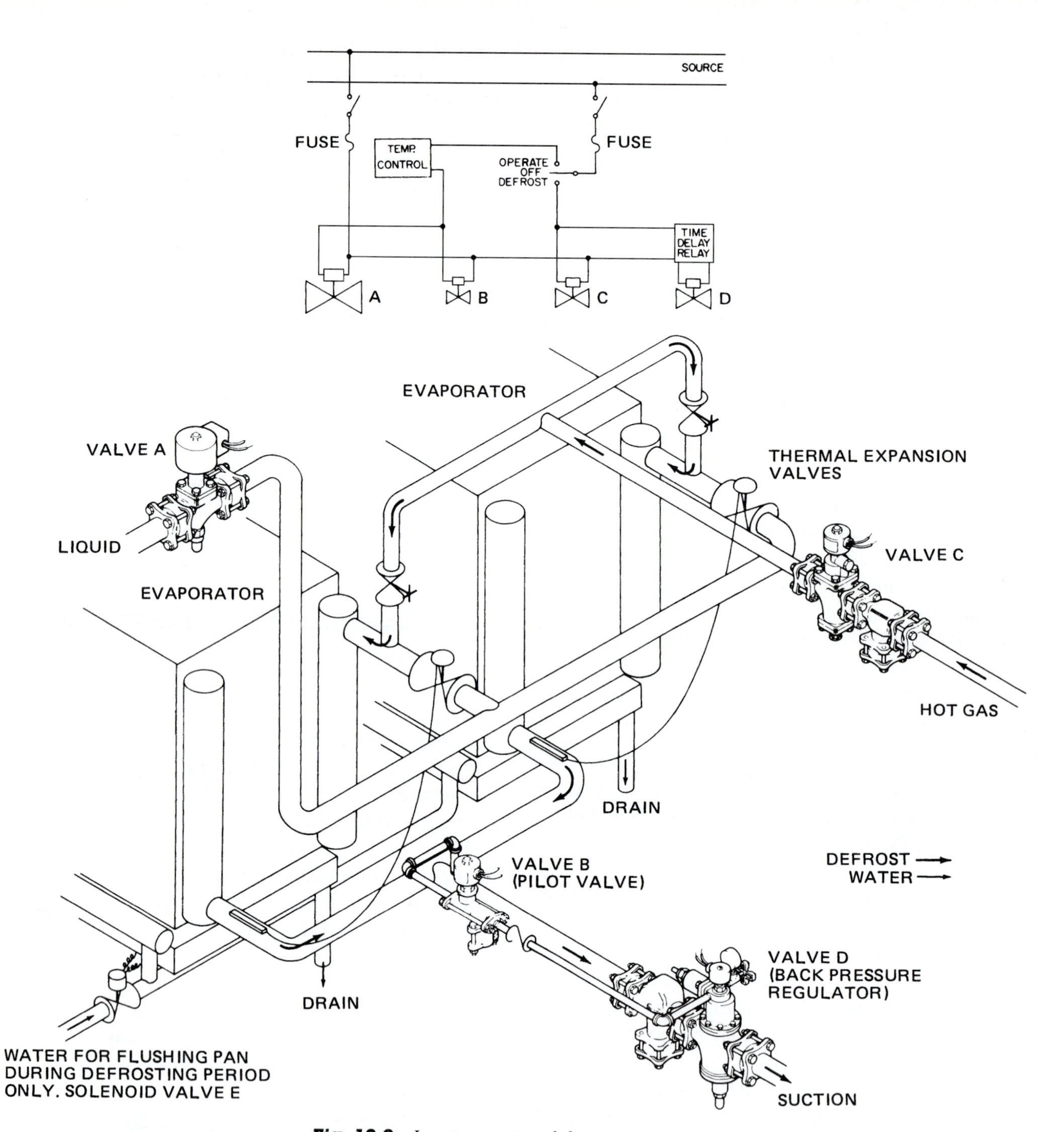

Fig. 10-9 *Low-temperature defrost system.* (Hubbell)

Flooded-Ceiling Evaporator—Liquid-Leg Shutoff (Bottom Hot-Gas Feed)

Figure 10-14 illustrates a flooded-ceiling evaporator. Upon initiation of the defrost sequence, the hot-gas solenoid (Number 1) is opened. Gas flows to gas-powered check valve, isolating the bottom of the surge tank from the evaporator. The hot gas flows through the pan coil and the in-line check valve into the evaporator. Excess gas pressure is dumped into the surge tank. It will bleed through valve A. During defrost, this valve has been deenergized to perform as a relief regulator set at approximately 70 psig.

Flooded-Ceiling Evaporator—Liquid-Leg Shutoff (Top Hot-Gas Feed)

Figure 10-15 shows a multiple flooded-evaporator system using input and output headers to connect the various evaporators and the surge drum. Note that, upon defrost,

Table 10-2 *Valve Sizing for Low-Temperature System*

Tons Refrigerant	Liquid Solenoid (in.)	Hot-Gas Solenoid (in.)	Back-Pressure Regulator (in.)	Pilot Solenoid (in.)	Defrost Water (in.)
3	1/2	1/2	1	1/4	1/2
5	1/2	1/2	1 1/4	1/4	3/4
7	1/2	1/2	1 1/4	1/4	3/4
10	1/2	1/2	1 1/2	1/4	1
12	1/2	1/2	1 1/2	1/4	1
15	1/2	1/2	2	1/4	1 1/4
20	1/2	1/2	2	1/4	1 1/4
25	1/2	1/2	2 1/2	1/4	1 1/2
30	1/2	1/2	2 1/2	1/4	1 1/2
35	1/2	1/2	3	1/4	2
40	3/4	3/4	3	1/4	2
45	3/4	3/4	3	1/4	2
50	3/4	3/4	3	1/4	2

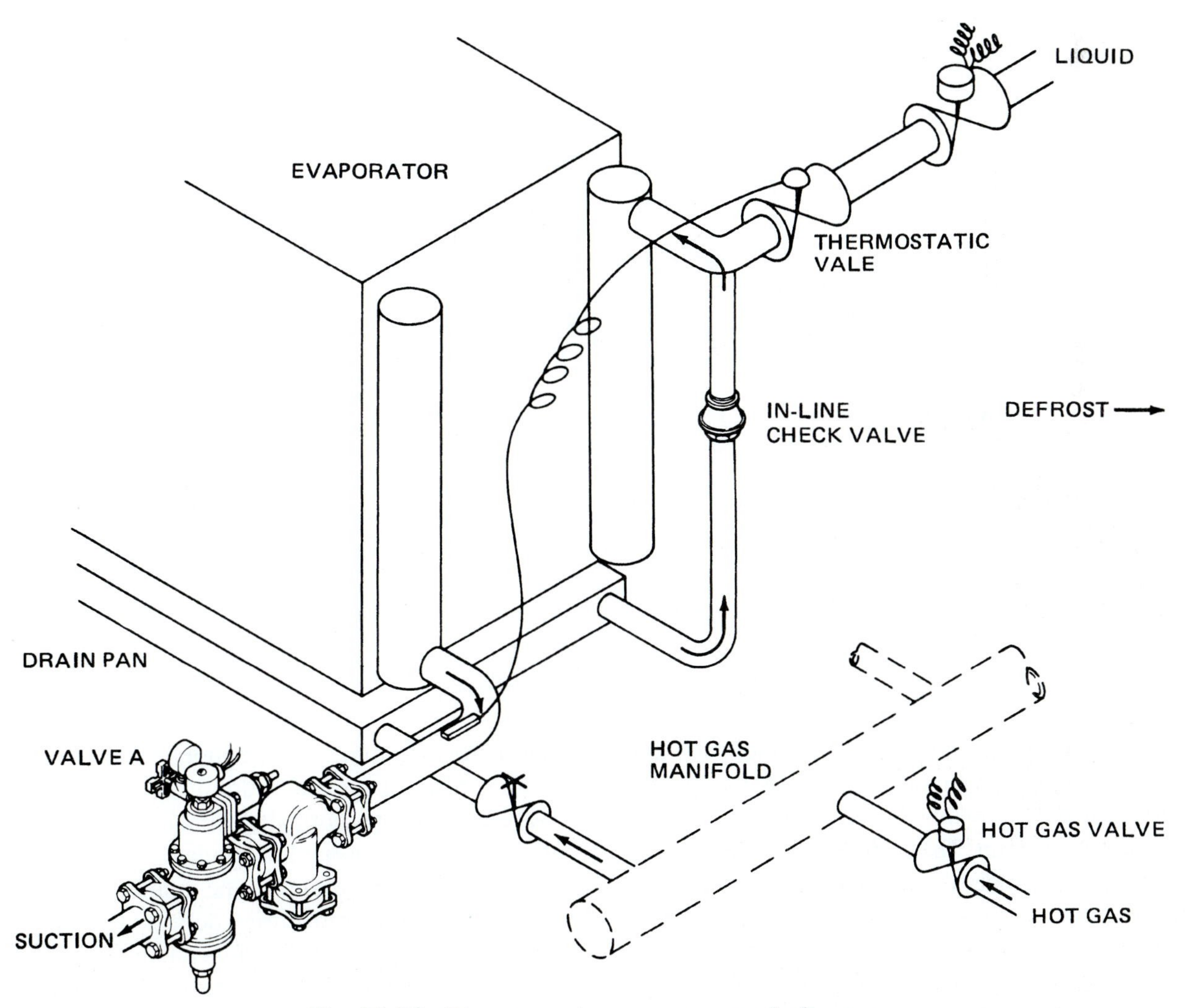

Fig. 10-10 *Direct-expansion evaporator—top feed.* (Hubbell)

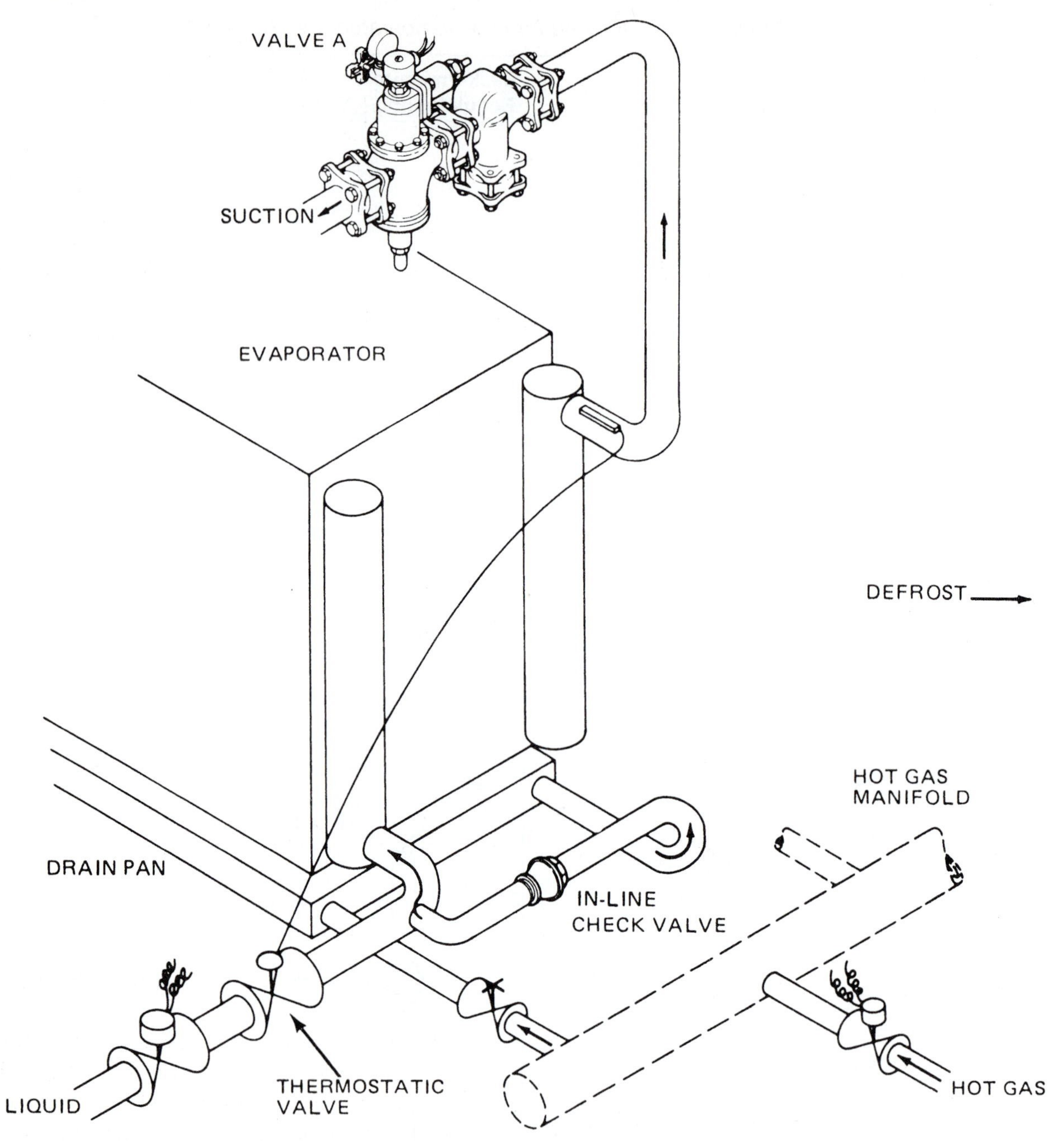

Fig. 10-11 *Direct-expansion evaporator—bottom feed.* (Hubbell)

the fluid and condensate, are purged from the evaporator and surge drum into the remote accumulator through the regulator, which is a reseating safety valve. This is usually set at about 70 psig. The accumulator must be sized to accept the refrigerant, plus hot-gas condensate.

Flooded-Ceiling Blower (Top Hot-Gas Feed)

Figure 10-16 shows a modification of the system shown in Fig. 10-15. In the system shown in Fig. 10-16, top-fed hot-defrost gas forces the evaporator fluid directly to the bottom of the large surge drum. The defrost regulator (valve A), which is normally open, is deenergized during the defrost to act as a relief regulator.

To minimize heating of the ammonia that accumulates in the surge drum during defrost, a thermostat bulb should be used to sense the temperature rise in the bottom header. This thermostat can be used to terminate the defrost cycle. Once again, the gas-powered check valve isolates the evaporator from the surge drum until the gas pressure is shut off.

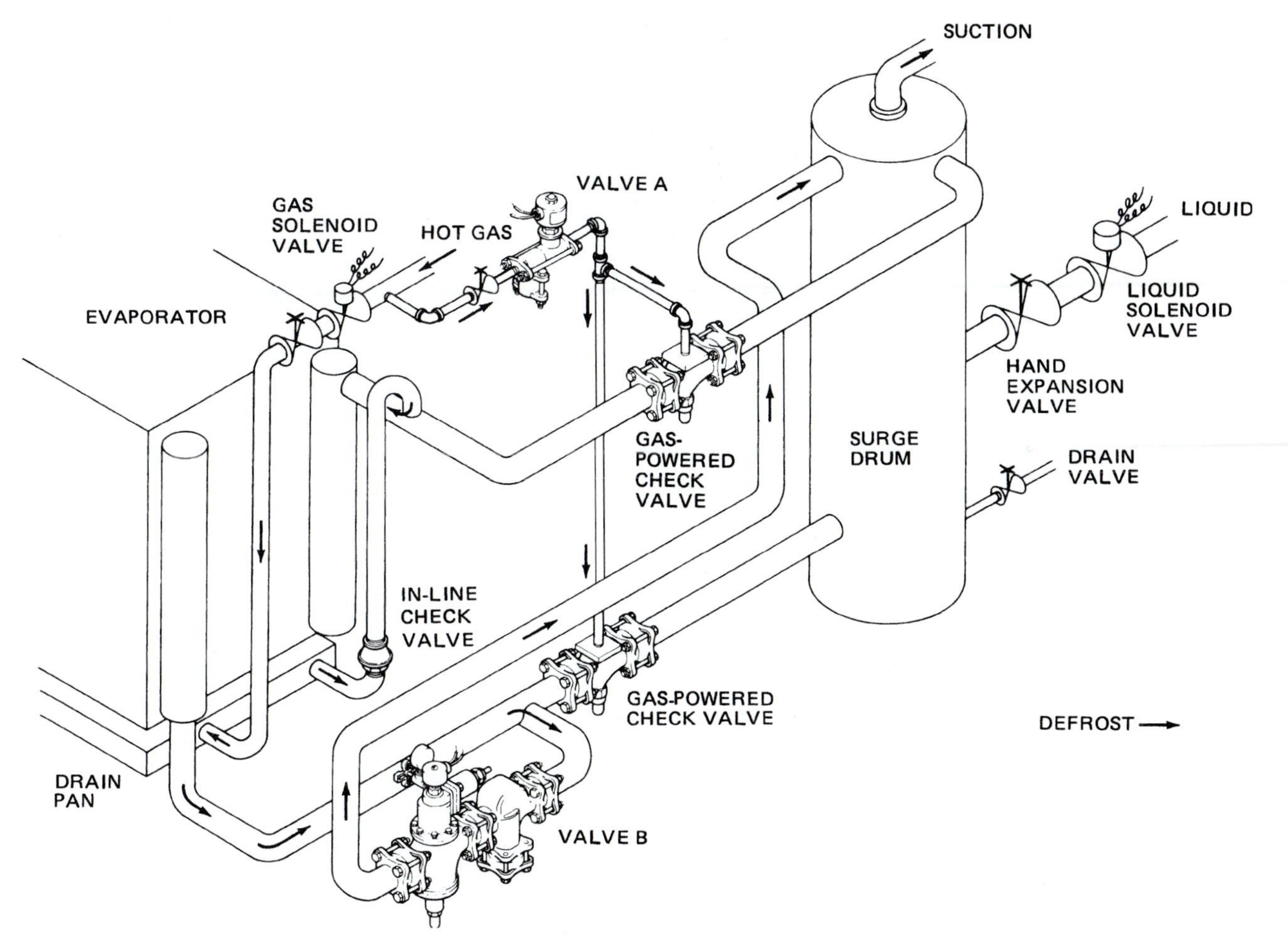

Fig. 10-12 *Gas- and liquid-leg shutoff—top feed.* (Hubbell)

Flooded-Ceiling Blower (Hot-Gas Feed through Surge Drum)

Figure 10-17 shows a simple defrost. It is a setup for the refrigeration system shown in Fig. 10-16. However, both the evaporator and surge drum are emptied during the defrost, necessitating the use of an ample suction accumulator to protect the compressor. In this system the pilot-solenoid valve in conjunction with the reverse-acting pressure regulator limits the system pressure. This permits the use of a simple solenoid valve and globe valve for rate control in the relief line.

Flooded Floor-Type Blower (Gas and Liquid-Leg Shutoff)

Figure 10-18 illustrates a flooded floor unit suitable for operation down to –70°F (–57°C).

The gas–pressure powered valve used in this circuit has a solenoid pilot operator. This provides positive action with gas or liquid loads at high or low temperatures and pressures.

To defrost a group of evaporators without affecting the temperatures of the common surge drum, the gas-powered valve is used in each end of the evaporator. A reseating safety valve is a relief regulator. It controls the defrost pressure to the relief-line accumulator. A check valve prevents back flow into the relief line. The in-line check valve prevents crossover between adjacent evaporators.

At high temperatures [above –25°F (–31°C)], use of the gas-powered check valve in place of the gas-powered solenoid valve is recommended.

Flooded Floor-Type Blower (Gas Leg Shutoff)

The system shown in Fig. 10-19 is similar to that shown in Fig. 10-18. However, a single gas-type, pressure-powered valve is used.

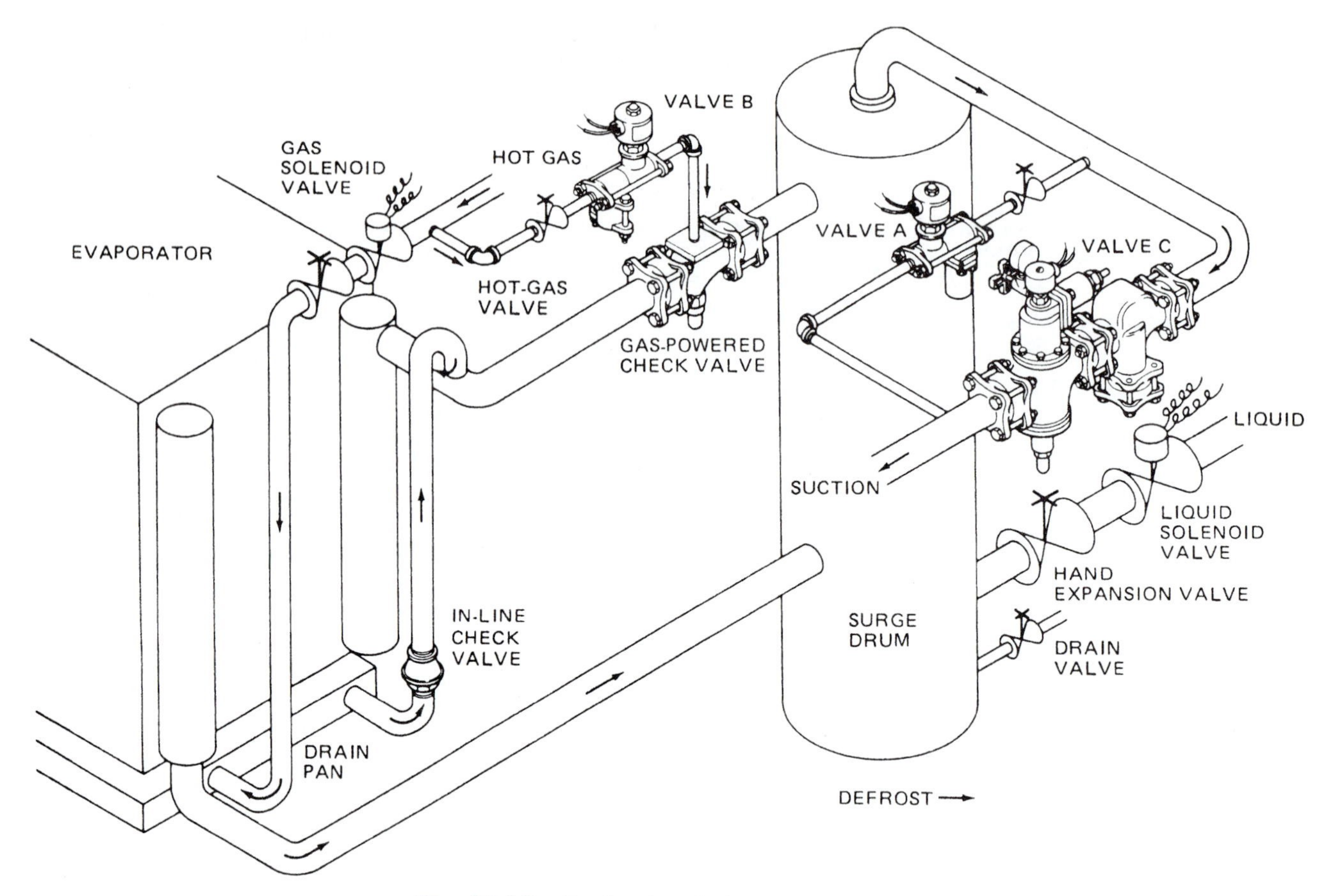

Fig. 10-13 *Gas-leg shutoff—bottom feed.* (Hubbell)

Overpressure at the surge drum is relieved by valve B, a defrost-relief regulator. This is normally wide open. It becomes a regulating valve when its solenoid is deenergized during defrost.

Defrost gas flows through the hot-gas solenoid when energized. It then flows through the glove valve and the in-line check valve to force the evaporator fluid into the surge drum.

An optional hot-gas thermostat bulb may be used to sense heating of the bottom of the evaporator. Thus, it can act as a backup for the timed defrost cycle.

LIQUID-RECIRCULATING SYSTEMS

Liquid–refrigerant recirculating systems are frequently fed by liquid flow upward through their evaporators. These systems are called *bottom fed*. This is accomplished by either mechanical or gas displacement recirculators during the refrigerant cycle. See Fig. 10-20.

In some systems, more than a single evaporator is fed from the same recirculator, as shown in Fig. 10-20. Then, a proper distribution of liquid between evaporators must be maintained to achieve efficient operation of each evaporator. This balance is usually accomplished by the insertion of adjustable globe valves or orifices into the liquid-feeder line. Similarly, adjustment of the globe valves or insertion of orifices is also often used properly to distribute hot gas during the defrost cycle.

Equalizing orifices or globe valves are not used if the hot gas used for defrosting is fed to the bottom of the evaporators as shown in Fig. 10-20. In such cases, most of the hot gas could flow through the circuits nearest the hot-gas supply line. The same would also happen in circuits where both vertical and horizontal headers are used, as in Fig. 10-21. The more remote circuits could remain full of cold liquid. Consequently, they would not defrost.

Supplying hot gas to the top of the evaporator forces liquid refrigerant down through the evaporator and out through a reseating safety-valve relief regulator into the suction line return to the accumulator. See Fig. 10-21. Reseating safety-valve relief regulators are usually set to relieve at 60 to 80 psig to provide rapid defrost.

The use of check valves is important in flooded liquid-recirculating systems fed by mechanical

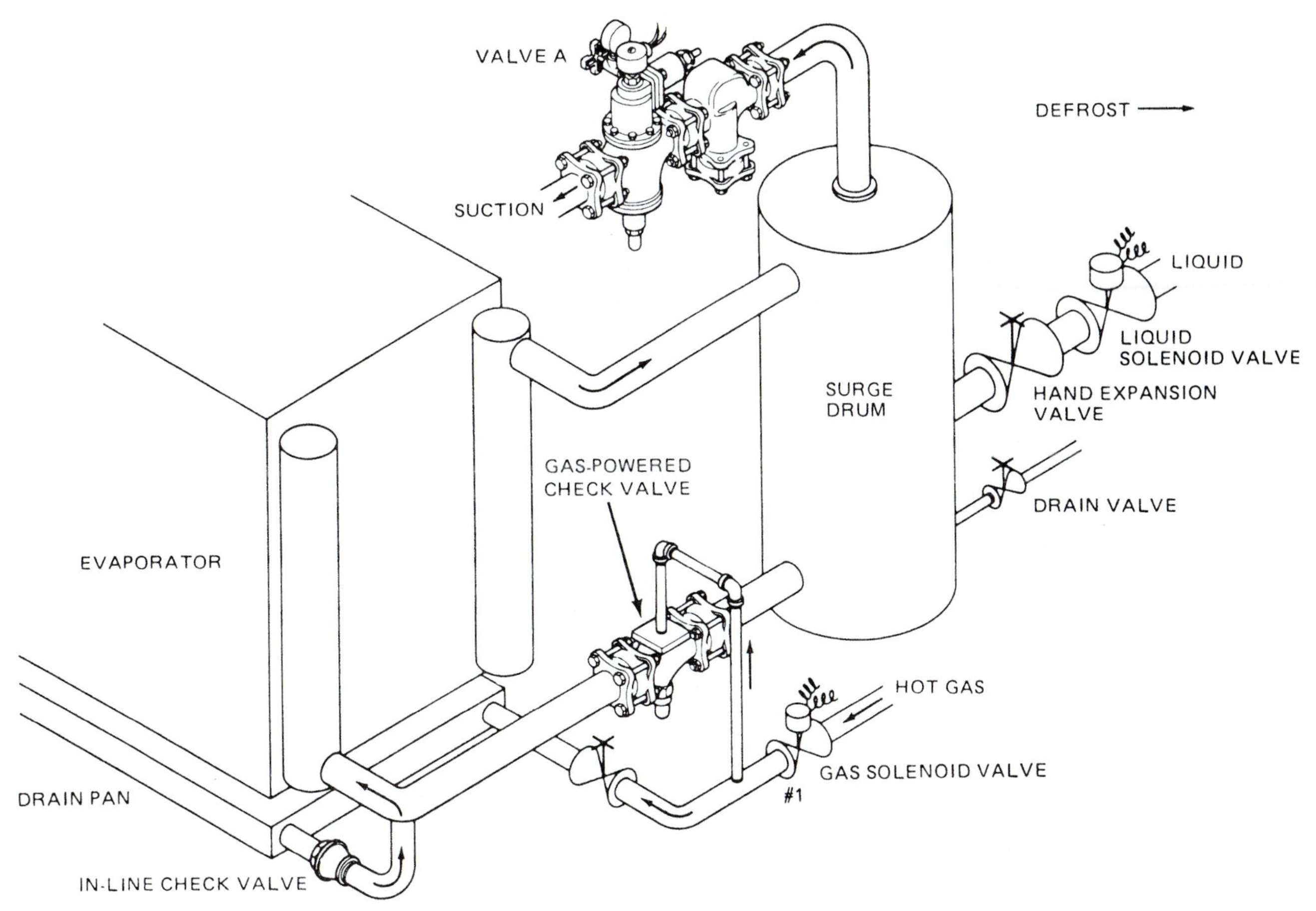

Fig. 10-14 *Ceiling evaporator, liquid-leg shutoff—bottom feed.* (Hubbell)

gas-displacement liquid recirculators. The check valves are used where the pressure of the hot gas used for defrost is higher than the system pressure. The reseating safety-check valve must be used to stop this gas at high pressure from flowing back into the liquid supply line.

Flooded Recirculator (Bottom Hot-Gas Feed)

The multiple system shown in Fig. 10-20 shows the check valve mounted in each of the liquid-refrigerant branch lines. A single solenoid valve is used in the main refrigerant line. The defrost gas is bottom fed.

Flooded Recirculator (Top-Gas Feed)

The system illustrated in Fig. 10-21 shows a check valve mounted directly at the outlet of each of the liquid-solenoid valves. The defrost gas is top fed. This system permits selective defrosting of each evaporator. A single accumulator is used to protect the compressor during defrost, as well as to accumulate both liquid refrigerant and defrost condensate. This protection is accomplished by using a differential pressure-regulator valve in an evaporator bypass circuit.

The differential pressure-regulator valve will open sufficiently to relieve excess pressure across the compressor inlet. The pressure will discharge as this excess pressure differential occurs. When the pressure differential is less than the regulator valve setting, the regulator will be tightly closed.

Low-Temperature Ceiling Blower

The low-temperature liquid recirculating system, illustrated in Fig. 10-22, uses several controls. During the cooling cycle, No. 1 pilot valve is opened and No. 2 pilot valve is closed, holding the gas-powered solenoid valve wide open. This allows flow of liquid through the energized liquid-solenoid valve from the recirculator and then through the circuit of the unit. The in-line check valve installed between the drain pan coil header

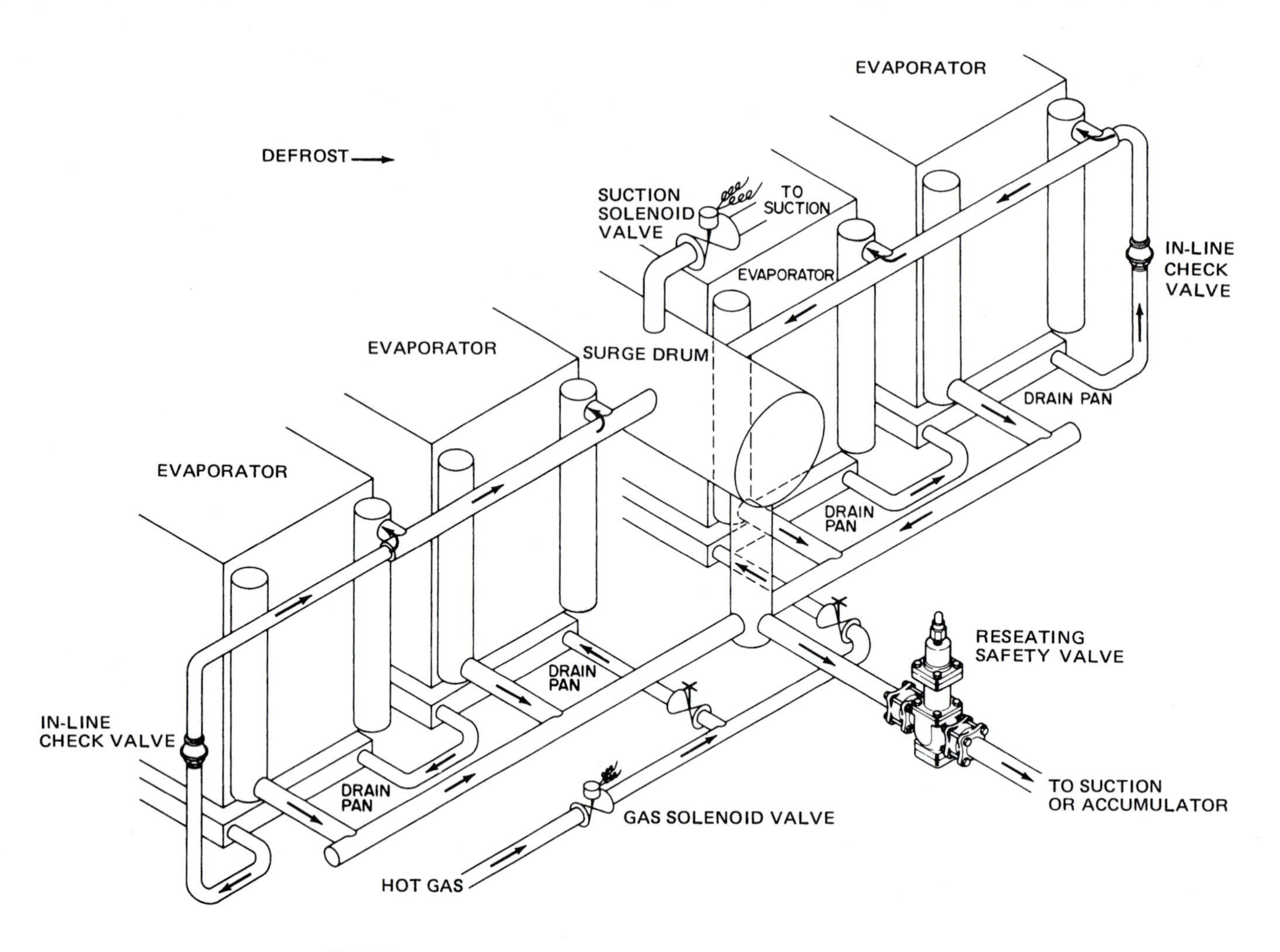

Fig. 10-15 *Ceiling evaporator, liquid-leg shutoff—top feed.* (Hubbell)

and suction line prevents drainage of liquid into the drain pan coil.

For defrost, the liquid-solenoid valve is closed. The No. 1 pilot solenoid is deenergized. The No. 2 solenoid is opened, closing the gas-powered solenoid valve tightly. The hot gas, solenoid is energized. This allows distribution of the hot gas through the drain pan coils, the in-line check valve, the top of the suction header, and the coil. The gas comes out the bottom of the liquid header.

Check valve A prevents the, flow of the high-pressure gas in the liquid line. Therefore, the gas is relieved through the safety-valve relief regulator (B). This is set to maintain pressure in the evaporator to promote rapid, or efficient defrost.

YEAR–ROUND AUTOMATIC CONSTANT LIQUID-PRESSURE CONTROL SYSTEM

The constant liquid control system is a means of increasing the efficiency of a refrigeration system that utilizes air-cooled, atmospheric, or evaporative condensers. See Fig. 10-23. This is accomplished by automatically maintaining a constant liquid pressure throughout the year to ensure efficient operation. Constant liquid pressure on thermal expansion valves, float controls, and other expansion devices results in efficient low-side operation. Hot gas defrosting, liquid recirculation, or other refrigerant control systems require constant liquid pressure for successful operation. Liquid pressure is reduced by cold weather and extremely low wet-bulb temperatures with low-refrigeration loads.

To compensate for a decrease in liquid measure, it is necessary automatically to throttle the discharge to a predetermined point and regulate the flow of discharge pressure to the liquid line coming from the condenser and going to the receiver. Thus, predetermined pressure is applied to the top of the liquid in the receiver. The constant liquid pressure control does this. In addition, when the compressor "start and stop" is controlled by pressurestats, the pressure-operated hot-gas flow control valve is a tight closing stop valve during stop periods. This permits efficient "start and stop" operation of the compressor by pressure control of the low side.

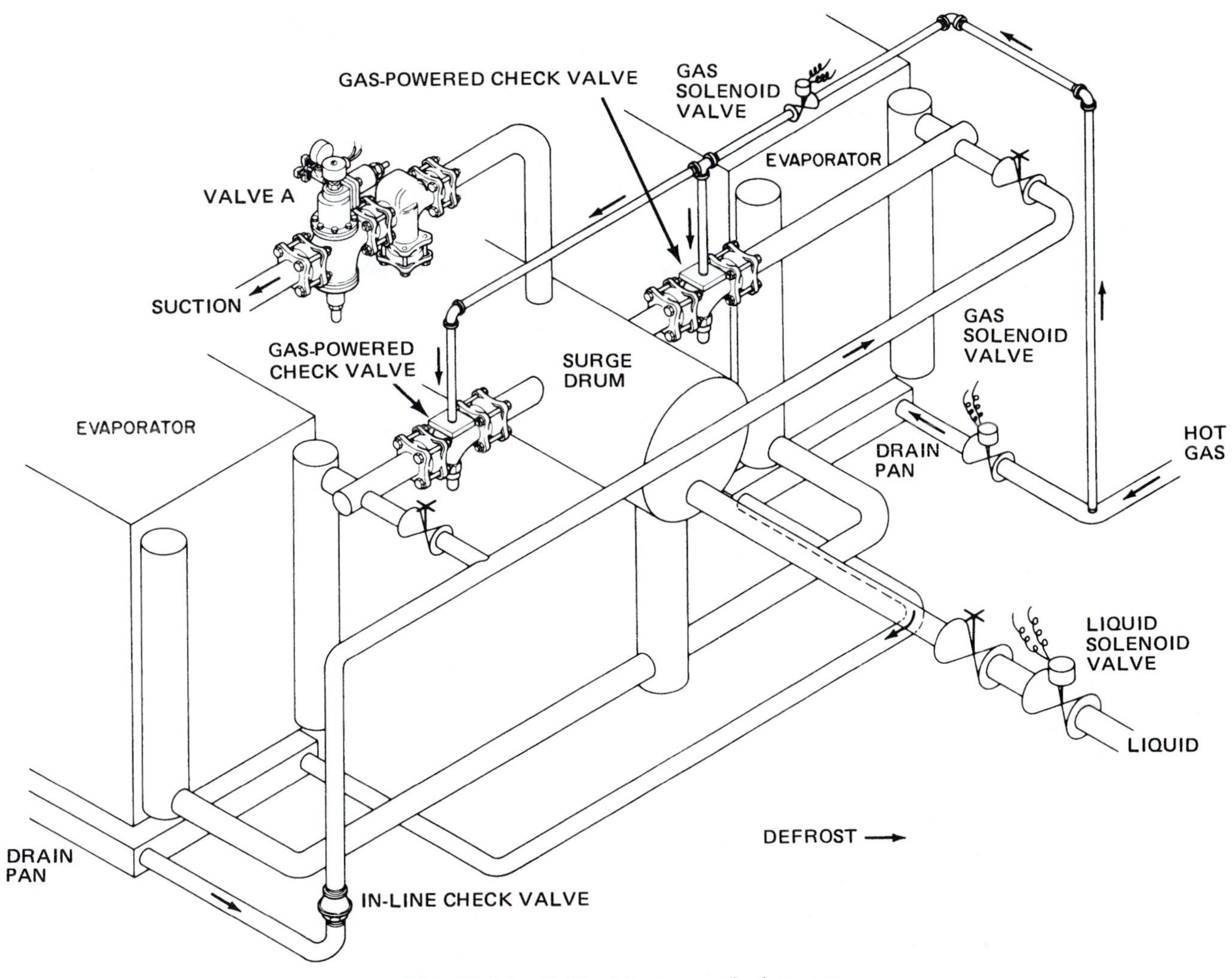

Fig. 10-16 *Ceiling blower—top feed.* (Hubbell)

The three valves in the system shown in Fig. 10-23 are:

- The reverse-acting pressure regulator
- The pressure-operated hot-gas flow-control valve
- The relief-check valve

The function of the control system is to maintain a constant liquid pressure (A). The reverse acting pressure regulator valve accomplishes this, which is a modulating-type valve. It maintains a constant predetermined pressure on the downstream side of the regulator. To maintain a constant pressure (A) it is necessary to maintain a discharge pressure (B) approximately 5 psi above (A). This is accomplished by the hot-gas control valve, which will maintain a constant pressure (B) on the upstream or inlet side of the regulator. Due to the design of the regulator, a constant supply of gas will be available at a predetermined pressure to supply the pressure regulator to maintain pressure (A). Excess hot gas is not required to maintain a fill flow into the condenser.

The relief-check valve prevents pressure (A) from causing backflow into the condenser. When the compressor shuts down, the hot-gas flow-control valve closes tightly and shuts off the discharge line. This prevents gas from flowing into the condenser.

The check valve actually prevents the backflow of liquid into the condenser. Thus, liquid cannot back up into the condenser in extremely cold weather. Sufficient low-side pressure will be maintained to start the compressor when refrigeration is required.

DUAL-PRESSURE REGULATOR

A dual-pressure regulator is shown in Fig. 10-24. It is used on a shell-and-tube cooler. The dual-pressure regulator is particularly adaptable for the control of shell-and-tube brine, or water coolers, which at intervals may be subjected to increased loads. Such an arrangement is shown in Fig. 10-24.

The high-pressure diaphragm is set at a suction pressure suitable for the normal load. The low-pressure

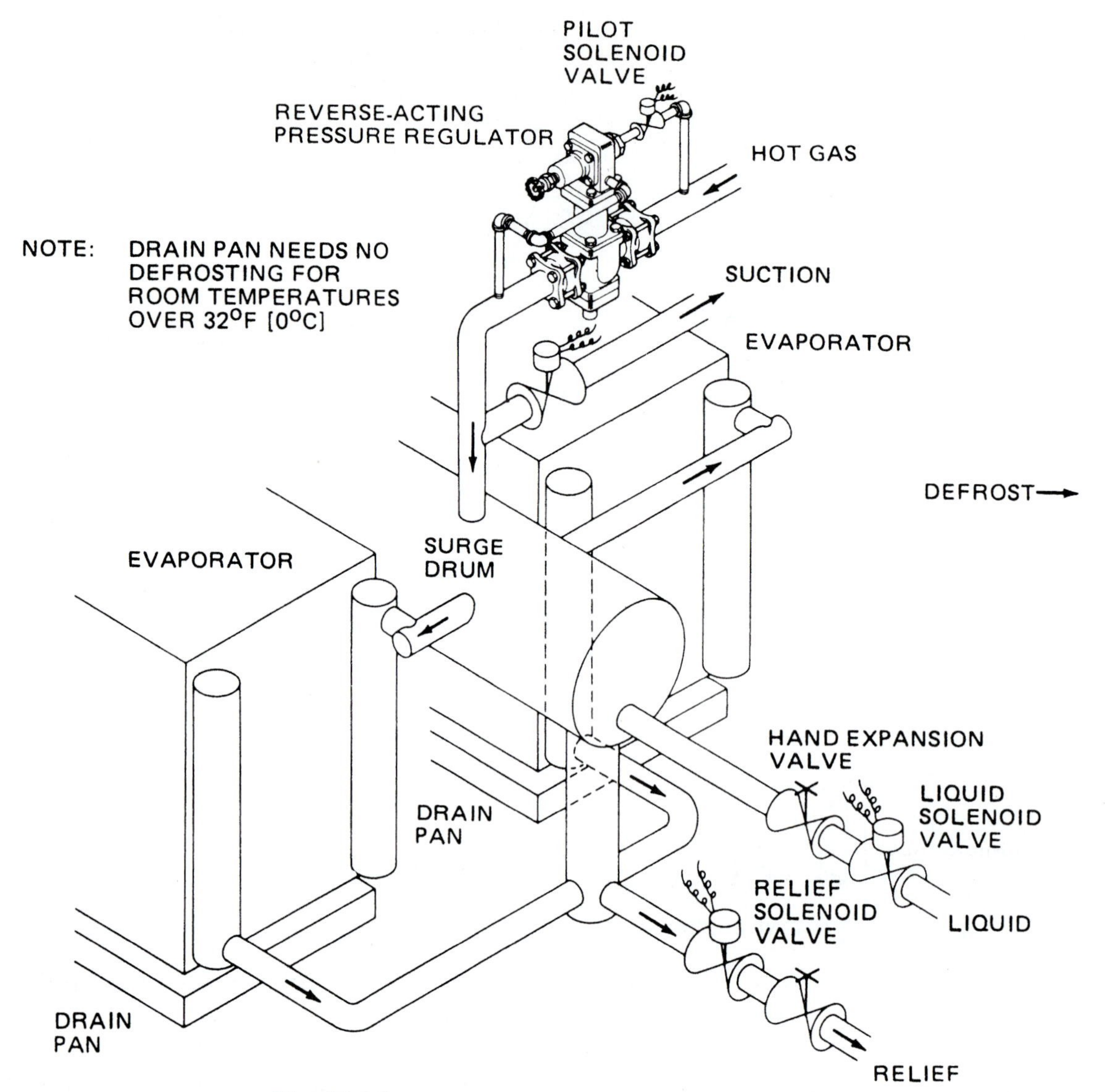

Fig. 10-17 *Ceiling blower—feed through surge drum.* (Hubbell)

diaphragm is set for a refrigerant temperature low enough to take care of any intermittent additional loads on the cooler.

In this case, a thermostat affects the transfer between low and high pressure. The remote bulb of the thermostat is located in the water or brine line leaving the cooler. A temperature increase at this bulb, indicating an increase in load, will cause the thermostat to open the electric pilot and transfer control of the cooler to the low-temperature diaphragm. Upon removal of the excess load, the thermostat will cause the electric pilot to close the low-pressure port. The cooler is then automatically transferred to the normal pressure for which the high-pressure diaphragm is set. The diaphragms may be set at any two evaporator pressures at which it is desirable to operate. Any electric switching device responsive to load change may be used to change from one evaporator pressure to the other.

VALVES AND CONTROLS FOR HOT-GAS DEFROST OF AMMONIA-TYPE EVAPORATORS

The following valves and controls are used in the hot-gas defrost systems of ammonia-type evaporators:

Hot-gas or pilot solenoid valve. The valve is a $^1/_8$ in. ported-solenoid valve. It is a direct-operated valve suitable as a liquid, suction, hot gas, or pilot valve at pressures up to 300 lb.

Suction, liquid, or gas-solenoid valve. The suction-solenoid valve is a one-piston, pilot-operated valve suitable for suction, liquid, or gas lines at pressures up to 300 lb. It is available with a $^9/_{16}$ or $^3/_4$ in. port.

Pilot-operated solenoid valve. The valve is a one-piston, pilot-operated solenoid valve used as a positive stop

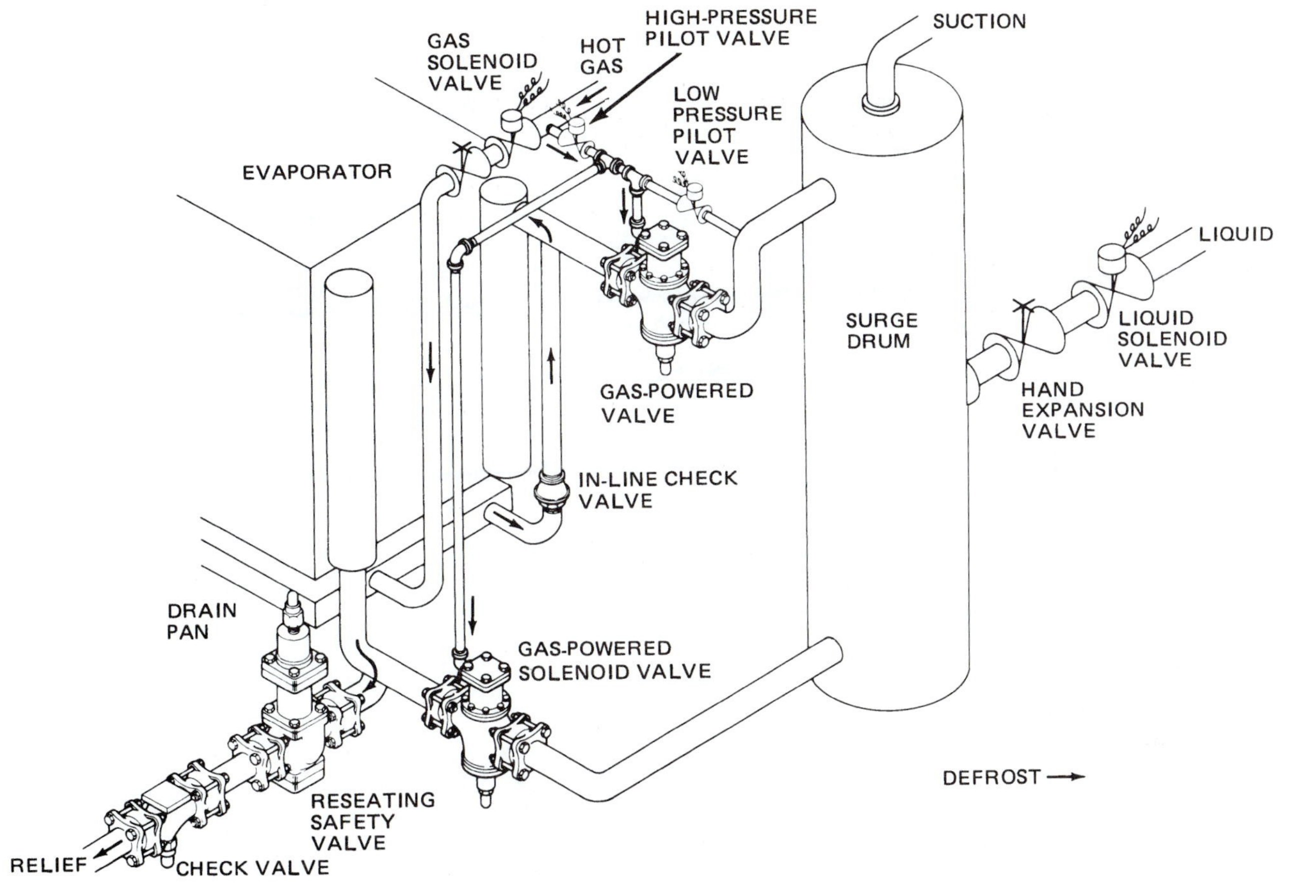

Fig. 10-18 *Floor blower—gas and liquid shutoff.* (Hubbell)

valve for applications above –30°F (–34°C) on gas or liquid.

Pilot-operated two-piston valve. The solenoid valve is a rugged, pilot-operated, two-piston valve with spring return for positive closing under the most adverse conditions. It is used for compressor unloader, suction, liquid, and hot-gas applications.

Gas-powered solenoid valve. The gas-powered solenoid valve is a power-piston type of valve that uses high pressure to force the valve open through the control of pilot valves. Because of the high power available to open these valves, heavy springs may be used to close the valves positively at temperatures down to –90°F (–68°C).

Dual-pressure regulator valve. The dual-pressure regulator valve is designed to operate at two predetermined pressures without resetting or adjustment. By merely opening and closing a pilot solenoid, either the low- or high-pressure setting is maintained.

Reseating safety valve. The reseating safety valve is generally used as a relief regulator to maintain a predetermined system pressure. The pressure maintained by the valve is adjustable manually.

Back-pressure regulator arranged for full capacity. The back-pressure regulator is normally used where pressure control of the evaporator is not required, as in a direct expansion system. A pilot solenoid is energized, allowing pressure to bypass the sensing chamber of the regulator holding the valve wide open. Deenergizing the pilot valve allows the valve to revert to its function as a back-pressure regulator maintaining a preset pressure upstream of the valve. The valve performs both as a suction solenoid and as a relief regulator.

Differential relief valve. The differential relief valve is a modulating regulator for liquid or gas use. It will maintain a constant preset pressure differential between the upstream and downstream side of a regulator.

Reverse-acting pressure regulator. The reverse-acting pressure regulator is used to maintain a constant predetermined pressure downstream of the valve. When complete shutoff of the regulator is required, a pilot valve is installed in the upstream feeder line. When the solenoid valve is closed, the regulator closes tightly. When the solenoid valve is open, the regulator is free to operate as the pressure demands. With

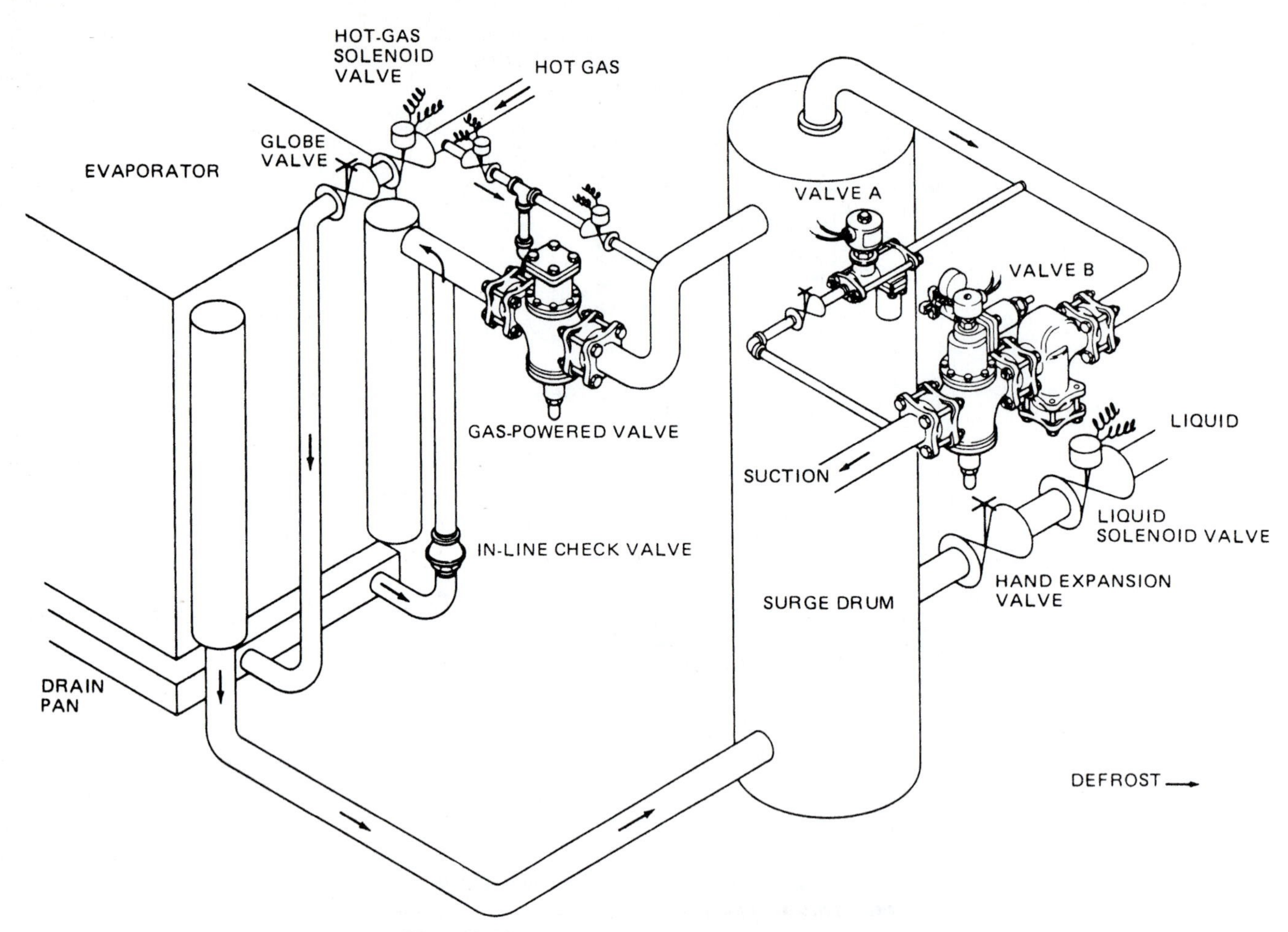

Fig. 10-19 *Floor blower—gas-leg shutoff.* (Hubbell)

the solenoid installed as described earlier, this becomes a combination of reverse-acting regulator and stop valve.

Gas-powered check valve. The gas-powered check valve is held in a normally open position by a strong spring. Gas pressure applied at the top of the valve closes the valve positively against the high-system pressures. A manual opening stem is standard.

Check valve. The check valve is a spring-loaded positive check valve with manual opening stem. It is used to prevent backup of relatively high pressure into lower pressure lines.

In-line check valve. The in-line check valve is used in multiple-branch liquid lines fed by a single solenoid valve. This check valve prevents circulation between evaporators during refrigeration. The in-line check valve is also used between drain pans and evaporators to prevent frosting of the drain pan during refrigeration.

These valves and controls are necessary. They cause defrosting operations to take place in large evaporators used in commercial jobs. Some manufacturing operations also call for large-capacity refrigeration equipment.

BACK-PRESSURE REGULATOR APPLICATIONS OF CONTROLS

In a refrigeration system designed to maintain a predetermined temperature at full load, any decrease in load would tend to lower below full-load temperature the temperature of the medium being cooled.

To maintain constant temperatures in applications having varying loads, means must be provided to change refrigerant temperature to meet varying load requirements.

Refrigerant temperature is a function of evaporator pressure. Thus, the most direct means of changing refrigerant temperature to meet varying load requirements is to vary the system pressure. This variation of system pressure is accomplished by adjusting the setting of a back-pressure regulator.

A number of back-pressure valve controls are available. Some of them are shown in the following sections.

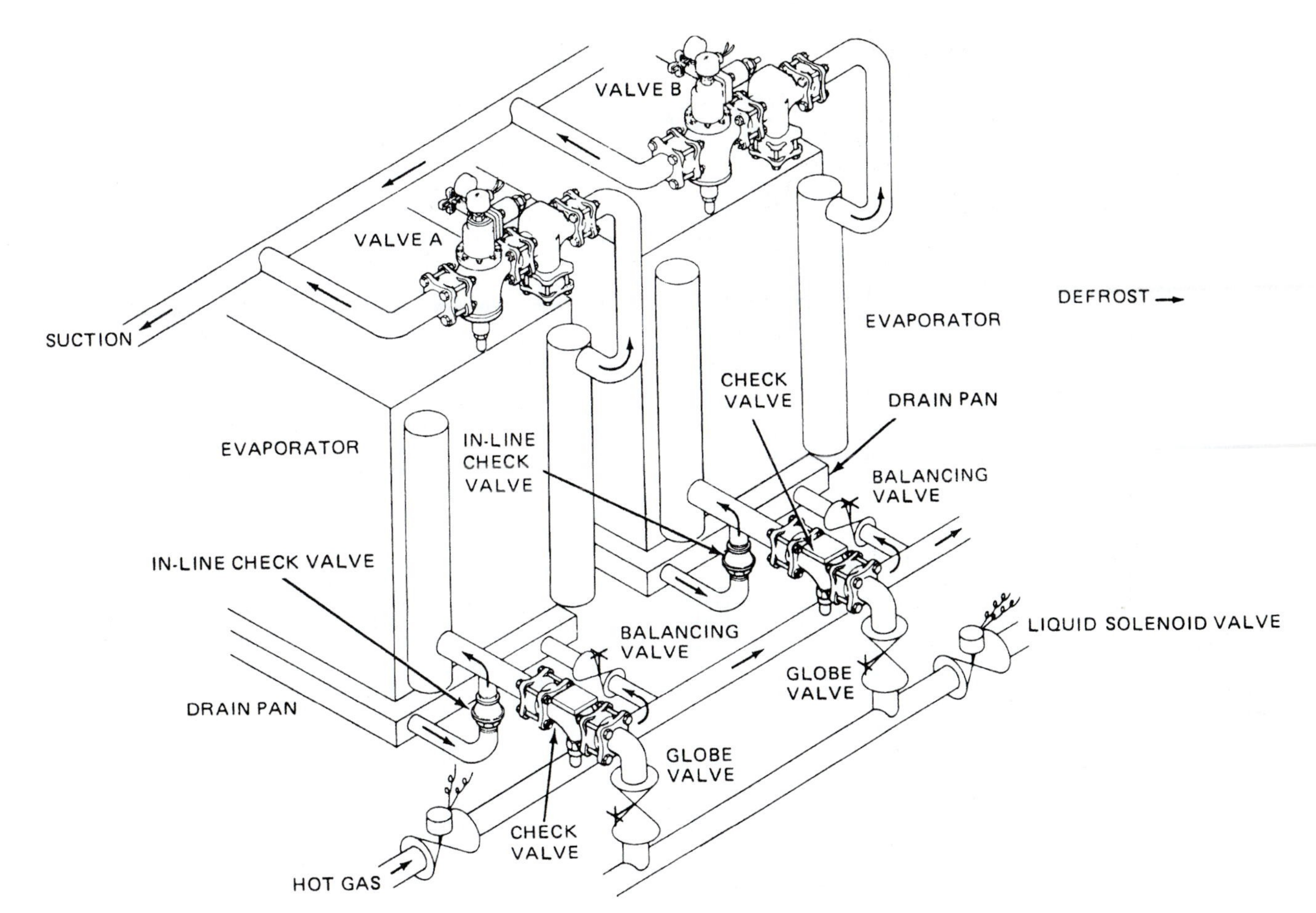

Fig. 10-20 *Flooded recirculator—bottom feed.* (Hubbell)

Refrigerant-Powered Compensating-Type Pilot Valve

The upper portion of the valve head is similar to a standard pressure-regulating head. On the lower portion of the head another diaphragm is connected to the main diaphragm by a push rod. As the thermal bulb warms, the liquid in it expands, pushing up on the rod and opening the regulator. Because this is accomplished by an outside power source, the pressure drop through the head is reduced considerably. The valve head will function in connection with the regulator on a $^1/_2$- to $^3/_4$-lb overall pressure drop. The point at which the modulation or compensation takes place may be adjusted by turning the adjusting stem. By turning the stem in, the product temperature is increased. By turning the stem out, the product temperature is decreased. The back-pressure valve will remain wide open, taking advantage of the line suction pressure until the product being cooled approaches the temperature at which modulation is to begin. The valve head will hold the temperature of the product up to and within $\pm^1/_2$°F (0.28°C) of the desired temperature. In the case of failure of the thermal element, the valve head can be used as a straight back-pressure valve by readjusting it to the predetermined suction pressure at which you desire the system to operate. See Fig. 10-25.

Air-Compensating Back-Pressure Regulator

A standard regulator is reset by manually turning the adjusting stem, which increases the spring pressure on top of the diaphragm. In an air-compensated regulator, a change of pressure on top of the diaphragm is accomplished by introducing air pressure into the airtight bonnet over the diaphragm. As this air pressure is increased, the setting of the regulator will be increased. This will produce like changes of evaporator pressure and refrigerant temperature. The variations in air pressure are produced by the temperature changes of the thermostatic remote bulb placed in the stream of the medium being cooled as it leaves the evaporator.

Temperature changes in the medium being cooled over the remote bulb of the thermostat will cause the thermostat to produce air pressures in the regulator

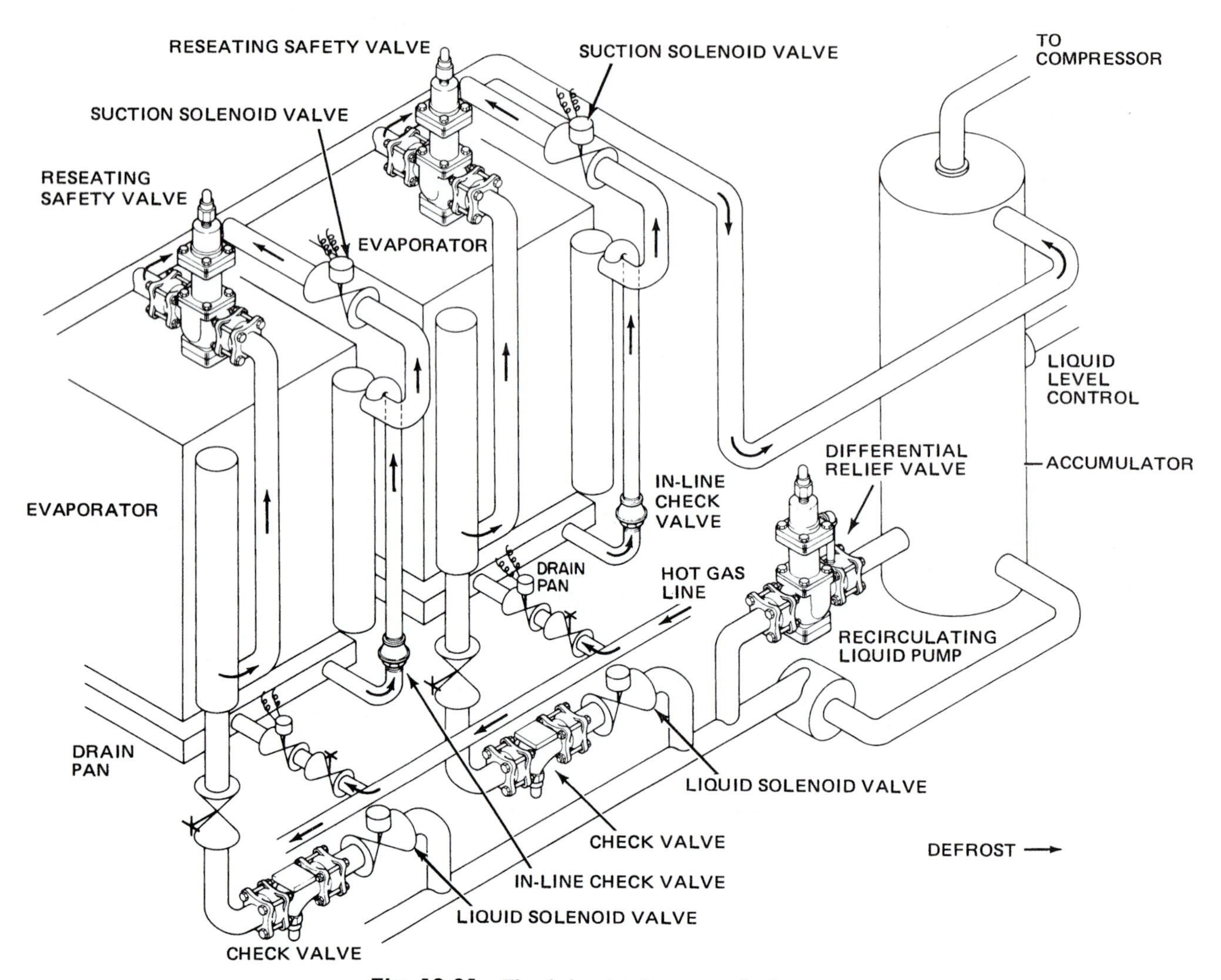

Fig. 10-21 *Flooded recirculator—top feed.* (Hubbell)

bonnet within a range of 0 to 15 lb. This will cause the regulator to change the evaporator suction pressure in a like amount. A more definite understanding of this operation is obtained by assuming certain working conditions for the purpose of illustration. In cases where a larger range of modulation is required, a three-to-one air relay may be installed. This will permit a 45-lb range of modulation. See Fig. 10-26.

Electric-Compensating Back-Pressure Regulator

A standard regulator is reset manually by turning the adjusting stem, usually found at the top of the regulator. In an electrically compensated regulator, turning the stem to obtain different refrigerant pressures and temperatures in the evaporator is accomplished by a small electric motor. This motor rotates the adjusting stem in accordance with temperature variations in a thermostatic bulb placed in the medium being cooled as it leaves the evaporator. The adjusting stem, spring, and controlling diaphragm have been separated from their positions at the top of the regulator. They have been placed in a small remote unit mounted on a common base with the motor and gear drive. This compensating unit may be located in any convenient place within 20 ft of the main regulator. The unit is connected to it by two small pipe-lines. These convey the pressure changes set up by the control diaphragm.

The total arc of rotation of the motor and the large gear on the motor, acting through the smaller pinion on the adjusting stem of the diaphragm unit, will rotate the stem about two turns. This is sufficient to cause the regulator to vary the evaporator pressure through a total range of about 13 lb. See Fig. 10-27.

VALVE TROUBLESHOOTING

Most of the problems in an evaporator system occur in the valves that make the defrost system operate properly.

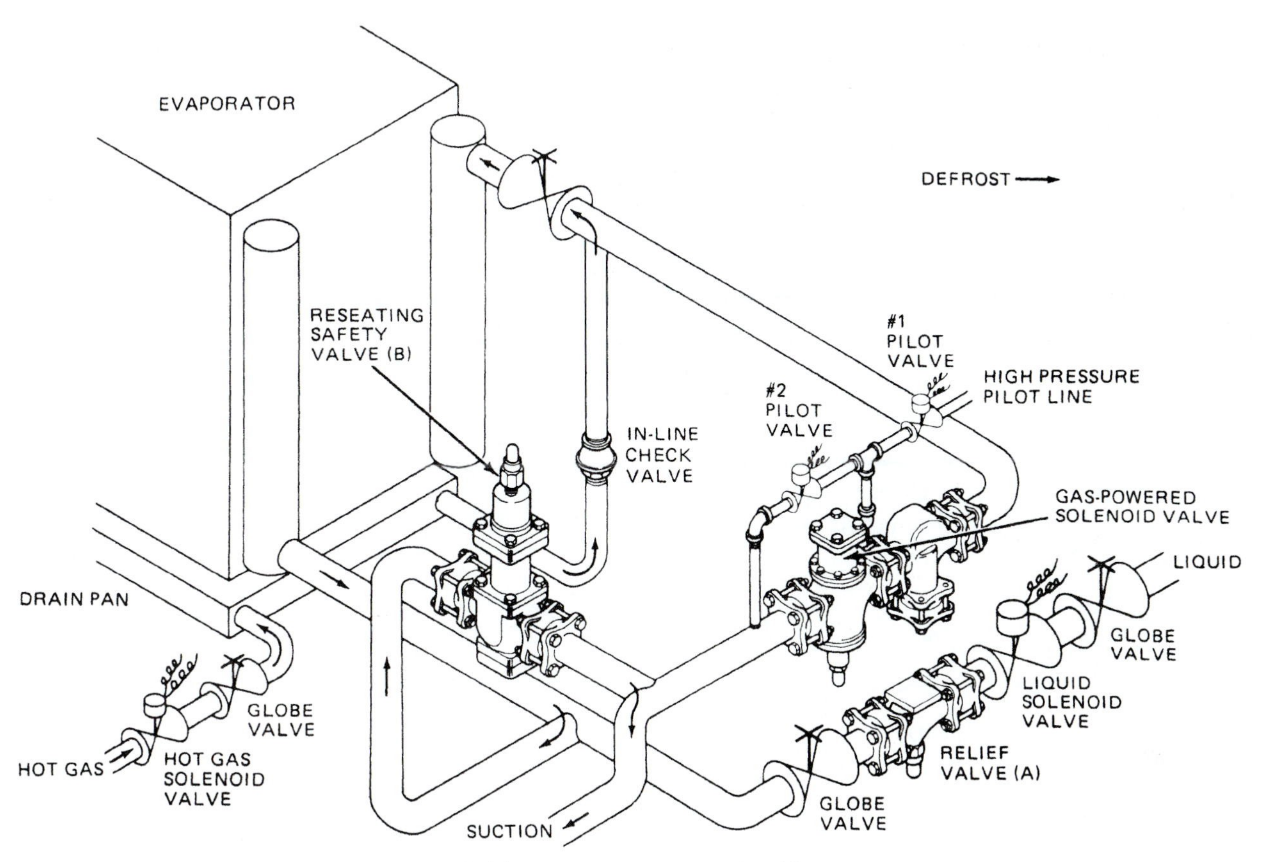

Fig. 10-22 *Low-temperature ceiling blower.* (Hubbell)

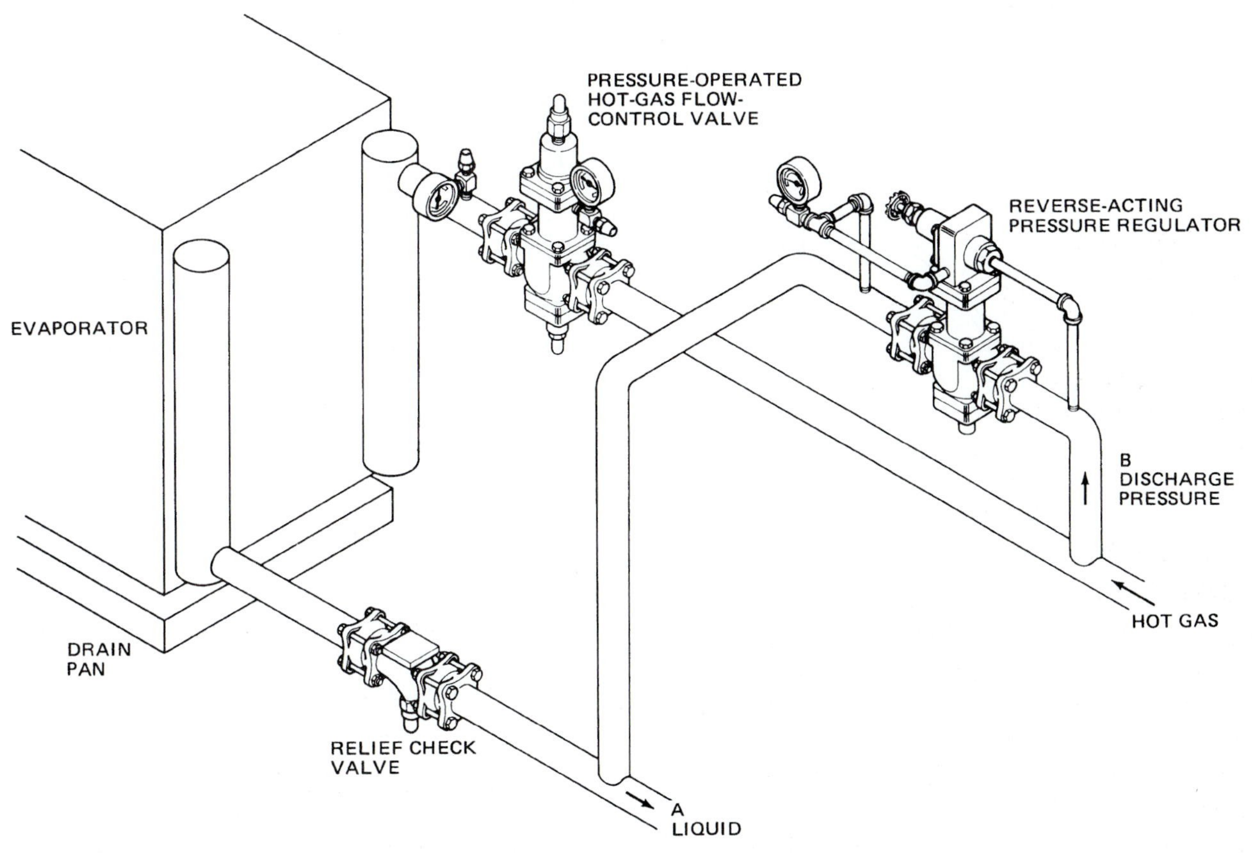

Fig. 10-23 *Year-round automatic control system.* (Hubbell)

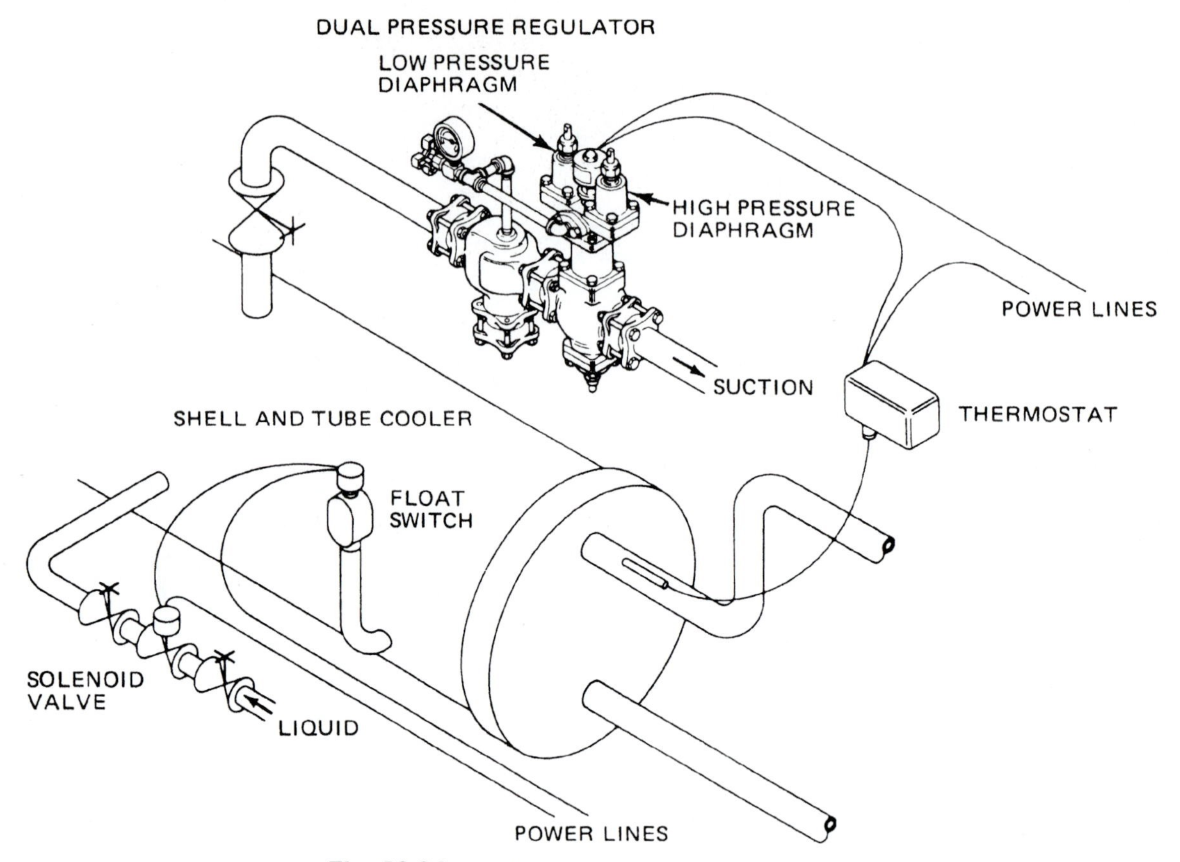

Fig. 10-24 *Dual–pressure regulator application.* (Hubbell)

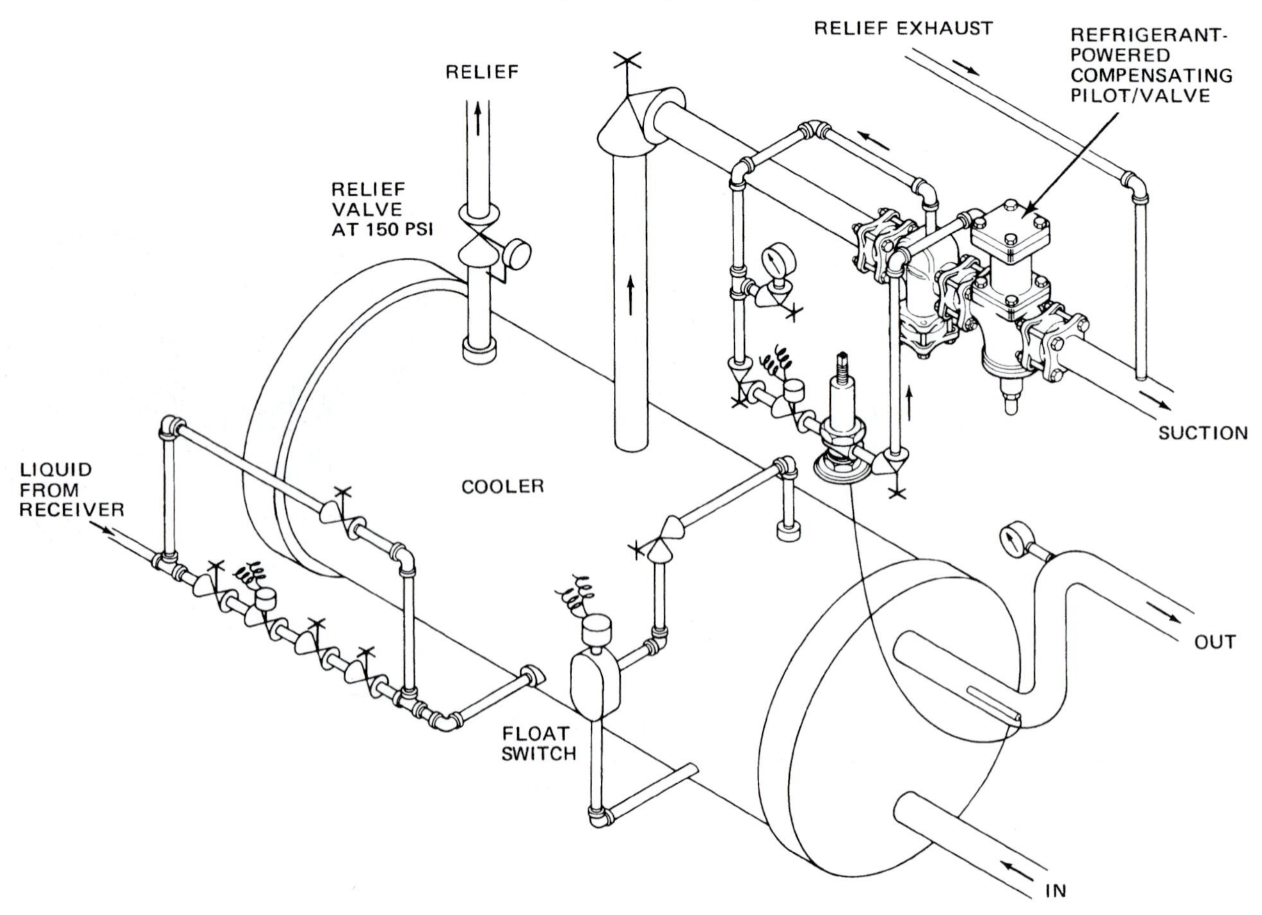

Fig. 10-25 *Thermal-compensating back-pressure regulator.* (Hubbell)

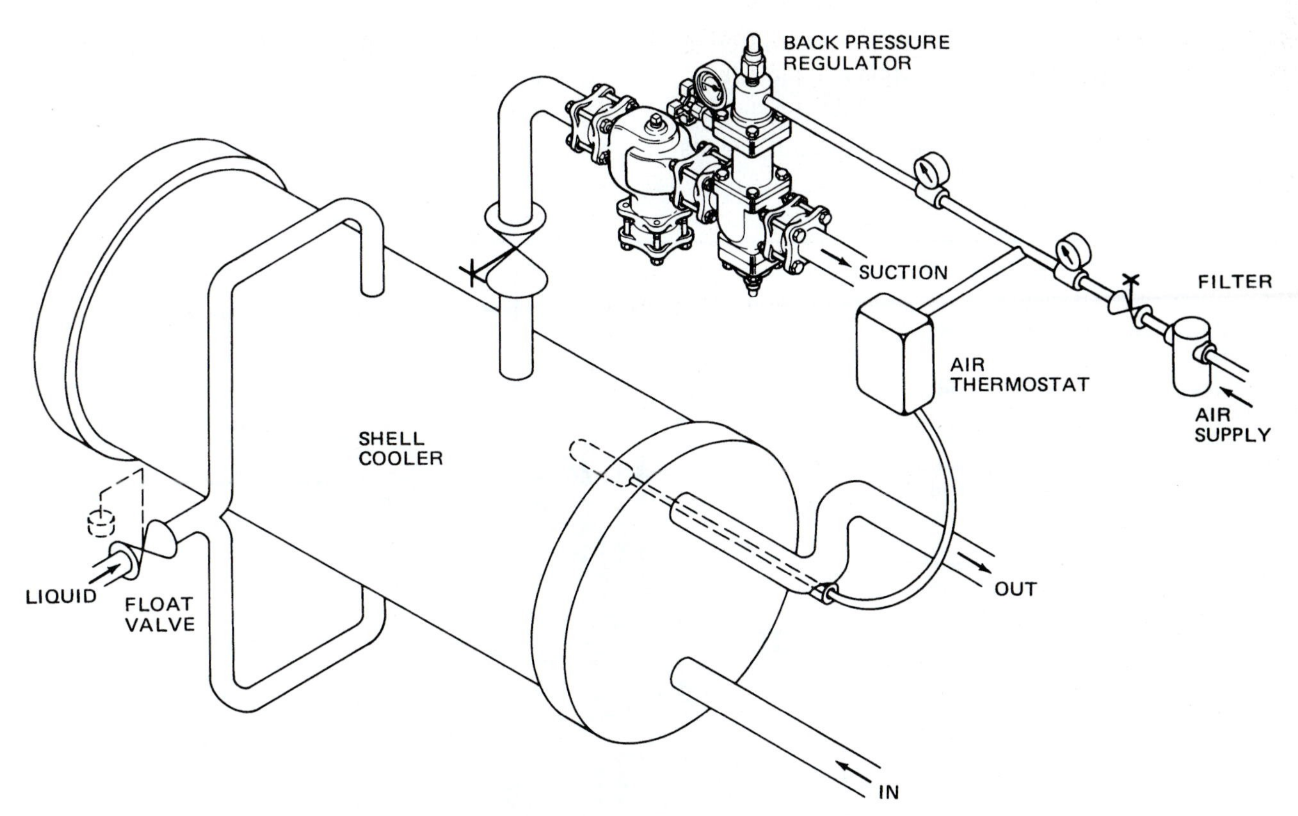

Fig. 10-26 *Air-compensating back-pressure regulator.* (Hubbell)

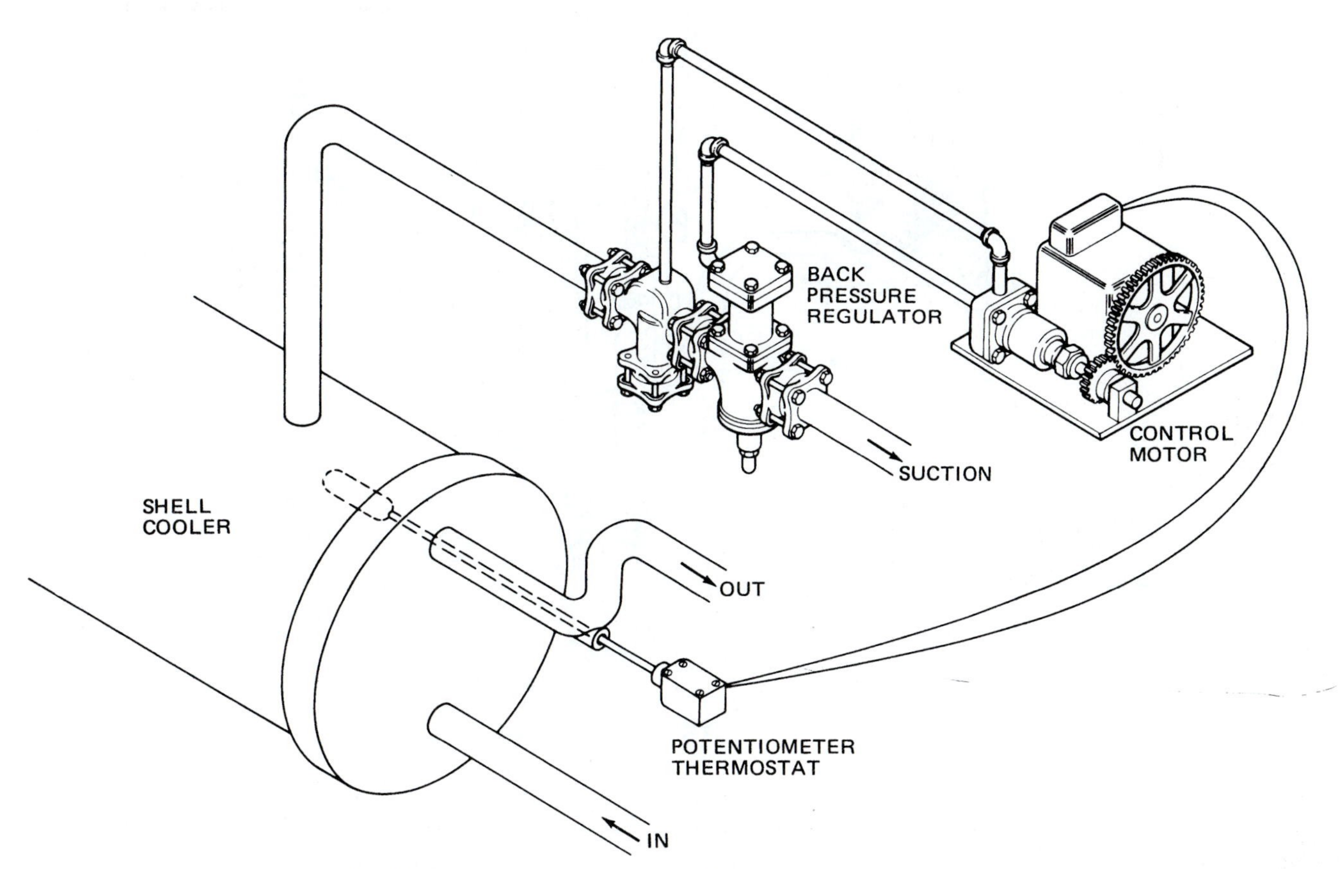

Fig. 10-27 *Electric-compensating back-pressure regulator.* (Hubbell)

Every valve has its own particular problems. A differential pressure-relief regulator valve is shown in Fig. 10-28.

A listing of its component parts should help you see the areas where trouble may occur. Table 10-3 lists possible causes and remedies.

The valve difficulties and remedies listed in Table 10-3 are for one particular type of valve. Manufacturers issue troubleshooting tables such as the one shown. These should be consulted when troubleshooting the valves of the evaporator system.

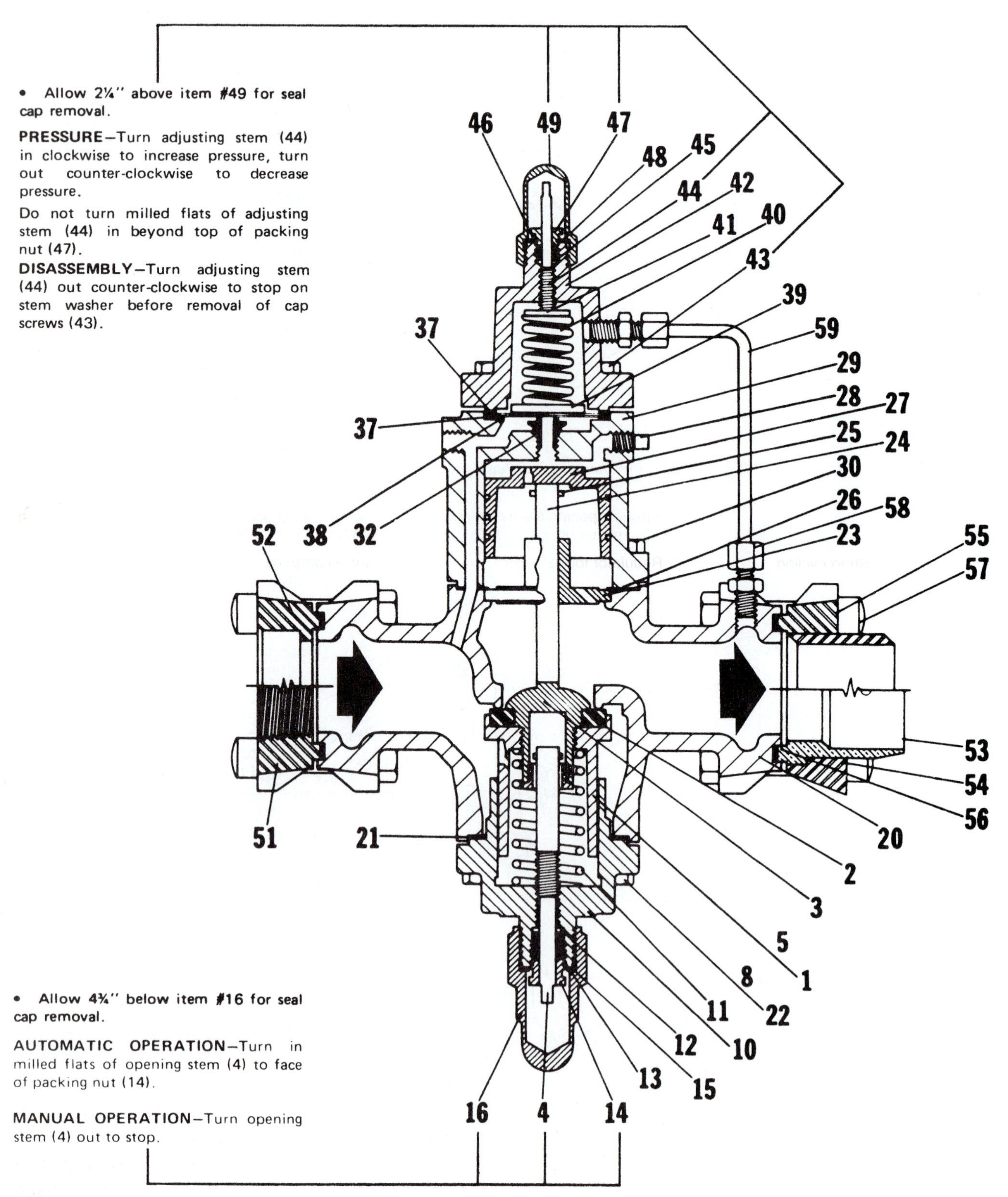

Fig. 10-28 *Differential pressure-relief regulators automatically maintain a preset differential between the upstream (inlet) and the downstream (outlet) side of the control valve.* (Hubbell)

1 DISC PISTON
2 SEAT DISC
3 SEAT DISC RETAINER
4 OPENING STEM
5 ROLL PIN
8 STEM RETAINING NUT
10 BOTTOM CAP
11 DISC PISTON SPRING
12 OPENING STEM WASHER
13 PACKING
14 PACKING NUT
15 SEAL CAP GASKET
16 SEAL CAP
BODY ASSEMBLY
20 BODY (SQUARE)
21 BOTTOM CAP GASKET
22 CAP SCREW
23 GUIDE PLATE
24 PUSH ROD
25 ROLL PIN
26 CYLINDER GASKET
CYLINDER ASSEMBLY
27 POWER PISTON
28 PIPE PLUG
29 CYLINDER
30 CAP SCREW
BONNET ASSEMBLY
32 PILOT SEAT BEAD
37 DIAPHRAGM GASKET
38 DIAPHRAGM
39 ADJUSTING SPRING PLATE
40 ADJUSTING SPRING
41 ADJUSTING SPRING GUIDE
42 BONNET
43 CAP SCREW
44 PRESSURE ADJUSTING STEM
45 PRESSURE ADJ. STEM WASHER
46 PACKING
47 PACKING NUT
48 SEAL CAP GASKET
49 SEAL CAP
50 NAME PLATE (NOT SHOWN)
51 THREADED FLANGE
52 SOCKET WELD FLANGE
53 O.D.S. FITTING
54 O.D.S. FLANGE
55 WELD NECK FLANGE
56 GASKET
57 BOLT & NUT
58 4 × 4 MALE CONNECTOR
59 0.250 DIA. TUBING

Fig. 10-28 *(Continued)*

Table 10-3 *Troubleshooting a Differential Pressure-Relief Regulator*

Symptom	Probable Cause	*Remedy
Erratic Operation. No adjustment. Regulator remains open.	Damaged pilot seat bead and/or diaphragms.	Replace.
	Dirt-binding power or disc pistons. Dirt lodged in seat disc or pilot seat bead area.	Clean, repair, and/or replace damaged items.
	Tubing sensing downstream pressure blocked.	Remove obstruction.
	Manual opening stem holding disc piston open.	Turn opening stem to automatic position.
Short cycling, hunting, or chattering.	Regulator too large for load conditions.	Install properly sized metered orifice control.
	Power piston bleed hole enlarged.	Replace or contact factory for sizing.
	O.D. of power piston worn, creating excessive clearance.	Replace piston.
Excessive pressure drop.	Regulator too small for load.	Replace with correctly sized regulator.
	Passage to sensing chamber blocked.	Remove obstructions.
	Strainer blocked.	Clean strainer—replace screen if damaged.
No adjustment over 90 psig.	Range spring rated at 2 to 90 psig.	Order range kit rated at 75 to 300 psig.
No adjustment under 2 psig.	Range spring rated at 2 to 90 psig.	Order range kit rated at 25 in. vacuum to 50 psig.

*If repair requires metal removal—replace part.

Noise in Hot-Gas Lines

Noise in hot-gas lines between interconnected compressors and evaporator condensers may be eliminated by the installation of mufflers. This noise may be particularly noticeable in large installations and is usually caused by the pulsations in gas flow caused by the reciprocating action of the compressors and velocity of the gas through the hot-gas line from the compressor. The proper location of a muffler is in a horizontal or down portion of the hot-gas line, immediately after leaving the compressor. It should never be installed in a riser. The problem of decreased system capacity due to excessive vertical lifts in the liquid line is usually solved by the installation of subcoolers. Check valves are used in the suction line of low-temperature fixtures when they are multiplied with high-temperature fixtures. Their use is most important when the condensing unit is regulated by low-pressure control.

REVIEW QUESTIONS

1. What is the purpose of an evaporator?
2. What is the purpose of a shell-and-tube chiller?
3. Describe the cooling cycle.
4. Describe the defrost cycle.
5. What is meant by direct expansion with top hot-gas feed?
6. What is a flooded-ceiling evaporator?
7. How are liquid refrigerant recirculating systems fed?
8. Where are equalizing orifices or globe valves used?
9. Why is the constant liquid control system used in refrigeration systems?
10. What is a dual-pressure regulator?

11
CHAPTER

Refrigerant: Flow Control

PERFORMANCE OBJECTIVES

After studying this chapter you should:

1. Know the various meter devices used for controlling refrigerants.
2. Know the fittings and hardware used in refrigerant control.
3. Know the sizes of refrigerant lines.
4. Know how driers, line strainers, and filters operate properly.
5. Know how a basic *thermostat-expansion valve* (TEV) operates.
6. Know how to install the TEV with multievaporators.
7. Know how to service crankcase pressure-regulating valves.
8. Know how to service and troubleshoot head pressure control valves.
9. Know how to service and troubleshoot discharge-bypass valves.
10. Know how to service and troubleshoot level-control valves.
11. Know why and how accumulators work.

Many devices are used in refrigeration and air-conditioning systems to control the flow of the refrigerant. Proper selection, installation, and maintenance hold the key to efficient performance under varying conditions.

METERING DEVICES

Metering devices divide the high side from the low side of the refrigeration system. Acting as a pressure control, metering devices allow the correct amount of refrigerant to pass into the evaporator.

Hand-Expansion Valve

Of the several types of metering devices, the hand-expansion valve is the simplest. See Fig. 11-1. Used only on manually controlled installations, the hand-expansion valve is merely a needle valve with a fine adjustment stem. When the machine is shut down, the hand-expansion valve must be closed to isolate the liquid line.

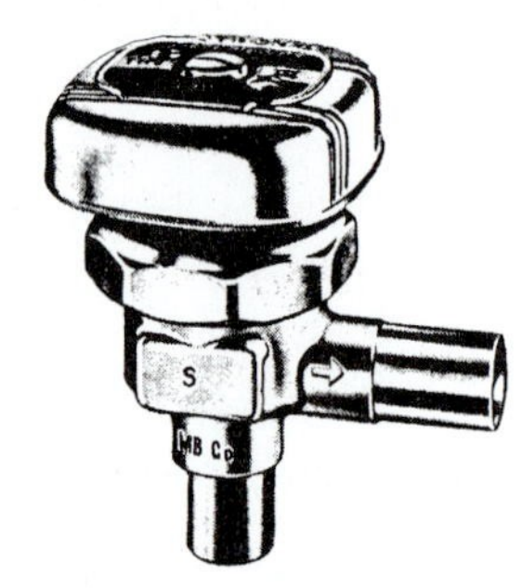

Fig. 11-1 *Hand-expansion valve.* (Mueller Brass)

Automatic-Expansion Valve

The automatic-expansion valve controls liquid flow by responding to the suction pressure of the unit acting on its diaphragm or bellows. See Fig. 11-2. When the valve opens, liquid refrigerant passes into the evaporator. The resulting increase in pressure in the evaporator closes the valve.

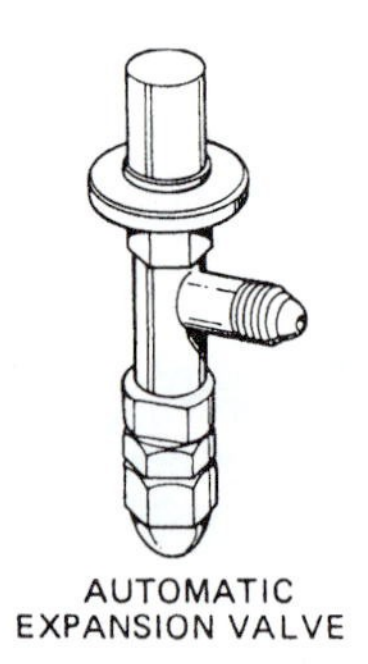

Fig. 11-2 *Automatic-expansion valve.* (Mueller Brass)

Meanwhile, the compressor is pulling gas away from the coils, reducing the pressure. This pressure reduction allows the expansion valve to open again. In operation, the valve never quite closes. The needle floats just off the seat and opens wide when the unit calls for refrigeration. When the machine is shut down; the pressure building up in the coils closes the expansion valve until the unit starts up.

Thermostatic-Expansion Valve

The TEV, used primarily in commercial refrigeration and in air conditioning, is a refinement of the automatic-expansion valve. See Fig. 11-3. A bellows or diaphragm responds to pressure from a remote bulb charged with a substance similar to the refrigerant in the system. The bulb is attached to the suction line near the evaporator outlet. It is connected to the expansion valve by a capillary tube.

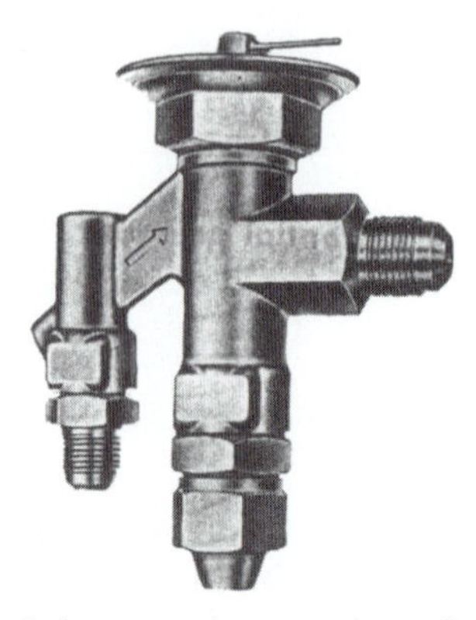

Fig. 11-3 *A thermostatic-expansion valve.* (Mueller Brass)

In operation, the TEV keeps the frost line of the unit at the desired location by reacting to the superheat of the suction gas. Superheat cannot be present until all liquid refrigerant in the evaporator has been vaporized. Thus, it is possible to obtain a range of evaporator temperatures by adjusting the superheat control of the TEV.

The prime importance of this type of metering device is its ability to prevent the flood-back of slugs of liquid through the suction line to the compressor. If this liquid returns to the compressor, it could damage it. The compressor is designed to pump vapors, not liquids.

Capillary Tubing

Small-bore capillary tubing is used as a metering device. It is used on everything, from the household refrigerator to the heat pump. Essentially, it is a carefully measured length of very small diameter tubing. It creates a predetermined pressure drop in the system. The capillary has no moving parts.

Because a capillary tube cannot stop the flow of refrigerant when the condensing unit stops, such a refrigeration unit will always equalize high-side and low-side pressures on the off-cycle. For this reason, it is important that the refrigerant charge be of such a quantity that it can be held on the low side of the system without damage to the compressor. In a charge of several pounds, this "critical charge" of refrigerant may have to be carefully weighed.

An accumulator, or enlarged chamber, is frequently provided on a capillary-tube system to prevent slugs of liquid refrigerant from being carried into the suction line.

Float Valve

A float valve, either high side or low side, can serve as a metering device. The high-side float, located in the liquid line, allows the liquid to flow into the low side when a sufficient amount of refrigerant has been condensed to move the float ball. No liquid remains in the receiver. A charge of refrigerant just sufficient to fill the coils is put into the system on installation. This type of float, formerly used extensively, is now limited to use in certain types of industrial and commercial systems.

The low-side float valve keeps the liquid level constant in the evaporator. It is used in flooded-type evaporators where the medium being cooled flows through tubes in a bath of refrigerant. The low-side float is more critical in operation than the high-side float and must be manufactured more precisely. A malfunction will cause the evaporator to fill during shutdown. This condition will result in serious pounding and probable compressor trouble on start-up.

Needle valves, either diaphragm or packed type, may be used as hand-expansion valves. As such, they are usually installed in a bypass line around an automatic-or TEV. They are placed in operation when the normal control is out of order or is removed for repairs.

FITTINGS AND HARDWARE

Modern refrigerants can escape through the most minute openings. Since porosity in a fitting could create such an opening, it is mandatory that porosity be eliminated from fittings and accessories that are to be used with refrigerants. See Fig. 11-4. One way to eliminate porosity in fittings is to either forge or draw them from brass rod. This creates a final grain structure that prevents the seepage of refrigerant due to porosity. The threads on fittings must be machined with some degree of accuracy to prevent leaks. Solder-type fittings should be made of wrought copper, brass rod, or brass forgings. See Fig. 11-5. This eliminates the possibility of leaks due to porosity of the metal. The tube is not weakened by the cutting of threads, as is the case with iron pipe. A soldered joint allows the use of a much lighter wall tube with complete safety and with significant cost savings.

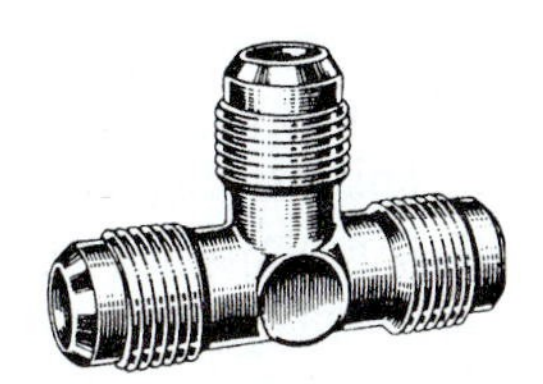

Fig. 11-4 *45°-flare fitting.* (Mueller Brass)

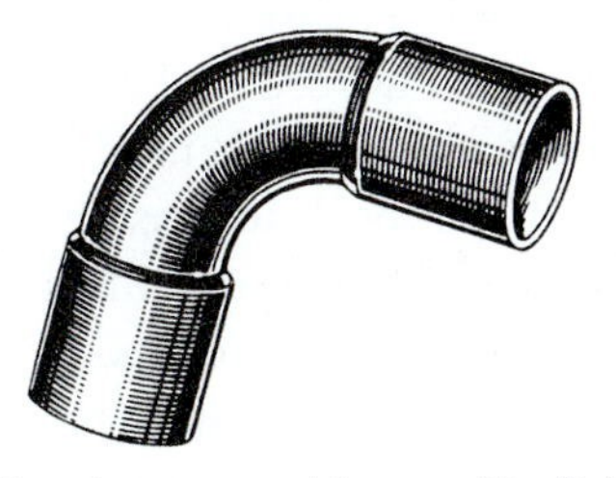

Fig. 11-5 *Wrought copper, solder-to-solder fitting.* (Mueller Brass)

One advantage of copper pipe over iron is the elimination of scale and corrosion. In service, a light coating of copper oxide forms on the outside of the copper tube. This coating prevents chemical attack. There is no "rusting out" of copper tube.

Copper Tubing

For flare-fitting applications, seamless soft copper tube is recommended. See Fig. 11-6. This tube is furnished with sealed ends. It is supplied in 50 ft lengths, in sizes

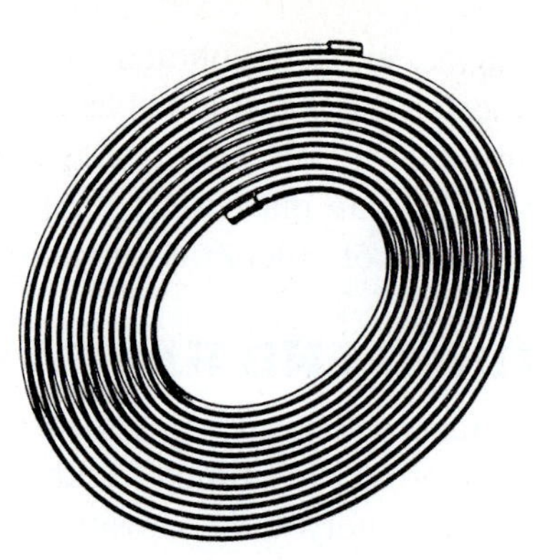

Fig. 11-6 *Refrigeration service tube in a 50-ft coil.* (Mueller Brass)

from $^{1}/_{8}$ through $^{3}/_{4}$ in. *outside diameter* (OD) for flaring and through OD for soldering.

The chief demand for this tube is in sizes from $^{1}/_{4}$ through $^{5}/_{8}$ in. OD. Sizes smaller than $^{1}/_{4}$ in. are seldom used in commercial refrigeration. To uncoil the tube without kinks, hold one free end against the floor, or on a bench, and uncoil along the floor or bench. The tube may be cut to length with a hacksaw or tube cutter. In either case, deburr the end before flaring. Bending is readily accomplished with either external or internal bending springs or lever-type bending tools.

ACR (*air-conditioning, refrigeration*) tube is frequently used. It is cleaned, degreased, dried, and end-sealed at the factory. This assures the user that he or she is installing a clean, trouble-free tube. Some tubing is available with an inert gas (nitrogen). See Fig. 11-7. The *nitrogenized*-ACR tube is purged, charged with clean, dry nitrogen, and then sealed with reusable plugs. After cutting the tube, the remaining length can easily be replugged. The remaining nitrogen limits excess oxides during succeeding brazing operations. It comes in 20 ft lengths. Type-L hard tube has from $^{3}/_{8}$ through $3^{1}/_{8}$ in. OD. Type-K tube is also available.

Where tubing will be exposed inside food compartments, tinned copper is recommended. Type-L, hard-temper copper tube is recommended for field installations using solder-type fittings. Type M is sufficiently strong for any pressures of the commonly used refrigerants. However, it is used chiefly in manufactured assemblies where external damage to the tube is not as likely as in field installations. For maximum protection against possible external damage to refrigerant lines, a few cities require the use of type-K copper tube.

Line

Correct line sizes are essential to obtain maximum efficiency from refrigeration equipment. In supermarkets, for example, the long lines running under the floor from the display cases to the machine room at the rear of the store must be fully engineered. Otherwise, problems of oil return, slugging, or erratic refrigeration are quite likely. Table 11-1 lists refrigerant-line sizes. When available, the manufacturer's recommendations must be followed regarding step-sizing, risers, traps, and the like. Available information on Refrigerant 502 claims performance at temperatures below 5°F (−15°C) when compared with Refrigerant 22. The newer refrigerants like 502a are said to be equivalent in performance to the older designated refrigerant.

Solder

Each solder is designed for a certain job. For instance, 50-50 solder, which consists of 50 percent tin and 50 percent lead, will not function well in some instances. In fact, 50-50 solder will deteriorate in some refrigerated food-storage compartments where normally wet-refrigerant lines and high carbon dioxide content are present. For this reason, No. 95 solder is recommended. It has 95 percent tin and 5 percent antimony. Number 122 solder (45 percent silver brazing alloy) is used for joints in refrigerant lines where 50-50 solder may deteriorate.

Suction Line P-Traps

For years, the P-trap was made by forming two or more fittings. It has now become available in one piece. See Fig. 11-8. The newer one-piece P-trap promotes efficient oil migration in refrigeration systems. This is increasingly important today. Many large food markets place their compressors and condensers on balconies or mezzanines. Such remote condensing units are

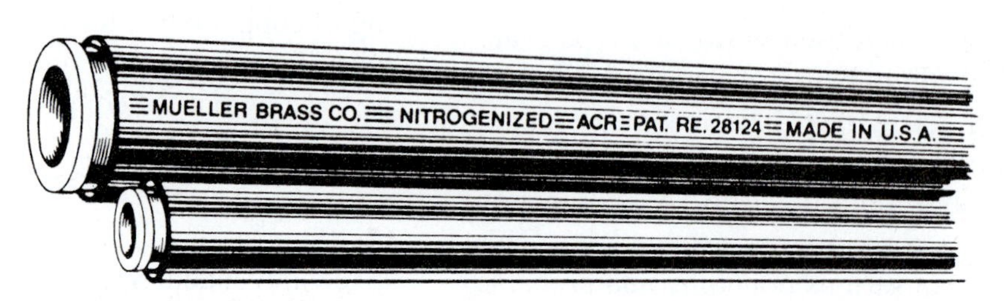

Fig. 11-7 *Nitrogenized ACR-copper tube.* (Mueller Brass)

Table 11-1 *Sizes of Refrigerant Lines*

	Refrigerant 12			Refrigerant 22			Refrigerant 40			Refrigerant 502		
		Suction Line			Suction Line			Suction Line			Suction Line	
Btu Per Hour	Liquid Line	5°F [–15°C]	40°F [4.4°C]	Liquid Line	5°F [–15°C]	40°F [4.4°C]	Liquid Line	5°F [–15°C]	40°F [4.4°C]	Liquid Line	5°F [–15°C]	40°F [4.4°C]
3,000	1/4	1/2	1/2	1/4	1/2	1/2	1/4	1/2	1/2	1/4	1/2	1/2
6,000	3/8	5/8	5/8	3/8	5/8	5/8	1/4	1/2	1/2	3/8	5/8	5/8
9,000	3/8	7/8	5/8	3/8	7/8	5/8	3/8	5/8	5/8	3/8	7/8	5/8
12,000	3/8	1 1/8	7/8	3/8	7/8	7/8	3/8	7/8	7/8	3/8	7/8	7/8
15,000	3/8	1 1/8	7/8	3/8	1 1/8	7/8	3/8	7/8	7/8	3/8	1 1/8	7/8
18,000	3/8	1 1/8	7/8	3/8	1 1/8	7/8	3/8	1 1/8	7/8	3/8	1 1/8	7/8
21,000	1/2	1 1/8	1 1/8	1/2	1 1/8	1 1/8	3/8	1 1/8	7/8	1/2	1 1/8	1 1/8
24,000	1/2	1 3/8	1 1/8	1/2	1 1/8	1 1/8	1/2	1 1/8	7/8	1/2	1 1/8	1 1/8
30,000	5/8	1 3/8	1 1/8	1/2	1 3/8	1 1/8	1/2	1 1/8	1 1/8	5/8	1 3/8	1 1/8
36,000	5/8	1 3/8	1 1/8	5/8	1 3/8	1 1/8	1/2	1 3/8	1 1/8	5/8	1 3/8	1 1/8
42,000	5/8	1 5/8	1 3/8	5/8	1 3/8	1 3/8	1/2	1 3/8	1 1/8	5/8	1 3/8	1 3/8
48,000	5/8	1 5/8	1 3/8	5/8	1 5/8	1 3/8	1/2	1 3/8	1 1/8	5/8	1 5/8	1 3/8
54,000	5/8	1 5/8	1 3/8	5/8	1 5/8	1 3/8	5/8	1 3/8	1 1/8	5/8	1 5/8	1 3/8
60,000	7/8	1 5/8	1 3/8	5/8	1 5/8	1 3/8	5/8	1 5/8	1 3/8	7/8	1 5/8	1 3/8
72,000	7/8	2 1/8	1 5/8	7/8	1 5/8	1 3/8	5/8	1 5/8	1 3/8	7/8	1 5/8	1 3/8
96,000	7/8	2 1/8	1 5/8	7/8	2 1/8	1 5/8	5/8	2 1/8	1 5/8	7/8	2 1/8	1 5/8
108,000	7/8	2 5/8	2 1/8	7/8	2 1/8	1 5/8	7/8	2 1/8	1 5/8	7/8	2 1/8	1 5/8
120,000	7/8	2 5/8	2 1/8	7/8	2 1/8	1 5/8	7/8	2 1/8	1 5/8	7/8	2 1/8	1 5/8
150,000	1 1/8	2 5/8	2 1/8	7/8	2 1/8	2 1/8	7/8	2 1/8	2 1/8	1 1/8	2 1/8	2 1/8
180,000	1 1/8	2 5/8	2 1/8	1 1/8	2 5/8	2 1/8	7/8	2 5/8	2 1/8	1 1/8	2 5/8	2 1/8
210,000	1 1/8	3 1/8	2 1/8	1 1/8	2 5/8	2 1/8	7/8	2 5/8	2 1/8	1 1/8	2 5/8	2 1/8
240,000	1 3/8	3 1/8	2 5/8	1 3/8	2 5/8	2 1/8	7/8	2 5/8	2 1/8	1 3/8	2 5/8	2 1/8
300,000	1 3/8	3 1/8	2 5/8	1 3/8	3 1/8	2 5/8	1 1/8	2 5/8	2 1/8	1 3/8	3 1/8	2 5/8
360,000	1 3/8	3 5/8	2 5/8	1 1/8	3 1/8	2 5/8	1 1/8	3 1/8	2 5/8	1 3/8	3 1/8	2 5/8
420,000	1 5/8	3 5/8	3 1/8	1 3/8	3 1/8	2 5/8	1 1/8	3 1/8	2 5/8	1 5/8	3 1/8	2 5/8
480,000	1 5/8	4 1/8	3 1/8	1 5/8	3 5/8	3 1/8	1 1/8	3 1/8	2 5/8	1 5/8	3 5/8	3 1/8
540,000	1 5/8	4 1/8	3 1/8	1 5/8	3 5/8	3 1/8	1 3/8	3 5/8	3 1/8	1 5/8	3 5/8	3 1/8
600,000	1 5/8	4 1/8	3 1/8	1 5/8	4 1/8	3 1/8	1 3/8	3 5/8	3 1/8	1 5/8	4 1/8	3 1/8

To convert Btu per hour to tons of refrigeration—divide by 12,000
Suction temperature, condensing medium, compressor design, and many other factors determine horsepower required for a ton of refrigerating capacity. Consult ASHRAE Handbook.
Mueller Brass

likely to have long horizontal suction lines or vertical risers exceeding 3 ft in height in the suction line. In such cases, the oil concentration in the circulating refrigerant may be expected to be above 0.6 percent. A low-vapor velocity may be encountered. This results in unsatisfactory oil return to the compressor. Tests have proven that with a P-trap installed, vapor velocity can fall as low as 160 ft/min and satisfactory oil return can still be achieved. The P-trap drains the oil from the horizontal runs approaching the risers. This oil, in turn, migrates up through the riser to the compressor in one of the three different forms: as a rippling oil film, as a mist, or as a transparent-colloidal dispersion in the vaporized refrigerant. The method of oil migration depends upon the vapor velocity in the suction line.

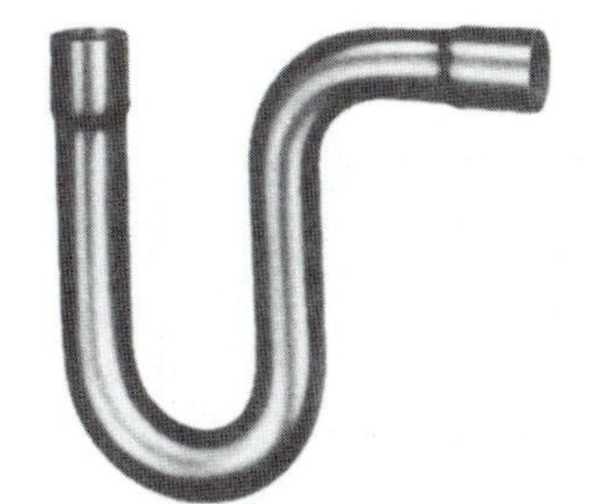

Fig. 11-8 *Suction line P-trap.* (Mueller Brass)

Compressor Valves

There are three types of compressor valves:

- Adjustable
- Double port
- Single port

Open and semihermetic compressors are usually fitted with compressor-service valves, one each at the suction and discharge ports. The service valve has no operating function. Nevertheless, it is indispensable when service is to be performed on any part of the refrigeration system. See Fig. 11-9.

Compressor-service valves are back-seating. They are constructed so that the stem forms a seal against a seat, whether the stem is full forward or full backward. Valve packing is depended upon only when the stem is

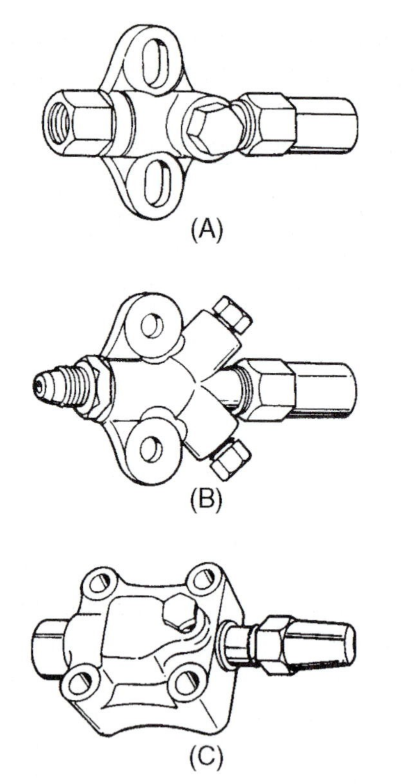

Fig. 11-9 *Compressor valves.* (A) *Adjustable compression valve.* (B) *Double-port compressor valve.* (C) *Single-port compressor valve.* (Mueller Brass)

in the intermediate position. In one style of construction, the front seat, including the one connection is threaded. Silver is brazed into the body after the stem has been assembled. When the valve is full open (normal position when the unit is running), the gage and charging port plug or cap may be removed without loss of refrigerant. A charging line or pressure gage may be attached to this side port. It is also possible to repack the valve without interruption of service.

Line Valves

Line valves are essential components of refrigerant systems. See Fig. 11-10. Installed in key locations, line valves make it possible to isolate any portion of a system or, in a multiple hookup, to separate one system from the rest. Local codes frequently specify the location of line valves in commercial and industrial refrigeration and air-conditioning systems.

There are two types of line valves—packed and packless. They must be designed to prevent refrigerant leakage. Since refrigerants are difficult to retain, packed valves are usually equipped with seal caps. Some seal caps are designed to be removed and used as wrenches for operating the valves.

In large packed valves, such as that shown in Fig. 11-11, O-rings are used as seals between the bonnets and valve bodies. They are available in either straight through ($^7/_8$ to $4^1/_8$ in.) or angle type ($7^7/_8$ through $3^1/_8$ in.) construction.

The packless design is often preferred for smaller valves. The packless-type valve is used to good advantage on

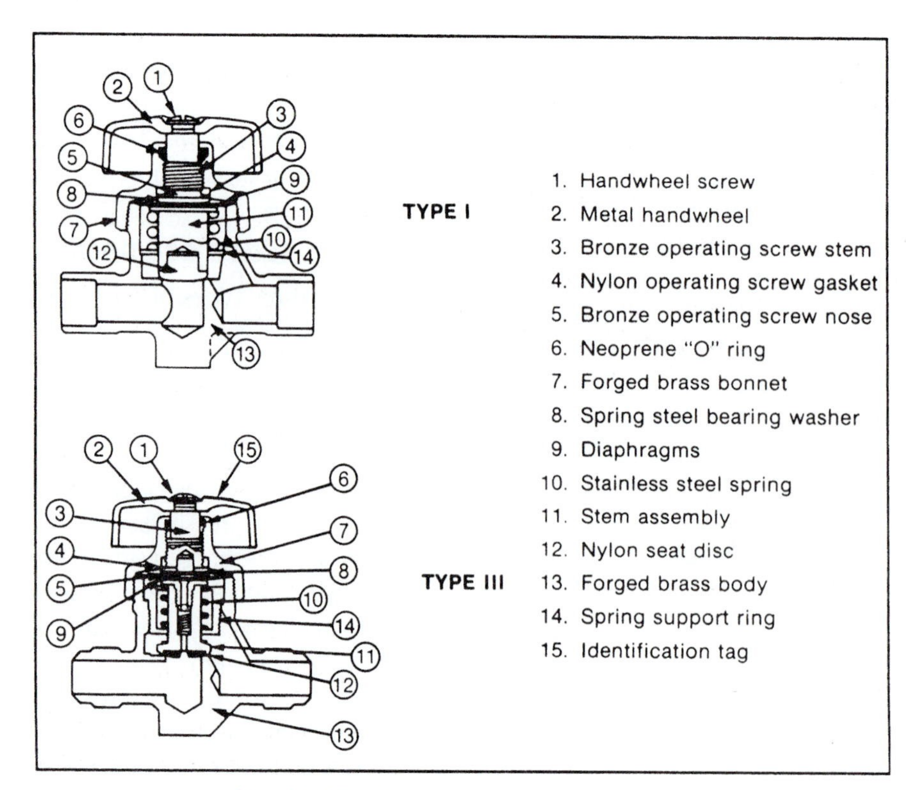

Fig. 11-10 *Packless-line valves.* (Mueller Brass)

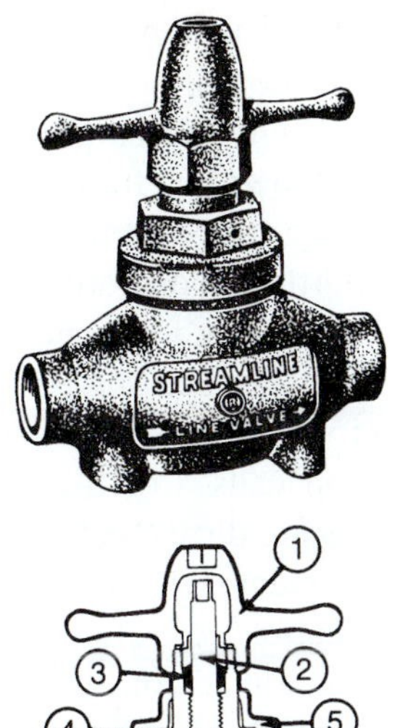

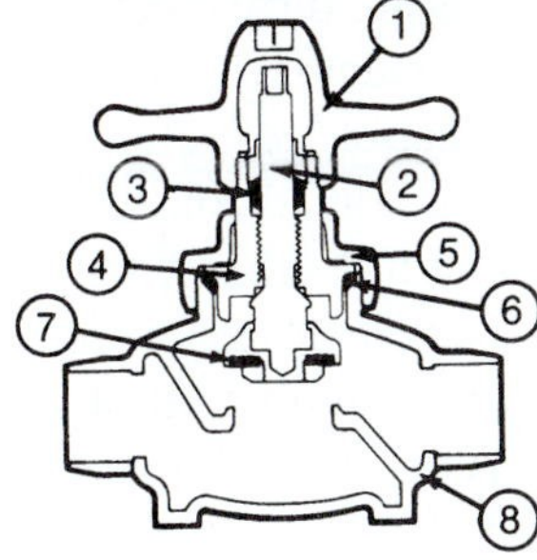

1. Cast bronze wing-seal cap
2. Bronze stem
3. Molded stem packing
4. Forged brass bonnet in sizes over 1⅛
5. Forged brass union collar
6. Neoprene "O" ring
7. Nylon seat disc
8. Cast bronze body

Fig. 11-11 *Packed-line valve.* (Mueller Brass)

charging boards. The valves contain triple diaphragms—one of phosphor bronze and two of stainless steel. These valves must be frost-proof. They must be designed for use where condensation is likely to occur. During the off-cycle, condensation may seep down the stem of a nonfrost-proof valve into the bonnet. There, it will alternately freeze and thaw. The eventual buildup of ice against the diaphragms may close the valve. Another factor that should be considered is whether or not the valve has back-seating. This prevents all pressure pulsations while the valve is open. The back-seating should allow inspection of the diaphragms without shutting down the system.

DRIERS, LINE STRAINERS, AND FILTERS

Of the three items to be examined and understood, one of the first to be considered is the drier.

Driers

Most authorities agree that moisture is the most detrimental material in a refrigeration system. A unit can stand only minute amounts of water. For this reason, most refrigeration and air-conditioning systems, both field and factory-assembled, contain driers. See Fig. 11-12.

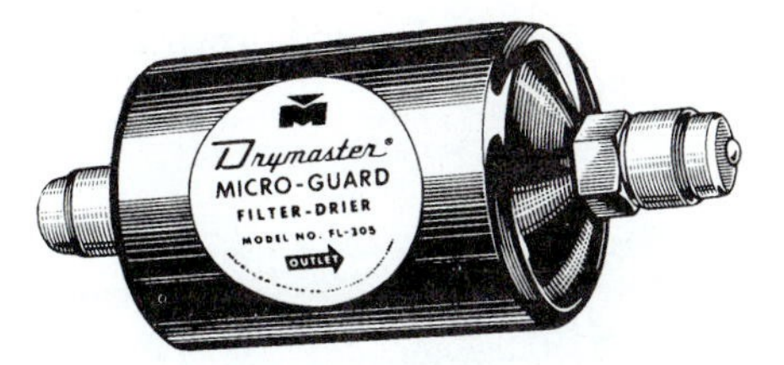

Fig. 11-12 *Filter drier.* (Mueller Brass)

Moisture Moisture or water is always present in refrigeration systems. Acceptable limits vary from one unit to another and from one refrigerant to another. Moisture is harmful even if freeze-ups do not occur. Moisture is an important factor in the formation of acids, sludge, and corrosion. To be safe, keep the moisture level as low as possible.

Moisture will react with today's halogen-type refrigerants to form harmful hydrochloric and hydrofluoric acids within the system. To minimize the possibility of freeze-up or corrosion, the following maximum safe limits of moisture should be observed:

> Note: These are halogen-type refrigerants.

Refrigerant 12	15 parts per million (15 ppm)
Refrigerant 22	60 parts per million (60 ppm)
Refrigerant 502	30 parts per million (30 ppm)

If the moisture exceeds these figures, corrosion is possible. Also, excess water may freeze at the metering device if the system operates below 32°F (0°C). Freeze-ups do not occur in air-conditioning systems where evaporator temperatures are normally above 40°F (4.4°C).

A drier charged with a moisture-removing substance and installed in the liquid line is the most practical way to remove moisture. With a drier of the proper size, excess water is stored in the drier. Here, it can neither react with the refrigerant nor travel through the system.

Many materials have been tried as desiccants, or drying agents. Today, the desiccant materials most commonly used are:

- Silica gel
- Activated alumina
- Calcium sulfate
- Zeolite-type materials

These are known as molecular sieves and micro traps. The total drier design considers not only drying and filtering, but also maintaining maximum refrigerant flow.

Filter driers must allow free flow of refrigerant. They must also prevent fine particles of the adsorbent or other foreign matter from passing through to the metering device, usually located downstream from the drier. See Fig. 11-13.

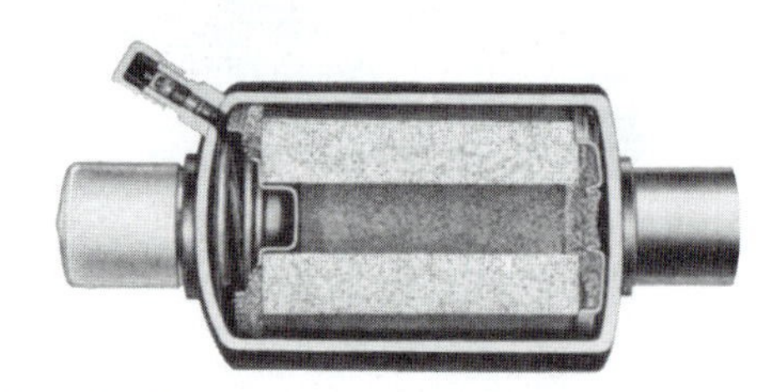

Fig. 11-13 *Suction-line filter drier.* (Mueller Brass)

Dirt Dirt, sludge, flux, and metallic particles are frequently found in refrigeration systems. Numerous metallic contaminants—cast-iron dust, rust, and scale, plus steel, copper, and brass filings—can damage cylinder walls and bearings. They can plug capillary tubes and TEV screens. These contaminants are catalytic and contribute to decomposition of the refrigerant-oil mixture at high temperatures.

Acids By themselves, Refrigerants 12 and 22 are very stable, even when heated to a high temperature. However, under some conditions, reactions occur that can result in the formation of acids. For example, at elevated temperatures, Refrigerant 12 will react with the oil to form hydrochloric and hydrofluoric acids. These acids are usually present as a gas in the system and are highly corrosive. Where an "acid acceptor," such as electrical insulation paper, is present, Refrigerant 22 will decompose at high temperatures to form hydrochloric acid. The reaction of refrigerants with water may cause hydrolysis and the formation of hydrochloric and hydrofluoric acids. In ordinary usage this reaction is negligible. However, in a very wet system operating at abnormally high temperatures, some hydrolysis may occur.

All of these reactions are increased by elevated temperature and are catalytic in effect. They result in the formation of corrosive compounds.

Another source of acidity in refrigeration systems is the organic acid formed from oil breakdown. Appreciable amounts of organic acid are found in the majority of oil samples analyzed in the laboratory. These acids will also corrode the metals in a system. Therefore, they must be removed.

Acid may be neutralized by the introduction of an alkali, but the chemical combination of the two creates further hazards. They release additional moisture and form a salt. Both of these are detrimental to the system.

Sludge and Varnish The utmost care may be taken in the design and fabrication of a system. Nonetheless, in operation, unusually high-discharge temperatures will cause the oil to break down and form sludge and varnish.

Temperatures may vary in different makes of compressors and under different operating conditions. Temperatures of 265°F (130°C) are not unusual at the discharge valve under normal operation. Temperatures well above 300°F (150°C) frequently occur under unusual conditions. Common causes of high temperatures in refrigeration systems are dirty condensers, noncondensable gases in the condenser, high compression ratio, high superheat of suction gas returned to the compressor, and fan failure on forced convection condensers.

In addition to high-discharge temperatures, certain catalytic metals contribute to oil-refrigerant mixture breakdown. The most significant of these is iron. It is used in all systems and is an active catalyst. Copper is a catalyst also, but its action is slower.

However, the end result is the same. The reaction causes sludges and other corrosive materials that will hinder the normal operation of compressor valves and control devices. In addition, air in a system will also accelerate oil deterioration.

Line Strainers and Filters

It is impossible to keep all foreign matter out of factory or field-assembled refrigeration systems. Core sand from the compressor casting, brazing oxides in piping or tubing, chips from cutting or baring, sawdust, and dirt are found in most refrigeration systems. This is especially so with field-assembled installations. Tubing, for example, may have been exposed to air-carried dirt for several days. Cleanliness is difficult to maintain in the field. All tubing to be used for refrigeration applications should be protected by capping or sealing. It should be recapped or sealed after each use.

Moving through the system with the flowing refrigerant, particles of foreign matter may score critical moving parts or clog orifices. To prevent such damage, strainers or filters are frequently installed in the system. See Fig. 11-14. A filter drier placed in the liquid line ahead of the metering device is the normal precaution. However, on multiple installations, it is usual to install a strainer upstream of each metering device just ahead of key valves and controls. To protect the compressor, most engineers also specify filters for the suction line.

The simplest strainer consists of a set of metal screens of the proper mesh. See Fig. 11-15. Adding felt pads or asbestos cloth creates a very effective filter, rather than a mere strainer. A cellulose fiber core as used in Fig. 11-13 is also very effective for this purpose.

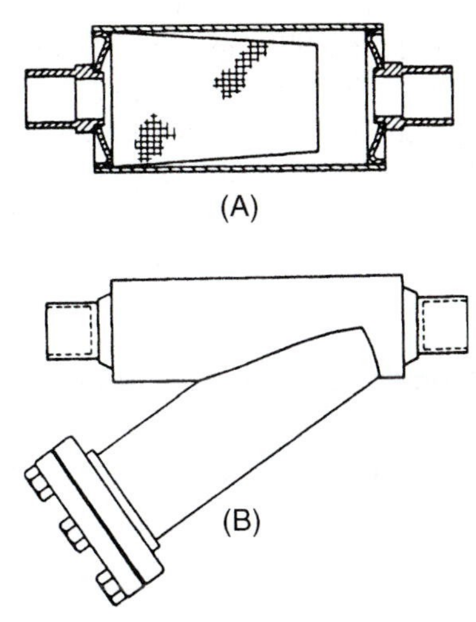

Fig. 11-14 *Strainers.* (A) *Non-cleanable strainer.* (B) *Y-type line strainer.* (Sporlan Valve)

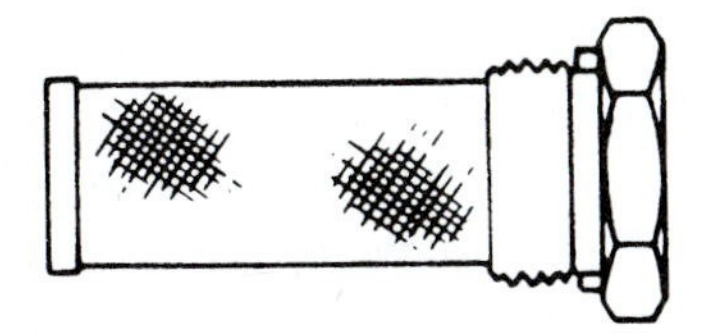

Fig. 11-15 *Strainer-filter, removable metal screen.* (Mueller Brass)

Suction-line strainers and filters are designed with sufficient flow capacity to prevent excessive pressure drop. Since determining the need for cleaning or replacement of a suction-line filter is related to pressure drop, some designs are offered with pressure taps. These permit pressure gage installation to determine the degree of pressure drop.

Strainers are made in several designs. They are also supplied in a wide range of sizes. Screen areas are large. In cartridge-type strainers, provision is made for removal of the screens or filters. See Fig. 11-16.

LIQUID INDICATORS

Liquid indicators are inserted in a refrigerant line to indicate the amount of refrigerant in a system. See Fig. 11-17. Proper operation of a refrigeration system depends upon there being the correct amount of refrigerant in the unit. Looking through the window of a liquid indicator is the simplest way to determine whether there is a refrigerant shortage. A shortage of refrigerant may be due to a leak in the system or due to failure to charge enough refrigerant into a unit after field service.

Liquid indicators normally disclose a shortage of refrigerant by the appearance of bubbles. Some use a special assembly that shows by the appearance of the word "FULL" that there is sufficient refrigerant at that point in the system.

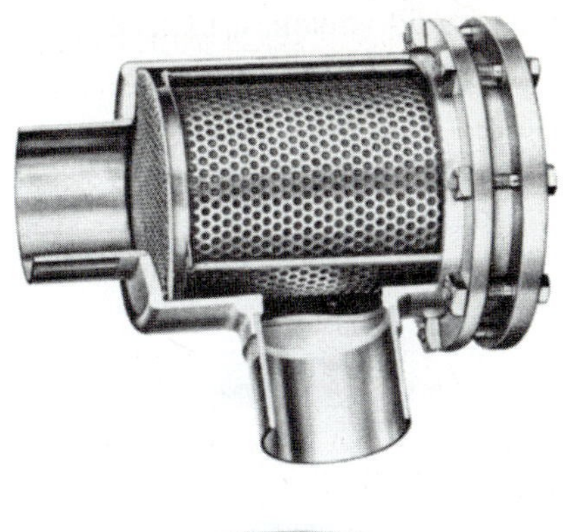

Fig. 11-16 *A strainer and its replacement-type filter.* (Sporlan Valve)

Liquid indicators are manufactured in single-port, double-port, and straight-through types. In the single- and double-port indicators, an internal compression bushing seals the glass firmly against the body. Assemblies are furnished with a protective dust cap or seal cap.

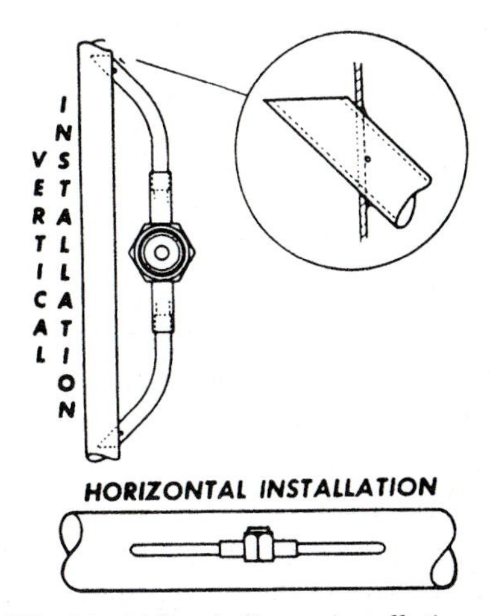

Fig. 11-17 *Liquid-line indicator installation.* (Sporlan Valve)

To observe the liquid stream, it is necessary to remove the cap. See Fig. 11-18.

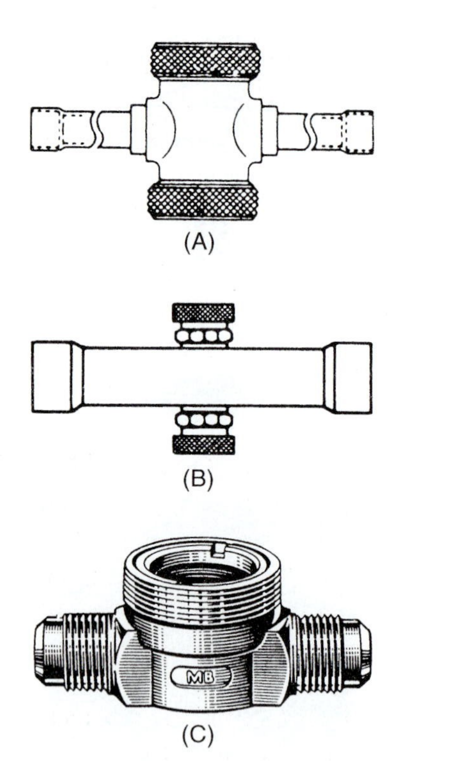

Fig. 11-18 *Liquid indicators.* (A) *and* (B) *Double port.* (C) *Single port.* *(Mueller Brass)*

A relatively new addition to the function of the liquid indicator is that of moisture detection. Special materials used in the ports of liquid-moisture indicators change color to indicate excessive moisture in the system.

Indicators with solder-type ends, but without extended ends, are normally furnished disassembled so the heating required for soldering will not damage glass or gaskets. In addition, single- and double-port types are supplied assembled with extended ends. These make it possible to solder without damaging the indicator, as long as normal precautions are observed.

Construction

The indicator is a porous filter paper impregnated with a chemical salt that is sensitive to moisture. The salt changes color according to the moisture content (relative saturation) of the refrigerant. The indicator changes color below moisture levels generally accepted as a safe operation range. This device is not suitable for use with ammonia or sulfur dioxide. However, it does have a full application with Refrigerants 11, 12, 22, 113, 114, 500, and 502. See Table 11-2. The indicator should be installed after the filter drier and ahead of the expansion device. Prior to installation, the indicator will be yellow, indicating a wet condition. This is a normal situation, since the air in contact with the element is above 0.5 percent relative humidity. This does not affect the operation or calibration of the indicator. As soon as it is installed in a system, the indicator element will begin to change according to the moisture content in the refrigerant. The action of the indicator element is completely reversible.

Table 11-2 *Moisture Content (In Parts Per Million)*

Unit Shows	Liquid Line Temp. (°F.)	Refrigerants 11 & 12			Refrigerant 22			Refrigerant 500			Refrigerants 502, 113, & 114		
		75°	**100°**	125°	75°	**100°**	125°	75°	**100°**	125°	75°	**100°**	125°
Green DRY		Below 5	Below **10**	Below 20	Below 30	Below **45**	Below 60	Below 40	Below **60**	Below 100	Below 10	Below **20**	Below 30
Chartreuse CAUTION		5-15	**10-30**	20-50	30-90	**45-130**	60-180	40-90	**60-150**	100-230	10-45	**20-65**	30-110
Yellow WET		Above 15	Above **30**	Above 50	Above 90	Above **130**	Above 180	Above 90	Above **150**	Above 230	Above 45	Above **65**	Above 110

Sporlan Valve

BOLD figures are for the average design conditions of refrigerant liquid lines operating at 100° F. Since the actual temperature is not critical, a satisfactory estimate can be made by comparing it to body temperature. If it feels cool to touch, use 75° F. If it feels warm, use 125° F. column figures.

The unit calibration information given above is based on detailed experimental data for Refrigerants 12, 22, 500, 502, and 113. The calibration information on other refrigerants and solvents was obtained from a comparison of their properties with 12, 22, 500, 502, and 113. For the less common liquids, the following moisture calibration is suggested.

Refrigerant 13......... use ...... "12" calibration
Refrigerant 21......... use ...... "22" calibration
Trichloroethylene.......use ...... "22" calibration
Perchloroethylene..........use ...... "113" calibration
Carbon Tetrachloride.......use........"12" calibration
Propane or Butane.........use......."500" calibration

AIR TEST—Recent tests on AIR show that the unit changes color in the range of 0.5 to 2.0%. In ordinary air lines this means that the unit will change color at dew points in the range of −40 to −60° F.

The element will change color as often as the moisture content of the system varies. Some change may take place rapidly at the start-up of a new system or after replacement of a drier on existing installations. However, the equipment should be operated for about 12 h to allow the system to reach equilibrium before deciding if the drier needs to be changed.

Installation

Indicators with $^1/_4$ through $1^1/_8$ in. *outside diameter flanged* (ODF) connections should not be disassembled in the field for brazing or any other purpose. The long fittings on sweat models are copper-plated steel and do not conduct heat as readily as copper fittings.

On indicators with $1^3/_8$, $1^5/_8$, and $2^1/_8$ in., ODF connections, the indicator cartridge must be removed from the brass-saddle fitting before brazing the indicator in the main liquid line. It is shipped hand tight for easy removal.

Bypass Installations

On systems having liquid lines larger than $2^1/_8$ in. OD, the indicator should be installed in a bypass line. During the operating cycle, this will provide sufficient flow to obtain a satisfactory reading for both moisture and liquid indication.

Best results will be obtained if the bypass line is parallel to the main liquid line and the takeoff and return tubes project into the main liquid line at 45°. Preformed $^1/_4$ and $^3/_8$ in. tubing is available. It can be used with either flare or sweat-type indicators.

Excess Oil and the Indicator

When a system is circulating an excessive amount of oil, the indicator may become saturated. This causes the indicator to appear brown or translucent and lose its ability to change color. However, this does not damage the indicator. Let the indicator unit remain in the line. The circulating refrigerant will remove excess oil and the indicator element will return to its proper color.

Alcohol

Do not install the color-changing indicator in a system that has methyl alcohol or a similar liquid-dehydrating agent. Remove the alcohol by using a filter and then install the indicator. Otherwise, the alcohol will damage the color indicator.

Leak Detectors

Die-type visual leak detectors will also mask the color-changing indicator. Here again, use a filter to remove all leak-detector color from the system before installing the indicator.

Liquid Water

Occasionally, it is possible for large quantities of water to enter a refrigeration system. An example would be a broken tube in a water-cooled condenser. If the free water contacts the indicator element; the element will be damaged. All moisture indicators are made of a chemical salt. These salts must be soluble in water to change color. If excessive water is present, the salts will dissolve. Permanent damage to the indicator will result. The indicator may remain yellow, or even turn white.

Hermetic-Motor Burnouts

After a hermetic-motor burnout, install a filter to remove the acid and sludge contamination. When the system has operated for 48 h, replace the filter. At the same time, install the color indicator for moisture.

The acid formed by the burnout may damage the indicator element of the color-changing unit. Thus, it should be installed only after the greater percentage of contaminants has been removed.

Hardware and Fittings

In assembling a unit in the factory or the field, strict standards of quality must be observed. Cleanliness is very important. The cleanliness of a part can determine the efficiency of a piece of equipment. Figure 11-19 illustrates some of the hardware and fittings.

THERMOSTATIC-EXPANSION VALVE (TEV)

Several different valves are used to control the flow of refrigerants. All refrigerants are relatively expensive. They will leak through fittings and tubing capable of retaining water at high pressures. A leak results in the loss of expensive refrigerant and in possible product loss, such as of frozen food. For this reason, all refrigerant lines and fittings must be absolutely seepage-proof.

Proper fittings and controls also have a bearing on the efficiency and capacity of a refrigerating machine. The capacity of a condensing unit depends, among other things, upon the suction pressure at which the unit operates. Normally, the higher the suction pressure, the greater the efficiency of the compressor.

Suction pressure at the compressor is governed by the design of the evaporator. The desired temperature in the medium being cooled and the pressure drop in the suction line from the evaporator to the compressor

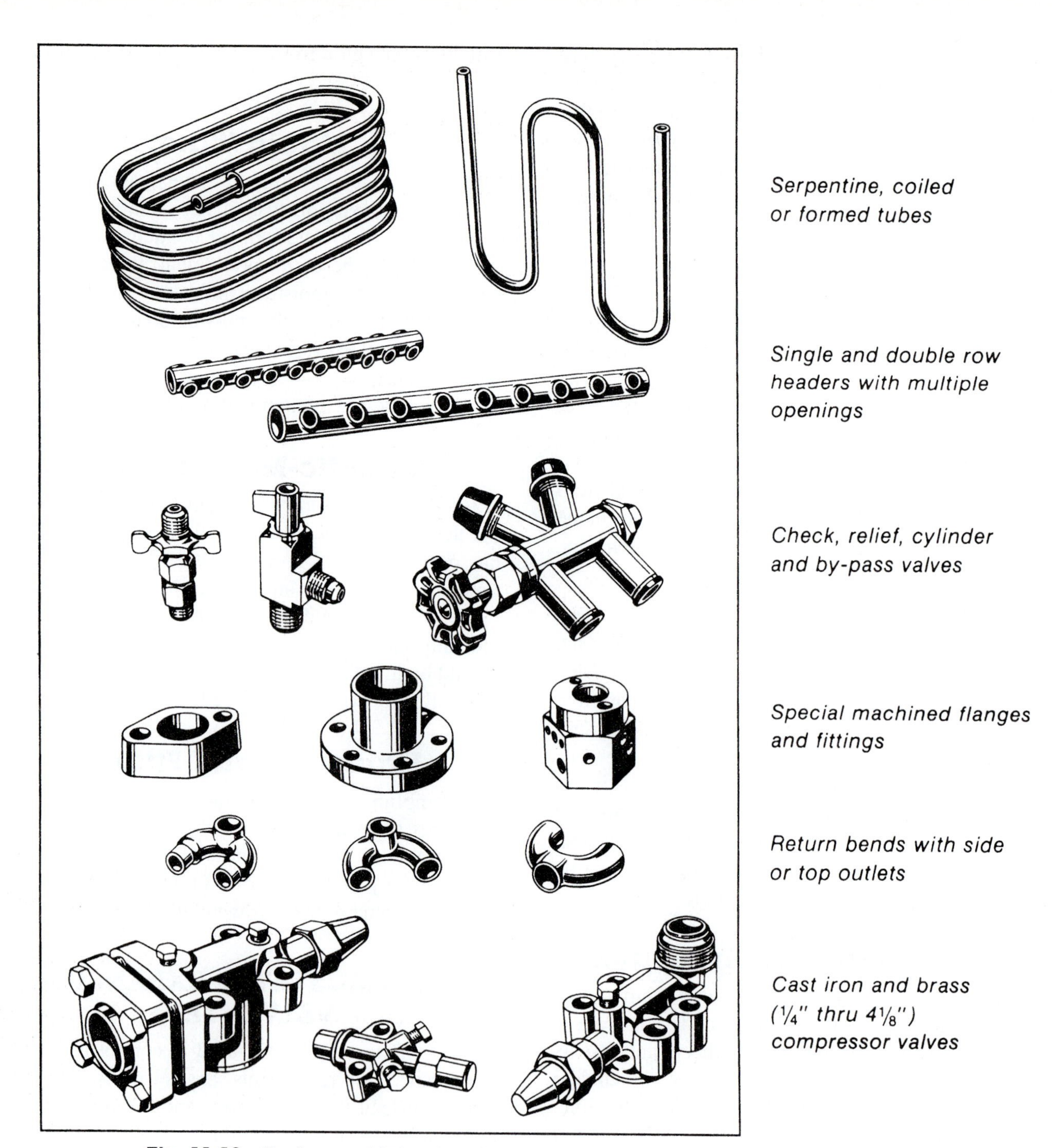

Fig. 11-19 *Hardware and fittings for refrigeration and air-conditioning installation.* (Mueller Brass)

also govern the design pressure. This pressure drop in the suction line can be kept to a minimum by use of ample line sizes, fittings, and accessories designed to eliminate restrictions. Pressure drop is also a factor in liquid lines between the receiver and the metering device. Excessive pressure drop will result in "flashing" or partial vaporization, of the liquid refrigerant before it reaches the metering device. The metering device is designed to handle liquid. It will not function properly if fed a mixture of vapor and liquid. Here, valves play an important role in controlling and metering the flow of liquid in the system.

The TEV uses the fluctuations of the pressure of the saturated refrigerant sealed inside the power element to control the flow of refrigerant through the valve. See Fig. 11-20.

Basically, TEV operation is determined by the following three fundamental pressures:

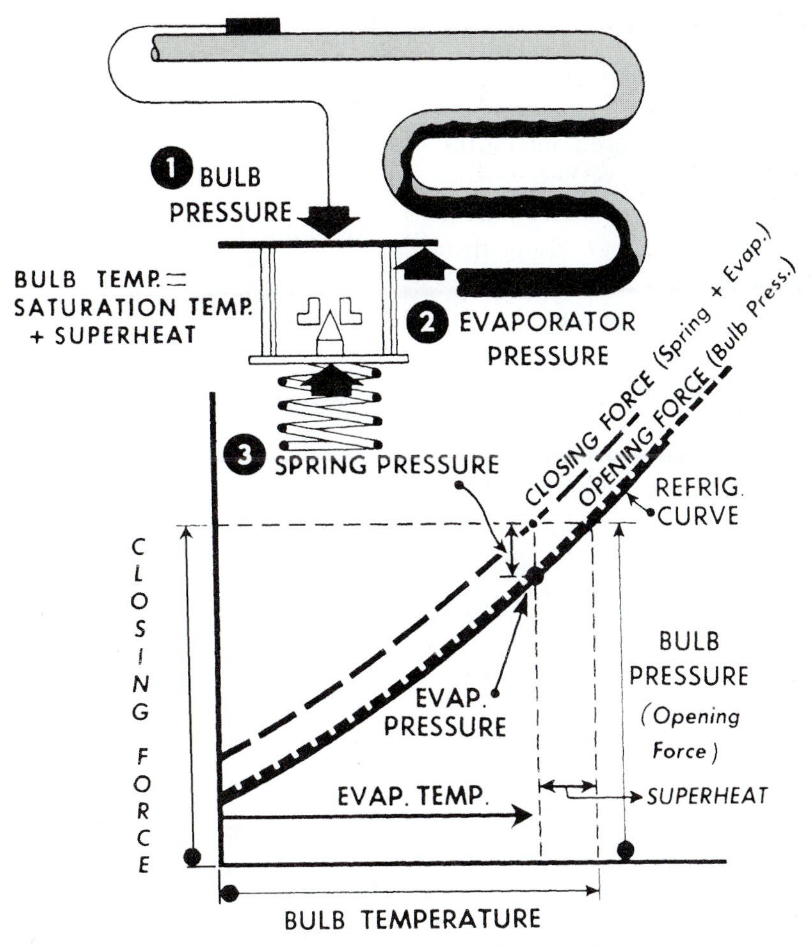

Fig. 11-20 *Basic thermostatic-expansion valve operation.* (Virginia Chemicals)

- Bulb pressure on one side of the diaphragm tends to open the valve.
- Evaporator pressure on the opposite side of the diaphragm tends to close the valve.
- Spring pressure is applied to the pin carrier and is transmitted through the push rods to the evaporator side of the diaphragm. This assists in closing the valve.

When the valve is modulating, bulb pressure is balanced by the evaporator pressure and spring pressure. When the same refrigerant is used in the thermostatic element and refrigeration system, each will exert the same pressure if their temperatures are identical. After evaporation of the liquid refrigerant in the evaporator, the suction gas is superheated. Its temperature will increase. However, the evaporator pressure, neglecting pressure drop, is unchanged. This warmer vapor flowing through the suction line increases the bulb temperature. Since the bulb contains both vapor and liquid refrigerant, its temperature and pressure increase. This higher bulb pressure acting on the top (bulb side) of the diaphragm is greater than the opposing evaporator pressure and spring pressure, which causes the valve pin to be moved away from the seat. The valve is opened until the spring pressure, combined with the evaporator pressure, is sufficient to balance the bulb pressure. See Fig. 11-20.

If the valve does not feed enough refrigerant, the evaporator pressure drops or the bulb temperature is increased by the warmer vapor leaving the evaporator (or both). The valve then opens. This admits more refrigerant until the three pressures are again in balance. Conversely, if the valve feeds too much refrigerant, the bulb temperature is decreased, or the evaporator pressure increases (or both). The spring pressure tends to close the valve until the three pressures are in balance.

With an increase in evaporator load, the liquid refrigerant evaporates at a faster rate and increases the evaporator pressure. The higher evaporator pressure results in a higher evaporator temperature and a correspondingly higher bulb temperature. The additional evaporator pressure (temperature) acts on the bottom of the diaphragm. The additional bulb pressure (temperature) acts on the top of the diaphragm. Thus, the two pressure increase on the diaphragm cancel each other. The valve easily adjusts to the new load condition with a negligible chance in superheat.

Valve Location

TEVs may be mounted in any position. However, they should be installed as close as possible to the evaporator inlet. If a refrigerant distributor is used, mount the distributor directly to the valve outlet for best performance. If a hand valve is located on the outlet side of the TEV, it should have a full-sized port. No restrictions should appear between the TEV and the evaporator, except a refrigerant distributor if one is used.

When the evaporator and TEV valve are located above the receiver, there is a static-pressure loss in the liquid line. This is due to the weight of the column of liquid refrigerant. This weight may be interpreted in terms of pressure loss in pounds per square inch, see Table 11-3. If the vertical lift is great enough, vapor, or flash gas, will form in the liquid line. This greatly reduces the capacity of the TEV. When an appreciable vertical lift is unavoidable, precautions should be taken to prevent the accompanying pressure loss from producing liquid-line vapor. This can be accomplished by providing enough subcooling to the liquid refrigerant, either in the condenser or after the liquid leaves the receiver. Subcooling is found by subtracting the actual liquid temperature from the condensing temperature (corresponding to the condensing pressure). The amount of subcooling necessary to prevent vapor formation in the liquid line is usually available in a table. See Table 11-4.

CAUTION: Ammonia valves should never be permitted to operate with vapor in the liquid line. This causes severe pin and seat erosion. It also will drastically reduce the life of the valve.

Bulb Location

The location of the bulb is extremely important. In some cases, it determines the success or failure of the refrigerating plant. For satisfactory expansion-valve control, good *thermal contact* between the bulb and suction line is essential. The bulb should be securely

Table 11-3 *Vertical Lift and Pressure Drop*

	Vertical Lift (F)					Average Pressure Drop Across Distributor
	20	40	60	80	100	
Refrigerant	Static Pressure Loss (psi)					
12	11	22	33	44	55	25 psi
22	10	20	30	40	50	35 psi
500	10	19	29	39	49	25 psi
502	10	21	31	41	52	35 psi
717 (Ammonia)	5	10	15	20	25	40 psi

Sporlan Valve

Table 11-4 *Pressure Loss and Required Subcooling for 100 and 130°F Condensing of Refrigerants*

Refrigerant	100°F [37.8°C] Condensing						130°F [54.4°C] Condensing					
	Pressure Loss (psi)						Pressure Loss (psi)					
	5	10	20	30	40	50	5	10	20	30	40	50
	Required Subcooling (°F)						Required Subcooling (°F)					
12	3	6	12	18	25	33	3	5	9	14	18	23
22	2	4	8	11	15	19	2	4	6	9	12	14
500	3	5	10	15	21	27	2	4	8	11	15	19
502	2	3	7	10	14	18	1	3	5	8	11	13
717 (Ammonia)	2	4	7	10	14	17	2	3	5	7	10	12

Sporlan Valve

fastened with two bulb straps to a clean, straight section of the suction line.

Application of the bulb to a horizontal run of suction line is preferred. If a vertical installation cannot be avoided, the bulb should be mounted so that the capillary tubing comes out at the top. On suction lines, OD and larger, the surface temperature may vary slightly around the circumference of the line. On these lines, it is generally recommended that the bulb be installed at a point midway on the side of the horizontal line and parallel to the direction of flow. On smaller lines the bulb may be mounted at any point around the circumference. However, locating the bulb on the bottom of the line is not recommended, since an oil-refrigerant mixture is generally present at that point. Certain conditions peculiar to a particular system may require a different bulb location than that normally recommended. In these cases, the proper bulb location may be determined by trial. Accepted principles of good suction-line piping should be followed to provide a bulb location that will give the best possible valve control. Never locate the bulb in a trap or pocket in the suction line. Liquid refrigerant or a mixture of liquid refrigerant and oil boiling out of the trap will falsely influence the temperature of the bulb and result in poor valve control.

Recommended suction-line piping includes a horizontal line leaving the evaporator to which the TEV bulb is attached. This line is pitched slightly downward. When a vertical riser follows, a short trap is placed immediately ahead of the vertical line. See Fig. 11-21. The trap will collect any liquid refrigerant or oil passing through the suction line and prevent it from influencing the bulb temperature.

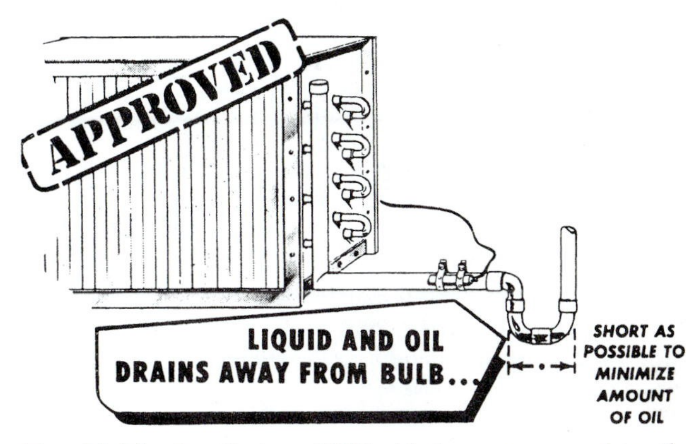

Fig. 11-21 *Installation of TEV with the compressor above the evaporator.* (Virginia Chemicals)

On multiple evaporator installations the piping should be arranged so that the flow from any valve cannot affect the bulb of another. Approved piping practices, including the proper use of traps, ensure individual control for each valve without the influence of refrigerant and oil flow from other evaporators. See Fig. 11-22.

For recommended suction-line piping when the evaporator is located above the compressor see Fig. 11-23. The vertical riser extending to the height of the evaporator

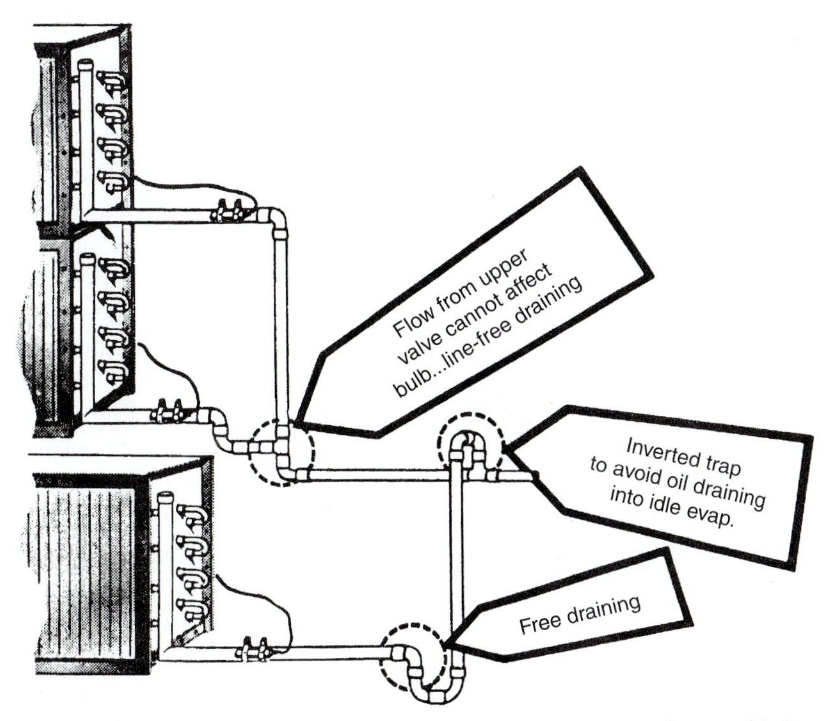

Fig. 11-22 *Installation of the TEV with multiple evaporators, above and below main suction line.* (Virginia Chemical)

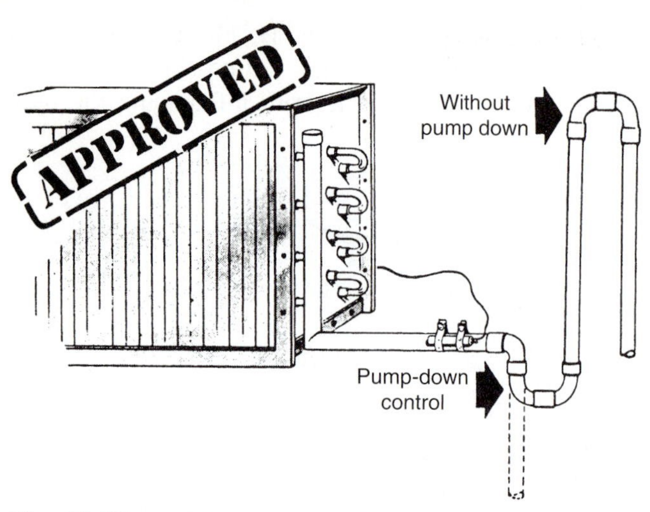

Fig. 11-23 *Installation of the TEV with the compressor below the evaporator.* (Virginia Chemicals)

prevents refrigerant from draining by gravity into the compressor during the off-cycle. When a pump-down control is used, the suction line may turn down without a trap.

On commercial and low-temperature applications, the bulb should be the same as the evaporator temperature during the off-cycle. This will ensure tight closing of the valve when the compressor stops. If bulb insulation is used on lines operating below 32°F (0°C), use nonwater-absorbing insulation to prevent water from freezing around the bulb.

On brine tanks and water coolers the bulb should be below the liquid surface. Here, it will be at the same temperature as the evaporator during the off-cycle. A solenoid valve must be used ahead of the TEV.

Some air-conditioning applications have TEVs equipped with charged elements. Here, the bulb may be located inside or outside the cooled space or duct. The valve body should not be located in the air stream leaving the evaporator. Avoid locating the bulb in the return air stream unless the bulb is well insulated.

External Equalizer

As the evaporating temperature drops, the maximum pressure drop that can be tolerated between the valve outlet and the bulb location without serious capacity loss for an internally equalized valve also decreases. This is shown in Table 11-4. There are, of course, applications that may satisfactorily employ the internal equalizer when higher pressure drop is present. This should usually be verified by laboratory tests. The general recommendations given in Table 11-4 are suitable for most field-installed systems. Use the external equalizer when pressure drop between the outlet and bulb locations exceeds values shown in Table 11-4. When the expansion valve is equipped with an external equalizer, it must be connected. Never cap an external equalizer. The valve may flood, starve, or regulate erratically. There is no operational disadvantage in using an external equalizer, even if the evaporator has a low pressure drop.

> NOTE: The external equalizer must be used on evaporators that use a pressure–drop type refrigerant distributor.

See Fig. 11-24. Generally, the external equalizer connection is in the suction line immediately downstream of the bulb. See Fig. 11-24. However, equipment manufacturers sometimes select other locations that are compatible with their specific design requirements.

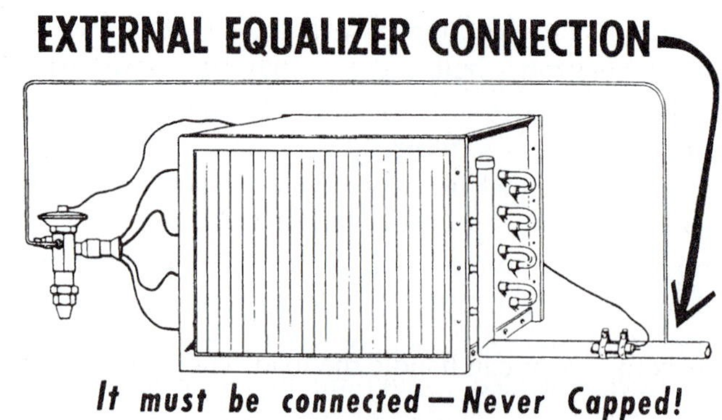

Fig. 11-24 *External equalizer connection.* (Virginia Chemicals)

Field Service

The TEV is erroneously considered by some to be a complex device. As a result, many valves are needlessly replaced when the cause of the system malfunction is not immediately recognized.

Actually, the TEV performs only one very simple function. It keeps the evaporator supplied with enough refrigerant to satisfy all load conditions. It is not a temperature control, suction-pressure control, a control to vary the compressor-running time, or a humidity control.

The effectiveness of the valve's performance is easily determined by measuring the superheat. See Fig. 11-25. Observing the frost on the suction line or considering only the suction pressure may be misleading. Checking the superheat is the first step in a simple and systematic analysis of TEV performance.

If insufficient refrigerant is being fed to the evaporator, the superheat will be high. If too much refrigerant is being fed to the evaporator, the superheat will be low. Although these symptoms may be attributed to improper TEV control, more frequently the origin of the trouble lies elsewhere.

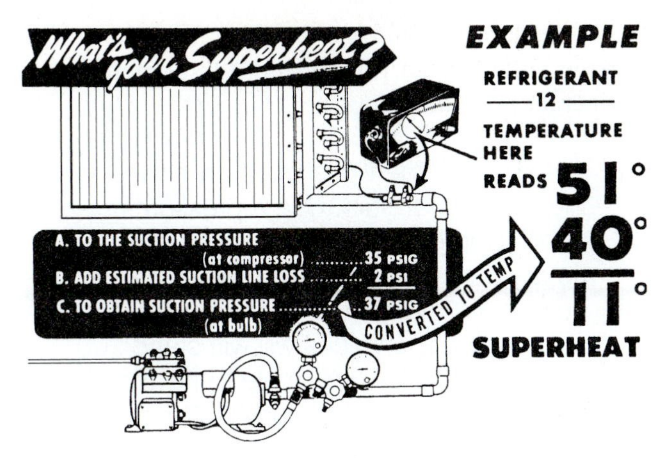

Fig. 11-25 *How to figure superheat?* (Virginia Chemicals)

CRANKCASE PRESSURE-REGULATING VALVES

Crankcase pressure-regulating valves are designed to prevent overloading of the compressor motor. They limit the crankcase pressure during and after a defrost cycle or after a normal shutdown period. When properly installed in the suction line, these valves automatically throttle the vapor flow from the evaporator until the compressor can handle the load. They are available in the range of 0 to 60 psig.

Operation of the Valve

Crankcase pressure-regulating valves (CROS) are sometimes called suction pressure-regulating valves. They are sensitive only to their outlet pressure. This would be the compressor crankcase or suction pressure. To indicate this trait, the designation describes the operation: close on rise of outlet pressure, or CRO. As shown in Fig. 11-26, the inlet pressure is exerted on the underside of the bellows and on top of the seat disc. Since the effective area of the bellows is equal to the area of the port, the inlet pressure cancels out and does not affect valve operation. The valve-outlet pressure acting on the bottom of the disc exerts a force in the closing direction. This force is opposed by the adjustable spring force. These are the operating forces of the CRO. The CRO's pressure setting is determined by the spring force. Thus, by increasing the spring force, the valve setting or the pressure at which the valve will close is increased.

As long as the valve-outlet pressure is greater than the valve-pressure setting, the valve will remain closed. As the outlet pressure is reduced, the valve will open and pass refrigerant vapor into the compressor. Further reduction of the outlet pressure will allow the valve to open to its rated position, where the rated pressure drop will exist across the valve port. An increase in the outlet pressure will cause the valve to throttle until the pressure setting is reached.

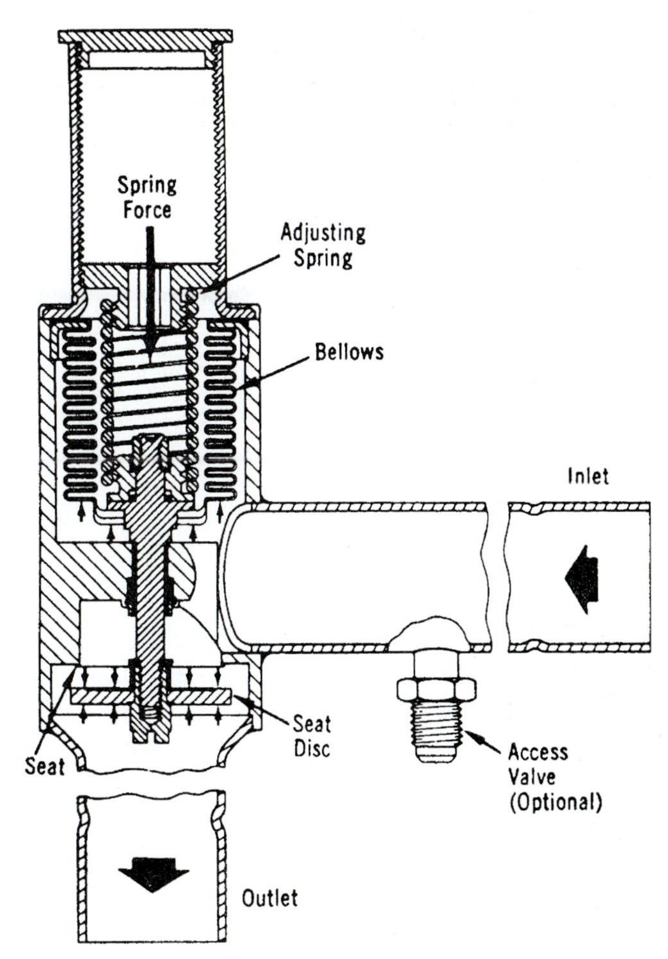

Fig. 11-26 *Crankcase pressure-regulating valve.* (Sporlan Valve)

The operation of a valve of this type is improved by an antichatter device built into the valve. Without this device, the CRO would be susceptible to compressor pulsations that greatly reduce the life of a bellows. This feature allows the CRO to function at low-load conditions without any chattering or other operation difficulties.

Valve Location

As Fig. 11-27 indicates, the CRO valve is applied in the suction line between the evaporator and the compressor. Normally, the CRO is installed downstream of any other controls or accessories. However, on some applications it may be advisable or necessary to locate other system components, such as an accumulator, downstream of the CRO. This is satisfactory as long as the CRO valve is applied only as a crankcase pressure-regulating valve. CRO valves are designed for application in the suction line only. They should not be applied in hot–gas bypass lines or any other refrigerant line of a system.

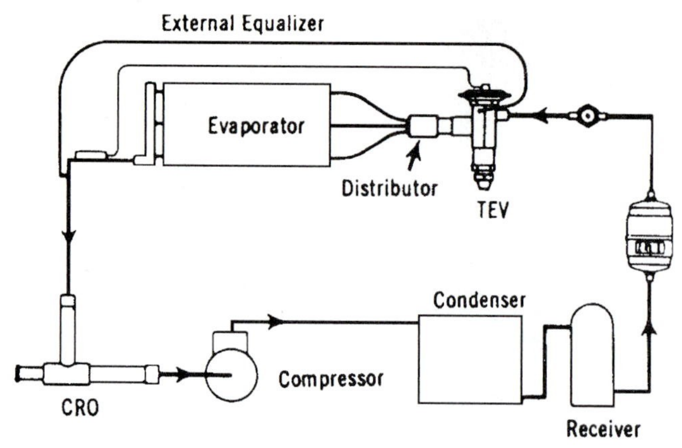

Fig. 11-27 *CRO valve applied in the suction line between the evaporator and the compressor.* (Sporlan Valve)

Strainer

Just as with any refrigerant flow-control device, the need for an inlet strainer is a function of system cleanliness and proper installation procedures. See Fig. 11-28. When the strainer is used, the tubing is inserted in the valve connection up to the tubing stop. Thus, the strainer has been locked in place. Moisture and particles too small for the inlet strainer are harmful to the system and must be removed. Therefore, it is recommended that a filter drier be installed according to the application recommendations.

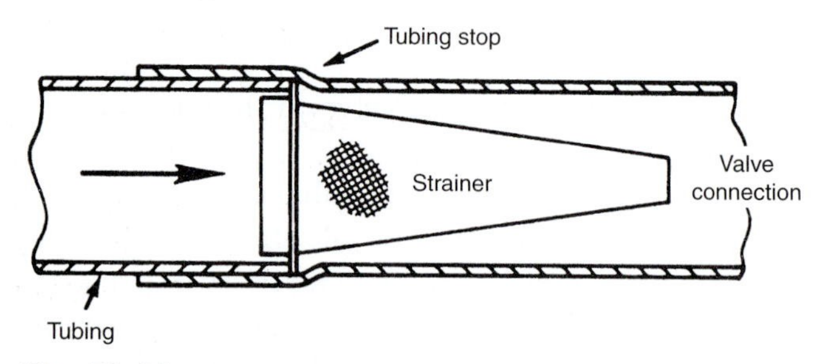

Fig. 11-28 *Strainer for cleanliness.* (Sporlan Valve)

Brazing Procedures

When installing CROs with solder connections, the internal parts must be protected by wrapping the valve with a *wet* cloth to keep the body temperature below 250°F (121°C). The tip of the torch should be large enough to avoid prolonged heating of the connections. Overheating can also be minimized by directing the flame away from the valve body.

Test and Operating Pressures

Excessive leak testing or operating pressures may damage these valves by reducing the life of the bellows. For leak detection, an inert gas such as nitrogen or CO_2 may be added to an idle system to supplement the refrigerant pressure.

> CAUTION: Inert gas must be added to the system carefully. Use a pressure regulator. Unregulated gas pressure can seriously damage the system and endanger human life. Never use oxygen or explosive gases. The values will withstand 200 to 300 psig. However, check the manufacturer's recommendations first.

Adjusting the Pressure

The standard setting by the factory for CROs in the 0/60 psig range is 30 psig. Since these valves are adjustable, the setting may be altered to suit the specific system requirements. CROs should be adjusted at start-up when the pressure in the evaporator is above the desired setting. The final valve setting should be below the maximum suction pressure recommended by the compressor or unit manufacturer.

The main purpose of the CRO is to prevent the compressor motor from overloading due to high-suction pressure. Thus, it is important to arrive at the correct pressure setting. The best way to see if the motor is overloaded is to check the current draw at start-up or after a defrost cycle. If overloading is evident, a suction gage should be put on the compressor. The CRO setting may be too high and may have to be adjusted. If the compressor is overloaded and the CRO valve is to be reset, the following procedure should be followed.

The unit should be shut off long enough for the system pressure to equalize. Observe the suction pressure as the unit is started, since this is the pressure the valve is controlling. If the setting is to be decreased, slowly adjust the valve in a counterclockwise direction approximately one-quarter turn for each 1 psi pressure change required. After a few moments of operation, the unit should be cycled off and the system pressure allowed to equalize again. Observe the suction pressure (valve setting) as the unit is started up. If the setting is still too high, the adjustment should be repeated. The proper size hex wrench is used to adjust these valves. A clockwise rotation increases the valve setting, while a counterclockwise rotation decreases the setting.

When CROs are installed in parallel, each should be adjusted the same amount. If one valve has been adjusted more than the other, best performance will occur if both are adjusted all the way in before resetting them an equal amount.

Service

Since CRO valves are hermetic and cannot be disassembled for inspection and cleaning, they are usually replaced if inoperative. If a CRO fails to open, close properly, or will not adjust, solder or other foreign material is probably lodged in the port. It is sometimes possible to dislodge these materials by turning the adjustment nut all the way in with the system running. If the CRO develops a refrigerant leak around the spring housing, it probably has been overheated during installation or the bellows have failed due to severe compressor pulsations. In either case, the valve must be replaced.

EVAPORATOR PRESSURE-REGULATING VALVES

Evaporator pressure-regulating valves offer an efficient means of balancing the system capacity and the load requirements during periods of low loads. They also are able to maintain different evaporator conditions on multitemperature systems. The main function of this valve is to prevent the evaporator pressure from falling below a predetermined value at which the valve has been set.

Control of evaporator pressure by cycling the compressor with a thermostat or some other method is quite adequate on most refrigeration systems. Control of the evaporator pressure also controls the saturation temperature. As the load drops off, the evaporating pressure starts to decrease and the system performance falls off. These valves automatically throttle the vapor flow from the evaporator. This maintains the desired minimum evaporator pressure. As the load increases, the evaporating pressure will increase above the valve setting and the valve will open further.

Operation

For any pressure-sensitive valve to modulate to a more closed or open position, a change in the operating pressure is required. The unit change in the valve stroke for a given change in the operating pressure is called the *valve gradient*. Every valve has a specific gradient designed into it for the best possible operation. Valve sensitivity and the valve's capacity rating are functions of the valve gradient. Thus, a relatively sensitive valve is needed when a great change in the evaporating temperature cannot be tolerated. Therefore, the valves have nominal ratings based on the 8 psi evaporator pressure change, rather than a full stroke.

Evaporator pressure-regulator valves respond only to variations in their inlet pressure (evaporator pressure). Thus, the designation for evaporator pressure-regulating valves is *opens on the rise of the inlet pressure* (ORI). See Fig. 11-29.

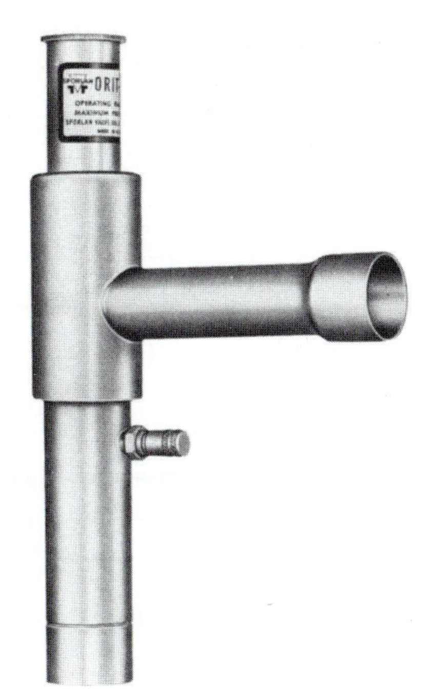

Fig. 11-29 *Evaporator pressure-regulating valve.* (Sporlan Valve)

Pressure at the outlet is exerted on the underside of the bellows and on top of the seat disc. The effective area of the bellows is equal to the area of the port. Thus, the outlet pressure cancels out and the inlet pressure acting on the bottom of the seat disc opposes the adjustable spring force. These two forces are the operating forces of the ORIT (The T added to the valve designation indicates an access valve on the inlet connection.). When the evaporator load changes, the ORIT opens or closes in response to the change in evaporator pressure. An increase in inlet pressure above the valve setting tends to open the valves. If the load drops, less refrigerant is boiled off in the evaporator and evaporator pressure will decrease. The decrease in evaporator pressure tends to move the ORIT to a more closed position. This, in turn, keeps the evaporator pressure up. The result is that the evaporator pressure changes as the load changes. The operation of a valve of this type is improved by an antichatter device built into the valve. Without this device, the OBIT would be susceptible to compressor pulsations that can reduce the life of bellows. This antichatter feature allows the ORIT to function at low-load conditions without chattering or other operating difficulties.

Type of System

The proper application of the evaporator pressure-regulating valve involves the consideration of several system factors.

One type of system is a single evaporator type, such as a water chiller. Here, the valve is used to prevent freeze-up at light loads. See Fig. 11-30.

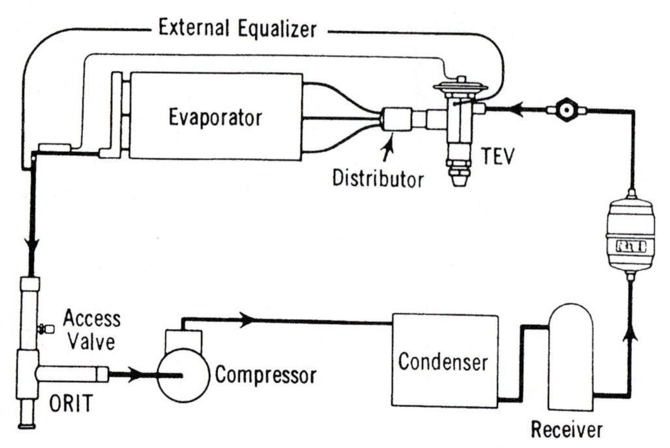

Fig. 11-30 *Valve location in a single-evaporator system.* (Sporlan Valve)

Another type of system is a multitemperature refrigeration system with evaporators operating at different temperatures. See Fig. 11-31. A valve may be required on one or more of the evaporators to maintain pressures higher than that of the common suction line. For example, if evaporator A in Fig. 11-31 is designed for 35°F (1.7°C) (72.6 psig on Refrigerant 502), evaporator B for 32°F (0°C) on the same refrigerant (68.2 psig), and other evaporators for 25°F (–3.9°C) (58.7 psig), the valves (ORIT) are used to maintain a pressure of 72.6 psig in evaporator A and 68.2 psig in evaporator B. However, some multitemperature systems may require an OBIT on each evaporator, depending on the type of product being refrigerated.

Valve Location

ORITs must be installed upstream of any other suction-line controls or accessories. These valves may be installed in the position most suited to the application. However, these valves should be located so that they do not act as an oil trap or so that solder cannot run into the internal parts during brazing in the suction line. Since these valves are hermetic, they cannot be disassembled to remove solder trapped in the internal parts. Installation of a filter drier and a strainer may be worth the expense to keep the system clean and operational. Brazing procedures are the same as for other valves of this type. The valve core of the access valve is shipped in an envelope attached to the access valve. If the access-valve connection is to be used as a reusable pressure tap to check the valve setting, the OBIT must be brazed in before the core is installed. This protects the synthetic material of the core. If the access valve is to be used as a permanent pressure tap, the core and access valve cap may be discarded.

Test and Operating Pressures

As with other pressure valves, it is possible to introduce nitrogen or CO_2 in an idle system to check for correct pressure settings.

The usual precautions for working with gases apply here. The standard factory setting for the 0/50 psig range is 30 psi. For the 30/100 psig range, it is 60 psig. Since these valves are adjustable, the setting may be altered to suit the system.

The main purpose of an OBIT valve is to keep the evaporator pressure above some given point at minimum load conditions. The valves are selected on the basis of the pressure drop at full-load conditions. Nevertheless, they should be adjusted to maintain the minimum allowable evaporator pressure under the actual minimum load conditions.

These valves can be adjusted by removing the cap and turning the adjustment screw with a hex wrench of the proper size. A clockwise rotation increases the valve

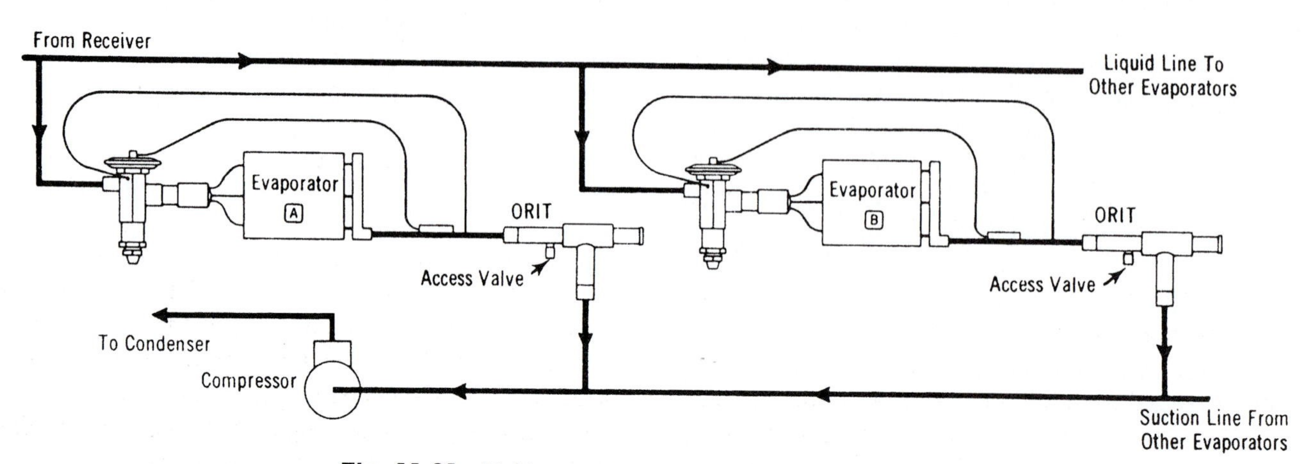

Fig. 11-31 *Multitemperature refrigeration system.* (Sporlan Valve)

setting, while a counterclockwise rotation decreases the setting. To obtain the desired setting, a pressure gage should be utilized on the inlet side of the valve. Thus, the effects of any adjustments can be observed.

When these valves are installed in parallel, each should be adjusted the same amount. If one valve has been adjusted more than the other, the best performance will occur if both are adjusted all the way before resetting them an equal amount.

Service

Since these valves are hermetic and cannot be disassembled for inspection and cleaning, they usually must be replaced if found defective or inoperative. It is possible sometimes to adjust the valve until the obstruction is dislodged. This usually works best when the system is running. If it leaks around the spring housing, it will have to be replaced. The bellows have been permanently damaged.

HEAD-PRESSURE CONTROL VALVES

Design of air-conditioning and refrigeration systems using air-cooled condensing units involves two main problems that must be solved if the system is to be operated reliably and economically. These problems are high-ambient and low-ambient operation. If the condensing unit is properly sized, it will operate satisfactorily during extreme ambient temperatures. However, most units will be required to operate at ambient temperatures below their design's dry-bulb temperature during most of the year. Thus, the solution to low-ambient operation is more complex.

Without good head-pressure control during low-ambient operation, the system can have running-cycle and off-cycle problems. Two running-cycle problems are of prime concern:

- The pressure differential across the TEV port affects the rate of refrigerant flow. Thus, low-head pressure generally causes insufficient refrigerant to be fed to the evaporator.
- Any system using hot gas for defrost or compressor capacity control must have a normal head pressure to operate properly. In either case, failure to have sufficient head pressure will result in low-suction pressure and/or iced-evaporator coils.

The primary off-cycle problem is the possible inability to get the system on-the-line if the refrigerant has migrated to the condenser. The evaporator pressure may not build up to the cut-in point of the low-pressure control. The compressor cannot start, even though refrigeration is required. Even if the evaporator pressure builds up to the cut-in setting, insufficient flow through the TEV will cause a low-suction pressure, which results in compressor cycling.

There are nonadjustable and adjustable methods of head-pressure control by valves. Each method uses two valves designed specifically for this type of application. Low-ambient conditions are encountered during fall-winter-spring operation on air-cooled systems, with the resultant drop in condensing pressure. Then, the valve's purpose is to hold back enough of the condensed liquid refrigerant to make part of the condenser surface inactive. This reduction of active condensing surface raises condensing pressure and sufficient liquid-line pressure for normal system operation.

Operation

The ORI head–pressure control valve is an inlet pressure-regulating valve. It responds to changes in condensing pressure only. The valve designation stands for ORI pressure. As shown in Fig. 11-32, the outlet pressure is exerted on the underside of the bellows and on top of the seat disc. Since the effective area of the bellows is equal to the area of the port, the outlet pressure cancels out. The inlet pressure acting on the bottom of the seat disc opposes the adjusting spring force. These two forces are the operating forces of the ORI.

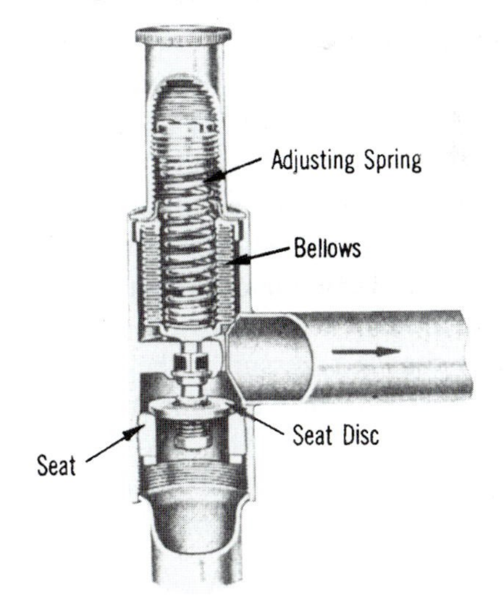

Fig. 11-32 *Head-pressure control valve.* (Sporlan Valve)

When the outdoor ambient temperature changes, the ORI opens or closes in response to the change in condensing pressure. An increase in inlet pressure above the valve setting tends to open the valve. If the ambient temperature drops, the condenser capacity is increased and the condensing pressure drops off. This causes the ORI to start to close or assume a throttling position.

ORO-Valve Operation

The ORO head-pressure control valve is an outlet pressure-regulating valve that responds to changes in receiver pressure. The valve designation stands for "opens on rise of outlet" pressure. See Fig. 11-33. The inlet and outlet pressures are exerted on the underside of the seat disc in an opening direction. Since the area of the port is small in relationship to the diaphragm area, the inlet pressure has little direct effect on the operation of the valve. The outlet or receiver pressure is the control pressure. The force on top of the diaphragm that opposes the control pressure is due to the air charge in the element. These two forces are the operating forces of the ORO.

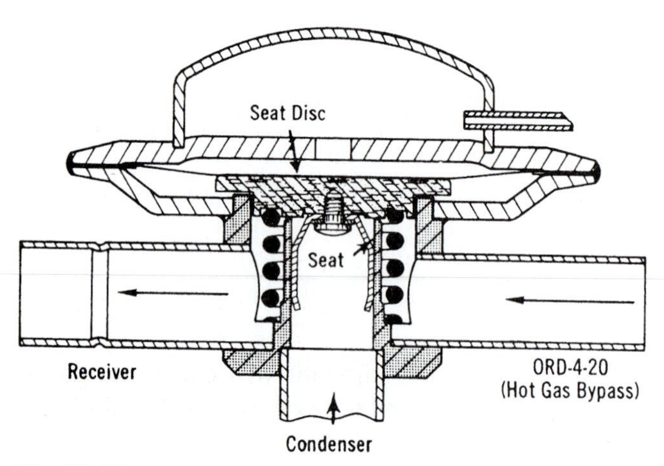

Fig. 11-33 *Head-pressure control valve that opens on rise of outlet pressure (ORO).* (Sporlan Valve)

When the outdoor ambient temperature changes, the condensing pressure changes. This causes the receiver pressure to fluctuate accordingly. As the receiver pressure decreases, the ORO throttles the flow of liquid from the condenser. As the receiver pressure increases, the valve modulates in an opening direction to maintain a nearly constant pressure in the receiver. Since the ambient temperature of the element affects the valve-pressure setting, the control pressure may change slightly when the ambient temperature changes. However, the valve and element temperature remain fairly constant.

ORD Valve Operation

The ORD valve is a pressure-differential valve. It responds to changes in the pressure difference across the valve. See Fig. 11-34. The valve designation stands for "opens on rise of differential" pressure. Therefore, the ORD is dependent on some other control valve or action for its operation. In this respect, it is used with either the ORI or ORO for head-pressure control.

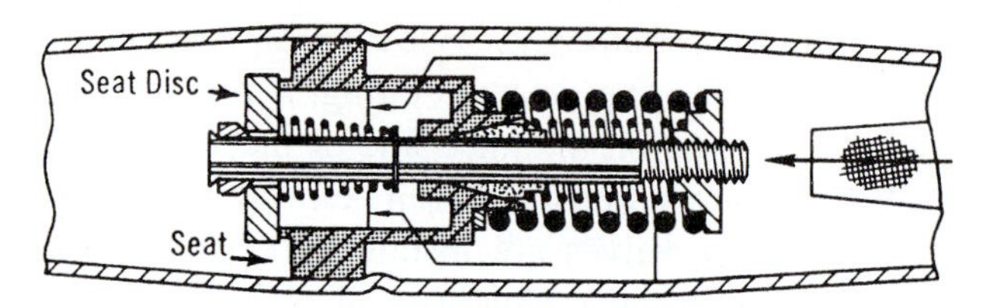

Fig. 11-34 *Head-pressure control valve that opens on rise of differential across the valve (ORD).* (Sporlan Valve)

As either the ORI or ORO valve starts to throttle the flow of liquid refrigerant from the condenser, a pressure differential is created across the ORD. When the differential reaches 20 psi, the ORD starts to open and bypasses hot gas to the liquid drain line. As the differential increases, the ORD opens further until its full stroke is reached at a differential of 30 psi. Due to its function in the control of head pressure, the full stroke can be utilized in selecting the ORD. While the capacity of the ORD increases, as the pressure differential increases, the rating point at 30 psi is considered a satisfactory maximum value.

The standard pressure setting for the ORD is 20 psig. For systems where the condenser pressure drop is higher than 10 or 12 psi, an ORD with a higher setting can be ordered.

Head-pressure control can be improved with an arrangement, such as that shown in Fig. 11-35. In this operation, a constant receiver pressure is maintained for normal system operation. The ORI is adjustable over a nominal range of 100 to 225 psig. Thus, the desired pressure can be maintained for all of the commonly used refrigerants—12, 22, and 502, as a well as the latest alternatives.

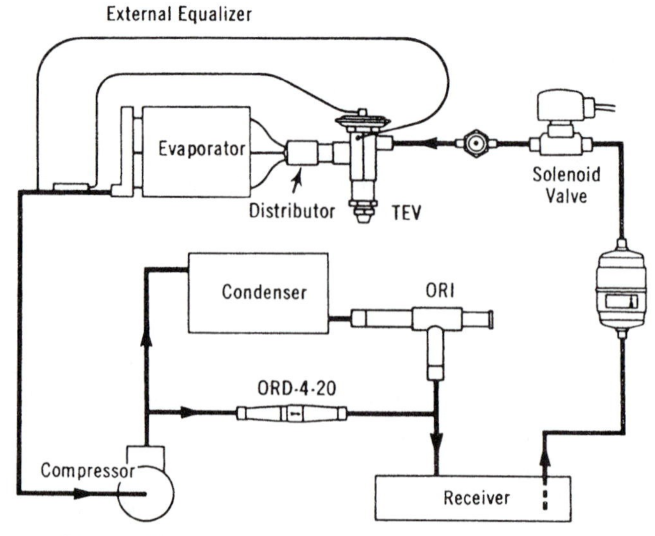

Fig. 11-35 *Adjustable ORI/ORD system.* (Sporlan Valve)

The ORI is located in the liquid drain line between the condenser and the receiver. The ORD is located in a hot-gas line bypassing the condenser. During periods of low-ambient temperature, the condensing pressure falls until it approaches the setting of the ORI valve. The ORI then throttles, restricting the flow of liquid from the condenser. This causes refrigerant to back up in the condenser, thus reducing the active condenser surface. This raises the condensing pressure. Since it is really receiver pressure that needs to be maintained, the bypass line with the ORD is required.

The ORD opens after the ORI has offered enough restriction to cause the differential between condensing pressure and receiver pressure to exceed 20 psi. The hot gas flowing through the ORD heats up the cold liquid being passed through the ORI. Thus, the liquid reaches the receiver warm and with sufficient pressure to assure proper expansion-valve operation. As long as sufficient refrigerant charge is in the system, the two valves modulate the flow automatically to maintain proper receiver pressure regardless of outside ambient temperature.

Installation

To insure proper performance, head-pressure control valves must be selected and applied correctly. These valves can be installed in either horizontal or vertical lines, if possible, the valves should be oriented so solder cannot run into the internal parts during brazing. Care should be taken to install the valves with the flow in the proper direction. The ORI and ORO valves *cannot* be installed in the discharge line for any reason.

In most cases the valves are located at the condensing unit. When the condenser is remote from the compressor, the usual location is near the compressor. In all cases it is important that some precautions be taken in mounting the valves. While the heaviest valve is approximately 2.5 lb (1.14 kg) in weight, it is suggested that they be adequately supported to prevent excessive stress on the connections. Since discharge lines are a possible source of vibrations that result from discharge gas pulses and inertia forces associated with the moving parts, fatigue in tubing, fittings, and connections may result. Pulsations are best handled by placing a good muffler as close to the compressor as possible.

Vibrations from moving parts of the compressor are best isolated by flexible loops or coils (discharge lines or smaller) or flexible metal hoses for larger lines. For best results, the hoses should be installed as close to the compressor shut-off valves as possible. The hoses should be mounted horizontally, and parallel to the crankshaft, or vertically. The hoses should *never* be mounted horizontally and 90° from the crankshaft. A rigid brace should be placed on the outlet end of the hose. This brace will prevent vibrations beyond the hose.

Brazing Procedures

Any of the commonly used brazing alloys for high-side usage are satisfactory. It is very important that the internal parts be protected by wrapping the valve with a wet cloth to keep the body temperature below 250°F (121°C). Also, when using high-temperature solders, the torch tip should be large enough to avoid prolonged heating of the copper connections. Always direct the flame away from the valve body.

Test and Operating Pressures

Excessive leak testing or operating pressures may damage these valves and reduce the life of the operating members. For leak detection, an inert dry gas, such as nitrogen or CO_2, can be added to an idle system to supplement the refrigerant pressure.

Remove the cap and adjust the adjustment screw with the proper wrench. Check the manufacturer's recommended pressures before making adjustments.

Refrigerant and charging procedures require that enough refrigerant be available for flooding the condenser at the lowest expected ambient temperature. There must still be enough charge in the system for proper operation.

A shortage of refrigerant will cause hot gas to enter the liquid line and the expansion valve. Refrigeration will cease.

The receiver must have sufficient capacity to hold at least all of the excess liquid refrigerant in the system. This is because such refrigerant will be returned to the receiver when high-ambient conditions prevail. If the receiver is too small, liquid refrigerant will be held back in the condenser during high-ambient condition. Excessively high-discharge pressures will be experienced.

CAUTION: All receivers must utilize a pressure-relief valve or device according to the applicable standards or codes.

Follow the manufacturer's recommendations for charging the system. Procedures may vary with different valve manufacturers.

Service

There are several possible causes for system malfunction with "refrigerant side" head-pressure control. These may be difficult to isolate from each other. As with any form of system troubleshooting, it is necessary to know the existing operating temperatures and pressures before system problems can be determined. Once the malfunction is established, it is easier to pinpoint the cause and then take suitable action. Table 11-5 lists

Table 11-5 *Troubleshooting Head-Pressure Control Valves*

Possible Cause	Remedy
Malfunction—Low Head Pressure	
1. Insufficient refrigerant charge to adequately flood condenser.	1. Add charge.
2. Low-pressure setting on ORI.	2. Increase setting.
3. ORI fails to close due to foreign material in valve.	3. Turn adjustment out so material passes through valve. If unsuccessful, replace ORI.
4. ORI fails to adjust properly.	4. See 3 above.
5. Wrong setting on ORO (e.g., 100 psig on Refrigerant 22 or 502 system).	5. Replace ORO with valve with correct setting.
6. ORO fails to close due to:	6. See below:
a. Foreign material in valve.	a. Cause ORO to open by raising condensing/receiver pressure above valve setting by cycling condenser fan. If foreign material does not pass through valve, replace ORO.
b. Loss of air charge in element.	b. Replace ORO.
7. ORD fails to open (on ORI/ORD system only) due to:	7. See below:
a. Less than 20 psi pressure drop across ORD.	a. Check ORI causes/remedies above: 2, 3, or 4.
b. Internal parts damaged by overheating when installed.	b. Replace ORD.
8. Refrigerant leak at adjustment housing of ORI.	8. Replace ORI.
Malfunction—High Head Pressure	
1. Dirty condenser coil.	1. Clean coil.
2. Air on condenser blocked off.	2. Clear area around unit.
3. Too much refrigerant charge.	3. Remove change until proper head pressure is maintained.
4. Undersized receiver.	4. Check receiver capacity against refrigerant required to maintain desired head pressure.
5. Noncondensibles (air) in system.	5. Purge from system.
6. High-pressure setting on ORI.	6. Decrease setting.
7. ORI or ORO restricted due to inlet strainer being plugged.	7. Open inlet connection to clean strainer.
8. ORI fails to adjust properly or to open due to foreign material in valve.	8. Turn adjustment out so material passes through valve. If unsuccessful, replace ORI.
9. Wrong setting on ORO (e.g., 180 psig on Refrigerant 12 system).	9. Replace ORO with valve with correct setting.
10. ORD fails to open due to internal parts being damaged by overheating when installed (only when used with ORO).	10. Replace ORD.
11. ORD bypassing hot gas when not required due to:	11. See below:
a. Internal parts damaged by overheating when installed.	a. Replace ORD.
b. Pressure drop across condenser coil, ORI or ORO, and connecting piping above 14 psi.	b. Reduce pressure drop (e.g., use larger ORI or ORI or ORO valves in parallel) or order ORD-4 with higher setting.

Sporlan Valve

the most common malfunctions, the possible causes, and the remedies.

Nonadjustable ORO/ORD System Operation

The nonadjustable ORO head-pressure control valve and the ORD pressure-differential valve offer the most economical system of refrigerant side–head pressure control. Just as the ORI/ORD system simplified this type of control, the ORO/ORD system offers the capability of locating the condenser and receiver on the same elevation. See Fig. 11-36. By making these two valves available either separately or brazed together, there is added flexibility in the piping layout. The operation of the ORO/ORD system is such that a nearly constant receiver pressure is maintained for normal operation. As the temperature of the ORO element decreases, the pressure setting decreases accordingly.

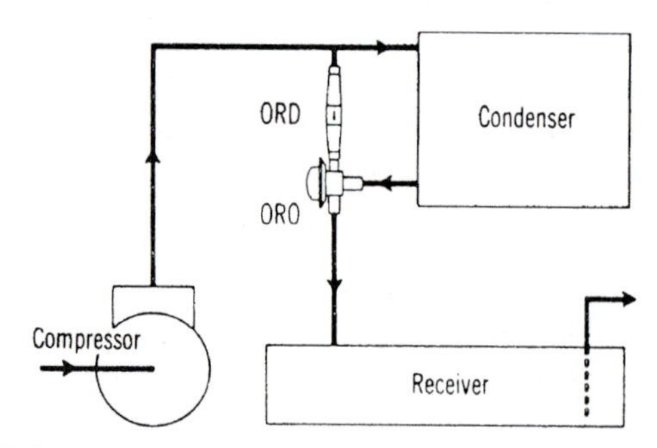

Fig. 11-36 *The ORO is located in the liquid drain line between the condenser and the receiver.* (Sporlan Valve)

However, by running the bypassed hot gas through the ORO the element temperature is adequately maintained so the ORO/ORD system functions well to ambient temperatures of –40°F (–40°C) and below. This third connection on the ORO also eliminates the need for a tee connection in the liquid drain line.

Note that in Fig. 11-36, the ORO is located in the liquid drain line between the condenser and the receiver, while the ORD is located in a hot-gas line bypassing the condenser. Other than the fact that the ORO operates in response to its outlet pressure (receiver pressure), the ORO/ORD operates in the same basic manner as the ORI/ORD system previously explained.

DISCHARGE BYPASS VALVES

On many air-conditioning and refrigeration systems it is desirable to limit the minimum evaporating pressure. This is so especially during periods of low load either to prevent coil icing or to avoid operating the compressor at lower suction pressure than it was designed for. Various methods of operation have been designed to achieve the result—integral cylinder unloading, gas engines with variable speed control, or multiple smaller systems. Compressor cylinder unloading is used extensively on larger systems. However, it is too costly on small equipment, usually 10 hp or below. Cycling the compressor with a low-pressure cutout control has had widespread usage, but is being reevaluated for three reasons:

- On-off control on air-conditioning systems is uncomfortable and does a poor job of humidity control.
- Compressor cycling reduces equipment life.
- In most cases, compressor cycling is uneconomical because of peak-load demand charges.

One solution to the problem is to bypass a portion of the hot discharge gas directly into the low side. This is done by the modulating control valve—commonly called a *discharge bypass valve* (DBV). This valve, which opens on a decrease in suction pressure, can be set to maintain automatically a desired minimum evaporating pressure, regardless of the decrease in evaporator load.

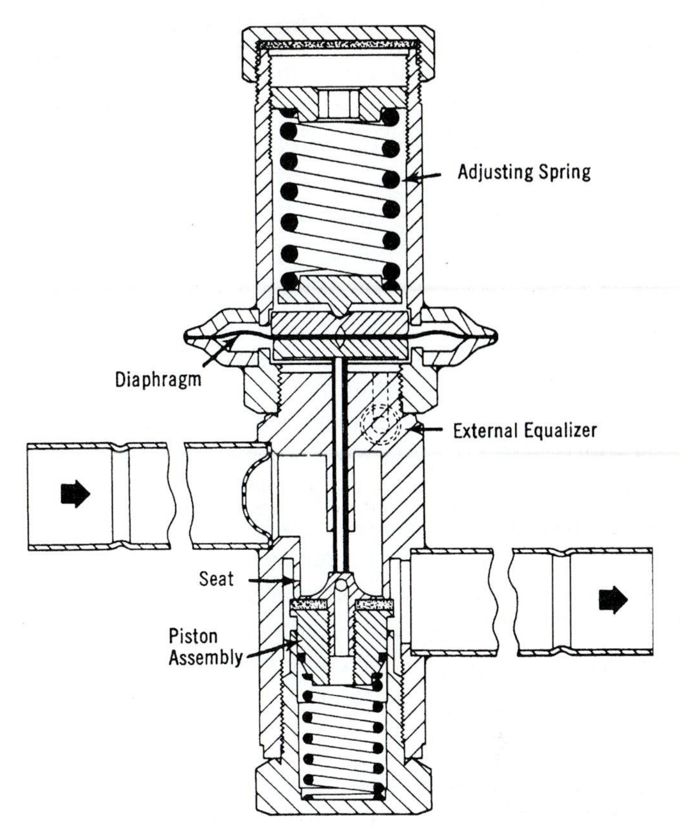

Fig. 11-37 *Discharge-bypass valve.* (Sporlan Valve)

Operation

DBVs respond to changes in downstream or suction pressure. See Fig. 11-37. When the evaporating pressure is above the valve setting, the valve remains closed. As the suction pressure drops below the valve setting, the valve responds and begins to open. As with all modulating-type valves, the size of the opening is proportional to the change in the variable being controlled. In this case, the variable is the suction pressure. As the suction pressure drops, the valve opens further until the limit of the valve stroke is reached. However, on normal applications there is no sufficient pressure change to open these valves to the limit of their stroke. The amount of pressure change available from the point at which it is desired to have the valve closed to the point at which it is to be open varies widely with the refrigerant used and the evaporating temperature. For this reason, DBVs are rated on the basis of allowable evaporator temperature change from closed position to rated opening. A 6°F (3.3°C) change is considered normal for most applications and is the basis of capacity ratings.

Application

DBVs provide an economical method of compressor capacity control in place of cylinder un-loaders or of handling unloading requirements below the last step of cylinder unloading.

On air-conditioning systems, the minimum allowable evaporating temperature that will avoid coil icing depends on evaporator design. The amount of air passing over the coil also determines the allowable evaporator minimum temperature. The refrigerant temperature may be below 32°F (0°C). However, coil icing will not usually occur with high air velocities, since the external surface temperature of the tube will be above 32°F (0°C). For most air-conditioning systems the minimum evaporating temperature should be 26 to 28°F (–3.3 to –2.2°C). DBVs are set in the factory. They start to open at an evaporating pressure equivalent to 32°F (0°C) saturation temperature. Therefore, evaporating temperature of 26°F (–3.3°C) is their rated capacity. However, since they are adjustable, these valves can be set to open at a higher evaporating temperature.

On refrigeration systems, DBV are used to prevent the suction pressure from going below the minimum value recommended by the compressor manufacturer. A typical application would be a low-temperature compressor designed for operation at a minimum evaporating temperature on Refrigerant 22 of –40°F (–40°C). The required evaporating temperature at normal load conditions is –30°F (–34°C). A DBV would be selected that would start to open at the pressure equivalent to –34°F (–36°C) and bypass enough hot gas at –40°F (–40°C) to prevent a further decrease in suction pressure. Valve settings are according to manufacturer's recommendations.

The DBV is applied in a branch line off the discharge line as close to the compressor as possible. The bypassed vapor can enter the low side at one of the following locations:

- To evaporator inlet with distributor
- To evaporator inlet without distributor
- To suction line

Figure 11-38 shows the bypass to evaporator inlet with a distributor. The primary advantage of this method is that the system TEV will respond to the increased superheat of the vapor, leaving the evaporator, and will provide the liquid required for desuperheating. The evaporator also serves as an excellent mixing chamber for the bypassed hot gas and the liquid-vapor mixture from the expansion valve. This ensures that dry vapor reaches the compressor suction. Oil return from the evaporator is also improved, since the velocity in the evaporator is kept high by the hot gas.

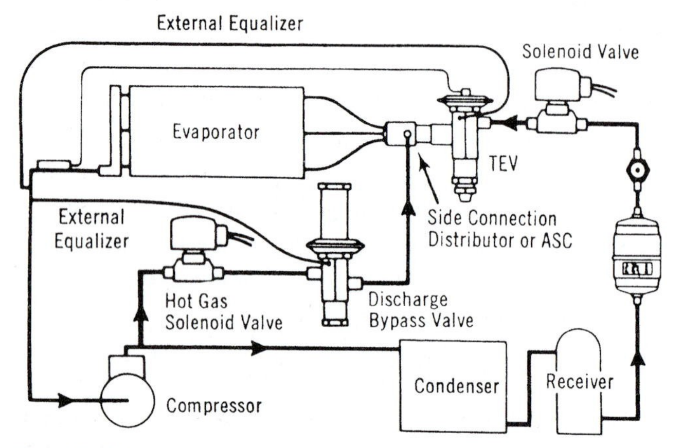

Fig. 11-38 *Connection arrangement for a discharge-bypass valve.* (Sporlan Valve)

Externally Equalized Bypass Valves

The primary function of the DBV is to maintain suction pressure. Thus, the compressor suction pressure is the control pressure. It must be exerted on the underside of the valve diaphragm. When the DBV is applied, as shown in Fig. 11-38, where there is an appreciable pressure drop between the valve outlet and the compressor suction, the externally equalized valve must be used. This is true because when the valve opens, a sudden rise in pressure occurs at the valve outlet. This creates a false control pressure, which would cause the internally equalized valve to close.

Many refrigeration systems and water chillers do not use refrigerant distributors but may require some method of compressor capacity control. This type of application provides the advantages discussed earlier.

Bypass to Evaporator Inlet without Distributor

On many applications, it may be necessary to bypass directly into the suction line. This is generally true of systems with multievaporators or remote-condensing units. It may also be true for existing systems where it is easier to connect to the suction line than the evaporator inlet. The latter situation involves systems fed by TEVs or capillary tubes. When hot gas is bypassed, temperature starts to increase. This can cause breakdown of the oil and refrigerant, possibly resulting in a compressor burnout. On close-coupled systems, this can be eliminated by locating the main expansion-valve bulb downstream of the bypass connection, as shown in Fig. 11-39.

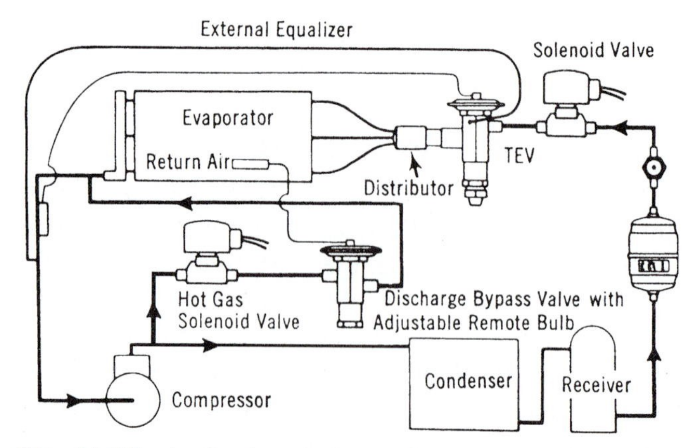

Fig. 11-39 *Application of a hot-gas bypass to an existing system with only minor piping changes.* (Sporlan Valve)

Installation

Bypass valves can be installed in horizontal or vertical lines, whichever best suits the application and permits easy accessibility to the valves. However, consideration should be given to locating these valves so that they do not act as oil traps. Also solder must not run into the internal parts during brazing.

The DBV should always be installed at the condensing unit rather than at the evaporator section. This will ensure the rated bypass capacity of the DBV. It will also eliminate the possibility of hot gas condensing in the bypass line. This is especially true on remote systems.

When externally equalized lines are used, the equalizer connection must be connected to the suction line where it will sense the desired operating pressure. See Fig. 11-40.

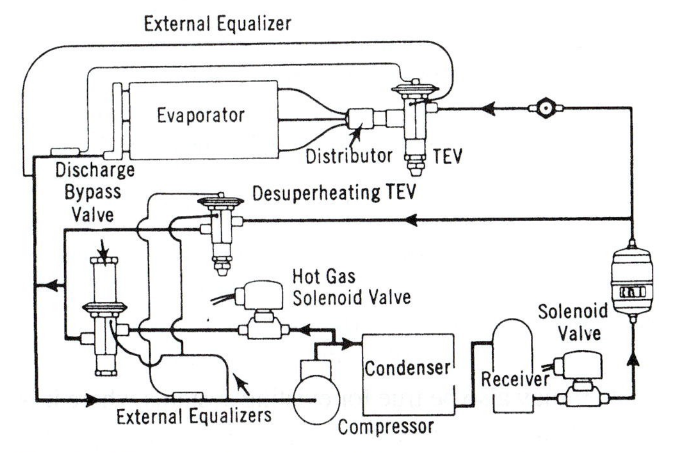

Fig. 11-40 *Externally equalized discharge-bypass valve.* (Sporlan Valve)

Since the DBV is applied in a bypass line between the discharge line and the low side of a system, the valve is subjected to compressor vibrations. Unless the valve, connecting fittings, and tubing are properly isolated from the vibrations, fatigue failures may occur. While the heaviest valve weighs only 3.5 lb (1.6 kg), it should be adequately supported to prevent excessive stress on the connections.

If the *remote-bulb* type bypass valve is used, the bulb must be located in a fairly constant ambient temperature because the element-bulb assembly is air charged. These valves are set at the factory in an 80°F (27°C) ambient temperature. Thus, any appreciable variation from this temperature will cause the pressure setting to vary from the factory setting. For a nonadjustable valve, the remote bulb may be located in an ambient of 80°F ±10°F (27°C ±5.5°C). The adjustable remote bulb model can be adjusted to operate in a temperature of 80°F ±30°F (27°C ±16.7°C). On many units the manufacturer will have altered the pressure setting to compensate for an ambient temperature appreciably different than 80°F (27°C). Therefore, on some units it may be necessary to consult with the equipment manufacturer for the proper opening pressure setting of the bypass valve.

There are numerous places on a system where the remote bulb can be located. Two possible locations are the return air stream and a structural member of the unit, if it is located in a conditioned space. Other locations, where the temperature is fairly constant but different than 80°F (27°C) are also available. These include the return water line on a chiller, the compressor suction line, or the main liquid line. As previously mentioned, the setting may have been altered.

A bulb strap with bolts and nuts is usually supplied with each remote-bulb type DBV. This strap is for use in fastening the bulb in place.

Special Considerations

If a DBV is applied on a system with an evaporator pressure regulating valve (ORIT or other type), the DBV may bypass into either the evaporator inlet or the suction line. The bypass will depend on the specific system. Valve function and the best piping method to protect the compressor should be the deciding factors. If the DBV is required on a system with a crankcase pressure-regulating valve (CRO or other type), the bypass valve can bypass to the low side of the evaporator inlet or the suction line without difficulties. The only decision necessary is whether an internally or externally equalized valve is required. This depends on where the hot gas enters the low side. The pressure setting of the DBV must be lower than the CRO setting for each valve to function properly.

The hot-gas solenoid valve is to be located upstream of the bypass valve. If the solenoid valve is installed downstream of the DBV, the oil and/or liquid refrigerant may be trapped between the two valves. Depending on the ambient temperature surrounding the valves and piping, this could be dangerous.

If the hot-gas solenoid valve is required for pump-down control, it should be wired in parallel with the liquid-solenoid valve so that it can be deenergized by a thermostat.

The hot-gas solenoid is sometimes used for protection against high superheat conditions because the compressor does not have an integral temperature-protection device. If this is done, the solenoid valve is wired in series with a bimetal thermostat fastened to the discharge line close to the compressor.

Testing and Operating Pressures

Excessive leak testing or operating pressures may damage these valves and reduce the life of the operating members. Since a high-side test pressure differential of approximately

350 psig or higher will force the DBV open, the maximum allowable test pressures for DBV are the same as for the high and low side of the system. If greater high-side test pressures than those given in the manufacturer's specifications are to be encountered, some method of isolating the DBV from these high pressures must be found.

Valve setting and adjustment must be done according to the manufacturer's recommendations. Proper instrumentation must be used to determine exactly when these valves are open.

Hot Gas

Hot gas may be required for other system functions besides bypass capacity control. Hot gas may be needed for defrost and head pressure control. Normally, these functions will not interfere with each other. However, compressor cycling on low-suction pressure may be experienced on system start-up when the DBV is operating and other functions require the hot gas also. For example, the head-pressure control requires hot gas to pressurize the receiver and liquid line to get the TEV operating properly. In this case, the DBV should be prevented from functioning by keeping the hot–gas solenoid valve closed until adequate liquid line or suction pressure is obtained.

Malfunctions

There are several reasons for system malfunctions. Possible causes of trouble, when hot gas bypass for capacity control is used, are listed in Table 11-6.

Valves are coded by the manufacturer. The part numbers given in Table 11-6 are those of the Sporlan Valve Company. Note that each letter and number has a meaning. The coded part numbers in Fig. 11-41 are given as examples. Similar codes are used by other valve manufacturers. To be informed of such codes, you will need the manufacturers' bulletins. A good file of such bulletins will enable you to quickly identify the various valve problems. These can be obtained on the internet. Just use the manufacturer's name and dot com and you will usually open their website. From their home page you will be able to navigate their various departments and offerings.

LEVEL CONTROL VALVES

Capillary tubes and float valves are used to control the refrigerant in a system.

Capillary Tubes

Capillary tubes are used to control pressure and temperature in a refrigeration unit. They are most commonly used in domestic refrigeration, milk coolers, ice-cream cabinets, and smaller units. Commercial refrigeration units use other devices. The capillary tube consists of a tube with a very small diameter. The length of the tube depends on the size of the unit to be served, the refrigerant used, and other physical considerations. To effect the necessary heat exchange, this tube is usually soldered to the suction line between the condenser and the evaporator. The capillary tube acts as a constant throttle or restrictor on the refrigerant. Its length and diameter offer sufficient frictional resistance to the flow of refrigerant to build up the head pressure needed to condense the gas.

If the condenser and evaporator were simply connected by a large tube, the pressure would rapidly adjust itself to the same value in both of them. A small diameter water pipe will hold back water, allowing a pressure to be built up behind the water column, but with a small rate of flow. Similarly, the small diameter capillary tube holds back the liquid refrigerant. This enables a high pressure to be built up in the condenser during the operation of the compressor. At the same time, this permits the refrigerant to flow slowly into the evaporator. See Fig. 11-42. A filter drier is usually inserted between the condenser and the capillary tube. This is necessary because the line or tube is so small that it is easily clogged.

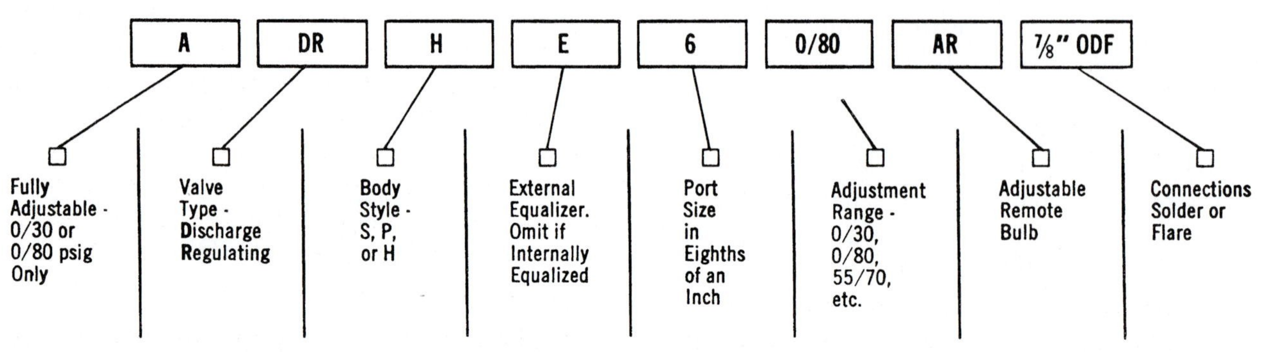

Fig. 11-41 *Codes used to identify the discharge-bypass valve.* (Sporlan Valve)

Table 11-6 *Troubleshooting Discharge-Bypass Valves*

Valve Type*	Malfunction	Cause	Remedy
		Fully Adjustable Models—ADR Type	
ADRS-2 ADRSE-2 ADRP-3 ADRPE-3	Failure to open	1. Dirt or foreign material in valve.	1. Disassemble valve and clean.
	Failure to close	1. Dirt or foreign material in valve. 2. Diaphragm failure. 3. Equalizer passageway plugged. 4. External equalizer not connected or equalizer line pinched shut.	1. Disassemble valve and clean. 2. Replace element only. 3. Disassemble valve and clean. 4. Connect or replace equalizer line.
ADRH-6 ADRHE-6	Failure to open	1. Dirt or foreign material in valve. 2. Equalizer passageway plugged. 3. External equalizer not connected or equalizer line pinched shut.	1. Disassemble valve and clean. 2. Disassemble valve and clean. 3. Connect or replace equalizer line.
	Failure to close	1. Dirt or foreign material in valve. 2. Diaphragm failure.	1. Disassemble valve and clean. 2. Replace element only.
		"LIMITED" ADJUSTABLE MODELS—DR–AR TYPE	
DRP-3-AR DRPE-3-AR	Failure to open	1. Dirt or foreign material in valve. 2. Diaphragm failure. 3. Air charge in element lost.	1. Disassemble valve and clean. 2. Replace element only. 3. Replace element only.
	Failure to close	1. Dirt or foreign material in valve. 2. Equalizer passageway plugged. 3. External equalizer not connected or equalizer line pinched shut.	1. Disassemble valve and clean. 2. Disassemble valve and clean. 3. Connect or replace equalizer line.
DRH-6-AR DRHE-6-AR	Failure to open	1. Dirt or foreign material in valve. 2. Diaphragm failure. 3. Equalizer passageway plugged. 4. External equalizer not connected or equalizer line pinched shut. 5. Air charge in element lost.	1. Disassemble valve and clean. 2. Replace element only. 3. Disassemble valve and clean. 4. Connect or replace equalizer line. 5. Replace element only.
	Failure to close	1. Dirt or foreign material in valve.	1. Disassemble valve and clean.
		NON-ADJUSTABLE MODELS—REMOTE BULB and DOME TYPE	
DRS-2 DRSE-2 DRP-3 DRPE-3	Failure to open	1. Dirt or foreign material in valve. 2. Diaphragm failure. 3. Air charge in element lost.	1. Disassemble valve and clean. 2. Replace element only. 3. Replace element only.
	Failure to close	1. Dirt or foreign material in valve. 2. Equalizer passageway plugged. 3. External equalizer not connected or equalizer line pinched shut.	1. Disassemble valve and clean. 2. Disassemble valve and clean. 3. Connect or replace equalizer line.
DRH-6 DRHE-6	Failure to open	1. Dirt or foreign material in valve. 2. Diaphragm failure. 3. Equalizer passageway plugged. 4. External equalizer not connected or equalizer line pinched shut. 5. Air charge in element lost.	1. Disassemble valve and clean. 2. Replace element only. 3. Disassemble valve and clean. 4. Connect or replace equalizer line. 5. Replace element only.
	Failure to close	1. Dirt or foreign material in valve.	1. Disassemble valve and clean.

*The model numbers are for Sporlan valves.

Capillary tubes may be cleaned and unplugged by the method suggested in Chap. 1. Replacement should be performed in the shop after discharging the unit. In replacing the capillary tube, make sure that the same length of tube is used. The bore or inner diameter should be exactly the same as the old tube. It is easy to check with the proper tool. This tool is described in Chap. 1.

Float Valve

A hollow float is sometimes used to control the level of refrigerant. See Fig. 11-43. The float is fastened to a lever arm. The arm is pivoted at a given point and connected to a needle that seats at the valve opening. If there is no liquid in the evaporator, the ball-lever arm rests on a stop and the needle is not seated, thus leaving the valve open. Once liquid refrigerant under pressure from the compressor enters the float chamber, the float rises with the liquid level until, at a predetermined level, the needle closes the needle-valve opening.

In some plants of large size and Freon-12 as a refrigerant, multiple ports are provided for handling the larger quantities of liquid.

Installation The following precautions must be observed before the installation of a float valve:

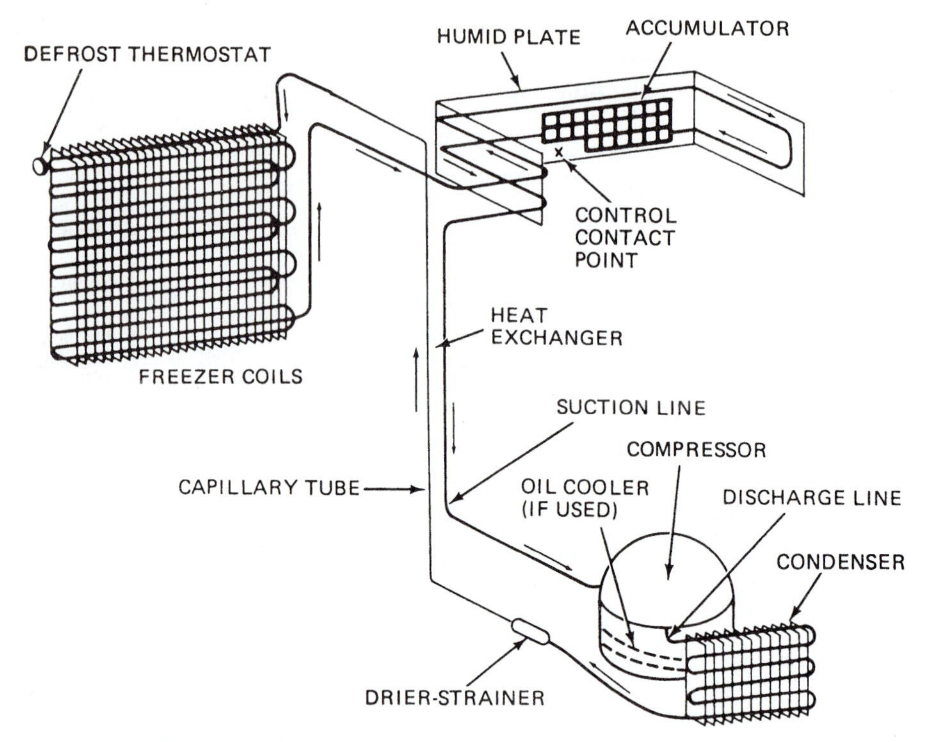

Fig. 11-42 *Refrigerant flow with a capillary tube in the line.*

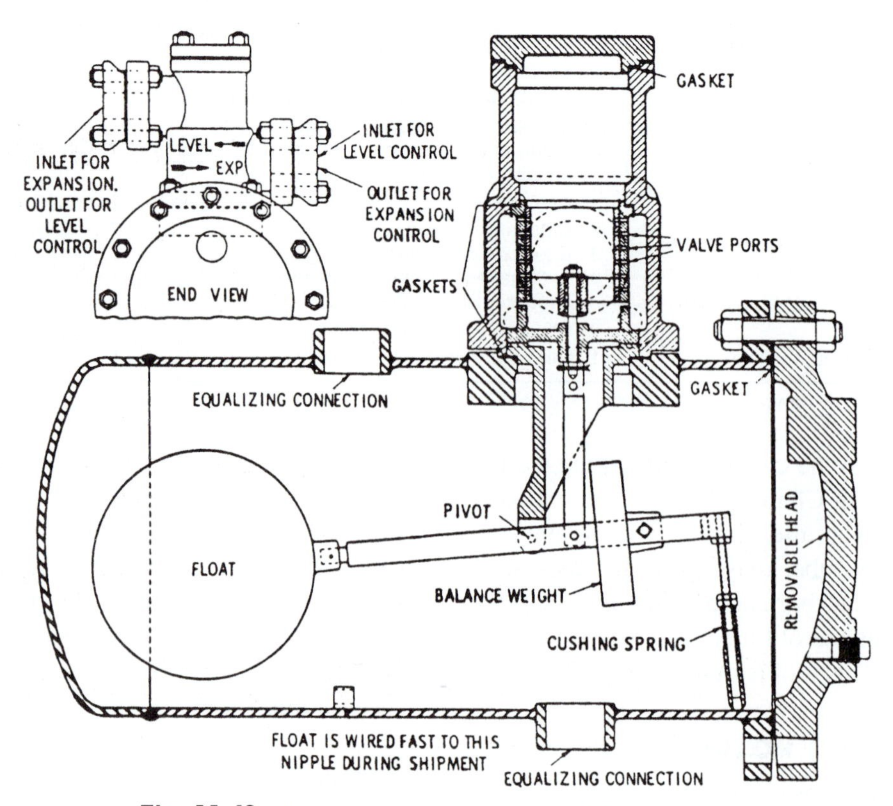

Fig. 11-43 *Interior construction of a typical float valve.* (Frick)

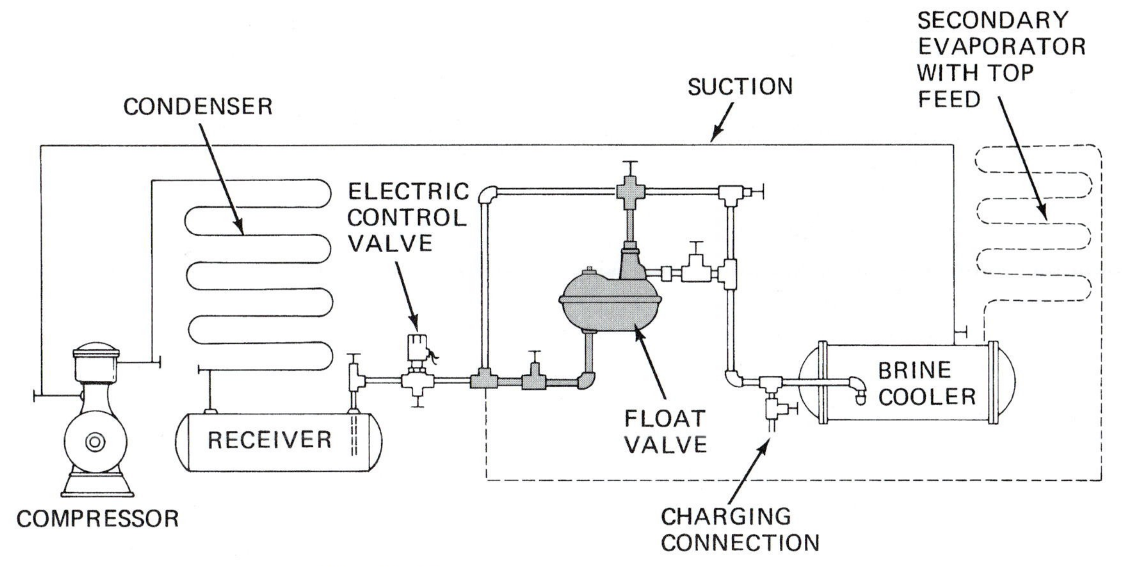

Fig. 11-44 *High-pressure float-control system.*

- Most float controls are designed for a maximum differential pressure of 200 lb.
- If the pressure will exceed 190 psi, there are stems and orifices of special size available for low-temperature use.
- In any application, keep the bottom equalizing line above the bottom of the evaporator to avoid oil logging.
- Make sure there are no traps in the equalizing line.
- The stems of a globe valve must be in a horizontal plane.
- Refrigerant flow must be kept to less than 100 ft/min where a bottom float equalizing connection is made to the header or accumulator return. That means the header and accumulator pipe must be properly sized.
- Accumulators of a small diameter with a velocity of over 50 ft/min are not suitable for accurate float application. However, the float may control within wider limits with higher velocities. The top equalizing connection must be connected to a point of practically zero gas velocity.
- In automatic plants, always provide a solenoid valve in the liquid line ahead of the float control. This solenoid valve is to close either when the temperatures are satisfactory or when the compressor stops.

Figure 11-44 illustrates the connections for a high-pressure float control. There have been new developments in the control of liquid level since the early days of refrigeration.

LEVEL-MASTER CONTROL

The level-master control is a positive liquid-level control device suitable for application to all flooded evaporators. See Fig. 11-45. The level-master control is a standard TEV with a level-master element. The combination provides a simple, economical, and highly effective

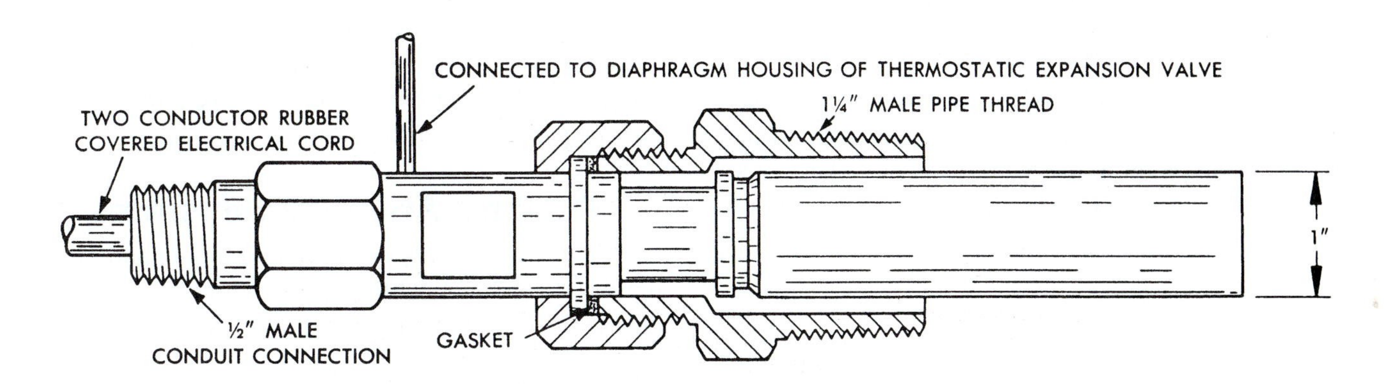

Fig. 11-45 *Level-master control.* (Sporlan Valve)

liquid-level control. The bulb of the conventional thermostatic element has been modified to an insert type of bulb that incorporates a low-wattage heater. A 15-W heater is supplied as standard. For applications below −60°F (−51°C) evaporating temperature, a special 25-W heater is needed.

The insert bulb is installed in the accumulator or surge drum at the point of the desired liquid level. As the level at the insert bulb drops, the electrically added heat increases the pressure within the thermostatic element and opens the valve. As the liquid level at the bulb rises, the electrical input is balanced by the heat transfer from the bulb to the liquid refrigerant. The level-master control either modulates or eventually shuts off. The evaporator pressure and spring assist in providing a positive closure.

Installation

The level-master control is applicable to any system that has been specifically designed for flooded operation. The valve is usually connected to feed into the surge drum above the liquid level. It can feed into the liquid leg or coil header.

The insert bulb can be installed directly in the shell, surge drum, or liquid leg on new or existing installations. Existing float systems can be easily converted by installing the level-master control insert bulb in the float chamber.

Electrical Connections

The heater is provided with a two-wire neoprene-covered cord 2 ft in length. It runs through a moisture-proof grommet and a 1/2 in. male-conduit connection affixed to the insert bulb assembly. See Fig. 11-46.

The heater circuit must be interrupted when refrigeration is not required. Wire the heater in parallel with the holding coil of the compressor-line starter or solenoid valve—not in series.

Hand Valves

On some installations, the valve is isolated from the surge drum by a hand valve. A 2- to 3-lb pressure drop from the valve outlet to the bulb location is likely. For such installations, an externally equalized valve is recommended.

Oil Return

All reciprocating compressors will allow some oil to pass into the discharge line along with the discharge gas. Mechanical oil separators are used extensively. However, they are never completely effective. The untrapped oil passes through the condenser, liquid line, expansion device, and into the evaporator.

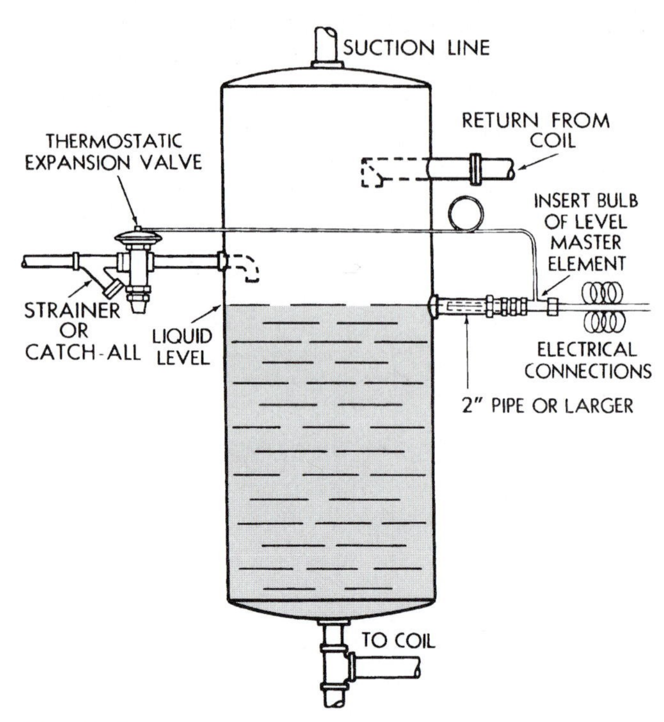

Fig. 11-46 *Installation of the level-master control.* (Sporlan Valve)

In a properly designed direct-expansion system, the refrigerant velocity in the evaporator tubes and the suction line is sufficiently high to ensure a continuous return of oil to the compressor crankcase. However, this is not characteristic of flooded systems. Here, the surge drum is designed for a relatively low-vapor velocity. This prevents entrainment of liquid-refrigerant droplets and consequent carry-over into the suction line. This design also prevents the return of any oil from the low side in the normal manner.

If oil is allowed to concentrate at the insert bulb location of the level-master control, overfeeding with possible flood-back can occur. The tendency to overfeed is due to the fact that the oil does not convey the heat from the low-wattage heater element away from the bulb as rapidly as does pure liquid refrigerant. The bulb pressure is higher than normal, and the valve remains in the open or partially open position.

Oil and Ammonia Systems

For all practical purposes, liquid ammonia and oil are immiscible (not capable of being mixed). Since the density of oil is greater than that of ammonia, it will fall to the bottom of any vessel containing such a mixture if the mixture is relatively placid. Therefore, the removal of oil from an ammonia system is a comparatively simple task. Generally, on systems equipped with a surge drum, the liquid leg is extended downward below the point where the liquid is fed off to the evaporator. A drain valve is provided to allow periodic manual draining. See Fig. 11-47.

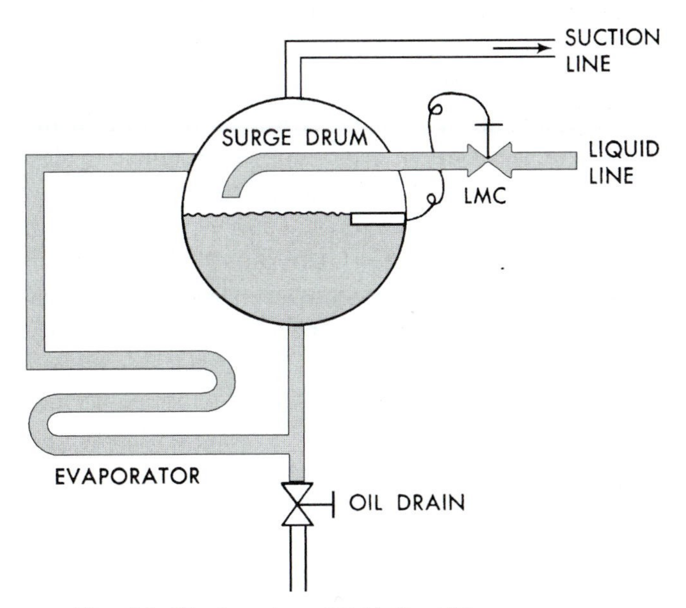

Fig. 11-47 *Location of LM in liquid line.* (Sporlan Valve)

For flooded chillers that do not use a surge drum, a sump with a drain valve is usually provided at the bottom of the chiller shell. These methods are quite satisfactory, except possibly on some low-temperature systems. Here, the drain leg or sump generally must be warmed prior to attempting to draw off the oil. The trapped oil becomes quite viscous at lower temperatures.

If oil is not drained from a flooded ammonia system, a reduction in the evaporator heat-transfer rate can occur due to an increase in the refrigerant-film resistance. Difficulty in maintaining the proper liquid level with any type of flooded control can also be expected.

With a float valve, you can expect the liquid level in the evaporator to increase with high concentration of oil in a remote float chamber. If a level-master control is used with the insert bulb installed in a remote chamber, oil concentration at the bulb can cause overfeeding with possible flood-back. The lower or liquid balance line must be free of traps and be free-draining into the surge drum or chiller, as shown in Fig. 11-48. The oil drain leg or sump must be located at the lowest point in the low side.

Oil and Halocarbon Systems

With halocarbon systems (Refrigerants 12, 22, 502, and so forth) the oil and refrigerant are miscible (capable of being mixed) under certain conditions. Oil is quite

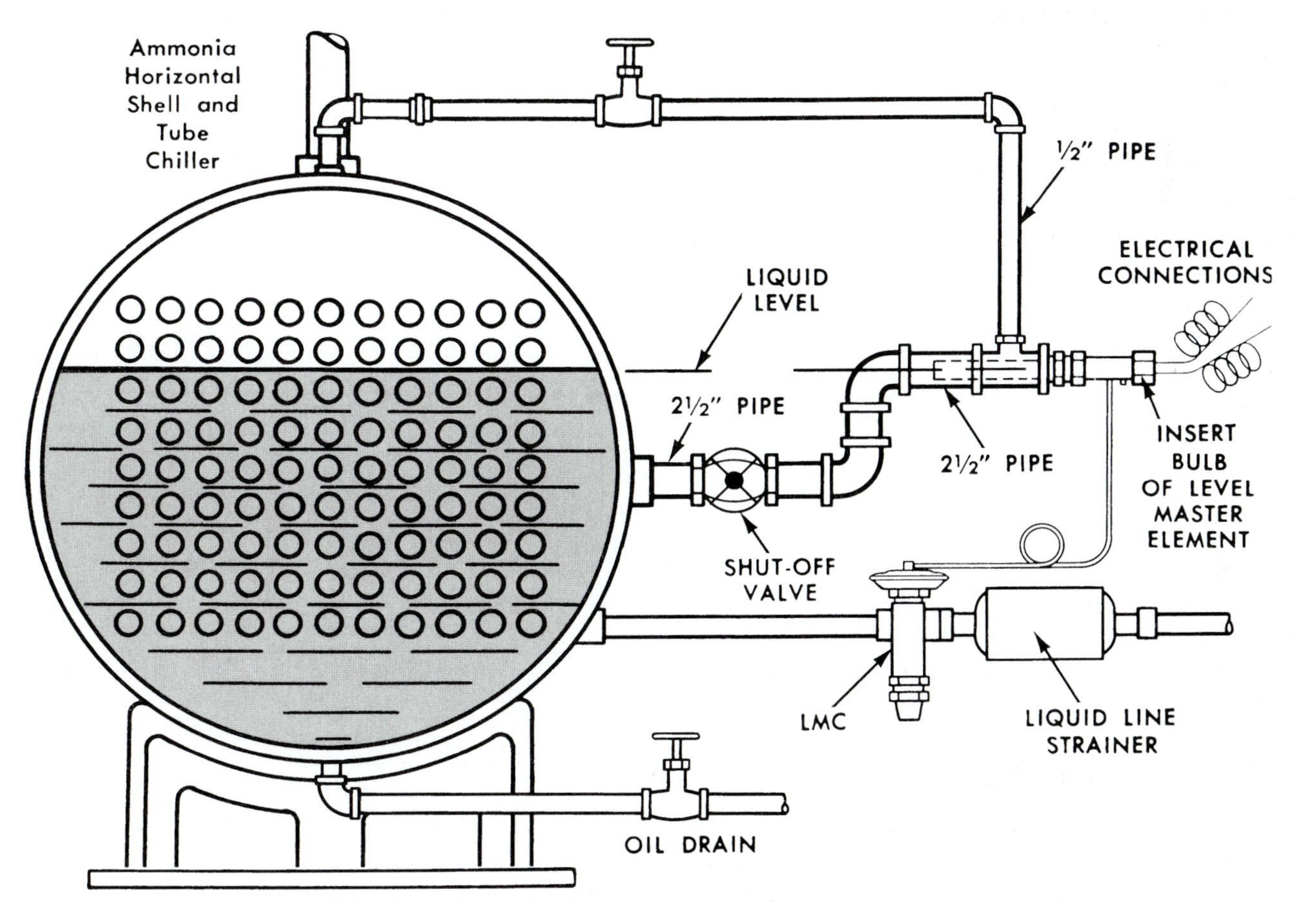

Fig. 11-48 *Level-master control with the bulb inserted in a remote chamber.* (Sporlan Valve)

soluble in liquid Refrigerant 12 and partially so in liquid Refrigerant 22 and 502. For example, for a 5 percent (by weight) solution of a typical *napthenic* (a petroleum-based oil) oil in liquid refrigerant, the oil will remain in solution down to about −75°F (−59°C) for Refrigerant 12, down to about 0°F (−18°C) for Refrigerant 22, and down to about 20°F (−7°C) for Refrigerant 502. Depending upon the type of oil and the percentage of oil present, these figures can vary. However, based on the foregoing, we can assume that for the majority of Refrigerant-12 systems the oil and refrigerant are completely miscible at all temperatures normally encountered. However, at temperatures below 0°F (−18°C) with Refrigerant 22 and a 5 percent oil concentration and temperatures below 20°F (−7°C) with Refrigerant 502 and a 5 percent oil concentration, a liquid-phase separation occurs. An oil-rich solution will appear at the top and a refrigerant-rich solution will lay at the bottom of any relatively placid remote-bulb chamber.

Oil in a halocarbon-flooded evaporator can produce many results. Oil as a contaminant will raise the boiling point of the liquid refrigerant. For example, with Refrigerant 12, the boiling point increases approximately 1°F (0.56°C) for each 5 percent of oil (by weight) in solution. As in an ammonia system, oil can foul the heat-transfer surface with a consequent loss in system capacity. Oil can produce foaming and possible carry-over of liquid into the suction line. Oil can also affect the liquid-level control. With a float valve you can normally expect the liquid level in the evaporator to decrease with increasing concentrations of oil in the float chamber. This is due to the difference in density between the lighter oil in the chamber and the lower balance leg and the heavier refrigerant/oil mixture in the evaporator. A lower column of dense mixture in the evaporator will balance a higher column of oil in the remote chamber and piping. This is similar to a "U-tube" manometer with a different fluid in each leg.

With the level-master control, the heat transfer rate at the bulb is decreased, producing overfeeding and possible flood-back. What can be done? First of all, the oil concentration must be kept as low as possible in the evaporator, surge drum, and remote insert bulb chamber (if one is used). With Refrigerant 12, since the oil/refrigerant mixture is homogenous, it can be drained from almost any location in the chiller, surge drum, or remote chamber that is below the liquid level. With Refrigerants 22 and 502, the drain must be located at or slightly below the surface of the liquid, since the oil-rich layer is at the top. There are many types of oil-return devices:

- Direct drain into the suction line
- Drain through a high-pressure, liquid-warmed heat exchanger
- Drain through a heat exchanger with the heat supplied by an electric heater

Draining directly into the suction line, as shown in Fig. 11-49, is the simplest method. However, the hazard of possible flood-back to the compressor remains.

Draining through a heat exchanger, as indicated in Fig. 11-50, is a popular method. The liquid refrigerant flood-back problems are minimized by using the warm liquid to vaporize the liquid refrigerant in the oil/refrigerant mixture.

The use of a heat exchanger with an insert electric heater, as shown in Fig. 11-51, is a variation of the preceding method.

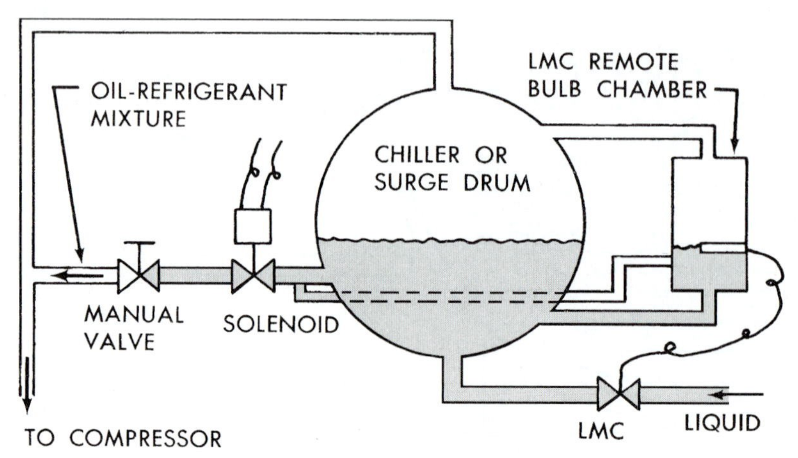

Fig. 11-49 *Direct drain of oil to the suction line is one of the three ways to recover oil in flooded systems. Heat from the environment or a liquid-suction heat exchanger is required to vaporize the liquid refrigerant so drained. Vapor velocity carries oil back to the compressor.* (Sporlan Valve)

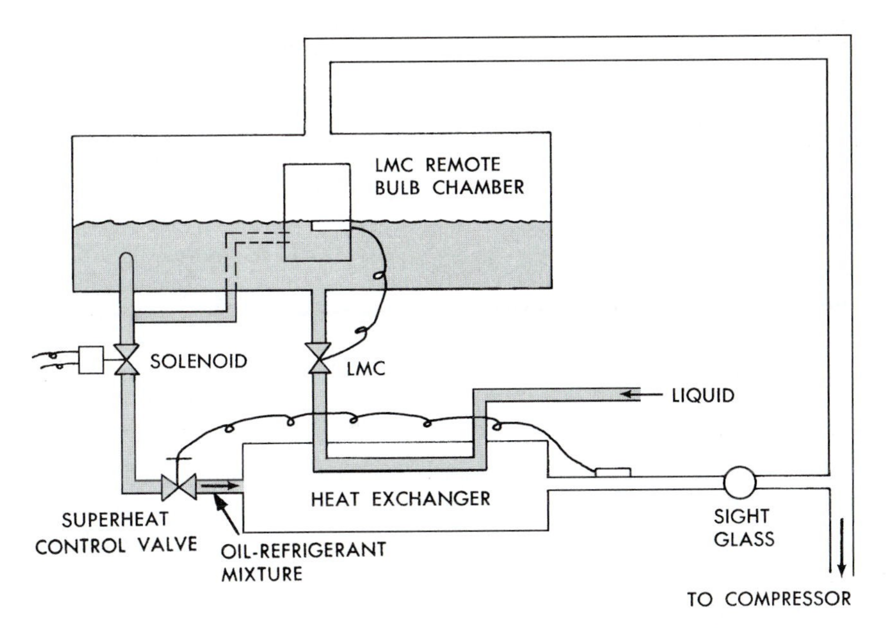

Fig. 11-50 *Oil return by draining oil-refrigerant mixture through a heat exchanger is shown here. Heat in incoming liquid vaporizes refrigerant to prevent return of liquid to the compressor. Liquid feed is controlled by a thermostatic- or hand-expansion valve.* (Sporlan Valve)

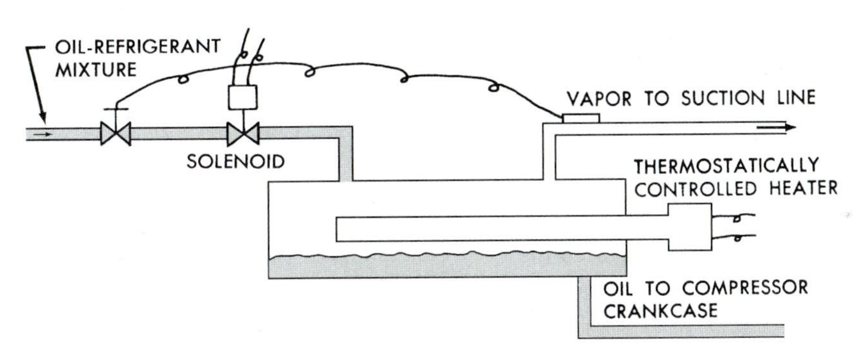

Fig. 11-51 *An electric heater may also be added to separate oil and refrigerant. This system is similar to that shown in Fig. 10-49, except that the heat required for vaporization is added electrically.* (Sporlan Valve)

In all of the return arrangements discussed, a solenoid valve should be installed in the drain line and arranged to close when the compressor is not in operation. Otherwise, liquid refrigerant could drain from the low side into the compressor crankcase during the off cycle.

If the insert bulb is installed directly into the surge drum or chiller, oil return is necessary only from this point. However, the insert bulb is sometimes located in a remote chamber that is tied to the surge drum or chiller with liquid and gas balance lines. Then oil return should be made from both locations, as shown in Fig. 11-49, 11-50, and 11-52.

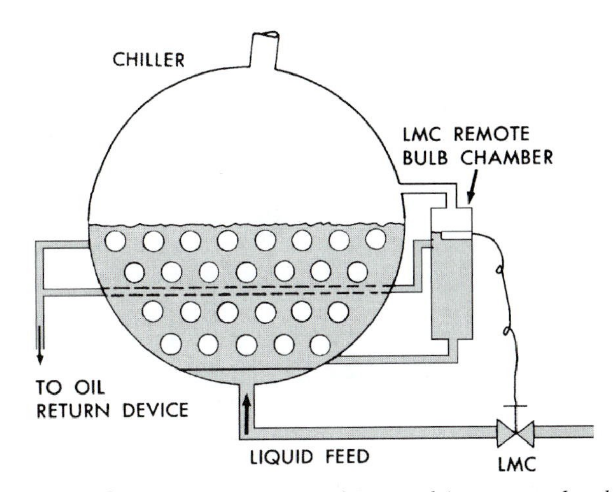

Fig. 11-52 *Level-master control inserted in remote chamber.* (Sporlan Valve)

Conclusions

The problem of returning oil from a flooded system is not highly complex. There are undoubtedly other methods in use today that are comparable to those outlined here. Regardless of how it is accomplished, oil return must be provided for proper operation of any flooded system. This is necessary not only with the level-master control, but also with a float or other type of level-control device.

OTHER TYPES OF VALVES

There are check valves, water valves and receiver valves.

Service Valves on Sealed Units

Hermetic-refrigeration systems, also called sealed units, normally have no service valves on the compressor. Instead, a charging plug or valve may be mounted on the compressor. A special tool is needed to operate the charging device, which varies on different makes.

A service engineer needs the correct valve-operating device for a unit. Thus, a kit is made that contains adapters and wrench ends to fit many makes of sealed units.

Essentially, the device is a body with a union connection and provisions for charging line and pressure-gage connections. The stem may be turned or pushed in or out of the body as required. Figure 11-53 shows a line-piercing valve. They are used for charging, testing, or purging those hermetic units not provided with a charging plug or valve. These valves may be permanently attached to the line without danger of refrigerant loss.

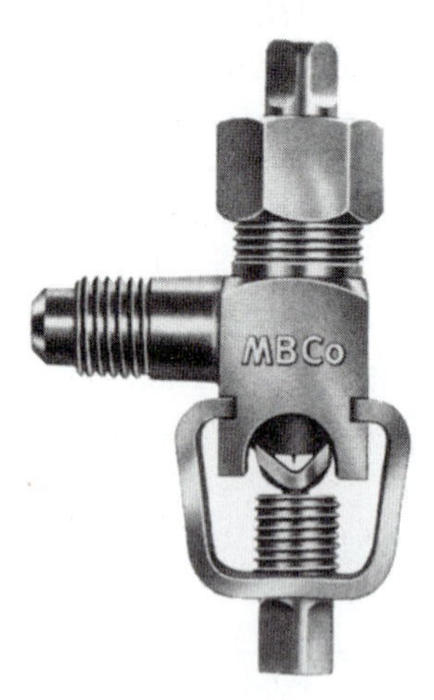

Fig. 11-53 *Line-piercing valve.* (Mueller Brass)

Water Valves

Manually operated valves are installed on water circuits associated with refrigeration systems—either on cooling towers or in secondary brine circuits. They are installed for convenience in servicing and for flexibility in operating conditions. These valves make it possible to recircuit, bypass, or shut off water flow as desired. See Fig. 11-54.

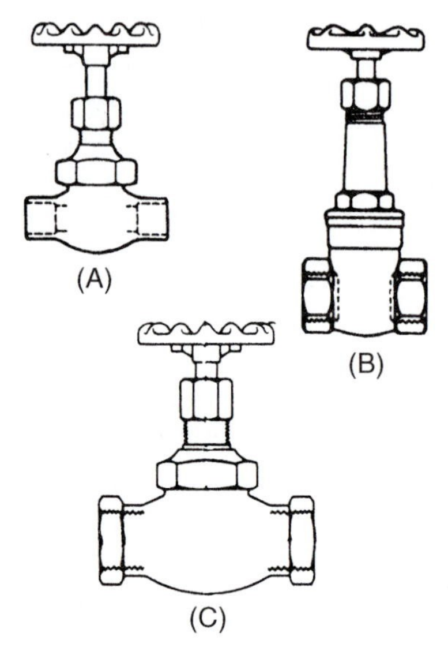

Fig. 11-54 *Water valves.* (A) *Stop valve.* (B) *Gate valve.* (C) *Globe valve.* (Mueller Brass)

These manually operated shutoff or flow-control valves are available in a wide variety of styles and sizes. Valve stems and body seats are accurately machined to close tolerances, ensuring easy and positive shutoff. They are made of nonporous cast bronze.

There are three main types (See Fig. 11-54):

- Stop valves
- Globe valves
- Gate valves

Check Valves

Some refrigeration systems are designed in which the refrigerant liquid or vapor flows to several components, but must never flow back through a given line. A check valve is needed in such installations. As its name implies, a check valve checks or prevents the flow of refrigerant in one direction, while allowing free flow in the other direction. For example, two evaporators might be controlled by a single condensing system. In this case, a check valve should be placed in the line from the lower temperature evaporator to prevent the suction gas from the higher temperature evaporator from entering the lower temperature evaporator. See Fig. 11-55.

Check valves are designed to eliminate chattering and to give maximum refrigerant flow when the unit is operating. If the spring tension is sufficient to overcome the weight of the valve disc, the check valve may be mounted in any position.

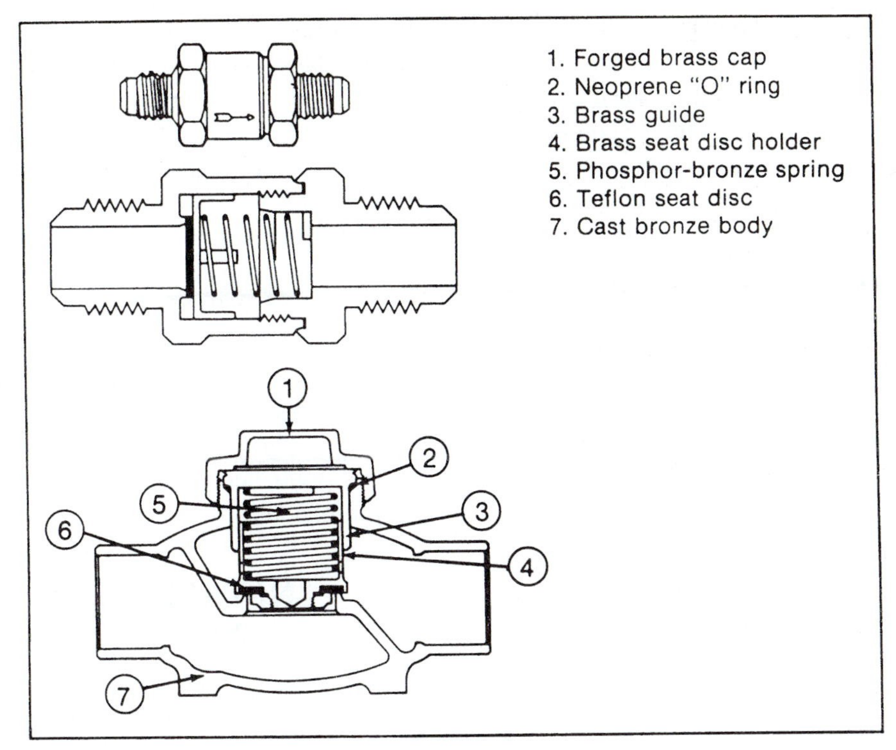

Fig. 11-55 *Check valves.* *(Mueller Brass)*

Receiver Valves

Receivers may be fitted with two valves—an inlet valve and an outlet valve. The outlet valve may have the inlet in the form of an ordinary connection, such as an elbow. An inlet valve permits closing the receiver should a leak develop between the compressor and the receiver. The receiver outlet valve is important when the system is "pumped down," when for reasons of service all the refrigerant is conveyed to the receiver for temporary storage. See Fig. 11-56.

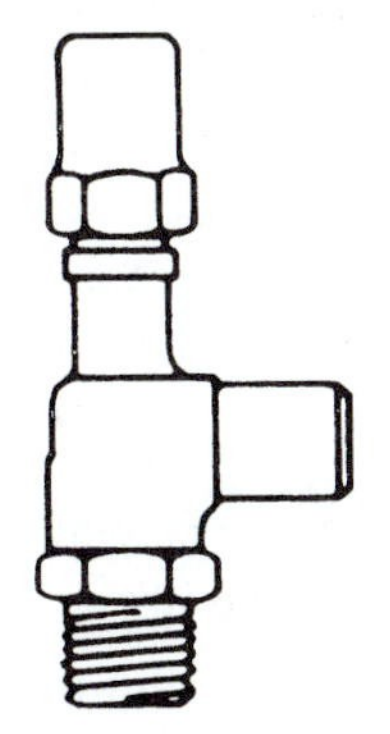

Fig. 11-56 *Receiver-angle valve.* *(Mueller Brass)*

ACCUMULATORS

Accumulators have been used for years on original equipment. More recently they have been field installed. The significance with respect to accumulator and system performance has never been clarified. Engineers have been forced to evaluate each model in terms of the system on which it is to be applied. Application in the field has been primarily based on choosing a model with fittings that will accommodate the suction line and be large enough to hold about half of the refrigerant charge.

There is no standard rating system for accumulators. The accuracy of rating data becomes a function of the type of equipment used to determine the ratings. Some data is now available to serve as a guide to those checking the use of an accumulator.

Purpose

The purpose of an accumulator is to prevent compressor damage due to slugging of refrigerant and oil. They provide a positive oil return at all rated conditions. They are designed to operate at –40°F (–40°C) evaporator temperature. Pressure drop is low across them. They act as a suction muffler. They can take suction-gas temperatures as low as 10°F (–12.2°C) at the accumulator. Most of them can withstand a working pressure of 300 psi and have fusible relief devices.

Compressors are designed to compress vapors, not liquids. Many systems, especially low-temperature systems, are subject to the return of excessive quantities of liquid refrigerant. This returned refrigerant dilutes the oil and washes out bearings. In some cases, it causes complete loss of oil in the crankcase. This results in broken valve reeds, and damage to pistons, rods, crankcase, and other moving parts. The accumulator acts as a reservoir to hold temporarily the excess oil-refrigerant mixture and return it at a rate the compressor can safely handle. Figure 11-57 shows the interior view.

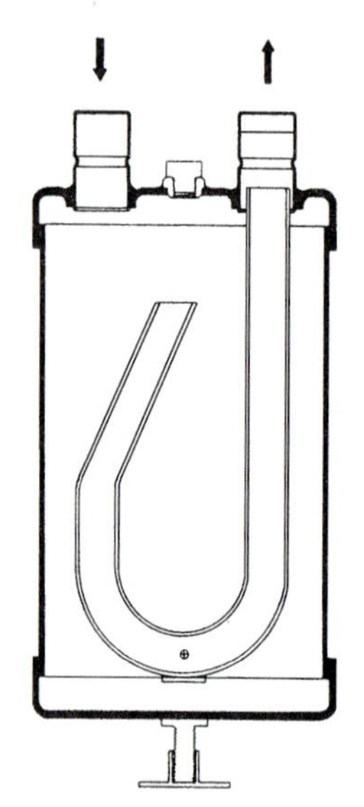

Fig. 11-57 *Suction-line accumulator. (Virginia Chemicals)*

Rating Data

The refrigerant-holding capacity of the accumulator is based on an average condition of 65 percent fill under running conditions.

Refrigerant-Holding Capacity It is obvious that directly on startup or after long off-cycles the amount held may fluctuate from empty to nearly full.

Minimum Evaporator Temperature and Minimum Temperature of Suction Gas at the Accumulator

The oil-refrigerant mixture in the suction line has been studied over the range of –50 to +40°F (–46 to +4°C). The value of –40°F (–40°C) was chosen as a minimum evaporator temperature because it appears adequate for commercial refrigeration. Yet, it is conservative enough to provide a margin of safety. More important is the requirement that the temperature of the suction gas at the accumulator be 10°F (–12°C) or higher. Particularly with refrigerants such as Freon 502, in the low-temperature range up to 0°F (–17.8°C), the oil and refrigerant separate into two layers, with the upper layer being the oil-rich layer. At these low temperatures, the oil-rich layer can become so viscous that it will not flow. When the refrigerant below the heavy oil layer leaves the accumulator, the very thick oil settles over the oil-return port and stops all oil return. This condition will occur regardless of accumulator design. If temperatures below 10°F (–12°C) at the accumulator are to be used, auxiliary heat must be added to keep the oil fluid.

Maximum recommended actual tonnage is based on pressure drop through the accumulator equivalent to an effect of 1°F (0.56°C) on evaporator temperature.

Minimum recommended actual tonnage is based on the minimum flow through the accumulator necessary to insure positive oil return.

For operating conditions outside the manufacturer's published ratings, contact the manufacturer for recommendations.

INSTALLATION OF THE ACCUMULATOR

Locate the accumulator as close to the compressor as possible. In systems employing reverse cycle, the accumulator must be installed between the reversing valve and the compressor. Proper inlet (from the evaporator) and outlet (to the compressor) must be observed. The accumulator must be installed vertically. Proper sizing of an accumulator may not necessarily result in the accumulator connections matching the suction-line size. This new technology must replace the dangerous and outmoded practice of matching the accumulator connections to the suction-line size. To accommodate mismatches, bushing down may be required.

The accumulator should not be installed in a bypass line or in suction lines that experience other than total refrigerant flow.

When installing an accumulator with solder connections, direct the torch away from the top access plug to prevent possible damage to the O-ring seal. When installing a model equipped with a fusible plug, a dummy plug should be inserted in place of the fusible plug until all brazing or soldering is complete.

REVIEW QUESTIONS

1. What is capillary tubing?
2. Why do some cities require the use of K-type copper tubing?
3. What is the composition of No. 95 solder?
4. Name two types of line valves.
5. What is the most detrimental material in a refrigeration system?
6. What happens to halogen-type refrigerants when they combine with water?

7. What happens to Refrigerants 12 and 22 when heated to a high temperature?
8. How are sludge and varnish formed in a refrigeration system?
9. What happens to a liquid indicator if the system has too much oil?
10. What does TEV stand for?
11. What is flash gas?
12. What is the purpose of the crankcase pressure-regulating valve?
13. What does CRO stand for?
14. Where are ORITs installed?
15. What is the purpose of an ORO head-pressure valve?
16. What is the function of a discharge-bypass valve?
17. How is a remote bulb used in a bypass valve?
18. List two possible locations for the mounting of the remote bulb.
19. How are capillary tubes used in a refrigeration unit?
20. What is an accumulator?
21. What is the refrigerant-holding capacity of the accumulator?

12

CHAPTER

Servicing and Safety

PERFORMANCE OBJECTIVES

After studying this chapter you will:

1. Know how to handle compressed gas cylinders.
2. Know how to work safely in lifting and repair work.
3. Know how to troubleshoot and service compressors.
4. Know what to do with compressor-motor burnout.
5. Know how to replace the filter drier.
6. Know how to repair the perimeter tube.
7. Know how to splice a power cord.
8. Know how to replace the evaporator-heat exchanger assembly.
9. Know how to add refrigerant.
10. Know how to test for refrigerant leaks.
11. Know the difference between capacitors used for start and run circuits.
12. Know how to field-test a hermetic compressor.
13. Know how to check PSC compressor-motor troubles and correct them.
14. Know how to test electrical components.

One of the most important parts of working around air-conditioning and refrigeration equipment is that of doing the job safely. The possibility of incorrect procedures being followed can make it very painful both physically and mentally. Some of the suggestions that follow should aid in your understanding of careful work habits and use of the proper tool for the job.

SAFETY

Safe practices are important in servicing refrigeration units. Such practices are common sense, but must be reinforced to make one aware of the problems that can result when a job is done incorrectly.

Handling Cylinders

Refrigeration and air-conditioning servicepersons must be able to handle compressed gases. Accidents occur when compressed gases are not handled properly.

Oxygen or Acetylene Must Never be Used to Pressurize a Refrigeration System Oxygen will explode when it comes in contact with oil. Acetylene will explode under pressure, except when properly dissolved in acetone as used in commercial acetylene cylinders.

Dry nitrogen or dry carbon dioxide are suitable gases for pressurizing refrigeration or air-conditioning systems for leak tests or system cleaning. However, the following specific restrictions must be observed:

Nitrogen (N_2). Commercial cylinders contain pressures in excess of 2000 lb/in.^{2} at normal room temperature.

Carbon dioxide (CO_2). Commercial cylinders contain pressures in excess of 800 lb/in.^{2} at normal room temperature.

Cylinders should be handled carefully. Do not drop them or bump them. Keep cylinders in a vertical position and securely fastened to prevent them from tipping over. Do not heat the cylinder with a torch, or other open flame. If heat is necessary to withdraw gas from the cylinder, apply heat by immersing the lower portion of the cylinder in warm water. Never heat a cylinder to a temperature over 110°F (43°C).

Pressurizing

Pressure Testing or Cleaning Refrigeration and Air-Conditioning Systems Can be Dangerous! Extreme caution must be used in the selection and use of pressurizing equipment. Follow these procedures:

- Never attempt to pressurize a system without first installing an appropriate pressure-regulating valve on the nitrogen or carbon dioxide cylinder discharge. This regulating valve should be equipped with two functioning pressure gages. One gage indicates cylinder pressure. The other gage indicates discharge or downstream pressure.
- Always install a pressure-relief valve or frangible-disc type pressure-relief device in the pressure-supply line. This device should have a discharge port of at least 1/2 in. *national pipe thread* (NPT) size. This valve or frangible-disc device should be set to release at 175 psig.
- A system can be pressurized up to a *maximum* of 150 psig for leak testing or purging. See Fig. 12-1.

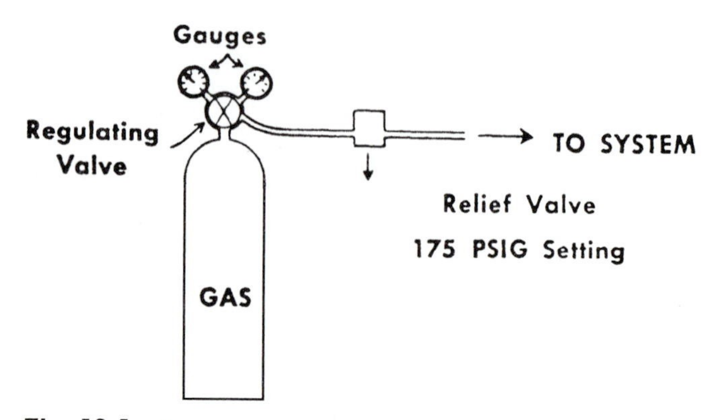

Fig. 12-1 *Pressurizing set-up for charging refrigeration systems.*

Tecumseh hermetic-type compressors are low-pressure housing compressors. The compressor housings (cans or domes) are not normally subjected to discharge pressures. They operate instead at relatively low-suction pressures. These Tecumseh compressors are generally installed on equipment where it is impractical to disconnect or isolate the compressor from the system during pressure testing. Therefore, do not exceed 150 psig when pressurizing such a complete system.

When flushing or purging a contaminated system, care must be taken to protect the eyes and skin from contact with acid-saturated refrigerant or oil mists. The eyes should be protected with goggles. All parts of the body should be protected by clothing to prevent injury by refrigerant. If contact with either skin or eyes occurs, flush the exposed area with cold water. Apply an ice pack if the burn is severe, and see a physician at once.

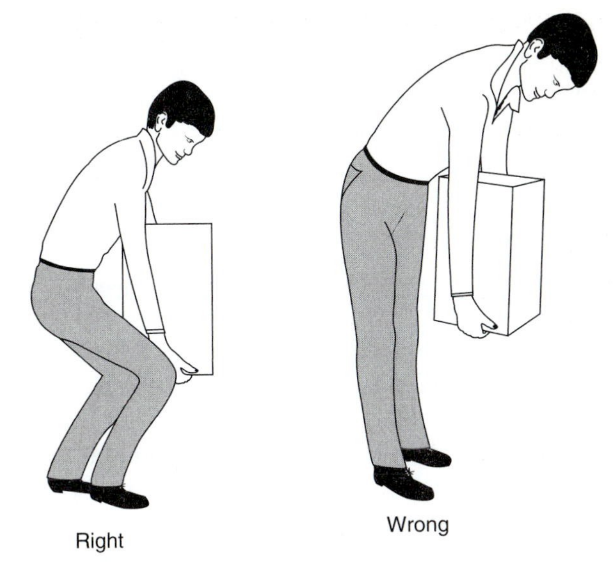

Fig. 12-2 *Safety first. Lift with the legs not the back.* (Tecumseh)

Working with Refrigerants

R-12 has effectively been replaced in modern air-conditioning equipment with R-134a or any of the approved substitutes and R-22 has some acceptable substitutes also. They are considered to be nontoxic and noninflammable. However, any gas under pressure can be hazardous. The latent energy in the pressure alone can cause damage. In working with R-12 and R-22 (or their substitutes), observe the same precautions that apply when working with other pressurized gases.

Never completely fill any refrigerant gas cylinder with liquid. Never fill more than 80 percent with liquid. This will allow for expansion under normal conditions.

Make sure an area is properly ventilated before purging Or evacuating a system that uses R-12, R-22 or their equivalents. In certain concentrations and in the presence of an open flame, such as a gas range or a gas water heater, R-12 and R-22 may break down and form a small amount of harmful phosgene gas. This gas was used in World War I as the poison gas designed for warfare.

Lifting

Lifting heavy objects can cause serious problems. Strains and sprains are often caused by improper lifting methods. Figure 12-2 indicates the right and the wrong way to lift heavy objects. In this case, a compressor is shown.

To avoid injury, learn to lift the safe way. Bend your knees, keep your back erect, and lift gradually with your leg muscles.

The material you are lifting may slip from your hands and injure your feet. To prevent foot injuries, wear the proper shoes.

Electrical Safety

Many Tecumseh single-phase compressors are installed in systems requiring off-cycle crankcase heating. This is designed to prevent refrigerant accumulation in the compressor housing. The power is on at all times. Even if the compressor is not running, power is applied to the compressor housing where the heating element is located.

Another popular system uses a run capacitor that is always connected to the compressor motor windings, even when the compressor is not running. Other devices are energized when the compressor is not running. That means there is electrical power applied to the unit even when the compressor is not running. This calls for an awareness of the situation and the proper safety procedures.

Be safe. Before you attempt to service any refrigeration system, make sure that the main circuit breaker is open and all power is off.

SERVICING THE REFRIGERATOR SECTION

The refrigerant cycle is a continuous cycle, which occurs whenever the compressor is operating. Liquid refrigerant is evaporated in the evaporator by the heat that enters the cabinet through the insulated walls and by product load and door openings. The refrigerant vapor passes from the evaporator, through the suction line, to the compressor dome, which is at suction pressure. From the top interior of the dome, the vapor passes down through a tube into the pump cylinder. The pressure and temperature of the vapor are raised in the cylinder by compression. The vapor is then forced through the discharge valve into the discharge line and the condenser. Air passing over the condenser surface removes heat from the high-pressure vapor, which then condenses to a liquid. The liquid refrigerant flows from the condenser to the evaporator through the small diameter liquid line (capillary tube). Before it enters the evaporator, it is subcooled in the heat exchanger by the low-temperature suction vapor in the suction line. See Fig. 12-3.

Sealed Compressor and Motor

All models are equipped with a compressor with internal-spring suspension. Some compressors have a plug-in magnetic starting relay, with a separate motor overload protector. Others have a built-in metallic motor overload protector. When ordering a replacement compressor, you should always give the refrigerator model number and serial number, and the compressor part number. Every manufacturer has a listing available to servicepersons.

Condenser

Side-by-side and top-freezer models with a vertical, natural draft, wire-tube type condenser have a water-evaporating coil connected in series with the condenser. The high-temperature, high-pressure, compressed refrigerant vapor passes first through the water-evaporating coil. There, part of the latent heat of evaporation and sensible heat of compression are released. See Fig. 12-3. The refrigerant then flows back through the oil-cooling coil in the compressor shell. There, additional heat is picked up from the oil. The refrigerant then flows back to the main condenser, where sufficient heat is released to the atmosphere. This results in the condensation of refrigerant from a high-pressure vapor to high-pressure liquid.

Filter Drier

A filter drier is located in the liquid line at the outlet of the condenser. Its purpose is to filter or trap minute particles of foreign materials and absorb any moisture in the system. Fine mesh screens filter out foreign particles. The desiccant absorbs the moisture.

Capillary Tube

The capillary tube is a small diameter liquid line connecting the condenser to the evaporator. Its resistance or,

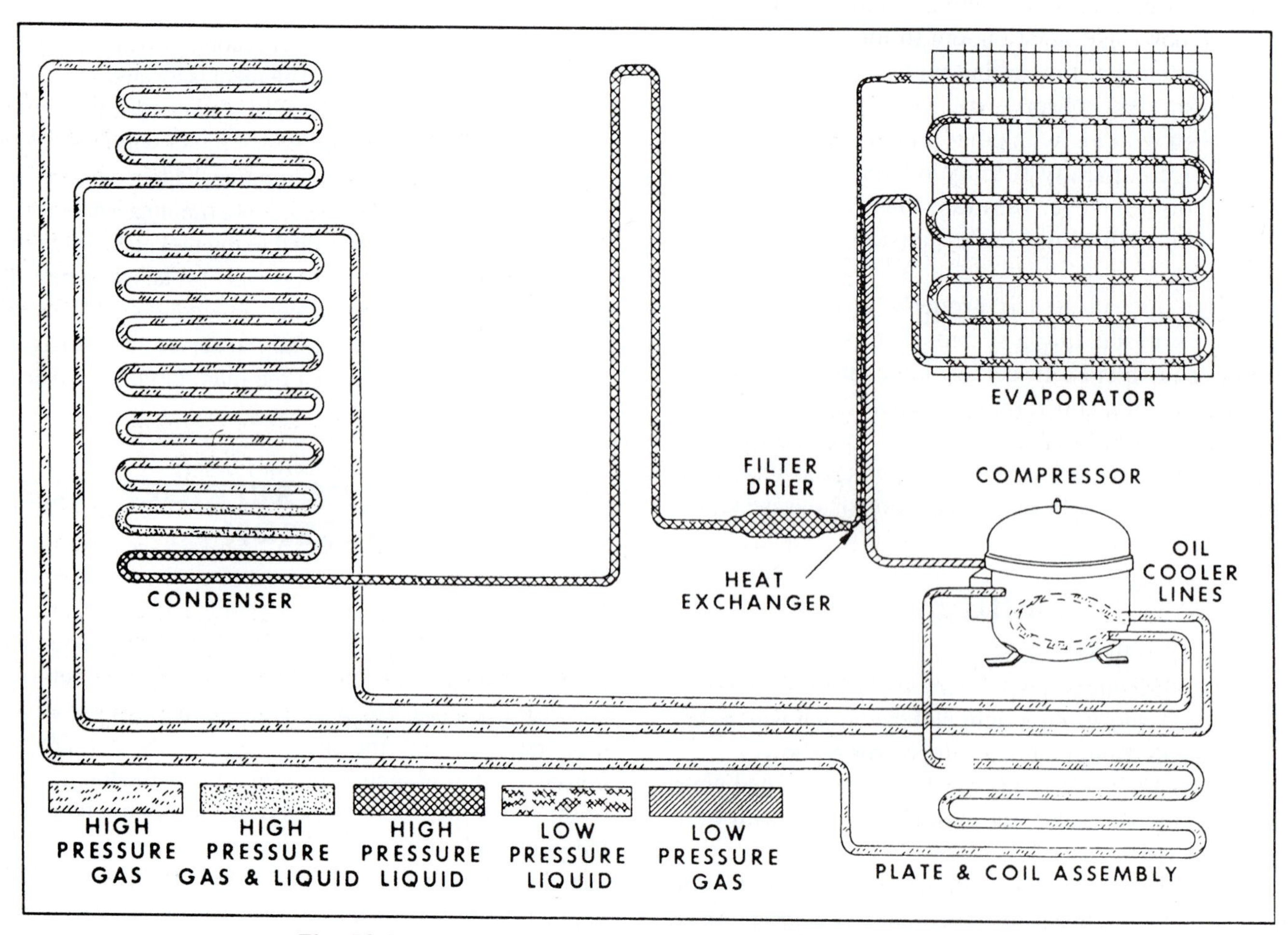

Fig. 12-3 *Refrigerating system with various pressures located.* (Kelvinator)

pressure drop due to the length of the tube and its small diameter, meters the refrigerant flow into the evaporator.

The capillary tube allows the high-side pressure to unload, or balance out, with the low-side pressure during the *off-cycle*. This permits the compressor to start under a no-load condition.

The design of the refrigerating system for capillary feed must be carefully engineered. The capillary feed must be matched to the compressor for the conditions under which the system is most likely to operate. Both the high side (condenser) and low side (evaporator) must be specifically designed for use with a capillary tube.

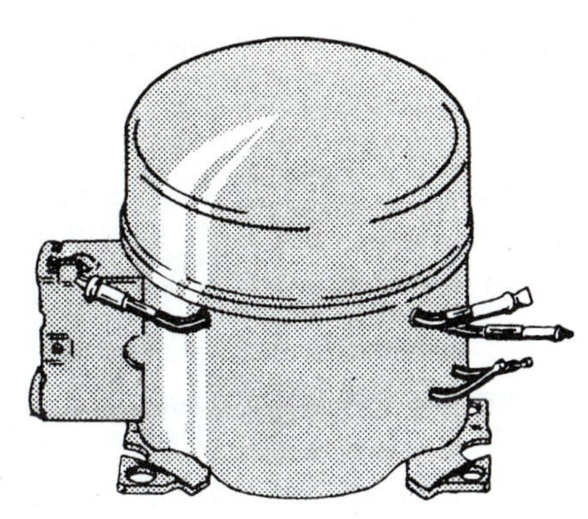

Fig. 12-4 *N-line replacement compressor for top-freezer models.* (Kelvinator)

Fig. 12-5 *A-line replacement compressor.* (Kelvinator)

Heat Exchanger

The heat exchanger is formed by soldering a portion of the capillary tube to the suction line. The purpose of the heat exchanger is to increase the over-all capacity and efficiency of the system. It does this by using the cold suction gas leaving the evaporator to cool the warm liquid refrigerant passing through the capillary tube to the evaporator. If the hot liquid refrigerant from the condenser were permitted to flow uncooled into the evaporator, part of the refrigerating effect of the refrigerant in the evaporator would have to be used to cool the incoming hot liquid down to evaporator temperature.

Freezer-Compartment and Provision-Compartment Assembly

Liquid refrigerant flows through the capillary and enters the freezer evaporator. Expansion and evaporation starts at this point. See Fig. 12-3.

COMPRESSOR REPLACEMENT

Replacement compressor packages are listed by the manufacturer. Check with the refrigerator manufacturer to be sure you have the proper replacement. Refer to the compressor number in the refrigerator under repair. Compare that number to the suggested replacement number. Replacement compressors are charged with oil and a holding charge of nitrogen. A replacement filter drier is packaged with each replacement compressor. It must be installed with the compressor. Figure 12-4 shows the N-line replacement compressor designed for top-freezer Kelvinator models. The A-line replacement compressor is shown in Fig. 12-5. It is used on Kelvinator chest or upright freezers.

The new relay-overload protector assembly supplied with the N-line replacement compressor should always be used. The new motor-overload protector supplied with the A-line replacement compressor should always be used. Transfer the relay, the relay cover, and the cover clamp from the original compressor to the replacement compressor. If a small quantity of refrigerant is used, a major portion of it will be absorbed by the oil in the compressor. It occurs when the refrigerator has been inoperative for a considerable length of time. When opening the system, use care to prevent oil from blowing out with the refrigerant.

TROUBLESHOOTING COMPRESSORS

There are several common compressor problems. Table 12-1 lists these problems and their solutions.

TROUBLESHOOTING REFRIGERATOR COMPONENTS

Compressor Will Not Run

Cause

1. Inoperative thermostat. Replace.
2. Service cord is pulled from the wall receptacle. Replace.
3. Service is pulled from the harness. Disconnect.

Table 12-1 Compressor-Troubleshooting and Service

Complaint	Possible Cause	Repair
Compressor will not start. There is no hum.	1. Line disconnect switch open. 2. Fuse removed or blown. 3. Overload-protector tripped. 4. Control stuck in open position. 5. Control off due to cold location. 6. Wiring improper or loose.	1. Close start or disconnect switch. 2. Replace fuse. 3. Refer to electrical section. 4. Repair or replace control. 5. Relocate control. 6. Check wiring against diagram.
Compressor will not start. It hums, but trips on overload protector.	1. Improperly wired. 2. Low voltage to unit. 3. Starting capacitor defective. 4. Relay failing to close. 5. Compressor motor has a winding open or shorted. 6. Internal mechanical trouble in compressor. 7. Liquid refrigerant in compressor.	1. Check wiring against diagram. 2. Determine reason and correct. 3. Determine reason and replace. 4. Determine reason and correct, replace if necessary. 5. Replace compressor. 6. Replace compressor. 7. Add crankcase heater and/or accumulator.
Compressor starts, but does not switch off of start winding.	1. Improperly wired. 2. Low voltage to unit. 3. Relay failing to open. 4. Run capacitor defective. 5. Excessively high-discharge pressure. 6. Compressor motor has a winding open or shorted. 7. Internal mechanical trouble in compressor (tight).	1. Check wiring against diagram. 2. Determine reason and correct. 3. Determine reason and replace if necessary. 4. Determine reason and replace. 5. Check discharge shut-off valve, possible overcharge, or insufficient cooling of condenser. 6. Replace compressor. 7. Replace compressor.
Compressor starts and runs, but short cycles on overload protector.	1. Additional current passing through the overload protector. 2. Low voltage to unit (or unbalanced if three-phase). 3. Overload-protector defective. 4. Run capacitor defective. 5. Excessive discharge pressure. 6. Suction pressure too high. 7. Compressor too hot—return gas hot. 8. Compressor motor has a winding shorted.	1. Check wiring against diagram. Check added fan motors, pumps, and the like, connected to wrong side of protector. 2. Determine reason and correct. 3. Check current, replace protector. 4. Determine reason and replace. 5. Check ventilation, restrictions in cooling medium, restrictions in refrigeration system. 6. Check for possibility of misapplication. Use stronger unit. 7. Check refrigerant charge. (Repair leak.) Add refrigerant, if necessary. 8. Replace compressor.
Unit runs, but short cycles on.	1. Overload protector. 2. Thermostat. 3. High pressure cut-out due to insufficient air or water supply, overcharge, or air in system. 4. Low pressure cut-out due to: a. Liquid-line solenoid leaking. b. Compressor valve leak. c. Undercharge. d. Restriction in expansion device.	1. Check current. Replace protector. 2. Differential set too close. Widen. 3. Check air or water supply to condenser. Reduce refrigerant charge, or purge. 4. a. Replace. b. Replace. c. Repair leak and add refrigerant. d. Replace expansion device.
Unit operates long or continuously.	1. Shortage of refrigerant. 2. Control contacts stuck or frozen closed. 3. Refrigerated or air conditioned space has excessive load or poor insulation. 4. System inadequate to handle load. 5. Evaporator coil iced. 6. Restriction in refrigeration system. 7. Dirty condenser. 8. Filter dirty.	1. Repair leak. Add charge. 2. Clean contacts or replace control. 3. Determine fault and correct. 4. Replace with larger system. 5. Defrost. 6. Determine location and remove. 7. Clean condenser. 8. Clean or replace.
Start capacitor open, shorted, or blown.	1. Relay contacts not operating properly. 2. Prolonged operation on start cycle due to: a. Low voltage to unit. b. Improper relay. c. Starting load too high. 3. Excessive short cycling. 4. Improper capacitor.	1. Clean contacts or replace relay if necessary. 2. a. Determine reason and correct. b. Replace. c. Correct by using pump-down arrangement if necessary. 3. Determine reason for short cycling as mentioned in previous complaint. 4. Determine correct size and replace.
Run capacitor open, shorted, or blown.	1. Improper capacitor. 2. Excessively high line voltage (110% of rated maximum).	1. Determine correct size and replace. 2. Determine reason and correct.

Table 12-1 *(Continued)*

Complaint	Possible Cause	Repair
Relay defective or burned out.	1. Incorrect relay. 2. Incorrect mounting angle. 3. Line voltage too high or too low. 4. Excessive short cycling. 5. Relay being influenced by loose vibrating mounting. 6. Incorrect run capacitor.	1. Check and replace. 2. Remount relay in correct position. 3. Determine reason and correct. 4. Determine reason and correct. 5. Remount rigidly. 6. Replace with proper capacitor.
Space temperature too high.	1. Control setting too high. 2. Expansion valve too small. 3. Cooling coils too small. 4. Inadequate air circulation.	1. Reset control. 2. Use larger valve. 3. Add surface or replace. 4. Improve air movement.
Suction-line frosted or sweating.	1. Expansion-valve oversized or passing excess refrigerant. 2. Expansion valve stuck open. 3. Evaporator fan not running. 4. Overcharge of refrigerant.	1. Readjust valve or replace with smaller valve. 2. Clean valve of foreign particles. Replace if necessary. 3. Determine reason and correct. 4. Correct charge.
Liquid-line frosted or sweating.	1. Restriction in dehydrator or strainer. 2. Liquid shutoff (king valve) partially closed.	1. Replace part. 2. Open valve fully.
Unit noisy.	1. Loose parts or mountings. 2. Tubing rattle. 3. Bent fan blade causing vibration. 4. Fan motor bearings worn.	1. Tighten. 2. Reform to be free of contact. 3. Replace blade. 4. Replace motor.

4. No voltage at the wall receptacle. House fuse blown.
5. Faulty cabinet wiring. Repair or replace.
6. Relay leads. Disconnect.
7. Relay loose or inoperative. Tighten or replace.
8. Compressor windings open. Replace compressor.
9. Compressor stuck. Replace.
10. Low voltage, causing compressor to cycle on overload. Voltage fluctuation should not exceed 10 percent, plus or minus, from the nominal rating of 115 V.

Compressor Runs, but There Is No Refrigeration

Cause

1. System is out of refrigerant. Check for leaks.
2. Compressor is not pumping. Replace.
3. Restricted filter drier. Replace.
4. Restricted capillary tube. Replace.
5. Moisture in the system. Pump down and recharge.

Compressor Short Cycles

Cause

1. Erratic thermostat operation. Replace.
2. Faulty relay. Replace.
3. Restricted airflow over the condenser. Remove restrictions.
4. Low voltage. Fluctuation exceeds 10 percent.
5. Inoperative condenser fan. Repair or replace.
6. Compressor draws excessive wattage. Replace.

Compressor Runs Too Much or 100 Percent

Cause

1. Erratic thermostat or thermostat is set too cold. Replace or reset to normal position.
2. Refrigerator exposed to unusual heat. Relocate.
3. Abnormally high room temperature. If outside temperature is cooler, open windows to lower temperature. Turn on fans to move the air.
4. Low pumping capacity compressor. Replace.
5. Door gaskets not sealing. Check with 100-W lamp.
6. System is undercharged or overcharged. Correct the charge.
7. Interior light stays on. Check door switch.
8. Noncondensable are in the system. Evacuate and recharge.
9. Capillary-tube kinked or partially restricted.
10. Filter drier or strainer partially restricted. Replace.
11. Excessive service load. Remove part of the load.
12. Restricted airflow over the condenser. Remove restriction.

Noise

Cause

1. Tubing vibrates. Adjust tubing.
2. Internal compressor noise. Replace.
3. Compressor vibrating on the cabinet frame. Adjust.
4. Loose water-evaporating pan. Tighten.
5. Rear machine compartment cover missing. Replace.
6. Compressor is operating at a high head pressure due to restricted airflow over the condenser. Reduce or remove restrictions.
7. Inoperative condenser fan motor. Check fuses, circuit breaker, and condition of fan motor. Replace, if necessary.

To Replace the Compressor

Following is the method for replacing the compressor on the Kelvinator refrigerator.

The Kelvinator refrigerator has been chosen since there were many of them made over a number of years. They are also readily available to training courses from any number of sources including those appliance shops that donate trade ins to schools for student work.

The following material applies to the servicing of the Kelvinator refrigerator.

1. Bleed the refrigerant slowly by cutting the process tube on the compressor with diagonal cutters. If the refrigerator has not been in operation for some time, oil may be discharged with the refrigerant. Use care when bleeding the refrigerant. Place a cloth over the process tube to prevent oil and refrigerant from splattering the room. Preferably, run the compressor, if operative, until the dome becomes warm. This will separate the refrigerant from the oil.

   CAUTION: Ventilate the room while purging, especially when open-flame cooking or baking is being done in the kitchen.

2. Use diagonal pliers. Cut the discharge, suction, and oil-cooler tubes. Crimp the tubes that remain on the compressor dome to prevent oil leakage during shipment.
3. Remove wire leads from the relay. Remove mounting-cap screws. Then, remove compressor from machine compartment.

   CAUTION: If original compressor showed signs of burnout, follow instructions for "cleaning system after burnout."

4. Remove the filter drier by cutting the "I" inlet tube 1 in. from the brazed connection. Use a file to score the capillary tube uniformly about 1 in. from the brazed joint at the filter drier. Break off the capillary tube.
5. Transfer the rubber mounts from the inoperative compressor to the replacement compressor. Set the replacement compressor in place and install.
6. Remove the line caps and bleed off the holding charge of nitrogen. Use a suitable tool. Cut the suction, discharge, and oil-cooler extension tubes to the required lengths. Swage the tubes as required. Join to tubes on cabinet for brazing.
7. Install the replacement filter drier (packaged with replacement compressor).
8. Braze the refrigerant tubes to the filter drier and compressor. Use silver solder (Easy Flo-45).

   CAUTION: Do not remove the ends of the filter drier until all tubes have been processed for installation of the filter drier.

9. Install the hand valve and the charging hose to the compressor copper process tube.

   NOTE: On "N"-line compressors, silver solder a 4 in. piece of $^1/_4$ in. *outside diameter* (OD) copper tubing into the steel process tube. Pressurize the system to 75 lb/in.^{2} with R-12 refrigerant or its substitute. Leak test all low-side joints. Operate the compressor for a few minutes. Then leak test all high-side joints. Discharge, and evacuate system with a vacuum pump.

10. Close the hand valve. Remove the vacuum pump. Connect the charging cylinder to the hand valve. Purge the charging hose between the charging cylinder and the hand valve. Open the valve on the charging cylinder and allow liquid refrigerant to fill the charging hose up to the hand valve.

    NOTE: Do not open the hand valve until the charging hose is full of liquid refrigerant and the amount of refrigerant in the charging cylinder has been recorded. Failure to follow this procedure results in an undercharged system.

11. Open the hand valve and charge the system. Then, close the hand valve. Refer to manufacturer's recommendations for the proper refrigerant charge.

12. Pinch off the copper process tube after the charge has been established. With the pinch-off tool on the process tube, remove the charging hand valve and the charging hose. Flatten the end of the tube and seal it with phos-copper. Then, remove pinch-off tool.

COMPRESSOR MOTOR BURNOUT

There are four major causes of motor burnout: low line voltage, loss of refrigerant, high head pressure, and moisture.

Low line voltage. When the motor winding in a motor gets too hot, the insulation melts and the winding short circuits. A blackened burned-out run winding is the result. Low line voltage causes the winding to get very hot because it is forced to carry more current at the same compressor load. When this current gets too high, or is carried for too many hours, the motor run winding fails. A burnout caused by low voltage is generally a slow burnout. This contaminates the system.

Loss of refrigerant. A second cause of motor burnout is loss of refrigerant. In a hermetically sealed motor compressor, the refrigerant vapor passes down around the motor windings. The cool refrigerant vapor keeps the motor operating at proper temperature. If there is a refrigerant leak, and there is no refrigerant to cool the motor, the windings become too hot. A burnout results. The overload protector may not always protect against this type of burnout since it requires the transfer of high heat from the motor through the refrigerant vapor to the compressor dome.

High head pressure. High head pressure is a third cause of motor burnout. With high head pressure, the motor load is increased. The increased current causes the winding to overheat and eventually fail. Poor circulation of air over the high-side condenser can cause motor failure for this reason.

Moisture. The fourth major cause of motor burnout is moisture. It takes very little moisture to cause problems. In the compressor dome, refrigerant is mixed with lubricating oil, and heat from the motor windings and compressor operation. If there is any air present, the oxygen can combine chemically with hydrogen in the refrigerant and oil to form water. Just one drop of water can cause problems. When water contacts the refrigerant and oil in the presence of heat, hydrochloric or hydrofluoric acid is formed. These acids destroy the insulation on the motor winding. When the winding short circuits, a momentary temperature of over 3000°F (1648°C) is created. Acids combine chemically with the insulation and oil in the compressor dome to create sludge. This quickly contaminates the refrigerating system. Sludge collects in various places throughout the system. It is very hard to dislodge. Purging the refrigerant charge or blowing refrigerant vapor through the system will not clean the system.

CLEANING SYSTEM AFTER BURNOUT

Remove the inoperative compressor and filter drier. Flush the high side and low side of the system with R-12 (or R-134a) liquid refrigerant. (Invert the refrigerant drum.)

Connect the high side and low side of the system. Also, connect the oil-cooler tubes. Then, evacuate the system using a vacuum pump. Never use the new replacement compressor for this purpose. It will quickly become contaminated. Break the vacuum with refrigerant. Repeat the process. Then, and this is extremely important, repeat the process a third time. Thus, there are three purges and three evacuations.

Remove the vacuum pump and install a new replacement compressor and filter drier. Follow the procedures previously outlined. Use silver solder (Easy Flo-45) or phos-copper to make brazed connections.

REPLACING THE FILTER DRIER

If the compressor is not to be changed, follow these procedures to replace the filter drier.

1. To replace the filter drier, move the refrigerator to a location where the rear of the machine compartment is accessible.
2. Remove the machine compartment sound deadener baffle.
3. Cut the copper process tube on the compressor and bleed the refrigerant. Retain as much length as possible. Remove the filter drier by cutting $^1/_4$ in. inlet tube 1 in. from the brazed connection. Use a file to score the capillary tube uniformly approximately 1 in. from the brazed joint at the filter drier. Then, break off the capillary tube.
4. Install the replacement filter drier with its inlet at the top. (Arrow indicates direction of flow.) See Fig. 12-6. Braze the refrigerant tubes to the filter drier. Use silver solder (Easy Flo-45).

CAUTION: Do not remove caps from the replacement filter drier until all the refrigerant tubes have been processed for installation of the filter drier.

Fig. 12-6 Replacement filter drier. (Kelvinator)

5. Install a process-tube adapter kit. See Fig. 12-7. If the adapter is not available, slip a 1/4 in. flare nut over the copper tube and flare the tube.

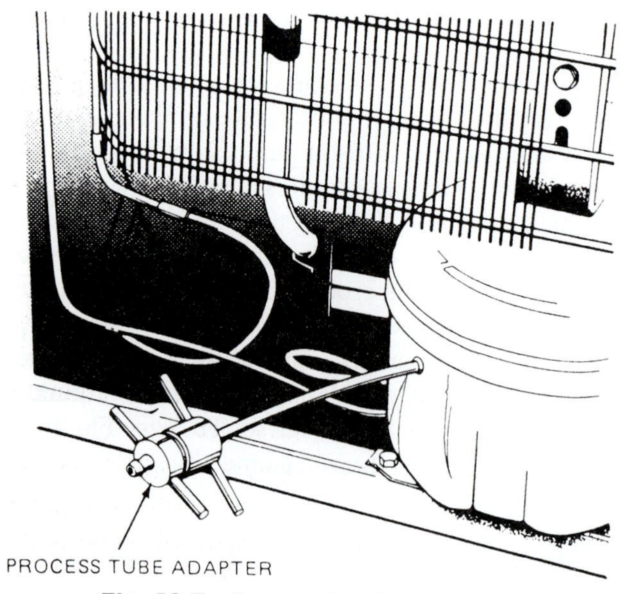

Fig. 12-7 Process-tube adapter. (Kelvinator)

6. Install a hand valve and a charging hose. Connect the hose to the vacuum pump and evacuate the system. See Fig. 12-8.
7. Shut off the vacuum pump and close the hand valve at the process-tube adapter. Remove the vacuum pump and charging hose at the hand valve.

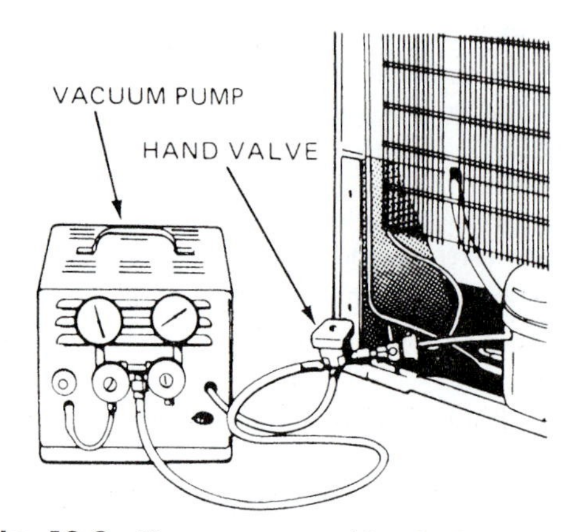

Fig. 12-8 Vacuum pump and hand valve. (Kelvinator)

8. Connect a drum of R-12 refrigerant, or a charging cylinder, of the hand valve at the process-tube adapter. See Fig. 12-9. Purge the charging hose between the drum or charging cylinder and the hand valve. Open the valve on the drum, or charging cylinder. Allow the liquid refrigerant to fill the charging hose up to the hand valve at the process tube.

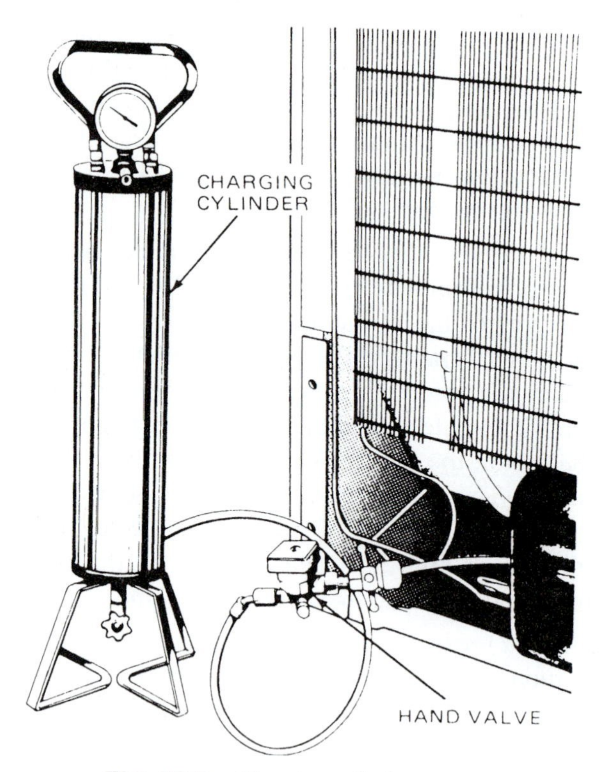

Fig. 12-9 Charging cylinder. (Kelvinator)

NOTE: Do not open the hand valve until the charging hose is full of liquid and the amount of refrigerant in the charging cylinder has been recorded. Failure to follow this procedure results in an undercharge of refrigerant.

9. Open the hand valve and charge the system. Then, close the hand valve. If the system is charged from a refrigerant drum, operate the system until it has cycled to determine if the charge is proper. Refer to manufacturer's tables for operating pressures.

10. After the charge is established, pinch off the copper process tube with a pinch-off tool. With a pinch-off tool on the process tube, remove the charging adapter and hand valve. Seal the end of the tube with phos-copper. Then remove the pinch-off tool. See Fig. 12-10.

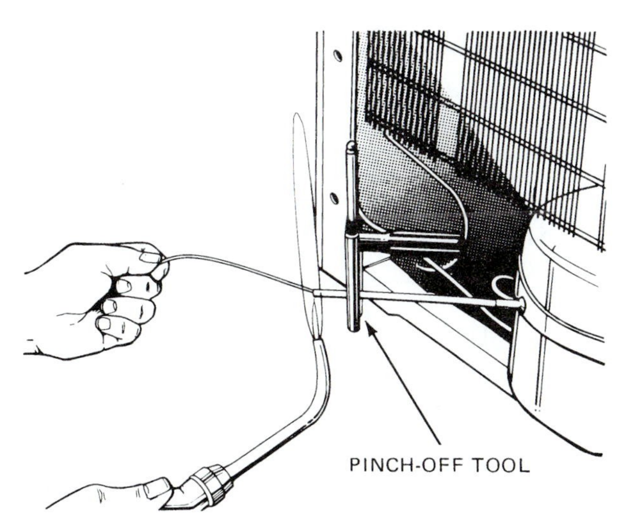

Fig. 12-10 *Brazing process tube with pinch-off tool in place.* (Kelvinator)

REPLACING THE CONDENSER

To replace the condenser, bleed off the refrigerant as outlined under "Replacing the Filter Drier." Cut the condenser and the filter-drier inlet and outlet lines. Remove the condenser and the filter drier. Install a new condenser. Braze the refrigerant lines. Use silver solder (Easy Flo-45). Then, follow procedure for "Replacing the Filter Drier."

REPLACING THE HEAT EXCHANGER

To replace the heat exchanger, follow the procedure given in "Replacing the Evaporator-Heat Exchanger Assembly."

REPAIRING THE PERIMETER TUBE (FIBERGLASS INSULATED)

Top-Freezer and Side-by-Side Models

A perimeter tube, which is part of the refrigerating system, extends across the top and down, both sides of the cabinet. Should a refrigerant leak develop in this tube, repairs are made as follows:

1. Use a tubing cutter. Cut and deburr the $^3/_{16}$ in. OD perimeter tube at A and B. See Fig. 12-11.
2. Use an ice pick to pierce a $^1/_{32}$ in. diameter hole through the wall of the plastic sleeve 1 to $^1/_2$ in. from the end. Thread one end of the nylon line through the pierced hole and piston. Tie a triple knot in the line. Loop the opposite end of the line around the plastic sleeve and tie. See Fig. 12-12.
3. Insert the piston into the perimeter tube A. Slide the plastic sleeve onto the perimeter tube A. Insert the $^3/_{16}$ in. OD copper tube. Flare the tube, and connect it to the refrigerant drum. See Fig. 12-13.
4. Open the valve on the refrigerant drum. Blow the piston and the nylon-line assembly through the perimeter tube.
5. Cut the nylon line that is looped around the plastic sleeve. Disconnect the refrigerant drum. Slide the

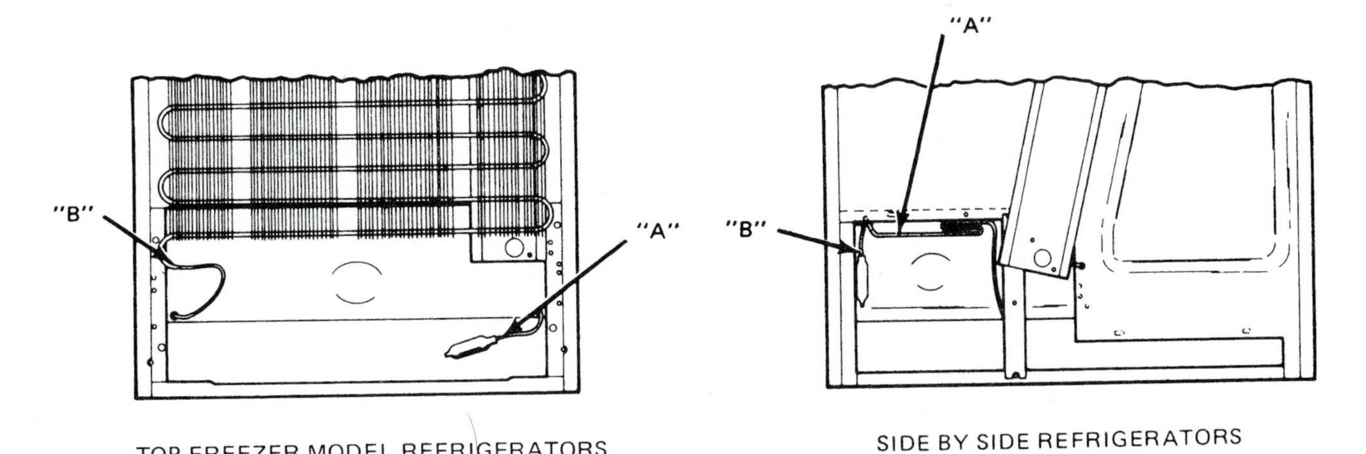

Fig. 12-11 *Repairing the perimeter tube. Cut and deburr the $^3/_{16}$ in. OD perimeter tube at "A" and "B."* (Kelvinator)

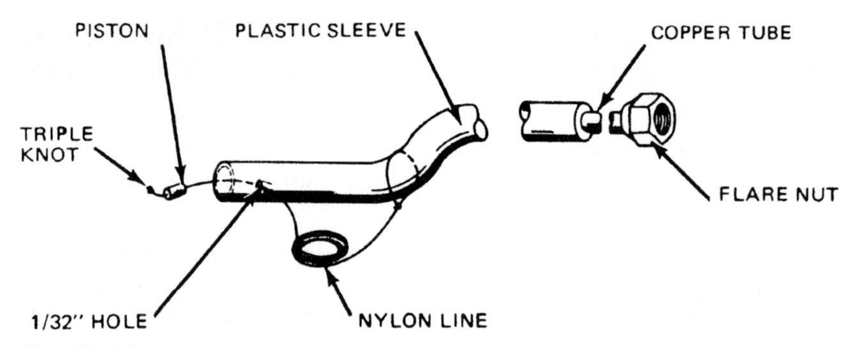

Fig. 12-12 *Repair tool made for repair of the perimeter tube.* (Kelvinator)

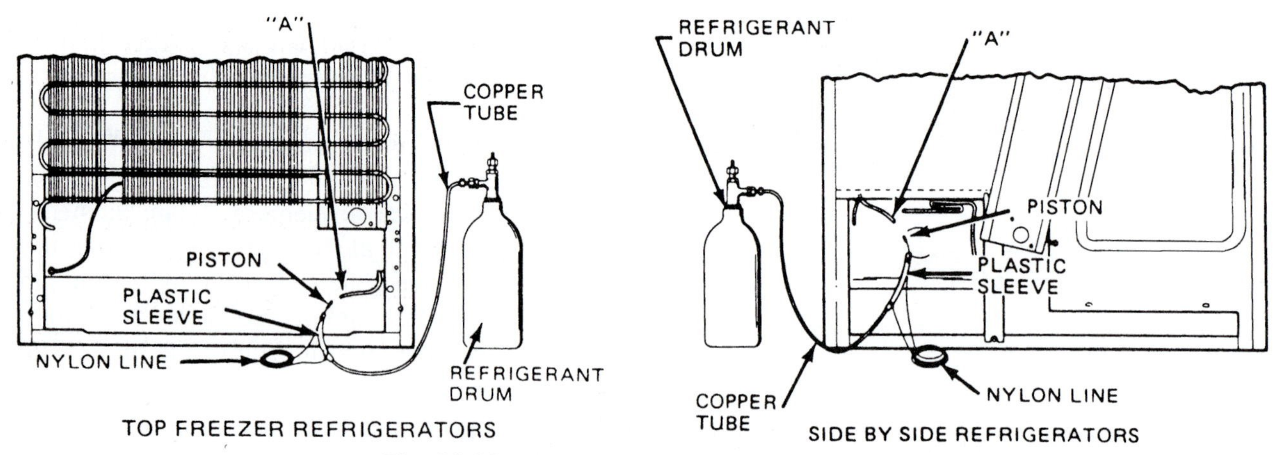

Fig. 12-13 *Repairing the perimeter tube.* (Kelvinator)

plastic sleeve off the perimeter tube. Tie the nylon line to the electric-drier coil (heater). Grasp the piston end of the nylon line and pull the heater into the perimeter tube until the heater-lead connector on the opposite end rests against the end of the perimeter tube A. Remove the nylon line from the heater.

6. Swing the condenser aside (top-freezer models). Remove the cabinet harness channel. Cut the power cord inside the channel about 3 in. from the harness restraining grommet. On freezers and models where the power cord is not in the harness channel, cut the power cord 3 in. inside the restraining strap.

   Strip $^5/_8$ in. of insulation off the power cord, as illustrated in Fig. 12-14. Splice one end each of the white wires and the ends of the ribbed power cord together with wire connectors. See Figs. 12-15, 12-16, 12-17, and 12-18. Connect the green (ground) wires together with a wire connector. Wrap the wire connectors with a piece of electrical tape. Remove the nylon line from the heater.

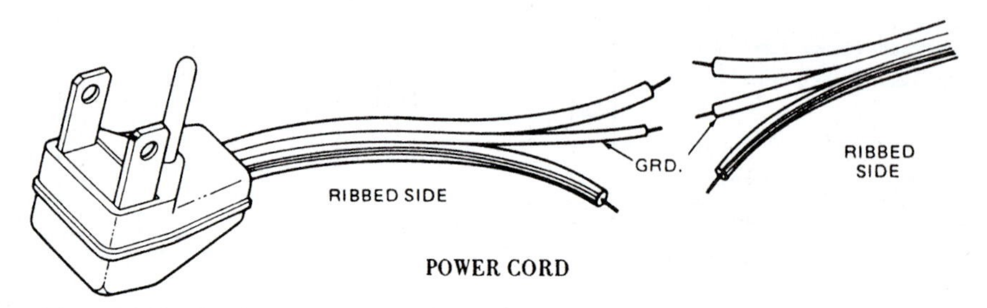

Fig. 12-14 *Splicing a new power-cord plug onto the existing power-cord line.* (Kelvinator)

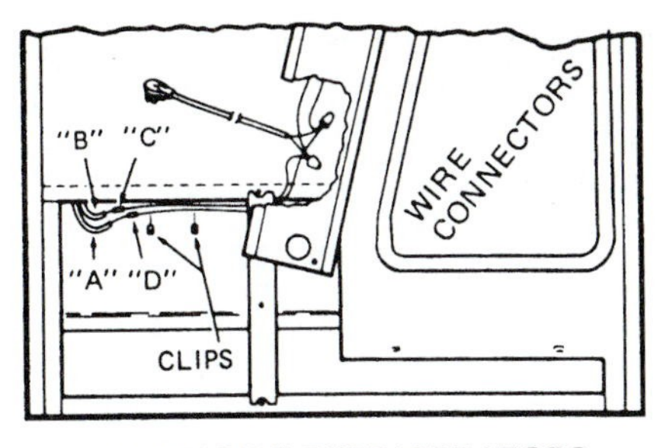

Fig. 12-15 *Power-cord location and splices, side-by-side models.* (Kelvinator)

CAUTION: Recheck your work to be sure the ribbed ends of power cord are spliced together to maintain polarity. Shape the perimeter tubes A and B as required for routing the heater leads along the flange of the cabinet to the harness channel.

7. Cut off the excess heater approximately 2 in. from the end of perimeter tube B.

NOTE: Exercise extreme care when stripping insulation to prevent damaging the heater-resistance wire.

Cut the black and white wires to their proper length. Splice the white wire to the heater lead C and the black wire to heater lead D with wire connectors. Secure the leads to the cabinet flange with clips. Reinstall the harness channel.

8. Score the capillary tube and remove the filter drier. Use a copper tube and the filter drier to connect the condenser to the capillary tube. See Fig. 12-19.

Foam-Insulated 12 and 14 ft³, Top-Freezer Models

1. Disconnect the service cord from the power supply.
2. Use a tubing cutter to cut and deburr the $^{3}/_{16}$ in. OD hot tube at A and B. See Fig. 12-11.
3. Mount the auxiliary condenser on the bottom of the existing condenser, using bolts, nuts, and spacers.

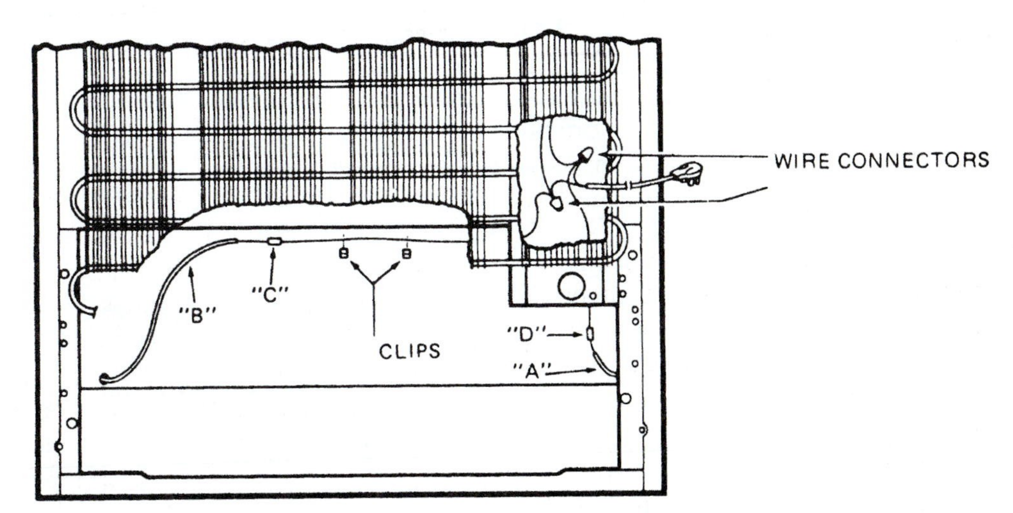

Fig. 12-16 *Power-cord splicing on top-freezer models.* (Kelvinator)

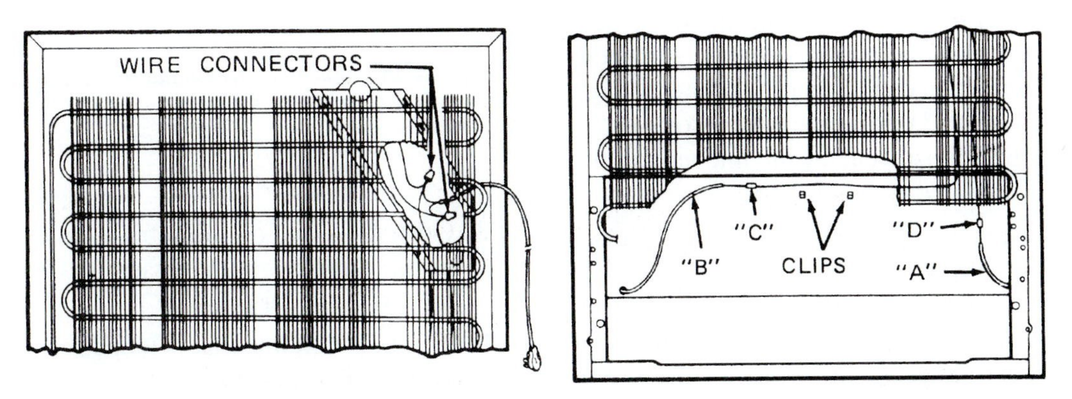

Fig. 12-17 *Another top-freezer refrigerator power-cord splice.* (Kelvinator)

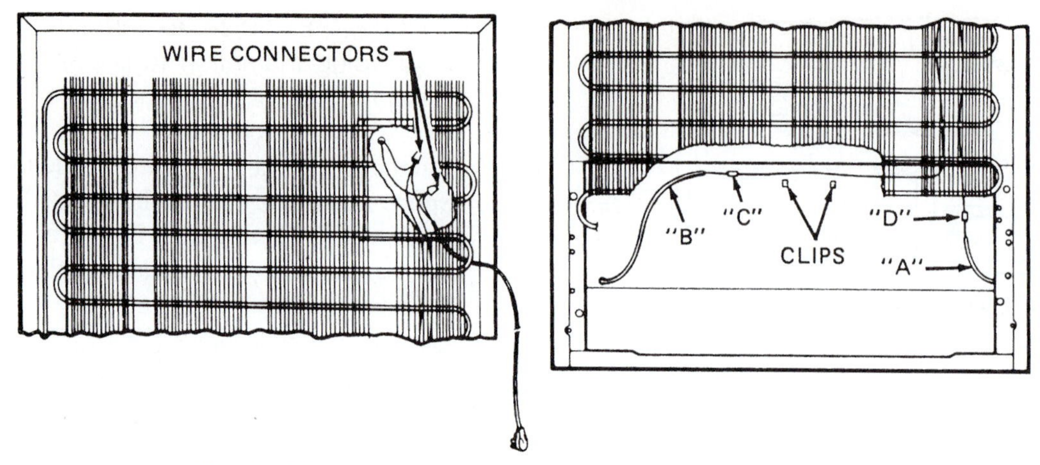

Fig. 12-18 *Top-freezer refrigerator-cord splice location.* (Kelvinator)

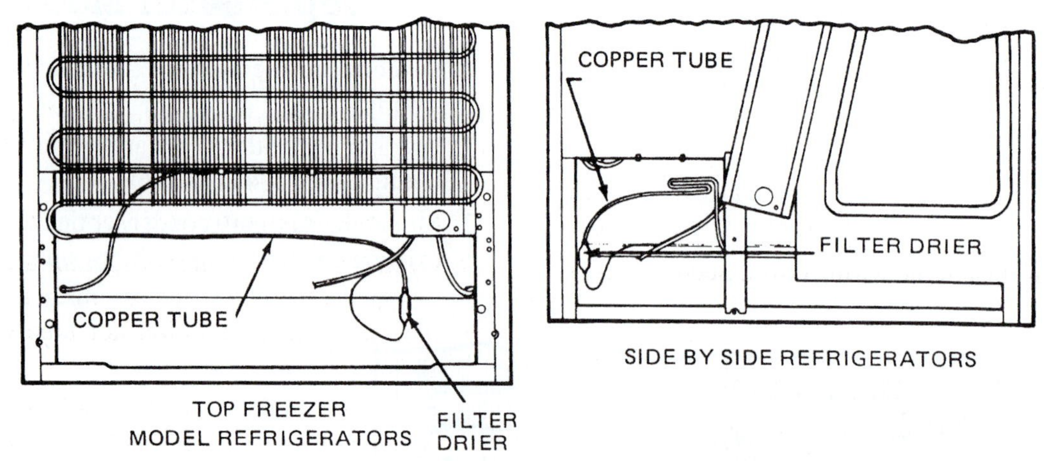

Fig. 12-19 *Side-by side and top-freezer model refrigerator filter-drier location.* (Kelvinator)

4. Use silver solder to connect the refrigerant lines that were removed from the hot tube, to the auxiliary-condenser lines.

   NOTE: Refrigerant must enter the top of the auxiliary condenser.

5. Install a replacement drier. Evacuate and recharge the system using the nameplate charge.
6. Use an ice pick to pierce a $^1/_{32}$ in. diameter hole through the wall of the plastic sleeve $1^1/_2$ in. from the end. Thread one end of the nylon line through the pierced hole and piston. Tie a triple knot in the line. Loop the opposite end of the line around the plastic sleeve and tie. See Fig. 12-12.
7. Insert the piston into the hot tube A. Slide the plastic sleeve onto the hot tube A. Insert the $^3/_8$ in. OD copper tube into the opposite end of the plastic sleeve. Install a flare nut on the copper tube and flare the tube. Connect to the refrigerant drum. See Fig. 12-20.

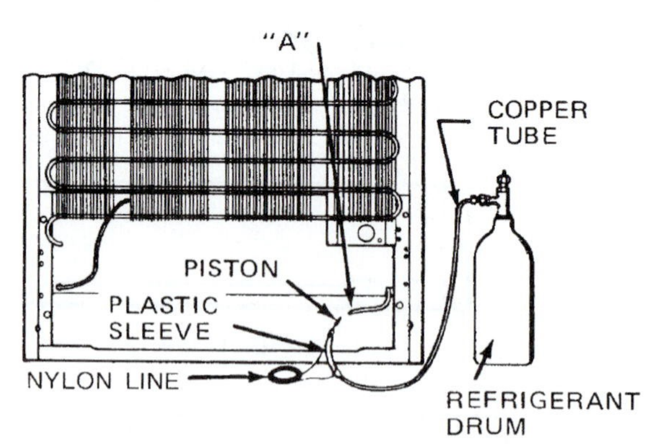

Fig. 12-20 *An older technique using vapor pressure from refrigerant drum to blow the piston and nylon assembly through the hot tube.* (Kelvinator)

8. Open the valve on the refrigerant drum. Blow the piston and nylon-line assembly through the hot tube. (Use vapor pressure.)
9. Cut the nylon line that is looped around the plastic sleeve. Disconnect the refrigerant drum. Slide the plastic sleeve off the hot tube. Tie the nylon line to

the electric heater. Grasp the piston end of the nylon line and pull the heater into the hot tube until the heater-lead connector on the opposite end rests against the end of hot tube A. Remove the nylon line from the heater.

10. Remove the wiring cover from the back of the cabinet. Disconnect the power cord from the cabinet harness. Connect the adapter harness to the cabinet harness and power cord. See Fig. 12-19.
11. Shape the hot tubes A and B as required for routing the heater leads along the flange of the cabinet to the suction line. Cut the excess heater approximately 2 in. from the end of the hot tube B and strip the insulation back 1 in.

NOTE: Exercise extreme care when stripping insulation to prevent damaging the heater-resistance wire.

12. Cut the adapter harness and heater lead to the proper length. Connect the adapter harness to the heater with connectors. Secure the leads to the suction line. Reinstall the wiring cover. See Fig. 12-19.

Foam-Insulated 19 ft³ Side-by-Side Models

1. Disconnect the service cord from the power supply.
2. Use a tubing cutter to cut and deburr the $^3/_{16}$ in. OD hot tubes A and B. See Fig. 12-11.
3. Mount the auxiliary condenser on the bottom of the existing condenser, using bolts, nuts, and spacers.
4. Use silver solder to connect the refrigerant lines removed from the hot tube to the auxiliary-condenser lines.

NOTE: Refrigerant must enter the top of the auxiliary condenser.

5. Install the replacement drier. Evacuate and recharge the system using the nameplate charge.
6. Remove the mullion to gain access to the mullion heater.
7. Cut 2 in. from the bottom of the mullion hot tube at the U-bend. Deburr the remaining tubes C and D. See Fig. 12-21.
8. Use an ice pick to pierce a $^1/_{32}$-in. diameter hole through the wall of the plastic sleeve $1^1/_2$ in. from the end. Thread one end of the nylon line through the pierced hole and piston. Tie a triple knot in the line. Loop the opposite end of the line around the plastic sleeve and tie. See Fig. 12-12.

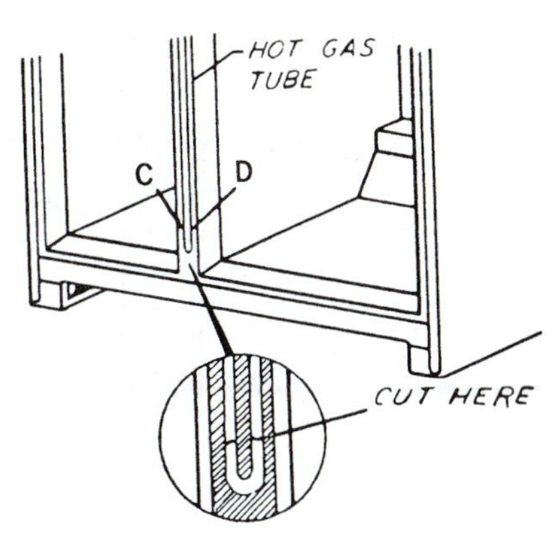

Fig. 12-21 *Where to cut the hot-gas tube for repair?* (Kelvinator)

9. Insert the piston into hot tube A. Slide the plastic sleeve onto hot tube A. Insert the $^3/_{16}$ in. OD copper tube into the opposite end of the plastic sleeve. Install a flare nut on the copper tube and flare tube. Connect to the refrigerant drum. See Fig. 12-20.
10. Open the valve on the refrigerant drum. Blow the piston and the nylon line assembly through the hot tube to open the mullion hot tube C. Use vapor pressure.
11. Cut the nylon line looped around the plastic sleeve. Disconnect the refrigerant drum. Slide the plastic sleeve off the hot tube. Tie the nylon line to the electric heater. Grasp the piston end of the nylon line. Pull the heater into the hot tube until the heater-lead connector on the opposite end rests against the end of hot tube A. Remove the nylon line from the piston.
12. Start at the remaining open mullion hot tube D. Repeat steps 8 through 11 to install the heater in the remaining part of the hot tube. Leave the free loop of wire at the end of the mullion heater.
13. Reinstall the mullion.
14. Remove the wiring cover from the back of the cabinet harness. Disconnect the power cord from the cabinet. Connect the adapter harness to the cabinet harness and the power cord. See Fig. 12-22.
15. Shape hot tubes A and B as required for routing the heater leads along the flange of the cabinet to the suction line. Cut the excess heater approximately 2 in. from the end of the hot tube B and strip the insulation back $^1/_2$ in.

NOTE: Exercise extreme care when stripping insulation to prevent damaging the heater-resistance wire.

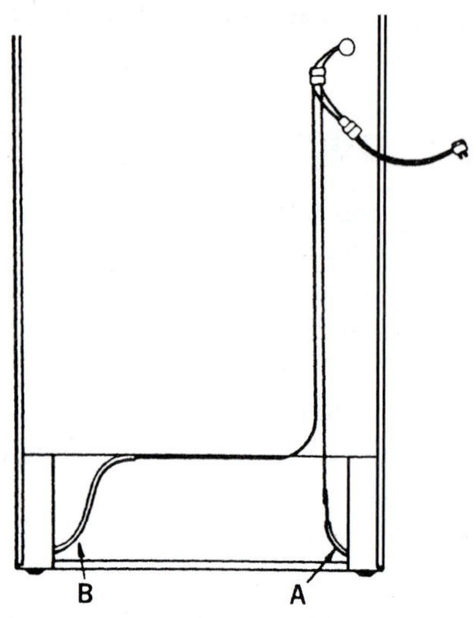

Fig. 12-22 *Location of the leads and power cord.* (Kelvinator)

16. Cut the adapter harness and the heater lead to their proper lengths. Connect the adapter harness to the heater with connectors. Secure the leads to the suction line. Reinstall the wiring cover. See Fig. 12-22.

REPLACING THE EVAPORATOR-HEAT EXCHANGER ASSEMBLY

Top-Freezer, No-Frost Models

When an evaporator-heat exchanger assembly develops a refrigerant leak and the compressor has operated after the refrigerant escaped, air and moisture have entered the system. To protect the system, it must be flushed with liquid refrigerant and evacuated. A replacement filter drier must be installed in conjunction with the evaporator-heat exchanger assembly.

To replace the evaporator-heat exchanger assembly, move the refrigerator to a location where the front and rear are accessible. Remove the machine compartment cover on models so equipped. Then, follow steps (1) and (4) in the discussion of "To Replace the Compressor." Use a tubing cutter to cut suction tube at the compressor. Swing the condenser aside. Disconnect the defrost heater and defrost termination thermostat leads from the cabinet harness. Bend the heat exchanger upward. Remove the sealer from the tubing-harness entrance hole. Remove the evaporator cover mounting screws and lay the cover on the bottom of the freezer compartment. Remove the evaporator mounting screws and RH and LH air barriers. Pull the evaporator-heat exchanger assembly forward out of the freezer compartment. Transfer the defrost heater and defrost termination thermostat to the replacement evaporator. Install the replacement evaporator-heat exchanger assembly. Install the evaporator RH and LH air barriers and evaporator cover. Press the sealer firmly into the tubing-harness entrance hole. Connect the defrost heater and termination thermostat leads to the harness. Swing the condenser into place and secure. Braze the suction tube to the compressor. Use silver solder (Easy Flo-45). Then, follow steps (4) through (10) discussed in "Replacing the Filter Drier." Install the machine compartment cover on models so equipped.

Side-by-Side Models

When an evaporator-heat exchanger assembly develops a refrigerant leak and the compressor has operated after the refrigerant escaped, air and moisture have entered the system. To protect the system, it must be flushed with liquid refrigerant and evacuated.

Proper procedure must be followed to make sure the flushing refrigerant does not escape into the atmosphere. Use the proper recovery techniques for reclaiming refrigerant. A replacement filter drier must be installed in conjunction with the evaporator-heat exchanger assembly.

To replace the evaporator-heat exchanger assembly, move the refrigerator to a location where the front and rear are accessible.

Remove the machine compartment cover. Then, follow steps (1) and (4) discussed in "To Replace the Compressor." Use a tubing cutter to cut the suction tube at the compressor. Disconnect the defrost termination thermostat leads from the cabinet harness. Bend the heat exchanger upward. Remove the sealer from the tubing-harness entrance hole. Remove the shelves, shelf supports, evaporator cover, and evaporator mounting screws. Pull the evaporator-heat exchanger assembly forward from the freezer compartment. Transfer the defrost heater and defrost termination thermostat to the replacement evaporator. Install the replacement evaporator-heat exchanger assembly. Connect the defrost heater and defrost termination thermostat leads to the cabinet harness. Install the sealer firmly into the tubing-harness entrance hole from the front and rear. Install the evaporator cover, shelf supports, and shelves. Braze the suction tube to the compressor. Use silver solder (Easy Flo-45). Then, follow steps (4) through (10) discussed in "Replacing the Filter Drier." Install the machine compartment cover.

ADDING REFRIGERANT

CAUTION: Always introduce refrigerant into the system in a vapor state. When the operation of a system indicates that it is short of refrigerant, it must be assumed that there is a leak in the system. Proceed to test the system with a leak detector. When the leak is located, it should be repaired. First, however, you must determine if the leak is repairable.

Unless the system has lost most of its refrigerant, the leak test can be made without the addition of extra refrigerant. If the system is completely out, sufficient refrigerant must be added to make a leak test. A new filter drier must be installed.

Low-Side Leak or Slight Undercharge

If a slight undercharge of refrigerant is indicated, without a leak being found, the charge can be corrected without changing the compressor.

In the case of a low-side refrigerant leak, resulting in complete loss of refrigerant, the compressor will run. However, there will be no refrigeration. Suction pressure will drop below atmospheric pressure. Air and moisture are drawn into the system, saturating the filter drier.

It is not necessary to replace the compressor. The leak should be repaired. The system should be flushed with liquid refrigerant. A replacement filter drier should be installed. The system should be excavated and recharged.

The system may have operated for a considerable length of time with no refrigerant and the leak may have occurred in the evaporator. In this case, excessive amounts of moisture may have entered the system. In such cases the compressor may need to be replaced to prevent repetitive service.

High-Side Leak or Slight Undercharge

It is not necessary to change a compressor when a leak is found in the system. If a slight undercharge of refrigerant is indicated, without a leak being found, the charge can be corrected without changing the compressor.

It is recommended that the system be flushed with liquid refrigerant and evacuated. A replacement filter drier should be installed to protect the system against moisture. Make sure the flushing refrigerate is captured and reprocessed.

Overcharge of Refrigerant

When the cabinet is pulled down to temperature, an indication of an overcharge is that the suction line will be colder than normal. The normal temperature of the suction line will be a few degrees cooler than room temperature. If its temperature is much lower than room temperature, the unit will run longer because the liquid is pulled into the heat exchanger. When the overcharge is excessive, the suction line will sweat or frost.

TESTING FOR REFRIGERANT LEAKS

If the system is diagnosed as short of refrigerant and has not been recently opened, there is probably a leak in the system. Adding refrigerant without first locating and repairing the leak will not permanently correct the difficulty. *The leak must be found.* Sufficient refrigerant may have escaped to make it impossible to leak test effectively. In such cases, add a $^{1}/_{4}$ in. line-piercing valve to the compressor process tube. Add sufficient refrigerant to increase the pressure to 75 lb/in.^{2}. Through this procedure, minute leaks are more easily detected before discharging the system and contaminating the surrounding air.

NOTE: The line-piercing valve (clamp-on type) should be used only for adding refrigerant and for test purposes. It must be removed from the system after it has served its purpose. Braze-on type line-piercing valves may be left on the process tube to evacuate the system and recharge after repairs are completed.

Various types of leak detectors are available. Liquid detectors (bubbles), halide torches, halogen-sensing electronic detectors, and electronic transistor pressure-sensing detectors are used.

You can sometimes spot a leak by the presence of oil around it. To be conclusive, however, use a leak detector.

Liquid detectors (bubbles) can be used to detect small leaks in the following manner. Brush liquid detector over the suspected area and watch for the formation of bubbles as the gas escapes. If the leak is slight, you may have to wait several minutes for a bubble to appear.

CAUTION: Use the bubble method only when you are sure that the system has positive pressure. Using it where a vacuum is present could pull liquid detector into the system.

When testing with the halide torch, be sure the room is free from refrigerant vapors. Watch the flame for the slightest change in color. A very faint green indicates a small leak. The flame will be unmistakably green to purple when large leaks are encountered. To simplify leak detection, keep the system pressurized to a minimum of 75 lb/in.^{2}.

For more sensitive testing, use an electronic leak detector. Halogen-sensing electronic detectors can detect minute refrigerant leaks, even though the surrounding air may contain small amounts of refrigerant.

An electronic, transistorized, pressure-sensing detector does not require that the system be pressurized with refrigerant. Dry air or nitrogen may be used to pressurize the system. The escaping pressure through a minute opening is detected.

In "urethane" froth foam-insulated models, R-11 refrigerant (halogenated hydrocarbons), is used as a blowing agent in the foaming process. Molecules of R-11 refrigerant are encased within the cellular formation of the insulation and between the inner and outer wall of the cabinet. Therefore, when checking for refrigerant leaks in these models with a halide torch or halogen electronic leak detector, a false indication of a refrigeration leak will be experienced where no leak actually exists. This process has now been modified and materials produced in recent years will not have this refrigerant-type agent used in the foam-making process.

However, if the equipment is older the false reading may be a problem, especially where the cellular formation is disturbed or broken by moving the refrigerant lines or probing into the older insulation. The electronic, transistorized, pressure-sensing detector is not affected by the presence of refrigerants in the air. Where refrigerant tubes are encased within the foam insulation, there is one continuous tube. All brazed joints are accessible for checking leaks.

A joint suspected of leaking can be enclosed in an envelope of cellophane film. Tightly tape both ends and any openings to make it gas tight. See Fig. 12-23. After about an hour, you can pierce one end of the film for your probe and pierce the other end for air to enter. If you get a response, the joint should be rebrazed. The component should be replaced if a leak is found at the aluminum-to-copper butt-weld joint.

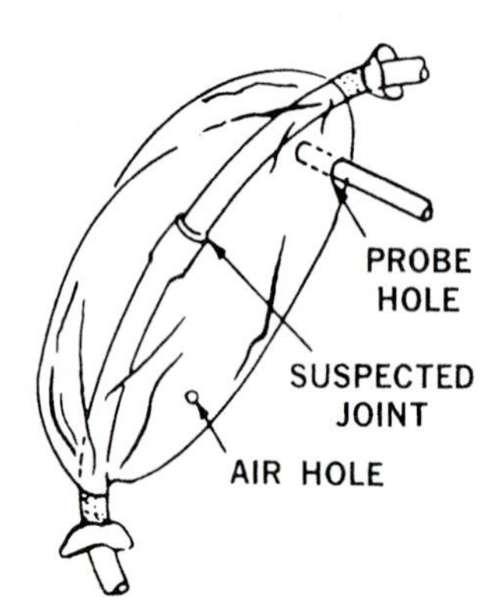

Fig. 12-23 *Leak-detection envelope.* *(Kelvinator)*

SERVICE DIAGNOSIS

To service refrigeration equipment properly, the serviceperson must possess the following:

- A thorough understanding of the theory of refrigeration
- A working knowledge of the purpose, design, and operation of the various mechanical parts of the refrigerator
- The ability to diagnose and correct any trouble that may develop

On The Initial Contact

Always allow the customer to explain the problem. Many times the trouble can be diagnosed more quickly through the customer's explanation. Most of all, do not jump to conclusions until you have evaluated the information obtained from the customer. Then, proceed with your diagnosis.

Before Starting a Test Procedure

Connect the refrigerator service cord to the power source, through a wattmeter, combined with a voltmeter. Then, make a visual inspection and operational check of the refrigerator to determine the following:

- Is the refrigerator properly leveled?
- If the refrigerator is a static condenser model, is it located for proper dissipation of heat from the condenser? Check recommended spacing from rear wall and clearance above cabinet.
- Feel the condenser. With the compressor in operation, the condenser should be hot with a gradual reduction in temperature from the top to the bottom of the condenser.
- Are door gaskets sealing on pilaster area?
- Does the door, PC, or FC actuate the light switch? (PC is the abbreviation for provisions compartment; FC is the abbreviation for freezer compartment.) Is FC fan guard in place? Is FC fan properly located on motor shaft?
- Is the thermostat thermal element properly positioned? The thermal element must not contact the evaporator.
- Observe the frost pattern on the evaporator.
- Check the thermostat knob setting.
- Check the air-damper control knob setting.
- Inscribe bracket opposite slotted shaft of defrost timer to determine if timer advances.
- Is the condenser fan motor operating?

NOTE: Condenser fan motor operates only when compressor operates.

- Are air ducts free of obstructions?

The service technician should inquire regarding the number of people in the family. This will help determine the service load and daily door openings. In addition, the service technician should know the room temperature.

After this phase of diagnosis is completed, a thorough operational check should be made of the refrigeration system. Any components not previously checked should be checked in the following order.

Thermostat Cut-Out and Cut-In Temperatures

To check the cut-out and cut-in temperatures of the thermostat, use a refrigeration tester or a recording meter. Attach a test bulb to the thermal element on the top-freezer and side-by-side models. Replace the evaporator cover before starting the test.

When using a refrigeration tester equipped with several bulbs, place a No. 2 bulb in FC air (center) and a No. 3 bulb in PC air (center).

Allow the system to operate through a complete cut-out and cut-in cycle, with the thermostat set on middle position.

For accurate reading or recording, the temperature of the thermostat thermal element must not be reduced more than 1°F (0.5°C) per minute through the final 5°F (2.7°C) prior to cut out.

Erratic operation of the thermostat will affect both the FC and PC air temperature.

Freezer- and Provision-Compartment Air Temperatures

Freezer- and provision-compartment temperatures are affected by the following:

- Improper door seal
- Frost accumulation on the PC and the FC evaporators
- Service load
- Ambient temperature
- Percentage of relative humidity
- Thermostat calibration (cut in and cut out)
- Location of the FC fan blade on motor shaft
- Compressor efficiency

From this, it is evident that temperatures are not always the same in every refrigerator, even under identical conditions. However, an average FC air temperature of 0 ±6°F (−17.8 ± 3.3°C) with a PC air temperature of 36 (2.2°C) to 42°F (5.6°C), should be obtained.

Freezer Compartment Too Warm
Cause:

1. Inoperative fan motor
2. Improperly position fan
3. Evaporator iced up
4. Defrost heater burned out
5. Inoperative defrost timer
6. Inoperative defrost temperature termination thermostat
7. Wire loose at defrost timer
8. Fan guard missing
9. Excessive service load
10. Abnormally low room temperatures
11. FC or PC door left open
12. Thermostat out of calibration
13. PC or FC door gasket not sealing; check with 100-W lamp
14. Thermostat thermal element touching the evaporator
15. Inoperative condenser fan motor
16. Shortage of refrigerant (side-by-side models only)
17. Restricted filter drier or capillary tube

Provision Compartment Too Warm
Cause

1. Inoperative fan motor
2. Improperly positioned fan
3. Fan guard missing
4. PC air inlet air duct restricted
5. PC to FC return air duct restricted
6. Air-flow control on *warmer* position
7. Thermostat out of calibration.
8. Thermostat knob set at warm setting
9. Thermostat thermal element touching the evaporator
10. Evaporator iced up
11. Inoperative defrost timer
12. Inoperative defrost heater
13. Inoperative defrost temperature termination thermostat
14. Loose wire at defrost timer
15. Excessive service load, resulting from too much food in compartment

16. Inoperative capacitor in the fan motor
17. PC or FC door left open
18. Inoperative or erratically opening FC and/or FC door switch
19. Shelves covered with foil wrap or paper, retarding air circulation
20. Restricted capillary tube or filter drier

Evaporator Blocked with Ice
Cause

1. Inoperative defrost timer
2. Defrost timer terminates too early
3. Defrost timer incorrectly wired; check wiring
4. Inoperative fan motor
5. Inoperative termination thermostat
6. Inoperative defrost heater
7. PC or FC door left open
8. FC drain plugged; clean
9. FC drain sump or drain-trough heater burned out; replace.

Line Voltage

It is essential to know the line voltage at the appliance. A voltage reading should be taken the instant the compressor starts and while the compressor is running. Line-voltage fluctuation should not exceed 10 percent, plus or minus, from nominal rating. Low voltage will cause overheating of the compressor motor windings. This will result in compressor cycling on thermal overload, or the compressor may fail to start.

Inadequate line-wire size and overloaded lines are the most common reasons for low voltage at the appliance.

Wattage

Wattage is a true measure of power. It is the measure of the rate at which electrical energy is consumed. Therefore, wattage readings are useful in determining compressor efficiency, proper refrigerant charge, and the presence of a restriction. They also help detect the malfunction of an electrical component.

Amperes, measured with an Amprobe, multiplied by the voltage is not a true measurement of power in an *alternating current* (AC) circuit. It gives only "volt-amperes" or "apparent power." This value must be multiplied by the power factor (phase angle), to obtain the true or actual (AC) power. The actual power is indicated by a wattmeter.

$$\text{Watts} = \text{volts} \times \text{amperes} \times \text{power factor}$$

Or

$$\text{Power factor} = \frac{\text{watts}}{\text{volts} \times \text{amperes (VA)}}$$

$$= \frac{\text{actual power}}{\text{apparent power}}$$

Thus, the power factor may be expressed as the ratio of the actual watts to the apparent watts. The apparent watts value is the product of the amperes and volts as indicated by an ammeter and a voltmeter. The power factor varies from 0 to 1.00 (or 100 percent). On resistance heaters, such as the drier coil and drain heater, the actual watts are equal to the amperes multiplied by the volts. Thus, the power factor is 100 percent or 1.00. On electric motors, because of their magnetic reaction, the actual watts are not equal to amperes multiplied by volts. This means the power factor is less than 1.00. For this reason, a wattmeter should be used.

Compressor Efficiency

A low-capacity pumping compressor causes excessive or continuous compressor operation, depending on the ambient temperature and service load. Recovery of cabinet temperature will be slow. If cycling does occur, wattage will generally be below normal. Condenser temperature will be near normal.

Refrigerant Shortage

A loss of refrigerant results in the following:

- Excessive or continuous compressor operation.
- Above-normal PC temperature.
- A partially frosted evaporator (depending on amount of refrigerant lost).
- Below-normal FC temperature.
- Low-suction pressure (vacuum).
- Low wattage. The condenser will be "warm to cool" again, depending on the amount of refrigerant lost.

When refrigerant is added, the frost pattern will improve. The suction and discharge pressures will rise. The condenser will become hot. The wattage will increase. The refrigerator should then be turned off and thoroughly leak tested.

It is not always necessary to change a compressor when a leak is found in the system. If a slight undercharge of refrigerant is indicated, without a leak being found, the charge can be corrected without changing the compressor.

It is recommended, however, that the filter drier be replaced to protect the system against moisture. This is essential if the refrigerant charge has leaked out or if moisture may have entered the system. Refer to "Replacing the Filter Drier."

Restrictions

Restrictions are classified as total or partial, as a result of foreign matter, oil, or moisture in the capillary tube or drier. A permanent or total restriction completely stops the flow of refrigerant through the system. The result is continuous compressor operation, low wattage, low-suction pressure (vacuum), and a cool condenser. A cool condenser indicates liquid refrigerant trapped in the condenser. A partial restriction results in the following:

- A partially frosted evaporator.
- Excessive or continuous compressor operation.
- Above-normal PC temperature.
- Below-normal FC temperature.
- Low wattage.
- Low-suction pressure (vacuum), depending on the amount of restriction.
- The lower (or outlet) one-half or one-third of the condenser will be cool. Such coolness indicates liquid refrigerant trapped in the condenser.

To make sure that the trouble is a partial restriction and not a shortage of refrigerant, cover the condenser.

Allow the compressor to operate to increase the discharge pressure and temperature. (Note the increase in wattage.) The increase in discharge pressure will force refrigerant through the restricted area. Frosting of the evaporator will occur. The frost pattern will not improve with a shortage of refrigerant.

A total or partial "moisture" restriction always occurs at the outlet of the capillary tube. If moisture is suspected, turn the refrigerator "off." Allow all system temperatures to rise above 32°F (0°C) or manually initiate a defrost cycle. If moisture is present, the restriction will be released. The system should be discharged. A replacement drier should be installed and the system evacuated and recharged.

Defrost-Timer Termination

Manually initiate a defrost cycle. Do this by turning the slotted shaft of the timer clockwise.

> NOTE: Rotate the timer shaft slowly into the defrost cycle or part of the defrost timer will be missed. If the length of the defrost cycle is not in accord with specifications, change the defrost timer.

Computing Percent Run Time

That period of operation between the cut-in and cut-out points is called the pull-down cycle, on-cycle, or running cycle.

The period of time between the cut-out and cut-in point is called the warm up cycle, or off-cycle.

A complete cycle of operation is equal to the on-cycle plus the off-cycle. Such a cycle usually is timed in minutes.

The percent running time is computed by the following formula:

$$\frac{\text{On-cycle time}}{\text{On-cycle time} + \text{off-cycle time}} \times 100$$

$$= \text{percent of running time}$$

START AND RUN CAPACITORS

One of the frequent problems with air conditioners and refrigerators has nothing to do with the refrigerant and the refrigeration cycle—other than the operation of the electric motor, which powers the whole operation. In order to start under load—in case the power to the refrigerator goes off and then comes on again with power loss on the line or some other cause—the motor must start under extreme load. The capacitor-start motor has the added torque necessary to do the job. In order to improve the power factor and electrical efficiency of the motor the inductance of the motor windings is balanced off with the introduction of the run capacitor. In some instances, like a refrigerator sitting for long periods of inactivity and then plugged in and expected to run, the capacitors will malfunction, in some cases due to becoming dried out. The electrolyte of the electrolytic capacitor has a tendency to become dry and therefore changes its electrical characteristics.

Capacitor Ratings

Never use a capacitor with a lower rating than specified on the original equipment. The voltage rating and the microfarad rating are important. A higher voltage rating than that specified is always usable. However, a voltage rating lower than that specified can cause damage. Make sure the capacitance marked on the capacitor in MFD, or microfarads, is as specified. Replace with a capacitor of the same size in µF, uF, MF, UF, or MFD. All these abbreviations are used to indicate microfarads.

Start Capacitor and Bleeder Resistors

The development of high power factor, low-current, single-phase compressor motors that require start and run capacitors used with potential type relays created electrical peculiarities. These did not exist in previous designs.

In some situations, relay contacts may weld together, causing compressor motor failure. This phenomenon occurs due to the high voltage in the start capacitor discharging (arcing) across the potential relay contacts. To eliminate this, start capacitors are equipped with bleeder resistors across the capacitor terminals. See Fig. 12-24.

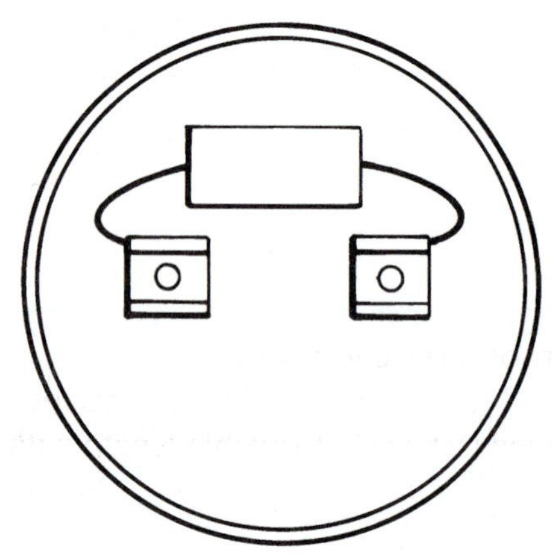

Fig. 12-24 *Bleeder resistor across the capacitor terminals.* *(Tecumseh)*

Bleeder resistor equipped capacitors may not be available. Then, a 2-W 15,000-Ω resistor can be soldered across the capacitor terminals.

If the relay solenoid opens, the start capacitor is left in the circuit too long. Normally, it is in the circuit less than 10 s. If it stays longer, it is subject to excessive heat buildup. It will spray its contents on the equipment nearest it. It is a good idea to mount the capacitor where it will cause little damage if it does malfunction.

Run Capacitors

The marked terminal of run capacitors should be connected to the "R" terminal of the compressor, and thus to L_2. Check the wiring diagram for the correct terminal.

The run capacitor is in the circuit whenever the compressor is running. It is an oil-filled electrolytic capacitor that can take continuous use. The start capacitor is a dry type. It has in it a substance that can react quickly if too long in the circuit. The oil-filled type is a wet electrolytic. It will take longer circuit use.

There are at least three ways of attaching leads to the terminals of electrolytic capacitors. See Fig. 12-25.

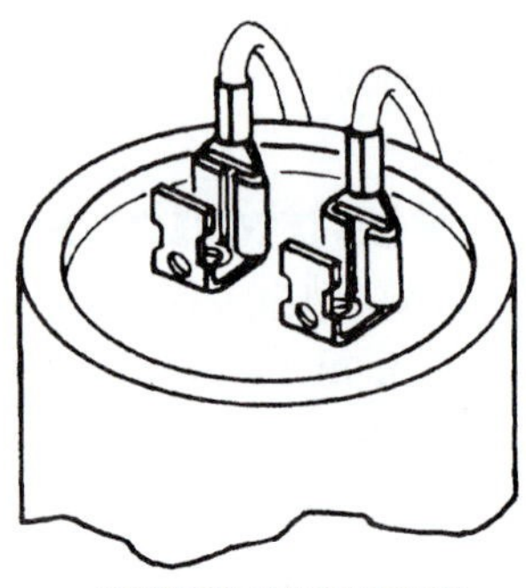

QUICK CONNECT

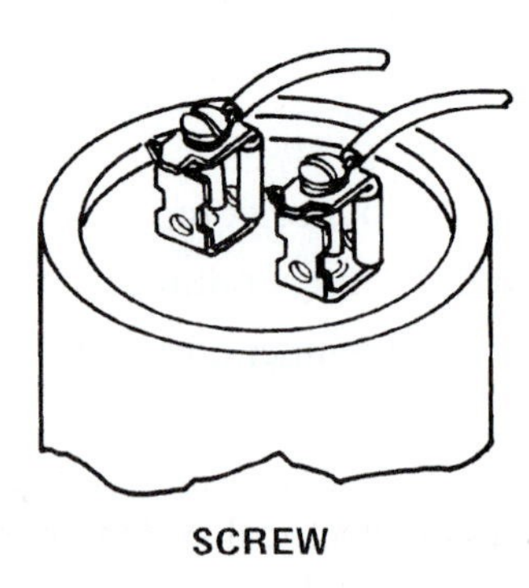

SCREW

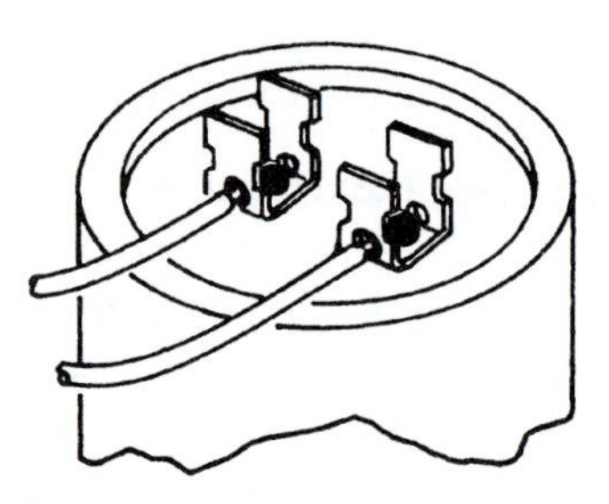

Fig. 12-25 *There are three ways to attach terminals to an electrolytic capacitor.* *(Tecumseh)*

PERMANENT SPLIT-CAPACITOR (PSC) COMPRESSOR MOTORS

The *permanent split-capacitor* (PSC) motor eliminates the need for potentially troublesome and costly extra electrical components. Start capacitors and potential motor starting are needed for capacitor-start motors.

Conditions that affect the PSC motor starting include the following:

- *Low voltage reduces motor starting and running torque.* Torque varies as the square of the voltage. Low voltage can prevent starting, cause slow starting, light flicker, and TV screen flip-flop. The minimum voltage required in starting a 230-V or 230/208-V PSC compressor is 200-V locked rotor (LRV) measured at the compressor terminals. This cannot be measured accurately after the compressor starts. It can be measured only when on locked rotor.

- *Circuit breaker or fuse trips.* Branch circuit fuses or circuit breakers sized too small will cause nuisance tripping, incorrectly diagnosed as compressor "no start."
- *Unequalized system pressure.* The maximum equalized pressure against which a PSC compressor is designed to start is 170 psig. System pressure may not be equalized within the 3-min design limitation due to improper refrigerant metering device, excessive refrigerant charge, and rapid cycling of room thermostat. Because of these, the compressor will not start.
- *Starting load too great.* A number of conditions can cause too great a starting load on a PSC compressor motor. System refrigerant charge may be excessive. Liquid refrigerant may have migrated to the compressor and formed a high liquid level in the crankcase.

Figure 12-26 shows how the run capacitor is wired into the compressor electrical system. Note how the run capacitor is in series with the start winding and in parallel with the main winding. The snap-acting disc opens the circuit when too much current is drawn through the windings. The heater provides the heat needed to cause quick action of the snap-acting disc.

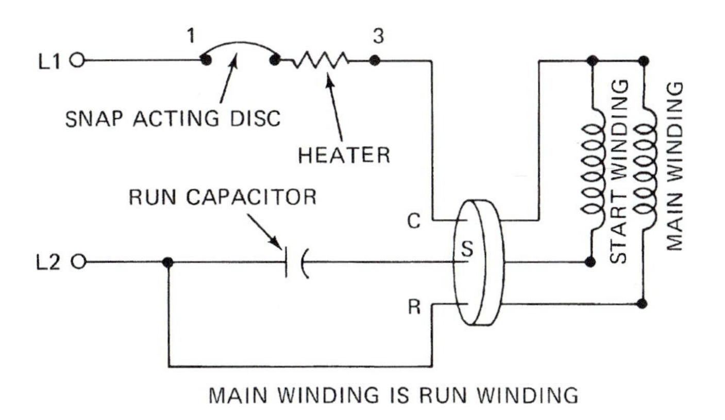

Fig. 12-26 *The PSC motor eliminates the need for potentially troublesome and costly extra electrical components, such as start capacitors and potential motor-starting relays.* (Tecumseh)

The charts shown in Fig. 12-27 give wire size, LRV, and circuit breaker or fuse-size requirements for compressors.

Table 12-2 lists PSC compressor motor troubles and corrections.

FIELD TESTING HERMETIC COMPRESSORS

Before a hermetic compressor is returned to the source, it must be tested. A replacement service charge is made on all in-warranty and out-of-warranty compressors and units that are returned when no defect is found. To avoid this unnecessary expense and loss of time, all compressors must be tested before they are returned for repair.

Most authorized wholesalers have a compressor test stand. The following equipment is needed:

- Variable high-voltage transformer (high potential) for checking for grounds. Capacitor check and analyzer for checking start and run capacitors for shorts, opens, intermittent, and capacitance. A suggested model (Mike-o-Meter) is manufactured by Sprague Products Company, North Adams, Massachusetts.
- Voltmeter capable of measuring the three voltages are shown below:
  a. 115-V, 60-Hz, single phase
  b. 230-V, 60-Hz, single phase
  c. 220-V, 60-Hz, three phase
- Two sets of service test cords.
  a. One 12-gage, two-conductor, stranded copper, insulated test cord of suitable length with alligator test clips on one end for testing a single-phase compressor.
  b. One 12-gage, three-conductor, stranded copper insulated test cord of suitable length with alligator clips on one end for testing a three-phase compressor.
- One ammeter having a range of 40 A adjustable from 0 to 10, 0 to 20, and 0 to 40 A for 220-V, single-phase usage.
- One bypass line, including connector tube clamps for discharge and suction line, 200-psi pressure gage and control valves as shown in Fig. 12-28.
- Hose adapters.
- Four start capacitors having the following ratings:
  a. 100μF at 125-V AC rating (for $^1/_2$, $^1/_3$, and $^3/_4$ hp, 120-V rated compressors).
  b. 50μF at 250-V AC rating (for $^1/_3$, $^1/_2$, and $^3/_4$ hp, 230/240-V rated compressors).
  c. 100 μF at 250-V AC rating (for 1, $1^1/_2$, $1^3/_4$, and 2 hp, 230/240-V rated compressors).
  d. 200 μF at 230/240-V AC rating (for 3, 4, and 5 hp, 230/240-V rated compressors).
- Continuity test cord and lamp as shown in Fig. 12-29. Electrical instrument requirements can be covered by a tester manufactured by Airserco Manufacturing Company.

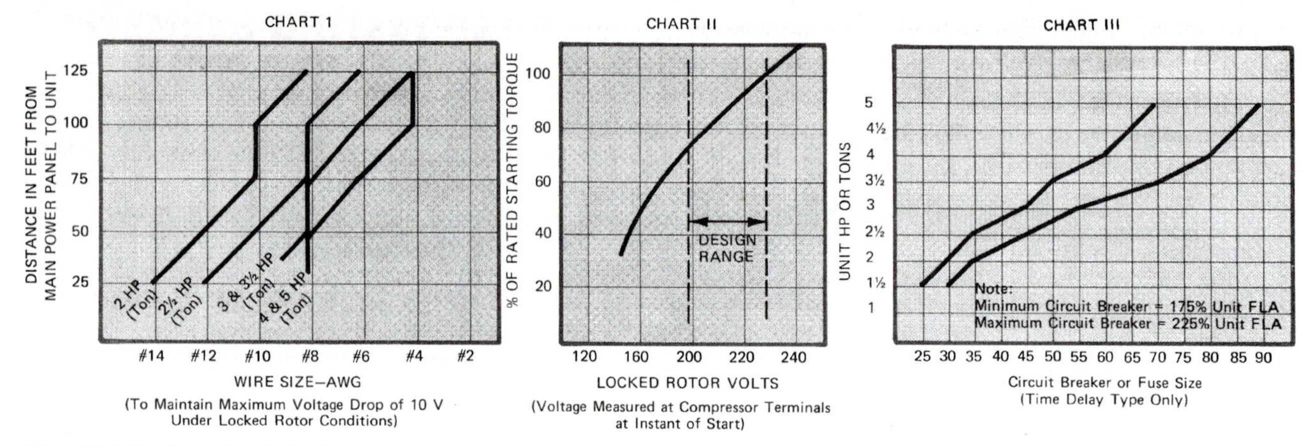

Fig. 12-27 *Wire size, locked rotor volts, and circuit breaker or fuse size requirements for compressors.* (Tecumseh)

Table 12-2 *PSC Compressor Motor Troubles and Corrections*

Causes	Corrections
Low Voltage	
1. Inadequate wire size.	1. Increase wire size.
2. Watt-hour meter too small.	2. Call utility company.
3. Power transformer too small or feeding too many homes.	3. Call utility company.
4. Input voltage too low.	4. Call utility company.
(Note: Starting torque varies as the square of the input voltage.)	
Branch Circuit Fuse or Circuit Breaker Tripping	
1. Rating too low.	1. Increase size to a minimum of 175% of unit FLA (Full Load Amperes) to a maximum of 225% of FLA.
System Pressure High or Not Equalized	
1. Pressure not equalizing within 3 min.	1. a. Check metering device (capillary tube or expansion valve). b. Check room thermostat for cycling rate. Off cycle should be at least 5 min. Also check for "chattering." c. Has some refrigerant dryer or some other possible restriction been added?
2. System pressure too high.	2. Make sure refrigerant charge is correct.
3. Excessive liquid in crankcase (split-system applications).	3. Add crankcase heater and suction line accumulator.
Miscellaneous	
1. Run capacitor open or shorted.	1. Replace with new, properly-sized capacitor.
2. Internal overload open.	2. Allow two hours to reset before changing compressor.

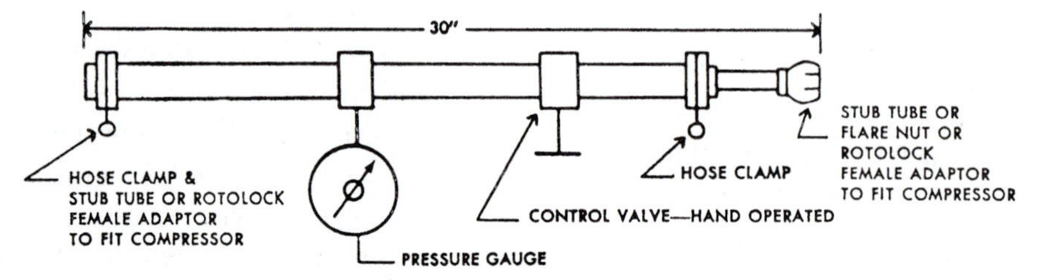

Fig. 12-28 *Compressor test stand.* (Tecumseh)

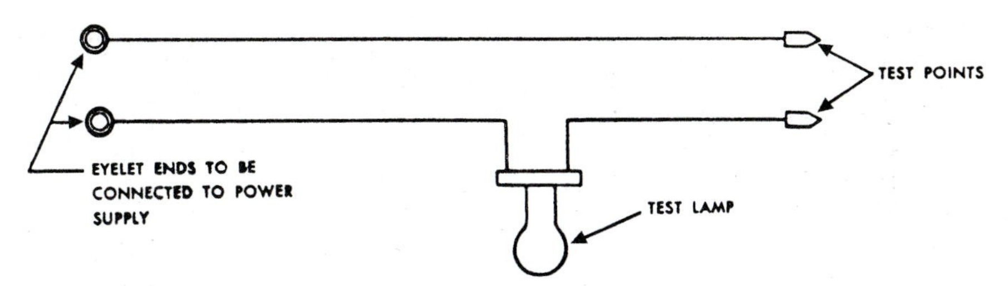

Fig. 12-29 *Continuity test cord and lamp.* (Tecumseh)

Warranty Test Procedure

Several checks must be made to make sure the compressor is operating correctly. The high-potential test is one of the tests to be performed. Following are the voltage requirements:

- Use 950 test volts for any compressor having rating up to $^1/_2$ hp.
- Use 1450 test volts for any compressor having a rating of $^1/_2$ hp and greater.

METHOD OF TESTING

With the transformer adjusted to the correct specified voltage, attach one lead of the tester as a ground by holding it against a nonpainted portion of the compressor housing. Touch the other lead for 1 s to any one of the compressor terminal posts. The high-potential ground tester will then indicate if a ground is present.

If a ground is indicated, do not check further. Remove the compressor and attach a tag to it. Note on the tag "grounded compressor." Return the compressor to the manufacturer for replacement.

A run test is made if the compressor is not showing a ground. The procedure is as follows:

1. Check the compressor assembly for correctness of wiring and for loose or broken terminals or joints. Change or repair where possible.
2. Install a bypass line from the suction to the discharge line as shown in Fig. 12-30. The control valve in the line is put in *full open* position.

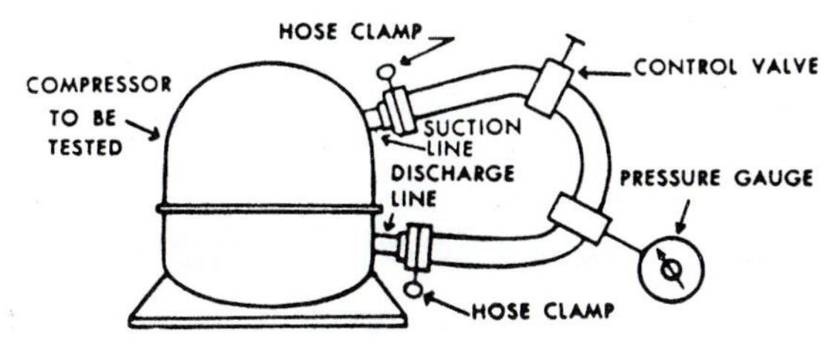

Fig. 12-30 *Bypass line from suction to discharge line.* (Tecumseh)

3. Remove all electrical components.
4. Place an ammeter in L_1 leg of the power supply.
5. Hook up the compressor motor to the power supply as follows:

   Single phase:

   a. L_1 to C (common) and L_2 to R (run) terminals.
   b. Using proper start capacitor from the manufacturer's chart, connect one capacitor terminal to R. Let the lead from the other capacitor terminal dangle loose. See Fig. 12-31.

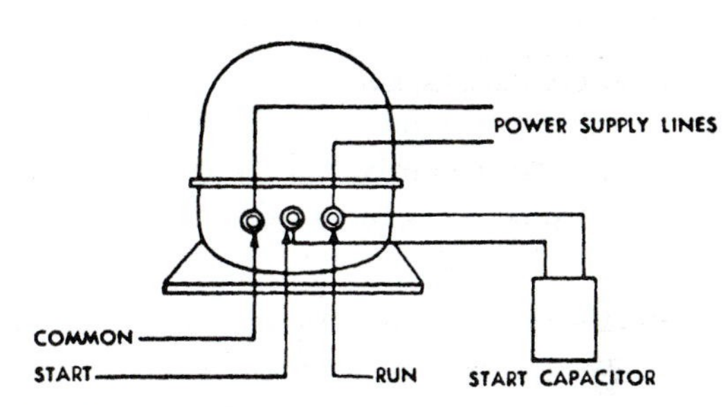

Fig. 12-31 *Location of start capacitor in a circuit.* (Tecumseh)

   c. Energize the compressor and momentarily touch the loose start capacitor lead to S.

   CAUTION: Do not touch the start capacitor lead to S for more than a few seconds or the start winding will burn.

   Three phase(3Φ)

   a. Connect L_1, L_2, and L_3 to the three compressor terminals.
   b. Energize the compressor.
6. Observe the ammeter.
   a. If the compressor does not run and draws *locked rotor amperes* (LRA-check the serial plate for this information), attach a tag noting that the compressor is "stuck." Return the compressor for replacement.

b. If the compressor runs, but is abnormally noisy, attach a tag noting that the compressor is "noisy." Return the compressor for replacement.

c. If the compressor does not run and draws no amperes, attach a tag noting that the compressor has "open winding." Return the compressor for replacement.

7. If the compressor runs normally, close the control valve in the bypass line until the gage reads approximately 175 psig. Read the amperes. If the current is more than the rated FLA, attach a tag noting that "high amperes" are the problem. Return the compressor for replacement.

8. If the current and sound are normal, close the control valve. Stop the compressor and clock the rate of pressure fall on the bypass line gage. The following pressures are for Tecumseh compressors. For other compressors, the manufacturer's recommendation must be checked.

   Twenty-five psig per minute initial rate of discharge pressure drop should not be exceeded on models AE, T, AT, AK, AJ, AR, AU, and ISM.

   Forty psig per minute initial rate of discharge pressure drop should not be exceeded on models AB, AH, B, C, P, and AP.

   Eighty psig per minute initial rate discharge pressure drop should not be exceeded on models F and PF.

   If pressure changes occur in excess of those indicated, tag the compressor and mark it "internal leak." Return the compressor to the manufacturer for replacement.

9. If the compressor tests normal, open the control valve in the bypass line. Remove the bypass, and immediately seal the discharge and suction tubes.

10. Check the resistance of the motor windings against the values furnished by the manufacturer.

11. If the compressor checks out normal, return it to the customer.

Resistance Checks

The run (main) windings of a single-phase hermetic compressor motor consist of large-diameter wire having very low resistance. The accurate measurement of run-windings resistance requires a digital ohmmeter. Less accurate meters are not sensitive enough to give correct readings.

Motors are sometimes diagnosed as grounded when the problem actually lies with the ohmmeter. It may not be sensitive to the normal 1- or 2-Ω run-winding resistance.

There is a test procedure that can circumvent this measurement problem. This technique also enables the serviceperson to determine the operating position of an internal line break overload if one is installed in the motor.

The procedure is as follows:

1. Remove all electrical connections from the compressor terminals.

2. Measure the resistance across the run and start terminals of the motor. This measures the combined resistance of the run and start windings. If the measured resistance approaches the value given in the manufacturer's specifications, the windings can be considered normal. If no resistance is read, the windings are open. The motor should be rejected.

3. If step 2 indicates no problems, measure the resistance of the start winding only by checking across the terminal S (start) and C (common). The ohms should approach the range given by the manufacturer. If no resistance is read, the internal common lead to the motor is open.

   If the motor has an external overload, then an open common lead means the motor is defective. If the motor has an internal line break overload, then the overload may be open because the motor is overly hot.

It is not uncommon for the overload to remain open for more than an hour if tripped by a hot motor. Motors cool slowly. The internal overload will not close until the compressor dome (and thus the motor iron) is cool to the touch—below 130°F (54°C).

4. If steps 2 and 3 indicate no problems, check each terminal in turn for a ground to the compressor housing. First, file a shiny spot to ensure a good electrical connection.

5. If steps 2, 3 and 4 indicate no problems, the problem may not be with the single-phase motor. Three-phase motors are more difficult to diagnose, since all three windings are run windings with low resistance. The internal overload, if present, is across all three windings. Here, the suggested procedure is as follows:

   a. Check for continuity between each of the three terminal pairs.

   b. If the circuit is open and an internal overload is present, be sure the motor is cool before rejecting the motor.

   c. If step a, indicates no problems, check for ground between each terminal and the housing.

Testing Electrical Components

The individual electrical components in a compressor may be tested as follows:

1. Check the rating and part number of each component to ensure that the component is correct for the compressor model. If the wrong component is being used, replace with the correct one.
2. Using an ohmmeter, check the external overload (at room temperature) for continuity across the terminals. If defective, replace with the proper component.
3. The start and run capacitor should be checked on the recommended capacitor tester according to the manufacturer's instructions. An alternate, but less precise method, uses a good ohmmeter. With the meter on the R × 1 range, continuity indicates a shorted capacitor. Replace. With the meter on the R × 100,000 scale, no needle deflection indicates an open capacitor. Replace.
4. It is difficult to check the starting relay without special equipment.

   Use the ohmmeter to make continuity checks.

   While being checked, current relays must be held upright in their normal operating position.

   Contacts should be open between terminals 1 (or L) and S. Therefore there should be no continuity.

   Terminals 2 (or L) and M should indicate continuity through the operating coil. (If there is no terminal 2, use terminal 1.)

   Test the potential relay as follows:

   Contacts should be closed between terminals 1 and 2. Therefore, there should be continuity.

   There should be continuity through the operating coil between terminals 2 and 5.
5. If all the above tests prove satisfactory, change the relay. The new relay will eliminate any faulty electrical characteristics, such as improper pickup and dropout. These cannot be determined in the previous tests.
6. As a final check, connect the new relay to the compressor and the capacitors previously checked. If the compressor fails to start at serial plate voltage, the compressor should be considered inoperative because of internal defects. It must be replaced. If a capacitor checker other than that recommended was used, try new capacitors before rejecting the compressor.

INSTALLING AN AIR-COOLED CONDENSING UNIT

There are some important steps to be taken in unpacking and installing a unit for the first time. General shipping and receiving instructions should be followed and the unit be checked out before attempting to put it into operation. A service manual is provided with each new unit package. Be sure to follow it completely in order that the warranty will cover any future problems. The following is a typical unit installation instructional guide. The following information and illustrations are provided by Rheem Manufacturing Company.

General Information

The information contained in this manual has been prepared to assist in the proper installation, operation, and maintenance of the air-conditioning system. Improper installation, or installation not made in accordance with these instructions, can result in unsatisfactory operation and/or dangerous conditions, and can cause the related warranty not to apply.

Read this manual and any instructions packaged with separate equipment required to make up the system prior to installation. Retain this manual for future reference.

To achieve unit design operating efficiency and capacity, the indoor cooling coils listed in the condensing unit specification sheet should be used.

Check the outside sheet metal for signs of damage or rough handling. See Fig. 12-32.

Checking Product Received

Upon receiving unit, inspect it for any shipping damage. Claims for damage, either apparent or concealed, should be filed immediately with the shipping company. Check condensing unit model number, electrical characteristics and accessories to determine if they are correct. Check system components (evaporator coil, condensing unit, evaporator, and blower) to make sure they are properly matched. Figure 12-33 shows the way to make sure the right model was delivered. The dimensions will be given in a sheet similar to that in Table 12-3.

Remove shipping bracket under compressor if supplied.

Corrosive Environment

The metal parts of this unit may be subject to rust or deterioration if exposed to a corrosive environment. This oxidation could shorten the equipment's useful life. Corrosive elements include salt spray, fog or mist in seacoast areas, sulfur or chlorine from lawn-watering systems, and various chemical contaminants from industries such as paper mills and petroleum refineries.

If the unit is to be installed in an area where contaminants are likely to be a problem, special attention should be given to the equipment location and exposure.

- Avoid having lawn sprinkler heads spray directly on the unit cabinet.

Fig. 12-32 *Air-cooled condensing unit.* (Rheem)

- In coastal areas, locate the unit on the side of the building away from the waterfront.
- Shielding provided by a fence or shrubs may give some protection.
- Elevating the unit off its slab or base enough to allow air circulation will help avoid holding water against the base-pan. Regular maintenance will reduce the buildup of contaminants and help to protect the unit's finish.

WARNING: Disconnect all power to unit before starting maintenance.

- Frequent washing of the cabinet, fan blade, and coil with fresh water will remove most of the salt or other contaminants that build up on the unit.
- Regular cleaning and waxing of the cabinet with a good automobile polish will provide some protection.
- A good liquid cleaner may be used several times a year to remove matter that will not wash off with water.

Several different types of protective coatings are offered in some areas. These coatings may provide some benefit. But the equipment manufacturer cannot verify the effectiveness of such coating materials.

Locating Unit

Consult local building codes or ordinances for special installation requirements. When selecting a site to locate the outdoor unit, consider the following:

- A minimum clearance of 24 in. on one side for service access, 12 in. for air inlets on all sides, and 60 in. for air discharge (unit top) is required.
- The unit must be located outdoors and cannot be connected to ductwork.
- Locate unit where operating sound will not disturb owner or neighbors.
- Locate unit, so roof runoff water does not pour directly on the unit. Provide gutter or other shielding at roof level.
- If a concrete pad is used, do not connect slab to building foundation or structure to prevent noise transmission.
- Do not obstruct openings in bottom of the unit.
- The length of refrigerant piping and wiring should be as short as possible to avoid capacity losses and increased operating costs.
- Locate the pad at a level sufficient above grade to prevent ground water from entering the unit.

Unit Mounting

If elevating the condensing unit, either on a flat roof or on a slab, observe the following guidelines:

- The base pan provided elevates the condenser coil $^3/_4$ in. above the base pad.
- If elevating a unit on a flat roof, use 4×4 in. stringers positioned to distribute unit weight evenly and prevent noise and vibration.

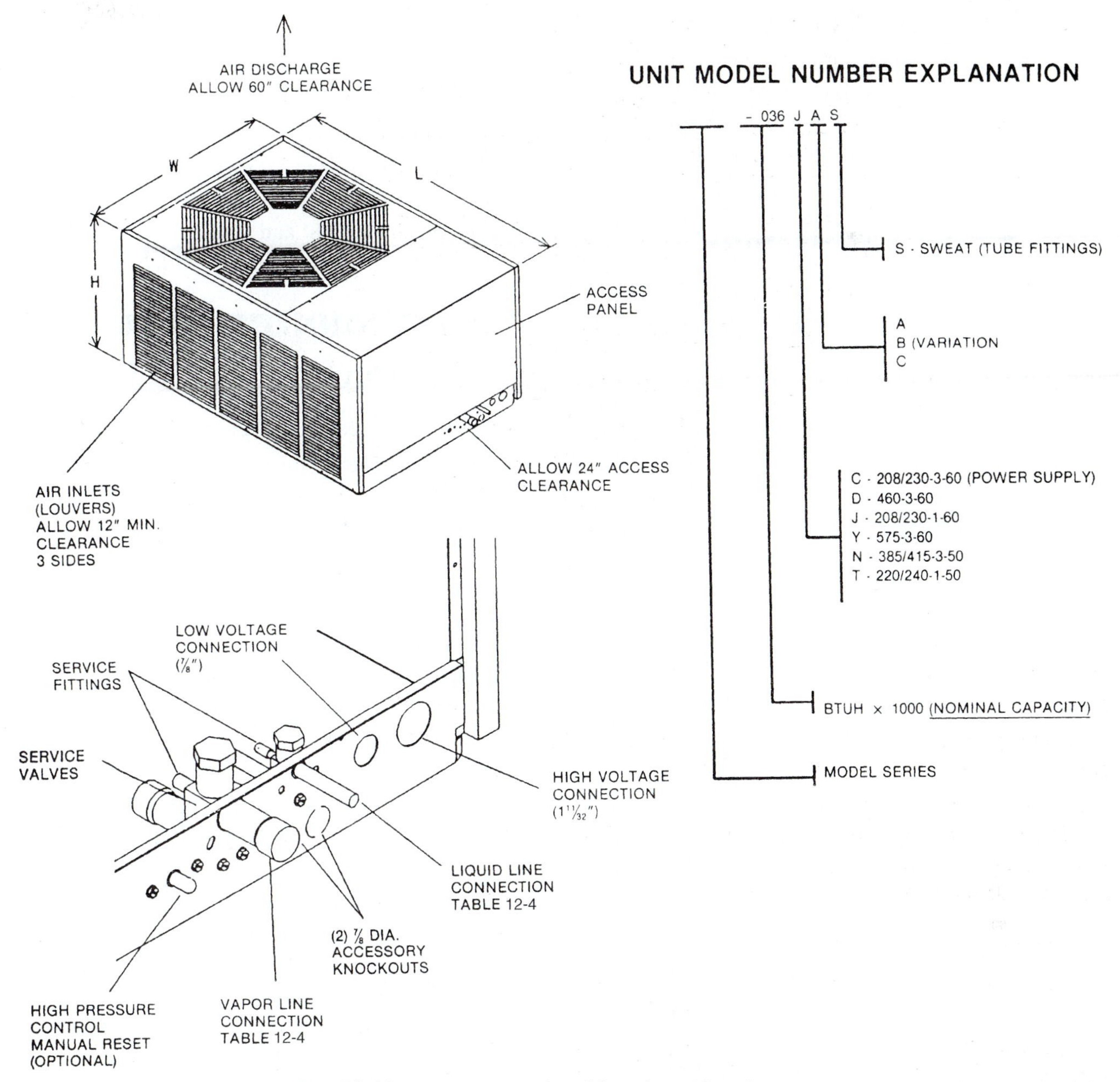

Fig. 12-33 *Dimensions and model number explanation.* (Rheem)

Table 12-3 *Dimensional Data*

Condensing unit model						
	(-)AKA-	018, 024	030	036, 042	048, 060	
	(-)ALB-		018, 024	030, 036	042, 048, 060	
	(-)AMA-				018, 024, 030	
					036, 042, 048	060
Length "H"		$16^3/_4$	$20^3/_4$	$20^3/_4$	$26^3/_4$	$34^3/_4$
Length "L"		$33^{11}/_{16}$	$33^{11}/_{16}$	$38^{11}/_{16}$	$42^9/_{18}$	43
Width "W"		$23^1/_4$	$23^1/_4$	$27^1/_8$	31	31

Refrigerant Connections

All units are factory charged with refrigerant, either R-22 or an approved substitute. All models are supplied with service valves. Keep tube ends sealed until connection is to be made. This is to prevent system contamination.

Replacement Units

To prevent failure of a new condensing unit, the existing evaporator tubing system must be cleaned or replaced. Care must be exercised that the expansions device is not plugged. Liquid-line filter driers are recommended on all units if compressor motor has failed. Test the oil for acid. If positive, a suction-line filter drier is mandatory.

Evaporator Coil

NOTE: The application evaporator temperature range is +32 to +53.5°F for unlisted condensing compressor units.

Tubing Connections

Coils have only a holding charge of dry nitrogen. Keep all tube ends sealed until connections are to be made.

Location

- Never locate the coil in return duct of a gas or oil furnace.
- Provide a service inlet to the coil for inspection and cleaning.
- Keep the coil pitched toward the drain connection.

CAUTION: When the coil is installed over a finished ceiling and/or living area, it is recommended that a secondary sheet-metal condensate pan be constructed and installed under entire unit. In some recently made units the drain or condensate pan is a molded-plastic one to aid in the remove of water to prevent the formation of mold and other stagnant-water problems.

Drain Lines

- Never connect a drain line that is smaller than the fitting provided on the coil. Pitch the line at least a quarter of an inch per foot away from the coil.
- Run the line to an open drain or outdoors.

Air Flow

Insure that the proper airflow is available as recommended for matched systems. See specifications sheet.

Duct System

In an uninsulated attic or crawl space, 1-in. fiberglass insulation with a vapor barrier skin is required. Make sure the ducts are tight and leak free. Duct must be of adequate size for required airflow with furnace and coil.

INTERCONNECTING TUBING

Suction and Liquid Lines

- Keep all lines sealed until connection is made.
- Connect the tubes to the evaporator coil first. Refer to the refrigerant line size charts, Tables 12-4 and 12-5, for correct piping sizes.
- Always use the shortest length possible with a minimum number of bends. Use a tube bender for short bends to insure that there are no kinks in the line. To prevent noise transmission, never position the suction or liquid lines in direct contact with the building structure. Also use an isolated or suspension type hanger when possible.

Maximum Length of Interconnecting Tubing

There is no fixed maximum length for interconnecting tubing. But if over 200 ft, add 3 oz of dry refrigerant oil for each 10 ft of line over 200 ft. Oil may be added not to exceed three fluid ounces for each 10 ft over 60 ft. Connecting lines should be kept as short as possible, as there is capacity loss for each foot of vapor-line tubing used. Table 12-4 shows multipliers to be used to determine capacity for various vapor-line lengths and diameters. The losses that occur due to lines being exposed to outdoor conditions are not included. Systems equipped with over 60 ft of lines must be equipped with crankcase heaters and the applications of a pump-down cycle or a suction accumulator.

Refrigerant charging for units using capillary or fixed orifice tubes must be by the suction-line superheat method. Units using thermostatic-expansion valves must be charged by liquid-pressure method. See "charging data" attached inside the service panel of the unit.

Condensing Unit Installed Below Evaporator

Maximum height of evaporator above the condensing unit is 40 ft. If over 20 ft, the liquid lines should be

Table 12-4 Refrigerant Line Size Information

<table>
<tr><td colspan="26">CONDENSING UNIT REFRIGERANT TUBING DATA</td></tr>
<tr><td rowspan="3">MODEL</td><td>(-)AKA-</td><td>018</td><td></td><td>024</td><td></td><td>030</td><td></td><td>036</td><td></td><td></td><td></td><td>042</td><td></td><td></td><td></td><td></td><td></td><td></td><td></td><td>048</td><td></td><td></td><td>060</td><td></td><td></td></tr>
<tr><td>(-)ALB-</td><td></td><td>018</td><td></td><td>024</td><td></td><td>030</td><td></td><td></td><td></td><td></td><td></td><td>036</td><td></td><td>042</td><td></td><td></td><td></td><td>048</td><td></td><td></td><td></td><td></td><td>060</td><td></td></tr>
<tr><td>(-)AMA-</td><td></td><td></td><td></td><td></td><td></td><td></td><td></td><td>018</td><td>024</td><td>030</td><td></td><td></td><td></td><td></td><td></td><td>036</td><td></td><td></td><td></td><td>042</td><td>048</td><td></td><td></td><td>060</td></tr>
<tr><td colspan="2">FACTORY CHARGE, OZ ①</td><td>46</td><td>61</td><td>51</td><td>63</td><td>68</td><td>74</td><td>75</td><td>88</td><td>85</td><td>96</td><td>79</td><td>78</td><td></td><td>100</td><td></td><td>96</td><td></td><td>192</td><td>113</td><td>100</td><td>172</td><td>112</td><td>173</td><td>192</td></tr>
<tr><td colspan="2">REFRIGERANT VAPOR CONNECTION SIZE ON UNIT</td><td>5/8" I.D. SWEAT</td><td colspan="10">3/4" I.D. SWEAT</td><td colspan="7">7/8" I.D. SWEAT</td><td colspan="6">1-1/8" I.D. SWEAT④</td></tr>
<tr><td colspan="2">REFRIGERANT LIQUID CONNECTION SIZE ON UNIT</td><td colspan="16">5/16" I.D. SWEAT③</td><td colspan="8">3/8" I.D. SWEAT</td></tr>
<tr><td colspan="26">RECOMMENDED VAPOR AND LIQUID LINE SIZES (O.D.) FOR VARIOUS LENGTH RUNS</td></tr>
<tr><td rowspan="2">LENGTHS UP TO 30 FEET</td><td>SUCT.</td><td colspan="2">5/8" O.D.</td><td colspan="9">3/4" O.D.</td><td colspan="7">7/8" O.D.</td><td colspan="6">1-1/8" O.D</td></tr>
<tr><td>LIQ.</td><td colspan="10">1/4" O.D.</td><td colspan="6">5/16" O.D.</td><td colspan="8">3/8" O.D.</td></tr>
<tr><td rowspan="2">LENGTHS 31 TO 45 FEET</td><td>SUCT.</td><td colspan="2">5/8" O.D.</td><td colspan="9">3/4" O.D.</td><td colspan="7">7/8" O.D.</td><td colspan="6">1-1/8" O.D.</td></tr>
<tr><td>LIQ.</td><td colspan="10">20'-1/4" O.D. BALANCE 5/16" O.D.</td><td colspan="6">25'-5/16" O.D. BALANCE 3/8" O.D.</td><td colspan="8">25'-3/8" O.D. BALANCE 1/2" O.D</td></tr>
<tr><td rowspan="2">LENGTHS 46 TO 60 FEET</td><td>SUCT.</td><td colspan="2">3/4" O.D.</td><td colspan="9">3/4" or 7/8" O.D. SEE CAPACITY MULTIPLIER ②</td><td colspan="7">7/8" or 1-1/8" O.D. SEE CAPACITY MULTIPLIER②</td><td colspan="6">1-1/8" O.D</td></tr>
<tr><td>LIQ.</td><td colspan="10">15'-1/4" O.D. BALANCE 5/16" O.D.</td><td colspan="6">20'-5/16" O.D. BALANCE 3/8" O.D.</td><td colspan="8">20 -3/8" O.D. BALANCE 1/2" O.D.</td></tr>
<tr><td rowspan="2">LENGTHS 61 TO 90 FEET</td><td>SUCT.</td><td colspan="2">3/4" O.D.</td><td colspan="9">3/4" or 7/8" O.D. SEE CAPACITY MULTIPLIER②</td><td colspan="7">7/8" or 1-1/8" O.D. SEE CAPACITY MULTIPLIER②</td><td colspan="6">1-1/8" O.D.</td></tr>
<tr><td>LIQ.</td><td colspan="10">5/16" O.D.</td><td colspan="6">10'-5/16" O.D. BALANCE 3/8" O.D.</td><td colspan="8">10'-3/8" O.D. BALANCE 1/2" O.D.</td></tr>
<tr><td rowspan="2">LENGTHS 91 TO 120 FEET</td><td>SUCT.</td><td colspan="2">3/4" O.D.</td><td colspan="9">3/4" or 7/8" O.D. SEE CAPACITY MULTIPLIER②</td><td colspan="7">7/8" or 1-1/8" O.D. SEE CAPACITY MULTIPLIER②</td><td colspan="6">1-1/8" O.D.</td></tr>
<tr><td>LIQ.</td><td colspan="10">80'-5/16" O.D. BALANCE 3/8" O.D.</td><td colspan="6">3/8" O.D.</td><td colspan="8">1/2" O.D.</td></tr>
<tr><td rowspan="2">LENGTHS 121 TO 150 FEET</td><td>SUCT.</td><td colspan="2">3/4" O.D.</td><td colspan="9">3/4" or 7/8" O.D. SEE CAPACITY MULTIPLIER②</td><td colspan="7">7/8" or 1-1/8" O.D. SEE CAPACITY MULTIPLIER②</td><td colspan="6">1-1/8" O.D.</td></tr>
<tr><td>LIQ.</td><td colspan="10">60'-5/16" O.D. BALANCE 3/8" O.D.</td><td colspan="6">100'-3/8" O.D. BALANCE 1/2" O.D.</td><td colspan="8">1/2" O.D.</td></tr>
</table>

CAPACITY MULTIPLIER FOR VARYING VAPOR LINE SIZES AND LENGTH RUNS②

VAPOR TUBING SIZE O.D.	5/8 3/4	5/8 3/4 7/8	3/4 7/8	3/4 7/8	3/4 7/8 1-1/8	3/4 7/8 1-1/8	7/8 1-1/8	7/8 1-1/8 1-3/8	7/8 1-1/8 1-3/8
30 FT. LENGTH	1.00 1.01	.98 1.00 1.01	1.00 1.01	1.00 1.01	.99 1.00 1.01	.99 1.00 1.01	1.00 1.01	.99 1.00 1.01	.99 1.00 1.01
60 FT. LENGTH	.98 1.00	.96 .98 1.00	.98 1.00	.98 1.00	.97 .99 1.01	.97 .99 1.01	.98 1.00	.97 1.00 1.01	.97 1.00 1.01
90 FT. LENGTH	.97 .99	.94 .97 .99	.97 99	.96 .99	.96 .98 1.00	.94 98 1.00	.96 1.00	96 99 1.00	.95 99 1.00
120 FT. LENGTH	.95 .99	.91 .95 .99	.95 .99	.94 .98	.94 .98 1.00	.92 .97 1.00	.94 .99	.94 .99 1.00	.93 .99 1.00
150 FT. LENGTH	.94 .98	.88 .94 .98	.94 .98	.93 .97	.92 .97 1.00	.89 .96 1.00	.92 .98	.93 .99 1.00	.91 .98 1.00

①Factory charge is sufficient for 25 Ft. of recommended liquid line and matching evaporator. For different lengths, adjust charge accordingly:

1/4" Liquid Line ± .3 oz per foot
5/16" Liquid Line ± .4 oz per foot
3/8" Liquid Line ± .6 oz per foot
1/2" Liquid Line ± 1.2 oz per foot

②Capacity multiplier × rated capacity = actual capacity
Example -024 with 60' of 3/4" O.D. Vapor Line
24.000 × 98 = 23.520 BTUH.

③Approx. 5/16" I.D. will accept 1/4" O.D. field liquid line.
④Requires adapter supplied with unit (packed inside).

increased to the next size larger than shown in Table 12-4. This is to prevent flashing in the liquid line. Liquid refrigerant in a vertical column will exert a downward pressure of 0.5 psi per foot of elevation, in effect adding to pressure drop when flow is up.

Condensing Unit Installed Above Evaporator

There is no absolute fixed limit as to how high the condensing unit may be above the evaporator, but if over 10 ft, close attention must be given to line sizing, to addition of oil, and to refrigerant charging. Oil traps are not required. The vertical portion of the

Table 12-5 Line Sizing with Condensing Unit Over 10 ft Above Evaporator

	Liquid Line Sizing (O.D.)	
Nominal Tons	Horizontal Run*	Vertical Run†
1-1/2	1/4″	100% - 3/16″
2	1/4″	80% - 1/4″ & 20% - 5/16″
2.1/2	1/4″	40% - 1/4″ & 60% - 5/16″
3	5/16″	20% - 1/4″ & 80% - 5/16″
3.1/2	5/16″	100% - 5/16″
4	3/8″	80% - 5/16″ & 20% - 3/8″
5	3/8″	40% - 5/16″ & 60% - 3/8″

*See Table 12-4 if horizontal run exceeds 30 ft.
†The smaller size tubing must be at the bottom of the run. The combination shown will result in approximately zero net pressure drop for vertical run.

liquid line must be sized, as shown in Table 12-5, for use with flow checks and expansion-valve coils. Erratic operating pressure can result if piping is not properly sized.

TUBING INSTALLATION

Observe the following when installing refrigerant tubing between the condensing unit and evaporator coil:

- Use clean, dehydrated, sealed refrigeration grade tubing.
- Always keep tubing sealed until tubing is in place and connections are to be made.
- If there is any question as to how clean the liquid and vapor lines are, blow out with dry nitrogen before connecting to the outdoor unit and indoor coil. Any debris in the line set will end up plugging the expansion device.
- As an added precaution, you may install a good filter drier in the liquid line.
- Vapor line and liquid line must not be in contact with each other. The vapor line must be insulated.
- If tubing has been cut, make sure ends are deburred while holding in a position to prevent chips from falling into tubing. Burrs such as those caused by tubing cutters can affect performance dramatically, particularly on small liquid-line sizes.
- For best operation, keep tubing run as short as possible with a minimum number of elbows or bends.
- Locations where the tubing will be exposed to mechanical damage should be avoided. If it is necessary to use such locations, copper tubing should be housed to prevent damage.
- If tubing is to be run underground, it must be run in a sealed watertight chase.
- Use care in routing tubing and do not kink or twist. Use a good tubing bender on the vapor line to prevent kinking.
- Route the tubing using temporary hangers, then straighten the tubing and install permanent hangers. Line must be adequately supported.
- The vapor line must be insulated to prevent dripping, sweating, and prevent performance losses. Rubatex and Armaflex are satisfactory insulations for this purpose. Use minimum $^1/_2$ in. insulation thickness, additional insulation may be required for long runs.
- Check Table 12-4 for correct liquid and vapor-line sizing for condensing unit size and length of run.

TUBING CONNECTIONS

- Be certain both refrigerant service valves at the condensing unit are closed (turn fully clockwise).
- All lines should be assembled with type "L" refrigerant tubing and not with copper water pipe. They should be brazed with the following alloys:
  - Copper to copper—5 percent silver alloy (no flux)
  - Copper to steel or brass—35 percent silver alloy (no flux)
- Clean the inside of fittings and the outside of the tubing with steel wool or sand cloth before soldering. Always keep chips, steel wool, dirt, and the like, out of the inside when cleaning. Assemble tubing part way into fitting. Apply flux all around the outside of the tubing and push tubing into the stop. This procedure will keep the flux from getting inside the system.
- Remove cap and Schrader core from service port to protect seals from heat damage.
- Wrap service valves with a wet rag. Before applying heat, to braze the tubing between outdoor unit and indoor coil. Flow dry nitrogen into a service port and through the tubing while brazing.

LEAK TESTING

- Pressurize line set and coil through service fittings with dry nitrogen to 150 psig maximum. Leak test all joints using liquid detergent. If a leak is found, recover pressure and repair.

WARNING: Do not use oxygen to purge lines of pressure system for leak test. Oxygen reacts violently with oil, which can cause an explosion resulting in severe personal injury or death.

- The indoor coil and tubing of sweat-type units must be evacuated before operating unit. See evacuation procedure.
- The brass valve is not a back-seating valve. Opening or closing valve does not close service port.

NOTE: $^3/_{16}$ and $^5/_{16}$ in. Allen wrenches are required for brass-service valves. Extreme caution must be exercised not to force valve stem against the retaining ring when opening the valves. System pressure could force the valve stem out of the valve body and possibly cause personal injury. In the event that the retaining ring is missing, do not attempt to open the valve.

FLOW-CHECK PISTON

The flow-check piston is a multipurpose device. With flow into the compression nut end from the liquid line, the piston acts as the expansion device with flow through the metering orifice in the center of the piston. The "O" ring on the end of the piston prevents refrigerant from bypassing the metering orifice. Flow from the metering orifice is centered into a distributor that serves to evenly distribute refrigerant to the evaporator circuits. See Fig. 12-34.

It is essential that the evaporator and condensing unit be properly matched. Use only matched components as shown in sales specification sheets. See Table 12-6 for the appropriate piston size for the evaporator and condensing-unit combination.

A piston size that is too small will cause starving and one that is too large will cause flooding. In any case, system performance and reliability will be unacceptable.

If a combination is used that requires a piston-size change (see Table 12-7), the combination cannot be used without changing to the correct size piston.

Change the piston in the distributor on the evaporator before installing the coil and charging the system following the procedure shown:

- Using a back-up wrench on the distributor body, loosen the compression nut to gain access to the piston.
- Using the wire provided with replacement pistons, run (hooked end) through hole in piston.
- Hook nose of piston and lift gently from distributor body.
- Replace piston with one of proper size (see Table 12-6). Install piston with gasket end of piston in distributor. Do not force piston into distributor.

> NOTE: With piston in distributor, seal end should be down and should not be seen looking in end of distributor. Pistons must be free to rotate and move up and down. Make sure piston is free to move in distributor body.

- Insure that distributor gasket is located properly in the distributor body.
- Replace compression nut using back-up wrench on distributor body. Torque compression nut with 8 to 10 lb/ft.
- Original piston size is stamped on outside of distributor body. Remove the new piston-size label from its poly bag. The new piston comes in the plastic bag. Install the new size label on the outside of the distributor tube.
- Check fittings for leaks after installation, evacuation, and charging are complete.

> CAUTION: Do not attempt to drill pistons to size in the field. Metering holes have a special champhered inlet and

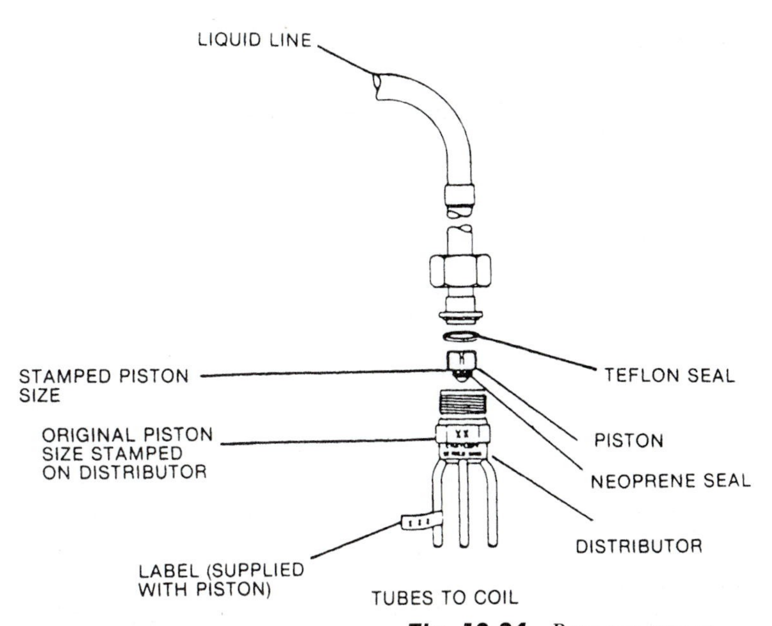

Fig. 12-34 *Base pan.* (Rheem)

Table 12-6 Condensing-Unit Approved Application Matches with Flow-check Piston Sizes Required

Condensing Unit Model and Size	Evaporator and Model Number and Size	Piston Size Required	*Coil Code		System Chg. OZ
			Elec. Furn.	HP AH	
(-)AKA-018	RCBA-2453	53	B	B	46
(-)AKA-024	RCBA-2457	57	C	D	51
(-)AKA-030	RCBA-3665	65	B	D	68
(-)AKA-036	RCBA-3673	73	C	C	75
(-)AKA-042	RCBA-4878	78	B	B	79
(-)AKA-048	RCBA-4876	76	C	D	113
(-)AKA-060	RCBA-6089	89	B	B	112

*Coil code in electric furnace or air-handler model number.

Table 12-7 Condensing-Unit Approved Application Matches with TXV and Piston Sizes Required

Condensing Unit Model and Size	Evaporator Model Number and Size	TXV Size (TON)	Piston Size	Coil Slabs
(-)ALB-018 (-)AMA-018	RCGA-24A1	1.5	120	4
(-)ALB-024 (-)AMA-024	RCGA-24A2	2.0	172	4
(-)ALB-030 (-)AMA-030	RCGA-36A1	2.5	157	6
(-)ALB-036 (-)AMA-036	RCGA-36A2	3.0	157	6
(-)ALB-042 (-)AMA-042	RCGA-48A1	4.0	172	8
(-)ALB-048 (-)AMA-048	RCGA-48A1	4.0	172	8
(-)ALB-060	RCGA-60A1	5.0	172	10
(-)AMA-060	RCGA-60A1	5.0	172	10

cannot be modified. WARNING: Do not replace the neoprene "O" ring on the piston with any type of seal. Contact the parts department for the exact replacement "O" ring.

EVACUATION PROCEDURE

Evacuation is the most important part of the entire service procedure. The life and efficiency of the equipment is dependent upon the thoroughness exercised by the serviceman when evacuating air and moisture from the system.

Air in a system causes high-condensing temperatures and pressure, resulting in increased power input and reduced performance.

Moisture chemically reacts with the refrigerant and oil to form corrosive hydrofluoric and hydrochloric acids. These attack motor windings and parts, causing breakdown.

After the system has been leak checked and proven sealed, connect the vacuum pump and evacuate system to 29.5 in. The vacuum pump must be connected to both the high and low sides of the system through adequate connections. Use the largest size connections available since restrictive service connections may make the process so slow as to be unacceptable. This may lead to false readings because of pressure drop through the fittings.

CAUTION: Compressors (especially scroll type) should never be used to evacuate the air-conditioning system. Vacuums this low can cause internal electrical arcing, resulting in a damaged or failed compressor.

- With the thermostat in the "OFF" position, turn the power "ON" to the furnace and the condensing unit.
- Before starting the condensing unit, allow 12 h time to elapse, giving crankcase heater (if provided) time to drive refrigerant from the compressor, thus preventing damage during start-up.
- Start the condensing unit and the furnace with the thermostat. Make sure the blower is operating.

CHECKING REFRIGERANT CHARGE

Charge for all systems should be checked against the "charging chart" inside the access panel cover. Before using the chart, the indoor conditions must be within 2°F (wet bulb) of desired comfort conditions and system must be run until operating conditions stabilize (15 to 30 min).

> CAUTION: Do not operate the compressor without charge in system.

Addition of R-22, or the refrigerant being used, will raise pressures (vapor, liquid, and discharge) and lower vapor temperature.

> CAUTION: If addition of Refrigerant raises both vapor pressure and temperature, unit is over-charged.

Charging by Superheat

Superheat charging method is used for charging systems when a flow-check piston or capillary tubes are used on the evaporator as a metering device.

Pressure reading and charging is accomplished using the service port located on the vapor-service valve (large valve) located on the base pan. See Fig. 12-35.

Vapor temperature readings must be taken on the vapor line going from the vapor-service valve (large valve) and the compressor. A remote temperature indicator is most convenient. If this is not available, a thermometer properly located and insulated can be used.

Measure and record the three values required. Find the intersection of vapor-line pressure and outdoor ambient on the charging chart. The vapor-line temperature should approximate the intersect value on the chart.

The most likely causes for the intersection of vapor pressure and ambient temperature in the open area to (left) or (right) of table values are:

- *Left.* Low charge, low
- *Right.* Overcharge, high

Charging by Liquid Pressure

Liquid-pressure method is used for charging systems in the cooling mode when an expansion valve is used on the evaporator. The service port on the liquid-service valve (small valve) is used for this purpose.

Measure and record the three values required. Find the intersection of outdoor ambient and indoor ambient (°F W.B.) on the "charging chart." The liquid-line pressure should approximate the intersection value on the chart.

Charging by Weight

For a new installation, evacuation of interconnecting tubing and evaporator coil is adequate; otherwise, evacuate the entire system. Use the factory charge shown in Table 12-4 of these instructions or unit data plate. Note that charge value includes charge required for 25 ft of standard size interconnecting liquid line. Calculate actual charge required with installed liquid-line size and length using:

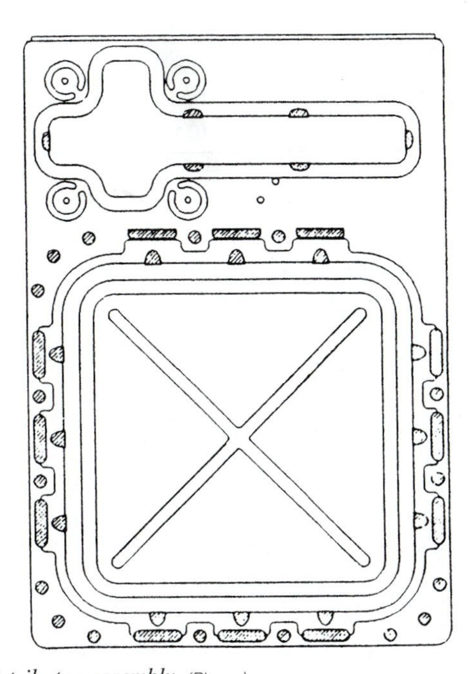

Fig. 12-35 *Piston and distributor assembly.* (Rheem)

- $^1/_4$ in. OD = 0.6 oz/ft
- $^5/_{16}$ in. OD = 0.4 oz/ft
- $^3/_8$ in. OD = 0.6 oz/ft
- $^1/_2$ in. OD = 1.2 oz/ft

With an accurate scale (±1 oz) or volumetric charging device, adjust charge difference between that shown on the unit data plate and that calculated for the new installation. If the entire system has been evacuated, add the total calculated charge.

FINAL LEAK TESTING

After the unit has been properly evacuated and charged, a halogen leak detector should be used to detect leaks in the system. All piping within the condensing unit, evaporator, and interconnecting tubing should be checked for leaks. If a leak is detected, the refrigerant should be recovered before repairing the leak. The Clean Air Act prohibits releasing refrigerant into the atmosphere.

SERVICE

Operation

Most single-phase units are operated PSC (no starting components). It is important that such systems be off for a minimum of 5 min before restarting to allow equalization of pressure. The thermostat should not be moved to cycle unit without waiting 5 min. To do so may cause the compressor to go off on an automatic overload device or blow a fuse. Poor electrical service can also cause nuisance tripping on overloads or blow fuses. This generally can be corrected by adding start components. Check with factory for recommended start components, if required. For PSC type operation, refrigerant metering must be done with fixed orifice, cap tubes, or bleed type expansion valves because of low-starting torque. If nonbleed expansion valve coils (supplied by factory) are used, start components are required.

Single-Pole Compressor Contactor (CC)

Single-pole contactors are used on all standard single-phase units through 5 tons. Caution must be exercised when servicing, as only one leg of the power supply is broken with the contactor. Two pole contactors are used on some three-phase units.

Compressor Crankcase Heat (CCH)

All heaters are located on the lower half of the compressor shell. Its purpose is to drive refrigerant from the compressor shell during long off-cycles, thus preventing damage to the compressor during starting.

At initial start-up or after extended shutdown periods, make sure the heater is energized for at least 12 h before the compressor is started. (Disconnect switch is ON and wall thermostat is OFF.)

NOTE: Crankcase heaters are not required for scroll compressors.

Hard Start Components (SC and SR)

They are available through parts department only. The start component kit includes *start capacitor* (SC) and *start relay* (SR). Start components are required with all nonbleed expansion-valve coils.

NOTE: Start components are not required for scroll compressors.

Time Delay Control (TDC)

The *time delay control* (TDC) is in the low-voltage control circuit. When the compressor shuts off due to a power failure or thermostat operation, this control keeps it off at least 5 min which allows the system pressure to equalize, thus not damaging the compressor or blowing fuses on start-up.

Low Ambient Control (LAC)

This component senses compressor head pressure and shuts the condenser fan off when the head pressure drops to approximately 175 psig. This allows the unit to build a sufficient head pressure at lower ambient in order to maintain system balance and obtain improved capacity. Low ambient control should be used on all equipment operated below 65°F ambient.

High- and Low-Pressure Controls (HPC or LPC)

These controls keep the compressor from operating in pressure ranges that can damage it. Both controls are in the low-voltage control circuit. *High-pressure control* (HPC) is a manual reset that operates near 450 psig. Do not reset arbitrarily without first determining what caused it to function. The *low-pressure control* (LPC) is an automatic reset that operates near 15 psig and resets near 40 psig.

NOTE: High- and low-pressure controls may be standard on some models.

ELECTRICAL WIRING

Field wiring must comply with the National Electric Code (C.E.C. in Canada) and any applicable local ordinance.

Power Wiring

It is important that proper electrical power is available at the condensing-unit contactor. Voltage should not vary more than 10 percent of that stamped on the rating plate when the unit tries to start. Interphase variation on the three-phase units must not be more than 3 percent.

Install a branch circuit disconnect within sight of the unit and of adequate size to handle the starting current. See Table 12-8.

Table 12-8 *Electrical and Physical Data*

AWG Copper Wire Size	AWG Aluminum Wire Size	Connector Type & Size (or equivalent)
12	10	T & B Wire Nut PT2
10	8	T & B Wire Nut PT3
8	8	Sherman Split Bolt TSP6
6	4	Sherman Split Bolt TSP4
4	2	Sherman Split Bolt TSP2

For branch circuit wiring (main power supply to unit disconnect), the minimum wire size for the length of run can be determined from Table 12-9 using the circuit ampacity found on the unit rating plate. From the unit disconnect to unit, the smallest wire size allowable in Table12-9 should be used.

Table 12-9 *Copper wire size-AWG*

Supply wire length (feet)	(1% Voltage Drop)							
200	6	4	4	4	3	3	2	2
150	8	6	6	4	4	4	3	3
100	10	8	8	6	6	6	4	4
50	14	12	10	10	8	8	6	6
Supply-circuit ampacity	15	20	25	30	35	40	45	50

Power wiring must be run in grounded, rain-tight conduit. Conduit must be run through the connector panel below the access cover and attached to the bottom of the control box. (See Fig. 12-33.)

Connect power wiring to contactor located in outdoor heat pump electrical box. (See wiring diagram attached to unit access panel.)

Check all electrical connections, including factory wiring within the unit and make sure all connections are tight.

Do not connect aluminum field wire to the contactor terminals.

Special Instruction for Power Wiring with Aluminum conductors

- Select the equivalent aluminum wire size from the following tabulation:
- Attach a length (6 in. or more) of recommended size copper wire to the unit contactor terminals L_1 and L_3 for single phase, L_2 and L_3 for three phase.
- Splice copper wire pigtails to aluminum wire with *Underwriters' Laboratories* (UL) recognized connectors for copper-aluminum splices. Follow these instructions very carefully to make a positive and lasting connection.
- Strip insulation from the aluminum conductor.
- Coat the stripped end of the aluminum wire with the recommended inhibitor, and wire brush aluminum surface through inhibitor. *Inhibitors*: Brundy-Pentex "A," Alcoa No. 2EJC; T & B-KPOR Shield.
- Clean and recoat aluminum conductor with inhibitor.
- Make the splice using the earlier- listed wire nuts or split-bolt connectors.
- Coat the entire connection with inhibitor and wrap with electrical-insulating tape.

GROUNDING WARNING: The unit must be permanently grounded.

A grounding lug is provided near contactor for a ground wire. Grounding may be accomplished by grounding the power wire conduit to the condensing unit. Make sure the conduit nut locking teeth have pierced the insulating paint film.

Control Wiring

If the low-voltage control wiring is run in conduit with the power supply, Class-I insulation is required. Class-II insulation is required if run separate. Low-voltage wiring may be run through the insulated bushing provided in the $^7/_8$-in. hole in the base panel, up to and attached to the pigtails from the bottom of the control box. Conduit can be run to the base panel if desired by removing the insulated bushing.

A thermostat and a 24-V, 20-VA minimum transformer are required for the control circuit of the condensing unit. The furnace or the air-handler transformer may be used if it has the proper voltage and current available. See the wiring diagram and Table 12-10 for reference.

Table 12-10 *Field Wire Sizes for 24-V Thermostat Circuits*

Thermostat Load-Amps	Solid Copper Wire-Awg.					
3.0	16	14	12	10	10	10
2.5	16	14	12	12	10	10
2.0	18	16	14	12	12	10
Length of Run-Feet*	50	100	150	200	250	300

*Wire length equals twice the run distance.
NOTE: Do not use control wiring smaller than No. 18 AWG between thermostat and outdoor unit.

START-UP AND PERFORMANCE

Even though the unit is factory charged with refrigerant, the charge must be checked. This is done using the charge table attached to the service panel and adjusted, if required. Allow a minimum of 30 min running. See the instructions on the unit service panel for marking the total charge on the unit rating plate.

TROUBLESHOOTING

A troubleshooting chart is provided in Table 12-11 for checking the unit before and after it has been installed. Keep in mind that you should disconnect all power to the unit before servicing.

Table 12-11 *Troubleshooting Chart*

Symptom	Possible Cause	Remedy
High head-low suction	a. Restriction in liquid line or capillary tube or filter drier	a. Remove or replace defective component
High head-high or normal suction	a. Dirty condenser coil	a. Clean coil
	b. Overcharged	b. Correct system charge
	c. Condenser fan not running	c. Repair or replace
Low head-high suction	a. Incorrect capillary tube	a. Replace evaporator assembly
	b. Defective compressor valves	b. Replace compressor
Unit will not run	a. Power off or loose electrical connection	a. Check for unit voltage at contactor in condensing unit
	b. Thermostat out of calibration-set too high	b. Reset
	c. Defective contactor	c. Check for 24 volts at contactor coil replace if contacts are open
	d. Blown fuses	d. Replace fuses
	e. Transformer defective	e. Check wiring-replace transformer
	f. High-pressure control open	f. Reset-also see high head pressure remedy. The high-pressure control opens at 430 PSI
	g. Compressor overload contacts open	g. If external overload-replace OL. If internal replace compressor NOTE: Wait at least 2 h for overload to reset
Condenser fan runs. compressor does not	a. Run or start capacitor defective	a. Replace
	b. Start relay defective	b. Replace
	c. Loose connection	c. Check for unit voltage at compressor-check and tighten all connections
	d. Compressor stuck, grounded or open motor winding, open internal overload.	d. Wait at least 2 h for overload to reset If still open, replace the compressor.
	e. Low-voltage condition	e. Add start kit components
Low suction-cool compressor Iced Evaporator Coil	a. Low-indoor airflow	a. Increase speed of blower or reduce restriction-replace air filter
	b. Operating unit at temperatures below 65° outdoor temperature.	b. Add low ambient kit
Compressor short cycles	a. Defective overload protector	a. Replace-check for correct voltage
Registers sweat	a. Low airflow	a. Increase speed of furnace blower or reduce restriction replace air filter.
High-suction pressure	a. Excessive load	a. Recheck load calculation
	b. Defective compressor	b. Replace
Insufficient cooling	a. Improperly sized unit	a. Recalculate load
	b. Improper airflow	b. Check-should be approximately 400 CFM per ton
	c. Incorrect refrigerant charge	c. Charge per procedure attached to unit service panel
	d. Incorrect voltage	d. At compressor terminals, voltage must be within 10% of nameplate volts when unit is operating

WARNING: Disconnect all power to unit before servicing. Contactor may break only one side of line.

REVIEW QUESTIONS

1. Why is oxygen not used to pressurize a refrigeration system?
2. Why can pressure testing of a refrigeration system be dangerous?
3. Why should you wear goggles when purging a contaminated system?
4. Why do you not fill a gas cylinder completely full?
5. What is meant by off-cycle crankcase heating?
6. Where is the filter drier located in the liquid line?
7. What is a heat exchanger?
8. What are the four major causes of motor burnout on a compressor?
9. How do you repair the perimeter tube if it develops a break?
10. What is the mullion on a refrigerator cabinet?
11. Why do you not use the bubble method for finding a leak when a vacuum is present?
12. What color of flame indicates a leak with a halide torch?
13. Why can you get a false reading of a refrigerant leak in a refrigerator when the insulation is urethane foam?
14. How do you test the potential relay?

13
CHAPTER

Freezers

PERFORMANCE OBJECTIVES

After studying this chapter you will:

1. Know how to install a chest type or an upright freezer.
2. Know how to identify and assemble freezer parts.
3. Know how to select the proper thermostat for a repair job.
4. Know how to troubleshoot freezers: upright models.
5. Know how to troubleshoot freezers: chest models.
6. Know how to work on a portable ice-cream freezer.

Freezers come in a variety of styles and shapes. Much depends on the use to which the freezer is to be put—whether it be for commercial purposes or for home use. Many homes now have upright and chest-type freezers for convenience in storing meat and fish as well as other foodstuffs. This chapter is concerned with the operation, maintenance, and repair of both types of freezes, commercial and noncommercial.

TYPES OF FREEZERS

There are two types of domestic freezers—the upright freezer and the chest-type freezer. The essential parts of the upright freezer are shown in Fig. 13-1.

Notice that the evaporator coils are built into the shelves as part of that unit. This means the shelves are not adjustable. Notice that the condenser coils (17) in Fig. 13-1 are welded to the outside of the cabinet. This prevents sweating and aids the dissipation of heat over a large surface. The primary convenience of this type of freezer is that the frozen food is visible, easily arranged, and easily removed.

The chest-type freezer provides a different storage arrangement. See Fig. 13-2. In some instances, the condenser coils are mounted on the back of the chest-type freezer.

The electrical diagram for a chest-type freezer is shown in Fig. 13-3. Note that the thermostat controls the on-off operation of the compressor. The light switch completes the circuit from one side of the power

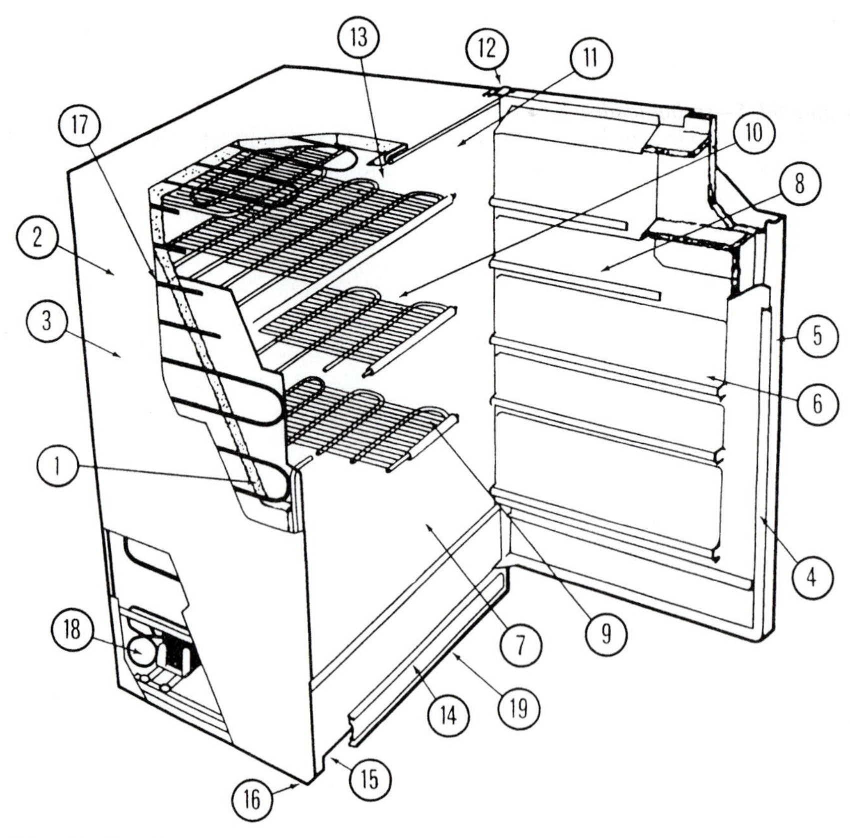

Fig. 13-1 *The parts of an upright home freezer: 1—polyurethane foam-insulation cabinets, 2—wrap-around steel cabinet, 3—baked-on enamel finish, 4—magnetic-door seal, 5—key-ejecting lock, 6—bookshelf door storage, 7—slide-out basket, 8—juice-can shelf, 9—steel shelves, 10—fast two-way freezing level, 11—temperature-control knob, 12—door stops, 13—interior light, 13— power-on light, 15—defrost water drain, 16—adjustable leveling legs, 17—coils welded to outer walls, 18—sealed compressor, and 19. RED.*

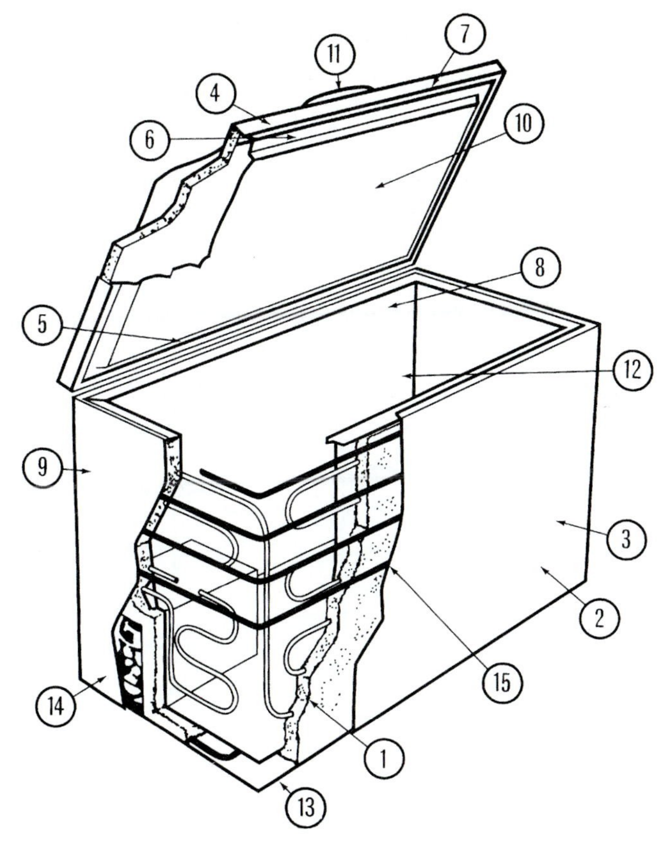

Fig. 13-2 The parts of a chest-type freezer: 1—polyurethane foam insulation, 2—wrap-around steel cabinet, 3—baked-on enamel finish, 4— self-adjusting lid, 5—spring-loaded hinges, 6—vinyl lid gasket, 7—safety lock and self-ejecting key, 8—lift-out wire baskets, 9—temperature-control knob, 10—automatic interior light, 11—power-on light, 12—vertical cabinet divider, 13—defrost water drain, 13—sealed compressor, and 15—wrap-around condenser.

cord to the other through the light. Some models have a mercury switch that operates when the lid is up. This type of freezer does not have automatic defrost. Defrosting must be done manually.

INSTALLING A FREEZER

It may be necessary to remove the freezer door for passage through narrow doors. With each freezer there is an instruction sheet explaining the step-by-step procedure for door removal. Screw-type levelers are used to adjust the level of the freezer. Upright freezer models use a screw-type leveler that can be moved up and down by turning to the left or right. See Fig. 13-4.

The cabinet must be level side-to-side with a very slight tilt toward the rear. This will aid in obtaining a tight door gasket seal. If the cabinet is tilted toward the front, the weight of the door, plus the door food load, will result in poor gasket seal.

> NOTE: Caution the user not to slam the door. If the door is slammed, the air pressure maybe sufficient to open the door slightly. Putting metal or wood shims between the floor and the freezer, as required, levels chest freezers.

Do not locate the freezer adjacent to a stove or other heat source. Avoid an area that is exposed to direct sunlight for long periods.

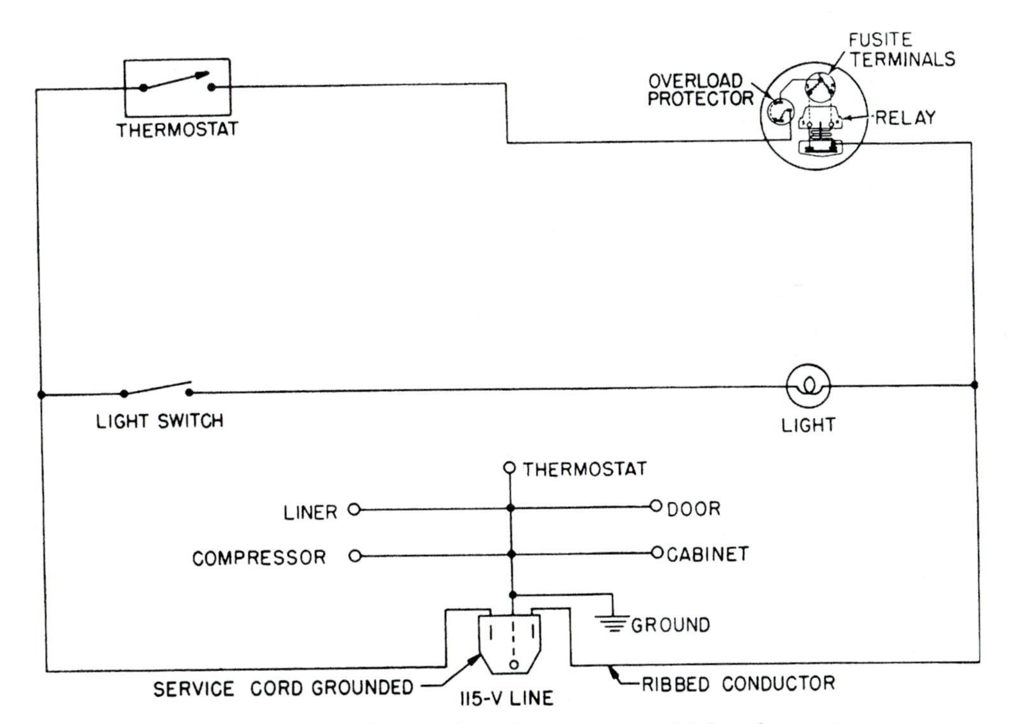

Fig. 13-3 Electrical schematic for a chest-type manual defrost freezer. (Kelvinator)

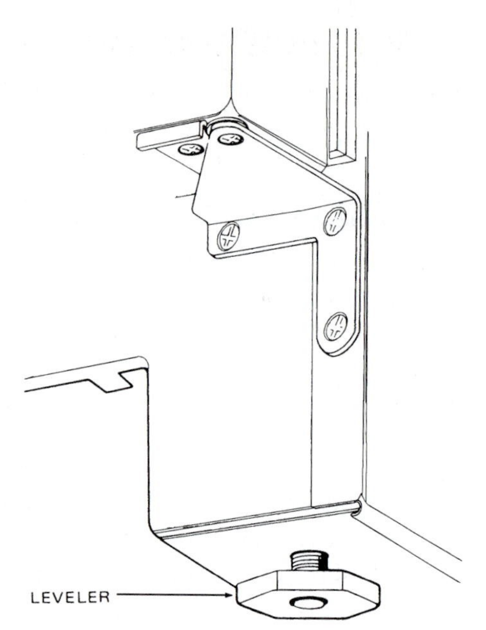

Fig. 13-4 *Adjustable screw for the upright model freezer.* (Kelvinator)

FREEZER COMPONENTS

Wrapped Condenser

The wrapped condenser incorporates a precooler condenser in series (through an oil cooler) with the main condenser. The condenser, made of 1/4 in. steel tubing, is clamped to the cabinet wrapper. Thermal mastic is applied to each pass for maximum heat dissipation.

A wrapped condenser depends on natural convection of room air for dissipation of heat. Restricted air circulation around the cabinet will cause high-operating temperatures and reduced capacity. The wrapped condenser reduces the possibility of moisture condensing on the cabinet shell during extremely humid weather. It also eliminates the need for periodic cleaning of the condenser. Figure 13-5 shows the condenser layout on an upright freezer.

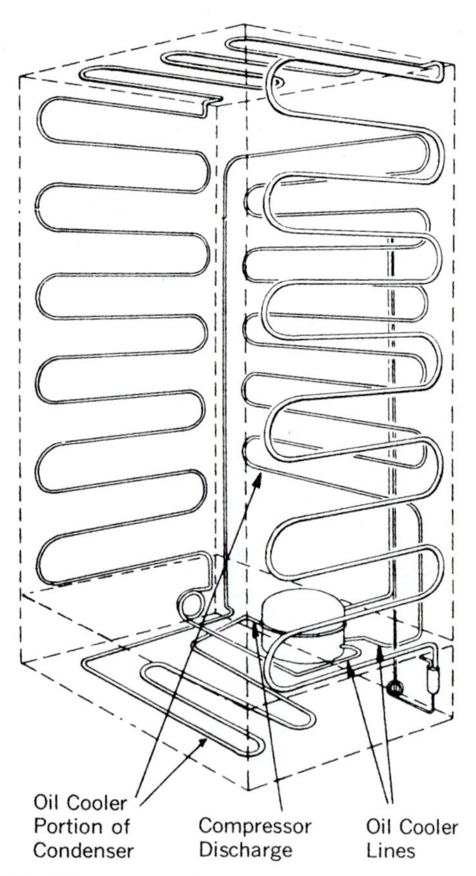

Fig. 13-5 *Wrapper condenser for the upright model freezer.* (Kelvinator)

Cold-Ban Trim

Upright models have four-piece sectional cold-ban trim strips that extend around the periphery of the freezer storage compartment. These trim strips are replaceable.

Starting at the lower corners, force the side trims toward the opposite side of the freezer as shown in Fig. 13-6.

Use a small flat screwdriver to release the cold-ban trim from the cabinet U-channel. Then, pull the trims down and out from the overlapping top trim strip. Remove the top and bottom cold trim strips by grasping one end and pulling the trim out of the cabinet U-channel.

Before installing replacement cold-ban trim, be sure the fiberglass filler insulation sections are in place. Install the bottom cold-ban trim. Squeeze one end of the trim and press the front flanges of the trim into the U-channel as shown in Fig. 13-7. Then, use the palm of the hand to press on the rear edge of trim, forcing the lock tabs on the trim over the flange of the freezer liner.

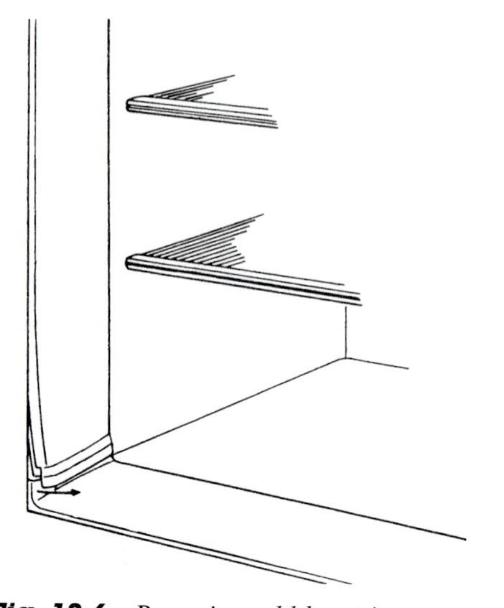

Fig. 13-6 *Removing cold-ban trim.* (Kelvinator)

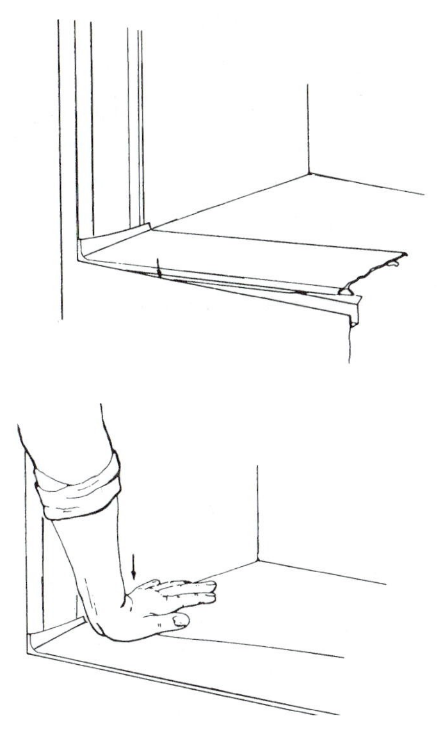

Fig. 13-7 *Installing cold-ban trim.* (Kelvinator)

Install the right-hand and the left-hand side cold-ban trims. Then install the top cold-ban trim.

Shelf Fronts

The lower portions of the door shelves are formed as a unit with the inner panel. The door shelf fronts are removable. To remove, push down on the end caps and tilt out slightly at the top. See Fig. 13-8. Slide the end caps up. Pull outward on the bottom to free the caps from the slots in the door inner panel.

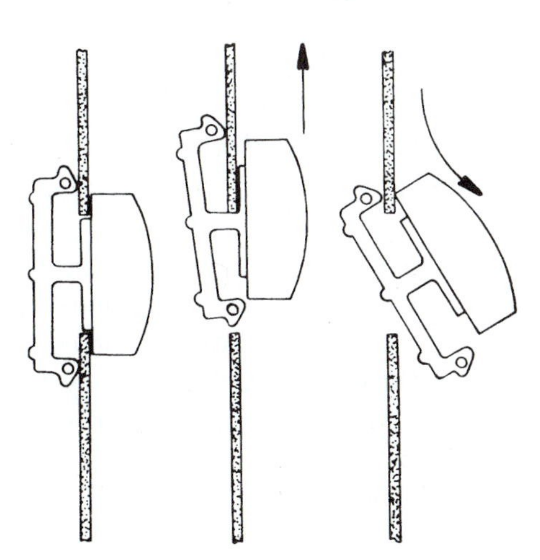

Fig. 13-8 *Removing shelf fronts in a chest-type freezer.* (Kelvinator)

Vacuum Release

Some models have a vacuum release in the bottom edge of the door. This speeds up equalization of the air pressures, permitting successive door openings.

Some models do not incorporate a vacuum-release device. If they have a good airtight gasket seal and the freezer is in operation, the door cannot be opened the second time when two door openings are required in quick succession. This is due to a difference in air pressure between the freezer interior and the room atmosphere. Opening the door the first time results in spillage of cold air from the freezer. This cold air is replaced by warm air. When the door is closed, this warm air is cooled, reducing its specific volume, thus creating a vacuum. Leaving the door closed for about 1 and $1^1/_2$ min will allow the air pressure to equalize.

Lock Assembly

Some models have a lock. The lock assembly is mounted in the door outer panel. It is held in place by the lock retainer. Figure 13-9 shows an exploded view of a typical lock assembly.

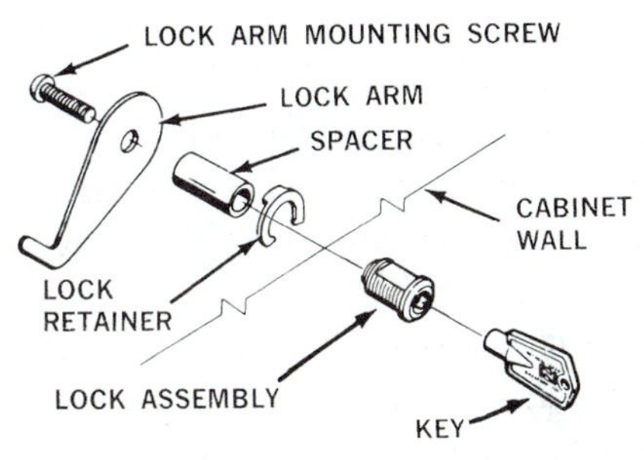

Fig. 13-9 *Door lock assembly.* (Kelvinator)

To replace the lock, remove the door inner panel. Then remove the lock retainer and lock assembly. Replace in the reverse order of removal. Remember that the lock key is self-ejecting.

Hinges

Chrome-plated steel hinges are used on all models. See Figs. 13-10 and 13-11. The hinge pins ride in nylon thimbles placed in the door panel piercing. Nylon spacers are placed over the hinge pins to form the weight-hearing surfaces.

To replace hinges, remove the top hinge screws and hinge. Then remove the door from the freezer.

Hinge Adjustment Some hinges have enlarged mounting holes to permit hinge adjustment. Shims maybe added or removed from behind the bottom hinge to eliminate

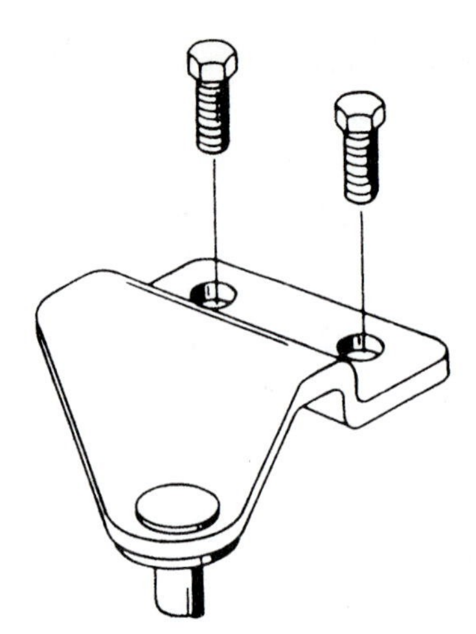

Fig. 13-10 *Top hinge.* (Kelvinator)

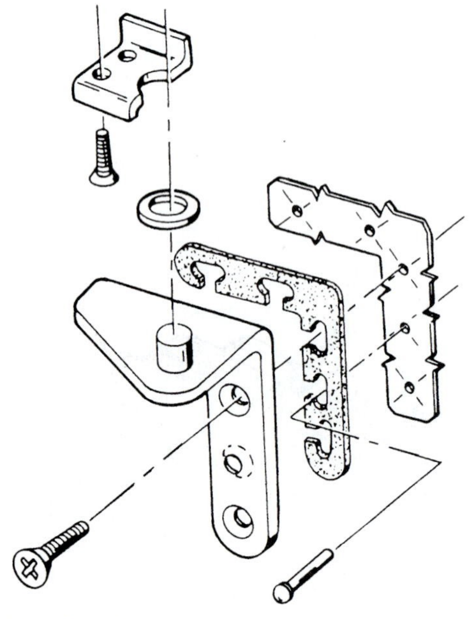

Fig. 13-11 *Bottom hinge assemble.* (Kelvinator)

hinge bind or to improve the door-gasket seal at the hinge side.

Chest models have spring-loaded hinges. These hinges incorporate a strong coil spring. The force of this spring counterbalances the weight of the lid and lifts the lid to the open position.

The hinge butt is fastened to the cabinet with four screws. These screws engage a fixed tapping plate inside the wrapper wall. The hinge leaf is fastened to the lid's outer panel with four screws that engage a fixed tapping plate. Pressed fiber shims are used under the hinge butt.

The hinge-butt holes are slotted vertically to allow adjustment of the lid and to secure proper gasket fit. The hinge-leaf holes are slotted horizontally to allow adjustment of the lid, either to the left or to the right.

Lid

Most chest-type freezers have flexible lids. Even when the lid is lifted at one corner, it will seal properly on closing under its own weight. The lid's outer panel is drawn from one piece of steel. The edge is turned back to form a flat flange. This gives strength and furnishes a plane surface for support of the gasket and the lid's inner panel. Tapping plates for the hinges are welded in place.

The inner lid panel, gasket, handle, lock assembly, and insulation may be seen in Fig. 13-12.

Thermostats

Freezers and refrigerators have the same theory of operation. The start relay for the compressor operates the same way as the start relay for a refrigerator. However, the thermostats are somewhat different.

On upright freezers the thermostat is mounted in the upper right-hand corner of the storage compartment in all manual defrost models. The thermostat knob in the manual defrost models is numbered 1 through 6 or *coldest* and *off*.

On all other models, remove the right-hand side coldban trim and filler insulation. Loosen and remove the thermostat thermal element clamp from beneath the refrigerated shelf. Straighten the thermal element and attach a 3-ft length of cord to the end of the thermal element. Use tape. Remove the light shield and thermostat knob. Disconnect the wire leads and remove the thermostat from the mounting bracket. Pull the thermal element out of the insulation.

> NOTE: Cord taped to the end of the thermal element feeds into the insulation cavity as the thermal element is pulled out. Remove the cord from the inoperative thermostat and tape it to the replacement thermostat thermal element. Pulling on the opposite end of the cord, thread the thermal element through the insulation and hole in the liner. Attach the thermal element to the refrigerated shelf. Mount the thermostat. Replace the light shield and knob.

On the chest models the thermostat is located on the left end of the cabinet near the top of the unit compartment. The dial is marked *off*, normal, and cold. To stop the compressor during a normal running cycle, pull the service cord from the electrical outlet or turn the thermostat to the *off* position.

To replace the thermostat, first disconnect the power cord from the electrical outlet, and then remove the knob. Remove the thermostat mounting screws. Pull the

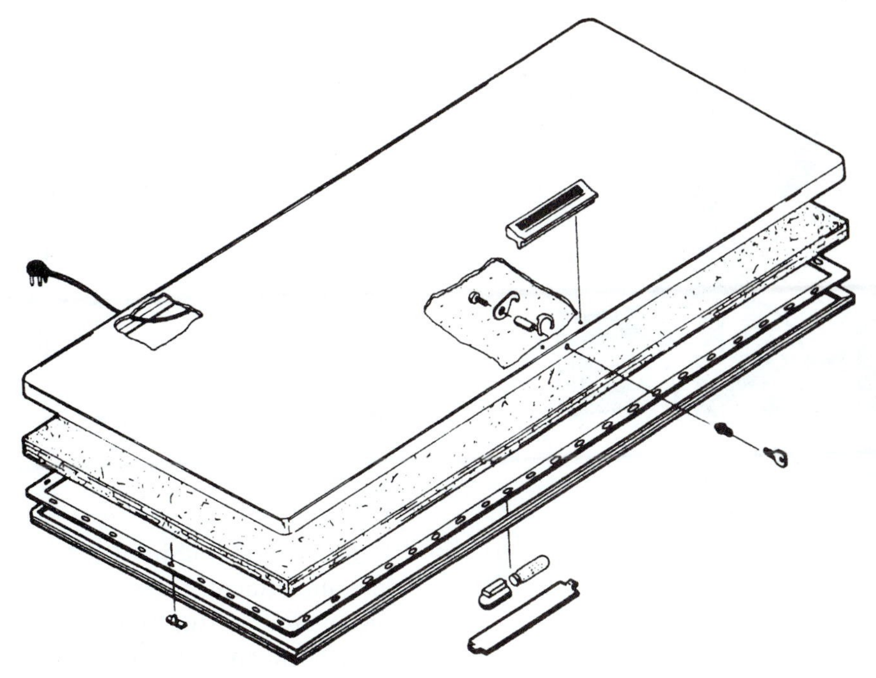

Fig. 13-12 *Freezer-lid construction for chest-type freezer.* (Kelvinator)

thermostat into view in the machine-compartment opening. Disconnect the wire leads from the thermostat terminals.

Remove the mastic sealer from around the thermal element where it enters the thermal-well trough in the cabinet outer wrapper in the machine compartment.

Before removing the thermal element from the thermal well, wrap a small piece of tape around the thermal element next to the opening of the thermal well.

Remove the thermostat from the machine compartment. Wrap a piece of tape on the new thermostat thermal element at the same location as the tape on the inoperative thermostat. Push the thermal element into the thermal well. To insure the correct length of thermal element in the well for positive contact, the tape on the thermostat thermal element should be at the entrance of the well. Replace the mastic sealer. Connect the wire leads. Install the thermostat mounting screws and knob.

Figures 13-13, 13-14, and 13-15 show three types of thermostats. The thermostats are set at the factory in accordance with the manufacturer's specifications for cut-in and cutout. No adjustment should be made unless it is absolutely proven that the thermostats are not in accordance with specifications.

If a higher or lower range than is obtainable by the selector knob is desired, adjust the range (altitude)-adjustment screw.

On GE thermostats, the range-adjustment screw is reached through the small hole in the face of the thermostat. See Fig. 13-13. Turn the screw to the left to lower the cutout and cut-in. Turn the screw to the right to raise the cutout and cut-in temperatures. Turning the range screw to the right makes altitude adjustments.

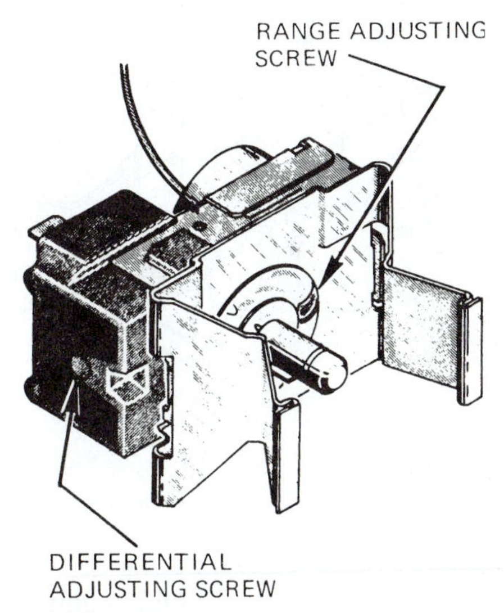

Fig. 13-13 *GE thermostat.* (Kelvinator)

The range-adjustment screw on the Ranco thermostats is located behind a removable cover. See Fig. 13-13. Turn the screw to the left to lower the cutout and cut-in temperatures, and to the right to raise cutout and cut-in temperatures. Turning the range screw to the right makes altitude adjustments.

Cutler-Hammer thermostats have cut-in and cutout temperature adjustment screws. See Fig. 13-15. Turn

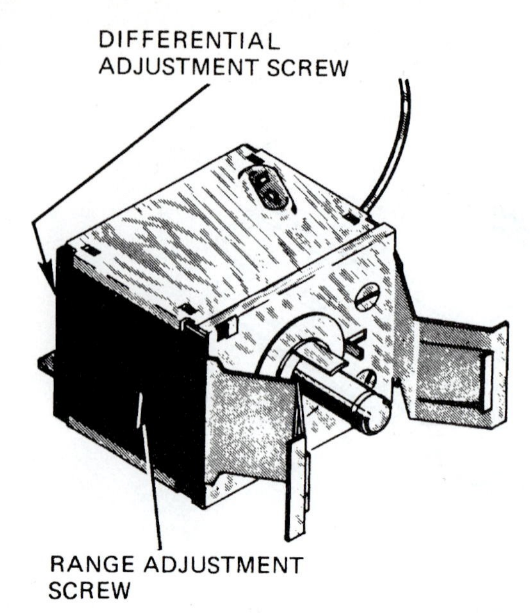

Fig. 13-14 *Ranco thermostat.* (Kelvinator)

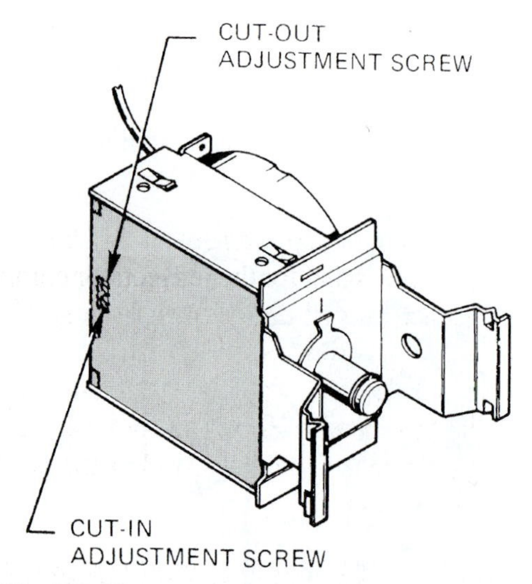

Fig. 13-15 *Cutler-Hammer thermostat.* (Kelvinator)

the screws to the left to raise the cutout and cut-in temperatures and to the right to lower the cut-in and cutout temperatures.

Both cutout and cut-in screws must be adjusted counterclockwise to compensate for altitudes above 1000 ft.

Drain System

Manual defrost models have a defrost water drain and tube assembly for draining defrost water into a shallow pan. The drain tube is located behind the removable front grille. See Fig. 13-16.

Chest models must be defrosted manually. A drain and tube assembly is located in the bottom left-hand corner of the storage compartment. See Fig. 13-17.

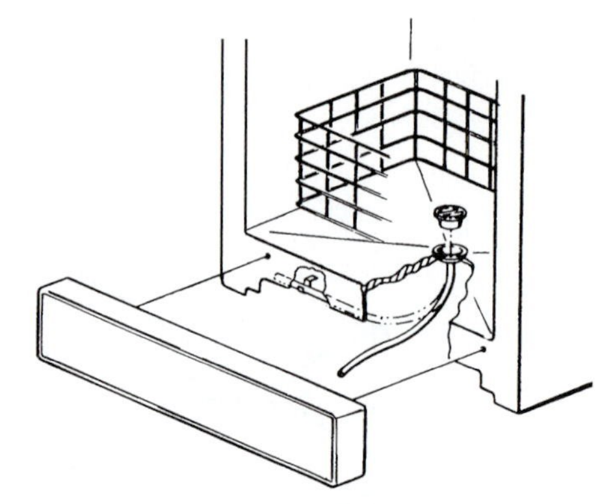

Fig. 13-16 *Drain system for manual-defrost model upright freezer.* (Kelvinator)

Remove the drain plug from the inside bottom of the compartment. Place a shallow pan under the drain tube in front of the freezer and remove the cap. An alternate method is to insert a male garden hose adapter into the drain tube and attach a garden hose. Remove the hose and adapter when defrosting is completed. Replace the drain plug and cap.

Wrapper Condenser

All compressors have internal spring suspension with four point external mounting. The compressors have plug-in magnetic starting relays. These mount directly over the compressor Fusite terminal assembly and a separate motor overload protector.

The wrapper condenser incorporates a precooler condenser in series with the main condenser. A wrapper condenser depends on the natural convection of the room air for dissipation of heat.

The high-temperature, high-pressure discharge refrigerant vapor is pumped into the precooler condenser. This is located on the back wall of the freezer where it releases part of its latent heat of vaporization and sensible heat of compression. From the precooler condenser, the refrigerant passes back to the machine compartment and through the cooler coil in the compressor dome (where additional heat is picked up from the oil). It then passes back to the main condenser, where additional heat is released to the atmosphere. This results in condensation of the refrigerant from the high-pressure vapor to the high-pressure liquid.

Ample condenser area is provided to keep the surface temperature of the cabinet only 10 to 15°F (5.5 to 8.3°C) above room temperatures. Heat released by the condenser helps reduce the possibility of moisture condensation on the cabinet surface in humid areas. The wrapper-type condenser eliminates service calls caused by plugged or dirty condensers. For the life of the

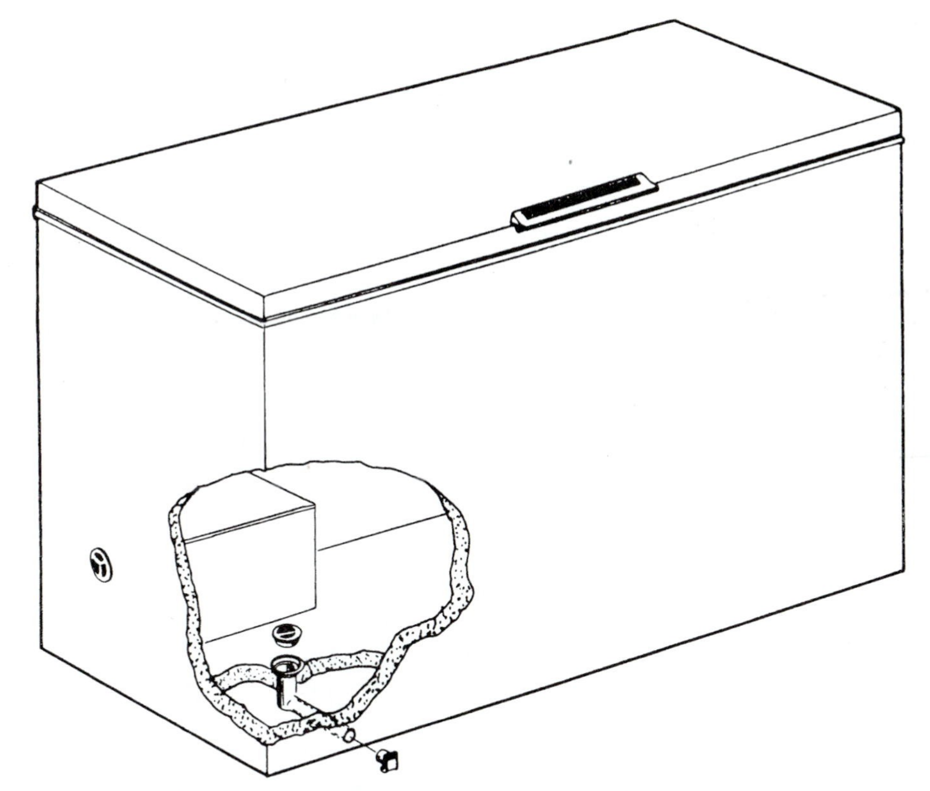

Fig. 13-17 *Drain system for a chest-model freezer.* (Kelvinator)

freezer, the wrapper condenser remains efficient. A filter drier is located in the liquid line at the outlet of the condenser.

Evaporator Coil

Liquid refrigerant flows through the capillary tube and into the evaporator coil where expansion and evaporation of the refrigerant takes place. The evaporator coil (low-side) is a pattern of zigzag passes of tubing. The evaporator coil is designed to produce adequate refrigeration and maintain uniform storage temperatures throughout the cabinet. See Figs. 13-18 and 13-19.

REPLACING THE COMPRESSOR

With a relatively small amount of refrigerant used in the freezer, the oil in the compressor will absorb a major portion of it when the freezer has been inoperative for a considerable length of time. When opening the system, use care to prevent the oil from blowing out with the refrigerant.

When replacing a compressor on a freezer that is in operation, disconnect the service cord from the electrical outlet. Allow the low side to warm up to room temperature before removing the compressor. Placing an electric lamp inside the cabinet will help raise the temperature.

The procedure for replacement of the freezer compressor is the same as that for the refrigerator compressor. Check the earlier part of the chapter for a step-by-step procedure.

The procedures for replacing the filter drier and heat exchanger, cleaning the capillary tube, replacing the evaporator, grid (or shelf), as well as the condenser are discussed in Chap. 13.

REPAIRING THE CONDENSER

When a refrigerant leak is found in any portion of the internal wrapper-type condenser (including the precooler) on models with removable liners, the condenser must be repaired. See Fig. 13-20. On the foam-insulated models, an external natural-draft condenser is installed on the rear of the freezer.

INSTALLING THE DRIER COIL

Install the drier coil in the U-channel as shown in Fig. 13-21. Secure the drier coil to the outer shell with short strips of mastic sealer spaced approximately 6 in. apart.

Thread the drier-coil wire leads down through the wiring harness grommet in the lower right-hand front corner of the freezer U-channel. Route the wire lead to the rear of the machine compartment and splice into the service cord.

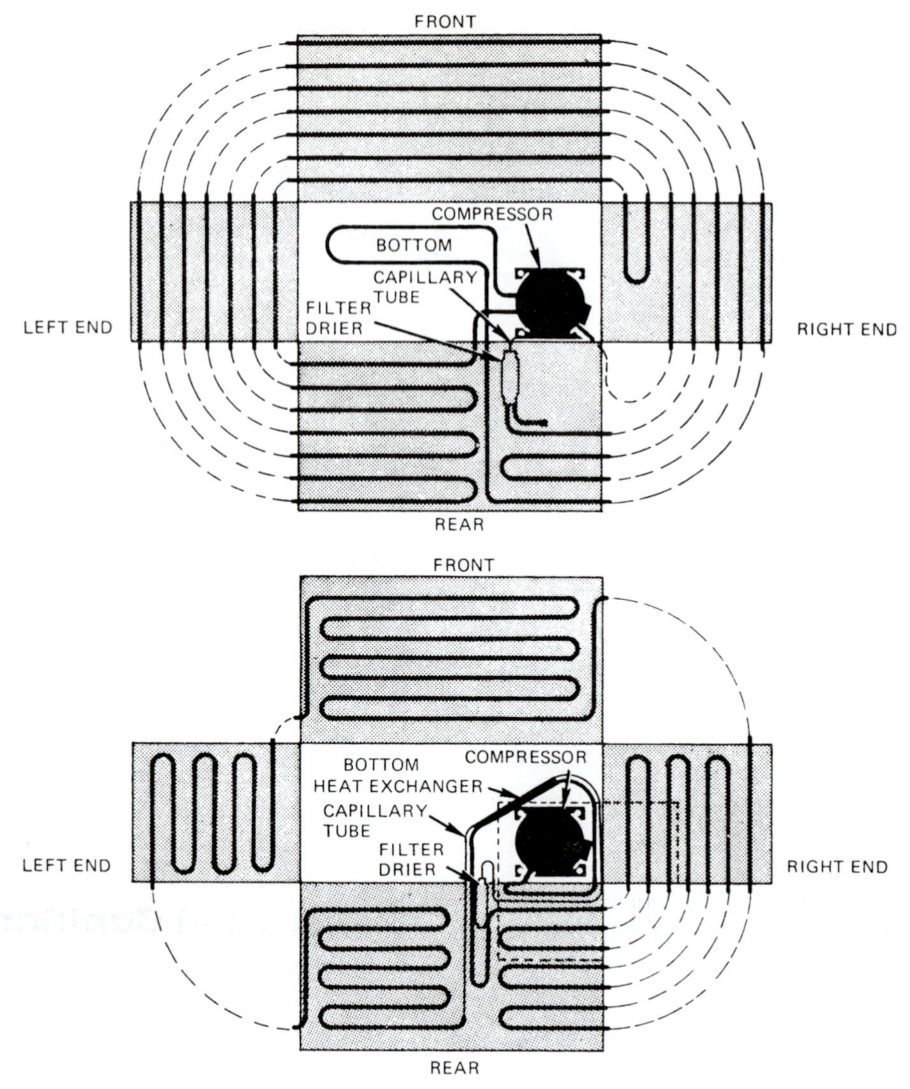

Fig. 13-18 *Refrigerant systems for chest-model freezer.* (Kelvinator)

Install a length of mastic sealer, supplied with the drier coil, on the lower front flange and up to 2 in. on each side of the freezer liner. Install the bottom cold-ban trim and press the rear edge firmly into the mastic sealer. Then, install the remaining cold-ban trim section.

All chest models have the wrapper-type condenser attached to the inner surface of the cabinet's outer shell. The condenser, encased with foam insulation, is not accessible for repair.

When there is evidence of an internal refrigerant leak, the evaporator (low side) and condenser (high side) should be disconnected from the system and individually pressurized and leak tested.

> CAUTION: Do not disturb or puncture the cellular formation of foam insulation when testing for leaks. R-11 refrigerant entrapped in the cellular formation will be released, indicating a refrigerant leak. (In older models it may be of interest and concern that R-11 was used to make the foam insulation. It is trapped inside the insulation and will be released when disturbed.)

A refrigerant leak found in any part of the internal wrapper-type condenser (including the precooler of the freezer) can be repaired by installing an external natural draft-type condenser on the rear of the freezer. Remember that the condenser outlet line must be positioned so that it has a gradual slope downward from the point it leaves the freezer to where it enters the drier. The filter drier must be vertical or at a 30° to 45° angle, with the outlet end down so that a liquid refrigerant seal is maintained.

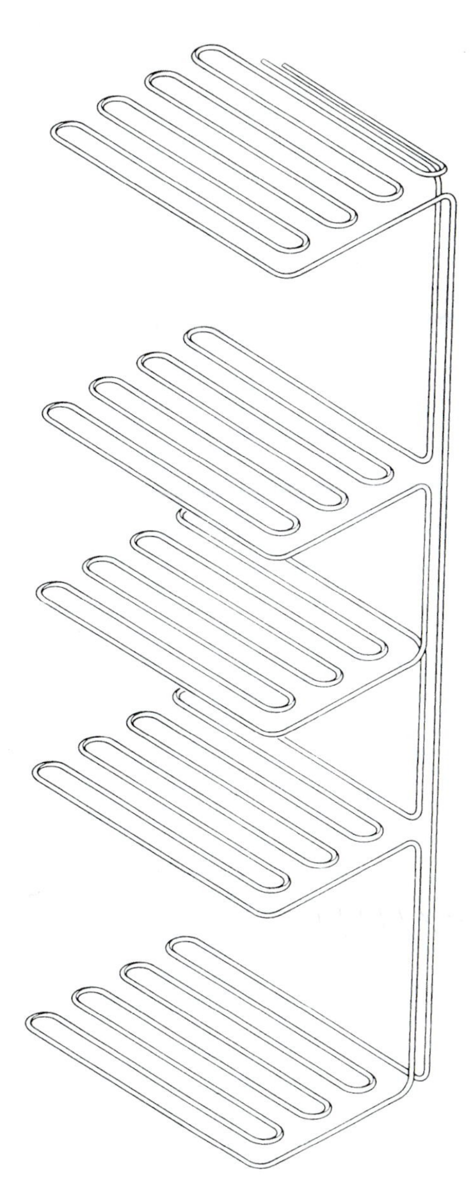

Fig. 13-19 *Evaporator for a manual defrost upright model.* (Kelvinator)

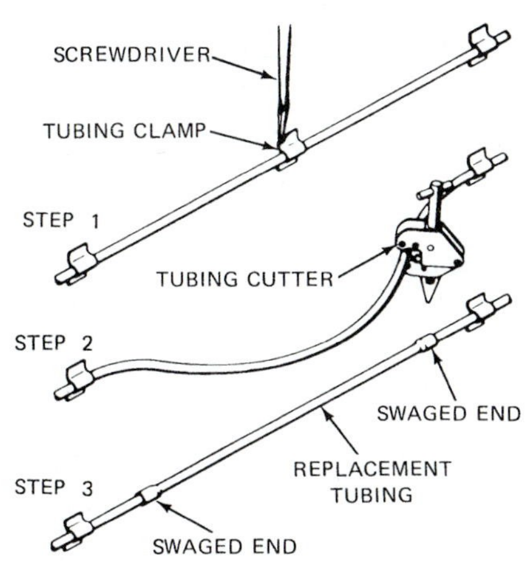

Fig. 13-20 *Repairing a condenser on the upright model. Step 1: Loosen the tubing within problems area. Step 2: Use tubing cutter to remove problem area. Cut in two places. Step 3: Swage the new pipe ends so they fit over the existing tubing.* (Kelvinator)

COMPLETE RECHARGE OF REFRIGERANT

In the case of a major refrigerant leak that is repairable, such as a broken or cracked refrigerant line in the machine compartment, the unit will run. However, there may be no refrigeration or only partial refrigeration. Suction pressure will drop below the atmospheric pressure. Thus, with a leak on the low side, air and moisture are drawn into the system, saturating the filter drier. If there is reason to believe that the system contains an appreciable amount of moisture, the compressor and drier filter should be replaced. The system should be cleaned with liquid refrigerant and evacuated, after the leak has been repaired.

OVERCHARGE OF REFRIGERANT

When the cabinet is pulled down to temperature, an indication of an overcharge is that the suction line will be cooler than normal. The normal temperature of the suction line will be a few degrees cooler than room temperature. If its temperature is much lower than room temperature, the unit will run longer because the liquid is pulled beyond the accumulator into the heat exchanger. When the overcharge is excessive, the suction line will sweat or frost.

Restricted Capillary Tube

The capillary tube is restricted when the flow of liquid refrigerant through the tube is completely or partially interrupted. Symptoms are similar to those of a system that has lost its refrigerant. However, the major part of the refrigerant charge will be pumped into the high side (condenser), the same as with a moisture restriction. The suction pressure will range slightly below normal to very low (2 to 20 in. vacuum), depending on the amount of restriction.

TESTING FOR REFRIGERANT LEAKS

If the system is diagnosed as short of refrigerant and has not been recently opened, there is probably a leak in the system. Adding refrigerant without first locating and repairing the leak, or replacing the faulty component, would not permanently correct the difficulty. The leak must *be found.*

Sufficient refrigerant may have escaped to make it impossible to leak test effectively. In such cases, add a $^1/_4$-in. line-piercing valve to the compressor process tube. Add sufficient refrigerant to increase the pressure to 75 lb/in.^{2}. Through this procedure, minute leaks are more easily detected before the refrigerant is discharged from the system and contaminates the surrounding air.

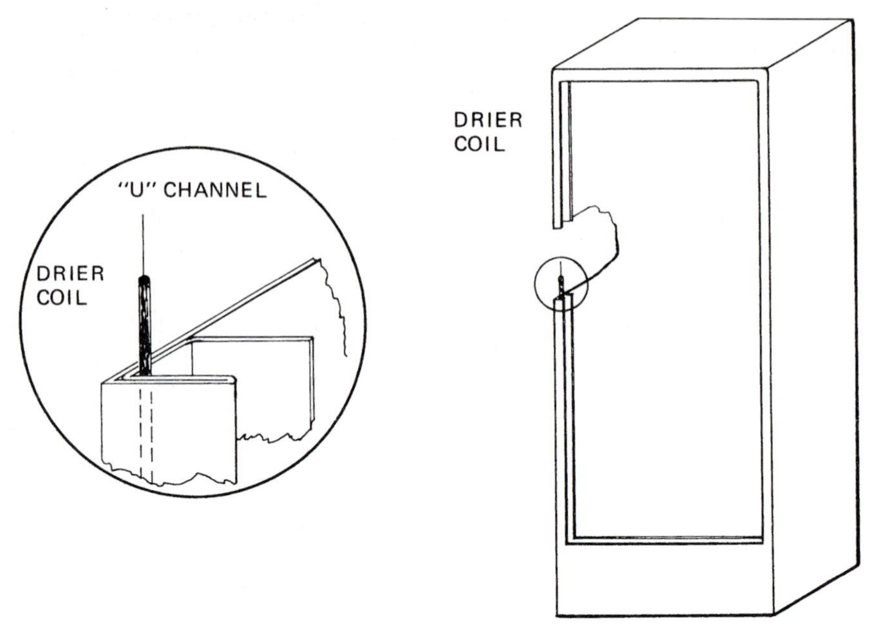

Fig. 13-21 *External condenser on an upright-model freezer.* (Kelvinator)

The line-piercing valve should be used for adding refrigerant and for test purposes only. It must be removed from the system after it has served its purpose. Braze a 4 in. piece of in. OD copper tube into the compressor process tube. Evacuate the system and recharge after repairs are completed.

TROUBLESHOOTING FREEZERS

Table 13-1 lists troubleshooting procedures for upright freezer models. Table 13-2 lists troubleshooting procedures for chest-type freezer models.

Portable Freezers

Many uses can be found for portable freezers. For example, much of the ice cream sold during the summer months is sold from refrigerated trucks. It is essential that these trucks have reliable portable freezing units. See Figs. 13-22 and 13-23.

The compressor is run by 110/120-V, 60-Hz AC. It runs during the night when the truck is out of service. A small fan circulates the cold air. The fan runs on the truck battery during the day. At night, it is plugged into line current (120-V, AC).

This unit uses a hermetically sealed compressor designed for use with R-22. About 10 or 12 oz of R-22 are used for a full charge.

Troubleshooting If the machine runs short of refrigerant, it should be allowed to warm to room temperature and checked for leaks with a halide torch. For finding small leaks, at least 90 lb of internal pressure are needed. It maybe necessary to add refrigerant to obtain this pressure. If so, connect the suction-line service opening to a drum of refrigerant (probably R-22), making sure the drum remains upright so that only gas will enter the unit.

Never connect in this manner a drum that is warmer than any part of the system. The gas will condense in the system, resulting in overcharge and waste of refrigerant.

If the unit is charged due to leaks, or any major repairs are made on the system, it is recommended that a new drier be installed. When replacing the original drier, be certain that the replacement drier has a good filter and strainer incorporated with the drying agent.

If a gas leak has allowed air to enter the system, the system must be evacuated or thoroughly cleaned with R-22. A new drier must be installed before charging. Air remaining in the system cannot be purged off. It permeates the complete system and is not trapped in the high side as in other systems using a liquid receiver.

Allow pressure to build to approximately 100 lb in the unit and shut off the charging valve immediately. After the unit is started, add refrigerant slowly until backpressure is between 10 and 16 lb, depending on the ambient temperature. (A high-ambient temperature will produce a higher head and back pressure.) The back pressure will then remain about the same until the eutectic (contents of the freezer) is completely frozen.

Table 13-1 *Troubleshooting Freezers: Upright Models*

Trouble	Probable Cause	Remedy or Repair
Product too cold.	Temperature selector knob set too cold.	Set warmer.
	Thermostat bulb contact bad.	If the bulb contact is bad, the bulb temperature will lag behind the cooling coil temperature. The unit will run longer and make the freezer too cold. See that the bulb makes good contact wih the bulb well.
	Thermostat is out of adjustment.	Readjust or change the thermostat.
Product too warm.	Thermostat selector knob set too warm.	Set cooler.
	Thermostat contact points dirty or burned.	Replace thermostat.
	Thermostat out of adjustment.	Readjust or change the thermostat.
	Loose electrical connection.	This may break the circuit periodically and cause the freezer to be come warm because of irregular or erratic operation. Check the circuit and repair or replace parts.
	Excessive service load or abnormally high room temperature.	Unload part of the contents. Move unit to a room with lower temperature or exhaust excess room heat.
	Restricted air circulation over wrapped condenser.	Allow 6 in. clearance above the top and $3^1/_2$ in. clearance at the sides and between the back of the cabinet and the wall.
	Excessive frost accumulation on the refrigerated shelves (manual defrost models).	Remove the frost.
	Compressor cycling on overload protector.	Check the protector and line voltage at the compressor.
Unit will not operate.	Service cord out of wall receptacle.	Plug in the service cord.
	Blown fuse in the feed circuit.	Check the wall receptacle with a test lamp for a live circuit. If the receptacle is dead, but the building has current, replace the fuse. Determine the cause of the overload or short circuit.
	Bad service cord plug, loose connection or broken wire.	If the wall receptacle is live, check the circuit and make necessary repairs.
	Inoperative thermostat.	Power element may have lost charge or points may be dirty. Check the points. Short out the thermostat. Repair or replace the thermostat.
	Inoperative relay.	Replace.
	Stuck or burned-out compressor.	Replace the compressor.
	Low voltage. Cycling on overload.	Call utility company, asking them to increase voltage to the house. Or, move unit to a separate household circuit.
	Inoperative overload protector.	Replace.
Unit runs all the time.	Thermostat out of adjustment.	Readjust or change the thermostat.
	Short refrigerant charge (up to 4 oz.). Cabinet temperatures abnormally low in lower section.	Not enough refrigerant to flood the evaporator coil at the outlet to cause the thermostat to cut-out. Recharge and test for leaks.
	Restricted air flow over the wrapper condenser.	Provide proper clearances around the cabinet.
	Inefficient compressor.	Replace.
Unit short cycles.	Thermostat erratic or out of adjustment.	Readjust or change the thermostat.
	Cycling on the relay.	This may be caused by low or high line voltage that varies more than 10% from the 115 V. It may also be caused by high-discharge pressures caused by air or noncondensable gases in the system. Correct either condition.
Unit runs too much.	Abnormal use of the cabinet.	Heavy usage requires more operation. Check the usage and correct or explain.
	Shortage of refrigerant.	Unit will run longer to remove the necessary amount of heat and it will operate at a lower than normal suction pressure. Put in the normal charge and check for leaks.
	Overcharge of refrigerant.	Excessively cold or frosted suction line results in lost refriger- ation effort. Unit must run longer to compensate for the loss. Purge off excessive charge.
	Restricted airflow over the condenser.	This can result if the cabinet is enclosed. This will obstruct the air flow around the cabinet shell. Restricted airflow can also be caused by air or noncondensable gases in the system. This results in a higher head pressure. The higher head pressure produces more reexpansion during the suction stroke of the compressor. Consequently, less suction vapor is taken. Increased running time must compensate for loss of efficiency. Correct the condition.
	High room or ambient temperature.	Any increase in temperature around the cabinet will increase the refrigeration load. This will result in longer running time to maintain cabinet temperature.
Too much frost on refrigerated surfaces—lowside.	Abnormally heavy usage in humid weather.	Do not leave the freezer door open any longer than neces- sary to load or remove products.
	Poor door-gasket seal.	This permits the entrance of moisture by migration, which freezes out of the air as frost on the refrigerated surfaces.

Table 13-2 *Troubleshooting Freezers: Chest Models*

Trouble	Probable cause	Remedy or repair
	Inoperative relay.	Replace relay.
	Stuck or burned-out compressor.	Replace compressor.
	Low voltage. Cycling on overlead.	Call utility company, asking them to increase voltage. Or, change unit to a different circuit in house.
	Inoperative overload protector.	Replace.
Unit runs all the time.	Thermostat out of adjustment.	Readjust or change the thermostat.
	Short refrigerant charge (up to 4 oz).	Cabinet temperatures abnormally low in the lower section. Not enough refrigerant to flood the evaporator coil at the outlet to cause the thermostat to cut-out. Recharge and test for leaks.
	Restricted airflow over the wrapper condenser.	Provide proper clearances around the cabinet.
	Inefficient compressor.	Replace compressor.
Unit short cycles.	Thermostat erratic or out of adjustment.	Readjust or change.
	Cycling on the relay.	This may be caused by low-or high-line voltage that varies more than 10% from 115 V. It may also be caused by high-discharge pressures caused by air or noncondensable gases in the system. Correct either condition.
Unit runs too much.	Abnormal use of the cabinet.	Heavy usage requires more operation. Check the usage and correct or explain.
	Shortage of refrigerant.	Unit will run longer to remove the necessary amount of heat and it will operate at a lower than normal suction pressure. Put in the normal charge and check for leaks.
	Overcharge of refrigerant.	Excessively cold or frosted suction line results in lost refrigerant effort. Unit must run longer to compensate for the loss. Purge off the excess charge.
	Restricted airflow over the condenser.	This can be the result of enclosing the cabinet. This will cause obstruction to the air flow around the cabinet shell. Inefficient compression can be caused by air or noncondensable gases in the system. This results in a higher head pressure. The higher head pressure produces more reexpansion during the suction stroke of the compressor and, consequently, less suction vapor is taken. Increased running time must compensate for the loss of efficiency. Correct the condition.
	High-room or-ambient temperature.	Any increase in temperaure around the cabinet location will increase the refrigeration load. This will result in longer running time to maintain the cabinet temperature.
Too much frost on refrigerated surfaces—lowside.	Abnormally heavy usage in humid weather.	Do not leave the freezer lid open any longer than necessary to load and unload.
	Poor gasket seal.	This permits the entrance of moisture by migration, which freezes out the air as frost on the refrigerated surfaces.

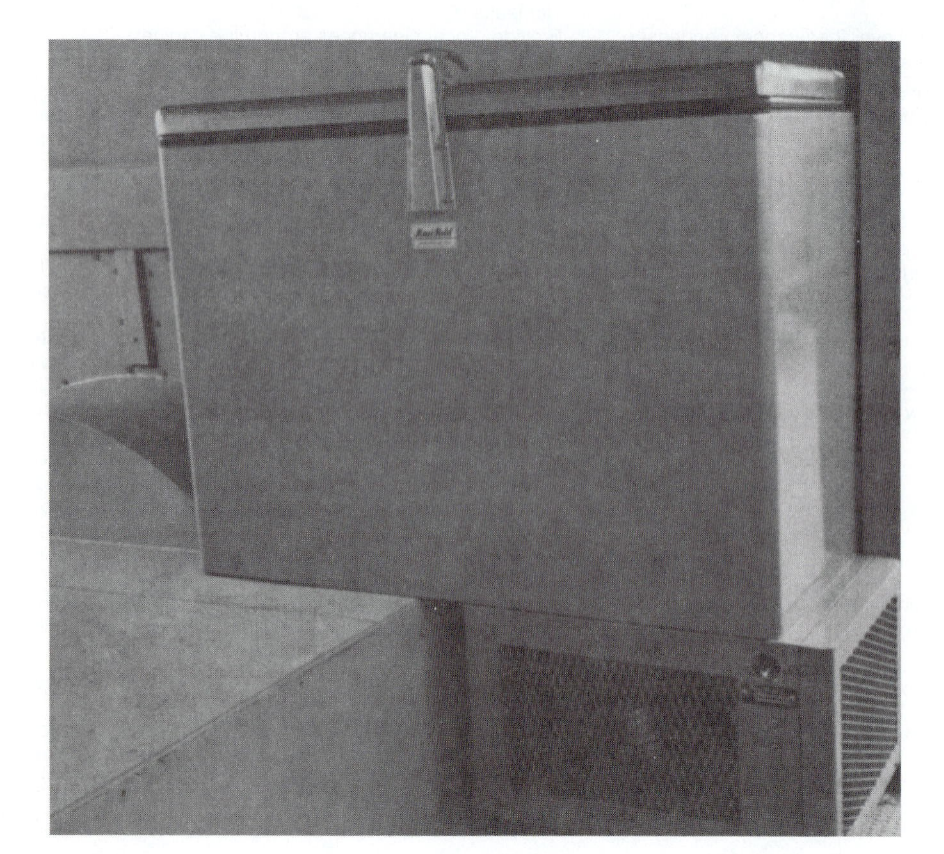

Fig. 13-22 *Portable freezer used in a truck to transport ice cream and milk.* (Kari-Kold)

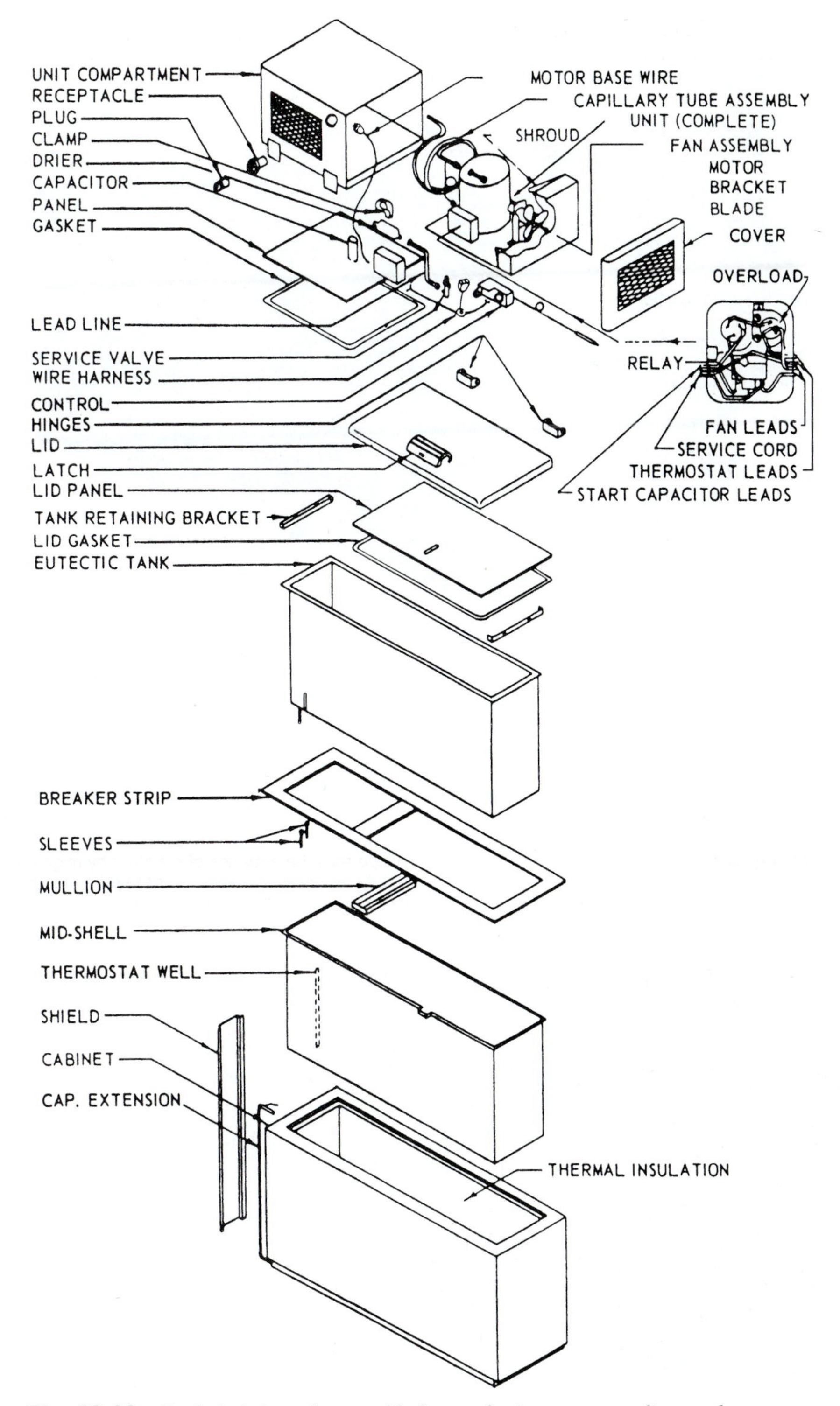

Fig. 13-23 *Exploded view of a portable freezer for ice-cream vending on the street or the beach.* (Kari-Kold)

The charge should be checked again when the cabinet is around −15°F (−26°C) or colder. Then, with the condensing unit running, the suction line should frost out of the cabinet about 6 to 8 in. The desired frost line can be obtained by adding or purging of refrigerant (make sure the purged refrigerant is captured and reclaimed) a little at a time, allowing time for the system to equalize. If the compressor will not start, but the condenser fan is

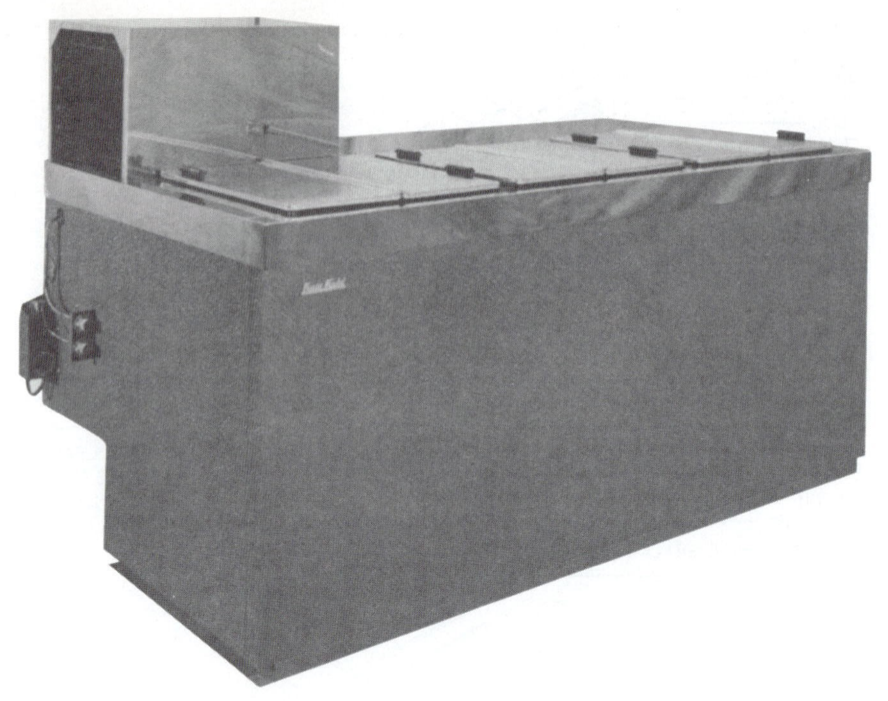

Fig. 13-24 *Ice-cream dispenser capable of being mounted on a truck and used to transport frozen goods.* (Kari-Kold)

running, check the head and back pressures. If the pressures are not equal, a capillary tube maybe clogged with moisture or foreign material. Heating the end of the capillary tube where it enters the cabinet will usually begin to equalize pressures if the restriction is due to moisture freezing. Evacuate the system, install a new drier, and recharge the system as described already. If the capillary tube is clogged with material other than frozen moisture, it should be replaced.

When the compressor does not start and the head and back pressures are approximately equal, check for trouble as follows.

Check the line voltage by holding the voltmeter leads on contacts of the motor base plug. Take a reading when the overload protector clicks in and the compressor is trying to start. This reading should be 100 V or more. If less, the trouble is probably in the supply line.

- Replace the capacitor, if the unit has one.
- Replace the relay and/or overload.

If, after these checks, the compressor will not start, the unit should be returned to the manufacturer.

Figure 13-24 shows an ice-cream vending unit. Most of the mechanical parts are located on top of the unit to prevent damage when the unit is handled frequently. This type of freezer, in various sizes, can be mounted in a variety of vehicles. The cabinet provides for economical operation that can pay for itself in dry ice savings alone. The unit is plugged in at night. The smaller units rely upon insulation to hold the cold air. Other units plug into the vehicle's battery.

Small trucks can be fitted with portable freezers. These are useful to dairies servicing school cafeterias and other large food-dispensing operations where milk must be kept cool and ready for servicing at a specific time. Some units can handle between 400 and 700 bottles or cartons of milk. They can be rolled into a cafeteria line.

REVIEW QUESTIONS

1. What are the two types of domestic freezers?
2. What is a wrapped condenser?
3. How do you remove the hinges on domestic freezers?
4. In what way do freezer and refrigerator thermostats differ?
5. How is the heat dissipated with a wrapped condenser?
6. What type of condenser is used when repairing a foam-insulated freezer?
7. Why must you be careful not to puncture the cellular formation of foam insulation when testing for leaks in a freezer?

8. When does the suction pressure drop below the atmospheric pressure in a defective freezer?
9. What indicates an overcharge of refrigerant in a freezer?
10. What symptoms are observed when the capillary tube is restricted and partially interrupts the flow of refrigerant?
11. How much internal pressure is needed in a freezer system to aid in locating small leaks?
12. When the compressor does not start and the head and back pressures are approximately equal, what should be checked first?

14
CHAPTER

Temperature, Psychrometrics, and Air Control

PERFORMANCE OBJECTIVES

After studying this chapter you should:

1. Know how to convert Fahrenheit to Celsius degrees.
2. Know the importance of the psychrometric chart and be able to read it.
3. Know how to obtain relative humidity.
4. Know how to use the manometer.
5. Know the difference between convection, conduction, and radiation.
6. Know how to place registers to control airflow and comfort in a room.
7. Know how to design a perimeter system of air control.
8. Know how to control air noise.
9. Understand room air motion.
10. Know how and when to install linear grilles.

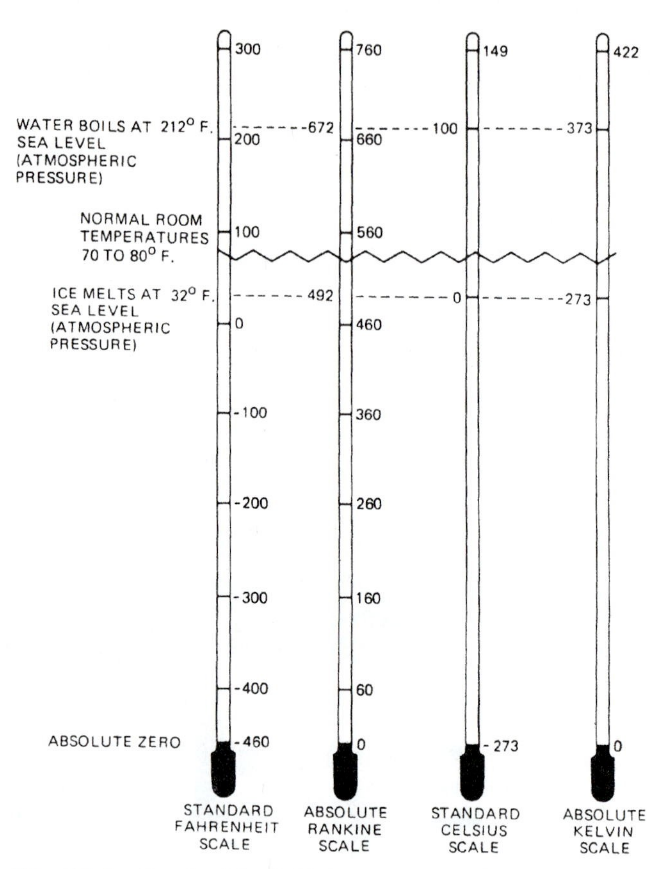

Fig. 14-1 *Standard dry-bulb thermometer scales.* (Johnson Controls)

TEMPERATURE

Temperature is defined as the thermal state of matter. Matter receives or gives up heat as it is contacted by another object. If no heat flows upon contact, there is no difference in temperature. Figure 14-1 shows the different types of dry-bulb thermometers. The centigrade scale is now referred to as degrees Celsius (°C). (Centi is metric for 100.) The Celsius scale is divided into 100 degrees, from the freezing point of water to the boiling point.

Degrees Fahrenheit

American industry and commerce still use the Fahrenheit scale for temperature measurement. However, the metric scale (degrees Celsius, °C) is becoming rapidly accepted. The Fahrenheit scale divides into 180 parts the temperature range from the freezing point of water to its boiling point. The Fahrenheit temperature scale measures water at its freezing point of 32° and its boiling point of 212°. The pressure reference is sea level, or 14.7 lb/in.^{2}.

Degrees Celsius

In laboratory work and in the metric system, the temperature is measured in degrees Celsius. It ranges from the freezing point of water (0°C) to its boiling point (100°C). Again the pressure reference is sea level.

Absolute Temperature

Absolute temperatures are measured from absolute zero. This is the point at which there is no heat. On the Fahrenheit scale, absolute zero is –460°. Temperatures on the absolute Fahrenheit scale (Rankine) can be found by adding 460° to the thermometer reading. On the Celsius scale, absolute zero is –273°. Any temperature on the absolute Celsius scale (Kelvin) can be found by adding 273° to the thermometer reading. See Fig. 14-1.

Absolute zero temperature is the base point for calculations of heat. For example, if air or steam is kept in a closed vessel, the air or steam pressure will change roughly in direct proportion to its absolute temperature. Thus, if 0°F air (460° absolute) is heated to 77°F (537° absolute), without increasing the volume, the pressure will increase to roughly 537/460 times the original pressure. A more formal statement of the important physical law involved states:

> At a constant temperature, as the absolute temperature of a perfect gas varies, its absolute pressure will vary directly. Or, at a constant pressure, as the absolute temperature of a perfect gas varies, the volume of the gas will vary directly.

This statement is known as the *perfect gas law*. It can be expressed mathematically by the following equation.

$$PV = TR$$

where P = absolute pressure
V = volume
T = absolute temperature
R = a constant, depending on the units selected

CONVERTING TEMPERATURES

It is sometimes necessary to convert from one temperature scale to another. In converting from the Fahrenheit to the Celsius scale, $5/9$°F is equal to 1°C. Or, 1°C is equal to 1°F. Equations facilitate converting from one scale to the other.

EXAMPLE:

Convert 77°F to °C. Use the following formula: °C = $5/9$ (°F–32)

$$\frac{5}{9}(77-32)$$

$$\frac{5}{9}(45)$$

$$\frac{5(45)}{9}$$

$$\frac{225}{9} = 25$$

$$77°F = 25°C$$

EXAMPLE:

Convert 25°C to °F. Use the following formula: °F = $9/5$ (°C) + 32

$$\frac{9}{5}(25) + 32$$

$$\frac{9(25)}{5} + 32$$

$$\frac{225}{5} = 45 + 32 = 77$$

$$25°C = 77°F$$

Temperature conversion tables are available. Using them, it is easy to convert temperatures from one temperature scale to another.

A calculator can be used for the above temperature conversions. If a calculator is used, the number 0.55555555 can be substituted for $5/9$. The number 1.8 can be substituted for $9/5$.

PSYCHROMETRICS

Psychrometry is the science and practice of air mixtures and their control. The science deals mainly with dry air, water vapor mixtures, with the specific heat of dry air and its volume. It also deals with the heat of water, heat of vaporization or condensation, and the specific heat of steam in reference to moisture mixed with dry air. Psychrometry is a specialized area of thermodynamics.

PRESSURES

All devices that measure pressure must be exposed to two pressures. The measurement is always the *difference* between two pressures, such as gage pressure and atmospheric pressure.

Gage Pressure

On an ordinary pressure gage, one side of the measuring element is exposed to the medium under pressure. The other side of the measuring element is exposed to the atmosphere. Atmospheric pressure varies with altitude and climatic conditions.

Thus, it is obvious that gage pressure readings will not represent a precise, definite value unless the atmospheric pressure is known. Gage pressures are usually designated as psig (pounds per square inch gage). Pressure values that include atmospheric pressure are designated psia (pounds per square inch absolute).

Atmospheric Pressure

A barometer is used to measure atmospheric pressure. A simple mercury barometer may be made with a glass tube slightly more than 30 in. in length. It should be sealed at one end and filled with mercury. The open end should be inverted in a container of mercury. See Fig. 14-2. The mercury will drop in the tube until the weight of the atmosphere on the surface of the mercury in the pan just supports the weight of the mercury column in the tube. At sea level and under certain average climatic conditions, the height of mercury in the tube will be 29.92 in. The space above the mercury in the tube will be an almost perfect vacuum, except for a slight amount of mercury vapor.

A mercury *manometer* is an accurate instrument for measuring pressure. The mercury is placed in a glass U-tube. With both ends open to the atmosphere, the mercury will stand at the same level in both sides of the tube. See Fig. 14-3. A scale is usually mounted on one of the tubes with its zero point at the mercury level.

PRESSURE MEASURING DEVICES

Low pressures, such as in an air-distribution duct, are measured with a manometer using water instead of mercury in the U-tube. See Fig. 14-3. The unit of

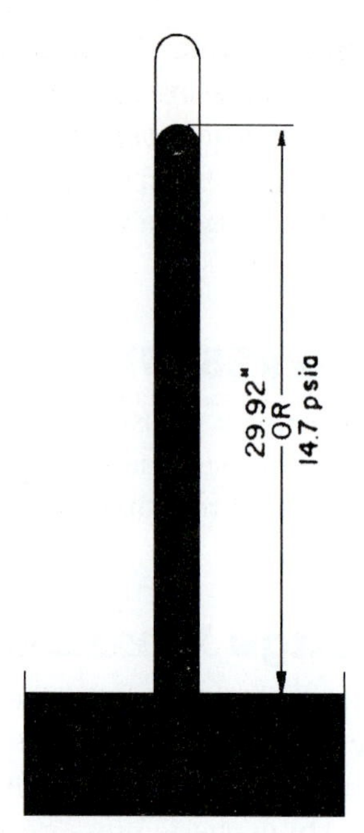

Fig. 14-2 *Simple mercury barometer.* (Johnson Controls)

measurement, 1 in. of water, is often abbreviated as 1-in. H_2O or 1-in. w.g. (water gage).

Water is much lighter than mercury. Thus, a water column is a more sensitive gage of pressure than a mercury column. Figure 14-4 shows a well-type manometer. It indicates 0 in. w.g. Figure 14-5 shows an incline manometer. It spreads a small range over a longer scale for accurate measurement of low pressures. Most manometers of this type use a red oily liquid in place of water to provide a more practical and useful instrument. Figure 14-6 shows a magnahelic mechanical manometer.

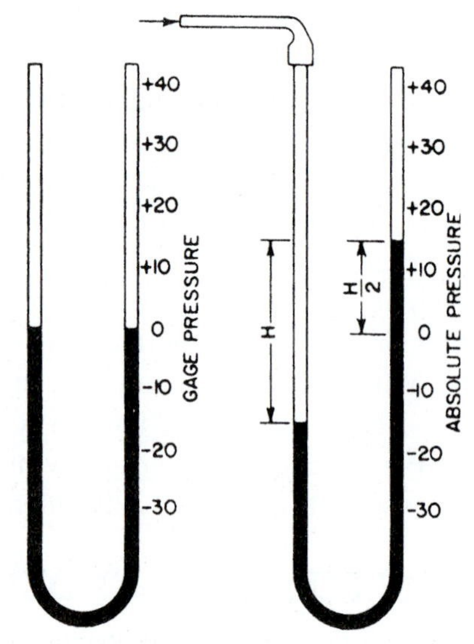

Fig. 14-3 *Mercury manometer at rest (left) and with pressure applied (right). H is the height from the top of one tube to the top of the other; H/2 is one-half of H.*

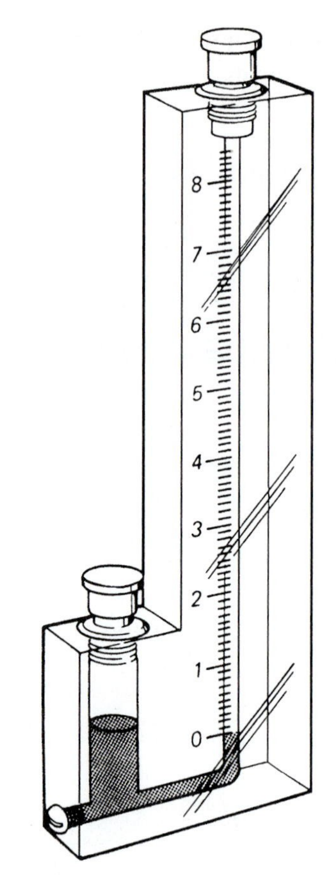

Fig. 14-4 *Well-type manometer.* (Johnson)

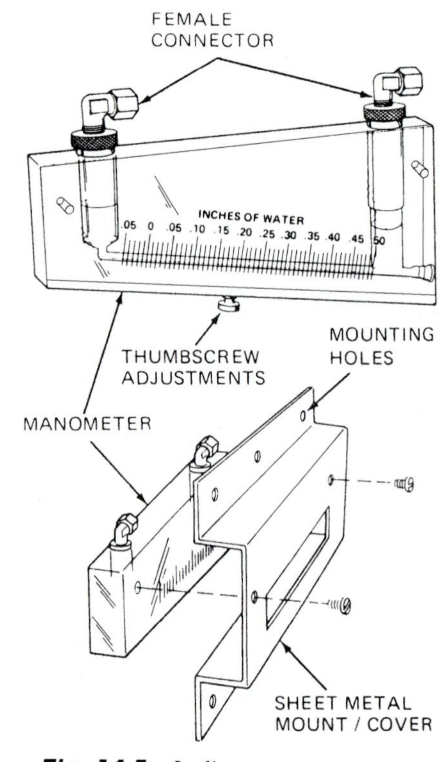

Fig. 14-5 *Incline manometer.* (Dwyer)

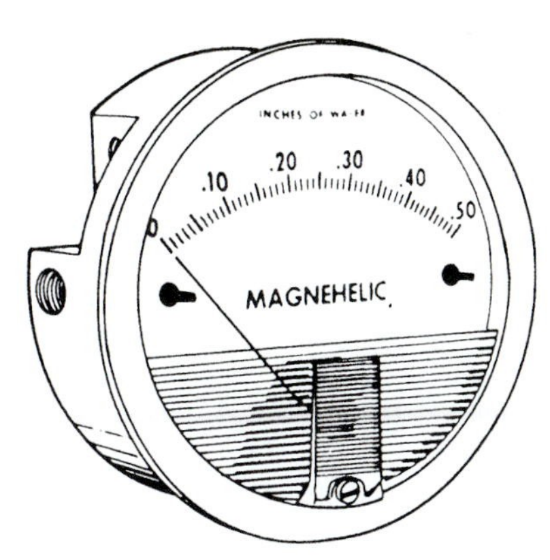

Fig. 14-6 *Magnahelic manometer.*

It is designed to eliminate the liquid in the water gage. It is calibrated in hundredths of an inch water gage for very sensitive measurement.

Figure 14-7 shows a standard air-pressure gage used for adjusting control instruments. It is calibrated in inches of water and psig. It is apparent from the scale that 1 psig is equal to 27.6-in. w.g.

Fig. 14-7 *Inches of water gage.* (Johnson)

Bourdon (spring) tube gages and metal diaphragm gages are also used for measuring pressure. These gages are satisfactory for most commercial uses. They are not as accurate as the manometer or barometer because of the mechanical methods involved. The Bourdon tube gage is discussed in Chap. 2.

Another unit of pressure measurement is the "atmosphere." Zero gage pressure is one atmosphere (14.7 lb/in.^{2}) at sea level. For rough calculations, one atmosphere can be considered 15 psig. A gage pressure of 15 psia is approximately two atmospheres. The volume of a perfect gas varies inversely with its pressure as long as the temperature remains constant. Thus, measuring pressures in atmospheres is convenient in some cases, as may be observed from the following example:

A 30-gal tank, open to the atmosphere, contains 30 gal of free air at a pressure of one atmosphere. If the tank is closed and air pumped in until the pressure equals two atmospheres, the tank will contain 60 gal of free air. The original 30 gal now occupies only one-half of the volume it originally occupied.

Conversion charts for pressures (psi to inches of Hg) may be found in various publications. The reference section of an engineering data book is a good source.

HYGROMETER

A *hygrometer* is described as an instrument (by www.wikipedia.org) used to measure the amount of moisture in the air. If a moist wick is placed over a thermometer bulb the evaporation of moisture from the wick will lower the thermometer reading (temperature). If the air surrounding a wet-bulb thermometer is dry, evaporation from the moist wick will be more rapid than if the bulb thermometer is wet. When the air is saturated no water will evaporate from the cloth wick and the temperature of the wet-bulb thermometer will be the same as the reading on the dry-bulb thermometer. However, if the air is not saturated, the water will evaporate from the wick, causing the tempeature reading to be lower. The accuracy of the wet-bulb thermometer depends on how fast air passes over the bulb. Speeds of 5000 ft/min (60 mph) are best, but it is dangerous to move a thermometer at that speed. Errors up to 15 percent can occur if the air movement is too slow or if there is too much radiant heat (sunlight for example) present. A wet-bulb temperature taken with air moving at about 1 to 2 m/sec is referred to as a screen temperature, whereas a temperature taken with air moving about 3.5 m/s or more is referred to as the sling temperature.

Properties of Air

Air is composed of nitrogen, oxygen, and small amounts of water vapor. Nitrogen makes up 77 percent, while oxygen accounts for 23 percent. Water vapor can account for 0 to 3 percent under certain conditions. Water vapor is measured in grains or, in some cases, pounds per pound of dry air. Seven thousand grains of water equal 1 lb.

Temperature determines the amount of water vapor that air can hold. Hotter temperatures mean that air has a greater capacity to hold water suspended. Water is condensed out of air as it is cooled. Outside, water condensation becomes rain. Inside, it becomes condensation on the window glass.

Thus, dry air acts somewhat like a sponge. It absorbs moisture. There are four properties of air that account for its behavior under varying conditions. These properties are dry-bulb temperature, wet-bulb temperature, dew-point temperature, and relative humidity.

Dry-Bulb Temperature Dry-bulb temperature is the air temperature that is determined by an ordinary thermometer. There are certain amounts of water vapor per pound of dry air. They can be plotted on a psychrometric chart. *Psychro* is a Greek term meaning "cold." A psychrometer is an instrument for measuring the aqueous vapor in the atmosphere. A difference between a wet-bulb thermometer and a dry-bulb thermometer is an indication of the dryness of the air. A psychrometer, then, is a hygrometer, which is a device for measuring water content in air. A psychrometric chart indicates the different values of temperature and water moisture in air.

Wet-Bulb Temperature Wet-bulb temperature reflects the cooling effect of evaporating water. A wet-bulb thermometer is the same as a dry-bulb thermometer, except that it has a wet cloth around the bulb. See Fig. 14-8A. The thermometer is swung around in the air. The temperature is read after this operation. The wet-bulb temperature is lower than the dry-bulb temperature. It is the lowest temperature that a water-wetted body will attain when exposed to an air current. The measurement is an indication of the moisture content of the air.

The Bacharach sling psychrometer, shown in Fig. 14-8B, is a compact sling type that determines the percent of relative humidity. It has a built-in side rule calculator that correlates wet- and dry-bulb temperatures to relative humidity. The dual range, high and low temperature, scales are designed for better resolution. The thermometers telescope into the handle for protection when not in use. They are available as either the red spirit-filled or the mercury-filled thermometers, and in °F or °C. There is a built-in water reservoir that holds sufficient water for several hours of testing. It is designed for portability and ease of use and ruggedness. Accuracy is within ±5 percent relative humidity. The thermometers are constructed of shock-resistant glass. The stems have deep-etched numbers and 1° scale divisions for easy reading.

The mercury-filled and spirit-filled psychrometers have a range of +25 to +120°F. They can be obtained in Celsius with a range in degrees Celsius of –5 to +50.

Dew-Point Temperature Dew-point temperature is the temperature below which moisture will condense out of air. The dew point of air is reached when the air contains all the moisture it can hold. The dry-bulb and wet-bulb temperatures are the same at this point. The air is said to be at 100 percent relative humidity when both thermometers read the same. Dew point is important when designing a humidifying system for human comfort. If the humidity is too high in a room, the moisture will condense and form on the windows.

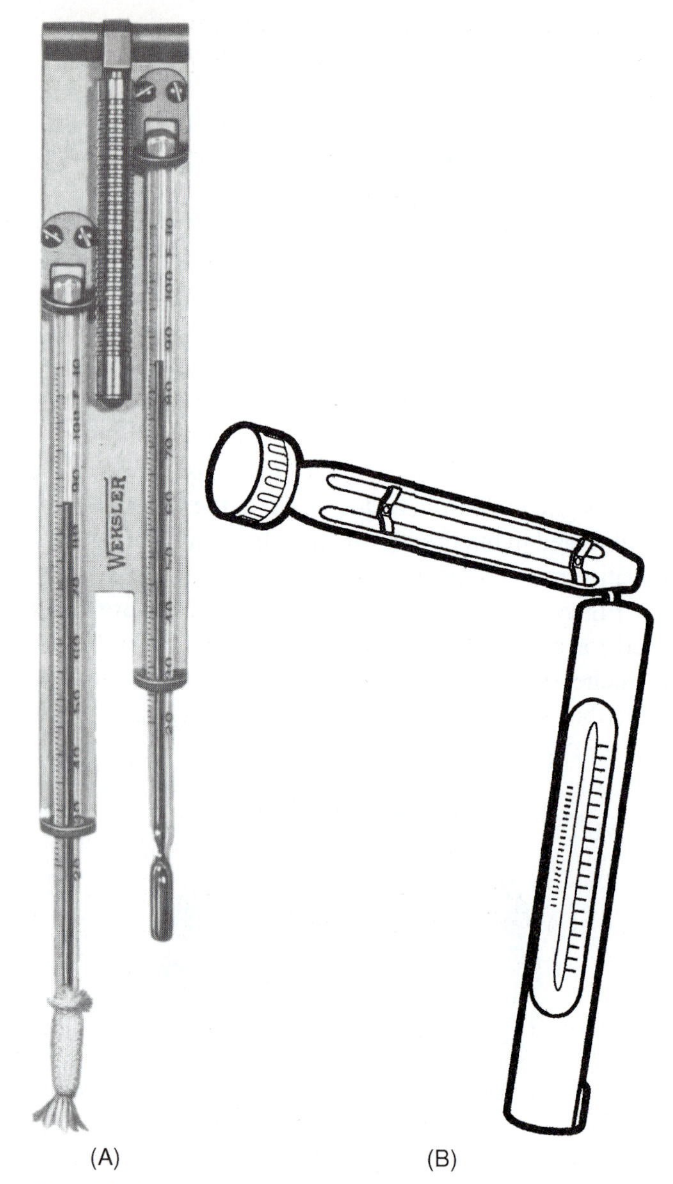

Fig. 14-8 *(A) Wet-bulb and dry-bulb thermometers mounted together. Note the knurled and ringed rod between the two at the top of the scales. It is used to hold the unit and twirl it in the air. (B) Bacharach sling psychrometer.* (Weksler)

Relative Humidity Relative humidity is a measure of how much moisture is present compared to how much moisture the air could hold at that temperature. Relative humidity (rh) is a measure of the percentage of humidity contained in the air, based on the saturation condition of the air. A reading of 70 percent means that the air contains 70 percent of the moisture it can hold. Relative humidity lines on the psychrometric chart are sweeping curves, as shown in Fig. 14-13.

To keep the home comfortable in winter it is sometimes necessary to add humidity. Hot air heat will in time remove most of the moisture in the living space. The addition of moisture is accomplished in a number of ways. Humidifiers are used to spray water into the air or large areas of water are made available to evaporate. Showers and running water also add moisture to a living space.

In summer, however, the amount of moisture per pound on the outside is greater than on the inside, especially when the room is air conditioned. This means the vapor pressure is greater on the outside than the inside. Under these conditions, moisture will enter the air-conditioned space by any available route. It will enter through cracks, around doors and windows, and through walls. In winter, the moisture moves the other way—from the inside to the outside.

The percentage of relative humidity is never more than 100 percent. When the air is not saturated, the dry-bulb temperature will always be higher than the wet-bulb temperature. The dew-point temperature will always be the lowest reading. Also, the greater the difference between the dew-point temperature and the dry-bulb temperature, the lower will be the percentage of relative humidity. The wet-bulb reading can never be higher than the dry-bulb reading. Nor can the dew-point reading be higher than the dry-bulb reading. Consider the following example:

Saturated air (100 percent humidity)

Temperature is:
Dry bulb 90°F
Wet bulb 90°F
Relative humidity 100 percent
Dew point 90°F

Unsaturated air (less than 100 percent humidity)

Temperature is:
Dry bulb 80°F
Wet bulb 75°F
Relative humidity 80 percent
Dew point 73°F

Temperature is:
Dry bulb 90°F
Wet bulb 75°F
Relative humidity 50 percent
Dew point 69°F

Manufacturers of humidifiers furnish a dial similar to the thermostat for controlling the humidity. A chart on the control tells what the humidity setting should be when the temperature outside is at a given point. Table 14-1 gives an example of what the settings should be.

The relationship between humidity, wet-bulb temperature, and dry-bulb temperature has much to do with the designing of air-conditioning systems. There are three methods of controlling the saturation of air.

1. Keep the dry-bulb temperature constant. Raise the wet-bulb temperature and the dew-point temperature to the dry-bulb temperature. Adding moisture to the air can do this. This, in turn, will raise the dew-point temperature to the dry-bulb temperature that automatically raises the wet-bulb temperature to the dry-bulb temperature.
2. Keep the wet-bulb temperature constant. Lower the dry-bulb temperature. Raise the dew-point temperature to the wet-bulb temperature. Cooling the dry-bulb temperature to the level of the wet-bulb temperature does this. The idea here is to do it without adding or removing any moisture. The dew point is automatically raised to the wet-bulb temperature.
3. Keep the dew-point temperature constant and the wet-bulb temperature at the dew-point temperature.

Table 14-1 *Permissible Relative Humidity (in the Winter)*

Outside Temperature		Brick Wall 12 in. Thick Plastered Inside	Single Glass	Double Glass
°F	°C			
		Percentage		
−20	−29.0	45	7	35
−10	−23.0	50	10	40
0	−17.8	60	18	45
10	−12.2	64	25	50
20	−6.7	67	30	55
30	−1.1	74	38	60
40	4.4	80	45	65
50	10.0	85	50	70
60	15.6	90	55	75

Cooling the dry-bulb and wet-bulb temperatures to the dew-point temperature can do this.

People and Moisture

People working inside a building or occupied space give off moisture as they work. They also give off heat. Such moisture and heat must be considered in determining air-conditioning requirements. Table 14-2 indicates some of the heat given off by the human body when working.

Table 14-2 *Activity-Heat Relationships*

Activity	Total Heat in Btu Per Hour
Person at rest	385
Person standing	430
Tailor	480
Clerk	600
Dancing	760
Waiter	1000
Walking	1400
Bowling	1500
Fast walking	2300
Severe exercise	2600

PSYCHROMETRIC CHART

The psychrometric chart holds much information. See Fig. 14-9. However, it is hard to read. It must be studied for some time. The dry-bulb temperature is located in one place and the wet bulb in another. If the two are known, it is easy to find the relative humidity and other factors relating to air being checked. Both customary and metric psychrometric charts are available.

Air contains different amounts of moisture at different temperatures. Table 14-3 shows the amounts of moisture that air can hold at various temperatures.

An explanation of the various quantities shown on a psychrometric chart will enable you to understand the chart. The different quantities on the chart are shown separately on the following charts. These charts will help you see how the psychrometric chart is constructed. See Fig. 14-10.

Across the bottom, the vertical lines are labeled from 25 to 110°F in increments of 5°F. These temperatures indicate the dry-bulb temperature. See Fig. 14-11.

The horizontal lines are labeled from 0 to 180°F. This span of numbers represents the grains of moisture per pound of dry air (when saturated). See Fig. 14-11.

The outside curving line on the left side of the graph indicates the wet-bulb, dew-point, or saturation temperature. See Fig. 14-12.

At 100 percent saturation the wet-bulb temperatures are the same as the dry-bulb and the dew-point temperatures. This means the wet-bulb lines start from the 100 percent saturation curve. Diagonal lines represent the wet-bulb temperatures. The point where the diagonal line of the wet bulb crosses the dry bulb's vertical line is the dew point. The temperature of the dew point will be found by running the horizontal line to the left and reading the temperature on the curve since the wet-bulb and dew-point temperatures are on the curve.

The curving lines within the graph indicate the percentage of relative humidity. These lines are labeled 10, 20 percent, and so on. See Fig. 14-13.

The pounds of water per pound of dry air are shown in the middle column of numbers on the right. See Fig. 14-14.

The grains of moisture per pound of dry air are shown in the left-hand column of the three columns of numbers on the right. See Fig. 14-14.

Table 14-3 shows that 1 lb of dry air will hold 19.1 grains of water at 25°F (–3.9°C). One pound of dry air will hold 415 grains of water at 110°F. It can be seen that the higher the temperature, the more moisture the air can hold. This is one point that should be remembered. To find the weight per grain, divide 1 by 7000 to get 0.00014 lb per grain. Therefore, on the chart 0.01 lb corresponds with about 70 grains. The volume of dry air (ft^3/lb) is represented by diagonal lines. See Fig. 14-15. The values are marked along the lines. They represent the cubic feet of the mixture of vapor and air per pound of dry air. The chart indicates that volume is affected by temperature relationships of the wet and dry-bulb readings. The lines are usually at intervals of $^1/_2$ ft^3/lb.

Enthalpy is the total amount of heat contained in the air above 0°F (–17.8°C). See Fig. 14-16. The lines on the chart that represent enthalpy are extensions of the wet-bulb lines. They are extended and labeled in Btu per pound. This value can be used to help determine the load on an air-conditioning unit.

AIR MOVEMENT

Convection, Conduction, and Radiation

Heat always passes from a warmer to a colder object or space. The action of refrigeration depends upon this natural law. The three methods by which heat can be transferred are convection, conduction, and radiation.

- *Convection* is heat transfer that takes place in liquids and gases. In convection, the molecules carry the heat from one point to another.
- *Conduction* is heat transfer that takes place chiefly in solids. In conduction, the heat passes from one

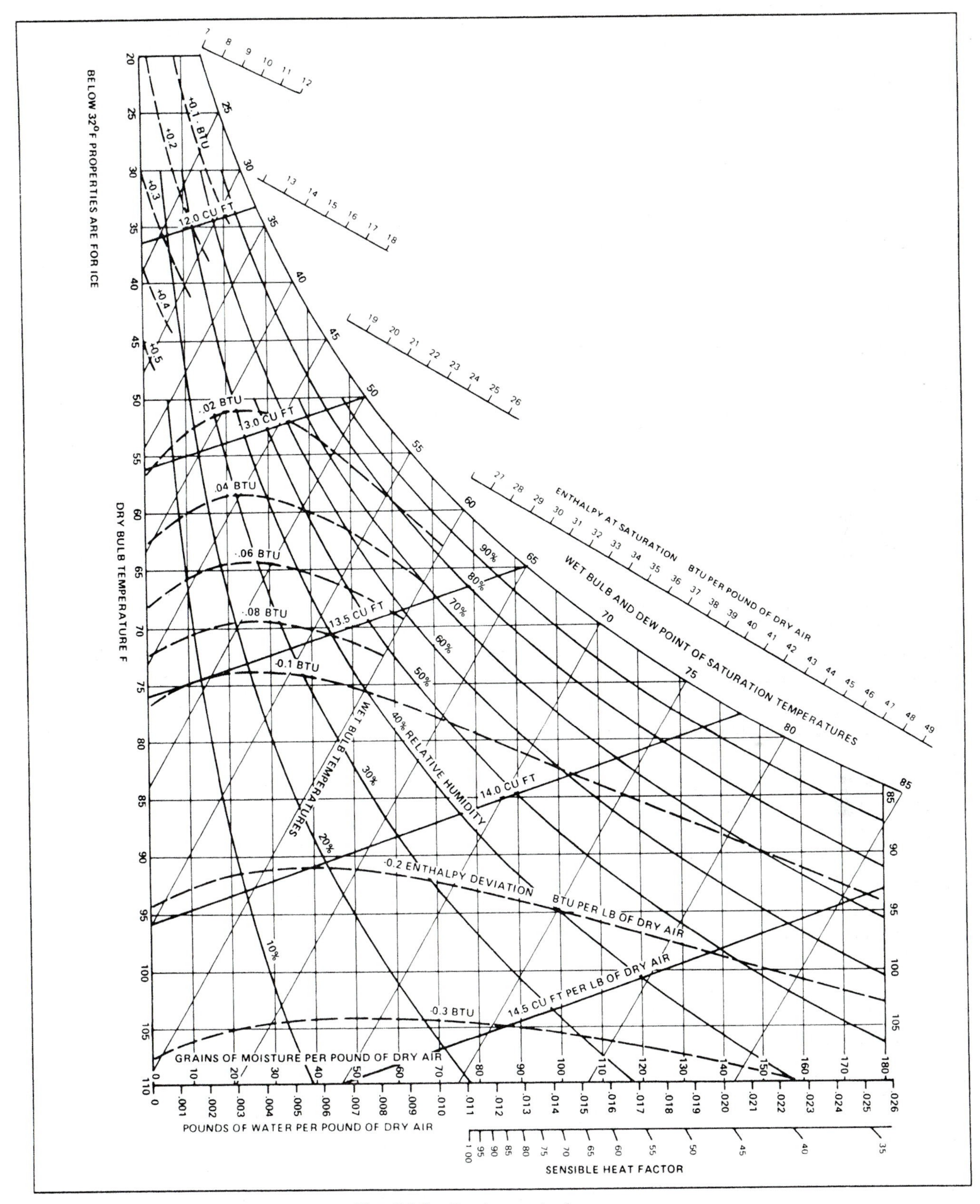

Fig. 14-9 *Psychrometric chart.* (Carrier)

Table 14-3 *Saturated Vapor per Pound of Dry Air (Barometer Reading at 29.92 in. per Square in.)*

Temperature		Weight in Grains
°F	°C	
25	−3.9	19.1
30	−1.1	24.1
35	1.7	29.9
40	4.4	36.4
45	7.2	44.2
50	10.0	53.5
55	12.8	64.4
60	15.6	77.3
65	18.3	92.6
70	21.1	110.5
75	23.9	131.4
80	26.7	155.8
85	29.4	184.4
90	32.2	217.6
95	35.0	256.3
100	37.8	301.3
105	40.6	354.0
110	43.3	415.0

molecule to another without any noticeable movement of the molecules.

- *Radiation* is heat transfer in waveform, such as light or radio waves. It takes place through a transparent medium such as air, without affecting that medium's temperature, volume, and pressure. Radiant heat is not apparent until it strikes an opaque surface, where it is absorbed. The presence of radiant heat is felt when it is absorbed by a substance or by your body.

Convection can be used to remove heat from an area. Then it can be used to cool. Air or water can be cooled in one plan and circulated through pipes of radiators in another location. In this way the cool water or air is used to remove heat.

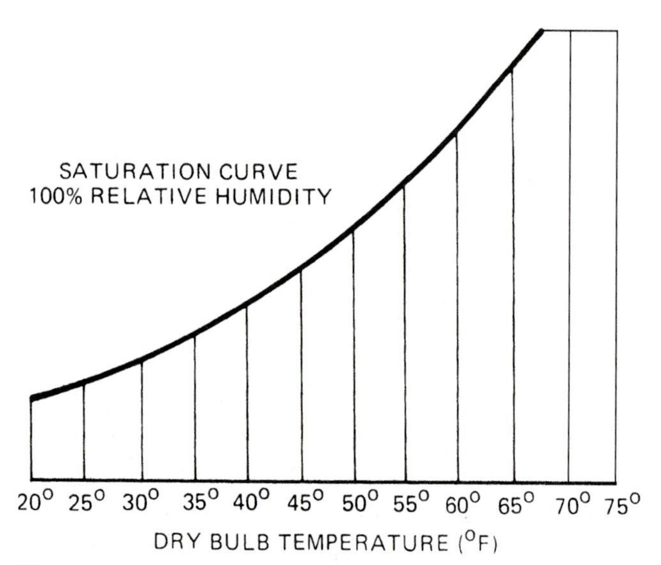

Fig. 14-10 *Temperature lines (dry bulb) on psychrometric chart. (Only a portion of the chart is shown).*

COMFORT CONDITIONS

The surface temperature of the average adult's skin is 80°F (26.7°C). The body can either gain or lose heat according to the surrounding air. If the surrounding air is hotter than the skin temperature, the body gains heat and the person may become uncomfortable. If the surrounding air is cooler than the skin temperature, then the body loses heat. Again, the person may become uncomfortable. If the temperature is much higher than the skin temperature or much cooler than the body temperature, then the person becomes uncomfortable. If the air is about 70°F (21.1°C) then the body feels comfortable. Skin temperature fluctuates with the temperature of the surface air. The total range of skin temperature is between 40 and 105°F (4.4 and 40.6°C). However, if the ambient temperature rises 10°F (5.5°C), the skin temperature rises only 3°F (1.7°C). Most of the time the normal temperature of the body ranges from 75 to 100°F (23.9 to 37.8°C). Both humidity and temperature affect the comfort of the human body. However, they are not the only factors that cause a person to be comfortable or uncomfortable. In heating or cooling a room, the air velocity, noise level, and temperature variation caused by the treated air must also be considered.

Velocity

When checking for room comfort, it is best to measure the velocity of the air at the distance of 4 to 72 in. from floor level. Velocity is measured with a velometer. See Fig. 14-19. Following is a range of air velocities and their characteristics.

- Slower than 15 feet per minute (fpm): stagnant air
- 20 to 50 fpm: acceptable air velocities
- 25 to 35 fpm: the best range for human comfort
- 35 to 50 fpm: comfortable for cooling purposes

Velocities of 50 fpm, or higher, call for a very high speed for the air entering the room. A velocity of about 750 fpm, or greater, is needed to create a velocity of 50 fpm or more inside the room. When velocities greater than 750 fpm are introduced, noise will also be present.

Sitting and standing levels must be considered when designing a cooling system for a room. People will tolerate cooler temperatures at the ankle level than at the sitting level, which is about 30 in. from the floor. Variations of 4°F (2.2°C) are acceptable between levels. This is also an acceptable level for temperature variations between rooms.

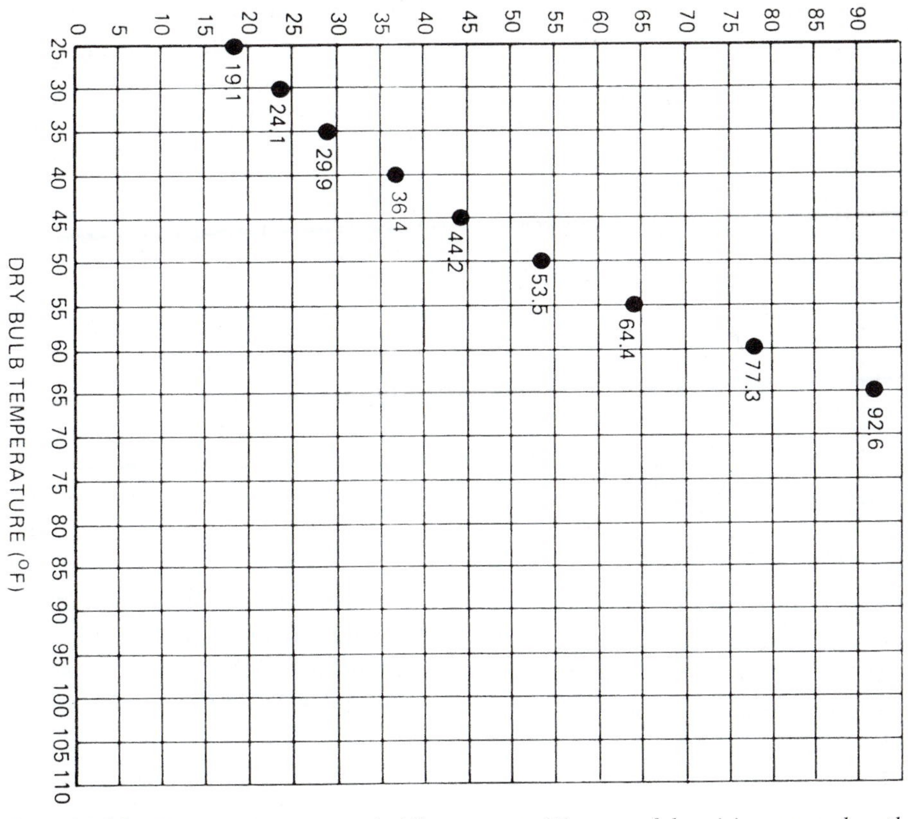

Fig. 14-11 *The moisture content in kilograms per kilogram of dry airis measured on the vertical column. Here, only a portion of the chart is shown.*

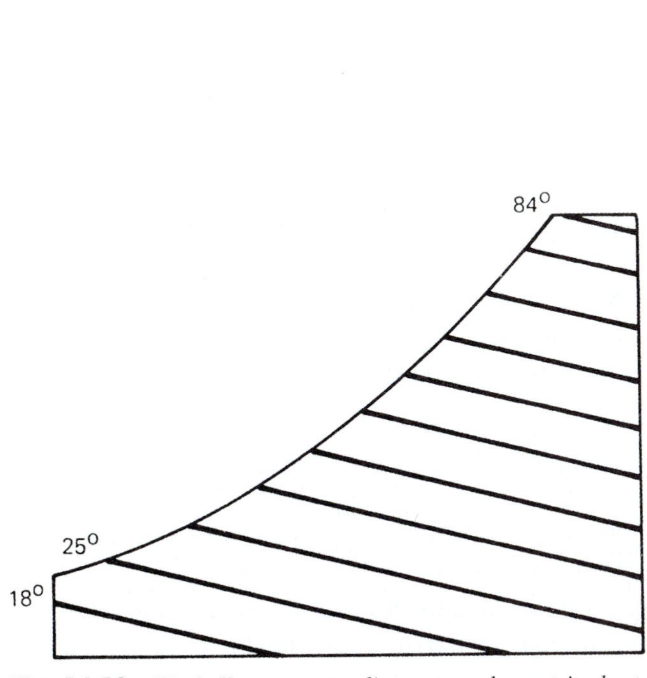

Fig. 14-12 *Wet-bulb temperature lines on psychrometric chart.*

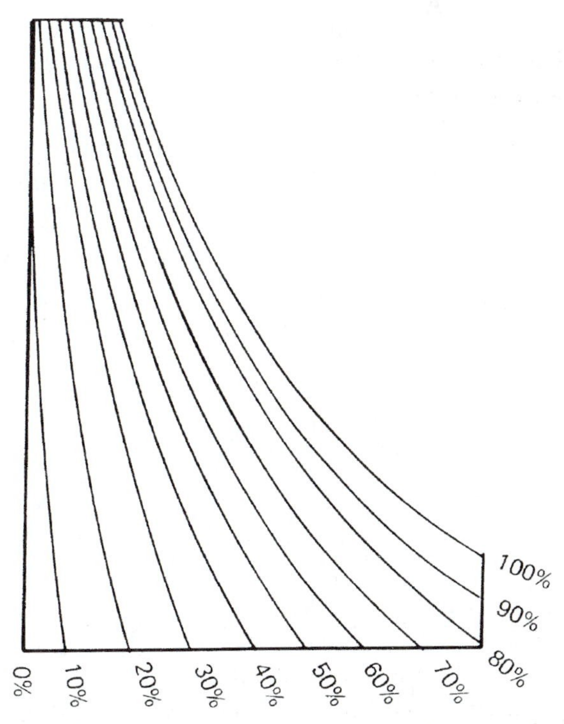

Fig. 14-13 *Relative humidity lines on a psychrometric chart.*

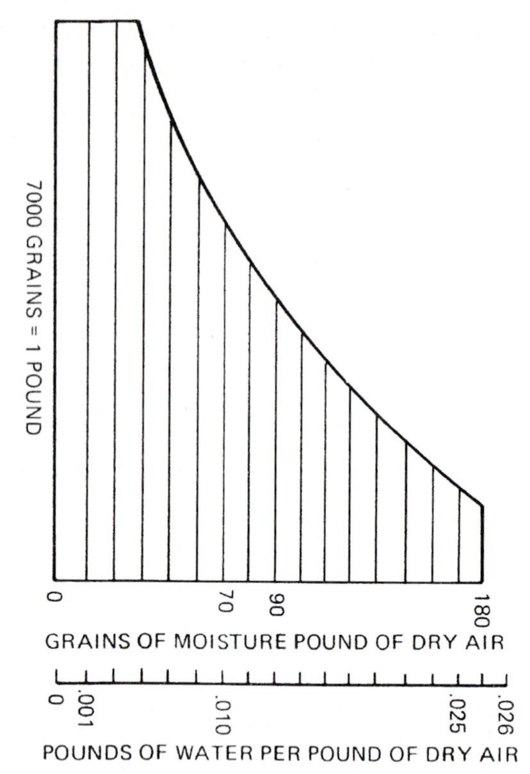

Fig. 14-14 *The sensible heat factor on a psychrometeric chart.*

To make sure that the air is properly distributed for comfort, it is necessary to look at the methods used to accomplish the job.

TERMINOLOGY

The following terms apply to the movement of air. They are frequently used in referring to air-conditioning systems.

- *Aspiration* is the induction of room air into the primary air stream. Aspiration helps eliminate stratification of air within the room. When outlets are properly located along exposed walls, aspiration also aids in absorbing undesirable currents from these walls and windows. See Fig. 14-17.

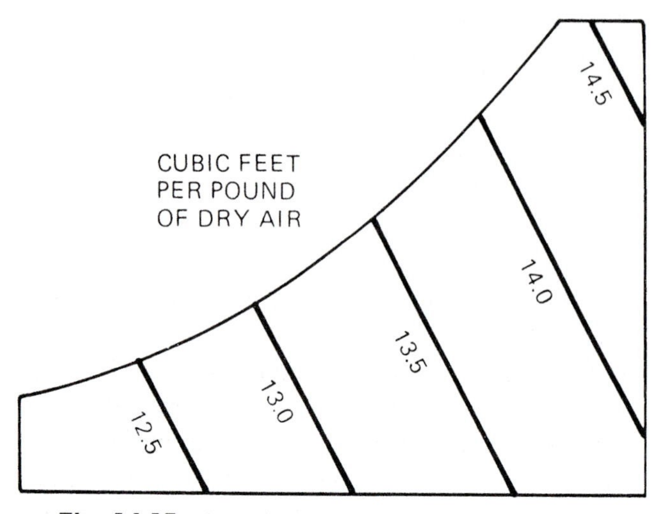

Fig. 14-15 *Air-volume lines on a psychrometric chart.*

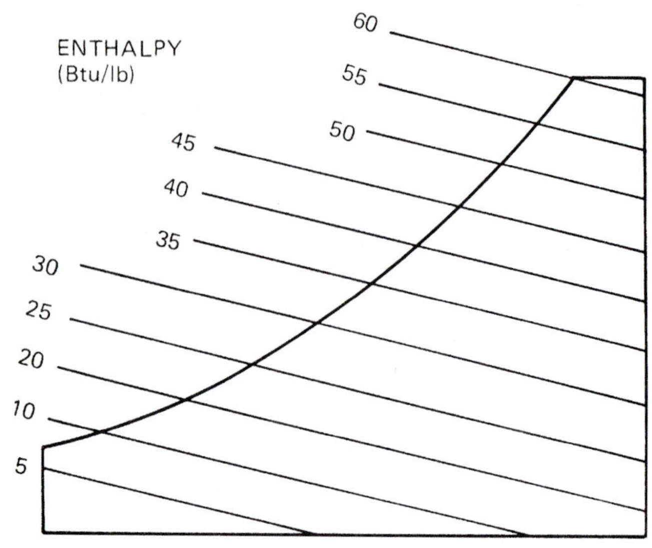

Fig. 14-16 *Enthalpy lines on a psychrometric chart.*

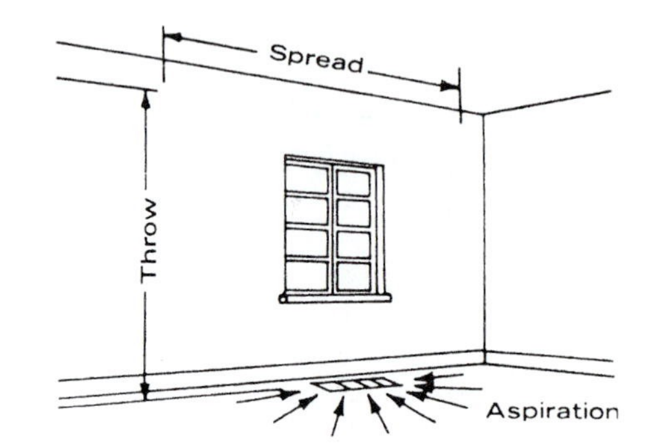

Fig. 14-17 *Aspiration, throw, and spread.* (Lima)

- *Cubic feet per minute* (cfm) is the measure of a volume of air. Air now in cubic feet per minute of a register or grille is computed by multiplying the face velocity times the free area in square feet.

EXAMPLE:

A resister with 144 in.^{2} (1 ft^2) of free area and a measured face velocity of 500 fpm would be delivering 500 cfm.

- *Decibels* (db) are units of measure of sound level. It is important to keep this noise at a minimum. In most catalogs for outlets, there is a line dividing the noise level of the registers or diffusers. Lower total pressure loss provides a quieter system.
- *Drop* is generally associated with cooling where air is discharged horizontally from high sidewall outlets. Since cool air has a natural tendency to drop, it will

fall progressively as the velocity decreases. Measured at the point of terminal velocity, drop is the distance in feet that the air has fallen below the level of the outlet. See Fig. 14-18.

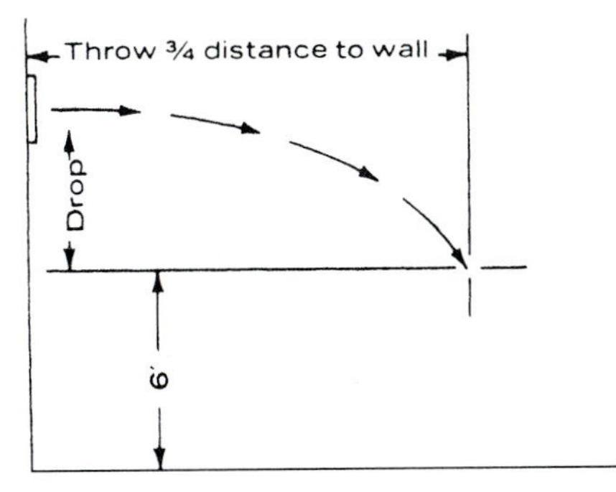

Fig. 14-18 *Drop.* (ARI)

- *Diffusers* are outlets that have a widespread, fan-shaped pattern of air,
- *Effective area* is the smallest net area of an outlet utilized by the air stream in passing through the outlet passages. It determines the maximum, or jet, velocity of the air in the outlet. In many outlets, the effective area occurs at the velocity measuring point and is equal to the outlet area. See Fig. 14-19.

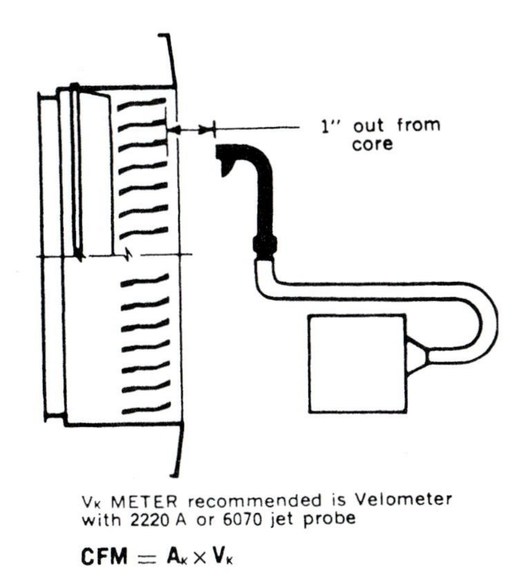

Fig. 14-19 *Air measurement at the grille.* (Lima)

- *Face velocity* is the average velocity of air passing through the face of an outlet or a return.
- *Feet per minute* (fpm) is the measure of the velocity of an air stream. This velocity can be measured with a velocity meter that is calibrated in fpm.
- *Free area* is the total area of the openings in the outlet or inlet through which air can pass. With gravity systems, free area is of prime importance. With forced air systems, free area is secondary to total pressure loss, except in sizing return air grilles.
- *Noise criteria* (NC) is an outlet sound rating in pressure level at a given condition of operation, based on established criteria and a specific room acoustic absorption value.
- *Occupied zone* is that interior area of a conditioned space that extends to within 6 in. of all room walls and to a height of 6 ft above the floor.
- *Outlet area* is the area of an outlet utilized by the air stream at the point of the outlet velocity as measured with an appropriate meter. The point of measurement and type of meter must be defined to determine cfm accurately.
- Outlet *velocity* (V_k) is the measured velocity at the started point with a specific meter.
- *Perimeter systems* are heating and cooling installations in which the diffusers are installed to blanket the outside walls. Returns are usually located at one or more centrally located places. High sidewall or ceiling returns are preferred, especially for cooling. Low returns are acceptable for heating. High sidewall or ceiling returns are highly recommended for combination heating and cooling installations.
- *Registers* are outlets that deliver air in a concentrated stream into the occupied zone.
- *Residual velocity* (V_R) is the average sustained velocity within the confines of the occupied zone, generally ranging from 20 to 70 fpm.
- *Sound power level* (L_w) is the total sound created by an outlet under a specified condition of operation.
- *Spread* is the measurement (in feet) of the maximum width of the air pattern at the point of terminal velocity. See Fig. 14-20.

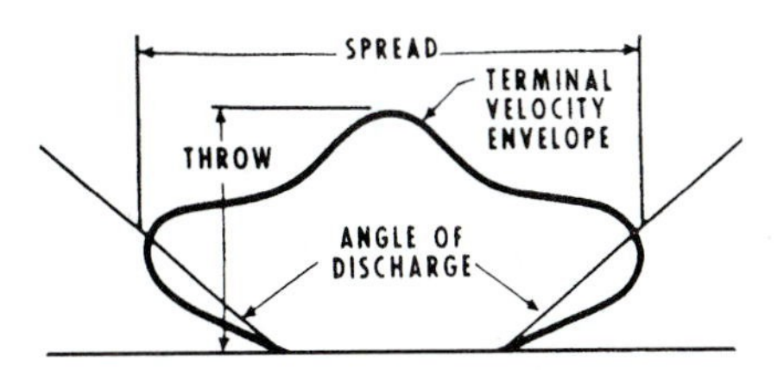

Fig. 14-20 *Typical air-stream pattern.* (Tuttle & Bailey)

- *Static pressure* (sp) is the outward force of air within a duct. This pressure is measured in inches of water. The static pressure within a duct is comparable to the air pressure within an automobile tire. A manometer measures static pressure. See Figs. 14-3 through 14-6.
- *Temperature differential* (ΔT) is the difference between primary supply and room air temperatures.

- *Terminal velocity* is the point at which the discharged air from an outlet decreases to a given speed, generally accepted as 50 fpm.
- *Throw* is the distance (measured in feet) that the air stream travels from the outlet to the point of terminal velocity. Throw is measured vertically from perimeter diffusers and horizontally from registers and ceiling diffusers. See Fig. 14-17.
- *Total pressure* (tp) is the sum of the static pressure and the velocity pressure. Total pressure is also known as impact pressure. This pressure is expressed in inches of water. The total pressure is directly associated with the sound level of an outlet. Therefore, any factor that increases the total pressure will also increase the sound level. Under sizing of outlets or increasing the speed of the blower will increase total pressure and the sound level.
- *Velocity pressure* (vp) is the forward moving force of air within a duct. This pressure is measured in inches of water. The velocity pressure is comparable to the rush of air from a punctured tire. A velometer is used to measure air velocity. See Fig. 14-19.

DESIGNING A PERIMETER SYSTEM

After the heat loss or heat gain has been calculated, the sum of these heat losses or heat gains will determine the size of the duct systems and the heating and cooling unit.

The three factors that ensure proper delivery and distribution of air within a room are location of outlet, type of outlet, and size of outlet. Supply outlets, if possible, should always be located to blanket every window and every outside wall. See Fig. 14-21. Thus, a register is recommended under each window.

The outlet selected should be a diffuser whose air pattern is fan shaped to blanket the exposed walls and windows.

The *American Society of Heating, Refrigeration, and Air-Conditioning Engineers* (ASHRAE) furnishes a chart with the locations and load factors needed for the climate of each major city in the United States. The chart should be followed carefully. The type of house, the construction materials, house location, room sizes, and exposure to sun and wind are important factors. With such information, you can determine how much heat will be dissipated. You can also determine how much heat and how much cooling will be dissipated in a building.

The *ASHRAE Handbook* of *Fundamentals* lists the information needed to compute the load factors. Calculate the heat loss or heat gain of the room; divide this figure by the number of outlets to be installed. From this you can determine the Btu/h required of each outlet. Refer to the performance data furnished by the manufacturer to determine the size the outlet should

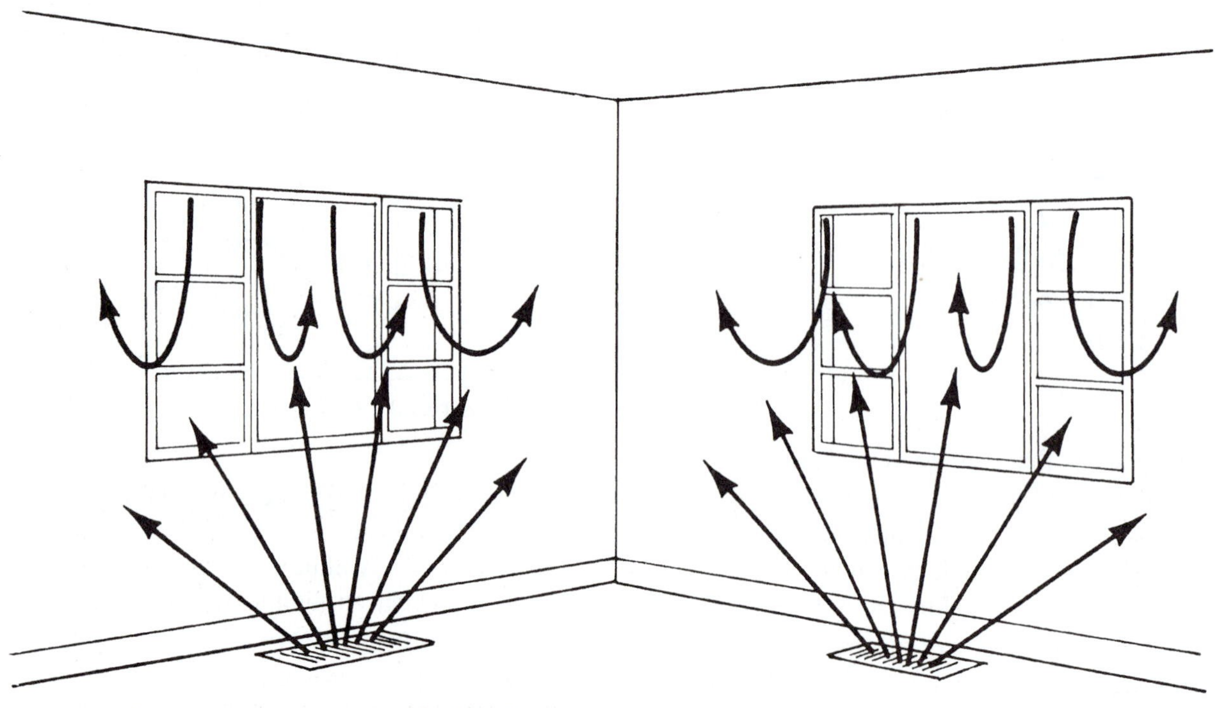

Fig.16-21 *Location of an outlet.* (Lima)

be. For residential application, the size selected should be large enough so that the Btu/h capacity on the chart falls to the side where the quiet zone is indicated. There is still a minimum vertical throw of 6 ft where cooling is involved.

Locating and Sizing Returns

Properly locating and sizing return air grilles are important. It is generally recommended that the returns be installed in high sidewall or the ceiling. They should be in one or more centrally located places. This depends upon the size and floor plan of the structure. Although such a design is preferred, low returns are acceptable for heating.

To minimize noise, care must be taken to size correctly the return air grille. The blower in the equipment to be used is rated in cfm by the manufacturer. This rating can usually be found in the specification sheets. Select the grille or grilles necessary to handle this cfm.

The grille or grilles selected should deliver the necessary cfm for the air to be conditioned. Thus, the proper size must be selected. The throw should reach approximately three-quarters of the distance from the outlet to the opposite wall. See Fig. 14-18. The face velocity should not exceed the recommended velocity for the application. See Table 14-4. The drop should be such that the air stream will not drop into the occupied zone. The occupied zone is generally thought of as 6 ft above floor level.

The sound caused by an air outlet in operation varies in direct proportion to the velocity of the air passing through it. Air velocity depends partially on outlet size. Table 14-5 lists recommendations for outlet velocities within safe sound limits for most applications.

Airflow Distribution

Bottom or side outlet openings in horizontal or vertical supply ducts should be equipped with adjustable flow equalizing devices. Figure 14-22 indicates the pronounced one-sided flow effect from an outlet opening. This is before the corrective effect of air-turning devices. A control grid is added in Fig. 14-23 to equalize flow in the takeoff collar. A Vectrol is added in Fig. 14-24 to turn air into the branch duct and provide volume control. Air-turning devices are recommended for installation at all outlet collars and branch duct connections.

Square unvaned elbows are also a source of poor duct distribution and high-pressure loss. Nonuniform flow in a main duct, occurring after an unvaned ell, severely limits the distribution of air into branch ducts in the vicinity of the ell. One side of the duct may be void, thus starving a branch duct. Conversely, all flow

Table 14-4 *Register or Grille Size Related to Air Capacities (in cfm)*

Register of Grille	Area in	Air Capacities in cfm										
Size	Ft²	250 fpm	300 fpm	400 fpm	500 fpm	600 fpm	700 fpm	750 fpm	800 fpm	900 fpm	1000 fpm	1250 fpm
8 × 4	.163	41	49	65	82	98	114	122	130	147	163	204
10 × 4	.206	52	62	82	103	124	144	155	165	185	206	258
10 × 6	.317	79	95	127	158	190	222	238	254	285	317	396
12 × 4	.249	62	75	100	125	149	174	187	199	224	249	311
12 × 5	.320	80	96	128	160	192	224	240	256	288	320	400
12 × 6	.383	96	115	153	192	230	268	287	306	345	383	479
14 × 4	.292	73	88	117	146	175	204	219	234	263	292	365
14 × 5	.375	94	113	150	188	225	263	281	300	338	375	469
14 × 6	.449	112	135	179	225	269	314	337	359	404	449	561
16 × 5	.431	108	129	172	216	259	302	323	345	388	431	539
16 × 6	.515	129	155	206	258	309	361	386	412	464	515	644
20 × 5	.541	135	162	216	271	325	379	406	433	487	541	676
20 × 6	.647	162	194	259	324	388	453	485	518	582	647	809
20 × 8	.874	219	262	350	437	524	612	656	699	787	874	1093
24 × 5	.652	162	195	261	326	391	456	489	522	587	652	815
24 × 6	.779	195	234	312	390	467	545	584	623	701	779	974
24 × 8	1.053	263	316	421	527	632	737	790	842	948	1053	1316
24 × 10	1.326	332	398	530	663	796	928	995	1061	1193	1326	1658
24 × 12	1.595	399	479	638	798	951	1117	1196	1276	1436	1595	1993
30 × 6	.978	245	293	391	489	587	685	734	782	880	978	1223
30 × 8	1.321	330	396	528	661	793	925	991	1057	1189	1371	1651
30 × 10	1.664	416	499	666	832	998	1165	1248	1331	1498	1664	2080
30 × 12	2.007	502	602	803	1004	1204	1405	1505	1606	1806	2007	2509
36 × 8	1.589	397	477	636	795	953	1112	1192	1271	1430	1589	1986
36 × 10	2.005	501	602	802	1003	1203	1404	1504	1604	1805	2005	2506
36 × 12	2.414	604	724	966	1207	1448	1690	1811	1931	2173	2414	3018

*Based on LIMA registers of the 100 Series.

Table 14-5 Outlet Velocity Ratings

Area	Rating (in. fpm)
Broadcast studios	500
Residences	500 to 750
Apartments	500 to 750
Churches	500 to 750
Hotel bedrooms	500 to 750
Legitimate theatres	500 to 1000
Private offices, acoustically treated	500 to 1000
Motion picture theatres	1000 to 1250
Private offices, not acoustically treated	1000 to 1250
General offices	1250 to 1500
Stores	1500
Industrial buildings	1500 to 2000

Fig. 14-22 *This flow path diagram shows the pronounced one-sided flow effect from an outlet opening before corrective effect of air-turning devices.* (Tuttle & Bailey)

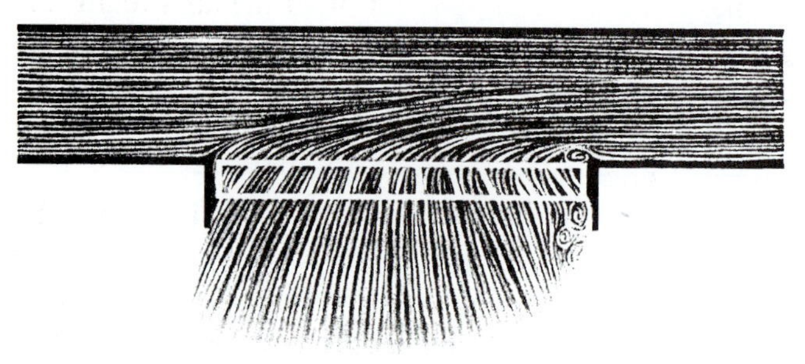

Fig. 14-23 *A control grid is added to equalize flow in the take-off collar.* (Tuttle & Bailey)

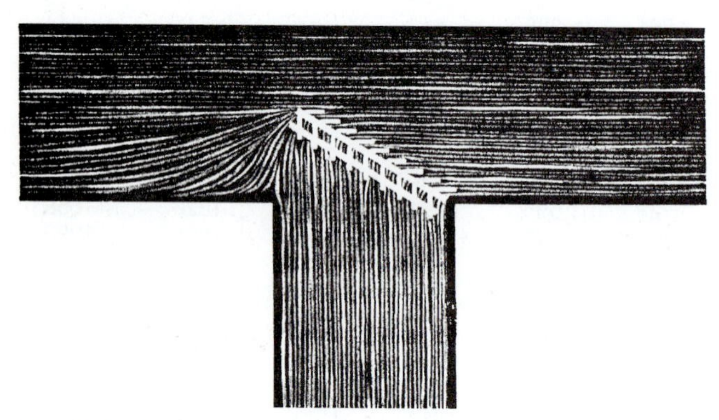

Fig. 14-24 *A Vectrol is added to turn air into the branch duct and provide volume control.* (Tuttle & Bailey)

may be stacked up on one side. This requires dampers to be excessively closed, resulting in higher sound levels.

Flow diagrams show the pronounced turbulence and piling up of airflow in an ell. See Fig. 14-25. Duc-turns reduce the pressure loss in square elbows as much as 80 percent. Their corrective effect is shown in Fig. 14-26.

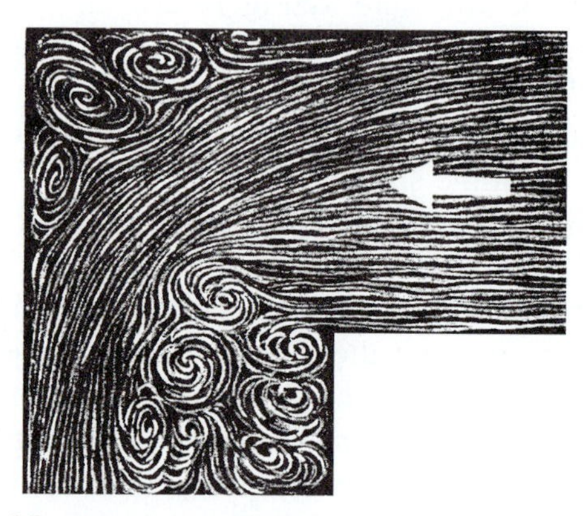

Fig. 14-25 *Note the turbulence and piling up of airflow in an ell.* (Tuttle & Bailey)

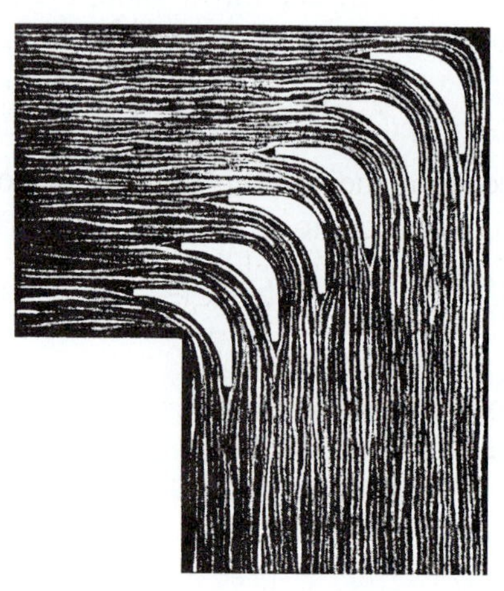

Fig. 14-26 *A ducturn reduces the pressure loss in square elbows by as much as 80 percent.* (Tuttle & Bailey)

SELECTION OF DIFFUSERS AND GRILLES

The selection of a linear diffuser or grille involves job-condition requirements, selection judgment, and performance data analysis.

Diffusers and grilles should be selected and sized according to the following characteristics:

- Type and style
- Function
- Air-volume requirement
- Throw requirement

- Pressure requirement
- Sound requirement

Air-Volume Requirement

The air volume per diffuser or grille is that which is necessary for the cooling, heating, or ventilation requirements of the area served by the unit. The air volume required, when related to throw, sound, or pressure-design limitations, determines the proper diffuser or grille size.

Generally, air volumes for internal zones of building spaces vary from 1 to 3 cfm/ft^2 of floor area. Exterior zones will require higher air volumes of 2.5 to 4 cfm/ft^2. In some cases, only the heating or cooling load of the exterior wall panel or glass surface is to be carried by the distribution center. Then, the air volume per linear foot of diffuser or grille will vary from 20 to 200 cfm, depending on heat-transfer coefficient, wall height, and infiltration rate.

Throw Requirement

Throw and occupied area air location are closely related. Both could be considered in the analysis of specific area requirements. The minimum-maximum throw for a given condition of aeration is based upon a terminal velocity at that distance from the diffuser. The residual room velocity is a function of throw to terminal velocity. Throw values are based on terminal velocities ranging from 75 to 150 fpm with corresponding residual room velocities of 75 to 150 fpm. The diffuser or grille location together with the air pattern selected, should generally direct the air path above the occupied zone. The air path then induces room air along its throw as it expands in cross section. This equalizes temperature and velocity within the stream. With the throw terminating in a partition or wall surface, the mixed air path further dissipates energy.

Ceiling mounted grilles and diffusers are recommended for vertical down pattern. Some locations in the room may need to be cooler than others. Also, some room locations may be harder to condition because of airflow problems. They are used in areas adjacent to perimeter wall locations that require localized spot conditioning. Ceiling heights of 12 ft or greater are needed. The throw for vertical projection is greatly affected by supply air temperature and proximity of wall surfaces.

Sidewall mounted diffusers and grilles have horizontal values based on a ceiling height of 8 to 10 ft. The diffuser or grille is mounted approximately 1 ft below the ceiling. For a given listed throw, the room air motion will increase or decrease inversely with the ceiling height. For a given air pattern setting and room air motion, the listed minimum-maximum throw value can be decreased by 1 ft for each 1 ft increase in ceiling height above 10 ft. Throw values are furnished by the manufacturer.

When sidewall grilles are installed remote from the ceiling (more than 3 ft away), reduce rated throw values by 20 percent.

Sill-mounted diffusers or grilles have throw values based on an 8- to 10-ft ceiling height. This is with the outlet installed in the top of a 30 in. high sill. For a given listed throw, the room air motion will change with the ceiling height. For a given air pattern setting and room air motion, the listed minimum-maximum throw value can be decreased by 2 ft for each 1 ft increase in ceiling height above 10 ft. Decrease 1 ft for each 1 ft decrease in sill height.

The minimum throw results in a room air motion higher than that obtained when utilizing the maximum throw. Thus, 50 fpm, rather than 35 fpm, is the air motion. The listed minimum throw indicates the minimum distance recommended. The minimum distance is from the diffuser to a wall or major obstruction such as a structural beam. The listed maximum throw is the recommended maximum distance to a wall or major obstruction. Throw values for sidewall grilles and ceiling diffusers and the occupied area velocity are based on flush ceiling construction providing an unobstructed air-stream path. The listed maximum throw times 1.3 is the complete throw of the air stream. This is where the terminal velocity equals the room air velocity. Rated occupied area velocities range from 25 to 35 fpm for maximum listed throws, and 35 to 50 fpm for minimum listed throw values.

Cooled-air drop or heated-air rise are of practical significance when supplying heated or cooled air from a sidewall grille. If the throw is such that the air stream prematurely enters the occupied zone, considerable draft may be experienced. This is due to incomplete mixing. The total airdrop must be considered when the wall grille is located at a distance from the ceiling. Cooled airdrop is controlled by spacing the wall grille from the ceiling and adjusting the grilles upward 15 in. Heated-air rise contributes significantly to temperature stratification in the upper part of the room.

The minimum separation between grille and ceiling must be 2 ft or more. The minimum mounting separation must be 2 ft or more. The minimum mounting height should be 7 ft.

Pressure Requirement

The diffuser or grille minimum pressure for a given air volume reflects itself in ultimate system fan horsepower requirements.

A diffuser or grille with a lower pressure rating requires less total energy than a unit with a higher pressure rating for a given air volume and effective area. Diffusers and grilles of a given size, having lower pressure ratings, usually have a lower sound level rating at a specified air volume.

Sound Requirement

Diffusers and grilles should be selected for the recommended noise criteria rating for a specific application. The data for each specific diffuser or grille type contains a *noise criteria* (NC) rating.

Table 14-6 lists recommended NC and area of application.

Air Noise High velocities in the duct or diffuser typically generate air noise. The flow turbulence in the duct and the excessive pressure reductions in the duct and diffuser system also generate noise. Such noise is most apparent directly under the diffuser. Room background levels of NC 35 and less provide little masking effect. Any noise source stands out above the background level and is easily detected.

Typically, air noise can be minimized by the following procedures:

- Limiting branch duct velocities to 1200 fpm
- Limiting static pressure in branch ducts adjacent to outlets to 0.15 in. H_2O
- Sizing diffusers to operate at outlet jet velocities up to 1200 fpm, (neck velocities limited to 500 to 900 fpm), and total pressures of 0.10 in. H_2O
- Using several small diffusers (and return grilles) instead of one or two large outlets or inlets that have a higher sound power
- Providing low-noise dampers in the branch duct where pressure drops of more than 0.20 in. of water must be taken
- Internally lining branch ducts near the fan to quiet this noise source
- Designing background sound levels in the room to be a minimum of NC 35 or NC 40.

CASING RADIATED NOISE

Casing noise differs from air noise in the way it is generated. Volume controllers and pressure- reducing dampers generate casing noise. Inside terminal boxes are sound baffles, absorbing blankets, and orifice restrictions to eliminate line of sight through the box. All these work to reduce the generated noise before the air and air noise discharge from the box into the outlet duct. During this process, the box casing is vibrated by the internal noise. This causes the casing to radiate noise through the suspended ceiling into the room. See Fig. 14-27.

Locating Terminal Boxes

In the past, terminal boxes and ductwork were separated from the room by dense ceilings. These ceilings prevented the system noise from radiating into the room. Plaster and taped sheetrock ceilings are examples of dense ceilings. Current architectural practice is to utilize lightweight (and low-cost) decorative suspended ceilings. These ceilings are not dense. They have only one-half the resistance to noise transmission that plaster and sheetrock ceilings have. Exposed tee-bar grid ceilings with 2×4 glass fiber pads, and perforated metal pan ceilings are examples. The end result is readily apparent. Casing radiated noise in lightweight modern buildings is a problem.

Table 14-6 *Recommended NC Criteria*

NC Curve	Communication Environment	Typical Occupancy
Below NC 25	Extremely quiet environment, suppressed speech is quite audible, suitable for acute pickup of all sounds.	Broadcasting studios, concert halls, music rooms.
NC 30	Very quiet office, suitable for large conferences; telephone use satisfactory.	Residences, theatres, libraries, executive offices, directors' rooms.
NC 35	Quiet office; satisfactory for conference at a 15 ft table; normal voice 10 to 30 ft telephone use satisfactory.	Private offices, schools, hotel rooms, courtrooms, churches, hospital rooms.
NC 40	Satisfactory for conferences at a 6 to 8 ft table; normal voice 6 to 12 ft; telephone use satisfactory.	General offices, labs, dining rooms.
NC 45	Satisfactory for conferences at a 4 to 5 ft; table; normal voice 3 to 6 ft; raised voice 6 to 12 ft; telephone use occasionally difficult.	Retail stores, cafeterias, lobby areas large drafting and engineering offices, reception areas.
Above NC 50	Unsatisfactory for conferences of more than two or three persons; normal voice 1 to 2 ft; raised voice 3 to 6 ft; telephone use slightly difficult.	Photocopy rooms, stenographic pools, print machine rooms, process areas.

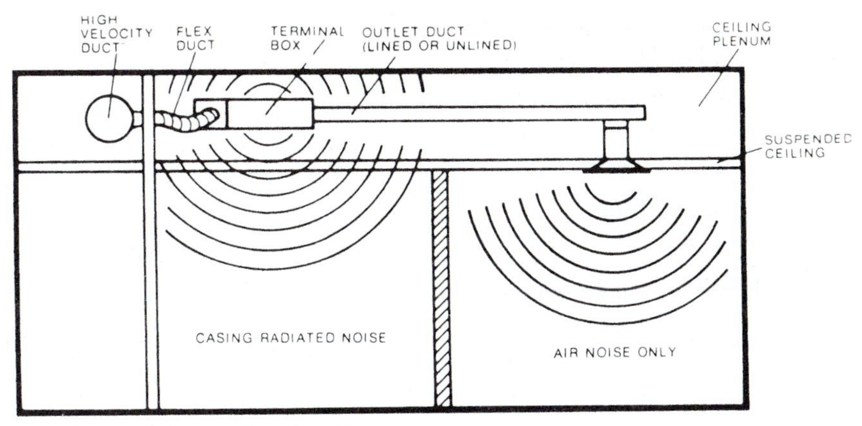

Fig. 14-27 *Casing noises.* (Tuttle & Bailey)

Controlling Casing Noise

Terminal boxes can sometimes be located over noisy areas (corridors, toilet areas, machine equipment rooms), rather than over quiet areas. In quiet areas casing noise can penetrate the suspended ceiling and become objectionable. Enclosures built around the terminal box (such as sheetrock or sheet lead over a glass fiber blanket wrapped around the box) can reduce the radiated noise to an acceptable level.

However, this method is cumbersome and limits access to the motor and volume controllers in the box. It depends upon field conditions for satisfactory performance, and is expensive. Limiting static pressure in the branch ducts minimizes casing noise. This technique, however, limits the flexibility of terminal box systems. It hardly classifies as a control.

Vortex Shedding

Product research in controlling casing noise has developed a new method of reducing radiated noise. The technique is known as *vortex shedding*. When applied to terminal boxes, casing radiated noise is dramatically lowered. *Casing radiation attenuation* (CRA) vortex shedders can be installed in all single- or dual-duct boxes up to 7000 cfm, both constant volume or variable volume, with or without reheat coils. CRA devices provide unique features and the following benefits:

- No change in terminal box size. Box is easier to install in tight ceiling plenums to insure minimum casing noise under all conditions.
- Factory-fabricated box and casing-noise eliminator, a one-piece assembly, reduces cost of installation. Only one box is hung. Only one duct connection is made.
- Quick-opening access door is provided in box. This assures easy and convenient access to all operating parts without having to cut and patch field-fabricated enclosures.
- Equipment is laboratory tested and performance rated. Engineering measurements are made in accordance with industry standards. Thus, on-job performance is insured. Quiet rooms result and owner satisfaction is assured.

RETURN GRILLES

Performance

Return air grilles are usually selected for the required air volume at a given sound level or pressure value. The intake air velocity at the face of the grille depends mainly on the grille size and the air volume.

The grille style and damper setting have a small effect on this intake velocity. The grille style, however, has a very great effect on the pressure drop. This, in turn, directly influences the sound level.

The intake velocity is evident only in the immediate vicinity of the return grille. It cannot influence room air distribution. Recent ASHRAE research projects have developed a scientific computerized method of relating intake grille velocities, measured 1 in. out from the grille face, to air volume. Grille measuring factors for straight, deflected bar, open, and partially closed dampers are in the engineering data furnished with the grille.

It still remains the function of the supply outlets to establish proper coverage, air motion, and thermal equilibrium. Because of this, the location of return grilles is not critical and their placement can be largely a matter of convenience. Specific locations in the ceiling may be desirable for local heat loads, or smoke exhaust, or a location in the perimeter sill or floor may be desirable for an exterior zone intake under a window wall section. It is not advisable to locate large centralized return grilles in an occupied area. The large mass of air moving through the grille can cause objectionable air motion for nearby occupants.

Return Grille Sound Requirement

Return air grilles should be selected for static pressures. These pressures will provide the required NC rating and conform to the return system performance characteristics. Fan-sound power is transmitted through the return air system as well as the supply system. Fan silencing may be necessary or desirable in the return side. This is particularly so if silencing is being considered on the supply side.

Transfer grilles venting into the ceiling plenum should be located remote from plenum noise source. The use of a lined sheet-metal elbow can reduce transmitted sound. Lined elbows on vent grilles and lined common ducts on ducted return grilles can minimize "cross talk" between private offices.

TYPES OF REGISTERS AND GRILLES

The spread of an unrestricted air stream is determined by the grille bar deflection. Grilles with vertical face bars at 0° deflection will have a maximum throw value. As the deflection setting of vertical bars is increased, the air stream covers a wider area and the throw decreases.

Registers are available with adjustable valves. An air leakage problem is eliminated if the register has a rubber gasket mounted around the grille. When it pulls up tightly against the wall, an airtight seal is made. This helps to eliminate noise. The damper has to be cam-operated so that it will stay open and not blow shut when the air comes through.

On some registers, a simple tool can be used to change the direction of the deflection bars. This means that adjusting the bars in the register can create a number of deflection patterns.

FIRE AND SMOKE DAMPERS

Ventilating, air conditioning, and heating ducts provide a path for fire and smoke, which can travel throughout a building. The ordinary types of dampers that are often installed in these ducts depend on gravity-close action or spring and level mechanisms. When their releases are activated, they are freed to drop inside the duct.

A fusible-link attachment to individual registers also helps control fire and smoke. Figure 14-28 shows a fusible-link type register. The link is available with

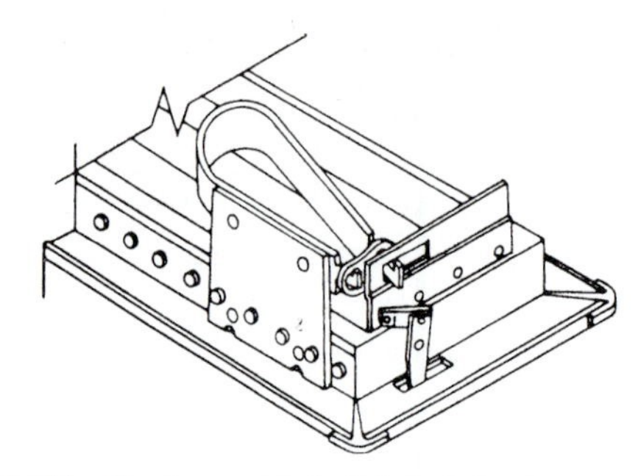

Fig. 14-28 *Register with fusible link for fire control.* (Lima)

melting points of 160°F (71.1°C) or 212°F (100°C). When the link melts, it releases a spring that forces the damper to a fully closed position. The attachment does not interfere with damper operation.

Smoke Dampers for High-Rise Buildings

Fire and smoke safety concepts in high-rise buildings are increasingly focusing on providing safety havens for personnel on each floor. This provision is to optimize air flow to or away from the fire floor or adjacent floors. Such systems require computer-actuated smoke dampers. Dampers are placed in supply and return ducts that are reliable. They must be closed tightly, and must offer minimum flow resistance when fully open.

CEILING SUPPLY GRILLES AND REGISTERS

Some ceiling grilles and registers have individually adjustable vanes. They are arranged to provide a one-way ceiling air pattern. They are recommended for applications in ceiling and sidewall locations for heating and cooling systems. They work best where the system has 0.75 to 1.75 cfm/ft^2 of room area. See Fig. 14-29.

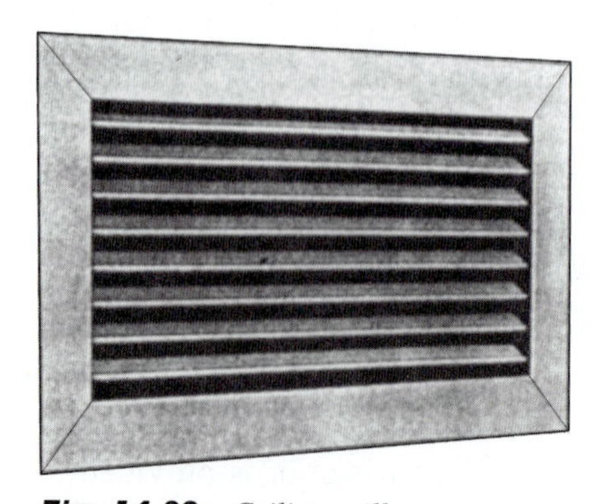

Fig. 14-29 *Ceiling grille.* (Tuttle & Bailey)

Some supply ceiling grilles and registers have individually adjustable curved vanes. They are arranged to provide a three-way ceiling air pattern. The vertical face vanes are a three-way diversion for air. A horizontal pattern with the face vanes also produces a three-way dispersion of air. These grilles and registers are recommended for applications in ceiling locations for heating and cooling systems handling 1.0 to 2.0 cfm/ft^2 of room area.

Figure 14-30 shows a grille with four-way vertical face vanes. Horizontal face vanes are also available. They, too, are adjustable individually for focusing an air stream in any direction. Both the three-way and four-way pattern grilles can be adjusted to a full or partial down blow position. The curved streamlined vanes are adjusted to a uniform partially closed position. This deflects the air path while retaining an effective area capacity of 35 percent of the neck area. In the full down blow position, grille effective area is increased by 75 percent.

Fig. 14-30 *Vertical face vanes in a four-way ceiling supply grille.* (Tuttle & Bailey)

Perforated adjustable diffusers for ceiling installation are recommended for heating and cooling. See Fig. 14-31. They are also recommended for jobs requiring on-the-job adjustment of air diffusion patterns.

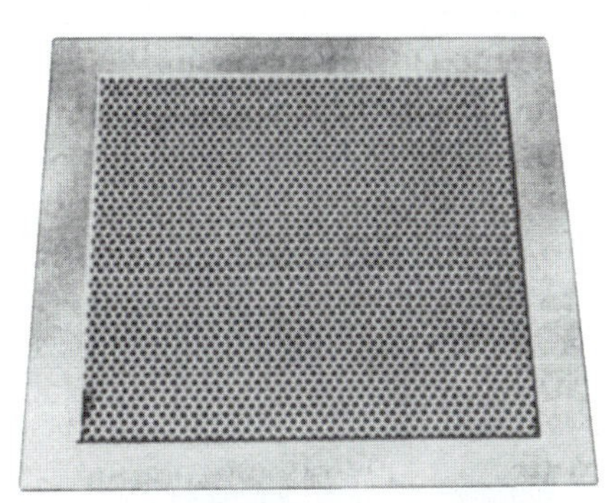

Fig. 14-31 *Perforated face adjustable diffuser for full flow and a deflector for ceiling installation.* (Tuttle & Bailey)

Full-flow square or round necks have expanded metal air-pattern deflectors. They are adjustable for four-, three-, two-, or one-way horizontal diffusion patterns. This can be done without change in the air volume, pressure, or sound levels. This deflector and diffuser have high-diffusion rates. The result is rapid temperature and velocity equalization of the mixed air mass well above the zone of occupancy.

They diffuse efficiently with 6 to 18 air changes per hour.

CEILING DIFFUSERS

There are other designs in ceiling diffusers. The type shown in Fig. 14-32 is often used in a supermarket or other large store. Here, it is difficult to mount other means of air distribution. These round diffusers with a flush face and fixed pattern are for ceiling installation. They are used for heating, ventilating, and cooling. They are compact and simple flush diffusers. High induction rates result in rapid temperature and velocity equalization of the mixed air mass. Mixing is done above the zone of occupancy.

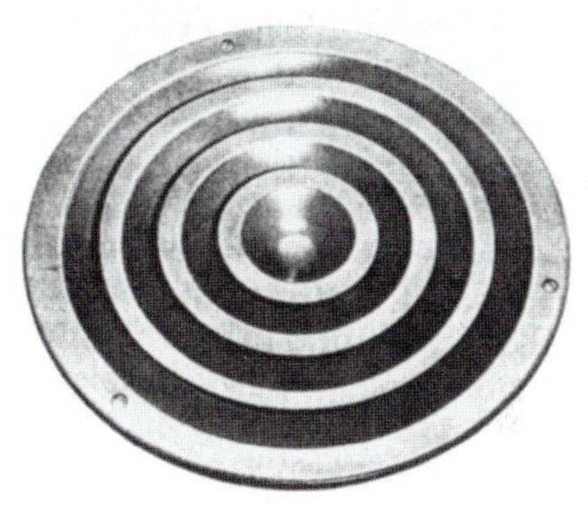

Fig. 14-32 *Round diffusers with flush face, fixed pattern for ceiling installation.* (Tuttle & Bailey)

Grids are used and sold as an accessory to these diffusers. The grid, see Fig. 14-33, is a multiblade device designed to insure uniform airflow in a diffuser collar. It is individually adjustable. The blades can be moved to control the air stream precisely.

For maximum effect, the control grid should be installed with the blades perpendicular to the direction of approaching airflow. Where short collars are encountered, a double bank of control grids is recommended. The upper grid is placed perpendicular to the branch duct flow. The lower grid is placed parallel to the branch duct flow. The control grid is attached to the duct collar by means of mounting straps. It is commonly used with volume dampers.

Fig. 14-33 *Control grid with multiblade devices to control airflow in a diffuser collar.* (Tuttle & Bailey)

Antismudge Rings

The antismudge ring is designed to cause the diffuser discharge air path to contact the ceiling in a thin-layered pattern. This minimizes local turbulence, the cause of distinct smudging. See Fig. 14-34.

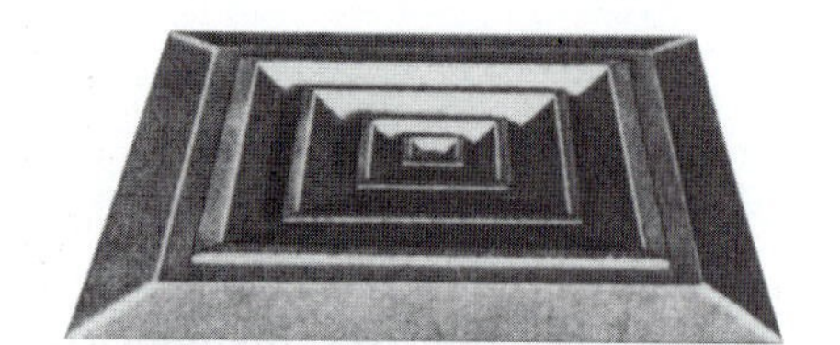

Fig. 14-34 *Antismudge ring.* (Tuttle & Bailey)

For the best effect, the antismudge ring must fit evenly against the ceiling surface. It is held in position against the ceiling by the diffuser margin. This eliminates any exposed screws.

Air-Channel Diffusers

Air-channel supply diffusers are designed for use with integrated air handling ceiling systems. They are adaptable to fit between open parallel tee bars. They fit within perforated or slotted ceiling runners. The appearance of the integrated ceiling remains unchanged regardless of the size of the unit. They are painted out to be invisible when viewing the ceiling. These high-capacity diffusers provide a greater air-handling capability. See Fig. 14-35.

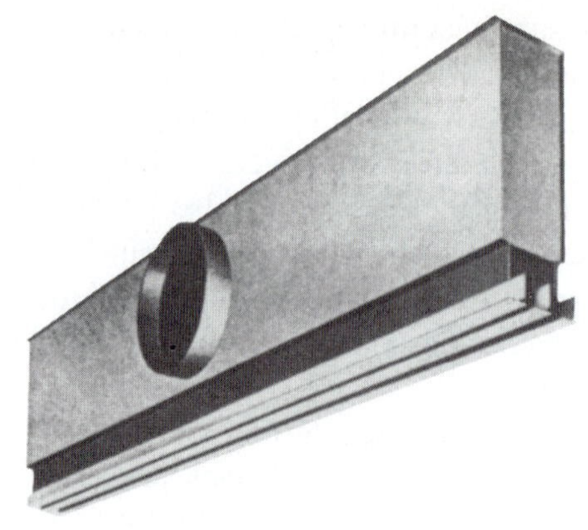

Fig. 14-35 *High-capacity air channel diffuser with fixed pattern for suspended grid ceilings.* (Tuttle & Bailey)

Luminaire Diffusers

The luminaire is a complete lighting unit. The luminaire diffuser fits close to the fluorescent lamp fixtures in the ceiling. The single-side diffuser with side inlet is designed to provide single-side concealed air distribution. See Fig. 14-36. These diffusers are designed with oval-shaped side inlets and inlet dampers. They provide effective single-point dampering.

Fig. 14-36 *Single-side diffuser with side inlet.* (Tuttle & Bailey)

Dual-side diffusers with side inlet are designed to provide concealed air distribution. Note the crossover from the oval side inlet to the other side of the diffuser. This type of unit handles more air and spreads it more evenly when used in large areas. See Fig. 14-37. This type of diffuser is also available with an insulation jacket when needed.

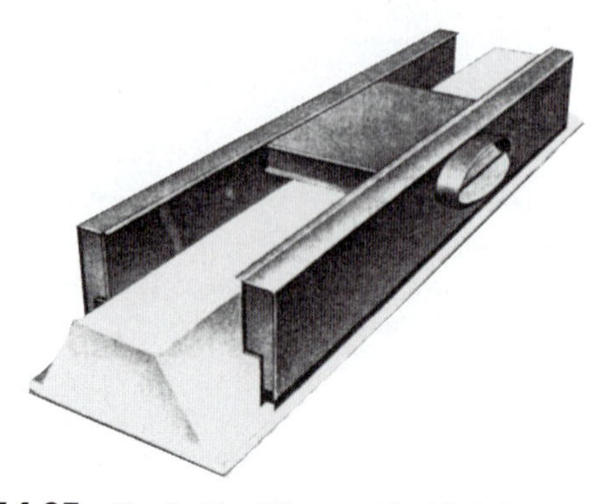

Fig. 14-37 *Dual-side diffuser with side inlet.* (Tuttle & Bailey)

Room Air Motion

Figure 14-38 illustrates the airflow from ceiling diffusers. The top view illustrates the motion from the diffuser. The side view shows how the temperature differential is very low. Note that the temperature is 68°F near the ceiling and sidewall, and 73°F on the opposite wall near the ceiling.

LINEAR GRILLES

Linear grilles are designed for installation in the sidewall, sill, floor, and ceiling. They are recommended for supplying heated, ventilated, or cooled air, and for returning or exhausting room air. See Fig. 14-39.

When installed in the sidewall near the ceiling, the linear grilles provide a horizontal pattern above the occupied zone. Core deflections of 15 and 30° direct the air path upward to overcome the drop effect resulting from cool primary air.

When installed in the top of a sill or enclosure, the linear grilles provide a vertical up-pattern. This is effective in overcoming uncomfortable cold downdrafts. It also offsets the radiant effect of glass surfaces. Core deflections of 0 and 15° directed toward the glass surface provide upward airflow to the ceiling, and along the ceiling toward the interior zone.

When installed in the ceiling, linear grilles provide a vertical downward air pattern. This pattern is effective in projection heating and in cooling the building perimeter from ceiling heights above 13 to 15 ft. Application of down flow primary air should be limited to insure against excessive drafts at the end of the throw. Core deflections of 0, 15, and 30° direct the air path angularly downward as required,

Debris screens can be integrally attached. See Fig. 14-40.

FANS AND MECHANICAL VENTILATION

Mechanical ventilation differs from natural ventilation mainly in that the air circulation is performed by mechanical means (such as fans or blowers). In natural ventilation, the air is caused to move by natural forces. In mechanical ventilation, the required air changes are effected partly by diffusion, but chiefly by positive currents put in motion by electrically operated fans or blowers, as shown in Fig. 14-41. Fresh air is usually circulated through registers connected with the outside and warmed as it passes over and through the intervening radiators.

Air Volume

The volume of air required is determined by the size of the space to be ventilated and the number of times per hour that the air in the space is to be changed. In many cases, existing local regulations or codes will govern the ventilating requirements. Some of these codes are based on a specified amount of air per person, and others on the air required per square foot of floor area.

Fans and Blowers

The various devices used to supply air circulation in air-conditioning applications are known as fans, blowers, exhausts, or propellers. The different types of fans may be classified with respect to their construction as follows:

- Propeller
- Tube axial
- Vane axial
- Centrifugal

A *propeller fan* consists essentially of a propeller or disk-type wheel within a mounting ring or plate and includes the driving-mechanism supports for either belt or direct drive. A *tube-axial fan* consists of a propeller or disk-type wheel within a cylinder and includes the driving-mechanism supports for either belt drive or direct connection. A *vane-axial fan* consists of a disk-type wheel within a cylinder and a set of air-guide vanes located before or after the wheel. It includes the driving-mechanism supports for either belt drive or direct connection. A *centrifugal fan* consists of a fan rotor or wheel within a scroll-type housing and includes the driving-mechanism supports for either belt drive or direct connection. Figure 14-42 shows the mounting arrangements.

Fan performance may be stated in various ways, with the air volume per unit time, total pressure, static pressure, speed, and power input being the most important. The terms, as defined by the National Association of Fan Manufacturers, are as follows:

- *Volume* handled by a fan is the number of cubic feet of air per minute expressed as fan-outlet conditions.
- *Total pressure* of a fan is the rise of pressure from fan inlet to fan outlet.
- *Velocity pressure* of a fan is the pressure corresponding to the average velocity determination from the volume of airflow at the fan-outlet area.
- *Static pressure* of a fan is the total pressure diminished by the fan-velocity pressure.

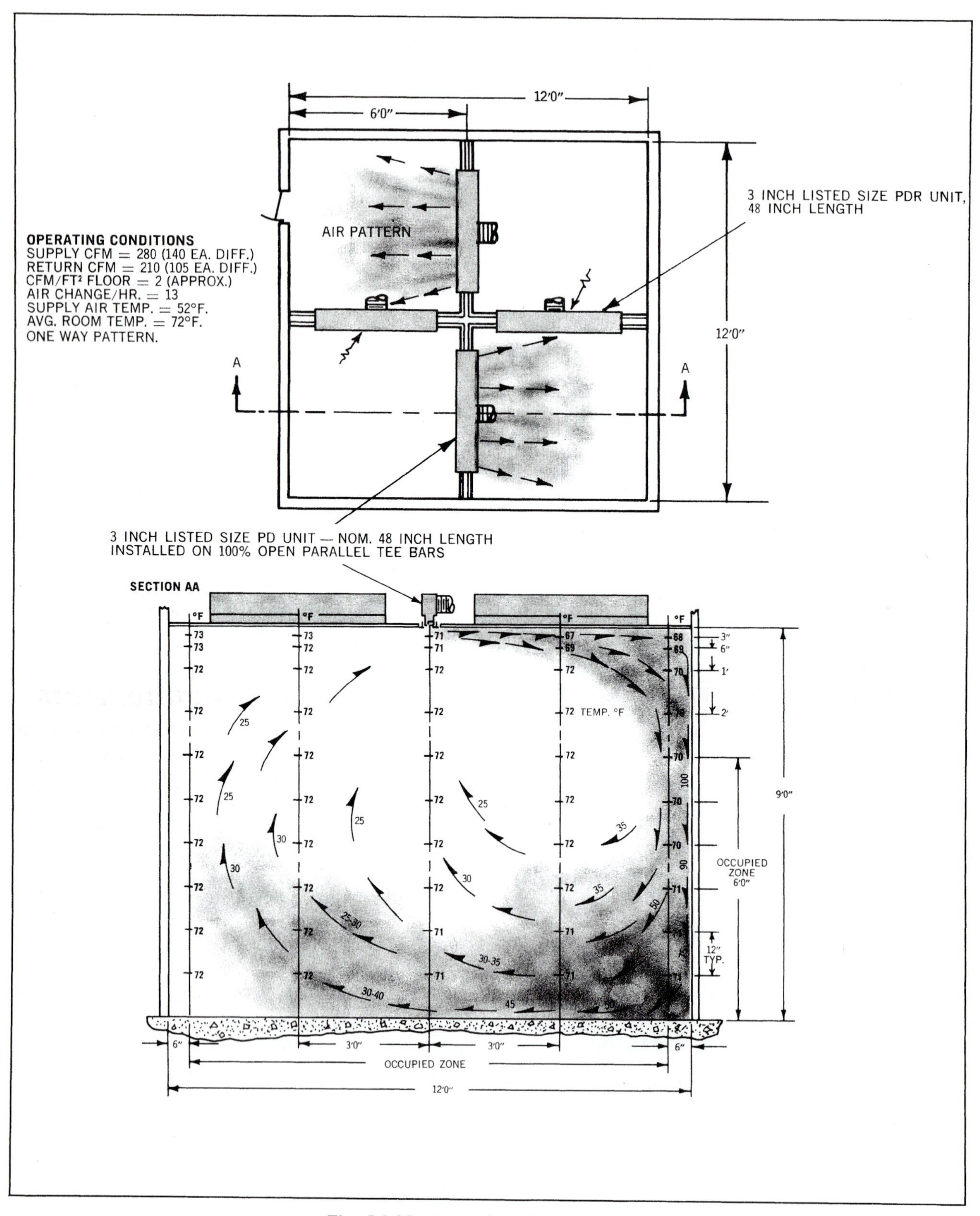

Fig. 14-38 Room air motion. (Tuttle & Bailey)

Fig. 14-39 *Linear grille with a hinged access door.* (Tuttle & Bailey)

Fig. 14-40 *Debris screen for linear grilles.* (Tuttle & Bailey)

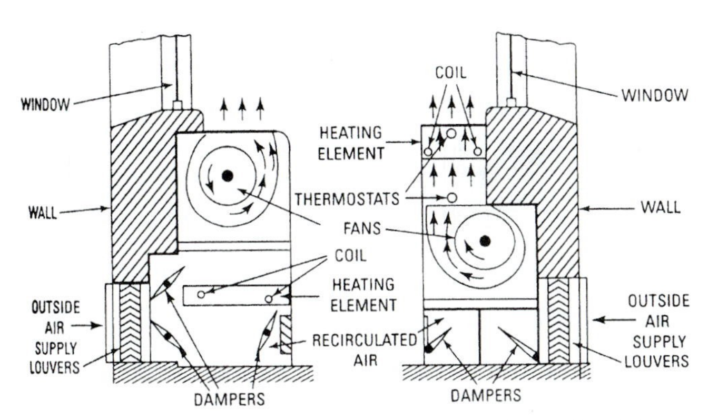

Fig. 14-41 *Typical mechanical ventilators for residential use. Note placement of fans and other details.*

- *Power output* of a fan is expressed in horsepower and is based on fan volume and the fan total pressure.
- *Power input* of a fan is expressed in horsepower and is measured as horsepower delivered to the fan shaft.
- *Mechanical efficiency* of a fan is the ratio of power output to power input.
- *Static efficiency* of a fan is the mechanical efficiency multiplied by the ratio of static pressure to the total pressure.
- *Fan-outlet area* is the inside area of the fan outlet.
- *Fan-inlet area* is the inside area of the inlet collar.

Air Volume

The volume of air required is determined by the size of the space to be ventilated and the number of times per hour that the air in the space is to be changed. Table 14-7 shows the recommended rate of air change for various types of spaces.

In many cases, existing local regulations or codes will govern the ventilating requirements. Some of these codes are based on a specified amount of air per person and on the air required per square foot of floor area. Table 14-7 should serve as a guide to average conditions. Where local codes or regulations are involved, they should be taken into consideration. If the number of persons occupying the space is larger than would be normal for such a space, the air should be changed more often than shown.

Horsepower Requirements

The horsepower required for any fan or blower varies directly as the cube of the speed, provided that the area of the discharge orifice remains unchanged. The horsepower requirements of a centrifugal fan generally decrease with a decrease in the area of the discharge orifice if the speed remains unchanged. The horsepower requirements of a propeller fan increase as the area of the discharge orifice decreases if the speed remains unchanged.

Fan Driving Methods

Whenever possible, the fan wheel should be directly connected to the motor shaft. This can usually be accomplished with small centrifugal fans and with propeller

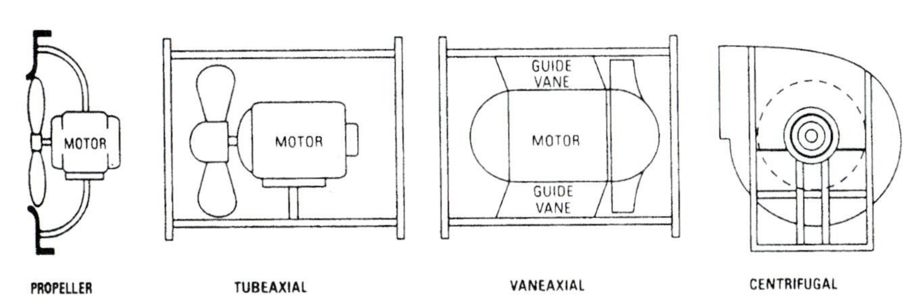

Fig. 14-42 *Fan classifications with proper mounting arrangement.*

Table 14-7 *Volume of Air Required*

Space to Be Ventilated	Air Changes per Hour	Minutes per Change
Auditoriums	6	10
Bakeries	20	3
Bowling alleys	12	5
Club rooms	12	5
Churches	6	10
Dining rooms (restaurants)	12	5
Factories	10	6
Foundries	20	3
Garages	12	5
Kitchens (restaurants)	30	2
Laundries	20	3
Machine shops	10	6
Offices	10	6
Projection booths	60	1
Recreation rooms	10	6
Sheet-metal shops	10	6
Ship holds	6	10
Stores	10	6
Toilets	20	3
Tunnels	6	10

fans up to about 60 in. in diameter. The deflection and the critical speed of the shaft, however, should be investigated to determine whether or not it is safe.

When selecting a motor for fan operation, it is advisable to select a standard motor one size larger than the fan requirements. It should be kept in mind, however, that direct-connected fans do not require as great a safety factor as that of belt-driven units. It is desirable to employ a belt drive when the required fan speed or horsepower is in doubt, since a change in pulley size is relatively inexpensive if an error is made. See Fig. 14-43.

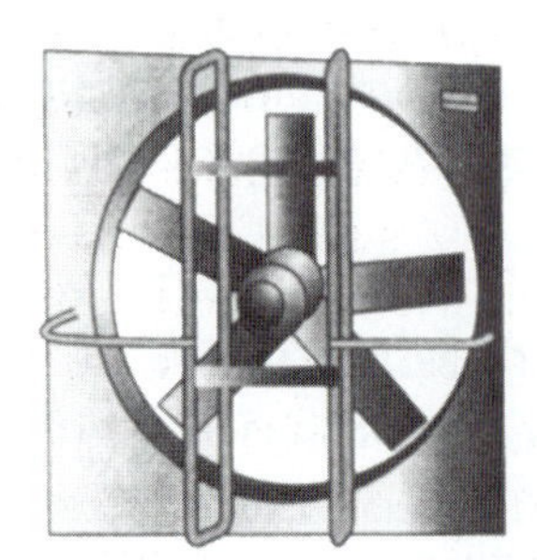
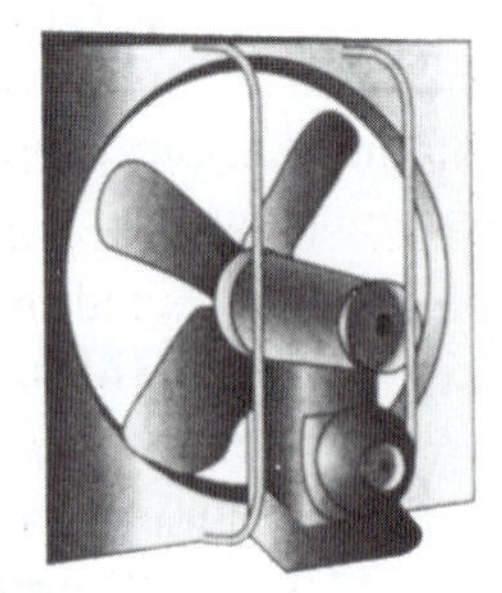

Fig. 14-43 *Various types of propeller fan drives and mounting arrangements.*

Directly connected small fans for single-phase AC motors of the split-phase, capacitor, or shaded-pole type usually drive various applications. The capacitor motor is more efficient electrically and is used in districts where there are current limitations. Such motors, however, are usually arranged to operate at one speed. With such a motor, if it is necessary to vary the air volume or pressure of the fan or blower, the throttling of air by a damper installation is usually made.

In large installations (such as when mechanical draft fans are required), various drive methods are used:

- A slip-ring motor to vary the speed
- A constant-speed, directly connected motor, which, by means of moveable guide vanes in the fan inlet, serves to regulate the pressure and air volume

Selecting A Fan

Most often, the service determines the type of fan to use. When operation occurs with little or no resistance, and particularly when no duct system is required, the propeller fan is commonly used because of its simplicity and economy in operation. When a duct system is involved, a centrifugal or axial type of fan is usually employed. In general, centrifugal and axial fans are comparable with respect to sound effect, but the axial fans are somewhat lighter and require considerably less space. The following information is usually required for proper fan selection:

- Capacity requirement in cubic feet per minute
- Static pressure or system resistance
- Type of application or service
- Mounting arrangement of system
- Sound level or use of space to be served
- Nature of load and available drive

The various fan manufacturers generally supply tables or characteristic curves that ordinarily show a wide range of operating particulars for each fan size. The tabulated data usually include static pressure, outlet velocity, revolutions per minute, brake horsepower, tip or peripheral speed, and so on.

Applications of Fans

The numerous applications of fans in the field of air conditioning and ventilation are well known, particularly to engineers and air-conditioning repair and maintenance personnel. The various fan applications are as follows:

- Attic fans
- Circulating fans

- Cooling-tower fans
- Exhaust fans
- Kitchen fans

Exhaust fans are found in all types of applications, according to the American Society of Heating and Ventilating Engineers. Wall fans are predominantly of the propeller type, since they operate against little or no resistance. They are listed in capacities from 1000 to 75,000 ft^3/min. They are sometimes incorporated in factory-built penthouses and roof caps or provided with matching automatic louvers. Hood exhaust fans involving duct work are predominantly centrifugal, especially in handling hot or corrosive fumes.

Spray-booth exhaust fans are frequently centrifugal, especially if built into self-contained booths. Tube-axial fans lend themselves particularly well to this application where the case of cleaning and of suspension in a section of ductwork is advantageous. For such applications, built-in cleanout doors are desirable.

Circulating fans are invariably propeller or disk-type units and are made in a vast variety of blade shapes and arrangements. They are designed for appearance as well as utility. *Cooling-tower fans* are predominantly the propeller type. However, axial types are also used for packed towers, and occasionally a centrifugal fan is used to supply draft. *Kitchen fans* for domestic use are small propeller fans arranged for window or wall mounting and with various useful fixtures. They are listed in capacity ranges from 300 to 800 ft^3/min.

Attic fans are used during the summer to draw large volumes of outside air through the house or building whenever the outside temperature is lower than that of the inside. It is in this manner that the relatively cool evening or night air is utilized to cool the interior in one or several rooms, depending on the location of the air-cooling unit. It should be clearly understood, however, that the attic fan is not strictly a piece of air-conditioning equipment since it only moves air and does not cool, clean, or dehumidify. Attic fans are used primarily because of their low cost and economy of operation, combined with their ability to produce comfort cooling by circulating air rather than conditioning it.

Operation of Fans

Fans may be centrally located in an attic or other suitable space (such as a hallway), and arranged to move air proportionately from several rooms. A local unit may be installed in a window to provide comfort cooling for one room only when desired. Attic fans are usually propeller types and should be selected for low velocities to prevent excessive noise. The fans should have sufficient capacity to provide at least 30 air changes per hour.

To decrease the noise associated with air-exchange equipment, the following rules should be observed:

- The equipment should be properly located to prevent noise from affecting the living area.
- The fans should be of the proper size and capacity to obtain reasonable operating speed.
- Equipment should be mounted on rubber or other resilient material to assist in preventing transmission of noise to the building.

If it is unavoidable to locate the attic air-exchange equipment above the bedrooms, it is essential that every precaution be taken to reduce the equipment noise to the lowest possible level. Since high-speed AC motors are usually quieter than low-speed ones, it is often preferable to use a high-speed motor connected to the fan by means of an endless V-belt, if the floor space available permits such an arrangement.

Installation of Attic Fans

Because of the low static pressures involved (usually less than $^1/_8$ in. of water), disk or propeller fans are generally used instead of the blower or housed types. It is important that the fans have quiet operating characteristics and sufficient capacity to give at least 30 air changes per hour. For example, a house with 10,000 ft^3 content would require a fan with a capacity of 300,000 ft^3/h or 5000 ft^3/min to provide 30 air changes per hour.

The two general types of attic fans in common use are *boxed-in fans* and *centrifugal fans*. The boxed-in fan is installed within the attic in a box or suitable housing located directly over a central ceiling grille or in a bulkhead enclosing an attic stair. This type of fan may also be connected by means of a direct system to individual room grilles. Outside cool air entering through the windows in the downstairs room is discharged into the attic space and escapes to the outside through louvers, dormer windows, or screened openings under the eaves.

Although an air-exchange installation of this type is rather simple, the actual decision about where to install the fan and where to provide the grilles for the passage of air up through the house should be left to a ventilating engineer. The installation of a multiblade centrifugal fan is shown in Fig. 14-44. At the suction side, the fan is connected to exhaust ducts leading to grilles, which are placed in the ceiling of the two bedrooms. The air exchange is accomplished by admitting fresh air through open windows and up through the suction

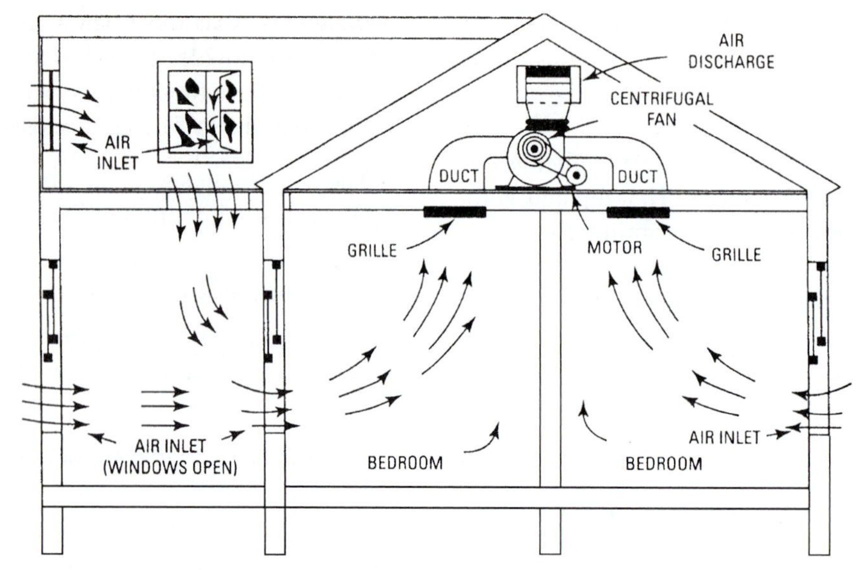

Fig. 14-44 *Installing a centrifugal fan in a one-family dwelling.*

side of the fan; the air is finally discharged through louvers as shown in Fig. 14-44.

Another installation is shown in Fig. 14-45. This fan is a centrifugal curved-blade type, mounted on a light angle-iron frame, which supports the fan wheel, shaft, and bearings. The air inlet in this installation is placed close to a circular opening, which is cut in an airtight board partition that serves to divide the attic space into a suction and discharge chamber. The air is admitted through open windows and doors and is then drawn up the attic stairway through the fan into the discharge chamber.

Routine Fan Operation

The routine of operation to secure the best and most efficient results with an attic fan is important. A typical operating routine might require that, in the late afternoon when the outdoor temperature begins to fall, the windows on the first floor and the grilles in the ceiling or the attic

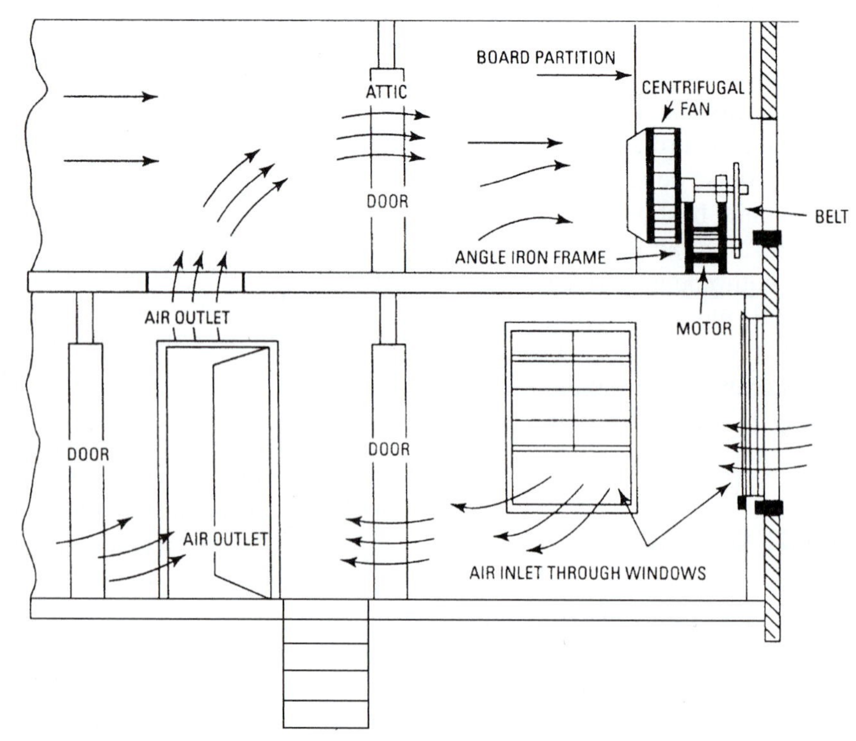

Fig. 14-45 *Typical attic installation of a belt-driven fan.*

floor be opened and the second-floor windows kept closed. This will place the principal cooling effect in the living rooms. Shortly before bedtime, the first-floor windows may be closed and those on the second floor opened to transfer the cooling effect to the bedrooms. A suitable time clock may be used to shut the motor off before arising time.

VENTILATION METHODS

Ventilation is produced by two basic methods: natural and mechanical. Open windows, vents, or drafts obtain natural ventilation, whereas mechanical ventilation is produced by the use of fans.

Thermal effect is possibly better known as *flue effect*. Flue effect is the draft in a stack or chimney that is produced within a building when the outdoor temperature is lower than the indoor temperature. This is caused by the difference in weight of the warm column of air within the building and the cooler air outside.

Air may be filtered two ways: dry filtering and wet filtering. Various air-cleaning equipment (such as filtering, washing, or combined filtering and washing devices) is used to purify the air. When designing the duct network, ample filter area must be included so that the air velocity passing through the filters is sufficient. Accuracy in estimating the resistance to the flow of air through the duct system is important in the selection of blower motors. Resistance should be kept as low as possible in the interest of economy. Ducts should be installed as short as possible.

Competent medical authorities have properly emphasized the effect of dust on health. Air-conditioning apparatus removes these contaminants from the air. The apparatus also provides the correct amount of moisture so that the respiratory tracts are not dehydrated, but are kept properly moist. Dust is more than just dry dirt. It is a complex, variable mixture of materials and, as a whole, is rather uninviting, especially the type found in and around human habitation. Dust contains fine particles of sand, soot, earth, rust, fiber, animal and vegetable refuse, hair, and chemicals.

REVIEW QUESTIONS

1. Define temperature.
2. What is the difference between absolute zero and zero on the Fahrenheit scale?
3. What is the formula for converting degrees Fahrenheit to degrees Celsius?
4. What is a manometer?
5. What does psig stand for?
6. What is the meaning of psia?
7. What is the difference between a wet-bulb thermometer and a dry-bulb thermometer?
8. Define dew point.
9. What is relative humidity?
10. What is a psychrometric chart?
11. How is a psychrometeric chart used in designing cooling systems?
12. What is enthalpy?
13. What is the purpose of ducturns?
14. Name at least three factors to be considered when selecting and sizing diffusers and grilles.
15. Name two ways to minimize air noise.
16. What is vortex shedding?
17. How do you minimize or eliminate cross talk between private offices due to the air-distribution system?
18. What is a fusible link?

15

CHAPTER

Comfort Air Conditioning

PERFORMANCE OBJECTIVES

After studying this chapter you will:

1. Know how to mount window units properly.
2. Know how to select the proper plug for operation of a window unit.
3. Know how to install a basement located furnace with air-conditioning ability.
4. Know how to install single-package rooftop units.
5. Understand the refrigerant piping for the rooftop unit.
6. Know how to select the proper copper tubing suction line for a unit.
7. Know how to select the proper copper tubing liquid line and size the installation.
8. Know how to troubleshoot mobile home units.
9. Know how to troubleshoot hermetic compressor type air-conditioning equipment.
10. Understand the newer types of ductless air conditioners and their controls.

Most of us think of window air conditioners when we think of motels and homes in areas where central air conditioning throughout the house is not affordable. These units require a lot of attention from the maintenance person inasmuch as they are operated by a number of different people, and children can reach the controls.

WINDOW UNITS

Room air conditioners are appliances that cool, dehumidify, filter, and circulate air. See Fig. 15-1. There are several manufacturers of window-mounted air-conditioning units. No attempt is being made here to represent all of them. Some general information is needed to help the repairperson make recommendations to those who request information on various units.

Today, energy conservation is important. Thus, most manufacturers specify the energy savings their unit will produce. The best way to determine the amount of energy consumed is to check the electrical requirements of the unit. Check this figure with the Btu generated by that power. Room air conditioners are usually rated according to the number of Btu they will produce in cooling. The smallest unit is about 5000 Btu. Thus, it is possible to determine which unit can produce a greater amount of cooling in terms of Btu per kilowatt hour.

Figure 15-2 shows a room air conditioner with the cover removed. The compressor, fan, evaporator, and condenser are visible. The three main parts of an air conditioner are the fan, filter, and cooling element. The fan pulls warm moist air into the unit. The air is moved through the filter, where dust particles are removed. Next, the air is passed over the cooling element, where it is cooled and dehumidified. The fan then returns the conditioned air to the room. This conditioned air has been cooled, dehumidified, and filtered.

Mounting

Window units are mounted in several different ways. Some are mounted in windows and some are mounted in a hole in the wall. The hole-in-the-wall mount is usually designed into the building at the time it is constructed. Different window types require that adapters

Fig. 15-1 *Window-mounted air-conditioning unit.* (Admiral)

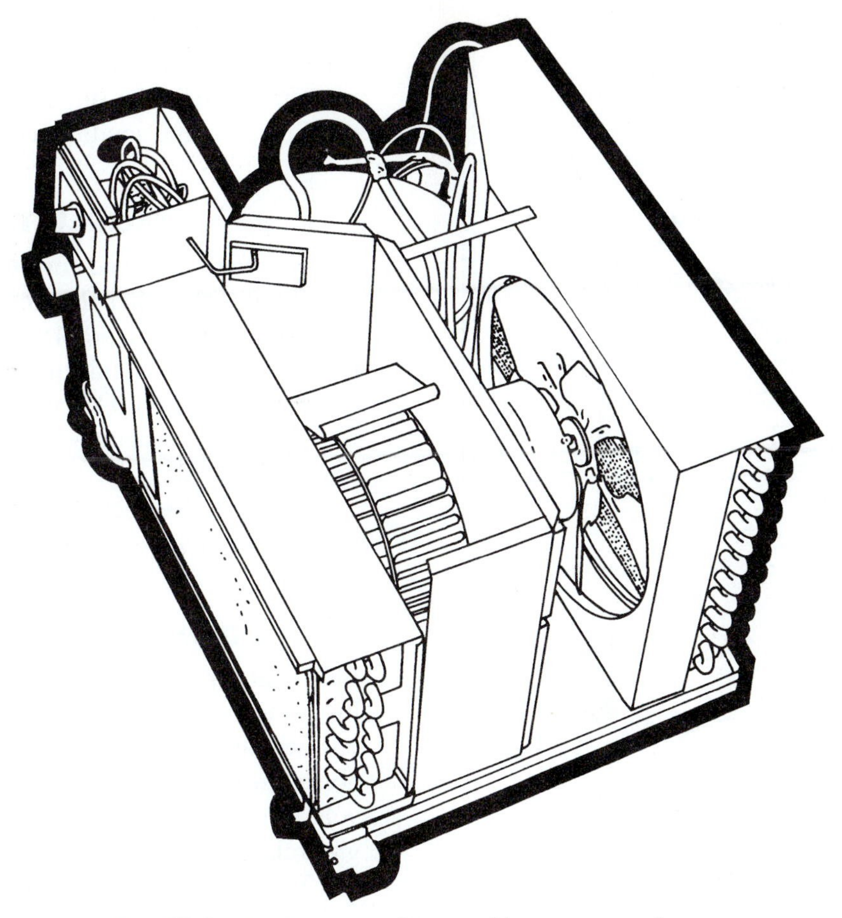

Fig. 15-2 *Window air conditioner with cover removed.* (Chrysler)

be fitted to the air-conditioning unit. Figure 15-3 shows a model with expanding side panels. A universal mounting kit is shown with flush mounting in Fig. 15-4. It indicates how the units will look inside and outside. Figures 15-5 and 15-6 show two mounting methods. Figure 15-7 illustrates a kit for mounting a unit in a sliding or casement window from $21^1/_4$ to 36 in. high. A Mylar slide-up weather tight wing is pulled up to fill the space around the air conditioner.

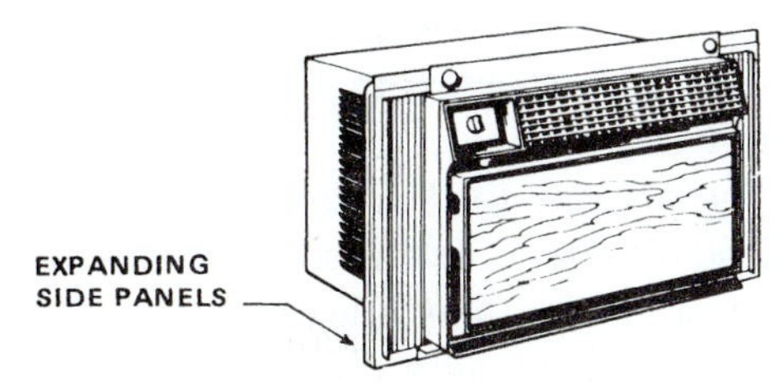

Fig. 15-3 *Window air conditioner with expanding side panels for window mounting.* (Chrysler)

The wall mounting of a small unit is shown in Fig. 15-8. A telescopic wall sleeve is available or an in-wall fixed sleeve is constructed when the house or office is built or remodeled.

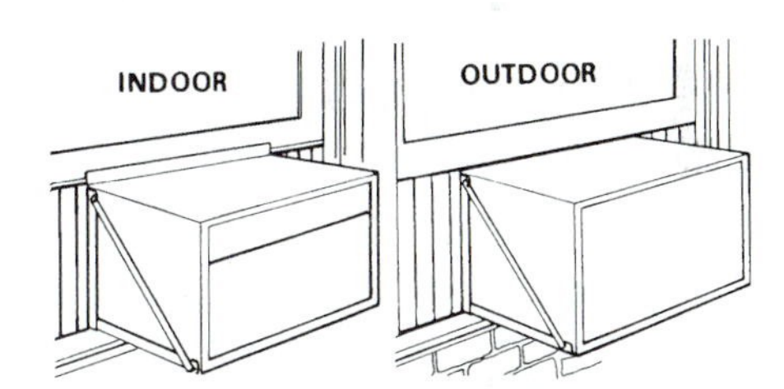

Fig. 15-4 *Universal mounting kit for flush mounting of window air conditioners.* (Chrysler)

The main advantage of the window unit is ease of installation. It is ready for use when plugged into the wall plug.

Electrical Plugs

Some units demand more current than can be safely furnished by a 120-V, 15-A circuit. Thus, a 230-240-V plug must be installed. Different plugs are used for different outlet sizes. Figure 15-9 shows several different plugs. Note the difference as in the arrangement of the slots. It is obvious that the plug must be fitted to the correct socket. The correct socket will have wiring of the proper size to handle the load. Plugs vary in design

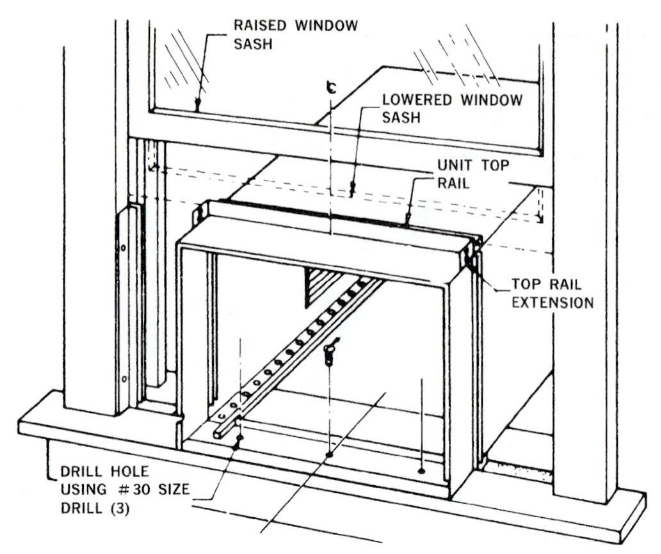

Fig. 15-5 *In-window mounting for an air conditioner.* (Chrysler)

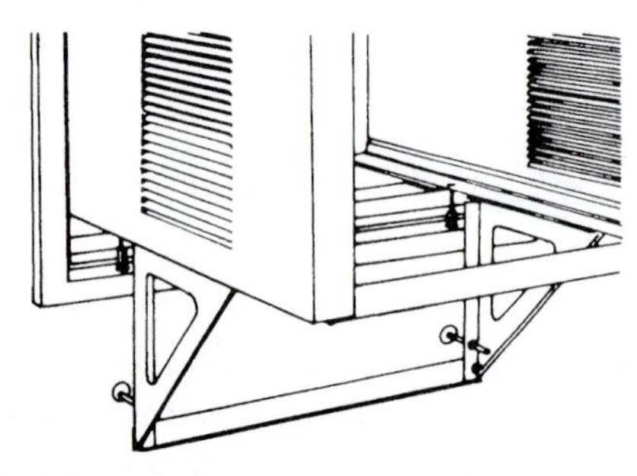

Fig. 15-6 *Another arrangement for in-window mounting of an air conditioner.* (Chrysler)

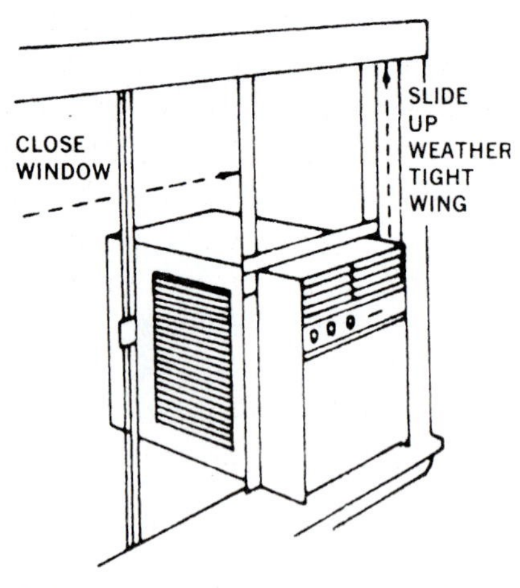

Fig. 15-7 *Mounting an air conditioner in a sliding or casement window.* (Chrysler)

because the power demands of the air conditioners will vary with the Btu rating. For example, one of the plugs shown in Fig. 15-9, requires 50 A. In most cases, this would require a special circuit for the air conditioner. In most cases with high current demands, the air conditioner is also capable of heating with an electric-heating coil installed for cold weather use.

Larger air conditioners do not come with plugs. They must be wired directly to the service. Wire of the correct size must be run to the junction box mounted on the side of the conditioner unit. See Fig. 15-10.

Maintenance

The units are designed for ease of maintenance. They usually require a filter change or cleaning at least once a year. Where dust is a problem, such maintenance should be more frequent. At this time, the condenser coil should be brushed with a soft brush and flushed with water. The filters should be vacuumed and then washed to remove dust. The outside of the case should be wiped clean with a soapy cloth. Needless to say, the cleaner the filter, the more efficient the unit.

Low-Voltage Operation

Electrical apparatus is designed to produce at full capacity at the voltage indicated on the rating plate. Motors operated at lower than rated voltage cannot provide full horsepower without shortening their service life. Low voltage can result in energy that is insufficient to energize relays and coils.

The *Air Conditioning and Refrigeration Institute* (ARI) certifies cooling units after testing them. The units are tested to make sure they will operate with 10 percent above or 10 percent below rated voltage. This does not mean that he units will operate continuously without damage to the motor. A large proportion of air-conditioning compressor burnouts can be traced to low voltage. That is because a hermetic compressor motor is entirely enclosed within the refrigerant cycle, it is very important that it is not abused either by overloading or low voltage. Both of these conditions can occur during peak load conditions.

A national survey has shown that the most common cause of compressor low voltage is the use of undersized conductors between the utility lines and the condensing unit. Low voltage becomes extremely important when it is necessary to plug into an existing circuit. The existing load on the circuit may be sufficient to load the circuit. In this case, the air-conditioning unit will result in too much load, blowing the fuse or tripping the circuit breaker.

However, in some cases, the fuse does not blow and the circuit breaker does not trip. This reduces the line voltage, since the wire for the circuit is too small to handle the current needed to operate all devices plugged into it. Check the circuit load before adding the air conditioner to the line. This will prevent damage to the unit.

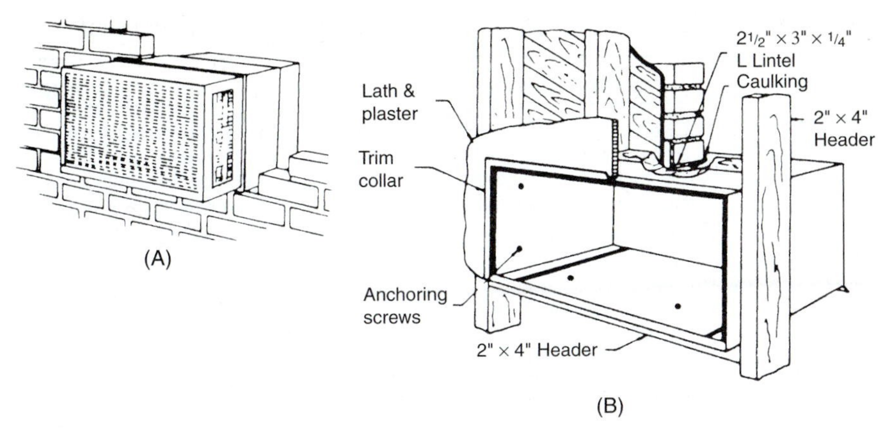

Fig. 15-8 *Two types of mounting in the wall. (A) Using a telescopic wall sleeve. (B) Using a fixed in-wall sleeve.* (Chrysler)

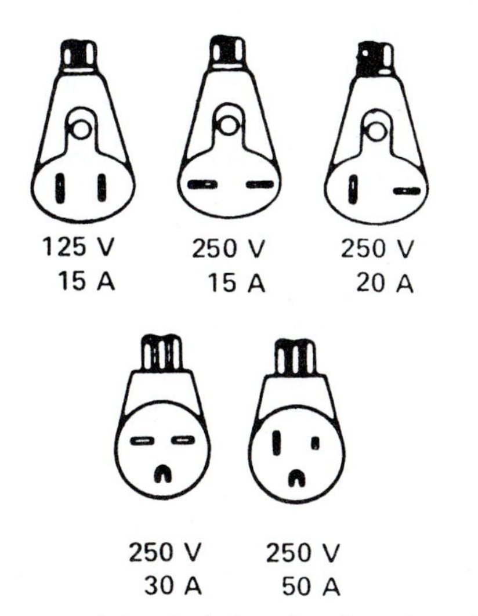

Fig. 15-9 *Types of electrical plugs found on air conditioners for home use.* (Chrysler)

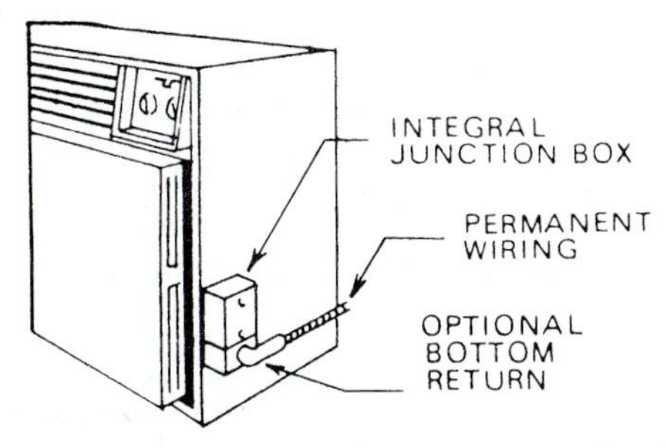

Fig. 15-10 *Wiring an air conditioner with a fixed electrical connection.* (Chrysler)

Troubleshooting

To troubleshoot this type of air-conditioning equipment, a troubleshooting table (Table 15-1) has been provided at the end of this chapter. The general troubleshooting procedures listed in the table are used for hermetically sealed compressors.

Evaporator Maintenance

If there is a low-temperature application of the evaporator with dry-coil-type units, then defrosting can be obtained. This is done by providing arrangements for the intake of outside air and the exhaust of this air to the outside during the defrosting period. Keep in mind that the outside temperature must be above 35°F when using this defrosting method. At very low temperatures, hot-gas defrosting may be employed. But, only when there is more than one evaporator connected to a condensing unit. This is so that the other unit can continue to provide refrigeration to the fixture.

When the operating conditions, such as design-fixture temperature, operating refrigerant temperature, and condensing-unit running time, will allow, the removal of frost on the evaporator may be accomplished automatically. Controls normally supplied with condensing units by most manufactures will do this. By adjusting these controls, the condensing unit can be cycled to suit the job at hand.

Automatic Defrosting

A variety of automatic defrosting that can be obtained on some installations using a time clock. The clock shuts the system down at suitable and convenient intervals. It returns the unit to normal operation afterward. As a general rule, this system is usable only when the fixture temperatures are around 32°F or higher. The fixture temperature air is used when the fixture can be quickly warmed from 32°F up to approximately 35°F in a reasonable period.

Table 15-1 *Troubleshooting Hermetic Compressor Type Air-Conditioning Equipment*

Trouble	Probable Cause	Remedy or Repair
Compressor will not start. (No hum.)	Open line circuit.	Check the wiring, fuses, and receptacle.
	Protector open.	Wait for reset. Check current drawn from line.
	Contacts open on control relay.	Check control, and check pressure readings.
	Open circuit in the motor stator.	Replace the stator or the whole compressor.
Compressor will not start. However, it hums intermittently. Cycles with the protector.	Not wired correctly.	Check wiring diagram and actual wiring.
	Line voltage low.	Check line voltage. Find where line voltage is dropped. Correct.
	Start capacitor open.	Replace start capacitor.
	Relay contacts do not close.	Check by manually operating. Replace if defective.
	Start winding open.	Check stator leads. Replace compressor if the leads are OK.
	Stator winding grounded. (Usually blows fuse.)	Check stator leads. Replace compressor if leads are OK.
	Discharge pressure too high.	Remove cause of excessive pressure. Discharge shutoff and receiver valves should be open.
.	Compressor too tight.	Check oil level. Correct the binding cause. If this cannot be done, replace compressor.
	Start capacitor weak.	Replace the capacitor.
Compressor starts. Motor will not speed up enough to have start winding drop out of circuit.	Line voltage low.	Increase the voltage.
	Wired incorrectly.	Rewire according to wiring diagram.
	Relay defective.	Check operation. If defective, replace.
	Run capacitor shorted.	Disconnect the run capacitor and check for short.
	Start and run windings shorted.	Check winding resistances. If incorrect, replace the compressor.
	Start capacitor weak.	Check capacitors. Replace those defective.
	Discharge pressure high.	Check discharge shutoff valves. Check pressure.
	Tight compressor.	Check oil level. Check binding. Replace if necessary.
Compressor starts and runs. However, it cycles on the protector.	Low line voltage.	Increase the voltage.
	Additional current being drawn through the protector.	Check to see if fans or pumps are wired to the wrong connector.
	Suction pressure is too high.	Check compressor. See if it is the right size for the job.
	Discharge pressure is too high.	Check ventilation. Check for over-charge. Also check for obstructions to air flow or refrigerant flow.
	Protector is weak.	Check current. Replace protector if it is not clicking out at right point.
	Run capacitor defective.	Check capacitance. Replace if found defective.
	Stator partially shorted or grounded.	Check resistance for a short to the frame. Replace if found shorted to ground (frame).
	Inadequate motor cooling.	Correct air flow.
	Compressor tight.	Check oil level. Check cause of binding.
	3ϕ line unbalanced.	Check each leg or phase. Correct if the voltages are not the same between legs
	Discharge valve leaks or is damaged.	Replace the valve plate.
Start capacitors burn out.	Short cycling.	Reduce the number of starts. They should not exceed 20 h.
	Prolonged operation with start winding in circuit.	Reduce the starting load. Install a crankcase pressure limit valve. Increase low voltage if this is found to be the condition. Replace the relay if it is found to be defective.
	Relay contacts sticking.	Clean the relay contacts. Or, replace the relay.
	Wrong relay or wrong relay setting.	Replace the relay.
	Wrong capacitor.	Check specifications for correct size capacitor. Be sure the MFD and WVDC are correct for this compressor.
	Working voltage of capacitor too low.	Replace with capacitor of correct WVDC.
	Water shorts out terminals of the capacitor.	Place capacitor so the terminals will not get wet.

Table 15-1 (Continued)

Trouble	Probable Cause	Remedy or Repair
Run capacitors burn out. They spew their contents over the surfaces of anything nearby. This problem can usually be identified with a visual check.	Excessive line voltage.	Reduce line voltage. It should not be over 10% of the motor rating.
	Light load with a high line voltage.	Reduce voltage if not within 10% overage limit.
	Voltage rating of capacitor too low.	Replace with capacitors of the correct WVDC.
	Capacitor terminals shorted by water.	Place capacitor so the terminals will not get wet.
Relays burn out.	Low-line voltage.	Increase voltage to within 10% of limit.
	High-line voltage.	Reduce voltage to within 10% of the motor rating.
	Wrong size capacitor.	Use correct size capacitor. The proper MFD rating should be installed.
	Short cycling.	Decrease the number of starts per hour.
	Relay vibrating.	Make sure you mount the relay rigidly.
	Wrong relay.	Use the recommended relay for the compressor motor.

Note: These are general problems that can be identified with any hermetic compressor. Problems with the electrical switches, valves, and tubing are located by using the knowledge you have acquired previously in the theory and applications sections of this book.

EVAPORATORS FOR ADD-ON RESIDENTIAL USE

One of the more efficient ways of adding whole-house air conditioning is by adding an evaporator coil in the furnace. The evaporator coil becomes an important part of the whole system. It can be added to the existing furnace to make a total air conditioning and heating package. There are two types of evaporators—down-flow and up-flow.

The down-flow evaporator is installed beneath a down-flow furnace. See Fig. 15-11. Lennox makes down-flow models in 3-, 4-, and 5-ton sizes with an inverted "A" coil. See Fig. 15-12. Condensate runs down the slanted side to the drain pan. This unit can be installed in a closet or utility room wherever the furnace is located. This type of unit is shipped factory assembled and tested.

Up-flow evaporators are installed on top of the furnace. They are used in basement installations and in closet installations. See Fig. 15-13. The adapter base and the coil in Fig. 15-14, are shown, as they would fit onto the top of an up-flow existing furnace. The plenum must be removed and replaced once the coil has been placed on top of the furnace.

In most cases, use of an add-on top evaporator means the fan motor must be changed to a higher horsepower rating. The evaporator in the plenum makes it more difficult to force air through the heating system. In some cases, the pulley size on the blower and the motor must be changed to increase the cfm (cubic feet per minute) moving past the evaporator.

Some motors have sealed bearings. Some blower assemblies, such as that shown in Fig. 15-15, have sealed bearings. However, some have sleeve bearings. In such cases, the owner should know that the motor and blower must be oiled periodically to operate efficiently.

Figure 15-16 shows how the evaporator coil sits on top of the furnace, making the up-flow type of air conditioning operate properly. The blower motor is located below the heater and plenum.

The evaporator is not useful unless it is connected to a compressor and condenser. These are usually located outside the house. Figure 15-17 shows the usual outdoor compressor and condenser unit. This unit is capable of furnishing 2.4 to 5 tons of air conditioning, ranging in capacities from 27,000 to 58,000 Btu. Note that this particular unit has a U-shaped condenser coil that forms three vertical sides. The extra surface area is designed to make the unit more efficient in heat transfer. The fan, which is thermostatically operated, has two-speeds. It changes to low speed when the outside temperature is below 75°F (23.9°C).

Like most compressors designed for residential use, this compressor is hermetically sealed. The following safety devices are built in: a suction-cooled overload protector, a pressure-relief valve, and a crankcase heater. Controls include high- and low-pressure switches. They automatically shut off the unit if discharge pressure becomes excessive or suction pressure falls too low.

In apartments where space is at a premium, a different type of unit is used. It differs only in size. See Fig. 15-18. These compact units have a blower, filter, and evaporator coil contained in a small package. They have electric-heat coils on top. In some cases, hot water is used to heat in the winter.

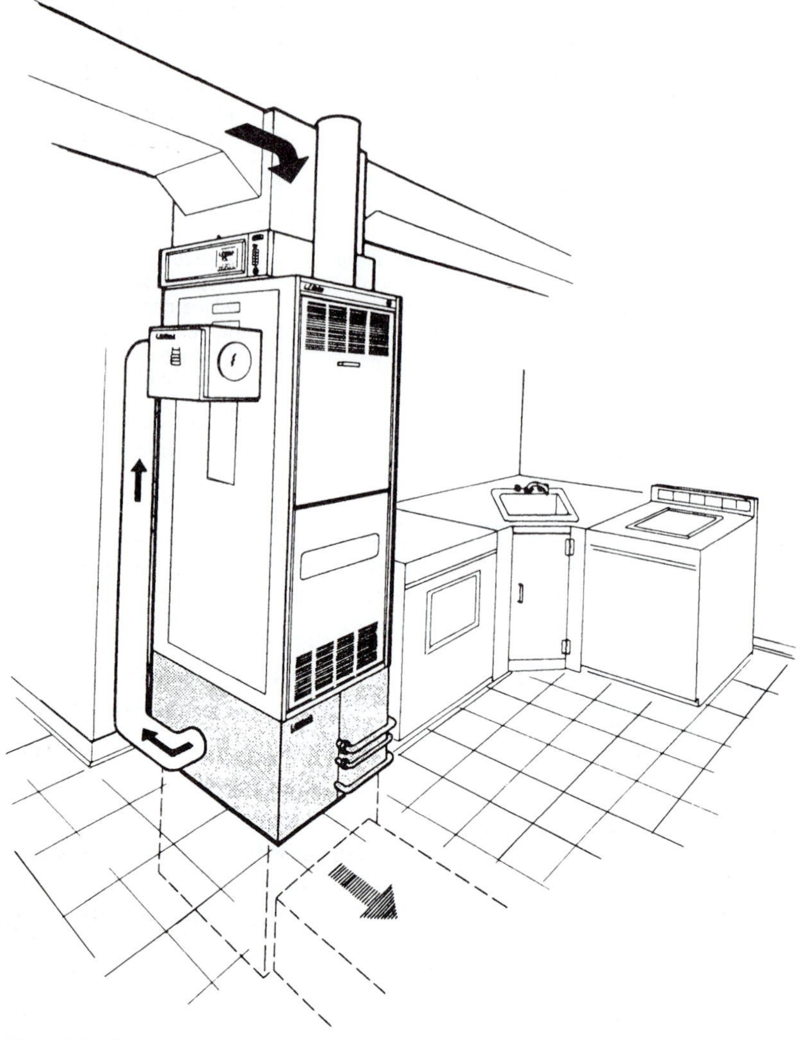

Fig. 15-11 *Utility room installation of a down-flow evaporator on an existing furnace.* (Lennox)

Fig. 15-12 *Evaporator units for add-on cooling for up-flow and down-flow furnaces.* (Lennox)

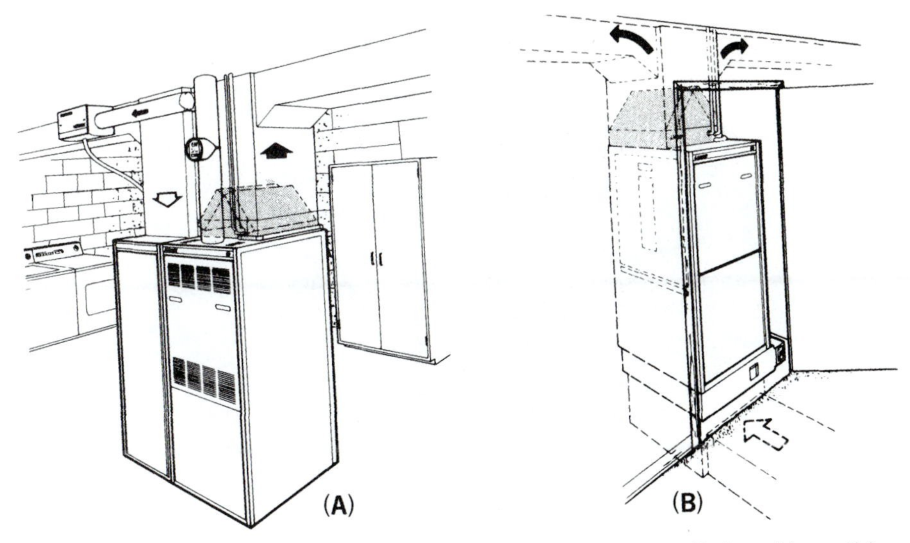

Fig. 15-13 *Installation of up-flow evaporators. (A) Basement installation with an oil furnace, return air cabinet, and automatic humidifier. (B) Closed installation of an evaporator coil with electric furnace and electronic air cleaner.* (Lennox)

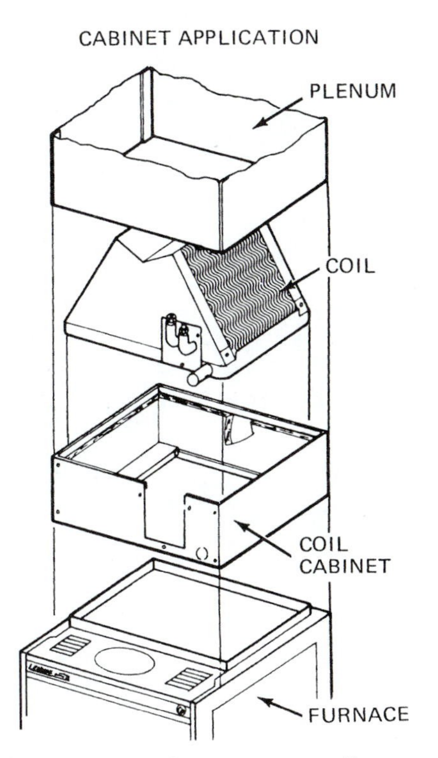

Fig. 15-14 *Installation of an evaporator coil on top of an existing furnace installation.* (Lennox)

Figure 15-19 shows the various ways in which these units may be mounted. The capacity is usually 18,000 to 28,000 Btu. Note location of the control box. This is important since most of the maintenance problems are caused by electrical, rather than refrigerant, malfunctions. This type of unit allows each apartment tenant to have his or her own controls.

Troubleshooting

To troubleshoot this type of air-conditioning equipment, a troubleshooting table (Table 15-1) has been provided. The general troubleshooting procedures listed there are used for hermetically sealed compressors.

REMOTE SYSTEMS

A remote system designed for home or commercial installation can be obtained with a complete package. It has the condensing unit, correct operating refrigerant charge, refrigerant lines, and evaporator unit.

The charge in the line makes it important to have the correct size of line for the installation. Metering control of the refrigerant in the system is accomplished by the exact sizing (bore and length) of the liquid line. The line must be used as delivered. It cannot be shortened.

Lennox has the *refrigerant flow control* (RFC) system. It is a very accurate means of metering refrigerant in the system. It must never be tampered with during installation. The whole principle of the RFC system involves matching the evaporator coil to the proper length and bore of the liquid line. This is believed by the manufacturer to be superior to the capillary-tube system. The RFC equalizes pressures almost instantly after the compressor stops. Therefore, it starts unloaded, eliminating the need for any extra controls. In addition, a precise amount of refrigerant charge is added to the system at the factory, resulting in trouble-free operation.

The condensing unit is shipped with a complete refrigerant charge. The condensing unit and evaporator

Fig. 15-15 *Motor-driven blower unit.* (Lennox)

Fig. 15-16 *Cutaway view of the furnace, blower, and evaporator coil on an air-conditioning and heating unit.* (Lennox)

Fig. 15-17 *A 2.5- to 5-ton condensing unit with a one-piece, wraparound, U-shaped condenser coil. This unit has a two-speed fan and sealed-in compressor for quite operation.* (Lennox)

are equipped with flared liquid and suction lines for quick connection. The compressor is hermetically sealed.

The unit may be built in and weatherproofed as a rigid part of the building structure. See Fig. 15-20. The condensing unit can be free-standing on rooftops or slabs at grade level.

Figure 15-21 shows the condensing unit designed for the apartment developer and volume builder. It comes in 1-, 1.5-, and 2-ton sizes. Cooling capacities

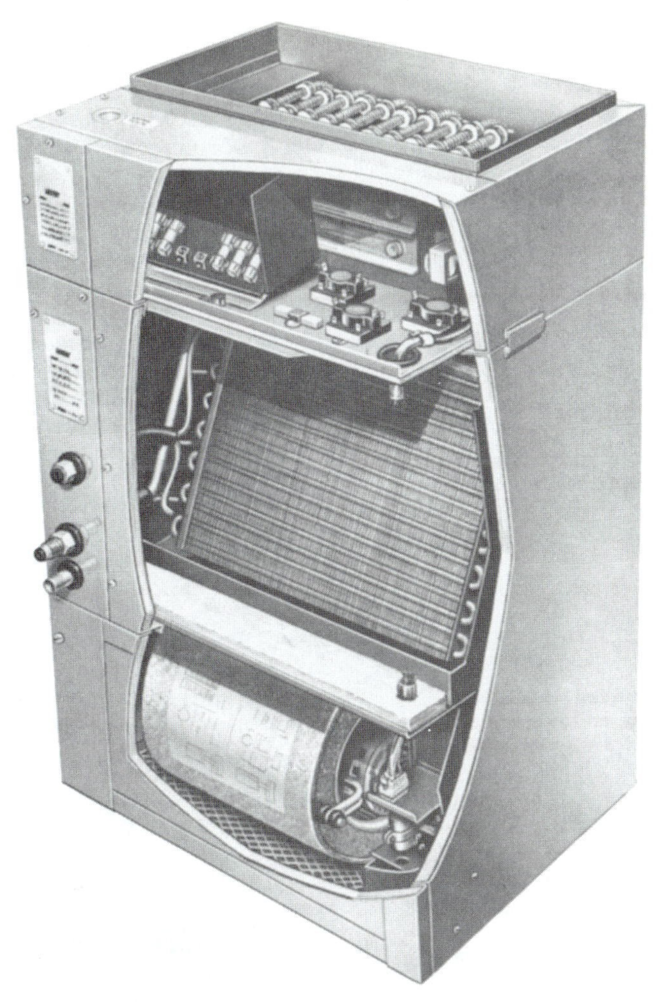

Fig. 15-18 *This unit fits in small closets or corners. It has the possibility of producing hot water or electric heat as well as cooling. Note the heating coils on the top of the unit. It comes in three sizes, 1.5, 2, and 2.5 tons.* (Lennox)

range from 17,000 through 28,000 Btu. An aluminum grille protects the condenser while offering low resistance to air discharge. The fan is mounted for less noise. It also reduces the possibility of air re-circulation back through the condenser when it is closely banked for multiple installations. When mounted at grade level, this also keeps hot air discharge from damaging grass and shrubs.

SINGLE-PACKAGE ROOFTOP UNITS

Single-rooftop units can be used for both heating and cooling for industrial and commercial installations. Figure 15-22 shows such a unit. It can provide up to 1.5 million Btu if water heat is used. It can also include optional equipment to supply heat, up to 546,000 Btu, using electricity. They can use oil, gas, or propane for heating fuels. These units require large amounts of energy to operate. It is possible to conserve energy by using more sensitive controls.

Highly sensitive controls monitor supply air. They send signals to the control module. It, in turn, cycles the mechanical equipment to match the output to the load condition.

An optional device for conserving energy is available. It has a "no load" band thermostat that has a built-in differential of 6°F (3.3°C). This gives the system the ability to "coast" between the normal control points without consuming any primary energy within the recommended comfort-setting range.

Another feature that is prevalent is a refrigerant heat-reclaim coil. It can reduce supermarket heating costs significantly. A reheat coil can be factory installed downstream from the evaporator coil. It will use the condenser heat to control humidity and prevent overcooling.

A unit of this size is designed for a large store or supermarket. Figure 15-23 shows how the rooftop model is mounted for efficient distribution of the cold air. Since cold air is heavy, it will settle quickly to floor level. Hot air rises and stays near the ceiling in a room. Thus, it is possible for this warmer air to increase the temperature of the cold air from the conditioner before it comes into contact with the room's occupants.

Smoke Detectors

Photocell detectors detect smoke within the system. They actuate the blower motor controls and other devices to perform the following:

1. Shut off the entire system.
2. Shut down the supply blower, close outside-air and return-air dampers and runs.
3. Run supply-air blower, open outside-air dampers, close return-air dampers, and stop return-air blower or exhaust fans.
4. Run supply-air blower, open outside-air dampers, close return-air dampers, and run return-air blower or exhaust fan. Actuation occurs when smoke within the unit exceeds a density that is sufficient to obscure light by a factor of 2 or 4 percent per foot. A key switch is used for testing. Two detectors are used. One is located in the return air section. The other is located in the blower section downstream from the air filters.

Firestats

Firestats are furnished as standard equipment. Manual reset types are mounted in the return air and supply air stream. They will shut off the unit completely when either firestat detects excessive air temperatures.

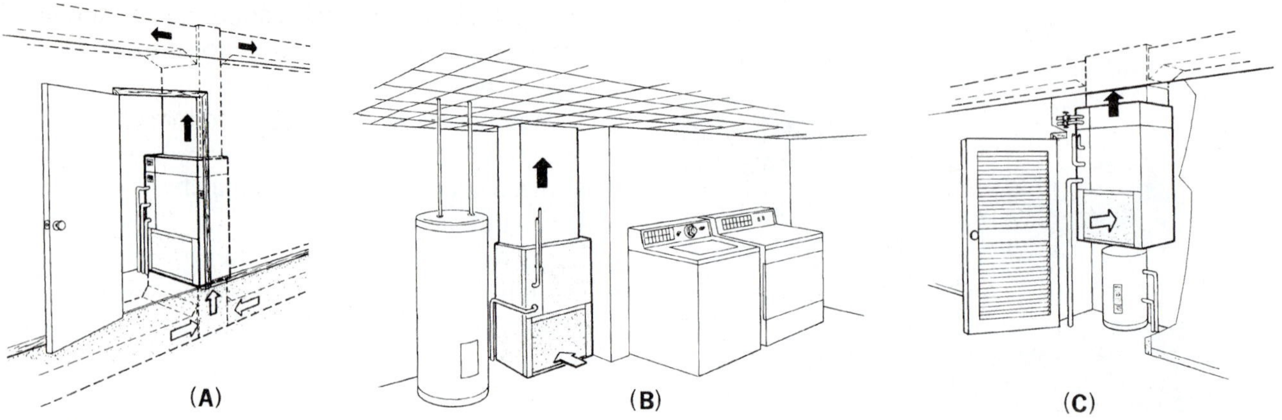

Fig. 15-19 *Typical installations of the blower, coil, and filter units. (A) A closet installation with electric-heat section. (B) A utility-room installation. (C) A wall-mounted installation with hot-water section.* (Lennox)

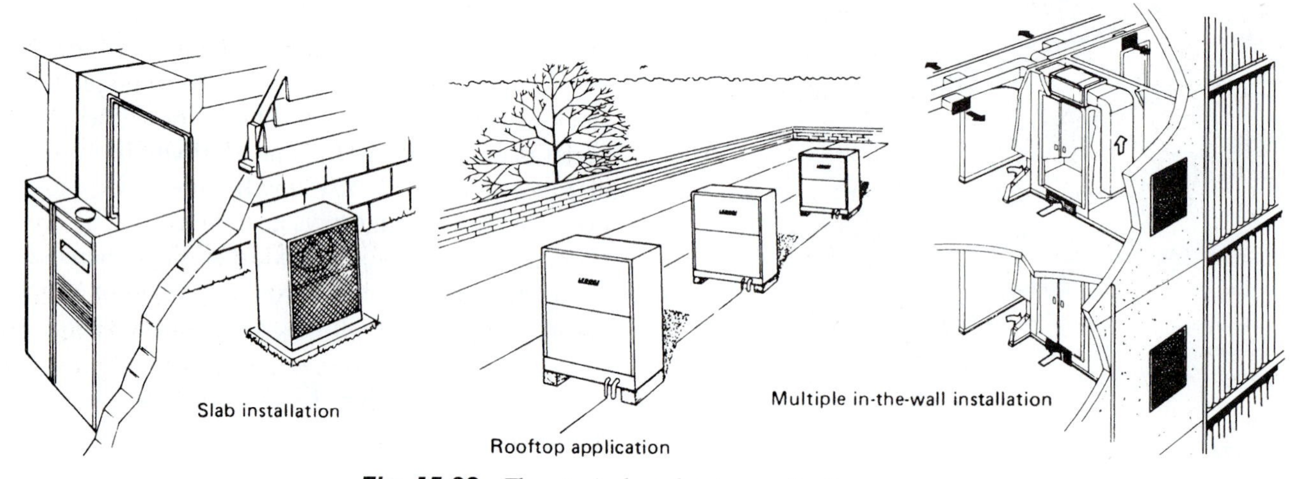

Fig. 15-20 *Three typical condensing applications.* (Lennox)

On this type of unit, the blowers are turned by one motor with a shaft connecting the two fans. There are three small fan motors and blades mounted to exhaust air from the unit. There are four condenser fans. The evaporator coil is slanted. There are two condenser coils mounted at angles. There are two compressors. The path for the return air is through the filters and evaporator coil back to the supply-air ductwork.

Return-Air Systems

Return-air systems are generally one of two types: the ducted return-air system or the open plenum return-air system ("sandwich space"). See Fig. 15-24. The ducted return-air system duct is lined with insulation, which greatly reduces noise.

The open-plenum system eliminates the cost of return-air returns or ducts and is extremely flexible. In a building with relocatable interior walls, it is much easier to change the location of a ceiling grille than reroute a ducted return system.

Acoustical Treatment

Insulating the supply duct reduces duct loss or gain and prevents condensation. Use $1^1/_2$-lb density on ducts that deliver air velocities up to 1500 fpm.

Three-pound density or neoprene-coated insulation is recommended for ducts that handle air at velocities greater than 1500 fpm. Insulation can be $^1/_2$ or 1 in. thick on the inside of the duct.

Where rooftop equipment uses the sandwich space for the return air system, a return air chamber should be connected to the air-inlet opening. Such an air chamber is shown in Fig. 15-24. This reduces air handling sound transmission through the thin ceiling panels. It should be sized not to exceed 1500-fpm return-air velocity. The duct can be fiberglass or a fiberglass-lined metal

Fig. 15-21 *Apartment house or residential condensing unit. It can be installed through the wall, free standing or at grade level, or on the roof.* (Lennox)

duct. A ceiling return-air grille should not be installed within 15 ft of the duct inlet.

Volume Dampers

Volume dampers are important to a good system design. Lengths of supply runs vary and usually have the same cubic-foot measurements. Therefore, balancing dampers should be used in each supply branch run. The installer must furnish and install the balancing dampers. Dampers should be installed between the supply air duct and the diffuser outlet.

There are several ways in which rooftop conditioners can be installed. Figure 15-25 shows three installation methods.

Refrigerant Piping

Figure 15-26 shows how the unit, just discussed, is hooked up for refrigerant flow. Note how the two compressors are hooked into the operation of the unit. Note also the location of the reheat condenser coil, if it is installed in this type of unit.

Troubleshooting

To troubleshoot this type of air-conditioning equipment, a troubleshooting table (Table 15-1) has been provided. The general troubleshooting procedures

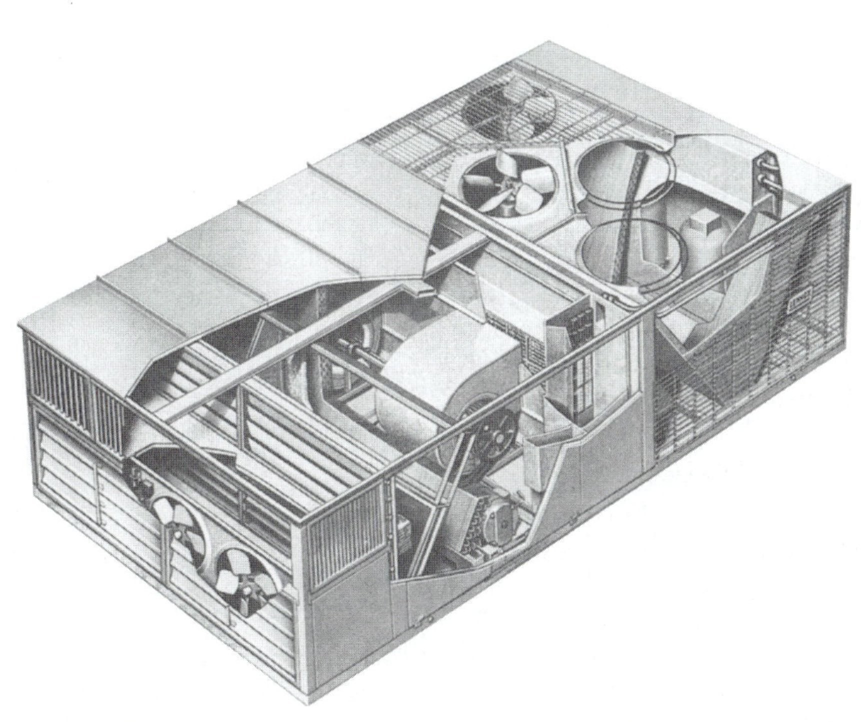

Fig. 15-22 *Single-zone rooftop system. This unit is used for industrial and commercial-market applications. Cooling capacity ranges from 8 through 60 tons.* (Lennox)

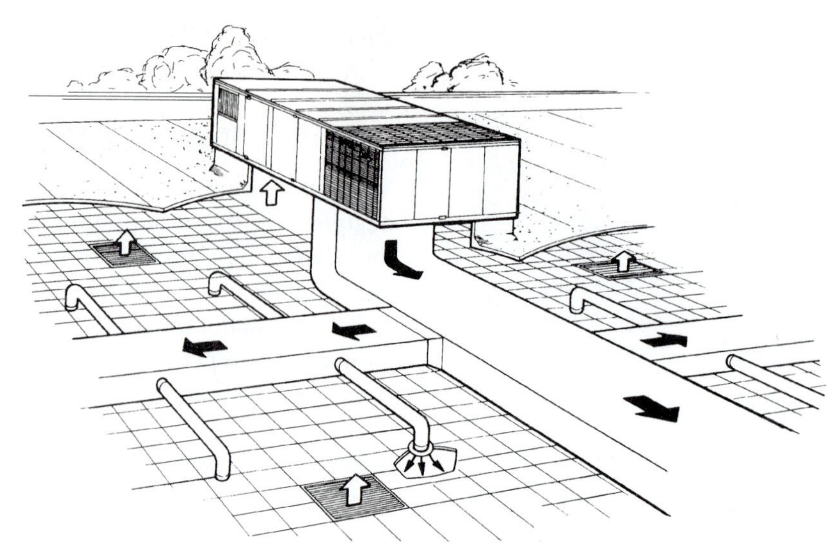

Fig. 15-23 *Typical rooftop installation of the single-zone system.* (Lennox)

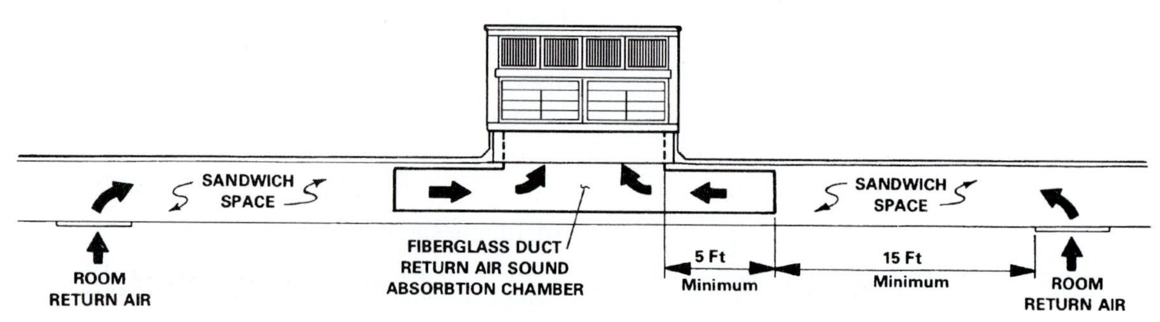

Fig. 15-24 *Return-air system for the rooftop unit.* (Lennox)

REFRIGERATION AND AIR CONDITIONING TECHNOLOGY

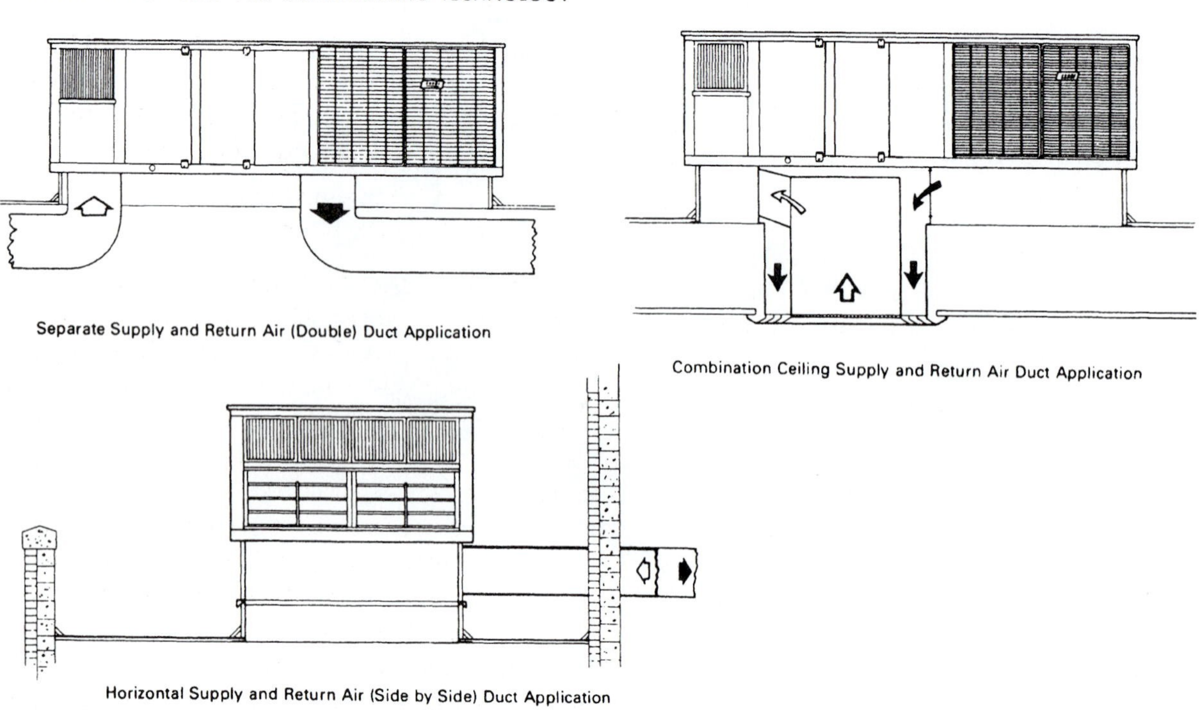

Fig. 15-25 *Choice of air patterns for the rooftop unit.* (Lennox)

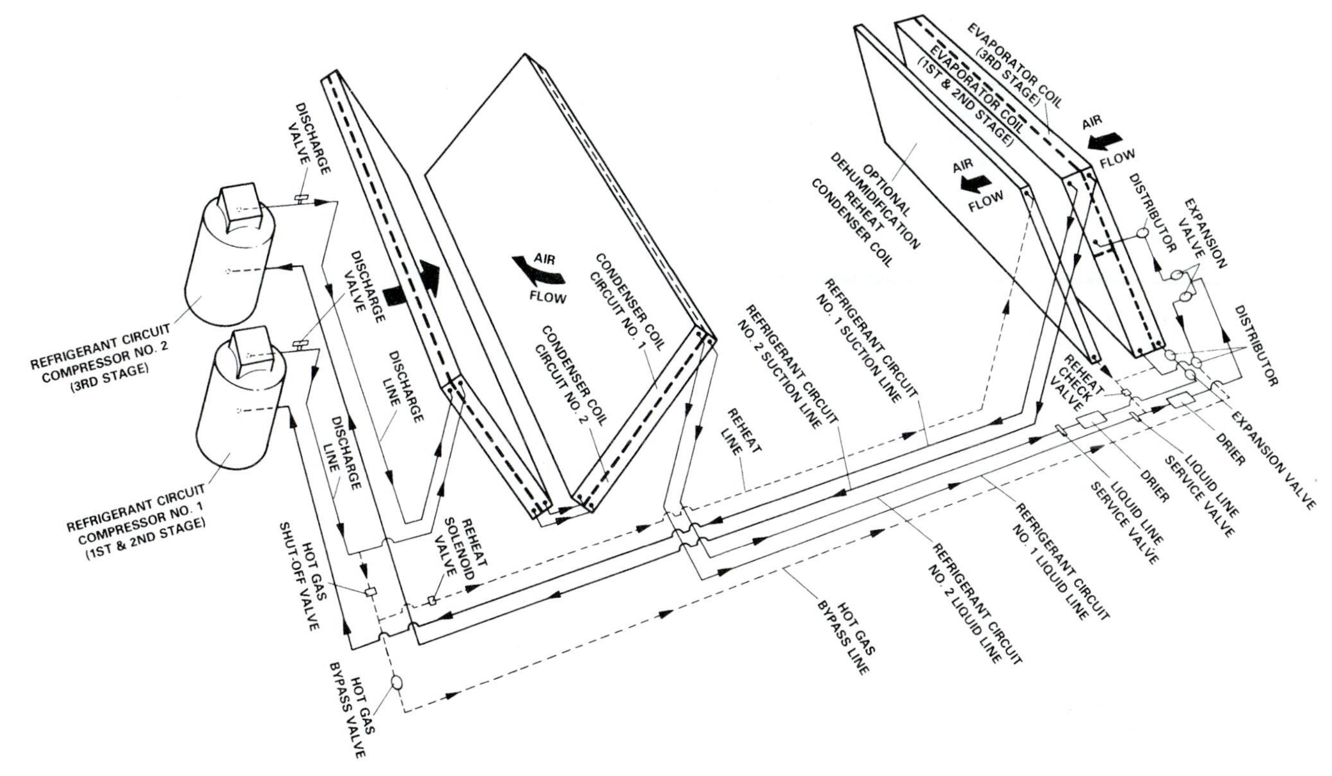

Fig. 15-26 *Refrigerant piping for the rooftop unit.* (Lennox)

listed in the table are used for hermetically sealed compressors.

REFRIGERANT PIPE SIZES

In some installations on rooftop or slab, the unit does not come self-contained. This means the condensing unit may be located on a slab and the evaporator coil in some other location. See Fig. 15-27. In this instance, it is necessary to make sure the refrigerant-piping design is correct.

The principal objectives of refrigerant piping are to:

- Insure proper feed to evaporators.
- Provide practical line sizes without excessive pressure drop because pressure losses decrease capacity and increase power requirements.
- Protect compressors by preventing excessive lubricating oil from being trapped in the system.
- Minimize the loss of lubricating oil from the compressor at all times.
- Prevent liquid refrigerant from entering the compressor during operation and shutdown. In general, larger pipe sizes have lower pressure drops, lower power requirements, and high capacity. The more economical smaller pipe sizes provide sufficient velocities to carry oil at all loads.

Liquid-Line Sizing

Liquid-line sizing presents less of a problem than suction-line sizing for the following reasons:

- The smaller liquid-line piping is much cheaper than suction-line piping.
- Compressor lubricating oil and fluorinated hydrocarbon refrigerants, such as R-22 in the liquid state, mix well enough that, in normal comfort air-conditioning uses, positive oil return is not a problem.
- Vertical risers, traps, and low velocities do not interfere with oil return in liquid lines.

Although liquid-line sizing offers more latitude than suction-line sizing, high-pressure drops should be avoided to prevent flash-gas formation in the liquid line. Flash gas interferes with expansion-valve operation. It also causes liquid-distribution problems where more than one evaporator coil is being used. Where applications requirements are such that flash gas is unavoidable, there are methods of making allowance for it. Liquid refrigerant pumps and separation tanks can be used. (See *ASHRAE Guide and Data Book* for details.) The acceptable pressure drop depends on the amount of subcooling the condenser unit offers and the inherent losses resulting from liquid lift, if present. It is advisable to have the liquid slightly subcooled when it reaches the expansion

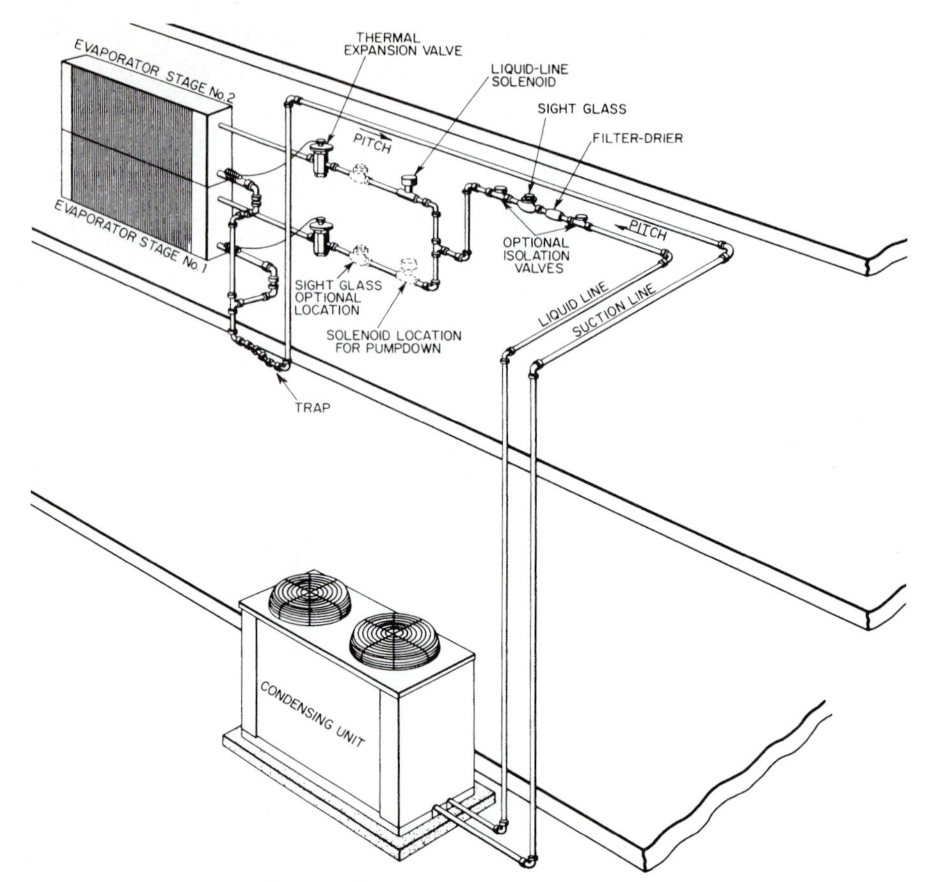

Fig.15-27 *Refrigerant pipe sizes.* (Bryant)

valve. This helps avoid flash-gas formation and provides stable operation of the expansion valve.

Suction-Line Sizing

The importance of suction-line design and sizing cannot be overemphasized. Lubricating oil does not mix well with the cold refrigerant vapor leaving the evaporator(s). It must be returned to the compressor either by entrainment with the refrigerant vapor or by gravity.

Traps and areas where oil may pool must be kept to a minimum. This is because large quantities of oil may become "lost" in the system. Piping should be level or with a slight pitch in the direction of the compressor.

Suction-line evaporator takeoffs should be designed so that oil cannot drain into idle coils. The common suction for multievaporator coils should be lower than the lowest evaporator outlet. Where an application requires that a common suction be above one or more of the coils on a multievaporator coil application, a suction riser with top loop connection is recommended.

Systems requiring a suction riser are more difficult to design. Sizing the pipe for minimum gas velocity at minimum system capacity (minimum displacement and suction temperature) may result in excessive pressure losses at full load. Excessive pressure losses in a suction riser may be compensated for by increasing pipe sizing in horizontal or down runs to reduce total system pressure losses. It can also be compensated for by using double suction risers. In comfort air conditioning, the use of double-suction risers is the *exception*, rather than the rule. However, where necessary, it proves a valuable tool. (See *ASHRAE Guide Book* for details.)

Sizing Procedure Use the following procedure for selecting the proper refrigerant pipe size:

- Determine the "measured" length of the straight pipe. Do this separately for the liquid line and the suction line.
- The fittings cause additional friction above that created by the measured length of straight pipe. To account for this extra friction, the equivalent method is used. This technique consists of adding to the measured length an additional length of straight pipe that has the same pressure loss as the fittings.

The pressure loss of the fittings depends upon the number, type, and diameter of the fittings used. The dependence of the pressure loss in the fitting on the diameter of the fittings presents a real problem. To determine the pipe diameter, it is necessary to know the total equivalent length of pipe, the sum of the measured straight pipe length, and the fitting losses in equivalent pipe length. On the other hand, since the fitting pressure losses do depend upon diameter, it is necessary to know the diameter of the pipe to determine the fitting losses in equivalent pipe length.

The result is a situation in which it seems that the solution is needed to solve the problem. Fortunately, if the pipe diameter can be estimated with reasonable accuracy, the fitting losses can be determined. In turn, the correct pipe diameter can be determined.

A pipe diameter selection based upon 1.5 times the measured straight pipe length can be used to obtain an estimate of the pipe diameter.

- In the majority of applications, the design conditions are a pressure loss equal to 2°F (1.1°C), a suction temperature of 40°F (4.4°C), and a condensing temperature of 105°F (40.6°C).
- Once the estimated pipe diameter is determined, the actual equivalent length (corrected if necessary) can be determined. Then the final pipe diameter can be determined.

For example, the following demonstrates the procedure that was just outlined. A Bryant unit—566B360RCU—is used for the example. This 360-size unit has two separate 180,000 Btu/h (15-ton) condensing sections. Each section is piped individually. The liquid and suction lines for each condensing section should be sized for 15 tons. This example covers only suction-line sizing. However, the same procedure is used for the liquid lines.

In the majority of comfort air-conditioning applications, the design conditions closely approximate 105°/40°/2°. The following design conditions are used to demonstrate the use of the correction factors:

Condensing temperature	110°F
Suction temperature	35°F
Maximum friction drop	2.5°F

The measured straight pipe length in suction or liquid line equals 100 ft. Number and type of fitting in each line:

Ten standard 45° elbows

Four gate valves

Four standard 90° elbows

The measured straight pipe length of the suction line is equal to 100 ft. Therefore, use 150 ft as a first approximation of the total equivalent length for a combination of 150 ft and 15 tons. Table 15-2 gives an estimated pipe diameter of $1^{5}/_{8}$ in.

Once the estimated pipe diameter is obtained, obtain the following fitting losses in equivalent lengths from Tables 15-3 and 15-4. Use the 2 in. pipe size for $1^{5}/_{8}$ in. pipe.

Ten standard 45° elbows:

$$10 \times 2.6 = 26 \text{ ft}$$

Four gate valves:

$$4 \times 2.3 = 9.2 \text{ ft}$$

Four standard 90° elbows:

$$4 \times 5.0 = \frac{20}{55.2} \text{ ft}$$

Actual total equivalent pipe length:

100 ft measured pipe length + 55.2 ft fitting losses = 155.2 ft

Correct the nominal tonnage for the 35°F suction and 110°F condensing temperatures. The factor is 1.13.

Table 15-2 *Copper Tubing Suction Line Sizes (in Inches) for Pressure Drop Corresponding to 2°F*

Equivalent Pipe Length (ft)	R-22 Refrigerant Systems (tons)															
	2	3	3.5	4	5	7.5	10	15	20	25	30	40	50	60	80	100
500	$1^{1}/_{8}$	$1^{3}/_{8}$	$1^{3}/_{8}$	$1^{3}/_{8}$	$1^{5}/_{8}$	$2^{1}/_{8}$	$2^{1}/_{8}$	$2^{5}/_{8}$	$2^{5}/_{8}$	$2^{5}/_{8}$	$3^{1}/_{8}$	$3^{1}/_{8}$	$3^{5}/_{8}$	$3^{5}/_{8}$	$4^{1}/_{8}$	$5^{1}/_{8}$
400	$1^{1}/_{8}$	$1^{3}/_{8}$	$1^{3}/_{8}$	$1^{3}/_{8}$	$1^{3}/_{8}$	$1^{5}/_{8}$	$2^{1}/_{8}$	$2^{1}/_{8}$	$2^{5}/_{8}$	$2^{5}/_{8}$	$3^{1}/_{8}$	$3^{1}/_{8}$	$3^{5}/_{8}$	$3^{5}/_{8}$	$4^{1}/_{8}$	$5^{1}/_{8}$
300	$1^{1}/_{8}$	$1^{1}/_{8}$	$1^{3}/_{8}$	$1^{3}/_{8}$	$1^{3}/_{8}$	$1^{5}/_{8}$	$2^{1}/_{8}$	$2^{1}/_{8}$	$2^{5}/_{8}$	$2^{5}/_{8}$	$2^{5}/_{8}$	$3^{1}/_{8}$	$3^{1}/_{8}$	$3^{5}/_{8}$	$3^{5}/_{8}$	$4^{1}/_{8}$
200	$^{7}/_{8}$	$1^{1}/_{8}$	$1^{1}/_{8}$	$1^{1}/_{8}$	$1^{3}/_{8}$	$1^{5}/_{8}$	$1^{5}/_{8}$	$2^{1}/_{8}$	$2^{1}/_{8}$	$2^{5}/_{8}$	$2^{5}/_{8}$	$2^{5}/_{8}$	$3^{1}/_{8}$	$3^{1}/_{8}$	$3^{5}/_{8}$	$3^{5}/_{8}$
150	$^{7}/_{8}$	$1^{1}/_{8}$	$1^{1}/_{8}$	$1^{1}/_{8}$	$1^{3}/_{8}$	$1^{3}/_{8}$	$1^{5}/_{8}$	$1^{5}/_{8}$	$2^{1}/_{8}$	$2^{1}/_{8}$	$2^{5}/_{8}$	$2^{5}/_{8}$	$2^{5}/_{8}$	$3^{1}/_{8}$	$3^{1}/_{8}$	$3^{5}/_{8}$
100	$^{7}/_{8}$	$^{7}/_{8}$	$1^{1}/_{8}$	$1^{1}/_{8}$	$1^{1}/_{8}$	$1^{3}/_{8}$	$1^{3}/_{8}$	$1^{5}/_{8}$	$2^{1}/_{8}$	$2^{1}/_{8}$	$2^{1}/_{8}$	$2^{5}/_{8}$	$2^{5}/_{8}$	$2^{5}/_{8}$	$3^{1}/_{8}$	$3^{5}/_{8}$
80	$^{7}/_{8}$	$^{7}/_{8}$	$1^{1}/_{8}$	$1^{1}/_{8}$	$1^{1}/_{8}$	$1^{3}/_{8}$	$1^{3}/_{8}$	$1^{5}/_{8}$	$2^{1}/_{8}$	$2^{1}/_{8}$	$2^{1}/_{8}$	$2^{1}/_{8}$	$2^{5}/_{8}$	$2^{5}/_{8}$	$3^{1}/_{8}$	$3^{1}/_{8}$
60	$^{7}/_{8}$	$^{7}/_{8}$	$^{7}/_{8}$	$^{7}/_{8}$	$1^{1}/_{8}$	$1^{1}/_{8}$	$1^{3}/_{8}$	$1^{5}/_{8}$	$1^{5}/_{8}$	$2^{1}/_{8}$	$2^{1}/_{8}$	$2^{1}/_{8}$	$2^{5}/_{8}$	$2^{5}/_{8}$	$2^{5}/_{8}$	$3^{1}/_{8}$
50	$^{7}/_{8}$	$^{7}/_{8}$	$^{7}/_{8}$	$^{7}/_{8}$	$1^{1}/_{8}$	$1^{1}/_{8}$	$1^{3}/_{8}$	$1^{3}/_{8}$	$1^{5}/_{8}$	$2^{1}/_{8}$	$2^{1}/_{8}$	$2^{1}/_{8}$	$2^{1}/_{8}$	$2^{5}/_{8}$	$2^{5}/_{8}$	$3^{1}/_{8}$
40	$^{7}/_{8}$	$^{7}/_{8}$	$^{7}/_{8}$	$^{7}/_{8}$	$^{7}/_{8}$	$1^{1}/_{8}$	$1^{1}/_{8}$	$1^{3}/_{8}$	$1^{5}/_{8}$	$1^{5}/_{8}$	$2^{1}/_{8}$	$2^{1}/_{8}$	$2^{1}/_{8}$	$2^{5}/_{8}$	$2^{5}/_{8}$	$2^{5}/_{8}$
30	$^{5}/_{8}$	$^{7}/_{8}$	$^{7}/_{8}$	$^{7}/_{8}$	$^{7}/_{8}$	$1^{1}/_{8}$	$1^{1}/_{8}$	$1^{3}/_{8}$	$1^{3}/_{8}$	$1^{5}/_{8}$	$1^{5}/_{8}$	$2^{1}/_{8}$	$2^{1}/_{8}$	$2^{1}/_{8}$	$2^{5}/_{8}$	$2^{5}/_{8}$
20	$^{5}/_{8}$	$^{7}/_{8}$	$^{7}/_{8}$	$^{7}/_{8}$	$^{7}/_{8}$	$1^{1}/_{8}$	$1^{1}/_{8}$	$1^{3}/_{8}$	$1^{3}/_{8}$	$1^{5}/_{8}$	$1^{5}/_{8}$	$2^{1}/_{8}$	$2^{1}/_{8}$	$2^{1}/_{8}$	$2^{1}/_{8}$	$2^{5}/_{8}$

Bryant

Table 15-3 *Valve Losses in Equivalent Feet of Pipe (Screwed, Welded, Flanged, and Flared Connections)*

Nominal Pipe Size (inches)	Globe and Lift Check	60°–Y	45°–Y	Angle	Gate	Swing check	Y—Type Strainer: Flanged End	Y—Type Strainer: Screwed End
3/8	17	8	6	6	0.6	5	—	—
1/2	18	9	7	7	0.7	6	—	3
3/4	22	11	9	9	0.9	8	—	4
1	29	15	12	12	1.0	10	—	5
1 1/4	38	20	15	15	1.5	14	—	9
1 1/2	43	24	18	18	1.8	16	—	10
2	55	30	24	24	2.3	20	27	14
2 1/2	69	35	29	29	2.8	25	28	20
3	84	43	35	35	3.2	30	42	40
3 1/2	100	50	41	41	4.0	35	48	—
4	120	58	47	47	4.5	40	60	—
5	140	71	58	58	6.0	50	80	—

Table 15-4 *Fitting Losses in Equivalent Feet of Pipe (Screwed, Welded, Flared, and Brazed Connections)*

Nominal Pipe Size (in.)	Smooth Bend Elbows						Smooth Bend Tees				Sudden Enlargement*** d/D			Sudden Contraction** d/D		
	90° Standard*	90° Long Radius	90° Street*	45° Standard*	45° Street*	180° Standard*	Flow-thru Branch	Straight-thru Flow: No Reduction	Straight-thru Flow: Reduced (d, 3/4 d)	Straight-thru Flow: Reduced (d, 1/2 d)	1/4	1/2	3/4	1/4	1/2	3/4
3/8	1.4	0.9	2.3	0.7	1.1	2.3	2.7	0.9	1.2	1.4	1.4	0.8	0.3	0.7	0.5	0.3
1/2	1.6	1.0	2.5	0.8	1.3	2.5	3.0	1.0	1.4	1.6	1.8	1.1	0.4	0.9	0.7	0.4
3/4	2.0	1.4	3.2	0.9	1.6	3.2	4.0	1.4	1.9	2.0	2.5	1.5	0.5	1.2	1.0	0.5
1	2.6	1.7	4.1	1.3	2.1	4.1	5.0	1.7	2.3	2.6	3.2	2.0	0.7	1.6	1.2	0.7
1 1/4	3.3	2.3	5.6	1.7	3.0	5.6	7.0	2.3	3.1	3.3	4.7	3.0	1.0	2.3	1.8	1.0
1 1/2	4.0	2.6	6.3	2.1	3.4	6.3	8.0	2.6	3.7	4.0	5.8	3.6	1.2	2.9	2.2	1.2
2	5.0	3.3	8.2	2.6	4.5	8.2	10.0	3.3	4.7	5.0	8.0	4.8	1.6	4.0	3.0	1.6
2 1/2	6.0	4.1	10.0	3.2	5.2	10.0	12.0	4.1	5.6	6.0	10.0	6.1	2.0	5.0	3.8	2.0
3	7.5	5.0	12.0	4.0	6.4	12.0	15.0	5.0	7.0	7.5	13.0	8.0	2.6	6.5	4.9	2.6
3 1/2	9.0	5.9	15.0	4.7	7.3	15.0	18.0	5.9	8.0	9.0	15.0	9.2	3.0	7.7	6.0	3.0
4	10.0	6.7	17.0	5.2	8.5	17.0	21.0	6.7	9.0	10.0	17.0	11.0	3.8	9.0	6.8	3.8
5	13.0	8.2	21.0	6.5	11.0	21.0	25.0	8.2	12.0	13.0	24.0	15.0	5.0	12.0	9.0	5.0

* R/D approximately equal to 1.
** R/D approximately equal to 1.5.
** Enter table for losses at smallest diameter "d."
Bryant

The factor is given in the equipment manufacturer's specifications for the unit. Instead of using 15.0 tons, use 15.0 tons (1.13 × 15 = 16.95) for the final pipe size selection.

In addition, the acceptable friction loss is 2.5°F, instead of 2.0°F. The correction factor from Table 15-5 is 0.8. Thus, use 124 ft (0.8 × 155.2) for the final pipe size selection. A combination of 124 ft and 15 tons (gives a final pipe size selection of 2 1/8 in. diameter. See Table 15-2. The 2 1/8 in. is used since it is given for 20 tons. Seventeen tons are more than fifteen tons, so you move to the next highest value in the table. It is better to have a larger pipe than a smaller one. You also use the 150 ft equivalent pipe length, since a 100 ft length would be too short and not allow for errors in the original estimates.

Troubleshooting

To troubleshoot this type of air-conditioning equipment, a troubleshooting table (Table 15-1) has been provided. The general procedures listed there are used for hermetically sealed compressors.

MOBILE HOMES

Some units are now available for mobile home installations. See Fig. 15-28. Such a unit will furnish from 2 to 4 tons of cooling. The unit is 3 by 3 ft and will occupy a very small area outside the mobile home. Electric heat can be added to provide a comfortable year-round condition. The noise problem is minimized by ducting the condenser exhaust fan upward.

Table 15-5 *Friction-Drop Correction Factors—Liquid and Suction Lines*

Friction Drop (°F)	0.5	1.0	1.5	2.0	2.5	3.0	4.0	5.0	6.0
Multiplier	4.0	2.0	1.3	1.0	0.8	0.7	0.5	0.4	0.3

Bryant

The cooling coil, blower, compressor, and all other refrigeration components are contained in a low-silhouette weatherproof cabinet.

If a mobile home owner decides to move, the heating and cooling unit can be disconnected from the mobile home, transported to a new home-site, and easily reconnected.

A flexible insulated duct with round flanges simplifies hookups to mobile home ductwork. Conventional metal ductwork can be attached if desired.

This type of unit has a relatively large cooling capacity. Thus, it can be used on smaller homes, vacation cottages, and other small buildings. It is delivered as a complete package. All that is needed is the electrical power source, a thermostat connection, and a hookup to the ductwork of the building or mobile home.

Troubleshooting

To troubleshoot this type of air-conditioning equipment, a troubleshooting table (Table 15-1) has been provided. The general troubleshooting procedures listed in the table are used for hermetically sealed compressors.

Each manufacturer publishes manuals for use with the equipment the manufacturer makes. As you get more involved in the troubleshooting of specific types of equipment you will build your own library of troubleshooting manuals. Many of these contain wiring and piping schematics.

Other chapters in this book detail the proper operation of this type of equipment. Knowing the details of equipment operation will help you use the manuals more effectively. Familiarity with trade magazines will lead you to articles on problems with specific equipment.

WALL-MOUNTED DUCTLESS AIR CONDITIONERS

Another type of air-conditioning equipment is the wall-mounted unit. This type of unit requires no ductwork and has a wireless remote controller. The compressor-condenser unit is mounted or placed outside the living quarters. There are models from 8800 Btu/h with a SEER rating of 10.7 up to 14,600 Btu/h with a SEER rating of 10.0.

Figure 15-29A shows the wall-mounted unit made by Mitsubishi Electric. Figure 15-29B shows the remote

Fig. 15-28 *Mobile home self-contained air-conditioning unit.* (Bryant)

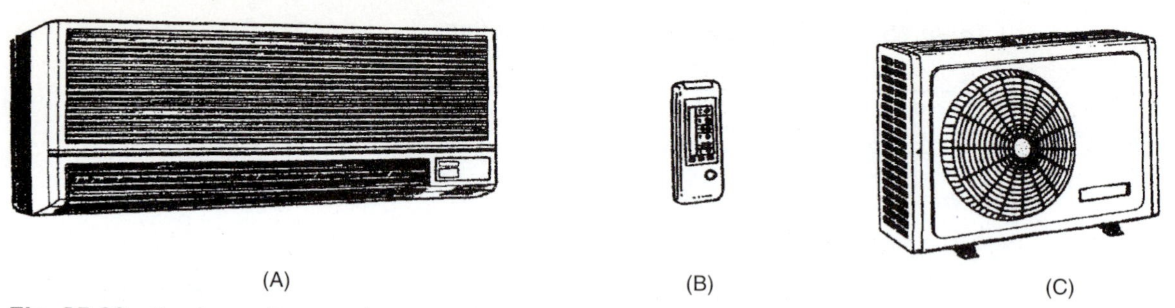

Fig. 15-29 *Ductless wall-mounted air-conditioner system. (A) Wall-mounted unit in room to be cooled. (B) LCD wireless remote controller. (C) Outside unit consisting of compressor-condenser unit.* (Mitsubishi)

wireless controller, which resembles the handheld TV remote. Figure 15-29C is the compressor-condenser unit. The system has a wireless remote controller that incorporates a number of features, which provide greater control and ease of use. It has a *liquid crystal display* (LCD) which indicates such information as mode, fan speed, and temperature selected as well as the timer setting and time remaining. It is also equipped with "I feel control" which is a feature unique to Mitsubishi Electric equipment. It allows the user to adjust the temperature exactly to the level he or she wants, simply by tapping the button that describes present conditions: too cool, too warm, or okay. The optimum temperature set this way is then memorized for immediate recall whenever the air conditioner is used again. And what's more, the new control has been made more compact and easy to handle than even before.

Fan Control Mode

A variety of microprocessor-controlled functions are available with this system. There is an automatic ON/OFF timer. A convenient timer function includes *auto start, auto stop*, both of which can be set between 1 to 12 h. Auto stop is the night setback mode. This mode raises the temperature setting by 1°F after 1 h, and by another 1° the next half an hour. When $1^1/_2$ h have passed, the air conditioner is running at a setting that is 2° higher, thus preventing excessive cooling.

The automatic fan speed control is shown in Fig. 15-30. When there is a large difference between

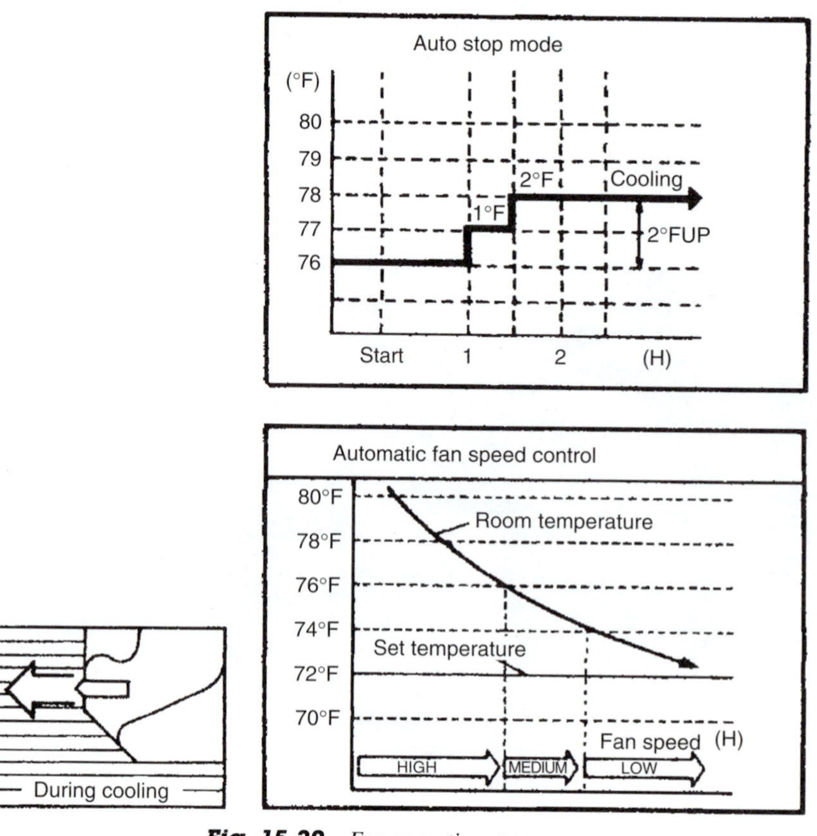

Fig. 15-30 *Fan operation.* (Mitsubishi)

the set temperature and the existing room temperature, the fan speed automatically adjusts to HIGH. As the temperature differential decreases, the fan speed shifts down automatically to MEDIUM and LOW. Operation is almost whisper quiet.

Restart Function

There is a restart function that restarts the equipment after a power outage. Operation resumes in the mode in which the equipment was running immediately before the outage.

Rotary Compressor

The compressor is of the rotary design. It is designed to be energy efficient for lower operating costs and longer service life. See Fig. 15-31.

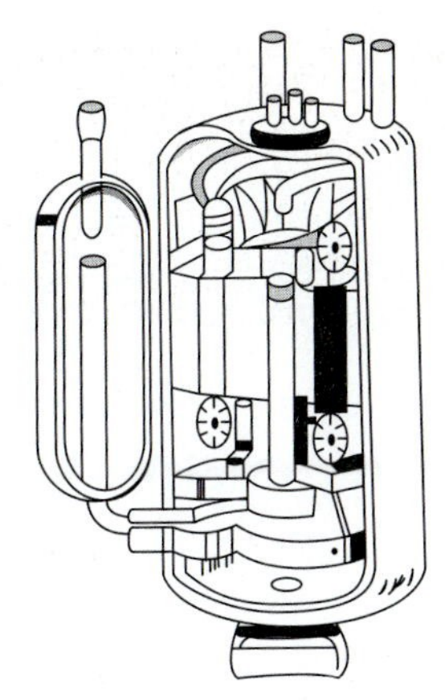

Fig. 15-31 *Rotary compressor. (Mitsubishi)*

REVIEW QUESTIONS

1. How often should a window unit's air filter be cleaned?
2. What causes a large percentage of motor burnouts in air conditioners?
3. Name two types of evaporators for residential use.
4. What is a plenum?
5. Where are single-rooftop units used?
6. Why are smoke detectors needed to work in conjunction with an air conditioner?
7. What is a firestat?
8. What are the two types of return air systems?
9. What is a volume damper?
10. What are the two reasons why liquid-line sizing presents less of a problem than suction-line sizing?

16
CHAPTER

Commercial Air-Conditioning Systems

PERFORMANCE OBJECTIVES

After studying this chapter you will:

1. Know how to troubleshoot commercial air-conditioning systems.
2. Know how to read the wiring diagram for a 208/230 V 60 Hz unit.
3. Know how to troubleshoot the direct multizone system.
4. Know how to troubleshoot the evaporative cooling system.
5. Know how to troubleshoot the absorption-type air-conditioning system.
6. Know how to troubleshoot the chilled water air-conditioning system.
7. Understand the operation of the chilled water system.
8. Understand the refrigeration cycle of a hermetic-type centrifugal chiller.
9. Be able to troubleshoot a console model (self-contained) air conditioner.
10. Know how to keep a service log used with an open-drive centrifugal refrigeration machine (chiller).

There are several types of commercial air-conditioning systems. This chapter discusses the following systems: expansion-valve air-conditioning systems, packaged cooling units, rooftop heating and cooling units, direct multizone systems, evaporative cooling systems, absorption-type air-conditioning systems, chilled water air conditioning, chillers, and console-type air-conditioning systems.

EXPANSION-VALVE AIR-CONDITIONING SYSTEM

This type of condensing unit can be installed in singles or multiples. Such units are used in residential, apartment, motel, and commercial applications. These units are applicable only to expansion-valve systems. The low height and upward discharge of air make it easy to conceal the unit among shrubs on a slab at ground level or out of sight on a roof.

Compressor

The compressor is hermetically sealed. Built-in protection devices protect from excessive current and temperature. It is suction cooled. Overload protection is by an internal pressure relief. A crankcase heater is the standard equipment on all of these units. It ensures proper compressor lubrication at all times. The crankcase heater is thermostatically controlled and temperature actuated to operate only when required. Rubber mounts help to reduce the noise associated with compressors.

Condenser

The condenser coil is U-shaped to provide a large surface area for heat exchange. The joints in the compressor are silver soldered. The compressor is tested for leaks at 450 to 500 psi. Refrigerant lines come precharged. This unit comes with a drier. Solid-state controls prevent compressor short-cycling and also allow time for suction and discharge pressure to equalize. This permits the compressor to start in an unloaded condition. An automatic reset control will shut off the compressor for 5 min.

EXPANSION-VALVE KIT

An expansion-valve kit has been developed to match the evaporator unit. The expansion valve is equipped with a bleed port. This permits pressures to equalize after the compressor stops. This means the compressor can restart in an unloaded condition. Flare fittings permit connections on the valve in a simple field installation.

Since single-phase models require large amounts of current, they can cause lights to blink when they start up. A *positive temperature coefficient* (PTC) kit can eliminate some of the start-up problems. It consists of a solid-state circuit with a ceramic thermistor. The thermistor provides extra starting torque to solve most compressor hard starting problems such as low voltage or light dimming. It switches itself out of the circuit after start-up.

A start kit consisting of a start capacitor and potential relay must be installed in some models when used with certain evaporator units and expansion-valve kits. The added capacitance is taken from the circuit when the coil energizes almost instantly.

However, it does help with the starting load. Since the coil is energized whenever the air-conditioner system is operating, it is possible for the coil to become open. This causes the start capacitor to "blow-up" and spread its contents inside the control housing. Replacement of the capacitor and the coil of the relay is necessary in order to repair the equipment for proper operation. Usually the entire relay must be replaced, since the coil is not always available separately.

Troubleshooting

To troubleshoot this type of air-conditioning equipment, refer to Table 16-1. The general troubleshooting procedures listed in the table are used for hermetically sealed compressors.

Table 16-1 *Troubleshooting a Console Model (Self-Contained) Air Conditioner*

Symptom and Possible Cause	Possible Remedy
Unit Fails to Start	
1. Start switch off. 2. Reset button out. 3. Power supply off. 4. Loose connection in wiring. 5. Valves closed.	1. Place start switch in start position. 2. Push reset button. 3. Check voltage at connection terminals. 4. Check external and internal wiring connections. 5. See that all valves are opened.
Motor Hums, but Fails to Start	
1. Motor is a single-phase on a three-phase circuit. 2. Belts too tight. 3. Not oil in bearings. Bearings tight from lack of lubrication.	1. Test for blown fuse and/or tripped circuit breaker. 2. See that the motor is floating freely on trunnion base. See that the belts are in the pulley groove and not binding. 3. Use proper oil for motor.
Unit Fails to Cool	
1. Thermostat set incorrectly. 2. Fan not running. 3. Coil frosted.	1. Check thermostat setting. 2. Check electrical circuit for fan motor. Determine if fan blade and motor shaft revolve freely. 3. Dirty filters restrict air flow through unit. Check for an obstruction at air grille. Fan not operating. Check fan operation. Attempting to operate unit at too low a coil temperature.
Unit Runs Continuously, but no Cooling	
1. Shortage of refrigerant.	1. Check liquid refrigerant level. Check for leaks. Repair and add refrigerant to proper level.
Noisy Compressor	
1. Thermostat differential too close.	1. Check differential setting of thermostat and adjust setting.
Unit Vibrates	
1. Unit is not level. 2. Shipping bolts not removed. 3. Belts jerking. 4. Unit suspension springs not balanced.	1. Level all sides. 2. Remove all shipping bolts and steel bandings. 3. Motor not floating freely. 4. Adjust unit suspension until unit ceases vibrating.
Condensate Leaks	
1. Drain lines not properly installed. 2. Slime formation in pan and drain lines. Slime sometimes. present on evaporator fins.	1. Drain pipe sizes, proper fall in drain line, traps, and possible obstruction due to foreign matter should be checked. 2. This formation is largely biological and usually complex in nature. Different localities produce different types. It is largely a local problem. Check Chap. 9 ("Cooling Water Problems") Periodic cleaning will tend to reduce the trouble, but will not eliminate it totally. Filtering air thoroughly will also help. However, at times some capacity must be sacrificed when doing this.
Unit Cycles too Often	
1. Too much vibration in unit. 2. Slugging oil. 3. Bearing knock. 4. Oil level low in crankcase.	1. Check for vibration point. 2. Low-suction pressure. 3. Liquid in crankcase. 4. Pump-down system. Add oil if too low.

Note: The aspects of troubleshooting detailed in Table 16-1 are as comprehensive as they can be within the limits of this text. Obviously, there is no substitute for experience. Working with air-conditioning and refrigeration systems will give you this experience. Thus, you will sharpen your troubleshooting skills.

PACKAGED COOLING UNITS

A 2- to 5-ton packaged unit is available from several manufacturers. Lennox makes a self-contained unit that can be mounted on a slab or on the rooftop. See Fig. 16-1. This one is designed for the residential replacement market. The compressor, control box, filter, condenser coil, and evaporator coil are in one package. The blower is also located in the package. The only component inside the building is the ductwork. Return air enters in the lower opening through the evaporator coil and is discharged out through the top opening.

One of the advantages of a unit of this type is its completeness. It comes ready to connect to the ductwork

Fig. 16-1 *Horizontal, single-package air conditioner.* (Lennox)

and the electrical outlets. The blower is located outside the house. The inside noise is that of the air moving through the ductwork. The air filter is also located outside. It is a vacuum-cleanable type with polyurethane coating. It is coated with oil to increase efficiency. If the filter is washed, it should be reoiled.

Up to 58,000 Btu can be added with an optional field-installed heating unit. If electric heat is desired, it is possible for this type of unit to heat and cool, using the same ductwork. Figure 16-2 shows typical unit installations.

ROOFTOP HEATING AND COOLING UNITS

Rooftop heating and cooling units are made by many manufacturers. Some units are delivered with a full refrigerant charge. This means there are no refrigerant lines to connect. This cuts labor costs and installation time. Since the unit is on the rooftop, no inside room has to be allocated for the heating and cooling equipment.

This unit uses gas for heating up to 112,500 Btu. The cooling can reach 60,000 Btu or 5 tons. Typical

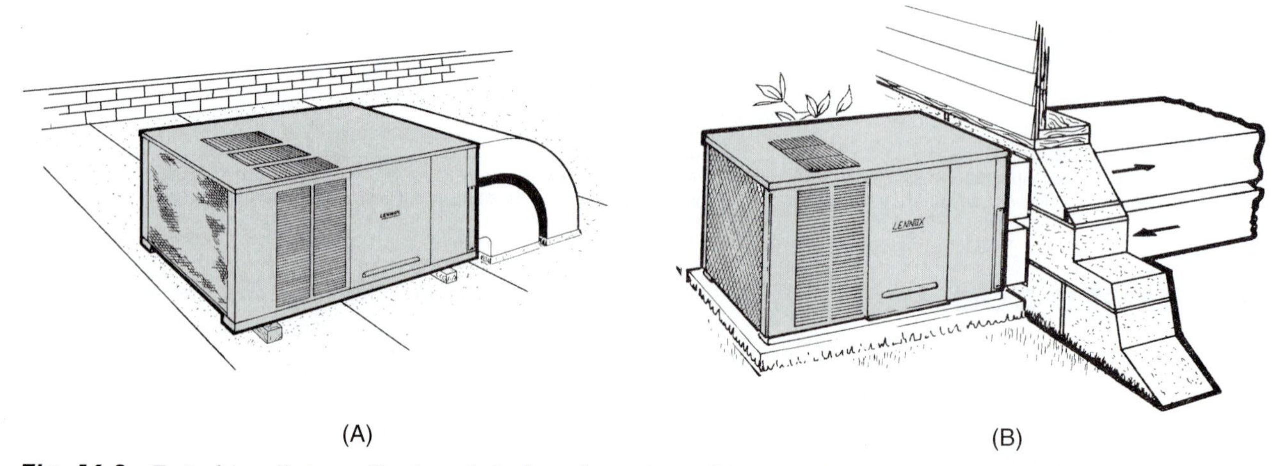

Fig. 16-2 *Typical installations of horizontal single-package air conditioner. (A) A rooftop installation. (B) A unit on a slab at grade level.* (Lennox)

installation systems are shown in Fig. 16-3. One of the advantages is that heat and cooling can be added rather quickly in the construction phase.

The installers can work in comfort and thus improve their efficiency during the construction phase of the building. The unit can be set in place on a slab at ground level. The duct, gas, and electrical connections can be made to it at that location.

Electrical

Since this unit has the ability to deliver 5 tons of air conditioning, it needs some type of electrical control to ensure that proper operation is obtained. Figure 16-4 shows the 230-V, 60-Hz, single-phase unit's electrical schematic. The same unit is manufactured with 208/230-V, 60-Hz, and three-phase wiring. The same unit can be operated on 460V, 60 Hz, three-phase. The motors and some controls must be changed to take the higher voltages. This does not mean that a unit that operates on 460 V will operate on 220 V when the supply is changed.

Certain parts must be changed. In some cases, relay coils and the compressor motor must be rated at the voltage present for operation. These units are factory

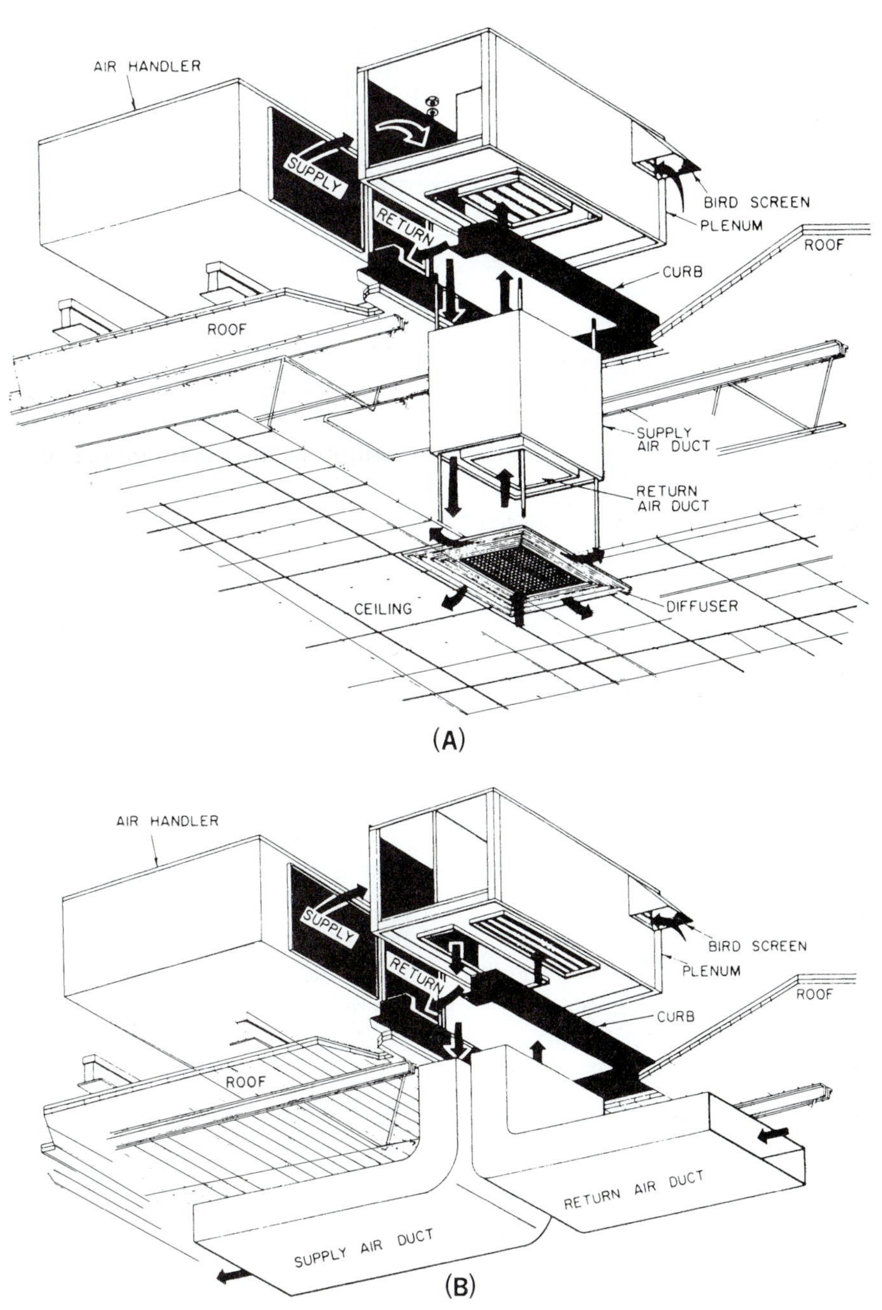

Fig. 16-3 *Typical system installations on the rooftop. (A) A concentric duct arrangement. (B) A side-by-side duct arrangement.* (Lennox)

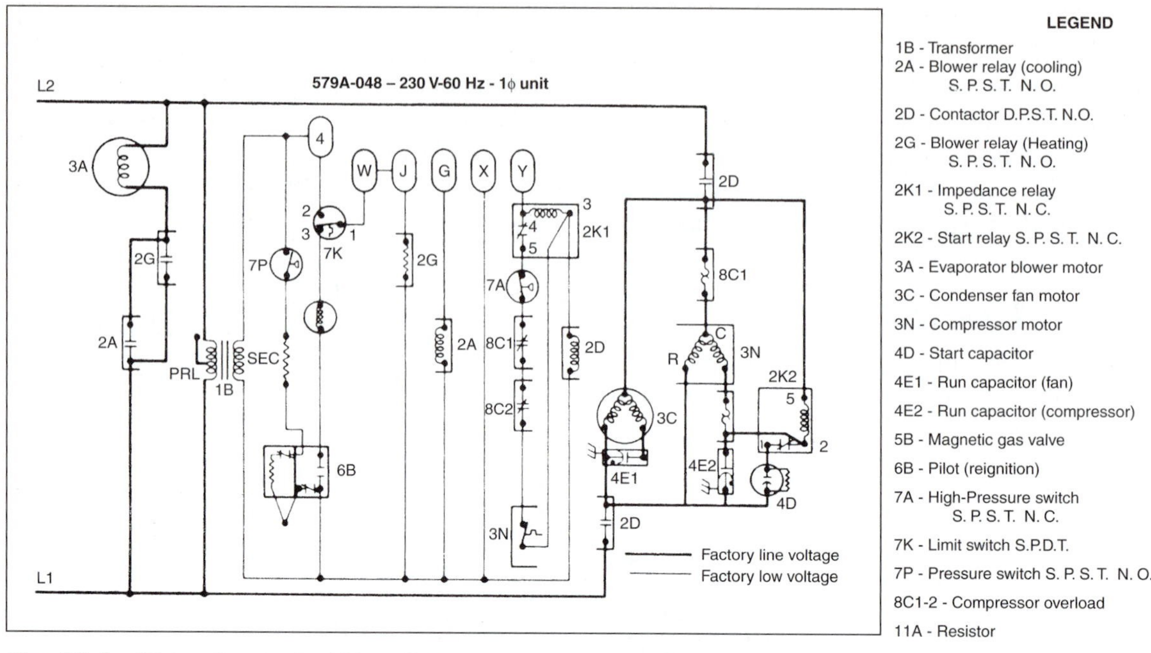

Fig. 16-4 *Wiring diagram for 230-V, 60-Hz, single-phase unit with its parts labeled.* (Bryant)

wired for both high voltage and low voltage. Connections to the thermostat, located inside the living space, are connected to terminals marked W, J, G, X, and Y.

Note the part of the circuit responsible for the gas heating. This part would be eliminated if the unit was used only for cooling. For instance, if cooling is called for, the contacts on 7K would be making contact between points 1 and 2, instead of point 3, as shown in Fig. 16-4.

Sequence of Operation

Only the cooling operations are shown in Fig. 16-5. This rooftop conditioner operates in the following manner. Most others operate in the same way. This sequence permits operation on the 208 to 230-V units. With thermostat-system switch and fan switch in auto position, the operation sequence is as follows.

When there is a demand for cooling by the thermostat, terminal R "makes" to terminals Y1 and G through the thermostat. This thermostat switching action electrically connects blower motor contactor (2M 1) and cooling relay (2A) across the 24-V secondary of the control transformer (1B). This causes the blower motor contactor (2M1) and cooling relay (2A) to become energized.

The contacts of the energized blower motor contactor close to energize blower motor (3E). This starts the blower motor instantly.

The contacts of cooling relay (2A) close to energize compressor contactor (2M2). As the compressor contactor pulls in, the compressor (3L) starts. At the same time the compressor contactor (2M2) is energized, the condenser fan motor (3C2) is placed in operation. The condenser fan motor (3Cl) does not start until the high-side pressure reaches 280 psig, at which point the low-ambient pressure switch (7P) closes to complete the line-voltage circuit to the fan motor (3C1) and starts this motor. When the compressor discharge pressure drops to 178 psig, the low-ambient pressure switch (7P) will reopen and the condenser fan motor (3C1) will stop. This provides high-side pressure for low-ambient operation down to 32°F (0°C). When the pressure builds back up to 280 psig, the low-ambient pressure switch (7P) will close again. This restarts the condenser fan motor (3C1).

During this time of operation, only one-half of the evaporator coil is being used. Should the indoor temperature continue to rise, the thermostat will make between R and Y2, at which time the liquid-line solenoid valve (5B) is energized and opens. This permits the refrigerant to flow to the other half of the evaporator coil. Keep in mind that two-stage cooling is not available on all models. Those with single-stage cooling will not have half of the evaporator coil operating.

As the conditioned space temperature drops, the second stage contacts, R to Y2, will open within the thermostat and close the liquid-line solenoid valve (5B). The unit will continue to operate at two-thirds capacity. As the temperature within the conditioned space continues to drop and reaches the thermostat setting, contacts R to Y1 and G will open. At this time,

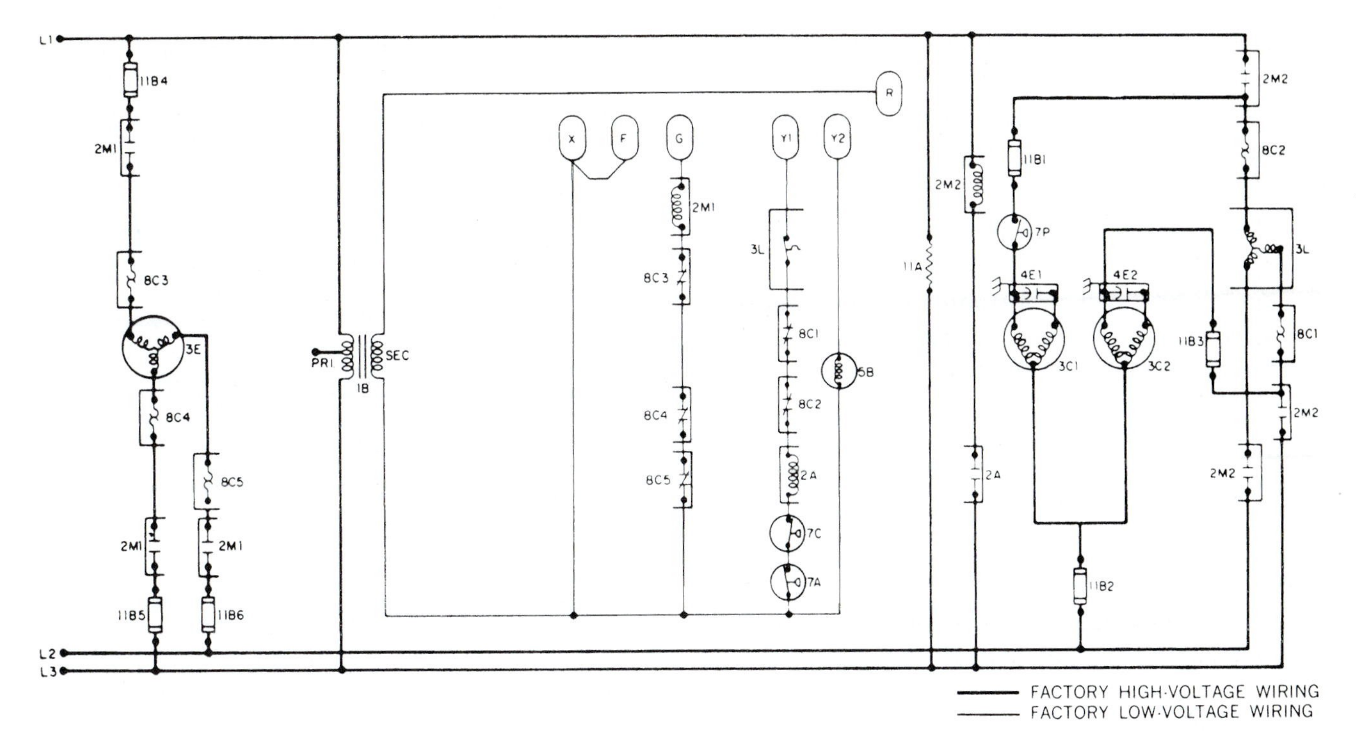

LEGEND

1B-Transformer (Tapped Primary)
2A-Cooling Relay N.O.
2M1-Blower Motor Contactor N.O.
2M2-Compressor Contactor N.O.
3C1 & 2-Fan Motors
3E-Blower Motor
3L-Compressor with Internal Overload
4E1 & 2-Run Capacitor
5B-Liquid Line Solenoid Valve
7A-High-Pressure Switch N.C.
7C-Low-Pressure Switch N.C.
7P-Low-Ambient Pressure Switch N.O.
8C1 & 2-Compressor Overloads
8C3, 4, & 5-Blower Motor Overload
11A-Crankcase Heater
11B1, 2, & 3-Fan Motor Fuses
11B4, 5, & 6-Blower Motor Fuses

Fig. 16-5 *Wiring diagram for a 208/230-V operation.* (Bryant)

compressor (3L), condenser fan motors (3C1 and 3C2) and the blower motor (3E) will stop. After all motors have stopped, the unit remains in standby position ready for the next call for cooling by the thermostat.

Compressor Safety Devices

Several safety devices protect the compressor in abnormal situations. For instance, the high pressure switch interrupts the compressor control circuit when the refrigerant high-side pressure becomes excessive. A low-pressure switch interrupts the compressor control circuit when the refrigerant low-side pressure becomes too low. The compressor is protected from overloads by current-operated circuit breakers. Thermal devices embedded in the windings of the compressor motor open the circuit when too much heat is generated by the windings. Some manufacturers place a 5-min delay device in series with the compressor motor. Thus, the motor cannot be restarted for 5 min after shutdown.

When any of the above safety devices are actuated, current in the Y1 leg is interrupted and shuts off the compressor and condenser fan motors. See Fig. 16-5.

Maintenance

Before performing any maintenance on the unit, make sure the main line switch is open or in the OFF position. Label the switch so that someone will not turn it on while you are working.
The components should be checked and serviced as follows:

- *Blower motor oil.* According to the manufacturer's recommendations, the rating plate will usually give lubrication instructions.
- *Electrical connections.* The electrical connections should be checked periodically and retightened.
- *Pulley alignment and belt tension.* Check the blower and motor pulley for alignment. Also check the belt

for proper tension. It should have approximately 1 in. of sag under normal finger pressure.

- *Blower bearings.* Blowers are equipped with prelubricated bearings and need no lubrication. If, however, there are a blower unit and blower motor without sealed bearings, a few drops of oil must be added occasionally.
- *Condenser and evaporator coils.* Coils should be inspected occasionally and cleaned as necessary. Be careful not to bend the soft aluminum fins.

WARNING: Make sure the main line disconnect switch is in the off position before cleaning the coils.

- *Filters.* System air filters should be inspected every two months for clogging because of dirt. When necessary, replace disposable-type filters.

Special Instructions

- Do not rapid-cycle the unit. Allow at least 5 min before turning on the unit after it has shut oil.
- If a general power failure occurs, the electrical power supply should be turned off at the unit disconnect switch until the electrical power supply has been restored.
- Air filters should be cleaned or replaced at regular intervals to ensure against restricted airflow across the cooling coil.
- During the off season, the main power supply may be left on or turned off. Leaving the power turned on will keep the compressor crankcase heaters energized.
- If power has been off during the winter, it must be turned on for at least 12 h before spring start-up of the unit. This allows the crankcase heaters to vaporize any liquid refrigerant that may be condensed in the compressor.

Troubleshooting

To troubleshoot this type of air-conditioning equipment, refer to Table 16-1. The general troubleshooting procedures listed in the table are used for hermetically sealed compressors.

DIRECT MULTIZONE SYSTEM

The direct multizone system unit is roof mounted and can be used for cooling and heating. See Fig. 16-6. It can use chilled water for cooling up to 550,000 Btu/h.

Air distribution is 12- or 16-zone multizone control at the unit or double duct with independent mixing dampers at each zone. Figure 16-7 shows the typical applications of such a unit with a zone distribution system using mixing dampers located at the unit. A double-duct distribution system with zone damper boxes can be used. Mixing dampers are remote from the unit. The net weight of the unit is 2525 lb or $1^1/_4$ tons.

Figure 16-8 shows the location of the parts inside the unit. Figure 16-9 shows how the refrigerant piping

Fig. 16-6 *Direct multizone system for rooftop mounting.* (Lennox)

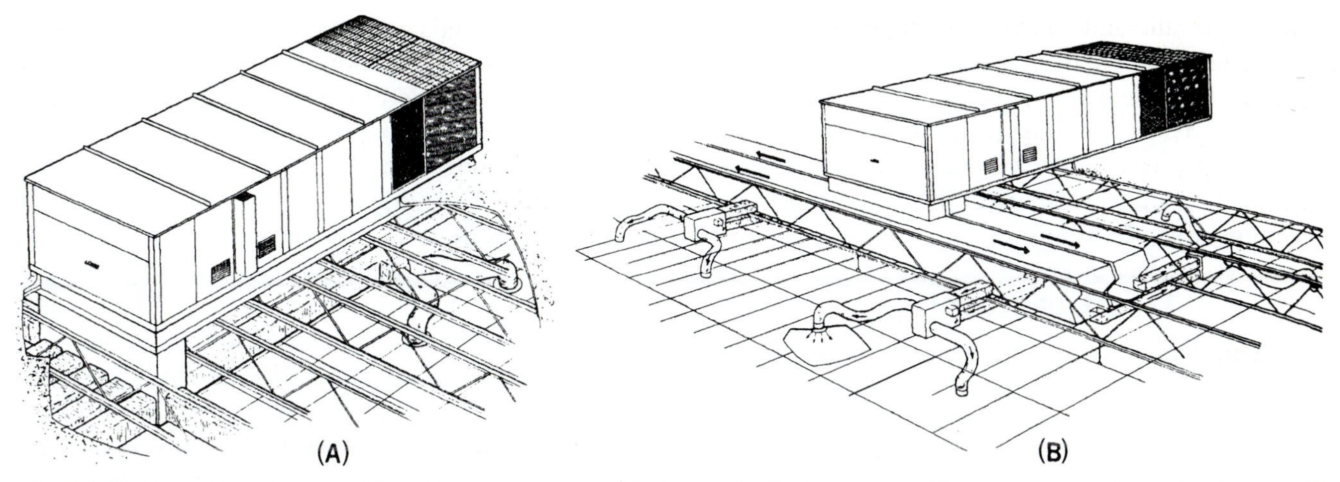

Fig. 16-7 *Typical installations of the multizone system unit. (A) A zone distribution system with mixing dampers located at the unit. (B) A double-duct distribution system with zone damper boxes and mixing dampers remote from the unit.* (Lennox)

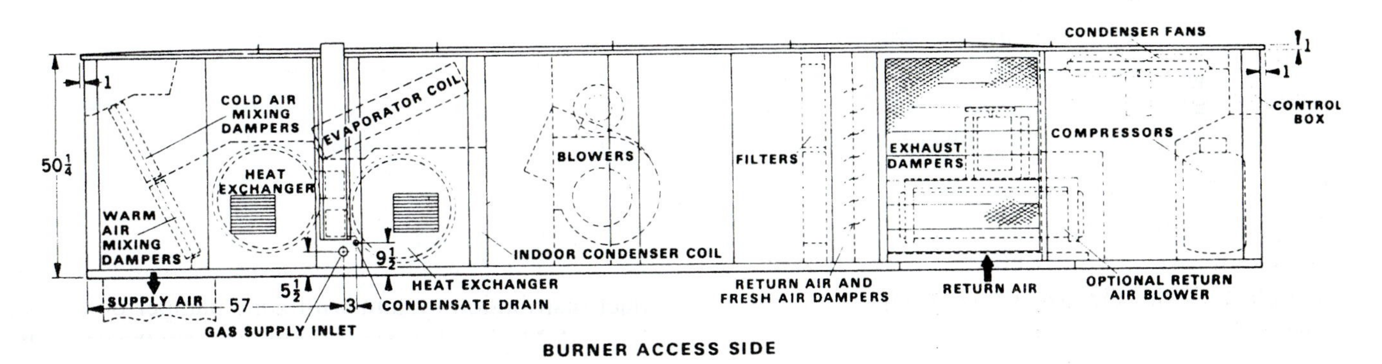

Fig. 16-8 *Location of the component parts to the multizone unit.* (Lennox)

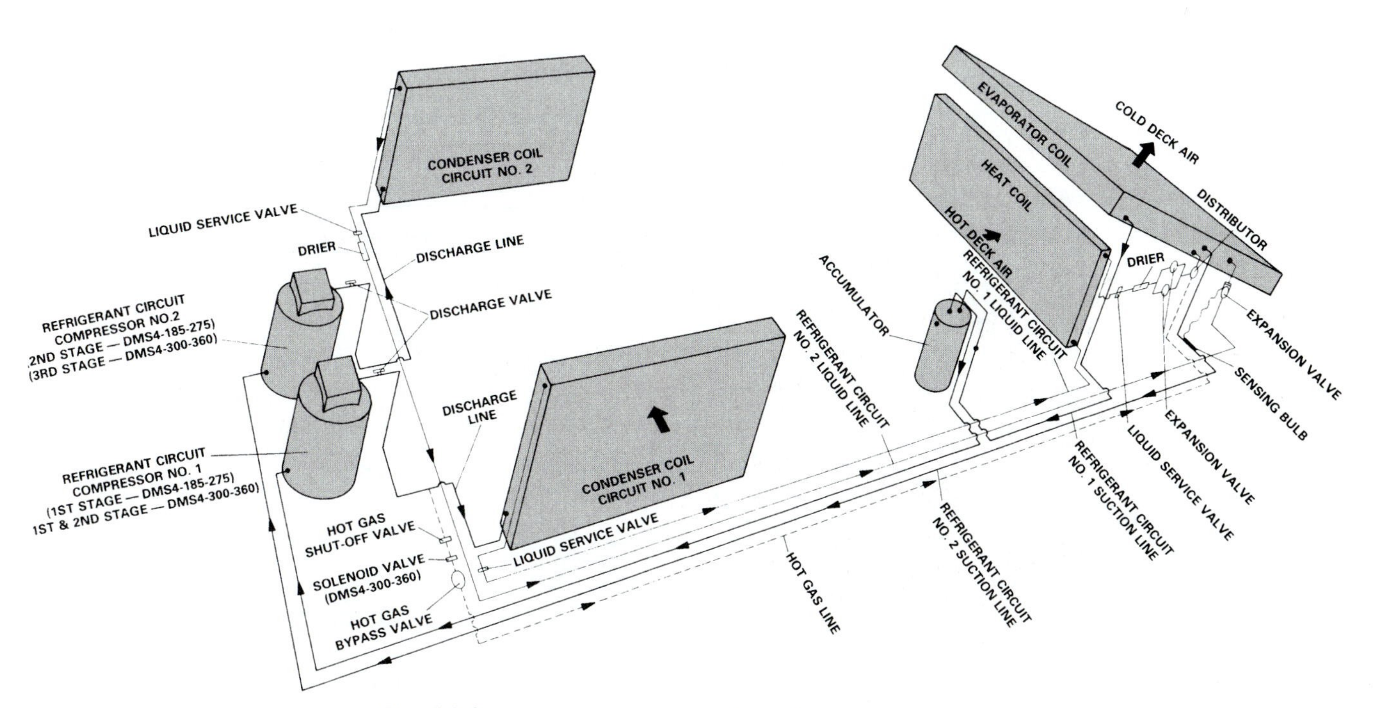

Fig. 16-9 *Refrigerant piping for the multizone system.* (Lennox)

is laid out for the unit when two compressors are used for cooling purposes. Note that this unit uses an accumulator. There are certain conditions under which the capacity of such a unit must be rated. These conditions are the temperature of the evaporator air, the condenser coil air temperature, the speed of the blower motor, and its volume of air delivered.

Troubleshooting

To troubleshoot this type of air-conditioning equipment, refer to Table 16-1. The general troubleshooting procedures listed in the table are used for hermetically sealed compressors.

EVAPORATIVE COOLING SYSTEM

In some locations, it is possible to use the cooling tower principles to condense the refrigerant. This method has the usual problems with water and tower fungi. Those problems are discussed in Chap. 10.

The condensing coil is cooled by air drawn in from outside the tower and forced upward over the coil. Water is pumped continuously to a distribution system and sprayed so that it drops in small droplets over the condensing coil. See Fig. 16-10. The water is reused since it cools as it drops through the moving air stream. In some systems the water is pumped up and into a trough. The water drips down over the condenser coils and cools them.

In some cases, the water moves through the tubes that surround the refrigerant-carrying tubes. The air stream then removes the heat and discharges it into the surrounding air. This means the cooling tower should be mounted outside a building. In some instances it is possible to mount the tower inside. However, a duct is then needed to carry the discharged air outside. As shown in Fig. 16-10B, the water is carried off and must be replaced as it, too, evaporates. The pan is filled to level when the float moves down and allows the water makeup valve to open. If the condenser temperature reaches or exceeds 100°F (37.8°C), the thermostat turns on the water and the fan.

Problems with this system center in the electrical control system and the water system. The controls, fan

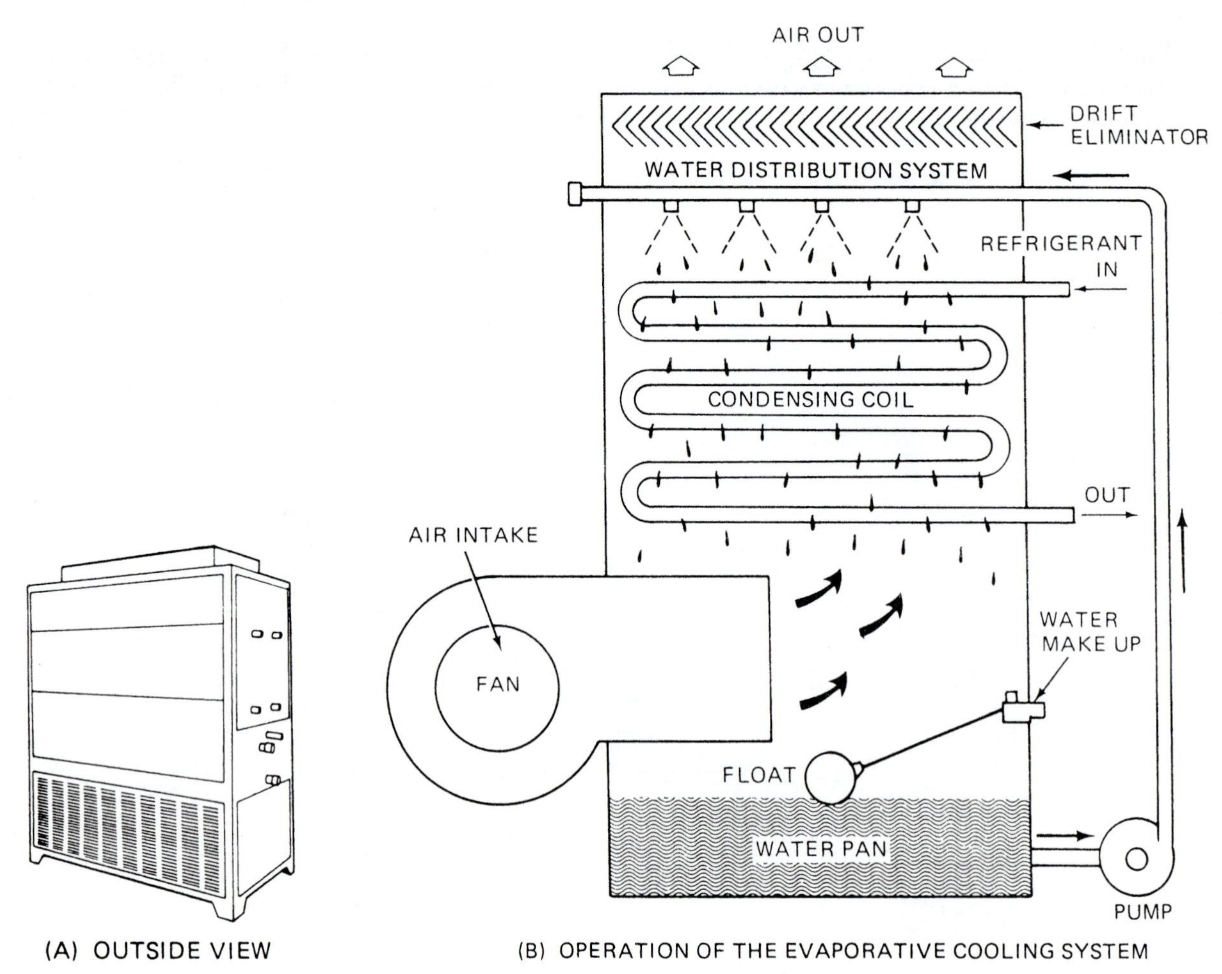

Fig. 16-10 *An evaporative cooling system. (A) Outside view. (B) Operation of the evaporative cooling system.*

motor, and pump motor are electrically operated. Thus, troubleshooting involves the usual electrical-circuit checks.

ABSORPTION-TYPE AIR-CONDITIONING SYSTEMS

A boiling refrigerant in an evaporator *absorbs* heat. The evaporator pressure must be low for boiling to take place. To produce the low pressure, it is necessary to remove the refrigerant as soon as the boiling refrigerant vaporizes. Vapors can be absorbed quickly by another liquid. However, the other liquid must be able to absorb the vapor when it is cool. It will then release the absorbed heat when it is heated.

Ammonia is one of the refrigerants most commonly used in the absorption-type air-conditioning systems. Ammonia vapors are absorbed quickly by large amounts of cool water. In fact, it can absorb vapor as quickly as a compressor.

High-pressure ammonia can be fed as a pure liquid through a metering device directly into an evaporator. See Fig. 16-11. Refrigeration takes place until the high-side liquid ammonia is exhausted or the water in the absorber tank is saturated. Once saturated, it no longer absorbs ammonia. If the ammonia tank and the absorber are large enough, these components can be used as part of an air-conditioning system.

A system can be devised to handle large installations. See Fig. 16-12. In this system, some of the ammonia is removed from the water. This leaves a weak water solution of ammonia. This solution flows by gravity to the absorber.

The water in the absorber absorbs the ammonia. Such absorption continues until ammonia represents 30 percent of the water-ammonia solution.

Such a strong (30 percent) solution of ammonia is called *strong aqua*. Aqua means *water*. The strong aqua is pumped up to the generator. The absorber operates at low-side evaporator pressure. That is why the pump is necessary. The generator has a high-side pressure.

Air is driven out of water by heat. Ammonia also can be driven out of water by applying heat. The high-temperature ammonia vapor rises and moves to the condenser. Weak condensed liquid flows back to the absorber through the force of gravity. In the condenser, the latent heat is removed from the ammonia vapor. Condensed ammonia liquid flows through the liquid receiver to the evaporator. In the evaporator, the ammonia boils at reduced pressure. Latent heat is absorbed. The liquid ammonia changes into a vapor. In changing to a vapor, the ammonia produces refrigeration. Ammonia is only one refrigerant used for this type of system. Lithium bromide and water also can be made into a refrigerant. Figure 16-13 shows a typical absorber system. Several manufacturers make packaged units for absorber systems.

CHILLED WATER AIR CONDITIONING

To produce air conditioning for large areas, such as department stores and office buildings, it is necessary to use another means of cooling the air. Chilled water is used to produce the cooling needed to reduce the interior temperature of offices and stores. To understand the function of the chilled water, it is necessary to look at the total system. See Fig. 16-14.

The refrigerating machine is the chiller. Water is supplied to the chiller. There, its temperature is reduced to about 48°F (8.9°C). The chilled water then flows to the coils in the fan coil unit. The fan coil unit is located

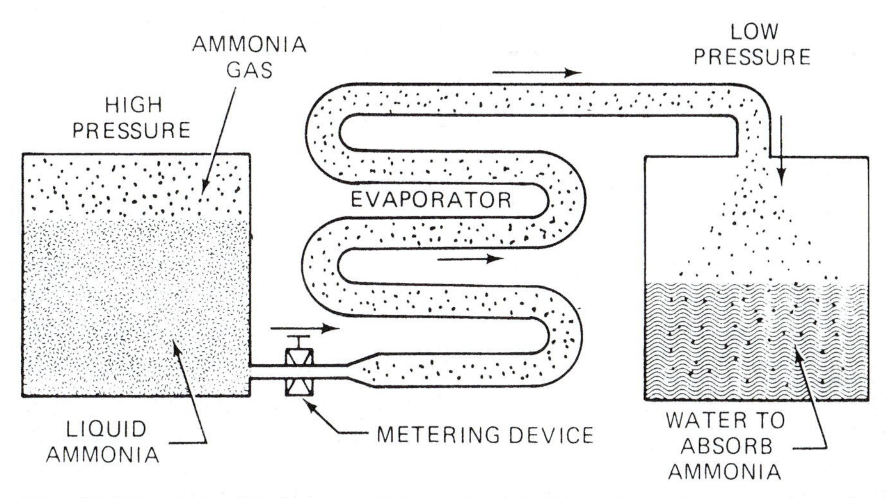

Fig. 16-11 *A simplified system of absorption of refrigeration using ammonia as the refrigerant.* (Arkla)

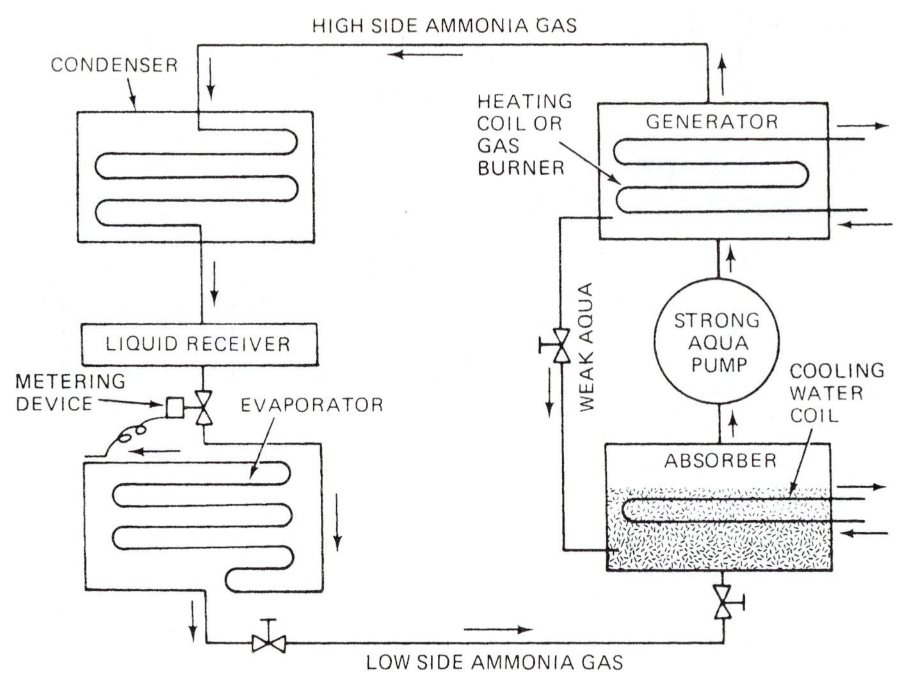

Fig. 16-12 *The ammonia absorption system of refrigeration used in large installations.* *(Arkla)*

in the space to be conditioned. In some cases, a central air-handling system is used. Pumps are used to move the water between the chiller and the air-handling equipment. The water is heated by the room air that is pulled over the chilled water coils. Thus, the water reaches a temperature of about 55°F (12.8°C). In some installations it reaches 58°F (14.4°C). The water absorbs about 10°F (5.5°C) of heat as it is exposed to the room air being drawn into the unit by blowers.

The heated water is then pumped back to the chiller. There, the water is chilled again by the machine removing the absorbed heat. Once chilled, the 48°F (8.9°C) water is again ready to be pumped back to the fan coil unit or the central air-handling system. This process of recirculation is repeated as needed to reduce the temperature of the space being conditioned.

A cooling tower is used to remove the heat to the outside of the building. Cooling towers were discussed in Chap. 9. The condenser water is cooled by the cooling tower. See Fig. 16-14.

Figure 16-15 shows the refrigeration cycle of a chiller.

Refrigerant Cycle

When the compressor starts, the impellers draw large quantities of refrigerant vapor from the cooler at a rate determined by the size of the guide-vane opening. This compressor suction reduces the pressure within the cooler. This causes the liquid refrigerant to boil vigorously at a fairly low temperature [typically 30 to 35°F (−1.1 to 1.7°C)].

The liquid refrigerant obtains the energy necessary for the change to vapor by removing heat from the water in the cooler tubes. The cold water can then be used for process chilling or air conditioning, as desired.

After removing heat from the water, the warm refrigerant vapor is compressed by the first-stage impeller. It then mixes with flash economizer gas and is further compressed by the second-stage impeller.

Compression raises the temperature of the refrigerant vapor above that of the water flowing through the condenser tubes. The compressed vapor is then discharged into the condenser at 95 to 105°F (35 to 40.6°C). Thus, the relatively cool condensing water [typically 75 to 85°F (23.9 to 29 4°C)] removes some of the heat, condensing the vapor into a liquid.

Liquid refrigerant then drains into a valve chamber with a liquid seal. This prevents gas from passing into the economizer. When the refrigerant level in the valve chamber reaches a preset level, the valve opens. This allows liquid to flow through spray pipes in the economizer.

Pressure in the economizer is midway between those of the condenser and cooler pressures. At this low pressure, some of the liquid refrigerant flashes to gas, cooling the remaining liquid.

Flash gas is piped through the compressor motor for cooling purposes. It is then mixed with gas already

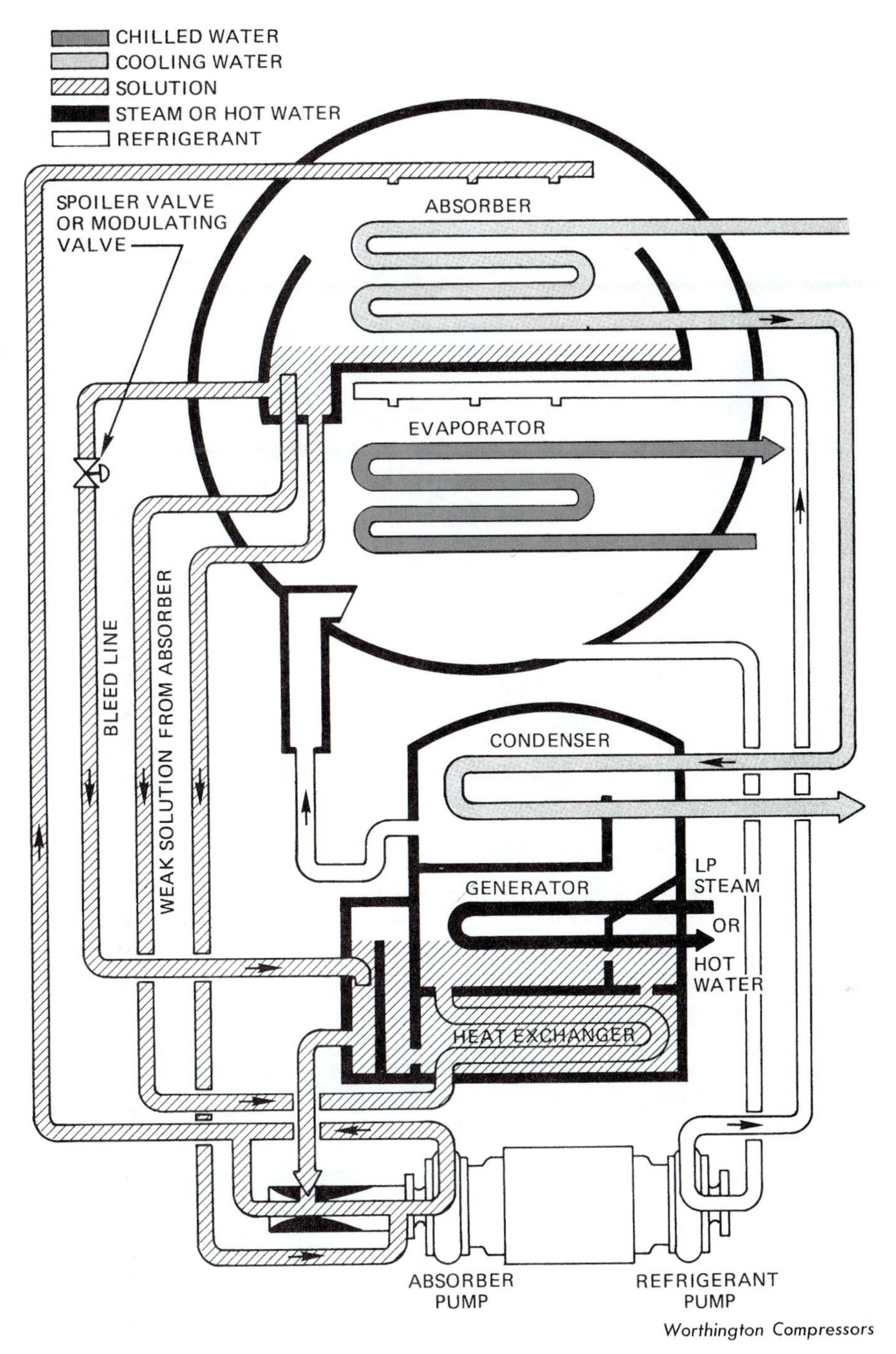

Fig. 16-13 *A typical absorption system used in commercial air-conditioning applications.* (Worthington Compressors)

compressed by the first-stage impeller. The cooled liquid refrigerant in the economizer is metered through a low-side valve chamber into the cooler. Pressure in the cooler is lower than economizer pressure. Thus, some of the liquid flashes, cooling the remainder to cooler (evaporator) temperature. The cycle is now complete.

Figure 16-16 shows a cutaway view of a chiller. This is how Fig. 16-15 looks in a packaged unit. These chillers are available from 425 to 2500 tons of refrigeration.

Such large systems must be operated by a person with specialized knowledge of the unit. The manufacturers publish training manuals that detail the operation, maintenance, and repair of the units. Some of the training manuals are as long as this textbook. Thus, there can be presented here only a brief discussion of the information you will need to operate and maintain such a cooling operation.

A typical chiller installation is shown in Fig. 16-17. Note the piping and wiring systems. As can be observed, the electrical system is rather complicated.

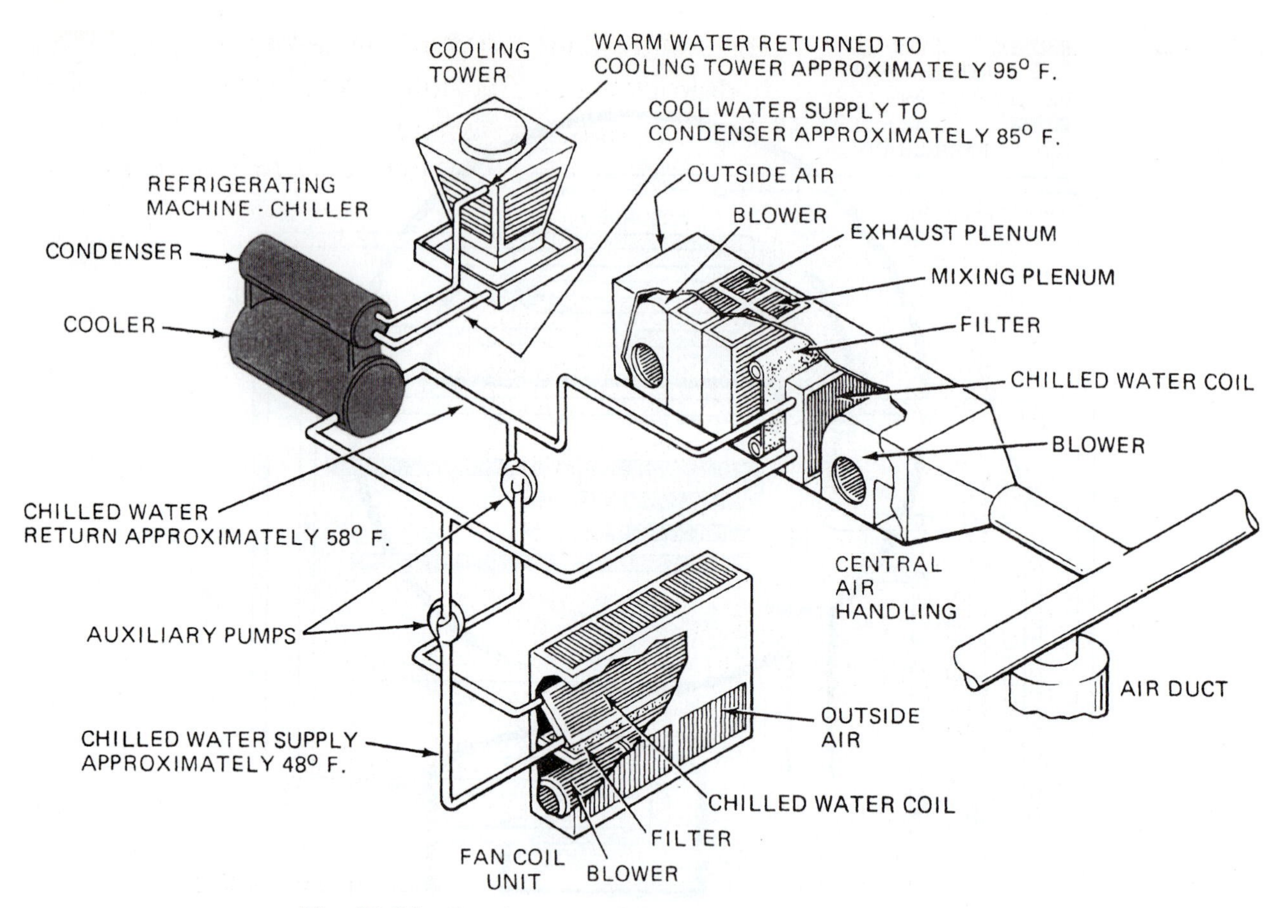

Fig. 16-14 *Complete air-conditioning system using chilled water.* (Carrier)

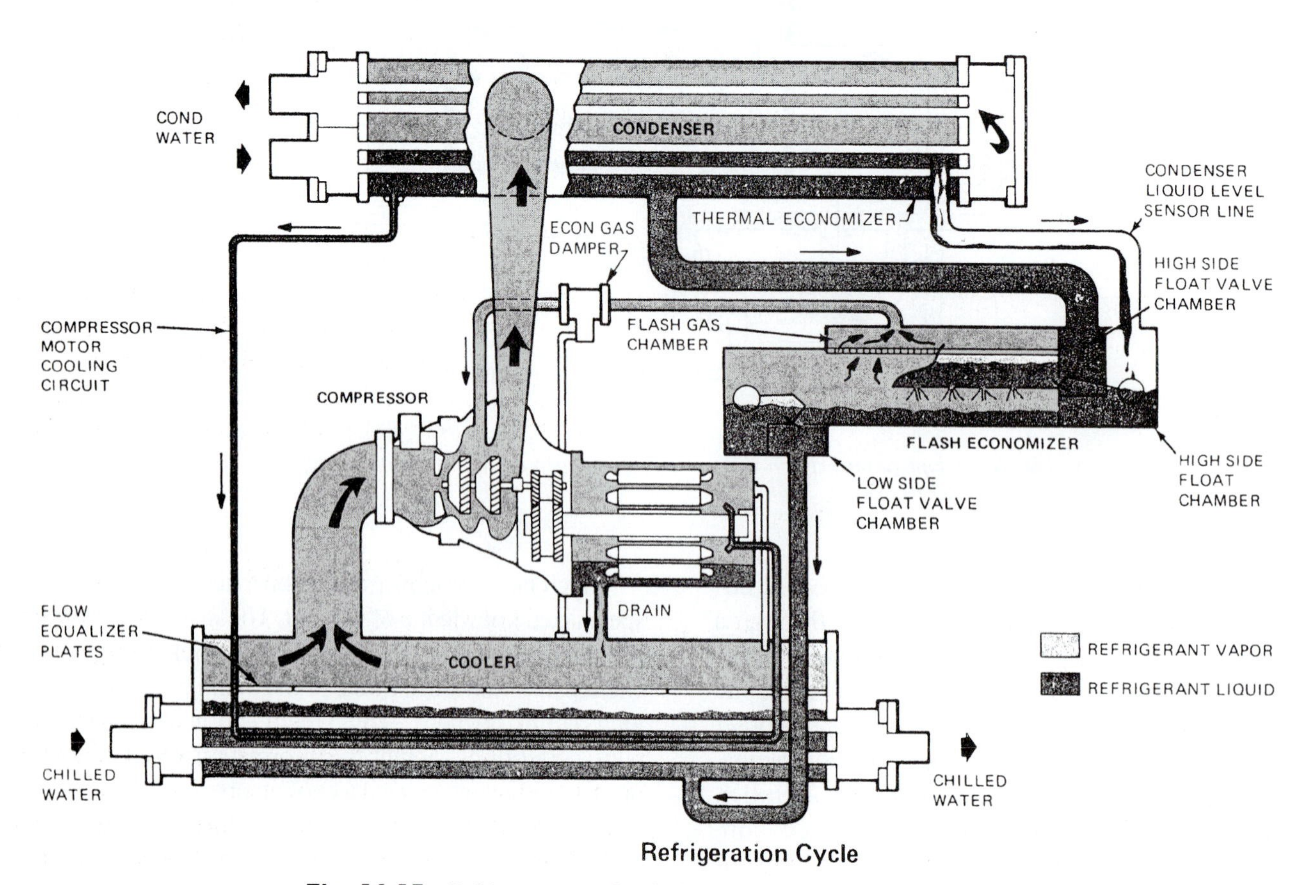

Fig. 16-15 *Refrigeration cycle of a hermetic-type centrifugal chiller.* (Carrier)

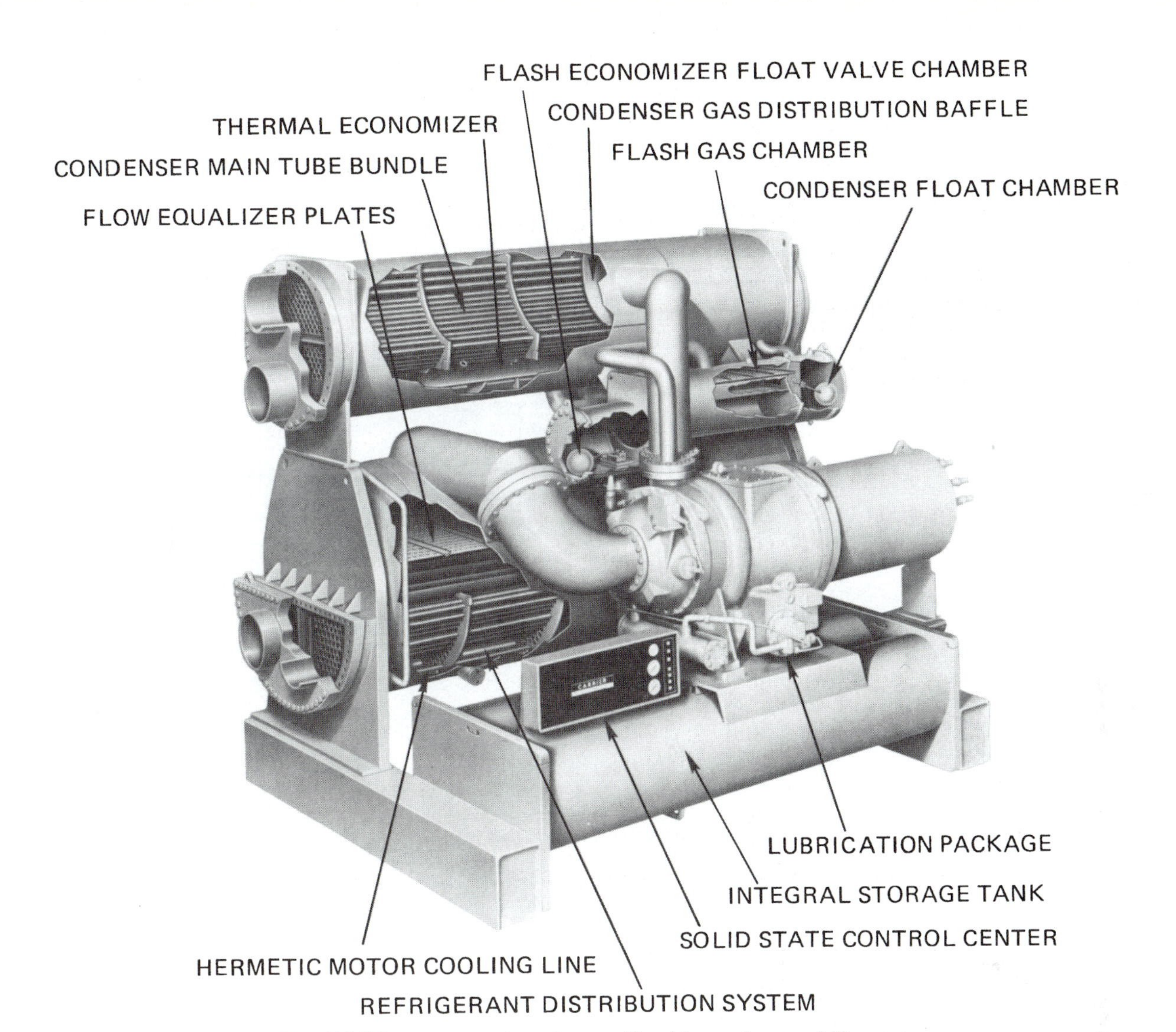

Fig. 16-16 *Cutaway view of a centrifugal hermetic-type chiller.* (Carrier)

Control System

Chiller safety controls are electronic. Chiller capacity controls may be either solid state (transistorized) or pneumatic (air-pressure controlled).

Chiller operating capacity is determined by the position of the guide vanes at the entrance to the compressor suction. As cooling needs change, the vanes open and close automatically.

A thermistor probe in the chilled water line constantly monitors chilled water temperature. The probe signals any temperature change to a capacity control module in the machine control center. The module, in turn, initiates a response from the guide-vane actuator. When chilled water temperature drops, the vane actuator causes the guide vanes to move toward the closed position. The chiller capacity decreases.

Conversely, a rise in chilled water temperature causes the guide vanes to open and increase chiller capacity. If the water temperature continues to rise, the vanes open further to compensate. Built-in safeguards prevent motor overloads. To minimize start-up demand, control interlocks keep the guide vanes closed (at minimum capacity position) until the compressor reaches run condition.

CHILLERS

Chillers are divided by type according to the compressors they use. Thus, they are reciprocating compressors and centrifugal compressors.

Reciprocating compressors may have single-acting or double-acting arrangements. They are made with from 1 to 16 cylinders. These cylinders may be arranged in a V-, W-, or radial pattern. Each cylinder arrangement is designed for a specific requirement. Reciprocating compressors have already been discussed in Chap. 7.

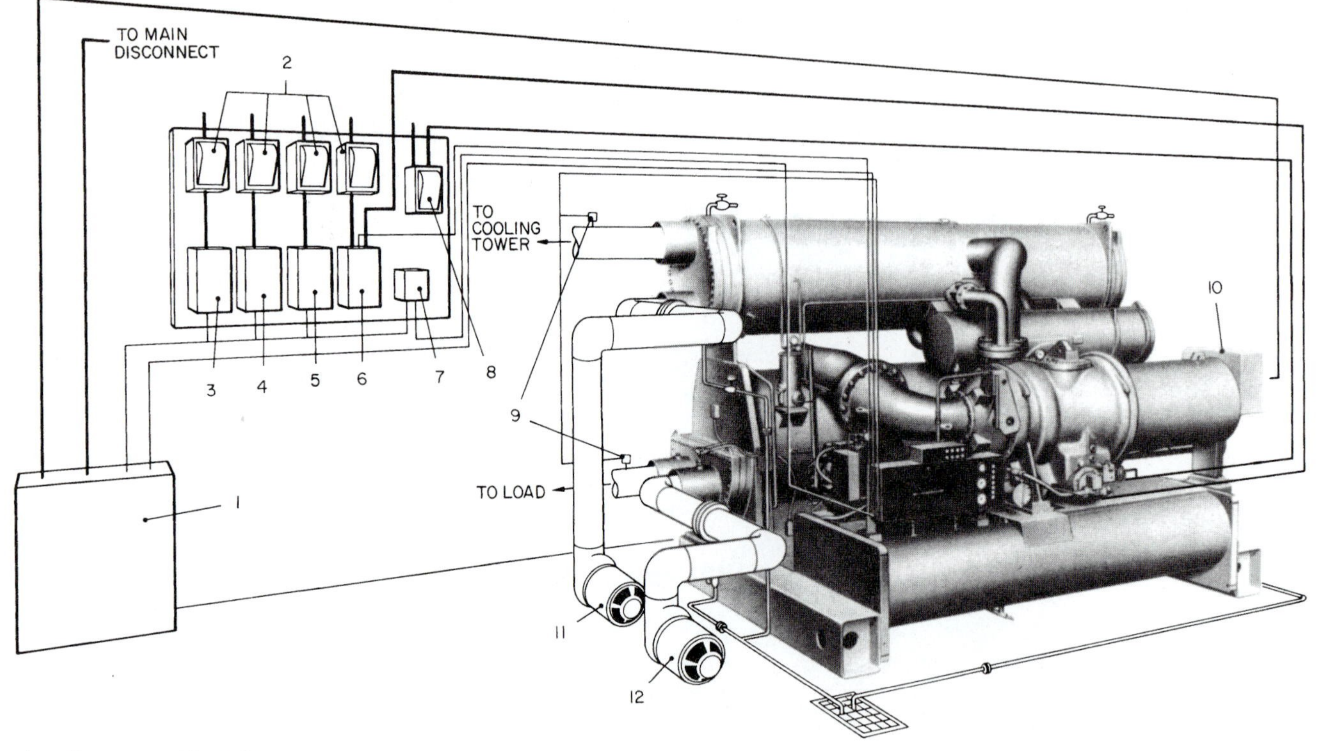

1 — Compressor Motor Starter
2 — Fused Disconnect
3 — Cooling Tower Fan Starter
4 — Condenser Water Pump Starter
5 — Cooler Water Pump Starter
6 — Lube Oil Pump Starter
7 — Pilot Relay
8 — Fused Disconnect for Oil Heater and Heater Controller (115 v)
9 — Water Flow Switches
10 — Compressor Motor Terminal Box
11 — Condenser Water Pump
12 — Cooler Water Pump

NOTES:
1. Separate 115-volt source for controls, unless transformer is furnished with compressor motor starter.
2. Wiring and piping shown do not include all details for a specific installation. Refer to certified electrical drawings.
3. Wiring must conform to applicable local and national codes.
4. Pipe per standard techniques.
→ 5. Oil cooler water source must be clean and noncorrosive. City water or chilled water may be used. Recommended flow condition is: 30 gpm water at 85 – 100 F. Vary gpm to obtain 100 F leaving water.
→ 6. Oil heater must be on separate circuit providing continuous service.

Fig. 16-17 *Typical piping and wiring diagram for the chiller in Fig. 16-16.* (Carrier)

There are two types of centrifugal compressors for chillers. They are the hermetic centrifugal compressor and the open-drive centrifugal compressor.

The open-drive centrifugal compressor is another type of chiller. Centrifugal compressors are used in units that produce 200 tons or more of refrigeration. Centrifugal compressors used in industrial applications may range from 200 to 10,000 tons. Flexible under varying loads, they are efficient at loads of less than 40 percent of their design capacity.

Large volumes of refrigerant are used in centrifugal compressors. They operate at relatively low pressures. The refrigerants used are R-11, R-12, R-113, R-114, and R-500 and several substitutes for some units.

Centrifugal compressors operate most efficiently with a high molecular weight, high specific-volume refrigerant.

Reciprocating Chillers

Reciprocating chillers are made in sizes up to 200 tons. They cannot handle large quantities of refrigerant. Thus, more than one compressor must be used. That is why the compressors are stacked on a large frame. Usually no more than two compressors are used in a refrigeration circuit. Thermostatic-expansion valves, discussed earlier, are used as metering devices.

Components Used with Chillers

Some components of the chiller system are huge. They are capable of handling large volumes of air. Such rugged air-handling equipment is necessary for this type of installation. See Fig. 16-18.

Air terminals are used to distribute the air when it reaches its destination. Figure 16-19 shows the whole

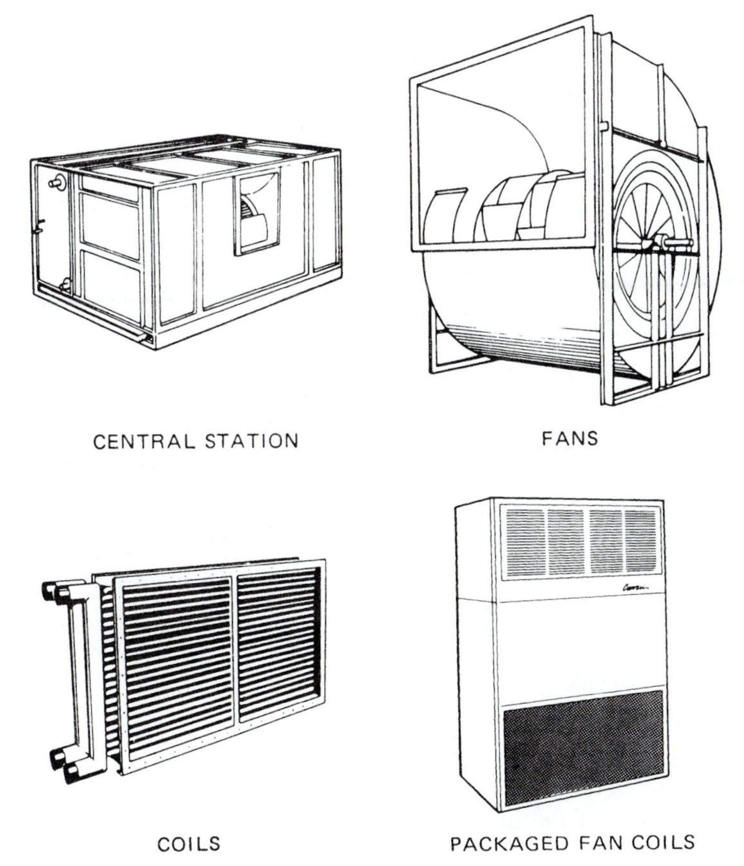

Fig. 16-18 *Air-handling equipment used with chiller system.* (Carrier)

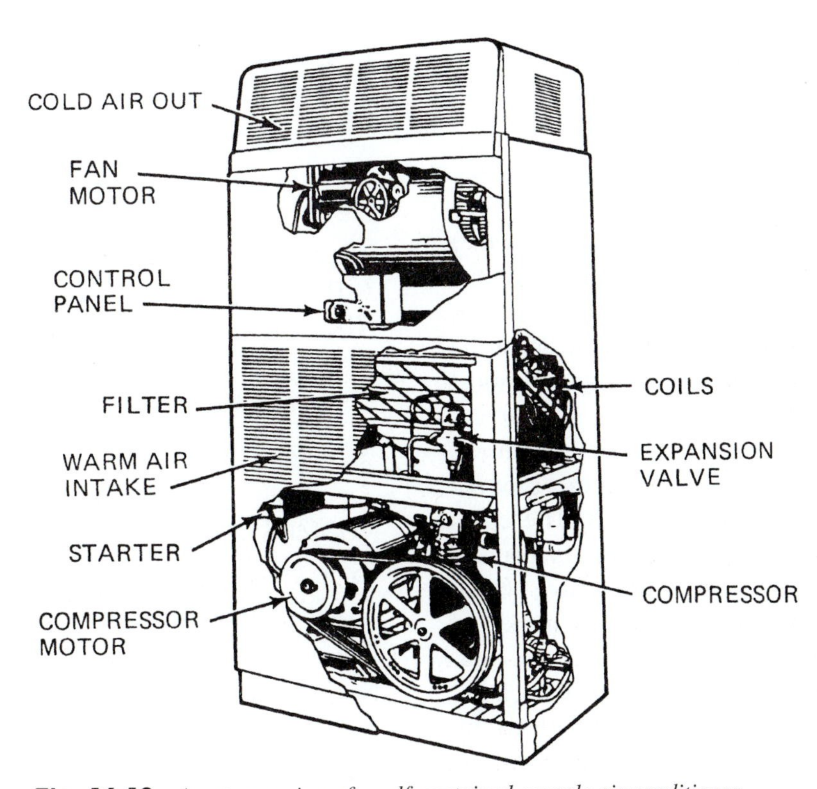

Fig. 16-19 *A cutaway view of a self-contained console air conditioner.*

system with the entire unit hooked up to furnish cool air to a room or building. Installation of this type of equipment requires a thorough knowledge of plumbing, electricity, and air-conditioning refrigeration. Trained specialists are needed to handle problems that arise from the operation of these large systems.

CONSOLE-TYPE AIR-CONDITIONING SYSTEMS

The console air conditioner is a self-contained unit. These units come in 2- to 10-hp sizes. They are used in small commercial buildings, restaurants, stores, and banks. They may be water cooled or air cooled.

Figure 16-19 shows an air-cooled console air conditioner. You should be able to vent to the outside the hot air produced by the compressor and the condenser.

There are also water-cooled console air conditioners. They will require connections to the local water supply as well as a water drain and condensate drain. See Fig. 16-20. Note the location of the parts in Fig. 16-19. Water is used to cool the compressor. In both models, the evaporator coil is mounted in the top of the unit. See Fig. 16-21. Air blown through the evaporator is cooled and directed to the space to be conditioned. In some areas, a water-cooled model is not feasible.

Since the evaporator coil also traps moisture from the air, this condensate must be drained. This dehumidifying action accounts for large amounts of water on humid days. If outside air is brought in, the condensate will be more visible than if inside air is recirculated.

Installation

The console air conditioner is produced by the factory ready for installation. It must be moved to a suitable location and hooked to electrical and plumbing connections. Once located and connected, it must be checked for level. Electrical and plumbing work must conform to local codes.

Check the unit for damage that may have occurred during shipping and installation. Note the type of compressor and the type of cooling used. The compressor is usually hermetic. Refrigerant is usually controlled by a thermostatic-expansion valve. Once installed, check the operating conditions. Check and record the temperature in and the temperature out. The difference in temperature is important. Check the electrical circuits so no overload is produced by adding the unit to the line. Record your observations for future use in servicing.

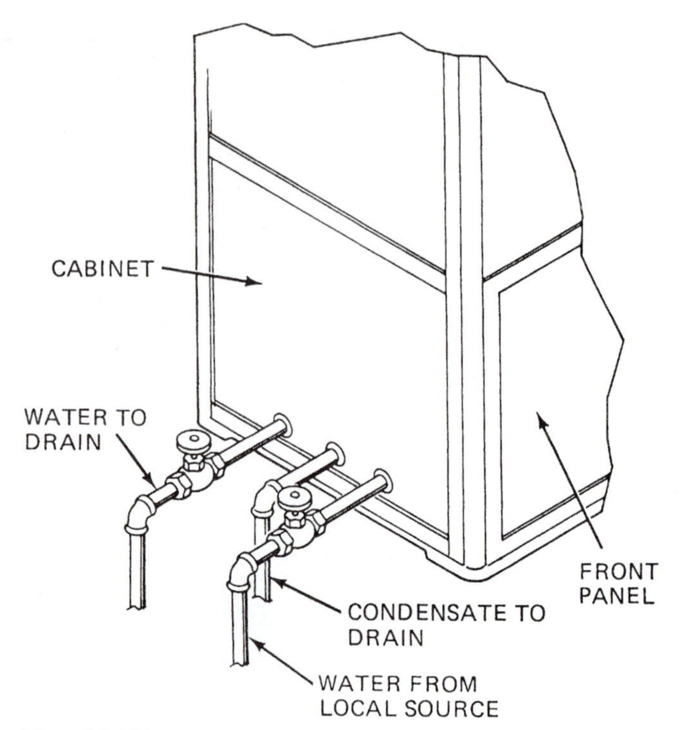

Fig. 16-20 *Plumbing connections for a water-cooled, self-contained console air conditioner.*

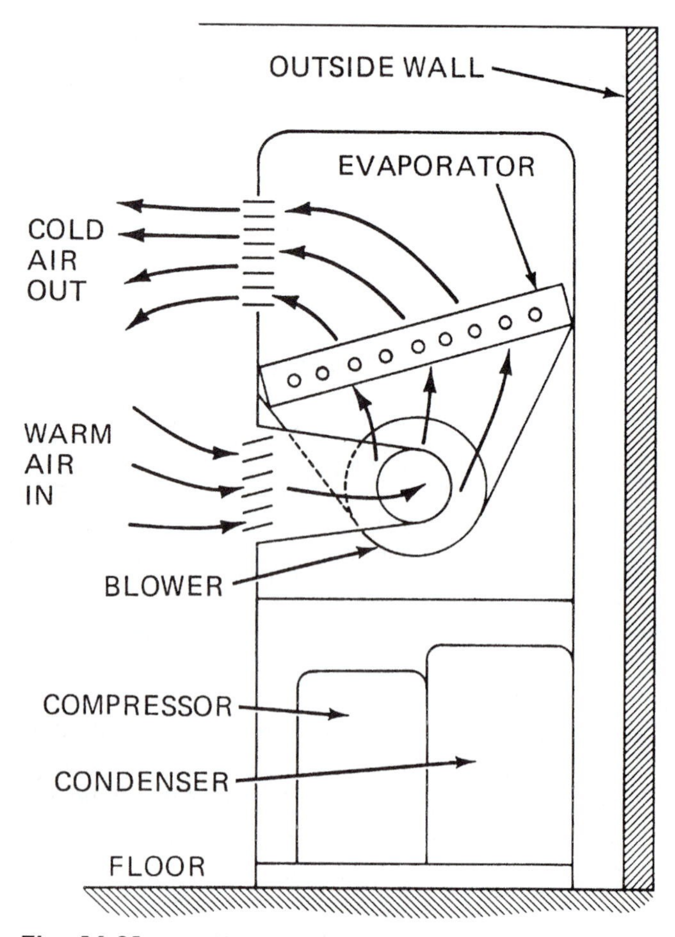

Fig. 16-21 *A self-contained console air conditioner, showing airflow over the evaporator*

Carrier Air Conditioning Company

A DIVISION OF CARRIER CORPORATION

REFRIGERATION LOG – OPEN DRIVE CENTRIFUGAL REFRIGERATION MACHINE

JOB NAME ______________

MACHINE SIZE ______________

MACHINE SERIAL NO. ______________

1	2	3	4	5	6	7	8	9	10	11	12	13	14	15	16	17	18	19	20	21	22	23	24
TIME	COOLER					CONDENSER				COMPRESSOR							GEAR OIL		PRIME MOVER*				OPERA-TOR
	GPM ____					GPM ____				DAM-PER POSI-TION	BEARING TEMP		OIL				PRES	TEMP	A	B	C	D	
	REFRIGERANT			BRINE TEMP		REFRIGERANT		WATER TEMP					Level	Reser-voir Temp.	Pressure								
	Pres	Temp	Shut-down Level	In	Out	Pres	Temp	In	Out		Thrust End	Seal End			Supply	Seal Housing							

REMARKS: ie, shutdown on safety controls, repairs, amount of oil or ref. added or removed. Check purge operation, frequency of purge starts, oil level, ref. level, discharge pressure, record water volume drained from purge.

DATE ______________

* ELECTRIC MOTOR	STEAM TURBINE	GAS ENGINE
A Volts	A Steam Pressure	A Gas Pressure
B Amps	B Gland Steam Pres	B Manifold Pres
C Bearing Temp.	C Bearing Temp.	C Jacket Water Temp.
D Oil Level	D Oil Pressure	D Oil Temp.

Fig. 16-22 *A service log used with an open-drive centrifugal refrigeration machine (chiller).* (Carrier)

Service

The unit is easily serviced since the component parts are located in one cabinet. Remove panels to gain access to the compressor, valves, blowers, filter, evaporator, and motors. A maintenance schedule should be set up and followed. Most maintenance consists of changing filters and checking pressures. The servicing of the refrigerating unit has already been described in detail in this chapter. The servicing of evaporators has also been described.

Cleaning the filters, cleaning the inside of the cabinet with a vacuum, and cleaning the evaporator fins are the normal service procedures. Water connections and electrical control devices should be checked for integrity. Clean the fan motor. Oil the bearings on the blowers and motors whenever specified by the manufacturer. If there are problems with a water-cooled condenser, refer to Chap. 8.

Scheduled maintenance is very important for all types of air-conditioning units. For some units, a log is kept to make sure the various components are checked periodically. Check the log first for any abnormal readings. See Fig. 16-22.

Troubleshooting

Table 16-1 lists basic troubleshooting procedures for the console model (self-contained) air conditioner.

REVIEW QUESTIONS

1. List three types of commercial air-conditioning systems.
2. What is the major advantage of a packaged cooling unit?
3. How often should the system air filters be inspected for dirt and clogging?

4. How long must power be on before starting the unit in the spring?
5. Where is the direct multizone system mounted?
6. What refrigerant is most commonly used in an absorption-type air-conditioning system?
7. What is the meaning of the term *strong aqua*?
8. What is a chiller?
9. How are cooling towers used to remove heat from a building?
10. How do you determine the operating capacity of a chiller?
11. Where is the hot air produced by the console model's compressor and condenser, vented?
12. What maintenance procedures are necessary for a console air-conditioning unit?

17

CHAPTER

Various Types of Air Conditioners and Heat Pumps

PERFORMANCE OBJECTIVES

After studying this chapter you will:

1. Know how gas is used to air condition.
2. Know how gas-fired chillers work.
3. Know how to use antifreeze to prevent damage to water in a chilled water cooler.
4. Know how to work with absorption refrigeration machines.
5. Understand how lithium-bromide water absorption cycle type cooling operates.
6. Know how solar air conditioners work.
7. Understand how heat pumps work.
8. Know how to troubleshoot heat pumps.
9. Know why a defrost cycle is needed in a heat-pump system.

There are a number of types of air conditioners made by a variety of manufacturers. Heat pumps are also available by a number of manufacturers. More of the heat pumps are being installed, especially in areas where there is not such a cold winter or rapid temperature swings.

GAS AIR CONDITIONING

There are three main types of gas air-conditioning cycles used today: compression, absorption, and dehumidification.

In both compression and absorption cycles, air temperature and humidity are tailored to meet variations in surrounding air conditions and changes in room occupancy. Both of these cycles evaporate and condense a refrigerant. They require energy for operation. Mechanical energy is used in the compression type. Heat energy is used in the absorption type. The dehumidification cycle is used primarily in industrial and commercial applications. Dehumidification reduces the moisture content of the air.

Absorption Cooling Cycle

The absorption type of air-conditioning equipment works on two basic principles: a salt solution absorbs water vapor and the evaporation of water causes cooling. In this particular discussion, the absorption cooling cycle is appropriate since it is used in gas-fired air conditioners.

Most gas-fueled air-conditioning equipment use a solution of lithium bromide (LiBr) in water. Lithium bromide is a colorless, saltlike compound that dissolves in water, even to a greater extent than does common salt.

A solution of lithium bromide and water can absorb still more water. Note that, in Fig. 17-1, a tank of absorbing solution (tank B) is connected with a tank of water (tank A). The air in the system is almost completely evacuated. The partial vacuum aids the evaporation process. Water vapor is drawn from the evaporator to the absorber. Evaporation of the water in the evaporator causes the water remaining in it to cool about 10°F (5.5°C). The evaporator effect in the evaporator is greatly hastened if the water is sprayed through several shower-bath sprinkler heads. A coil of pipe through which a material such as water passes can be placed within the shower of evaporating water. The water entering the coil of pipe at 55°F (12.8°C) will be cooled to about 45°F (7.2°C).

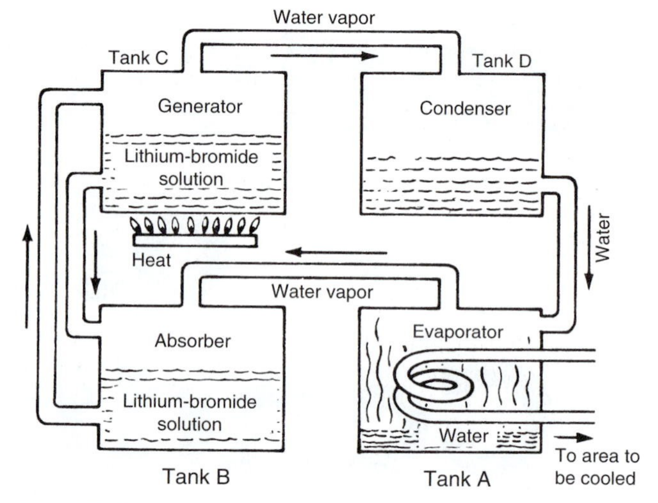

Fig. 17-1 *Lithium-bromide system of refrigeration.*

Since the absorber (B), shown in Fig. 17-1, continually receives water, it would soon overflow if the excess water that comes to it as water vapor was not removed. To avoid overflow, the solution that has absorbed water is pumped to a generator (C).

In the generator, the solution is heated directly by a natural gas flame. A steam coil may heat it indirectly. The steam is made in a gas-fueled boiler.

When the solution is heated, some of the water evaporates and passes into the condenser (D). The concentrated solution, that remains is sprayed back into the absorber (B). Here, it again absorbs water vapor that comes from the evaporator.

Water vapor in the condenser (D) is cooled by a separate coil of pipe through which water passes. The condensed water is returned to the evaporator (A).

Careful engineering is needed to make the system work well and economically. Attention must be given to temperatures, pressures, and heat transfer in all parts of the system. Practical machines with very few moving parts have now been developed.

Absorption units may also use ammonia as the refrigerant. In such system, heat from natural gas is used to boil an ammonia-water solution. The operation of the lithium-bromide cycle discussed earlier generally applies. In a system using ammonia, the temperature of the evaporator can go below the freezing point of water. Ammonia is referred to as R-717.

Ammonia Refrigerant in a Gas-Fired System

Ammonia is also used as a refrigerant in a gas-fired system. As the ammonia is moved through the system, it changes state becoming strong and weak vapor and a liquid. Chilled water is used as a circulation coolant. Very few electrical pumps are needed. Fans are still needed to remove the collected heat. A pump is needed for circulating the ammonia. A pump is needed for circulating the chilled water.

GAS-FIRED CHILLERS

Chillers operate on natural gas or propane gas. See Fig. 17-2. Gas is used for the major job of cooling. Electricity is used for the smaller energy requirements of fans, motors, and controls. This means electrical power requirements are only about 20 percent of those of a completely electrical unit.

Gas units are available in 3-, 4-, and 5-ton capacities. They use heat as a catalyst. They have no compressor. This means they have fewer moving parts than other types of systems. They are designed for outside installation. They cool by circulating, the refrigerant, which is plain tap water, through a matching coil. The coil is added to a new or existing furnace in the house. As the chilled water produced by the unit circulates through the coil, it absorbs heat from the conditioned space. The water, bearing the absorbed heat, is then returned to the unit outdoors, where the heat is dissipated to the outside air. Table 17-1 shows the amount of permanent antifreeze required when the outside temperature is below freezing. A defoaming agent also must be added.

Table 17-2 shows the specifications of Arkla 3-, 4-, and 5-ton units. Note that the refrigerant is R-717 (ammonia). Also, note the amount of gas needed to produce 3 tons of air conditioning—79,000 Btu.

Gas-fired units may be connected in 5-ton multiples to provide up to 30 tons of air conditioning. Figure 17-3 shows how they are doubled up to provide 10 tons. For some units, it takes 250,000 Btu of gas input per hour to provide 120,000 Btu/h of cooling. That means 48 percent efficiency, if the electrical energy needed is not accounted for in the figuring. For the unit referred to, the operating voltage is 230 V, with 60-Hz, single-phase operation. Wiring size is not too large,

Fig. 17-2 *Gas-fired air-conditioning unit.* (Arkla)

Table 17-1 Antifreeze Needed to Prevent Damage to Water in a Chilled Water Cooler

Lowest Expected Outdoor Temp. °F (Freezing Point of Mixture)	Permanent Antifreeze Percentage by Volume (%)
25	10
15	20
5	30
0	33
–5	35
–10	40
–20	45

since there is a maximum of 8 A drawn for the condenser fan motors and 33 A for the solution pumps. Normal running current for the solution pump motors is only 5 A. Locked rotor current of 33 A occurs only if the motor is stuck or jammed so it cannot start. The start currents may also reach this 33-A level under some load conditions.

The chilled water system uses stainless steel to prevent problems with rust and other ferrous metal piping problems. There is only one electrical, one gas, and one chilled water supply and return connection for each unit.

CHILLER-HEATER

Some gas-fired units furnish cooling for the summer and heat for the winter. The user changes the functions simply by changing the settings of a room thermostat. The "all-year" units are designed for outdoor installation. They operate on either natural or propane gas.

Changeover Sequence for Chilled Water Operation

When the thermostat calls for cooling, the hot water pump is off. The chilled water pump moves water from the chiller tank and pumps it up a "candy cane" shaped loop and out to the air handler. See Fig. 17-4. As the water returns to the chiller tank to be cooled again, it passes through the water reservoir.

The water does not flow through the tubes of the hot water generator as it returns to the chiller tank. The water in the generator is dormant because it is plugged by a check ball. This is held in place by the pressure from the discharge side of the chilled water pump.

Changeover Sequence for Hot Water Operation

When the thermostat is set for heating, the chilled water pump is off. The pressure from the hot water pump moves the check ball to seal off the water in the chiller tank. See Fig. 17-5. Now the water in the chiller system is dormant. The hot water pump circulates the water from the hot water generator through the air handler and back to the generator through the water reservoir. During the heating cycle the reservoir also serves as a place to relieve air from the system. The tube from the top of the reservoir passes through the chiller tank and runs up to the distribution pan, which is open to atmospheric pressure.

Self-Leveling Feature

Self-leveling of water between the chiller tank and the water reservoir during the heating cycle is another unique feature of this system.

Should the water level in the reservoir drop below normal, a vacuum is created in the top of the reservoir. The vacuum causes a negative pressure that acts as a suction to draw water up the air release tube, refilling the reservoir. See Fig. 17-6. As the water level in the distribution tube falls below the level of the water in the chiller tank, the second check ball is forced away from the seat to allow the water level to return to normal. See Fig. 17-7. It should be remembered that antifreeze and a defoaming agent are necessary for this water system.

ABSORPTION REFRIGERATION MACHINE

The absorption refrigeration machine is used primarily in air-conditioning applications. Chilled water is the output of the machine. The chilled water is then used to cool. This particular machine is available for capacities of 100 through 600 tons. These units are small, relatively lightweight, and vibration free. They can be located wherever a source of steam or very hot water is available. Lithium bromide, a salt solution, is used as the absorbent.

Absorption Operation Cycle

Figure 17-8 is a schematic diagram of an absorption cold generator. Note that the evaporator, absorber, concentrator, and condenser are enclosed in a single casing. The heat exchanger is located externally below the main shell.

- *Evaporator.* The evaporator pump circulates the refrigerant (water) from the refrigerant pump into the spray trees. To utilize the maximum surface for evaporation, the refrigerant is sprayed over the evaporator tubes. As the spray contacts the relatively

Table 17-2 Specifications for 3-, 4-, and 5-Ton Gas-Fired Air-Conditioning Units

SPECIFICATIONS	MODEL		
	3 Ton ACB 36-00	4 Ton ACB 48-00	5 Ton ACB 60-00
Performance Ratings			
Gas input, Btu/h	79,000	100,000	125,000
Delivered capacity, Btu/h*	36,000	48,000	60,000
Condenser entering air temperature	95° F.	95° F.	95° F.
Condenser air flow, cfm, approx.	4,000	6,000	6,000
Chilled water** entering temperature	55° F.	55° F.	55° F.
Chilled water** leaving temperature	45° F.	45° F.	45° F.
Chilled water** flow, gpm	7.2	9.6	12.0
maximum permissible flow, gpm	12.0	12.0	16.0
Allowable friction loss for piping and coil, feet of water	27	26	25
Maximum vertical distance, top of coil above unit base			
with rigid piping, feet	25	25	25
with flexible piping, feet	15	15	15
Do not use ferrous metal piping or tubing.			
Electrical Ratings			
Required voltage, 60 Hz, 1 phase	115	230	230†
Condenser fan motor horsepower	1/3	1/3	1/2
full load/locked rotor amperes, nominal	4.6/9.9	3.5/7.2	3.5/7.2
Pump drive motor horsepower	1/2	1/2	3/4
full load/locked rotor amperes, nominal	7.4/44.8	3.6/21.8	5.1/30.5
Operating wattage draw, typical	875	1,000	1,275
Branch circuit ampacity, minimum	13.85	8.00	9.88
Number and size of time delay fuses	1-20 amperes	2-15 amperes	2-15 amperes
Physical Data			
Refrigerant type	717	717	717
Unit chilled water** volume, gallons	3.0	3.0	5.0
Operating weight, pounds	550	750	775
Shipping weight, pounds	590	825	850
Inlet chilled water connection size, FPT	3/4"	1"	1"
Outlet chilled water connection size, FPT	3/4"	1"	1"
Gas inlet connection size, FPT	1/2"	1/2"	1/2"
Electric entrance knockouts, diameter	7/8"	7/8"	7/8"

*See tables below for additional capacity data.

**"Chilled water" is a solution of good quality tap water and 10% by volume of permanent antifreeze, with sufficient defoaming agent added to prevent foaming.

†ACB 60-00 equipped for 208-volt operation available on special order.

Once Through or Process Applications

Only water or glycol-water mixtures shall be circulated through the unit. In process application where cooling of other solutions is desired, they should be circulated through a secondary heat exchanger.

Refrigeration Capacity

The capacity of the air-cooled chillers varies with ambient air temperature and leaving chilled water temperature. Capacity characteristics of the units are shown in the tables below.

CAPACITY ACB36-00 IN BTU/H

Leaving Chilled Water Temperature (° F.)	Air temperature entering condenser (° F.)			
	90	95	100	105
48	36,720	36,180	34,920	32,652
46	36,650	36,036	34,200	31,320
45	36,576	36,000	33,840	30,600
44	36,500	35,820	33,720	29,592
42	36,360	35,532	32,220	27,432
40	36,000	35,028	30,600	23,400

CAPACITY ACB48-00 IN BTU/H

Leaving Chilled Water Temperature (° F.)	Air temperature entering condenser (° F.)			
	90	95	100	105
48	48,960	48,240	46,560	43,536
46	48,912	48,048	45,600	41,760
45	48,768	48,000	45,120	40,800
44	48,624	47,760	44,496	39,456
42	48,480	47,328	42,960	36,576
40	48,000	46,704	41,300	31,200

CAPACITY ACB60-00 IN BTU/H

Leaving Chilled Water Temperature (° F.)	Air temperature entering condenser (° F.)			
	90	95	100	105
48	61,200	60,300	58,200	54,420
46	61,140	60,060	57,000	52,200
45	60,960	60,000	56,400	51,000
44	60,780	59,700	55,620	49,320
42	60,600	59,160	53,700	45,720
40	60,000	58,380	51,000	39,000

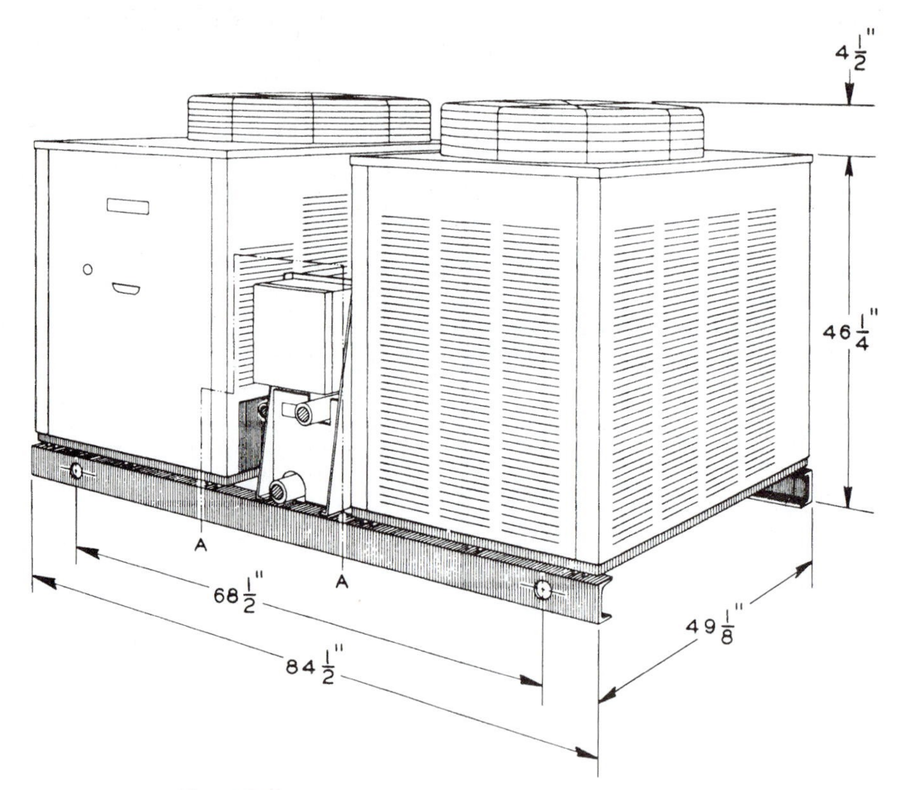

Fig. 17-3 *Ten-ton gas-fired air-conditioning unit.* (Arkla)

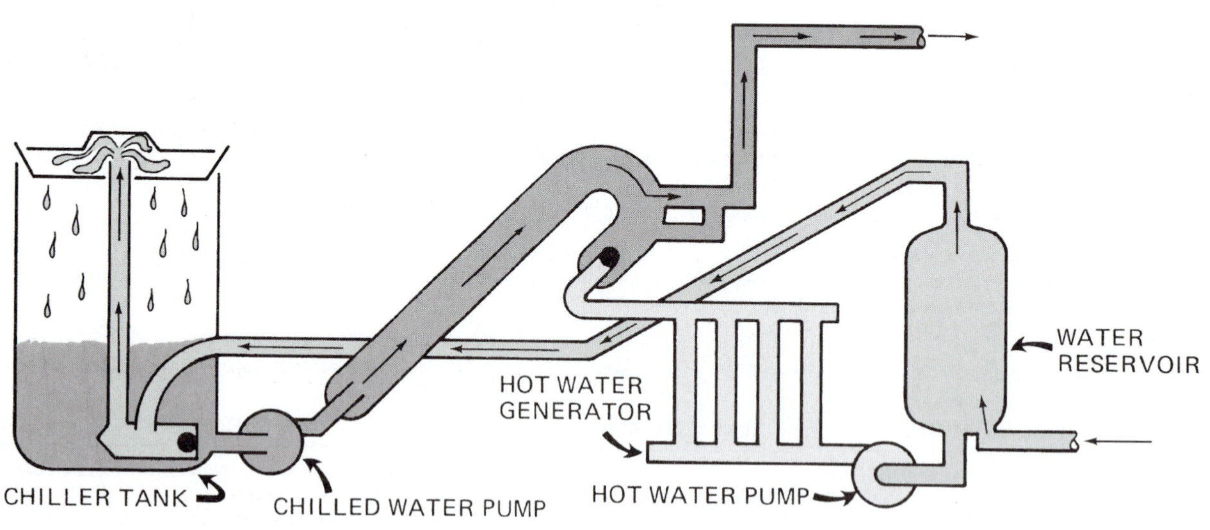

Fig. 17-4 *The cooling cycle in a chiller heater.* (Arkla)

warm surface of the tubes carrying the water to be chilled, a vapor is created. In this manner heat is extracted from the tube surface, chilling the fluid in the tubes. The vapor created in this process passes through eliminators to the absorber.

- *Absorber.* The lithium-bromide solution (under proper conditions) keeps the pressure in the absorber section low enough to pull the refrigerant vapor from the high-pressure evaporator. As the vapor flows into the absorber, it mixes with the absorbent solution being sprayed over the tube bundle.
- *Heat exchanger.* The heat exchanger is used only as an economizer. The cool diluted solution from the concentrator pump is heated by the hotter concentrated solution moving from the concentrator to the absorber steam or hot water (heating medium) is

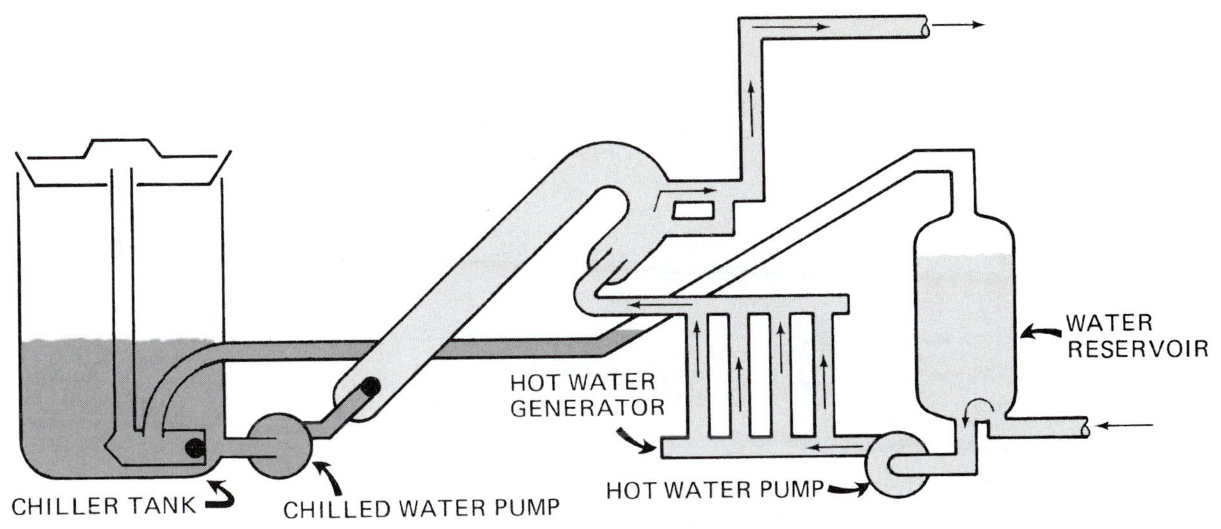

Fig. 17-5 *The heating cycle in a chiller heater.* (Arkla)

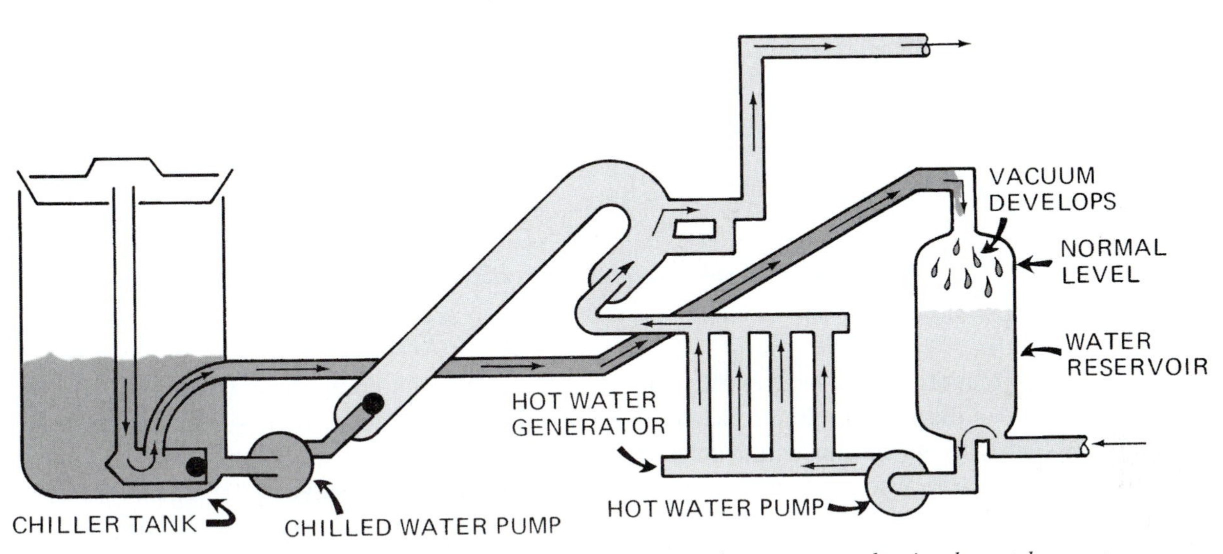

Fig. 17-6 *Self-leveling feature in a chiller-heater. Vacuum draws water up the air-release tube.* (Arkla)

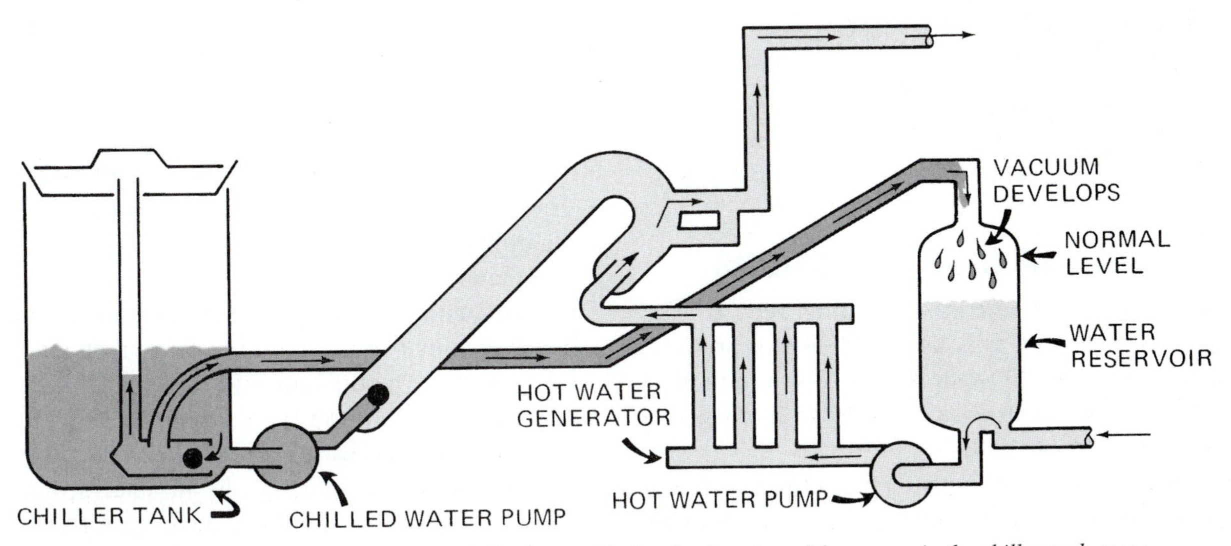

Fig. 17-7 *Self-leveling feature in a chiller heater. Notice the direction of the arrows in the chiller tank.* (Arkla)

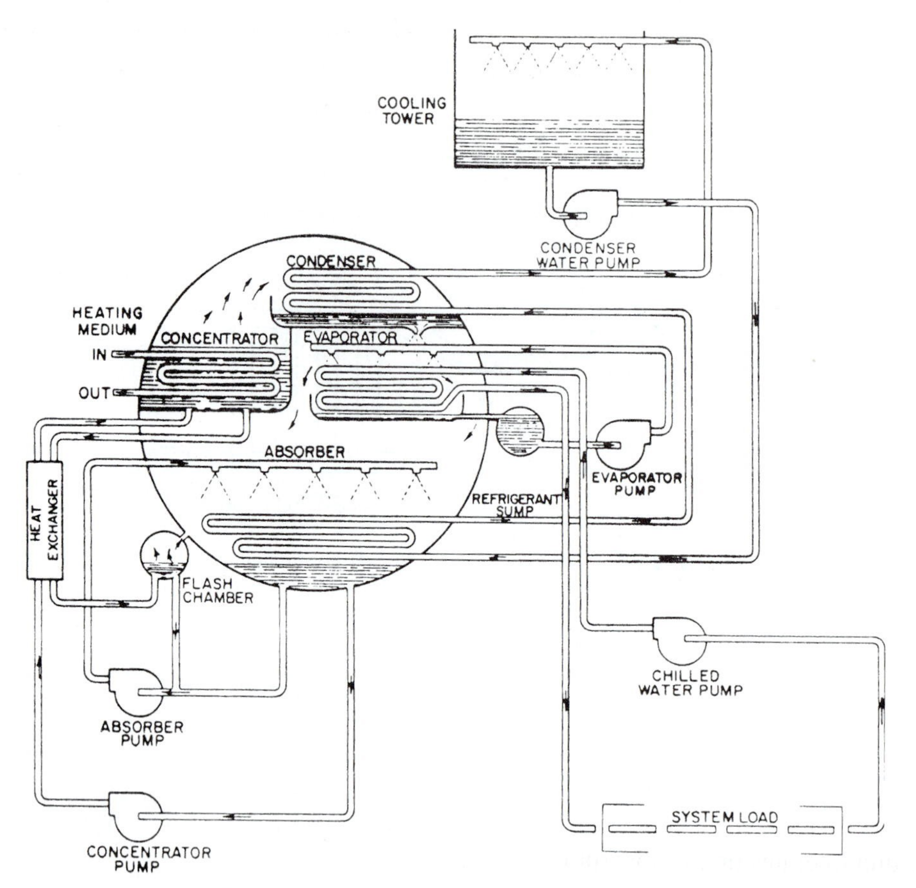

Fig. 17-8 *Schematic of an absorption cold generator.* (Johnson)

conserved. The heat transfer in the heat exchanger brings the temperature of the diluted solution closer to the boiling point. It also brings the concentrated solution temperature closer to the absorber temperature.

- *Concentrator.* Steam or high-temperature water entering the concentrator is controlled to boil off the same quantity of refrigerant picked up by the absorber. The refrigerant vapor is given up by boiling the solution in the concentrator. The vapor passes through eliminators to the tube surface of the condenser.
- *Condenser.* The refrigerant vapor from the concentrator is condensed on the tube surface of the condenser and falls into the pan below the tube bundle.

SOLAR AIR CONDITIONERS

Harnessing the sun's energy is nothing new. As far back as 1878, the sun was used to power a printing plant. In Egypt, in 1913, solar energy was used to produce steam to operate an engine that drove an irrigation pump.

In the United States, solar energy powered the phone of a Georgia cotton grower in 1955. Even then, the costs involved in harnessing solar energy were astronomical in comparison to the costs for the abundantly available fossil fuels—coal, oil, and natural gas. Thus, the research programs languished until the energy crisis of the early 1970s. Then, fossil-fuel shortages, environmental concerns, and the rising costs of energy, reawakened interest in solar energy. It was known that solar-heated water could provide the power for space heating and water heating. However, could it effectively cool a home? Could the costs of using solar energy be brought within the reach of the average person?

History of Solar Cooling

Prior to 1972, little, if any, research had been done regarding the use of solar energy for cooling. Therefore, no air-conditioning equipment specially designed for use with solar energy was available. Currently, however, there are cooling systems that lend themselves to easy modification for use with solar energy. One is the absorption-type system manufactured by Arkla Industries. The other is the *Rankine cycle*.

The Rankine cycle needs an intermediate step. This involves replacing the electric motor in the conventional vapor compression refrigeration cycle with a turbine or using solar cells to produce electricity. In either case, making modifications for the Rankine cycle is more costly than making modifications of the absorption system.

The National Science Foundation has worked on a cooling system specially designed for use with solar energy by giving grant money to researchers. Work is being done on improving both the residential and medium-tonnage range.

Systems of Solar Cooling

There are two systems used in solar cooling: the *direct system* and the *indirect system.* The direct system of application uses the absorption cooling system. It provides higher firing water temperatures directly from the storage tanks to the unit's generator. See Fig. 17-9.

The indirect system is a closed system in which a heat exchanger transfers the heat from the solar-heated water storage tanks. This allows the use of an antifreeze fluid. See Fig. 17-10.

Lithium-Bromide Water Absorption Cycle

The Arkla-Solaire unit operates on the absorption principle. See Fig. 17-11. It uses solar-heated water as the energy source. Lithium bromide and water are used as the absorbent/refrigerant solution. The refrigeration tonnage is delivered through a chilled water circuit that flows between the unit's evaporator and a standard fan-coil assembly located inside the conditioned space. The heat from the conditioned space is dissipated externally at the cooling tower.

The four main components of the Solaire cooling unit are the generator, condenser, evaporator, and absorber.

When the solar-heated water enters the tubes inside the generator, the heat from the hot water vaporizes the refrigerant (water), separating it from the absorbent (lithium bromide).

The vaporized refrigerant vapor then flows to the absorber. There, it again liquefies and combines with the circulating solution. The reunited lithium bromide and water solution then passes to the liquid heat exchanger. There, it is reheated before being returned to the generator.

Figure 17-12 shows a medium-tonnage air-conditioning unit specially designed for use with solar energy. This is a nominally rated 25-ton unit that can operate with a firing water temperature as low as 190°F (87.8°C).

There is also a 3-ton absorption unit that operates with solar-heated water. It provides full capacity with a firing water temperature of 210°F (98.9°C).

Solar Cooling Research Centers

Figure 17-13 pictures a *Solar House*, which was conceived by the Copper Development Association. Inc. a number of years ago. Almost everything in this innovative (for

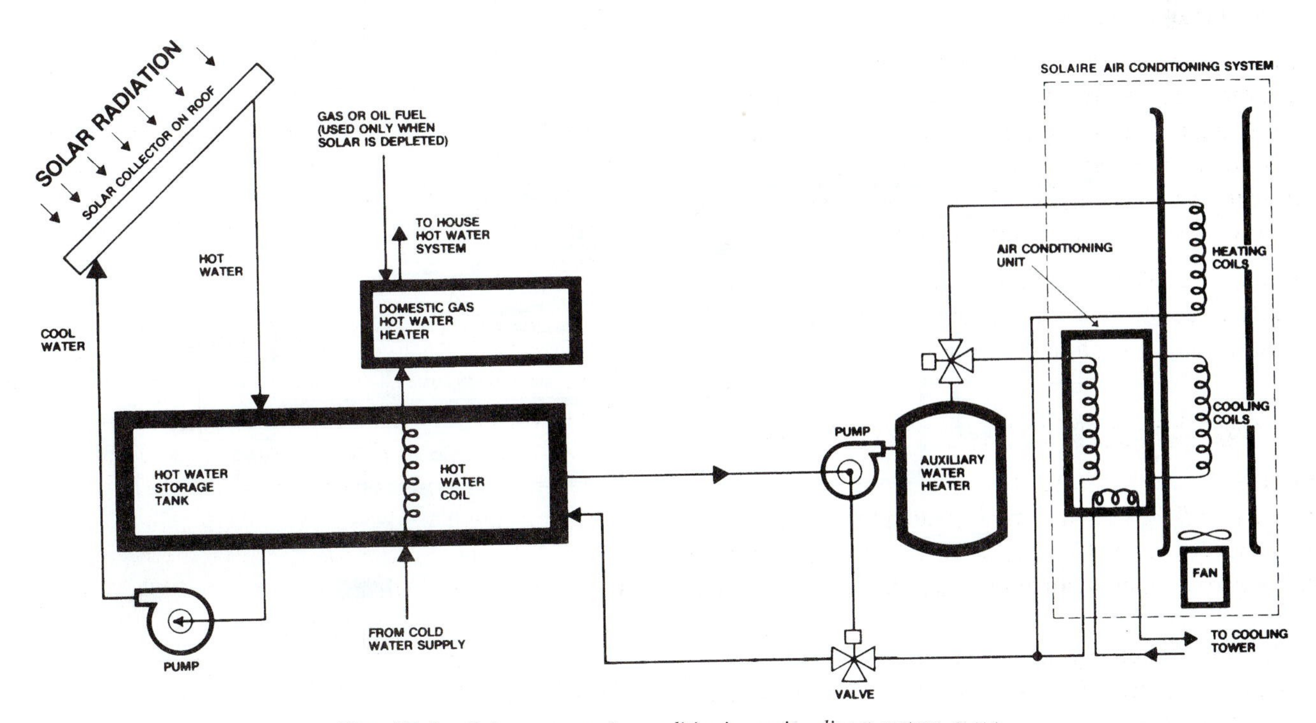

Fig. 17-9 *Solar-energy air-conditioning unit—direct system.* (Arkla)

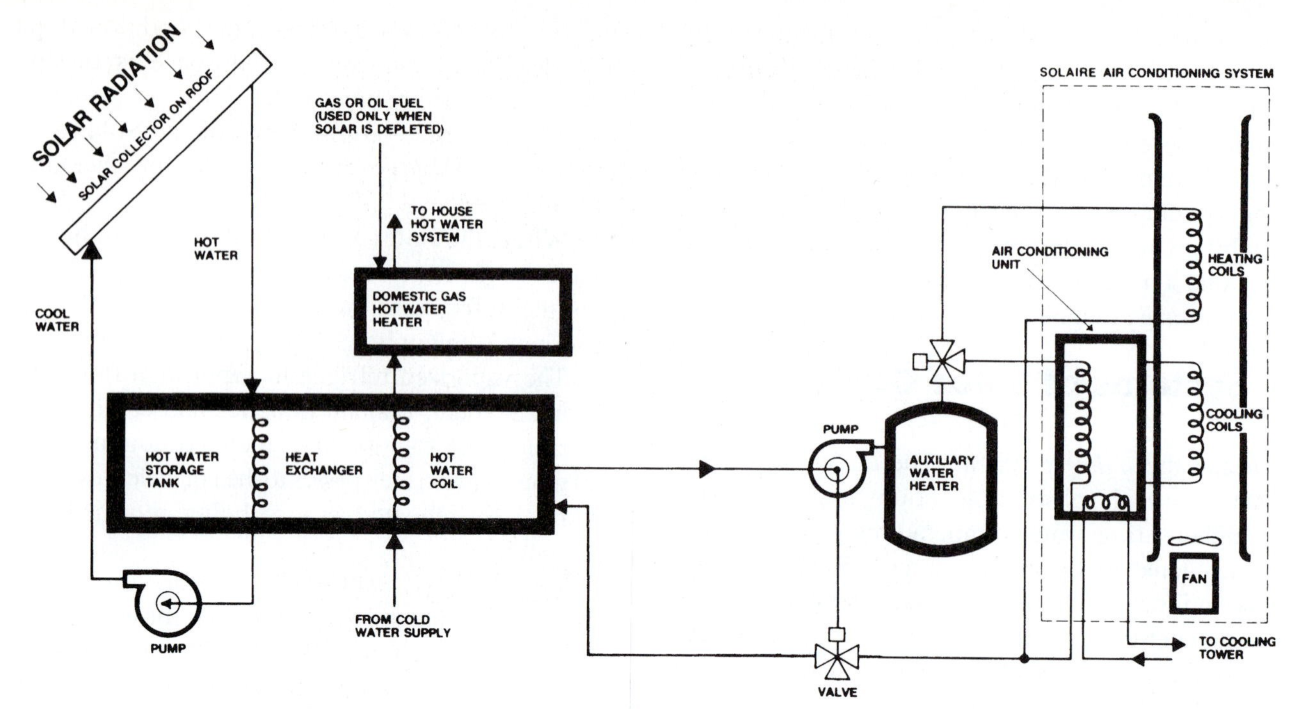

Fig. 17-10 *Solar-energy air-conditioning unit—indirect system.* (Arkla)

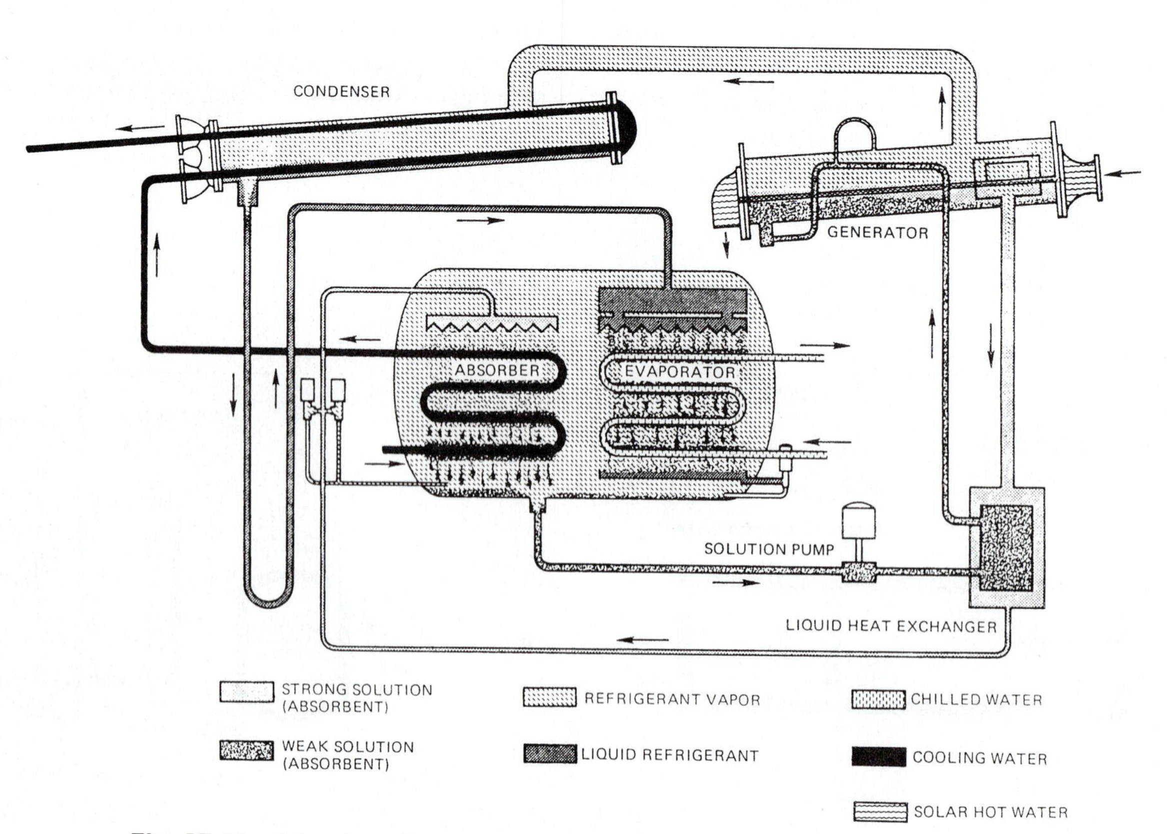

Fig. 17-11 *Solar air conditioning using lithium bromide and water absorption cycle.* (Arkla)

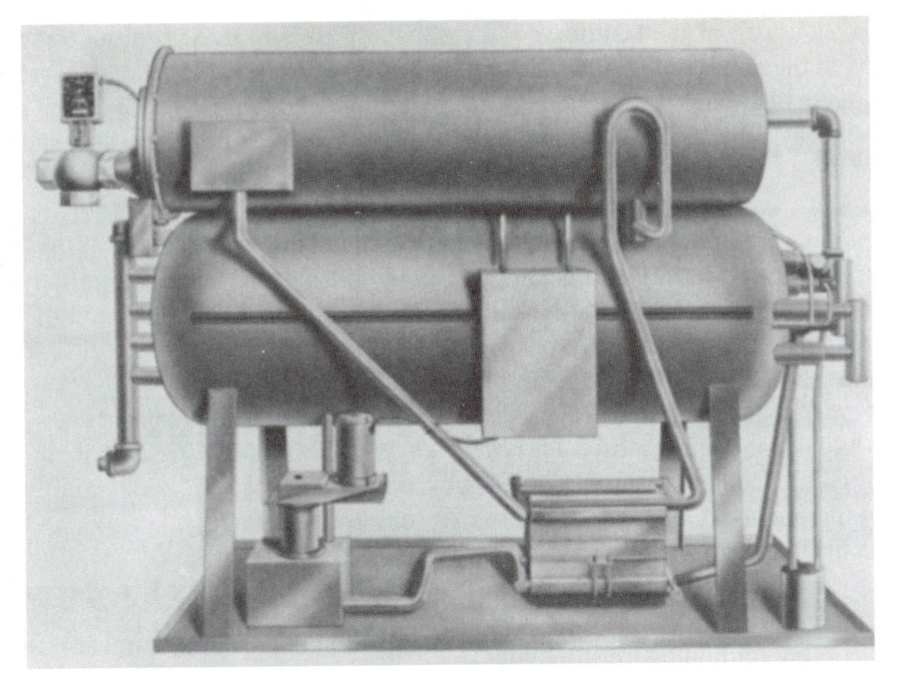

Fig. 17-12 *A medium-tonnage air-conditioning unit specially designed for use with solar energy. This is a 25-ton unit that operates with a firing temperature as low as 190°F.* (Arkla)

Fig. 17-13 *Artist's rendering of a solar house in Tucson, Arizona. The four-bedroom home is cooled by two units fired by solar-heated water.* (Arkla)

its time) four-bedroom house from the heating, cooling, and sound systems to the door chimes and kitchen clock can be run on stored solar energy. Coupled with the solar energy collector system, the heating/cooling system is probably the most challenging technological innovation in the house.

The climate control system consists of two 3-ton Solaire units for cooling and two duct coils for heating. As designed, it was anticipated so that the solar-heated water would be able to operate the cooling cycle nearly 75 percent of the time. It should supply 100 percent of the heating requirements. A back-up water heater was installed near the 3000-gal hot water thermal energy storage. The backup unit functioned automatically. That is, if the water temperature dropped too low to operate the climate control system. The house, fully computerized, can analyze hard data on solar energy in a normal home environment.

Solar heating and cooling for homes are still under development and engineering study in a number of locations. For instance, one of those was a Colorado State University solar research project that studied a residential-type structure. It was designed as a laboratory for testing and evaluating the performance of solar equipment designed for heating and cooling. The solar cooling system was a 3-ton system which used lithium-bromide absorption, modified for using hot water as the heat source.

The Marshall Space Flight Center Solar House was a simulated residence that used three surplus office trailers with a freestanding roof. See Fig. 17-14. It had the effective areas and the heating/cooling load equivalent of an average three-bedroom house. A 3500-gal water tank was used as a heat storage reservoir. A lithium-bromide type of air-conditioning system was used.

One of the first schools heated and cooled by solar energy is shown in Fig. 17-15. It used a lithium-bromide system for cooling. The control system's components included 10,000 ft^2 of collector area and 45,000 gal of fluid for thermal storage. This experimental project was designed to provide solar heating and cooling, and domestic hot water for the 32,000 ft^2, one-story school.

Take a closer look at the solar energy collectors shown in Fig. 17-16. This is an absorption air conditioner. It is used for both cooling and heating a portion of a training facility.

HEAT PUMPS

A heat pump is a mechanical refrigeration system. It is arranged and controlled to utilize the condenser heat. The condenser heat is wasted or dissipated into the air when a condenser-compressor are mounted outside a building being air conditioned. By utilizing the heat generated by the condenser can be used for some useful purpose. This useful purpose is in most instances, space heating. Heat-pump systems may be classified as:

- Package, or built-up
- Air-to-air
- Water-to-air
- Water-to-water

Fig. 17-14 *This research solar-energy house is cooled by a lithium-bromide air conditioner. It uses solar-heated water.* (Arkla)

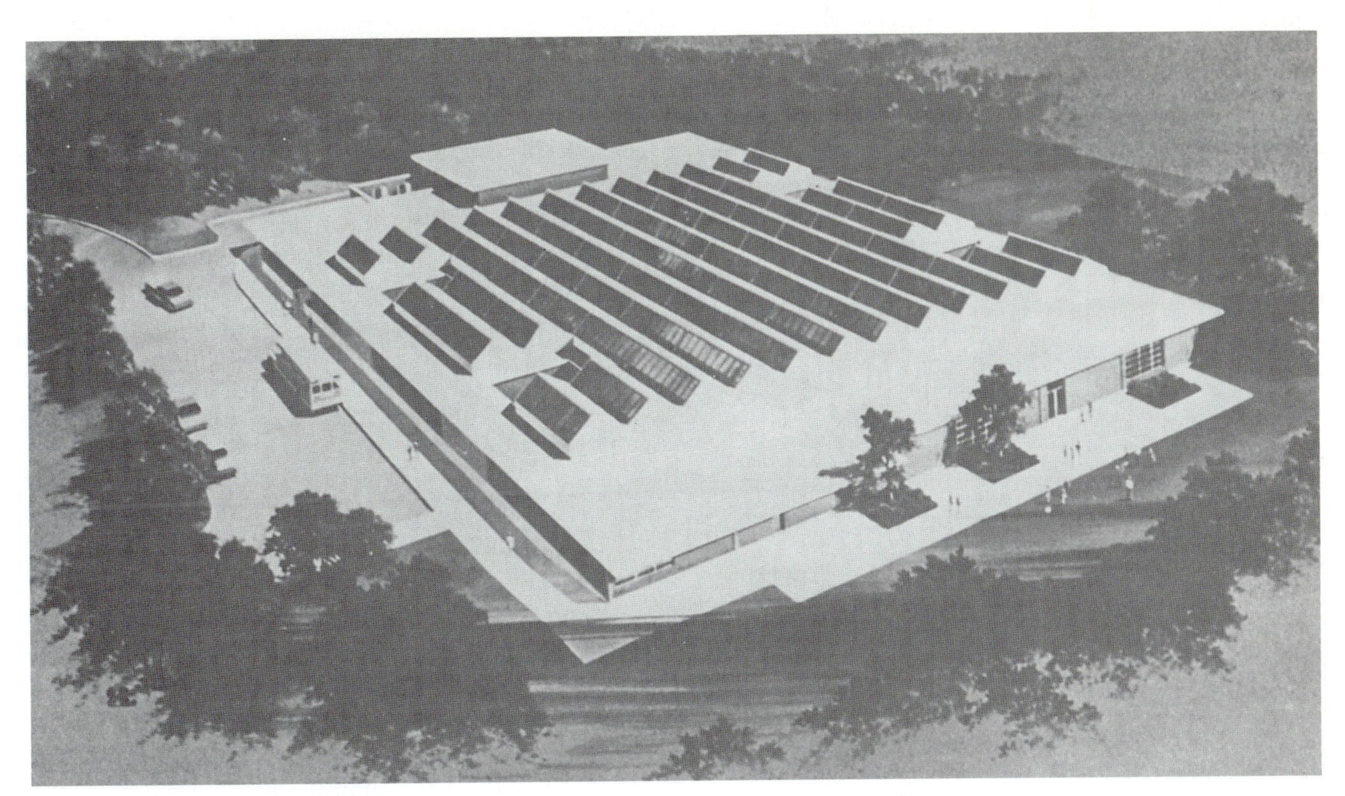

Fig. 17-15 *This school in Atlanta, Georgia is cooled by solar energy.* (Arkla)

Fig. 17-16 *An absorption air conditioner is used to heat this training facility. Solar water heats and cools the house located in Morton Grove, Illinois.* (Arkla)

Earth coupled systems are also used as a variation of the water-to-water concept. Keep in mind that the heat pump is primarily a central air conditioner. It can also act as a heating system. During the cooling season the heat pump performs exactly like a central air conditioner. It removes heat from the indoor air and discharges it outside. See Fig. 17-17.

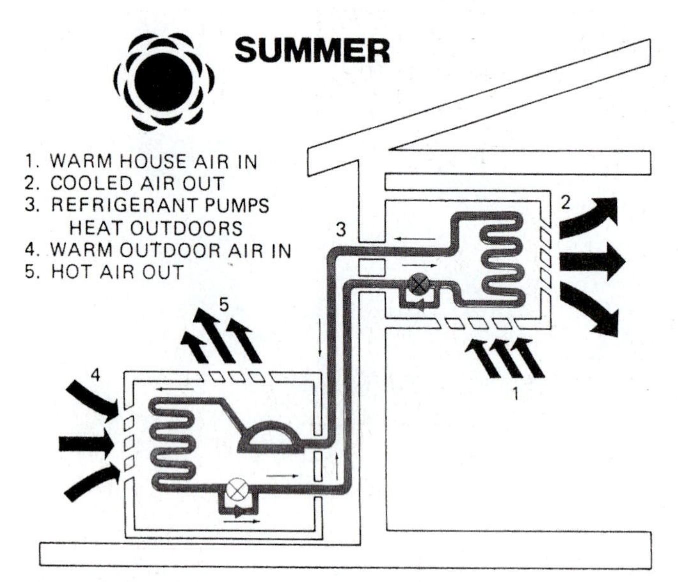

Fig. 17-17 *Operation of a heat pump during the summer.* (GE)

During the heating season, the heat pump reverses its function. It changes from a cooling system to a heating system. It then removes the available heat from the outdoor air and discharges it inside the house. See Fig. 17-18.

There is heat in outdoor air, even at 0°F (−17.8°C). In fact, heat is available in outdoor air down to −460°F (−273°C).

Since the heat pump is a refrigeration machine, it needs only enough electrical power to run a compressor, an outdoor fan, and an indoor blower. The result is a heating system with a seasonal efficiency of better than 150 percent. This means that for every kilowatt of electric power used, the heat pump will produce more than 1.5 kW of heat energy. Only the heat pump can give this level of efficiency.

Heat pumps are available in all sizes for apartments, homes, and commercial applications. Heat pumps are not new. General Electric has been selling them since 1952. There are now various types of units on the market.

One unit, the Fuelmaster, works with a heat pump. It can be used with gas, oil, and electric furnaces. See Fig. 17-19. As can be seen from the illustration, the heat pump resembles a compressor-condenser unit. However, the control box is different. See Fig. 17-20. The control box has relays and terminal strips factory installed and wired. The heat pump delay and defrost limit control are included in the unit.

Operation

On mild temperature heating days, the heat pump handles all heating needs. When the outdoor temperature reaches the "balance point" of the home (heat loss equals heat pump heating capacity), the two-stage indoor thermostat activates the furnace (secondary heat source). When the furnace fires, a heat relay deenergizes the heat pump.

When the second stage (furnace) need is satisfied and plenum temperature has cooled to 90 to 100°F (32.2 to 37.8°C), the heat pump delay turns the heat pump back on. It controls the conditioned space until the second stage (full heat) operation is required again.

When outdoor temperature drops below the setting of the low-temperature compressor monitor (field-installed option) the control shuts out the heat pump. The furnace handles all of the heating need. The low-temperature compressor monitor is standard on models dated 1974 and after.

During the cooling season the heat pump operates in its normal cooling mode. It uses the furnace blower as the primary air mover. See Fig. 17-21.

Defrost

During a defrost cycle, the heat pump switches from heating to cooling. To prevent cool air from being circulated when heating is needed, the control automatically turns on the furnace to compensate for the heat pump defrost cycle. (Most modern heat pump systems do the same thing with strip heating.) When supply air temperature climbs above 110 to 120°F (43.3 to 48.9°C), the defrost limit control turns off the furnace and keeps indoor air from getting too warm.

After a defrost cycle, the air temperature downstream of the coil may go above the 115°F (46.1°C) closing point of the heat pump delay. Then, the compressor will stop until the heat exchanger has cooled to 90 to 100°F (32.2 to 37.8°C), as it does during normal cycling operation between furnace and heat pump.

Outdoor Thermostat

In a straight heat pump or supplementary electric heater application, at least one outdoor thermostat is required to cycle the heaters as the outdoor temperature

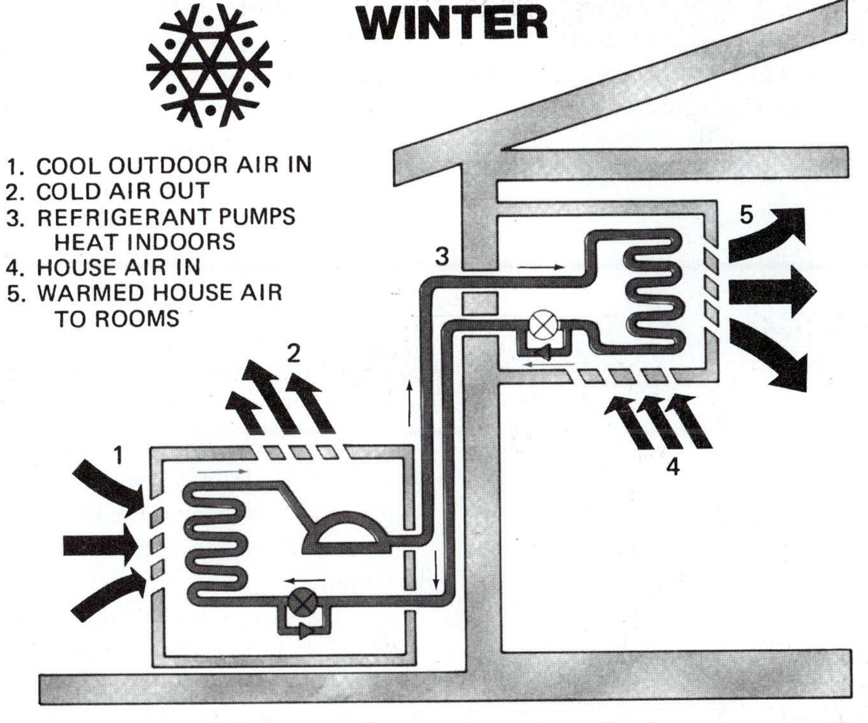

Fig. 17-18 *Operation of a heat pump during the winter.* (GE)

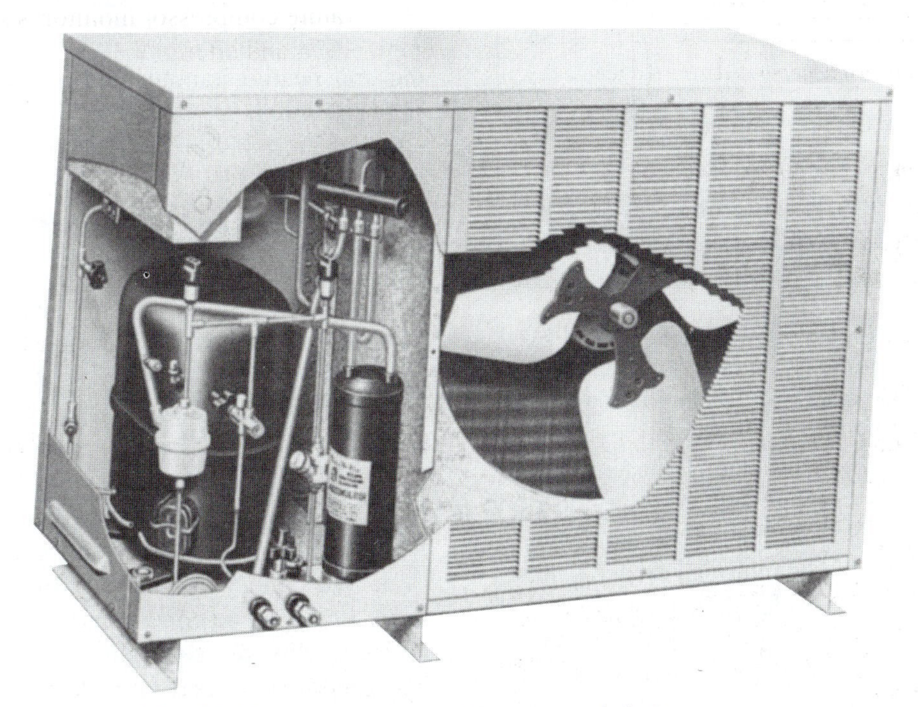

Fig. 17-19 *Outside unit for a heat pump made by Lennox.* (Lennox)

drops. In the Fuelmaster system, the indoor thermostat controls the supplemental heat source (furnace). The outdoor thermostat is not required. Since the furnace is serving as the secondary heat source, the Fuelmaster system does not require the home rewiring usually associated with supplemental electric strip heating.

Special Requirements of Heat Pump Systems

The installation, maintenance, and operating efficiency of the heat pump system are like those of no other comfort system. A heat pump system requires the same air

Fig. 17-20 *Control box for a heat pump add-on unit made by Lennox.* (Lennox)

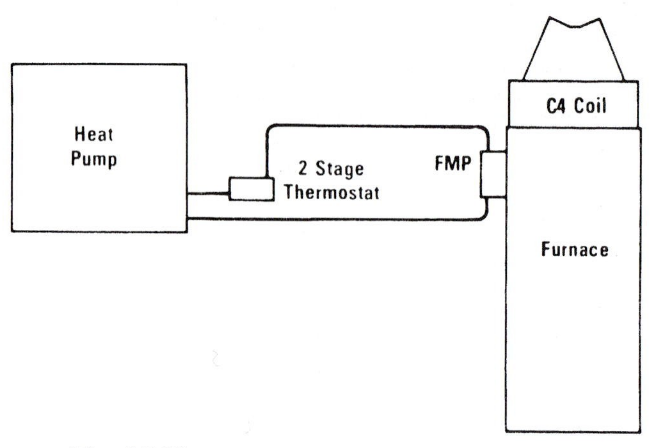

Fig. 17-21 *Typical heat pump components.* (Lennox)

quantity for heating and cooling. Because of this, the air moving capability of an existing furnace is extremely important. It should be carefully checked before a heat pump is added. Heating and load calculations must be accurate. System design and installation must be precise.

The air-distribution system and diffuser location are equally important. Supply ducts must be properly sized and insulated. Adequate return air is also a must.

Heating supply air is cooler than with other systems. This is quite noticeable to homeowners accustomed to gas or oil heat. This makes diffuser location and system balancing critical. Typical installations of heat pumps are shown in Fig. 17-22.

Sizing Equipment

Home insulation, exposure, design and construction, climate, and living habits determine the efficiency of a heat pump system. Each heat pump installation is unique. Each job must be calculated carefully. Remember, there are no *rules of thumb* for heat pump sizing.

To determine the most economical operating cost, size the system for its cooling load. Use supplemental heating (second stage) to make up the difference between heat pump heating capacity and building heat loss. Sizing the heat pump to handle the entire heat loss will result in oversized cooling capacity that will adversely affect dehumidification during cooling.

Defrost Cycle

During a heat pump heating cycle, the outdoor coil absorbs heat from outdoor air. To do this the coil must be cooler than the air.

When air temperature falls below 40°F (4.4°C), coil surface temperature is below freezing. Moisture in the air freezes on the coil. Frost or ice builds up, reducing the air passage through the coil and cutting heat pump output.

The heat pump defrost cycle removes this buildup. The system is reversed to a cooling cycle, which heats the coil and melts the ice. Supplemental heat is used to counteract the cooling effect of the cycle change.

Balance Point

The outdoor temperature at which heat loss of a building and heating output of a heat pump are equal is called the *balance point*. It is the lowest temperature at which the heat pump alone can handle the heating load.

Using the Heat Pump

In the last fifty years, several large office buildings and small college campuses have been constructed using water-to-water heat pumps. Their capacities were up to several hundred tons. These systems usually use well water. That means two wells are used. One is used for supply and one for disposal. A possible arrangement is shown in Fig. 17-23.

The supply and disposal wells are manually selected. Well water and return water are mixed, for both evaporator and condenser. This is done on a temperature basis. Under some conditions, this system can become an internal source heat pump. That is, when the exterior-zone heating and interior-zone cooling loads are in balance, or nearly so, little or no well water is needed.

Internal source heat pumps without wells are used where there is sufficient internal cooling load to supply the net heating requirements under all conditions. Excess heat can be disposed of through cooling towers.

A problem with these systems is related to a high electrical load for the pumping system. A variety of variable-flow piping schemes have been devised to overcome this problem.

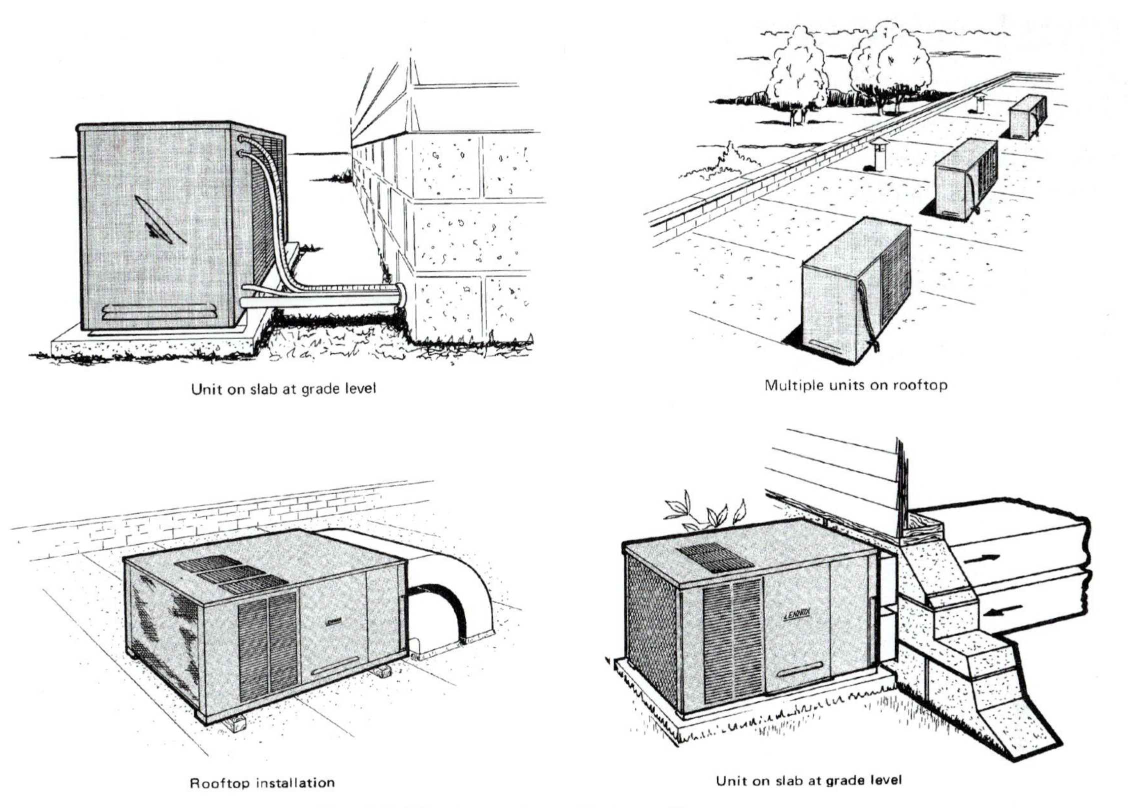

Fig. 17-22 *Typical installations of heat pumps.* (Lennox)

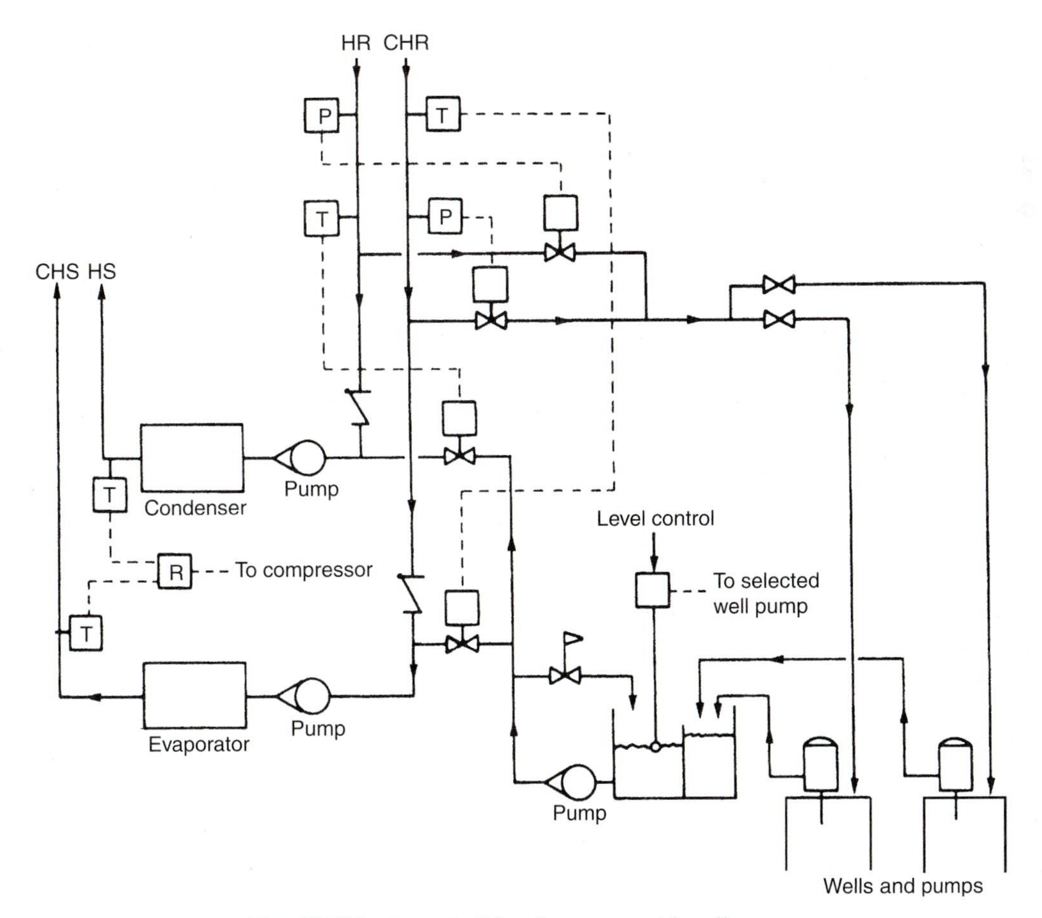

Fig. 17-23 *Large building heat pump with well water source.*

REVIEW QUESTIONS

1. What are the three main types of gas air-conditioning cycles used today?
2. What are the two basic principles of operation of the absorption-type airconditioner?
3. What type of refrigerant does an absorption-type unit use?
4. What type of refrigerant is used in a gas-fired system?
5. How can problems from rust and other ferrous metals be prevented in a chilled water system?
6. What is a concentrator?
7. What two types of systems are used in solar cooling?
8. What determines the efficiency of a heat pump system?

18

CHAPTER

Estimating Load and Insulating Pipes

PERFORMANCE OBJECTIVES

After studying this chapter you will:

1. Know how to figure refrigeration and air-conditioning load.
2. Know how air doors are used and installed.
3. Know how insulation is properly used to aid in the load factor.
4. Know how insulated pipes can improve efficiency of a system.

REFRIGERATION AND AIR-CONDITIONING LOAD

The load for a refrigeration unit comes from many sources. This load comes from several heat sources, the more common being the following:

- Heat from the outside leaks through doors and windows or conducted through the insulated walls.
- Transparent materials allow heat to penetrate them. This occurs when windows are used in a refrigerated space.
- Open doors and windows may allow heat to enter a refrigerated place. Cracks around the doors and windows also allow heat to enter the refrigerated space.
- The materials (products) stored in the refrigerated space give off heat. To lower the product temperature, it is necessary to lower the heat content of the materials.
- People who occupy a cooled space give off heat. This must be considered when figuring any load for a given refrigeration unit.
- Equipment inside the refrigerated space may give off heat. For example, motors, electric lights, electronic equipment, steam tables, urns, hair dryers, and similar items give off heat. To obtain an accurate figure, it is necessary to consider all of the heat sources. This will, in turn, determine how long the equipment must run to maintain a given temperature in the space being cooled.

RUNNING TIME

The time necessary for the cooling equipment to maintain or come down to a certain temperature is called the *running* time. The time used for calculations is 24 h. Equipment capacity is rated in Btu per hour. Therefore, 24 h times the Btu per hour will produce the normal capacity of refrigeration equipment. A quick way to determine this is to use the following formula:

$$\text{Required Btu/h equipment capacity} = \frac{\text{Total cooling load in Btu for 24 h}}{\text{Desired running time}}$$

Most equipment cannot run for 24 h, since defrosting consumes some of the time. That is why the total cooling load for 24 h is divided by the desired running time.

Moisture taken from the stored product causes frost to form on the evaporator coils or surface. This frost must be removed to maintain the efficiency of the unit. The defrost cycle must be determined by the amount of frost that will form on the coils. The refrigeration process stops while defrosting is being accomplished.

A system may be stopped long enough for the frost to melt. This is not the desired method in most cases. It may allow the product's temperature to rise, thus spoiling the product. To speed up the defrosting process, a heating element is usually introduced into the system. The heating element melts the frost rapidly. The water is drained to the outside of the unit. The off-cycle type of defrost is time consuming. It usually takes 8 h. This means the loss of 8 h of refrigeration or a total running time of 16 h for the equipment. (It takes 8 h for defrosting and about 8 h to bring the temperature back to the previous point.) Where heated defrost methods are used, it usually takes about 6 h on the average. That produces an equipment running time of 18 h. This is usually taken as average for calculations. Several heating methods are used. Electric heating elements have been mounted near the evaporator surfaces, or hot gas may be recirculated to produce the same effect. This method is explored in detail in Chap. 10.

In air-conditioning equipment, the temperature of the coil rarely gets below 40°F (4.4°C). Thus, no frost accumulates. This means that in most cases a defrost method is not necessary in air-conditioning equipment. Air-conditioning equipment is designed for continuous running. The running time is determined by the Btu actually needed to cool a room.

CALCULATING COOLING LOAD

The individual loads should be figured first. These are then totaled. This produces the load to be used for figuring running time and equipment design characteristics.

The four sources of load are:

- Wall gain load
- Air change load
- Product load
- Miscellaneous load

Wall Gain Load

Heat that leaks through the wall is the *wall gain load.* This heat comes from outside the refrigerated area. There is no perfect insulation. Thus, there is always some heat leakage through the walls. Heat always moves toward a less-heated (cooled) area. So, if the inside of a space is cooler than the outside, there is always movement of heat from the warmer to the cooler area. Insulation is used to slow down this heat movement. Air conditioners and commercial refrigeration systems are always subject to wall gain or heat gain from outside the cooled area.

Air Change Load

The *air change load* originates when the door is opened to a refrigerated space. The warm moist air that enters the area must be cooled to the inside temperature. This cooling presents a load to the equipment.

In some cases, this is not a factor. In the case of chillers, there is no opening through which air can pass. Thus, this type of load does not exist. However, this is not the case with air-conditioning equipment. The cracks around doors and windows also add to the load in an air-conditioned space. In some cases air is introduced from outside to improve ventilation. This is especially true in air-conditioning systems. The outside air must be cooled. Therefore, it presents a load to the air conditioner.

The introduction of air for ventilation is the ventilation load. The air that leaks around doors and windows is the infiltration load. Every air-conditioning system must deal with this type of load.

Most commercial refrigerators have well-fitted door gaskets. Thus, they have little infiltration load. Here air changes are the result of opening and closing the door(s).

Product Load

Any material stored in a refrigerated space must be brought down to the temperature of the inside space if it is not already to that temperature. In some cases, the temperature of the product is lower than the inside temperature. This means it can also add to the cooling process. However, in most cases the temperature of such a product is not taken into consideration. The refrigeration process it contributes to is gradual and, usually, is slowly diminishing.

Once the product is cooled to the temperature of the refrigerated space it is no longer a part of the load.

Fruits and vegetables give off respiration heat the entire time they are in storage. They give off heat even though they reach the temperature of the storage area. There is no further decrease in their temperature, however.

In sonic instances a product will give off heat all the time it is stored. In this case it is best to place it in a chiller first, then transfer it to cold storage.

Air conditioning has no product load as such. However, there is often a *pull-down load.* This is thought of as a *product load.*

Miscellaneous Loads

Heat from electrical equipment, electric lights, and people working in a refrigerated place is thought of as a *miscellaneous load.*

In an air-conditioned space there is no miscellaneous load. It has been taken care of previously. The people in the air-conditioned space make up the major part of the load. In fact, human occupancy is the primary load on most air-conditioned spaces. There are exceptions in which electrical equipment, or other types of equipment, is the entire load. This would occur when the air-conditioning system was designed solely for cooling some type of equipment for better operation.

CALCULATING HEAT LEAKAGE

It is difficult to estimate some loads with accuracy. Heat leakage can be estimated with some degree of accuracy. Heat leakage through walls, floors, and ceilings depends on the insulating material and its thickness. The formula for determining heat leakage is:

$$H = kA\,(t_1 - t_2)$$

where H = heat leakage
k = heat transfer coefficient in Btu per square foot per degree Fahrenheit
A = area in square feet
$t_1 - t_2$ = temperature gradient through the wall, which is expressed in degrees Fahrenheit.

In this example, the wall is made of 6 in. of concrete and 4 in. of cork. What is the heat transfer coefficient of the wall?

Assume that the concrete wall is an outside wall. It is exposed to air circulation from the environment. The transfer of heat through the air film next to a concrete and cork wall is about 4.2 Btu/ft^2/h for each degree difference in temperature between the inside and outside. This information can be found in handbooks for engineers. The resistance is the reciprocal of this, or

$$\frac{1}{4.2} = 0.24 \text{ (0.238095 rounded to 0.24)}$$

The coefficient for concrete is about 8 Btu/ft^2/h/° difference per inch of thickness. Or, 1 over 8 is equal to 0.125. The resistance is 0.125.

The resistance of a 6 in. wall is 6 times greater (6×0.125), or 0.75. The coefficient for cork is about 0.31 Btu/ft^2/h/° difference per inch of thickness, or

$$\frac{1}{0.31} = 3.225806$$

This means it has a resistance of $^1/_{0.31}$ or 3.225806.

The resistance for a 4 in. wall is 4×3.225806 or 12.903224. The inside wall contacts the still air. Experience has shown that an average value of the coefficient for the film is about 1.4. The resistance then is $^1/_{1.4}$ or 0.714286. The total resistance for this wall is $0.24 + 0.75 + 12.9 + 0.71 = 14.60$.

Therefore, the overall coefficient is $^1/_{14.60}$ or 0.068493. It is apparent that the principal resistance is offered by the cork. By using the same method, it is possible to obtain the overall coefficient for any type of wall.

Once you have found the heat transfer coefficient, the heat leakage can be found by the use of the formula. Suppose you have a room 20 by 20 ft and 8 ft high. The walls of the room are of 6 in. concrete and 4 in. cork. The perimeter of the room is 4 (sides) $\times$ 20 or 80 ft. The total wall surface is 8×80 ft, or 640 ft^2. If the outside temperature is 75°F and the inside temperature is 30°F, the heat leakage through the walls is found by using the following formula:

$$H = 0.068493 \times 640(75 - 30) = 1972.5984 \text{ Btu/h}$$

To find the total heat leakage, you must also figure the heat leakage from the floor and the ceiling. Once the ceiling and floor leakages have been added, then the refrigeration needed to cool the product must be added to obtain the total load on the refrigeration system.

CALCULATING PRODUCT COOLING LOAD

Product cooling load can be figured also. Heat emitted from the product to be cooled can be calculated. The amount of product per locker should be known. Most tables indicate that the average locker user will place an average of 2 to 2.5 lb of product (meat) per day in the locker storage compartment. This means that, in a chill room, having a 300-locker installation, the daily load would be 300×2.5, or 750 lb. The initial temperature of the meat may be as high as 95°F. The final temperature can be assumed to be 36°F.

Various kinds of products will vary in terms of specific heat. However, an average specific heat of 0.7 is generally used as a value for making calculations. For a 300-locker unit, the daily product cooling load would be $0.7 \times 750(95 - 36)$, or 30,975 Btu per day.

Heat change loads are caused by the entrance of warm air when the doors are opened. This load is difficult to estimate accurately. It is affected by room usage, interior volume, whether or not the room is entered through an outside door, the size of the door, and how many times the door is opened.

EXAMPLE:

When the temperature of meat reaches 35°F, the meat is moved to the cold storage room. During preparation for storage, the meat may warm up to 40 or 50°F. That means the meat must be cooled to 32°F before it will begin to freeze. The average heat of fusion amounts to about 90 Btu/lb for meat when the latent heat of fusion is being removed. In this case, the heat of fusion is the heat that is removed at the freezing point before the meat is frozen. After freezing the meat is subcooled to the quick-freeze temperature. It then has a specific heat on the average of 0.4.

It is possible to calculate the amount of heat removed from a pound of meat. Assume the final temperature of the quick freeze to be −10°F. It is possible to obtain the amount of heat removed.

Cooling the meat to the freezing point, remove $0.7(45 - 28)$ [0.7 = specific heat; 45 = average of 40 and 50; 28 = temperature just below freezing point] =	11.9 Btu
Freezing the meat (latent heat of fusion) =	90.0 Btu
Subcooling the meat to −10°F $= 0.4(28 + 10) =$	15.2 Btu
Total per pound	117.1Btu

If you allow 2.5 lb of meat per locker, the quick freezer in a 300-locker installation should have a capacity to freeze 750 lb of meat per day (300×2.5). In this case, the product load would be 87,825 Btu per day (750×117.1).

Since miscellaneous loads cannot be accurately calculated, locker and freezer doors should be opened no more than is necessary. This will keep the load, due to such openings, to a minimum. Experience indicates that no more than 15 to 20 percent of the leakage load is caused by such openings.

Capacity of the Machines Used in the System

The capacity of any refrigerating compressor depends on its running speed and the number and size of its cylinders. The efficiency of the compressor must be

Table 18-1 *Usage Factors in Btu per 24 Hours per Cubic Foot Interior Capacity*

Inside Volume	Type of Service	Temperature Difference (Room Temperature Minus Refrigerator Temperature)								
		40°F	45°F	50°F	55°F	60°F	65°F	70°F	75°F	80°F
15 ft³	Normal	108	122	135	149	162	176	189	203	216
	Heavy	134	151	168	184	201	218	235	251	268
50 ft³	Normal	97	109	121	133	145	157	169	182	194
	Heavy	124	140	155	171	186	202	217	233	248
100 ft³	Normal	85	96	107	117	128	138	149	160	170
	Heavy	114	128	143	157	171	185	200	214	228
200 ft³	Normal	74	83	93	102	111	120	130	139	148
	Heavy	104	117	130	143	156	169	182	195	208
300 ft³	Normal	68	77	85	94	102	111	119	218	136
	Heavy	98	110	123	135	147	159	172	184	196
400 ft³	Normal	65	73	81	89	97	105	113	122	130
	Heavy	95	107	119	130	142	154	166	178	190

considered. The number of hours of operation per day, and the suction and discharge pressure play an important role in the capacity of the machine. Capacity rating is usually based on conditions standardized by *American Society of Heating, Refrigeration, and Air Conditioning Engineers* (ASHRAE) and *Air Conditioning and Refrigeration Institute* (ARI). These standards call for compressor suction of 5°F and 19.6 lb gage pressure and discharge pressure of 86°F and 154.5 lb gage pressure.

In the case of smaller coolers (1600 ft³), there is a shorter method for figuring the load. If these are used for general purpose cooling and storage, the product load is difficult to determine. It may vary. In these cases, an average is used. The wall gain load and the usage or service load are used to determine the total load.

Calculate the wall gain as previously shown. The usage load will equal the interior volume multiplied by a usage factor.

Usage factors are available from, tables developed by, ASHRAE. See Table 18-1. These factors vary with the difference in temperatures between the outside and inside of the cooler. An allowance is made for normal and heavy usage. There is no safety factor in this equation. To find the average daily load, divide the total loading factor by the desired operating time for the equipment. Equipment is selected from information supplied by manufacturers of the cooling units.

When determining the miscellaneous load, you will need a constant that has been found to be very reliable. This constant is that 1 W of electrical energy produces 3.415 Btu. Thus, a 25-W bulb will generate 85.375 Btu (25 × 3.415).

When you have determined the load requirements, you will then need to check the manufacturer's recommendations for a particular unit to match the condensing unit to the load. For instance, Table 18-2 shows that a 1-hp unit provides 8190 Btu/h. A 2-hp unit provides 16,150 Btu/h. For example, if a load calls for 16,000 Btu, then a 2-hp unit will suffice. Some allowance should be made for the load factor changing under maximum load conditions. It is, of course, wise to know exactly what the cooler will be used for before choosing the condensing unit.

Table 18-2 *Capacity of Typical Air-Cooled Condensing Units*

Condensing Unit (hp)	Capacity per Hour (Btu)
1/3	2,460
1/2	4,010
3/4	5,820
1	8,190
1 1/2	12,050
2	16,150

The size of the condensing unit can also be found by dividing the Btu needed to cool the load by 12,000. This will give the horsepower rating of the condensing unit. This works because 12,000 Btu/hp is an industry standard. A ton of ice melting in 1 h will remove 12,000 Btu of heat from the area in which it is located.

AIR DOORS

One of the ways to minimize temperature losses and prevent warm moist air entering is by using an air door. An air door also provides protection against insects, dust, dirt, and fumes. It provides an invisible barrier affording people an unobstructed view of the work area. It ensures a constant interior temperature by preventing the entry of hot or cold air. See Fig. 18-1.

An air door can be used to seal in cold air and save energy by preventing excessive operation of the refrigeration system.

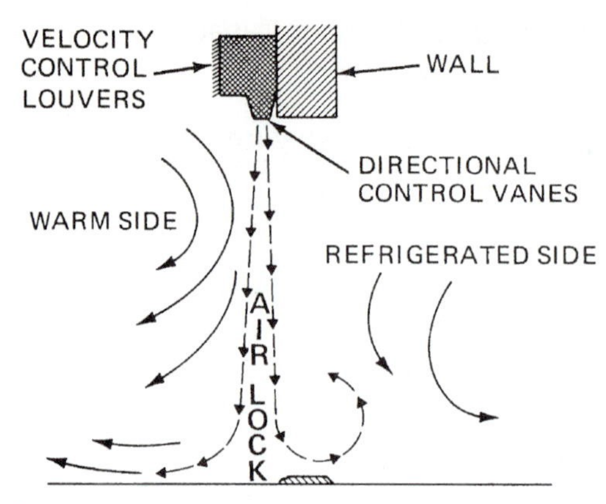

Fig. 18-1 *Using an air door to keep cold air inside a refrigerated area.* (Virginia Chemicals)

Refrigeration coolers and freezers use the air door to maintain interior temperature. They stop the entry of warm moist air when doors are opened to the cooler. They lessen the frequency of expensive defrosting and prevent the refrigeration system from overloading. It is easier for personnel to see who is coming in or going out of the cooler. Thus, accident prevention is a good by-product of the air-door installation.

The main component of an air door is a fan mounted in a unit. The fan controls the volume of air directed downward. This air seals off an area from any temperature change for short periods. Air doors are used in meat packing plants, food processing plants, supermarkets, commissaries, restaurants, hospitals, cold storage plants, and breweries.

INSULATION

Insulation is needed to prevent the penetration of heat through a wall or air hole into a cooled space. There are several insulation materials such as wood, plastic, concrete, and brick. Each has its application. However, more effective materials are constantly being developed and made available.

Sheet Insulation

Vascocel is an expanded, closed cell, sponge rubber that is made in a continuous sheet form of 36 in. wide. It comes in a wide range of thickness: $^3/_8$, $^1/_2$, and $^3/_4$ in.). This material is designed primarily for insulating oversize pipes, large tanks and vessels, and other similar medium- and low-temperature areas. Because of its availability on continuous rolls, this material lends itself ideally to application on large air ducts and irregular shapes.

This material is similar to its companion product, Vascocel tubing. It may be cut and worked with ordinary hand tools such as scissors or a knife. The sheet stock is easily applied to clean, dry surfaces with an adhesive. See Fig. 18-2. The *k* factor (heat transfer coefficient) of this material is 0.23. It has some advantages over other materials. It is resistant to water penetration, water absorption, and physical abrasion.

Tubing Insulation

Insulation tape is a special synthetic rubber and cork compound designed to prevent condensation on pipes and tubing. It is usually soft and pliable. Thus, it can be molded to fit around fittings and connections. There are many uses for this type of insulation. It can be used on hot or cold pipes or tubing. It is used in residential buildings, air-conditioning units, and commercial installations. It comes in 2 in. wide rolls that are 30 ft long. The tape is thick. If stored or used in temperatures under 90°F (32.2°C), the lifetime is indefinite.

Foam insulation tape is made specifically for wrapping cold pipes to prevent pipe sweat. See Fig. 18-3. It can be used to hold down heat loss on hot pipes below 180°F (82.2°C). It can be cut in pieces and easily molded around fittings and valves. It adheres to itself and clean metal surfaces. It is wrapped over pipes with about $^1/_4$ in. overlap on each successive lap. Remember one precaution: *Never wrap two or more parallel runs of tubing or pipe together, leaving air voids under the tape.* Fill the voids between the pipes with Permagum before wrapping. This will prevent moisture from collecting in the air spaces. This foam insulation tape has a unicellular composition. The *k* factor is 0.26 at 75°F (23.9°C).

Permagum is a nonhardening, water-resistant sealing compound. It is formulated to be nonstaining, nonbleeding, and to have excellent adhesion to most clean surfaces. It comes in containers in either slugs or cords. See Fig. 18-4.

This sealer is used to seal metal to metal joints in air conditioners, freezers, and coolers. It can seal metal to wood joints and set plastic and glass windows in wood or metal frames. It can be used to seal electrical or wire entries in air-conditioning installations or in freezers. It can be worked into various spaces. It comes with a paper backing so that it will not stick to itself.

Extrusions are simple to apply. Unroll the desired length and smooth it into place. It is soft and pliable. The bulk slug material can be formed and applied by hand or with tools such as a putty knife.

Sealing compounds are sometimes needed to seal a joint or an entry location. These compounds can be purchased in small units in white, nonstaining compositions.

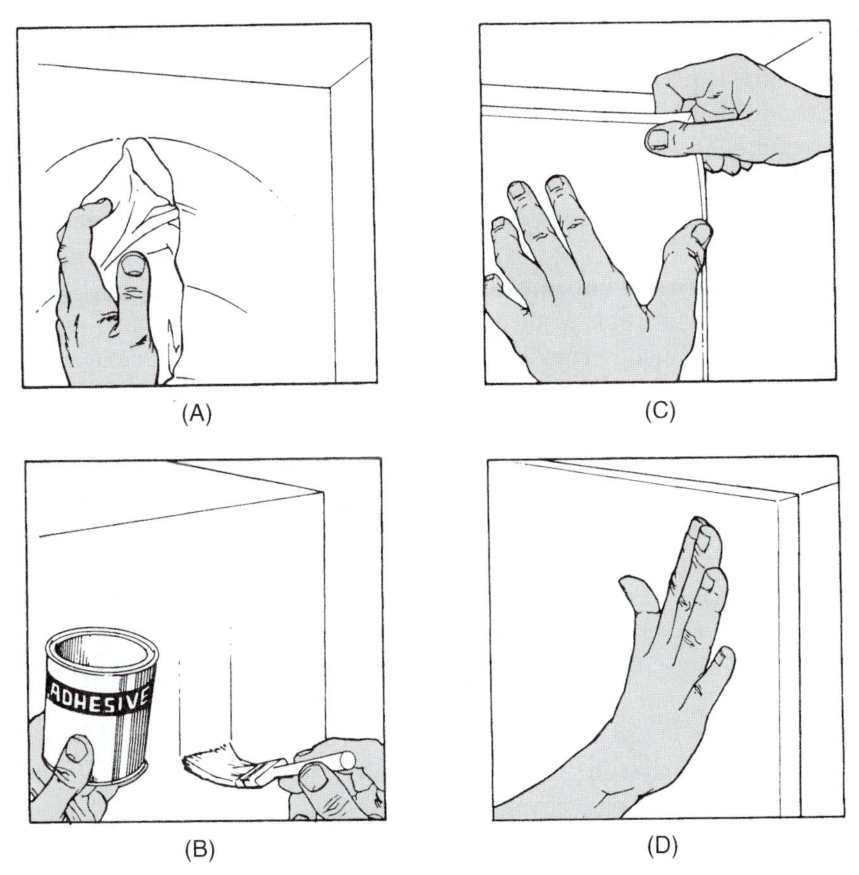

Fig. 18-2 *Installing sheet insulation. (A) Prepare the surfaces for application of the sheet insulation by wiping with a soft, dry cloth to remove any dust or foreign matter. Use a solvent to remove grease or oil. (B) Apply the adhesive in a thin, even coat to the surface to be insulated. (C) Position the sheet of insulation over the surface and then simply smooth it in place. The adhesive is a contact type. The sheet must be correctly positioned before it contacts the surface. (D) Check for adhesion of ends and edges. The surface can be painted.*

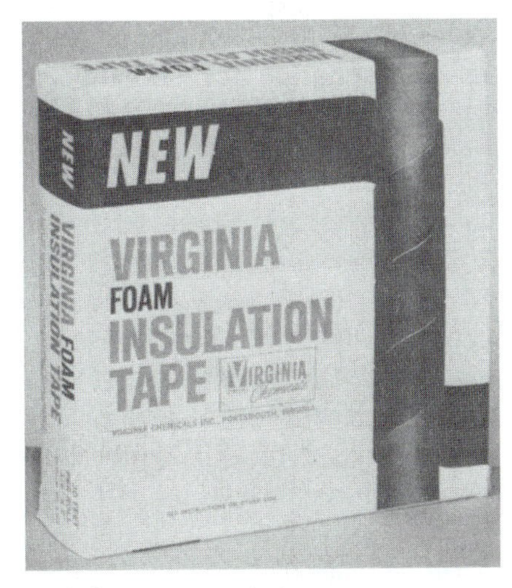

Fig. 18-3 *Foam insulation tape.* (Virginia Chemicals)

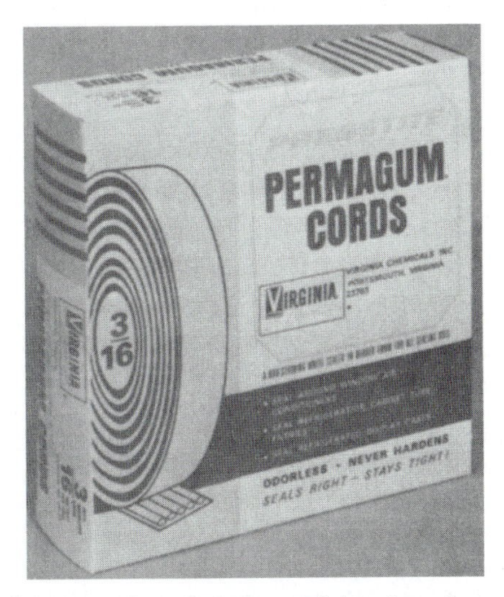

Fig. 18-4 *Slugs of insulation material and cords are workable into locations where sheet material cannot fit.* (Virginia Chemicals)

Pipe Insulation

Pipe fittings are insulated for a number of reasons. Methods of insulating three different fittings are shown in Fig. 18-5. In most cases it is advisable to clean all joints and waterproof them with cement. A mixture of hot crude paraffin and granulated cork can be used to fill the cracks around the fittings.

Figure 18-6 shows a piece of rock cork insulation. It is molded from a mixture of rock wool and waterproof binder. Rock wool is made from limestone that has been melted at about 3000°F (1649°C). It is then blown into fibers by high-pressure steam. Asphaltum is the binder used to hold it into a moldable form. This insulation has approximately the same insulation qualities as cork. It can be made waterproof when coated with asphalt. Some more modern materials have been developed to give the same or better insulation qualities. The Vascocel tubing can be used in the insulation of pipes. Pipe wraps are available to give good insulation and prevent dripping, heat loss, or heat gain.

Figure 18-7 shows a fitting insulated with preshrunk wool felt. This is a built-up thickness of pipe covering made of two layers of hair felt. The inside portion is covered with plastic cement before the insulation material is applied. After the application, waterproof tape and plastic cement should be added for protection against moisture infiltration. This type of insulation is used primarily on pipes located inside a building. If the pipe is located outside, another type of insulation should be used.

REFRIGERATION PIPING

The use of various materials for insulation purposes in the refrigeration field over the years has resulted in some equipment still operational today. It is this equipment that service people are most often called to repair or maintain. It is therefore necessary for the present day repairperson to be acquainted with the older types of insulations that may be encountered during the workday.

The success of any refrigeration plant depends largely on the proper design of the refrigeration piping and a thorough understanding of the necessary accessories and their functions in the system. In sizing refrigerant lines, it is necessary to consider the optimum sizes with respect to economics, friction losses, and oil return. It is desirable to have line sizes as small as possible from the cost standpoint. On the other hand, suction- and discharge-line pressure drops cause a loss of compressor capacity, and excessive liquid-line pressure drops may cause flashing of the liquid refrigerant with consequent faulty expansion-valve operation.

Refrigerant piping systems, to operate successfully, should satisfy the following:

- Proper refrigerant feed to the evaporators should be ensured.
- Refrigerant lines should be of sufficient size to prevent an excessive pressure drop.
- An excessive amount of lubricating oil should be prevented from being trapped in any part of the system.
- Liquid refrigerant should be prevented from entering the compressor at all times.

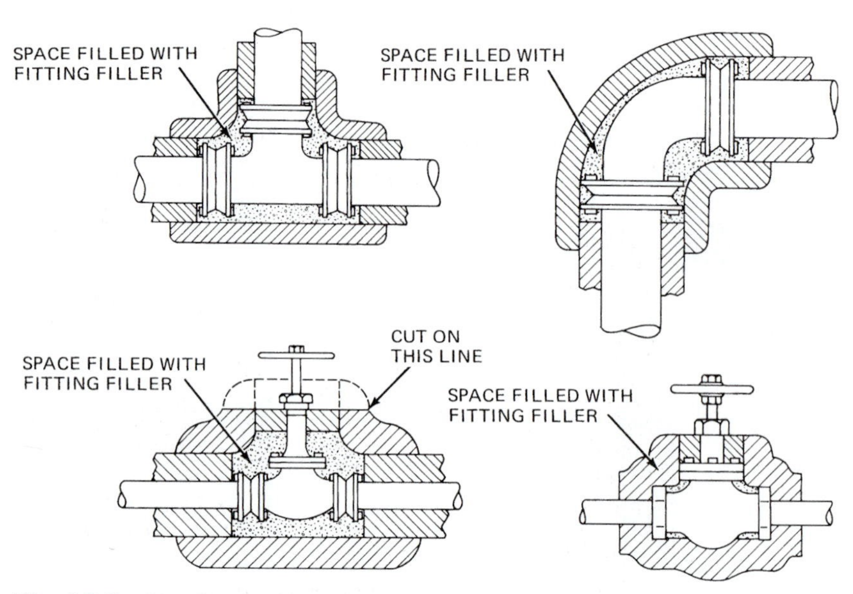

Fig. 18-5 *Pipe-fittings covered with cork-type insulation. On one of the valves the top section can be removed if the packing needs replacing.*

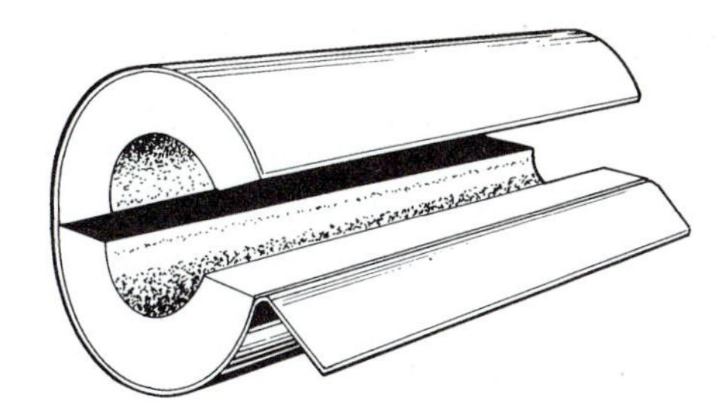

Fig. 18-6 *Pipe insulation.*

PRESSURE-DROP CONSIDERATIONS

Pressure drop in liquid lines is not as critical as it is in the suction and discharge lines. The important thing to remember is that the pressure drop should not be so great as to cause gas formation in the liquid line and/or insufficient liquid pressure at the liquid-feed device. A system should normally be designed so that the pressure drop, due to friction in the liquid line, is not greater than that corresponding to 1 to 2° change in saturation temperature. Friction pressure drops in the liquid line include the drop in accessories, such as the solenoid valve, strainer-drier, and hand valves, as well as in the actual pipe and fittings from the receiver outlet to the refrigerant feed device at the evaporator.

Friction pressure drop in the suction line means a loss in system capacity because it forces the compressor to operate at a lower suction pressure to maintain the desired evaporating temperature in the coil. It is usually standard practice to size the suction line to have a pressure drop due to friction not any greater than the equivalent of a 1 to 2° change in saturation temperature.

LIQUID REFRIGERANT LINES

The liquid lines do not generally present any design problems. Refrigeration oil is sufficiently miscible with commonly used refrigerants in the liquid form to assure adequate mixture and positive oil return. The following factors should be considered when designating liquid lines.

The liquid lines, including the interconnected valves and accessories, must be of sufficient size to prevent excessive pressure drops.

When interconnecting condensing units with condenser receivers or evaporative condensers, the liquid lines from each unit should be brought into a common liquid line.

Each unit should join the common liquid line as far below the receivers as possible, with a minimum of 2 ft preferred. The common liquid line should rise to the ceiling of the machine room. The added heat of liquid is provided to prevent, as far as possible, hot gas blowing back from the receivers.

All liquid lines from the receivers to the common line should have equal pressure drops in order to provide, as nearly as possible, equal liquid flow and prevent the blowing of gas.

Remove all liquid-line filters from the condensing units, and install them in parallel in the common liquid line at the ceiling level.

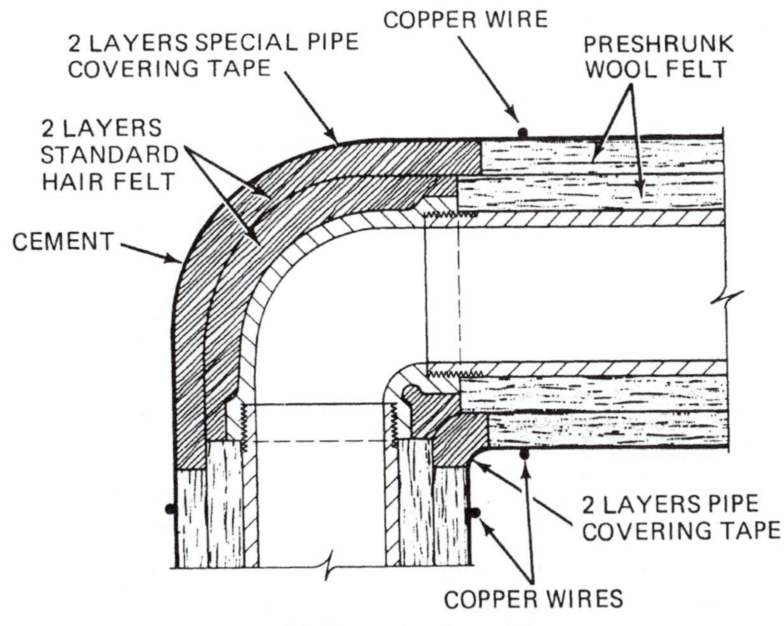

Fig. 18-7 *Insulated pipe fittings.*

Hot gas blowing from the receivers can be condensed in reasonable quantities by liquid subcoolers, as specified for the regular condensing units, having a minimum lift of 60 ft at 80°F condensing medium temperature.

Interconnect all the liquid receivers of the evaporative condensers above the liquid level to equalize the gas pressure.

The common and interconnecting liquid line should have an area equal to the sum of the areas of the individual lines. Install a hand shut-off valve in the liquid line from each receiver. Where a reduction in pipe size is necessary in order to provide sufficient gas velocity to entrain oil up the vertical risers at partial loads, greater pressure drops will be imposed at full load. These can usually be compensated for by oversizing the horizontal and down comer lines to keep the total pressure drop within the desired limits.

INTERCONNECTION OF SUCTION LINES

When designing suction lines, the following important considerations should be observed:

- The lines should be of sufficient capacity to prevent any considerable pressure drop at full load.
- In multiple-unit installations, all suction lines should be brought to a common manifold at the compressor.
- The pressure drop between each compressor and main suction line should be the same in order to ensure a proportionate amount of refrigerant gas to each compressor, as well as a proper return of oil to each compressor.
- Equal pipe lengths, sizes, and spacing should be provided.
- All manifolds should be level.
- The inlet and outlet pipes should be staggered.
- Never connect branch lines at a cross or tee.
- A common manifold should have an area equal to the sum of the areas of the individual suction lines.
- The suction lines should be designed so as to prevent liquid from draining into the compressor during shutdown of the refrigeration system.

DISCHARGE LINES

The hot-gas loop accomplishes two functions: it prevents gas that may condense in the hot-gas line from draining back into the heads of the compressor during the *off* cycles, and it prevents oil leaving one compressor from draining down into the head of an idle machine.

It is important to reduce the pressure loss in hot-gas lines because losses in these lines increase the required compressor horsepower per ton of refrigeration and decrease the compressor capacity. The pressure drop is kept at a minimum by sizing the lines generously to avoid friction losses, but still making sure that refrigerant line velocities are sufficient to entrain and carry along oil at all load conditions. In addition, the following pointers should be observed:

- The compressor hot-gas discharge lines should be connected as shown in Fig. 18-8.

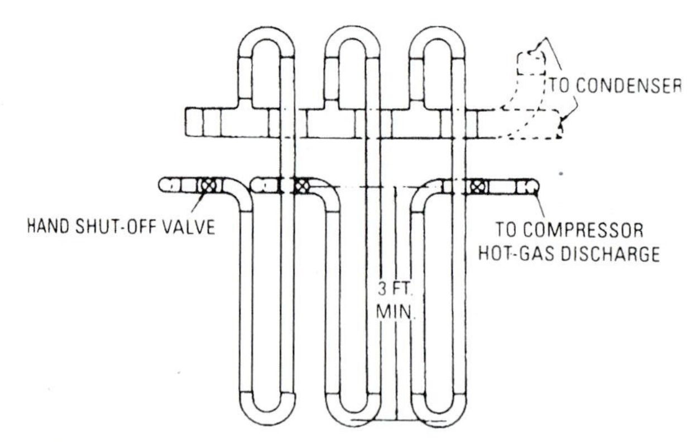

Fig. 18-8 *One way to connect hot-gas discharge lines.*

- The maximum length of the risers to the horizontal manifold should not exceed 6 ft.
- The manifold size should be at least equal to the size of the common hot-gas line to the evaporative condenser.
- If water-cooled condensers are interconnected, the hot-gas manifolds should be at least equal to the size of the discharge of the largest compressor.
- If evaporative condensers are interconnected, a single gas line should be run to the evaporative condensers, and the same type of manifold provided at the compressors should be installed.
- Always stagger and install the piping at the condensers.
- When the condensers are above the compressors, install a loop having a minimum depth of 3 ft in the hot-gas main line.
- Install a hand shut-off valve in the hot-gas line at each compressor.

WATER VALVES

The water-regulating valve is the control used with water-cooled condensers. When installing water valves, the following should be observed:

- The condenser water for interconnected compressor condensers should be applied from a common water line.
- Single automatic water valves or multiple valves in parallel (Fig. 18-9) should be installed in the common water line.

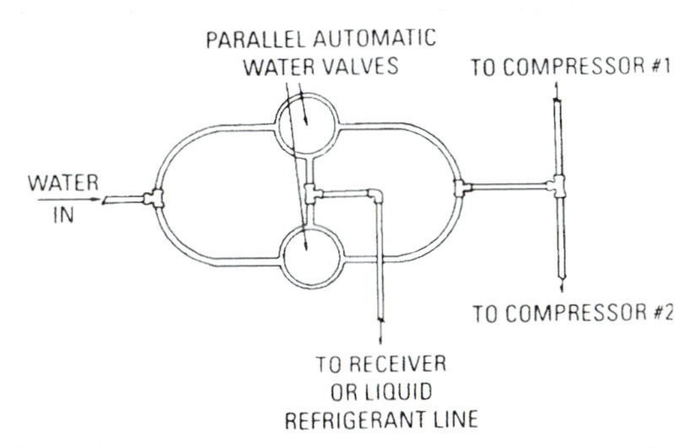

Fig. 18-9 *A method of interconnecting water valves.*

- Pressure-control tubing from the water valves should be connected to a common line, which, in turn, should be connected to one of the receivers or to the common liquid line.

MULTIPLE-UNIT INSTALLATION

Multiple compressors operating in parallel must be carefully piped to ensure proper operation. The suction piping at parallel compressors should be designed so that all compressors run at the same suction pressure and oil is returned in equal proportions to the running compressors. All suction lines should be brought into a common suction header in order to return the oil to each crankcase as uniformly as possible.

The suction header should be run above the level of the compressor suction inlets so that oil can drain into the compressors by gravity. The header should not be below the compressor suction inlets because it can become an oil trap. Branch suction lines to the compressors should be taken off from the side of the header. Care should be taken to make sure that the return mains from the evaporators are not connected into the suction header so as to form crosses with the branch suction lines to the compressors. The suction header should be run full size along its entire length. The horizontal takeoffs to the various compressors should be the same size as the suction header. No reduction should be made in the branch suction lines to the compressors until the vertical drop is reached.

Figure 18-10 shows the suction and hot-gas header arrangements for two compressors operating in parallel.

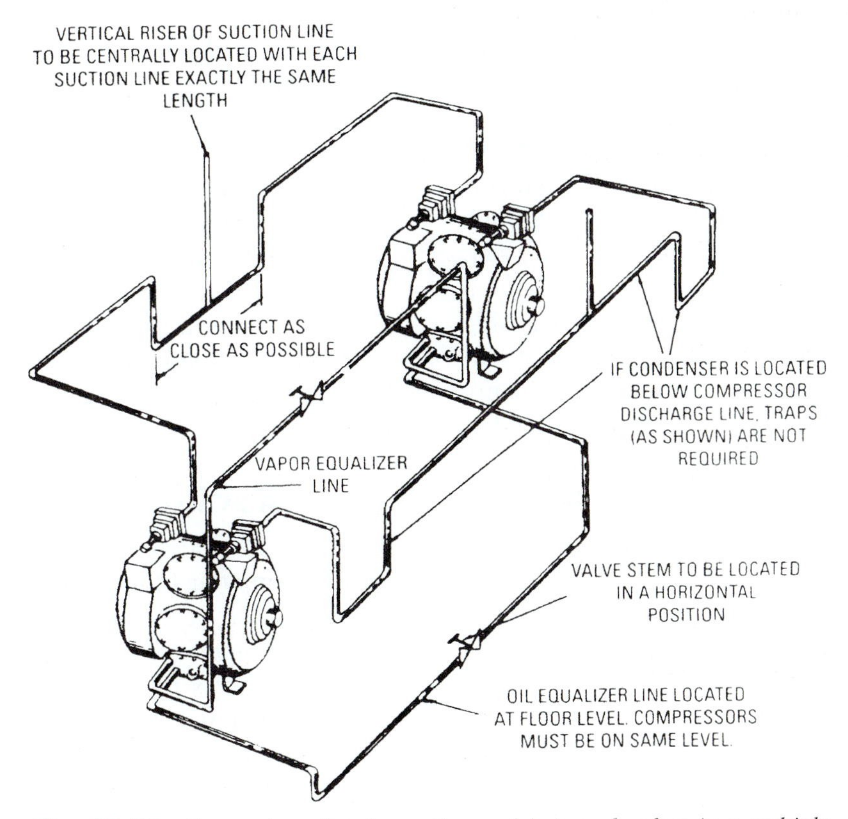

Fig. 18-10 *Connections for the suction and hot-gas headers in a multiple compressor installation.*

Takeoffs to each compressor from the common suction header should be horizontal and from the side, to ensure equal distribution of oil and prevent accumulating liquid refrigerant in an idle compressor in case of slop-over.

Piping Insulation

Insulation is required for refrigeration piping to prevent moisture condensation and prevent heat gain from the surrounding air. The desirable properties of insulation are that it should have a low coefficient of heat transmission, be easy to apply, have a high degree of permanency, and provide protection against air and moisture infiltration. Finally, it should have a reasonable installation cost.

The type and thickness of insulation used depends on the temperature difference between the surface of the pipe and the surrounding air, and also on the relative humidity of the air. It should be clearly understood that although a system is designed to operate at a high-suction temperature, it is quite difficult to prevent colder temperatures occurring from time to time. This may be due to a carrying over of some liquid from the evaporator or the operation of an evaporator pressure valve. Interchangers are preferable to insulation, in this case.

One of the safest pipe insulations available is molded cork or rock cork of the proper thickness. Hair-felt insulation may be used, but great care must be taken to have it properly sealed. For temperatures above 40°F, wool felt or a similar insulation may be used, but here again success depends on the proper seal against air and moisture infiltration.

Liquid refrigerant lines carry much higher temperature refrigerant than suction lines; and if this temperature is above the temperature of the space through which they pass, no insulation is usually necessary. However, if there is danger of the liquid lines going below the surrounding air temperatures and causing condensation, they should be insulated when condensation will be objectionable. If they must unavoidably pass through highly heated spaces, such as those adjacent to steam pipes, through boiler rooms, then the liquid lines should also be insulated to ensure a solid column of liquid to the expansion valve.

There were four types of insulation in use before the discovery of modern insulation materials. Those you may encounter were in general use for refrigerator piping. They are namely:

- Cork
- Rock cork
- Wool felt with waterproof jacket
- Hair felt with waterproof jacket

Cork Insulation

Cork pipe covering is prepared by pressing dried and granulated cork in metal molds. The natural resins in the cork bind the entire mass into its new shape. In the case of the cheaper cork, an artificial binder is used. The cork may be molded to fit pipe and fittings, or it may be made into flat boards of varying sizes and thickness. Cork has a low thermal conductivity. The natural binder in the material itself makes cork highly water-resistant, and its structure ensures a low capillarity. It can be made practically impervious to water by surfacing with odorless asphalt.

All fittings in the piping, as well as the pipe itself, should be thoroughly insulated to prevent heat gain to protect the pipe insulation from moisture infiltration and deterioration, and eliminate condensation problems. Molded cork covering, made especially for this purpose, is available for all common types of fittings. Each covering should be the same in every respect as the pipe insulation, with the exception of the shape, and should be formed so that it joins to the pipe insulation with a break. Typical cork fitting covers are furnished in three standard thicknesses for ice water, brine, and special brine.

To secure maximum efficiency and long life from cork covering, it must be correctly applied and serviced, as well as properly selected. Hence, it is essential that the manufacturer's recommendations and instructions be followed in detail. The following general information is a summary of the data that are of general interest.

All pipelines should be thoroughly cleaned, dried, and free from all leaks. It is also advisable to paint the piping with waterproof paint before applying the insulation, although this is not recommended by all manufacturers. All joints should be sealed with waterproof cement when applied. Fitting insulation should be applied in substantially the same manner, with the addition of a mixture of hot crude paraffin and granulated cork used to fill the space between the fittings shown in Fig. 18-5.

Rock-Cork Insulation

Rock-cork insulation is manufactured commercially by molding a mixture of rock wool and a waterproof binder into any shape or thickness desired. The rock wool is made from limestone melted at about 3000°F and then blown into fibers by high-pressure steam. It is mixed with an asphalt binder and molded into the various commercial forms. The heat conductivity is about the same as cork, and the installed price may be less. Because of its mineral composition, it is odorless, vermin-proof, and free from decay. Like cork, it can be made completely waterproof by surfacing with odorless asphaltum. The pipe covering fabricated from rock wool and a binder is premolded in single-layer sections, 36-in. long, to fit all standard pipe sizes and is usually furnished with a factory-applied waterproof jacket.

When pipelines are insulated with rock-cork covering, the fittings are generally insulated with built-up rock wool impregnated with asphalt. This material is generally supplied in felted form, having a nominal thickness of about 1 in. and a width of about 18 in. It can be readily adapted to any type of fitting and is efficient as an insulator when properly applied.

Before applying the formed rock-cork insulation, it is first necessary to thoroughly clean and dry the piping and then paint it with waterproof asphalt paint. The straight lengths of piping are next covered with the insulation, which has the two longitudinal joints and one end joint of each section coated with plastic cement. The sections are butted tightly together with the longitudinal joints at the top and bottom and temporarily held in place by staples. The plastic cement should coat that part of the exterior area of each section to be covered by the waterproof lap and the lap pressed smoothly into it. The end joints should be sealed with a waterproof fabric embedded in a coat of the plastic cement. Each section should then be secured permanently in place with three to six loops of copper-plated annealed steel wire.

Wool-Felt Insulation

Wool felt is a relatively inexpensive type of pipe insulation and is made up of successive layers of waterproof wool felt that are indented in the manufacturing process to form air spaces. The inner layer is a waterproof asphalt-saturated felt, while the outside layer is an integral waterproof jacket. This insulating material is satisfactory when it can be kept airtight and moisture free. If air is allowed to penetrate, condensation will take place in the wool felt, and it will quickly deteriorate. Thus, it is advisable to use it only where temperatures above 40°F are encountered and when it is perfectly sealed. Under all conditions, it should carry the manufacturer's guarantee for the duty that it is to perform.

After all the piping is thoroughly cleaned and dried, the sectional covering is usually applied directly to the pipe with the outer layer slipped back and turned so that all joints are staggered. The joints should be sealed with plastic cement, and the flap of the waterproof jacket should be sealed in place with the same material. Staples and copper-clad steel wire should be provided to permanently hold the insulation in place, and then the circular joints should be covered with at least two layers of waterproof tape to which plastic cement is applied.

Pipe fittings should be insulated with at least two layers of hair felt (Fig. 18-7) built up to the thickness of the pipe covering; but before the felt is placed around the fittings, the exposed ends of the pipe insulation should be coated with plastic cement.

After the felt is in place, two layers of waterproof tape and plastic cement should be applied for protection from moisture infiltration.

Insulation of this type is designed for installation in buildings where it is normally protected against outside weather conditions. When outside pipes are to be insulated, one of the better types of pipe covering should be used. In all cases, the manufacturer's recommendations should be followed during the application.

Hair-Felt Insulation

Hair-felt insulation is usually made from pure cattle hair that has been especially prepared and cleaned. It is a very good insulator against heat, having a low thermal conductivity. Its installed cost is somewhat lower than cork; but it is more difficult to install and seal properly, and hence its use must be considered a hazard with the average type of workmanship. Prior to installation, the piping should be cleaned and dried and then prepared by applying a thickness of waterproof paper or tape wound spirally, over which the hair felt of approximately 1-in. thickness is spirally wound for the desired length of pipe. It is then tightly bound with jute twine, wrapped with a sealing tape to make it entirely airtight, and finally painted with waterproof paint. If more than one thickness of hair felt is desired, it should be built up in layers with tarpaper between. When it is necessary to make joints around fittings, the termination of the hair felt should be tapered down with sealing tape and the insulation applied to the fittings should overlap this taper, thus ensuring a permanently tight fit.

The important point to remember is that this type of insulation must be carefully sealed against any air or moisture infiltration, and even then difficulty may occur after it has been installed. At any point where air infiltration (or "breathing," as it is called) is permitted to occur, condensation will start and travel great distances along the pipe, even undermining the insulation that is properly sealed.

There are several other types of pipe insulation available, but they are not used extensively. These include various types of wrapped and felt insulation, but they are seldom applied with success. Whatever insulation is used, it should be critically examined to see whether it will provide the protection and permanency required of it; otherwise it should never be considered. Although all refrigerant piping, joints, and fittings should be covered, it is not advisable to do so until the system has been thoroughly leak tested and operated for a time.

Pressure drop in the various parts of commercial refrigeration systems, due to pipe friction, and the proper dimensioning to obtain the best operating results are important items when installation of equipment is made.

By careful observation of the foregoing detailed description of refrigeration piping and methods of installation, the piping problem will be greatly simplified and result in proper system operation.

REVIEW QUESTIONS

1. Why is off-cycle defrost not used in commercial units?
2. Why does frost not form on an air-conditioning unit?
3. Explain wall gain load, air change load, and product load.
4. How do you calculate heat leakage?
5. What are miscellaneous loads?
6. What does the capacity of a refrigerating compressor depend on?
7. What is the purpose of an air door?
8. What is the main component of an air door?
9. What is the basic purpose of Vascocel?
10. Why is tubing insulation needed in a refrigerating system?

19

CHAPTER

Installing and Controlling Electrical Power for Air-Conditioning Units

PERFORMANCE OBJECTIVES

After studying this chapter you will:

1. Understand how to find the correct wire size for a refrigeration unit.
2. Know the permissible maximum voltage drops in a service line.
3. Know the effects of voltage variations on AC motors.
4. Know how to calculate starting current values and inrush voltage drops.
5. Know the code limitations on amperes per conductor.
6. Know how to select the correct circuit protection.
7. Know the different types of fuses and circuit breakers.

In order to operate properly, the correct voltage, amperage, and wattage must be furnished to the air-conditioning unit. A number of devices have to be utilized to keep the voltage source clean and available for efficient operation of the unit. It all starts with the choosing of the correct wire from the power pole to the house or store or school where the air-conditioning unit is used.

CHOOSING WIRE SIZE

There are two criteria for choosing wire size for installation of air-conditioning or refrigeration equipment. The size of the electrical conductor wire recommended for a given appliance circuit depends upon two things: limitation on voltage loss and minimum wire size.

Limiting Voltage Loss

Proper operation of an electrical device must be under the conditions for which it was designed. The wire size selected must be low in resistance per foot of length. This will assure that the full load "line loss" of the total length of the circuit does not cause low voltage of the appliance terminals. Since the length of electrical feeders varies with each installation, wire sizing to avoid excessive voltage loss becomes the responsibility of the installing contractor. The *National Electrical Code* (NEC) or local code should be followed.

Minimum Wire Size

To avoid field wiring being damaged by tensile stress or overheating, national and local codes establish minimum wire sizes. The maximum amperage permitted for a given conductor limits internal heat generation so that temperature will not damage its insulation. This assumes proper fusing that will limit the maximum current flow so that the conductor will always be protected.

Wire size and voltage loss go hand in hand, so to speak. The larger the wire, the more current it can handle without voltage loss along the lines. Each conductor or wire has resistance. This resistance, measured in ohms per unit of wire length, increases as the cross-sectional area of the wire decreases. The size of the wire is indicated by gage number. The higher the gage number, the smaller the wire. *American Wire Gage* (AWG) is the standard used for wire size. Each gage number has a resistance value in ohms per foot of wire length. The resistance of aluminum wire is 64 percent greater than that for copper of the same gage number.

Wire Selection

The wire size recommended for actual use should be the heavier of the two indicated by the procedures that follow.

Local approval is usually necessary for any installation that has large current draws. The data presented here are based on the NEC. Much of the detail has been omitted in the interest of simplification. Thus, there may be areas of incompleteness not covered by a footnote or reference. In all cases it is recognized that final approval must come from the authority having local jurisdiction. The NEC sets forth minimum standards. It is an effort to establish some standard for safe operation of equipment.

WIRE SIZE AND LOW VOLTAGE

The voltage at which a motor or device should operate is stamped on the nameplate. This voltage indicates that the full capacity of the device is being utilized when that particular voltage is available. Motors operated at lower than rated voltage are unable to provide full horsepower without jeopardizing their service life. Electric heating units lose capacity even more rapidly at reduced voltages.

Low voltage can result in insufficient spark for oil burner ignition, reluctant starting of motors, and overheating of motors handling normal loads. Thus, it is not uncommon to protect electrical devices by selecting relays that will not close load circuits if the voltage is more than 15 percent below rating.

Air Conditioning and Refrigeration Institute (ARI) certified that cooling units are tested to assure they will start and run at 10 percent above and 10 percent below their rating plate voltage. However, this does not imply that continuous operation at these voltages will not affect their capacity, performance, and anticipated service life. A large proportion of air-conditioning compressor burnouts can be traced to low voltage. Because the motor of a hermetic compressor is entirely enclosed within the refrigerant cycle, it is important that it not be abused either by overloading or undervoltage. Both of these can occur during peak load conditions. A national survey has shown that the most common cause of compressor low voltage is the use of *undersized* conductors between the utility lines and the condensing unit.

The size of the wire selected must be one that, under full-load conditions, will deliver acceptable voltages to the appliance terminals. The NEC requires that the conductors be sized to limit voltage drop between the outdoor-pole service tap and the appliance terminals; not in excess of 5 percent of rated voltage under full-load conditions. This loss may be subdivided, with 3 percent permissible in service drops, feeders, meters, and overcurrent protectors at the distribution panel and the appliance. See Fig. 19-1.

In a 240-V service, the wire size selected for an individual appliance circuit should cause no more than 4.8-V drop under full-load conditions. Even with this 5 percent limitation on voltage drop, the voltage at the equipment terminals is still very apt to be below the rating plate values. See Table 19-1.

Table 19-1 *Permissible Maximum Voltage Drops*

For a line voltage of:	120	208	240	480
Feeders to distribution panel (3%)	3.6	6.24	7.2	14.4
Branch circuit to appliance (2%)	2.4	4.16	4.8	9.6
Total voltage drop fully loaded	6	10.4	12	24
Resultant* voltage at appliance	114	197.6	228	456

*Assumes full-rated voltage where feeders connect to utility lines. If utility voltage runs low, the overall voltage drop should be further reduced so as to make available at the appliance terminals a voltage as close as practical to that specified on the appliance rating plate.

Voltage Drop Calculations

Just as friction creates pressure loss in water flow through pipes, so does electrical resistance create voltage drop as current flows through a conductor. The drop increases with the length of the conductor (in feet), the current flow (in amperes), and the ohms of resistance per foot of wire. This relationship may be expressed as follows:

$$\text{Voltage drop} = \text{amperes} \times \text{ohms/foot} \times \text{length of conductor}$$

Figure 19-2 illustrates how voltage drop per 100 ft of copper conductor will increase with the amount of current drawn through the conductor. The wire size is indicated on the straight line. Match the amperes with the wire size. Then follow over to the left column to determine the voltage drop. For instance, Fig. 19-2 shows that there will be a 2.04-V drop per 100 ft of copper conductor for 20 A of current through a No. 10 wire.

THE EFFECTS OF VOLTAGE VARIATIONS ON AC MOTORS

Motors will run at the voltage variations already mentioned. This does not imply such operation will comply

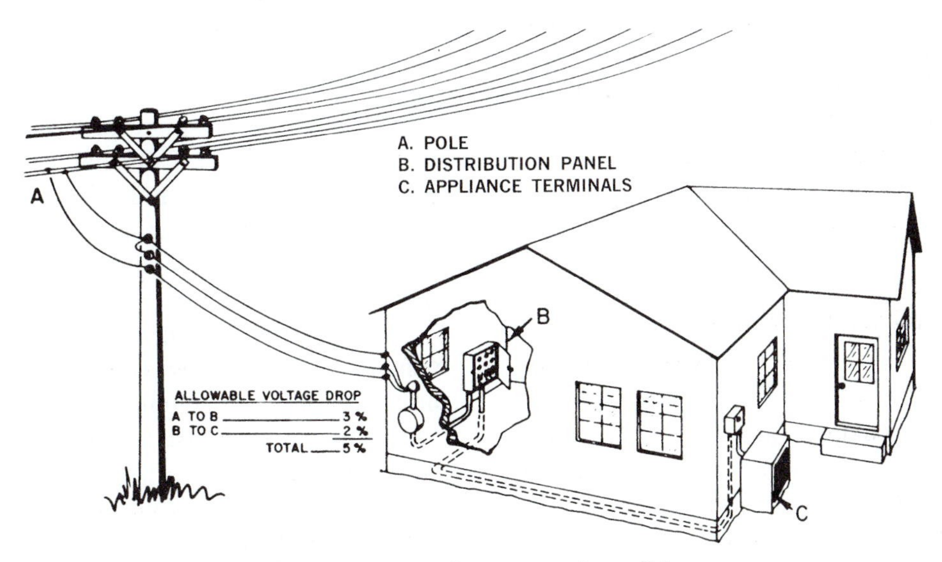

Fig. 19-1 *Voltage drops from post to air conditioner.* (Bryant)

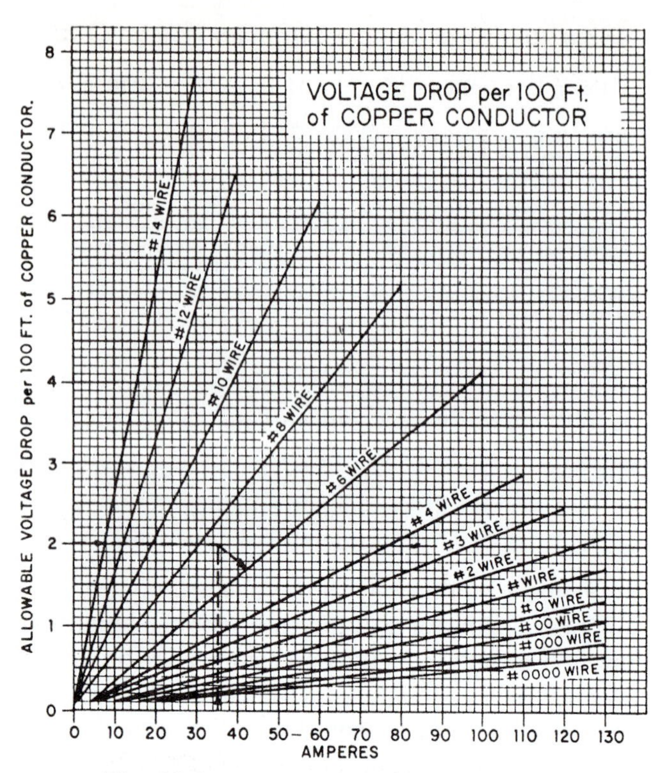

Fig. 19-2 *Conductor voltage drop per 100 ft.*

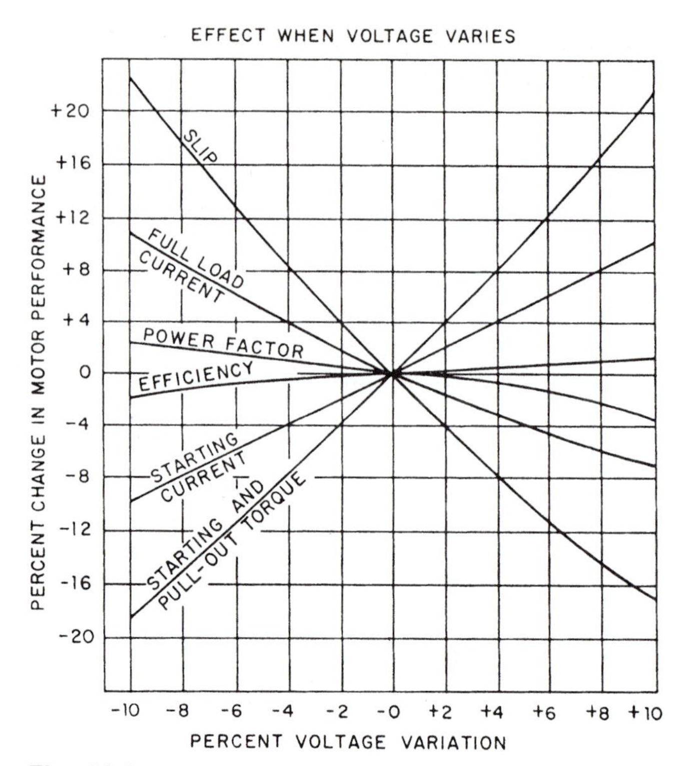

Fig. 19-3 *General effects of voltage variations on induction motor characteristics.*

with industry standards of capacity, temperature rise, or normally anticipated service life. Figure 19-3 shows general effects. Such effects are not guaranteed for specific motors.

The temperature rise and performance characteristics of motors sealed within hermetic compressor shells constitute a special case. These motors are cooled by return suction gas of varying quantity and temperature. Thus, Fig. 19-3 is not necessarily applicable to this specialized type of equipment.

The chart shows the approximate effect of voltage variations on motor characteristics. The reference base of voltage and frequency is understood to be as that shown on the nameplate of the motor.

Some of the terms used in the chart are explained here:

Normal slip = synchronous speed—the rating plate speed.

Slip in the graph indicates the change in normal slip.

Synchronous speeds for 60-Hz motors are:

Two pole 3600 r/min or rpm

Four pole 1800 r/min or rpm

Six pole 1200 r/min or rpm

Eight pole 900 r/min or rpm

Table 19-2 indicates the voltage drop that may be anticipated for various ampere flow rates through copper conductors of different gage size. Figure 19-2 provides the same data in graphic form.

These data are applicable to both single-phase and three-phase circuits. In each case, the wire length equals twice the distance from the power distribution panel to the appliance terminals, measured along the path of the conductors. This is twice the distance between B and C in Fig. 19-1, measured along the path of the conductors. For motorized appliances, particularly those that start under loaded conditions, the voltage at the appliance terminals should not drop more than 10 percent below rating plate values unless approved by the manufacturer. Thus, the voltage drop permissible in the load leads must anticipate any reduction below rated voltage that may be suffered under full-load conditions at the point of power source connection (point A in Fig. 19-1).

Troublesome voltage losses may also occur elsewhere if electrical joints or splices are mechanically imperfect and create unanticipated resistance. Such connections may exist in the distribution panel, the meter socket, or even where outdoor power drops are clamped to the feeder lines on poles. Where there is a wide variation between no-load voltage and operating voltage, sources of voltage drop can be determined by taking voltmeter readings at various points in the circuit. These points might be ahead of the meter, after the circuit disconnect switch, at the appliance terminals, and at other locations.

SELECTING PROPER WIRE SIZE

To provide adequate voltage at the appliance terminals, anticipate the minimum voltage that may exist at the distribution panel. Then determine the allowable voltage drop acceptable in the appliance circuit. This should not exceed 2 percent of rated voltage. It should, for example, not exceed 4.1 V for 208-V service. Table 19-1 shows voltage drops for 120-, 208-, 240-, and 408-V service.

Determine the length of feed conductor. This is twice the length of the wire path from the source to the appliance. In Fig. 19-1, this is two times the distance from B to C measured along the path of the wire. If it is single phase or three phase, consider two conductors in establishing the total length of the circuit.

Determine the allowable drop per 100 ft of conductor.

Example:

If for a 230-V installation, a 4.6-V drop is permissible, and the wire path is 115 ft from the distribution panel to the appliance (this makes 230 ft of conductor), then the allowable drop per 100 ft will be:

$$4.6 \text{ V} \times \frac{100}{230} \text{ ft} = 2.0 \text{ V/100 ft}$$

Using either Table 19-2 or Fig. 19-2, determine the gage wire required. When using the graph, select the gage number closest below and to the right.

Example:

The full-load value is 35 A. The allowable voltage drop is 2.0 V/100 ft. See Fig. 19-2.

Table Solution (Table 19-2):

Select No. 6 wire. This results in a drop of 1.44 V/100 ft.

Graph Solution (Fig. 19-2):

Intersection lies between No. 6 wire and No. 8 wire. Select the larger of the two, in this case it would be No. 6.

UNACCEPTABLE MOTOR VOLTAGES

Occasionally, it becomes necessary to determine causes of unacceptable voltage conditions at motor terminals. Often this is necessary where excessive voltage

Table 19-2 *Voltage Drop per 100 Feet of Copper Conductor of Wire Gage*

Amperes*	No. 14	No. 12	No. 10	No. 8	No. 6	No. 4	No. 3	No. 2	No. 1	No. 0	No. 00	No. 000	No. 0000
5	1.29	0.81	0.51	0.32	0.21	0.13	0.11						
10	2.57	1.62	1.02	0.64	0.41	0.26	0.21	0.16	0.13	0.10			
15	3.86	2.43	1.53	0.96	0.62	0.39	0.31	0.24	0.19	0.15	0.12	0.10	
20	5.14	3.24	2.04	1.28	0.82	0.52	0.41	0.32	0.26	0.20	0.16	0.13	0.10
25	6.43	4.05	2.55	1.60	1.03	0.65	0.51	0.41	0.32	0.26	0.20	0.16	0.13
30	7.71	4.86	3.06	1.92	1.23	0.78	0.62	0.49	0.39	0.31	0.24	0.19	0.15
35		5.67	3.57	2.24	1.44	0.91	0.72	0.57	0.45	0.36	0.28	0.22	0.18
40		6.48	4.08	2.56	1.64	1.04	0.82	0.65	0.52	0.41	0.32	0.26	0.20
45			4.59	2.88	1.85	1.17	0.92	0.73	0.58	0.46	0.36	0.29	0.23
50			5.10	3.20	2.05	1.30	1.03	0.81	0.65	0.51	0.41	0.32	0.26
60			6.12	3.84	2.46	1.56	1.23	0.97	0.77	0.61	0.49	0.38	0.31
70				4.48	2.87	1.82	1.44	1.13	0.90	0.71	0.57	0.45	0.36
80				5.12	3.28	2.08	1.64	1.30	1.03	0.82	0.65	0.51	0.41
90					3.69	2.34	1.85	1.46	1.16	0.92	0.73	0.58	0.46
100					4.10	2.59	2.05	1.62	1.29	1.02	0.81	0.64	0.51
110						2.85	2.26	1.78	1.42	1.12	0.89	0.70	0.56
120							2.46	1.94	1.55	1.22	0.97	0.77	0.61
130								2.10	1.68	1.33	1.05	0.83	0.66
140									1.81	1.43	1.13	0.90	0.71
150										1.53	1.22	0.96	0.77
ohms/100 ft													
copper	0.257	0.162	0.1018	0.064	0.041	0.0259	0.0205	0.0162	0.0128	0.0102	0.0081	0.0064	0.0051
aluminum	4.22	0.266	0.167	0.105	0.0674	0.0424	0.0336	0.0266	0.0129	0.0168	0.0133	0.0105	0.0084

*To determine voltage drop for aluminum conductors, enter the chart using 1.64 × actual amperes.
The conductor's lengths is twice the length of the branch leads, whether single or three phase.
Since resistance varies with temperature, it may be necessary to correct for wire temperature under load conditions if the ambient materially exceeds 80°F. If so, increase ampere values using the multiplier 1.0 + 0.002 × (ambient temperautre –80°F).
Example: If current flow and environment result in conductors reaching 140°F under load conditions, the appliance ampere ratings should be increased by the multiplying factor 1.0 + 0.002 (140° – 80°) = 1.0 + 0.12 = 1.12.

Table 19-3 *Range of Locked Rotor Amperes per Motor Horsepower*

NEMA Code Letter	115	208		230		460	
	1ϕ	1ϕ	3ϕ	1ϕ	3ϕ	1ϕ	3ϕ
A	0–27.4	0–15.1	0–9.1	0–13.7	0–7.9	0–6.9	0–4.0
B	27.5–30.9	15.2–17.0	9.2–9.8	13.8–15.5	8.0–9.0	7.0–7.7	4.1–4.5
C	31.0–34.8	17.1–19.4	9.9–11.2	15.6–17.4	9.1–10.1	7.8–8.7	4.6–5.0
D	34.9–39.2	19.5–21.6	11.3–12.5	17.5–19.6	10.2–11.3	8.8–9.8	5.1–5.7
E	39.3–43.5	21.7–24.0	12.6–13.9	19.7–21.7	11.4–12.5	9.9–10.9	5.8–6.3
F	43.6–48.7	24.1–26.9	14.0–15.5	21.8–24.4	12.6–14.1	11.0–12.2	6.4–7.0
G	48.8–54.8	27.0–30.3	15.6–17.5	24.5–27.4	14.2–15.8	12.3–13.7	7.1–7.9
H	54.9–61.7	30.4–33.7	17.6–19.5	27.5–30.6	15.9–17.7	13.8–15.3	8.0–8.8
J	61.8–69.6	33.8–38.4	19.6–22.2	30.7–34.8	17.8–20.1	15.4–17.4	8.9–10.1
K	69.7–78.4	38.5–43.3	22.3–25.0	34.9–39.2	20.2–22.6	17.5–19.6	10.2–11.3
L	78.5–87.1	43.4–48.0	25.1–27.7	39.3–43.2	22.7–25.2	19.7–21.8	11.4–12.6
M	87.2–97.4	48.1–53.8	27.8–31.1	43.3–48.7	25.3–28.7	21.9–24.4	12.7–14.1
N	97.5–109	53.9–60.0	31.2–34.6	48.7–54.5	28.3–31.5	24.5–27.3	14.2–15.8
P	110–122	60.1–67.2	34.7–38.8	54.6–61.0	31.6–35.2	27.4–30.5	15.9–17.6
R	123–139	67.3–76.8	38.9–44.4	61.1–69.6	35.3–40.2	30.6–34.8	17.7–20.1
S	140–157	76.9–86.5	44.5–50.0	69.7–78.4	40.3–45.3	34.9–39.2	20.2–22.6
T	158–174	86.6–96.0	50.1–55.5	78.5–87.0	45.4–50.2	39.3–43.5	22.7–25.1
U	175–195	96.1–108	55.6–56.4	87.1–97.5	50.3–56.3	44.5–48.8	25.2–28.2
V	196 and up	109 and up	56.5 and up	97.6 and up	56.4 and up	48.9 and up	28.3 and up

Note: Locked rotor amperes appear on rating plates of hermetic compressors.
The NEMA code letter appears on the motor rating plate.
Multiply above values by motor horsepower.

drops are encountered as motors start. During this brief interval, the starting inrush current may approximate a motor's locked-rotor amperage rating.

Table 19-3 shows the range of *locked-rotor amperes* (LRA) per motor horsepower. LRA appear on the rating plates of hermetic compressors. Depending on the type of motor, its locked-rotor amperage may be two to six times its rated full-load current. Motor-starting torque varies as the square of the voltage. Thus, only 81 percent of the anticipated torque is available if the voltage drops to 90 percent of the rating during the starting period.

The full-load amperage value must be considered in choosing the proper wire size and making sure the motor has acceptable voltages. These are shown in Table 19-4.

Table 19-4 *Approximate Full-Load Amperage Values for AC Motors*

Motor	Single Phase*		Three-Phase, Squirrel Cage Induction		
HP	115 V	230 V	230 V	460 V	575 V
1/6	4.4	2.2			
1/4	5.8	2.9			
1/3	7.2	3.6			
1/2	9.8	4.9	2	1.0	0.8
3/4	13.8	6.9	2.8	1.4	1.1
1	16	8	3.6	1.8	1.4
1 1/2	20	10	5.2	2.6	2.1
2	24	12	6.8	3.4	2.7
3	34	17	9.6	4.8	3.9
5	56	28	15.2	7.6	6.1
7 1/2			22	11.0	9.0
10			28	14.0	11.0
15			42	21.0	17.0
20			54	27.0	22.0
25			68	34.0	27.0

*Does not include shaded pole.

CALCULATING STARTING CURRENT VALUES AND INRUSH VOLTAGE DROPS

Single-Phase Current

Wire size and inrush voltage drop can be calculated. The following formula can be used for single-phase current.

Example:

If a single-phase 230-V condensing unit, rated at 22-A full load and having a starting current of 91 A is located 125 ft from the distribution panel and so utilizes 250 ft of the No. 10 copper wire. The voltage drop expected during full-load operation is calculated as follows:

Refer to the lower lines of Table 19-2. Note that the resistance of No. 10 copper wire is 0.1018 Ω/100 ft.

$$\text{Voltage drop} = 22\ \text{A} \times 0.10018 \times \frac{250}{100}\ \text{ft} = 5.6\ \text{V}$$

(Note that 5.6 V exceeds the 2 percent loss factor, which is 4.6 V.) If the full 3 percent loss (6.9 V) allowed ahead of the meter is present, then the voltage at the load terminal of the meter will be 223.1 V (230 – 6.9 = 223.1). Subtract the voltage drop calculated in the foregoing equation and there will be only 217.5 V at the unit terminals during full-load operation.

Thus,

$$223.1 - 5.6 = 217.5\ \text{V}$$

With a total loss of 5.4 percent, (230 – 217.5 = 12.5 V or 5.47 percent), it is common practice to move to the next largest wire size. Therefore, for this circuit, AWG No. 8 wire should be used instead of No. 10.

Insofar as motor starting and relay operation are concerned, the critical period is during the initial instant of start-up when the inrush current closely approximates the locked-rotor value. For the equipment described in the above example, the voltage drop experienced at 91-A flow for No. 10 wire is again excessive, indicating the wisdom of using No. 8 wire.

For No. 10 wire:

$$\text{Inrush voltage drop} = 91\ \text{A} \times \frac{0.1018\ \Omega}{100\ \text{ft}} \times 250\ \text{ft}$$
$$= 23.1595\ \text{V or } 23.2\ \text{V}$$

For No. 8 wire:

$$\text{Inrush voltage drop} = 91\text{A} \times \frac{0.064\ \Omega}{100\ \text{ft}} \times 250\ \text{ft}$$
$$= 14.56\ \text{V}$$

For a 230-V circuit, the 23.2 V slightly exceeds a 10 percent drop between the meter and the appliance. To this must be added the voltage drop incurred in the lead-in wires from the outdoor power line. This total must then be deducted from the power line voltage on the poles, which may be less than 230 V during utility peak load periods. Although the inrush current may exist for only an instant, this may be long enough to cause a starting relay to open, thus cutting off current to the motor. Without current flow, the voltage at the unit immediately rises enough to reclose the relay, so there is another attempt to start the motor. While the unit may get underway after the second or third attempt, such "chattering relay" operation is not good for the relay, the capacitors, or the motor.

For electrical loads such as lighting, resistance heating, and cooking, inrush current may be considered the equivalent of normal current flow. In the case of rotating machinery, it is only during that initial period or rotation that the start-up current exceeds that of final operation. The same is true of relays during the instance of "pull-in."

Three-Phase Circuits

Calculating the inrush voltage drop for three-phase circuits is the same as calculating the drop for single-phase circuits. Again, the value for circuit length equals twice the length of an individual conductor. Because more conductors are involved, the normal current and the starting current per conductor are smaller for a motor of a given size. Thus, lighter wire may be used.

Example:

Using the same wire length as in the single-phase example and the lower values of 13.7-A full load and 61-A starting inrush per conductor for the three-phase rating of the same size compressor, the use of No. 10 conductor results in:

$$\text{Normal voltage drop} = 13.7\ \text{A} \times \frac{0.1018\ \Omega}{100\ \text{ft}} \times 250\ \text{ft}$$
$$= 3.5\ \text{V}$$

$$\text{Inrush voltage drop} = 61\ \text{A} \times \frac{0.1018\ \Omega}{100\ \text{ft}} \times 250\ \text{ft}$$
$$= 15.5\ \text{V}$$

Inrush Voltage Drop

The actual inrush current through an appliance usually is somewhat less than the total of locked-rotor current values. Locked-rotor current is measured with rated voltage at the appliance terminals. Because voltage drop in the feed lines reduces the voltage available at the terminals, less than rated voltage can be anticipated across the electrical components. Consequently, inrush currents and voltage drops are somewhat less. This fact is illustrated in the following, which is based on the same installation as that in the previous single-phase examples. However, here the actual locked-rotor current of 101 A is used. The formula can be found in the column on the right.

CODE LIMITATIONS ON AMPERES PER CONDUCTOR

Varied mechanical conditions are encountered in field wiring. Thus, the NEC places certain limitations on the smallness of conductors installed in the field. Such limitations apply regardless of conductor length. They assure the following:

- That the wire itself has ample strength to withstand the stress of pulling it through long conduits and chases. With specific exceptions, no wire lighter than No. 14 copper is permitted for field wiring of line-voltage power circuits.
- By stipulating the maximum amperage permissible for each wire gage, self-generated heat can be limited to avoid temperature damage to wire insulation. If wiring is installed in areas of high-ambient temperature, the amperage rating may need to be reduced.
- By stipulating the maximum amperage of overload protectors for circuits, current flow is limited to safe values for the conductor used. Some equipment has momentary starting currents that trip-out overload protectors sized on the basis of full-load current. Here heavier fusing is permissible, but only under specific circumstances. Current flow limitations for each gage protect wire insulation from damage due to overheating.

HEAT GENERATED WITHIN CONDUCTORS

Heat generation due to current flow through the wire is important for the following two reasons:

- Temperature rise increases the resistance of the wire and, therefore, the voltage drop in the circuit. Under most conditions of circuit usage, this added resistance generates additional heat in the wires. Finally, a temperature is reached where heat dissipation from the conductors equals the heat that they generate. It is desirable to keep this equilibrium temperature low. The number of Btu generated can be found by both of the following formulas [Eqs. (19.3) and (19.4)].

Indicated locked-rotor impedance

$$= \frac{\text{rated voltage}}{\text{locked-rotor amperes}}$$

$$= 230\text{ V}/101\text{ A} = 2.28\ \Omega$$

Resistance of No. 10 leads

$$= \text{ohms}/100\text{ ft} \times \text{length of wire (in ft)}$$

$$= 0.1018\ \Omega/100\text{ ft} \times 250\text{ ft} = 0.25\ \Omega$$

Total indicated load and conductor impedance

$$= (2.28 + 0.25) = 2.53\ \Omega$$

$$\text{Inrush current} = \frac{\text{Distribution panel voltage}}{\text{total indicated impedance}}$$

$$\text{Inrush current} = \frac{230\text{ V}}{2.53\ \Omega} = 91\text{ A}$$

Calculated inrush voltage drop

$$= 91.0 \times \frac{0.1018\ \Omega}{100\text{ ft}} \times 250\text{ ft} = 23.2\text{ V}$$

Locked-rotor voltage drop

$$= 101 \times \frac{0.1018\ \Omega}{100\text{ ft}} \times 250\text{ ft} = 25.7\text{ V}$$

$$\text{Btu generated} = \text{amperes}^2 \times \text{resistance in ohms} \times 3.4313^*$$

$$\text{Btu generated} = \text{amperes} \times \text{voltage drop} \times 3.4313^*$$

- Temperature also damages wire insulation. The degree of damage is dependent upon the insulation's ability to withstand temperature under varying degrees of exposure, age, moisture, corrosive environment, mechanical abuse, and thickness.

Estimating the probable operating temperature of a conductor and its insulation is difficult. The rate of heat dissipation from the wiring surfaces varies with the ambient temperature, the proximity of other heat-generating conductors, the heat conductivity of the insulation and jacket material, the availability of cooling air, and other factors. Freestanding individual conductors dissipate heat more effectively. However, the typical situation of two or three conductors, each carrying equal

*Conversion factor

current and enclosed in a common jacket, cable, or conduit, anticipates limitations as set forth by the NEC.

CIRCUIT PROTECTION

Circuits supplying power to appliances must incorporate some means for automatically disconnecting the circuit from the power source, should there be abnormal current flow due to accidental grounding, equipment overload, or short circuits. Such overload devices should operate promptly enough to limit the buildup of damaging temperatures in conductors or in the electrical components of an appliance. However, devices selected to protect circuits feeding motors must be slow enough to permit the momentary inrush of heavy starting current. They must then disconnect the circuit if the motor does not start promptly, as can happen under low voltage conditions.

Devices heavy enough to carry continuously the motor-starting current do not provide the overload protection desired. Likewise, heavily fused branch circuits do not adequately protect the low amperage components that cumulatively require the heavy fusing. For this reason some literature lists maximum allowable fuse sizes for equipment. While electrical components of factory-built appliances are individually safeguarded, the field combining of two or more units on one circuit may create a problem more complex than that normally encountered. Remember that the final authority is the local electrical inspector.

Standard Rule

With a few exceptions, the ampere capacity of an overload protector cannot exceed the ampacity values listed by wire size by the NEC. (Check the NEC for these exceptions.) If the allowable ampacity of a conductor does not match the rating of a standard size fuse or nonadjustable trip-circuit breaker, the device with the next largest capacity should be used. Some of the standard sizes of fuses and nonadjustable trip-circuit breakers are: 15, 20, 25, 30, 35, 40, 45, 50, 60, 70, 80, 90, 100, 110, 125, 150, 175, 200, 225, 250, and 300 A.

FUSES

One-Time Single-Element Fuses

If a current of more than rated load is continued sufficiently long, the fuse link becomes overheated. This causes the center portion to melt. The melted portion drops away. However, due to the short gap, the circuit is not immediately broken. An arc continues and burns the metal at each end until the arc is stopped because of the very high increase in resistance. The material surrounding the link tends to break the arc mechanically. The center portion melts first, because it is farthest from the terminals that have the highest heat conductivity. See Fig. 19-4.

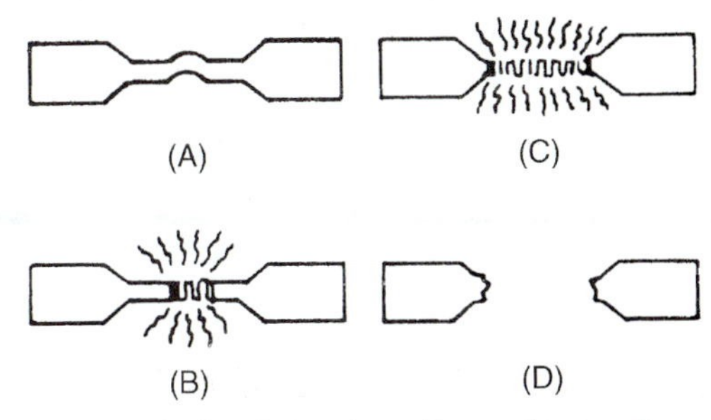

Fig. 19-4 *Illustration of how a fuse works.*

Fuses will carry a 10 percent overload indefinitely under laboratory-controlled conditions. However, they will blow promptly if materially overloaded. They will stand 150 percent of the rated amperes for the following time periods:

- 1 min (fuse is 30 A or less)
- 2 min (fuse of 31 to 60 A)
- 4 min (fuse of 61 to 100 A)

Time-Delay Two-element Fuses

Two-element fuses use the burn out link described previously. They also use a low-temperature soldered connection that will open under overload. This soldered joint has mass, so it does not heat quickly enough to melt if a heavy load is imposed for only a short time. However, a small but continuous overload will soften the solder so that the electrical contact can be broken.

With this type of protection against light overloads, the fusible link can be made heavier, yet blow quickly to protect against heavy overloads. This results in fewer nuisance burn outs and equipment shutdowns. Two types of dual-element fuses are shown in Fig. 19-5.

TYPES OF FUSES

In addition to the fuses just described, there are three general categories based on *shape* and *size*.

- The *automotive glass* (AG) fuse consists of a glass cylinder with metallic end caps between which is connected a slender metal element that melts on current overload. This fuse has a length of $1^{5}/_{16}$ in. and a diameter of $^{1}/_{4}$ in. It is available only for low amperages. While used in specific appliances, it is not used to protect permanently installed wiring.
- Cartridge fuses are like AG fuses. However, they are larger. The cylindrical tube is fiber, rather than glass.

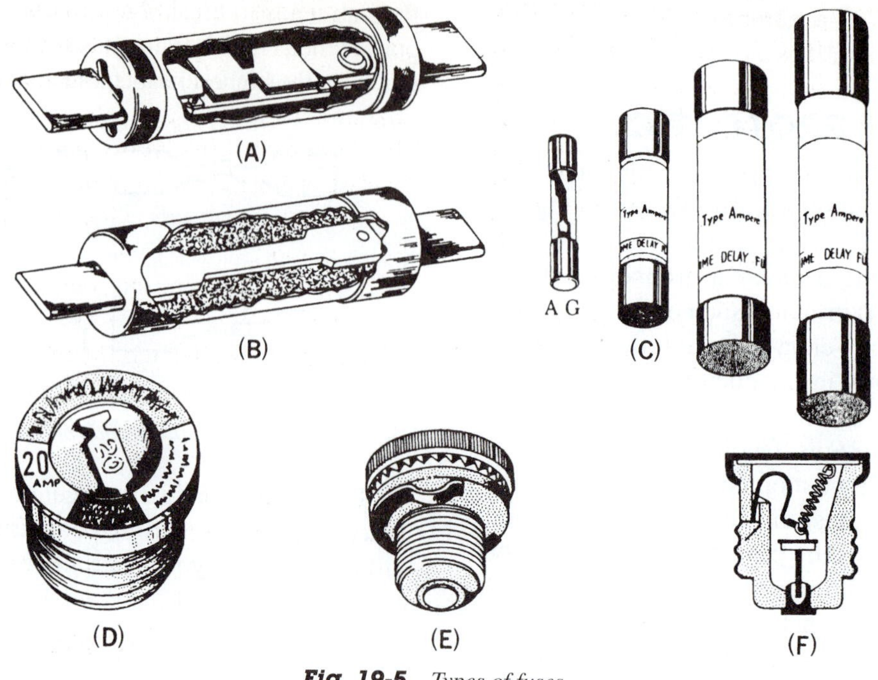

Fig. 19-5 *Types of fuses.*

The metallic end pieces may be formed as lugs, blades, or cylinders to meet a variety of fuse box socket requirements. The internal metal fusible link may be enclosed in sand or powder to quench the burnout arc.

Cartridge fuses are made in a variety of dimensions, based on amperage and voltage. Blade-type terminals are common above 60 A. Fuses used to break 600-V arcs are longer than those for lower voltages. Fuses are available in many capacities other than the listed standard capacities, particularly in the two-element, time-delay types. Often, they are so dimensioned as to not be interchangeable with fuses of other capacities.

- Plug fuses are limited in maximum capacity to 30 A. They are designed for use in circuits of not more than 150 V above ground. Two-element time-delay types are available to fit standard screw lamp sockets. They are also available with nonstandard threads made especially for various amperage ratings.

THERMOSTATS

The thermostat (or temperature control) stops and starts the compressor in response to room temperature requirements. Each thermostat has a charged power element containing either a volatile liquid or an active vapor charge. The temperature-sensitive part of this element (thermostat feeler bulb) is located in the return air stream. As the return air temperature rises, the pressure of the liquid or vapor inside the bulb increases This closes the electrical contacts and starts the compressor. As the return air temperature drops, the reduced temperature of the feeler bulb causes the contacts to open and stops the compressor.

The advent of transistors and the semiconductor chips or integrated circuits produced a more accurate method of monitoring and adjusting temperatures within a system. The microprocessor makes use of the semiconductor and chip's abilities to compare temperatures. It can also program on and off cycles, as well as monitor the duration of each cycle. This leads to more accurate temperature control.

Figure 19-6 shows a microprocessor-based thermostat. As you can see from the front of the control panel, you can adjust the program to do many things. It can also save energy, whether it is operating the furnace for heat or the air-conditioning unit for cooling. These units usually come with a battery so that the memory can retain whatever is programmed into it. The battery is also a backup for the clock so that the program is retained even if the line power is interrupted.

Thermostat As a Control Switch

The control switch (thermostat) may be located in the room to be cooled or heated, depending upon the particular switch selection point. The control switch (heat, off, cool, and auto) is of the sliding type and normally

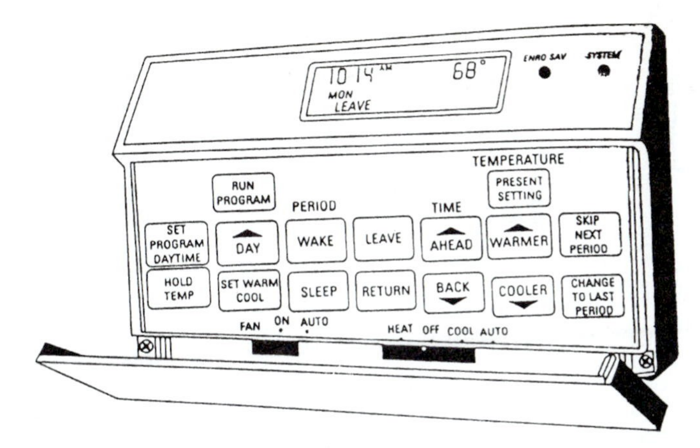

Fig. 19-6 *Microprocessor thermostat used for residential temperature and cooling control.*

has four positions, marked HEAT, OFF, COOL, and AUTO. The thermostat is taken from its socket and programmed according to the manufacturer's directions. Then it is activated by plugging it into the wall socket and replacing a couple of screws to hold it in place.

To operate the unit as a ventilator, the switch on the left is marked "Fan" with an "On" and "Auto" choice to select the FAN operation. When a thermostat is installed for automatic cooling, the compressor and fans will cycle according to the dial requirements.

Figure 19-7 shows the electrical circuitry for a home heat-cool thermostat. Keep in mind that the thermostat should always be on an inside partition, never on an outside wall. Do not mount the instrument on a part of the wall that has steam or hot water pipes or warm air ducts behind it. The location should be such that direct sunshine or fireplace radiation cannot strike the thermostat. Be careful that the spot selected is not likely to have a floor lamp near it or a table lamp under it. Do not locate the thermostat where heat from kitchen appliances can affect it. Do not locate it on a wall that has cold unused room on the other side.

After a thermostat has been mounted, it is wise to fill the stud space behind the instrument with insulating material. This is to prevent any circulation of cold air. Furthermore, the hole behind the thermostat for the wires should be sealed so that air cannot emerge from the stud space and blow across the thermostat element. It is quite common to find considerable air motion through this hole caused by a chimney effect in the stud space.

Service

Servicemen who have a good knowledge of refrigeration and air conditioning will be able to competently service air conditioners. Most air-conditioning units of present design contain compressors of the hermetic or sealed type. The only parts that can be serviced in the field are the relay, control switch, fan, fan motor, start and run capacitors, air filters, and cabinet parts. The refrigerating system (consisting of the cooling unit, condensers, compressors, and connecting lines) generally

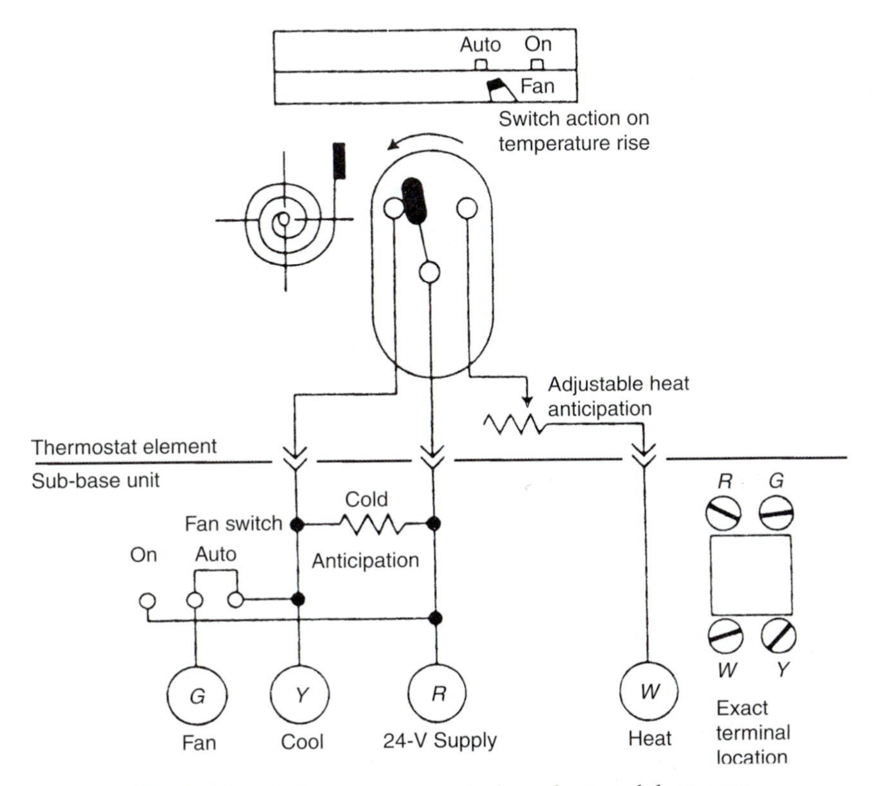

Fig. 19-7 *Electrical circuitry of a home heat-cool thermostat.*

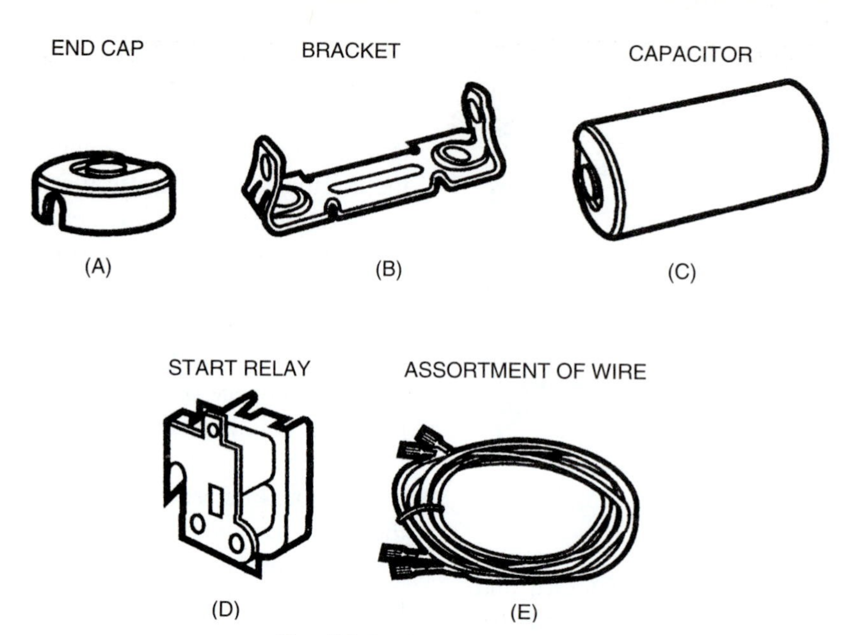

Fig. 19-8 *Starter kits.* (Carrier)

cannot be serviced in the field. Most servicemen will find the newer electronic type thermostats easy to program when following the manufacturer's instructions included with every thermostat and usually given to the homeowner at closing on the house. See App. 3 for more information on programmable thermostats.

Start Kits

There are kits of components to use with the *permanent start capacitor* (PSC) that need additional starting torque. These kits are available for most of the popular brands of compressors, such as Copeland, Tecumseh, Bristol, and Carlyle. See Fig. 19-8.

The kit consists of the end cap (Fig. 19-8A), bracket (Fig. 19-8B), capacitor (Fig. 19-8C), start relay (Fig. 19-8D), and an assortment of wire of four different lengths and three colors (Fig. 19-8E).

Table 19-5 shows the specifics of each part with the capacitor voltage rating and microfarad range. Note the dropout volts for the start relay.

Table 19-5 *Start Capacitors and Relays*

START CAPACITOR		START RELAY				
MFD RANGE	VOLTS	CONTACT RATING	MAX OPER. VOLTS	PICK UP VOLTS	DROPOUT VOLTS	PART NO.
21-25	330	50 AMP	336	162-175	55-115	P296-0007
43-53	320	35 AMP	336	162-175	40-90	P296-0006
43-53	320	50 AMP	395	180-195	40-105	P296-0001
72-88	330	35 AMP	336	162-175	40-90	P296-0012
88-108	330	35 AMP	336	162-175	40-90	P296-0005
88-108	330	50 AMP	420	195-224	60-123	P296-0002
88-108	330	50 AMP	395	180-195	40-105	P296-0004
88-108	330	50 AMP	336	162-175	55-115	P296-0009
108-130	330	50 AMP	420	195-224	70-140	P296-C010
135-155	320	50 AMP	420	195-224	60-123	P296-0003
145-175	330	50 AMP	395	180-195	40-105	P296-0008
189-227	330	50 AMP	420	262-290	60-121	P296-0011

SINGLE-PHASE LINE MONITORS

The single-phase line monitor is a very low cost, highly accurate, rugged module that was designed to protect single-phase devices from over or under voltage, rapid short-cycling, and power interruptions. See Fig. 19-9. Its small design is easy to mount to continuously monitor incoming line voltage for errors. When line power is appropriate, the module closes a set of *normally open* contacts and lights a green *light emitting diodes* (LED). When the incoming power is outside of the user-selected parameters, the *normally closed* contacts will close and a red LED will illuminate, indicating current fault conditions. This unit also interrogates the line during fault conditions to reduce nuisance trips from transients or compressor start-ups. The unit trips if power is abnormal for 66 percent of interrogation time. The time delay on make is 0 to 10 min. A green LED shows when power is on and valid and the relay is energized. The red LED rapid flash when the unit currently detects high/low voltage situation. Red LED blinking shows the power is current valid and waiting for the end of a delay. The wiring diagram in Figs. 19-10 and 19-11 shows the pictorial wiring of the unit.

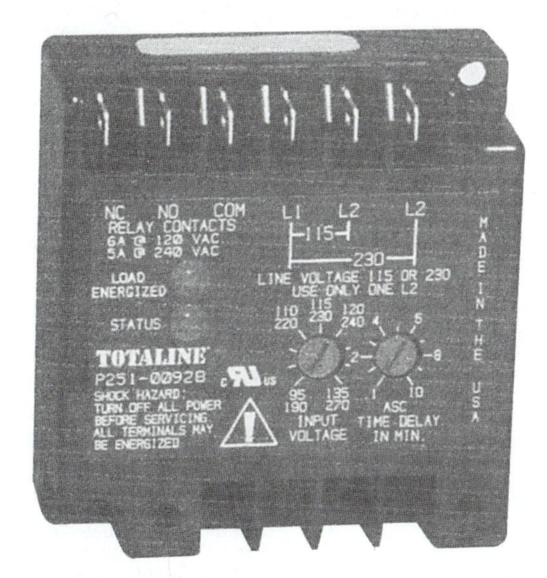

Fig. 19-9 *Single-phase line monitor.* (Carrier)

The unit is epoxy encapsulated for use in extreme environmental conditions. The power-loss detection or trip takes place within 45 ms. There is a user selectable *antishort cycle* (ASC) delay of 0 to 10 min. This unit also trips if the power is abnormal for 55 percent of interrogation period. The relay contacts can handle 6 A. The unit proper requires only 21 to 31 mA for power operation. Table 19-6 presents a competetive cross reference guide to various manufacturers' units.

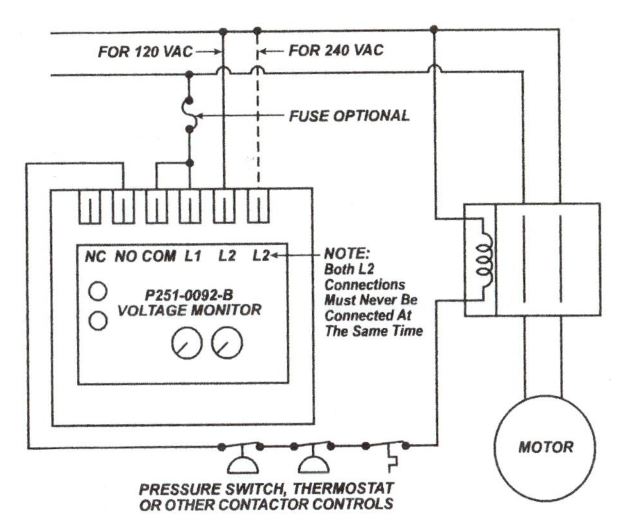

Fig. 19-10 *Wiring diagram for single-phase line monitor.* (Carrier)

TIME DELAYS

Time delays also come in epoxy encapsulated units. They will meet requirements for numerous applications. They are a simple two-wire hook-up usually, and will work with the anticipator-type thermostat. One of those shown in Fig. 19-12 is specifically designed to control circulating fan in heat pumps, airconditioners, and forced air systems. The table contained in the figure indicates the specifics of the units and their variety.

HEAD PRESSURE CONTROL

The solid-state, epoxy encapsulated head pressure control was designed specifically for HVAC and refrigeration applications. The unit is a temperature-sensitive fan motor speed control that regulates head pressure at low ambients by varying the air volume through the condenser. Its unique construction permits exceptionally quick response times to change in liquid-line temperatures. See Fig. 19-13. The dimensions of the unit are shown in Fig. 19-14. It provides full torque to the motor during start-up to ensure proper fan rotation and lubrication of bearings. It also features high temperature bypass and applies full voltage to the condenser fan under normal operating conditions. It also determines the minimum rpm level at which the condenser fan should operate. The Setpoint Hysteresis (Deadband) prevents system oscillation, which may occur with small temperature fluctuations in the liquid line. This unit can be used with a wide variety of motors.

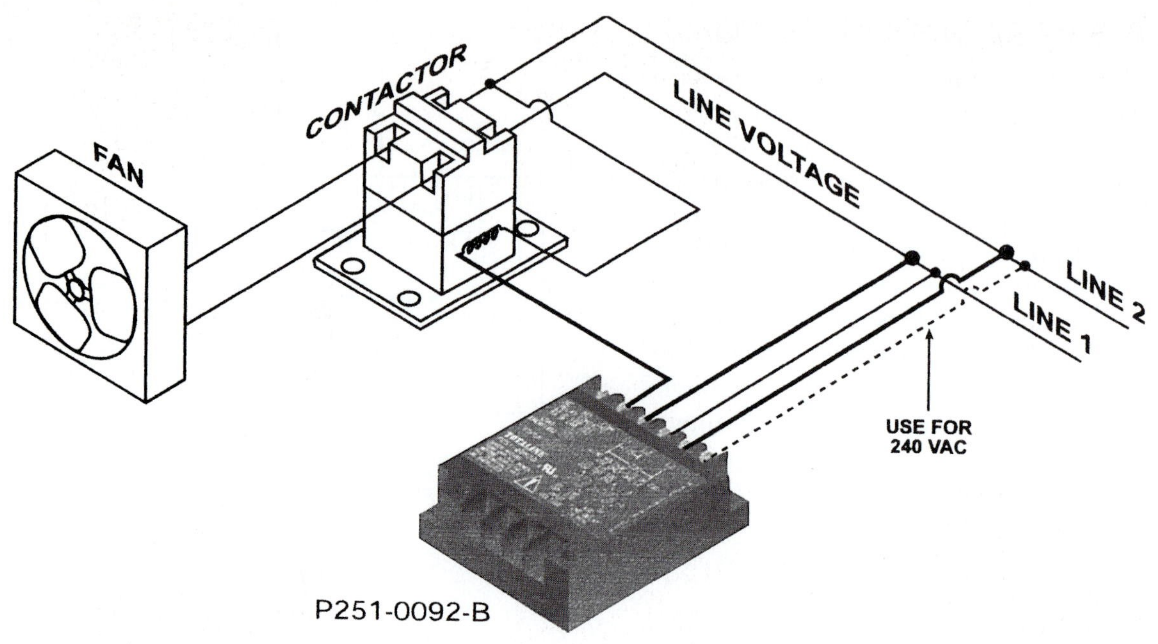

Fig. 19-11 *System diagram for single-phase line monitor.* (Carrier)

Table 19-6 *Competitive Cross-Reference Guide*

	P251-0083C Model Totaline	800/800A Series Hoffman	E31 Series Ranco	P66 Johnson Controls	Totaline Advantage
Hard start	Field adjustable 1-5 s	Factory fixed at 5 s	NO	NO	Installer may adjust the "Hard start" period to satisfy different fan size and job requirements.
Low temperature cutoff	Field adjustable 30°F–70°F (min, span)	Fixed at 50°F	NO	NO	Installer may determine the minimum rpm level at which condenser fan should operate.
High temperature bypass	YES	YES	NO	NO	Applies full voltage to the condenser fan under normal operating conditions.
Setpoint hysteresis (Deadband)	Fixed at 3°F	Fixed at 3°F	NO	NO	Prevents system oscillation, which may occur with small temperature fluctuations in the liquid line
24 VAC supply	YES	NO	NO	YES	A 24-VAC supply provides control from the low-voltage side via a low-voltage sensor. This eliminates costly high-voltage wiring and allows for easier and safer installation.
Surface mount sensor	YES	YES	YES	NO	Helps to prevent system penetration and is easier to install.
Independent adjustment	YES	NO	NO	NO	Variable adjustment is quick, easy and independent... simplifying field calibration to a wide variety of motors.
Multiple voltage operation	One model covers 120–600 VAC!	NO Three models required	NO 240 VAC only	NO Two models required	Reduces inventory. One model is easier, less costly to stock and helps reduce the chance of miswiring in the field.

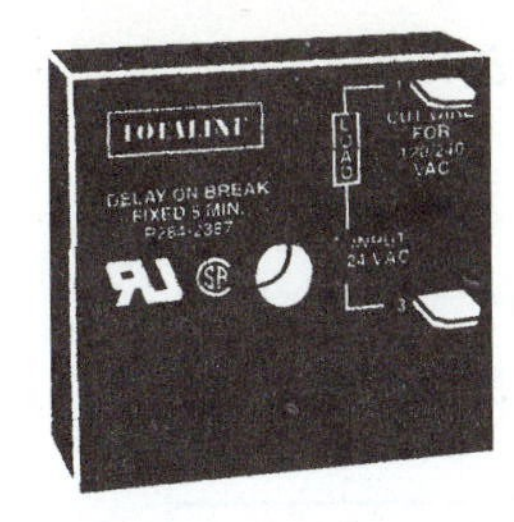

2387

- Simple 2 wire hook-up
- Works with anticipator type thermostat
- 24 VAC

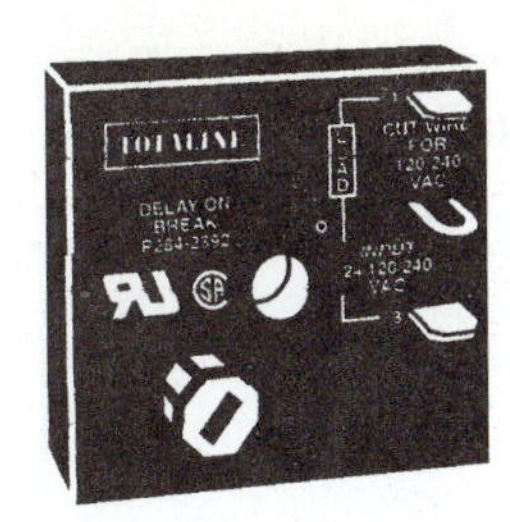

2392

- Simple 2 wire hook-up
- Works with anticipator type thermostat
- Provides jumper wire for 120/240 VAC operation
- 24/120/240 VAC

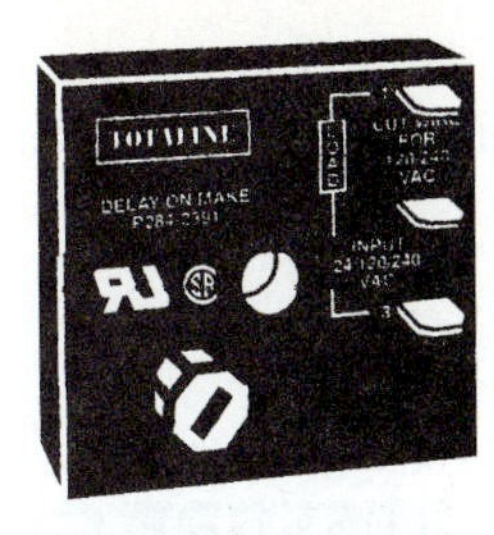

2391

- Simple 2 wire hook-up
- Works with anticipator type thermostat
- Provides jumper wire for 120/240 VAC operation
- 24/120/240 VAC

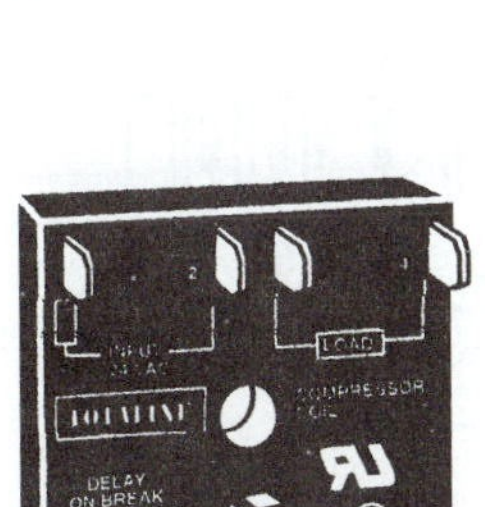

2390

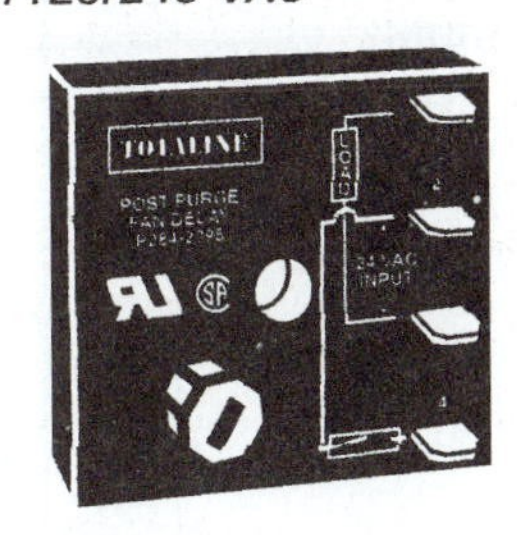

2395

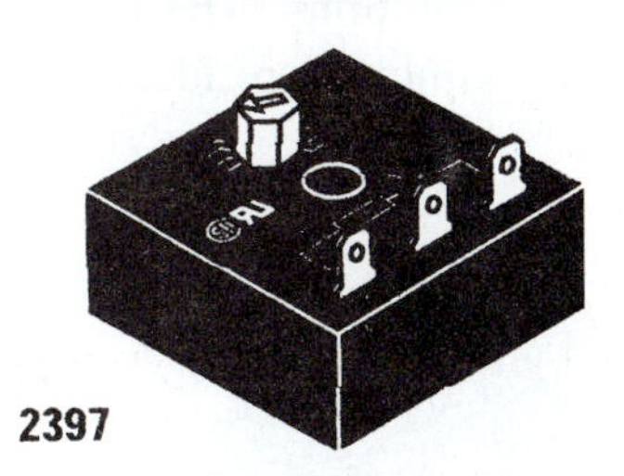

2397

PART NO.	FIXED	ADJUSTABLE	TYPE OF DELAY	USAGE
P284-2387	5 minute	-	On break	Delay period starts at end of cycle
P284-2392	-	0 - 10 minutes	On break	Delay period starts at end of cycle
P284-2390	-	0 - 10 minutes	On break	Delay period starts at end of cycle plus brownout protection
P284-2395	-	0 - 10 minutes	Post purge	Fan remains on at end of cycle
P284-2391	-	0 - 10 minutes	On make	Delay period starts at beginning of cycle
P284-2397	-	0 - 10 minutes	Bypass timer	Delay period starts at beginning of cycle

Fig. 19-12 *Time delay modules.* (Carrier)

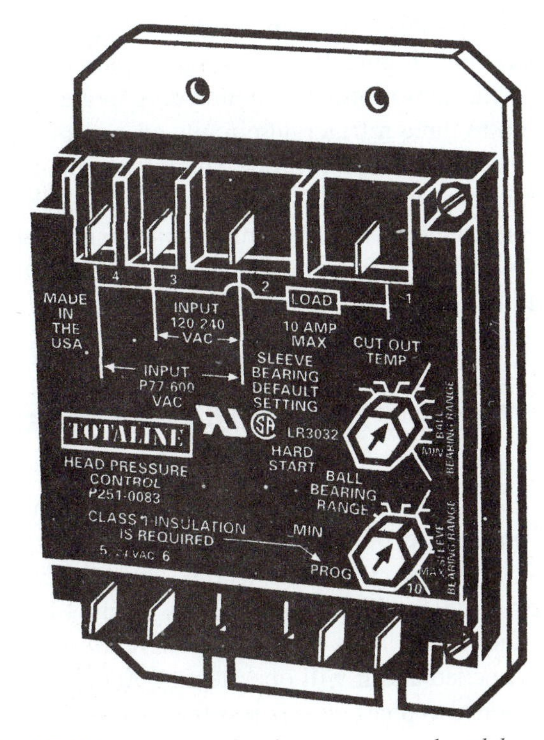

Fig. 19-13 *Solid-state head pressure control module.* (Carrier)

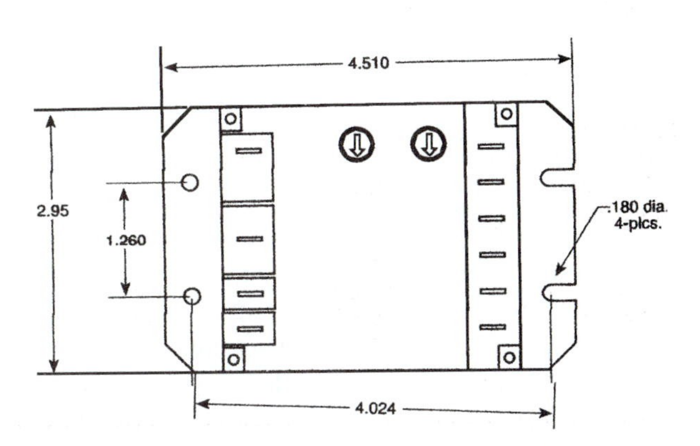

Fig. 19-14 *Dimension for head pressure control module.* (Carrier)

Table 19-6 shows the head pressure control specifics and compares the Carrier or Totaline with the Hoffmann, Ranco, and Johnson Controls models of the same unit.

PRESSURE CONTROLS

Line-Voltage Head Pressure Controls

Preventing evaporator freeze-ups, liquid-slugged compressors, and low pressure cutouts can occur during low ambient conditions. The dual line-voltage input head pressure controls are ideal for refrigeration. See Fig. 19-15.

Fig. 19-15 *Line-voltage head pressure controls.* (Carrier)

Connection Diagrams

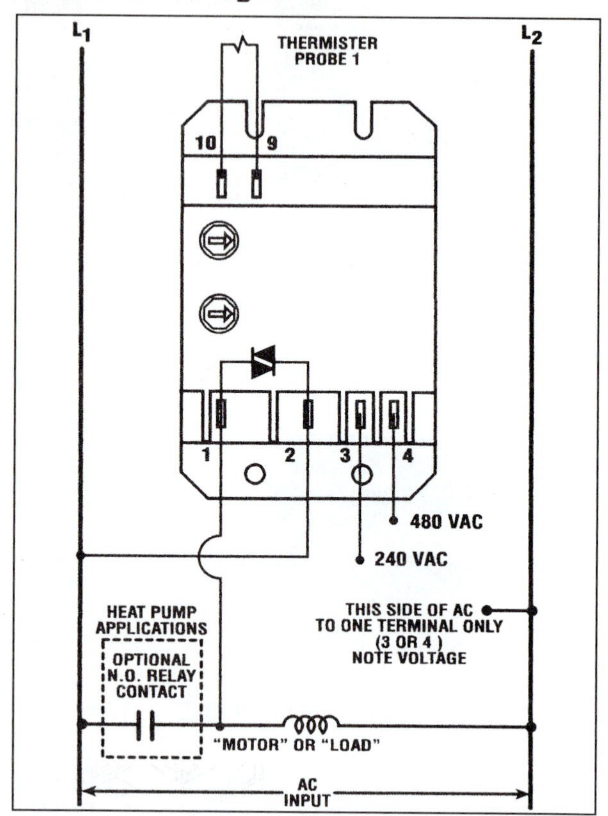

Fig. 19-16 *Connection diagrams for Fig. 21-14.* (Carrier)

The unit controls temperature sensitive motor fan speed. They have been designed to regulate head pressure at low ambients by varying the amount of airflow through the condenser and help to ensure sufficient pressure across the expansion valve. This unit has a dual-voltage 240/480 input transformer as part of the package. The user is able to select and adjust set points on the unit: hard start has 0.1 to 5 s adjustment possible, low temperature cutoff adjustment has adjustment for minimum pressure.

Basically, the unit prevents evaporator freeze-ups, low pressure cutouts, and liquid-slugged compressors. It provides full torque to the motor during start-up to help ensure proper fan rotation and lubrication of bearings.

It is possible to set the low temperature cutoff at the minimum rpm level at which the condenser fan should operate. And, the high temperature bypass applies full voltage to the condenser fan under normal conditions. Note the schematic drawing for connectons in Fig. 19-16. There are two models available, one operates on 120/240 V and the other on 208/480-V AC.

Table 19-7 shows the probe resistance versus various temperatures. Figure 19-17 gives a graphic illustration of how the sensor probe is mounted. A single unit controls up to three refrigerant circuits on a single remote condenser. The sensor probe should be mounted up several bends into the condenser (upper $^1/_3$ as shown at the left), to more closely monitor condensing temperature. On lower efficiency systems the sensor may be placed directly on the liquid line.

Three-Phase Line-Voltage Monitor

The three-phase line monitor protects against phase loss, phase reversal, and phase unbalance. It is ideally suited to protect scroll and screw compressors from reverse rotation. See Fig. 19-18. A bright LED indicates faults and loads energized. There are no adjustments needed and it will operate on 190 to 600-V AC. Its response to phase loss is less than 1 s, with an automatic reset from a fault condition.

Table 19-7 *Probe Resistance vs. Temperature*

P251-0084			P251-0085			P251-0086		
°F	°C	Resistance K Ohms	°F	°C	Resistance K Ohms	°F	°C	Resistance K Ohms
32	0	22.8	59	15	28.3	59	15	42.4
50	10	15.9	68	20	22.5	68	20	33.7
68	20	11.4	77	25	18.0	77	25	27.0
77	25	9.7	86	30	14.5	86	30	21.8
86	30	7.6	95	35	11.8	95	35	17.6
104	40	4.6	11	40	9.6	104	40	14.4
22	50	2.8	113	45	7.9	113	45	11.8
40	60	1.8	22	50	6.5	122	50	9.7
58	70	1.2	131	55	5.4	131	55	81

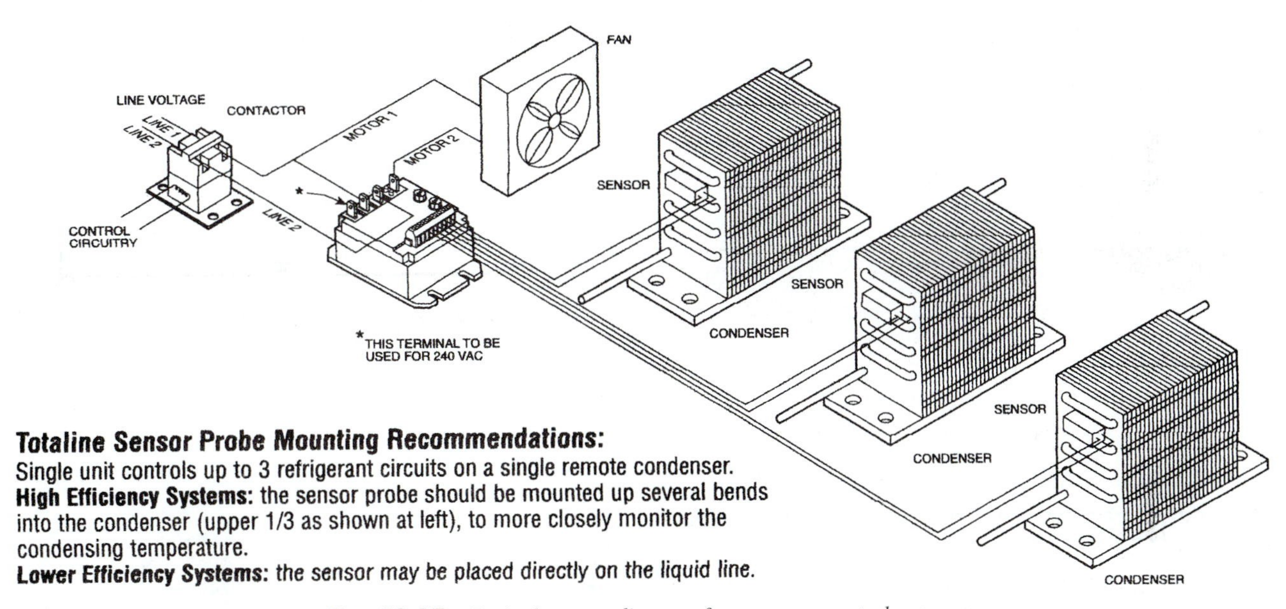

Fig. 19-17 *Typical system diagram for pressure controls. (Carrier)*

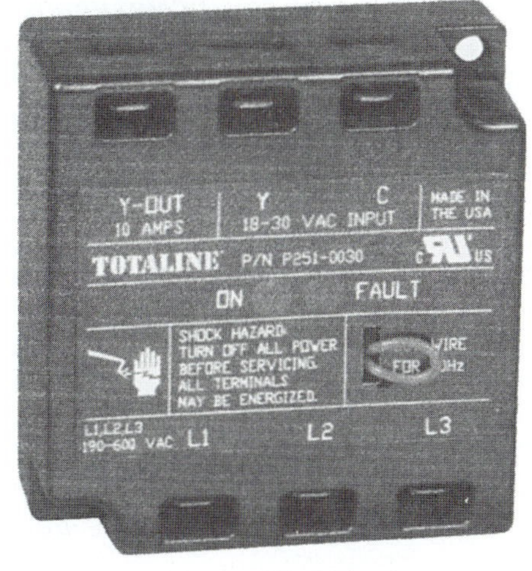

Fig. 19-18 *Three-phase line-voltage monitor. (Carrier)*

If all the three phases are relattively equal and in proper sequence, the normally open contacts (Y-Y-OUT) will close when 24 V is applied betweeen C and Y. If the phases are out of sequence or one is missing, the contacts will never close. If a phase is lost while the unit is energized, the contacts will open immediately, and remain open until the error is corrected.

An illuminated green LED indicates that the output is energized and a red LED indicates that there is a fault and the output is deenergized. See Fig. 19-19.

Figure 19-20 shows a graphic presentation of a typical wiring diagram, while Fig. 19-21 shows typical part-winding start wiring.

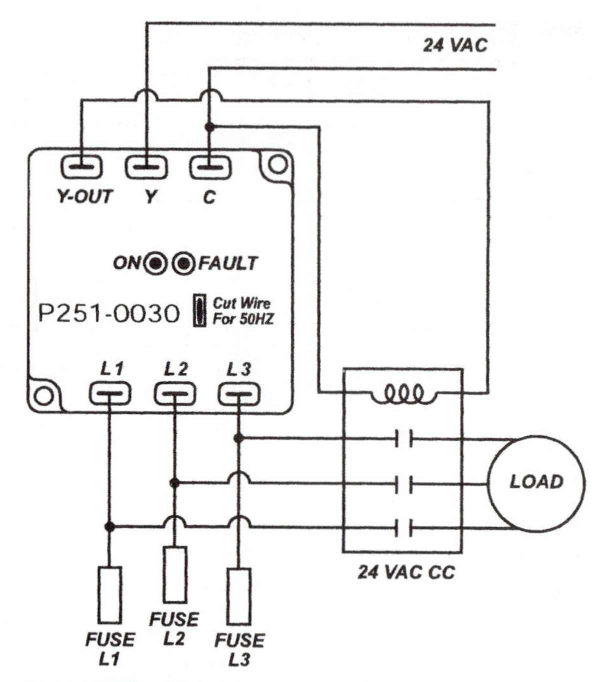

Fig. 19-19 *Wiring diagram for a three-phase line-voltage monitor.* (Carrier)

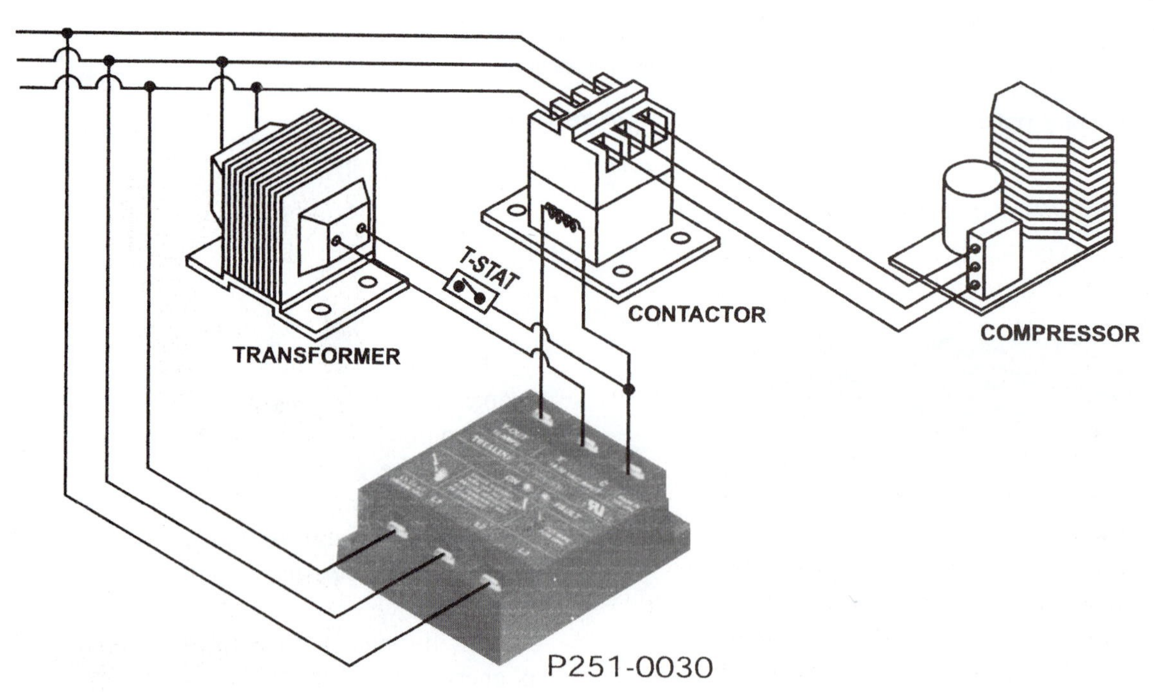

Fig. 19-20 *Typical wiring diagram for a three-phase line-voltage monitor.* (Carrier)

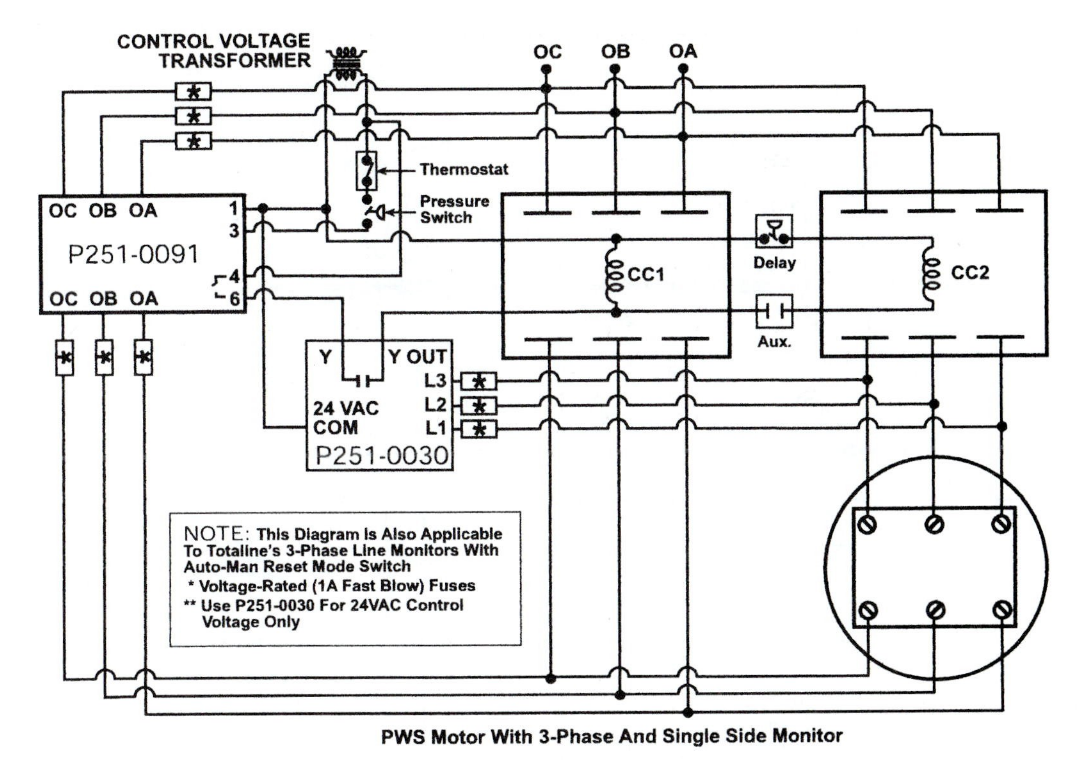

Fig. 19-21 *Typical part winding start wiring diagram.* (Carrier)

Current Sensing

The rapid start current-sensing unit is able to engage the hard-start capacitor for precisely the correct amount of time. This is done by monitoring the compressor current upon start-up. It ensures maximum starting torque without the risk of supplying too much current into the start winding.

The unit is a timed safety circuit that operates in the event the motor fails to start within 600 ms or 0.6 s. See Fig. 19-22.

This easy to install two-wire unit is solid state circuitry. It boosts starting torque and disengages upon start. It can fit into tight locations. Various wiring configurations and dimensions are shown in Fig. 19-23. Table 19-8 shows some rapid start comparisons to illustrate further the operation and properties of the current-sensing device.

Fig. 19-22 *Current sensor.* (Carrier)

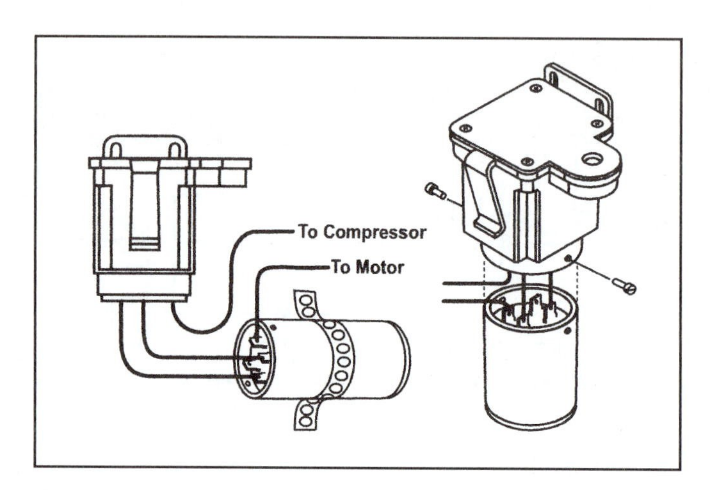

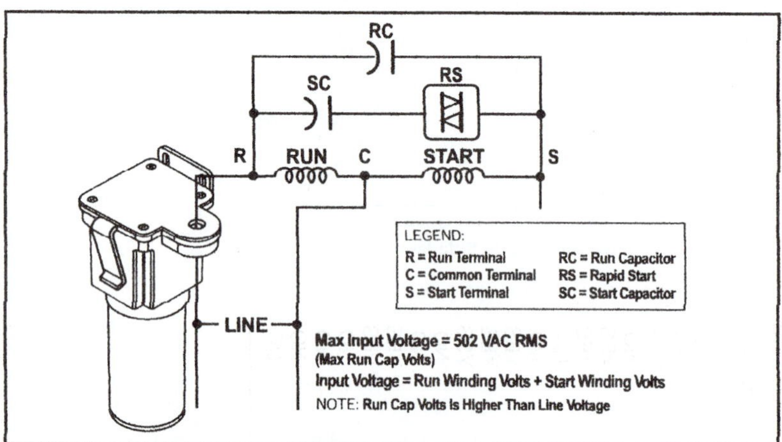

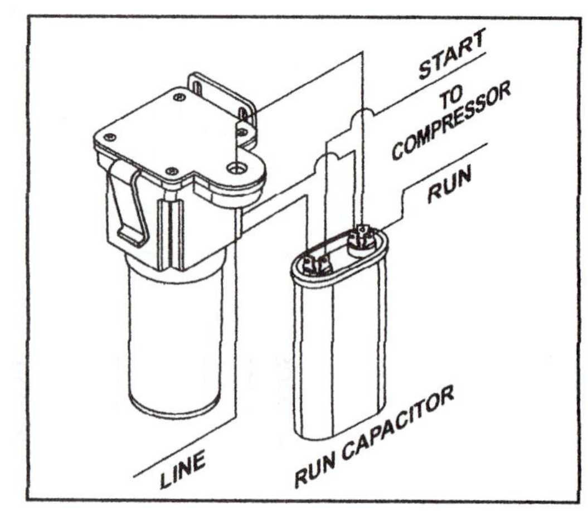

Assorted Wiring Configurations

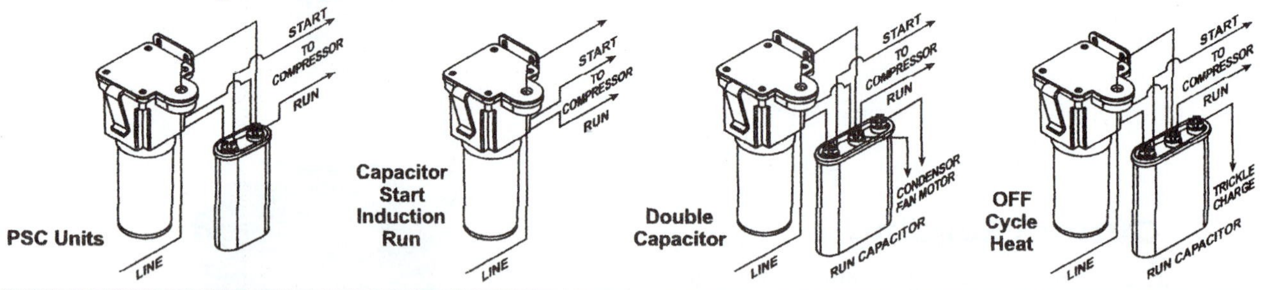

Dimensions

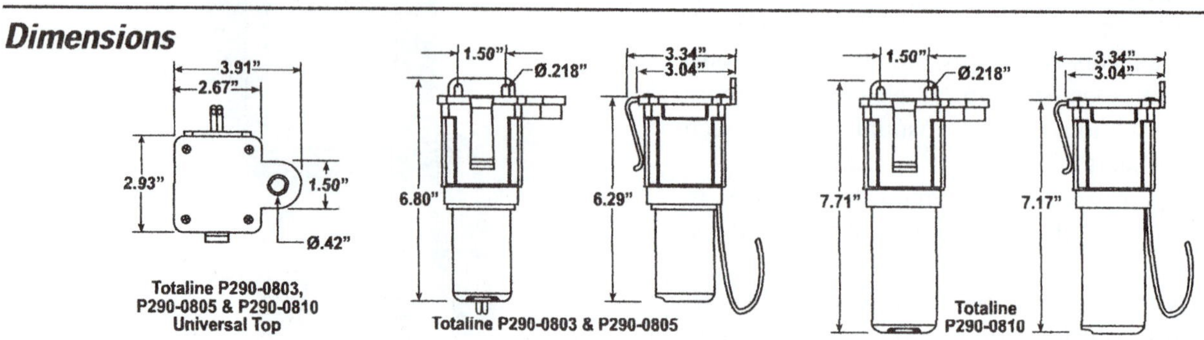

Fig. 19-23 *Wiring diagrams for current sensors along with dimensions.* (Carrier)

Table 19-8 *Rapid Start Comparisons (Carrier)*

	Hard Start			Soft Start	
	Differential Current Relay	Potential		PTCR Devices	Timing Devices
Totaline Rapid Start Comparison	Totaline Rapid Start	Kickstart	Conventional Three Wire Relay Capacitor Kit	GEMLINE HS600 & HS650 MARS 32701 & 32702 ROBERTSHAW 600-052 & 600-057 SUPCO SPP5, SPP6 SPP7 WATSCO WSX-5, WSX-6	SUPCO SPP8 WATSCO WSX-1
Self adjusting	YES	NO	NO	NO	NO
Uses current differential technology	YES	NO	NO	NO	NO
Use potential motor start relay	NO	YES	YES	NO	NO
Two wires, non polarized	YES	YES	NO	YES	YES
Recycles instantly	YES	YES	YES	NO	NO
Senses whether motor started or not	YES	YES	YES	NO	NO
Replaces three wire relay and capacitor kit	YES	YES	N/A	NO	NO
UL and CSA recognized	PENDING	YES	NO	NO	NO
Approved by compressor manufacturers	YES	YES	YES	NO	NO
Approved by equipment manufacturers	YES	YES	YES	NO	NO
Used by OEM manufacturers	YES	YES	YES	NO	NO
Safety cut-off	YES	NO	NO	NO	NO
True power factor starting	YES	NO	NO	NO	NO
Factory calibration	NOT REQUIRED	YES	YES	YES	YES
Voltage sensitive	NO	YES	YES	YES	YES
PTCR device	NO	NO	NO	YEs	NO
Timing circuit device	NO	NO	NO	YES	YES
Affected by ambient temperature	NO	NO	NO	YES	YES
Stays in circuit to long at start up	NO	NO	NO	YES	YES

REVIEW QUESTIONS

1. What are the two criteria for choosing wire size for installation in air-conditioning and refrigeration equipment?
2. What are some of the results of low voltage to a refrigerating system?
3. What is the most common cause of compressor low voltage?
4. What is synchronous speed?
5. How do you choose the proper size of wire for a job?
6. What does the abbreviation LRA mean in reference to motors?
7. How much torque is available from a motor if the voltage drops to 90 percent of its rated value?
8. What is the minimum number of wires needed to wire a three-phase compressor?
9. What causes heat generation in wire conductors?
10. What is the purpose of a fuse in a circuit?
11. What is the difference between a cartridge fuse and a plug fuse?

20
CHAPTER

Air-Conditioning and Refrigeration Careers

PERFORMANCE OBJECTIVES

After studying this chapter you will:

1. Know the industries that employ air-conditioning and refrigeration mechanics.
2. Know what job qualifications are needed for work in the air-conditioning/refrigeration field.
3. Know the various sources of information for those in the field.
4. Know the opportunities for teaching in the field.

The field of air conditioning and refrigeration offers a variety of career opportunities. Air-conditioning and refrigeration mechanics install and service air-conditioning and refrigeration equipment. There are a number of different types of equipment that require service. Some systems are complex, but are easily broken down into smaller units for repair purposes. Some mechanics specialize in a particular part of a system.

INDUSTRIES THAT EMPLOY AIR-CONDITIONING AND REFRIGERATION MECHANICS

Approximately 259,000 persons work as air-conditioning and refrigeration mechanics. Cooling contractors employ most of these mechanics. Food chains, school systems, manufacturers, and other organizations use the services of air-conditioning and refrigeration mechanics. Large air-conditioning systems use many mechanics to keep the equipment operational. However, not all mechanics work for other people. About one in every seven mechanics is self-employed.

Manufacturers use refrigeration equipment for a variety of processes. Meat packers and chemical makers use refrigeration in some form. Temperature control is very important for many manufacturing processes.

Mechanics work in homes, office buildings, and factories. They work anywhere there is air conditioning or refrigeration to be installed or serviced. They bring to the job sites the tools and parts they need. During the repair season, mechanics may do considerable driving. Radio communications and cell phones may be used to dispatch them to the jobs. If major repairs are needed, mechanics will transport parts or inoperative machinery to a repair shop.

Mechanics work in buildings that are often uncomfortable. This is because the air-conditioning or refrigeration system has failed. The mechanic may have to work in a cramped position in an attic, a basement, or a crawl space.

Rooftop units are a common practice in keeping smaller installations cool. Many of the systems have at least one unit on the roof of the building. Cooling towers are usually mounted on top of a building. Thus, the mechanic may be called upon to work in high places. In summer, the rooftop may be very hot. This trade does require some hazardous work. For instance, there are the dangers of electrical shock, torch burns, muscle strains, and other injuries from handling heavy equipment.

System installation calls for work with motors, compressors, condensing units, evaporators, and other components. These devices must be installed properly. This calls for the mechanic to be able to follow the designer's specification. In most instances, blueprints are used to indicate where and how the equipment is to be installed. The ability to read blueprints is essential for the air-conditioning and refrigeration mechanic. Such ability will help ensure that the ductwork, refrigerant lines, and electrical service are properly connected. See Fig. 20-1.

After making the connections, it is then necessary to charge the system with the proper refrigerant. Proper operation must be assured before the mechanic is through with the job.

Equipment installation is but one of the jobs the mechanic must perform. If the equipment fails, then the mechanic must diagnose the cause of the trouble and make the proper repairs. The mechanic must:

- Find defects
- Inspect parts
- Be able to know if thermostats and relays are working correctly

During the winter, air-conditioning mechanics inspect parts such as relays and thermostats. They also perform required maintenance. Overhauling may be included if compressors need attention or recharging. They may also adjust the airflow ducts for the change of seasons.

Air-conditioning and refrigeration mechanics use a number of special tools. They also use more common tools such as hammers, wrenches, metal snips, electric drills, pipe cutters and benders, and acetylene torches. Air ducts and refrigeration lines require more specialized tools. Voltammeters are also part of the mechanic's toolbox. Electrical circuits and refrigeration lines must be tested and checked. Testing of electrical components is also required. A good background in electricity is necessary for any mechanic. See Fig. 20-2.

Cooling systems sometimes are installed or repaired by other craft workers. For example, on a large air-conditioning installation job, especially where people are covered by union contracts, sheet-metal specialists might do ductwork. Electricians will do electrical work. The installation of piping will be done by pipe

Fig. 20-1 *The air-conditioning and refrigeration mechanic must be familiar with blueprint symbols. Reading blueprints is essential to the proper installation of systems.*

fitters. However, in small towns or small companies, the AC man will probably have to do all these specialized skills.

Job Qualifications

Most air-conditioning and refrigeration mechanics start as helpers. They acquire their skills by working for several years with experienced mechanics. New people usually begin by assisting experienced mechanics. They do the simple jobs at first. They may carry materials or insulate refrigerant lines. In time, they do more difficult jobs such as cutting and soldering pipes and sheet metal and checking electrical circuits. In four or five years the new mechanics are capable of making all types of repairs and installations.

The armed forces operate their own technical schools. They range from six months to two years depending on the specialty and degree of skill needed for the *military occupational specialty* (MOS).

Civilian apprenticeship programs are run by unions and air-conditioning contractors. In addition to on-the-job training, apprentices receive 144 h of classroom instruction each year. This is in related subjects. Such subjects include the use and care of tools, safety practices, blueprint reading, and air-conditioning theory. Applicants for apprenticeships must have a high-school diploma. They are given a mechanical aptitude test. Apprenticeship programs last for four to five years.

Many high schools, private vocational schools, and junior colleges offer programs in air conditioning and

Fig. 20-2 *The air-conditioning and refrigeration mechanic must be able to test effectiveness of a component. This demands skill in using a variety of instruments.*

refrigeration. Students study air-conditioning and refrigeration theory and equipment design and construction. They also learn the basics of installation, maintenance, and repair. Employers, may prefer to hire graduates of these programs because they require less on-the-job training.

High-school graduates are preferred as helpers. If they have mechanical aptitude and have had courses in mathematics, mechanical drawing, electricity, physics, and blueprint reading, they have a better opportunity to be hired. Good physical condition is also necessary. Mechanics sometimes have to lift and move heavy equipment.

To keep up with technological change and to expand their skill, experienced mechanics may take courses, offered by a number of sources. The *Refrigeration Service Engineers Society* (RSES) and the *Air-Conditioning Contractors of America* (ACCA) offer updating courses for mechanics. There are a number of trade magazines, that help keep the mechanic up-to-gate in the latest equipment and troubleshooting procedures.

Mechanics can advance. They can become supervisors. Some open their own contracting businesses. However, it is becoming difficult for single-person operations to operate successfully.

The Future

Employment of air-conditioning and refrigeration mechanics is expected to increase. The increase is expected to be about as fast as the average of all occupations during the past 10 years. Many openings will occur as experienced mechanics transfer to other fields of work. As experienced mechanics retire or die they have to be replaced. The numbers will vary from year to year. However, in the United States, as people move toward the sunbelt there will be more jobs for air-conditioning mechanics. See Fig. 20-3.

Opportunities for air-conditioning and refrigeration mechanics are expected to follow trends in residential and commercial construction. Even during periods of slow growth, many mechanics will be needed to service existing systems. Installations of new energy-saving air-conditioning systems will also create new jobs. In addition, more refrigeration equipment will be needed in the production, storage, and marketing of food and

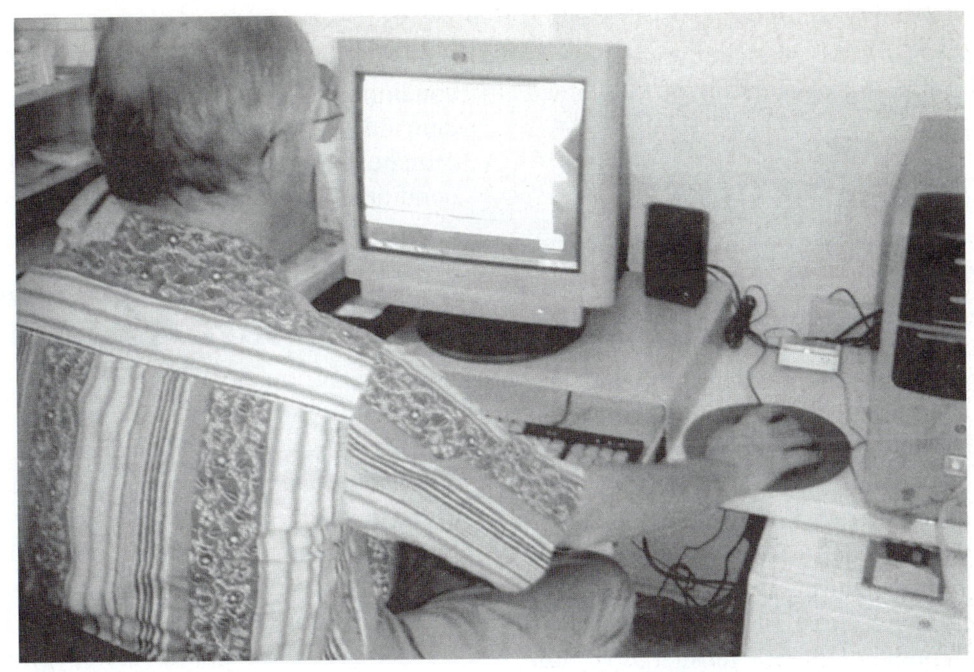

Fig. 20-3 *The temperature in data processing centers must be carefully controlled if the data-processing equipment is to operate properly. Such equipment often generates a great deal of heat. Thus, air conditioning is essential for the equipment to function properly. This operator works in an air-conditioned room.*

other perishables. Because these trades have attracted many people, the beginner mechanics may face competition for jobs as helpers or apprentices.

Pay and Benefits

Most mechanics work for hourly wages. These skilled air-conditioning and refrigeration specialists receive pay frequently higher than those who work in similar specialties.

Apprentices receive a percentage of the wage paid to the experienced mechanics. Their percentage is about 40 percent at the beginning of their training. They receive 80 percent during their fourth year.

Mechanics usually work a 40-h week. However, during seasonal peaks they often work overtime or irregular hours. Most employers try to provide a full workweek for the entire year. By doing this, they have mechanics when they need them most during the summer months. However, they may temporarily reduce hours or lay off some mechanics when seasonal peaks end. In most shops that service air-conditioning and refrigeration equipment, employment is largely stable through out the year.

Median hourly earnings of heating, air-conditioning, and refrigeration mechanics and installers were $16.78 in 2002. The middle 50 percent earned between $12.95 and $21.37 an hour. The lowest 10 percent earned less than $10.34, and the top 10 percent earned more than $26.20. Median hourly earnings in the industries employing the largest numbers of heating, air-conditioning, and refrigeration mechanics and installers in 2002 were as follows:

Hardware, and plumbing and heating equipment and supplies merchant wholesalers	$18.78
Commercial and industrial machinery and equipment (except automotive and electronic) repair and maintenance	$17.16
Direct selling establishments	$17.14
Elementary and secondary schools	$16.80
Building equipment contactors	$16.03

Apprentices usually begin at about 40 to 50 per cent of the wage rate paid to experienced workers. As they gain experience and improve their skills, they receive periodic increases until they reach the wage rate of experienced workers.

Heating, air-conditioning, and refrigeration mechanics and installers enjoy a variety of employer-sponsored benefits. In addition to typical benefits, such as health insurance and pension plans, some employers pay for work-related training and provide uniforms, company vans, and tools.

About 20 percent of heating, air-conditioning, and refrigeration mechanics and installers are members of a union. The unions to which the greatest numbers of mechanics and installers belong to are the *Sheet Metal*

Workers' International Association (SMWIA) and the United Association of Journeymen and Apprentices of the Plumbing and Pipe-fitting Industry of the United States and Canada.

TEACHING AS A CAREER

A person interested in passing on to others his or her knowledge of air conditioning might be to teach in vocational schools. The public schools also offer classes in air conditioning and refrigeration. Teachers of such courses (vocational education) often come from the trade itself. Once they have secured a position in the school, they may have to take certain college-level courses. These courses will deal with teaching methods and other subjects related to education.

Private trade schools are usually in need of good people with experience in the trade. They are needed to organize and teach apprentices. These schools may be sponsored by air-conditioning and refrigeration contractors or by unions. See Fig. 20-4.

Pay and benefits, are the same as for any other teacher in the public or the private schools. Working conditionings are similar throughout the country. The demand varies with the temperature. Therefore, the climate has much to do with the demand for air-conditioning and refrigeration specialists.

SOURCES OF ADDITIONAL INFORMATION

For more information about opportunities for training, certification, and employment in this trade, contact local vocational and technical schools; local heating, air-conditioning, and refrigeration contractors; a local of the unions or organizations previously mentioned; a local joint union-management apprenticeship committee; or the nearest office of the state employment service or apprenticeship agency.

For information on career opportunities, training, and technician certification, contact:

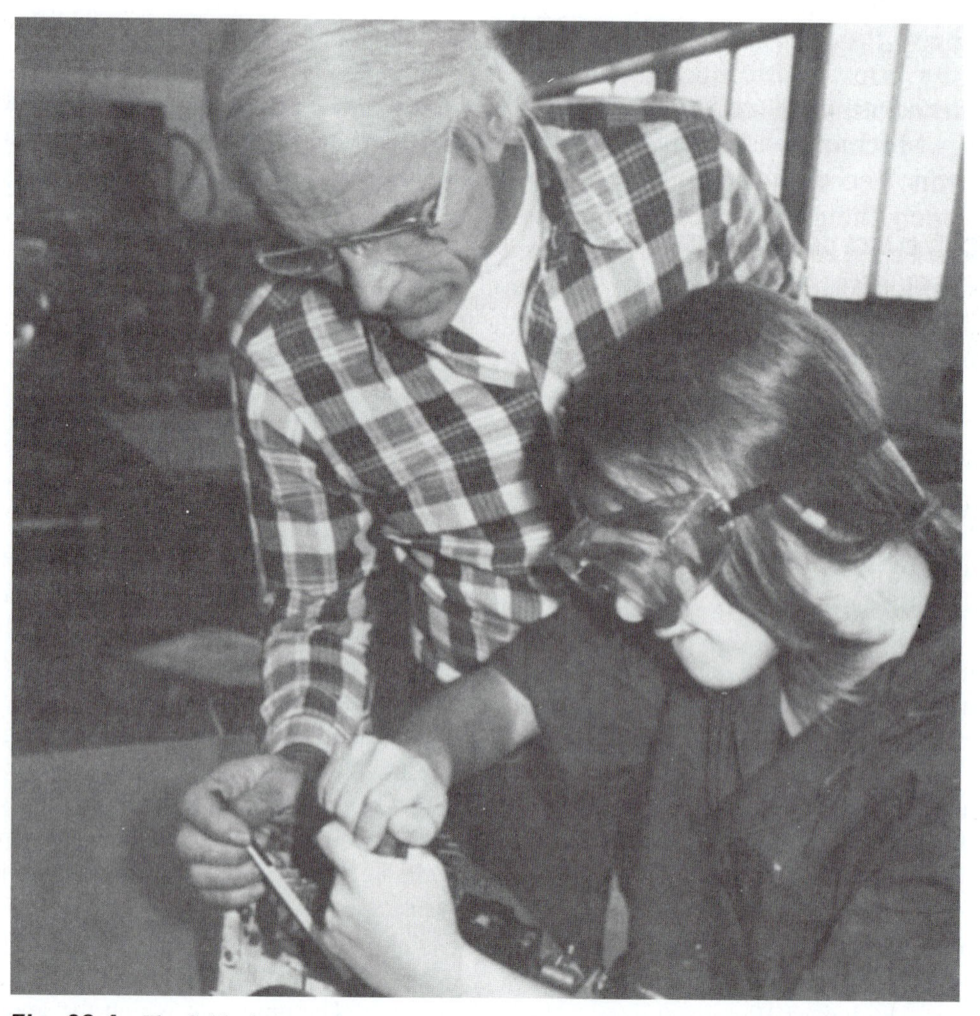

Fig. 20-4 *The field of air-conditioning and refrigeration offers many opportunities to qualified teachers. Such teachers can work in private trade schools, as well as public schools.*

Air-Conditioning Contractors of America (ACCA), 2800 Shirlington Rd., Suite 300, Arlington, VA 22206. Available at http://www.acca.org

Refrigeration Service Engineers Society (RSES), 1666 Rand Rd., Des Plaines, IL 60016-3552. Available at http://www.rses.org

Plumbing-Heating-Cooling Contractors (PHCC), 180 S. Washington St., P.O. Box 6808, Falls Church, VA 22046. Available at http://www.phccweb.org

Sheet Metal and Air-Conditioning Contractors' National Association, 4201 Lafayette Center Dr., Chantilly, VA 20151-1209. Available at http://www.smacna.org

For information on technician testing and certification, contact:

North American Technician Excellence (NATE), 4100 North Fairfax Dr., Suite 210, Arlington, VA 22203. Available at http://www.natex.org

For information on career opportunities and training, contact:

Associated Builders and Contractors, Workforce Development Department, 4250 North Fairfax Dr., 9th Floor, Arlington, VA 22203.

Home Builders Institute, 1201 15th St. NW., 6th Floor, Washington, DC 20005-2800. Available at http://www.hbi.org

Mechanical Contractors Association of America, 1385 Piccard Dr., Rockville, MD 20850-4329. Available at http://www.mcaa.org

Air-Conditioning and Refrigeration Institute, 4100 North Fairfax Dr., Suite 200, Arlington, VA 22203. Available at http://www.coolcareers.org or http://www.ari.org

There are more than 500 occupations registered by the U.S. Department of Labor's National Apprenticeship System. For more information on the Labor Department's registered apprenticeship system and links to state apprenticeship programs, check their Web site, (http://www.doleta.gov).

REVIEW QUESTIONS

1. How many hours of classroom instruction are received in an air-conditioning and refrigeration apprenticeship program?
2. What business employs most air-conditioning and refrigeration mechanics?
3. What percentage of the wage paid to experienced workers are paid to apprentices?
4. What does the future promise for the air-conditioning and refrigeration mechanic?
5. How many people work in this industry today?
6. What are some of the professional organizations that exist for service to the air-conditioning and refrigeration technician?

APPENDIX

Some New Refrigerants

Recommendations for R-12 Retrofit Products

Closest Match/Easiest (top) → **Poorest Match/Most Difficult** (bottom)

R-12 AC	R-500 AC	R-12 small equipment Higher T	R-12 small equipment Lower T	R-12 larger equipment Higher T	R-12 larger equipment Lower T
R-414B	R-409A	R-414B	R-409A	R-414B	R-409A
R-416A	R-401B	R-416A	R-401A	R-409A	R-401A
R-401A	R-401A	R-401A	R-414B	R-401A	R-414B
R-409A	R-414B	R-409A	R-416A	R-416A	R-416A
R-134a	R-134a				R-134a
	R-416A	R-134a	R-134a	R-134a	

"R-12" Refrigerants: Property Comparison

Refrigerant	Components	Composition	Glide	Lube	Pressure Match –20	10	40	90P
R-12	(pure)	100	0	M	0.6	14.6	37	100
R-134a	(pure)	100	0	P	4″v	12	35	104
R-401A	22/152a/124	53/13/34	8	MAP	1	16	42	116
R-401B	22/152a/124	61/11/28	8	AP	2	19	46	124
R-409A	22/124/142b	60/25/15	13	MAP	0	16	40	115
R-414B	22/600a/124/142b	50/1.5/39/9.5	13	MAP	1	16	41	113
R-416A	134a/600/124	59/2/39	3	P	7.5″v	8	28	88
Freezone	134a/142b	80/20	4	P	6″v	15	31	93

Note: M—Mineral oil; A—Alkyl benzene; P—Polyolester.

"R-502" Refrigerants: Property Comparison

Refrigerant	Components	Composition	Glide	Lube	Pressure Match –20	10	40	90P
R-502	22/115	49/51	0	MA	15	41	81	187
Retrofit Blends								
R-402A	125/290/22	60/2/38	2.5	M+AP	19	48	93	215
R-402B	125/290/22	38/2/60	2.5	M+AP	15	42	83	198
R-408A	125/143a/22	7/46/47	1	M+AP	14	38	77	186
HFC Blends								
R-404A	125/143a/134a	44/52/4	1.5	P	16	48	84	202
R-507	125/143a	50/50	0	P	18	46	89	210

Note: M—Mineral oil; A—Alkyl benzene; P—Polyolester.

Thermodynamic Properties of R-401A

Temp [F]	Pressure Liquid [psia]	Pressure Vapor [psia]	Density Liquid [lb/ft³]	Density Vapor [lb/ft³]	Enthalpy Liquid [Btu/lb]	Enthalpy Vapor [Btu/lb]	Entropy Liquid [Btu/R-lb]	Entropy Vapor [Btu/R-lb]
–60	6.5	4.7	88.18	0.1049	–5.371	94.93	–0.01309	0.2418
–55	7.5	5.5	87.71	0.1215	–4.035	95.60	–0.00977	0.2402
–50	8.7	6.4	87.24	0.1401	–2.694	96.26	–0.00648	0.2386
–45	9.9	7.4	86.77	0.1610	–1.350	96.93	–0.00323	0.2372
–40	11.4	8.6	86.29	0.1842	0.000	97.59	0.00000	0.2358
–35	12.9	9.9	85.82	0.2101	1.354	98.25	0.00320	0.2345
–30	14.7	11.3	85.33	0.2386	2.714	98.91	0.00637	0.2333
–25	16.6	12.9	84.85	0.2701	4.078	99.56	0.00952	0.2321
–20	18.7	14.7	84.36	0.3048	5.449	100.2	0.01265	0.2310
–15	21.0	16.6	83.86	0.3429	6.825	100.9	0.01575	0.2299
–10	23.6	18.8	83.37	0.3846	8.207	101.5	0.01882	0.2289
–5	26.4	21.2	82.86	0.4302	9.595	102.1	0.02188	0.2279
0	29.4	23.8	82.36	0.4799	10.99	102.8	0.02492	0.2269
5	32.7	26.6	81.84	0.5340	12.39	103.4	0.02793	0.2261
10	36.2	29.7	81.33	0.5927	13.80	104.0	0.03093	0.2252
15	40.1	33.1	80.80	0.6563	15.21	104.6	0.03391	0.2244
20	44.2	36.7	80.27	0.7251	16.64	105.2	0.03687	0.2236
25	48.7	40.7	79.74	0.7995	18.07	105.8	0.03982	0.2229
30	53.5	45.0	79.20	0.8798	19.51	106.4	0.04275	0.2221
35	58.6	49.6	78.65	0.9662	20.95	107.0	0.04566	0.2214
40	64.2	54.6	78.10	1.059	22.41	107.6	0.04857	0.2208
45	70.1	59.9	77.54	1.159	23.88	108.2	0.05145	0.2201
50	76.4	65.6	76.97	1.267	25.35	108.7	0.05433	0.2195
55	83.1	71.8	76.39	1.382	26.83	109.3	0.05720	0.2189
60	90.2	78.3	75.81	1.505	28.33	109.8	0.06005	0.2183
65	97.8	85.3	75.21	1.637	29.83	110.4	0.06290	0.2178
70	105.9	92.8	74.61	1.779	31.35	110.9	0.06573	0.2172
75	114.5	100.7	74.00	1.930	32.87	111.4	0.06856	0.2167
80	123.5	109.2	73.37	2.092	34.41	111.9	0.07138	0.2162
85	133.1	118.1	72.74	2.265	35.96	112.4	0.07420	0.2156
90	143.2	127.6	72.09	2.449	37.52	112.8	0.07701	0.2151
95	153.9	137.7	71.43	2.647	39.10	113.3	0.07981	0.2146
100	165.2	148.3	70.76	2.858	40.69	113.7	0.08261	0.2141
105	177.0	159.6	70.08	3.083	42.30	114.1	0.08541	0.2136
110	189.5	171.4	69.38	3.324	43.92	114.5	0.08822	0.2131
115	202.6	183.9	68.66	3.581	45.56	114.9	0.09102	0.2126
120	216.3	197.1	67.93	3.857	47.21	115.2	0.09382	0.2120
125	230.7	211.0	67.17	4.152	48.89	115.6	0.09663	0.2115
130	245.8	225.6	66.40	4.468	50.58	115.9	0.09945	0.2110
135	261.7	240.9	65.60	4.807	52.30	116.2	0.1023	0.2104
140	278.2	257.1	64.77	5.171	54.04	116.4	0.1051	0.2098
145	295.5	274.0	63.92	5.564	55.81	116.6	0.1080	0.2092
150	313.6	291.7	63.04	5.987	57.61	116.8	0.1108	0.2085
155	332.6	310.3	62.12	6.444	59.43	116.9	0.1137	0.2078

Thermodynamic Properties of R-134a

Temp [F]	Pressure [psia]	Density (L) [lb/ft³]	Density (V) [lb/ft³]	Enthalpy (L) [Btu/lb]	Enthalpy (V) [Btu/lb]	Entropy (L) [Btu/R-lb]	Entropy (V) [Btu/R-lb]
−60	4.0	90.49	0.09689	−5.957	94.13	−0.01452	0.2359
−55	4.7	90.00	0.1127	−4.476	94.89	−0.01085	0.2347
−50	5.5	89.50	0.1305	−2.989	95.65	−0.00720	0.2336
−45	6.4	89.00	0.1505	−1.498	96.41	−0.00358	0.2325
−40	7.4	88.50	0.1729	0.000	97.17	0.00000	0.2315
−35	8.6	88.00	0.1978	1.503	97.92	0.00356	0.2306
−30	9.9	87.49	0.2256	3.013	98.68	0.00708	0.2297
−25	11.3	86.98	0.2563	4.529	99.43	0.01058	0.2289
−20	12.9	86.47	0.2903	6.051	100.2	0.01406	0.2282
−15	15.3	85.95	0.3277	7.580	100.9	0.01751	0.2274
−10	16.6	85.43	0.3689	9.115	101.7	0.02093	0.2268
−5	18.8	84.90	0.4140	10.66	102.4	0.02433	0.2262
0	21.2	84.37	0.4634	12.21	103.2	0.02771	0.2256
5	23.8	83.83	0.5173	13.76	103.9	0.03107	0.2250
10	26.6	83.29	0.5761	15.33	104.6	0.03440	0.2245
15	29.7	82.74	0.6401	16.90	105.3	0.03772	0.2240
20	33.1	82.19	0.7095	18.48	106.1	0.04101	0.2236
25	36.8	81.63	0.7848	20.07	106.8	0.04429	0.2232
30	40.8	81.06	0.8663	21.67	107.5	0.04755	0.2228
35	45.1	80.49	0.9544	23.27	108.2	0.05079	0.2224
40	49.7	79.90	1.050	24.89	108.9	0.05402	0.2221
45	54.8	79.32	1.152	26.51	109.5	0.05724	0.2217
50	60.2	78.72	1.263	28.15	110.2	0.06044	0.2214
55	65.9	78.11	1.382	29.80	110.9	0.06362	0.2212
60	72.2	77.50	1.510	31.45	111.5	0.06680	0.2209
65	78.8	76.87	1.647	33.12	112.2	0.06996	0.2206
70	85.8	76.24	1.795	34.80	112.8	0.07311	0.2204
75	93.5	75.59	1.953	36.49	113.4	0.07626	0.2201
80	101.4	74.94	2.123	38.20	114.0	0.07939	0.2199
85	109.9	74.27	2.305	39.91	114.6	0.08252	0.2197
90	119.0	73.58	2.501	41.65	115.2	0.08565	0.2194
95	128.6	72.88	2.710	43.39	115.7	0.08877	0.2192
100	138.9	72.17	2.935	45.15	116.3	0.09188	0.2190
105	149.7	71.44	3.176	46.93	116.8	0.09500	0.2187
110	161.1	70.69	3.435	48.73	117.3	0.09811	0.2185
115	173.1	69.93	3.713	50.55	117.8	0.1012	0.2183
120	185.9	69.14	4.012	52.38	118.3	0.1044	0.2180
125	199.3	68.32	4.333	54.24	118.7	0.1075	0.2177
130	213.4	67.49	4.679	56.12	119.1	0.1106	0.2174
135	228.3	66.62	5.052	58.02	119.5	0.1138	0.2171
140	243.9	65.73	5.455	59.95	119.8	0.1169	0.2167
145	260.4	64.80	5.892	61.92	120.1	0.1201	0.2163
150	277.6	63.83	6.366	63.91	120.4	0.1233	0.2159
155	295.7	62.82	6.882	65.94	120.6	0.1265	0.2154
160	314.7	61.76	7.447	68.00	120.7	0.1298	0.2149

Physical Properties of Refrigerants	R-134a
Environmental classification	HFC
Molecular weight	102.3
Boiling point (1 atm, F)	–14.9
Critical pressure (psia)	588.3
Critical temperature (F)	213.8
Critical density (lb./ft^3)	32.0
Liquid density (70 F, lb./ft^3)	76.2
Vapor density (bp, lb./ft^3)	0.328
Heat of vaporization (bp, BTU/lb.)	93.3
Specific heat liquid (70 F, BTU/lb. F)	0.3366
Specific heat vapor (1 atm, 70 F, BTU/lb. F)	0.2021
Ozone depletion potential (CFC 11 = 1.0)	0
Global warming potential (CO_2 = 1.0)	1320
ASHRAE Standard 34 safety rating	A1

Available in the following sizes:

R-134a

012R134a 12-oz cans
30R134a 30-lb cylinder
A30R134a 30-lb auto AC
50R134a 50-lb cylinder
125R134a 125-lb cylinder*
1000R134a ½-ton cylinder*
2000R134a ton cylinder*

* Deposit Required

Pressure-Temp Chart

Temp (F)	R-134a (psig)
–40	14.8
–35	12.5
–30	9.9
–25	6.9
–20	3.7
–15	0.6
–10	1.9
–5	4.0
0	6.5
5	9.1
10	11.9
15	15.0
20	18.4
25	22.1
30	26.1
35	30.4
40	35.0
45	40.1
50	45.5
55	51.3
60	57.5
65	64.1
70	71.2
75	78.8
80	86.8
85	95.4
90	104
95	114
100	124
105	135
110	147
115	159
120	171
125	185
130	199
135	214
140	229
145	246
150	263

Physical Properties of Refrigerants	R-401A	R-401B
Environmental classification	HCFC	HCFC
Molecular weight	94.4	92.8
Boiling point (1 atm, F)	–29.9	–32.3
Critical pressure (psia)	669	679.1
Critical temperature (F)	221	218.3
Critical density (lb./ft^3)	30.9	31.1
Liquid density (70 F, lb./ft^3)	74.6	74.6
Vapor density (bp, lb./ft^3)	0.306	0.303
Heat of vaporization (bp, BTU/lb.)	97.5	98.2
Specific heat liquid (70 F, BTU/lb. F)	0.3037	0.3027
Specific heat vapor (1 atm, 70 F, BTU/lb. F)	0.1755	0.1725
Ozone depletion potential (CFC 11 = 1.0)	0.037	0.039
Global warming potential (CO_2 = 1.0)	1163	1267
ASHRAE Standard 34 safety rating	A1	A1
Temperature glide (F) (see section II)	8	8

Available in the following sizes:

R-401A
30R401A 30-lb cylinder
125R401A 125-lb cylinder*
1700R401A 1-ton cylinder*

R-401B
30R401B 30-lb cylinder
125R401B 125-lb cylinder*

* Deposit Required

Pressure-Temp Chart

Temp (F)	R-401A Liquid (psig)	R-401A Vapor (psig)	R-401B Liquid (psig)	R-401B Vapor (psig)
–40	8.1	13.2	6.5	11.8
–35	5.1	10.7	3.3	9.1
–30	1.7	7.9	0.2	6.1
–25	1.0	4.8	2.1	2.8
–20	3.0	1.4	4.3	0.5
–15	5.2	1.2	6.6	2.5
–10	7.7	3.3	9.2	4.7
–5	10.3	5.5	12.0	7.1
0	13.2	8.0	15.1	9.7
5	16.3	10.7	18.4	12.6
10	19.7	13.7	22.0	15.8
15	23.4	16.9	25.9	19.2
20	27.4	20.4	30.1	23.0
25	31.7	24.2	34.6	27.0
30	36.4	28.3	39.5	31.4
35	41.3	32.8	44.8	36.1
40	46.6	37.6	50.4	41.1
45	52.4	42.7	56.4	46.6
50	58.5	48.2	62.8	52.4
55	65.0	54.1	69.6	58.7
60	71.9	60.4	76.9	65.4
65	79.3	67.2	84.7	72.5
70	87.1	74.4	92.9	80.1
75	95.4	82.1	102	88.2
80	104	90.2	111	96.8
85	114	98.9	121	106
90	123	108	131	116
95	134	118	142	126
100	145	128	153	137
105	156	139	166	148
110	169	151	178	160
115	181	163	192	173
120	195	176	206	187
125	209	189	220	201
130	224	203	236	216
135	239	218	252	231
140	255	234	269	248
145	272	250	287	265
150	290	267	305	283

Physical Properties of Refrigerants	R-402A	R-402B
Environmental classification	HCFC	HCFC
Molecular weight	101.6	94.7
Boiling point (1 atm, F)	–56.5	–52.9
Critical pressure (psia)	600	645
Critical temperature (F)	168	180.7
Critical density (lb./ft^3)	33.8	33.1
Liquid density (70 F, lb./ft^3)	72.61	72.81
Vapor density (bp, lb./ft^3)	0.356	0.328
Heat of vaporization (bp, BTU/lb.)	83.58	90.42
Specific heat liquid (70 F, BTU/lb. F)	0.3254	0.317
Specific heat vapor (1 atm, 70 F, BTU/lb. F)	0.1811	0.1741
Ozone depletion potential (CFC 11 = 1.0)	0.019	0.03
Global warming potential (CO_2 = 1.0)	2746	2379
ASHRAE Standard 34 safety rating	A1	A1
Temperature Glide (see section II)	2.5	2.5

Available in the following sizes:

R-402A
27R402A 27-lb cylinder
110R402A 110-lb cylinder*

R-402B
13R402B 13-lb cylinder

* Deposit Required

Pressure-Temp Chart

Temp (F)	R402A (psig)	R402B (psig)
–40	6.3	3.6
–35	9.1	6.0
–30	12.1	9.0
–25	15.4	12.0
–20	18.9	15.4
–15	22.9	18.6
–10	27.1	22.6
–5	31.7	27.0
0	36.7	31.0
5	42.1	36.0
10	48.0	42.0
15	54.2	47.0
20	60.9	54.0
25	68.1	60.0
30	75.8	67.0
35	84.0	75.0
40	92.8	83.4
45	102	91.6
50	112	100
55	123	110
60	134	120
65	146	133
70	158	143
75	171	155
80	185	170
85	200	183
90	215	198
95	232	213
100	249	230
105	267	247
110	286	262
115	305	283
120	326	303
125	347	323
130	370	345
135	393	—
140	418	—
145	443	—
150	470	—

APPENDIX

Electrical and Electronic Symbols Used in Schematics

Table B-1 Useful Mechanical, Electrical, and Heat Equivalents

Unit	Equivalent Value in Other Units	
1 kWh	1000	watt-hours
	1.34	horsepower-hours
	2,654,200	pound-feet
	3,600,000	joules
	3412	heat units
	367,000	kilogram-meters
	0.235	pound carbon oxidized with perfect efficiency
	3.53	pounds water evaporated from and at 212°F
	22.75	pounds water raised from 62 to 212°F
1 hp-h	0.746	kilowatt-hour
	1,980,000	pound-feet
	2545	heat-units
	273,740	kilogram-meters
	0.175	pound carbon oxidized with perfect efficiency
	2.64	pounds water evaporated from and at 212°F
	17.0	pounds water raised from 62 to 212°F
1 kW	1000	watts
	1.34	horsepower
	2,654,200	pound-feet per hour
	44,240	pound-feet per minute
	737.3	pound-feet per second
	3412	heat units per hour
	56.9	heat units per minute
	0.948	heat unit per second
	0.2275	pound carbon oxidized per hour
	3.53	pounds water evaporated per hour from and at 212°F
1 hp	746	watts
	0.746	kilowatt
	33,000	pound-feet per minute
	550	pound-feet per second
	2545	heat units per hour
	42.4	heat units per minute
	0.707	heat unit per second
	0.175	pound carbon oxidized per hour
	2.64	pounds water evaporated per hour from and at 212°F

(Calcium Chloride Institute)

Diagrams are more useful if you know what the symbols mean. The schematic diagram of an electrical circuit aids in being able to troubleshoot. They are also useful in making it possible to understand what happens in a given arrangement of symbols.

These symbols are part of ARI Standard 130-88. ARI is the abbreviation for the *Air Conditioning and Refrigeration Institute.*

In some instances notes are added near the symbol for a special purpose. For instance, if IEC shows up near the symbol it means the symbol has been recommended by the *International Electro-technical Commission.* The following symbols are not necessarily in alphabetical order.

FUNDAMENTAL ITEMS

Resistor

Resistor, general

Tapped Resistor

Variable resistor: with adjustable contact

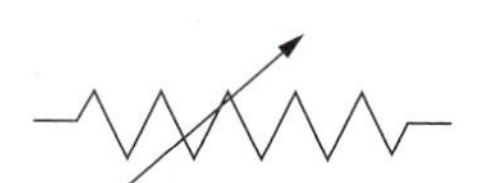

Thermistor

General

t*

NOTE: The asterisk is not part of the symbol. It means to indicate an appropriate value.

With Independent Integral Heater

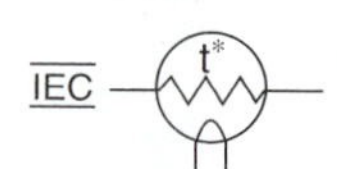

Capacitor

General If it is necessary to identify the capacitor electrodes, the curved element shall represent the outside electrode.

IEC

Variable

Battery The long line is always positive, but polarity may be indicated in addition.

−
+

Multicell

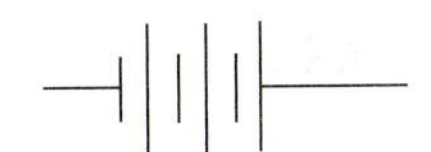

Temperature-Measuring Thermocouple

IEC

Thermopile for pumping heat

COLD

HOT

Spark Gap, Igniter Gap

Transmission Path

Transmission Path: Conductors, Cables, Wiring ··· Factory Wired

Field Installed or Sales Option, if specified

Crossing of Paths or Conductors Not Connected
The crossing is not necessarily at a 90° angle.

IEC

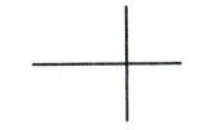

Junction of Paths or Conductors Junction (Connection)

IEC

Application: Junction of Connected Paths, Conductors, or Wires

IEC

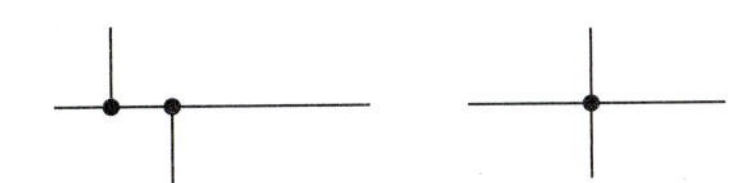

Terminal block

*Terminal number

Assembled Conductors: Cable. Shielded Single Conductor

Shielded Cable (5-Conductor Shown)

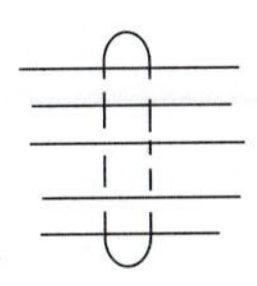

Shielded Cable With Shield Grounded (2-Conductor Shown)

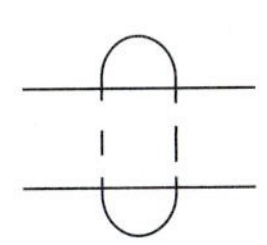

Cable (2-Conductor Shown)

IEC

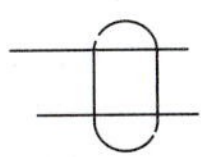

Ribbon Cable

Twisted Cable (pair, triple, and the like)

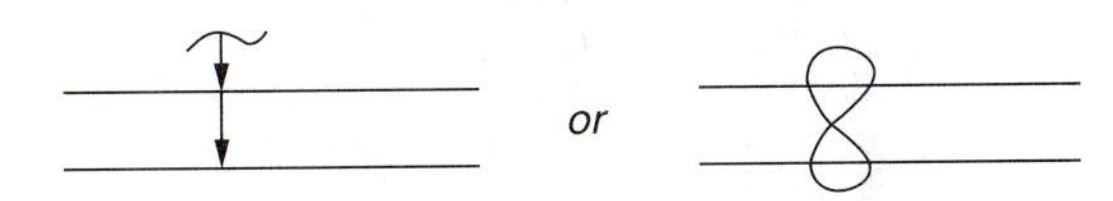

Circuit Return

Ground (1) A direct conducting connection to the earth or body of water that is a part thereof. (2) A conducting connection to a structure (chassis) that serves a function similar to that of an earth ground (that is, a structure such as a frame of an air, space, or, land vehicle that is not conductively connected to earth).

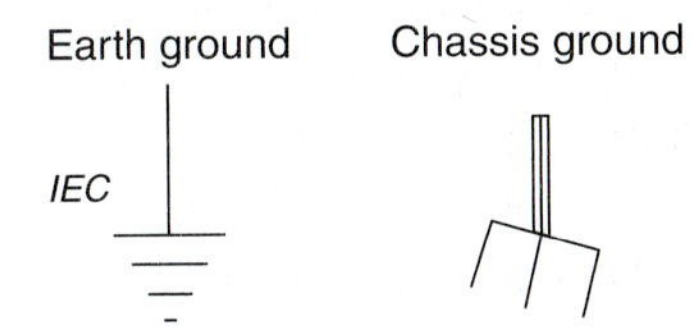

Normally Closed Contact (break)

Normally Open Contact (make)

Operating Coil (Relay coil)

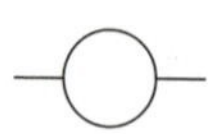

Solenoid Coil

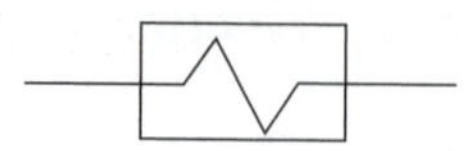

Switch Fundamental symbols for contacts, mechanical connections, and so forth, may be used for switch symbols.

The standard method of showing switches is in a position with no operating force applied. For switches that may be in any one of two or more positions with no operating force applied and for switches actuated by some mechanical device (as in air-pressure, liquid-level, rate-of flow, and so forth, switches), a clarifying note may be necessary to explain the point at which the switch functions.

Pushbutton, Momentary or Spring-Return

Normally Open, Circuit Closing (make)

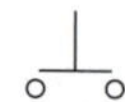

Normally closed, Circuit Opening (break)

Two-Circuit (dual)

Two Circuit, Maintained or Not Spring-Return

Maintained (Locking) Switch

Toggle Switch Single Throw

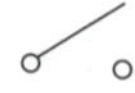

Application: 3 Disconnect Switch

Transfer, 2-Position—Double Throw

IEC

Transfer 3-Position

off IEC

Transfer, 2-Position

IEC

Transfer 3-Position

Selector or Multiposition Switch The position in which the switch is shown may be indicated by a note or designation of switch position.

General (for Power or Control Diagrams) Any number of transmission paths may be shown.

Segmental Contact

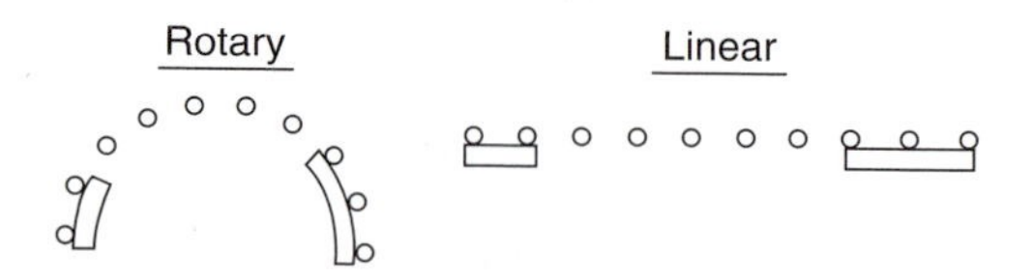

Slide

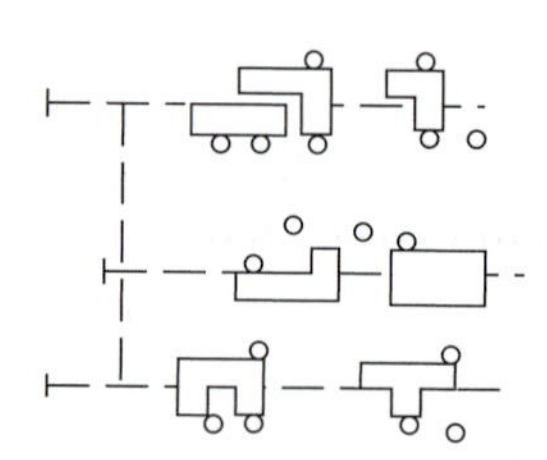

Master or Control

Detached contacts shown elsewhere on diagram

Contact	Indicator position		
	A	B	C
1–2			X
3–4	X		
5–6			X
7–8	X		

X-indicates contacts closed

Limit Switch, Directly Actuated, Spring Returned

Normally Open

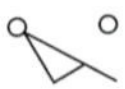

Normally Open—Held Closed

Normally Open Switch with Time-Delay Closing (NOTC)

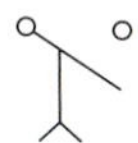

Normally Closed Switch With Time-Delay Opening (NCTO)

With Single Heater (Single Phase)

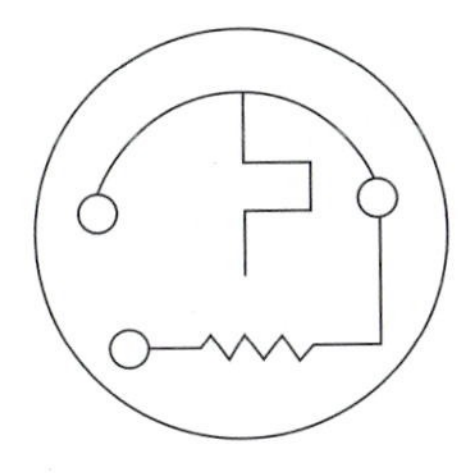

With Heaters (Three Phase)

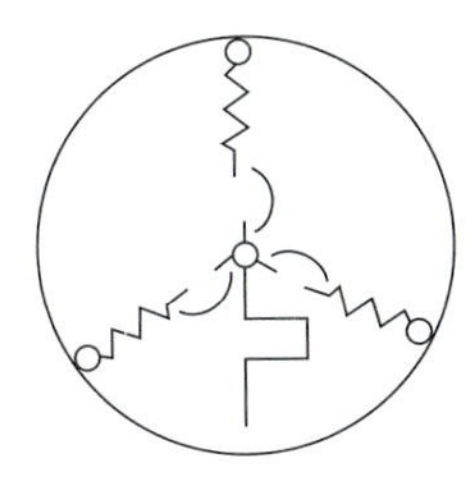

Humidity Actuated Switch

Closes on Rising Humidity

Opens on Rising Humidity

CONNECTORS

Connector, Disconnecting Device

The connector symbol is not an arrowhead. It is larger and the lines are drawn at a 90° angle.

Female Contact

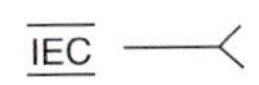

Male Contact

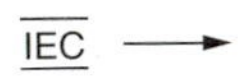

Separable Connectors (engaged)

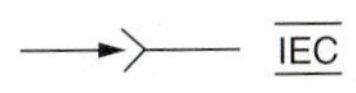

Application: Engaged 4-Conductor

Connectors The Plug has one male and three female contacts.

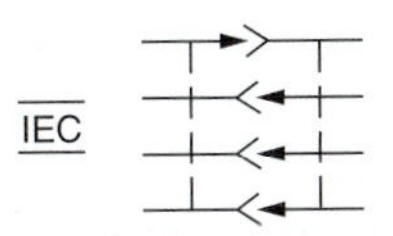

TRANSFORMERS, INDUCTORS, WINDINGS

Transformer

Current Transformer

Magnetic Core Transformer (nonsaturating)

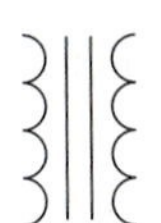

With Taps—Single Phase

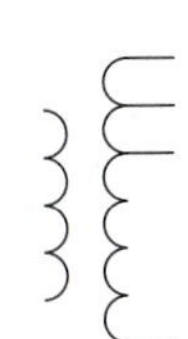

Autotransformer, Single Phase

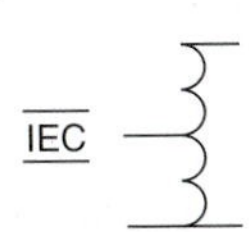

Adjustable Autotransformer

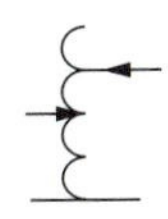

SEMICONDUCTOR DEVICES

Semiconductor Device, Transistor, Diode

In general, the angle at which a lead is brought to a symbol element has no significance. IEC

Orientation, including a mirror-image presentation, does not change the meaning of a symbol. IEC

The elements of the symbol must be drawn in such an order as to show clearly the operating function of the device. IEC

Element Symbols

Rectifying Junction or Junction Which Influence a Depletion Layer Arrowheads (—▸|—) shall be half the length of the arrow away from the semiconductor base region. IEC

The equilateral (—▸|—) triangle shall be filled and shall touch the semiconductor base-region symbol IEC

The triangle points in the direction of the forward (easy) current as indicated by a direct current ammeter, unless otherwise noted adjacent to the symbol. Electron flow is in the opposite direction.

Special Property Indicators

If necessary, a special function or property essential for circuit operation shall be indicated by a supplementary symbol included as part of the symbol.

Typical Applications: Two-Terminal Devices

Semiconductor Diode: Semiconductor Rectifier diode

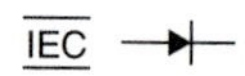

Breakdown Diode: Overvoltage Absorber

Unidirectional Diode; Voltage Regulator; Zener Diode

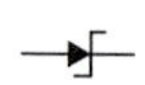

Bidirectional Diode

IEC

Unidirectional Negative-Resistance Breakdown Diode; Trigger Diac

NPN-type

IEC

PNP-type

IEC

Bidirectional Negative-Resistance Breakdown Diode; Trigger Diac

NPN-type

PNP-type

Photodiode

Photosensitive Type

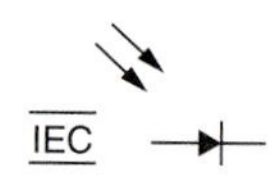

Photoemissive Type

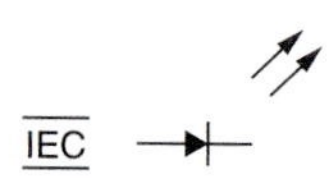

Phototransistor (NPN-type) (without external base-region connection)

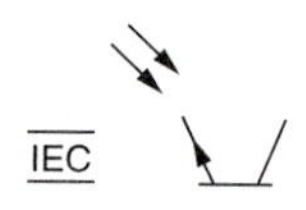

Typical Applications: Three (or more) Terminal Devices

PNP Transistor (also PNIP transistor, if omitting the intrinsic region will not result in ambiguity)

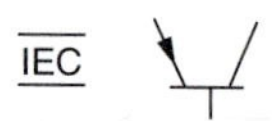

Application: PNP transistor with One Electrode Connected to Envelope

NPN Transistor (also NPIN transistor, if omitting the intrinsic region will not result in ambiguity)

Unijunction Transistor with N-Type Base

UnijunctionTtransistor with P-Type Base

Field-Effect Transistor with N-Channel Junction Gate

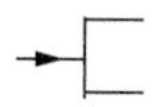

Field-Effect Transistor with P-Channel Junction Gate

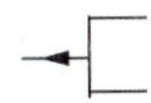

Thyristor, Reverse-Blocking Triode-Type, N-type Gate; Semiconductor-Controlled Rectifier, N-Type Gate

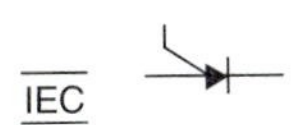

Thyristor, Reverse-Blocking Triode-, P-Type Gate; Semiconductor-Controlled Rectifier, P-Type Gate

Thyristor, Reverse-Blocking Tetrode-Type; Semiconductor-Controlled Switch

Thyristor, Bidirectional Triode-type; Triac; Gated Switch

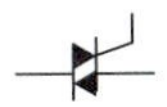

Phototransistor (PNF-Type)

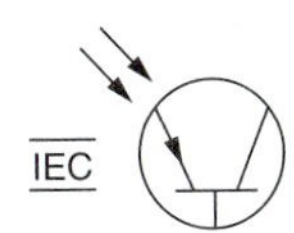

Photon-Coupled Isolator

NOTE: T is the transmitter; R is the receiver. The letters are for explanation and are not part of the symbol. Explanatory information should be added to explain circuit operation.

General

Complete Isolator (single-package type)

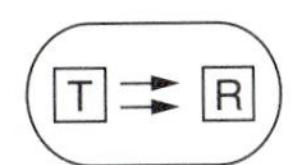

Application: Incandescent Lamp and Symmetrical Photoconductive Transducer

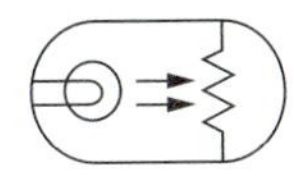

Application: Photoemissive Diode and Phototransistor

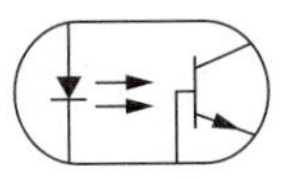

Field-Effect Transistor with N-Channel MOS Gate

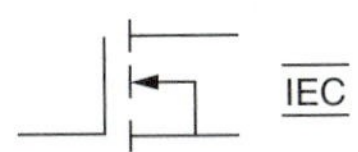

Field-Effect Transistor with P-Channel MOS Gate

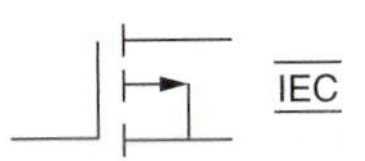

Thyristor, Gate Turn-Off Type

CIRCUIT PROTECTORS

Fuse

General

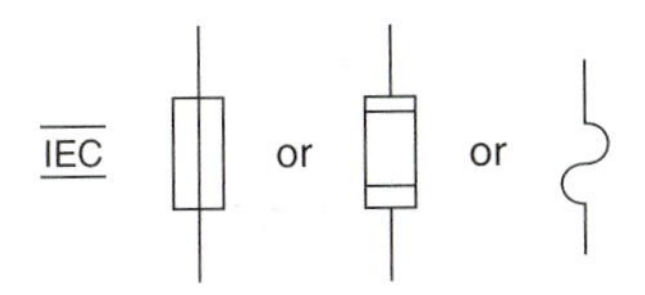

Circuit Breaker

General

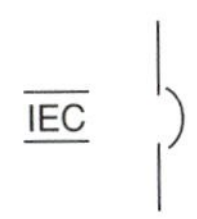

Application: Three-Pole Circuit Breaker with Thermal-Overload Device in all Three Poles

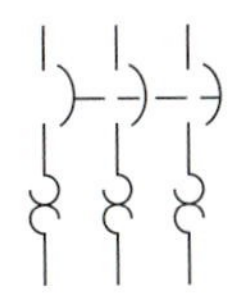

Application: Three-Pole Circuit Breaker with Magnetic-Overload Device in all Three Poles

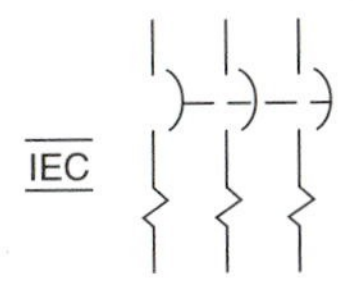

ACOUSTIC DEVICES

Audible-Signaling Device

Bell, Electrical

If specific identification is required, the abbreviation AC or DC may be added within the square.

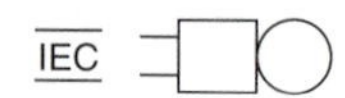

Horn, Electrical

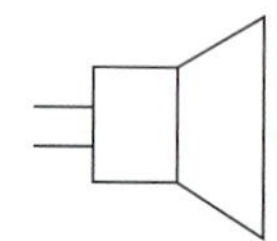

LAMPS AND VISUAL SIGNALING DEVICES

Indicating, Pilot, Signaling, or Switchboard Light

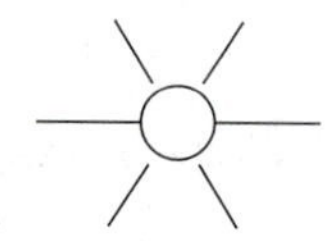

To indicate the characteristic, insert the specified letter or letters inside the symbol.

A	Amber
B	Blue
C	Clear
F	Fluorescent
G	Green
NE	Neon
O	Orange
OP	Opalescent
P	Purple
R	Red
W	White
Y	Yellow

Application: Green Signal Light

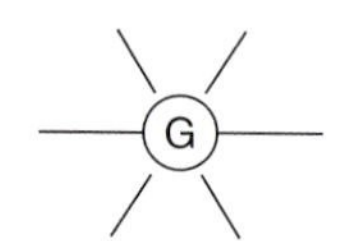

ROTATING MACHINERY

Rotating Machine

Generator (General)

IEC (G) Single phase

(G) Three phase

Motor (General)

IEC (M) Single phase

(M) Three phase

Application: Alternating-Current Motors

Two Lead Type

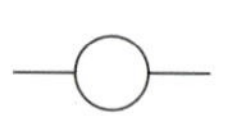

External Capacitor Type

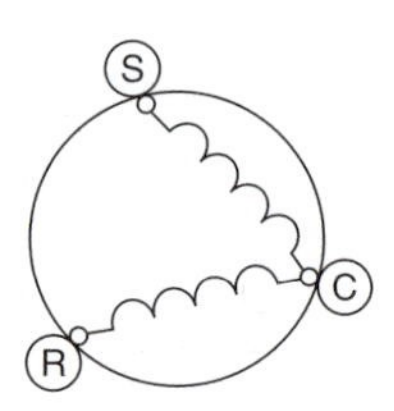

Polyphase Type

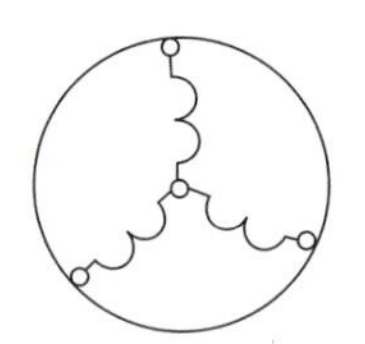

Application: Single Phase with Internal Line Break Protector

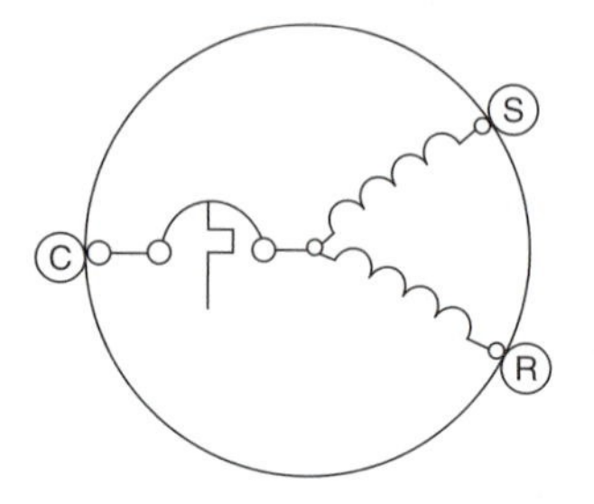

Application: Three-Phase with Internal Line Break Protector

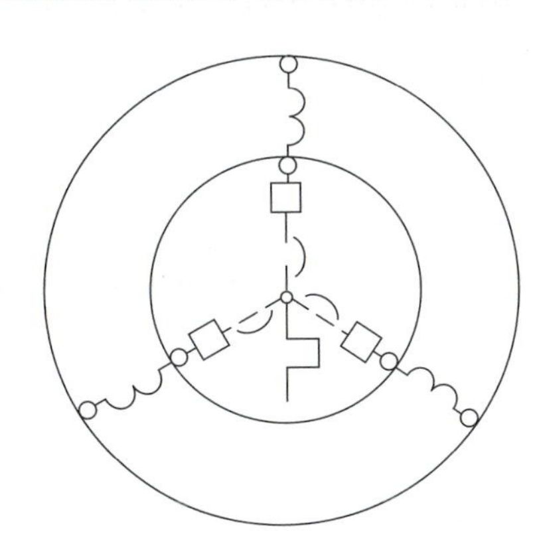

Overloads (Current)

Thermal

Service Trip

Remote Trip

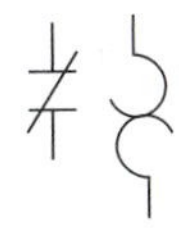

Magnetic

Series Trip

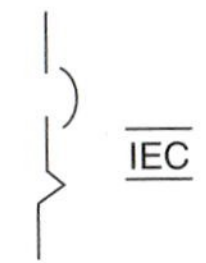

Remote Trip

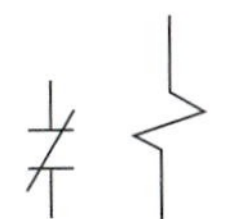

Overload Coils

Thermal

Magnetic

Application: Bimetallic (Thermal)

No Heater

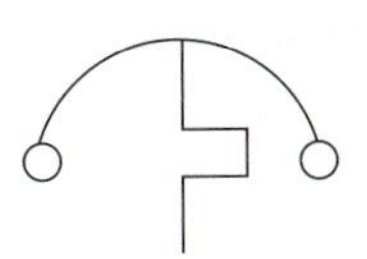

APPENDIX

Programming Thermostats*

NOTE: A variety of thermostats are available; those shown here are from one manufacturer only. See Internet for more complete coverage and homeowner's guides.

*All the figures are taken from http://www.totaline.com.

1
2
3
4
5
6
7
8
71

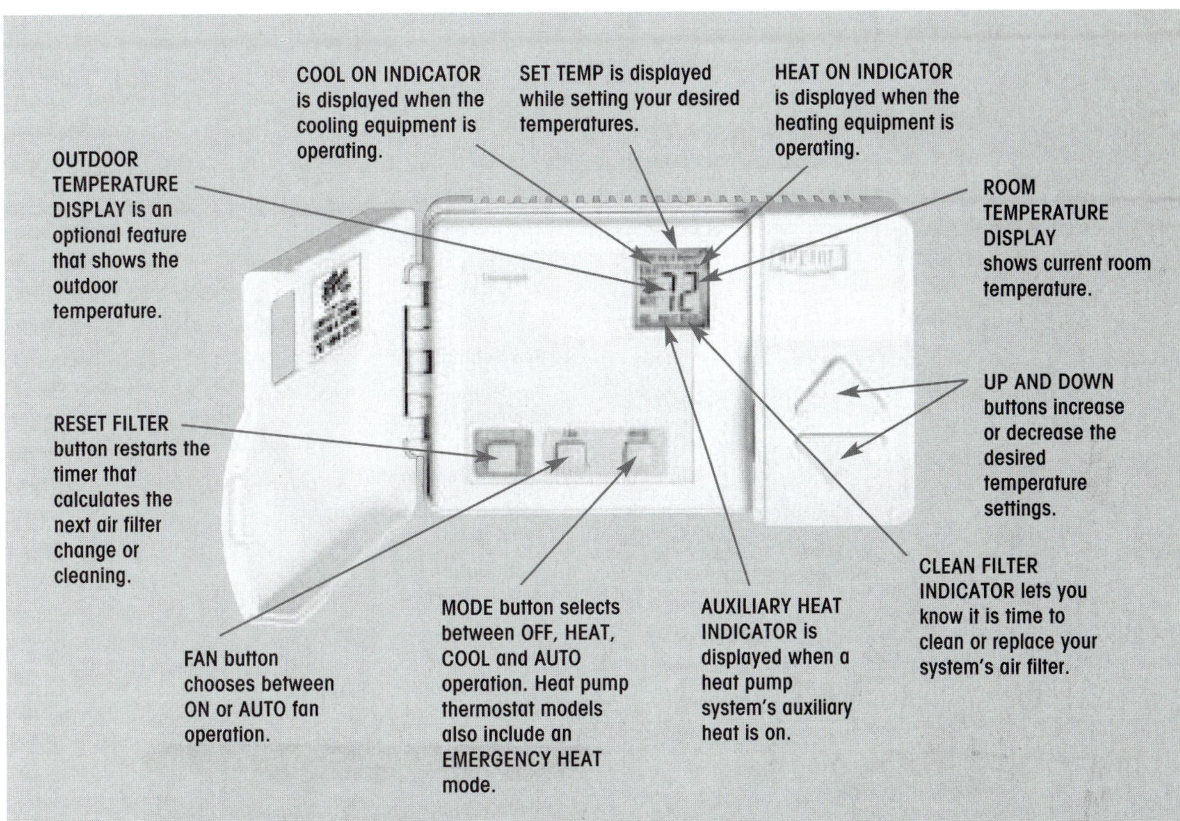

NOTE: Not all messages displayed in above illustration will appear at once in any situation.

Comfort At Your Command.

This is no ordinary thermostat. Bryant listened to the needs of homeowners nationwide and delivered a product to meet those needs. The result is a thermostat that interacts with people as effectively as it does with your heating and cooling system. It's a simple, yet powerful control that puts comfort at your command.

Making Life Easier.

Take a few minutes to review the features and functions listed above. Bryant gives you control over your comfort with simple instructions, responsive push buttons and an easy-to-read backlit display. Once set, this thermostat reliably monitors indoor temperatures and responsively meets your comfort demands.

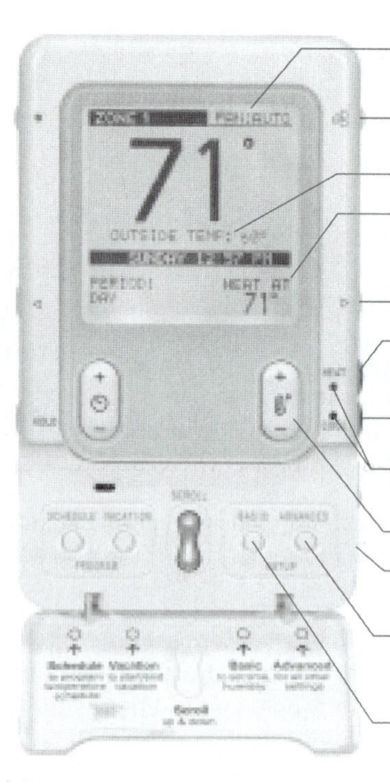

Fan Setting

Fan Button chooses “high,” “medium,” “low,” or “auto” fan mode.

Outside Temperature

Desired Heating/Cooling Temperature

Right Button provides system status.

Heat Button selects heating operation.

Cool Button selects cooling operation.

Heat/Cool LEDs indicate heating or cooling operation.

Temp (+\-) Button

Off Button turns the system on and off.

Advanced Setup Button provides access to customizable features.

Basic Setup Button provides access to current day, time and desired humidity level.

HOW TO DETERMINE A MODEL NUMBER FOR TOTALINE, CARRIER, AND BRYANT THERMOSTATS

Gold Series

All of these thermostats have a door with Up and Down buttons on the right hand side.

The **Standard Residential** stats have the following buttons from left to right:
Reset Filter – Fan - Mode
There are 3 different models in this group and the same Owners Manual covers all 3.

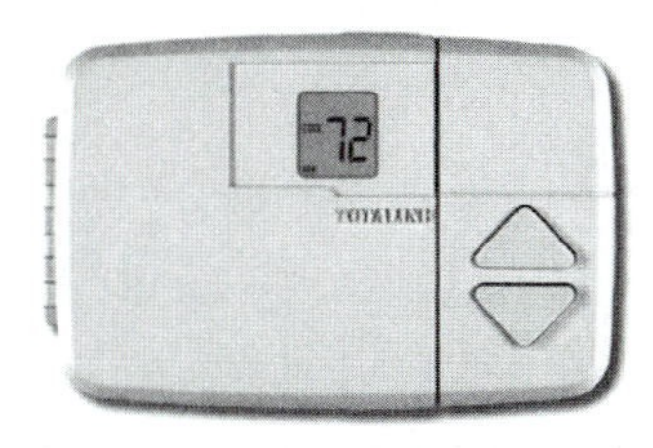

Old Version

Totaline: P274-0100, 0200, 0300
Carrier: TSTATCCNAC01-B, NHP01-B, N2S01-B
Bryant: TSTATBBNAC01-B, NHP01-B, N2S01-B

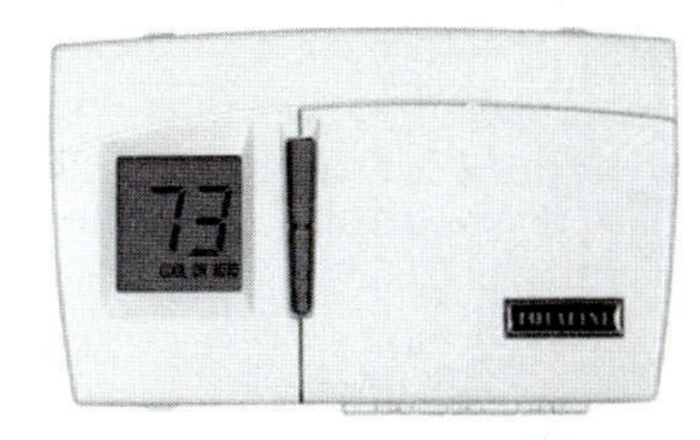

New Version As Of 11/1/04

Totaline: P274-0100-C, 0200-C, 0300-C
Carrier: TSTATCCNAC01-C, NHP01-C, N2S01-C
Bryant: TSTATBBNAC01-C, NHP01-C, N2S01-C

The **Programmable Residential** stats have the following buttons from left to right:
Top Row: **Copy Previous Day – Program – Mode**
Middle row: **Change Day – End – Fan**
Bottom Row: **Set Time/Temp – Reset Filter - Hold**
There are 3 different models in this group and the same Owners Manual covers all 3.

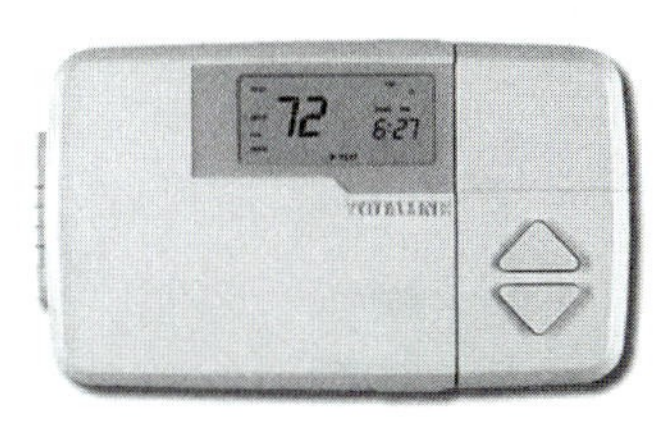

Totaline: P274-1100, 1200, 1300
Carrier: TSTATCCPAC01-B, PHP01-B, P2S01-B
Bryant: TSTATBBPAC01-B, PHP01-B, P2S01-B

In addition to these three, Carrier and Bryant also have 2 other models.
The Duel Fuel model has the same appearance and buttons as the others, so you would have the customer pull the stat apart in order to read the model # on the back of the circuit board. These stats open like a door from left to right.
Carrier: TSTATCCPDF01-B
Bryant: TSTATBBPDF01-B

The Thermidistat also has the same appearance, but the following buttons:
Top Row: **Copy Previous Day – Program – Mode**
Middle row: **Change Day – Humidity – Fan**
Bottom Row: **Set Time/Temp – Vacation – Hold/End**

Carrier: TSTATCCPDF01-B
Bryant: TSTATBBPDF01-B

The **Commercial** stat has the same appearance as the Residential stats, but has slightly different buttons.
Top Row: **Copy Previous Day – Program – Mode**
Middle row: **Change Day – End – Fan**
Bottom Row: **Set Time/Temp – Occupied <Reset Filter>Hold**

Signature Series

The **Standard Residential** stats do not have a door and contain an Up and Down button on the right side and two buttons that read **Mode** and **Fan** under the display.
There are 2 Carrier/Bryant models and 4 Totaline models. One manual is used for all models.

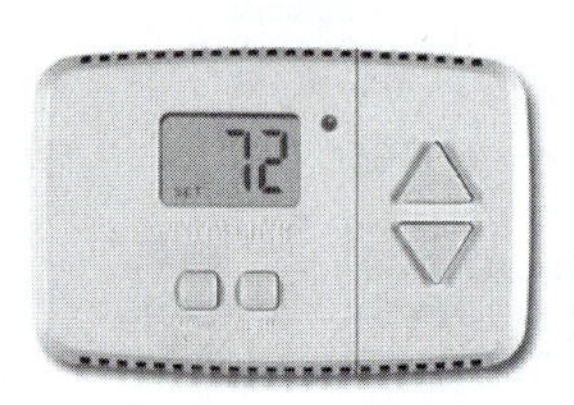

Totaline: P374-0000, 0100, 0200, 0300
Carrier: TSTATCCBAC01-B, TSTATCCBHP01-B
Bryant: TSTATBBBAC01-B, TSTATBBBHP01-B

The **Programmable Residential** stats have 3 different styles.
The first style has a door with 3 buttons on the outside. There are 4 different versions.

Version #1: **Fan – Emergency Heat – Backlight – Program – Set Clock – Mode**
These functions refer to the following models:
Totaline: P374-1000
Carrier: TSTATCCPS101
Bryant: TSTATBBPS101

Version #2: **Fan – Outside – Vacation – Program – Set Clock**
These functions refer to the following models:
Totaline: P374-1100
Carrier: TSTATCCPS701
Bryant: TSTATBBPS701

Version #3: **Fan – Outside – Vacation – Program – Set Clock**
While these functions are the same as Version #2, check to see if the customer has a duel fuel system … ie. A heat pump and a gas furnace.
Totaline Only: P374-1500

Version #4: "INTELLISTAT" **Fan – Outside – Humidity – Program – Set Clock - Mode**
Totaline Only: P374-1600

The second style is a flushmount or "flatstat". It has 4 buttons under the display, plus a raised round bubble where the sensor is enclosed. There are 2 different versions that both have the same buttons.
Mode – Fan – Up - Down

1-Day or Non-Programmable:
Totaline: P374-1000FM
Carrier: TSTATCCPF101
Bryant: TSTATBBPF101

7-Day Programmable:
Totaline: P374-1100FM
Carrier: TSTATCCPF701
Bryant: TSTATBBPF701

The third style has the Up/Down buttons on the right and 6 buttons under the display. There are two versions but the manual is the same for both. These are not available in Totaline branding.
Top Row: **Mode – Fan**
Bottom Row: **Program – Time/Temp – Day – Hold/End**

Carrier: TSTATCCSAC01 and TSTATCCSHP01
Bryant: TSTATBBSAC01 and TSTATBBSHP01

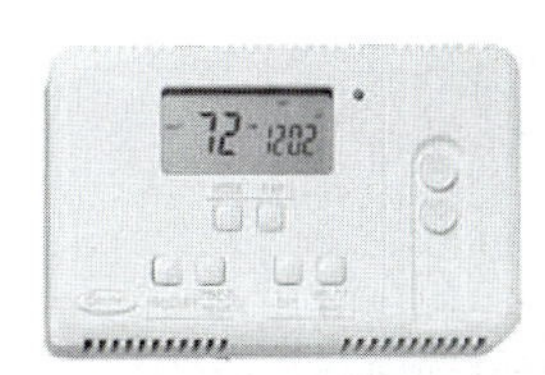

The **Commercial** stats have 2 different styles and several different versions.
The first style has a door with 3 buttons on the outside. There are 4 different versions.

Version #1: **Mode – Fan – Emerg. Heat – Backlight – Reset Filter**
Totaline: P374-2100
Carrier: 33CS071-01
Bryant: TSTATBB071-01

Version #2: **Mode – Fan – Holiday – Program – Set Clock**
Totaline: P374-2200 or P374-2200LA (Light Activated)
Carrier: 33CS220-01 or 33CS220-LA
Bryant: TSTATBB220-01 or TSTATBB220-LA

Version #3: **Mode – Fan – Holiday – Program – Set Clock**
While these buttons are the same as Version #2, there is a difference, please call.
Totaline: P374-2300 or P374-2300LA (Light Activated)
Carrier: 33CS250-01 or 33CS250-LA
Bryant: TSTATBB250-01 or TSTATBB250-LA

The second style is a flushmount or "flatstat". It has 4 buttons under the display, plus a raised round bubble where the sensor is enclosed. There are 2 different versions that both have the same buttons. This is not available in the Bryant branding.
Mode – Override – Up - Down

Standard:
Totaline: P374-2200FM
Carrier: 33CS220-FS

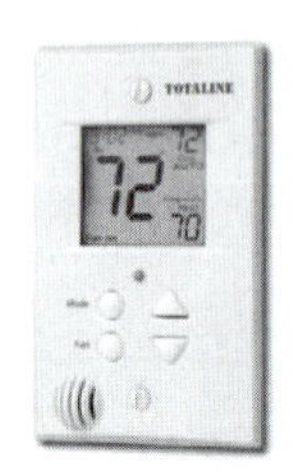

Delux:
Totaline: P374-2300FM
Carrier: 33CS250-FS

Star Series

The **Standard Residential** stats have 2 different styles.
The first style is a battery operated stat that has the Up/Down buttons plus slide switches on the bottom – and the side if a Heat Pump model (P474-0140).

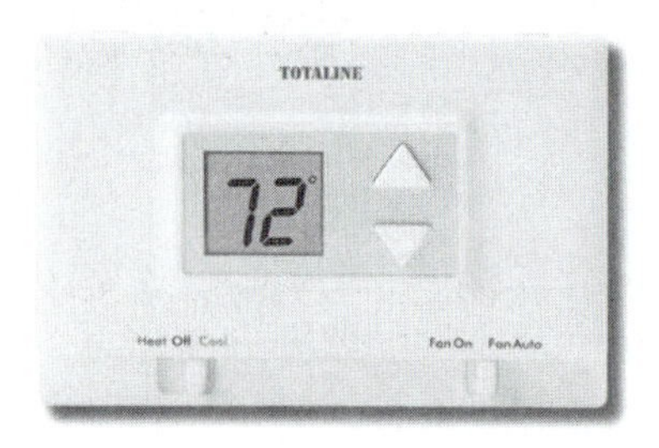

Totaline: P474-0130
Carrier: TSTATCCNQ001
Bryant: TSTATBBNQ001

Totaline Only: P474-0140

The second style has two buttons on each side of the display. There are two different versions, but they both have the same buttons. Have the consumer pull the stat off of the backplate. The part # is located on the circuit board.
Left side: **Mode – Fan** Right side: **Up - Down**

Totaline: P474-0100 or P474-0220
Carrier: TSTATCCNB001 or TSTATCCN2W01
Bryant: TSTATBBNB001 or TSTATBBN2W01

The **Programmable Residential** stats have 4 different styles.
The first style has two buttons on each side of the display. It looks exactly like the above picture, but is 1-day programmable. Have the customer pull the body of the stat off of the back plate. The part # is located on the circuit board.

Totaline: P474-1010
Carrier: TSTATCCBP101
Bryant: TSTATBBPB101

The second style is a battery operated stat that has the Up/Down buttons plus slide switches on the bottom and the side.

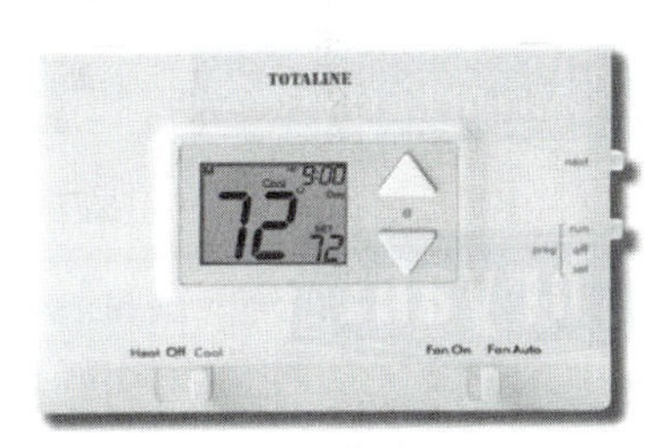

Totaline: P474-1035
Carrier: TSTATCCPQ501
Bryant: TSTATBBPQ501

The third style has 4 buttons all located under the display. There are two different versions so the customer needs to pull the body of the stat off of the back plate. The part # is printed on the circuit board. **Mode – Fan – Down - Up**

Totaline: P474-1020 or P474-1050
Carrier: TSTATCCP2W01 or TSTATCCPB501
Bryant: TSTATBBP2W01 or TSTATBBPB501

There are also 2 Commercial versions of this stat. One version has the same buttons as the Residential model. Again, the customer needs to pull the body of the stat off to determine the model #. The second version has one different button: **Mode – Override – Down- Up**

Totaline: P474-2050 and P474-2150
Carrier: 33CSN2-WC and 33CSSP2-WC
Bryant: Not Available

The fourth style is the wireless stat. It consists of two parts, the transmitter and the receiver. The transmitter has 4 buttons located vertically under the display: **Up – Down – Mode – Fan**

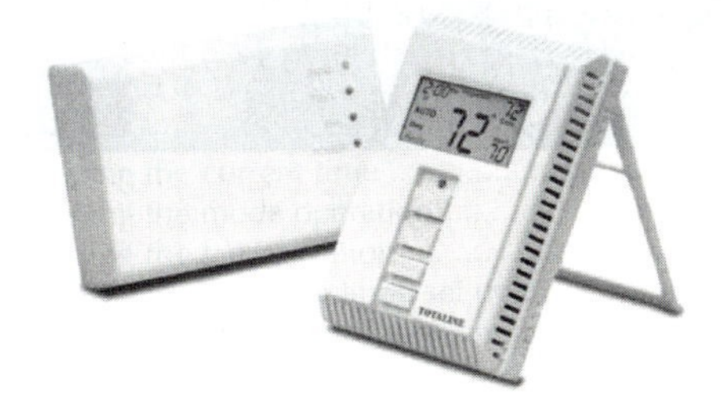

Totaline: P474-1100RF and P474-1100REC
Carrier: TSTATCCPRF01 and TSTATCCREC01
Bryant: TSTATBBPRF01 and TSTATBBREC01

The Commercial version of the wireless stat looks exactly the same, but one button changes:
Up – Down – Mode – Override

Totaline: P474-2300RF and P474-2300REC
Carrier: 33CS250-RC and 33CS250-RE
Bryant: Not Available

SETTING UP THE THERMIDISTAT CONTROL 9

e. Humidify setting only displayed — humidify setting does not change according to outdoor temperature.

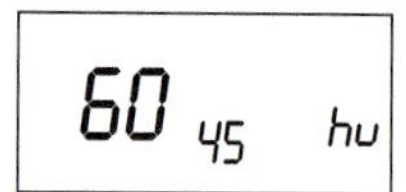

END

4 Press the END button to exit the humidify mode.

Setting the dehumidification set point

NOTE: This function is for use with variable-speed equipment only.

SET DHUM

1 Press the SET DHUM button to enter the dehumidify mode.

The current indoor humidity (large number) and dehumidify set point (small number) are displayed along with the dehumidify indicator (dhu).

SETTING UP THE THERMIDISTAT CONTROL 10

2 With the dehumidify indicator (dhu) displayed, press the UP or DOWN button to adjust the dehumidify set point. Dehumidify levels can be set from 50% to 90%. Or, to turn dehumidification off, press the MODE button until "OF" appears on the display.

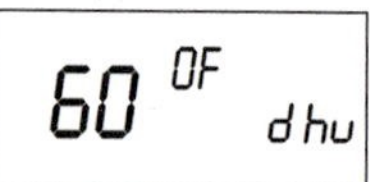

(See suggested settings on page 26.)

END

3 Press the END button to exit.

Setting the "cool to dehumidify" function

This setting allows a standard comfort system to provide moderate dehumidification by running the air conditioner. The function can also be used with variable-speed equipment.

NOTE: While in the "cool to dehumidify" mode, the indoor air temperature will not drop more than 3° below the cooling set point with a dehumidification demand.

SETTING UP THE THERMIDISTAT CONTROL 11

SET DHUM

1 Press the SET DHUM button. "dhu" is displayed.

2 Press the UP or DOWN button to raise or lower the dehumidify set point.

Dehumidification can be set from 50% to 90%.

MODE

3 Press the MODE button until the COOL icon is displayed.

END

4 Press the END button to exit.

OPERATING THE THERMIDISTAT CONTROL 12

Checking current temperature

The Thermidistat Control will display the current temperature.

To view your current temperature set points, press the UP or DOWN button once. The heating and cooling set points will be displayed.

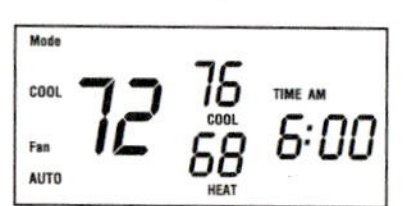

Checking the outdoor temperature and indoor humidity

1 Press the UP and DOWN buttons simultaneously.

2 The outdoor temperature will appear on the display. Then, the indoor humidity will be displayed.

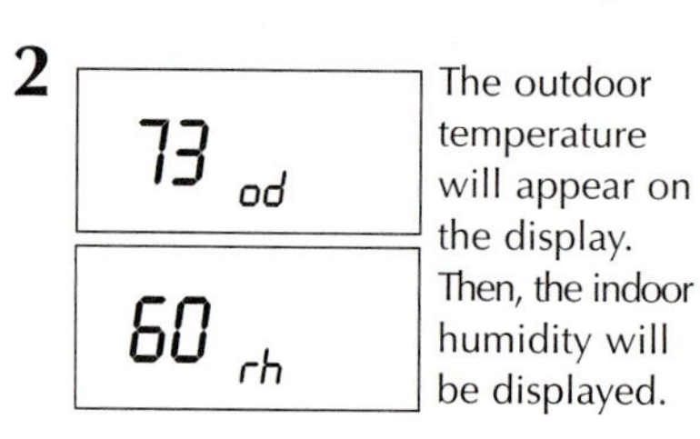

OPERATING THE THERMIDISTAT CONTROL 13

NOTE: If 2 dashes (--) appear, your Thermidistat Control does not include the outdoor air temperature sensor or the sensor is not working properly. Check with your dealer if you are unsure.

Checking current humidification and dehumidification set points

SET HUM

1 Press the SET HUM button.

The current indoor humidity (large number) is displayed along with the humidify set point (small number).

60 45 hu

END

2 Press the END button.

OPERATING THE THERMIDISTAT CONTROL 14

SET DHUM

3 Press the SET DHUM button. The current indoor humidity (large number) is displayed along with the dehumidify set point (small number).

COOL 60 65 dhu

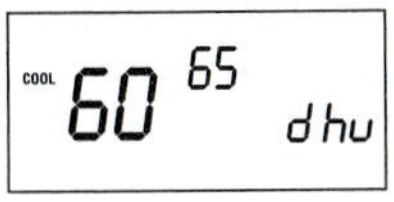

END

4 Press the END button.

Clean filter feature

Your Thermidistat Control reminds you when it's time to change or clean your filter by displaying the CLEAN FILTER indicator.

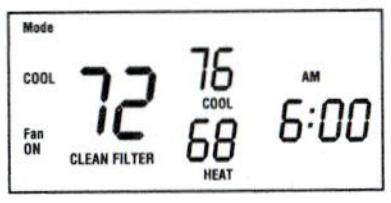

Mode COOL Fan ON 72 CLEAN FILTER 76 COOL 68 HEAT AM 6:00

VACATION

1 Press the VACATION and END buttons simultaneously after you have changed or cleaned your filter to restart the timer.

END

OPERATING THE THERMIDISTAT CONTROL 15

Vacation feature setup

The vacation feature allows a separate set of temperature and humidity set points to be stored for vacation and recalled with a single button press.

The vacation feature is preprogrammed for you with vacation settings for temperature and humidity. (Heat 60°, cool 85°, hu 10%, dhu 75%) If these are okay, skip ahead to "vacation feature operation." If you wish to enter new settings, continue with this section.

VACATION

1 Press the VACATION button to display the vacation temperature settings.

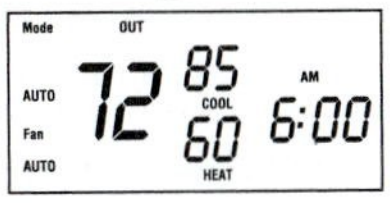

Mode OUT AUTO Fan AUTO 72 85 COOL 60 HEAT AM 6:00

The OUT indicator is displayed.

OPERATING THE THERMIDISTAT CONTROL 16

2 To change the cooling set point:

SET COOL

a.) Press the SET COOL button. COOL will flash on the display.

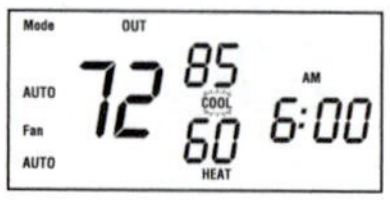

b.) Press the UP or DOWN button to adjust the setting.

END

c.) Press the END button to end.

3 To change the heating set point:

SET HEAT

a.) Press the SET HEAT button. HEAT will flash on the display.

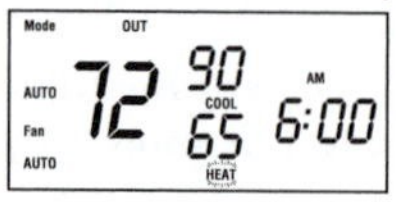

b.) Press the UP or DOWN button to adjust the setting.

END

c.) Press the END button to end.

OPERATING THE THERMIDISTAT CONTROL 17

4 To change the dehumidification set point:

SET DHUM

a.) Press the SET DHUM button. The "dhu" indicator will be displayed.

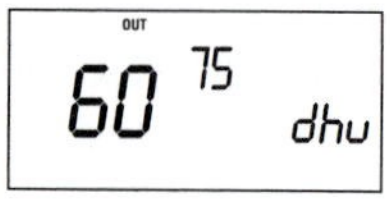

b.) Press the UP or DOWN button to adjust the setting.

MODE

c.) Press the MODE button to choose the dehumidification mode.

END

c.) Press the END button to end.

OPERATING THE THERMIDISTAT CONTROL 18

5 To change the humidification set point:

SET HUM

a.) Press the SET HUM button. The "hu" indicator will be displayed.

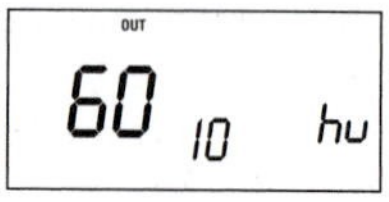

b.) Press the UP or DOWN button to adjust the setting.

MODE

c.) Press the MODE button to turn the humidification feature off (OF).

END

c.) Press the END button to end.

NOTE: In dehumidify, you may enter a set point, choose "COOL" to dehumidify, or turn dehumidification off (OF). In humidify, you may enter a setting or turn humidification off (OF).

OPERATING THE THERMIDISTAT CONTROL 19

Vacation feature operation

VACATION

1 ○ Press the VACATION button when you are ready to leave. Be sure you have properly selected the mode (HEAT, COOL, AUTO).

The OUT indicator is displayed, and your system will automatically follow your vacation temperature and humidity settings.

VACATION

2 ○ Press the VACATION button when you return to resume normal operation.

AUTO CHANGEOVER 20

Your Thermidistat Control provides complete, automatic control over heating and cooling with auto changeover. Auto changeover means your system will automatically heat or cool as needed to maintain your temperature set points.

Auto changeover makes life easier because you no longer have to manually switch the thermostat between heating or cooling operation. Just set your heating and cooling set points and let the Thermidistat Control do the rest!

NOTE: If Auto Changeover mode is not necessary in your area of the country, your installer may disable the AUTO mode.

WHAT IF... 21

AUXILIARY HEAT indicator is displayed...

The AUXILIARY HEAT indicator appears on the heat pump version of the Thermidistat Control only. It is displayed when your system is operating on auxiliary heat.

NOTE: This indicator does not reflect a problem with your system.

CLEAN FILTER indicator is displayed ...

The CLEAN FILTER indicator tells you when to clean or replace your system's air filter. Press the VACATION and END buttons simultaneously after cleaning or replacing the filter to turn off the indicator and restart the timer.

NOTE: This indicator does not reflect a problem with your system.

WHAT IF... 22

OUT Indicator is displayed ...

The OUT indicator reminds you that your system is in vacation mode. This function automatically adjusts the temperature and humidity settings to levels appropriate for when you're away. Press the VACATION button to resume normal system operation.

NOTE: This indicator does not reflect a problem with your system.

EQUIPMENT ON Indicator is displayed ...

When the cooling equipment is operating, the word COOL preceded by a small triangle is displayed below the cooling set point. When the heating equipment is operating, the word HEAT preceded by a small triangle is displayed below the heating set point. If the equipment turn on is being delayed, the triangle and the word will flash.

NOTE: This indicator does not reflect a problem with your system.

WHAT IF ... 23

You have a power outage ...

An internal power source eliminates the need to re-enter your settings into the Thermidistat Control after power outages. The comfort settings you have entered will be maintained indefinitely. The clock will run for 8 hours.

You have a system error message ...

The display may appear as follows:

--, E3, E4, E5, or E6

-- indicates a problem with the indoor air temperature sensor

E3 indicates a problem with the outdoor air temperature sensor

E4, E5, or E6 indicates an internal failure.

WARRANTY 24

This Thermidistat Control includes a 1-year limited warranty. For detailed warranty information, please refer to the All Product Limited Warranty Card included in your information packet. This Thermidistat Control is also eligible for manufacturer's extended system warranties. Ask your dealer for details on extended warranties for longer-term protection.

COMMON TERMS AND WHERE TO FIND THEM 25

Auxiliary Heat ....Pg. 21

Most heat pump systems require a supplemental heating source, called auxiliary heat, to maintain your comfort when outdoor temperatures fall significantly. Your Thermidistat Control lets you know when your home is being warmed with supplemental heat.

Clean Filter ....Pg. 14

Your system's air filter will require regular cleaning to reduce the dirt and dust in the system and your indoor air. The CLEAN FILTER indicator lets you know when it's time to clean the filter.

COMMON TERMS AND WHERE TO FIND THEM 26

Dehumidification Set Point ....Pg. 9

The amount of moisture to be removed from your home. You can check your actual humidity level and your desired dehumidification set point by pressing the SET DHUM button.

Suggested settings: 50% – 60% suggested depending on installation, area of the country, and your heating and cooling equipment.

Emergency Heat ....Pg. 4

This indicates that auxiliary heat is being used without the heat pump.

End ....Pg. 1

The END button returns the Thermidistat Control to normal operation.

COMMON TERMS AND WHERE TO FIND THEM 27

Fan ....Pg. 5

Your system's fan can run continuously or only as called for during heating or cooling. Continuous operation helps with air circulation and cleaning. Automatic operation provides energy savings. Press the FAN button to make your choice.

Humidification Set Point ....Pg. 6

The amount of moisture desired in your home to be supplied by the humidifier. You can check the actual humidity level and your desired humidification set point by pressing the SET HUM button.

Suggested settings:

Outdoor Temp (F)	-20	-10	0	10	20
Suggested Hum Set Point	15	20	25	30	35

COMMON TERMS AND WHERE TO FIND THEM 28

Mode ....Pgs. 3-5

Mode refers to the type of operation your system is set up to perform. Mode settings include: OFF, HEAT, COOL, and AUTO. Heat pump systems also include EMERGENCY HEAT (EHEAT).

Outdoor Temperature ....Pg. 12

Your Thermidistat Control not only measures the indoor temperature, but it may also be equipped to measure and display the outdoor temperature as well. Press the UP and DOWN buttons simultaneously to read the outdoor temperature display.

COMMON TERMS AND WHERE TO FIND THEM 29

Power Outage ....**Pg. 22**
Complete loss of electricity. Your Thermidistat Control has an internal power source that allows the clock to continue to run for 8 hours or more without electricity. Settings are stored indefinitely without the aid of batteries.

Reset Filter ....**Pg. 14**
The reset filter function turns off the CLEAN FILTER indicator and restarts the timer. Press the VACATION and END buttons simultaneously after you've cleaned and replaced the system's air filter.

COMMON TERMS AND WHERE TO FIND THEM 30

Set Time ....**Pg. 2**
This function allows you to set the proper time. Press the SET TIME button to activate.

Temperature Sensor ....**Pg. 12**
Temperature sensors measure the current indoor or outdoor temperatures which are displayed on the Thermidistat Control.

Temperature Set Points ....**Pg. 1**
These are the desired heating and cooling set points entered into the Thermidistat Control. The actual room temperature will automatically be displayed, but you can check the desired temperature for the current mode by pressing either the UP or DOWN button.

7310 West Morris Street, Indianapolis, IN 46231

Printed on recycled paper.

Form: OM17-25 Replaces: OM17-22 Printed in the U.S.A. 9-98 Catalog No. 13TS-TA11

COMMON TERMS AND WHERE TO FIND THEM 31

Time ....**Pg. 2**
The current time is displayed continuously on the display.

Up and Down Buttons ....**Pg. 1**
These buttons are used to set the clock and enter temperature and humidity information.

APPENDIX

Tools of the Trade (Plus Frequently Asked Questions with Answers)

There are a number of distributors and manufacturers of refrigeration and air-conditioning equipment and refrigerants. The *Internet* is an ideal place to look for the latest tools, equipment, and refrigerant sources. The following are but two of the many. www.mastercool.com has an online catalog for refrigeration and A/C service tools. Another online catalog is found at www.yellowjacket.com, which also has leak detectors, fin straighteners, and other equipment for the tradesman.

Mastercool, Inc.
1 Aspen Drive
Randolph, NJ 07869
www.mastercool.com

Ritchie Engineering Company, Inc.
10950 Hampshire Avenue S.
Bloomington, MN 55438-2623
www.yellowjacket.com

These are examples of the questions and answers available on the Web site for those interested in furthering their careers and being able to respond to customer questions and problems. (Used through the courtesy of Ritchie Engineering, Inc.)

R-100 Recovery system

(A)

Refrigerant recovery cylinders

(B)

Fig. D-1 *(A) Recovery units. (B) Recovery unit cylinders.* (Ritchie Engineering, available at www.yellowjacket.com)

FREQUENTLY ASKED QUESTIONS

Q. **For what refrigerants are the R 60a, R 70a, and R 80a rated**? See Fig. D-1 for recovery units.

A. They are tested by *Underwriter's Laboratories, Inc.* (UL) to ARI 740-98 and approved for medium pressure refrigerants R-12, R-134a, R-401C, R406A, R-500; medium high pressure refrigerants R-401A, R-409A, R-401B, R-412A, R-411A, R-407D, R-22, R411B, R-502, R-407C, R-402B, R-408A, R-509; and high pressure refrigerants R-407A, R-404A, R-402A, R-507, R407B, R-410A.

Q. **Why should I purchase a recovery System?**

A. With the Yellow Jacket name on a hose, you know you have got the genuine item for performance backed by more than 50 years. Now, you will also find the name on refrigerant recovery systems that are based on RRTI- and RST-proven designs. RRTI was one of the original recovery companies and helped DuPont design its original unit. With the purchase of RST in 1998, Ritchie Engineering combined Yellow Jacket standards of manufacturing and testing with the RST track record of tough reliability.

Q. **Is *American Refrigeration Institute* (ARI) the only testing agency?**

A. No. ARI is only a certifying agency that hires another agency to perform the actual testing. UL is also *Enviromental Protection Agency* (EPA) approved as a testing and certifying agency. Yellow Jacket Systems are UL tested for performance. Some Yellow Jacket Systems are tested to CSA, CUL, CE, and TUV safety standards which go beyond the ARI performance standards.

Q. Can I compare systems by comparing their ARI or UL ratings?

A. Yes. ARI and UL test standards should be the same. And remember that manufacturers can change the conditions under which they test their own machines to give the appearance of enhanced performance. Only ARI and UL test results provide consistent benchmarks and controls on which to make objective comparisons.

Q. How dependable are Yellow Jacket refrigerant recovery systems?

A. Yellow Jacket recovery systems get pushed to the limits: day in and day out in dirty conditions, on roof tops, and sometimes in freezing or high ambient temperatures. Yellow Jacket equipment has been tested at thousands of cycles, and are backed with the experience of units in the field since 1992.

Q. Is pumping liquid the fastest way to move refrigerant?

A. Yes, and the R 50a, R 60a, R 70a, R 80a, and R 100 monitor liquid flow at a rate safe for the compressor. In Yellow Jacket lab testing, over 80,000 lb of virgin liquid R-22 were continuously and successfully pumped. That is over 2,500 standard 30-lb tanks, or the equivalent of refrigerant in over 25,000 typical central AC systems.

Q. What is the push/pull recovery method?

A. Many technicians use this recovery method, particularly on large jobs. By switching the R 50a, R 60a, R 70a, R 80a, or R 100 discharge valve to "recover" during push/pull recovery, the condenser is bypassed, increasing the "push" pressure and speeding the recovery.

Q. Why do the R 50a through 100 feature a built-in filter?

A. Every recovery machine requires an in-line filter to protect the machine against the particles and "gunk" that can be found in failed refrigeration systems. For your convenience, the R 50a, R 60a, R 70a, R 80a, and R 100 series incorporate a built-in 200-mesh filter that you can clean, and if necessary, replace. The filter traps 150-micron particles and protects against the dirtiest systems to maximize service life. In case of a burn-out, an acid-core filter/drier is mandatory. The Yellow Jacket filter is built-in to prevent breaking off like some competitive units with external filters.

Q. What is auto purge and how does it work?

A. At the end of each cycle, several ounces of refrigerant can be left in the recovery machine to possibly contaminate the next job, or be illegally vented. Many competitive recovery machines require switching hoses, tuning the unit off and on, or other time-consuming procedures. The R 50a, R 60a, R 70a, R 80a, and R 100 can be quickly purged "on the fly" by simply closing the inlet valve, and switching the discharge valve to purge. In a few seconds, all residual refrigerant is purged and you are finished. Purging is completed without switching off the recovery unit.

Q. Can increased airflow benefit recovery cylinder pressure?

A. Yes. For reliable performance in the high ambient temperatures, Yellow Jacket units are engineered with a larger condenser and more aggressive fan blade with a greater pitch. This allows the unit to run cooler and keeps the refrigerant cooler in the recovery cylinder.

Q. Can I service a Yellow Jacket System in the field?

A. Although Yellow Jacket Systems feature either a full one or optional two year warranty, there are times when a unit will need a tune-up. The service manual with every unit includes a wide variety of information such as tips to speed recovery, troubleshooting guides, and parts listings. On the side of every unit, you will find hook-up instructions, a quick start guide, and simple tips for troubleshooting. And if ever in doubt, just call 1-800-769-8370 and ask for customer service.

All service and repair parts are readily available through your nearest Yellow Jacket wholesaler.

Q. Are there any recovery systems certified for R-410A?

A. ARI 740-98 has been written but not yet enacted by the EPA. The Yellow Jacket R 60a, R 70a, and R 80a series have also been UL tested and certified for high-pressure gases such as R-410A that are covered under ARI 740-98.

Q. What features should I demand in a system to be used for R-410A?

A. Look for the following three features as a minimum:

- High-volume airflow through an oversized condenser to keep the unit running cooler and help eliminate cut-outs in high ambient temperatures
- Single automatic internal high-pressure switch for simple operation

- *Constant pressure regulator* (CPR) valve rated to 600 psi for safety and it eliminates the need to monitor and regulate the unit during recovery

Q. What is a CPR valve?

A. The CPR valve is the feature that makes the Yellow Jacket R 70a and R 80a the first truly automatic recovery systems for every refrigerant. The single 600 psi rated high-pressure switch covers all refrigerants and eliminates the need for a control panel with two selector switches for R-410A.

The CPR valve automatically reduces the pressure of the refrigerant being recovered. Regardless of which refrigerant, input is automatically regulated through a small orifice that allows refrigerant to flash into vapor for compression without slugging the compressor. Throttling is not required.

Under normal conditions, you could turn the machine to the "liquid" and "recover" settings. The machine will complete the job while you work elsewhere.

Q. With the Yellow Jacket R 70a design, do I have to manually reset a pressure switch between medium and high-pressure gasses?

A. No. Some competitive machines require you to choose between medium and high-pressure gas settings before recovery. You will see the switch on their control panels. With the Yellow Jacket R 70a, the single internal automatic high-pressure switch makes the choice for you. That is why only R 70a is truly automatic.

Frequently Asked Questions About Pumps

1. How can I select the right pump cfm?

A. See Fig. D-2 for a sampling of pumps. The following guidelines are for domestic through commercial applications.

System(tons)	Pump (cfm)
10–15	2.0
15–30	4.0
30–45	6.0
45–60	8.0
60–90	12.0
90–130	18.0
Over 130	24.0

2. Can I use a vacuum pump for recovery?

A. A vacuum pump removes water vapor and is *not* for refrigerant recovery. Connecting a vacuum pump to a pressurized line will damage the pump and vent refrigerant to the atmosphere, which is a crime.

Large capacity supervac™ pumps 12, 18, 24 cfm

Fig. D-2 *Vacuum pumps.* (Ritchie Engineering, available at www.yellowjacket.com)

3. How much of a vacuum should I pull?

A. A properly evacuated system is at 2500 microns or less. This is $^1/_{10}$ of 1 in. and impossible to detect without an electronic vacuum gauge. For most refrigeration systems, *American Society of Heating, Refrigerating, and Air-Conditioning Engineers* (ASHRAE) recommends pulling vacuum to 1000 microns or less. Most system manufacturers recommend pulling to an even lower number of microns.

4. Do I connect an electronic vacuum gauge to the system or pump?

A. To monitor evacuation progress, connect it to the system with a vacuum/charge valve.

5. Why does the gauge micron reading rise after the system is isolated from the vacuum pump?

A. This indicates that water molecules are still detaching from the system's interior surfaces. The rate of rise indicates the level of system contamination and if evacuation should continue.

6. Why does frost form on the system exterior during evacuation?

A. Because ice has formed inside. Use a heat gun to thaw all spots. This helps molecules move off system walls more quickly toward the pump.

7. How can I speed evacuation?

A.

- Use clean vacuum pump oil. Milky oil is water saturated and limits pump efficiency.

- Remove valve cores from both high and low fittings with a vacuum/charge valve tool to reduce time through this orifice by at least 20 percent.
- Evacuate both high and low sides at the same time. Use short, 3/8-in. diameter and larger hoses.

SuperEvac Systems can reduce evacuation time by over 50 to 60 percent. SuperEvac pumps are rated at 15 microns (or less) to pull a vacuum quickly. Large inlet allows you to connect a large diameter hose. With large oil capacity, SuperEvac pumps can remove more moisture from systems between oil changes.

8. What hose construction is best for evacuation?

A. Stainless steel. There is no permeation and outgassing.

Frequently Asked Questions About Fluorescent Leak Scanners

Q. Does the *ultra violet* (UV) scanner light work better than an electronic leak detector? See Fig. A4-3.

A. No one detection system is better for all situations. But, with a UV lamp you can scan a system more quickly and moving air is never a problem. Solutions also leave a telltale mark at every leak site. Multiple leaks are found more quickly.

Q. How effective are new *light emitting diode* (LED) type UV lights? See Fig. A4-4.

A. LEDs are small, compact lights for use in close range. Most effective at 6-in. range. The model with two blue UV and three UV bulbs has a slightly greater range. Higher power Yellow Jacket lights are available.

Q. Can LED bulbs be replaced?

A. No. The average life is 110,000 h.

Q. Are RediBeam lamps as effective as the System II lamps?

A. The RediBeam lamp has slightly less power to provide lightweight portability. But with the patented reflector and filter technologies, the RediBeam 100-W bulb produces sufficient UV light for pinpointing leaks.

Q. Does the solution mix completely in the system?

A. Solutions are combinations of compatible refrigeration oil and fluorescent material designed to mix completely with the oil type in the system.

UV LEAK DETECTION TOOLS

53515 MACH IV FLEXIBLE UV LIGHT

The MACH IV flexible UV light has 4 TRUE UV LEDS delivering a brilliant fluorescent glow. Gets into tight areas easily.

53012 UV SWIVEL HEAD LIGHT

- 12V/100 WATT
- 180° Swivel Head Gets Into Tight Places
- Instant ON
- Heavy Duty Metal Construction
- Includes UV Enhancing Safety Glasses

53312 UV MINI LIGHT

- 12V/50 WATT
- High Intensity
- Compact & Lightweight
- Instant ON
- 16 ft. (5m) Cord
- Includes UV Enhancing Safety Glasses

RECHARGEABLE UV LIGHT

53411
- High Intensity 12V/50 WATT
- Cordless with Rechargeable Battery Cartridge and Charger
- Compact & Portable
- Includes UV Enhancing Safety Glasses

53412
- Same as 53411 (less Battery Charger and glasses)

53413 BATTERY CARTRIDGE

- 12V/50 WATT
- This powerful rechargeable battery cartridge holds a charge equal to 30 minutes of continuous use

BATTERY CHARGER

53414 110V/60 HZ
53414-220 220V/50-60 HZ

LIGHT PART#	LIGHT BULB PART#	LENS PART#
53012	53012-B	53012-L
53112	53012-B	53110-L
53312	53312-B	53312-L
53411	53312-B	53312-L
53515	53515-B	–

ACCESSORIES

53314 **DYE REMOVER** - 4 oz
53315 **SERVICE LABELS** - 25 per pack. Bright yellow label indicating that the system has been charged with UV Dye.
92398 **UV ENHANCING SAFETY GLASSES** This is a must for protection against ultraviolet light during leak detection.

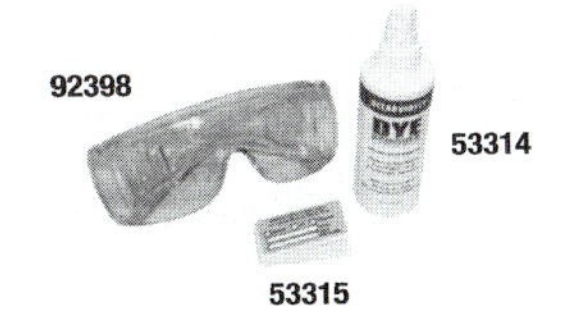

53809 MINI DYE INJECTOR

"Cartridge Type" Dye Injector (10 Appl.) Concentrated Dye.

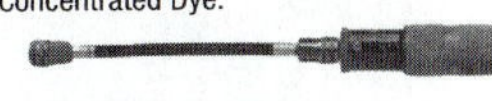

53223 "CARTRIDGE TYPE" UNIVERSAL DYE/OIL INJECTOR

Adding dye with the new 53223 is fast and easy, simply connect the injector to the low side of the A/C system and twist the handle to the next application line. The replaceable cartridge provides 25 applications of universal dye that is compatible with R12/22/502 and R134a systems. The injector hose comes complete with a R134a coupler and auto shut-off valve adapter for 1/4"FL systems.

"REFILLABLE" UNIVERSAL DYE/OIL INJECTOR

To inject the oil or dye, simply connect the injector to the low side of the A/C system and twist the handle until you reach the desired amount. The injector hose comes complete with a R134a coupler and auto shut-off valve adapter for 1/4"FL systems.

53123 **REFILLABLE UNIVERSAL DYE INJECTOR** with 2 oz Bottle of Concentrated A/C Dye (25 Appl.)
53123-A **REFILLABLE UNIVERSAL DYE/OIL INJECTOR** (without dye)
53134 **REFILLABLE DYE INJECTOR** with (R134a 13mm Connection) and 2 oz Bottle of Concentrated A/C Dye (25 Appl.)
53322 **REFILLABLE DYE INJECTOR** with (1/4" Auto Shut-off Valve Connection) and 2 oz Bottle of Concentrated A/C Dye (25 Appl.)

ULTRAVIOLET DYES

Standard Universal A/C Dyes

Pack 1/4 oz - **92699** 1 Application Standard Universal Dye Six 1/4 oz Packs (6 pcs)
8 oz Bottle - **92708** 32 Applications R12/22/502 Standard Universal Dye
32 oz Bottle - **92732** 128 Applications R12/22/502 Standard Universal Dye

Concentrated Universal Dyes

Replaceable Cartridge - **53825** 25 Applications Universal Dye Cartridge
Replaceable Cartridge - **53810** 10 Applications Concentrated Dye Cartridge
2 oz Bottle - **53625** 25 Applications Universal Dye

Fig. A4-3 *Leak scanners (Flourescent type).* (Mastercool, available at www.yellowjacket.com)

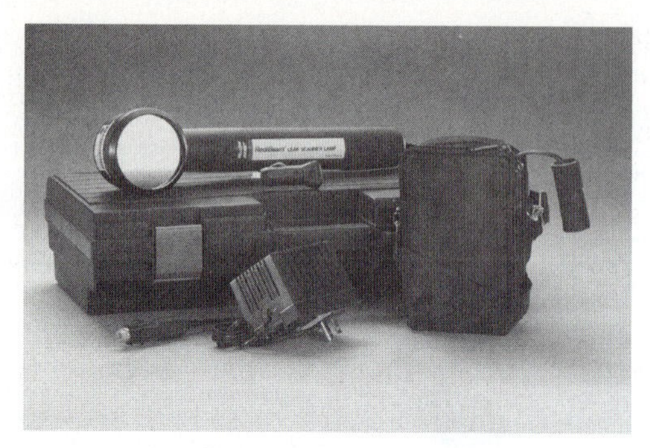

Universal UV system

Fig. A4-4 *LED-ultraviolet lights.* (Ritchie Engineering, available at www.yellowjacket.com)

Q. How are solutions different?

A. Solutions are available with mineral, alkylbenzene, PAG, or polyol ester base stock to match oil in the system.

Q. What is universal solution?

A. It is made from polyol ester and mixes well with newer oils. It also works in mineral oil systems, but can be harder to see.

Q. What is the lowest operating temperature?

A. It is –40°F for all solutions. Alkylbenzene in alkylbenzene systems to –100°F.

Q. Does solution stay in the system?

A. Yes. When future leaks develop, just scan for the sources. In over six years of testing, the fluorescent color retained contrast. When the oil is changed in the system, scanner solution must be added to the new oil.

Q. Is the solution safe?

A. Solutions were tested for three years before introduction and have been performance proven in the field since 1989. Results have shown the solutions safe for technicians, the environment and all equipment when used as directed. Solutions are pure and do not contain lead, chromium, or *chlorofluorocarbon* (CFC) products.

Presently, solutions are approved and used by major manufacturers of compressors, refrigerant, and equipment.

Q. How do I determine oil type in system?

A. Many times the oil is known due to the type of refrigerant or equipment application. Systems should be marked with the, kind of oil used. Always tag system when oil type is changed.

Q. In a system with a mix of mineral and alkylbenzene oil, which scanner solution should be used?

A. Base your choice of solution on whatever oil is present in the larger quantity. If you do not know which oil is in greater quantity, assume it to be alkylbenzene.

Q. How do I add solution to the system?

A. In addition to adding solution using injectors or mist infuser, you have other possibilities.

If you do not want to add more gas to the system, connect the injector between the high and low side allowing system pressure to do the job. Or, remove some oil from the system, then add solution to the oil and pump back in.

Q. How is the solution different from visible colored dyes?

A. Unlike colored dyes, Yellow Jacket fluorescent solutions mix completely with refrigerant and oil and do not settle out. Lubrication, cooling capacity, and unit life are not affected; and there is no threat to valves or plugging of filters. The solutions will also work in a system containing dytel.

Q. How do I test the system?

A. Put solution into a running system to be mixed with oil and carried throughout the system. Nitrogen charging for test purposes will not work since nitrogen will not carry the oil. To confirm solution in the system, shine the lamp into the system's sight glass. Another way is to connect a hose and a sight glass between the high and low sides, and monitor flow with the lamp. The most common reason for inadequate fluorescence is insufficient solution in the system.

Q. Can you tell me more about bulbs?

A. 115-V systems are sold with self-ballasted bulbs in the 150-W range. Bulbs operate in the 365 nanometer long range UV area and produce the light necessary to activate the fluorescing material in the solution. A filter on the front of the lamp allows only "B" band rays to come through. "B" band rays are not harmful.

Q. What is the most effective way to perform an acid test?

A. Scanner solution affects the color of the oil slightly. Use a two-step acid-test kit which factors out the solution in the oil, giving a reliable result.

Q. Can fluorescent product be used in nonrefrigerant applications?

A. Yes, in many applications.

ACCUPROBE LEAK DETECTOR WITH HEATED SENSOR TIP

This is the only tool you need for fast, easy, and certain leak detection. The heated sensor tip of the Yellow Jacket Accuprobe leak detector positively identifies the leak source for all refrigerants. That includes CFCs such as R-12 and R-502, *hydrochlorofluorocarbons* (HCFCs) as small as 0.03 oz/year, and *hydrofluorocarbon* (HFC) leaks of 0.06 oz/year, even R-404A, and R-410A. See Fig. A4-5.

Frequency of flashes in the tip and audible beeping increases the closer you get to the leak source. You zero-in and the exclusive Smart Alarm LED shows how big or small the leak on a scale of 1 to 9 is. Maximum value helps you determine if the leak needs immediate repair.

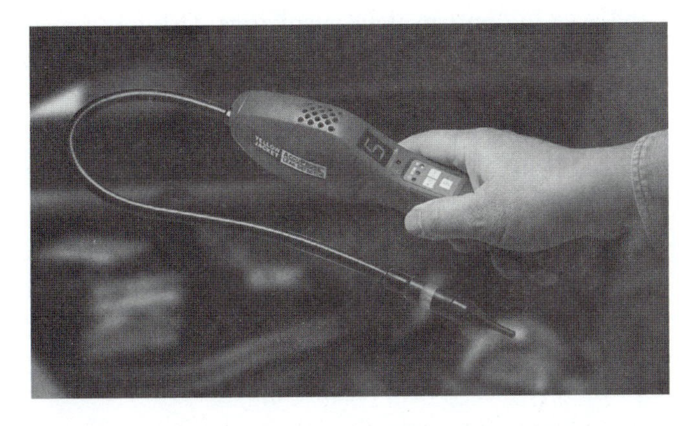

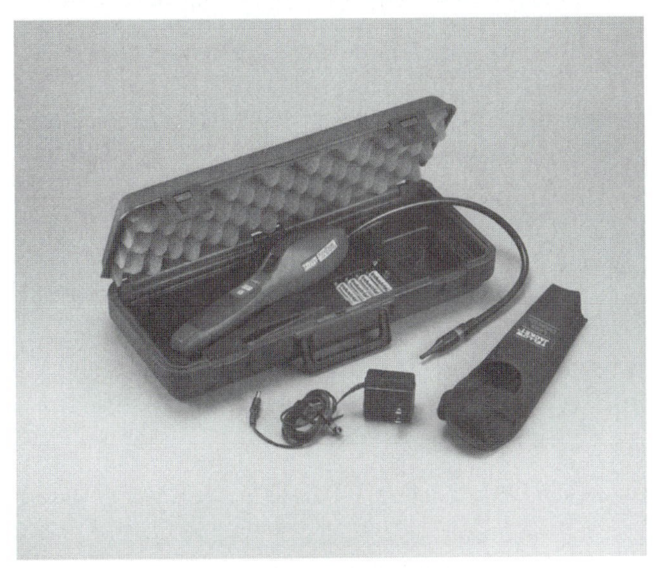

Fig. A4-5 *Accu Probe leak detector.* (Ritchie Engineering, available at www.yellowjacket.com)

Service life of the replaceable sensor is more than 300 h with minimal cleaning and no adjustments. Replaceable filters help keep out moisture and dust that can trigger false sensing and alarms.

Three sensitivity levels include ultrahigh to detect leaks that could be missed with other detection systems.

Additional features and benefits:

- Detect all HFCs including R-134a, R-404A, R-410A, and R407C and R-507; all HCFCs such as R-22 and all CFCs
  - Audible beeping can be muted.
  - Extended flexible probe for easy access in hard-to-reach areas.
- Sensor not poisoned by large amount of refrigerant and does not need recalibration
  - Sensor failure report mode
  - Weighs only about 15 oz for handling comfort and ease
  - Carrying holster and hard, protective case included
  - Bottle of nonozone depleting chemical included for use as a leak standard to verify proper functioning of sensor and electronic circuitry

TECHNOLOGY COMPARISON-HEATED SENSOR OR NEGATIVE CORONA?

Heated Sensor Leak Detectors

When the heated sensing element is exposed to refrigerant, an electrochemical reaction changes the electrical resistance within the element causing an alarm. The sensor is refrigerant specific with superior sensitivity to all HFCs, HCFCs, and CFCs and minimal chance of false alarms. When exposed to large amounts of refrigerant which could poison other systems, the heated sensor clears quickly and does not need recalibration before reuse. See Fig. A4-6.

Negative Corona Leak Detectors

In the sensor of an old-style corona detector, high voltage applied to a pointed electrode creates a corona. When refrigerant breaks the corona arc, the degree of breakage generates the level of the alarm. This technology has good sensitivity to R-12 and R-22, but only fair for R-134a, and poor for R-41 OA, R-404A, and R-407C. Sensitivity decreases with exposure to dirt, oils, and water. And false alarms can be triggered by dust, dirt specks, soap bubbles, humidity, smoke, small variations in the electrode emission, high levels of hydrocarbon vapors, and other nonrefrigerant variables. See Fig. A4-7.

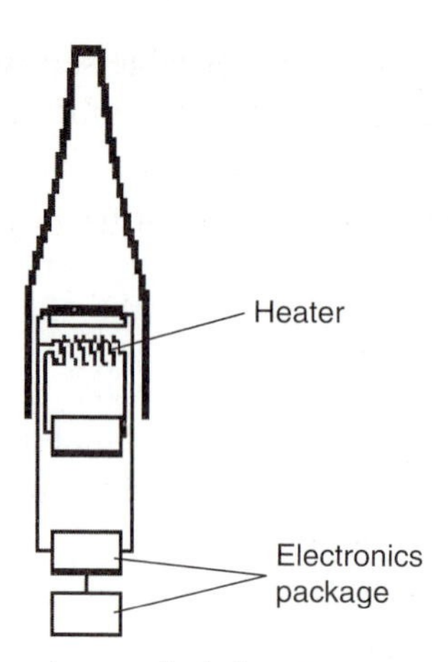

Fig. A4-6 *Heated sensor leak detector.* (Ritchie Engineering, available at www.yellowjacket.com)

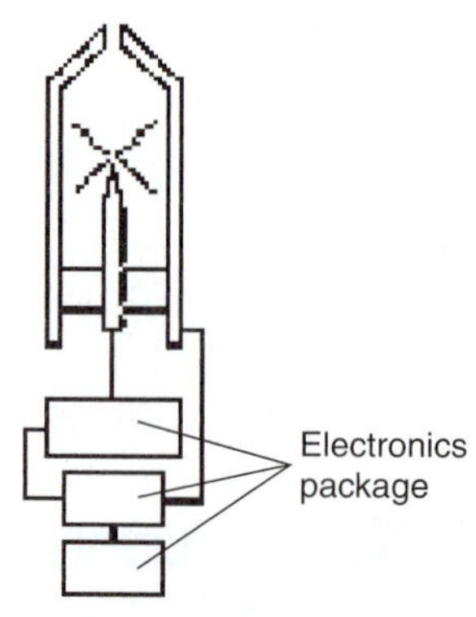

Fig. A4-7 *Negative corona leak detector.* (Ritchie Engineering, available at www.yellowjacket.com)

TIPS FOR DETECTING SYSTEM LEAKS

1. Inspect entire A/C system for signs of oil leakage, corrosion cracks, or other damage. Follow the system in a continuous path so no potential leaks are missed.
2. Make sure there is enough refrigerant in a system (about 15 percent of system capacity or 50 psi/min) to generate pressure to detect leaks.
3. Check all service access port fittings. Check seals in caps.
4. Move detector probe at 1 in/s within $^1/_4$ in. of suspected leak area.
5. Refrigerant is heavier than air, so position probe below test point.
6. Minimize air movement in area to make it easier to pinpoint the leak.
7. Verify an apparent leak by blowing air into the suspected leak.
8. When checking for evaporator leaks, check for gas in condensate drain tube.
9. Use heated sensor type detector for difficult-to-detect R-134a, R-410A, R-407C, and R-404A.

NEW COMBUSTIBLE GAS DETECTOR—WITH ULTRASENSITIVE, LONG-LIFE SENSOR

Hand-held precision equipment detects all hydrocarbon and other combustible gases including propane, methane, butane, industrial solvents, and more. See Fig. A4-8.

- Sensitivity, bar graph, and beeping to signal how much and how close.
- Unit is preset at normal sensitivity, but you can switch to high or low. After warm-up you will hear a slow beeping. Frequency increases when a leak is detected until an alarm sounds when moving into high gas concentration. The illuminated bar graph indicates leak size.
- If no leak is detected in an area you suspect, select high sensitivity. This will detect even low levels in the area to confirm your suspicions. Use low sensitivity as you move the tip over more defined areas, and you will be alerted when the tip encounters the concentration at the leak source.
- Ultrasensitive sensor detects less than 5 ppm methane and better than 2 ppm for propane. They perform equally well on a complete list of detectable gases including acetylene, butane, and isobutane.
- Automatic calibration and zeroing.
- Long-life sensor easily replaced after full service life.

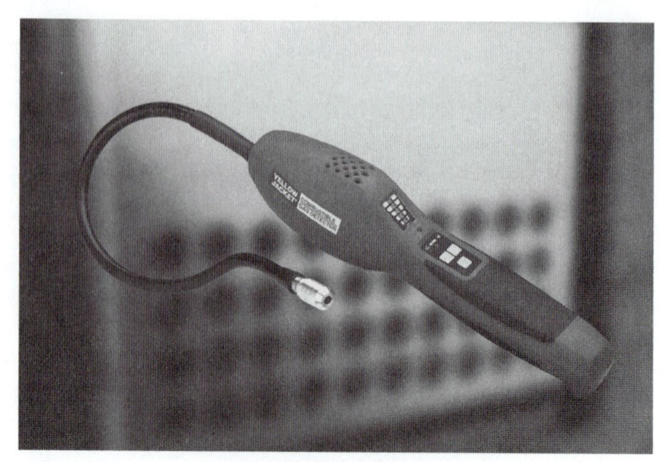

Combustible gas detector

Fig. A4-8 *Gas detectors.* (Ritchie Engineering, available at www.yellowjacket.com)

Applications

- Gas lines/pipes
- Propane filling stations
- Gas heaters
- Combustion appliances
- Hydrocarbon refrigerant
- Heat exchangers
- Marine bilges
- Manholes
- Air quality
- Arson residue(accelerants)

FREQUENTLY ASKED QUESTIONS ABOUT FIXED MONITORING SYSTEMS

Q. Are calibrated leak testers available to confirm that the monitor is calibrated correctly? See Fig. A4-9.

A. The Yellow Jacket calibrated leak is a nonreactive mixture of R-134a or NH_3 and CO_2. The nonreturnable cylinders contain 10 L of test gas. The cylinders require a reusable control valve and flow indicator. Test gases can be ordered for 100 ppm or 1000 ppm mixtures.

Q. What refrigerants will the leak monitors detect?

A. Leak monitors will detect most CFC, HFC, and HCFCs such as R-11, R-12, R-13, R-22, R-113, R-123, R-134a, R-404A, R407C, R-410A, R-500, R-502, and R-507. Yellow Jacket also has leak monitors available for ammonia and hydrocarbon-based refrigerants.

Q. Can the leak monitor be calibrated for specific applications?

A. Yes, the Yellow Jacket leak monitor can be calibrated for specific applications. Contact customer service for your specific need.

Q. If the unit goes into alarm, can it switch on the fan? Can it turn off the system at the same time?

A. The leak monitor has a pair of dry, normally open/normally closed contacts that can handle 10 A at 115 V. When the sensor indicates a gas presence higher than the set point, it opens the closed contacts and closes the open contacts which will turn equipment on or off.

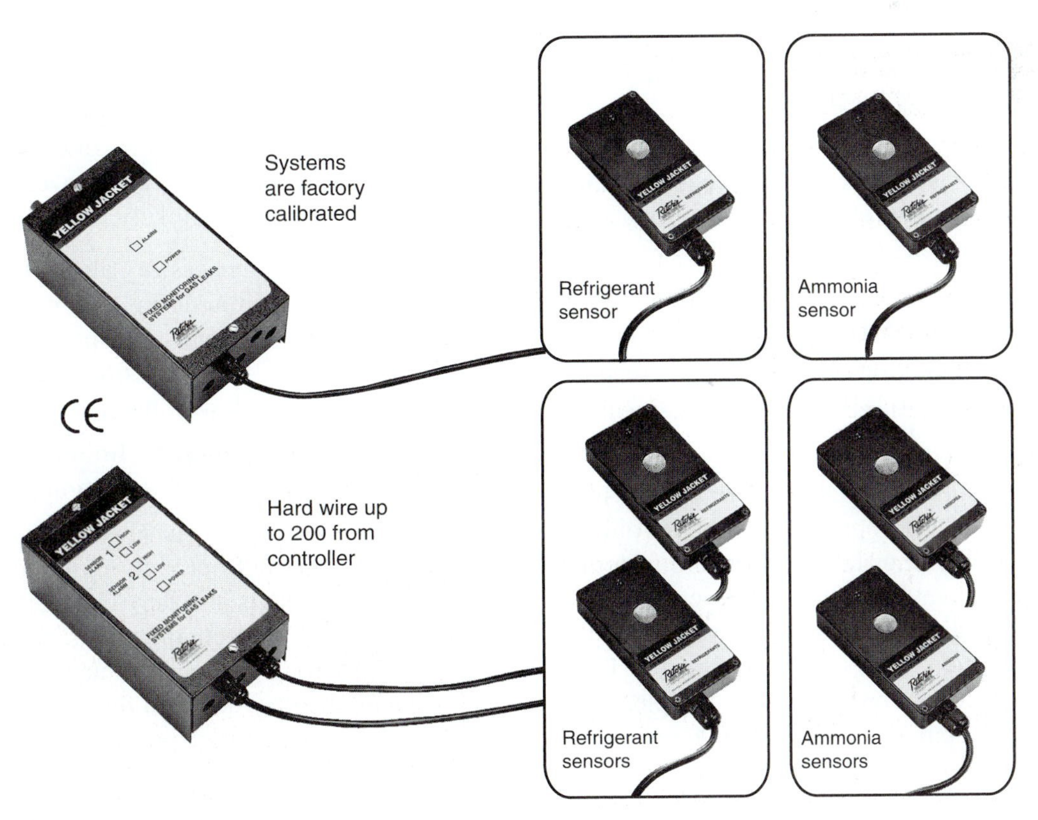

Fig. A4-9 *Fixed monitoring systems.* (Ritchie Engineering, available at www.yellowjacket.com)

Q. After a unit goes into alarm and the contacts close, what can it be connected to?

A. The open contacts can shut the system down, call a phone number, turn on a fan, or emergency light.

Q. How does the sensor work?

A. When the sintered metal oxide surface within the sensor absorbs gas molecules, electrical resistance is reduced in the surface allowing electrons to flow more easily. The system controller reads this increase in conductivity and signals an alarm. Metal oxide technology is proven for stability and performance.

Q. What is the detection sensitivity level of Yellow Jacket fixed monitors?

A. The dual sensitivity system has a low alarm level of about 100 ppm and a high level of about 1000 ppm for most CFC, HFC, and HCFC products. The high level for R-123 is an exception at about 300 ppm. Ammonia detection levels are about 100 ppm low and about 150 ppm high. The alarm level of all Yellow Jacket single-level systems is about 100 ppm.

Detection levels are preset at the factory to cover most situations. If necessary, however, you can order a custom level, or adjust the set point on site.

Q. What gas concentration must be detected?

A. Depends on the refrigerant. For a more thorough answer, terms established by U.S. agencies must first be understood:

- IDLH—immediately dangerous to life and health
- TWA—time weighted average concentration value over an 8-h work day or a 40-h work week (OSHA or NIOSH levels)
- STEL—short term exposure level measured over 15 min (NIOSH)
- Ceiling concentration—should not be exceeded in a working day (OSHA)

Obviously, the first consideration is IDLH. For most refrigerants, the IDLH is relatively high (e.g., R-12 is 15,000 ppm), and such a concentration would be unusual in a typical refrigerant leak situation. Leak detection, however, is still an immediate condition, so the STEL should be the next consideration, followed then by the 8-h TWA or ceiling concentration. R-22, for example, has a STEL of 1250 ppm and a TWA of 1000 ppm. (The TWA for most refrigerants is 1000 ppm.)

The draft UL standard for leak monitors requires gas detection at 50 percent of the IDLH. In other words, R-12 with a IDLH value of 15,000 ppm must be detected at 7500 ppm. As with most refrigerants, the TWA is 1000 ppm.

All of the foregoing suggests that for most CFC, HFC and HCFCs, detection at 1000 ppm provides a necessary safety margin for repair personnel. Ammonia with a significantly lower IDLH of 300 ppm and a TWA of 25 ppm requires detection at 150 ppm to comply with 50 percent IDLH requirements. R-123 has a TWA of 50 ppm and an IDLH of 1000 ppm, therefore detection at 100 ppm provides a good margin of safety. A monitor with a detection threshold of about 100 ppm for any gas provides an early warning so that repairs can be made quickly. This can save refrigerant, money, and the environment.

Q. How frequently should the system be calibrated?

A. Factory calibration should be adequate for 5 to 8 years. Routine calibration is unnecessary when used with intended refrigerants. Yellow Jacket sensors can not be poisoned, show negligible drift, and are stable long term. You should, however, routinely check performance.

Q. Can there be a false alarm?

A. For monitoring mixtures, the semiconductor must be able to respond to molecularly similar gases. With such sensitivity, false alarms can be possible. Engineered features in Yellow Jacket monitors help minimize false alarms:

- The two-level system waits about 30 s until it is "certain" that gas is present before signalling.
- At about 100 to 1000-ppm calibration level, false alarms are unlikely.

To prevent an unnecessary alarm, turn off the unit or disable the siren during maintenance involving refrigerants or solvents. Temperature, humidity, or transient gases may occasionally cause an alarm. If in doubt, check with the manufacturer.

Q. What are alternative technologies for monitoring and detecting refrigeration gas leaks?

A. Infrared technology is sensitive down to 1 ppm. This level is not normally required for refrigeration gases and is also very expensive compared to semiconductor technology. As an infrared beam passes through an air sample, each substance in the air absorbs the beam differently. Variations in the beam indicate the presence of a particular substance. The technique is very gas specific and in a room of mixed refrigerants, more than one system would be required. To get over this problem, newer models work on a broad band principle so they can see a range of gases. As a result they do not generally operate below 50 ppm and can experience false alarms.

Electrochemical cells can be used for ammonia. These cells are very accurate, but are expensive, and are normally used to detect low levels (less than 500 ppm), and perform for about two years.

With air sampling transport systems, tubing extends from the area(s) to be monitored back to a central controller/ sensor.

Micropumps move air through the system eliminating a number of on-site sensors, but there may be problems:

- Air in the area of concern is sampled at intervals rather than monitored continuously. This can slow the response to changing conditions.
- Dirt can be sucked into the tubes, blocking filters.
- Gases can be absorbed by the tube or leak out of the tube providing a concentration at the sensor lower than in the monitored area.
- Gases can leak into the tube in transit rather than the area monitored. The reading would be misleading.

NEW AND OLD TOOLS, OR CATALOG SHOPPING AND UPDATING

Refer to Fig. A4-10 for the following tools and supplies.

The Mastercool Company is indicative of the supply house supplies provided for those working in the refrigeration and air-conditioning field.

Some of the equipment you should be aware of as you continue to work in the field are shown in their catalog. A few of them are shown here as an example of some of the latest devices available to make your work day more efficient.

A convenient way to categorize the tools you work with are shown in the following example of a listing of available tools. This listing may change in time as the requirements for handling new refrigerants are brought about by accrediting agencies and standards writers.

- *Leak detection* relies on electronic detectors as well as the older types that have been around for years. Ultraviolet rays have now been utilized to more accurately identify and locate leaks. There are various dyes and injectors that need examining for keeping up. The combustible gas leak detector should also be examined as gases other than refrigerants are encountered on the job.

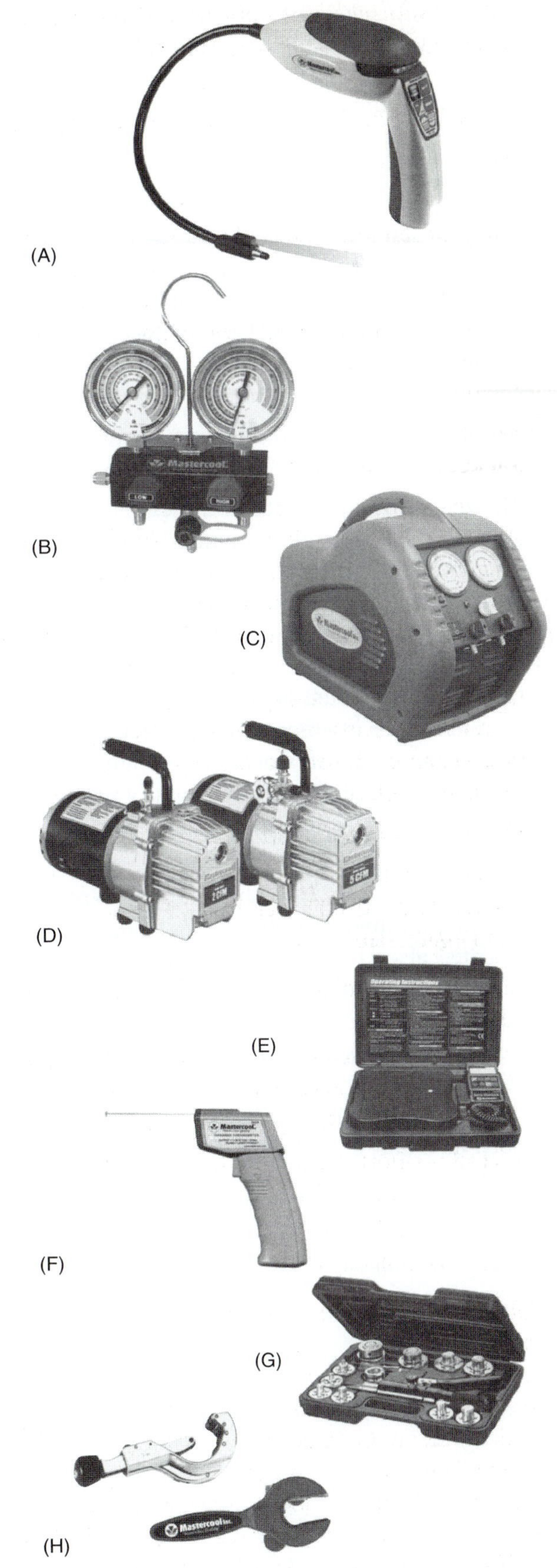

Fig. A4-10 *(A) Electronic leak detector. (B) Manifold and gauges. (C) Recovery unit. (D) Vacuum pumps. (E) Refrigerant scales. (F) Laser thermometer. (G) Hydraulic tools. (H) Tube cutters.* (Mastercool, available at www.yellowjacket.com)

- *Manifold/gages and hoses* are another of the categories most often recognized as essential to the technician working in the field and in-house. Hoses can stand some examination since they have been constantly improving through the years. And, there is always the chance a hose will/or has ruptured or leaked or deteriorated. Newer hoses are usually designed for a longer life than previously.
- Another category for classifying devices, tools, and other equipment is the *recovery equipment*, now so necessary to keep within the letter of the law and protect the environment.
- *Vacuum pumps* now have the rotary vane to produce deep vacuums. There are a number of pumps, oils, and accessories that fall into this classification process.
- *Refrigerant scales* have certified scales and programmable scales. The charging program allows the user to program desired quantities, and before the charge is complete, an alarm will sound allowing ample time to turn off the refrigerant supply. There are now features such as pause/charge: empty/full tank allows the user to know the amount of refrigerant left in the tank at any time. There is a repeat function that allows the user to charge to the previously stored amount. The scales are multilingual and have memory that allows programming for any number of vehicles or refrigerant applications.
- *Specialty hydraulic tools*, such as the tube expanding tool kit and the hydraulic flaring and swaging tool, are updated also. The new features are a hand-held hydraulic press that accurately flares and swages copper tubing. Once the die and adapter are secured in the fixture, a few pumps of the handle and you are done. This tool really takes the work out of swaging and flaring, especially on larger tube sizes. The kit includes dies and adapters for flaring and swaging copper tubing sizes from $^1/_4$ to $^7/_8$ in.
- *Tube cutters* have carbide steel cutting wheels for cutting hard and soft copper, aluminum, brass, and thin wall steel as well as stainless steel.
- *Charging station,* a lightweight durable steel frame cart, contains all the necessary tools to quickly and conveniently charge the A/C system. No need for different charging cylinders with units that have a rugged die cast electronic scale. Simply place the refrigerant cylinder on the scale and charge.
- *Electronic tank heater blanket* speeds up recharge time. It also assures total discharge of refrigerant from 30 lb and 50-lb tanks of 125°/55°C and maximum pressure of 185 psi (R134a) and 170 psi for R-12. They are available for use with 120 or 240 V.
- *Air content analyzer*: when an A/C system leaks, refrigerant is lost and air enters the system. Your refrigerant recycler cannot tell the difference between refrigerant and air—it cycles both from partially filled systems. You end up with an unknown quantity of efficiency robbing air in your supply tank. Excess actual pressure in your supply tank indicates the pressure of air, also called "noncondensable gases" or NCGs. When you release the excess pressure, you are also releasing air. The result is purer refrigerant which will work more efficiently. This one can be left on the supply tank for regular monitoring or it can be removed to check all your tanks.
- *Thermometers, valve core tools, and accessories:* valve core remover/installer controls refrigerant flow $^1/_4$ turn of the valve lever. Lever position also gives visual indication of whether valve is opened or closed. The infrared thermometer with laser has a back-kit LCD display and an expanded temperature range of –20 to 500°C or –4 to 932°F. An alkaline battery furnishes power for up to 15 h.

Glossary

absolute humidity The weight of water vapor per unit volume; grains per cubic foot; or grams per cubic meter.

absolute pressure The sum of gage pressure and atmospheric pressure. Thus, for example, if the gage pressure is 154 psi, the absolute pressure will be 154 + 14.7, or 168.7 psi.

absolute zero A temperature equal to –459.6°F or –273°C. At this temperature the volume of an ideal gas maintained at a constant pressure becomes zero.

absorption The action of a material in extracting one or more substances present in the atmosphere or a mixture of gases or liquids accompanied by physical change, chemical change, or both.

acceleration The time rate of change of velocity. It is the derivative of velocity with respect to time.

accumulator A shell placed in a suction line for separating the liquid entrained in the suction gas. A storage tank at the evaporator exit or suction line used to prevent floodbacks to the compressor.

acrolein A warning agent often used with methyl chloride to call attention to the escape of refrigerant. The material has a compelling, pungent odor and causes irritation of the throat and eyes. Acrolein reacts with sulfur dioxide to form a sludge.

ACR tube A copper tube usually hard drawn and sold to the trade cleaned and sealed with nitrogen inside to prevent oxidation. Identified by its actual *outside diameter* (OD).

activated alumina A form of aluminum oxide (Al_2O_3) that absorbs moisture readily and is used as a drying agent.

adiabatic Referring to a change in gas conditions where no heat is added or removed except in the form of work.

adiabatic process Any thermodynamic process taking place in a closed system without the addition or removal of heat.

adsorbent A sorbent that changes physically, chemically, or both during the sorption process.

aeration Exposing a substance or area to air circulation.

agitation A condition in which a device causes circulation in a tank containing fluid.

air, ambient Generally speaking, the air surrounding an object.

air changes A method of expressing the amount of air leakage into or out of a building or room in terms of the number of building volumes or room volumes exchanged per unit of time.

air circulation Natural or imparted motion of air.

air cleaner A device designed for the purpose of removing airborne impurities such as dust, gases, vapors, fumes, and smoke. An air cleaner includes air washers, air filters, electrostatic precipitors, and charcoal filters.

air conditioner An assembly of equipment for the control of at least the first three items enumerated in the definition of *air conditioning*.

air conditioner, room A factory-made assembly designed as a unit for mounting in a window, through a wall, or as a console. It is designed for free delivery of conditioned air to an enclosed space without ducts.

air conditioning The simultaneous control of all, or at least the first three, of the following factors affecting the physical and chemical conditions of the atmosphere within a structure—temperature, humidity, motion, distribution, dust, bacteria, odors, toxic gases, and ionization—most of which affect human health or comfort.

air-conditioning system, central fan A mechanical indirect system of heating, ventilating, or air conditioning in which the air is treated or handled by equipment located outside the rooms served, usually at a central location and conveyed to and from the rooms by means of a fan and a system of distributing ducts.

air-conditioning system, year round An air-conditioning system that ventilates, heats, and humidifies in winter, and cools and dehumidifies in summer to provide the desired degree of air motion and cleanliness.

air-conditioning unit A piece of equipment designed as a specific air-treating combination, consisting of a means for ventilation, air circulation, air cleaning, and heat transfer with a control means for maintaining temperature and humidity within prescribed limits.

air cooler A factory-assembled unit including elements, whereby the temperature of air passing through the unit is reduced.

air cooler, spray type A forced-circulation air cooler, wherein the coil surface capacity is augmented by a liquid spray during the period of operation.

air cooling A reduction in air temperature due to the removal of heat as a result of contact with a medium held at a temperature lower than that of the air.

air diffuser A circular, square, or rectangular air-distribution outlet, generally located in the ceiling, and comprised of deflecting members discharging supply air in various directions and planes, arranged to promote mixing of primary air with secondary room air.

air, dry In psychrometry, air unmixed with or containing no water vapor.

air infiltration The in-leakage of air through cracks, crevices, doors, windows, or other openings caused by wind pressure or temperature difference.

air, recirculated Return air passed through the conditioner before being again supplied to the conditioned space.

air, return Air returned from conditioned or refrigerated space.

air, saturated Moist air in which the partial pressure of the water vapor is equal to the vapor pressure of water at the existing temperature. This occurs when dry air and saturated water vapor coexist at the same dry-bulb temperature.

air, standard Air with a density of 0.075 lb/ft^3 and an absolute viscosity of 1.22×10 lb mass/ft-s. This is substantially equivalent to dry air at 70°F and 29.92 in. Hg barometer.

air washer An enclosure in which air is forced through a spray of water in order to cleanse, humidify, or precool the air.

ambient temperature The temperature of the medium surrounding an object. In a domestic system having an air-cooled condenser, it is the temperature of the air entering the condenser.

ammonia machine An abbreviation for a compression-refrigerating machine using ammonia as a refrigerant. Similarly, Freon machine, sulfur dioxide machine, and so forth.

ampere Unit used to measure electrical current. It is equal to 1 C of electrons flowing past a point in 1 s. A coulomb is 6.28×10^{18} electrons.

analyzer A device used in the high side of an absorption system for increasing the concentration of vapor entering the rectifier or condenser.

anemometer An instrument for measuring the velocity of air in motion.

antifreeze, liquid A substance added to the refrigerant to prevent formation of ice crystals at the expansion valve. Antifreeze agents in general do not prevent corrosion due to moisture. The use of a liquid should be a temporary measure where large quantities of water are involved, unless a drier is used to reduce the moisture content. Ice crystals may form when moisture is present below the corrosion limits, and in such instances, a suitable noncorrosive antifreeze liquid is often of value. Materials such as alcohol are corrosive and, if used, should be allowed to remain in the machine for a limited time only.

atmospheric condenser A condenser operated with water that is exposed to the atmosphere.

atmospheric pressure The pressure exerted by the atmosphere in all directions as indicated by a barometer. Standard atmospheric pressure is considered to be 14.695 psi (pounds per square inch), which is equivalent to 29.92 in. Hg (inches of mercury).

atomize To reduce to a fine spray.

automatic air conditioning An air-conditioning system that regulates itself to maintain a definite set of conditions by means of automatic controls and valves usually responsive to temperature or pressure.

automatic expansion valve A pressure-actuated device that regulates the flow of refrigerant from the liquid line into the evaporator to maintain a constant evaporator pressure.

baffle A partition used to divert the flow of air or a fluid.

balanced pressure The same pressure in a system or container that exists outside the system or container.

barometer An instrument for measuring atmospheric pressure.

blast heater A set of heat-transfer coils or sections used to heat air that is drawn or forced through it by a fan.

bleeder A pipe sometimes attached to a condenser to bleed off liquid refrigerant parallel to the main flow.

boiler A closed vessel in which liquid is heated or vaporized.

boiler horsepower The equivalent evaporation of 34.5 lb of water per hour from and at 212°F, which is equal to a heat output of $970.3 \times 34.5 = 33{,}475$ Btu.

boiling point The temperature at which a liquid is vaporized upon the addition of heat, dependent on the refrigerant and the absolute pressure at the surface of the liquid and vapor.

bore The inside diameter of a cylinder.

Bourdon tube Tube of elastic metal bent into circular shape that is found inside a pressure gage.

brine Any liquid cooled by a refrigerant and used for transmission of heat without a change in its state.

brine system A system whereby brine cooled by a refrigerating system is circulated through pipes to the point where the refrigeration is needed.

British thermal unit (Btu) The amount of heat required to raise the temperature of 1 lb of water 1°F. It is also the measure of the amount of heat removed in cooling 1 lb of water 1°F and is so used as a measure of refrigerating effect.

butane A hydrocarbon, flammable refrigerant used to a limited extent in small units.

calcium chloride A chemical having the formula $CaCl_2$, which, in granular form, is used as a drier. This material is soluble in water, and in the presence of large quantities of moisture may dissolve and plug up the drier unit or even pass into the system beyond the drier.

calcium sulfate A solid chemical of the formula $CaSO_4$, which may be used as a drying agent.

calibration The process of dividing and numbering the scale of an instrument; also of correcting and determining the error of an existing scale.

calorie Heat required to raise the temperature of 1 g of water 1°C (actually, from 4 to 5°C). Mean calorie is equal to one-hundredth part of the heat required to raise 1 g of water from 0 to 100°C.

capacitor An electrical device that has the ability to store an electrical charge. It is used to start motors, among other purposes.

capacity, refrigerating The ability of a refrigerating system, or part thereof, to remove heat. Expressed as a rate of heat removal, it is usually measured in Btu/h or tons/24 h.

capacity reducer In a compressor, a device, such as a clearance pocket, movable cylinder head, or suction bypass, by which compressor capacity can be adjusted without otherwise changing the operating conditions.

capillarity The action by which the surface of a liquid in contact with a solid (as in a slender tube) is raised or lowered.

capillary tube In refrigeration practice, a tube of small internal diameter used as a liquid refrigerant-flow control or expansion device between high and low sides; also used to transmit pressure from the sensitive bulb of some temperature controls to the operating element.

carbon dioxide ice Compressed solid CO_2; dry ice.

Celsius A thermometric system in which the freezing point of water is called 0°C and its boiling point 100°C at normal pressure. This system is used in the scientific community for research work and also by most European countries and Canada. This book has the Celsius value of each Fahrenheit temperature in parenthesis.

centrifugal compressor A compressor employing centrifugal force for compression.

centrifuge A device for separating liquids of different densities by centrifugal action.

change of air Introduction of new, cleansed, or recirculated air to a conditioned space, measured by the number of complete changes per unit time.

change of state Change from one state to another, as from a liquid to a solid, from a liquid to a gas, and so forth.

charge The amount of refrigerant in a system.

chimney effect The tendency of air or gas in a duct or other vertical passage to rise when heated due to its lower density compared with that of the surrounding air or gas. In buildings, the tendency toward displacement, caused by the difference in temperature, of internal heated air by unheated outside air due to the difference in density of outside and inside air.

clearance Space in a cylinder not occupied by a piston at the end of the compression stroke or volume of gas remaining in a cylinder at the same point, measured in percentage of piston displacement.

coefficient of expansion The fractional increase in length or volume of a material per degree rise in temperature.

coefficient of performance (heat pump) Ratio of heating effect produced to the energy supplied, each expressed in the same thermal units.

coil Any heating or cooling element made of pipe or tubing connected in series.

cold storage A trade or process of preserving perishables on a large scale by refrigeration.

comfort chart A chart showing effective temperatures with dry-bulb temperatures and humidities (and sometimes air motion) by which the effects of various air conditions on human comfort maybe compared.

compression system A refrigerating system in which the pressure-imposing element is mechanically operated.

compressor That part of a mechanical refrigerating system, which receives the refrigerant vapor at low pressure and compresses it into a smaller volume at higher pressure.

compressor, centrifugal A nonpositive displacement compressor that depends on centrifugal effect, at least in part, for pressure rise.

compressor displacement Compressor volume in cubic inches found by multiplying piston area by stroke by the number of cylinders.

Displacement in cubic feet per minute

$$= \frac{\pi \times r^2 \times L \times \text{rpm} \times n}{1728}$$

compressor, open-type A compressor with a shaft or other moving part, extending through a casing, to be driven by an outside source of power, thus requiring a stuffing box, shaft seals, or equivalent rubbing contact between a fixed and moving part.

compressor, reciprocating A positive-displacement compressor with a piston or pistons moving in a straight line but alternately in opposite directions.

compressor, rotary One in which compression is attained in a cylinder by rotation of a positive-displacement member.

compressor booster A compressor for very low pressures, usually discharging into the suction line of another compressor.

condenser A heat-transfer device that receives high-pressure vapor at temperatures above that of the cooling medium, such as air or water, to which the condenser passes latent heat from the refrigerant, causing the refrigerant vapor to liquefy.

condensing The process of giving up latent heat of vaporization in order to liquefy a vapor.

condensing unit A specific refrigerating machine combination, for a given refrigerant, consisting of one or more power-driven compressors, condensers, liquid receivers (when required), and the regularly furnished accessories.

condensing unit, sealed A mechanical condensing unit, in which the compressor and compressor motor are enclosed in the same housing, with no external shaft or shaft seal, the compressor motor operating in the refrigerant atmosphere.

conduction, thermal Passage of heat from one point to another by transmission of molecular energy from particle to particle through a conductor.

conductivity, thermal The ability of a material to pass heat from one point to another, generally expressed in terms of Btu per hour per square foot of material per inch of thickness per degree temperature difference.

conductor, electrical A material that will pass an electric current as part of an electrical system.

connecting rod A device connecting the piston to a crank and used to change rotating motion into reciprocating motion, or vice versa, as from a rotating crankshaft to a reciprocating piston.

constant-pressure valve A valve of the throttling type, responsive to pressure, located in the suction line of an evaporator to maintain a desired constant pressure in the evaporator higher than the main suction-line pressure.

constant-temperature valve A valve of the throttling type, responsive to the temperature of a thermostatic bulb. This valve is located in the suction line of an evaporator to reduce the refrigerating effect on the coil to just maintain a desired minimum temperature.

control Any device for regulation of a system or component in normal operation either manual or automatic. If automatic, the implication is that it is responsive to changes of temperature, pressure, or any other property whose magnitude is to be regulated.

control, high-pressure A pressure-responsive device (usually an electric switch) actuated directly by the refrigerant-vapor pressure on the high side of a refrigerating system (usually compressor-head pressure).

control, low-pressure An electric switch, responsive to pressure, connected into the low-pressure part of a refrigerating system (usually closes at high pressure and opens at low pressure).

control, temperature An electric switch or relay that is responsive to the temperature change of a thermostatic bulb or element.

convection The circulatory motion that occurs in a fluid at a nonuniform temperature, owing to the variation of its density and the action of gravity.

convection, forced Convection resulting from forced circulation of a fluid as by a fan, jet, or pump.

cooling tower, water An enclosed device for evaporative cooling water by contact with air.

cooling unit A specific air-treating combination consisting of a means for air circulation and cooling within prescribed temperature limits.

cooling water Water used for condensation of refrigerant. Condenser water.

copper plating Formation of a film of copper, usually on compressor walls, pistons, or discharge valves caused by moisture in a methyl chloride system.

corrosive Having a chemically destructive effect on metals (occasionally on other materials).

counter-flow In the heat exchange between two fluids, the opposite direction of flow, the coldest portion of one meeting the coldest portion of the other.

critical pressure The vapor pressure corresponding to the critical temperature.

critical temperature The temperature above which a vapor cannot be liquefied, regardless of pressure.

critical velocity The velocity above which fluid flow is turbulent.

Crohydrate An eutectic brine mixture of water and any salt mixed in proportions to give the lowest freezing temperature.

cycle A complete course of operation of working fluid back to a starting point measured in thermodynamic terms. Also used in general for any repeated process in a system.

cycle, defrosting That portion of a refrigeration operation, which permits the cooling unit to defrost.

cycle, refrigeration A complete course of operation of a refrigerant back to the starting point measured in thermodynamic terms. Also used in general for any repeated process for any system.

Dalton's law of partial pressure Each constituent of a mixture of gases behaves thermodynamically as if it alone occupied the space. The sum of the individual pressures of the constituents equals the total pressure of the mixture.

defrosting The removal of accumulated ice from a cooling unit.

degree day A unit based on temperature difference and time used to specify the nominal heating load in winter. For one day there exist as many degree-days as there are degrees Fahrenheit difference in temperature between the average outside air temperature, taken over a 24-h period, and a temperature of 65°F.

dehumidifier An air cooler used for lowering the moisture content of the air passing through it. An absorption or adsorption device for removing moisture from the air.

dehumidify To remove water vapor from the atmosphere or to remove water or liquid from stored goods.

dehydrator A device used to remove moisture from the refrigerant.

density The mass or weight per unit of volume.

dew point, air The temperature at which a specified sample of air, with no moisture added or removed, is completely saturated. The temperature at which the air, on being cooled, gives up moisture or dew.

differential (of a control) The difference between the cut-in and cutout temperature. A valve that opens at one pressure and closes at another. This allows a system to adjust itself with a minimum of overcorrection.

direct connected Driver and driven, as motor and compressor, positively connected in line to operate at the same speed.

direct expansion A system in which the evaporator is located in the material or space refrigerated or in the air-circulating passages communicating with such space.

discharge gas Hot, high-pressure vapor refrigerant, which has just left the compressor.

displacement, actual The volume of gas at the compressor inlet actually moved in a given time.

displacement, theoretical The total volume displaced by all the pistons of a compressor for every stroke during a definite interval (usually measured in cubic feet per minute).

drier Synonymous with dehydrator.

dry-type evaporator An evaporator of the continuous-tube type where the refrigerant from a pressure-reducing device is fed into one end and the suction line connected to the outlet end.

duct A passageway made of sheet metal or other suitable material, not necessarily leaktight, used for conveying air or other gas at low pressure.

dust An air suspension (aerosol) of solid particles of earthy material, as differentiated from smoke.

economizer A reservoir or chamber wherein energy or material from a process is reclaimed for further useful purpose.

efficiency, mechanical The ratio of the output of a machine to the input in equivalent units.

efficiency, volumetric The ratio of the volume of gas actually pumped by a compressor or pump to the theoretical displacement of the compressor.

ejector A device that utilizes static pressure to build up a high fluid velocity in a restricted area to obtain a lower static pressure at that point so that fluid from another source maybe drawn in.

element, bimetallic An element formed of two metals having different coefficients of thermal expansion, such as used in temperature-indicating and controlling devices.

emulsion A relatively stable suspension of small, but not colloidal, particles of a substance in a liquid.

engine Prime mover; device for transforming fuel or heat energy into mechanical energy.

enthalpy The total heat content of a substance, compared to a standard value 32°F or 0°C for water vapor. A measure of the energy content of a system per unit mass.

entropy The ratio of the heat added to a substance to the absolute temperature at which it is added.

equalizer A piping arrangement to maintain a common liquid level or pressure between two or more chambers.

eutectic solution A solution of such concentration as to have a constant freezing point at the lowest freezing temperature for the solution.

evaporative condenser A refrigerant condenser utilizing the evaporation of water by air at the condenser surface as a means of dissipating heat.

evaporative cooling The process of cooling by means of the evaporation of water in air.

evaporator A device in which the refrigerant evaporates while absorbing heat.

expansion valve, automatic A device that regulates the flow of refrigerant from the liquid line into the evaporator to maintain a constant evaporator pressure.

expansion valve, thermostatic A device that regulates the flow of refrigerant into an evaporator so as to maintain an evaporation temperature in a definite relationship to the temperature of a thermostatic bulb.

extended surface The evaporator or condenser surface that is not a primary surface. Fins or other surfaces that transmit heat from or to a primary surface, which is part of the refrigerant container.

external equalizer In a thermostatic expansion valve, a tube connection from the chamber containing the pressure-actuated element of the valve to the outlet of the evaporator coil. A device to compensate for excessive pressure drop through the coil.

Fahrenheit A thermometric system in which 32°F denotes the freezing point of water and 212°F the boiling point under normal pressure.

fan An air-moving device comprising a wheel, or blade, and housing or orifice plate.

fan, centrifugal A fan rotor or wheel within a scroll-type housing and including driving-mechanism supports for either belt-drive or direct connection.

fan, propeller A propeller or disk-type wheel within a mounting ring or plate and including driving-mechanism supports for either belt-drive or direct connection.

fan, tube-axial A disk-type wheel within a cylinder, a set of air-guide vanes located either before or after the wheel, and driving-mechanism supports for either belt-drive or direct connection.

filter A device to remove solid material from a fluid by a straining action.

flammability The ability of a material to burn.

flare fitting A type of connector for soft tubing that involves the flaring of the tube to provide a mechanical seal.

flash gas The gas resulting from the instantaneous evaporation of the refrigerant in a pressure-reducing device to cool the refrigerant to the evaporation temperature obtained at the reduced pressure.

float valve Valve actuated by a float immersed in a liquid container.

flooded system A system in which the refrigerant enters into a header from a pressure-reducing valve and the evaporator maintains a liquid level. Opposed to dry evaporator.

fluid A gas or liquid.

foaming Formation of a foam or froth of oil refrigerant due to rapid boiling out of the refrigerant dissolved in the oil when the pressure is suddenly reduced. This occurs when the compressor operates; and if large quantities of refrigerant have been dissolved, large quantities of oil may "boil" out and be carried through the refrigerant lines.

freezeup Failure of a refrigeration unit to operate normally due to formation of ice at the expansion valve. The valve maybe frozen closed or open, causing improper refrigeration in either case.

freezing point The temperature at which a liquid will solidify upon the removal of heat.

Freon-12 The common name for dichlorodifluoromethane (CCl_2F_2).

frostback The flooding of liquid from an evaporator into the suction line, accompanied by frost formation on the suction line in most cases.

furnace That part of a boiler or warm-up heating plant in which combustion takes place. Also a complete heating unit for transferring heat from fuel being burned to the air supplied to a heating system.

fusible plug A safety plug used in vessels containing refrigerant. The plug is designed to melt at high temperatures (usually about 165°F) to prevent excessive pressure from bursting the vessel.

gage An instrument used for measuring various pressures or liquid levels. (Sometimes spelled gauge).

gas The vapor state of a material.

generator A basic component of any absorption-refrigeration system.

gravity, specific The density of a standard material usually compared to that of water or air.

grille A perforated or louvered covering for an air passage, usually installed in a sidewall, ceiling, or floor.

halide torch A leak tester generally using alcohol and burning with a blue flame; when the sampling tube draws in halocarbon refrigerant vapor, the color of the flame changes to bright green. Gas given off by the burning halocarbon is phosgene, a deadly gas used in World War I in Europe against Allied troops (can be deadly if breathed in a closed or confined area).

halogen An element from the halogen group that consists of chlorine, fluorine, bromine, and iodine. Two halogens may be present in chlorofluorocarbon refrigerants.

heat Basic form of energy that may be partially converted into other forms and into which all other forms may be entirely converted.

heat of fusion Latent heat involved in changing between the solid and the liquid states.

heat, sensible Heat that is associated with a change in temperature; specific heat exchange of temperature, in contrast to a heat interchange in which a change of state (latent heat) occurs.

heat, specific The ratio of the quantity of heat required to raise the temperature of a given mass of any substance 1° to the quantity required to raise the temperature of an equal mass of a standard substance (usually water at 59°) 1°.

heat of vaporization Latent heat involved in the change between liquid and vapor states.

heat pump A refrigerating system employed to transfer heat into a space or substance. The condenser provides the heat, while the evaporator is arranged to pick up heat from air, water, and so forth. By shifting the flow of the refrigerant, a heat-pump system may also be used to cool the space.

heating system Any of the several heating methods usually termed according to the method used in its generation, such as steam heating, warm-air heating, and the like.

heating system, electric Heating produced by the rise of temperature caused by the passage of an electric current through a conductor having a high resistance to the current flow. Residence electric-heating systems generally consist of one or several resistance units installed in a frame or casing, the degree of heating being thermostatically controlled.

heating system, steam A heating system in which heat is transferred from a boiler or other source to the heating units by steam at, above, or below atmospheric pressure.

heating system, vacuum A two-pipe steam heating system equipped with the necessary accessory apparatus to permit operating the system below atmospheric pressure.

heating system, warm-air A warm-air heating plant consisting of a heating unit (fuel burning furnace) enclosed in a casing from which the heated air is distributed to various rooms of the building through ducts.

hermetically sealed unit A refrigerating unit containing the motor and compressor in a sealed container.

high-pressure cutout A control device connected into the high-pressure part of a refrigerating system to stop the machine when the pressure becomes excessive.

high side That part of the refrigerating system containing the high-pressure refrigerant. Also the term used to refer to the condensing unit consisting of the motor, compressor, condenser, and receiver mounted on a single base.

high-side float valve A float valve that floats in high-pressure liquid. Opens on an increase in liquid level.

hold over In an evaporator, the ability to stay cold after heat removal from the evaporator stops.

horsepower A unit of power. Work done at the rate of 33,000 lb-ft/min, or 550 lb-ft/s.

humidifier A device to add moisture to the air.

humidify To add water vapor to the atmosphere; to add water vapor or moisture to any material.

humidistat A control device actuated by changes in humidity and used for automatic control of relative humidity.

humidity, absolute The definite amount of water contained in a definite quantity of air (usually measured in grains of water per pound or per cubic foot of air).

humidity, relative The ratio of the water-vapor pressure of air compared to the vapor pressure it would have if saturated at its dry-bulb temperature. Very near to the ratio of the amount of moisture contained in air compared to what it could hold at the existing temperature.

humidity, specific The weight of vapor associated with 1 lb of dry air; also termed *humidity ratio*.

hydrocarbons A series of chemicals of similar chemical nature, ranging from methane (the main constituent of natural gas) through butane, octane, and so forth, to heavy lubricating oils. All are more or less flammable. Butane and isobutane have been used to a limited extent as refrigerants.

hydrolysis Reaction of a material, such as Freon-12 or methyl chloride, with water. Acid materials in general are formed.

hydrostatic pressure The pressure due to liquid in a container that contains no gas space.

hygrometer An instrument used to measure moisture in the air.

hygroscope See humidistat.

ice-melting equivalent The amount of heat (144 Btu) absorbed by 1 lb of ice at 32°F in liquefying to water at 32°F.

indirect cooling system See brine system.

infiltration The leakage of air into a building or space.

insulation A material of low heat conductivity.

irritant refrigerant Any refrigerant that has an irritating effect on the eyes, nose, throat, or lungs.

isobutane A hydrocarbon refrigerant used to a limited extent. It is flammable.

kilowatt Unit of electrical power equal to 1000 W, or 1.34 hp, approximately.

lag of temperature control The delay in action of a temperature-responsive element due to the time required for the temperature of the element to reach the surrounding temperature.

latent heat The quantity of heat that may be added to a substance during a change of state without causing a temperature change.

latent heat of evaporation The quantity of heat required changing 1 lb of liquid into a vapor with no change in temperature. Reversible.

leak detector A device used to detect refrigerant leaks in a refrigerating system.

liquid The state of a material in which its top surface in a vessel will become horizontal. Distinguished from solid or vapor forms.

liquid line The tube or pipe that carries the refrigerant liquid from the condenser or receiver of a refrigerating system to a pressure-reducing device.

liquid receiver That part of the condensing unit that stores the liquid refrigerant.

load The required rate of heat removal.

low-pressure control An electric switch and pressure-responsive element connected into the suction side of a refrigerating unit to control the operation of the system.

low side That part of a refrigerating system, which normally operates under low pressure, as opposed to the high side. Also used to refer to the evaporator.

low-side float A valve operated by the low-pressure liquid, which opens at a low level and closes at a high level.

main A pipe or duct for distributing to or collecting conditioned air from various branches.

manometer A U-shaped liquid-filled tube for measuring pressure differences.

mechanical efficiency The ratio of work done by a machine to the work done on it or energy used by it.

mechanical equivalent of heat An energy-conversion ratio of 778.18 lb-ft = 1 Btu.

methyl chloride A refrigerant having the chemical formula CH_3Cl.

micron (μ) A unit of length; the thousandth part of 1 mm or the millionth part of a meter.

Mollier chart A graphical representation of thermal properties of fluids, with total heat and entropy as coordinates.

motor A device for transforming electrical energy into mechanical energy.

motor capacitor A device designed to improve the starting ability of single-phase induction motors.

noncondensables Foreign gases mixed with a refrigerant, which cannot be condensed into liquid form at the temperatures and pressures at which the refrigerant condenses.

oil trap A device to separate oil from the high-pressure vapor from the compressor. Usually contains a float valve to return the oil to the compressor crankcase.

output Net refrigeration produced by the system.

ozone The O_3 form of oxygen, sometimes used in air conditioning or cold-storage rooms to eliminate odors, can be toxic in concentrations of 0.5 ppm and over.

packing The stuffing around a shaft to prevent fluid leakage between the shaft and parts around the shaft.

packless valve A valve that does not use packing to prevent leaks around the valve stem. Flexible material is usually used to seal against leaks and still permit valve movement.

performance factor The ratio of the heat moved by a refrigerating system to heat equivalent of the energy used. Varies with conditions.

phosphorous pentoxide An efficient drier material that becomes gummy reacting with moisture and hence is not used alone as a drying agent.

pour point, oil The temperature below, which the oil surface will not change when the oil container is tilted.

power The rate of doing work measured in horsepower, watts, kilowatts, and so forth.

power factor, electrical devices The ratio of watts to volt-amperes in an alternating current circuit.

pressure The force exerted per unit of area.

pressure drop Loss in pressure, as from one end of a refrigerant line to the other, due to friction, static head, and the like.

pressure gage See Gage.

pressure-relief valve A valve or rupture member designed to relieve excessive pressure automatically.

psychrometric chart A chart used to determine the specific volume, heat content, dew point, relative humidity, absolute humidity, and wet- and dry-bulb temperatures, knowing any two independent items of those mentioned.

purging The act of blowing out refrigerant gas from a refrigerant containing vessel usually for the purpose of removing noncondensables.

pyrometer An instrument for the measurement of high temperatures.

radiation The passage of heat from one object to another without warming the space between. The heat is passed by wave motion similar to light.

refrigerant The medium of heat transfer in a refrigerating system that picks up heat by evaporating at a low temperature and gives up heat by condensing at a higher temperature.

refrigerating system A combination of parts in which a refrigerant is circulated for the purpose of extracting heat.

relative humidity The ratio of the water-vapor pressure of air compared to the vapor pressure it would have if saturated at its dry-bulb temperature. Very nearly the ratio of the amount of moisture contained in air compared to what it could hold at the existing temperature.

relief valve A valve designed to open at excessively high pressures to allow the refrigerant to escape.

resistance, electrical The opposition to electric-current flow, measured in ohms.

resistance, thermal The reciprocal of thermal conductivity.

room cooler A cooling element for a room. In air conditioning, a device for conditioning small volumes of air for comfort.

rotary compressor A compressor in which compression is attained in a cylinder by rotation of a semiradial member.

running time Usually indicates percent of time a refrigerant compressor operates.

saturated vapor Vapor not superheated but of 100 percent quality, that is, containing no unvaporized liquid.

seal, shaft A mechanical system of parts for preventing gas leakage between a rotating shaft and a stationary crankcase.

sealed unit See hermetically sealed nit.

shell and tube Pertaining to heat exchangers in which a coil of tubing or pipe is contained in a shell or container. The pipe is provided with openings to allow the passage of a fluid through it, while the shell is also provided with an inlet and outlet for a fluid flow.

silica gel A drier material having the formula SiO_2.

sludge A decomposition product formed in a refrigerant due to impurities in the oil or due to moisture. Sludges may be gummy or hard.

soda lime A material used for removing moisture. Not recommended for refrigeration use.

solenoid valve A valve opened by a magnetic effect of an electric current through a solenoid coil.

solid The state of matter in which a force can be exerted in a downward direction only when not confined. As distinguished from fluids.

solubility The ability of one material to enter into solution with another.

solution The homogeneous mixture of two or more materials.

specific gravity The weight of a volume of a material compared to the weight of the same volume of water.

specific heat The quantity of heat required to raise the temperature of a definite mass of a material to a definite amount compared to that required to raise the temperature of the same mass of water the same amount. May be expressed as Btu/ pound/ degrees Fahrenheit.

specific volume The volume of a definite weight of a material. Usually expressed in cubic feet per pound. The reciprocal of density.

spray pond An arrangement for lowering the temperature of water by evaporative cooling of the water in contact with outside air. The water to be cooled is sprayed by nozzles into the space above a body of previously cooled water and allowed to fall by gravity into it.

steam Water in the vapor phase.

steam trap A device for allowing the passage of condensate, or air and condensate, and preventing the passage of steam.

subcooled Cooled below the condensing temperature corresponding to the existing pressure.

sublimation The change from a solid to a vapor state without an intermediate liquid state.

suction line The tube or pipe that carries refrigerant vapor from the evaporator to the compressor inlet.

suction pressure Pressure on the suction side of the compressor.

superheater A heat exchanger used on flooded evaporators, wherein hot liquid on its way to enter the evaporator is cooled by supplying heat to dry and superheat the wet vapor leaving the evaporator.

sweating Condensation of moisture from the air on surfaces below the dew-point temperature.

system A heating or refrigerating scheme or machine, usually confined to those parts in contact with the heating or refrigerating medium.

temperature Heat level or pressure. The thermal state of a body with respect to its ability to pick up heat from or pass heat to another body.

thermal conductivity The ability of a material to conduct heat from one point to another. Indicated in terms of Btu/per hour per square foot per inches of thickness per degrees Fahrenheit.

thermocouple A device consisting of two electrical conductors having two junctions—one at a point whose temperature is to be measured, and the other at a known temperature. The temperature between the two junctions is determined by the material characteristics and the electrical potential setup.

thermodynamics The science of the mechanics of heat.

thermometer A device for indicating temperature.

thermostat A temperature-actuated switch.

ton of refrigeration Refrigeration equivalent to the melting of 1 ton of ice per 24 h. 288,000 Btu/day, 12,000 Btu/h, or 200 Btu/min.

total heat The total heat added to a refrigerant above an arbitrary starting point to bring it to a given set of conditions (usually expressed in Btu/pound). For instance, in a super-heated gas, the combined heat added to the liquid necessary to raise its temperature from an arbitrary starting point to the evaporation temperature to complete evaporation, and to raise the temperature to the final temperature where the gas is superheated.

total pressure In fluid flow, the sum of static pressure and velocity pressure.

turbulent flow Fluid flow in which the fluid moves transversely as well as in the direction of the tube or pipe axis, as opposed to streamline or viscous flow.

unit heater A direct-heating, factory-made, encased assembly including a heating element, fan, motor, and directional outlet.

unit system A system that can be removed from the user's premises without disconnecting refrigerant-containing parts, water connection, or fixed electrical connections.

unloader A device in a compressor for equalizing high- and low-side pressures when the compressor stops and for a brief period after it starts so as to decrease the starting load on the motor.

vacuum A pressure below atmospheric, usually measured in inches of mercury below atmospheric pressure.

valve In refrigeration, a device for regulation of a liquid, air, or gas.

vapor A gas, particularly one near to equilibrium with the liquid phase of the substance, which does not follow the gas laws. Frequently used instead of gas for a refrigerant and, in general, for any gas below the critical temperature.

viscosity The property of a fluid to resist flow or change of shape.

water cooler Evaporator for cooling water in an indirect refrigerating system.

wax A material that may separate when oil/refrigerant mixtures are cooled. Wax may plug the expansion valve and reduce heat transfer of the coil.

wet-bulb depression Different between dry- and wet-bulb temperatures.

wet compression A system of refrigeration, in which some liquid refrigerant is mixed with vapor entering the compressor so as to cause discharge vapors from the compressor to tend to be saturated rather than superheated.

xylene A flammable solvent, similar to kerosene, used for dissolving or loosening sludges, and for cleaning compressors and lines.

zero, absolute, of pressure The pressure existing in a vessel that is entirely empty. The lowest possible pressure. Perfect vacuum.

zero, absolute, of temperature The temperature at which a body has no heat in it (–459.6°F or –273.1°C).

zone, comfort (average) The range of effective temperature during which the majority of adults feel comfortable.

Index

C

D

E

F

G

H

N

O

P

R

S

T

U

V

W

X

Z